Geometry
Common Core

Randall I. Charles
Basia Hall
Dan Kennedy
Laurie E. Bass
Art Johnson
Stuart J. Murphy
Grant Wiggins

PEARSON

Boston, Massachusetts • Chandler, Arizona • Glenview, Illinois • Upper Saddle River, New Jersey

ISBN-13: 978-0-13-318582-9
ISBN-10: 0-13-318582-6
5 6 7 8 9 10 V063 15 14 13 12

Contents *in Brief*

Welcome to *Pearson Geometry Common Core Edition* student book. Throughout this textbook, you will find content that has been developed to cover many of the Standards for Mathematical Content and all of the Standards for Mathematical Practice from the Common Core State Standards for Mathematics. The End-of-Course Assessment provides students with practice with all of the Standards for Mathematical Content listed on pages xvi to xix.

Series *Authors*

Randall I. Charles, Ph.D., is Professor Emeritus in the Department of Mathematics and Computer Science at San Jose State University, San Jose, California. He began his career as a high school mathematics teacher, and he was a mathematics supervisor for five years. Dr. Charles has been a member of several NCTM committees and is the former Vice President of the National Council of Supervisors of Mathematics. Much of his writing and research has been in the area of problem solving. He has authored more than 75 mathematics textbooks for kindergarten through college.

Dan Kennedy, Ph.D., is a classroom teacher and the Lupton Distinguished Professor of Mathematics at the Baylor School in Chattanooga, Tennessee. A frequent speaker at professional meetings on the subject of mathematics education reform, Dr. Kennedy has conducted more than 50 workshops and institutes for high school teachers. He is coauthor of textbooks in calculus and precalculus, and from 1990 to 1994 he chaired the College Board's AP Calculus Development Committee. He is a 1992 Tandy Technology Scholar and a 1995 Presidential Award winner.

Basia Hall currently serves as Manager of Instructional Programs for the Houston Independent School District. With 33 years of teaching experience, Ms. Hall has served as a department chair, instructional supervisor, school improvement facilitator, and professional development trainer. She has developed curricula for Algebra 1, Geometry, and Algebra 2 and co-developed the Texas state mathematics standards. A 1992 Presidential Awardee, Ms. Hall is past president of the Texas Association of Supervisors of Mathematics and is a state representative for the National Council of Supervisors of Mathematics (NCSM).

Consulting *Authors*

Stuart J. Murphy is a visual learning author and consultant. He is a champion of developing visual learning skills and using related strategies to help children become more successful students. He is the author of MathStart, a series of children's books that presents mathematical concepts in the context of stories. A graduate of the Rhode Island School of Design, he has worked extensively in educational publishing and has been on the authorship teams of a number of elementary and high school mathematics programs. He is a frequent presenter at meetings of the National Council of Teachers of Mathematics, the International Reading Association, and other professional organizations.

Grant Wiggins, Ed.D., is the President of Authentic Education in Hopewell, New Jersey. He earned his Ed.D. from Harvard University and his B.A. from St. John's College in Annapolis. Dr. Wiggins consults with schools, districts, and state education departments on a variety of reform matters; organizes conferences and workshops; and develops print materials and Web resources on curricular change. He is perhaps best known for being the coauthor, with Jay McTighe, of *Understanding by Design*[1] and *The Understanding by Design Handbook*, the award-winning and highly successful materials on curriculum published by ASCD. His work has been supported by the Pew Charitable Trusts, the Geraldine R. Dodge Foundation, and the National Science Foundation.

[1] ASCD, publisher of the "Understanding by Design Handbook" co-authored by Grant Wiggins and registered owner of the trademark "Understanding by Design", has not authorized or sponsored this work and is in no way affiliated with Pearson or its products.

Program *Authors*

Geometry

Laurie E. Bass is a classroom teacher at the 9–12 division of the Ethical Culture Fieldston School in Riverdale, New York. A classroom teacher for more than 30 years, Ms. Bass has a wide base of teaching experience, ranging from Grade 6 through Advanced Placement Calculus. She was the recipient of a 2000 Honorable Mention for the Radio Shack National Teacher Awards. She has been a contributing writer for a number of publications, including software-based activities for the Algebra 1 classroom. Among her areas of special interest are cooperative learning for high school students and geometry exploration on the computer. Ms. Bass is a frequent presenter at local, regional, and national conferences.

Art Johnson, Ed.D., is a professor of mathematics education at Boston University. He is a mathematics educator with 32 years of public school teaching experience, a frequent speaker and workshop leader, and the recipient of a number of awards: the Tandy Prize for Teaching Excellence, the Presidential Award for Excellence in Mathematics Teaching, and New Hampshire Teacher of the Year. He was also profiled by the Disney Corporation in the American Teacher of the Year Program. Dr. Johnson has contributed 18 articles to NCTM journals and has authored over 50 books on various aspects of mathematics.

Algebra 1 and Algebra 2

Allan E. Bellman, Ph.D., is a Lecturer/Supervisor in the School of Education at the University of California, Davis. Before coming to Davis, he was a mathematics teacher for 31 years in Montgomery County, Maryland. He has been an instructor for both the Woodrow Wilson National Fellowship Foundation and the T^3 program. He has been involved in the development of many products from Texas Instruments. Dr. Bellman has a particular expertise in the use of technology in education and speaks frequently on this topic. He was a 1992 Tandy Technology Scholar and has twice been listed in Who's Who Among America's Teachers.

Sadie Chavis Bragg, Ed.D., is Senior Vice President of Academic Affairs at the Borough of Manhattan Community College of the City University of New York. A former professor of mathematics, she is a past president of the American Mathematical Association of Two-Year Colleges (AMATYC), co-director of the AMATYC project to revise the standards for introductory college mathematics before calculus, and an active member of the Benjamin Banneker Association. Dr. Bragg has coauthored more than 50 mathematics textbooks for kindergarten through college.

William G. Handlin, Sr., is a classroom teacher and Department Chairman of Technology Applications at Spring Woods High School in Houston, Texas. Awarded Life Membership in the Texas Congress of Parents and Teachers for his contributions to the well-being of children, Mr. Handlin is also a frequent workshop and seminar leader in professional meetings throughout the world.

Reviewers *National*

Tammy Baumann
K-12 Mathematics Coordinator
School District of the City
 of Erie
Erie, Pennsylvania

Sandy Cowgill
Mathematics Department Chair
Muncie Central High School
Muncie, Indiana

Sheryl Ezze
Mathematics Chairperson
DeWitt High School
Lansing, Michigan

Dennis Griebel
Mathematics Coordinator
Cherry Creek School District
Aurora, Colorado

Bill Harrington
Secondary Mathematics
 Coordinator
State College School District
State College, Pennsylvania

Michael Herzog
Mathematics Teacher
Tucson Small School Project
Tucson, Arizona

Camilla Horton
Secondary Instruction Support
Memphis School District
Memphis, Tennessee

Gary Kubina
Mathematics Consultant
Mobile County School System
Mobile, Alabama

Sharon Liston
Mathematics Department Chair
Moore Public Schools
Oklahoma City, Oklahoma

Ann Marie Palmeri Monahan
Mathematics Supervisor
Bayonne Public Schools
Bayonne, New Jersey

Indika Morris
Mathematics Department Chair
Queen Creek School District
Queen Creek, Arizona

Jennifer Petersen
K-12 Mathematics Curriculum
 Facilitator
Springfield Public Schools
Springfield, Missouri

Tammy Popp
Mathematics Teacher
Mehlville School District
St. Louis, Missouri

Mickey Porter
Mathematics Teacher
Dayton Public Schools
Dayton, Ohio

Steven Sachs
Mathematics Department Chair
Lawrence North High School
Indianapolis, Indiana

John Staley
Secondary Mathematics
 Coordinator
Office of Mathematics, PK-12
Baltimore, Maryland

Robert Thomas, Ph.D.
Mathematics Teacher
Yuma Union High School
 District #70
Yuma, Arizona

Linda Ussery
Mathematics Consultant
Alabama Department of
 Education
Tuscumbia, Alabama

Denise Vizzini
Mathematics Teacher
Clarksburg High School
Montgomery County,
 Maryland

Marcia White
Mathematics Specialist
Academic Operations,
 Technology and Innovations
Memphis City Schools
Memphis, Tennessee

Merrie Wolf
Mathematics Department Chair
Tulsa Public Schools
Tulsa, Oklahoma

From the *Authors*

Welcome

Math is a powerful tool with far-reaching applications throughout your life. We have designed a unique and engaging program that will enable you to tap into the power of mathematics and mathematical reasoning. This award-winning program has been developed to align fully to the Common Core State Standards.

Developing mathematical understanding and problem-solving abilities is an ongoing process—a journey both inside and outside the classroom. This course is designed to help you make sense of the mathematics you encounter in and out of class each day and to help you develop mathematical proficiency.

You will learn important mathematical principles. You will also learn how the principles are connected to one another and to what you already know. You will learn to solve problems and learn the reasoning that lies behind your solutions. You will also develop the key mathematical practices of the Common Core State Standards.

Each chapter begins with the "big ideas" of the chapter and some essential questions that you will learn to answer. Through this question-and-answer process you will develop your ability to analyze problems independently and solve them in different applications.

Your skills and confidence will increase through practice and review. Work the problems so you understand the concepts and methods presented and the thinking behind them. Then do the exercises. Ask yourself how new concepts relate to old ones. Make the connections!

Everyone needs help sometimes. You will find that this program has built-in opportunities, both in this text and online, to get help whenever you need it.

This course will also help you succeed on the tests you take in class and on other tests like the SAT, ACT, and state exams. The practice exercises in each lesson will prepare you for the format and content of such tests. No surprises!

The problem-solving and reasoning habits and skills you develop in this program will serve you in all your studies and in your daily life. They will prepare you for future success not only as a student, but also as a member of a changing technological society.

Best wishes,

PowerGeometry.com

Welcome to Geometry. *Pearson Geometry Common Core Edition* is part of a blended digital and print environment for the study of high school mathematics. Take some time to look through the features of our mathematics program, starting with **PowerGeometry.com,** the site of the digital features of the program.

Hi, I'm Darius. My friends and I will be showing you the great features of the Pearson Geometry Common Core Edition program.

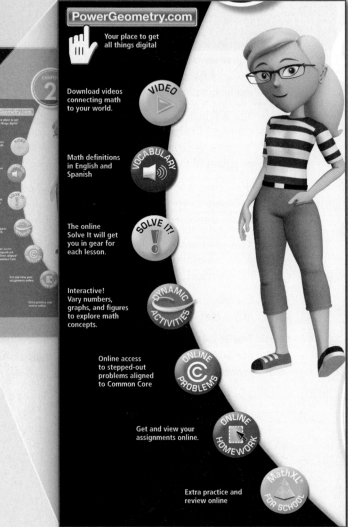

PowerGeometry.com

Your place to get all things digital

Download videos connecting math to your world.

Math definitions in English and Spanish

The online Solve It will get you in gear for each lesson.

Interactive! Vary numbers, graphs, and figures to explore math concepts.

Online access to stepped-out problems aligned to Common Core

Get and view your assignments online.

Extra practice and review online

In each chapter opener, you will be invited to visit the **PowerGeometry.com** site to access these online features. Look for these buttons throughout the lessons.

Big *Ideas*

We start with **Big Ideas.** Each chapter is organized around Big Ideas that convey the key mathematics concepts you will be studying in the program. Take a look at the Big Ideas on pages xx and xxi.

The Common Core State Standards have a similar organizing structure. They begin with **Conceptual Categories,** such as Algebra or Functions. Within each category are **domains** and **clusters.**

Common Core State Standards

Reasoning and Proof
You can observe patterns to make a conjecture; you can prove it is true using given information, definitions, properties, postulates, and theorems.

BIG ideas

Reasoning and Proof
Essential Question How can you make a conjecture and prove that it is true?

The **Big Ideas** are organizing ideas for all of the lessons in the program. At the beginning of each chapter, we'll tell you which Big Ideas you'll be studying. We'll also present an **Essential Question** for each Big Idea.

In the **Chapter Review** at the end of the chapter, you'll find the answers to the Essential Question for each Big Idea. We'll also remind you of the lesson(s) where you studied the concepts that support the Big Ideas.

Exploring *Concepts*

The lessons offer many opportunities to explore concepts in different contexts and through different media.

Hi, I'm Serena. I never have to power down when I am in math class now.

For each chapter, there is a video that you can access at **PowerGeometry.com.** The video presents concepts in a real-life context. And you can contribute your own math video.

Here's another cool feature. Each lesson opens with a **Solve It,** a problem that helps you connect what you know to an important concept in the lesson. Do you notice how the Solve It frame looks like it comes from a computer? That's because all of the Solve Its can be found at **PowerGeometry.com.**

The **Standards for Mathematical Practice** describe processes, practices, and habits of mind of mathematically proficient students. Many of the features in *Pearson Geometry Common Core Edition* help you become more proficient in math.

Developing Mathematical Proficiency

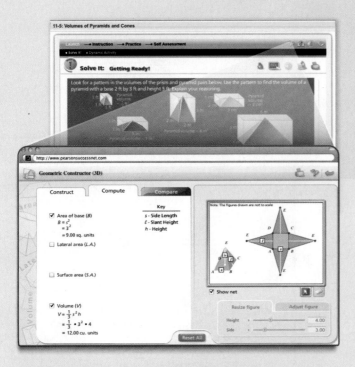

© **Content Standard**
Extends G.SRT.5 Use congruence . . . criteria for triangles to solve problems and prove relationships in geometric figures.

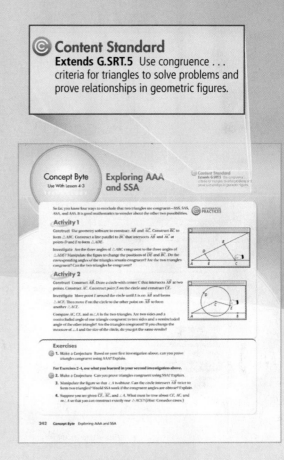

Want to do some more exploring? Try the **Math Tools** at **PowerGeometry.com.**

Click on this icon 🔧 to access these tools: Graphing Utility, Number Line, Algebra Tiles, and 2D and 3D Geometric Constructor. With the Math Tools, you can continue to explore the concepts presented in the lesson.

Try a **Concept Byte!** In a Concept Byte, you might explore technology, do a hands-on activity, or try a challenging extension.

The text in the top right corner of the first page of a lesson or Concept Byte tells you the **Standard for Mathematical Content** that you will be studying.

Thinking *Mathematically*

Mathematical reasoning is the key to solving problems and making sense of math. Throughout the program you'll learn strategies to develop mathematical reasoning habits.

Hello, I'm Tyler. These Plan boxes will help me figure out where to start.

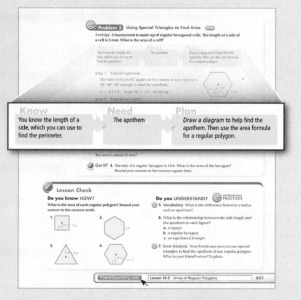

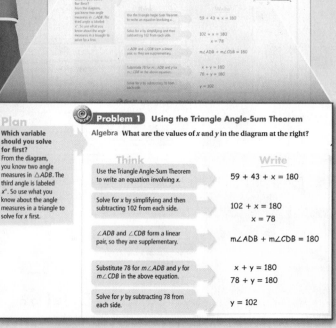

Plan

Problem 1 Using the Triangle Angle-Sum Theorem

Which variable should you solve for first?
From the diagram, you know two angle measures in △ADB. The third angle is labeled $x°$. So use what you know about the angle measures in a triangle to solve for x first.

Algebra What are the values of x and y in the diagram at the right?

Think	Write
Use the Triangle Angle-Sum Theorem to write an equation involving x.	$59 + 43 + x = 180$
Solve for x by simplifying and then subtracting 102 from each side.	$102 + x = 180$ $x = 78$
∠ADB and ∠CDB form a linear pair, so they are supplementary.	$m∠ADB + m∠CDB = 180$
Substitute 78 for $m∠ADB$ and y for $m∠CDB$ in the above equation.	$x + y = 180$ $78 + y = 180$
Solve for y by subtracting 78 from each side.	$y = 102$

The worked-out problems include call-outs that reveal the strategies and reasoning behind the solution. Look for the boxes labeled **Plan** and **Think.**

The **Think-Write** problems model the thinking behind each step of a solution.

Other worked-out problems model a problem-solving plan that includes the steps of stating what you **Know,** identifying what you **Need,** and developing a **Plan.**

The **Standards for Mathematical Practice** emphasize sense-making, reasoning, and critical reasoning. Many features in *Pearson Geometry Common Core Edition* provide opportunities for you to develop these skills and dispositions.

Standards for Mathematical Practice

Standards for Mathematical Practice

1. Make sense of problems and persevere in solving them.
2. Reason abstractly and quantitatively.
3. Construct viable arguments and critique the reasoning of others.
4. Model with mathematics.
5. Use appropriate tools strategically.
6. Attend to precision.
7. Look for and make use of structure.
8. Look for and express regularity in repeated reasoning.

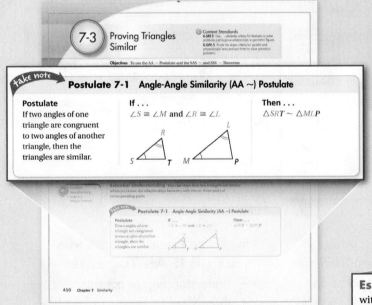

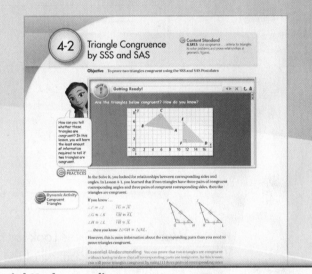

A **Take Note** box highlights important concepts in a lesson. You can use these boxes to review concepts throughout the year.

Essential Understanding You can prove that two triangles are congruent without having to show that *all* corresponding parts are congruent. In this lesson, you will prove triangles congruent by using (1) three pairs of corresponding sides and (2) two pairs of corresponding sides and one pair of corresponding angles.

Part of thinking mathematically is making sense of the concepts that are being presented. The **Essential Understandings** help you build a framework for the Big Ideas.

Practice *Makes Perfect*

Ask any professional and you'll be told that the one requirement for becoming an expert is practice, practice, practice.

Hello, I'm Anya. I can leave my book at school and still get my homework done. All of the lessons are at PowerGeometry.com

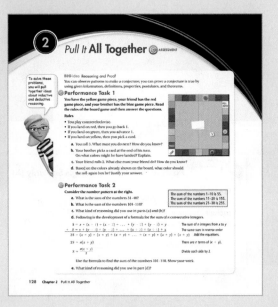

Want more practice? Look for this icon in your book. Check out all of the opportunities in **MathXL® for School.** Your teacher can assign you some practice exercises or you can choose some on your own. And you'll know right away if you got the right answer!

But the best practice occurs when you **Pull It All Together** — understanding of concepts, mathematical thinking, and problem solving — to solve interesting problems. And look: there are those Big Ideas again.

Acing *the Test*

Doing well on tests, whether they are chapter tests or state assessments, depends on a good understanding of math concepts, skill at solving problems, and, just as important, good test-taking skills.

All of these opportunities for practice help you prepare for assessments throughout the year, including the assessments to measure your proficiency with the Common Core State Standards.

Standards for Mathematical Practice

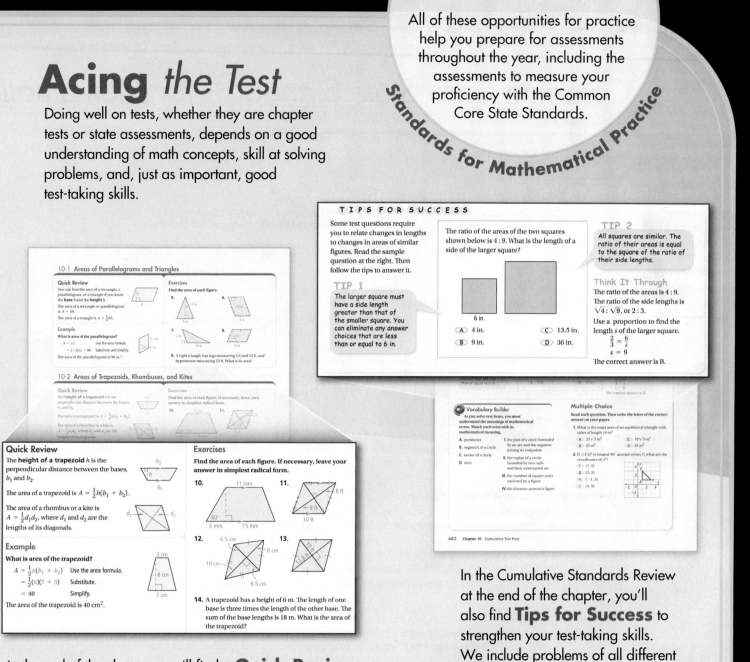

At the end of the chapter, you'll find a **Quick Review** of the concepts in the chapter and a few examples and exercises so you can check your skill at solving problems related to the concepts.

In the Cumulative Standards Review at the end of the chapter, you'll also find **Tips for Success** to strengthen your test-taking skills. We include problems of all different formats and types so you can feel comfortable with any test item on your state assessment.

Common Core *State Standards for Mathematics*
Geometry

Hi, I'm Max. Here is a list of the Common Core State Standards that you will study this year. Mastering these topics will help you be ready for the Common Assessment

Number and Quantity

Quantities

Reason quantitatively and use units to solve problems

N.Q.1 Use units as a way to understand problems and to guide the solution of multi-step problems; choose and interpret units consistently in formulas; choose and interpret the scale and the origin in graphs and data displays.

Geometry

Congruence

Experiment with transformations in the plane

G.CO.1 Know precise definitions of angle, circle, perpendicular line, parallel line, and line segment, based on the undefined notions of point, line, distance along a line, and distance around a circular arc.

G.CO.2 Represent transformations in the plane using, e.g., transparencies and geometry software; describe transformations as functions that take points in the plane as inputs and give other points as outputs. Compare transformations that preserve distance and angle to those that do not (e.g., translation versus horizontal stretch).

G.CO.3 Given a rectangle, parallelogram, trapezoid, or regular polygon, describe the rotations and reflections that carry it onto itself.

G.CO.4 Develop definitions of rotations, reflections, and translations in terms of angles, circles, perpendicular lines, parallel lines, and line segments.

G.CO.5 Given a geometric figure and a rotation, reflection, or translation, draw the transformed figure using, e.g., graph paper, tracing paper, or geometry software. Specify a sequence of transformations that will carry a given figure onto another.

Understand congruence in terms of rigid motions

G.CO.6 Use geometric descriptions of rigid motions to transform figures and to predict the effect of a given rigid motion on a given figure; given two figures, use the definition of congruence in terms of rigid motions to decide if they are congruent.

G.CO.7 Use the definition of congruence in terms of rigid motions to show that two triangles are congruent if and only if corresponding pairs of sides and corresponding pairs of angles are congruent.

G.CO.8 Explain how the criteria for triangle congruence (ASA, SAS, and SSS) follow from the definition of congruence in terms of rigid motions.

Prove geometric theorems

G.CO.9 Prove theorems about lines and angles. *Theorems include: vertical angles are congruent; when a transversal crosses parallel lines, alternate interior angles are congruent and corresponding angles are congruent; points on a perpendicular bisector of a line segment are exactly those equidistant from the segment's endpoints.*

G.CO.10 Prove theorems about triangles. *Theorems include: measures of interior angles of a triangle sum to 180°; base angles of isosceles triangles are congruent; the segment joining midpoints of two sides of a triangle is parallel to the third side and half the length; the medians of a triangle meet at a point.*

G.CO.11 Prove theorems about parallelograms. *Theorems include: opposite sides are congruent, opposite angles are congruent, the diagonals of a parallelogram bisect each other and its converse, rectangles are parallelograms with congruent diagonals.*

Make geometric constructions

G.CO.12 Make formal geometric constructions with a variety of tools and methods (compass and straightedge, string, reflective devices, paper folding, dynamic geometric software, etc.). *Copying a segment; copying an angle; bisecting a segment; bisecting an angle; constructing perpendicular lines, including the perpendicular bisector of a line segment; and constructing a line parallel to a given line through a point not on the line.*

G.CO.13 Construct an equilateral triangle, a square, and a regular hexagon inscribed in a circle.

Similarity, Right Triangles, and Trigonometry

Understand similarity in terms of similarity transformations

G.SRT.1   Verify experimentally the properties of dilations given by a center and a scale factor:

G.SRT.1.a   A dilation takes a line not passing through the center of the dilation to a parallel line, and leaves a line passing through the center unchanged.

G.SRT.1.b   The dilation of a line segment is longer or shorter in the ratio given by the scale factor.

G.SRT.2   Given two figures, use the definition of similarity in terms of similarity transformations to decide if they are similar; explain using similarity transformations the meaning of similarity for triangles as the equality of all corresponding pairs of angles and the proportionality of all corresponding pairs of side

G.SRT.3   Use the properties of similarity transformations to establish the AA criterion for two triangles to be similar.

Prove theorems involving similarity

G.SRT.4   Prove theorems about triangles. *Theorems include: a line parallel to one side of a triangle divides the other two proportionally and its converse; the Pythagorean Theorem proved using triangle similarity.*

G.SRT.5   Use congruence and similarity criteria for triangles to solve problems and to prove relationships in geometric figures.

Define trigonometric ratios and solve problems involving right triangles

G.SRT.6   Understand that by similarity, side ratios in right triangles are properties of the angles in the triangle, leading to definitions of trigonometric ratios for acute angles.

G.SRT.7   Explain and use the relationship between the sine and cosine of complementary angles.

G.SRT.8   Use trigonometric ratios and the Pythagorean Theorem to solve right triangles in applied problems.

Apply trigonometry to general triangles

G.SRT.9   (+) Derive the formula $A = 1/2\ ab \sin(C)$ for the area of a triangle by drawing an auxiliary line from a vertex perpendicular to the opposite side.

G.SRT.10   (+) Prove the Laws of Sines and Cosines and use them to solve problems.

G.SRT.11   (+) Understand and apply the Law of Sines and the Law of Cosines to find unknown measurements in right and non-right triangles (e.g., surveying problems, resultant forces).

Circles

Understand and apply theorems about circles

G.C.1   Prove that all circles are similar.

G.C.2   Identify and describe relationships among inscribed angles, radii, and chords. *Include the relationship between central, inscribed, and circumscribed angles; inscribed angles on a diameter are right angles; the radius of a circle is perpendicular to the tangent where the radius intersects the circle.*

G.C.3   Construct the inscribed and circumscribed circles of a triangle, and prove properties of angles for a quadrilateral inscribed in a circle.

G.C.4   (+) Construct a tangent line from a point outside a given circle to the circle.

Find arc lengths and areas of sectors of circles

G.C.5   Derive using similarity the fact that the length of the arc intercepted by an angle is proportional to the radius, and define the radian measure of the angle as the constant of proportionality; derive the formula for the area of a sector.

Look at the domains in bold and the cluster to get a good idea of the topics you'll study this year.

Expressing Geometric Properties with Equations

Translate between the geometric description and the equation for a conic section

G.GPE.1   Derive the equation of a circle of given center and radius using the Pythagorean Theorem; complete the square to find the center and radius of a circle given by an equation.

G.GPE.2   Derive the equation of a parabola given a focus and directrix.

Use coordinates to prove simple geometric theorems algebraically

G.GPE.4   Use coordinates to prove simple geometric theorems algebraically.

G.GPE.5   Prove the slope criteria for parallel and perpendicular lines and use them to solve geometric problems (e.g., find the equation of a line parallel or perpendicular to a given line that passes through a given point).

G.GPE.6   Find the point on a directed line segment between two given points that partitions the segment in a given ratio.

G.GPE.7   Use coordinates to compute perimeters of polygons and areas of triangles and rectangles, e.g., using the distance formula.

Geometric Measurement and Dimension

Explain volume formulas and use them to solve problems

G.GMD.1   Give an informal argument for the formulas for the circumference of a circle, area of a circle, volume of a cylinder, pyramid, and cone. *Use dissection arguments, Cavalieri's principle, and informal limit arguments.*

G.GMD.3   Use volume formulas for cylinders, pyramids, cones, and spheres to solve problems.

Visualize relationships between two-dimensional and three-dimensional objects

G.GMD.4   Identify the shapes of two-dimensional cross-sections of three-dimensional objects, and identify three-dimensional objects generated by rotations of two-dimensional objects.

Modeling with Geometry

Apply geometric concepts in modeling situations

G.MG.1   Use geometric shapes, their measures, and their properties to describe objects (e.g., modeling a tree trunk or a human torso as a cylinder).

G.MG.2   Apply concepts of density based on area and volume in modeling situations (e.g., persons per square mile, BTUs per cubic foot).

G.MG.3   Apply geometric methods to solve design problems (e.g., designing an object or structure to satisfy physical constraints or minimize cost; working with typographic grid systems based on ratios).

Statistics and Probability

Conditional Probability and the Rules of Probability

Understand independence and conditional probability and use them to interpret data

S.CP.1   Describe events as subsets of a sample space (the set of outcomes) using characteristics (or categories) of the outcomes, or as unions, intersections, or complements of other events ("or," "and," "not").

S.CP.2   Understand that two events A and B are independent if the probability of A and B occurring together is the product of their probabilities, and use this characterization to determine if they are independent.

S.CP.3   Understand the conditional probability of A given B as $P(A$ and $B)/P(B)$, and interpret independence of A and B as saying that the conditional probability of A given B is the same as the probability of A, and the conditional probability of B given A is the same as the probability of B.

S.CP.4   Construct and interpret two-way frequency tables of data when two categories are associated with each object being classified. Use the two-way table as a sample space to decide if events are independent and to approximate conditional probabilities

S.CP.5   Recognize and explain the concepts of conditional probability and independence in everyday language and everyday situations.

Use the rules of probability to compute probabilities of compound events in a uniform probability model

S.CP.6   Find the conditional probability of A given B as the fraction of B's outcomes that also belong to A, and interpret the answer in terms of the model.

S.CP.7   Apply the Addition Rule, $P(A$ or $B) = P(A) + P(B) - P(A$ and $B)$, and interpret the answer in terms of the model.

S.CP.8   (+) Apply the general Multiplication Rule in a uniform probability model, $P(A$ and $B) = P(A)P(B|A) = P(B)P(A|B)$, and interpret the answer in terms of the model.

S.CP.9   (+) Use permutations and combinations to compute probabilities of compound events and solve problems.

Using Probability to Make Decisions

Use probability to evaluate outcomes of decisions

S.MD.6   (+) Use probabilities to make fair decisions (e.g., drawing by lots, using a random number generator).

S.MD.7   (+) Analyze decisions and strategies using probability concepts (e.g., product testing, medical testing, pulling a hockey goalie at the end of a game).

BIGideas

Stay connected! These Big Ideas will help you understand how the math you study in high school fits together.

These Big Ideas are the organizing ideas for the study of important areas of mathematics: algebra, geometry, and statistics.

Algebra

Properties

- In the transition from arithmetic to algebra, attention shifts from arithmetic operations (addition, subtraction, multiplication, and division) to use of the *properties* of these operations.
- All of the facts of arithmetic and algebra follow from certain properties.

Variable

- Quantities are used to form expressions, equations, and inequalities.
- An expression refers to a quantity but does not make a statement about it. An equation (or an inequality) is a statement about the quantities it mentions.
- Using variables in place of numbers in equations (or inequalities) allows the statement of relationships among numbers that are unknown or unspecified.

Equivalence

- A single quantity may be represented by many different expressions.
- The facts about a quantity may be expressed by many different equations (or inequalities).

Solving Equations & Inequalities

- Solving an equation is the process of rewriting the equation to make what it says about its variable(s) as simple as possible.
- Properties of numbers and equality can be used to transform an equation (or inequality) into equivalent, simpler equations (or inequalities) in order to find solutions.
- Useful information about equations and inequalities (including solutions) can be found by analyzing graphs or tables.
- The numbers and types of solutions vary predictably, based on the type of equation.

Proportionality

- Two quantities are *proportional* if they have the same ratio in each instance where they are measured together.
- Two quantities are *inversely proportional* if they have the same product in each instance where they are measured together.

Function

- A function is a relationship between variables in which each value of the input variable is associated with a unique value of the output variable.
- Functions can be represented in a variety of ways, such as graphs, tables, equations, or words. Each representation is particularly useful in certain situations.
- Some important families of functions are developed through transformations of the simplest form of the function.
- New functions can be made from other functions by applying arithmetic operations or by applying one function to the output of another.

Modeling

- Many real-world mathematical problems can be represented algebraically. These representations can lead to algebraic solutions.
- A function that models a real-world situation can be used to make estimates or predictions about future occurrences.

Statistics and Probability

Data Collection and Analysis
- Sampling techniques are used to gather data from real-world situations. If the data are representative of the larger population, inferences can be made about that population.
- Biased sampling techniques yield data unlikely to be representative of the larger population.
- Sets of numerical data are described using measures of central tendency and dispersion.

Data Representation
- The most appropriate data representations depend on the type of data—quantitative or qualitative, and univariate or bivariate.
- Line plots, box plots, and histograms are different ways to show distribution of data over a possible range of values.

Probability
- Probability expresses the likelihood that a particular event will occur.
- Data can be used to calculate an experimental probability, and mathematical properties can be used to determine a theoretical probability.
- Either experimental or theoretical probability can be used to make predictions or decisions about future events.
- Various counting methods can be used to develop theoretical probabilities.

Geometry

Visualization
- Visualization can help you see the relationships between two figures and connect properties of real objects with two-dimensional drawings of these objects.

Transformations
- Transformations are mathematical functions that model relationships with figures.
- Transformations may be described geometrically or by coordinates.
- Symmetries of figures may be defined and classified by transformations.

Measurement
- Some attributes of geometric figures, such as length, area, volume, and angle measure, are measurable. Units are used to describe these attributes.

Reasoning & Proof
- Definitions establish meanings and remove possible misunderstanding.
- Other truths are more complex and difficult to see. It is often possible to verify complex truths by reasoning from simpler ones using deductive reasoning.

Similarity
- Two geometric figures are similar when corresponding lengths are proportional and corresponding angles are congruent.
- Areas of similar figures are proportional to the squares of their corresponding lengths.
- Volumes of similar figures are proportional to the cubes of their corresponding lengths.

Coordinate Geometry
- A coordinate system on a line is a number line on which points are labeled, corresponding to the real numbers.
- A coordinate system in a plane is formed by two perpendicular number lines, called the x- and y-axes, and the quadrants they form. The coordinate plane can be used to graph many functions.
- It is possible to verify some complex truths using deductive reasoning in combination with the distance, midpoint, and slope formulas.

1

Tools of Geometry

Numbers
Quantities
Reason quantitatively and use units to solve problems

Geometry
Congruence
Experiment with transformations in the plane
Prove geometric theorems
Make geometric constructions

Chapters 1 & 2

Reasoning and Proof

Visual See It!

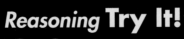

Reasoning Try It!

Practice Do It!

3

Parallel and Perpendicular Lines

Chapters 3 & 4

Geometry
Congruence
   Experiment with transformations in the plane
   Prove geometric theorems
   Make geometric constructions
Similarity, Right Triangles, and Trigonometry
   Prove theorems involving similarity

Geometry continued
Expressing Geometric Properties with Equations
   Use coordinates to prove simple geometric theorems algebraically
Modeling with Geometry
   Apply geometric concepts in modeling situations

4 Congruent Triangles

Visual **See It!**

Reasoning **Try It!**

Practice **Do It!**

5 Relationships Within Triangles

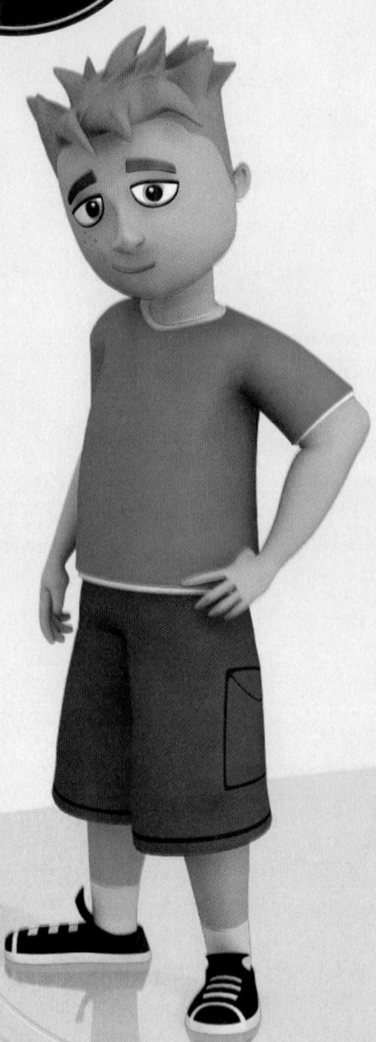

Chapters 5 & 6

Geometry

Congruence

Prove geometric theorems

Make geometric constructions

Similarity, Right Triangles, and Trigonometry

Prove theorems involving similarity

Geometry continued

Circles

Understand and apply theorems about circles

Expressing Geometric Properties with Equations

Use coordinates to prove simple geometric theorems algebraically

6 Polygons and Quadrilaterals

Visual See It!

Reasoning Try It!

Practice Do It!

7

Similarity

Chapters 7 & 8

Geometry

Similarity, Right Triangles, and Trigonometry

Prove theorems involving similarity

Define trigonometric ratios and solve problems involving right triangles

Apply trigonometry to general triangles

Geometry continued

Expressing Geometric Properties with Equations

Use coordinates to prove simple geometric theorems algebraically

Modeling with Geometry

Apply geometric concepts in modeling situations

8

Right Triangles and Trigonometry

Visual **See It!**

Reasoning **Try It!**

Practice **Do It!**

9 Transformations

Chapters 9 & 10

Geometry

Congruence
Experiment with transformations in the plane
Understand congruence in terms of rigid motions
Make geometric constructions

Similarity, Right Triangles, and Trigonometry
Understand similarity in terms of similarity transformations
Apply trigonometry to general triangles

Geometry continued

Circles
Understand and apply theorems about circles
Find arc lengths and areas of sectors of circles

Expressing Geometric Properties with Equations
Use coordinates to prove simple geometric theorems algebraically

Modeling with Geometry
Apply geometric concepts in modeling situations

Area

Visual See It!

Reasoning Try It!

Practice Do It!

11

Surface Area and Volume

Chapters 11 & 12

Geometry

Circles

Understand and apply theorems about circles

Expressing Geometric Properties with Equations

Translate between the geometric description and the equation for a conic section

Geometry continued

Geometric Measurement and Dimension

Explain volume formulas and use them to solve problems

Visualize relationships between two-dimensional and three-dimensional objects

Modeling with Geometry

Apply geometric concepts in modeling situations

12 Circles

Visual See It!

Reasoning Try It!

Practice Do It!

13

Probability

Chapter 13

Probability

Conditional Probability and the Rules of Probability

Understand independence and conditional probability and use them to interpret data

Use the rules of probability to compute probabilities of compound events in a uniform probability model

Using Probability to Make Decisions

Use probability to evaluate outcomes of decisions

Entry-Level Assessment

Multiple Choice

Read each question. Then write the letter of the correct answer on your paper.

1. What is the solution to $5a - 15 + 9a = 3a + 29$?

(A) $a = \frac{14}{11}$ (C) $a = 7$

(B) $a = 4$ (D) $a = 44$

2. A bag contains 4 blue marbles, 6 green marbles, and 2 red marbles. You select one ball at random from the bag. What is $P(\text{red})$?

(F) $\frac{1}{6}$ (H) $\frac{1}{2}$

(G) $\frac{1}{5}$ (I) $\frac{5}{6}$

3. You select one green marble from the full bag in Exercise 2. What is the probability that the next marble you select will be blue?

(A) $\frac{1}{3}$ (C) $\frac{4}{7}$

(B) $\frac{1}{5}$ (D) $\frac{4}{11}$

4. In the diagram below, the perimeter of the triangle is equal to the perimeter of the square. What is the length of a side of the square?

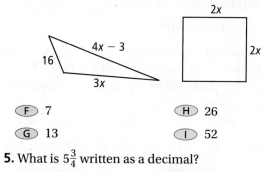

(F) 7   (H) 26

(G) 13   (I) 52

5. What is $5\frac{3}{4}$ written as a decimal?

(A) 3.75   (C) 5.75

(B) 5.25   (D) 20.3

6. Maria gave one half of her jelly beans to Carole. Carole gave one third of those to Austin. Austin gave one fourth of those to Tony. If Tony received two jelly beans, how many did Maria start with?

(F) 8   (H) 48

(G) 24   (I) 96

7. What is the ratio $0.6 : 2.4$ written in simplest form?

(A) $1 : 4$ (C) $4 : 1$

(B) $3 : 4$ (D) $6 : 24$

8. What is the slope of the line through $(-4, 2)$ and $(5, 8)$?

(F) $\frac{1}{6}$ (H) $\frac{3}{2}$

(G) $\frac{2}{3}$ (I) 6

9. What is the solution to the system of equations?

$$y = x - 2$$
$$2x + 2y = 4$$

(A) $(2, 0)$ (C) $(-2, 0)$

(B) $(0, -2)$ (D) $(0, 2)$

10. Which is the graph of a line with a slope of 3 and a y-intercept of -2?

(F) 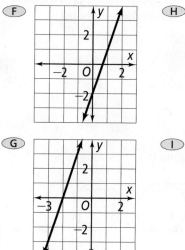 (H)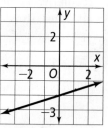

(G) 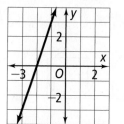 (I)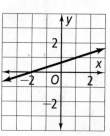

11. Which of the following is equivalent to $(-21)^2$?

(A) -441 (C) 42

(B) -42 (D) 441

12. What is the simplified form of $\sqrt{45a^5}$?

(F) $3a^2\sqrt{5a}$ (H) $5a\sqrt{3a^2}$

(G) $a^2\sqrt{45a}$ (I) $9a^2\sqrt{5a}$

13. What is the next term in the pattern?

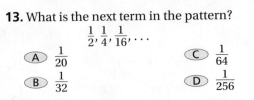

$$\frac{1}{2}, \frac{1}{4}, \frac{1}{16}, \cdots$$

- Ⓐ $\frac{1}{20}$
- Ⓒ $\frac{1}{64}$
- Ⓑ $\frac{1}{32}$
- Ⓓ $\frac{1}{256}$

14. What is the area of $\triangle ABC$, to the nearest tenth?

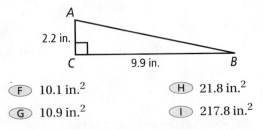

2.2 in.

9.9 in.

- Ⓕ 10.1 in.^{2}
- Ⓗ 21.8 in.^{2}
- Ⓖ 10.9 in.^{2}
- Ⓘ 217.8 in.^{2}

15. What is the value of the expression $-x(y - 8)^2$ for $x = -2$ and $y = 5$?

- Ⓐ -18
- Ⓒ 6
- Ⓑ -6
- Ⓓ 18

16. An athletic club has 248 members. Of these, 164 lift weights and 208 perform cardiovascular exercises regularly. All members do at least one of these activities. How many members do both?

- Ⓕ 40
- Ⓗ 84
- Ⓖ 44
- Ⓘ 124

17. What is the solution to $y - 7 > 3 + 2y$?

- Ⓐ $y < -10$
- Ⓒ $y > -\frac{10}{3}$
- Ⓑ $y > 4$
- Ⓓ $y < -4$

18. What is the next figure in the sequence below?

- Ⓕ a circle inside a square
- Ⓖ a square inside a circle inside a square
- Ⓗ a circle inside a square inside a circle inside a square
- Ⓘ a square inside a circle inside a square inside a circle

19. What is the ratio $18b^2$ to $45b$ written in simplest form?

- Ⓐ 18 to 45
- Ⓒ b to 2.5
- Ⓑ $2b^2$ to $5b$
- Ⓓ $2b$ to 5

20. A farmer leans a 12-ft ladder against a barn. The base of the ladder is 3 ft from the barn. To the nearest tenth, how high on the barn does the ladder reach?

- Ⓕ 9.2 ft
- Ⓗ 11.6 ft
- Ⓖ 10.8 ft
- Ⓘ 13.4 ft

21. A map has a scale of 1 in. : 25 mi. Two cities are 175 mi apart. How far apart are they on the map?

- Ⓐ 3 in.
- Ⓒ 6 in.
- Ⓑ 5 in.
- Ⓓ 7 in.

22. What is the equation of the line that is parallel to the line $y = 5x + 2$ and passes through the point $(1, -3)$?

- Ⓕ $y = -5x + 2$
- Ⓗ $y = \frac{1}{5}x - 8$
- Ⓖ $y = 5x + 8$
- Ⓘ $y = 5x - 8$

23. The graph below shows the distance and time of your car trip. What does the slope of the line mean?

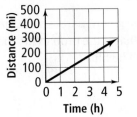

- Ⓐ You traveled 0.017 mi/h.
- Ⓑ You traveled for 5 h.
- Ⓒ You traveled 60 mi/h.
- Ⓓ You traveled 300 mi.

24. You are building a rectangular dog pen with an area of 90 ft^2. You want the length of the pen to be 3 ft longer than twice its width. Which equation can you use to find the width w of the pen?

- Ⓕ $90 = w(w + 3)$
- Ⓗ $90 = 2w(w + 3)$
- Ⓖ $90 = w(2w + 3)$
- Ⓘ $90 = (2 + w)(w + 3)$

25. The formula for the surface area of a sphere is $A = 4\pi r^2$. What is the formula solved for r?

- Ⓐ $r = \frac{A}{2\sqrt{\pi}}$
- Ⓒ $r = \frac{1}{2}\sqrt{\frac{A}{\pi}}$
- Ⓑ $r = \frac{A}{2\pi}$
- Ⓓ $r = 2\sqrt{\frac{A}{\pi}}$

CHAPTER 1

Get Ready!

Skills Handbook, p. 889

Squaring Numbers

Simplify.

1. 3^2 **2.** 4^2 **3.** 11^2

Skills Handbook, p. 890

Simplifying Expressions

Simplify each expression. Use 3.14 for π.

4. $2 \cdot 7.5 + 2 \cdot 11$ **5.** $\pi(5)^2$ **6.** $\sqrt{5^2 + 12^2}$

Skills Handbook, p. 890

Evaluating Expressions

Evaluate the following expressions for $a = 4$ and $b = -2$.

7. $\frac{a + b}{2}$ **8.** $\frac{a - 7}{3 - b}$ **9.** $\sqrt{(7 - a)^2 + (2 - b)^2}$

Skills Handbook, p. 892

Finding Absolute Value

Simplify each absolute value expression.

10. $|-8|$ **11.** $|2 - 6|$ **12.** $|-5 - (-8)|$

Skills Handbook, p. 894

Solving Equations

Algebra Solve each equation.

13. $2x + 7 = 13$ **14.** $5x - 12 = 2x + 6$ **15.** $2(x + 3) - 1 = 7x$

Looking Ahead Vocabulary

16. A child can *construct* models of buildings by stacking and arranging colored blocks. What might the term *construction* mean in geometry?

17. The *Mid*-Autumn Festival, celebrated in China, falls exactly in the middle of autumn, according to the Chinese lunar calendar. What would you expect a *midpoint* to be in geometry?

18. Artists often use long streaks to show *rays* of light coming from the sun. A ray is also a geometric figure. What do you think the properties of a *ray* are?

19. You and your friend work with each other. In other words, you and your friend are *co*-workers. What might the term *collinear* mean in geometry?

Tools of Geometry

 **DOMAINS**
• Congruence

You can see geometry everywhere, including at a concert. The laser beams are straight lines, and they intersect at points to form angles.

Points, lines, and planes are the building blocks of geometry, which you'll learn about in this chapter.

Vocabulary

English/Spanish Vocabulary Audio Online:

English	Spanish
angle bisector, *p. 37*	bisectriz de un ángulo
congruent segments, *p. 22*	segmentos congruentes
construction, *p. 43*	construcción
isometric drawing, *p. 5*	dibujo isométrico
linear pair, *p. 36*	par lineal
net, *p. 4*	plantilla
orthographic drawing, *p. 6*	dibujo ortográfico
perpendicular bisector, *p. 44*	mediatriz
postulate, *p. 13*	postulado
segment bisector, *p. 22*	bisectriz de un segmento
supplementary angles, *p. 34*	ángulos suplementarios
vertical angles, *p. 34*	ángulos verticales

My Math Video

00:04:04

VIDEO ▶

BIG ideas

1 Visualization
Essential Question How can you represent a three-dimensional figure with a two-dimensional drawing?

2 Reasoning
Essential Question What are the building blocks of geometry?

3 Measurement
Essential Question How can you describe the attributes of a segment or angle?

Chapter Preview

1-1 Nets and Drawings for Visualizing Geometry
1-2 Points, Lines, and Planes
1-3 Measuring Segments
1-4 Measuring Angles
1-5 Exploring Angle Pairs
1-6 Basic Constructions
1-7 Midpoint and Distance in the Coordinate Plane
1-8 Perimeter, Circumference, and Area

Nets and Drawings for Visualizing Geometry

© **Content Standard**
Prepares for G.CO.1 Know precise definitions of angle, circle, perpendicular line, parallel line, and line segment, based on the undefined notions of point, line, distance along a line, and distance around a circular arc.

Objective   To make nets and drawings of three-dimensional figures

SOLVE IT!

Getting Ready!

When you shine a flashlight on an object, you can see a shadow on the opposite wall. What shape would you expect the shadows in the diagram to have? Explain your reasoning.

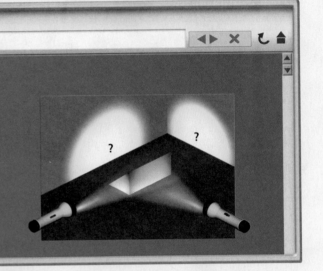

Try to visualize what the figure might look like from different perspectives.

© **MATHEMATICAL PRACTICES**

In the Solve It, you had to "see" the projection of one side of an object onto a flat surface. Visualizing figures is a key skill that you will develop in geometry.

Essential Understanding   You can represent a three-dimensional object with a two-dimensional figure using special drawing techniques.

A **net** is a two-dimensional diagram that you can fold to form a three-dimensional figure. A net shows all of the surfaces of a figure in one view.

Lesson Vocabulary
• net
• isometric drawing
• orthographic drawing

Think

How can you see the 3-D figure?
Visualize folding the net at the seams so that the edges join together. Track the letter positions by seeing one surface move in relation to another.

© **Problem 1**   **Identifying a Solid From a Net**

The net at the right folds into the cube shown beside it. Which letters will be on the top and front of the cube?

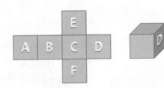

A, C, E, and F all share an edge with D when you fold the net, but only two of those sides are visible in the cube shown.

A wraps around and joins with D to become the back of the cube. B becomes the left side. F folds back to become the bottom.

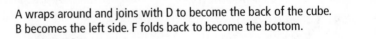

E folds down to become the top of the cube. C becomes the front.

Got It? **1.** The net in Problem 1 folds into the cube shown at the right. Which letters will be on the top and right side of the cube?

Packaging designers use nets to design boxes and other containers like the box in Problem 2.

Problem 2 **Drawing a Net From a Solid** STEM

Package Design What is a net for the graham cracker box to the right? Label the net with its dimensions.

Think

How can you see the net?
Visualize opening the top and bottom flaps of the box. Separate one of the side seams. Then unfold and flatten the box completely.

20 cm

6 cm

14 cm

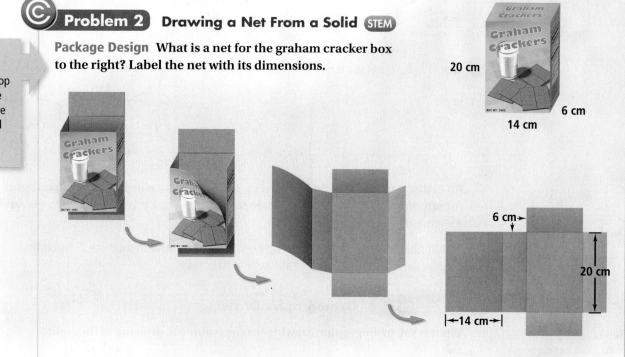

6 cm

20 cm

14 cm

Got It? **2. a.** What is a net for the figure at the right? Label the net with its dimensions.

b. Reasoning Is there another possible net for the figure in part (a)? If so, draw it.

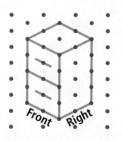

10 cm

10 cm

4 cm

7 cm

An **isometric drawing** shows a corner view of a three-dimensional figure. It allows you to see the top, front, and side of the figure. You can draw an isometric drawing on isometric dot paper. The simple drawing of a file cabinet at the right is an isometric drawing.

A net shows a three-dimensional figure as a folded-out flat surface. An isometric drawing shows a three-dimensional figure using slanted lines to represent depth.

Front   Right

Plan

Is there more than one way to make an isometric drawing?
Yes. You can start with any edge of the structure. Use that edge as a reference to draw the other edges.

© **Problem 3** Isometric Drawing

What is an isometric drawing of the cube structure at the right?

Step 1
Draw the front edges.

Step 2
Draw the right edges.

Step 3
Draw the back edges.

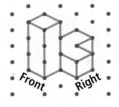

✓ **Got It?** **3.** What is an isometric drawing of this cube structure?

An **orthographic drawing** is another way to represent a three-dimensional figure. An orthographic drawing shows three separate views: a top view, a front view, and a right-side view.

Although an orthographic drawing may take more time to analyze, it provides unique information about the shape of a structure.

Plan

How can you determine the three views?
Rotate the structure in your head so that you can "see" each of the three sides straight on.

© **Problem 4** Orthographic Drawing

What is the orthographic drawing for the isometric drawing at the right?

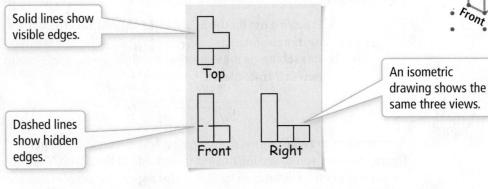

Solid lines show visible edges.

Dashed lines show hidden edges.

An isometric drawing shows the same three views.

✓ **Got It?** **4.** What is the orthographic drawing for this isometric drawing?

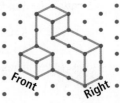

Lesson Check

Do you know HOW?

1. What is a net for the figure below? Label the net with its dimensions.

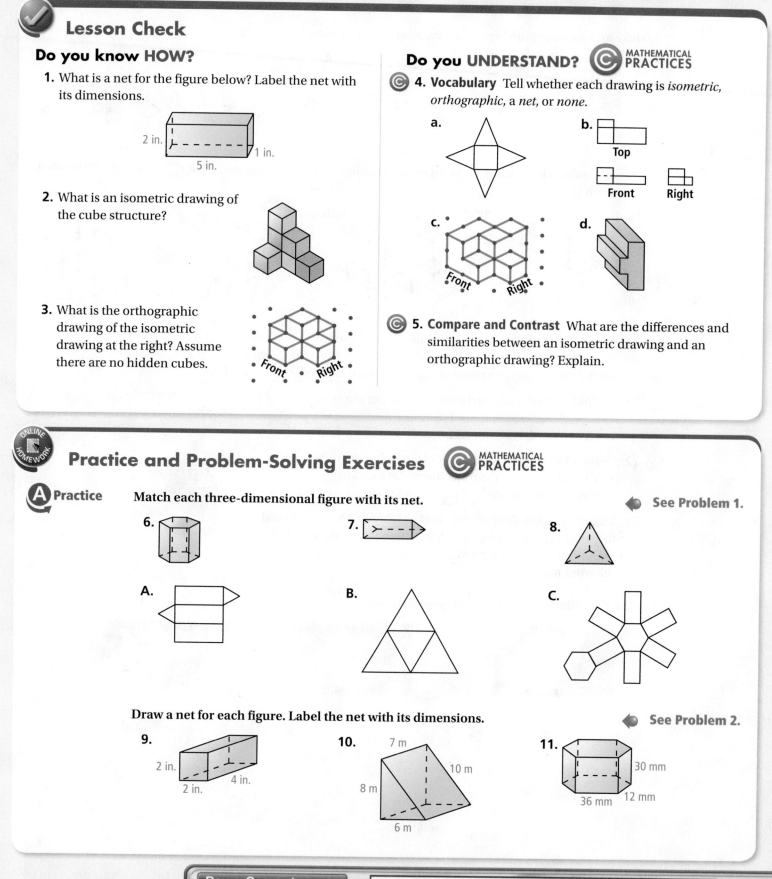

2 in.   1 in.   5 in.

2. What is an isometric drawing of the cube structure?

3. What is the orthographic drawing of the isometric drawing at the right? Assume there are no hidden cubes.

Front   Right

Do you UNDERSTAND?   MATHEMATICAL PRACTICES

4. Vocabulary Tell whether each drawing is *isometric*, *orthographic*, a *net*, or *none*.

a.

b.
Top
Front   Right

c.
Front   Right

d.

5. Compare and Contrast What are the differences and similarities between an isometric drawing and an orthographic drawing? Explain.

Practice and Problem-Solving Exercises   MATHEMATICAL PRACTICES

A Practice   Match each three-dimensional figure with its net.   See Problem 1.

6.

7.

8.

A.

B.

C.

Draw a net for each figure. Label the net with its dimensions.   See Problem 2.

9.
2 in.   4 in.   2 in.

10.
7 m   10 m   8 m   6 m

11.
30 mm   36 mm   12 mm

Make an isometric drawing of each cube structure on isometric dot paper.

● See Problem 3.

12.   **13.** 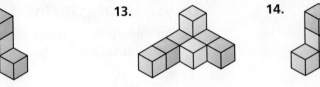 **14.** **15.**

For each isometric drawing, make an orthographic drawing. Assume there are no hidden cubes.

● See Problem 4.

16.  **17.** **18.** 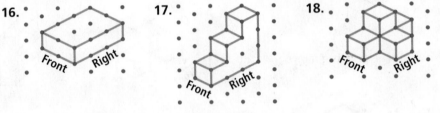 **19.**

Ⓑ Apply

© 20. Multiple Representations There are eight different nets for the solid shown at the right. Draw as many of them as you can. (*Hint*: Two nets are the same if you can rotate or flip one to match the other.)

© 21. a. Open-Ended Make an isometric drawing of a structure that you can build using 8 cubes.
  b. Make an orthographic drawing of this structure.

© 22. Think About a Plan Draw a net of the can at the right.
  • What shape are the top and bottom of the can?
  • If you uncurl the body of the can, what shape do you get?

23. History In 1525, German printmaker Albrecht Dürer first used the word *net* to describe a printed pattern that folds up into a three-dimensional shape. Why do you think he chose to use the word *net*?

STEM Manufacturing Match the package with its net.

24. 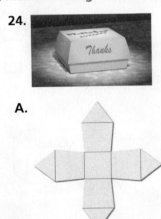 **25.** **26.**

A. 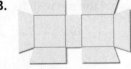 **B.** 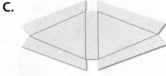 **C.**

27. Error Analysis Miquela and Gina drew orthographic drawings for the cube structure at the right. Who is correct?

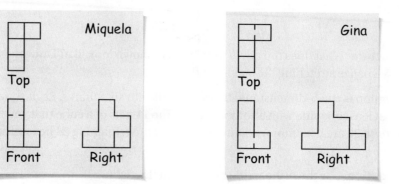

Miquela

Top

Front   Right

Gina

Top

Front   Right

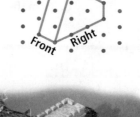

Front   Right

Make an orthographic drawing for each isometric drawing.

28.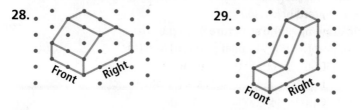
Front   Right

29.
Front   Right

30.
Front   Right

31. Fort Use the diagram of the fort at the right.
    a. Make an isometric drawing of the fort.
    b. Make an orthographic drawing of the fort.

STEM 32. Aerial Photography Another perspective in aerial photography is the "bird's-eye view," which shows an object from directly overhead. What type of drawing that you have studied in this lesson is a bird's-eye view?

33. Writing Photographs of buildings are typically not taken from a bird's-eye view. Describe a situation in which you would want a photo showing a bird's-eye view.

Visualization Think about how each net can be folded to form a cube. What is the color of the face that will be opposite the red face?

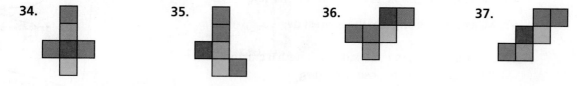

34.   **35.**   **36.**   **37.**

38. Multiple Representations There are 11 different nets for a cube. Four of them are shown above.
    a. Draw the other seven nets.
    b. Writing Suppose you want to make 100 cubes for an art project. Which of the 11 nets would you use? Explain why.

Challenge

39. The net at the right folds into a cube. Sketch the cube so that its front face is shaded as shown below.

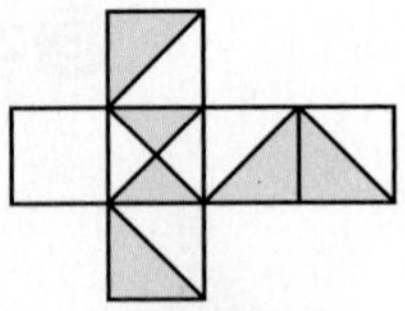

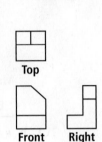

40. Architecture What does the net of the staircase shown look like? Draw the net. (*Hint*: Visualize stretching the stairs out flat.)

41. A hexomino is a two-dimensional figure formed with six squares. Each square shares at least one side with another square. The 11 nets of a cube that you found in Exercise 38 are hexominoes. Draw as many of the remaining 24 hexominoes as you can.

© 42. Visualization Use the orthographic drawing at the right.
   a. Make an isometric drawing of the structure.
   b. Make an isometric drawing of the structure from part (a) after it has been turned on its base 90° counterclockwise.
   c. Make an orthographic drawing of the structure from part (b).
   d. Turn the structure from part (a) 180°. Repeat parts (b) and (c).

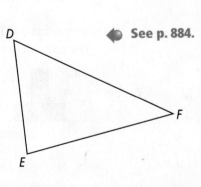

Top

Front   Right

Standardized Test Prep

SAT/ACT

43. How many possible nets does the solid at the right have?

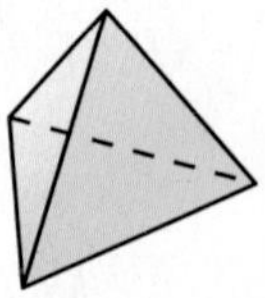

Ⓐ 1   Ⓑ 2   Ⓒ 3   Ⓓ 4

44. Solve $10a - 5b = 25$ for b.

Ⓕ $b = 10a + 25$ Ⓖ $b = 10a - 25$ Ⓗ $b = 2a + 5$ Ⓘ $b = 2a - 5$

Short Response

45. Graph the equation $x + 2y = -3$. Label the x- and y-intercepts.

Mixed Review

For Exercises 46 and 47, use the diagram at the right.

◀ See p. 884.

46. Measure *DE* and *EF* to the nearest millimeter.

47. Measure each angle to the nearest degree.

48. Draw a triangle that has sides of length 6 cm and 5 cm with a 90° angle between those two sides.

Get Ready! To prepare for Lesson 1-2, do Exercises 49–51.

Coordinate Geometry Graph the points on the coordinate plane.

◀ See p. 893.

49. $(0, 0), (2, 2), (0, 3)$ **50.** $(1, 2), (-4, 3), (-5, 0)$ **51.** $(-4, -5), (0, -1), (3, -2)$

Points, Lines, and Planes

© Content Standard

G.CO.1 Know precise definitions of angle, circle, perpendicular line, parallel line, and line segment, based on the undefined notions of point, line, distance along a line, and distance around a circular arc.

Objective To understand basic terms and postulates of geometry

SOLVE IT!

Getting Ready!

Make the figure at the right with a pencil and a piece of paper. Is the figure possible with a straight arrow and a solid board? Explain.

Does how the arrow goes through the board make sense?

MATHEMATICAL PRACTICES In this lesson, you will learn basic geometric facts to help you justify your answer to the Solve It.

Essential Understanding Geometry is a mathematical system built on accepted facts, basic terms, and definitions.

In geometry, some words such as *point*, *line*, and *plane* are undefined. Undefined terms are the basic ideas that you can use to build the definitions of all other figures in geometry. Although you cannot define undefined terms, it is important to have a general description of their meanings.

Lesson Vocabulary
- point
- line
- plane
- collinear points
- coplanar
- space
- segment
- ray
- opposite rays
- postulate
- axiom
- intersection

take note

Key Concept   Undefined Terms

Term Description	How to Name It	Diagram
A **point** indicates a location and has no size.	You can represent a point by a dot and name it by a capital letter, such as A.	A
A **line** is represented by a straight path that extends in two opposite directions without end and has no thickness. A line contains infinitely many points.	You can name a line by any two points on the line, such as $\overleftrightarrow{AB}$ (read "line AB") or $\overleftrightarrow{BA}$, or by a single lowercase letter, such as line ℓ.	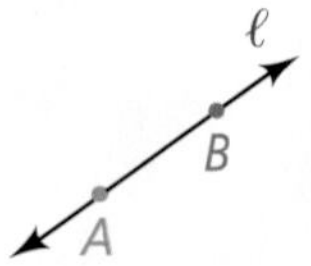 ℓ B A
A **plane** is represented by a flat surface that extends without end and has no thickness. A plane contains infinitely many lines.	You can name a plane by a capital letter, such as plane P, or by at least three points in the plane that do not all lie on the same line, such as plane ABC.	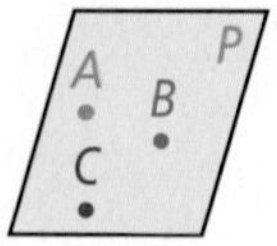 P A B C

Points that lie on the same line are **collinear points.** Points and lines that lie in the same plane are **coplanar.** All the points of a line are coplanar.

Problem 1 **Naming Points, Lines and Planes**

Think

Why can figures have more than one name?
Lines and planes are made up of many points. You can choose any two points on a line and any three or more noncollinear points in a plane for the name.

A What are two other ways to name $\overleftrightarrow{QT}$?

Two other ways to name $\overleftrightarrow{QT}$ are $\overleftrightarrow{TQ}$ and line m.

B What are two other ways to name plane P?

Two other ways to name plane P are plane RQV and plane RSV.

C What are the names of three collinear points? What are the names of four coplanar points?

Points R, Q, and S are collinear. Points R, Q, S, and V are coplanar.

Got It? **1. a.** What are two other ways to name $\overleftrightarrow{RS}$?
      b. What are two more ways to name plane P?
      c. What are the names of three other collinear points?
      d. What are two points that are *not* coplanar with points R, S, and V?

The terms *point, line,* and *plane* are not defined because their definitions would require terms that also need defining. You can, however, use undefined terms to define other terms. A geometric figure is a set of points. **Space** is the set of all points in three dimensions. Similarly, the definitions for *segment* and *ray* are based on points and lines.

take note

Key Concept **Defined Terms**

Definition	How to Name It	Diagram
A **segment** is part of a line that consists of two endpoints and all points between them.	You can name a segment by its two endpoints, such as $\overline{AB}$ (read "segment AB") or $\overline{BA}$.	A B
A **ray** is part of a line that consists of one endpoint and all the points of the line on one side of the endpoint.	You can name a ray by its endpoint and another point on the ray, such as $\overrightarrow{AB}$ (read "ray AB"). The order of points indicates the ray's direction.	A B
Opposite rays are two rays that share the same endpoint and form a line.	You can name opposite rays by their shared endpoint and any other point on each ray, such as $\overrightarrow{CA}$ and $\overrightarrow{CB}$.	A C B

Plan

How do you make sure you name all the rays?
Each point on the line is an endpoint for a ray. At each point, follow the line both left and right to see if you can find a second point to name the ray.

Problem 2 Naming Segments and Rays

A What are the names of the segments in the figure at the right?

The three segments are $\overline{DE}$ or $\overline{ED}$, $\overline{EF}$ or $\overline{FE}$, and $\overline{DF}$ or $\overline{FD}$.

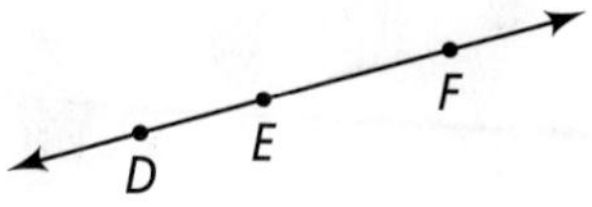

B What are the names of the rays in the figure?

The four rays are $\overrightarrow{DE}$ or $\overrightarrow{DF}$, $\overrightarrow{ED}$, $\overrightarrow{EF}$, and $\overrightarrow{FD}$ or $\overrightarrow{FE}$.

C Which of the rays in part (B) are opposite rays?

The opposite rays are $\overrightarrow{ED}$ and $\overrightarrow{EF}$.

Got It? 2. Reasoning $\overrightarrow{EF}$ and $\overrightarrow{FE}$ form a line. Are they opposite rays? Explain.

A **postulate** or **axiom** is an accepted statement of fact. Postulates, like undefined terms, are basic building blocks of the logical system in geometry. You will use logical reasoning to prove general concepts in this book.

You have used some of the following geometry postulates in algebra. For example, you used Postulate 1-1 when you graphed equations such as $y = 2x + 8$. You graphed two points and drew the line through the points.

take note

Postulate 1-1

Through any two points there is exactly one line.

Line t passes through points A and B. Line t is the only line that passes through both points.

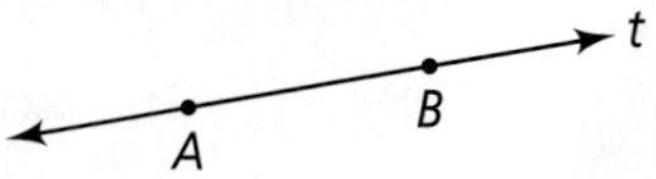

When you have two or more geometric figures, their **intersection** is the set of points the figures have in common.

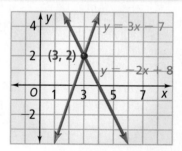

In algebra, one way to solve a system of two equations is to graph them. The graphs of the two lines $y = -2x + 8$ and $y = 3x - 7$ intersect in a single point (3, 2). So the solution is (3, 2). This illustrates Postulate 1-2.

take note

Postulate 1-2

If two distinct lines intersect, then they intersect in exactly one point.

$\overleftrightarrow{AE}$ and $\overleftrightarrow{DB}$ intersect in point C.

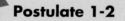

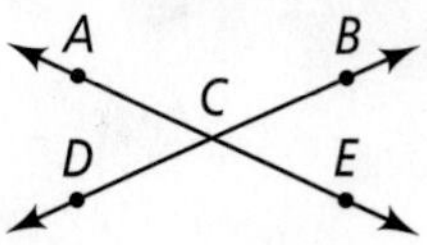

There is a similar postulate about the intersection of planes.

Postulate 1-3

If two distinct planes intersect, then they intersect in exactly one line.

Plane *RST* and plane *WST* intersect in $\overleftrightarrow{ST}$.

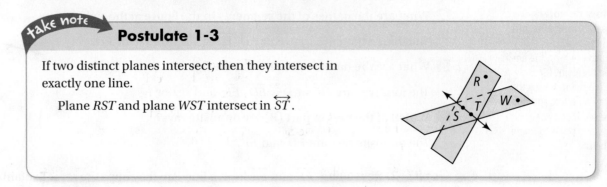

When you know two points that two planes have in common, Postulates 1-1 and 1-3 tell you that the line through those points is the intersection of the planes.

Problem 3 Finding the Intersection of Two Planes

Each surface of the box at the right represents part of a plane. What is the intersection of plane *ADC* and plane *BFG*?

Know

Plane *ADC* and plane *BFG*

Need

The intersection of the two planes

Plan

Find the points that the planes have in common.

Think

Is the intersection a segment?
No. The intersection of the sides of the box is a segment, but planes continue without end. The intersection is a line.

Focus on plane *ADC* and plane *BFG* to see where they intersect.

You can see that both planes contain point *B* and point *C*.

The planes intersect in $\overleftrightarrow{BC}$.

Got It? 3. a. What are the names of two planes that intersect in $\overleftrightarrow{BF}$?
b. Reasoning Why do you only need to find two common points to name the intersection of two distinct planes?

When you name a plane from a figure like the box in Problem 3, list the corner points in consecutive order. For example, plane *ADCB* and plane *ABCD* are also names for the plane on the top of the box. Plane *ACBD* is not.

Photographers use three-legged tripods to make sure that a camera is steady. The feet of the tripod all touch the floor at the same time. You can think of the feet as points and the floor as a plane. As long as the feet do not all lie in one line, they will lie in exactly one plane.

This illustrates Postulate 1-4.

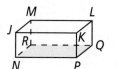

Postulate 1-4

Through any three noncollinear points there is exactly one plane.

Points *Q*, *R*, and *S* are noncollinear. Plane *P* is the only plane that contains them.

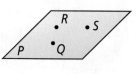

© **Problem 4** **Using Postulate 1-4**

Plan

How can you find the plane?
Try to draw all the lines that contain two of the three given points. You will begin to see a plane form.

Use the figure at the right.

Ⓐ **What plane contains points *N*, *P*, and *Q*? Shade the plane.**

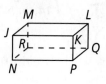

The plane on the bottom of the figure contains points *N*, *P*, and *Q*.

Ⓑ **What plane contains points *J*, *M*, and *Q*? Shade the plane.**

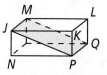

The plane that passes at a slant through the figure contains points *J*, *M*, and *Q*.

© ✓ **Got It?** **4. a.** What plane contains points *L*, *M*, and *N*? Copy the figure in Problem 4 and shade the plane.

 b. Reasoning What is the name of a line that is coplanar with $\overleftrightarrow{JK}$ and $\overleftrightarrow{KL}$?

Lesson Check

Do you know HOW?

Use the figure at the right.

1. What are two other names for $\overleftrightarrow{XY}$?

2. What are the opposite rays?

3. What is the intersection of the two planes?

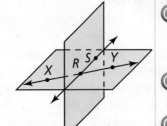

Do you UNDERSTAND?

MATHEMATICAL PRACTICES

4. **Vocabulary** A segment has endpoints R and S. What are two names for the segment?

5. Are $\overrightarrow{AB}$ and $\overrightarrow{BA}$ the same ray? Explain.

6. **Reasoning** Why do you use two arrowheads when drawing or naming a line such as $\overleftrightarrow{EF}$?

7. **Compare and Contrast** How is naming a ray similar to naming a line? How is it different?

Practice and Problem-Solving Exercises

MATHEMATICAL PRACTICES

 Practice

Use the figure at the right for Exercises 8–11.

8. What are two other ways to name $\overleftrightarrow{EF}$?

9. What are two other ways to name plane C?

10. Name three collinear points.

11. Name four coplanar points.

◀ See Problem 1.

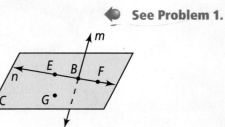

Use the figure at the right for Exercises 12–14.

12. Name the segments in the figure.

13. Name the rays in the figure.

14. **a.** Name the pair of opposite rays with endpoint T.
    b. Name another pair of opposite rays.

◀ See Problem 2.

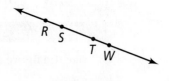

Use the figure at the right for Exercises 15–26.

Name the intersection of each pair of planes.

◀ See Problem 3.

15. planes QRS and RSW

16. planes UXV and WVS

17. planes XWV and UVR

18. planes TXW and TQU

Name two planes that intersect in the given line.

19. $\overleftrightarrow{QU}$

20. $\overleftrightarrow{TS}$

21. $\overleftrightarrow{XT}$

22. $\overleftrightarrow{VW}$

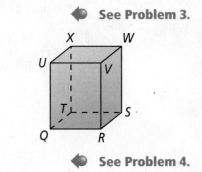

Copy the figure. Shade the plane that contains the given points.

◀ See Problem 4.

23. R, V, W

24. U, V, W

25. U, X, S

26. T, U, V

Postulate 1-4 states that any three noncollinear points lie in exactly one plane. Find the plane that contains the first three points listed. Then determine whether the fourth point is in that plane. Write *coplanar* or *noncoplanar* to describe the points.

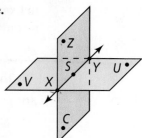

27. *Z, S, Y, C*   **28.** *S, U, V, Y*

29. *X, Y, Z, U*   **30.** *X, S, V, U*

31. *X, Z, S, V*   **32.** *S, V, C, Y*

If possible, draw a figure to fit each description. Otherwise, write *not possible*.

33. four points that are collinear   **34.** two points that are noncollinear

35. three points that are noncollinear   **36.** three points that are noncoplanar

ⓒ 37. Open-Ended Draw a figure with points *B, C, D, E, F,* and *G* that shows $\overleftrightarrow{CD}$, $\overleftrightarrow{BG}$, and $\overleftrightarrow{EF}$, with one of the points on all three lines.

ⓒ 38. Think About a Plan Your friend drew the diagram at the right to prove to you that two planes can intersect in exactly one point. Describe your friend's error.
- How do you describe a plane?
- What does it mean for two planes to intersect each other?
- Can you define an endpoint of a plane?

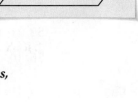

ⓒ 39. Reasoning If one ray contains another ray, are they the same ray? Explain.

For Exercises 40–45, determine whether each statement is *always, sometimes,* or *never* true.

40. $\overleftrightarrow{TQ}$ and $\overleftrightarrow{QT}$ are the same line.

41. $\overrightarrow{JK}$ and $\overrightarrow{JL}$ are the same ray.

42. Intersecting lines are coplanar.

43. Four points are coplanar.

44. A plane containing two points of a line contains the entire line.

45. Two distinct lines intersect in more than one point.

ⓒ 46. Use the diagram at the right. How many planes contain each line and point?
- **a.** $\overleftrightarrow{EF}$ and point *G*   **b.** $\overleftrightarrow{PH}$ and point *E*
- **c.** $\overleftrightarrow{FG}$ and point *P*   **d.** $\overleftrightarrow{EP}$ and point *G*
- **e. Reasoning** What do you think is true of a line and a point not on the line? Explain. (*Hint:* Use two of the postulates you learned in this lesson.)

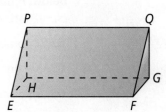

In Exercises 47–49, sketch a figure for the given information. Then state the postulate that your figure illustrates.

47. $\overleftrightarrow{AB}$ and $\overleftrightarrow{EF}$ intersect in point C.

48. The noncollinear points A, B, and C are all contained in plane N.

49. Planes LNP and MVK intersect in $\overleftrightarrow{NM}$.

STEM 50. Telecommunications A cell phone tower at point A receives a cell phone signal from the southeast. A cell phone tower at point B receives a signal from the same cell phone from due west. Trace the diagram at the right and find the location of the cell phone. Describe how Postulates 1-1 and 1-2 help you locate the phone.

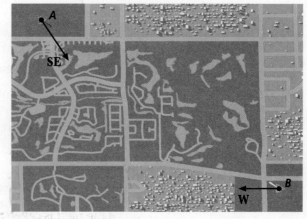

51. Estimation You can represent the hands on a clock at 6:00 as opposite rays. Estimate the other 11 times on a clock that you can represent as opposite rays.

52. Open-Ended What are some basic words in English that are difficult to define?

Coordinate Geometry Graph the points and state whether they are collinear.

53. $(1, 1), (4, 4), (-3, -3)$ **54.** $(2, 4), (4, 6), (0, 2)$ **55.** $(0, 0), (-5, 1), (6, -2)$

56. $(0, 0), (8, 10), (4, 6)$ **57.** $(0, 0), (0, 3), (0, -10)$ **58.** $(-2, -6), (1, -2), (4, 1)$

Ⓒ Challenge

59. How many planes contain the same three collinear points? Explain.

60. How many planes contain a given line? Explain.

61. a. Writing Suppose two points are in plane P. Explain why the line containing the points is also in plane P.

   b. Reasoning Suppose two lines intersect. How many planes do you think contain both lines? Use the diagram at the right and your answer to part (a) to explain your answer.

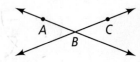

Probability Suppose you pick points at random from A, B, C, and D shown below. Find the probability that the number of points given meets the condition stated.

62. 2 points, collinear

63. 3 points, collinear

64. 3 points, coplanar

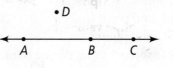

Standardized Test Prep

SAT/ACT

65. Which geometric term is undefined?

   Ⓐ segment   Ⓒ ray

   Ⓑ collinear   Ⓓ plane

66. Which diagram is a net of the figure shown at the right?

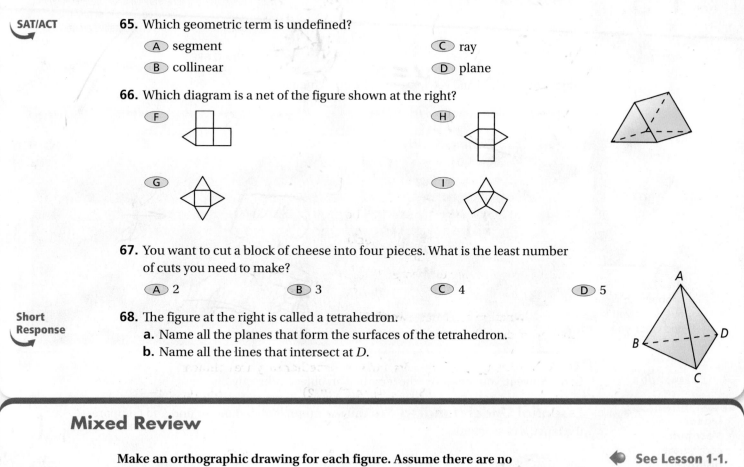

   Ⓕ   Ⓗ

   Ⓖ   Ⓘ

67. You want to cut a block of cheese into four pieces. What is the least number of cuts you need to make?

   Ⓐ 2   Ⓑ 3   Ⓒ 4   Ⓓ 5

Short Response

68. The figure at the right is called a tetrahedron.
   a. Name all the planes that form the surfaces of the tetrahedron.
   b. Name all the lines that intersect at D.

Mixed Review

Make an orthographic drawing for each figure. Assume there are no hidden cubes.

◀ See Lesson 1-1.

69.   Front Right   **70.**   Front Right   **71.**   Front Right

Simplify each ratio.

◀ See p. 891.

72. 30 to 12   **73.** $\dfrac{15x}{35x}$ **74.** $\dfrac{n^2 + n}{4n}$

Get Ready!   To prepare for Lesson 1-3, do Exercises 75–80.

Simplify each absolute value expression.

◀ See p. 892.

75. $|-6|$ **76.** $|3.5|$ **77.** $|7 - 10|$

Algebra Solve each equation.

◀ See p. 894.

78. $x + 2x - 6 = 6$ **79.** $3x + 9 + 5x = 81$ **80.** $w - 2 = -4 + 7w$

© **Content Standard**

G.CO.1 Know precise definitions of angle, circle, perpendicular line, parallel line, and line segment, based on the undefined notions of point, line, distance along a line, and distance around a circular arc.

Also G.GPE.6

Objective   To find and compare lengths of segments

SOLVE IT!

Getting Ready!   ◄▶ ✕ ↻ ⬆

On a freshwater fishing trip, you catch the fish below. By law, you must release any fish between 15 and 19 in. long. You need to measure your fish, but the front of the ruler on the boat is worn away. Can you keep your fish? Explain how you found your answer.

Analyze the problem to figure out what you know and what you need to find next.

© **MATHEMATICAL PRACTICES** In the Solve It, you measured the length of an object indirectly.

Lesson Vocabulary
• coordinate
• distance
• congruent segments
• midpoint
• segment bisector

Essential Understanding   You can use number operations to find and compare the lengths of segments.

take note

Postulate 1-5   Ruler Postulate

Every point on a line can be paired with a real number. This makes a one-to-one correspondence between the points on the line and the real numbers. The real number that corresponds to a point is called the **coordinate** of the point.

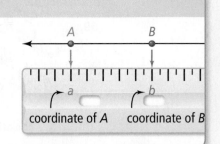

coordinate of *A*   coordinate of *B*

The Ruler Postulate allows you to measure lengths of segments using a given unit and to find distances between points on a number line. Consider $\overleftrightarrow{AB}$ at the right. The **distance** between points A and B is the absolute value of the difference of their coordinates, or $|a - b|$. This value is also AB, or the length of $\overline{AB}$.

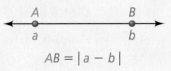

$$AB = |a - b|$$

What are you trying to find?
ST represents the length of $\overline{ST}$, so you are trying to find the distance between points *S* and *T*.

Problem 1 **Measuring Segment Lengths**

What is *ST*?

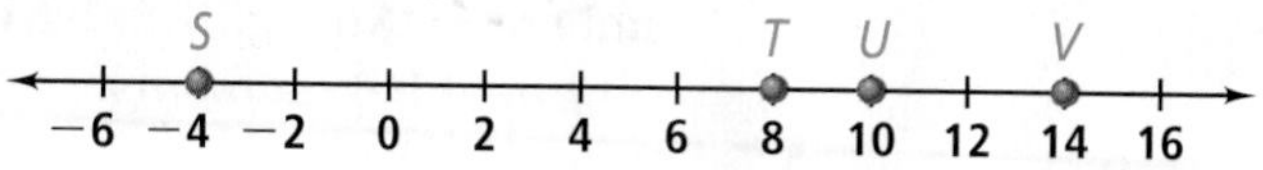

The coordinate of *S* is −4.

The coordinate of *T* is 8.

Ruler Postulate

$ST = |-4 - 8|$ Definition of distance

$= |-12|$ Subtract.

$= 12$ Find the absolute value.

Got It? **1.** What are *UV* and *SV* on the number line above?

---

take note

Postulate 1-6 **Segment Addition Postulate**

If three points *A*, *B*, and *C* are collinear and *B* is between *A* and *C*, then $AB + BC = AC$.

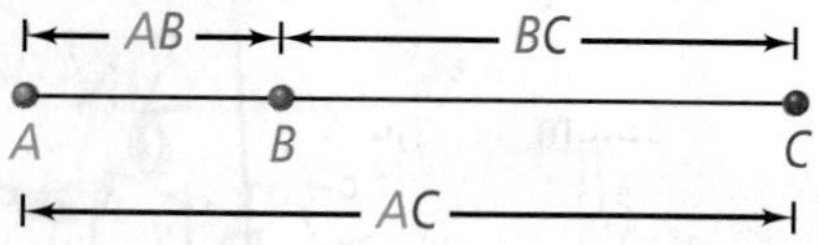

Problem 2 **Using the Segment Addition Postulate**

Algebra If *EG* = 59, what are *EF* and *FG*?

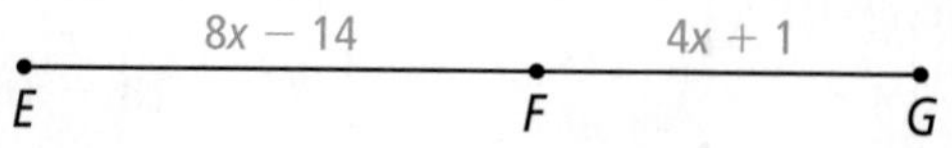

Know	Need	Plan
$EG = 59$ $EF = 8x - 14$ $FG = 4x + 1$	*EF* and *FG*	Use the Segment Addition Postulate to *write an equation.*

$EF + FG = EG$ Segment Addition Postulate

$(8x - 14) + (4x + 1) = 59$ Substitute.

$12x - 13 = 59$ Combine like terms.

$12x = 72$ Add 13 to each side.

$x = 6$ Divide each side by 12.

Use the value of *x* to find *EF* and *FG*.

$EF = 8x - 14 = 8(6) - 14 = 48 - 14 = 34$

$FG = 4x + 1 = 4(6) + 1 = 24 + 1 = 25$

Substitute 6 for *x*.

Got It? **2.** In the diagram, *JL* = 120. What are *JK* and *KL*?

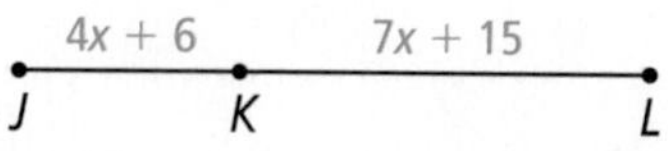

When numerical expressions have the same value, you say that they are equal (=). Similarly, if two segments have the same length, then the segments are **congruent (≅) segments.**

This means that if $AB = CD$, then $\overline{AB} \cong \overline{CD}$. You can also say that if $\overline{AB} \cong \overline{CD}$, then $AB = CD$.

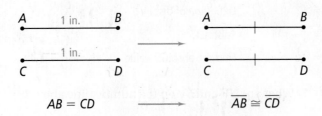

$$AB = CD \longrightarrow \overline{AB} \cong \overline{CD}$$

As illustrated above, you can mark segments alike to show that they are congruent. If there is more than one set of congruent segments, you can indicate each set with the same number of marks.

Plan

Problem 3 **Comparing Segment Lengths**

How do you know if segments are congruent?
Congruent segments have the same length. So find and compare the lengths of $\overline{AC}$ and $\overline{BD}$.

Are $\overline{AC}$ and $\overline{BD}$ congruent?

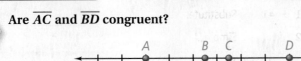

$AC = |-2 - 5| = |-7| = 7$

 Definition of distance

$BD = |3 - 10| = |-7| = 7$

Yes. $AC = BD$, so $\overline{AC} \cong \overline{BD}$.

Got It? **3. a.** Use the diagram above. Is $\overline{AB}$ congruent to $\overline{DE}$?
   b. Reasoning To find AC in Problem 3, suppose you subtract -2 from 5. Do you get the same result? Why?

The **midpoint** of a segment is a point that divides the segment into two congruent segments. A point, line, ray, or other segment that intersects a segment at its midpoint is said to *bisect* the segment. That point, line, ray, or segment is called a **segment bisector.**

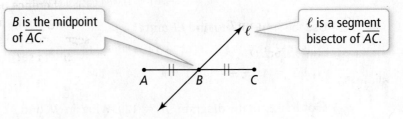

How can you use algebra to solve the problem?
The lengths of the congruent segments are given as algebraic expressions. You can set the expressions equal to each other.

Problem 4 Using the Midpoint

Algebra Q is the midpoint of $\overline{PR}$. What are PQ, QR, and PR?

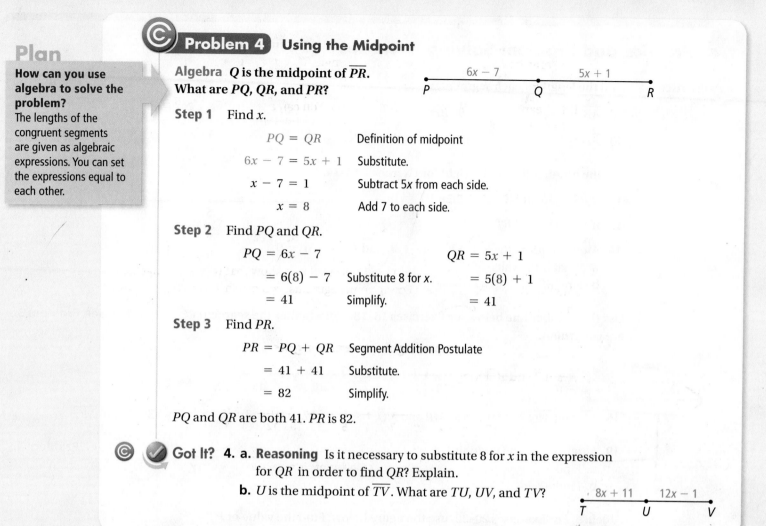

Step 1 Find x.

$PQ = QR$	Definition of midpoint
$6x - 7 = 5x + 1$	Substitute.
$x - 7 = 1$	Subtract $5x$ from each side.
$x = 8$	Add 7 to each side.

Step 2 Find PQ and QR.

$PQ = 6x - 7$ $\qquad\qquad$ $QR = 5x + 1$
$\quad = 6(8) - 7$ Substitute 8 for x. $\quad = 5(8) + 1$
$\quad = 41$ Simplify. $\quad = 41$

Step 3 Find PR.

$PR = PQ + QR$ Segment Addition Postulate
$\quad = 41 + 41$ Substitute.
$\quad = 82$ Simplify.

PQ and QR are both 41. PR is 82.

Got It? 4. a. Reasoning Is it necessary to substitute 8 for x in the expression for QR in order to find QR? Explain.

b. U is the midpoint of $\overline{TV}$. What are TU, UV, and TV?

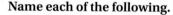
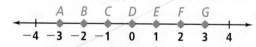

Lesson Check

Do you know HOW?

Name each of the following.

A B C D E F G
–4 –3 –2 –1 0 1 2 3 4

1. The point on $\overrightarrow{DA}$ that is 2 units from D

2. Two points that are 3 units from D

3. The coordinate of the midpoint of $\overline{AG}$

4. A segment congruent to $\overline{AC}$

Do you UNDERSTAND? MATHEMATICAL PRACTICES

5. Vocabulary Name two segment bisectors of $\overline{PR}$.

6. Compare and Contrast Describe the difference between saying that two segments are *congruent* and saying that two segments have *equal length*. When would you use each phrase?

7. Error Analysis You and your friend live 5 mi apart. He says that it is 5 mi from his house to your house and –5 mi from your house to his house. What is the error in his argument?

ℓ
P Q R S T
2 3 4 5 6

Practice and Problem-Solving Exercises

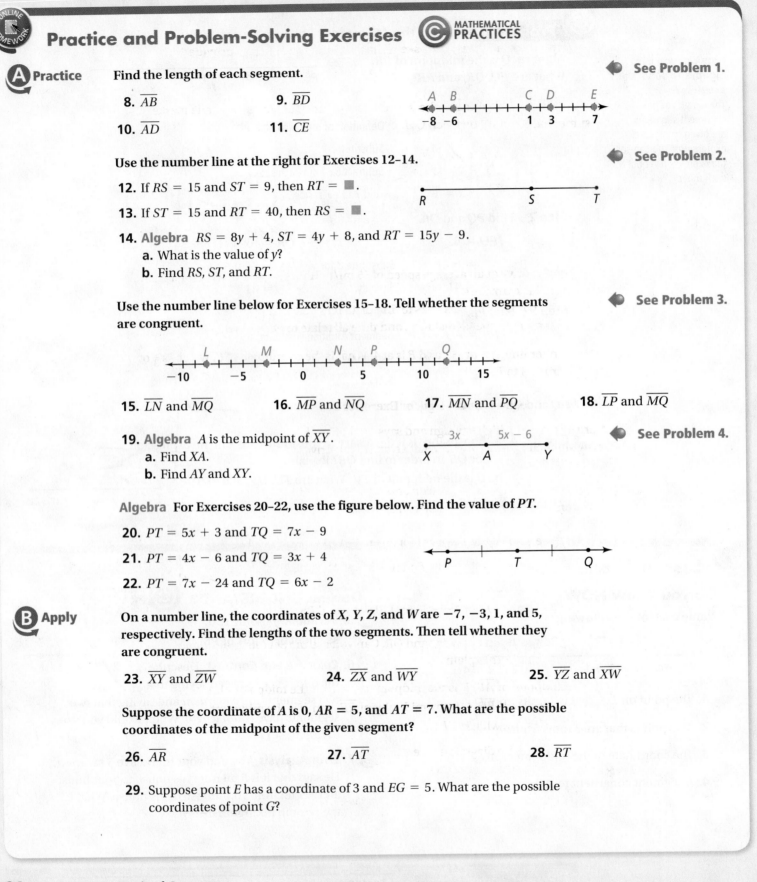

MATHEMATICAL PRACTICES

A Practice

Find the length of each segment.

See Problem 1.

8. $\overline{AB}$

9. $\overline{BD}$

10. $\overline{AD}$

11. $\overline{CE}$

Use the number line at the right for Exercises 12–14.

See Problem 2.

12. If $RS = 15$ and $ST = 9$, then $RT = \blacksquare$.

13. If $ST = 15$ and $RT = 40$, then $RS = \blacksquare$.

14. Algebra $RS = 8y + 4$, $ST = 4y + 8$, and $RT = 15y - 9$.
    a. What is the value of y?
    b. Find RS, ST, and RT.

Use the number line below for Exercises 15–18. Tell whether the segments are congruent.

See Problem 3.

15. $\overline{LN}$ and $\overline{MQ}$ **16.** $\overline{MP}$ and $\overline{NQ}$ **17.** $\overline{MN}$ and $\overline{PQ}$ **18.** $\overline{LP}$ and $\overline{MQ}$

19. Algebra A is the midpoint of $\overline{XY}$.
    a. Find XA.
    b. Find AY and XY.

See Problem 4.

Algebra For Exercises 20–22, use the figure below. Find the value of PT.

20. $PT = 5x + 3$ and $TQ = 7x - 9$

21. $PT = 4x - 6$ and $TQ = 3x + 4$

22. $PT = 7x - 24$ and $TQ = 6x - 2$

B Apply

On a number line, the coordinates of X, Y, Z, and W are -7, -3, 1, and 5, respectively. Find the lengths of the two segments. Then tell whether they are congruent.

23. $\overline{XY}$ and $\overline{ZW}$ **24.** $\overline{ZX}$ and $\overline{WY}$ **25.** $\overline{YZ}$ and $\overline{XW}$

Suppose the coordinate of A is 0, $AR = 5$, and $AT = 7$. What are the possible coordinates of the midpoint of the given segment?

26. $\overline{AR}$ **27.** $\overline{AT}$ **28.** $\overline{RT}$

29. Suppose point E has a coordinate of 3 and $EG = 5$. What are the possible coordinates of point G?

Visualization Without using your ruler, sketch a segment with the given length. Use your ruler to see how well your sketch approximates the length provided.

30. 3 cm   **31.** 3 in.   **32.** 6 in.   **33.** 10 cm   **34.** 65 mm

35. Think About a Plan The numbers labeled on the map of Florida are mile markers. Assume that Route 10 between Quincy and Jacksonville is straight.

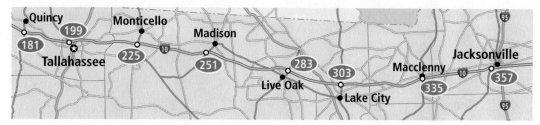

Suppose you drive at an average speed of 55 mi/h. How long will it take to get from Live Oak to Jacksonville?

- How can you use mile markers to find distances between points?
- How do average speed, distance, and time all relate to each other?

36. On a number line, A is at -2 and B is at 4. What is the coordinate of C, which is $\frac{2}{3}$ of the way from A to B?

Error Analysis Use the highway sign for Exercises 37 and 38.

37. A driver reads the highway sign and says, "It's 145 miles from Mitchell to Watertown." What error did the driver make? Explain.

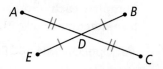

38. Your friend reads the highway sign and says, "It's 71 miles to Watertown." Is your friend correct? Explain.

Algebra Use the diagram at the right for Exercises 39 and 40.

39. If $AD = 12$ and $AC = 4y - 36$, find the value of y. Then find AC and DC.

40. If $ED = x + 4$ and $DB = 3x - 8$, find ED, DB, and EB.

41. Writing Suppose you know PQ and QR. Can you use the Segment Addition Postulate to find PR? Explain.

Challenge

42. C is the midpoint of $\overline{AB}$, D is the midpoint of $\overline{AC}$, E is the midpoint of $\overline{AD}$, F is the midpoint of $\overline{ED}$, G is the midpoint of $\overline{EF}$, and H is the midpoint of $\overline{DB}$. If $DC = 16$, what is GH?

43. a. Algebra Use the diagram at the right. What algebraic expression represents GK?
    b. If $GK = 30$, what are GH and JK?

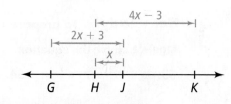

Standardized Test Prep

SAT/ACT

44. Points X, Y, and Z are collinear and Y is between X and Z. Which statement must be true?

    Ⓐ $XY = YZ$ Ⓒ $XY + XZ = YZ$

    Ⓑ $XZ - XY = YZ$ Ⓓ $XZ = XY - YZ$

45. Which is the top view of an orthographic drawing of the figure at the right?

    Ⓕ   Ⓗ

    Ⓖ   Ⓘ

46. Which statement is true based on the diagram?

    Ⓐ $\overline{BC} \cong \overline{CE}$ Ⓒ $AC + BD = AD$

    Ⓑ $BD < CD$ Ⓓ $AC + CD = AD$

Extended Response

47. Make an orthographic drawing of the structure at the right.

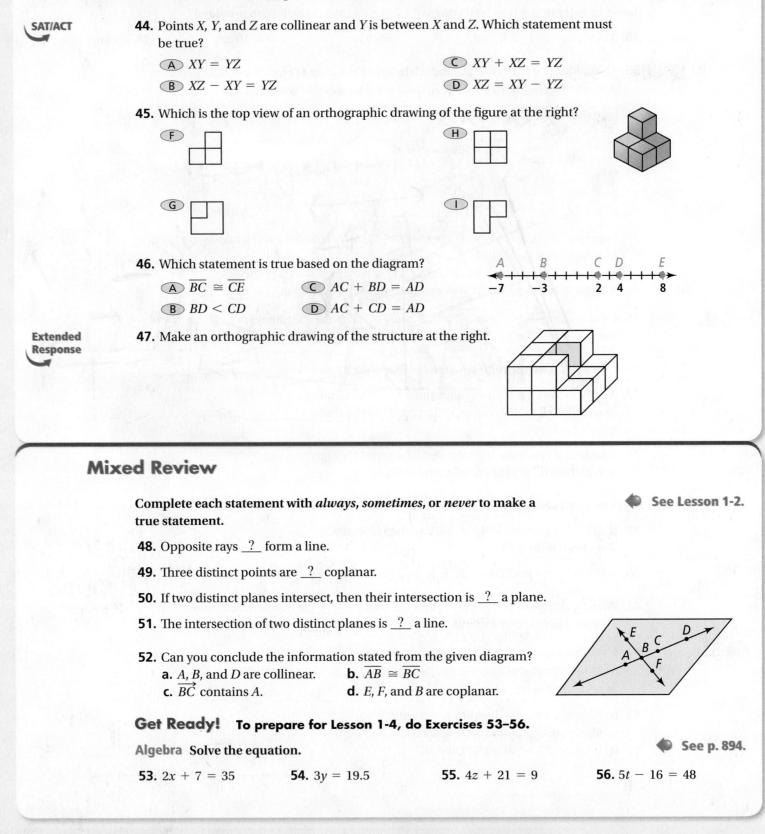

Mixed Review

Complete each statement with *always*, *sometimes*, or *never* to make a true statement.

    ◀ See Lesson 1-2.

48. Opposite rays _？_ form a line.

49. Three distinct points are _？_ coplanar.

50. If two distinct planes intersect, then their intersection is _？_ a plane.

51. The intersection of two distinct planes is _？_ a line.

52. Can you conclude the information stated from the given diagram?

    a. A, B, and D are collinear.   **b.** $\overline{AB} \cong \overline{BC}$

    c. $\overrightarrow{BC}$ contains A.   **d.** E, F, and B are coplanar.

Get Ready!   **To prepare for Lesson 1-4, do Exercises 53–56.**

Algebra Solve the equation.

    ◀ See p. 894.

53. $2x + 7 = 35$ **54.** $3y = 19.5$ **55.** $4z + 21 = 9$ **56.** $5t - 16 = 48$

© Content Standard
G.CO.1 Know precise definitions of angle, circle, perpendicular line, parallel line, and line segment, based on the undefined notions of point, line, distance along a line, and distance around a circular arc.

Objective   To find and compare the measures of angles

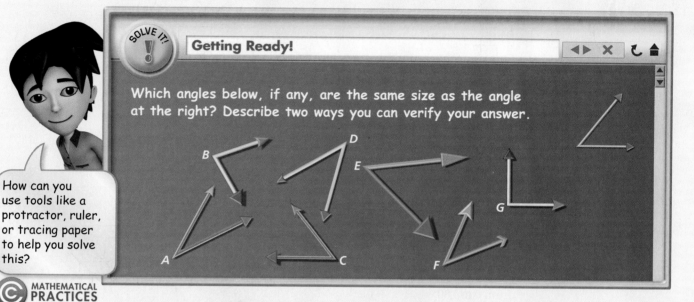

Getting Ready!

Which angles below, if any, are the same size as the angle at the right? Describe two ways you can verify your answer.

How can you use tools like a protractor, ruler, or tracing paper to help you solve this?

MATHEMATICAL PRACTICES

In this lesson, you will learn to describe and measure angles like the ones in the Solve It.

Essential Understanding   You can use number operations to find and compare the measures of angles.

Lesson Vocabulary
- angle
- sides of an angle
- vertex of an angle
- measure of an angle
- acute angle
- right angle
- obtuse angle
- straight angle
- congruent angles

take note

Key Concept   Angle

Definition	**How to Name It**	**Diagram**
An **angle** is formed by two rays with the same endpoint. The rays are the **sides** of the angle. The endpoint is the **vertex** of the angle.	You can name an angle by • its vertex, $\angle A$ • a point on each ray and the vertex, $\angle BAC$ or $\angle CAB$ • a number, $\angle 1$	The sides of the angle are $\overrightarrow{AB}$ and $\overrightarrow{AC}$. The vertex is A.

When you name angles using three points, the vertex must go in the middle.

The *interior* of an angle is the region containing all of the points between the two sides of the angle. The *exterior* of an angle is the region containing all of the points outside of the angle.

exterior

interior

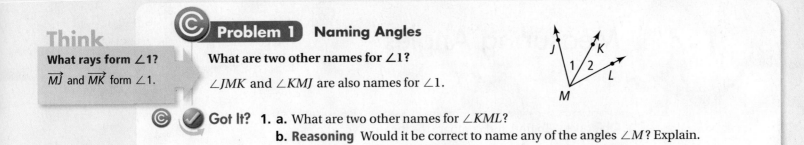

Problem 1 Naming Angles

Think

What rays form $\angle 1$?

$\overrightarrow{MJ}$ and $\overrightarrow{MK}$ form $\angle 1$.

What are two other names for $\angle 1$?

$\angle JMK$ and $\angle KMJ$ are also names for $\angle 1$.

Got It? **1. a.** What are two other names for $\angle KML$?

b. Reasoning Would it be correct to name any of the angles $\angle M$? Explain.

One way to measure the size of an angle is in degrees. To indicate the measure of an angle, write a lowercase *m* in front of the angle symbol. In the diagram, the measure of $\angle A$ is 62. You write this as $m\angle A = 62$. In this book, you will work only with degree measures.

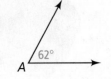

A circle has 360°, so 1 degree is $\frac{1}{360}$ of a circle. A protractor forms half a circle and measures angles from 0° to 180°.

take note

Postulate 1-7 Protractor Postulate

Consider $\overrightarrow{OB}$ and a point A on one side of $\overrightarrow{OB}$. Every ray of the form $\overrightarrow{OA}$ can be paired one to one with a real number from 0 to 180.

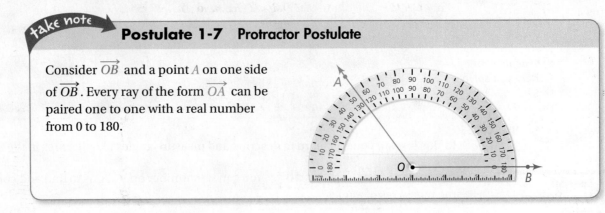

The Protractor Postulate allows you to find the measure of an angle. Consider the diagram below. The **measure** of $\angle COD$ is the absolute value of the difference of the real numbers paired with $\overrightarrow{OC}$ and $\overrightarrow{OD}$. That is, if $\overrightarrow{OC}$ corresponds with c, and $\overrightarrow{OD}$ corresponds with d, then $m\angle COD = |c - d|$.

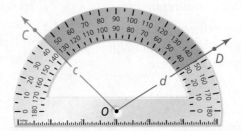

Notice that the Protractor Postulate and the calculation of an angle measure are very similar to the Ruler Postulate and the calculation of a segment length.

You can classify angles according to their measures.

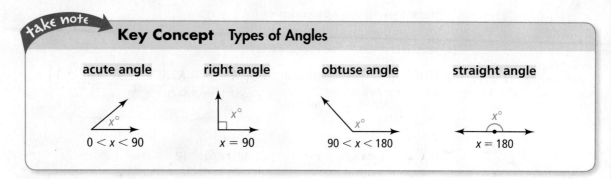

take note

Key Concept   Types of Angles

| acute angle | right angle | obtuse angle | straight angle |

The symbol ⌐ in the diagram above indicates a right angle.

© **Problem 2**   **Measuring and Classifying Angles**

**What are the measures of ∠LKN, ∠JKL, and ∠JKN? Classify each angle
as *acute*, *right*, *obtuse*, or *straight*.**

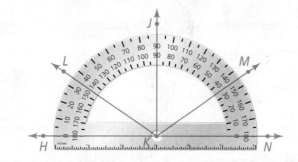

Think

**Do the classifications
make sense?**
Yes. In each case, the
classification agrees
with what you see in the
diagram.

Use the definition of the measure of an angle to calculate each measure.

$m\angle LKN = |145 - 0| = 145$; ∠LKN is obtuse.

$m\angle JKL = |90 - 145| = |-55| = 55$; ∠JKL is acute.

$m\angle JKN = |90 - 0| = 90$; ∠JKN is right.

✓ **Got It?** **2.** What are the measures of ∠LKH, ∠HKN, and ∠MKH? Classify
each angle as *acute*, *right*, *obtuse*, or *straight*.

Angles with the same measure are **congruent angles.** This means
that if $m\angle A = m\angle B$, then $\angle A \cong \angle B$. You can also say that if
$\angle A \cong \angle B$, then $m\angle A = m\angle B$.

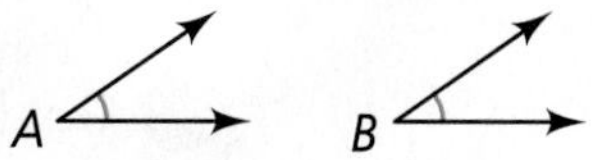

You can mark angles with arcs to show that they are congruent.
If there is more than one set of congruent angles, each set is
marked with the same number of arcs.

$m\angle A = m\angle B$
$\angle A \cong \angle B$

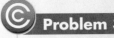

Think

Look at the diagram.
What do the angle
marks tell you?
The angle marks tell
you which angles are
congruent.

© Problem 3 Using Congruent Angles

Sports Synchronized swimmers form angles
with their bodies, as shown in the photo.
If $m\angle GHJ = 90$, what is $m\angle KLM$?

$\angle GHJ \cong \angle KLM$ because they
both have two arcs.
So, $m\angle GHJ = m\angle KLM = 90$.

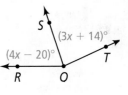

✓ Got It? 3. Use the photo in Problem 3.
If $m\angle ABC = 49$,
what is $m\angle DEF$?

The Angle Addition Postulate is similar to the Segment Addition Postulate.

take note

Postulate 1-8 Angle Addition Postulate

If point B is in the interior of $\angle AOC$,
then $m\angle AOB + m\angle BOC = m\angle AOC$.

Plan

How can you use the
expressions in the
diagram?
The algebraic expressions
represent the measures
of the smaller angles,
so they add up to the
measure of the larger
angle.

© Problem 4 Using the Angle Addition Postulate

Algebra If $m\angle RQT = 155$, what are $m\angle RQS$ and $m\angle TQS$?

$m\angle RQS + m\angle TQS = m\angle RQT$	Angle Addition Postulate
$(4x - 20) + (3x + 14) = 155$	Substitute.
$7x - 6 = 155$	Combine like terms.
$7x = 161$	Add 6 to each side.
$x = 23$	Divide each side by 7.

$m\angle RQS = 4x - 20 = 4(23) - 20 = 92 - 20 = 72$
$m\angle TQS = 3x + 14 = 3(23) + 14 = 69 + 14 = 83$

Substitute 23 for x.

✓ Got It? 4. $\angle DEF$ is a straight angle. What are $m\angle DEC$
and $m\angle CEF$?

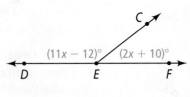

Lesson Check

Do you know HOW?

Use the diagram for Exercises 1–3.

1. What are two other names for ∠1?

2. Algebra If $m\angle ABD = 85$, what is an expression to represent $m\angle ABC$?

3. Classify ∠ABC.

Do you UNDERSTAND? MATHEMATICAL PRACTICES

4. Vocabulary How many sides can two congruent angles share? Explain.

5. Error Analysis Your classmate concludes from the diagram below that ∠JKL ≅ ∠LKM. Is your classmate correct? Explain.

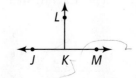

Practice and Problem-Solving Exercises MATHEMATICAL PRACTICES

Practice

Name each shaded angle in three different ways.

See Problem 1.

6.

7.

8.

Use the diagram below. Find the measure of each angle. Then classify the angle as *acute*, *right*, *obtuse*, or *straight*.

See Problem 2.

9. ∠EAF

10. ∠DAF

11. ∠BAE

12. ∠BAC

13. ∠CAE

14. ∠DAE

Draw a figure that fits each description.

15. an obtuse angle, ∠RST

16. an acute angle, ∠GHJ

17. a straight angle, ∠KLM

Use the diagram below. Complete each statement.

See Problem 3.

18. ∠CBJ ≅ ▪

19. ∠FJH ≅ ▪

20. If $m\angle EFD = 75$, then $m\angle JAB = $ ▪.

21. If $m\angle GHF = 130$, then $m\angle JBC = $ ▪.

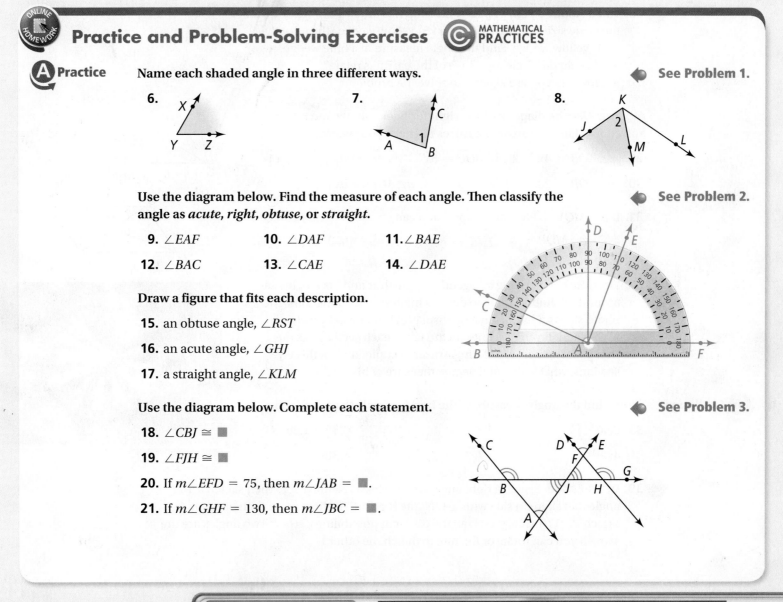

22. If $m\angle ABD = 79$, what are $m\angle ABC$ and $m\angle DBC$?

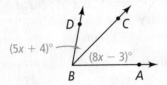

$(5x + 4)°$
$(8x - 3)°$

23. $\angle RQT$ is a straight angle. What are $m\angle RQS$ and $m\angle TQS$?

◀ **See Problem 4.**

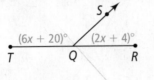

$(6x + 20)°$ $(2x + 4)°$

 Apply

Use a protractor. Measure and classify each angle.

24.   **25.** 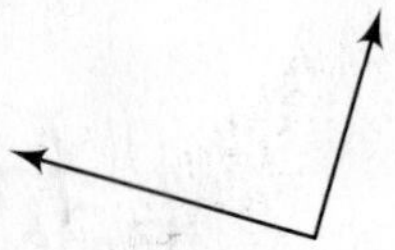   **26.**    **27.**

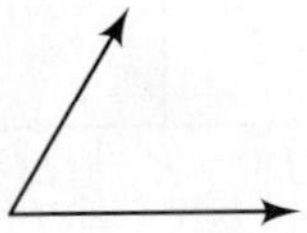

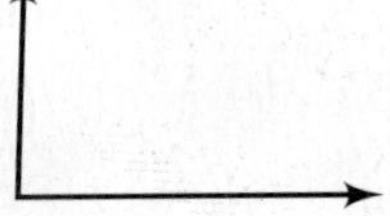

28. Think About a Plan A pair of earrings has blue wedges that are all the same size. One earring has a 25° yellow wedge. The other has a 14° yellow wedge. Find the angle measure of a blue wedge.
  - How do the angle measures of the earrings relate?
  - How can you use algebra to solve the problem?

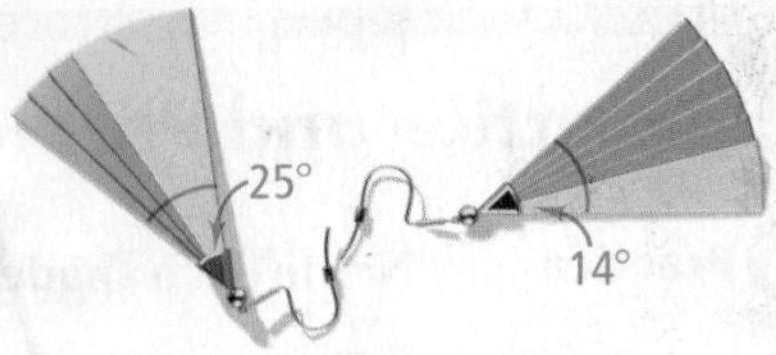

25°

14°

Algebra Use the diagram at the right for Exercises 29 and 30. Solve for x. Find the angle measures to check your work.

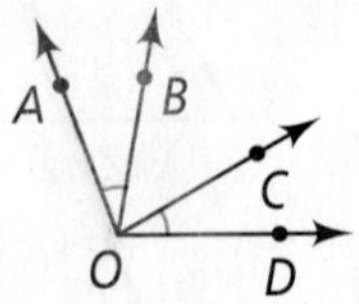

29. $m\angle AOB = 4x - 2, m\angle BOC = 5x + 10, m\angle COD = 2x + 14$

30. $m\angle AOB = 28, m\angle BOC = 3x - 2, m\angle AOD = 6x$

31. If $m\angle MQV = 90$, which expression can you use to find $m\angle VQP$?

  (A) $m\angle MQP - 90$
  (C) $m\angle MQP + 90$
  (B) $90 - m\angle MQV$
  (D) $90 + m\angle VQP$

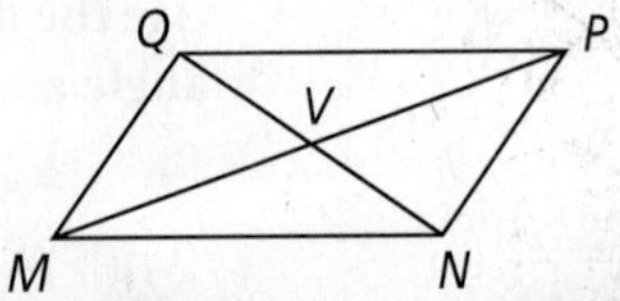

32. Literature According to legend, King Arthur and his knights sat around the Round Table to discuss matters of the kingdom. The photo shows a round table on display at Winchester Castle, in England. From the center of the table, each section has the same degree measure. If King Arthur occupied two of these sections, what is the total degree measure of his section?

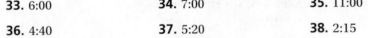

Challenge   **Time** Find the angle measure of the hands of a clock at each time.

33. 6:00   **34.** 7:00   **35.** 11:00

36. 4:40   **37.** 5:20   **38.** 2:15

39. Open-Ended Sketch a right angle with vertex V. Name it $\angle 1$. Then sketch a 135° angle that shares a side with $\angle 1$. Name it $\angle PVB$. Is there more than one way to sketch $\angle PVB$? If so, sketch all the different possibilities. (*Hint:* Two angles are the same if you can rotate or flip one to match the other.)

40. Technology Your classmate constructs an angle. Then he constructs a ray from the vertex of the angle to a point in the interior of the angle. He measures all the angles formed. Then he moves the interior ray as shown below. What postulate do the two pictures support?

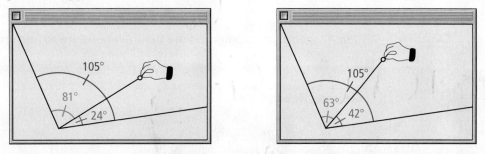

Standardized Test Prep

SAT/ACT

41. Which diagram shows the figure you can fold from the net at the right?

Ⓐ

Ⓑ

Ⓒ

Ⓓ

42. $\overline{XY}$ has endpoints $X = -72$ and $Y = 43$. What is XY?

Ⓕ -115 Ⓖ -29 Ⓗ 29 Ⓘ 115

Short Response

43. Use the figure at the right.
    a. What is the value of x?
    b. What is AC?

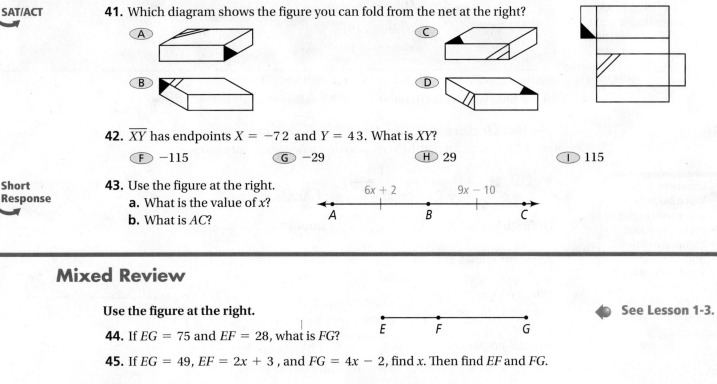

Mixed Review

Use the figure at the right.

See Lesson 1-3.

44. If $EG = 75$ and $EF = 28$, what is FG?

45. If $EG = 49$, $EF = 2x + 3$, and $FG = 4x - 2$, find x. Then find EF and FG.

Get Ready! **To prepare for Lesson 1-5, do Exercises 46–49.**

Algebra **Write and solve an equation to find the number(s).**

See p. 894.

46. Twice a number added to 4 is 28.

47. A number subtracted from 90 is three times that number.

48. The sum of two numbers is 180. One number is 5 times the other.

49. If $m\angle WXZ = 180$ and $m\angle ZXY = 115$, what is the measure of $\angle WXY$?

See Lesson 1-4.

Exploring Angle Pairs

© Content Standard
Prepares for G.CO.1 Know precise definitions of angle, circle, perpendicular line, parallel line, and line segment, based on the undefined notions of point, line, distance along a line, and distance around a circular arc.

Objective To identify special angle pairs and use their relationships to find angle measures

SOLVE IT!

Getting Ready!

The five game pieces at the right form a square to fit back in the box. Two of the shapes are already in place. Where do the remaining pieces go? How do you know? Make a sketch of the completed puzzle.

It might help if you make a sketch of the pieces and cut them out.

MATHEMATICAL PRACTICES

In this lesson, you will learn how to describe different kinds of angle pairs.

Essential Understanding Special angle pairs can help you identify geometric relationships. You can use these angle pairs to find angle measures.

Lesson Vocabulary
- adjacent angles
- vertical angles
- complementary angles
- supplementary angles
- linear pair
- angle bisector

take note

Key Concept   Types of Angle Pairs

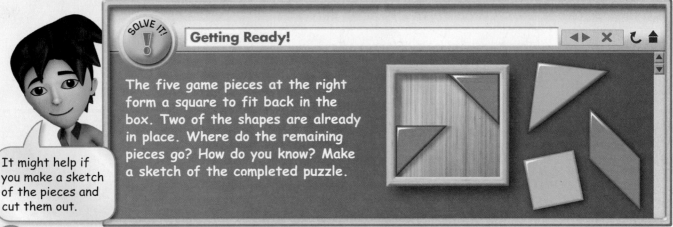

Definition	Example
Adjacent angles are two coplanar angles with a common side, a common vertex, and no common interior points.	$\angle 1$ and $\angle 2$, $\angle 3$ and $\angle 4$
Vertical angles are two angles whose sides are opposite rays.	$\angle 1$ and $\angle 2$, $\angle 3$ and $\angle 4$
Complementary angles are two angles whose measures have a sum of 90. Each angle is called the *complement* of the other.	$\angle 1$ and $\angle 2$, $\angle A$ and $\angle B$
Supplementary angles are two angles whose measures have a sum of 180. Each angle is called the *supplement* of the other.	$\angle 3$ and $\angle 4$, $\angle B$ and $\angle C$

What should you look for in the diagram?
For part (A), check whether the angle pair matches every part of the definition of adjacent angles.

Problem 1 Identifying Angle Pairs

Use the diagram at the right. Is the statement true? Explain.

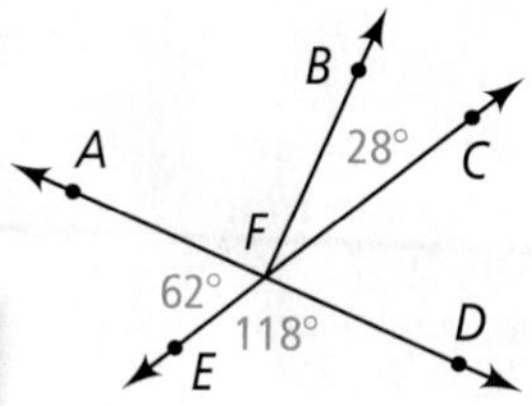

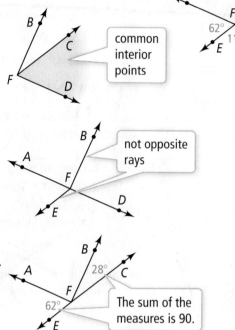

A ∠BFD and ∠CFD are adjacent angles.

No. They have a common side ($\overrightarrow{FD}$) and a common vertex (F), but they also have common interior points. So ∠BFD and ∠CFD are not adjacent.

common interior points

B ∠AFB and ∠EFD are vertical angles.

No. $\overrightarrow{FA}$ and $\overrightarrow{FD}$ are opposite rays, but $\overrightarrow{FE}$ and $\overrightarrow{FB}$ are not. So ∠AFB and ∠EFD are not vertical angles.

not opposite rays

C ∠AFE and ∠BFC are complementary.

Yes. $m\angle AFE + m\angle BFC = 62 + 28 = 90$. The sum of the angle measures is 90, so ∠AFE and ∠BFC are complementary.

The sum of the measures is 90.

Got It? **1.** Use the diagram in Problem 1. Is the statement true? Explain.
    a. ∠AFE and ∠CFD are vertical angles.
    b. ∠BFC and ∠DFE are supplementary.
    c. ∠BFD and ∠AFB are adjacent angles.

take note

Concept Summary Finding Information From a Diagram

There are some relationships you can assume to be true from a diagram that has no marks or measures. There are other relationships you cannot assume directly. For example, you *can* conclude the following from an unmarked diagram.

- Angles are adjacent.
- Angles are adjacent and supplementary.
- Angles are vertical angles.

You *cannot* conclude the following from an unmarked diagram.

- Angles or segments are congruent.
- An angle is a right angle.
- Angles are complementary.

Think

How can you get information from a diagram?
Look for relationships between angles. For example, look for congruent angles and adjacent angles.

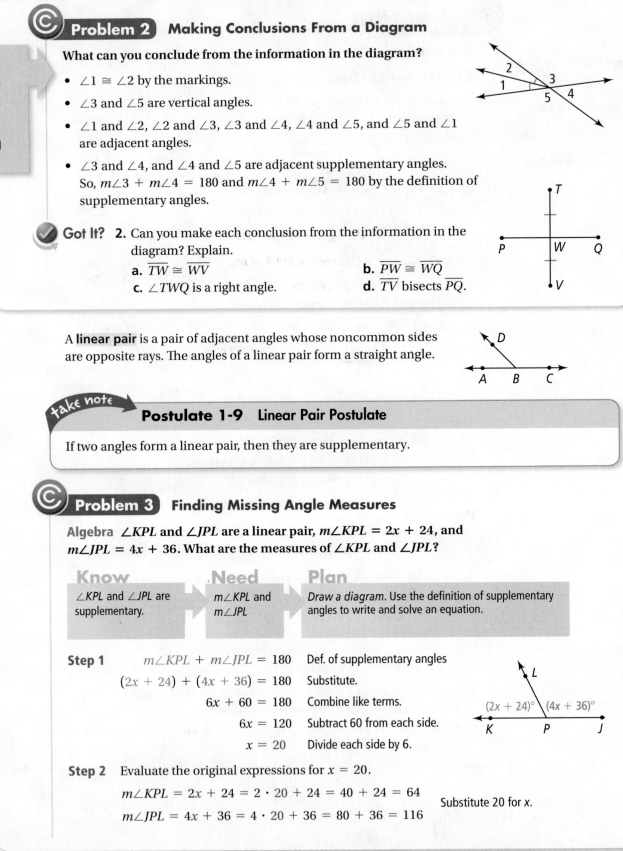

Problem 2 **Making Conclusions From a Diagram**

What can you conclude from the information in the diagram?

- ∠1 ≅ ∠2 by the markings.

- ∠3 and ∠5 are vertical angles.

- ∠1 and ∠2, ∠2 and ∠3, ∠3 and ∠4, ∠4 and ∠5, and ∠5 and ∠1 are adjacent angles.

- ∠3 and ∠4, and ∠4 and ∠5 are adjacent supplementary angles. So, $m\angle 3 + m\angle 4 = 180$ and $m\angle 4 + m\angle 5 = 180$ by the definition of supplementary angles.

Got It? **2.** Can you make each conclusion from the information in the diagram? Explain.

a. $\overline{TW} \cong \overline{WV}$ b. $\overline{PW} \cong \overline{WQ}$

c. ∠TWQ is a right angle.   d. $\overline{TV}$ bisects $\overline{PQ}$.

A **linear pair** is a pair of adjacent angles whose noncommon sides are opposite rays. The angles of a linear pair form a straight angle.

take note

Postulate 1-9 **Linear Pair Postulate**

If two angles form a linear pair, then they are supplementary.

Problem 3 **Finding Missing Angle Measures**

Algebra ∠KPL and ∠JPL are a linear pair, $m\angle KPL = 2x + 24$, and $m\angle JPL = 4x + 36$. What are the measures of ∠KPL and ∠JPL?

Know	Need	Plan
∠KPL and ∠JPL are supplementary.	$m\angle KPL$ and $m\angle JPL$	*Draw a diagram.* Use the definition of supplementary angles to write and solve an equation.

Step 1 $m\angle KPL + m\angle JPL = 180$ Def. of supplementary angles

$(2x + 24) + (4x + 36) = 180$ Substitute.

$6x + 60 = 180$ Combine like terms.

$6x = 120$ Subtract 60 from each side.

$x = 20$ Divide each side by 6.

Step 2   Evaluate the original expressions for $x = 20$.

$m\angle KPL = 2x + 24 = 2 \cdot 20 + 24 = 40 + 24 = 64$

Substitute 20 for x.

$m\angle JPL = 4x + 36 = 4 \cdot 20 + 36 = 80 + 36 = 116$

b. $\angle ADB$ and $\angle BDC$ are a linear pair. $m\angle ADB = 3x + 14$ and $m\angle BDC = 5x - 2$. What are $m\angle ADB$ and $m\angle BDC$?

An **angle bisector** is a ray that divides an angle into two congruent angles. Its endpoint is at the angle vertex. Within the ray, a segment with the same endpoint is also an angle bisector. The ray or segment bisects the angle. In the diagram, $\overrightarrow{AY}$ is the angle bisector of $\angle XAZ$, so $\angle XAY \cong \angle YAZ$.

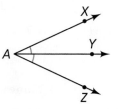

Problem 4 **Using an Angle Bisector to Find Angle Measures**

Multiple Choice $\overrightarrow{AC}$ bisects $\angle DAB$. If $m\angle DAC = 58$, what is $m\angle DAB$?

A. 29   B. 58   C. 87   D. 116

Draw a diagram.

$m\angle CAB = m\angle DAC$	Definition of angle bisector
$= 58$	Substitute.
$m\angle DAB = m\angle CAB + m\angle DAC$	Angle Addition Postulate
$= 58 + 58$	Substitute.
$= 116$	Simplify.

The measure of $\angle DAB$ is 116. The correct choice is D.

Got It? **4.** $\overrightarrow{KM}$ bisects $\angle JKL$. If $m\angle JKL = 72$, what is $m\angle JKM$?

Lesson Check

Do you know HOW?

Name a pair of the following types of angle pairs.

1. vertical angles

2. complementary angles

3. linear pair

4. $\overrightarrow{PB}$ bisects $\angle RPT$ so that $m\angle RPB = x + 2$ and $m\angle TPB = 2x - 6$. What is $m\angle RPT$?

Do you UNDERSTAND? 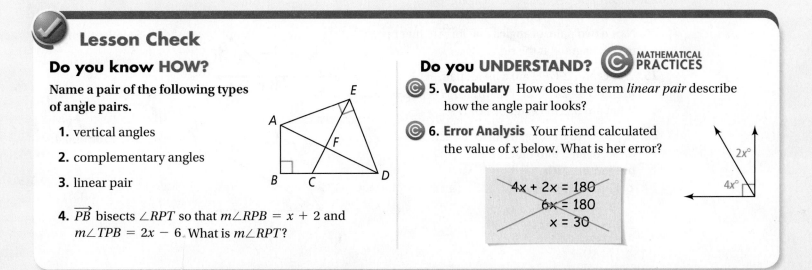 **MATHEMATICAL PRACTICES**

5. Vocabulary How does the term *linear pair* describe how the angle pair looks?

6. Error Analysis Your friend calculated the value of x below. What is her error?

$4x + 2x = 180$
$6x = 180$
$x = 30$

Practice and Problem-Solving Exercises

MATHEMATICAL PRACTICES

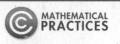

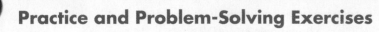

A Practice

Use the diagram at the right. Is each statement true? Explain.

See Problem 1.

7. $\angle 1$ and $\angle 5$ are adjacent angles.

8. $\angle 3$ and $\angle 5$ are vertical angles.

9. $\angle 3$ and $\angle 4$ are complementary.

10. $\angle 1$ and $\angle 2$ are supplementary.

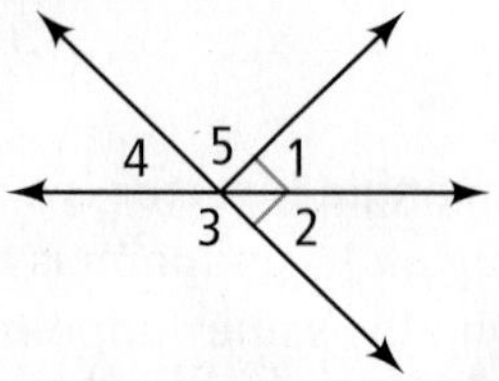

Name an angle or angles in the diagram described by each of the following.

11. supplementary to $\angle AOD$

12. adjacent and congruent to $\angle AOE$

13. supplementary to $\angle EOA$

14. complementary to $\angle EOD$

15. a pair of vertical angles

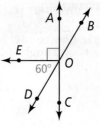

For Exercises 16–23, can you make each conclusion from the information in the diagram? Explain.

See Problem 2.

16. $\angle J \cong \angle D$

17. $\angle JAC \cong \angle DAC$

18. $m\angle JCA = m\angle DCA$

19. $m\angle JCA + m\angle ACD = 180$

20. $\overline{AJ} \cong \overline{AD}$

21. C is the midpoint of $\overline{JD}$.

22. $\angle JAE$ and $\angle EAF$ are adjacent and supplementary.

23. $\angle EAF$ and $\angle JAD$ are vertical angles.

See Problem 3.

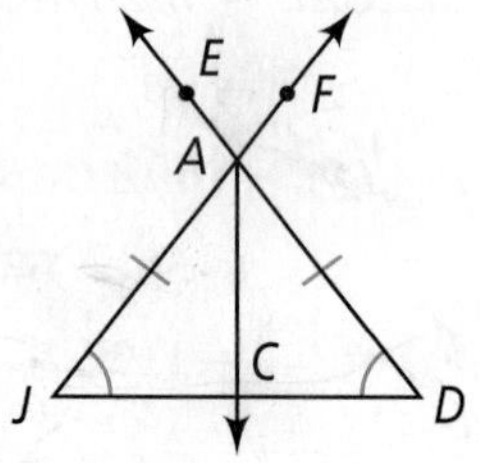

24. Name two pairs of angles that form a linear pair in the diagram at the right.

25. $\angle EFG$ and $\angle GFH$ are a linear pair, $m\angle EFG = 2n + 21$, and $m\angle GFH = 4n + 15$. What are $m\angle EFG$ and $m\angle GFH$?

See Problem 4.

26. Algebra In the diagram, $\overrightarrow{GH}$ bisects $\angle FGI$.

a. Solve for x and find $m\angle FGH$.

b. Find $m\angle HGI$.

c. Find $m\angle FGI$.

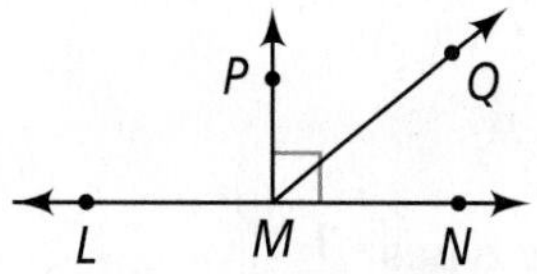

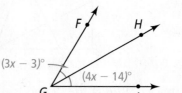

Algebra $\overrightarrow{BD}$ bisects $\angle ABC$. Solve for x and find $m\angle ABC$.

27. $m\angle ABD = 5x$, $m\angle DBC = 3x + 10$

28. $m\angle ABC = 4x - 12$, $m\angle ABD = 24$

29. $m\angle ABD = 4x - 16$, $m\angle CBD = 2x + 6$

30. $m\angle ABD = 3x + 20$, $m\angle CBD = 6x - 16$

Algebra Find the measure of each angle in the angle pair described.

31. Think About a Plan The measure of one angle is twice the measure of its supplement.
- How many angles are there? What is their relationship?
- How can you use algebra, such as using the variable x, to help you?

32. The measure of one angle is 20 less than the measure of its complement.

In the diagram at the right, $m\angle ACB = 65$. Find each of the following.

33. $m\angle ACD$ **34.** $m\angle BCD$

35. $m\angle ECD$ **36.** $m\angle ACE$

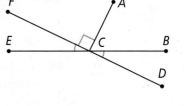

37. Algebra $\angle RQS$ and $\angle TQS$ are a linear pair where $m\angle RQS = 2x + 4$ and $m\angle TQS = 6x + 20$.
- **a.** Solve for x.
- **b.** Find $m\angle RQS$ and $m\angle TQS$.
- **c.** Show how you can check your answer.

38. Writing In the diagram at the right, are $\angle 1$ and $\angle 2$ adjacent? Justify your reasoning.

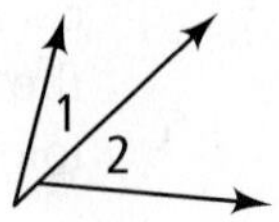

39. Reasoning When $\overrightarrow{BX}$ bisects $\angle ABC$, $\angle ABX \cong \angle CBX$. One student claims there is always a related equation $m\angle ABX = \frac{1}{2} m\angle ABC$. Another student claims the related equation is $2m\angle ABX = m\angle ABC$. Who is correct? Explain.

40. Optics A beam of light and a mirror can be used to study the behavior of light. Light that strikes the mirror is reflected so that the angle of reflection and the angle of incidence are congruent. In the diagram, $\angle ABC$ has a measure of 41.
- **a.** Name the angle of reflection and find its measure.
- **b.** Find $m\angle ABD$.
- **c.** Find $m\angle ABE$ and $m\angle DBF$.

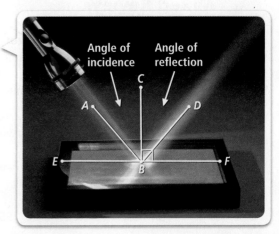

41. Reasoning Describe all situations where vertical angles are also supplementary.

Challenge Name all of the angle(s) in the diagram described by the following.

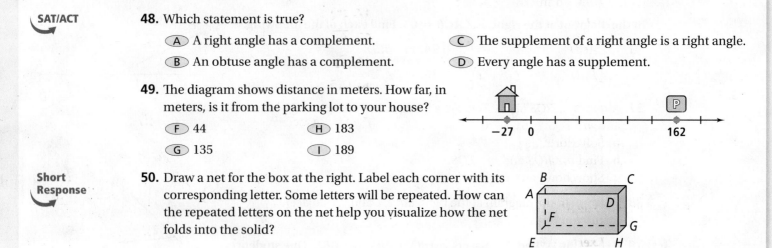

42. supplementary to ∠*JQM*

43. adjacent and congruent to ∠*KMQ*

44. a linear pair with ∠*LMQ*

45. complementary to ∠*NMR*

46. Coordinate Geometry The *x*- and *y*-axes of the coordinate plane form four right angles. The interior of each of the right angles is a quadrant of the coordinate plane. What is the equation for the line that contains the angle bisector of Quadrants I and III?

47. $\overrightarrow{XC}$ bisects ∠*AXB*, $\overrightarrow{XD}$ bisects ∠*AXC*, $\overrightarrow{XE}$ bisects ∠*AXD*, $\overrightarrow{XF}$ bisects ∠*EXD*, $\overrightarrow{XG}$ bisects ∠*EXF*, and $\overrightarrow{XH}$ bisects ∠*DXB*. If $m∠DXC = 16$, find $m∠GXH$.

Standardized Test Prep

SAT/ACT

48. Which statement is true?

  Ⓐ A right angle has a complement.

  Ⓒ The supplement of a right angle is a right angle.

  Ⓑ An obtuse angle has a complement.

  Ⓓ Every angle has a supplement.

49. The diagram shows distance in meters. How far, in meters, is it from the parking lot to your house?

  Ⓕ 44

  Ⓗ 183

  Ⓖ 135

  Ⓘ 189

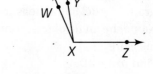

Short Response

50. Draw a net for the box at the right. Label each corner with its corresponding letter. Some letters will be repeated. How can the repeated letters on the net help you visualize how the net folds into the solid?

Mixed Review

Use the diagram at the right.

                                        ◗ See Lesson 1-4.

51. What is the acute angle?

52. What are the obtuse angles?

53. If $m∠WXZ = 150$, $m∠WXY = 8x - 1$, and $m∠ZXY = 17x + 26$, what is $m∠WXY$?

Get Ready! To prepare for Lesson 1-6, do Exercises 54–59.

Sketch each figure.

54. $\overrightarrow{GH}$ **55.** $\overline{CD}$ **56.** $\overleftrightarrow{AB}$ ◗ See Lesson 1-2.

57. acute ∠*ABC*   **58.** right ∠*PST*   **59.** straight ∠*XYZ*   ◗ See Lesson 1-4.

1 Mid-Chapter Quiz

MathXL® for School
Go to PowerGeometry.com

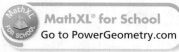

Do you know HOW?

Draw a net for each figure.

1.

2.

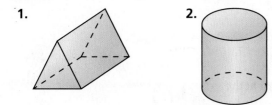

Determine whether the given points are coplanar. If *yes*, name the plane. If *no*, explain.

3. A, E, F, and B

4. D, C, E, and F

5. H, G, F, and B

6. A, E, B, and C

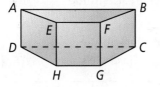

7. Use the figure from Exercises 3–6. Name the intersection of each pair of planes.

a. plane $AEFB$ and plane $CBFG$

b. plane $EFGH$ and plane $AEHD$

Use the figure below for Exercises 8–15.

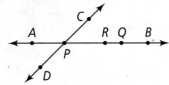

8. Give two other names for $\overleftrightarrow{AB}$.

9. Give two other names for $\overrightarrow{PR}$.

10. Give two other names for $\angle CPR$.

11. Name three collinear points.

12. Name two opposite rays.

13. Name three segments.

14. Name two angles that form a linear pair.

15. Name a pair of vertical angles.

16. a. Algebra Find the value of x in the diagram below.

b. Classify $\angle ABC$ and $\angle CBD$ as *acute*, *right*, or *obtuse*.

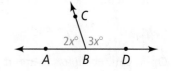

Find the length of each segment.

17. $\overline{PQ}$

18. $\overline{RS}$

19. $\overline{ST}$

20. $\overline{QT}$

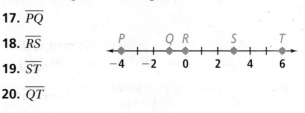

Use the figure below for Exercises 21–23.

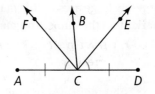

21. Algebra If $AC = 4x + 5$ and $DC = 3x + 8$, find AD.

22. If $m\angle FCD = 130$ and $m\angle BCD = 95$, find $m\angle FCB$.

23. If $m\angle FCA = 50$, find $m\angle FCE$.

Do you UNDERSTAND?

24. Error Analysis Suppose $PQ = QR$. Your friend says that Q is always the midpoint of $\overline{PR}$. Is he correct? Explain.

25. Reasoning Determine whether the following situation is possible. Explain your reasoning. Include a sketch.

Collinear points C, F, and G lie in plane M. $\overleftrightarrow{AB}$ intersects plane M at C. $\overleftrightarrow{AB}$ and $\overleftrightarrow{GF}$ do not intersect.

Concept Byte

Use With Lesson 1-6

A C T I V I T Y

Compass Designs

© **Content Standard**
Prepares for G.CO.12 Make formal geometric constructions with a variety of tools and methods (. . . dynamic geometric software . . .)

In Lesson 1-6, you will use a compass to construct geometric figures. You can construct figures to show geometric relationships, to suggest new relationships, or simply to make interesting geometric designs.

Activity

Step 1 Open your compass to about 2 in. Make a circle and mark the point at the center of the circle. Keep the opening of your compass fixed. Place the compass point on the circle. With the pencil end, make a small arc to intersect the circle.

Step 2 Place the compass point on the circle at the arc. Mark another arc. Continue around the circle this way to draw four more arcs—six in all.

Step 3 Place your compass point on an arc you marked on the circle. Place the pencil end at the next arc. Draw a large arc that passes through the circle's center and continues to another point on the circle.

Step 4 Draw six large arcs in this manner, each centered at one of the six points marked on the circle. You may choose to color your design.

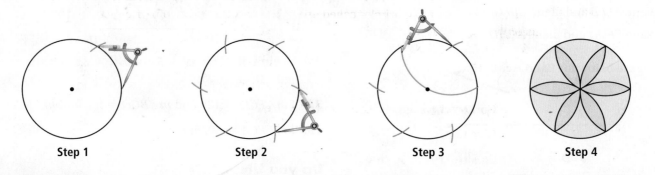

Step 1   Step 2   Step 3   Step 4

Exercises

1. In Step 2, did your sixth mark on the circle land precisely on the point where you first placed your compass on the circle?
   a. Survey the class to find out how many did.
   b. Explain why your sixth mark may not have landed on your starting point.

2. Extend your design by using one of the six points on the circle as the center for a new circle. Repeat Steps 1–4 with this circle. Repeat several times to make interlocking circles

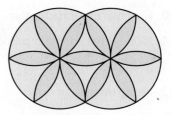

Basic Constructions

Content Standards
G.CO.12 Make formal geometric constructions with a variety of tools and methods (compass and straightedge . . .).
Also G.CO.1

Objective To make basic constructions using a straightedge and a compass

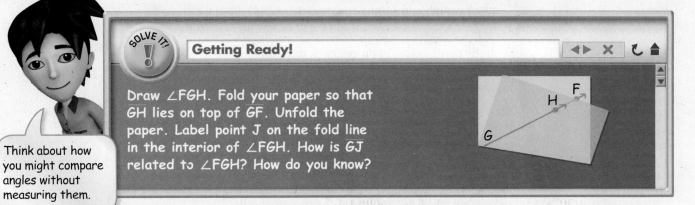

SOLVE IT!

Getting Ready!

Draw ∠FGH. Fold your paper so that $\overrightarrow{GH}$ lies on top of $\overrightarrow{GF}$. Unfold the paper. Label point J on the fold line in the interior of ∠FGH. How is $\overrightarrow{GJ}$ related to ∠FGH? How do you know?

Think about how you might compare angles without measuring them.

MATHEMATICAL PRACTICES

In this lesson, you will learn another way to construct figures like the one above.

Lesson Vocabulary
• straightedge
• compass
• construction
• perpendicular lines
• perpendicular bisector

Essential Understanding You can use special geometric tools to make a figure that is congruent to an original figure without measuring. This method is more accurate than sketching and drawing.

A **straightedge** is a ruler with no markings on it. A **compass** is a geometric tool used to draw circles and parts of circles called *arcs*. A **construction** is a geometric figure drawn using a straightedge and a compass.

Problem 1 Constructing Congruent Segments

Construct a segment congruent to a given segment.

Given: $\overline{AB}$

Construct: $\overline{CD}$ so that $\overline{CD} \cong \overline{AB}$

Think

Why must the compass setting stay the same?
Using the same compass setting keeps segments congruent. It guarantees that the lengths of $\overline{AB}$ and $\overline{CD}$ are exactly the same.

Step 1 Draw a ray with endpoint C.

Step 2 Open the compass to the length of $\overline{AB}$.

Step 3 With the same compass setting, put the compass point on point C. Draw an arc that intersects the ray. Label the point of intersection D.

$\overline{CD} \cong \overline{AB}$

✓ **Got It?** **1.** Use a straightedge to draw $\overline{XY}$. Then construct $\overline{RS}$ so that $RS = 2XY$.

Problem 2 Constructing Congruent Angles

Construct an angle congruent to a given angle.

Given: $\angle A$

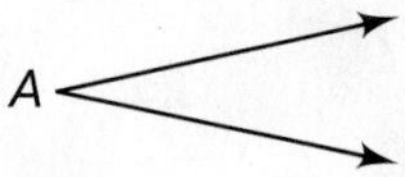

Construct: $\angle S$ **so that** $\angle S \cong \angle A$

Step 1
Draw a ray with endpoint S.

Step 2
With the compass point on vertex A, draw an arc that intersects the sides of $\angle A$. Label the points of intersection B and C.

Step 3
With the same compass setting, put the compass point on point S. Draw an arc and label its point of intersection with the ray as R.

Step 4
Open the compass to the length BC. Keeping the same compass setting, put the compass point on R. Draw an arc to locate point T.

Step 5
Draw $\overrightarrow{ST}$.

$\angle S \cong \angle A$

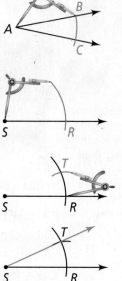

Think

Why do you need points like B and C?
B and C are reference points on the original angle. You can construct a congruent angle by locating corresponding points R and T on your new angle.

Got It? 2. a. Construct $\angle F$ so that $m\angle F = 2m\angle B$.

b. Reasoning How is constructing a congruent angle similar to constructing a congruent segment?

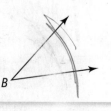

Dynamic Activity
Constructing Congruent Segments and Angles

Perpendicular lines are two lines that intersect to form right angles. The symbol $\perp$ means "is perpendicular to." In the diagram at the right, $\overleftrightarrow{AB} \perp \overleftrightarrow{CD}$ and $\overleftrightarrow{CD} \perp \overleftrightarrow{AB}$.

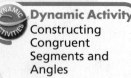

A **perpendicular bisector** of a segment is a line, segment, or ray that is perpendicular to the segment at its midpoint. In the diagram at the right, $\overleftrightarrow{EF}$ is the perpendicular bisector of $\overline{GH}$. The perpendicular bisector bisects the segment into two congruent segments. The construction in Problem 3 will show you how this works. You will justify the steps for this construction in Chapter 4, as well as for the other constructions in this lesson.

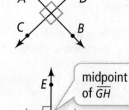
midpoint of $\overline{GH}$

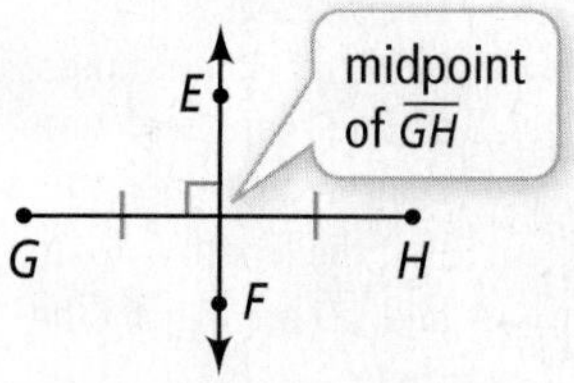

Problem 3   **Constructing the Perpendicular Bisector**

Construct the perpendicular bisector of a segment.

Given: $\overline{AB}$

Construct: $\overleftrightarrow{XY}$ so that $\overleftrightarrow{XY}$ is the perpendicular bisector of $\overline{AB}$

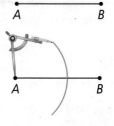

Step 1

Put the compass point on point A and draw a long arc as shown. Be sure the opening is greater than $\frac{1}{2}AB$.

Think

Why must the compass opening be greater than $\frac{1}{2}AB$?

If the opening is less than $\frac{1}{2}AB$, the two arcs will not intersect in Step 2.

Step 2

With the same compass setting, put the compass point on point B and draw another long arc. Label the points where the two arcs intersect as X and Y.

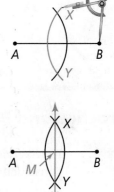

Step 3

Draw $\overleftrightarrow{XY}$. Label the point of intersection of $\overline{AB}$ and $\overleftrightarrow{XY}$ as M, the midpoint of $\overline{AB}$.

$\overleftrightarrow{XY} \perp \overline{AB}$ at midpoint M, so $\overleftrightarrow{XY}$ is the perpendicular bisector of $\overline{AB}$.

Got It?   **3.** Draw $\overline{ST}$. Construct its perpendicular bisector.

Problem 4   **Constructing the Angle Bisector**

Construct the bisector of an angle.

Given: $\angle A$

Construct: $\overrightarrow{AD}$, the bisector of $\angle A$

Step 1

Put the compass point on vertex A. Draw an arc that intersects the sides of $\angle A$. Label the points of intersection B and C.

Think

Why must the arcs intersect?

The arcs need to intersect so that you have a point through which to draw a ray.

Step 2

Put the compass point on point C and draw an arc. With the same compass setting, draw an arc using point B. Be sure the arcs intersect. Label the point where the two arcs intersect as D.

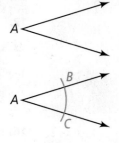

Step 3

Draw $\overrightarrow{AD}$.

$\overrightarrow{AD}$ is the bisector of $\angle CAB$.

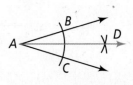

Got It?   **4.** Draw obtuse $\angle XYZ$. Then construct its bisector $\overrightarrow{YP}$.

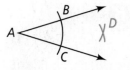

Lesson Check

Do you know HOW?

For Exercises 1 and 2, draw $\overline{PQ}$. Use your drawing as the original figure for each construction.

P ●————————————● Q

1. Construct a segment congruent to $\overline{PQ}$.

2. Construct the perpendicular bisector of $\overline{PQ}$.

3. Draw an obtuse $\angle JKL$. Construct its bisector.

Do you UNDERSTAND? © MATHEMATICAL PRACTICES

© 4. **Vocabulary** What two tools do you use to make constructions?

© 5. **Compare and Contrast** Describe the difference in accuracy between sketching a figure, drawing a figure with a ruler and protractor, and constructing a figure. Explain.

© 6. **Error Analysis** Your friend constructs $\overleftrightarrow{XY}$ so that it is perpendicular to and contains the midpoint of $\overline{AB}$. He claims that $\overline{AB}$ is the perpendicular bisector of $\overleftrightarrow{XY}$. What is his error?

Practice and Problem-Solving Exercises © MATHEMATICAL PRACTICES

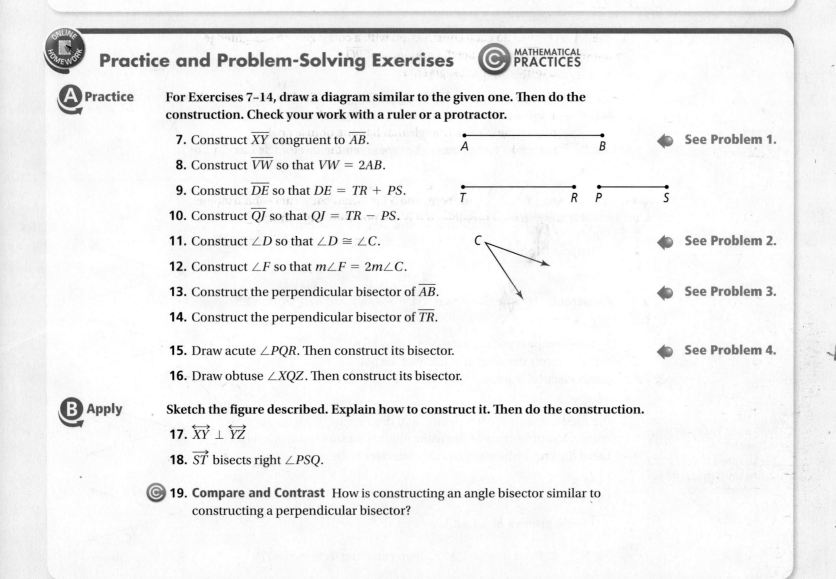

A Practice

For Exercises 7–14, draw a diagram similar to the given one. Then do the construction. Check your work with a ruler or a protractor.

7. Construct $\overline{XY}$ congruent to $\overline{AB}$.

8. Construct $\overline{VW}$ so that $VW = 2AB$.

9. Construct $\overline{DE}$ so that $DE = TR + PS$.

10. Construct $\overline{QJ}$ so that $QJ = TR - PS$.

11. Construct $\angle D$ so that $\angle D \cong \angle C$.

12. Construct $\angle F$ so that $m\angle F = 2m\angle C$.

13. Construct the perpendicular bisector of $\overline{AB}$.

14. Construct the perpendicular bisector of $\overline{TR}$.

15. Draw acute $\angle PQR$. Then construct its bisector.

16. Draw obtuse $\angle XQZ$. Then construct its bisector.

◀ See Problem 1.

◀ See Problem 2.

◀ See Problem 3.

◀ See Problem 4.

B Apply

Sketch the figure described. Explain how to construct it. Then do the construction.

17. $\overleftrightarrow{XY} \perp \overleftrightarrow{YZ}$

18. $\overrightarrow{ST}$ bisects right $\angle PSQ$.

© 19. **Compare and Contrast** How is constructing an angle bisector similar to constructing a perpendicular bisector?

20. Think About a Plan Draw an $\angle A$. Construct an angle whose measure is $\frac{1}{4}m\angle A$.
- How is the angle you need to construct related to the angle bisector of $\angle A$?
- How can you use previous constructions to help you?

21. Answer the questions about a segment in a plane. Explain each answer.
- **a.** How many midpoints does the segment have?
- **b.** How many bisectors does it have?
- **c.** How many lines in the plane are its perpendicular bisectors?
- **d.** How many lines in space are its perpendicular bisectors?

For Exercises 22–24, copy $\angle 1$ and $\angle 2$. Construct each angle described.

22. $\angle B; m\angle B = m\angle 1 + m\angle 2$

23. $\angle C; m\angle C = m\angle 1 - m\angle 2$

24. $\angle D; m\angle D = 2m\angle 2$

25. Writing Explain how to do each construction with a compass and straightedge.
- **a.** Draw a segment $\overline{PQ}$. Construct the midpoint of $\overline{PQ}$.
- **b.** Divide $\overline{PQ}$ into four congruent segments.

26. a. Draw a large triangle with three acute angles. Construct the bisectors of the three angles. What appears to be true about the three angle bisectors?
- **b.** Repeat the constructions with a triangle that has one obtuse angle.
- **c. Make a Conjecture** What appears to be true about the three angle bisectors of any triangle?

Use a ruler to draw segments of 2 cm, 4 cm, and 5 cm. Then construct each triangle with the given side measures, if possible. If it is not possible, explain why not.

27. 4 cm, 4 cm, and 5 cm

28. 2 cm, 5 cm, and 5 cm

29. 2 cm, 2 cm, and 5 cm

30. 2 cm, 2 cm, and 4 cm

31. a. Draw a segment, $\overline{XY}$. Construct a triangle with sides congruent to $\overline{XY}$.
- **b.** Measure the angles of the triangle.
- **c. Writing** Describe how to construct a 60° angle using what you know. Then describe how to construct a 30° angle.

32. Which steps best describe how to construct the pattern at the right?

- (A) Use a straightedge to draw the segment and then a compass to draw five half circles.

- (B) Use a straightedge to draw the segment and then a compass to draw six half circles.

- (C) Use a compass to draw five half circles and then a straightedge to join their ends.

- (D) Use a compass to draw six half circles and then a straightedge to join their ends.

33. Study the figures. Complete the definition of a line perpendicular to a plane: A line is perpendicular to a plane if it is __?__ to every line in the plane that __?__ .

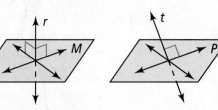

Line $r \perp$ plane M.

Line t is not $\perp$ plane P.

34. a. Use your compass to draw a circle. Locate three points A, B, and C on the circle.

b. Draw $\overline{AB}$ and $\overline{BC}$. Then construct the perpendicular bisectors of $\overline{AB}$ and $\overline{BC}$.

c. Reasoning Label the intersection of the two perpendicular bisectors as point O. What do you think is true about point O?

35. Two triangles are *congruent* if each side and each angle of one triangle is congruent to a side or angle of the other triangle. In Chapter 4, you will learn that if each side of one triangle is congruent to a side of the other triangle, then you can conclude that the triangles are congruent without finding the angles. Explain how you can use congruent triangles to justify the angle bisector construction.

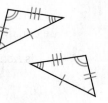

Standardized Test Prep

36. What must you do to construct the midpoint of a segment?

Ⓐ Measure half its length.　　　　Ⓒ Measure twice its length.

Ⓑ Construct an angle bisector.　　Ⓓ Construct a perpendicular bisector.

37. Given the diagram at the right, what is NOT a reasonable name for the angle?

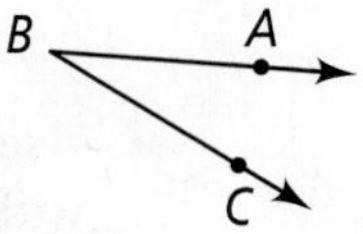

Ⓕ $\angle ABC$　　　　　　　　Ⓗ $\angle CBA$

Ⓖ $\angle B$　　　　　　　　　Ⓘ $\angle ACB$

38. M is the midpoint of $\overline{XY}$. Find the value of x. Show your work.

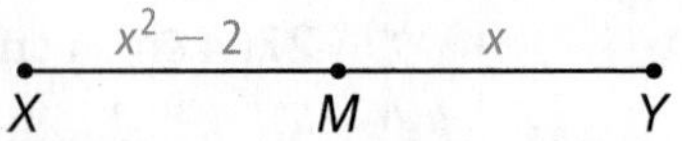

$$\overset{\;\;\;\;\;\;\;\;\;\;\;\;\;\;x^2 - 2 \;\;\;\;\;\;\;\;\;\;\;\;\;\; x}{\underset{X \;\;\;\;\;\;\;\;\;\;\;\;\;\;\;\; M \;\;\;\;\;\;\;\;\;\;\;\;\;\; Y}{\bullet \underline{\;\;\;\;\;\;\;\;\;\;\;\;\;\;\;\;} \bullet \underline{\;\;\;\;\;\;\;\;\;\;\;\;\;\;\;\;} \bullet}}$$

Mixed Review

39. $\angle DEF$ is the supplement of $\angle DEG$ with $m\angle DEG = 64$. What is $m\angle DEF$?

◀ **See Lesson 1-5.**

40. $m\angle TUV = 100$ and $m\angle VUW = 80$. Are $\angle TUV$ and $\angle VUW$ a linear pair? Explain.

Find the length of each segment.

◀ **See Lesson 1-3.**

41. $\overline{AC}$　　　　　　**42.** $\overline{AD}$

43. $\overline{CD}$　　　　　　**44.** $\overline{BC}$

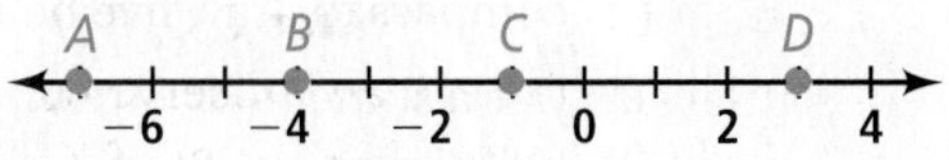

Get Ready!　**To prepare for Lesson 1-7, do Exercises 45–47.**

Algebra Evaluate each expression for $a = 6$ and $b = -8$.

◀ **See p. 890.**

45. $(a - b)^2$　　　　**46.** $\sqrt{a^2 + b^2}$　　　　**47.** $\dfrac{a + b}{2}$

Exploring Constructions

© **Content Standard**
G.CO.12 Make formal geometric constructions with a variety of tools and methods (compass and straightedge . . .).

© **MATHEMATICAL PRACTICES**

You can use Draw tools or Construct tools in geometry software to make points, lines, and planes. A figure made by Draw has no constraints. When you manipulate, or try to change, a figure made by Draw, it moves or changes size freely. A figure made by Construct is related to an existing object. When you manipulate the existing object, the constructed object moves or resizes accordingly.

In this Activity, you will explore the difference between Draw and Construct.

Activity

Draw $\overline{AB}$ and Construct the perpendicular bisector $\overleftrightarrow{DC}$. Then Draw $\overline{EF}$ and Construct G, any point on $\overline{EF}$. Draw $\overleftrightarrow{HG}$.

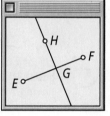
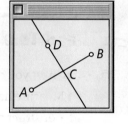

1. Find EG, GF, and $m\angle HGF$. Try to drag G so that $EG = GF$. Try to drag H so that $m\angle HGF = 90$. Were you able to draw the perpendicular bisector of $\overline{EF}$? Explain.

2. Drag A and B. Observe AC, CB, and $m\angle DCB$. Is $\overleftrightarrow{DC}$ always the perpendicular bisector of $\overline{AB}$ no matter how you manipulate the figure?

3. Drag E and F. Observe EG, GF, and $m\angle HGF$. How is the relationship between $\overleftrightarrow{EF}$ and $\overleftrightarrow{HG}$ different from the relationship between $\overline{AB}$ and $\overleftrightarrow{DC}$?

4. Write a description of the general difference between Draw and Construct. Then use your description to explain why the relationship between $\overleftrightarrow{EF}$ and $\overleftrightarrow{HG}$ differs from the relationship between $\overline{AB}$ and $\overleftrightarrow{DC}$.

Exercises

5. **a.** Draw $\angle NOP$. Draw $\overrightarrow{OQ}$ in the interior of $\angle NOP$. Drag Q until $m\angle NOQ = m\angle QOP$.

   b. Manipulate the figure and observe the different angle measures. Is $\overrightarrow{OQ}$ always the angle bisector of $\angle NOP$?

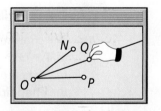

6. **a.** Draw $\angle JKL$.

   b. Construct its angle bisector, $\overrightarrow{KM}$.

   c. Manipulate the figure and observe the different angle measures. Is $\overrightarrow{KM}$ always the angle bisector of $\angle JKL$?

   d. How can you manipulate the figure on the screen so that it shows a right angle? Justify your answer.

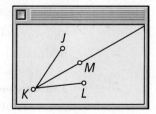

1-7 Midpoint and Distance in the Coordinate Plane

© Content Standards

Prepares for G.GPE.4 Use coordinates to prove simple geometric theorems algebraically.

Prepares for G.GPE.7 Use coordinates to compute perimeters of polygons and areas of triangles and rectangles.

Also G.GPE.6

Objectives To find the midpoint of a segment
To find the distance between two points in the coordinate plane

SOLVE IT!

Getting Ready!

In a video game, two ancient structures shoot light beams toward each other to form a time portal. The portal forms exactly halfway between the two structures. Your character is on the grid shown as a blue dot. How do you direct your character to the portal? Explain how you found your answer.

LEFT [   ]   RIGHT [   ]
UP [   ]   DOWN [   ]

Try drawing the situation on graph paper if you are having trouble visualizing it.

© MATHEMATICAL PRACTICES In this lesson, you will learn how to find midpoints and distance on a grid like the one in the Solve It.

Dynamic Activity
Finding Distance

Essential Understanding You can use formulas to find the midpoint and length of any segment in the coordinate plane.

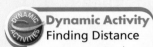

Key Concept Midpoint Formulas

Description	Formula	Diagram
On a Number Line The coordinate of the midpoint is the *average* or *mean* of the coordinates of the endpoints.	The coordinate of the midpoint M of $\overline{AB}$ is $\frac{a + b}{2}$.	
In the Coordinate Plane The coordinates of the midpoint are the average of the *x*-coordinates and the average of the *y*-coordinates of the endpoints.	Given $\overline{AB}$ where $A(x_1, y_1)$ and $B(x_2, y_2)$, the coordinates of the midpoint of $\overline{AB}$ are $M\left(\frac{x_1 + x_2}{2}, \frac{y_1 + y_2}{2}\right)$.	

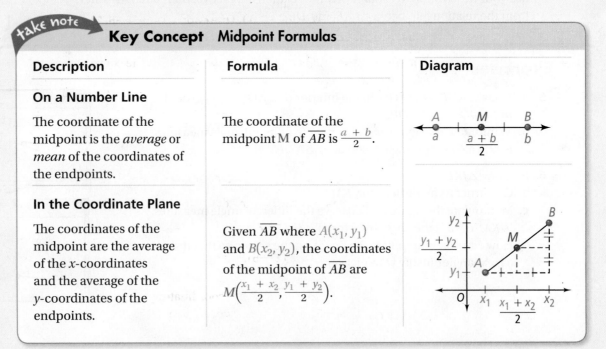

© **Problem 1** **Finding the Midpoint**

A $\overline{AB}$ has endpoints at -4 and 9. What is the coordinate of its midpoint?

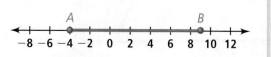

Let $a = -4$ and $b = 9$.

$$M = \frac{a + b}{2} = \frac{-4 + 9}{2} = \frac{5}{2} = 2.5$$

The coordinate of the midpoint of $\overline{AB}$ is 2.5.

B $\overline{EF}$ has endpoints $E(7, 5)$ and $F(2, -4)$. What are the coordinates of its midpoint M?

Let $E(7, 5)$ be (x_1, y_1) and $F(2, -4)$ be (x_2, y_2).

$$x\text{-coordinate of } M = \frac{x_1 + x_2}{2} = \frac{7 + 2}{2} = \frac{9}{2} = 4.5$$

$$y\text{-coordinate of } M = \frac{y_1 + y_2}{2} = \frac{5 + (-4)}{2} = \frac{1}{2} = 0.5$$

The coordinates of the midpoint of $\overline{EF}$ are $M(4.5, 0.5)$.

✓ **Got It?** **1. a.** $\overline{JK}$ has endpoints at -12 and 4 on a number line. What is the coordinate of its midpoint?

   b. What is the midpoint of $\overline{RS}$ with endpoints $R(5, -10)$ and $S(3, 6)$?

When you know the midpoint and an endpoint of a segment, you can use the Midpoint Formula to find the other endpoint.

© **Problem 2** **Finding an Endpoint**

The midpoint of $\overline{CD}$ is $M(-2, 1)$. One endpoint is $C(-5, 7)$. What are the coordinates of the other endpoint D?

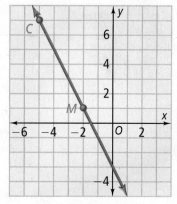

Let $M(-2, 1)$ be (x, y) and $C(-5, 7)$ be (x_1, y_1). Let the coordinates of D be (x_2, y_2).

$$(-2, 1) = \left(\frac{-5 + x_2}{2}, \frac{7 + y_2}{2} \right)$$

x		y
$-2 = \frac{-5 + x_2}{2}$	Use the Midpoint Formula.	$1 = \frac{7 + y_2}{2}$
$-4 = -5 + x_2$	Multiply each side by 2.	$2 = 7 + y_2$
$1 = x_2$	Simplify.	$-5 = y_2$

The coordinates of D are $(1, -5)$.

✓ **Got It?** **2.** The midpoint of $\overline{AB}$ has coordinates $(4, -9)$. Endpoint A has coordinates $(-3, -5)$. What are the coordinates of B?

In Lesson 1-3, you learned how to find the distance between two points on a number line. To find the distance between two points in a coordinate plane, you can use the Distance Formula.

Key Concept   Distance Formula

The distance between two points $A(x_1, y_1)$ and $B(x_2, y_2)$ is

$$d = \sqrt{(x_2 - x_1)^2 + (y_2 - y_1)^2}.$$

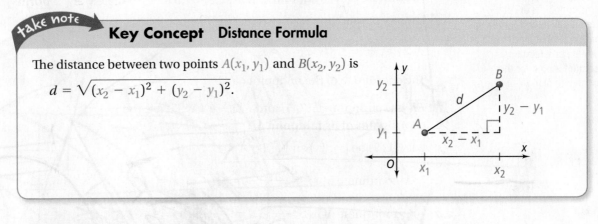

The Distance Formula is based on the *Pythagorean Theorem*, which you will study later in this book. When you use the Distance Formula, you are really finding the length of a side of a right triangle. You will verify the Distance Formula in Chapter 8.

$$a^2 + b^2 = c^2$$

Problem 3   Finding Distance   GRIDDED RESPONSE

What is the distance between $U(-7, 5)$ and $V(4, -3)$? Round to the nearest tenth.

Let $U(-7, 5)$ be (x_1, y_1) and $V(4, -3)$ be (x_2, y_2).

Think

What part of a right triangle is $\overline{UV}$?
$\overline{UV}$ is the hypotenuse of a right triangle with legs of length 11 and 8.

$d = \sqrt{(x_2 - x_1)^2 + (y_2 - y_1)^2}$	Use the Distance Formula.
$= \sqrt{(4 - (-7))^2 + (-3 - 5)^2}$	Substitute.
$= \sqrt{(11)^2 + (-8)^2}$	Simplify within the parentheses.
$= \sqrt{121 + 64}$	Simplify.
$= \sqrt{185}$	
$185 \;\sqrt{}\; 13.60147051$	Use a calculator.

To the nearest tenth, $UV = 13.6$.

Got It? 3. a. $\overline{SR}$ has endpoints $S(-2, 14)$ and $R(3, -1)$. What is SR to the nearest tenth?
   b. Reasoning In Problem 3, suppose you let $V(4, -3)$ be (x_1, y_1) and $U(-7, 5)$ be (x_2, y_2). Do you get the same result? Why?

Problem 4 **Finding Distance**

Recreation On a zip-line course, you are harnessed to a cable that travels through the treetops. You start at Platform *A* and zip to each of the other platforms. How far do you travel from Platform *B* to Platform *C*? Each grid unit represents 5 m.

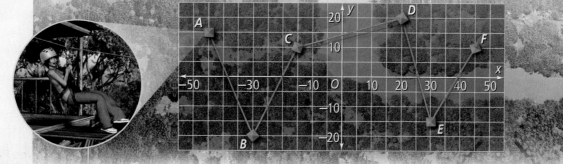

Think

Where's the right triangle?
The lengths of the legs of the right triangle are 15 and 30. There are two possibilities:

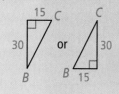

Let Platform $B(-30, -20)$ be (x_1, y_1) and Platform $C(-15, 10)$ be (x_2, y_2).

$d = \sqrt{(x_2 - x_1)^2 + (y_2 - y_1)^2}$ Use the Distance Formula.

$= \sqrt{(-15 - (-30))^2 + (10 - (-20))^2}$ Substitute.

$= \sqrt{15^2 + 30^2} = \sqrt{225 + 900} = \sqrt{1125}$ Simplify.

1125 = 33.54101966   Use a calculator.

You travel about 33.5 m from Platform *B* to Platform *C*.

Got It? **4.** How far do you travel from Platform *D* to Platform *E*?

Lesson Check

Do you know HOW?

1. $\overline{RS}$ has endpoints $R(2, 4)$ and $S(-1, 7)$. What are the coordinates of its midpoint *M*?

2. The midpoint of $\overline{BC}$ is $(5, -2)$. One endpoint is $B(3, 4)$. What are the coordinates of endpoint *C*?

3. What is the distance between points $K(-9, 8)$ and $L(-6, 0)$?

Do you UNDERSTAND? MATHEMATICAL PRACTICES

4. Reasoning How does the Distance Formula ensure that the distance between two different points is positive?

5. Error Analysis Your friend calculates the distance between points $Q(1, 5)$ and $R(3, 8)$. What is his error?

$d = \sqrt{(1 - 8)^2 + (5 - 3)^2}$
$= \sqrt{(-7)^2 + 2^2}$
$= \sqrt{49 + 4}$
$= \sqrt{53} \approx 7.3$

Practice and Problem-Solving Exercises

MATHEMATICAL PRACTICES

A Practice

Find the coordinate of the midpoint of the segment with the given endpoints.   ◀ **See Problem 1.**

6. 2 and 4   **7.** −9 and 6   **8.** 2 and −5   **9.** −8 and −12

Find the coordinates of the midpoint of $\overline{HX}$.

10. $H(0, 0)$, $X(8, 4)$ **11.** $H(-1, 3)$, $X(7, -1)$ **12.** $H(13, 8)$, $X(-6, -6)$

13. $H(7, 10)$, $X(5, -8)$ **14.** $H(-6.3, 5.2)$, $X(1.8, -1)$ **15.** $H\left(5\frac{1}{2}, -4\frac{3}{4}\right)$, $X\left(2\frac{1}{4}, -1\frac{1}{4}\right)$

The coordinates of point T are given. The midpoint of $\overline{ST}$ is $(5, -8)$. Find the   ◀ **See Problem 2.**
coordinates of point S.

16. $T(0, 4)$ **17.** $T(5, -15)$ **18.** $T(10, 18)$

19. $T(-2, 8)$ **20.** $T(1, 12)$ **21.** $T(4.5, -2.5)$

Find the distance between each pair of points. If necessary, round to the   ◀ **See Problem 3.**
nearest tenth.

22. $J(2, -1)$, $K(2, 5)$ **23.** $L(10, 14)$, $M(-8, 14)$ **24.** $N(-1, -11)$, $P(-1, -3)$

25. $A(0, 3)$, $B(0, 12)$ **26.** $C(12, 6)$, $D(-8, 18)$ **27.** $E(6, -2)$, $F(-2, 4)$

28. $Q(12, -12)$, $T(5, 12)$ **29.** $R(0, 5)$, $S(12, 3)$ **30.** $X(-3, -4)$, $Y(5, 5)$

Maps For Exercises 31–35, use the map below. Find the distance between the   ◀ **See Problem 4.**
cities to the nearest tenth.

31. Augusta and Brookline

32. Brookline and Charleston

33. Brookline and Davenport

34. Everett and Fairfield

35. List the cities in the order of least to
greatest distance from Augusta.

B Apply

Find (a) PQ to the nearest tenth and (b) the coordinates of the midpoint of $\overline{PQ}$.

36. $P(3, 2)$, $Q(6, 6)$ **37.** $P(0, -2)$, $Q(3, 3)$ **38.** $P(-4, -2)$, $Q(1, 3)$

39. $P(-5, 2)$, $Q(0, 4)$ **40.** $P(-3, -1)$, $Q(5, -7)$ **41.** $P(-5, -3)$, $Q(-3, -5)$

42. $P(-4, -5)$, $Q(-1, 1)$ **43.** $P(2, 3)$, $Q(4, -2)$ **44.** $P(4, 2)$, $Q(3, 0)$

© **45. Think About a Plan** An airplane at $T(80, 20)$ needs to fly to both $U(20, 60)$ and
$V(110, 85)$. What is the shortest possible distance for the trip? Explain.
- What type of information do you need to find the shortest distance?
- How can you use a diagram to help you?

46. Reasoning The endpoints of $\overline{AB}$ are $A(-2, -3)$ and $B(3, 2)$. Point C lies on AB and is $\frac{2}{5}$ of the way from A to B. What are the coordinates of Point C? Explain how you found your answer.

47. Do you use the Midpoint Formula or the Distance Formula to find the following?
  a. Given points K and P, find the distance from K to the midpoint of $\overline{KP}$.
  b. Given point K and the midpoint of $\overline{KP}$, find KP.

For each graph, find (a) AB to the nearest tenth and (b) the coordinates of the midpoint of $\overline{AB}$.

48.
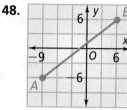

49.
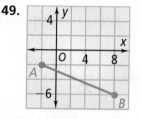

50.
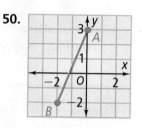

51. Coordinate Geometry Graph the points $A(2, 1)$, $B(6, -1)$, $C(8, 7)$, and $D(4, 9)$. Draw parallelogram $ABCD$, and diagonals $\overline{AC}$ and $\overline{BD}$.
  a. Find the midpoints of $\overline{AC}$ and $\overline{BD}$.
  b. What appears to be true about the diagonals of a parallelogram?

Travel The units of the subway map at the right are in miles. Suppose the routes between stations are straight. Find the distance you would travel between each pair of stations to the nearest tenth of a mile.

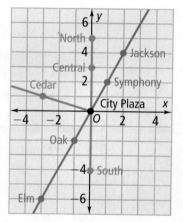

52. Oak Station and Jackson Station

53. Central Station and South Station

54. Elm Station and Symphony Station

55. Cedar Station and City Plaza Station

56. Maple Station is located 6 mi west and 2 mi north of City Plaza. What is the distance between Cedar Station and Maple Station?

57. Open-Ended Point $H(2, 2)$ is the midpoint of many segments.
  a. Find the coordinates of the endpoints of four noncollinear segments that have point H as their midpoint.
  b. You know that a segment with midpoint H has length 8. How many possible noncollinear segments match this description? Explain.

Challenge

58. Points $P(-4, 6)$, $Q(2, 4)$, and R are collinear. One of the points is the midpoint of the segment formed by the other two points.
  a. What are the possible coordinates of R?
  b. Reasoning $RQ = \sqrt{160}$. Does this information affect your answer to part (a)? Explain.

Geometry in 3 Dimensions You can use three coordinates (x, y, z) to locate points in three dimensions.

59. Point P has coordinates $(6, -3, 9)$ as shown at the right. Give the coordinates of points A, B, C, D, E, F, and G.

Distance in 3 Dimensions In a three-dimensional coordinate system, you can find the distance between two points (x_1, y_1, z_1) and (x_2, y_2, z_2) with this extension of the Distance Formula.

$$d = \sqrt{(x_2 - x_1)^2 + (y_2 - y_1)^2 + (z_2 - z_1)^2}$$

Find the distance between each pair of points to the nearest tenth.

60. $P(2, 3, 4)$, $Q(-2, 4, 9)$ **61.** $T(0, 12, 15)$, $V(-8, 20, 12)$

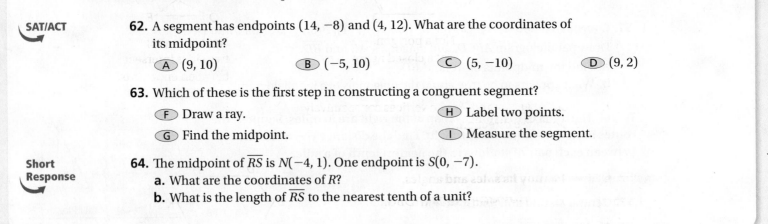

Standardized Test Prep

62. A segment has endpoints $(14, -8)$ and $(4, 12)$. What are the coordinates of its midpoint?

   Ⓐ $(9, 10)$ Ⓑ $(-5, 10)$ Ⓒ $(5, -10)$ Ⓓ $(9, 2)$

63. Which of these is the first step in constructing a congruent segment?

   Ⓕ Draw a ray.   Ⓗ Label two points.

   Ⓖ Find the midpoint.   Ⓘ Measure the segment.

Short Response

64. The midpoint of $\overline{RS}$ is $N(-4, 1)$. One endpoint is $S(0, -7)$.
   a. What are the coordinates of R?
   b. What is the length of $\overline{RS}$ to the nearest tenth of a unit?

Mixed Review

Use a straightedge and a compass.   ◀ See Lesson 1-6.

65. Draw $\overline{AB}$. Construct $\overline{PQ}$ so that $PQ = 2AB$.

66. Draw an acute $\angle RTS$. Construct the bisector of $\angle RTS$.

Use the diagram at the right.   ◀ See Lesson 1-4.

67. Name $\angle 1$ two other ways.

68. If $m\angle PQR = 60$, what is $m\angle RQS$?

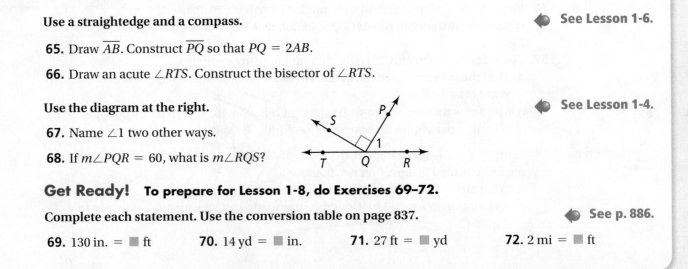

Get Ready!   To prepare for Lesson 1-8, do Exercises 69–72.

Complete each statement. Use the conversion table on page 837.   ◀ See p. 886.

69. 130 in. = ▪ ft   **70.** 14 yd = ▪ in.   **71.** 27 ft = ▪ yd   **72.** 2 mi = ▪ ft

Classifying Polygons

© **Content Standard**
Prepares for G.MG.1 Use geometric shapes, their measures, and their properties to describe objects.

In geometry, a figure that lies in a plane is called a *plane figure*.

A **polygon** is a closed plane figure formed by three or more segments. Each segment intersects exactly two other segments at their endpoints. No two segments with a common endpoint are collinear. Each segment is called a *side*. Each endpoint of a side is a *vertex*.

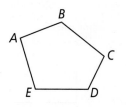

A polygon

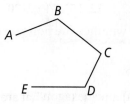

Not a polygon;
not a closed figure

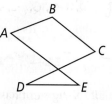

Not a polygon;
two sides intersect
between endpoints.

To name a polygon, start at any vertex and list the vertices consecutively in a clockwise or counterclockwise direction.

Example 1

Name the polygon. Then identify its sides and angles.

Two names for this polygon are *DHKMGB* and *MKHDBG*.

sides: $\overline{DH}$, $\overline{HK}$, $\overline{KM}$, $\overline{MG}$, $\overline{GB}$, $\overline{BD}$

angles: $\angle D$, $\angle H$, $\angle K$, $\angle M$, $\angle G$, $\angle B$

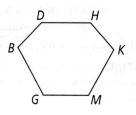

You can classify a polygon by its number of sides. The tables below show the names of some common polygons.

Names of Common Polygons

Sides	Name	Sides	Name
3	Triangle, or trigon	9	Nonagon, or enneagon
4	Quadrilateral, or tetragon	10	Decagon
5	Pentagon	11	Hendecagon
6	Hexagon	12	Dodecagon
7	Heptagon	⋮	⋮
8	Octagon	*n*	*n*-gon

You can also classify a polygon as concave or convex, using the diagonals of the polygon. A **diagonal** is a segment that connects two nonconsecutive vertices.

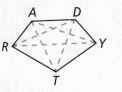

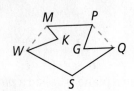

A **convex polygon** has no diagonal with points outside the polygon.

A **concave polygon** has at least one diagonal with points outside the polygon.

In this textbook, a polygon is convex unless otherwise stated.

Example 2

Classify the polygon by its number of sides. Tell whether the polygon is *convex* or *concave*.

The polygon has six sides. Therefore, it is a hexagon.

No diagonal of the hexagon contains points outside the hexagon. The hexagon is convex.

Exercises

Is the figure a polygon? If not, explain why.

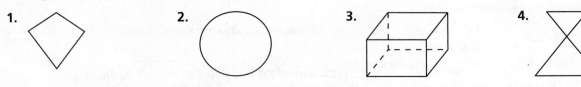

1.

2.

3.

4.

Name the polygon. Then identify its sides and angles.

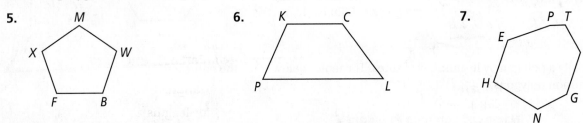

5.
M
X   W
F   B

6.
K   C
P   L

7.
P T
E   A
H
G
N

Classify the polygon by its number of sides. Tell whether the polygon is *convex* or *concave*.

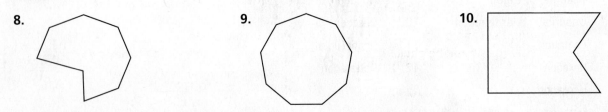

8.

9.

10.

Perimeter, Circumference, and Area

© **Content Standard**
N.Q.1 Use units as a way to understand problems and to guide the solution of multi-step problems; choose and interpret units consistently in formulas . . .

Objectives   To find the perimeter or circumference of basic shapes
To find the area of basic shapes

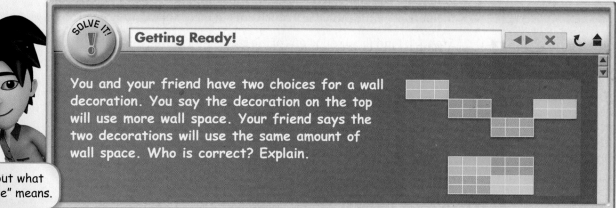

SOLVE IT!

Getting Ready!

You and your friend have two choices for a wall decoration. You say the decoration on the top will use more wall space. Your friend says the two decorations will use the same amount of wall space. Who is correct? Explain.

Think about what "wall space" means.

© MATHEMATICAL PRACTICES

In the Solve It, you considered various ideas of what it means to take up space on a flat surface.

Essential Understanding   Perimeter and area are two different ways of measuring geometric figures.

The **perimeter** P of a polygon is the sum of the lengths of its sides. The **area** A of a polygon is the number of square units it encloses. For figures such as squares, rectangles, triangles, and circles, you can use formulas for perimeter (or *circumference C* for circles) and area.

Dynamic Activity
Perimeter, Circumference, and Area

Lesson Vocabulary
• perimeter
• area

take note

Key Concept   **Perimeter, Circumference, and Area**

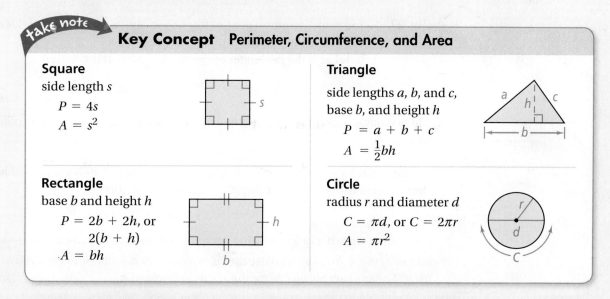

Square
side length s
$P = 4s$
$A = s^2$

Triangle
side lengths a, b, and c,
base b, and height h
$P = a + b + c$
$A = \frac{1}{2}bh$

Rectangle
base b and height h
$P = 2b + 2h$, or
$2(b + h)$
$A = bh$

Circle
radius r and diameter d
$C = \pi d$, or $C = 2\pi r$
$A = \pi r^2$

The units of measurement for perimeter and circumference include inches, feet, yards, miles, centimeters, and meters. When measuring area, use square units such as square inches (in.^{2}), square feet (ft^2), square yards (yd^2), square miles (mi^2), square centimeters (cm^2), and square meters (m^2).

Problem 1 **Finding the Perimeter of a Rectangle**

Landscaping The botany club members are designing a rectangular garden for the courtyard of your school. They plan to place edging on the outside of the path. How much edging material will they need?

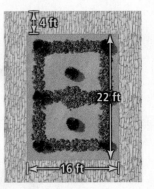

Step 1 Find the dimensions of the garden, including the path.

Plan

Why should you *draw a diagram*?
A diagram can help you see the larger rectangle formed by the garden and the path, and which lengths to add together.

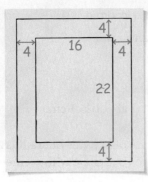

For a rectangle, "length" and "width" are sometimes used in place of "base" and "height."

Width of the garden and path
$$= 4 + 16 + 4 = 24$$

Length of the garden and path
$$= 4 + 22 + 4 = 30$$

Step 2 Find the perimeter of the garden including the path.

$P = 2b + 2h$	Use the formula for the perimeter of a rectangle.
$= 2(24) + 2(30)$	Substitute 24 for b and 30 for h.
$= 48 + 60$	Simplify.
$= 108$	

You will need 108 ft of edging material.

Got It? **1.** You want to frame a picture that is 5 in. by 7 in. with a 1-in.-wide frame.
   a. What is the perimeter of the picture?
   b. What is the perimeter of the outside edge of the frame?

You can name a circle with the symbol ⊙. For example, the circle with center A is written ⊙A.

The formulas for a circle involve the special number *pi* (π). Pi is the ratio of any circle's circumference to its diameter. Since π is an irrational number,

$$\pi = 3.1415926\ldots,$$

you cannot write it as a terminating decimal. For an approximate answer, you can use 3.14 or $\frac{22}{7}$ for π. You can also use the ⊙ key on your calculator to get a rounded decimal for π. For an exact answer, leave the result in terms of π.

Problem 2 Finding Circumference

What is the circumference of the circle in terms of π? What is the circumference of the circle to the nearest tenth?

A $\odot M$

$C = \pi d$	Use the formula for circumference of a circle.
$= \pi(15)$	This is the exact answer.
≈ 47.1238898	Use a calculator.

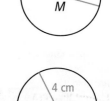

The circumference of $\odot M$ is 15π in., or about 47.1 in.

B $\odot T$

$C = 2\pi r$	Use the formula for circumference of a circle.
$= 2\pi(4)$	This is the exact answer.
$= 8\pi$	Simplify.
≈ 25.13274123	Use a calculator.

The circumference of $\odot T$ is 8π cm, or about 25.1 cm.

Got It? 2. a. What is the circumference of a circle with radius 24 m in terms of π?
　　　　　b. What is the circumference of a circle with diameter 24 m to the nearest tenth?

Problem 3 Finding Perimeter in the Coordinate Plane

Coordinate Geometry What is the perimeter of $\triangle EFG$?

Step 1 Find the length of each side.

$$EF = |6 - (-2)| = 8$$
$$FG = |3 - (-3)| = 6$$
　　　　　　　　　　　　Use the Ruler Postulate.

$$EG = \sqrt{(3 - (-3))^2 + (6 - (-2))^2}$$　Use the Distance Formula.

$$= \sqrt{6^2 + 8^2}$$　Simplify within the parentheses.

$$= \sqrt{36 + 64}$$　Simplify.

$$= \sqrt{100}$$

$$= 10$$

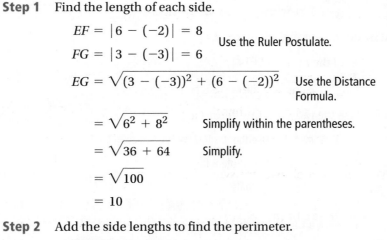

Step 2 Add the side lengths to find the perimeter.

$$EF + FG + EG = 8 + 6 + 10 = 24$$

The perimeter of $\triangle EFG$ is 24 units.

Got It? 3. Graph quadrilateral $JKLM$ with vertices $J(-3, -3)$, $K(1, -3)$, $L(1, 4)$, and $M(-3, 1)$. What is the perimeter of $JKLM$?

To find area, you should use the same unit for both dimensions.

Problem 4 Finding Area of a Rectangle

Banners You want to make a rectangular banner similar to the one at the right. The banner shown is $2\frac{1}{2}$ ft wide and 5 ft high. To the nearest square yard, how much material do you need?

Think

How can you check your conversion?
Yards are longer than feet, so the number you get in yards should be less than the given number in feet. Since $\frac{5}{6} < 2\frac{1}{2}$, the conversion checks.

Step 1 Convert the dimensions of the banner to yards. Use the conversion factor $\frac{1\text{ yd}}{3\text{ ft}}$.

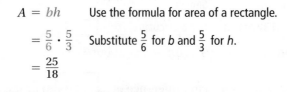

Width: $\frac{5}{2}\cancel{ft} \cdot \frac{1\text{ yd}}{3\cancel{ft}} = \frac{5}{6}\text{ yd}$ $2\frac{1}{2}\text{ ft} = \frac{5}{2}\text{ ft}$

Height: $5\cancel{ft} \cdot \frac{1\text{ yd}}{3\cancel{ft}} = \frac{5}{3}\text{ yd}$

Step 2 Find the area of the banner.

$A = bh$ Use the formula for area of a rectangle.

$= \frac{5}{6} \cdot \frac{5}{3}$ Substitute $\frac{5}{6}$ for b and $\frac{5}{3}$ for h.

$= \frac{25}{18}$

The area of the banner is $\frac{25}{18}$, or $1\frac{7}{18}$ square yards (yd^2). You need 2 yd^2 of material.

Got It? **4.** You are designing a poster that will be 3 yd wide and 8 ft high. How much paper do you need to make the poster? Give your answer in square feet.

Problem 5 Finding Area of a Circle

Plan

What are you given?
In circle problems, make it a habit to note whether you are given the radius or the diameter. In this case, you are given the diameter.

What is the area of $\odot K$ in terms of π?

Step 1 Find the radius of $\odot K$.

$r = \frac{16}{2}$, or 8   The radius is half the diameter.

Step 2 Use the radius to find the area.

$A = \pi r^2$ Use the formula for area of a circle.

$= \pi(8)^2$ Substitute 8 for r.

$= 64\pi$ Simplify.

The area of $\odot K$ is 64π m^2.

Got It? **5.** The diameter of a circle is 14 ft.
  a. What is the area of the circle in terms of π?
  b. What is the area of the circle using an approximation of π?
  c. Reasoning Which approximation of π did you use in part (b)? Why?

The following postulate is useful in finding areas of figures with irregular shapes.

take note

Postulate 1-10 Area Addition Postulate

The area of a region is the sum of the areas of its nonoverlapping parts.

Problem 6 Finding Area of an Irregular Shape

Multiple Choice What is the area of the figure at the right? All angles are right angles.

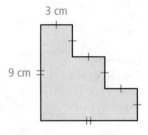
3 cm
9 cm

Ⓐ 27 cm^2

Ⓒ 45 cm^2

Ⓑ 36 cm^2

Ⓓ 54 cm^2

Step 1 Separate the figure into rectangles.

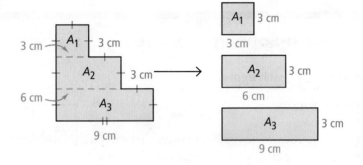

Think

What is another way to find the area?
Extend the figure to form a square. Then subtract the areas of basic shapes from the area of the square.

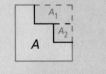

$A = A_{\text{square}} - A_1 - A_2$

Step 2 Find A_1, A_2, and A_3.

Area $= bh$ Use the formula for the area of a rectangle.

$A_1 = 3 \cdot 3 = 9$ Substitute for the base and height.

$A_2 = 6 \cdot 3 = 18$

$A_3 = 9 \cdot 3 = 27$

Step 3 Find the total area of the figure.

Total Area $= A_1 + A_2 + A_3$ Use the Area Addition Postulate.

$= 9 + 18 + 27$

$= 54$

The area of the figure is 54 cm^2. The correct choice is D.

Got It? 6. a. Reasoning What is another way to separate the figure in Problem 6?

b. What is the area of the figure at the right?

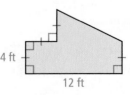
4 ft
12 ft

Lesson Check

Do you know HOW?

1. What is the perimeter and area of a rectangle with base 3 in. and height 7 in.?

2. What is the circumference and area of each circle to the nearest tenth?
   a. $r = 9$ in.   **b.** $d = 7.3$ m

3. What is the perimeter and area of the figure at the right?

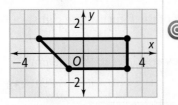

Do you UNDERSTAND? ⓒ MATHEMATICAL PRACTICES

ⓒ **4. Writing** Describe a real-world situation in which you would need to find a perimeter. Then describe a situation in which you would need to find an area.

ⓒ **5. Compare and Contrast** Your friend can't remember whether $2\pi r$ computes the circumference or the area of a circle. How would you help your friend? Explain.

ⓒ **6. Error Analysis** A classmate finds the area of a circle with radius 30 in. to be 900 in.². What error did your classmate make?

Practice and Problem-Solving Exercises ⓒ MATHEMATICAL PRACTICES

Ⓐ Practice   Find the perimeter of each figure.   ◀ See Problem 1.

7.

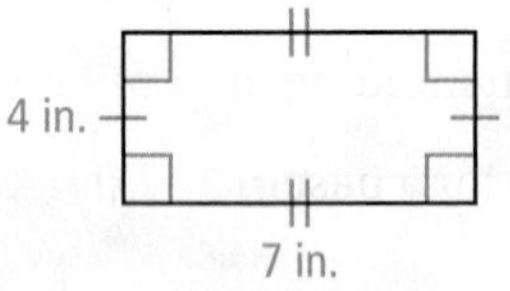

4 in.

7 in.

8.

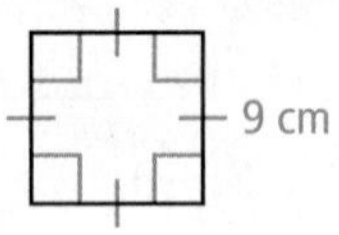

9 cm

9. Fencing A garden that is 5 ft by 6 ft has a walkway 2 ft wide around it. What is the amount of fencing needed to surround the walkway?

Find the circumference of ⊙C in terms of π.   ◀ See Problem 2.

10.

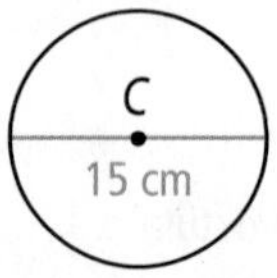

C

15 cm

11.

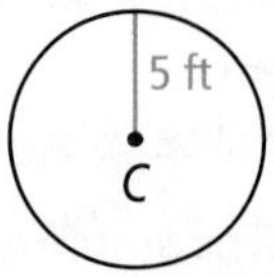

5 ft

C

12.

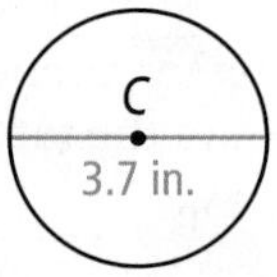

C

3.7 in.

13.

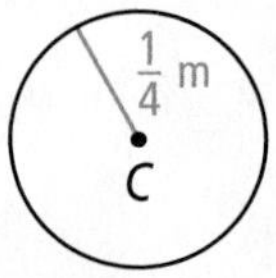

$\frac{1}{4}$ m

C

Coordinate Geometry Graph each figure in the coordinate plane. Find each perimeter.   ◀ See Problem 3.

14. $X(0, 2)$, $Y(4, -1)$, $Z(-2, -1)$

15. $A(-4, -1)$, $B(4, 5)$, $C(4, -2)$

16. $L(0, 1)$, $M(3, 5)$, $N(5, 5)$, $P(5, 1)$

17. $S(-5, 3)$, $T(7, -2)$, $U(7, -6)$, $V(-5, -6)$

Find the area of each rectangle with the given base and height.   ◀ See Problem 4.

18. 4 ft, 4 in.   **19.** 30 in., 4 yd   **20.** 2 ft 3 in., 6 in.   **21.** 40 cm, 2 m

22. Roads What is the area of a section of pavement that is 20 ft wide and 100 yd long? Give your answer in square feet.

Find the area of each circle in terms of π.

See Problem 5.

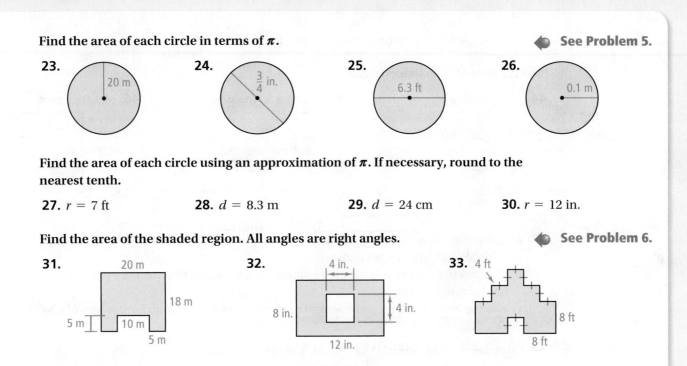

23. 20 m

24. $\frac{3}{4}$ in.

25. 6.3 ft

26. 0.1 m

Find the area of each circle using an approximation of π. If necessary, round to the nearest tenth.

27. $r = 7$ ft   **28.** $d = 8.3$ m   **29.** $d = 24$ cm   **30.** $r = 12$ in.

Find the area of the shaded region. All angles are right angles.

See Problem 6.

31. 20 m, 18 m, 5 m, 10 m, 5 m

32. 4 in., 8 in., 4 in., 12 in.

33. 4 ft, 8 ft, 8 ft

B **Apply**

Home Maintenance To determine how much of each item to buy, tell whether you need to know area or perimeter. Explain your choice.

34. wallpaper for a bedroom

35. crown molding for a ceiling

36. fencing for a backyard

37. paint for a basement floor

© 38. Think About a Plan A light-year unit describes the distance that one photon of light travels in one year. The Milky Way galaxy has a diameter of about 100,000 light-years. The distance to Earth from the center of the Milky Way galaxy is about 30,000 light-years. How many more light-years does a star on the outermost edge of the Milky Way travel in one full revolution around the galaxy compared to Earth?
- What do you know about the shape of each orbital path?
- Are you looking for circumference or area?
- How do you compare the paths using algebraic expressions?

39. a. What is the area of a square with sides 12 in. long? 1 ft long?
    b. How many square inches are in a square foot?

© 40. a. Count squares at the right to find the area of the polygon outlined in blue.
    b. Use a formula to find the area of each square outlined in red.
    c. Writing How does the sum of your results in part (b) compare to your result in part (a)? Which postulate does this support?

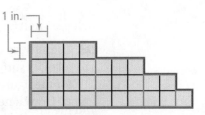

1 in.

41. The area of an 11-cm-wide rectangle is 176 cm². What is its length?

42. A square and a rectangle have equal areas. The rectangle is 64 cm by 81 cm. What is the perimeter of the square?

43. A rectangle has perimeter 40 cm and base 12 cm. What is its area?

Find the area of each shaded figure.

44. compact disc

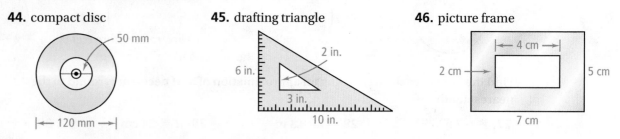

50 mm

120 mm

45. drafting triangle

2 in.

6 in.

3 in.

10 in.

46. picture frame

4 cm

2 cm

5 cm

7 cm

47. a. Reasoning Can you use the formula for the perimeter of a rectangle to find the perimeter of any square? Explain.
  b. Can you use the formula for the perimeter of a square to find the perimeter of any rectangle? Explain.
  c. Use the formula for the perimeter of a square to write a formula for the area of a square in terms of its perimeter.

48. Estimation On an art trip to England, a student sketches the floor plan of the main body of Salisbury Cathedral. The shape of the floor plan is called the building's "footprint." The student estimates the dimensions of the cathedral on her sketch at the right. Use the student's lengths to estimate the area of Salisbury Cathedral's footprint.

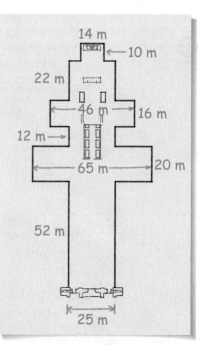

14 m
10 m
22 m
46 m
16 m
12 m
65 m
20 m
52 m
25 m

49. Coordinate Geometry The endpoints of a diameter of a circle are $A(2, 1)$ and $B(5, 5)$. Find the area of the circle in terms of π.

50. Algebra A rectangle has a base of x units. The area is $(4x^2 - 2x)$ square units. What is the height of the rectangle in terms of x?

Ⓐ $(4 - x)$ units   Ⓒ $(4x^3 - 2x^2)$ units

Ⓑ $(x - 2)$ units   Ⓓ $(4x - 2)$ units

Coordinate Geometry Graph each rectangle in the coordinate plane. Find its perimeter and area.

51. $A(-3, 2)$, $B(-2, 2)$, $C(-2, -2)$, $D(-3, -2)$

52. $A(-2, -6)$, $B(-2, -3)$, $C(3, -3)$, $D(3, -6)$

53. The surface area of a three-dimensional figure is the sum of the areas of all of its surfaces. You can find the surface area by finding the area of a net for the figure.

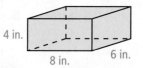

4 in.

8 in.

6 in.

  a. Draw a net for the solid shown. Label the dimensions.
  b. What is the area of the net? What is the surface area of the solid?

54. Coordinate Geometry On graph paper, draw polygon $ABCDEFG$ with vertices $A(1, 1)$, $B(10, 1)$, $C(10, 8)$, $D(7, 5)$, $E(4, 5)$, $F(4, 8)$, and $G(1, 8)$. Find the perimeter and the area of the polygon.

55. Pet Care You want to adopt a puppy from your local animal shelter. First, you plan to build an outdoor playpen along the side of your house, as shown on the right. You want to lay down special dog grass for the pen's floor. If dog grass costs $1.70 per square foot, how much will you spend?

56. A rectangular garden has an 8-ft walkway around it. How many more feet is the outer perimeter of the walkway than the perimeter of the garden?

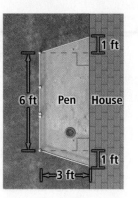

 Challenge   **Algebra** **Find the area of each figure.**

57. a rectangle with side lengths $\frac{2a}{5b}$ units and $\frac{3b}{8}$ units

58. a square with perimeter $10n$ units

59. a triangle with base $(5x - 2y)$ units and height $(4x + 3y)$ units

Standardized Test Prep

SAT/ACT

60. An athletic field is a 100 yd-by-40 yd rectangle with a semicircle at each of the short sides. A running track 10 yd wide surrounds the field. Find the perimeter of the outside of the running track to the nearest tenth of a yard.

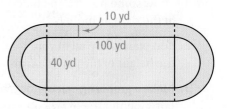

61. A square garden has a 4-ft walkway around it. The garden has a perimeter of 260 ft. What is the area of the walkway in square feet?

62. $A(4, -1)$ and $B(-2, 3)$ are points in a coordinate plane. M is the midpoint of $\overline{AB}$. What is the length of $\overline{MB}$ to the nearest tenth of a unit?

63. Find CD to the nearest tenth if point C is at $(12, -8)$ and point D is at $(5, 19)$.

Mixed Review

Find (a) AB to the nearest tenth and (b) the midpoint coordinates of $\overline{AB}$.   ◀ See Lesson 1-7.

64. $A(4, 1), B(7, 9)$ **65.** $A(0, 3), B(-3, 8)$ **66.** $A(-1, 1), B(-4, -5)$

$\overleftrightarrow{BG}$ **is the perpendicular bisector of** $\overline{WR}$ **at point** K.   ◀ See Lesson 1-6.

67. What is $m\angle BKR$?

68. Name two congruent segments.

Get Ready!   **To prepare for Lesson 2-1, do Exercise 69.**

69. a. Copy and extend this list to show the first 10 perfect squares.   ◀ See p. 889.
$$1^2 = 1, 2^2 = 4, 3^2 = 9, 4^2 = 16, \ldots$$

  b. Which do you think describes the square of any odd number?
     It is odd.   It is even.

Comparing Perimeters and Areas

© Content Standard
Prepares for G.MG.2 Apply concepts of density based on area and volume in modeling situations.

You can use a graphing calculator or spreadsheet software to find maximum and minimum values. These values help you solve real-world problems where you want to minimize or maximize a quantity such as cost or time. In this Activity, you will find minimum and maximum values for area and perimeter problems.

© MATHEMATICAL PRACTICES

Activity

You have 32 yd of fencing. You want to make a rectangular horse pen with maximum area.

1. Draw some possible rectangular pens and find their areas. Use the examples at the right as models.

2. You plan to use all of your fencing. Let X represent the base of the pen. What is the height of the pen in terms of X? What is the area of the pen in terms of X?

3. Make a graphing calculator table to find area. Again, let X represent the base. For Y_1, enter the expression you wrote for the height in Question 2. For Y_2, enter the expression you wrote for the area in Question 2. Set the table so that X starts at 4 and changes by 1. Scroll down the table.
   a. What value of X gives you the maximum area?
   b. What is the maximum area?

4. Use your calculator to graph Y_2. Describe the shape of the graph. Trace on the graph to find the coordinates of the highest point. What is the relationship, if any, between the coordinates of the highest point on the graph and your answers to Question 3? Explain.

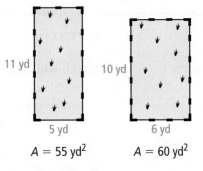

11 yd 5 yd
$A = 55 \text{ yd}^2$

10 yd 6 yd
$A = 60 \text{ yd}^2$

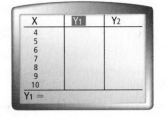

Exercises

5. For a fixed perimeter, what rectangular shape will result in a maximum area?

6. Consider that the pen is not limited to polygon shapes. What is the area of a circular pen with circumference 32 yd? How does this result compare to the maximum area you found in the Activity?

7. You plan to make a rectangular garden with an area of 900 ft^2. You want to use a minimum amount of fencing to keep the cost low.
   a. List some possible dimensions for the garden. Find the perimeter of each.
   b. Make a graphing calculator table. Use integer values of the base b, and the corresponding values of the height h, to find values for P, the perimeter. What dimensions will give you a garden with the minimum perimeter?

Pull It **All Together** © ASSESSMENT

To solve these problems, you will pull together many concepts and skills that you learned in this chapter. They are the basic tools used to study geometry.

BIG idea Visualization

You can draw and construct figures to understand geometric relationships.

© Performance Task 1

Graph $\triangle ABC$ with $A(4, 7)$, $B(0,0)$, and $C(8, 1)$.

a. Which sides of $\triangle ABC$ are congruent? How do you know?

b. Construct the bisector of $\angle B$. Mark the intersection of the ray and $\overline{AC}$ as D.

c. What do you notice about AD and CD?

BIG idea Measurement

Measures of segments and angles give you ways to describe geometric figures.

© Performance Task 2

Using the design below, you plan to make a rectangular wall hanging to decorate your bedroom. The green rectangle is 16 ft by 10 ft, the blue square is 7 ft by 7 ft, the orange triangles have a height of 1.73 ft and a side length of 2 ft, and the yellow circle has a diameter of 3 ft. You plan to glue the pieces together and then outline each piece with black cord.

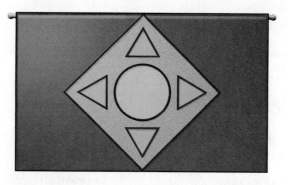

a. How much black cord does the design use? Show your work.

b. How much area of each color is showing in the design? Show your work.

c. At a craft store, fabric comes in bolts that are a fixed number of inches wide. You can only buy fabric by the whole yard. If each bolt is 48 in. wide, what is the least amount of each color that you need to buy? Explain.

Connecting **BIG** ideas and Answering the Essential Questions

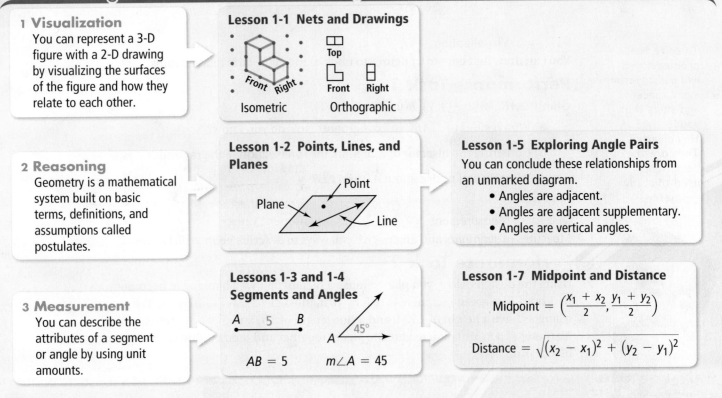

1 Visualization
You can represent a 3-D figure with a 2-D drawing by visualizing the surfaces of the figure and how they relate to each other.

Lesson 1-1 Nets and Drawings

Top

Front Right

Isometric

Front   Right

Orthographic

2 Reasoning
Geometry is a mathematical system built on basic terms, definitions, and assumptions called postulates.

Lesson 1-2 Points, Lines, and Planes

Point

Plane

Line

Lesson 1-5 Exploring Angle Pairs

You can conclude these relationships from an unmarked diagram.
- Angles are adjacent.
- Angles are adjacent supplementary.
- Angles are vertical angles.

3 Measurement
You can describe the attributes of a segment or angle by using unit amounts.

Lessons 1-3 and 1-4 Segments and Angles

A 5 B

$AB = 5$

A 45°

$m\angle A = 45$

Lesson 1-7 Midpoint and Distance

$$\text{Midpoint} = \left(\frac{x_1 + x_2}{2}, \frac{y_1 + y_2}{2} \right)$$

$$\text{Distance} = \sqrt{(x_2 - x_1)^2 + (y_2 - y_1)^2}$$

Chapter Vocabulary

- acute, right, obtuse, straight angles (p. 29)
- adjacent angles (p. 34)
- angle bisector (p. 37)
- collinear points, coplanar (p. 12)
- complementary angles (p. 34)
- congruent angles (p. 29)
- congruent segments (p. 22)
- construction (p. 43)
- isometric drawing (p. 5)
- linear pair (p. 36)
- measure of an angle (p. 28)
- net (p. 4)
- orthographic drawing (p. 6)
- perpendicular bisector (p. 44)
- perpendicular lines (p. 44)
- point, line, plane (p. 11)
- postulate, axiom (p. 13)
- ray, opposite rays (p. 12)
- segment (p. 12)
- segment bisector (p. 22)
- space (p. 12)
- supplementary angles (p. 34)
- vertex of an angle (p. 27)
- vertical angles (p. 34)

Choose the correct term to complete each sentence.

1. A ray that divides an angle into two congruent angles is a(n) __?__.

2. __?__ are two lines that intersect to form right angles.

3. A(n) __?__ is a two-dimensional diagram that you can fold to form a 3-D figure.

4. __?__ are two angles with measures that have a sum of 90.

1-1 Nets and Drawings for Visualizing Geometry

Quick Review

A **net** is a two-dimensional pattern that you can fold to form a three-dimensional figure. A net shows all surfaces of a figure in one view.

An **isometric drawing** shows a corner view of a three-dimensional object. It allows you to see the top, front, and side of the object in one view.

An **orthographic drawing** shows three separate views of a three-dimensional object: a top view, a front view, and a right-side view.

Example

Draw a net for the solid at the right.

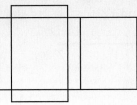

Exercises

5. The net below is for a number cube. What are the three sums of the numbers on opposite surfaces of the cube?

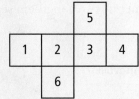

6. Make an orthographic drawing for the isometric drawing at the right. Assume there are no hidden cubes.

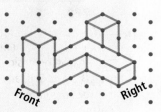

1-2 Points, Lines, and Planes

Quick Review

A **point** indicates a location and has no size.

A **line** is represented by a straight path that extends in two opposite directions without end and has no thickness.

A **plane** is represented by a flat surface that extends without end and has no thickness.

Points that lie on the same line are **collinear points**.

Points and lines in the same plane are **coplanar**.

Segments and **rays** are parts of lines.

Example

Name all the segments and rays in the figure.

Segments: $\overline{AB}$, $\overline{AC}$, $\overline{BC}$, and $\overline{BD}$

Rays: $\overrightarrow{BA}$, $\overrightarrow{CA}$ or $\overrightarrow{CB}$, $\overrightarrow{AC}$ or $\overrightarrow{AB}$, $\overrightarrow{BC}$, and $\overrightarrow{BD}$

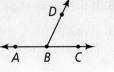

Exercises

Use the figure below for Exercises 7–9.

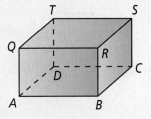

7. Name two intersecting lines.

8. Name the intersection of planes $QRBA$ and $TSRQ$.

9. Name three noncollinear points.

Determine whether the statement is *true* or *false*. Explain your reasoning.

10. Two points are always collinear.

11. $\overleftrightarrow{LM}$ and $\overrightarrow{ML}$ are the same ray.

1-3 Measuring Segments

Quick Review

The **distance** between two points is the length of the segment connecting those points. Segments with the same length are **congruent segments**. A **midpoint** of a segment divides the segment into two congruent segments.

Example

Are $\overline{AB}$ and $\overline{CD}$ congruent?

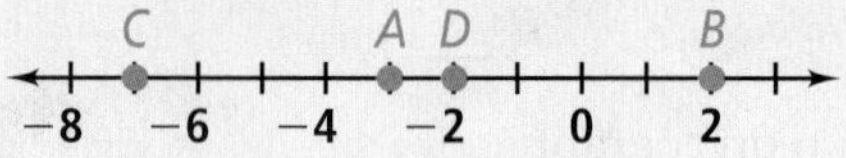

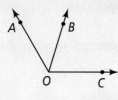

$AB = |-3 - 2| = |-5| = 5$

$CD = |-7 - (-2)| = |-5| = 5$

$AB = CD$, so $\overline{AB} \cong \overline{CD}$.

Exercises

For Exercises 12 and 13, use the number line below.

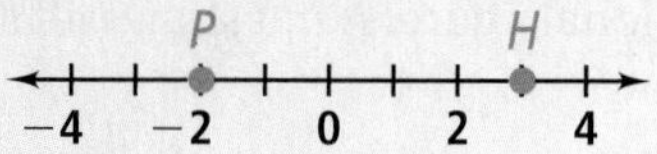

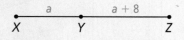

12. Find two possible coordinates of Q such that $PQ = 5$.

13. Use the number line above. Find the coordinate of the midpoint of $\overline{PH}$.

14. Find the value of m.

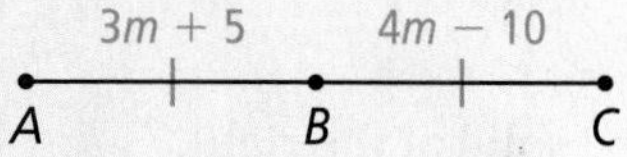

$$\underset{A}{\bullet} \quad \overset{3m + 5}{\underset{B}{\bullet}} \quad \overset{4m - 10}{\underset{C}{\bullet}}$$

15. If $XZ = 50$, what are XY and YZ?

$$\underset{X}{\bullet} \quad \overset{a}{\underset{Y}{\bullet}} \quad \overset{a + 8}{\underset{Z}{\bullet}}$$

1-4 Measuring Angles

Quick Review

Two rays with the same endpoint form an **angle**. The endpoint is the **vertex** of the angle. You can classify angles as acute, right, obtuse, or straight. Angles with the same measure are **congruent angles**.

Example

If $m\angle AOB = 47$ and $m\angle BOC = 73$, find $m\angle AOC$.

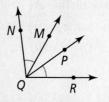

$m\angle AOC = m\angle AOB + m\angle BOC$

$\qquad = 47 + 73$

$\qquad = 120$

Exercises

Classify each angle as *acute, right, obtuse,* or *straight*.

16. **17.**

Use the diagram below for Exercises 18 and 19.

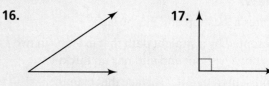

18. If $m\angle MQR = 61$ and $m\angle MQP = 25$, find $m\angle PQR$.

19. If $m\angle NQM = 2x + 8$ and $m\angle PQR = x + 22$, find the value of x.

1-5 Exploring Angle Pairs

Quick Review

Some pairs of angles have special names.

- **Adjacent angles:** coplanar angles with a common side, a common vertex, and no common interior points
- **Vertical angles:** sides are opposite rays
- **Complementary angles:** measures have a sum of 90
- **Supplementary angles:** measures have a sum of 180
- **Linear pair:** adjacent angles with noncommon sides as opposite rays

Angles of a linear pair are supplementary.

Example

Are ∠ACE and ∠BCD vertical angles? Explain.

No. They have only one set of sides with opposite rays.

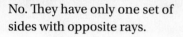

Exercises

Name a pair of each of the following.

20. complementary angles

21. supplementary angles

22. vertical angles

23. linear pair

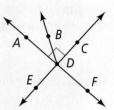

Find the value of x.

24.

$(3x + 31)°$ $(2x - 6)°$

25.

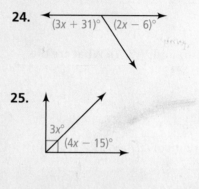

$3x°$
$(4x - 15)°$

1-6 Basic Constructions

Quick Review

Construction is the process of making geometric figures using a **compass** and a **straightedge**. Four basic constructions involve congruent segments, congruent angles, and bisectors of segments and angles.

Example

Construct $\overline{AB}$ congruent to $\overline{EF}$.

E ● ———— ● F

Step 1

Draw a ray with endpoint A.

A ●———————→

Step 2

Open the compass to the length of $\overline{EF}$. Keep that compass setting and put the compass point on point A. Draw an arc that intersects the ray. Label the point of intersection B.

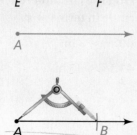

Exercises

26. Use a protractor to draw a 73° angle. Then construct an angle congruent to it.

27. Use a protractor to draw a 60° angle. Then construct the bisector of the angle.

28. Sketch $\overline{LM}$ on paper. Construct a line segment congruent to $\overline{LM}$. Then construct the perpendicular bisector of your line segment.

L ●———————————● M

29. a. Sketch ∠B on paper. Construct an angle congruent to ∠B.

    b. Construct the bisector of your angle from part (a).

1-7 Midpoint and Distance in the Coordinate Plane

Quick Review

You can find the coordinates of the midpoint M of $\overline{AB}$ with endpoints $A(x_1, y_1)$ and $B(x_2, y_2)$ using the **Midpoint Formula.**

$$M\left(\frac{x_1 + x_2}{2}, \frac{y_1 + y_2}{2}\right)$$

You can find the distance d between two points $A(x_1, y_1)$ and $B(x_2, y_2)$ using the **Distance Formula.**

$$d = \sqrt{(x_2 - x_1)^2 + (y_2 - y_1)^2}$$

Example

$\overline{GH}$ has endpoints $G(-11, 6)$ and $H(3, 4)$. What are the coordinates of its midpoint M?

$$x\text{-coordinate} = \frac{-11 + 3}{2} = -4$$

$$y\text{-coordinate} = \frac{6 + 4}{2} = 5$$

The coordinates of the midpoint of $\overline{GH}$ are $M(-4, 5)$.

Exercises

Find the distance between the points to the nearest tenth.

30. $A(-1, 5), B(0, 4)$

31. $C(-1, -1), D(6, 2)$

32. $E(-7, 0), F(5, 8)$

$\overline{AB}$ has endpoints $A(-3, 2)$ and $B(3, -2)$.

33. Find the coordinates of the midpoint of $\overline{AB}$.

34. Find AB to the nearest tenth.

M is the midpoint of $\overline{JK}$. Find the coordinates of K.

35. $J(-8, 4), M(-1, 1)$

36. $J(9, -5), M(5, -2)$

37. $J(0, 11), M(-3, 2)$

1-8 Perimeter, Circumference, and Area

Quick Review

The perimeter P of a polygon is the sum of the lengths of its sides. Circles have a circumference C. The area A of a polygon or a circle is the number of square units it encloses.

Square: $P = 4s$; $A = s^2$

Rectangle: $P = 2b + 2h$; $A = bh$

Triangle: $P = a + b + c$; $A = \frac{1}{2}bh$

Circle: $C = \pi d$ or $C = 2\pi r$; $A = \pi r^2$

Example

Find the perimeter and area of a rectangle with $b = 12$ m and $h = 8$ m.

$$P = 2b + 2h \qquad\qquad A = bh$$
$$= 2(12) + 2(8) \qquad\quad = 12 \cdot 8$$
$$= 40 \qquad\qquad\qquad = 96$$

The perimeter is 40 m and the area is 96 m^2.

Exercises

Find the perimeter and area of each figure.

38.

8 cm

39.

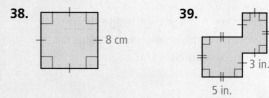

3 in.

5 in.

Find the circumference and the area for each circle in terms of π.

40. $r = 3$ in.

41. $d = 15$ m

Chapter Test

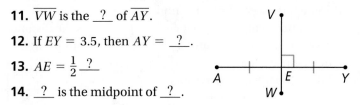

Do you know HOW?

1. Draw a net for a cube.

2. Draw an obtuse $\angle ABC$. Use a compass and a straightedge to bisect the angle.

Use the figure for Exercises 3–6.

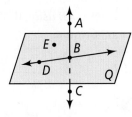

3. Name three collinear points.

4. Name four coplanar points.

5. What is the intersection of $\overleftrightarrow{AC}$ and plane Q?

6. How many planes contain the given line and point?

   a. $\overleftrightarrow{DB}$ and point A

   b. $\overleftrightarrow{BD}$ and point E

   c. $\overleftrightarrow{AC}$ and point D

   d. $\overleftrightarrow{EB}$ and point C

7. The running track at the right is a rectangle with a half circle on each end. $\overline{FI}$ and $\overline{GH}$ are diameters. Find the area inside the track to the nearest tenth.

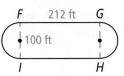

8. Algebra $M(x, y)$ is the midpoint of $\overline{CD}$ with endpoints $C(5, 9)$ and $D(17, 29)$.

   a. Find the values of x and y.

   b. Show that $MC = MD$.

9. Algebra If $JK = 48$, find the value of x.

$$\underset{4x-15}{\overset{J \qquad\qquad H \qquad\qquad K}{\bullet\!\!-\!\!-\!\!-\!\!-\!\!-\!\!-\!\!-\!\!-\!\!\bullet\!\!-\!\!-\!\!-\!\!-\!\!-\!\!-\!\!\bullet}}$$

10. To the nearest tenth, find the perimeter of $\triangle ABC$ with vertices $A(-2, -2)$, $B(0, 5)$, and $C(3, -1)$.

Use the figure to complete each statement.

11. $\overline{VW}$ is the ___?___ of $\overline{AY}$.

12. If $EY = 3.5$, then $AY = $ ___?___.

13. $AE = \frac{1}{2}$ ___?___

14. ___?___ is the midpoint of ___?___.

For the given dimensions, find the area of each figure. If necessary, round to the nearest hundredth.

15. rectangle with base 4 m and height 2 cm

16. square with side length 3.5 in.

17. circle with diameter 9 cm

Algebra Find the value of the variable.

18. $m\angle BDK = 3x + 4$, $m\angle JDR = 5x - 10$

19. $m\angle BDJ = 7y + 2$, $m\angle JDR = 2y + 7$

Do you UNDERSTAND?

Determine whether each statement is *always*, *sometimes*, or *never* true.

20. $\overrightarrow{LJ}$ and $\overrightarrow{TJ}$ are opposite rays.

21. Angles that form a linear pair are supplementary.

22. The intersection of two planes is a point.

23. Complementary angles are congruent.

24. Writing Explain why it is useful to have more than one way to name an angle.

25. You have 30 yd² of carpet. You want to install carpeting in a room that is 20 ft long and 15 ft wide. Do you have enough carpet? Explain.

TIPS FOR SUCCESS

Some questions ask you to find a distance using coordinate geometry. Read the sample question at the right. Then follow the tips to answer it.

What is the distance from the midpoint of $\overline{AB}$ to endpoint B?

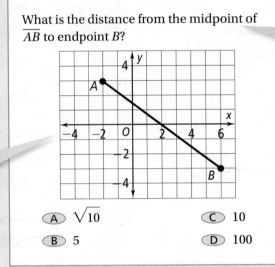

A $\sqrt{10}$

B   5

C   10

D   100

TIP 1

The midpoint divides the segment into two congruent segments that are each half of the total length.

Think It Through

Find AB using the Distance Formula.

$$AB = \sqrt{(-2 - 6)^2 + [3 - (-3)]^2}$$
$$= \sqrt{100}$$
$$= 10$$

The distance from the midpoint of $\overline{AB}$ to endpoint B is $\frac{1}{2}AB$, or 5. The correct answer is B.

TIP 2

Use the Distance Formula to find the length of the segment.

Vocabulary Builder

As you solve test items, you must understand the meanings of mathematical terms. Match each term with its mathematical meaning.

A. segment

B. angle bisector

C. construction

D. net

E. congruent angles

I. angles with the same measure

II. a two-dimensional diagram of a three-dimensional figure

III. the part of a line consisting of two endpoints and all points between

IV. a ray that divides an angle into two congruent angles

V. a geometric figure made using a straightedge and compass

Multiple Choice

Read each question. Then write the letter of the correct answer on your paper.

1. Points A, B, C, D, and E are collinear. A is to the right of B, E is to the right of D, and B is to the left of C. Which of the following is NOT a possible arrangement of the points from left to right?

   A D, B, A, E, C
   B D, B, A, C, E
   C B, D, E, C, A
   D B, A, E, C, D

2. A square and a rectangle have equal area. If the rectangle is 36 cm by 25 cm, what is the perimeter of the square?

   F   30 cm
   G   60 cm
   H   120 cm
   I   900 cm

3. Which construction requires drawing only one arc with a compass?

  Ⓐ constructing congruent segments

  Ⓑ constructing congruent angles

  Ⓒ constructing the perpendicular bisector

  Ⓓ constructing the angle bisector

4. Rick paints the four walls in a room that is 12 ft long and 10 ft wide. The ceiling in the room is 8 ft from the floor. The doorway is 3 ft by 7 ft, and the window is 6 ft by 5 ft. If Rick does NOT paint the doorway or window, what is the approximate area that he paints?

  Ⓕ 301 ft²   Ⓗ 331 ft²

  Ⓖ 322 ft²   Ⓘ 352 ft²

5. If ∠A and ∠B are supplementary angles, what angle relationship between ∠A and ∠B CANNOT be true?

  Ⓐ ∠A and ∠B are right angles.

  Ⓑ ∠A and ∠B are adjacent angles.

  Ⓒ ∠A and ∠B are complementary angles.

  Ⓓ ∠A and ∠B are congruent angles.

6. A net for a small rectangular gift box is shown below. What is the total area of the net?

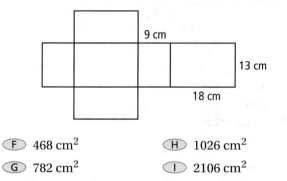

  Ⓕ 468 cm²   Ⓗ 1026 cm²

  Ⓖ 782 cm²   Ⓘ 2106 cm²

7. The measure of an angle is 12 less than twice the measure of its supplement. What is the measure of the angle?

  Ⓐ 28   Ⓒ 64

  Ⓑ 34   Ⓓ 116

8. Given: ∠A

What is the second step in constructing the angle bisector of ∠A?

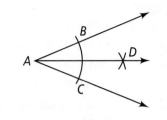

  Ⓕ Draw $\overrightarrow{AD}$.

  Ⓖ From points B and C, use the same compass setting to draw arcs that intersect at D.

  Ⓗ Draw a line segment connecting points B and C.

  Ⓘ From point A, draw an arc that intersects the sides of the angle at points B and C.

9. What is the distance, to the nearest tenth, from point D to point E through points C, A, B, and F?

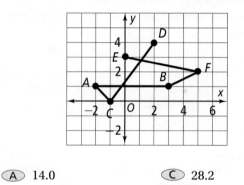

  Ⓐ 14.0   Ⓒ 28.2

  Ⓑ 18.7   Ⓓ 34.4

10. Which postulate most closely resembles the Angle Addition Postulate?

  Ⓕ Ruler Postulate

  Ⓖ Protractor Postulate

  Ⓗ Segment Addition Postulate

  Ⓘ Area Addition Postulate

11. What is the length of the segment with endpoints A(1, 7) and B(−3, −1)?

  Ⓐ $\sqrt{40}$ Ⓒ $\sqrt{80}$

  Ⓑ 8   Ⓓ 40

12. The measure of an angle is 78 less than the measure of its complement. What is the measure of the angle?

13. The face of a circular game token has an area of 10π cm². What is the diameter of the game token? Round to the nearest hundredth of a centimeter.

14. The measure of an angle is one third the measure of its supplement. What is the measure of the angle?

15. Bill's bike wheels have a 26-in. diameter. The odometer on his bike counts the number of times a wheel rotates during a trip. If the odometer counts 200 rotations during the trip from Bill's house to school, how many feet does Bill travel? Round to the nearest foot.

16. Y is the midpoint of $\overline{XZ}$. What is the value of b?

$$\overset{2b-1}{\underset{X \qquad\qquad Y}{\rule{3cm}{0.4pt}}} \overset{26-4b}{\underset{\qquad\qquad Z}{\rule{3cm}{0.4pt}}}$$

17. A rectangular garden has dimensions 6 ft by 17 ft. A second rectangular garden has dimensions 4 yd by 9 yd. What is the area in square feet of the larger of the two gardens?

18. The sum of the measures of a complement and a supplement of an angle is 200. What is the measure of the angle?

19. What is the area in square units of a rectangle with vertices $(-2, 5)$, $(3, 5)$, $(3, -1)$, and $(-2, -1)$?

20. $\overline{AB}$ has endpoints $A(-4, 5)$ and $B(3, 5)$. What is the x-coordinate of a point C such that B is the midpoint of $\overline{AC}$?

Short Response

21. The two blocks of cheese shown below are cut into slices of equal thickness. If the cheese sells at the same cost per slice, which type of cheese slice is the better deal? Explain your reasoning.

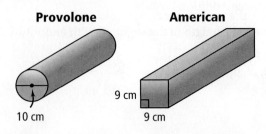

Provolone

10 cm

American

9 cm
9 cm

22. The midpoint of $\overline{GI}$ is $(2, -1)$. One endpoint is $G(-1, -3)$. What are the coordinates of endpoint I?

23. Copy the graph below. Connect the midpoints of the sides of the square consecutively. What is the perimeter of the new square? Show your work.

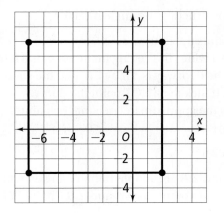

Extended Response

24. Make an orthographic drawing of the three-dimensional drawing. How might an orthographic drawing of the figure be more useful than a net?

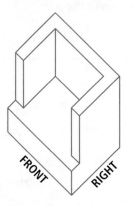

FRONT   RIGHT

25. A packaging company wants to fit 6 energy-drink cans in a cardboard box, as shown below. The bottom of each can is a circle with an area of 9π cm².

a. What is the total area taken up by the bottoms of the cans? Round to the nearest hundredth.

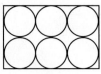

b. Will the cans fit in a box with length 16 cm and height 12 cm? Explain.

Get Ready!

Skills Handbook, p. 882

◀ Evaluating Expressions

Algebra **Evaluate each expression for the given value of x.**

1. $9x - 13$ for $x = 7$ **2.** $90 - 3x$ for $x = 31$ **3.** $\frac{1}{2}x + 14$ for $x = 23$

Skills Handbook, p. 886

◀ Solving Equations

Algebra **Solve each equation.**

4. $2x - 17 = 4$ **5.** $3x + 8 = 53$

6. $(10x + 5) + (6x - 1) = 180$ **7.** $14x = 2(5x + 14)$

8. $2(x + 4) = x + 13$ **9.** $7x + 5 = 5x + 17$

10. $(x + 21) + (2x + 9) = 90$ **11.** $2(3x - 4) + 10 = 5(x + 4)$

Lessons 1-3 through 1-5

◀ Segments and Angles

Use the figure at the right.

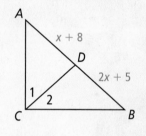

12. Name $\angle 1$ in two other ways.

13. If D is the midpoint of $\overline{AB}$, find the value of x.

14. If $\angle ACB$ is a right angle, $m\angle 1 = 4y$, and $m\angle 2 = 2y + 18$, find $m\angle 1$ and $m\angle 2$.

15. Name a pair of angles that form a linear pair.

16. Name a pair of adjacent angles that are not supplementary.

17. If $m\angle ADC + m\angle BDC = 180$, name the straight angle.

Looking Ahead Vocabulary

18. A scientist often makes an assumption, or *hypothesis*, about a scientific problem. Then the scientist uses experiments to test the *hypothesis* to see if it is true. How might a *hypothesis* in geometry be similar? How might it be different?

19. The *conclusion* of a novel answers questions raised by the story. How do you think the term *conclusion* applies in geometry?

20. A detective uses *deductive reasoning* to solve a case by gathering, combining, and analyzing clues. How might you use *deductive reasoning* in geometry?

CHAPTER 2

Reasoning and Proof

PowerGeometry.com

Your place to get all things digital

ownload videos
onnecting math
your world.

VIDEO

ath definitions
English and
panish

VOCABULARY

he online
olve It will get
ou in gear for
ach lesson.

SOLVE IT!

nteractive!
ary numbers,
raphs, and figures
o explore math
oncepts.

DYNAMIC ACTIVITIES

Online access to stepped-out problems aligned to Common Core

ONLINE PROBLEMS

Get and view your assignments online.

ONLINE HOMEWORK

Extra practice and review online

MathXL® FOR SCHOOL

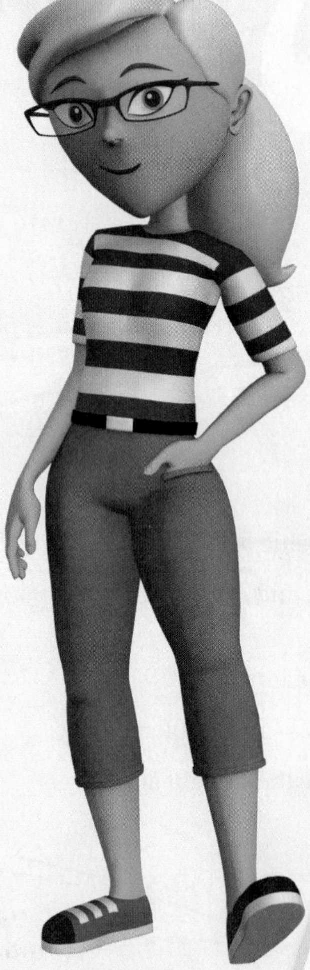

DOMAINS

- Construct Viable Arguments
- Attend to precision

Look at all the dominoes! Whether a domino falls depends on the dominoes before it. In a proof, each statement follows logically from the previous statements.

You'll learn about logical reasoning in this chapter.

Vocabulary

English/Spanish Vocabulary Audio Online:

English	Spanish
biconditional, *p. 98*	bicondicional
conclusion, *p. 89*	conclusión
conditional, *p. 89*	condicional
conjecture, *p. 83*	conjetura
contrapositive, *p. 91*	contrapositivo
converse, *p. 91*	recíproco
deductive reasoning, *p. 106*	razonamiento deductivo
hypothesis, *p. 89*	hipótesis
inductive reasoning, *p. 82*	razonamiento inductivo
inverse, *p. 91*	inverso
negation, *p. 91*	negación
theorem, *p. 120*	teorema

BIG ideas

Reasoning and Proof

Essential Question How can you make a conjecture and prove that it is true?

Chapter Preview

Patterns and Inductive Reasoning

© **Content Standards**
Prepares for G.CO.9 Prove theorems about lines and angles.
Prepares for G.CO.10 Prove theorems about triangles.
Prepares for G.CO.11 Prove theorems about parallelograms.

Objective To use inductive reasoning to make conjectures

Getting Ready!

Fold a piece of paper in half. When you unfold it, the paper is divided into two rectangles. Refold the paper, and then fold it in half again. This time when you unfold it, there are four rectangles. How many rectangles would you get if you folded a piece of paper in half eight times? Explain.

See if you can find a pattern to help you solve this problem.

© **MATHEMATICAL PRACTICES** In the Solve It, you may have used inductive reasoning. **Inductive reasoning** is reasoning based on patterns you observe.

Essential Understanding You can observe patterns in some number sequences and some sequences of geometric figures to discover relationships.

© **Problem 1** **Finding and Using a Pattern**

Look for a pattern. What are the next two terms in each sequence?

Plan

How do you look for a pattern in a sequence?
Look for a relationship between terms. Test that the relationship is consistent throughout the sequence.

A 3, 9, 27, 81, . . .

Each term is three times the previous term. The next two terms are
$81 \times 3 = 243$ and $243 \times 3 = 729$.

B

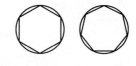

Each circle contains a polygon that has one more side than the preceding polygon. The next two circles contain a six-sided and a seven-sided polygon.

Lesson Vocabulary
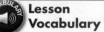
• inductive reasoning
• conjecture
• counterexample

 Got It? **1.** What are the next two terms in each sequence?
  a. 45, 40, 35, 30, . . .   **b.**

You may want to find the tenth or the one-hundreth term in a sequence. In this case, rather than find every previous term, you can look for a pattern and make a conjecture. A **conjecture** is a conclusion you reach using inductive reasoning.

 Problem 2 **Using Inductive Reasoning**

Plan

Do you need to draw a circle with 20 diameters?
No. *Solve a simpler problem* by finding the number of regions formed by 1, 2, and 3 diameters. Then look for a pattern.

Look at the circles. What conjecture can you make about the number of regions 20 diameters form?

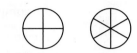

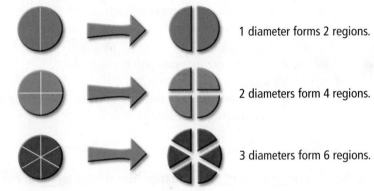

1 diameter forms 2 regions.

2 diameters form 4 regions.

3 diameters form 6 regions.

Each circle has twice as many regions as diameters. Twenty diameters form 20 · 2, or 40 regions.

Got It? **2.** What conjecture can you make about the twenty-first term in R, W, B, R, W, B, . . . ?

It is important to gather enough data before you make a conjecture. For example, you do not have enough information about the sequence 1, 3, . . . to make a reasonable conjecture. The next term could be 3 · 3 = 9 or 3 + 2 = 5.

 **Problem 3** **Collecting Information to Make a Conjecture**

Plan

What's the first step?
Start by gathering data. You can organize your data by *making a table*.

What conjecture can you make about the sum of the first 30 even numbers?

Find the first few sums and look for a pattern.

Number of Terms	Sum		
1	2	= 2	= 1 · 2
2	2 + 4	= 6	= 2 · 3
3	2 + 4 + 6	= 12	= 3 · 4
4	2 + 4 + 6 + 8	= 20	= 4 · 5

Each sum is the product of the number of terms and the number of terms plus one.

You can conclude that the sum of the first 30 even numbers is 30 · 31, or 930.

Got It? **3.** What conjecture can you make about the sum of the first 30 odd numbers?

Problem 4 Making a Prediction

Sales Sales of backpacks at a nationwide company decreased over a period of six consecutive months. What conjecture can you make about the number of backpacks the company will sell in May?

The points seem to fall on a line. The graph shows the number of sales decreasing by about 500 backpacks each month. By inductive reasoning, you can estimate that the company will sell approximately 8000 backpacks in May.

Backpacks Sold

Got It? **4. a.** What conjecture can you make about backpack sales in June?

b. Reasoning Is it reasonable to use this graph to make a conjecture about sales in August? Explain.

Not all conjectures turn out to be true. You should test your conjecture multiple times. You can prove that a conjecture is false by finding *one* counterexample. A **counterexample** is an example that shows that a conjecture is incorrect.

Problem 5 Finding a Counterexample

What is a counterexample for each conjecture?

A **If the name of a month starts with the letter J, it is a summer month.**

Counterexample: January starts with J and it is a winter month.

B **You can connect any three points to form a triangle.**

Counterexample: If the three points lie on a line, you cannot form a triangle.

These three points support the conjecture . . .   . . . but these three points are a counterexample to the conjecture.

Think

What numbers should you *guess-and-check*?
Try positive numbers, negative numbers, fractions, and special cases like zero.

C **When you multiply a number by 2, the product is greater than the original number.**

The conjecture is true for positive numbers, but it is false for negative numbers and zero.

Counterexample: $-4 \cdot 2 = -8$ and $-8 \ngtr -4$.

Got It? **5.** What is a counterexample for each conjecture?

a. If a flower is red, it is a rose.

b. One and only one plane exists through any three points.

c. When you multiply a number by 3, the product is divisible by 6.

Lesson Check

Do you know HOW?

What are the next two terms in each sequence?

1. 7, 13, 19, 25, . . .

2.

3. What is a counterexample for the following conjecture?

All four-sided figures are squares.

Do you UNDERSTAND?

④ 4. Vocabulary How does the word *counter* help you understand the term *counterexample*?

④ 5. Compare and Contrast Clay thinks the next term in the sequence 2, 4, . . . is 6. Given the same pattern, Ott thinks the next term is 8, and Stacie thinks the next term is 7. What conjecture is each person making? Is there enough information to decide who is correct?

Practice and Problem-Solving Exercises ⓒ MATHEMATICAL PRACTICES

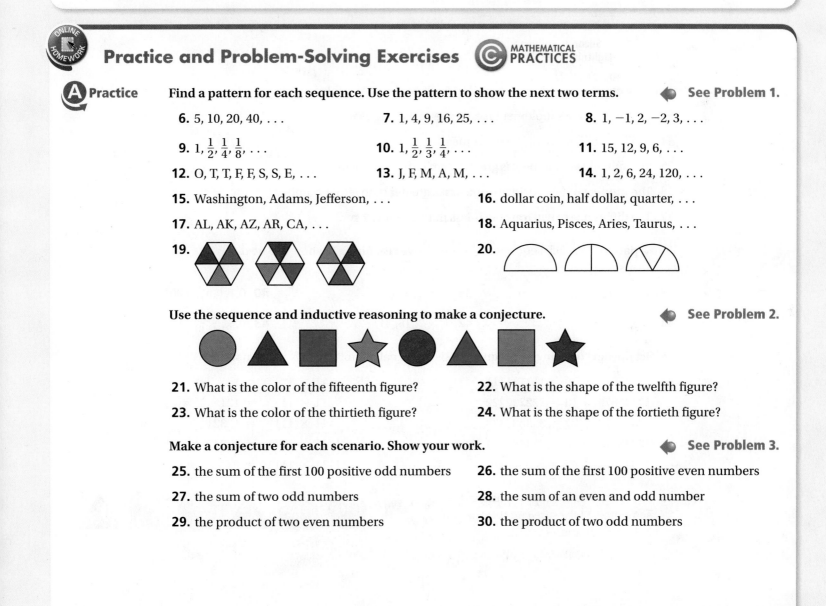

Ⓐ Practice

Find a pattern for each sequence. Use the pattern to show the next two terms. ◀ **See Problem 1.**

6. 5, 10, 20, 40, . . .

7. 1, 4, 9, 16, 25, . . .

8. 1, −1, 2, −2, 3, . . .

9. 1, $\frac{1}{2}$, $\frac{1}{4}$, $\frac{1}{8}$, . . .

10. 1, $\frac{1}{2}$, $\frac{1}{3}$, $\frac{1}{4}$, . . .

11. 15, 12, 9, 6, . . .

12. O, T, T, F, F, S, S, E, . . .

13. J, F, M, A, M, . . .

14. 1, 2, 6, 24, 120, . . .

15. Washington, Adams, Jefferson, . . .

16. dollar coin, half dollar, quarter, . . .

17. AL, AK, AZ, AR, CA, . . .

18. Aquarius, Pisces, Aries, Taurus, . . .

19.

20.

Use the sequence and inductive reasoning to make a conjecture. ◀ **See Problem 2.**

21. What is the color of the fifteenth figure?

22. What is the shape of the twelfth figure?

23. What is the color of the thirtieth figure?

24. What is the shape of the fortieth figure?

Make a conjecture for each scenario. Show your work. ◀ **See Problem 3.**

25. the sum of the first 100 positive odd numbers

26. the sum of the first 100 positive even numbers

27. the sum of two odd numbers

28. the sum of an even and odd number

29. the product of two even numbers

30. the product of two odd numbers

STEM **Weather** Use inductive reasoning to make a prediction about the weather.  **See Problem 4.**

31. Lightning travels much faster than thunder, so you see lightning before you hear thunder. If you count 5 s between the lightning and thunder, how far away is the storm?

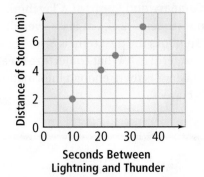

32. The speed at which a cricket chirps is affected by the temperature. If you hear 20 cricket chirps in 14 s, what is the temperature?

Number of Chirps per 14 Seconds	Temperature (°F)
5	45
10	55
15	65

Find one counterexample to show that each conjecture is false.  **See Problem 5.**

33. ∠1 and ∠2 are supplementary, so one of the angles is acute.

34. △*ABC* is a right triangle, so ∠*A* measures 90.

35. The sum of two numbers is greater than either number.

36. The product of two positive numbers is greater than either number.

37. The difference of two integers is less than either integer.

B **Apply**

Find a pattern for each sequence. Use inductive reasoning to show the next two terms.

38. 1, 3, 7, 13, 21, . . .

39. 1, 2, 5, 6, 9, . . .

40. 0.1, 0.01, 0.001, . . .

41. 2, 6, 7, 21, 22, 66, 67, . . .

42. 1, 3, 7, 15, 31, . . .

43. $0, \frac{1}{2}, \frac{3}{4}, \frac{7}{8}, \frac{15}{16}, \ldots$

Predict the next term in each sequence. Use your calculator to verify your answer.

44. 12345679 × 9 = 111111111
12345679 × 18 = 222222222
12345679 × 27 = 333333333
12345679 × 36 = 444444444
12345679 × 45 = ■

45.   1 × 1   = 1
   11 × 11   = 121
  111 × 111 = 12321
 1111 × 1111 = 1234321
11111 × 11111 = ■

© **46. Patterns** Draw the next figure in the sequence. Make sure you think about color and shape.

Draw the next figure in each sequence.

47.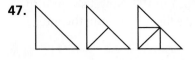

48.

© **49. Reasoning** Find the perimeter when 100 triangles are put together in the pattern shown. Assume that all triangle sides are 1 cm long.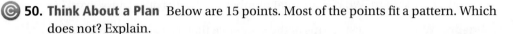

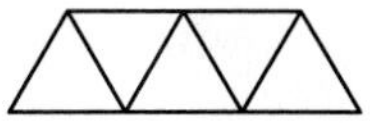

© **50. Think About a Plan** Below are 15 points. Most of the points fit a pattern. Which does not? Explain.

$A(6, -2)$ $B(6, 5)$ $C(8, 0)$ $D(8, 7)$ $E(10, 2)$ $F(10, 6)$ $G(11, 4)$ $H(12, 3)$
$I(4, 0)$ $J(7, 6)$ $K(5, 6)$ $L(4, 7)$ $M(2, 2)$ $N(1, 4)$ $O(2, 6)$

- How can you draw a diagram to help you find a pattern?
- What pattern do the majority of the points fit?

© **51. Language** Look for a pattern in the Chinese number system.
  a. What is the Chinese name for the numbers 43, 67, and 84?
  b. Reasoning Do you think that the Chinese number system is base 10? Explain.

© **52. Open-Ended** Write two different number sequences that begin with the same two numbers.

© **53. Error Analysis** For each of the past four years, Paulo has grown 2 in. every year. He is now 16 years old and is 5 ft 10 in. tall. He figures that when he is 22 years old he will be 6 ft 10 in. tall. What would you tell Paulo about his conjecture?

Chinese Number System

Number	Chinese Word	Number	Chinese Word
1	yī	9	jǐu
2	èr	10	shí
3	sān	11	shí-yī
4	sì	12	shí-èr
5	wǔ	⋮	⋮
6	lìu	20	èr-shí
7	qī	21	èr-shí-yī
8	bā	⋮	⋮
		30	sān-shí

STEM **54. Bird Migration** During bird migration, volunteers get up early on Bird Day to record the number of bird species they observe in their community during a 24-h period. Results are posted online to help scientists and students track the migration.
  a. Make a graph of the data.
  b. Use the graph and inductive reasoning to make a conjecture about the number of bird species the volunteers in this community will observe in 2015.

Bird Count

Year	Number of Species
2004	70
2005	83
2006	80
2007	85
2008	90

© **55. Writing** Describe a real-life situation in which you recently used inductive reasoning.

56. History When he was in the third grade, German mathematician Karl Gauss (1777–1855) took ten seconds to sum the integers from 1 to 100. Now it's your turn. Find a fast way to sum the integers from 1 to 100. Find a fast way to sum the integers from 1 to n. (*Hint:* Use patterns.)

57. Chess The small squares on a chessboard can be combined to form larger squares. For example, there are sixty-four 1×1 squares and one 8×8 square. Use inductive reasoning to determine how many 2×2 squares, 3×3 squares, and so on, are on a chessboard. What is the total number of squares on a chessboard?

58. a. Algebra Write the first six terms of the sequence that starts with 1, and for which the difference between consecutive terms is first 2, and then 3, 4, 5, and 6.

 b. Evaluate $\frac{n^2 + n}{2}$ for $n = 1, 2, 3, 4, 5$, and 6. Compare the sequence you get with your answer for part (a).

 c. Examine the diagram at the right and explain how it illustrates a value of $\frac{n^2 + n}{2}$.

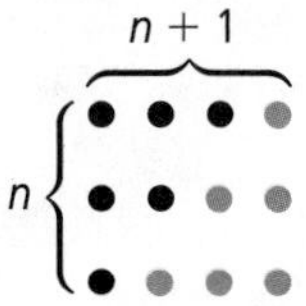

 d. Draw a similar diagram to represent $\frac{n^2 + n}{2}$ for $n = 5$.

Standardized Test Prep

SAT/ACT

59. What is the next term in the sequence 1, 1, 2, 3, 5, 8, 13, . . . ?

 Ⓐ 17 Ⓑ 20 Ⓒ 21 Ⓓ 24

60. A horse trainer wants to build three adjacent rectangular corrals as shown at the right. The area of each corral is 7200ft². If the length of each corral is 120 ft, how much fencing does the horse trainer need to buy in order to build the corrals?

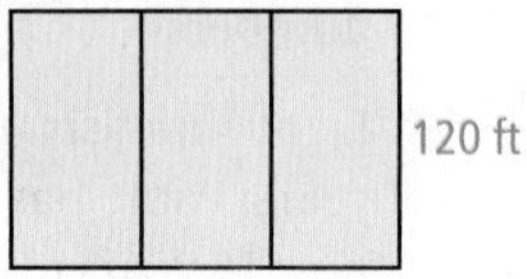

120 ft

 Ⓕ 300 ft Ⓖ 360 ft Ⓗ 560 ft Ⓘ 840 ft

Short Response

61. The coordinates x, y, a, and b are all positive integers. Could the points (x, y) and (a, b) have a midpoint in Quadrant III? Explain.

Mixed Review

62. What is the area of a circle with radius 4 in.? Leave your answer in terms of π. ◀ **See Lesson 1-8.**

63. What is the perimeter of a rectangle with side lengths 3 m and 7 m?

64. Solve for x if B is the midpoint of $\overline{AC}$. ◀ **See Lesson 1-3.**

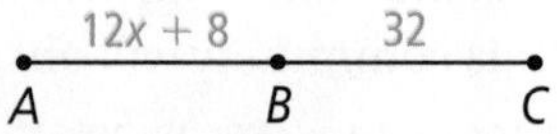

$$\underset{A}{\bullet}\overset{12x + 8}{}\underset{B}{\bullet}\overset{32}{}\underset{C}{\bullet}$$

Get Ready! **To prepare for Lesson 2-2, do Exercises 65 and 66.**

Tell whether each conjecture is *true* or *false*. Explain. ◀ **See Lesson 2-1.**

65. The sum of two even numbers is even. **66.** The sum of three odd numbers is odd.

Conditional Statements

© **Content Standards**
Prepares for G.CO.9 Prove theorems about lines and angles.
Prepares for G.CO.10 Prove theorems about triangles.
Prepares for G.CO.11 Prove theorems about parallelograms.

Objectives To recognize conditional statements and their parts
To write converses, inverses, and contrapositives of conditionals

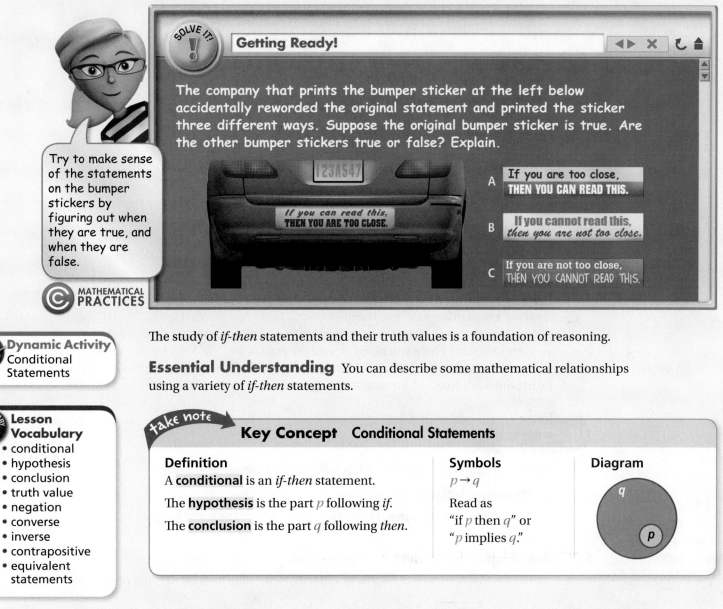

Try to make sense of the statements on the bumper stickers by figuring out when they are true, and when they are false.

© **MATHEMATICAL PRACTICES**

SOLVE IT!

Getting Ready!

The company that prints the bumper sticker at the left below accidentally reworded the original statement and printed the sticker three different ways. Suppose the original bumper sticker is true. Are the other bumper stickers true or false? Explain.

If you can read this, **THEN YOU ARE TOO CLOSE.**

A **If you are too close,** **THEN YOU CAN READ THIS.**

B **If you cannot read this,** *then you are not too close.*

C **If you are not too close,** THEN YOU CANNOT READ THIS.

Dynamic Activity
Conditional Statements

The study of *if-then* statements and their truth values is a foundation of reasoning.

Essential Understanding You can describe some mathematical relationships using a variety of *if-then* statements.

Lesson Vocabulary
• conditional
• hypothesis
• conclusion
• truth value
• negation
• converse
• inverse
• contrapositive
• equivalent statements

take note

Key Concept   Conditional Statements

Definition	Symbols	Diagram
A **conditional** is an *if-then* statement.	$p \rightarrow q$	
The **hypothesis** is the part p following *if*.	Read as	
The **conclusion** is the part q following *then*.	"if p then q" or "p implies q."	

The Venn diagram above illustrates how the set of things that satisfy the hypothesis lies inside the set of things that satisfy the conclusion.

What would a Venn diagram look like?
A robin is a kind of bird, so the set of robins (R) should be inside the set of birds (B).

© **Problem 1** Identifying the Hypothesis and the Conclusion

What are the hypothesis and the conclusion of the conditional?
    If an animal is a robin, then the animal is a bird.

Hypothesis (p): An animal is a robin.

Conclusion (q): The animal is a bird.

✓ **Got It? 1.** What are the hypothesis and the conclusion of the conditional?
                    If an angle measures 130, then the angle is obtuse.

Which part of the statement is the hypothesis (*p*)?
For two angles to be vertical, they must share a vertex. So the set of vertical angles (*p*) is inside the set of angles that share a vertex (*q*).

© **Problem 2** Writing a Conditional

How can you write the following statement as a conditional?
    Vertical angles share a vertex.

Step 1   Identify the hypothesis and the conclusion.

                Vertical angles share a vertex.

Step 2   Write the conditional.

                If two angles are vertical, then they share a vertex.

✓ **Got It? 2.** How can you write "Dolphins are mammals" as a conditional?

The **truth value** of a conditional is either *true* or *false*. To show that a conditional is true, show that every time the hypothesis is true, the conclusion is also true. A counterexample can help you determine whether a conditional with a true hypothesis is true. To show that the conditional is false, if you find one counterexample for which the hypothesis is true and the conclusion is false, then the truth value of the conditional is false.

© **Problem 3** Finding the Truth Value of a Conditional

Is the conditional *true* or *false*? If it is false, find a counterexample.

🅐 **If a woman is Hungarian, then she is European.**

        The conditional is true. Hungary is a European nation, so Hungarians are European.

How do you find a counterexample?
Find an example where the hypothesis is true, but the conclusion is false. For part (B), find a number divisible by 3 that is not odd.

🅑 **If a number is divisible by 3, then it is odd.**

        The conditional is false. The number 12 is divisible by 3, but it is not odd.

✓ **Got It? 3.** Is the conditional *true* or *false*? If it is false, find a counterexample.
                a. If a month has 28 days, then it is February.
                b. If two angles form a linear pair, then they are supplementary.

The **negation** of a statement p is the opposite of the statement. The symbol is $\sim p$ and is read "not p." The negation of the statement "The sky is blue" is "The sky is *not* blue." You can use negations to write statements related to a conditional. Every conditional has three related conditional statements.

take note

Key Concept   Related Conditional Statements

Statement	How to Write It	Example	Symbols	How to Read It
Conditional	Use the given hypothesis and conclusion.	If $m\angle A = 15$, then $\angle A$ is acute.	$p \rightarrow q$	If p, then q.
Converse	Exchange the hypothesis and the conclusion.	If $\angle A$ is acute, then $m\angle A = 15$.	$q \rightarrow p$	If q, then p.
Inverse	Negate both the hypothesis and the conclusion of the conditional.	If $m\angle A \neq 15$, then $\angle A$ is not acute.	$\sim p \rightarrow \sim q$	If not p, then not q.
Contrapositive	Negate both the hypothesis and the conclusion of the converse.	If $\angle A$ is not acute, then $m\angle A \neq 15$.	$\sim q \rightarrow \sim p$	If not q, then not p.

Below are the truth values of the related statements above. **Equivalent statements** have the same truth value.

Statement	Example	Truth Value
Conditional	If $m\angle A = 15$, then $\angle A$ is acute.	True
Converse	If $\angle A$ is acute, then $m\angle A = 15$.	False
Inverse	If $m\angle A \neq 15$, then $\angle A$ is not acute.	False
Contrapositive	If $\angle A$ is not acute, then $m\angle A \neq 15$.	True

A conditional and its contrapositive are equivalent statements. They are either both true or both false. The converse and inverse of a statement are also equivalent statements.

Problem 4   Writing and Finding Truth Values of Statements

What are the converse, inverse, and contrapositive of the following conditional? What are the truth values of each? If a statement is false, give a counterexample.
If the figure is a square, then the figure is a quadrilateral.

Think

Identify the hypothesis and the conclusion.

To write the converse, switch the hypothesis and the conclusion. Write $q \rightarrow p$.

To write the inverse, negate both the hypothesis and the conclusion of the conditional. Write $\sim p \rightarrow \sim q$.

To write the contrapositive, negate both the hypothesis and the conclusion of the converse. Write $\sim q \rightarrow \sim p$.

Write

p: The figure is a square.
q: The figure is a quadrilateral.
Converse: If the figure is a quadrilateral, **then** the figure is a square.
The converse is false.
Counterexample:
A rectangle that is not a square.
Inverse: If the figure is not a square, **then** the figure is not a quadrilateral.
The inverse is false.
Counterexamples:

Contrapositive: If the figure is not a quadrilateral, **then** the figure is not a square.
The contrapositive is true.

Got It?   4. What are the converse, inverse, and contrapositive of the conditional statement below? What are the truth values of each? If a statement is false, give a counterexample.
If a vegetable is a carrot, then it contains beta carotene.

Lesson Check

Do you know HOW?

1. What are the hypothesis and the conclusion of the following statement? Write it as a conditional.
Residents of Key West live in Florida.

2. What are the converse, inverse, and contrapositive of the statement? Which statements are true?
If a figure is a rectangle with sides 2 cm and 3 cm, then it has a perimeter of 10 cm.

Do you UNDERSTAND? MATHEMATICAL PRACTICES

3. Error Analysis Your classmate rewrote the statement "You jog every Sunday" as the following conditional. What is your classmate's error? Correct it.
If you jog, then it is Sunday.

4. Reasoning Suppose a conditional statement and its converse are both true. What are the truth values of the contrapositive and inverse? How do you know?

Practice and Problem-Solving Exercises © MATHEMATICAL PRACTICES

A Practice

Identify the hypothesis and conclusion of each conditional. See Problem 1.

5. If you are an American citizen, then you have the right to vote.

6. If a figure is a rectangle, then it has four sides.

7. If you want to be healthy, then you should eat vegetables.

Write each sentence as a conditional. See Problem 2.

8. Hank Aaron broke Babe Ruth's home-run record.

9. Algebra $3x - 7 = 14$ implies that $3x = 21$.

10. Thanksgiving in the United States falls on the fourth Thursday of November.

11. A counterexample shows that a conjecture is false.

12. Coordinate Geometry A point in the first quadrant has two positive coordinates.

Write a conditional statement that each Venn diagram illustrates.

13.   **14.**   **15.**

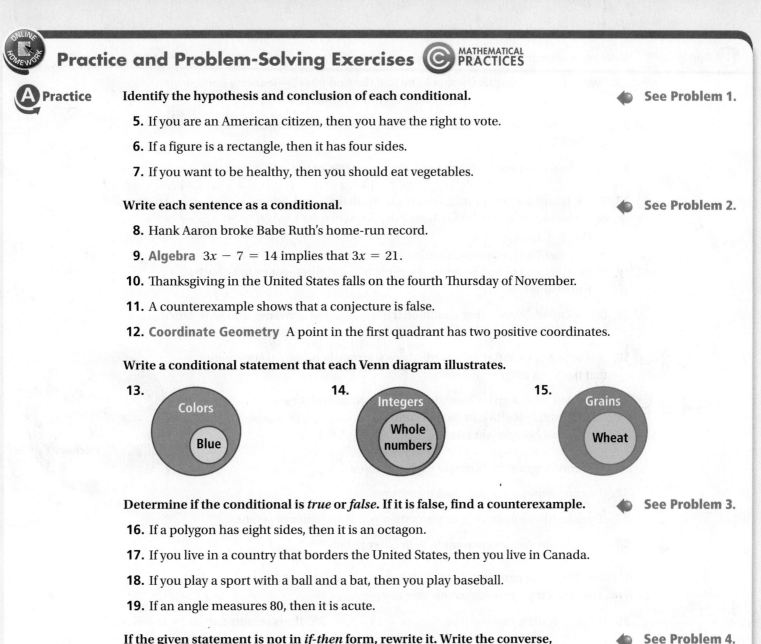

Determine if the conditional is *true* or *false*. If it is false, find a counterexample. See Problem 3.

16. If a polygon has eight sides, then it is an octagon.

17. If you live in a country that borders the United States, then you live in Canada.

18. If you play a sport with a ball and a bat, then you play baseball.

19. If an angle measures 80, then it is acute.

If the given statement is not in *if-then* form, rewrite it. Write the converse, inverse, and contrapositive of the given conditional statement. Determine the truth value of all four statements. If a statement is false, give a counterexample. See Problem 4.

20. If you are a quarterback, then you play football.

21. Pianists are musicians.

22. Algebra If $4x + 8 = 28$, then $x = 5$.

23. Odd natural numbers less than 8 are prime.

24. Two lines that lie in the same plane are coplanar.

Write each statement as a conditional.

25. "We're half the people; we should be half the Congress." —Jeanette Rankin, former U.S. congresswoman, calling for more women in office

26. "Anyone who has never made a mistake has never tried anything new." —Albert Einstein

27. Probability An event with probability 1 is certain to occur.

28. Think About a Plan Your classmate claims that the conditional and contrapositive of the following statement are both true. Is he correct? Explain.
   If $x = 2$, then $x^2 = 4$.
   • Can you find a counterexample of the conditional?
   • Do you need to find a counterexample of the contrapositive to know its truth value?

29. Open-Ended Write a true conditional that has a true converse, and write a true conditional that has a false converse.

30. Multiple Representations Write three separate conditional statements that the Venn diagram illustrates.

31. Error Analysis A given conditional is true. Natalie claims its contrapositive is also true. Sean claims its contrapositive is false. Who is correct and how do you know?

Draw a Venn diagram to illustrate each statement.

32. If an angle measures 100, then it is obtuse.

33. If you are the captain of your team, then you are a junior or senior.

34. Peace Corps volunteers want to help other people.

Algebra Write the converse of each statement. If the converse is true, write *true*. If it is not true, provide a counterexample.

35. If $x = -6$, then $|x| = 6$.

36. If y is negative, then $-y$ is positive.

37. If $x < 0$, then $x^3 < 0$.

38. If $x < 0$, then $x^2 > 0$.

39. Advertising Advertisements often suggest conditional statements. What conditional does the ad at the right imply?

Write each postulate as a conditional statement.

40. Two intersecting lines meet in exactly one point.

41. Two congruent figures have equal areas.

42. Through any two points there is exactly one line.

Challenge Write a statement beginning with *all*, *some*, or *no* to match each Venn diagram.

43.

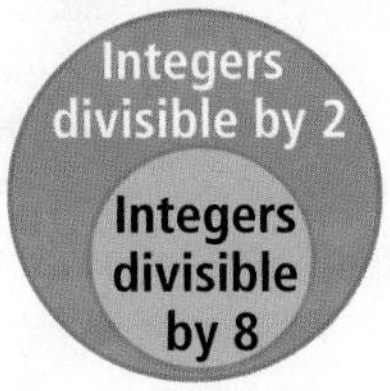

Integers
divisible by 2

Integers
divisible
by 8

44.

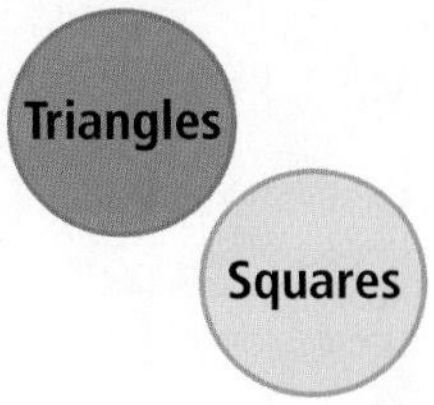

Triangles

Squares

45.

Students

Musicians

46. Let *a* represent an integer. Consider the five statements *r*, *s*, *t*, *u*, and *v*.

    r: a is even.   *s: a* is odd.   *t:* 2*a* is even.   *u:* 2*a* is odd.   *v:* 2*a* + 1 is odd.

How many statements of the form $p \rightarrow q$ can you make from these statements?
Decide which are true, and provide a counterexample if they are false.

Standardized Test Prep

SAT/ACT

47. Which conditional and its converse are both true?

  Ⓐ If $x = 1$, then $2x = 2$.   Ⓒ If $x = 3$, then $x^2 = 6$.

  Ⓑ If $x = 2$, then $x^2 = 4$.   Ⓓ If $x^2 = 4$, then $x = 2$.

48. What is the midpoint of the segment with endpoints $(-3, 7)$ and $(9, 5)$?

  Ⓕ $(6, 12)$ Ⓖ $(2, 4)$ Ⓗ $(3, 6)$ Ⓘ $(6, 6)$

49. Which is the best description of the figure at the right?

  Ⓐ convex pentagon   Ⓒ convex polygon

  Ⓑ concave octagon   Ⓓ concave pentagon

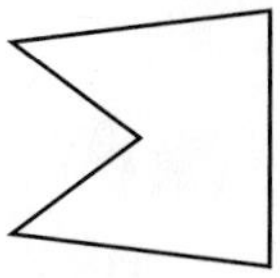

Short Response

50. Describe how to form the Fibonacci sequence 1, 1, 2, 3, 5, 8, 13 . . .

Mixed Review

Find a counterexample to show that each statement is false.   ◀ See Lesson 2-1.

51. You can connect any four points to form a rectangle.

52. The square of a number is always greater than the number.

Find the perimeter of each rectangle with the given base and height.   ◀ See Lesson 1-8.

53. 6 in., 12 in.   **54.** 3.5 cm, 7 cm   **55.** $1\frac{3}{4}$ yd, 18 in.   **56.** 11 m, 60 cm

Get Ready!   **To prepare for Lesson 2-3, do Exercises 57 and 58.**

Write the converse of each statement. Then determine the truth value of the original statement and of the converse.   ◀ See Lesson 2-2.

57. If today is September 30, then tomorrow is October 1.

58. If $\overline{AB}$ is the perpendicular bisector of $\overline{CD}$, then $\overline{AB}$ and $\overline{CD}$ are perpendicular.

Concept Byte

Use With Lesson 2-2

ACTIVITY

Logic and Truth Tables

© **Content Standards**

Prepares for G.CO.9 Prove theorems about lines and angles.
Prepares for G.CO.10 Prove theorems about triangles.
Prepares for G.CO.11 Prove theorems about parallelograms.

A **compound statement** combines two or more statements.

take note

Key Concepts Compound Statements

Compound Statement	How to Form It	Example	Symbols
conjunction	Connect two or more statements with *and*.	You will eat a sandwich and you will drink juice.	$s \wedge j$ You say "s and j."
disjunction	Connect two or more statements with *or*.	You will eat a sandwich or you will drink juice.	$s \vee j$ You say "s or j."

A conjunction $s \wedge j$ is true only when both s and j are true.

A disjunction $s \vee j$ is false only when both s and j are false.

Activity 1

For Exercises 1–4, use the statements below to construct the following compound statements.

> $s:$ We will go to the beach.
> $j:$ We will go out to dinner.
> $t:$ We will go to the movies.

1. $s \wedge j$ **2.** $s \vee j$ **3.** $s \vee (j \wedge t)$ **4.** $(s \vee j) \wedge t$

5. Write three of your own statements and label them $s, j,$ and t. Repeat Exercises 1–4 using your own statements.

For Exercises 6–9, use the statements below to determine the truth value of the compound statement.

> $x:$ Emperor penguins are black and white.
> $y:$ Polar bears are a threatened species.
> $z:$ Penguins wear tuxedos.

6. $x \wedge y$ **7.** $x \vee y$ **8.** $x \wedge z$ **9.** $x \vee z$

A **truth table** lists all the possible combinations of truth values for two or more statements.

Example	p	q	$p \rightarrow q$	$p \wedge q$	$p \vee q$
p: Ohio is a state. q: There are 50 states.	T	T	T	T	T
p: Georgia is a state. q: Miami is a state.	T	F	F	F	T
p: $2 + 2 = 5$ q: $2 \cdot 2 = 4$	F	T	T	F	T
p: $2 + 1 = 4$ q: Dolphins are big fish.	F	F	T	F	F

Activity 2

To find the possible truth values of a complex statement such as $(s \wedge j) \vee \sim t$, you can make a truth table like the one below. You start with columns for the single statements and add columns to the right. Each column builds toward the final statement. The table below starts with columns for s, j, and t and builds to $(s \wedge j) \vee \sim t$. Copy the table and work with a partner to fill in the blanks.

s	j	t	$\sim t$	$s \wedge j$	$(s \wedge j) \vee \sim t$
T	T	T	F	T	**20.** ?
T	T	F	**13.** ?	T	T
T	F	T	F	F	**21.** ?
T	F	F	T	**17.** ?	T
F	T	T	**14.** ?	F	**22.** ?
F	T	F	**15.** ?	F	**23.** ?
F	F	T	F	**18.** ?	F
10. ?	**11.** ?	**12.** ?	**16.** ?	**19.** ?	**24.** ?

25. Make a truth table for the statement $(\sim p \vee q) \wedge \sim r$.

26. a. Make a truth table for $\sim(p \wedge q)$. Make another truth table for $\sim p \vee \sim q$.
   b. Make a truth table for $\sim(p \vee q)$. Make another for $\sim p \wedge \sim q$.
   c. DeMorgan's Law states that $\sim(p \wedge q) = \sim p \vee \sim q$ and that $\sim(p \vee q) = \sim p \wedge \sim q$. How do the truth tables you made in parts (a) and (b) show that DeMorgan's Law is true?

2-3 Biconditionals and Definitions

Content Standards
Prepares for G.CO.9 Prove theorems about lines and angles.
Prepares for G.CO.10 Prove theorems about triangles.
Prepares for G.CO.11 Prove theorems about parallelograms.

Objective To write biconditionals and recognize good definitions

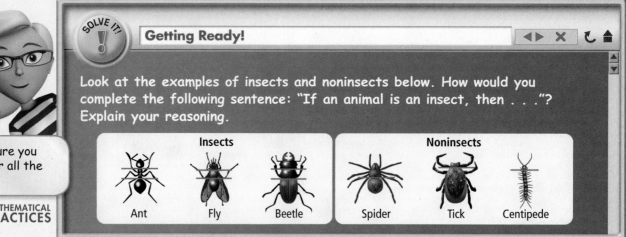

SOLVE IT!

Getting Ready!

Look at the examples of insects and noninsects below. How would you complete the following sentence: "If an animal is an insect, then . . ."? Explain your reasoning.

Make sure you consider all the data.

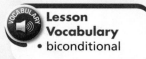
MATHEMATICAL PRACTICES

Insects	Noninsects
Ant Fly Beetle	Spider Tick Centipede

Lesson Vocabulary
• biconditional

In the Solve It, you used conditional statements. A **biconditional** is a single true statement that combines a true conditional and its true converse. You can write a biconditional by joining the two parts of each conditional with the phrase *if and only if*.

Essential Understanding A definition is good if it can be written as a biconditional.

Problem 1 Writing a Biconditional

What is the converse of the following true conditional? If the converse is also true, rewrite the statements as a biconditional.

> If the sum of the measures of two angles is 180, then the two angles are supplementary.

Think

How else can you write the biconditional?
You can also write the biconditional as "The sum of the measures of two angles is 180 if and only if the two angles are supplementary."

Converse: If two angles are supplementary, then the sum of the measures of the two angles is 180.
The converse is true. You can form a true biconditional by joining the true conditional and the true converse with the phrase *if and only if*.

Biconditional: Two angles are supplementary if and only if the sum of the measures of the two angles is 180.

 Got It? **1.** What is the converse of the following true conditional? If the converse is also true, rewrite the statements as a biconditional.
   If two angles have equal measure, then the angles are congruent.

Dynamic Activity
Biconditional
Statements

Key Concept **Biconditional Statements**

A biconditional combines $p \rightarrow q$ and $q \rightarrow p$ as $p \leftrightarrow q$.

| **Example** | **Symbols** | **How to Read It** |
| A point is a midpoint if and only if it divides a segment into two congruent segments. | $p \leftrightarrow q$ | "p if and only if q" |

You can write a biconditional as two conditionals that are converses.

 Problem 2 **Identifying the Conditionals in a Biconditional**

What are the two conditional statements that form this biconditional?
  A ray is an angle bisector if and only if it divides an angle into two congruent angles.

Let p and q represent the following:

p: A ray is an angle bisector.

q: A ray divides an angle into two congruent angles.

$p \rightarrow q$: If a ray is an angle bisector, **then** it divides an angle into two congruent angles.

$q \rightarrow p$: If a ray divides an angle into two congruent angles, **then** it is an angle bisector.

Got It? **2.** What are the two conditionals that form this biconditional?
        Two numbers are reciprocals if and only if their product is 1.

Plan

How can you separate the biconditional into two parts?
Identify the part before and the part after the phrase *if and only if*.

As you learned in Lesson 1-2, undefined terms such as *point, line,* and *plane* are the building blocks of geometry. You understand the meanings of these terms intuitively. Then you use them to define other terms such as *segment*.

A good definition is a statement that can help you identify or classify an object. A good definition has several important components.

✔ A good definition uses clearly understood terms. These terms should be commonly understood or already defined.

✔ A good definition is precise. Good definitions avoid words such as *large, sort of,* and *almost*.

✔ A good definition is reversible. That means you can write a good definition as a true biconditional.

Problem 3 **Writing a Definition as a Biconditional**

Plan

How do you determine whether a definition is reversible?
Write the definition as a conditional and the converse of the conditional. If both are true, the definition is reversible.

Is this definition of *quadrilateral* reversible? If yes, write it as a true biconditional.
    Definition: A quadrilateral is a polygon with four sides.

Think

Write a conditional.

Write the converse.

The conditional and its converse are both true. The definition is reversible. Write the conditional and its converse as a true biconditional.

Write

Conditional: If a figure is a quadrilateral, then it is a polygon with four sides.

Converse: If a figure is a polygon with four sides, then it is a quadrilateral.

Biconditional: A figure is a quadrilateral if and only if it is a polygon with four sides.

Got It? **3.** Is this definition of *straight angle* reversible? If yes, write it as a true biconditional.
    A straight angle is an angle that measures 180.

One way to show that a statement is *not* a good definition is to find a counterexample.

Problem 4 **Identifying Good Definitions**

Plan

How can you eliminate answer choices?
You can eliminate an answer choice if the definition fails to meet any one of the components of a good definition.

Multiple Choice **Which of the following is a good definition?**

Ⓐ A fish is an animal that swims.

Ⓒ Giraffes are animals with very long necks.

Ⓑ Rectangles have four corners.

Ⓓ A penny is a coin worth one cent.

Choice A is not reversible. A whale is a counterexample. A whale is an animal that swims, but it is a mammal, not a fish. In Choice B, *corners* is not clearly defined. All quadrilaterals have four corners. In Choice C, *very long* is not precise. Also, Choice C is not reversible because ostriches also have long necks. Choice D is a good definition. It is reversible, and all of the terms in the definition are clearly defined and precise. The answer is D.

Got It? **4. a.** Is the following statement a good definition? Explain.
        A square is a figure with four right angles.
    b. Reasoning How can you rewrite the statement "Obtuse angles have greater measures than acute angles" so that it is a good definition?

Lesson Check

Do you know HOW?

1. How can you write the following statement as two true conditionals?

   Collinear points are points that lie on the same line.

2. How can you combine the following statements as a biconditional?

   If this month is June, then next month is July.
   If next month is July, then this month is June.

3. Write the following definition as a biconditional.

   Vertical angles are two angles whose sides are opposite rays.

Do you UNDERSTAND? MATHEMATICAL PRACTICES

4. **Vocabulary** Explain how the term *biconditional* is fitting for a statement composed of *two* conditionals.

5. **Error Analysis** Why is the following statement a poor definition?

   Elephants are gigantic animals.

6. **Compare and Contrast** Which of the following statements is a better definition of a linear pair? Explain.

   A linear pair is a pair of supplementary angles.

   A linear pair is a pair of adjacent angles with noncommon sides that are opposite rays.

Practice and Problem-Solving Exercises MATHEMATICAL PRACTICES

A Practice

Each conditional statement below is true. Write its converse. If the converse is also true, combine the statements as a biconditional.

◀ See Problem 1.

7. If two segments have the same length, then they are congruent.

8. **Algebra** If $x = 12$, then $2x - 5 = 19$.

9. If a number is divisible by 20, then it is even.

10. **Algebra** If $x = 3$, then $|x| = 3$.

11. In the United States, if it is July 4, then it is Independence Day.

12. If $p \rightarrow q$ is true, then $\sim q \rightarrow \sim p$ is true.

Write the two statements that form each biconditional.

◀ See Problem 2.

13. A line bisects a segment if and only if the line intersects the segment only at its midpoint.

14. An integer is divisible by 100 if and only if its last two digits are zeros.

15. You live in Washington, D. C., if and only if you live in the capital of the United States.

16. A polygon is a triangle if and only if it has exactly three sides.

17. An angle is a right angle if and only if it measures 90.

18. **Algebra** $x^2 = 144$ if and only if $x = 12$ or $x = -12$.

Test each statement below to see if it is reversible. If so, write it as a true biconditional. If not, write *not reversible*.

◀ **See Problem 3.**

19. A perpendicular bisector of a segment is a line, segment, or ray that is perpendicular to a segment at its midpoint.

20. Complementary angles are two angles with measures that have a sum of 90.

21. A Tarheel is a person who was born in North Carolina.

22. A rectangle is a four-sided figure with at least one right angle.

23. Two angles that form a linear pair are adjacent.

Is each statement below a good definition? If not, explain.

◀ **See Problem 4.**

24. A cat is an animal with whiskers.

25. The red wolf is an endangered animal.

26. A segment is part of a line.

27. A compass is a geometric tool.

28. Opposite rays are two rays that share the same endpoint.

29. Perpendicular lines are two lines that intersect to form right angles.

 Apply

Ⓒ **30. Think About a Plan** Is the following a good definition? Explain.

A ligament is a band of tough tissue connecting bones or holding organs in place.

- Can you write the statement as two true conditionals?
- Are the two true conditionals converses of each other?

Ⓒ **31. Reasoning** Is the following a good definition? Explain.
An obtuse angle is an angle with measure greater than 90.

Ⓒ **32. Open-Ended** Choose a definition from a dictionary or from a glossary. Explain what makes the statement a good definition.

Ⓒ **33. Error Analysis** Your friend defines a right angle as an angle that is greater than an acute angle. Use a biconditional to show that this is not a good definition.

34. Which conditional and its converse form a true biconditional?

  Ⓐ If $x > 0$, then $|x| > 0$.

  Ⓒ If $x^3 = 5$, then $x = 125$.

  Ⓑ If $x = 3$, then $x^2 = 9$.

  Ⓓ If $x = 19$, then $2x - 3 = 35$.

Write each statement as a biconditional.

35. Points in Quadrant III have two negative coordinates.

36. When the sum of the digits of an integer is divisible by 9, the integer is divisible by 9 and vice versa.

37. The whole numbers are the nonnegative integers.

38. A hexagon is a six-sided polygon.

Language For Exercises 39–42, use the chart below. Decide whether the description of each letter is a good definition. If not, provide a counterexample by giving another letter that could fit the definition.

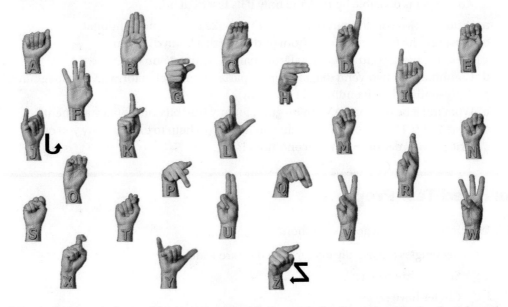

39. The letter *D* is formed by pointing straight up with the finger beside the thumb and folding the other fingers and the thumb so that they all touch.

40. The letter *K* is formed by making a *V* with the two fingers beside the thumb.

41. You have formed the letter *I* if and only if the smallest finger is sticking up and the other fingers are folded into the palm of your hand with your thumb folded over them and your hand is held still.

42. You form the letter *B* by holding all four fingers tightly together and pointing them straight up while your thumb is folded into the palm of your hand.

Reading Math Let statements *p, q, r,* and *s* be as follows:

p: ∠A and ∠B are a linear pair.
q: ∠A and ∠B are supplementary angles.
r: ∠A and ∠B are adjacent angles.
s: ∠A and ∠B are adjacent and supplementary angles.

Substitute for *p, q, r,* and *s,* and write each statement the way you would read it.

43. $p \rightarrow q$ **44.** $p \rightarrow r$ **45.** $p \rightarrow s$ **46.** $p \leftrightarrow s$

Challenge **47. Writing** Use the figures to write a good definition of a *line* in spherical geometry.

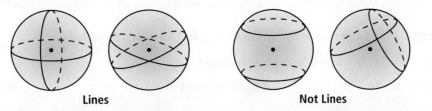

Lines   Not Lines

48. Multiple Representations You have illustrated true conditional statements with Venn diagrams. You can do the same thing with true biconditionals. Consider the following statement.

An integer is divisible by 10 if and only if its last digit is 0.

a. Write the two conditional statements that make up this biconditional.
b. Illustrate the first conditional from part (a) with a Venn diagram.
c. Illustrate the second conditional from part (a) with a Venn diagram.
d. Combine your two Venn diagrams from parts (b) and (c) to form a Venn diagram representing the biconditional statement.
e. What must be true of the Venn diagram for any true biconditional statement?
f. Reasoning How does your conclusion in part (e) help to explain why you can write a good definition as a biconditional?

Standardized Test Prep

SAT/ACT

49. Which statement is a good definition?

(A) Rectangles are usually longer than they are wide.

(B) Squares are convex.

(C) Circles have no corners.

(D) Triangles are three-sided polygons.

50. What is the exact area of a circle with a diameter of 6 cm?

(F) 28.27 cm   (G) 9π m^2 (H) 36π cm^2 (I) 9π cm^2

Extended Response

51. Consider this true conditional statement.

If you want to buy milk, then you go to the store.

a. Write the converse and determine whether it is *true* or *false*.
b. If the converse is false, give a counterexample to show that it is false. If the converse is true, combine the original statement and its converse as a biconditional.

Mixed Review

Write the converse of each statement.   ◀ See Lesson 2-2.

52. If you do not sleep enough, then your grades suffer.

53. If you are in the school chorus, then you have a good voice.

54. Reasoning What is the truth value of the contrapositive of a true conditional?

Get Ready!   **To prepare for Lesson 2-4, do Exercises 55–57.**

What are the next two terms in each sequence?   ◀ See Lesson 2-1.

55. 100, 90, 80, 70, . . .   **56.** 2500, 500, 100, 20, . . .   **57.** 1, 2, 0, 3, −1, . . .

Do you know HOW?

Use inductive reasoning to describe the pattern of each sequence. Then find the next two terms.

1. 1, 12, 123, 1234, . . .

2. 3, 4.5, 6.75, 10.125, . . .

3. 2, 3, 5, 7, 11, 13, . . .

Draw the next figure in each sequence.

4.

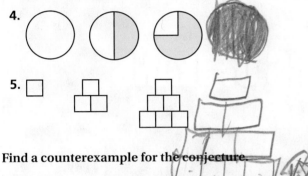

5.

Find a counterexample for the conjecture.

6. Three coplanar lines always make a triangle.

7. All balls are spheres.

8. When it rains, it pours.

Identify the hypothesis and the conclusion of the conditional statements.

9. If the traffic light is red, then you must stop.

10. If $x > 5$, then $x^2 > 25$.

11. If you leave your house, then you must lock the door.

Rewrite the statements as conditional statements.

12. Roses are beautiful flowers.

13. Apples grow on trees.

14. Quadrilaterals have four sides.

15. The world's largest trees are giant sequoias.

For Exercises 16–19, write the converse, inverse, and contrapositive of each conditional statement. Determine the truth value of each statement. If it is false, provide a counterexample.

16. If a figure is a circle with radius r, then its circumference is $2\pi r$.

17. If an integer ends with 0, then it is divisible by 2.

18. If you win the league championship game, then you win the league trophy.

19. If a triangle has one right angle, then the other two angles are complementary.

20. Write the two conditionals that make up this biconditional: An angle is an acute angle if and only if its measure is between 0 and 90.

For Exercises 21–23, rewrite the definition as a biconditional.

21. Points that lie on the same line are collinear.

22. Figures with three sides are triangles.

23. The moon is the largest satellite of Earth.

24. Which of the following is a good definition?

Ⓐ Grass is green.

Ⓑ Dinosaurs are extinct.

Ⓒ A pound weighs less than a kilogram.

Ⓓ A yard is a unit of measure exactly 3 ft long.

Do you UNDERSTAND?

ⓒ 25. Open-Ended Describe a situation where you used a pattern to reach a conjecture.

26. How does the word *induce* relate to the term *inductive reasoning*?

ⓒ 27. Error Analysis Why is the following not a good definition? How could you improve it?
Rain is water.

2-4 Deductive Reasoning

Content Standards
Prepares for G.CO.9 Prove theorems about lines and angles.
Prepares for G.CO.10 Prove theorems about triangles.
Prepares for G.CO.11 Prove theorems about parallelograms.

Objective To use the Law of Detachment and the Law of Syllogism

SOLVE IT!

Use each step to write an expression for the process as a whole.

MATHEMATICAL PRACTICES

Getting Ready!

You want to use the coupon to buy three different pairs of jeans. You have narrowed your choices to four pairs. The costs of the different pairs are $24.99, $39.99, $40.99, and $50.00. If you spend as little as possible, what is the average amount per pair of jeans that you will pay? Explain.

BUY TWO PAIRS OF JEANS
Get a THIRD Free*

*Free jeans must be of equal or lesser value.

ISBN 0-13-134022-0

Lesson Vocabulary
• deductive reasoning
• Law of Detachment
• Law of Syllogism

In the Solve It, you drew a conclusion based on several facts. You used deductive reasoning. **Deductive reasoning** (sometimes called logical reasoning) is the process of reasoning logically from given statements or facts to a conclusion.

Essential Understanding Given true statements, you can use deductive reasoning to make a valid or true conclusion.

take note

Property Law of Detachment

Law	**Symbols**
If the hypothesis of a true conditional is true, then the conclusion is true.	If $\quad p \rightarrow q \quad$ is true and $\quad p \quad$ is true, then $\quad q \quad$ is true.

To use the Law of Detachment, identify the hypothesis of the given true conditional. If the second given statement matches the hypothesis of the conditional, then you can make a valid conclusion.

Problem 1   Using the Law of Detachment

What can you conclude from the given true statements?

A Given: **If a student gets an A on a final exam, then the student will pass the course.**
Felicia got an A on her history final exam.

If a student gets an A on a final exam, then the student will pass the course.
Felicia got an A on her history final exam.

The second statement matches the hypothesis of the given conditional. By the Law of Detachment, you can make a conclusion.

You conclude: Felicia will pass her history course.

B Given: **If a ray divides an angle into two congruent angles, then the ray is an angle bisector.**
$\overrightarrow{RS}$ **divides** $\angle ARB$ **so that** $\angle ARS \cong \angle SRB$**.**

If a ray divides an angle into two congruent angles, then the ray is an angle bisector.
$\overrightarrow{RS}$ divides $\angle ARB$ so that $\angle ARS \cong \angle SRB$.

The second statement matches the hypothesis of the given conditional. By the Law of Detachment, you can make a conclusion.

You conclude: $\overrightarrow{RS}$ is an angle bisector.

Think

In part (C), the second statement is not a subset of the hypothesis. Instead, it is a subset of the conditional's conclusion.

C Given: **If two angles are adjacent, then they share a common vertex.**
$\angle 1$ **and** $\angle 2$ **share a common vertex.**

If two angles are adjacent, then they share a common vertex.
$\angle 1$ and $\angle 2$ share a common vertex.

The information in the second statement about $\angle 1$ and $\angle 2$ does not tell you if the angles are adjacent. The second statement does not match the hypothesis of the given conditional, so you cannot use the Law of Detachment. $\angle 1$ and $\angle 2$ could be vertical angles, since vertical angles also share a common vertex. You cannot make a conclusion.

Got It?   **1.** What can you conclude from the given information?
     a. If there is lightning, then it is not safe to be out in the open.
        Marla sees lightning from the soccer field.
     b. If a figure is a square, then its sides have equal length.
        Figure *ABCD* has sides of equal length.

Another law of deductive reasoning is the Law of Syllogism. The **Law of Syllogism** allows you to state a conclusion from two true conditional statements when the conclusion of one statement is the hypothesis of the other statement.

take note

Property   Law of Syllogism

Symbols

If $\quad p \rightarrow q \quad$ is true
and $\quad\quad q \rightarrow r \quad$ is true,
then $\quad p \;\rightarrow\; r \quad$ is true.

Example

If it is July, **then** you are on summer vacation.

If you are on summer vacation, **then** you work at a smoothie shop.

You conclude: If it is July, **then** you work at a smoothie shop.

© **Problem 2** **Using the Law of Syllogism**

Plan

When can you use the Law of Syllogism?
You can use the Law of Syllogism when the conclusion of one statement is the hypothesis of the other.

What can you conclude from the given information?

A **Given:**   If a figure is a square, then the figure is a rectangle.
If a figure is a rectangle, then the figure has four sides.

If a figure is a square, **then** the figure is a rectangle.
If a figure is a rectangle, **then** the figure has four sides.

The conclusion of the first statement is the hypothesis of the second statement, so you can use the Law of Syllogism to make a conclusion.

You conclude: If a figure is a square, **then** the figure has four sides.

B **Given:**   If you do gymnastics, then you are flexible.
If you do ballet, then you are flexible.

If you do gymnastics, **then** you are flexible.
If you do ballet, **then** you are flexible.

The statements have the same conclusion. Neither conclusion is the hypothesis of the other statement, so you cannot use the Law of Syllogism. You cannot make a conclusion.

✓ **Got It?** **2.** What can you conclude from the given information? What is your reasoning?
  a. If a whole number ends in 0, then it is divisible by 10.
   If a whole number is divisible by 10, then it is divisible by 5.
  b. If $\overrightarrow{AB}$ and $\overrightarrow{AD}$ are opposite rays, then the two rays form a straight angle.
   If two rays are opposite rays, then the two rays form a straight angle.

You can use the Law of Syllogism and the Law of Detachment together to make conclusions.

Problem 3 Using the Laws of Syllogism and Detachment

What can you conclude from the given information?

Given: If you live in Accra, then you live in Ghana.
If you live in Ghana, then you live in Africa. Aissa lives in Accra.

If you live in Accra, then you live in Ghana.
If you live in Ghana, then you live in Africa.
Aissa lives in Accra.

You can use the first two statements and the Law of Syllogism to conclude:
If you live in Accra, then you live in Africa.

You can use this new conditional statement, the fact that Aissa lives in Accra, and the Law of Detachment to make a conclusion.

You conclude: Aissa lives in Africa.

Think

Does the conclusion make sense?
Accra is a city in Ghana, which is an African nation. So if a person lives in Accra, then that person lives in Africa. The conclusion makes sense.

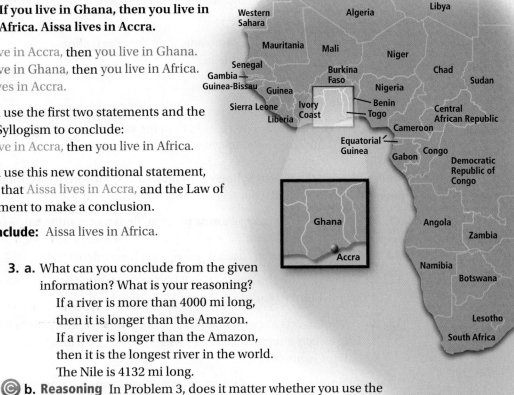

Got It? 3. a. What can you conclude from the given information? What is your reasoning?
If a river is more than 4000 mi long, then it is longer than the Amazon.
If a river is longer than the Amazon, then it is the longest river in the world.
The Nile is 4132 mi long.

b. Reasoning In Problem 3, does it matter whether you use the Law of Syllogism or the Law of Detachment first? Explain.

Lesson Check

Do You Know HOW?

If possible, make a conclusion from the given true statements. What reasoning did you use?

1. If it is Tuesday, then you will go bowling.
You go bowling.

2. If a figure is a three-sided polygon, then it is a triangle.
Figure *ABC* is a three-sided polygon.

3. If it is Saturday, then you walk to work.
If you walk to work, then you wear sneakers.

Do You UNDERSTAND? MATHEMATICAL PRACTICES

4. Error Analysis What is the error in the reasoning below?

> Birds that weigh more than 50 pounds cannot fly. A kiwi cannot fly. So, a kiwi weighs more than 50 pounds.

5. Compare and Contrast How is deductive reasoning different from inductive reasoning?

Practice and Problem-Solving Exercises

MATHEMATICAL PRACTICES

A Practice

If possible, use the Law of Detachment to make a conclusion. If it is not possible to make a conclusion, tell why.

See Problem 1.

6. If a doctor suspects her patient has a broken bone, then she should take an X-ray.
Dr. Ngemba suspects Lilly has a broken arm.

7. If a rectangle has side lengths 3 cm and 4 cm, then it has area 12 cm^2.
Rectangle *ABCD* has area 12 cm^2.

8. If three points are on the same line, then they are collinear.
Points *X*, *Y*, and *Z* are on line *m*.

9. If an angle is obtuse, then it is not acute.
$\angle XYZ$ is not obtuse.

10. If a student wants to go to college, then the student must study hard.
Rashid wants to go to Pennsylvania State University.

If possible, use the Law of Syllogism to make a conclusion. If it is not possible to make a conclusion, tell why.

See Problem 2.

STEM 11. Ecology If an animal is a Florida panther, then its scientific name is *Puma concolor coryi*.
If an animal is a *Puma concolor coryi*, then it is endangered.

12. If a whole number ends in 6, then it is divisible by 2.
If a whole number ends in 4, then it is divisible by 2.

13. If a line intersects a segment at its midpoint, then the line bisects the segment.
If a line bisects a segment, then it divides the segment into two congruent segments.

14. If you improve your vocabulary, then you will improve your score on a standardized test.
If you read often, then you will improve your vocabulary.

Use the Law of Detachment and the Law of Syllogism to make conclusions from the following statements. If it is not possible to make a conclusion, tell why.

See Problem 3.

15. If a mountain is the highest in Alaska, then it is the highest in the United States.
If an Alaskan mountain is more than 20,300 ft high, then it is the highest in Alaska.
Alaska's Mount McKinley is 20,320 ft high.

16. If you live in the Bronx, then you live in New York.
Tracy lives in the Bronx.
If you live in New York, then you live in the eleventh state to enter the Union.

17. If you are studying botany, then you are studying biology.
If you are studying biology, then you are studying a science.
Shanti is taking science this year.

© **18. Think About a Plan** If it is the night of your weekly basketball game, your family eats at your favorite restaurant. When your family eats at your favorite restaurant, you always get chicken fingers. If it is Tuesday, then it is the night of your weekly basketball game. How much do you pay for chicken fingers after your game? Use the specials board at the right to decide. Explain your reasoning.
- How can you reorder and rewrite the sentences to help you?
- How can you use the Law of Syllogism to answer the question?

> Monday
>   salads $4.99
> Tuesday
>   chicken fingers $5.99
> Wednesday
>   burgers $6.99

Beverages For Exercises 19–24, assume that the following statements are true.

A. If Maria is drinking juice, then it is breakfast time.
B. If it is lunchtime, then Kira is drinking milk and nothing else.
C. If it is mealtime, then Curtis is drinking water and nothing else.
D. If it is breakfast time, then Julio is drinking juice and nothing else.
E. Maria is drinking juice.

Use only the information given above. For each statement, write *must be true*, *may be true*, or *is not true*. Explain your reasoning.

19. Julio is drinking juice.　　**20.** Curtis is drinking water.　　**21.** Kira is drinking milk.

22. Curtis is drinking juice.　　**23.** Maria is drinking water.　　**24.** Julio is drinking milk.

STEM **25. Physics** Quarks are subatomic particles identified by electric charge and rest energy. The table shows how to categorize quarks by their flavors. Show how the Law of Detachment and the table are used to identify the flavor of a quark with a charge of $-\frac{1}{3} e$ and rest energy 540 *MeV*.

Rest Energy and Charge of Quarks						
Rest Energy (*MeV*)	360	360	1500	540	173,000	5000
Electric Charge (*e*)	$+\frac{2}{3}$	$-\frac{1}{3}$	$+\frac{2}{3}$	$-\frac{1}{3}$	$+\frac{2}{3}$	$-\frac{1}{3}$
Flavor	Up	Down	Charmed	Strange	Top	Bottom

Write the first statement as a conditional. If possible, use the Law of Detachment to make a conclusion. If it is not possible to make a conclusion, tell why.

26. All national parks are interesting.
Mammoth Cave is a national park.

27. All squares are rectangles.
ABCD is a square.

28. The temperature is always above 32°F in Key West, Florida.
The temperature is 62°F.

29. Every high school student likes art.
Ling likes art.

© **30. Writing** Give an example of a rule used in your school that could be written as a conditional. Explain how the Law of Detachment is used in applying that rule.

31. Reasoning Use the following algorithm: Choose an integer. Multiply the integer by 3. Add 6 to the product. Divide the sum by 3.

a. Complete the algorithm for four different integers. Look for a pattern in the chosen integers and in the corresponding answers. Make a conjecture that relates the chosen integers to the answers.

b. Let the variable x represent the chosen integer. Apply the algorithm to x. Simplify the resulting expression.

c. How does your answer to part (b) confirm your conjecture in part (a)? Describe how inductive and deductive reasoning are exhibited in parts (a) and (b).

STEM **32. Biology** Consider the following given statements and conclusion.

Given: If an animal is a fish, then it has gills.
A turtle does not have gills.
You conclude: A turtle is not a fish.

a. Make a Venn diagram to illustrate the given information.

b. Use the Venn diagram to help explain why the argument uses good reasoning.

Standardized Test Prep

SAT/ACT

33. What can you conclude from the given true statements?
If you wake up late, then you miss the bus.
If you miss the bus, then you are late for school.

Ⓐ If you are late for school, then you missed the bus.

Ⓑ If you wake up late, then you are late for school.

Ⓒ If you miss the bus, then you woke up late.

Ⓓ If you are late for school, then you woke up late.

Short Response

34. Claire reads anything Andrea reads. Ben reads what Claire reads, and Claire reads what Ben reads. Andrea reads whatever Dion reads.

a. Claire is reading *Hamlet*. Who else, if anyone, must also be reading *Hamlet*?

b. Exactly three people are reading *King Lear*. Who are they? Explain.

Mixed Review

35. Write the following definition as a biconditional.
Inductive reasoning is reasoning based on patterns you observe.

◀ **See Lesson 2-3.**

Get Ready! **To Prepare for Lesson 2-5, do Exercises 36–39.**

Use the figure at the right.

◀ **See Lessons 1-4 and 1-5.**

36. Name ∠1 in two other ways.

37. Name ∠2 in two other ways.

38. If ∠1 ≅ ∠2, name the bisector of ∠*AOC*.

39. Classify ∠*AOC*.

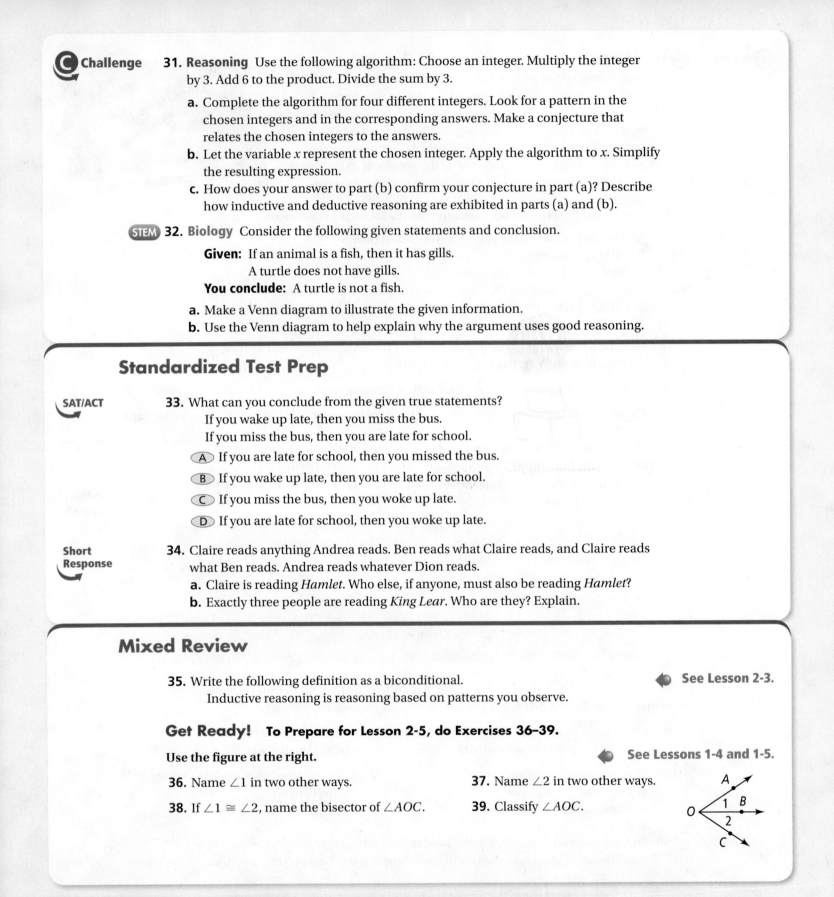

Reasoning in Algebra and Geometry

Content Standards
Prepares for G.CO.9 Prove theorems about lines and angles.
Prepares for G.CO.10 Prove theorems about triangles.
Prepares for G.CO.11 Prove theorems about parallelograms.

Objective To connect reasoning in algebra and geometry

SOLVE IT!

Getting Ready!

Follow the steps of the brainteaser using your age. Then try it using a family member's age. What do you notice? Explain how the brainteaser works.

- Write down your age.
- Multiply it by 10.
- Add 8 to the product.
- Double that answer and then subtract 16.
- Finally, divide the result by 2.

Think about how each step is related to the steps before it.

MATHEMATICAL PRACTICES

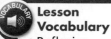

Lesson Vocabulary
- Reflexive Property
- Symmetric Property
- Transitive Property
- proof
- two-column proof

In the Solve It, you logically examined a series of steps. In this lesson, you will apply logical reasoning to algebraic and geometric situations.

Essential Understanding Algebraic properties of equality are used in geometry. They will help you solve problems and justify each step you take.

In geometry you accept postulates and properties as true. Some of the properties that you accept as true are the properties of equality from algebra.

take note

Key Concept Properties of Equality

Let a, b, and c be any real numbers.

Addition Property	If $a = b$, then $a + c = b + c$.
Subtraction Property	If $a = b$, then $a - c = b - c$.
Multiplication Property	If $a = b$, then $a \cdot c = b \cdot c$.
Division Property	If $a = b$ and $c \neq 0$, then $\frac{a}{c} = \frac{b}{c}$.
Reflexive Property	$a = a$
Symmetric Property	If $a = b$, then $b = a$.
Transitive Property	If $a = b$ and $b = c$, then $a = c$.
Substitution Property	If $a = b$, then b can replace a in any expression.

Key Concept   The Distributive Property

Use multiplication to distribute a to each term of the sum or difference within the parentheses.

Sum: Difference:

$$a(b + c) = a(b + c) = ab + ac$$ $$a(b - c) = a(b - c) = ab - ac$$

You use deductive reasoning when you solve an equation. You can justify each step with a postulate, a property, or a definition. For example, you can use the Distributive Property to justify combining like terms. If you think of the Distributive Property as $ab + ac = a(b + c)$ or $ab + ac = (b + c)a$, then $2x + x = (2 + 1)x = 3x$.

Problem 1   Justifying Steps When Solving an Equation

Algebra **What is the value of x? Justify each step.**

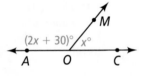

$\angle AOM$ and $\angle MOC$ are supplementary.	$\angle$s that form a linear pair are supplementary.
$m\angle AOM + m\angle MOC = 180$	Definition of supplementary $\angle$s
$(2x + 30) + x = 180$	Substitution Property
$3x + 30 = 180$	Distributive Property
$3x = 150$	Subtraction Property of Equality
$x = 50$	Division Property of Equality

Got It? **1.** What is the value of x? Justify each step.
    Given: $\overrightarrow{AB}$ bisects $\angle RAN$.

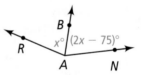

Some properties of equality have corresponding properties of congruence.

Key Concept   Properties of Congruence

Reflexive Property	$\overline{AB} \cong \overline{AB}$ $\angle A \cong \angle A$
Symmetric Property	If $\overline{AB} \cong \overline{CD}$, then $\overline{CD} \cong \overline{AB}$. If $\angle A \cong \angle B$, then $\angle B \cong \angle A$.
Transitive Property	If $\overline{AB} \cong \overline{CD}$ and $\overline{CD} \cong \overline{EF}$, then $\overline{AB} \cong \overline{EF}$. If $\angle A \cong \angle B$ and $\angle B \cong \angle C$, then $\angle A \cong \angle C$. If $\angle B \cong \angle A$ and $\angle B \cong \angle C$, then $\angle A \cong \angle C$.

Think

Is the justification a property of equality or congruence?
Numbers are equal (=) and you can perform operations on them, so (A) and (C) are properties of equality. Figures and their corresponding parts are congruent ($\cong$), so (B) is a property of congruence.

Problem 2 Using Properties of Equality and Congruence

What is the name of the property of equality or congruence that justifies going from the first statement to the second statement?

A $2x + 9 = 19$
$2x = 10$ — Subtraction Property of Equality

B $\angle O \cong \angle W$ and $\angle W \cong \angle L$
$\angle O \cong \angle L$ — Transitive Property of Congruence

C $m\angle E = m\angle T$
$m\angle T = m\angle E$ — Symmetric Property of Equality

Got It? 2. For parts (a)–(c), what is the name of the property of equality or congruence that justifies going from the first statement to the second statement?

a. $\overline{AR} \cong \overline{TY}$
$\overline{TY} \cong \overline{AR}$

b. $3(x + 5) = 9$
$3x + 15 = 9$

c. $\frac{1}{4}x = 7$
$x = 28$

d. Reasoning What property justifies the statement $m\angle R = m\angle R$?

A **proof** is a convincing argument that uses deductive reasoning. A proof logically shows why a conjecture is true. A **two-column proof** lists each statement on the left. The justification, or the reason for each statement, is on the right. Each statement must follow logically from the steps before it. The diagram below shows the setup for a two-column proof. You will find the complete proof in Problem 3.

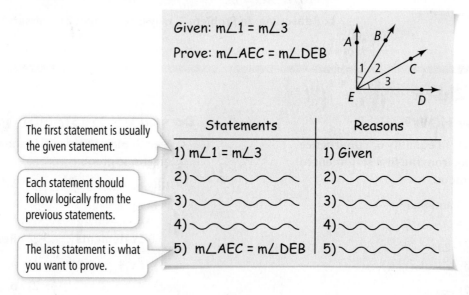

Given: $m\angle 1 = m\angle 3$

Prove: $m\angle AEC = m\angle DEB$

Statements	Reasons
1) $m\angle 1 = m\angle 3$	1) Given
2)	2)
3)	3)
4)	4)
5) $m\angle AEC = m\angle DEB$	5)

The first statement is usually the given statement.

Each statement should follow logically from the previous statements.

The last statement is what you want to prove.

Write a two-column proof.

Given: $m\angle 1 = m\angle 3$

Prove: $m\angle AEC = m\angle DEB$

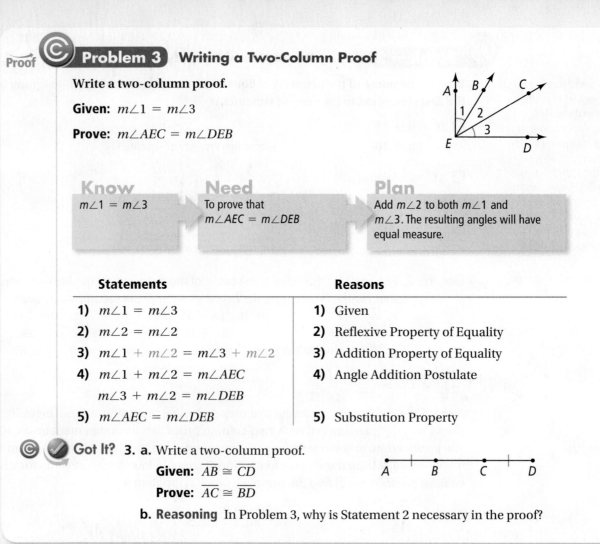

Know	Need	Plan
$m\angle 1 = m\angle 3$	To prove that $m\angle AEC = m\angle DEB$	Add $m\angle 2$ to both $m\angle 1$ and $m\angle 3$. The resulting angles will have equal measure.

Statements	**Reasons**
1) $m\angle 1 = m\angle 3$	1) Given
2) $m\angle 2 = m\angle 2$	2) Reflexive Property of Equality
3) $m\angle 1 + m\angle 2 = m\angle 3 + m\angle 2$	3) Addition Property of Equality
4) $m\angle 1 + m\angle 2 = m\angle AEC$ $m\angle 3 + m\angle 2 = m\angle DEB$	4) Angle Addition Postulate
5) $m\angle AEC = m\angle DEB$	5) Substitution Property

Got It? **3. a.** Write a two-column proof.

Given: $\overline{AB} \cong \overline{CD}$

Prove: $\overline{AC} \cong \overline{BD}$

b. Reasoning In Problem 3, why is Statement 2 necessary in the proof?

Lesson Check

Do you know HOW?

Name the property of equality or congruence that justifies going from the first statement to the second statement.

1. $m\angle A = m\angle S$ and $m\angle S = m\angle K$
   $m\angle A = m\angle K$

2. $3x + x + 7 = 23$
   $4x + 7 = 23$

3. $4x + 5 = 17$
   $4x = 12$

Do you UNDERSTAND? MATHEMATICAL PRACTICES

4. Developing Proof Fill in the reasons for this algebraic proof.

Given: $5x + 1 = 21$

Prove: $x = 4$

Statements	**Reasons**
1) $5x + 1 = 21$	1) a. ?
2) $5x = 20$	2) b. ?
3) $x = 4$	3) c. ?

Practice and Problem-Solving Exercises

MATHEMATICAL PRACTICES

A Practice

Algebra **Fill in the reason that justifies each step.**

See Problem 1.

5.

$\frac{1}{2}x - 5 = 10$	Given
$2\left(\frac{1}{2}x - 5\right) = 20$	**a.** ?
$x - 10 = 20$	**b.** ?
$x = 30$	**c.** ?

6.

$5(x + 3) = -4$	Given
$5x + 15 = -4$	**a.** ?
$5x = -19$	**b.** ?
$x = -\frac{19}{5}$	**c.** ?

7.

$\angle CDE$ and $\angle EDF$ are supplementary.	∠s that form a linear pair are supplementary
$m\angle CDE + m\angle EDF = 180$	**a.** ?
$x + (3x + 20) = 180$	**b.** ?
$4x + 20 = 180$	**c.** ?
$4x = 160$	**d.** ?
$x = 40$	**e.** ?

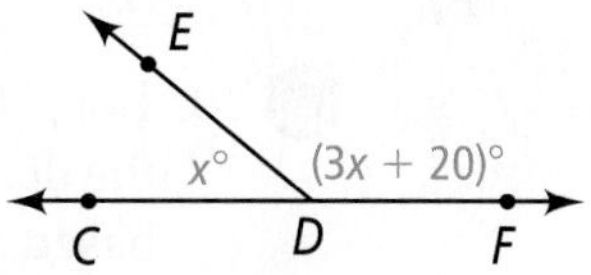

8.

$XY = 42$	Given
$XZ + ZY = XY$	**a.** ?
$3(n + 4) + 3n = 42$	**b.** ?
$3n + 12 + 3n = 42$	**c.** ?
$6n + 12 = 42$	**d.** ?
$6n = 30$	**e.** ?
$n = 5$	**f.** ?

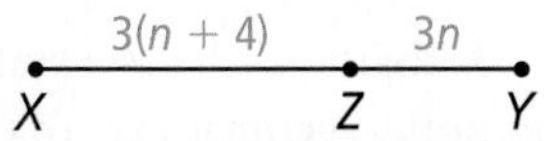

Name the property of equality or congruence that justifies going from the first statement to the second statement.

See Problem 2.

9. $2x + 1 = 7$
$2x = 6$

10. $5x = 20$
$x = 4$

11. $\overline{ST} \cong \overline{QR}$
$\overline{QR} \cong \overline{ST}$

12. $AB - BC = 12$
$AB = 12 + BC$

13. Developing Proof Fill in the missing statements or reasons for the following two-column proof.

See Problem 3.

Given: C is the midpoint of $\overline{AD}$.
Prove: $x = 6$

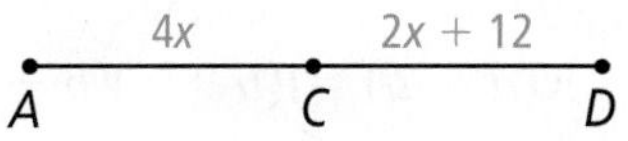

Statements	Reasons
1) C is the midpoint of $\overline{AD}$.	**1) a.** ?
2) $\overline{AC} \cong \overline{CD}$	**2) b.** ?
3) $AC = CD$	**3)** ≅ segments have equal length.
4) $4x = 2x + 12$	**4) c.** ?
5) d. ?	**5)** Subtraction Property of Equality
6) $x = 6$	**6) e.** ?

Use the given property to complete each statement.

14. Symmetric Property of Equality
If $AB = YU$, then __?__.

15. Symmetric Property of Congruence
If $\angle H \cong \angle K$, then __?__ $\cong \angle H$.

16. Reflexive Property of Congruence
$\angle POR \cong$ __?__

17. Distributive Property
$3(x - 1) = 3x -$ __?__

18. Substitution Property
If $LM = 7$ and $EF + LM = NP$,
then __?__ $= NP$.

19. Transitive Property of Congruence
If $\angle XYZ \cong \angle AOB$ and
$\angle AOB \cong \angle WYT$, then __?__.

20. Think About a Plan A very important part in writing proofs is analyzing the diagram for key information. What true statements can you make based on the diagram at the right?
- What theorems or definitions relate to the geometric figures in the diagram?
- What types of markings show relationships between parts of geometric figures?

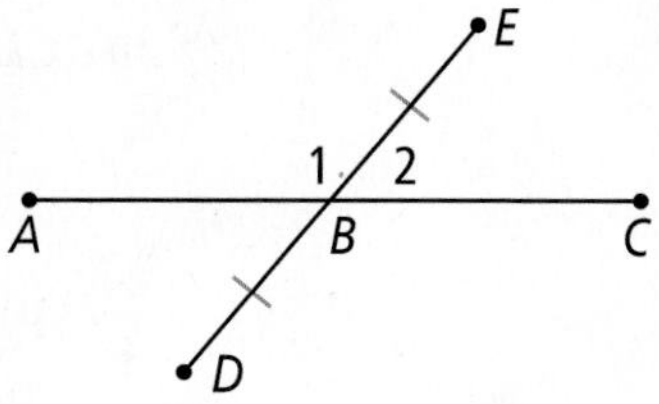

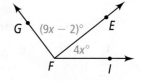

21. Writing Explain why the statements $\overline{LR} \cong \overline{RL}$ and $\angle CBA \cong \angle ABC$ are both true by the Reflexive Property of Congruence.

22. Reasoning Complete the following statement. Describe the reasoning that supports your answer.

The Transitive Property of Falling Dominoes: If Domino A causes Domino B to fall, and Domino B causes Domino C to fall, then Domino A causes Domino __?__ to fall.

Write a two-column proof.

23. Given: $KM = 35$
Proof **Prove:** $KL = 15$

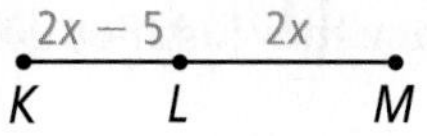

$\overset{2x - 5 \qquad 2x}{\underset{K \qquad\quad L \qquad\qquad M}{\rule{4cm}{0.4pt}}}$

24. Given: $m\angle GFI = 128$
Proof **Prove:** $m\angle EFI = 40$

25. Error Analysis The steps below "show" that $1 = 2$. Describe the error.

$a = b$	Given
$ab = b^2$	Multiplication Property of Equality
$ab - a^2 = b^2 - a^2$	Subtraction Property of Equality
$a(b - a) = (b + a)(b - a)$	Distributive Property
$a = b + a$	Division Property of Equality
$a = a + a$	Substitution Property
$a = 2a$	Simplify.
$1 = 2$	Division Property of Equality

Relationships Consider the following relationships among people. Tell whether each relationship is *reflexive, symmetric, transitive,* or *none of these.* Explain.

> **Sample:** The relationship "is younger than" is not reflexive because Sue is not younger than herself. It is not symmetric because if Sue is younger than Fred, then Fred is not younger than Sue. It is transitive because if Sue is younger than Fred and Fred is younger than Alana, then Sue is younger than Alana.

26. has the same birthday as

27. is taller than

28. lives in a different state than

Standardized Test Prep

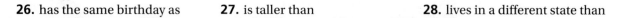

GRIDDED RESPONSE

SAT/ACT

29. You are typing a one-page essay for your English class. You set 1-in. margins on all sides of the page as shown in the figure at the right. How many square inches of the page will contain your essay?

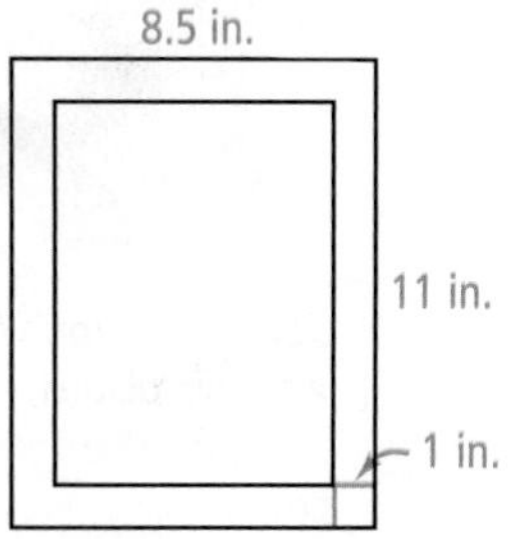

30. Given $2(m\angle A) + 17 = 45$ and $m\angle B = 2(m\angle A)$, what is $m\angle B$?

31. A circular flowerbed has circumference 14π m. What is its area in square meters? Use 3.14 for π.

32. The measure of the supplement of $\angle 1$ is 98. What is $m\angle 1$?

33. What is the next term in the sequence 2, 4, 8, 14, 22, 32, 44, . . . ?

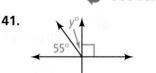

8.5 in.

11 in.

1 in.

Mixed Review

Ⓒ **34. Reasoning** Use logical reasoning to draw a conclusion.

See Lesson 2-4.

If a student is having difficulty in class, then that student's teacher is concerned.

Walt is having difficulty in science class.

Use the diagram at the right. Find each measure.

See Lesson 1-4.

35. $m\angle AOC$

36. $m\angle DOB$

37. $m\angle AOD$

38. $m\angle BOE$

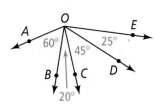

Get Ready! **To prepare for Lesson 2-6, do Exercises 39–41.**

Find the value of each variable.

See Lesson 1-5.

39.

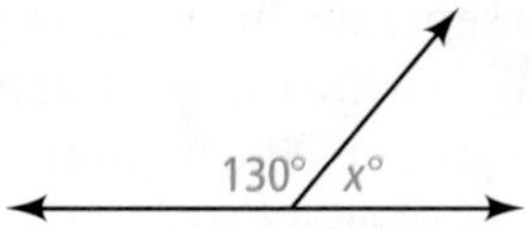

130° $x°$

40.

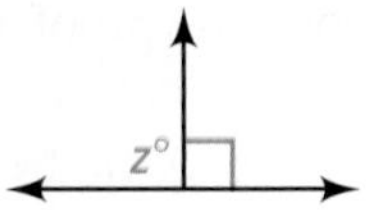

$z°$

41.

$y°$

55°

Proving Angles Congruent

© **Content Standard**
G.CO.9 Prove theorems about lines and angles. Theorems include: . . . vertical angles are congruent . . .

Objective   To prove and apply theorems about angles

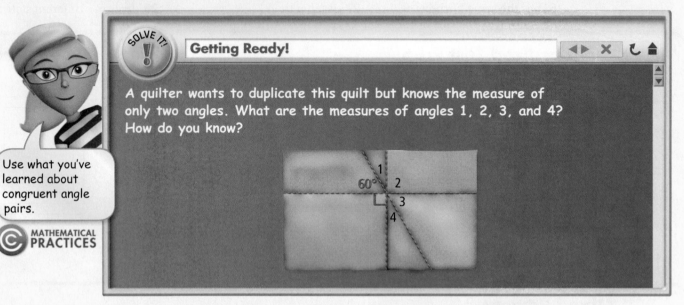

Use what you've learned about congruent angle pairs.

© **MATHEMATICAL PRACTICES**

Getting Ready!

A quilter wants to duplicate this quilt but knows the measure of only two angles. What are the measures of angles 1, 2, 3, and 4? How do you know?

In the Solve It, you may have noticed a relationship between vertical angles. You can prove that this relationship is always true using deductive reasoning. A **theorem** is a conjecture or statement that you prove true.

Essential Understanding   You can use given information, definitions, properties, postulates, and previously proven theorems as reasons in a proof.

Lesson Vocabulary
• theorem
• paragraph proof

take note

Theorem 2-1   Vertical Angles Theorem

Vertical angles are congruent.

$\angle 1 \cong \angle 3$ and $\angle 2 \cong \angle 4$

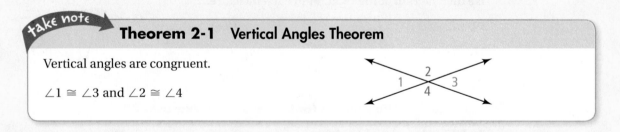

When you are writing a geometric proof, it may help to separate the theorem you want to prove into a hypothesis and conclusion. Another way to write the Vertical Angles Theorem is "If two angles are vertical, then they are congruent." The hypothesis becomes the given statement, and the conclusion becomes what you want to prove. A two-column proof of the Vertical Angles Theorem follows.

Proof **Proof of Theorem 2-1: Vertical Angles Theorem**

Given: $\angle 1$ and $\angle 3$ are vertical angles.

Prove: $\angle 1 \cong \angle 3$

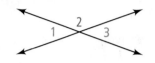

Statements	Reasons
1) $\angle 1$ and $\angle 3$ are vertical angles.	1) Given
2) $\angle 1$ and $\angle 2$ are supplementary. $\angle 2$ and $\angle 3$ are supplementary.	2) $\angle$ that form a linear pair are supplementary.
3) $m\angle 1 + m\angle 2 = 180$ $m\angle 2 + m\angle 3 = 180$	3) The sum of the measures of supplementary $\angle$ is 180.
4) $m\angle 1 + m\angle 2 = m\angle 2 + m\angle 3$	4) Transitive Property of Equality
5) $m\angle 1 = m\angle 3$	5) Subtraction Property of Equality
6) $\angle 1 \cong \angle 3$	6) $\angle$ with the same measure are $\cong$.

Plan

How do you get started?
Look for a relationship in the diagram that allows you to *write an equation* with the variable.

ⓒ Problem 1 **Using the Vertical Angles Theorem** `GRIDDED RESPONSE`

What is the value of x?

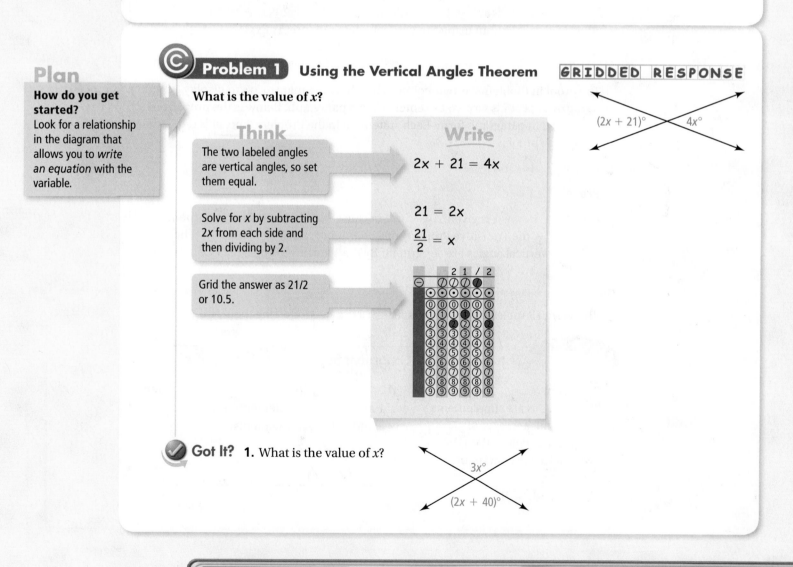

Think

The two labeled angles are vertical angles, so set them equal.

Solve for x by subtracting $2x$ from each side and then dividing by 2.

Grid the answer as 21/2 or 10.5.

Write

$2x + 21 = 4x$

$21 = 2x$

$\dfrac{21}{2} = x$

Got It? **1.** What is the value of x?

Proof (C) **Problem 2** **Proof Using the Vertical Angles Theorem**

Given: ∠1 ≅ ∠4

Prove: ∠2 ≅ ∠3

Think

Why does the Transitive Property work for statements 3 and 5?
In each case, an angle is congruent to two other angles, so the two angles are congruent to each other.

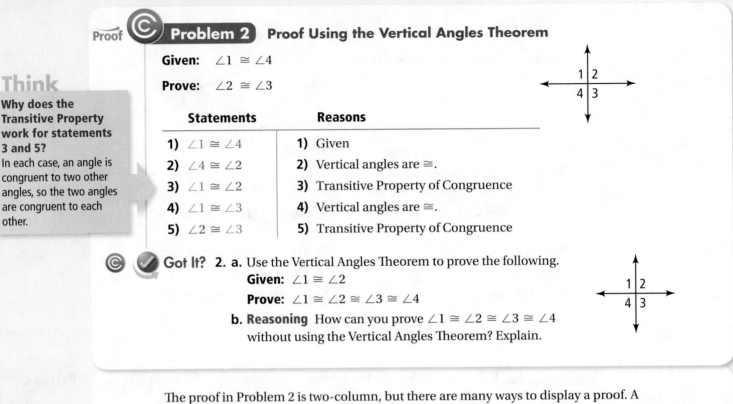

Statements	Reasons
1) ∠1 ≅ ∠4	**1)** Given
2) ∠4 ≅ ∠2	**2)** Vertical angles are ≅.
3) ∠1 ≅ ∠2	**3)** Transitive Property of Congruence
4) ∠1 ≅ ∠3	**4)** Vertical angles are ≅.
5) ∠2 ≅ ∠3	**5)** Transitive Property of Congruence

(C) ✓ **Got It?** **2. a.** Use the Vertical Angles Theorem to prove the following.

Given: ∠1 ≅ ∠2

Prove: ∠1 ≅ ∠2 ≅ ∠3 ≅ ∠4

b. Reasoning How can you prove ∠1 ≅ ∠2 ≅ ∠3 ≅ ∠4 without using the Vertical Angles Theorem? Explain.

The proof in Problem 2 is two-column, but there are many ways to display a proof. A **paragraph proof** is written as sentences in a paragraph. Below is the proof from Problem 2 in paragraph form. Each statement in the Problem 2 proof is red in the paragraph proof.

Proof **Given:** ∠1 ≅ ∠4

Prove: ∠2 ≅ ∠3

Proof: ∠1 ≅ ∠4 is given. ∠4 ≅ ∠2 because vertical angles are congruent. By the Transitive Property of Congruence, ∠1 ≅ ∠2. ∠1 ≅ ∠3 because vertical angles are congruent. By the Transitive Property of Congruence, ∠2 ≅ ∠3.

The Vertical Angles Theorem is a special case of the following theorem.

take note **Theorem 2-2** **Congruent Supplements Theorem**

Theorem	**If . . .**	**Then . . .**
If two angles are supplements of the same angle (or of congruent angles), then the two angles are congruent.	∠1 and ∠3 are supplements and ∠2 and ∠3 are supplements	∠1 ≅ ∠2

You will prove Theorem 2-2 in Problem 3.

Proof **Problem 3** **Writing a Paragraph Proof**

Given: ∠1 and ∠3 are supplementary.
∠2 and ∠3 are supplementary.

Prove: ∠1 ≅ ∠2

Proof: ∠1 and ∠3 are supplementary because it is given. So $m\angle 1 + m\angle 3 = 180$ by the definition of supplementary angles. ∠2 and ∠3 are supplementary because it is given, so $m\angle 2 + m\angle 3 = 180$ by the same definition. By the Transitive Property of Equality, $m\angle 1 + m\angle 3 = m\angle 2 + m\angle 3$. Subtract $m\angle 3$ from each side. By the Subtraction Property of Equality, $m\angle 1 = m\angle 2$. Angles with the same measure are congruent, so ∠1 ≅ ∠2.

Got It? **3.** Write a paragraph proof for the Vertical Angles Theorem.

The following theorems are similar to the Congruent Supplements Theorem.

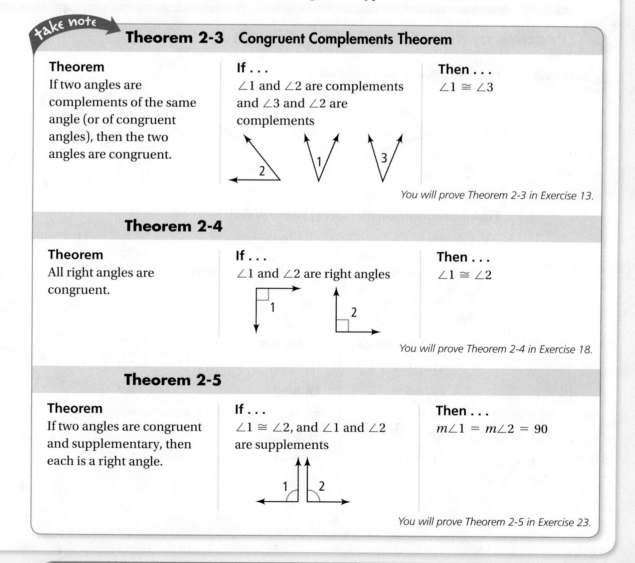

take note

Theorem 2-3 Congruent Complements Theorem

Theorem	**If . . .**	**Then . . .**
If two angles are complements of the same angle (or of congruent angles), then the two angles are congruent.	∠1 and ∠2 are complements and ∠3 and ∠2 are complements	∠1 ≅ ∠3

You will prove Theorem 2-3 in Exercise 13.

Theorem 2-4

Theorem	**If . . .**	**Then . . .**
All right angles are congruent.	∠1 and ∠2 are right angles	∠1 ≅ ∠2

You will prove Theorem 2-4 in Exercise 18.

Theorem 2-5

Theorem	**If . . .**	**Then . . .**
If two angles are congruent and supplementary, then each is a right angle.	∠1 ≅ ∠2, and ∠1 and ∠2 are supplements	$m\angle 1 = m\angle 2 = 90$

You will prove Theorem 2-5 in Exercise 23.

Lesson Check

Do you know HOW?

1. What are the measures of ∠1, ∠2, and ∠3?

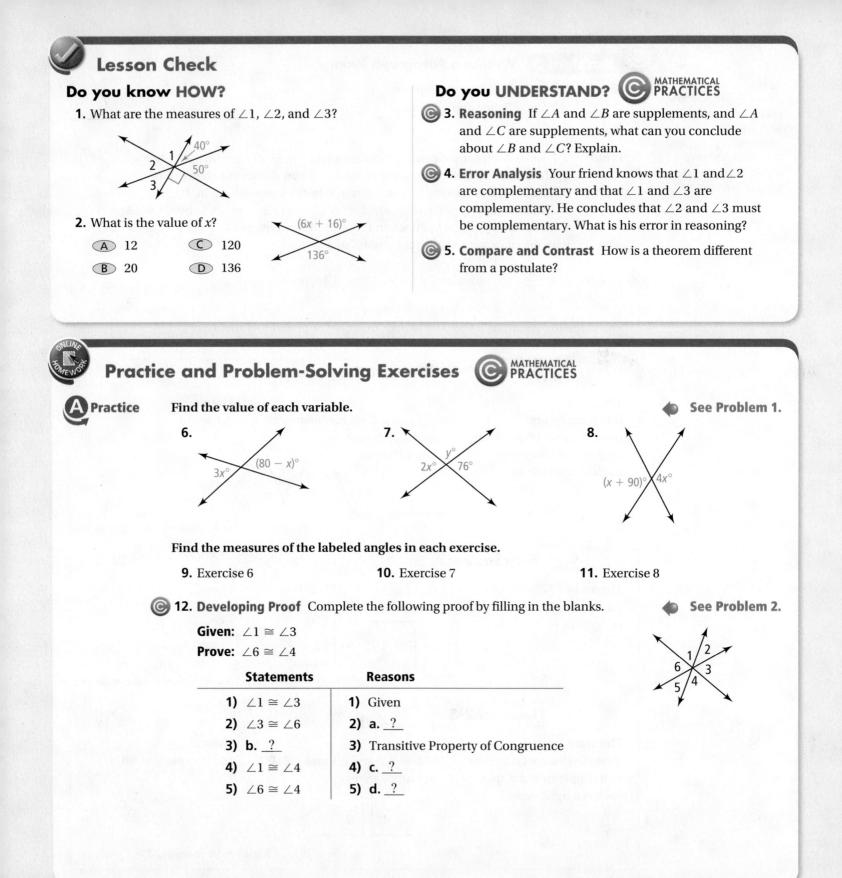

2. What is the value of x?

- Ⓐ 12
- Ⓑ 20
- Ⓒ 120
- Ⓓ 136

Do you UNDERSTAND? Ⓒ MATHEMATICAL PRACTICES

Ⓒ **3. Reasoning** If ∠A and ∠B are supplements, and ∠A and ∠C are supplements, what can you conclude about ∠B and ∠C? Explain.

Ⓒ **4. Error Analysis** Your friend knows that ∠1 and∠2 are complementary and that ∠1 and ∠3 are complementary. He concludes that ∠2 and ∠3 must be complementary. What is his error in reasoning?

Ⓒ **5. Compare and Contrast** How is a theorem different from a postulate?

Practice and Problem-Solving Exercises Ⓒ MATHEMATICAL PRACTICES

Ⓐ **Practice**   Find the value of each variable.   ◀ See Problem 1.

6.   **7.**   **8.**

$3x°$ $(80 − x)°$ $2x°$ $y°$ $76°$ $(x + 90)°$ $4x°$

Find the measures of the labeled angles in each exercise.

9. Exercise 6   **10.** Exercise 7   **11.** Exercise 8

Ⓒ **12. Developing Proof** Complete the following proof by filling in the blanks.   ◀ See Problem 2.

Given: ∠1 ≅ ∠3

Prove: ∠6 ≅ ∠4

Statements	Reasons
1) ∠1 ≅ ∠3	**1)** Given
2) ∠3 ≅ ∠6	**2) a.** ?
3) b. ?	**3)** Transitive Property of Congruence
4) ∠1 ≅ ∠4	**4) c.** ?
5) ∠6 ≅ ∠4	**5) d.** ?

See Problem 3.

13. Developing Proof Fill in the blanks to complete this proof of the Congruent Complements Theorem (Theorem 2-3).

If two angles are complements of the same angle, then the two angles are congruent.

Given: $\angle 1$ and $\angle 2$ are complementary.
$\angle 3$ and $\angle 2$ are complementary.

Prove: $\angle 1 \cong \angle 3$

Proof: $\angle 1$ and $\angle 2$ are complementary and $\angle 3$ and $\angle 2$ are complementary because it is given. By the definition of complementary angles, $m\angle 1 + m\angle 2 = $ **a.** _?_ and $m\angle 3 + m\angle 2 = $ **b.** _?_ . Then $m\angle 1 + m\angle 2 = m\angle 3 + m\angle 2$ by the Transitive Property of Equality. Subtract $m\angle 2$ from each side. By the Subtraction Property of Equality, you get $m\angle 1 = $ **c.** _?_ . Angles with the same measure are **d.** _?_ , so $\angle 1 \cong \angle 3$.

B Apply

14. Think About a Plan What is the measure of the angle formed by Park St. and 116th St.?
• Can you make a connection between the angle you need to find and the labeled angle?
• How are angles that form a right angle related?

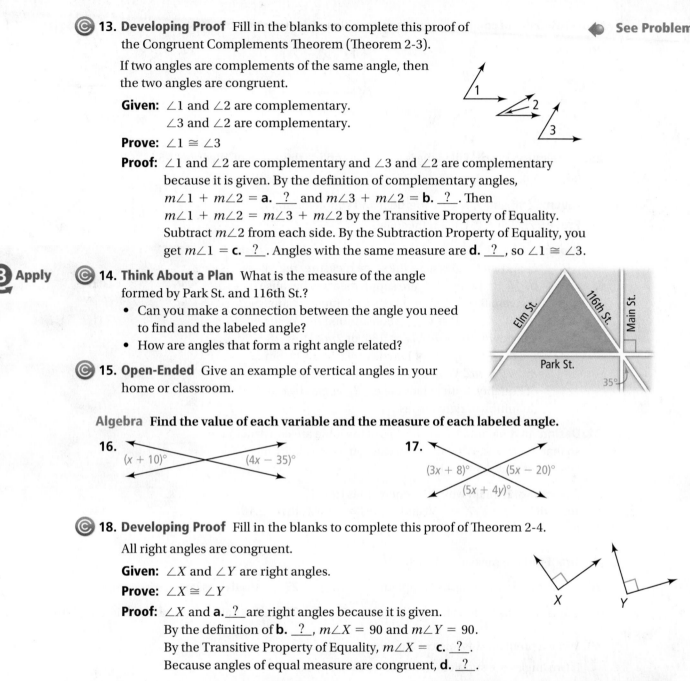

15. Open-Ended Give an example of vertical angles in your home or classroom.

Algebra Find the value of each variable and the measure of each labeled angle.

16. $(x + 10)°$ $(4x - 35)°$

17. $(3x + 8)°$ $(5x - 20)°$ $(5x + 4y)°$

18. Developing Proof Fill in the blanks to complete this proof of Theorem 2-4.

All right angles are congruent.

Given: $\angle X$ and $\angle Y$ are right angles.

Prove: $\angle X \cong \angle Y$

Proof: $\angle X$ and **a.** _?_ are right angles because it is given. By the definition of **b.** _?_ , $m\angle X = 90$ and $m\angle Y = 90$. By the Transitive Property of Equality, $m\angle X = $ **c.** _?_ . Because angles of equal measure are congruent, **d.** _?_ .

19. Miniature Golf In the game of miniature golf, the ball bounces off the wall at the same angle it hit the wall. (This is the angle formed by the path of the ball and the line perpendicular to the wall at the point of contact.) In the diagram, the ball hits the wall at a 40° angle. Using Theorem 2-3, what are the values of x and y?

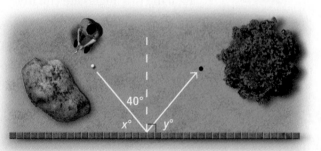

Name two pairs of congruent angles in each figure. Justify your answers.

20.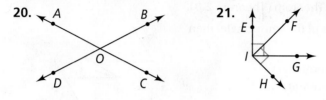

21.

22.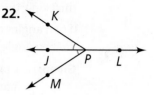

23. **Developing Proof** Fill in the blanks to complete this proof of Theorem 2-5.

If two angles are congruent and supplementary, then each is a right angle.

Given: ∠W and ∠V are congruent and supplementary.
Prove: ∠W and ∠V are right angles.

Proof: ∠W and ∠V are congruent because **a.** __?__ . Because congruent angles have the same measure, $m\angle W =$ **b.** __?__ . ∠W and ∠V are supplementary because it is given. By the definition of supplementary angles, $m\angle W + m\angle V =$ **c.** __?__ . Substituting $m\angle W$ for $m\angle V$, you get $m\angle W + m\angle W = 180$, or $2m\angle W = 180$. By the **d.** __?__ Property of Equality, $m\angle W = 90$. Since $m\angle W = m\angle V$, $m\angle V = 90$ by the Transitive Property of Equality. Both angles are **e.** __?__ angles by the definition of right angles.

24. **Design** In the photograph, the legs of the table are constructed so that ∠1 ≅ ∠2. What theorem can you use to justify the statement that ∠3 ≅ ∠4?

25. **Reasoning** Explain why this statement is true: If $m\angle ABC + m\angle XYZ = 180$ and ∠ABC ≅ ∠XYZ, then ∠ABC and ∠XYZ are right angles.

Algebra Find the measure of each angle.

26. ∠A is twice as large as its complement, ∠B.

27. ∠A is half as large as its complement, ∠B.

28. ∠A is twice as large as its supplement, ∠B.

29. ∠A is half as large as twice its supplement, ∠B.

30. Write a proof for this form of Theorem 2-2.

    Proof If two angles are supplements of congruent angles, then the two angles are congruent.

    Given: ∠1 and ∠2 are supplementary.
          ∠3 and ∠4 are supplementary.
          ∠2 ≅ ∠4
    Prove: ∠1 ≅ ∠3

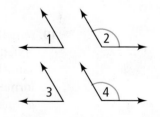

Challenge

31. **Coordinate Geometry** ∠DOE contains points $D(2, 3)$, $O(0, 0)$, and $E(5, 1)$. Find the coordinates of a point F so that $\overrightarrow{OF}$ is a side of an angle that is adjacent and supplementary to ∠DOE.

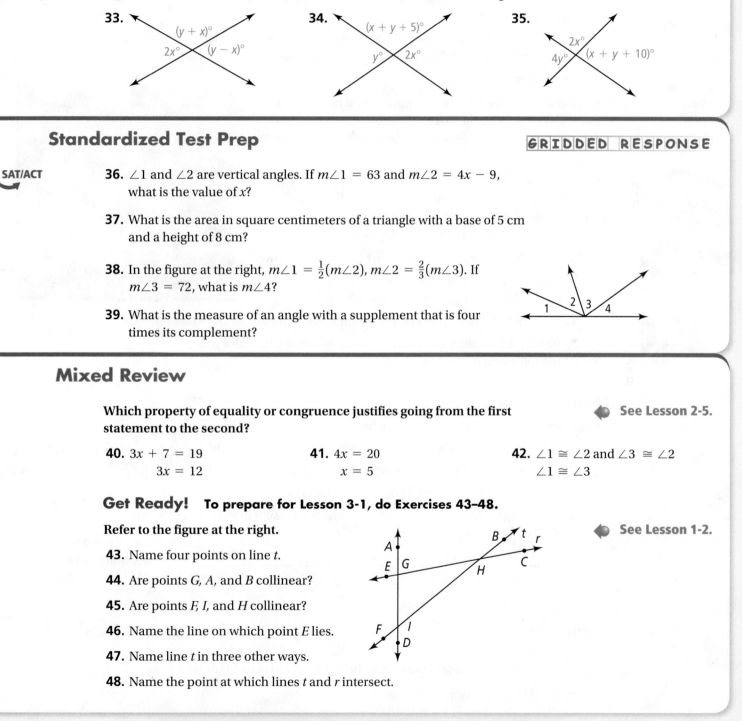

32. Coordinate Geometry $\angle AOX$ contains points $A(1, 3)$, $O(0, 0)$, and $X(4, 0)$.
- **a.** Find the coordinates of a point B so that $\angle BOA$ and $\angle AOX$ are adjacent complementary angles.
- **b.** Find the coordinates of a point C so that $\overrightarrow{OC}$ is a side of a different angle that is adjacent and complementary to $\angle AOX$.

Algebra Find the value of each variable and the measure of each angle.

33. $(y + x)°$ $2x°$ $(y - x)°$

34. $(x + y + 5)°$ $y°$ $2x°$

35. $2x°$ $4y°$ $(x + y + 10)°$

Standardized Test Prep

GRIDDED RESPONSE

SAT/ACT

36. $\angle 1$ and $\angle 2$ are vertical angles. If $m\angle 1 = 63$ and $m\angle 2 = 4x - 9$, what is the value of x?

37. What is the area in square centimeters of a triangle with a base of 5 cm and a height of 8 cm?

38. In the figure at the right, $m\angle 1 = \frac{1}{2}(m\angle 2)$, $m\angle 2 = \frac{2}{3}(m\angle 3)$. If $m\angle 3 = 72$, what is $m\angle 4$?

39. What is the measure of an angle with a supplement that is four times its complement?

Mixed Review

Which property of equality or congruence justifies going from the first statement to the second?

See Lesson 2-5.

40. $3x + 7 = 19$
$3x = 12$

41. $4x = 20$
$x = 5$

42. $\angle 1 \cong \angle 2$ and $\angle 3 \cong \angle 2$
$\angle 1 \cong \angle 3$

Get Ready! **To prepare for Lesson 3-1, do Exercises 43–48.**

Refer to the figure at the right.

See Lesson 1-2.

43. Name four points on line t.

44. Are points G, A, and B collinear?

45. Are points F, I, and H collinear?

46. Name the line on which point E lies.

47. Name line t in three other ways.

48. Name the point at which lines t and r intersect.

Pull It All Together © ASSESSMENT

To solve these problems, you will pull together ideas about inductive and deductive reasoning.

BIG idea Reasoning and Proof

You can observe patterns to make a conjecture; you can prove a conjecture is true by using given information, definitions, properties, postulates, and theorems.

© Performance Task 1

You have the yellow game piece, your friend has the red game piece, and your brother has the blue game piece. Read the rules of the board game and then answer the questions.

Rules

- You play counterclockwise.
- If you land on red, then you go back 1.
- If you land on green, then you advance 1.
- If you land on yellow, then you pick a card.

ROLL AGAIN

3

FINISH

START

  a. You roll 3. What must you do next? How do you know?

  b. Your brother picks a card at the end of his turn. On what colors might he have landed? Explain.

  c. Your friend rolls 2. What else must your friend do? How do you know?

  d. Based on the colors already shown on the board, what color should the roll-again box be? Justify your answer.

© Performance Task 2

Consider the number pattern at the right.

  a. What is the sum of the numbers 31–40?

  b. What is the sum of the numbers 101–110?

  c. What kind of reasoning did you use in parts (a) and (b)?

> The sum of the numbers 1–10 is 55.
> The sum of the numbers 11–20 is 155.
> The sum of the numbers 21–30 is 255.

  d. Following is the development of a formula for the sum of *n* consecutive integers.

$$S = x + (x + 1) + (x + 2) + \ldots + (y - 2) + (y - 1) + y \qquad \text{The sum of } n \text{ integers from } x \text{ to } y$$
$$\underline{+ \quad S = y + (y - 1) + (y - 2) + \ldots + (x + 2) + (x + 1) + x} \qquad \text{The same sum in reverse order}$$
$$2S = (x + y) + (x + y) + (x + y) + \ldots + (x + y) + (x + y) + (x + y) \qquad \text{Add the equations.}$$

$$2S = n(x + y) \qquad \text{There are } n \text{ terms of } (x + y).$$

$$S = \frac{n(x + y)}{2} \qquad \text{Divide each side by 2.}$$

  Use the formula to find the sum of the numbers 101–110. Show your work.

  e. What kind of reasoning did you use in part (d)?

Connecting BIG ideas and Answering the Essential Questions

Reasoning and Proof
You can observe patterns to make a conjecture; you can prove it is true using given information, definitions, properties, postulates, and theorems.

Inductive Reasoning (Lesson 2-1)
Inductive reasoning is the process of making conjectures based on patterns.

$$3 \quad 9 \quad 27 \quad 81$$
$$\times 3 \quad \times 3 \quad \times 3$$

Deductive Reasoning (Lessons 2-2, 2-3, 2-4)
Deductive reasoning is the process of making logical conclusions from given statements or facts.

Law of Detachment
If $p \rightarrow q$ is true and p is true, then q is true.

Law of Syllogism
If $p \rightarrow q$ is true and $q \rightarrow r$ is true, then $p \rightarrow r$ is true.

Proofs (Lessons 2-5 and 2-6)
A two-column proof lists the statements to the left and the corresponding reasons to the right. In a paragraph proof, the statements and reasons are written as sentences.

Statements	Reasons
• given information	• definitions
• information from a diagram	• properties
• logical reasoning	• postulates
• statement to prove	• previously proven theorems

Chapter Vocabulary

- biconditional (p. 98)
- conclusion (p. 89)
- conditional (p. 89)
- conjecture (p. 83)
- contrapositive (p. 91)
- converse (p. 91)
- counterexample (p.84)
- deductive reasoning (p. 106)
- equivalent statements (p. 91)
- hypothesis (p. 89)
- inductive reasoning (p. 82)
- inverse (p. 91)
- Law of Detachment (p. 106)
- Law of Syllogism (p. 108)
- negation (p. 91)
- paragraph proof (p. 122)
- proof (p. 115)
- theorem (p. 120)
- truth value (p. 90)
- two-column proof (p. 115)

Choose the correct vocabulary term to complete each sentence.

1. The part of a conditional that follows "then" is the ? .

2. Reasoning logically from given statements to a conclusion is ? .

3. A conditional has a(n) ? of true or false.

4. The ? of a conditional switches the hypothesis and conclusion.

5. When a conditional and its converse are true, you can write them as a single true statement called a(n) ? .

6. A statement that you prove true is a(n) ? .

7. The part of a conditional that follows "if" is the ? .

2-1 Patterns and Inductive Reasoning

Quick Review

You use **inductive reasoning** when you make conclusions based on patterns you observe. A **conjecture** is a conclusion you reach using inductive reasoning. A **counterexample** is an example that shows a conjecture is incorrect.

Example

Describe the pattern. What are the next two terms in the sequence?

$$1, -3, 9, -27, \ldots$$

Each term is -3 times the previous term. The next two terms are $-27 \times (-3) = 81$ and $81 \times (-3) = -243$.

Exercises

Find a pattern for each sequence. Describe the pattern and use it to show the next two terms.

8. 1000, 100, 10, . . .

9. 5, −5, 5, −5, . . .

10. 34, 27, 20, 13, . . .

11. 6, 24, 96, 384, . . .

Find a counterexample to show that each conjecture is false.

12. The product of any integer and 2 is greater than 2.

13. The city of Portland is in Oregon.

2-2 Conditional Statements

Quick Review

A **conditional** is an *if-then* statement. The symbolic form of a conditional is $p \to q$, where p is the **hypothesis** and q is the **conclusion**.

- To find the **converse**, switch the hypothesis and conclusion of the conditional ($q \to p$).

- To find the **inverse**, negate the hypothesis and the conclusion of the conditional ($\sim p \to \sim q$).

- To find the **contrapositive**, negate the hypothesis and the conclusion of the converse ($\sim q \to \sim p$).

Example

What is the converse of the conditional statement below? What is its truth value?

   If you are a teenager, then you are younger than 20.

Converse: If you are younger than 20, then you are a teenager.

A 7-year-old is not a teenager. The converse is false.

Exercises

Rewrite each sentence as a conditional statement.

14. All motorcyclists wear helmets.

15. Two nonparallel lines intersect in one point.

16. Angles that form a linear pair are supplementary.

17. School is closed on certain holidays.

Write the converse, inverse, and contrapositive of the given conditional. Then determine the truth value of each statement.

18. If an angle is obtuse, then its measure is greater than 90 and less than 180.

19. If a figure is a square, then it has four sides.

20. If you play the tuba, then you play an instrument.

21. If you baby-sit, then you are busy on Saturday night.

2-3 Biconditionals and Definitions

Quick Review

When a conditional and its converse are true, you can combine them as a true **biconditional** using the phrase *if and only if*. The symbolic form of a biconditional is $p \leftrightarrow q$. You can write a good **definition** as a true biconditional.

Example

Is the following definition reversible? If yes, write it as a true biconditional.

    A hexagon is a polygon with exactly six sides.

Yes. The conditional is true: If a figure is a hexagon, then it is a polygon with exactly six sides. Its converse is also true: If a figure is a polygon with exactly six sides, then it is a hexagon.

Biconditional: A figure is a hexagon *if and only if* it is a polygon with exactly six sides.

Exercises

Determine whether each statement is a good definition. If not, explain.

22. A newspaper has articles you read.

23. A linear pair is a pair of adjacent angles whose noncommon sides are opposite rays.

24. An angle is a geometric figure.

25. Write the following definition as a biconditional.

    An oxymoron is a phrase that contains contradictory terms.

26. Write the following biconditional as two statements, a conditional and its converse.

    Two angles are complementary if and only if the sum of their measures is 90.

2-4 Deductive Reasoning

Quick Review

Deductive reasoning is the process of reasoning logically from given statements to a conclusion.

Law of Detachment: If $p \rightarrow q$ is true and p is true, then q is true.

Law of Syllogism: If $p \rightarrow q$ and $q \rightarrow r$ are true, then $p \rightarrow r$ is true.

Example

What can you conclude from the given information? Given: If you play hockey, then you are on the team. If you are on the team, then you are a varsity athlete.

The conclusion of the first statement matches the hypothesis of the second statement. Use the Law of Syllogism to conclude: If you play hockey, then you are a varsity athlete.

Exercises

Use the Law of Detachment to make a conclusion.

27. If you practice tennis every day, then you will become a better player. Colin practices tennis every day.

28. $\angle 1$ and $\angle 2$ are supplementary. If two angles are supplementary, then the sum of their measures is 180.

Use the Law of Syllogism to make a conclusion.

29. If two angles are vertical, then they are congruent. If two angles are congruent, then their measures are equal.

30. If your father buys new gardening gloves, then he will work in his garden. If he works in his garden, then he will plant tomatoes.

2-5 Reasoning in Algebra and Geometry

Quick Review

You use deductive reasoning and properties to solve equations and justify your reasoning.

A **proof** is a convincing argument that uses deductive reasoning. A **two-column proof** lists each statement on the left and the justification for each statement on the right.

Example

What is the name of the property that justifies going from the first line to the second line?

$\angle A \cong \angle B$ and $\angle B \cong \angle C$

$\angle A \cong \angle C$

Transitive Property of Congruence

Exercises

31. Algebra Fill in the reason that justifies each step.

Given: $QS = 42$

Prove: $x = 13$

Statements	Reasons
1) $QS = 42$	**1)** a. ?
2) $QR + RS = QS$	**2)** b. ?
3) $(x + 3) + 2x = 42$	**3)** c. ?
4) $3x + 3 = 42$	**4)** d. ?
5) $3x = 39$	**5)** e. ?
6) $x = 13$	**6)** f. ?

Use the given property to complete the statement.

32. Division Property of Equality
If $2(AX) = 2(BY)$, then $AX = \underline{\ ?\ }$.

33. Distributive Property: $3p - 6q = 3(\underline{\ ?\ })$

2-6 Proving Angles Congruent

Quick Review

A statement that you prove true is a **theorem**. A proof written as a paragraph is a **paragraph proof**. In geometry, each statement in a proof is justified by given information, a property, postulate, definition, or theorem.

Example

Write a paragraph proof.

Given: $\angle 1 \cong \angle 4$

Prove: $\angle 2 \cong \angle 3$

$\angle 1 \cong \angle 4$ because it is given. $\angle 1 \cong \angle 2$ because vertical angles are congruent. $\angle 4 \cong \angle 2$ by the Transitive Property of Congruence. $\angle 4 \cong \angle 3$ because vertical angles are congruent. $\angle 2 \cong \angle 3$ by the Transitive Property of Congruence.

Exercises

Use the diagram for Exercises 34–37.

34. Find the value of y.

35. Find $m\angle AEC$.

36. Find $m\angle BED$.

37. Find $m\angle AEB$.

38. Given: $\angle 1$ and $\angle 2$ are complementary.
$\angle 3$ and $\angle 4$ are complementary.
$\angle 2 \cong \angle 4$

Prove: $\angle 1 \cong \angle 3$

Do you know HOW?

Use inductive reasoning to describe each pattern and find the next two terms of each sequence.

1. $-16, 8, -4, 2, \ldots$

2. $1, 4, 9, 16, 25, \ldots$

For Exercises 3 and 4, find a counterexample.

3. All snakes are poisonous.

4. If two angles are complementary, then they are not congruent.

5. Identify the hypothesis and conclusion:
If $x + 9 = 11$, then $x = 2$.

6. Write "all puppies are cute" as a conditional.

Write the converse, inverse, and contrapositive for each statement. Determine the truth value of each.

7. If a figure is a square, then it has at least two right angles.

8. If a square has side length 3 m, then its perimeter is 12 m.

© **Writing** Explain why each statement is not a good definition.

9. A pen is a writing instrument.

10. Supplementary angles are angles that form a straight line.

11. Vertical angles are angles that are congruent.

Name the property that justifies each statement.

12. If $UV = KL$ and $KL = 6$, then $UV = 6$.

13. If $m\angle 1 + m\angle 2 = m\angle 4 + m\angle 2$, then $m\angle 1 = m\angle 4$.

14. $\angle ABC \cong \angle ABC$

15. If $\angle DEF \cong \angle HJK$, then $\angle HJK \cong \angle DEF$.

16. The measure of an angle is 52 more than the measure of its complement. What is the measure of the angle?

17. Rewrite this biconditional as two conditionals.
A fish is a bluegill if and only if it is a bluish, freshwater sunfish.

For each diagram, state two pairs of angles that are congruent. Justify your answers.

18.   **19.**

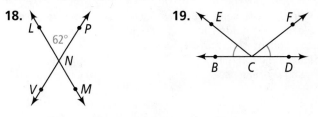

Use the Law of Detachment and the Law of Syllogism to make any possible conclusion. Write *not possible* if you cannot make any conclusion.

20. People who live in glass houses should not throw stones. Emily should not throw stones.

21. James wants to be a chemical engineer. If a student wants to be a chemical engineer, then that student must graduate from college.

Do you UNDERSTAND?

© **22. Open-Ended** Write two different sequences whose first three terms are 1, 2, 4. Describe each pattern.

© **23. Developing Proof** Complete this proof by filling in the blanks.

Given: $\angle FED$ and $\angle DEW$ are complementary.

Prove: $\angle FEW$ is a right angle.

$\angle FED$ and $\angle DEW$ are complementary because it is given. By the Definition of Complementary Angles, $m\angle FED + m\angle DEW =$ **a.** ? .
$m\angle FED + m\angle DEW = m\angle FEW$ by the **b.** ? .
$90 = m\angle FEW$ by the **c.** ? Property of Equality.
Then $\angle FEW$ is a right angle by the **d.** ? .

TIPS FOR SUCCESS

Some questions ask you to extend a pattern. Read the sample question at the right. Then follow the tips to answer it.

TIP 1

Look for a relationship between consecutive figures. Make sure the relationship holds for each pair of consecutive figures, not just the first two figures.

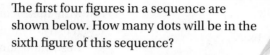

The first four figures in a sequence are shown below. How many dots will be in the sixth figure of this sequence?

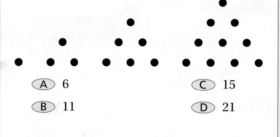

Ⓐ 6   Ⓒ 15

Ⓑ 11   Ⓓ 21

TIP 2

Use the relationship between the figures to extend the pattern.

Think It Through

The second figure has 2 more dots than the first figure, the third figure has 3 more dots than the second figure, and the fourth figure has 4 more dots than the third figure. So, the fifth figure will have 5 more dots than the fourth figure, or $10 + 5 = 15$ dots. The sixth figure will have 6 more dots than the fifth figure, or $15 + 6 = 21$ dots. The correct answer is D.

Vocabulary Builder

As you solve test items, you must understand the meanings of mathematical terms. Choose the correct term to complete each sentence.

I. Reasoning that is based on patterns you observe is called (*inductive, deductive*) reasoning.

II. The (*Law of Syllogism, Law of Detachment*) allows you to state a conclusion from two true conditional statements when the conclusion of one statement is the hypothesis of the other statement.

III. A conditional, or *if-then*, statement has two parts. The part following *if* is the (*conclusion, hypothesis*).

IV. The (*Reflexive Property, Symmetric Property*) says that if $a = b$, then $b = a$.

V. The (*inverse, converse*) of a conditional negates both the hypothesis and the conclusion.

Multiple Choice

Read each question. Then write the letter of the correct answer on your paper.

1. Which pair of angles must be congruent?

Ⓐ supplementary angles

Ⓑ complementary angles

Ⓒ adjacent angles

Ⓓ vertical angles

2. Which of the following best defines a postulate?

Ⓕ a statement accepted without proof

Ⓖ a conclusion reached using inductive reasoning

Ⓗ an example that proves a conjecture false

Ⓘ a statement that you prove true

3. What is the second step in constructing ∠S, an angle congruent to ∠A?

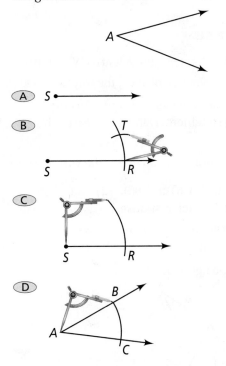

Ⓐ S •————————▶

Ⓑ (figure with T, S, R)

Ⓒ (figure with S, R)

Ⓓ (figure with A, B, C)

4. What is the converse of the following statement?

If a whole number has 0 as its last digit, then the number is evenly divisible by 10.

Ⓕ If a number is evenly divisible by 10, then it is a whole number.

Ⓖ If a whole number is divisible by 10, then it is an even number.

Ⓗ If a whole number is evenly divisible by 10, then it has 0 as its last digit.

Ⓘ If a whole number has 0 as its last digit, then it must be evenly divisible by 10.

5. The sum of the measures of the complement and the supplement of an angle is 114. What is the measure of the angle?

Ⓐ 12   Ⓒ 78

Ⓑ 66   Ⓓ 102

6. Which counterexample shows that the following conjecture is false?

Every perfect square number has exactly three factors.

Ⓕ The factors of 2 are 1, 2.

Ⓖ The factors of 4 are 1, 2, 4.

Ⓗ The factors of 8 are 1, 2, 4, 8.

Ⓘ The factors of 16 are 1, 2, 4, 8, 16.

7. How many rays are in the next two terms in the sequence?

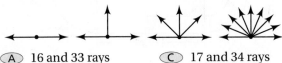

Ⓐ 16 and 33 rays   Ⓒ 17 and 34 rays

Ⓑ 17 and 33 rays   Ⓓ 18 and 34 rays

8. Which of the statements could be a conclusion based on the following information?

If a polygon is a pentagon, then it has one more side than a quadrilateral. If a polygon has one more side than a quadrilateral, then it has two more sides than a triangle.

Ⓕ If a polygon is a pentagon, then it has many sides.

Ⓖ If a polygon has two more sides than a quadrilateral, then it is a hexagon.

Ⓗ If a polygon has more sides than a triangle, then it is a pentagon.

Ⓘ If a polygon is a pentagon, then it has two more sides than a triangle.

9. The radius of each of the circular sections in the dumbbell-shaped table below is 3 ft. The rectangular portion has an area of 32 ft² and the length is twice the width. What is the area of the entire table to the nearest tenth?

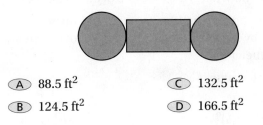

Ⓐ 88.5 ft²   Ⓒ 132.5 ft²

Ⓑ 124.5 ft²   Ⓓ 166.5 ft²

10. An athletic field is a rectangle, 120 yd by 60 yd, with a semicircle at each of the ends. A running track 15 yd wide surrounds the field. How many yards of fencing do you need to surround the outside edge of the track? Round your answer to the nearest tenth of a yard. Use 3.14 for π.

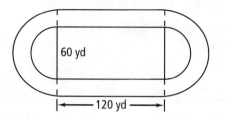

11. What is the next number in the pattern?

$$1, -4, 9, -16, \ldots$$

12. The base of a rectangle is 7 cm less than three times its height. If the base is 5 cm, what is the area of the rectangle in square centimeters?

13. $m\angle BZD = 107$
$m\angle FZE = 2x + 5$
$m\angle CZD = x$
What is the measure of $\angle CZD$?

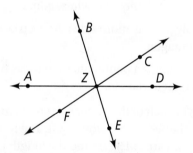

14. The measure of an angle is three more than twice its supplement. What is the measure of the angle?

15. A square and rectangle have equal area. The rectangle is 32 cm by 18 cm. What is the perimeter of the square in centimeters?

16. What is the value of x?

17. What is the y-coordinate of the midpoint of a segment with endpoints $(0, -4)$ and $(-4, 7)$?

Short Response

18. On a number line, P is at -4 and R is at 8. What is the coordinate of the point Q, which is $\frac{3}{4}$ the way from R to P?

19. Write the converse, inverse, and contrapositive of the following statement. Determine the truth value of each.

If you live in Oregon, then you live in the United States.

20. $\overline{AB}$ has endpoints $A(3, 6)$ and $B(9, -2)$ and midpoint M. Justify each response.
   a. What are the coordinates of M?
   b. What is AB?

21. Draw a net for the figure below.

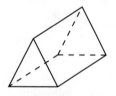

Extended Response

22. The sequence below lists the first eight powers of 7.

$$7^1, 7^2, 7^3, 7^4, 7^5, 7^6, 7^7, 7^8, \ldots$$

   a. Make a table that lists the digit in the ones place for each of the first eight powers of 7. For example, $7^4 = 2401$. The digit 1 is in the ones place.
   b. What digit is in the ones place of 7^{34}? Explain your reasoning.

23. Write a proof.

   Given: $\angle 1$ and $\angle 2$ are supplementary.

   Prove: $\angle 1$ and $\angle 2$ are right angles.

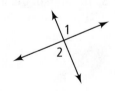

Get Ready!

CHAPTER 3

Lesson 1-5 ◀ **Identifying Angle Pairs**

Identify all pairs of each type of angles in the diagram.

1. linear pair

2. complementary angles

3. vertical angles

4. supplementary angles

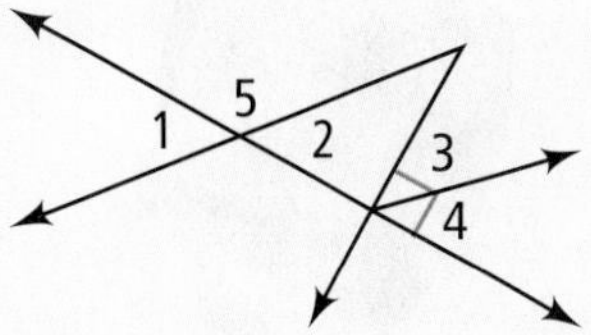

Lesson 2-5 ◀ **Justifying Statements**

Name the property that justifies each statement.

5. If $3x = 6$, then $x = 2$.

6. If $\angle 1 \cong \angle 2$ and $\angle 2 \cong \angle 3$, then $\angle 1 \cong \angle 3$.

Skills
Handbook,
p. 894
◀ **Solving Equations**

Algebra Solve each equation.

7. $3x + 11 = 7x - 5$ **8.** $(x - 4) + 52 = 109$ **9.** $(2x + 5) + (3x - 10) = 70$

Lesson 1-7 ◀ **Finding Distances in the Coordinate Plane**

Find the distance between the points.

10. $(1, 3)$ and $(5, 0)$ **11.** $(-4, 2)$ and $(4, 4)$ **12.** $(3, -1)$ and $(7, -2)$

🔊 Looking Ahead Vocabulary

13. The core of an apple is in the *interior* of the apple. The peel is on the *exterior*. How can the terms *interior* and *exterior* apply to geometric figures?

14. A ship sailing from the United States to Europe makes a transatlantic voyage. What does the prefix *trans-* mean in this situation? A *transversal* is a special type of line in geometry. What might a *transversal* do? Explain.

15. People in many jobs use *flow*charts to describe the logical steps of a particular process. How do you think you can use a *flow proof* in geometry?

Parallel and Perpendicular Lines

 DOMAINS
- Congruence
- Expressing Geometric Properties with Equations
- Modeling with Geometry

Check out the gymnast on the parallel bars! Why do you think they are called parallel bars?

You'll learn about properties of parallel lines in this chapter.

 Vocabulary

English/Spanish Vocabulary Audio Online:

English	Spanish
alternate exterior angles, *p. 142*	ángulos alternos externos
alternate interior angles, *p. 142*	ángulos alternos internos
corresponding angles, *p. 142*	ángulos correspondientes
exterior angle of a polygon, *p. 173*	ángulo exterior de un polígono
parallel lines, *p. 140*	rectas paralelas
same-side interior angles, *p. 142*	ángulos internos del mismo lado
skew lines, *p. 140*	rectas cruzadas
transversal, *p. 141*	transversal

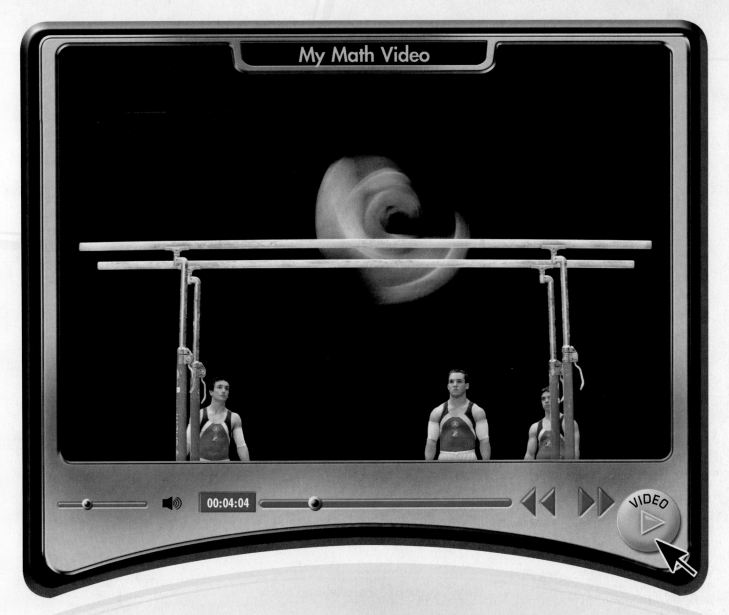

My Math Video

00:04:04

VIDEO ▶

BIG ideas

1 **Reasoning and Proof**
   Essential Question How do you prove that two lines are parallel?

2 **Measurement**
   Essential Question What is the sum of the measures of the angles of a triangle?

3 **Coordinate Geometry**
   Essential Question How do you write an equation of a line in the coordinate plane?

Chapter Preview

3-1 **Lines and Angles**
3-2 **Properties of Parallel Lines**
3-3 **Proving Lines Parallel**
3-4 **Parallel and Perpendicular Lines**
3-5 **Parallel Lines and Triangles**
3-6 **Constructing Parallel and Perpendicular Lines**
3-7 **Equations of Lines in the Coordinate Plane**
3-8 **Slopes of Parallel and Perpendicular Lines**

3-1 Lines and Angles

Content Standards

G.CO.1 Know precise definitions of . . . parallel line.
Prepares for G.CO.9 Prove theorems about lines and angles.

Objectives To identify relationships between figures in space
To identify angles formed by two lines and a transversal

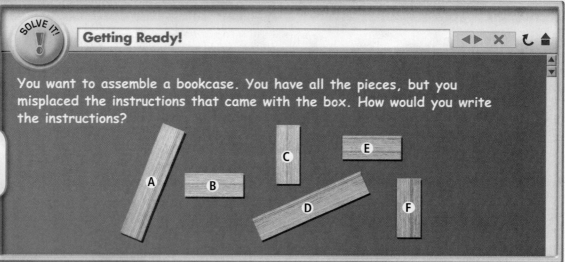

SOLVE IT!

Getting Ready!

You want to assemble a bookcase. You have all the pieces, but you misplaced the instructions that came with the box. How would you write the instructions?

Try visualizing how the bookcase looks in two dimensions.

MATHEMATICAL PRACTICES

In the Solve It, you used relationships among planes in space to write the instructions. In Chapter 1, you learned about intersecting lines and planes. In this lesson, you will explore relationships of nonintersecting lines and planes.

Essential Understanding Not all lines and not all planes intersect.

Lesson Vocabulary

- parallel lines
- skew lines
- parallel planes
- transversal
- alternate interior angles
- same-side interior angles
- corresponding angles
- alternate exterior angles

take note

Key Concept Parallel and Skew

Definition	Symbols	Diagram
Parallel lines are coplanar lines that do not intersect. The symbol ∥ means "is parallel to."	$\overleftrightarrow{AE} \parallel \overleftrightarrow{BF}$ $\overleftrightarrow{AD} \parallel \overleftrightarrow{BC}$	Use arrows to show $\overleftrightarrow{AE} \parallel \overleftrightarrow{BF}$ and $\overleftrightarrow{AD} \parallel \overleftrightarrow{BC}$.
Skew lines are noncoplanar; they are not parallel and do not intersect.	$\overleftrightarrow{AB}$ and $\overleftrightarrow{CG}$ are skew.	
Parallel planes are planes that do not intersect.	plane $ABCD$ ∥ plane $EFGH$	

A line and a plane that do not intersect are parallel. Segments and rays can also be parallel or skew. They are parallel if they lie in parallel lines and skew if they lie in skew lines.

 **Problem 1** **Identifying Nonintersecting Lines and Planes**

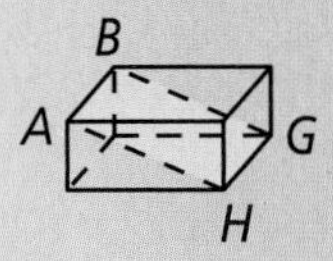

Think

Parallel lines are coplanar. Which planes contain $\overline{AB}$?
Planes *ABCD, ABFE,* and *ABGH* contain $\overline{AB}$. You need to visualize plane *ABGH*.

In the figure, assume that lines and planes that appear to be parallel are parallel.

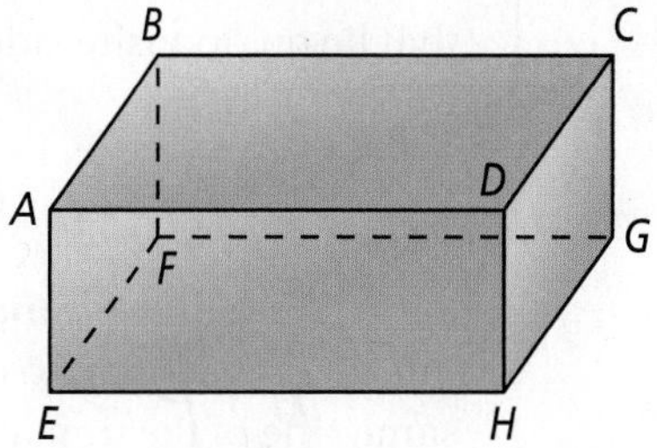

A Which segments are parallel to $\overline{AB}$?

$\overline{EF}$, $\overline{DC}$, and $\overline{HG}$

B Which segments are skew to $\overline{CD}$?

$\overline{BF}$, $\overline{AE}$, $\overline{EH}$, and $\overline{FG}$

C What are two pairs of parallel planes?

plane *ABCD* ∥ plane *EFGH*

plane *DCG* ∥ plane *ABF*

D What are two segments parallel to plane *BCGF*?

$\overline{AD}$ and $\overline{DH}$

Got It? **1.** Use the figure in Problem 1.

a. Which segments are parallel to $\overline{AD}$?

b. **Reasoning** Explain why $\overline{FE}$ and $\overline{CD}$ are *not* skew.

c. What is another pair of parallel planes?

d. What are two segments parallel to plane *DCGH*?

Essential Understanding When a line intersects two or more lines, the angles formed at the intersection points create special angle pairs.

A **transversal** is a line that intersects two or more coplanar lines at distinct points. The diagram below shows the eight angles formed by a transversal *t* and two lines ℓ and *m*.

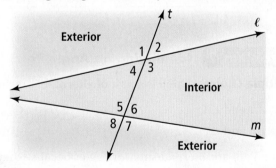

Notice that angles 3, 4, 5, and 6 lie between ℓ and *m*. They are *interior* angles. Angles 1, 2, 7, and 8 lie outside of ℓ and *m*. They are *exterior* angles.

Pairs of the eight angles have special names as suggested by their positions.

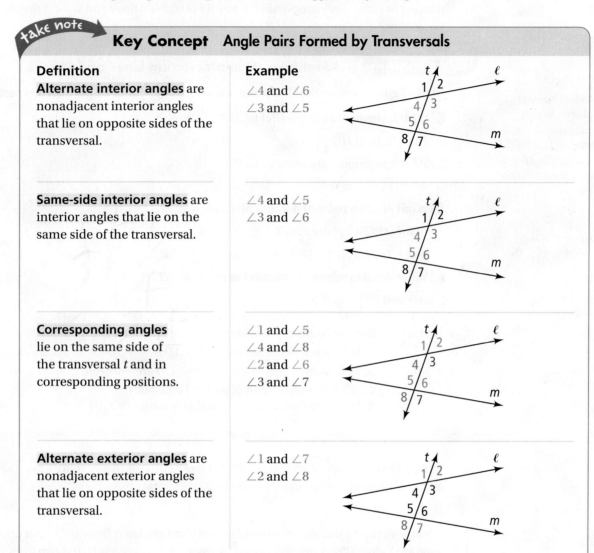

take note

Key Concept Angle Pairs Formed by Transversals

Definition

Alternate interior angles are nonadjacent interior angles that lie on opposite sides of the transversal.

Example

∠4 and ∠6
∠3 and ∠5

Same-side interior angles are interior angles that lie on the same side of the transversal.

∠4 and ∠5
∠3 and ∠6

Corresponding angles lie on the same side of the transversal *t* and in corresponding positions.

∠1 and ∠5
∠4 and ∠8
∠2 and ∠6
∠3 and ∠7

Alternate exterior angles are nonadjacent exterior angles that lie on opposite sides of the transversal.

∠1 and ∠7
∠2 and ∠8

Problem 2 **Identifying an Angle Pair**

Multiple Choice Which is a pair of alternate interior angles?

Ⓐ ∠1 and ∠3

Ⓒ ∠2 and ∠6

Ⓑ ∠6 and ∠7

Ⓓ ∠4 and ∠8

Think

Which choices can you eliminate?
You need a pair of interior angles. ∠1, ∠4, and ∠8 are exterior angles. You can eliminate choices A and D.

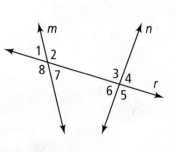

∠2 and ∠6 are alternate interior angles because they lie on opposite sides of the transversal *r* and in between *m* and *n*. The correct answer is C.

Got It? **2.** Use the figure in Problem 2. What are three pairs of corresponding angles?

Architecture The photo below shows the Royal Ontario Museum in Toronto, Canada. Are angles 2 and 4 *alternate interior angles, same-side interior angles, corresponding angles,* or *alternate exterior angles*?

Think

How do the positions of ∠2 and ∠4 compare?
∠2 and ∠4 are both interior angles and they lie on opposite sides of a line.

Angles 2 and 4 are alternate interior angles.

Got It? **3.** In Problem 3, are angles 1 and 3 *alternate interior angles, same-side interior angles, corresponding angles,* or *alternate exterior angles*?

Lesson Check

Do you know HOW?

Name one pair each of the segments, planes, or angles. Lines and planes that appear to be parallel are parallel.

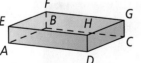

Exercises 1–3

1. parallel segments

2. skew segments

3. parallel planes

4. alternate interior

5. same-side interior

6. corresponding

7. alternate exterior

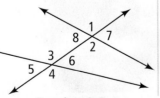

Exercises 4–7

Do you UNDERSTAND? MATHEMATICAL PRACTICES

8. Vocabulary Why is the word *coplanar* included in the definition for parallel lines?

9. Vocabulary How does the phrase *alternate interior angles* describe the positions of the two angles?

10. Error Analysis In the figure at the right, lines and planes that appear to be parallel are parallel. Carly says $\overline{AB} \parallel \overline{HG}$. Juan says $\overline{AB}$ and $\overline{HG}$ are skew. Who is correct? Explain.

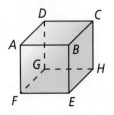

Practice and Problem-Solving Exercises

 MATHEMATICAL PRACTICES

See Problem 1.

A Practice Use the diagram to name each of the following. Assume that lines and planes that appear to be parallel are parallel.

11. a pair of parallel planes

12. all lines that are parallel to $\overleftrightarrow{AB}$

13. all lines that are parallel to $\overleftrightarrow{DH}$

14. two lines that are skew to $\overleftrightarrow{EJ}$

15. all lines that are parallel to plane *JFAE*

16. a plane parallel to $\overleftrightarrow{LH}$

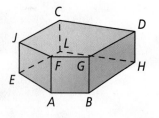

Identify all pairs of each type of angles in the diagram. Name the two lines and the transversal that form each pair.

See Problem 2.

17. corresponding angles

18. alternate interior angles

19. same-side interior angles

20. alternate exterior angles

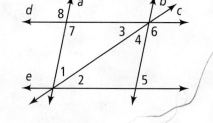

Are the angles labeled in the same color *alternate interior angles, same-side interior angles, corresponding angles,* or *alternate exterior angles*?

See Problem 3.

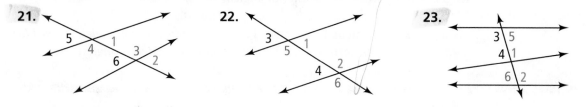

21.

22.

23.

24. **Aviation** The photo shows an overhead view of airport runways. Are ∠1 and ∠2 *alternate interior angles, same-side interior angles, corresponding angles,* or *alternate exterior angles*?

How many pairs of each type of angles do two lines and a transversal form?

25. alternate interior angles

26. corresponding angles

27. alternate exterior angles

28. vertical angles

29. Recreation You and a friend are driving go-karts on two different tracks. As you drive on a straight section heading east, your friend passes above you on a straight section heading south. Are these sections of the two tracks *parallel, skew,* or *neither*? Explain.

In Exercises 30–35, describe the statement as *true* or *false*. If false, explain. Assume that lines and planes that appear to be parallel are parallel.

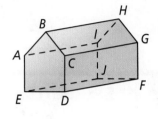

30. $\overleftrightarrow{CB} \parallel \overleftrightarrow{HG}$

31. $\overleftrightarrow{ED} \parallel \overleftrightarrow{HG}$

32. plane *AED* ∥ plane *FGH*

33. plane *ABH* ∥ plane *CDF*

34. $\overleftrightarrow{AB}$ and $\overleftrightarrow{HG}$ are skew lines.

35. $\overleftrightarrow{AE}$ and $\overleftrightarrow{BC}$ are skew lines.

36. Think About a Plan A rectangular rug covers the floor in a living room. One of the walls in the same living room is painted blue. Are the rug and the blue wall parallel? Explain.
  • Can you visualize the rug and the wall as geometric figures?
  • What must be true for these geometric figures to be parallel?

In Exercises 37–42, determine whether each statement is *always, sometimes,* or *never* true.

37. Two parallel lines are coplanar.

38. Two skew lines are coplanar.

39. Two planes that do not intersect are parallel.

40. Two lines that lie in parallel planes are parallel.

41. Two lines in intersecting planes are skew.

42. A line and a plane that do not intersect are skew.

43. a. Writing Describe the three ways in which two lines may be related.
   b. Give examples from the real world to illustrate each of the relationships you described in part (a).

44. Open-Ended The letter Z illustrates alternate interior angles. Find at least two other letters that illustrate pairs of angles presented in this lesson. Draw the letters. Then mark and describe the angles.

45. a. Reasoning Suppose two parallel planes *A* and *B* are each intersected by a third plane *C*. Make a conjecture about the intersection of planes *A* and *C* and the intersection of planes *B* and *C*.
   b. Find examples in your classroom to illustrate your conjecture in part (a).

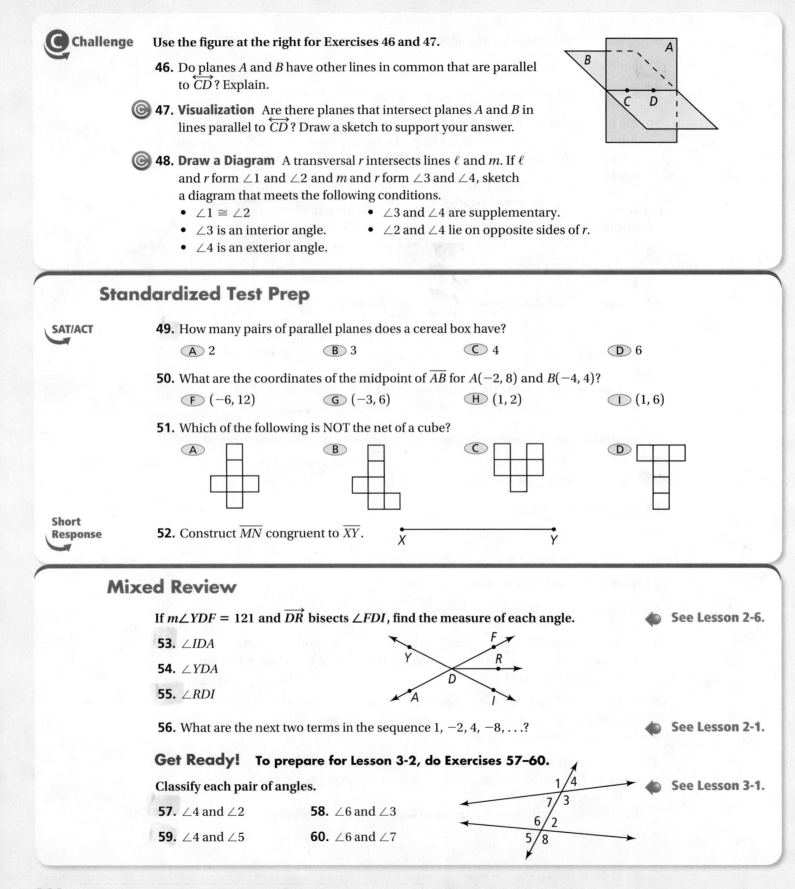

C **Challenge**   Use the figure at the right for Exercises 46 and 47.

    46. Do planes *A* and *B* have other lines in common that are parallel to $\overleftrightarrow{CD}$? Explain.

    Ⓒ **47. Visualization** Are there planes that intersect planes *A* and *B* in lines parallel to $\overleftrightarrow{CD}$? Draw a sketch to support your answer.

    Ⓒ **48. Draw a Diagram** A transversal *r* intersects lines ℓ and *m*. If ℓ and *r* form $\angle 1$ and $\angle 2$ and *m* and *r* form $\angle 3$ and $\angle 4$, sketch a diagram that meets the following conditions.
- $\angle 1 \cong \angle 2$
- $\angle 3$ is an interior angle.
- $\angle 4$ is an exterior angle.
- $\angle 3$ and $\angle 4$ are supplementary.
- $\angle 2$ and $\angle 4$ lie on opposite sides of *r*.

Standardized Test Prep

SAT/ACT

    49. How many pairs of parallel planes does a cereal box have?

      Ⓐ 2   Ⓑ 3   Ⓒ 4   Ⓓ 6

    50. What are the coordinates of the midpoint of $\overline{AB}$ for $A(-2, 8)$ and $B(-4, 4)$?

      Ⓕ $(-6, 12)$ Ⓖ $(-3, 6)$ Ⓗ $(1, 2)$ Ⓘ $(1, 6)$

    51. Which of the following is NOT the net of a cube?

      Ⓐ   Ⓑ   Ⓒ   Ⓓ

Short Response

    52. Construct $\overline{MN}$ congruent to $\overline{XY}$.

Mixed Review

    If $m\angle YDF = 121$ and $\overrightarrow{DR}$ bisects $\angle FDI$, find the measure of each angle.   ◀ **See Lesson 2-6.**

    53. $\angle IDA$

    54. $\angle YDA$

    55. $\angle RDI$

    56. What are the next two terms in the sequence 1, −2, 4, −8, . . .?   ◀ **See Lesson 2-1.**

    Get Ready!   **To prepare for Lesson 3-2, do Exercises 57–60.**

    Classify each pair of angles.   ◀ **See Lesson 3-1.**

    57. $\angle 4$ and $\angle 2$ **58.** $\angle 6$ and $\angle 3$

    59. $\angle 4$ and $\angle 5$ **60.** $\angle 6$ and $\angle 7$

Parallel Lines and Related Angles

© **Content Standards**
Prepares for G.CO.9 Prove theorems about lines and angles. Theorems include: . . . when a transversal crosses parallel lines, alternate interior angles are congruent and corresponding angles are congruent . . .
Also Prepares for G.CO.12

Activity

Use geometry software to construct two parallel lines. Check that the lines remain parallel as you manipulate them. Construct a point on each line. Then construct the transversal through these two points.

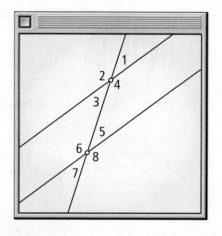

1. Measure each of the eight angles formed by the parallel lines and the transversal. Record the measurements.

2. Manipulate the lines. Record the new measurements.

3. When a transversal intersects parallel lines, what are the relationships among the angle pairs formed? Make as many conjectures as possible.

Exercises

4. Construct three or more parallel lines. Then construct a line that intersects all the parallel lines.
   a. What relationships can you find among the angles formed?
   b. How many different angle measures are there?

5. Construct two parallel lines and a transversal perpendicular to one of the parallel lines. What angle does the transversal form with the second line?

6. Construct two lines and a transversal, making sure that the two lines are *not* parallel. Locate a pair of alternate interior angles. Manipulate the lines so that these angles have the same measure.
   a. Make a conjecture about the relationship between the two lines.
   b. How is this conjecture different from the conjecture(s) you made in the Activity?

7. Again, construct two lines and a transversal, making sure that the two lines are *not* parallel. Locate a pair of same-side interior angles. Manipulate the lines so that these angles are supplementary.
   a. Make a conjecture about the relationship between the two lines.
   b. How is this conjecture different from the conjecture(s) you made in the Activity?

8. Construct perpendicular lines *a* and *b*. At a point that is not the intersection of *a* and *b*, construct line *c* perpendicular to line *a*. Make a conjecture about lines *b* and *c*.

Properties of Parallel Lines

Content Standard

G.CO.9 Prove theorems about lines and angles. Theorems include: . . . when a transversal crosses parallel lines, alternate interior angles are congruent . . .

Objectives To prove theorems about parallel lines
To use properties of parallel lines to find angle measures

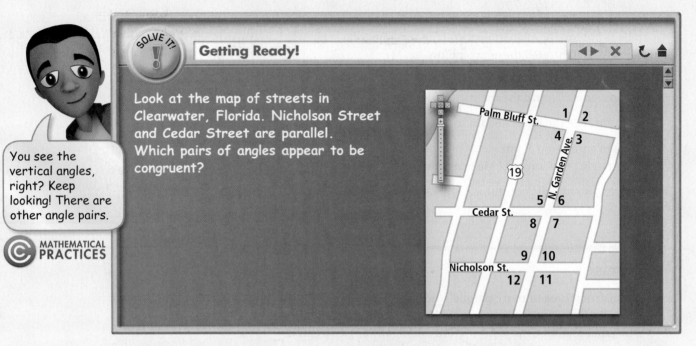

SOLVE IT!

Getting Ready!

Look at the map of streets in Clearwater, Florida. Nicholson Street and Cedar Street are parallel. Which pairs of angles appear to be congruent?

You see the vertical angles, right? Keep looking! There are other angle pairs.

MATHEMATICAL PRACTICES

In the Solve It, you identified several pairs of angles that appear congruent. You already know the relationship between vertical angles. In this lesson, you will explore the relationships between the angles you learned about in Lesson 3-1 when they are formed by *parallel* lines and a transversal.

Essential Understanding The special angle pairs formed by parallel lines and a transversal are congruent, supplementary, or both.

take note

Postulate 3-1 Same-Side Interior Angles Postulate

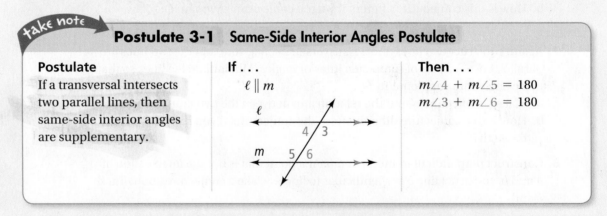

Postulate

If a transversal intersects two parallel lines, then same-side interior angles are supplementary.

If . . .

$\ell \parallel m$

Then . . .

$m\angle 4 + m\angle 5 = 180$
$m\angle 3 + m\angle 6 = 180$

Think

How do you find angles supplementary to a given angle?
Look for angles whose measure is the difference of 180 and the given angle measure.

Problem 1 **Identifying Supplementary Angles**

The measure of ∠3 is 55. Which angles are supplementary to ∠3? How do you know?

By definition, a straight angle measures 180.

If $m\angle a + m\angle b = 180$, then ∠a and ∠b are supplementary by definition of supplementary angles.

$180 - 55 = 125$, so any angle x, where $m\angle x = 125$, is supplementary to ∠3.

$m\angle 4 = 125$ by the definition of a straight angle.

$m\angle 8 = 125$ by the Same-Side Interior Angles Postulate.

$m\angle 6 = m\angle 8$ by the Vertical Angles Theorem, so $m\angle 6 = 125$.

$m\angle 2 = m\angle 4$ by the Vertical Angles Theorem, so $m\angle 2 = 125$.

Got It? **1. Reasoning** If you know the measure of one of the angles, can you always find the measures of all 8 angles when two parallel lines are cut by a transversal? Explain.

You can use the Same-Side Interior Angles Postulate to prove other angle relationships.

take note

Theorem 3-1   Alternate Interior Angles Theorem

Theorem	**If . . .**	**Then . . .**
If a transversal intersects two parallel lines, then alternate interior angles are congruent.	$\ell \parallel m$	$\angle 4 \cong \angle 6$ $\angle 3 \cong \angle 5$

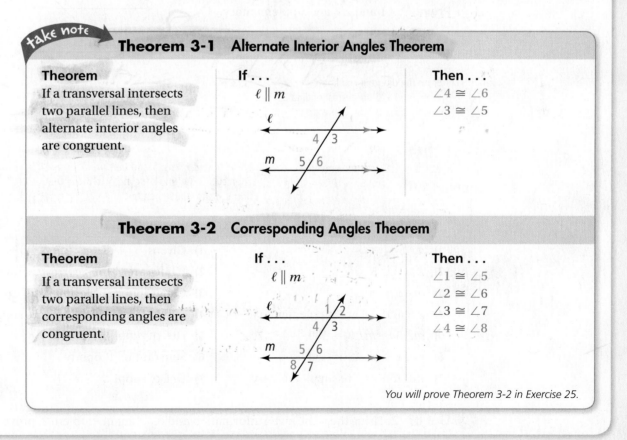

Theorem 3-2   Corresponding Angles Theorem

Theorem	**If . . .**	**Then . . .**
If a transversal intersects two parallel lines, then corresponding angles are congruent.	$\ell \parallel m$	$\angle 1 \cong \angle 5$ $\angle 2 \cong \angle 6$ $\angle 3 \cong \angle 7$ $\angle 4 \cong \angle 8$

You will prove Theorem 3-2 in Exercise 25.

Proof of Theorem 3-1: Alternate Interior Angles Theorem

Given: $\ell \parallel m$

Prove: $\angle 4 \cong \angle 6$

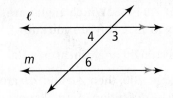

Statement	Reasons
1) $\ell \parallel m$	1) Given
2) $m\angle 3 + m\angle 4 = 180$	2) Supplementary Angles
3) $m\angle 3 + m\angle 6 = 180$	3) Same-Side Interior Angles Postulate
4) $m\angle 3 + m\angle 4 = m\angle 3 + m\angle 6$	4) Transitive Property of Equality
5) $m\angle 4 = m\angle 6$	5) Subtraction Property of Equality
6) $\angle 4 \cong \angle 6$	6) Definition of Congruence

Proof ©️ **Problem 2** **Proving an Angle Relationship**

Given: $a \parallel b$

Prove: $\angle 1$ and $\angle 8$ are supplementary.

Know

- $a \parallel b$
From the diagram you know
- $\angle 1$ and $\angle 5$ are corresponding
- $\angle 5$ and $\angle 8$ form a linear pair

Need

$\angle 1$ and $\angle 8$ are supplementary, or $m\angle 1 + m\angle 8 = 180$.

Plan

Show that $\angle 1 \cong \angle 5$ and that $m\angle 5 + m\angle 8 = 180$. Then substitute $m\angle 1$ for $m\angle 5$ to prove that $\angle 1$ and $\angle 8$ are supplementary.

Statements	Reasons
1) $a \parallel b$	1) Given
2) $\angle 1 \cong \angle 5$	2) If lines are $\parallel$, then corresp. ⩘ are ≅.
3) $m\angle 1 = m\angle 5$	3) Congruent ⩘ have equal measures.
4) $\angle 5$ and $\angle 8$ are supplementary.	4) ⩘ that form a linear pair are suppl.
5) $m\angle 5 + m\angle 8 = 180$	5) Def. of suppl. ⩘
6) $m\angle 1 + m\angle 8 = 180$	6) Substitution Property
7) $\angle 1$ and $\angle 8$ are supplementary.	7) Def. of suppl. ⩘

Got It? **2.** Using the same given information and diagram in Problem 2, prove that $\angle 1 \cong \angle 7$.

In the diagram for Problem 2, $\angle 1$ and $\angle 7$ are alternate exterior angles. In Got It 2, you proved the following theorem.

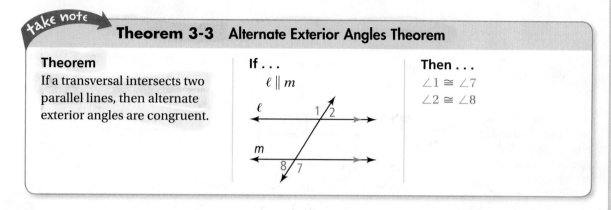

take note

Theorem 3-3 Alternate Exterior Angles Theorem

Theorem	**If . . .**	**Then . . .**
If a transversal intersects two parallel lines, then alternate exterior angles are congruent.	$\ell \parallel m$	$\angle 1 \cong \angle 7$ $\angle 2 \cong \angle 8$

If you know the measure of one of the angles formed by two parallel lines and a transversal, you can use theorems and postulates to find the measures of the other angles.

© **Problem 3** **Finding Measures of Angles**

What are the measures of $\angle 3$ and $\angle 4$? Which theorem or postulate justifies each answer?

Think

How do $\angle 3$ and $\angle 4$ relate to the given 105° angle?
$\angle 3$ and the given angle are alternate interior angles. $\angle 4$ and the given angle are same-side interior angles.

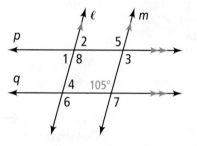

Since $p \parallel q$, $m\angle 3 = 105$ by the Alternate Interior Angles Theorem.

Since $\ell \parallel m$, $m\angle 4 + 105 = 180$ by the Same-Side Interior Angles Postulate. So, $m\angle 4 = 180 - 105 = 75$.

✓ **Got It?** **3.** Use the diagram in Problem 3. What is the measure of each angle? Justify each answer.

a. $\angle 1$	**b.** $\angle 2$
c. $\angle 5$	**d.** $\angle 6$
e. $\angle 7$	**f.** $\angle 8$

You can combine theorems and postulates with your knowledge of algebra to find angle measures.

 **Problem 4** **Finding an Angle Measure**

Algebra What is the value of y?

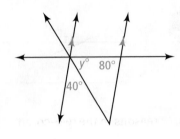

By the Angle Addition Postulate, $y + 40$ is the measure of an interior angle.

$$(y + 40) + 80 = 180 \qquad \text{Same-side interior } \angle\!\!\!\diagup \text{ of } \| \text{ lines are suppl.}$$
$$y + 120 = 180 \qquad \text{Simplify.}$$
$$y = 60 \qquad \text{Subtract 120 from each side.}$$

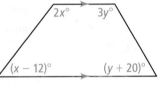

Got It? **4. a.** In the figure at the right, what are the values of x and y?

     b. What are the measures of the four angles in the figure?

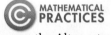

Lesson Check

Do you know HOW?

Use the diagram for Exercises 1–4.

1. Identify four pairs of congruent angles. (Exclude vertical angle pairs.)

2. Identify two pairs of supplementary angles. (Exclude linear pairs.)

3. If $m\angle 1 = 70$, what is $m\angle 8$?

4. If $m\angle 4 = 70$ and $m\angle 7 = 2x$, what is the value of x?

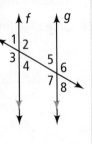

Do you UNDERSTAND? **MATHEMATICAL PRACTICES**

5. **Compare and Contrast** How are the Alternate Interior Angles Theorem and the Alternate Exterior Angles Theorem alike? How are they different?

6. In Problem 2, you proved that $\angle 1$ and $\angle 8$, in the diagram below, are supplementary. What is a good name for this pair of angles? Explain.

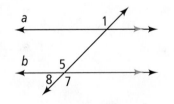

Practice and Problem-Solving Exercises

 MATHEMATICAL PRACTICES

(A) Practice   Identify all the numbered angles that are congruent to the given angle. Justify your answers.

◀ See Problem 1.

7.

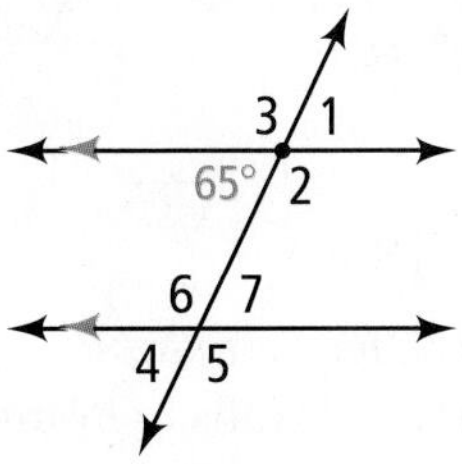

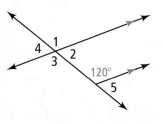

3 / 1
65° / 2
6 / 7
4 / 5

8.

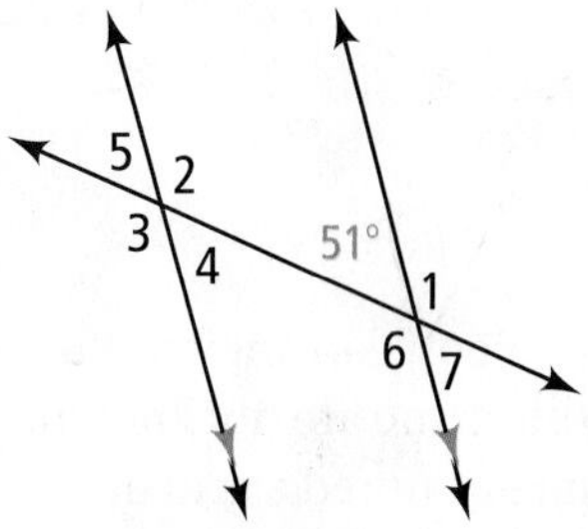

5 / 2
3 / 4
51°
6 / 7 / 1

9.

4 / 1
3 / 2
120°
5

(C) 10. Developing Proof Supply the missing reasons in the two-column proof.

◀ See Problem 2.

Given: $a \parallel b$, $c \parallel d$
Prove: $\angle 1 \cong \angle 3$

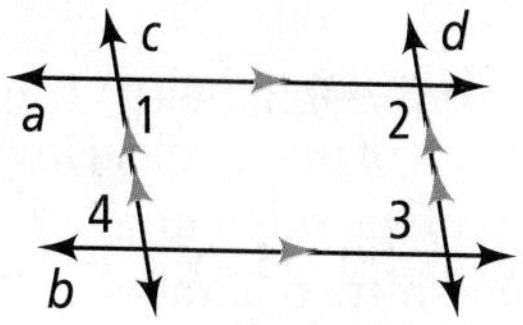

c   d
a   1   2
4   3
b

Statements	Reasons
1) $a \parallel b$	**1)** Given
2) $\angle 3$ and $\angle 2$ are supplementary.	**2) a.** _?_
3) $c \parallel d$	**3)** Given
4) $\angle 1$ and $\angle 2$ are supplementary.	**4) b.** _?_
5) $\angle 1 \cong \angle 3$	**5) c.** _?_

11. Write a two-column proof for Exercise 10 that does not use $\angle 2$.

Proof

Find $m\angle 1$ and $m\angle 2$. Justify each answer.

◀ See Problem 3.

12.

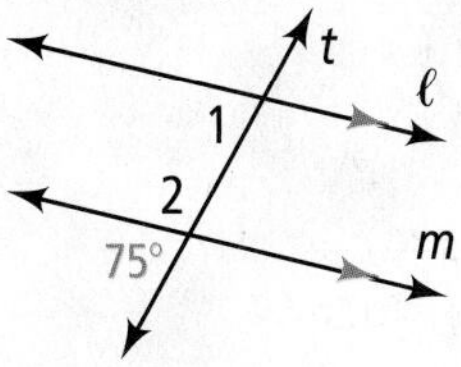

t
ℓ
1
2
75°
m

13.

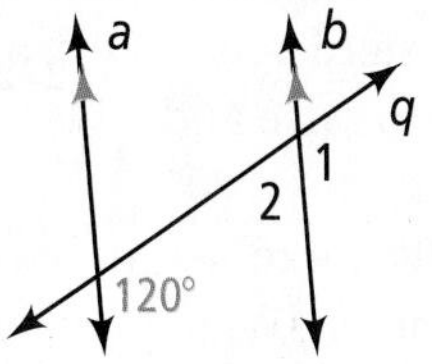

a   b
q
2 / 1
120°

14.

A   B
80°   70°
1   2
D   C

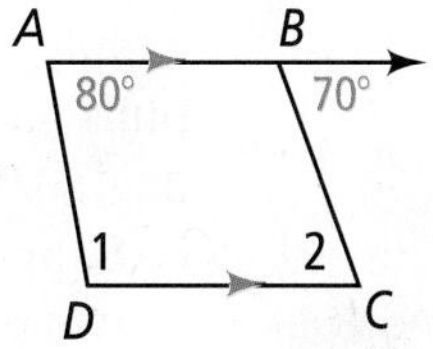

Algebra Find the value of x. Then find the measure of each labeled angle.

◀ See Problem 4.

15.

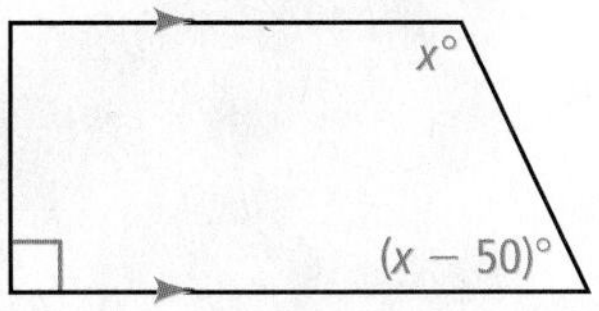

$x°$
$(x - 50)°$

16.

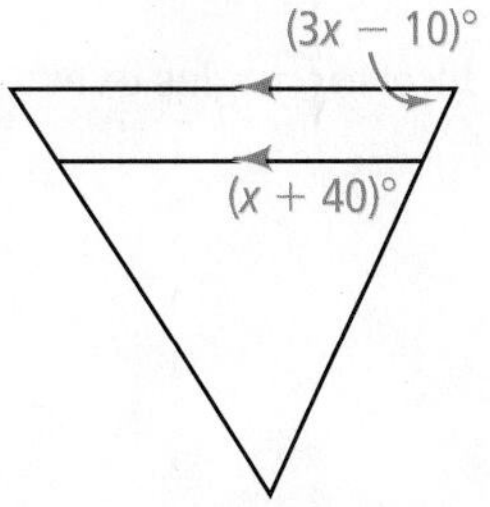

$(3x - 10)°$
$(x + 40)°$

17.

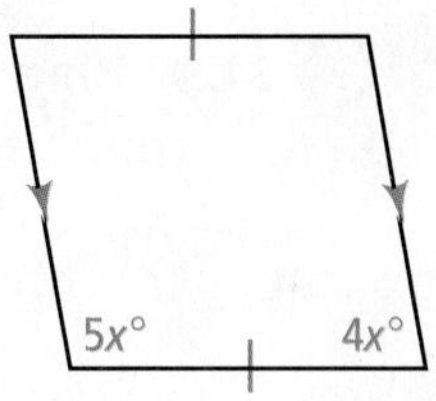

$5x°$ $4x°$

Algebra Find the values of the variables.

18.

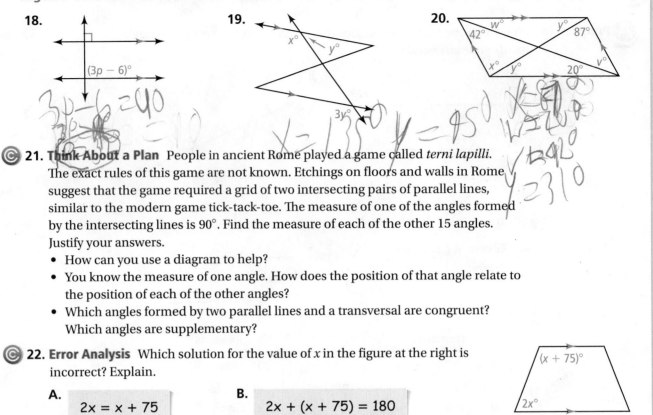

$(3p - 6)°$

19.

$x°$

$y°$

$3y°$

20.

$w°$ $y°$ $87°$

$42°$

$x°$ $y°$ $20°$ $v°$

© **21. Think About a Plan** People in ancient Rome played a game called *terni lapilli*. The exact rules of this game are not known. Etchings on floors and walls in Rome suggest that the game required a grid of two intersecting pairs of parallel lines, similar to the modern game tick-tack-toe. The measure of one of the angles formed by the intersecting lines is 90°. Find the measure of each of the other 15 angles. Justify your answers.

- How can you use a diagram to help?
- You know the measure of one angle. How does the position of that angle relate to the position of each of the other angles?
- Which angles formed by two parallel lines and a transversal are congruent? Which angles are supplementary?

© **22. Error Analysis** Which solution for the value of *x* in the figure at the right is incorrect? Explain.

A.

$2x = x + 75$
$x = 75$

B.

$2x + (x + 75) = 180$
$3x + 75 = 180$
$3x = 105$
$x = 35$

$(x + 75)°$

$2x°$

23. Outdoor Recreation Campers often use a "bear bag" at night to avoid attracting animals to their food supply. In the bear bag system at the right, a camper pulls one end of the rope to raise and lower the food bag.

   a. Suppose a camper pulls the rope taut between the two parallel trees, as shown. What is $m\angle 1$?

   b. Are $\angle 1$ and the given angle *alternate interior angles, same-side interior angles,* or *corresponding angles*?

© **24. Writing** Are same-side interior angles ever congruent? Explain.

63°

1

25. Write a two-column proof to prove the **Proof** Corresponding Angles Theorem (Theorem 3-2).

Given: $\ell \parallel m$

Prove: $\angle 2$ and $\angle 6$ are congruent.

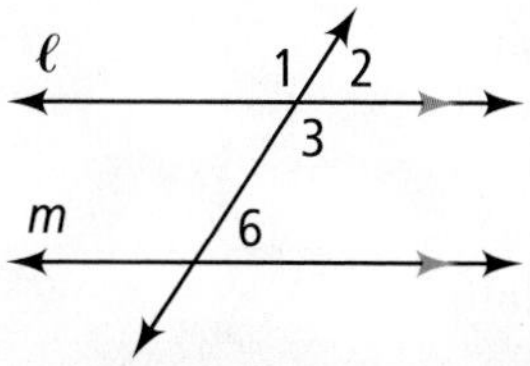

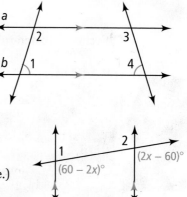

26. Write a two-column proof.
Proof **Given:** $a \parallel b$, $\angle 1 \cong \angle 4$

Prove: $\angle 2 \cong \angle 3$

C Challenge Use the diagram at the right for Exercises 27 and 28.

27. Algebra Suppose the measures of $\angle 1$ and $\angle 2$ are in a 4 : 11 ratio. Find their measures. (Diagram is not to scale.)

28. Error Analysis The diagram contains contradictory information. What is it? Why is it contradictory?

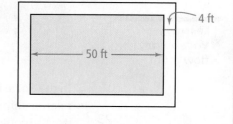

Standardized Test Prep

GRIDDED RESPONSE

SAT/ACT

29. $\angle 1$ and $\angle 2$ are same-side interior angles formed by two parallel lines and a transversal. If $m\angle 1 = 115$, what is $m\angle 2$?

30. The rectangular swimming pool shown at the right has an area of 1500 ft^2. A rectangular walkway surrounds the pool. How many feet of fencing do you need to surround the walkway?

31. The measure of an angle is two times the measure of its complement. What is the measure of the angle?

32. $\angle 1$ and $\angle 2$ are vertical angles. If $m\angle 1 = 4x$ and $m\angle 2 = 56$, what is the value of x?

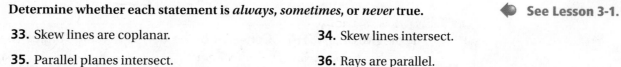

Mixed Review

Determine whether each statement is *always*, *sometimes*, or *never* true.

➤ See Lesson 3-1.

33. Skew lines are coplanar.

34. Skew lines intersect.

35. Parallel planes intersect.

36. Rays are parallel.

Get Ready! **To prepare for Lesson 3-3, do Exercises 37–39.**

Write the converse and determine its truth value.

➤ See Lesson 2-2.

37. If a triangle is a right triangle, then it has a 90° angle.

38. If two angles are vertical angles, then they are congruent.

39. If two angles are same-side interior angles, then they are supplementary.

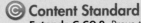

Content Standard
Extends G.CO.9 Prove theorems about lines and angles. Theorems include: . . . when a transversal crosses parallel lines, alternate interior angles are congruent and corresponding angles are congruent . . .

Objective To determine whether two lines are parallel

How can you use theorems you already know to solve this maze problem?

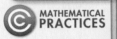
MATHEMATICAL
PRACTICES

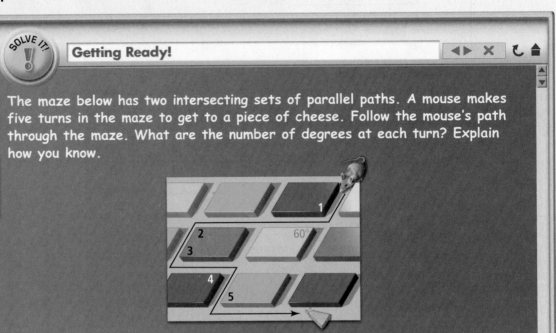

Getting Ready!

The maze below has two intersecting sets of parallel paths. A mouse makes five turns in the maze to get to a piece of cheese. Follow the mouse's path through the maze. What are the number of degrees at each turn? Explain how you know.

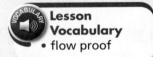
Lesson Vocabulary
• flow proof

In the Solve It, you used parallel lines to find congruent and supplementary relationships of special angle pairs. In this lesson you will do the converse. You will use the congruent and supplementary relationships of the special angle pairs to prove lines parallel.

Essential Understanding You can use certain angle pairs to decide whether two lines are parallel.

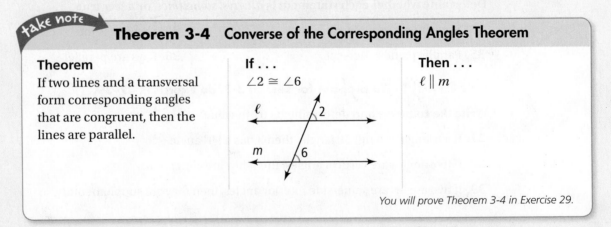

take note

Theorem 3-4 Converse of the Corresponding Angles Theorem

Theorem	**If . . .**	**Then . . .**
If two lines and a transversal form corresponding angles that are congruent, then the lines are parallel.	$\angle 2 \cong \angle 6$	$\ell \parallel m$

You will prove Theorem 3-4 in Exercise 29.

Problem 1 **Identifying Parallel Lines**

Which lines are parallel if ∠1 ≅ ∠2? Justify your answer.

∠1 and ∠2 are corresponding angles. If ∠1 ≅ ∠2, then $a \parallel b$ by the Converse of the Corresponding Angles Theorem.

Got It? **1.** Which lines are parallel if ∠6 ≅ ∠7? Justify your answer.

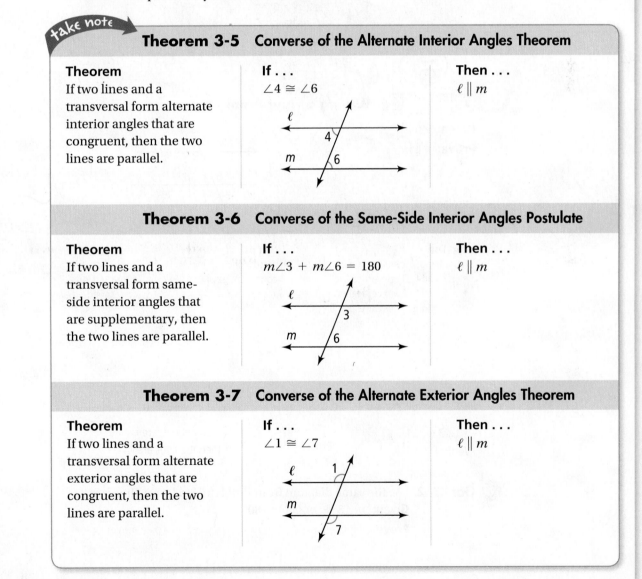

In Lesson 3-2 you proved theorems based on the Corresponding Angles Theorem. You can use the Converse of the Corresponding Angles Theorem to prove converses of the theorems and postulate you learned in Lesson 3-2.

take note

Theorem 3-5   Converse of the Alternate Interior Angles Theorem

Theorem	**If . . .**	**Then . . .**
If two lines and a transversal form alternate interior angles that are congruent, then the two lines are parallel.	∠4 ≅ ∠6	$\ell \parallel m$

Theorem 3-6   Converse of the Same-Side Interior Angles Postulate

Theorem	**If . . .**	**Then . . .**
If two lines and a transversal form same-side interior angles that are supplementary, then the two lines are parallel.	$m\angle 3 + m\angle 6 = 180$	$\ell \parallel m$

Theorem 3-7   Converse of the Alternate Exterior Angles Theorem

Theorem	**If . . .**	**Then . . .**
If two lines and a transversal form alternate exterior angles that are congruent, then the two lines are parallel.	∠1 ≅ ∠7	$\ell \parallel m$

The proof of the Converse of the Alternate Interior Angles Theorem below looks different than any proof you have seen so far in this course. You know two forms of proof—paragraph and two-column. In a third form, called **flow proof**, arrows show the logical connections between the statements. Reasons are written below the statements.

Proof **Proof of Theorem 3-5: Converse of the Alternate Interior Angles Theorem**

Given: $\angle 4 \cong \angle 6$
Prove: $\ell \parallel m$

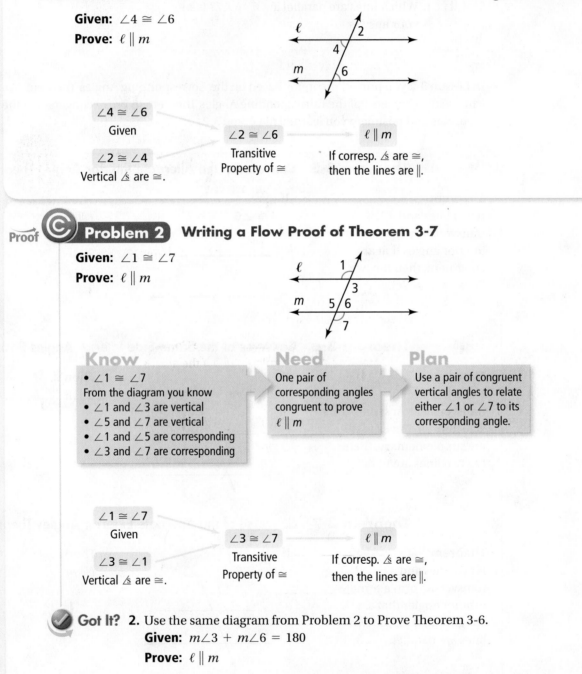

| $\angle 4 \cong \angle 6$ |
| Given |

| $\angle 2 \cong \angle 4$ |
| Vertical $\angle\!s$ are $\cong$. |

| $\angle 2 \cong \angle 6$ |
| Transitive Property of $\cong$ |

| $\ell \parallel m$ |
| If corresp. $\angle\!s$ are $\cong$, then the lines are $\parallel$. |

Proof © **Problem 2** **Writing a Flow Proof of Theorem 3-7**

Given: $\angle 1 \cong \angle 7$
Prove: $\ell \parallel m$

Know
- $\angle 1 \cong \angle 7$
From the diagram you know
- $\angle 1$ and $\angle 3$ are vertical
- $\angle 5$ and $\angle 7$ are vertical
- $\angle 1$ and $\angle 5$ are corresponding
- $\angle 3$ and $\angle 7$ are corresponding

Need
One pair of corresponding angles congruent to prove $\ell \parallel m$

Plan
Use a pair of congruent vertical angles to relate either $\angle 1$ or $\angle 7$ to its corresponding angle.

| $\angle 1 \cong \angle 7$ |
| Given |

| $\angle 3 \cong \angle 1$ |
| Vertical $\angle\!s$ are $\cong$. |

| $\angle 3 \cong \angle 7$ |
| Transitive Property of $\cong$ |

| $\ell \parallel m$ |
| If corresp. $\angle\!s$ are $\cong$, then the lines are $\parallel$. |

Got It? **2.** Use the same diagram from Problem 2 to Prove Theorem 3-6.
Given: $m\angle 3 + m\angle 6 = 180$
Prove: $\ell \parallel m$

The four theorems you have just learned provide you with four ways to determine if two lines are parallel.

© **Problem 3** **Determining Whether Lines are Parallel**

Think

How do ∠1 and ∠2 relate to each other in the diagram?
∠1 and ∠2 are both exterior angles and they lie on opposite sides of *t*.

The fence gate at the right is made up of pieces of wood arranged in various directions. Suppose ∠1 ≅ ∠2. Are lines *r* and *s* parallel? Explain.

Yes, *r* ∥ *s*. ∠1 and ∠2 are alternate exterior angles. If two lines and a transversal form congruent alternate exterior angles, then the lines are parallel (Converse of the Alternate Exterior Angles Theorem).

Got It? **3.** In Problem 3, what is another way to explain why *r* ∥ *s*? Justify your answer.

You can use algebra along with the postulates and theorems from Lesson 3-2 and Lesson 3-3 to help you solve problems involving parallel lines.

© **Problem 4** **Using Algebra**

Think

Work backward.
Think about what must be true of the given angles for *a* and *b* to be parallel.

Algebra What is the value of *x* for which *a* ∥ *b*?

The two angles are same-side interior angles. By the Converse of the Same-Side Interior Angles Postulate, *a* ∥ *b* if the angles are supplementary.

$$(2x + 9) + 111 = 180 \quad \text{Def. of supplementary angles}$$
$$2x + 120 = 180 \quad \text{Simplify.}$$
$$2x = 60 \quad \text{Subtract 120 from each side.}$$
$$x = 30 \quad \text{Divide each side by 2.}$$

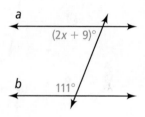

Got It? **4.** What is the value of *w* for which *c* ∥ *d*?

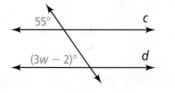

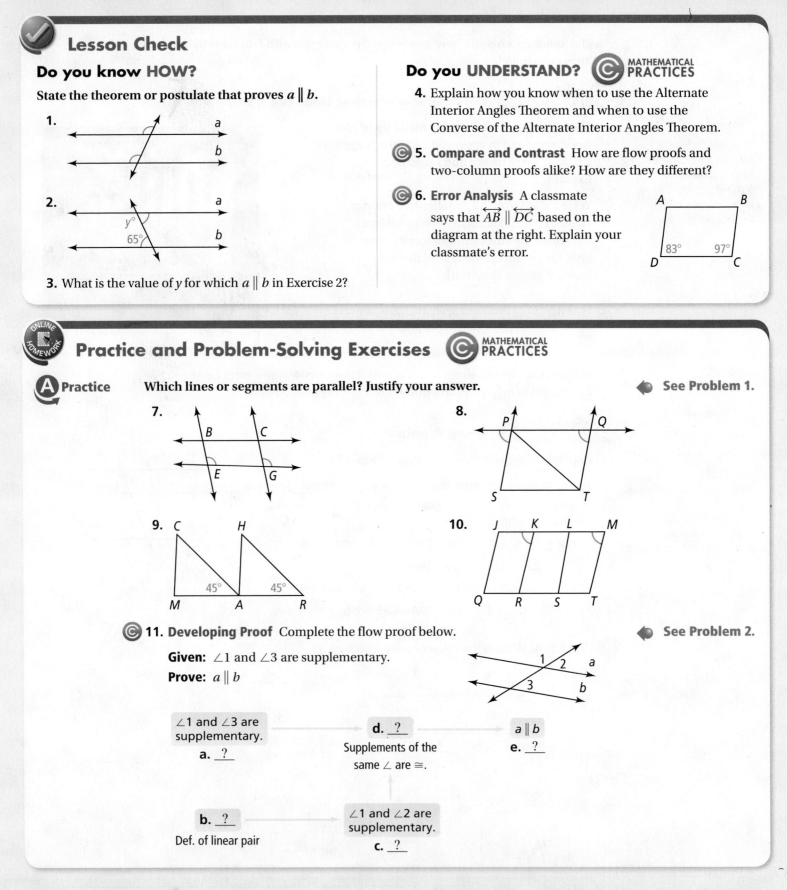

Lesson Check

Do you know HOW?

State the theorem or postulate that proves $a \parallel b$.

1.

    a

    b

2.

    a

    $y°$

    $65°$ *b*

3. What is the value of y for which $a \parallel b$ in Exercise 2?

Do you UNDERSTAND? MATHEMATICAL PRACTICES

4. Explain how you know when to use the Alternate Interior Angles Theorem and when to use the Converse of the Alternate Interior Angles Theorem.

5. Compare and Contrast How are flow proofs and two-column proofs alike? How are they different?

6. Error Analysis A classmate says that $\overleftrightarrow{AB} \parallel \overleftrightarrow{DC}$ based on the diagram at the right. Explain your classmate's error.

    A   *B*

    $83°$ $97°$

    D   *C*

Practice and Problem-Solving Exercises MATHEMATICAL PRACTICES

A Practice   Which lines or segments are parallel? Justify your answer.   **See Problem 1.**

7.

8.

9.

10.

11. Developing Proof Complete the flow proof below.   **See Problem 2.**

Given: $\angle 1$ and $\angle 3$ are supplementary.

Prove: $a \parallel b$

$\angle 1$ and $\angle 3$ are supplementary.

a. ?

d. ?

Supplements of the same $\angle$ are $\cong$.

$a \parallel b$

e. ?

b. ?

Def. of linear pair

$\angle 1$ and $\angle 2$ are supplementary.

c. ?

12. Parking Two workers paint lines for angled parking spaces. One worker paints a line so that $m\angle 1 = 65$. The other worker paints a line so that $m\angle 2 = 65$. Are their lines parallel? Explain.

See Problem 3.

Algebra Find the value of x for which $\ell \parallel m$.

See Problem 4.

13.

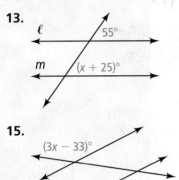

14.

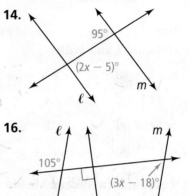

15.

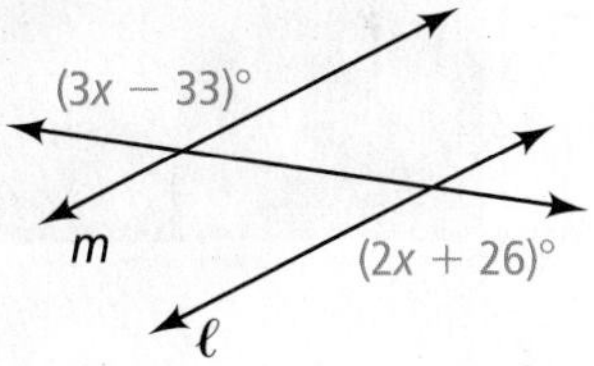

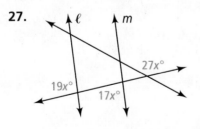

16.

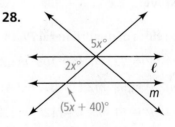

B Apply

C Developing Proof Use the given information to determine which lines, if any, are parallel. Justify each conclusion with a theorem or postulate.

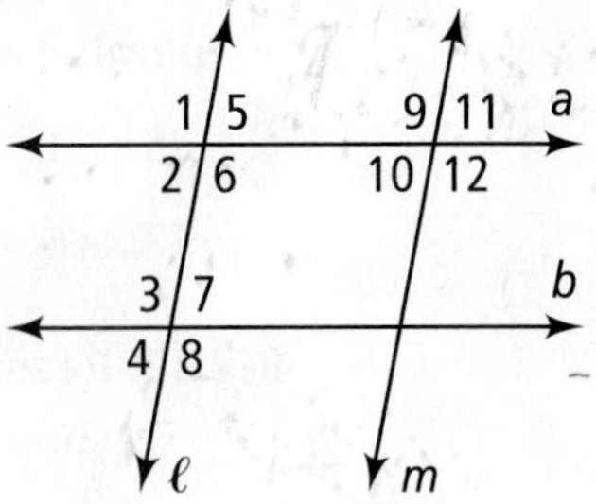

17. $\angle 2$ is supplementary to $\angle 3$.

18. $\angle 1 \cong \angle 3$.

19. $\angle 6$ is supplementary to $\angle 7$.

20. $\angle 9 \cong \angle 12$

21. $m\angle 7 = 65$, $m\angle 9 = 115$

22. $\angle 2 \cong \angle 10$

23. $\angle 1 \cong \angle 8$

24. $\angle 8 \cong \angle 6$

25. $\angle 11 \cong \angle 7$

26. $\angle 5 \cong \angle 10$

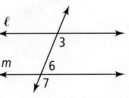

Algebra Find the value of x for which $\ell \parallel m$.

27.

28.

29. Prove the Converse of the Corresponding Angles Theorem (Theorem 3-4).

Proof

Given: $\angle 3 \cong \angle 7$

Prove: $\ell \parallel m$

30. Think About a Plan If the rowing crew at the right strokes in unison, the oars sweep out angles of equal measure. Explain why the oars on each side of the shell stay parallel.
- What type of information do you need to prove lines parallel?
- How do the positions of the angles of equal measure relate?

Algebra Determine the value of x for which $r \parallel s$. Then find $m\angle 1$ and $m\angle 2$.

31. $m\angle 1 = 80 - x$, $m\angle 2 = 90 - 2x$

32. $m\angle 1 = 60 - 2x$, $m\angle 2 = 70 - 4x$

33. $m\angle 1 = 40 - 4x$, $m\angle 2 = 50 - 8x$

34. $m\angle 1 = 20 - 8x$, $m\angle 2 = 30 - 16x$

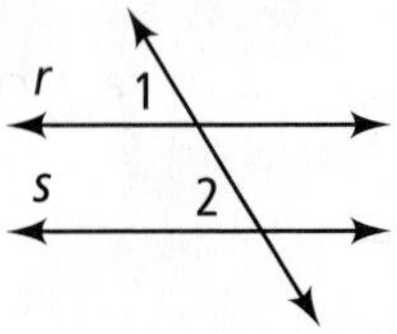

Use the diagram at the right below for Exercises 35–41.

Open-Ended Use the given information. State another fact about one of the given angles that will guarantee two lines are parallel. Tell which lines will be parallel and why.

35. $\angle 1 \cong \angle 3$

36. $m\angle 8 = 110$, $m\angle 9 = 70$

37. $\angle 5 \cong \angle 11$

38. $\angle 11$ and $\angle 12$ are supplementary.

39. Reasoning If $\angle 1 \cong \angle 7$, what theorem or postulate can you use to show that $\ell \parallel n$?

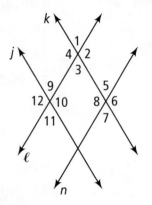

Write a flow proof.

40. Given: $\ell \parallel n$, $\angle 12 \cong \angle 8$
Proof **Prove:** $j \parallel k$

41. Given: $j \parallel k$, $m\angle 8 + m\angle 9 = 180$
Proof **Prove:** $\ell \parallel n$

Challenge

Which sides of quadrilateral *PLAN* must be parallel? Explain.

42. $m\angle P = 72$, $m\angle L = 108$, $m\angle A = 72$, $m\angle N = 108$

43. $m\angle P = 59$, $m\angle L = 37$, $m\angle A = 143$, $m\angle N = 121$

44. $m\angle P = 67$, $m\angle L = 120$, $m\angle A = 73$, $m\angle N = 100$

45. $m\angle P = 56$, $m\angle L = 124$, $m\angle A = 124$, $m\angle N = 56$

46. Write a two-column proof to prove the following: If a transversal intersects two
Proof parallel lines, then the bisectors of two corresponding angles are parallel. (*Hint:* Start by drawing and marking a diagram.)

Standardized Test Prep

SAT/ACT

Use the diagram for Exercises 47 and 48.

47. For what value of x is $c \parallel d$?

    Ⓐ 21   Ⓒ 43

    Ⓑ 23   Ⓓ 53

48. If $c \parallel d$, what is $m\angle 1$?

    Ⓕ 24   Ⓗ 136

    Ⓖ 44   Ⓘ 146

49. Which of the following is always a valid conclusion for the hypothesis?
If two angles are congruent, then __?__ .

    Ⓐ they are right angles   Ⓒ they have the same measure

    Ⓑ they share a vertex   Ⓓ they are acute angles

50. What is the value of x in the diagram at the right?

    Ⓕ $1.\overline{6}$ Ⓗ 17

    Ⓖ 10   Ⓘ 19

Short Response

51. Draw a pentagon. Is your pentagon convex or concave? Explain.

Mixed Review

Find $m\angle 1$ and $m\angle 2$. Justify each answer.   ◀ **See Lesson 3-2.**

52.

53.

Get Ready!   **To prepare for Lesson 3-4, do Exercises 54–57.**

Determine whether each statement is *always, sometimes,* or *never* true.   ◀ **See Lessons 1-6 and 3-1.**

54. Perpendicular lines meet at right angles.

55. Two lines in intersecting planes are perpendicular.

56. Two lines in the same plane are parallel.

57. Two lines in parallel planes are perpendicular.

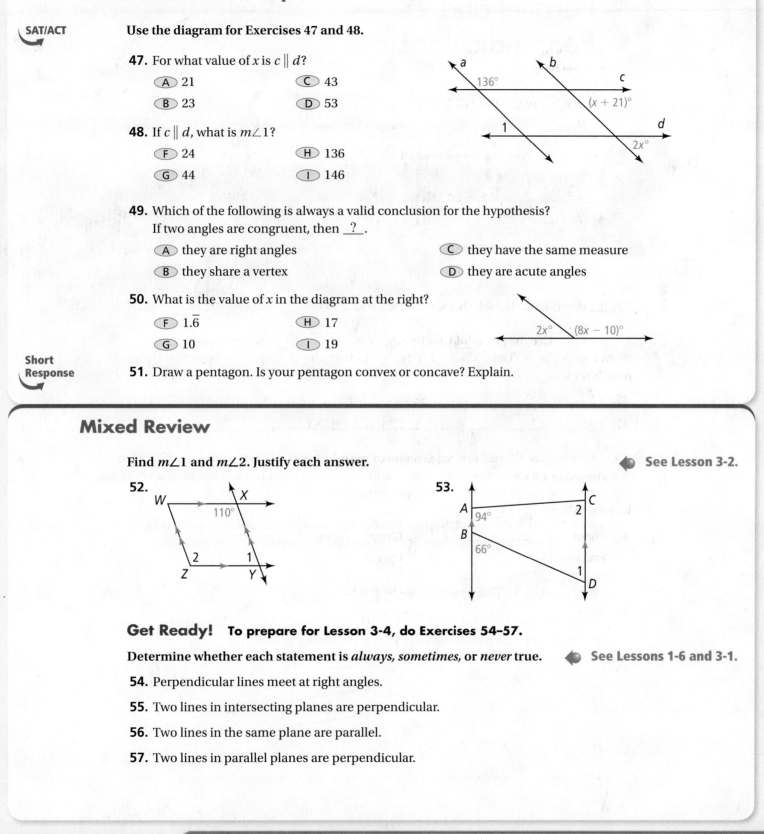

3-4 Parallel and Perpendicular Lines

© Content Standard
G.MG.3 Apply geometric methods to solve design problems.

Objective To relate parallel and perpendicular lines

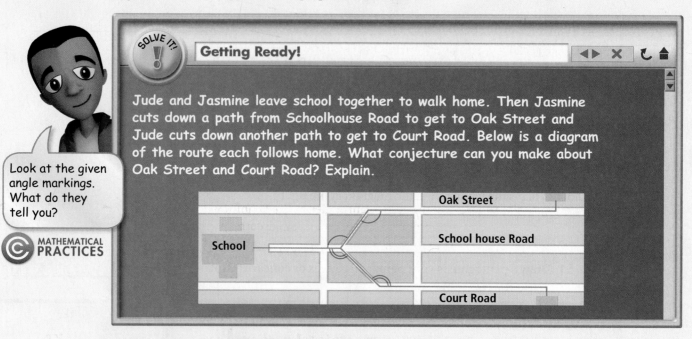

Look at the given angle markings. What do they tell you?

© MATHEMATICAL PRACTICES

In the Solve It, you likely made your conjecture about Oak Street and Court Road based on their relationships to Schoolhouse Road. In this lesson you will use similar reasoning to prove that lines are parallel or perpendicular.

Essential Understanding You can use the relationships of two lines to a third line to decide whether the two lines are parallel or perpendicular to each other.

take note

Theorem 3-8

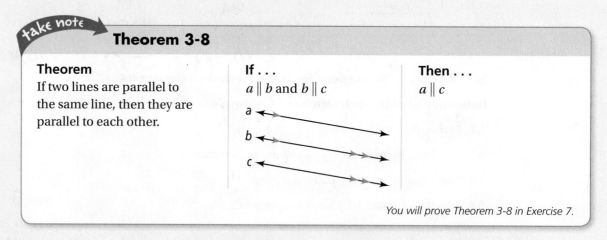

Theorem	If . . .	Then . . .
If two lines are parallel to the same line, then they are parallel to each other.	$a \parallel b$ and $b \parallel c$	$a \parallel c$

You will prove Theorem 3-8 in Exercise 7.

take note

Theorem 3-9

Theorem	If . . .	Then . . .
In a plane, if two lines are perpendicular to the same line, then they are parallel to each other.	$m \perp t$ and $n \perp t$ 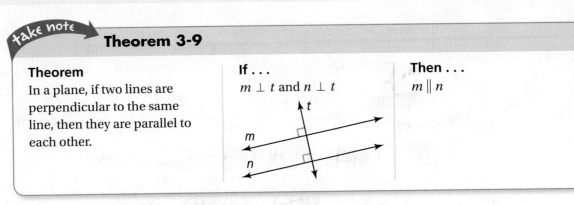	$m \parallel n$

Notice that Theorem 3-9 includes the phrase *in a plane*. In Exercise 17, you will consider why this phrase is necessary.

Proof Proof of Theorem 3-9

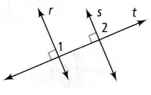

Given: In a plane, $r \perp t$ and $s \perp t$.

Prove: $r \parallel s$

Proof: $\angle 1$ and $\angle 2$ are right angles by the definition of perpendicular. So, $\angle 1 \cong \angle 2$. Since corresponding angles are congruent, $r \parallel s$.

© Problem 1 Solving a Problem With Parallel Lines STEM

Carpentry A carpenter plans to install molding on the sides and the top of a doorway. The carpenter cuts the ends of the top piece and one end of each of the side pieces at 45° angles as shown. Will the side pieces of molding be parallel? Explain.

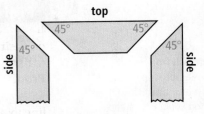

Know	Need	Plan
The angles at the connecting ends are 45°.	Determine whether the side pieces of molding are parallel.	Visualize fitting the pieces together to form new angles. Use information about the new angles to decide whether the sides are parallel.

Yes, the sides are parallel. When the pieces fit together, they form 45° + 45°, or 90°, angles. So, each side is perpendicular to the top. If two lines (the sides) are perpendicular to the same line (the top), then they are parallel to each other.

Got It? **1.** Can you assemble the pieces at the right to form a picture frame with opposite sides parallel? Explain.

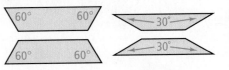

Theorems 3-8 and 3-9 give conditions that allow you to conclude that lines are parallel. The Perpendicular Transversal Theorem below provides a way for you to conclude that lines are perpendicular.

take note

Theorem 3-10   Perpendicular Transversal Theorem

Theorem	**If . . .**	**Then . . .**
In a plane, if a line is perpendicular to one of two parallel lines, then it is also perpendicular to the other.	$n \perp \ell$ and $\ell \parallel m$ 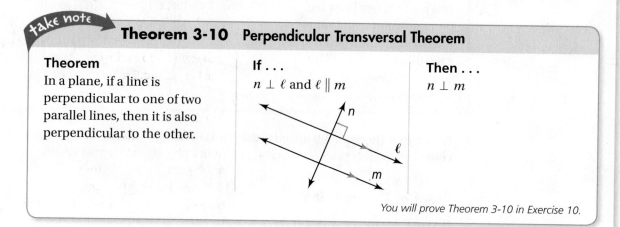	$n \perp m$

You will prove Theorem 3-10 in Exercise 10.

The Perpendicular Transversal Theorem states that the lines must be *in a plane*. The diagram at the right shows why. In the rectangular solid, $\overleftrightarrow{AC}$ and $\overleftrightarrow{BD}$ are parallel. $\overleftrightarrow{EC}$ is perpendicular to $\overleftrightarrow{AC}$, but it is not perpendicular to $\overleftrightarrow{BD}$. In fact, $\overleftrightarrow{EC}$ and $\overleftrightarrow{BD}$ are skew because they are not in the same plane.

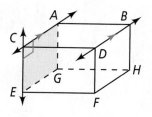

Proof © **Problem 2**   **Proving a Relationship Between Two Lines**

Think

Which line has a relationship to both lines *c* and *d*?
You know $c \perp b$ and $b \perp d$. So, line *b* relates to both lines *c* and *d*.

Given: In a plane, $c \perp b$, $b \perp d$, and $d \perp a$.

Prove: $c \perp a$

Proof: Lines *c* and *d* are both perpendicular to line *b*, so $c \parallel d$ because two lines perpendicular to the same line are parallel. It is given that $d \perp a$. Therefore, $c \perp a$ because a line that is perpendicular to one of two parallel lines is also perpendicular to the other (Perpendicular Transversal Theorem).

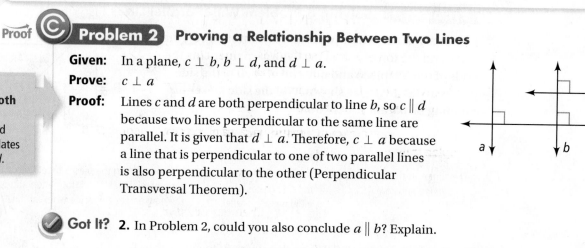

✓ **Got It?** **2.** In Problem 2, could you also conclude $a \parallel b$? Explain.

Lesson Check

Do you know HOW?

1. Main Street intersects Avenue A and Avenue B. Avenue A is parallel to Avenue B. Avenue A is also perpendicular to Main Street. How are Avenue B and Main Street related? Explain.

2. In the diagram below, lines a, b, and c are coplanar. What conclusion can you make about lines a and b? Explain.

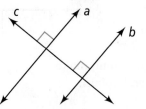

Do you UNDERSTAND? MATHEMATICAL PRACTICES

3. Explain why the phrase *in a plane* is not necessary in Theorem 3-8.

4. Which theorem or postulate from earlier in the chapter supports the conclusion in Theorem 3-9? In the Perpendicular Transversal Theorem? Explain.

5. **Error Analysis** Shiro sketched coplanar lines m, n, and r on his homework paper. He claims that it shows that lines m and n are parallel. What other information do you need about line r in order for Shiro's claim to be true? Explain.

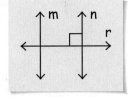

Practice and Problem-Solving Exercises MATHEMATICAL PRACTICES

A Practice

6. A carpenter is building a trellis for vines to grow on. The completed trellis will have two sets of diagonal pieces of wood that overlap each other.
   a. If pieces A, B, and C must be parallel, what must be true of $\angle 1$, $\angle 2$, and $\angle 3$?
   b. The carpenter attaches piece D so that it is perpendicular to piece A. If your answer to part (a) is true, is piece D perpendicular to pieces B and C? Justify your answer.

   See Problem 1.

7. **Developing Proof** Copy and complete this paragraph proof of Theorem 3-8 for three coplanar lines.

   See Problem 2.

   Given: $\ell \parallel k$ and $m \parallel k$

   Prove: $\ell \parallel m$

   Proof: Since $\ell \parallel k$, $\angle 2 \cong \angle 1$ by the **a.** _?_ Theorem. Since $m \parallel k$, **b.** _?_ $\cong$ **c.** _?_ for the same reason. By the Transitive Property of Congruence, $\angle 2 \cong \angle 3$. By the **d.** _?_ Theorem, $\ell \parallel m$.

8. Write a paragraph proof.
   Given: In a plane, $a \perp b$, $b \perp c$, and $c \parallel d$.
   Prove: $a \parallel d$

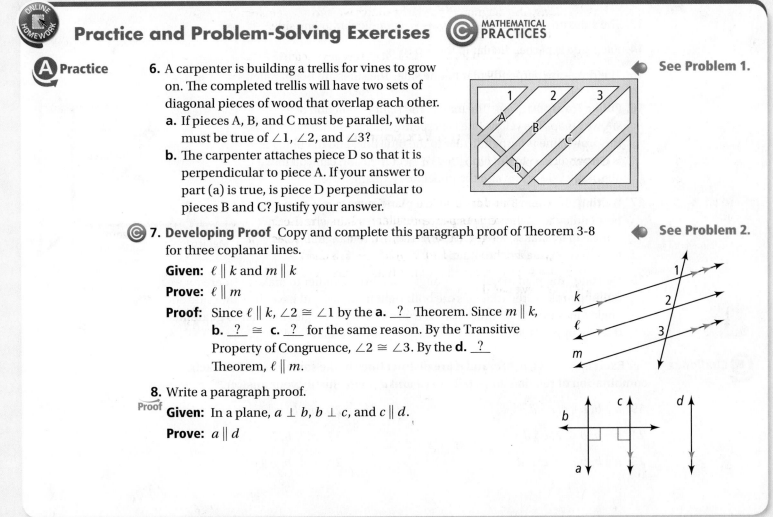

B **Apply**

© 9. Think About a Plan One traditional type of log cabin is a single rectangular room. Suppose you begin building a log cabin by placing four logs in the shape of a rectangle. What should you measure to guarantee that the logs on opposite walls are parallel? Explain.
- What type of information do you need to prove lines parallel?
- How can you use a diagram to help you?
- What do you know about the angles of the geometric shape?

10. Prove the Perpendicular Transversal Theorem (Theorem 3-10): In a
Proof plane, if a line is perpendicular to one of two parallel lines, then it is also perpendicular to the other.

Given: In a plane, $a \perp b$ and $b \parallel c$.

Prove: $a \perp c$

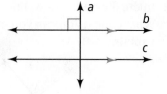

The following statements describe a ladder. Based only on the statement, make a conclusion about the rungs, one side, or both sides of the ladder. Explain.

11. The rungs are each perpendicular to one side.

12. The rungs are parallel and the top rung is perpendicular to one side.

13. The sides are parallel. The rungs are perpendicular to one side.

14. Each side is perpendicular to the top rung.

15. The rungs are perpendicular to one side. The sides are not parallel.

16. Public Transportation The map at the right is a section of a subway map. The yellow line is perpendicular to the brown line, the brown line is perpendicular to the blue line, and the blue line is perpendicular to the pink line. What conclusion can you make about the yellow line and the pink line? Explain.

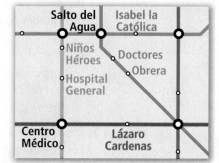

© 17. Writing Theorem 3-8 states that in a plane, two lines perpendicular to the same line are parallel. Explain why the phrase *in a plane* is needed. (*Hint:* Refer to a rectangular solid to help you visualize the situation.)

18. Quilting You plan to sew two triangles of fabric together to make a square for a quilting project. The triangles are both right triangles and have the same side and angle measures. What must also be true about the triangles in order to guarantee that the opposite sides of the fabric square are parallel? Explain.

© Challenge For Exercises 19–24, *a, b, c,* and *d* are distinct lines in the same plane. For each combination of relationships, tell how *a* and *d* relate. Justify your answer.

19. $a \parallel b, b \parallel c, c \parallel d$

20. $a \parallel b, b \parallel c, c \perp d$

21. $a \parallel b, b \perp c, c \parallel d$

22. $a \perp b, b \parallel c, c \parallel d$

23. $a \parallel b, b \perp c, c \perp d$

24. $a \perp b, b \parallel c, c \perp d$

25. Reasoning Review the reflexive, symmetric, and transitive properties for congruence in Lesson 2-5. Write reflexive, symmetric, and transitive statements for "is parallel to" (∥). Tell whether each statement is *true* or *false*. Justify your answer.

26. Reasoning Repeat Exercise 25 for "is perpendicular to" (⊥).

Standardized Test Prep

SAT/ACT

27. In a plane, line *e* is parallel to line *f*, line *f* is parallel to line *g*, and line *h* is perpendicular to line *e*. Which of the following MUST be true?

    Ⓐ *e* ∥ *g*   Ⓑ *h* ∥ *f*   Ⓒ *g* ∥ *h*   Ⓓ *e* ∥ *h*

28. Which point lies nearest to (5, 2) in the coordinate plane?

    Ⓕ (−1, 3)   Ⓖ (0, −2)   Ⓗ (4, −5)   Ⓘ (4, 10)

29. Which of the following is NOT a reason for proving two lines parallel.

    Ⓐ The lines are both ⊥ to the same line.   Ⓒ Vertical angles are congruent.

    Ⓑ Corresponding angles are congruent.   Ⓓ The lines are both ∥ to the same line.

Short Response

30. The diameter of a circle is the same length as the side of a square. The perimeter of the square is 16 cm. Find the diameter of the circle. Then find the circumference of the circle in terms of π.

Mixed Review

Algebra Determine the value of *x* for which *a* ∥ *b*.   ◀ **See Lesson 3-3.**

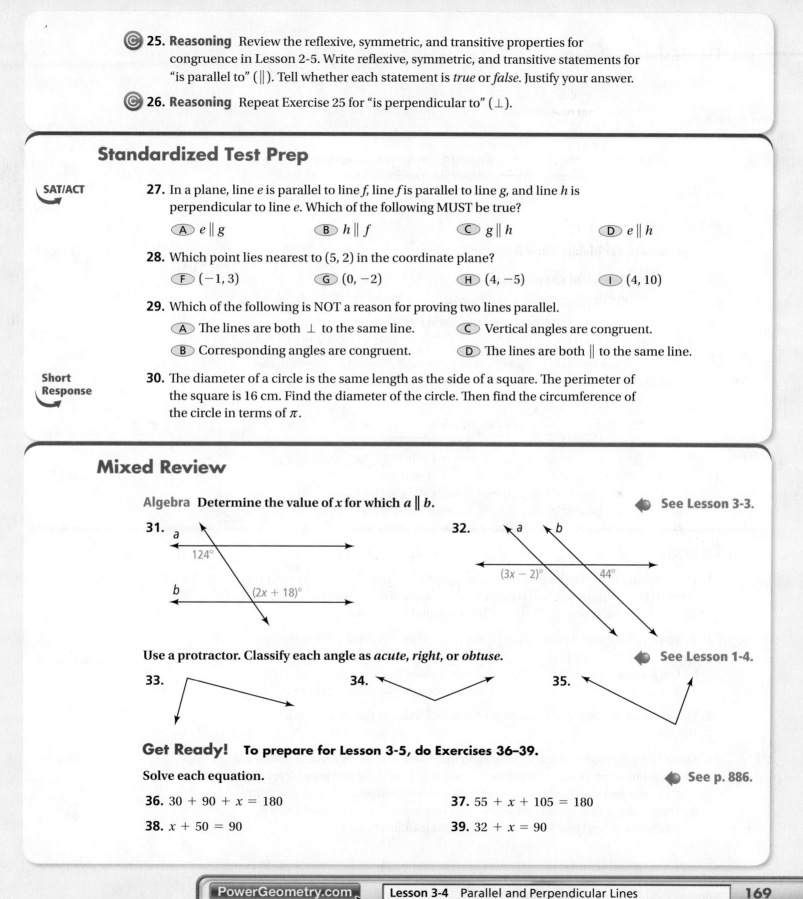

31.
a
124°
b
(2*x* + 18)°

32.
a *b*
(3*x* − 2)°   44°

Use a protractor. Classify each angle as *acute*, *right*, or *obtuse*.   ◀ **See Lesson 1-4.**

33.   **34.**   **35.**

Get Ready!   **To prepare for Lesson 3-5, do Exercises 36–39.**

Solve each equation.   ◀ **See p. 886.**

36. 30 + 90 + *x* = 180   **37.** 55 + *x* + 105 = 180

38. *x* + 50 = 90   **39.** 32 + *x* = 90

Concept Byte

Use With Lesson 3-4

ACTIVITY

Perpendicular Lines and Planes

© **Content Standard**

Extends G.CO.1 Know precise definitions of angle, circle, perpendicular line, parallel line, and line segment, based on the undefined notions of point, line, distance along a line, and distance around a circular arc.

As you saw in Chapter 1, you can use a polygon to represent a plane in space. You can sketch overlapping polygons to suggest how two perpendicular planes intersect in a line.

Activity

Draw perpendicular planes A and B intersecting in $\overleftrightarrow{CD}$.

Step 1 Draw plane A and $\overleftrightarrow{CD}$ in plane A.

Step 2 Draw two segments that are perpendicular to $\overleftrightarrow{CD}$. One segment should pass through point C. The other segment should pass through point D. The segments represent two lines in plane B that are perpendicular to plane A.

Step 3 Connect the segment endpoints to draw plane B. Plane B is perpendicular to plane A because plane B contains lines perpendicular to plane A.

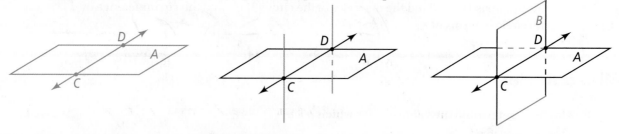

Exercises

1. Draw a plane in space. Then draw two lines that are in the plane and intersect at point A. Draw a third line that is perpendicular to each of the two lines at point A. What is the relationship between the third line and the plane?

2. **a.** Draw a plane and a point in the plane. Draw a line perpendicular to the plane at that point. Can you draw more than one perpendicular line?
   b. Draw a line and a point on the line. Draw a plane that is perpendicular to the line at that point. Can you draw more than one perpendicular plane?

3. Draw two planes perpendicular to the same line. What is the relationship between the planes?

4. Draw line ℓ through plane P at point A, so that line ℓ is perpendicular to plane P.
   a. Draw line m perpendicular to line ℓ at point A. How do m and plane P relate? Does this relationship hold true for every line perpendicular to line ℓ at point A?
   b. Draw a plane Q that contains line ℓ. How do planes P and Q relate? Does this relationship hold true for every plane Q that contains line ℓ?

3-5 Parallel Lines and Triangles

© **Content Standard**
G.CO.10 Prove theorems about triangles . . .
measures of interior angles of a triangle sum to 180°.

Objectives To use parallel lines to prove a theorem about triangles
To find measures of angles of triangles

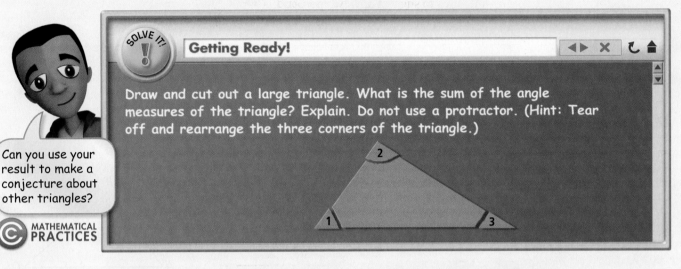

Can you use your result to make a conjecture about other triangles?

MATHEMATICAL PRACTICES

Dynamic Activity
Triangle Theorems Activity

Lesson Vocabulary
• auxiliary line
• exterior angle of a polygon
• remote interior angles

In the Solve It, you may have discovered that you can rearrange the corners of the triangle to form a straight angle. You can do this for any triangle.

Essential Understanding The sum of the angle measures of a triangle is always the same.

The Solve It suggests an important theorem about triangles. To prove this theorem, you will need to use parallel lines.

take note

Postulate 3-2 Parallel Postulate

Through a point not on a line, there is one and only one line parallel to the given line.

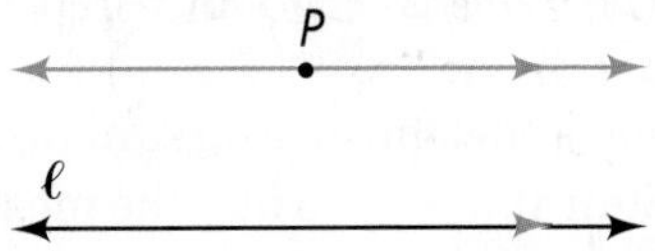

P

ℓ

There is exactly one line through *P* parallel to *ℓ*.

take note

Theorem 3-11 Triangle Angle-Sum Theorem

The sum of the measures of the angles of a triangle is 180.

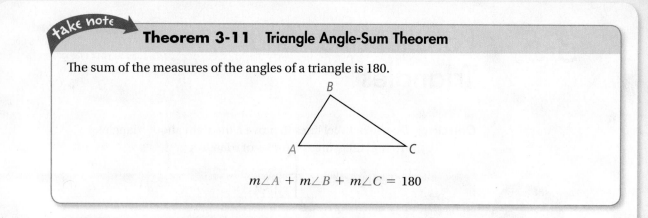

$$m\angle A + m\angle B + m\angle C = 180$$

The proof of the Triangle Angle-Sum Theorem requires an *auxiliary line*. An **auxiliary line** is a line that you add to a diagram to help explain relationships in proofs. The red line in the diagram below is an auxiliary line.

Proof **Proof of Theorem 3-11: Triangle Angle-Sum Theorem**

Given: $\triangle ABC$
Prove: $m\angle A + m\angle 2 + m\angle C = 180$

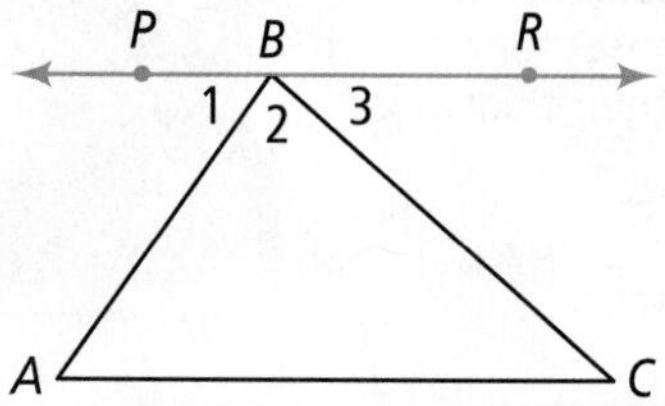

Statements	**Reasons**
1) Draw $\overleftrightarrow{PR}$ through B, parallel to $\overline{AC}$.	1) Parallel Postulate
2) $\angle PBC$ and $\angle 3$ are supplementary.	2) $\angle$s that form a linear pair are suppl.
3) $m\angle PBC + m\angle 3 = 180$	3) Definition of suppl. $\angle$s
4) $m\angle PBC = m\angle 1 + m\angle 2$	4) Angle Addition Postulate
5) $m\angle 1 + m\angle 2 + m\angle 3 = 180$	5) Substitution Property
6) $\angle 1 \cong \angle A$ and $\angle 3 \cong \angle C$	6) If lines are ∥, then alternate interior $\angle$s are $\cong$.
7) $m\angle 1 = m\angle A$ and $m\angle 3 = m\angle C$	7) Congruent $\angle$s have equal measure.
8) $m\angle A + m\angle 2 + m\angle C = 180$	8) Substitution Property

When you know the measures of two angles of a triangle, you can use the Triangle Angle-Sum Theorem to find the measure of the third angle.

Problem 1 **Using the Triangle Angle-Sum Theorem**

Algebra What are the values of *x* and *y* in the diagram at the right?

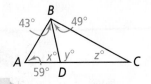

Plan

Which variable should you solve for first?
From the diagram, you know two angle measures in △*ADB*. The third angle is labeled *x*°. So use what you know about the angle measures in a triangle to solve for *x* first.

Think	Write
Use the Triangle Angle-Sum Theorem to write an equation involving *x*.	$59 + 43 + x = 180$
Solve for *x* by simplifying and then subtracting 102 from each side.	$102 + x = 180$ $x = 78$
∠*ADB* and ∠*CDB* form a linear pair, so they are supplementary.	$m\angle ADB + m\angle CDB = 180$
Substitute 78 for *m*∠*ADB* and *y* for *m*∠*CDB* in the above equation.	$x + y = 180$ $78 + y = 180$
Solve for *y* by subtracting 78 from each side.	$y = 102$

Got It? **1.** Use the diagram in Problem 1. What is the value of *z*?

An **exterior angle of a polygon** is an angle formed by a side and an extension of an adjacent side. For each exterior angle of a triangle, the two nonadjacent interior angles are its **remote interior angles.** In each triangle below, ∠1 is an exterior angle and ∠2 and ∠3 are its remote interior angles.

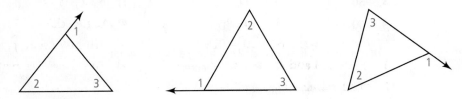

The theorem below states the relationship between an exterior angle and its two remote interior angles.

take note

Theorem 3-12 **Triangle Exterior Angle Theorem**

The measure of each exterior angle of a triangle equals the sum of the measures of its two remote interior angles.

$$m\angle 1 = m\angle 2 + m\angle 3$$

You will prove Theorem 3-12 in Exercise 33.

You can use the Triangle Exterior Angle Theorem to find angle measures.

Plan

What information can you get from the diagram?
The diagram shows you which angles are interior or exterior.

Ⓒ **Problem 2** **Using the Triangle Exterior Angle Theorem**

Ⓐ **What is the measure of ∠1?**

$m\angle 1 = 80 + 18$ Triangle Exterior Angle Theorem

$m\angle 1 = 98$ Simplify.

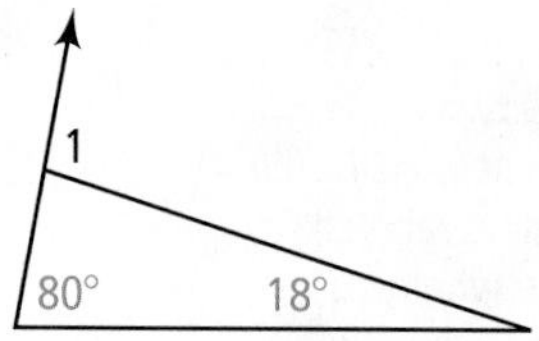

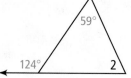

Ⓑ **What is the measure of ∠2?**

$124 = 59 + m\angle 2$ Triangle Exterior Angle Theorem

$65 = m\angle 2$ Subtract 59 from each side.

✔ **Got It?** **2.** Two angles of a triangle measure 53. What is the measure of an exterior angle at each vertex of the triangle?

Plan

How can you apply your skills from Problem 2 here?
Look at the diagram. Notice that you have a triangle and information about interior and exterior angles.

Ⓒ **Problem 3** **Applying the Triangle Theorems**

Multiple Choice When radar tracks an object, the reflection of signals off the ground can result in clutter. Clutter causes the receiver to confuse the real object with its reflection, called a ghost. At the right, there is a radar receiver at A, an airplane at B, and the airplane's ghost at D. What is the value of x?

Ⓐ 30   Ⓒ 70

Ⓑ 50   Ⓓ 80

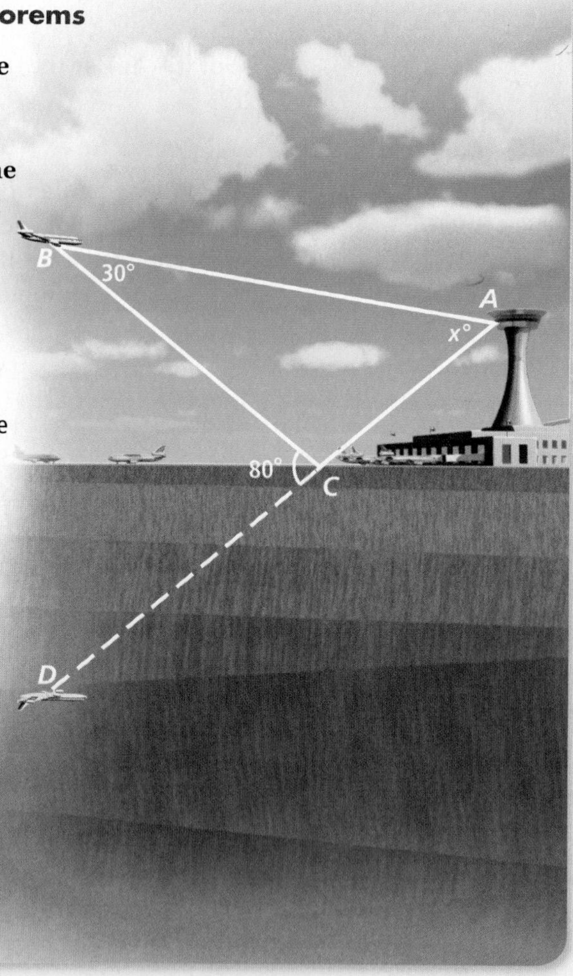

$m\angle A + m\angle B = m\angle BCD$ Triangle Exterior Angle Theorem

$x + 30 = 80$ Substitute.

$x = 50$ Subtract 30 from each side.

The value of x is 50. The correct answer is B.

Ⓒ ✔ **Got It?** **3. Reasoning** In Problem 3, can you find $m\angle A$ without using the Triangle Exterior Angle Theorem? Explain.

Lesson Check

Do you know HOW?

Find the measure of the third angle of a triangle given the measures of two angles.

1. 34 and 88

2. 45 and 90

3. 10 and 102

4. x and 50

In a triangle, $\angle 1$ is an exterior angle and $\angle 2$ and $\angle 3$ are its remote interior angles. Find the missing angle measure.

5. $m\angle 2 = 24$ and $m\angle 3 = 106$

6. $m\angle 1 = 70$ and $m\angle 2 = 32$

Do you UNDERSTAND? MATHEMATICAL PRACTICES

7. Explain how the Triangle Exterior Angle Theorem makes sense based on the Triangle Angle-Sum Theorem.

© **8. Error Analysis** The measures of the interior angles of a triangle are 30, x, and $3x$. Which of the following methods for solving for x is incorrect? Explain.

A.
$$x + 3x = 30$$
$$4x = 30$$
$$x = 7.5$$

B.
$$x + 3x + 30 = 180$$
$$4x + 30 = 180$$
$$4x = 150$$
$$x = 37.5$$

Practice and Problem-Solving Exercises MATHEMATICAL PRACTICES

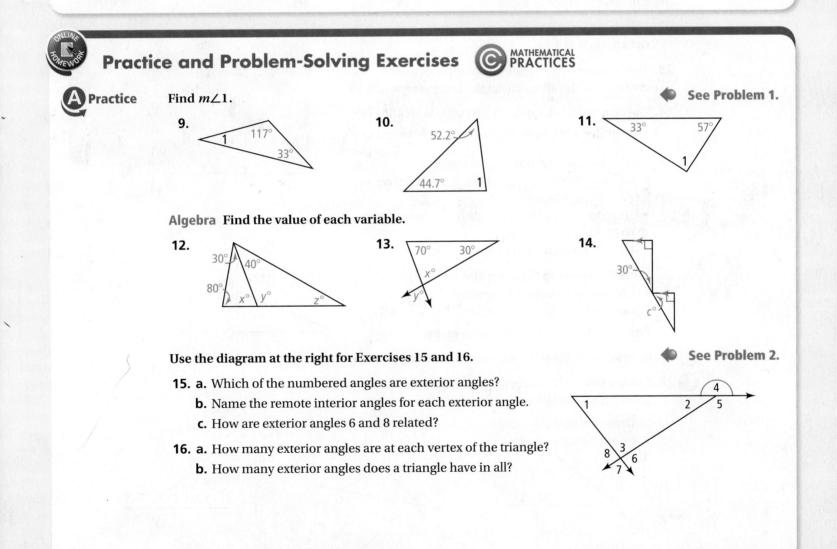

Ⓐ **Practice** Find $m\angle 1$.

See Problem 1.

9. 117° 1 33°

10. 52.2° 44.7° 1

11. 33° 57° 1

Algebra Find the value of each variable.

12. 30° 40° 80° $x°$ $y°$ $z°$

13. 70° 30° $x°$ $y°$

14. 30° $c°$

Use the diagram at the right for Exercises 15 and 16.

See Problem 2.

15. a. Which of the numbered angles are exterior angles?
   b. Name the remote interior angles for each exterior angle.
   c. How are exterior angles 6 and 8 related?

16. a. How many exterior angles are at each vertex of the triangle?
   b. How many exterior angles does a triangle have in all?

Algebra Find each missing angle measure.

17.

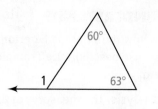

60°

1   63°

18.

128.5°

13°

2

19.

45°

3 4

47°

20. A ramp forms the angles shown at the right. What are the values of *a* and *b*?

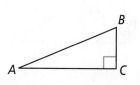

a°   b°   72°

See Problem 3.

21. A lounge chair has different settings that change the angles formed by its parts. Suppose $m\angle 2 = 71$ and $m\angle 3 = 43$. Find $m\angle 1$.

B **Apply**

Algebra Use the given information to find the unknown angle measures in the triangle.

22. The ratio of the angle measures of the acute angles in a right triangle is 1 : 2.

23. The measure of one angle of a triangle is 40. The measures of the other two angles are in a ratio of 3 : 4.

24. The measure of one angle of a triangle is 108. The measures of the other two angles are in a ratio of 1 : 5.

© **25. Think About a Plan** The angle measures of △*RST* are represented by $2x$, $x + 14$, and $x - 38$. What are the angle measures of △*RST*?
 • How can you use the Triangle Angle-Sum Theorem to write an equation?
 • How can you check your answer?

26. Prove the following theorem: The acute angles of
Proof a right triangle are complementary.
 Given: △*ABC* with right angle *C*
 Prove: $\angle A$ and $\angle B$ are complementary.

B

A   C

© **27. Reasoning** What is the measure of each angle of an equiangular triangle? Explain.

© **28. Draw a Diagram** Which diagram below correctly represents the following description? Explain your reasoning.

 Draw any triangle. Label it △*ABC*. Extend two sides of the triangle to form two exterior angles at vertex *A*.

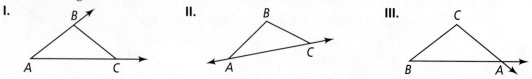

I.

B

A   C

II.

B

A   C

III.

C

B   A

Find the values of the variables and the measures of the angles.

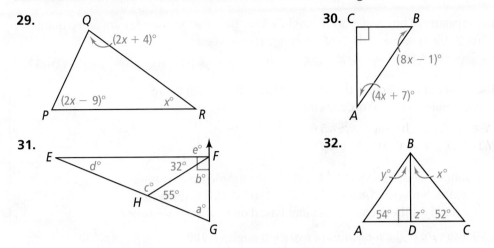

29.

Q

(2x + 4)°

(2x − 9)°   x°
P   R

30. C   B

(8x − 1)°

(4x + 7)°
A

31.

E   e° F
     d°   32°
          b°
       c° 55°
       H
          a°
             G

32.   B

y°   x°

54° z° 52°
A   D   C

33. Prove the Triangle Exterior Angle Theorem (Theorem 3-12).

Proof

The measure of each exterior angle of a triangle equals the sum of the measures of its two remote interior angles.

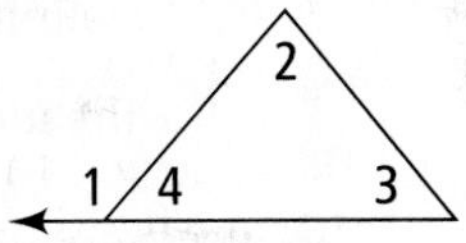

Given: ∠1 is an exterior angle of the triangle.

Prove: $m\angle 1 = m\angle 2 + m\angle 3$

2

1 / 4   3

© **34. Reasoning** Two angles of a triangle measure 64 and 48. What is the measure of the largest exterior angle of the triangle? Explain.

35. Algebra A right triangle has exterior angles at each of its acute angles with measures in the ratio 13 : 14. Find the measures of the two acute angles of the right triangle.

Ⓒ **Challenge**

Probability In Exercises 36–40, you know only the given information about the measures of the angles of a triangle. Find the probability that the triangle is equiangular.

36. Each is a multiple of 30.

37. Each is a multiple of 20.

38. Each is a multiple of 60.

39. Each is a multiple of 12.

40. One angle is obtuse.

41. In the figure at the right, $\overline{CD} \perp \overline{AB}$ and $\overline{CD}$ bisects ∠ACB. Find $m\angle DBF$.

42. If the remote interior angles of an exterior angle of a triangle are congruent, what can you conclude about the bisector of the exterior angle? Justify your answer.

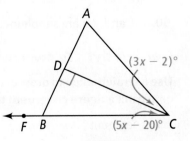

A

(3x − 2)°
D

F B   (5x − 20)° C

SAT/ACT

43. The measure of one angle of a triangle is 115. The other two angles are congruent. What is the measure of each of the congruent angles?

    Ⓐ 32.5   Ⓑ 57.5   Ⓒ 65   Ⓓ 115

44. The center of the circle at the right is at the origin. What is the approximate length of its diameter?

    Ⓕ 2   Ⓗ 5.6

    Ⓖ 2.8   Ⓘ 8

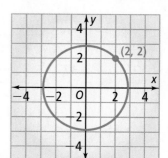

45. One statement in a proof is "∠1 and ∠2 are supplementary angles." The next statement is "$m\angle 1 + m\angle 2 = 180$." Which is the best justification for the second statement based on the first statement?

    Ⓐ The sum of the measures of two right angles is 180.

    Ⓑ Angles that form a linear pair are supplementary.

    Ⓒ Definition of supplementary angles

    Ⓓ The measure of a straight angle is 180.

Extended Response

46. $\triangle ABC$ is an obtuse triangle with $m\angle A = 21$ and $\angle C$ is acute.
    a. What is $m\angle B + m\angle C$? Explain.
    b. What is the range of whole numbers for $m\angle C$? Explain.
    c. What is the range of whole numbers for $m\angle B$? Explain.

Mixed Review

Use the diagram at the right for Exercises 47 and 48.

◀ **See Lesson 3-4.**

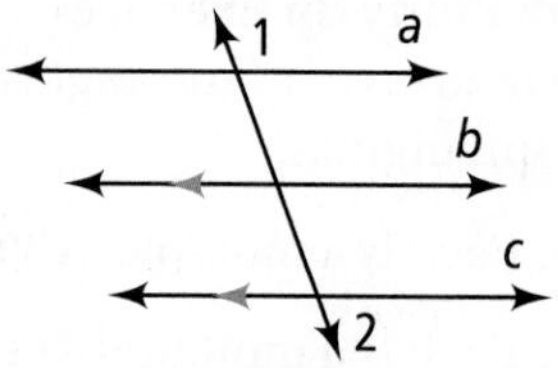

47. If ∠1 and ∠2 are supplementary, what can you conclude about lines a and c? Justify your answer.

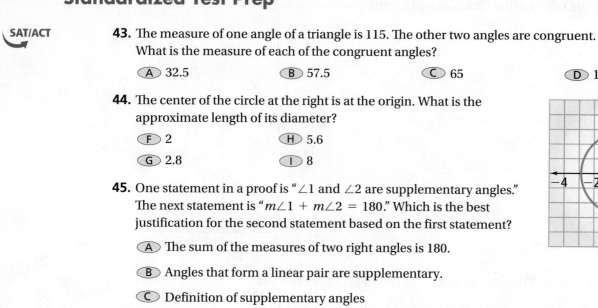

48. If $a \parallel c$, what can you conclude about lines a and b? Justify your answer.

49. $\angle ABC$ and $\angle CBD$ form a linear pair. If $m\angle ABC = 3x + 20$ and $m\angle CBD = x + 32$, find the value of x.

◀ **See Lesson 1-5.**

50. ∠1 and ∠2 are supplementary. If $\angle 1 \cong \angle 2$, find $m\angle 1$ and $m\angle 2$. Explain.

Get Ready!   To prepare for Lesson 3-6, do Exercises 51–53.

Use a straightedge to draw each figure. Then use a straightedge and compass to construct a figure congruent to it.

◀ **See Lesson 1-6.**

51. a segment   **52.** an obtuse angle   **53.** an acute angle

Exploring Spherical Geometry

© **Content Standard**

Extends G.CO.1 Know precise definitions of angle, circle, perpendicular line, parallel line, and line segment, based on the undefined notions of point, line, distance along a line, and distance around a circular arc.

Euclid was a Greek mathematician who identified many of the definitions, postulates, and theorems of high school geometry. Euclidean geometry is the geometry of flat planes, straight lines, and points.

In spherical geometry, the curved surface of a sphere is studied. A "line" is a great circle. A *great circle* is the intersection of a sphere and a plane that contains the center of the sphere.

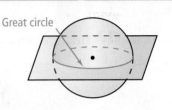
Great circle

Activity 1

You can use latitude and longitude to identify positions on Earth. Look at the latitude and longitude markings on the globe.

1. Think about "slicing" the globe with a plane at each latitude. Do any of your "slices" contain the center of the globe?

2. Think about "slicing" the globe with a plane at each longitude. Do any of your "slices" contain the center of the globe?

3. Which latitudes, if any, suggest great circles? Which longitudes, if any, suggest great circles? Explain.

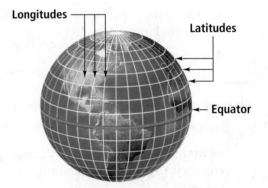
Longitudes
Latitudes
← Equator

You learned in Lesson 3–5 that through any point not on a line, there is one and only one line parallel to the given line (Parallel Postulate). That statement is not true in spherical geometry. In spherical geometry,

> *through a point not on a line, there is no line parallel to the given line.*

Since lines are great circles in spherical geometry, two lines always intersect. In fact, any two lines on a sphere intersect at *two* points, as shown at the right.

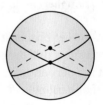

One result of the Parallel Postulate in Euclidean geometry is the Triangle Angle-Sum Theorem. The spherical geometry Parallel Postulate gives a very different result.

Activity 2

Hold a string taut between any two points on a sphere. The string forms a "segment" that is part of a great circle. Connect three such segments to form a triangle on the sphere.

Below are examples of triangles on a sphere.

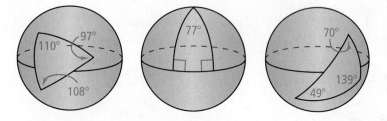

4. What is the sum of the angle measures in the first triangle? The second triangle? The third triangle?

5. How are these results different from the Triangle Angle-Sum Theorem in Euclidean geometry? Explain.

Exercises

For Exercises 6 and 7, draw a sketch to illustrate each property of spherical geometry. Explain how each property compares to what is true in Euclidean geometry.

6. There are pairs of points on a sphere through which you can draw more than one line.

7. Two equiangular triangles can have different angle measures.

8. For each of the following properties of Euclidean geometry, draw a counterexample to show that the property is *not* true in spherical geometry.
a. Two lines that are perpendicular to the same line do not intersect.
b. If two angles of one triangle are congruent to two angles of another triangle, then the third angles are congruent.

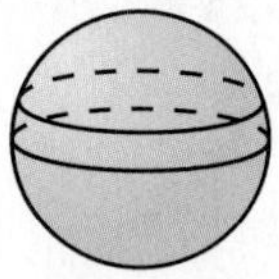

9. a. The figure at the right appears to show parallel lines on a sphere. Explain why this is not the case.
b. Explain why a piece of the top circle in the figure is *not* a line segment. (*Hint:* What must be true of line segments in spherical geometry?)

10. In Euclidean geometry, vertical angles are congruent. Does this seem to be true in spherical geometry? Explain. Make figures on a globe, ball, or balloon to support your answer.

Do you know HOW?

Identify the following. Lines and planes that appear to be parallel are parallel.

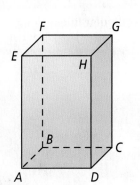

1. all segments parallel to $\overline{HG}$

2. a plane parallel to plane *EFB*

3. all segments skew to $\overline{EA}$

4. all segments parallel to plane *ABCD*

Use the diagram below for Exercises 5–14.

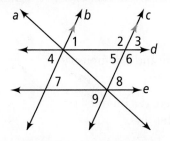

Name two pairs of each angle type.

5. corresponding angles

6. alternate interior angles

7. same-side interior angles

State the theorem or postulate that justifies each statement.

8. $\angle 7 \cong \angle 9$ **9.** $\angle 4 \cong \angle 5$

10. $m\angle 1 + m\angle 2 = 180$

Complete each statement.

11. If $\angle 5 \cong \angle 9$, then $\underline{\ ?\ } \parallel \underline{\ ?\ }$.

12. If $\angle 4 \cong \underline{\ ?\ }$, then $d \parallel e$.

13. If $e \perp b$, then $e \perp \underline{\ ?\ }$.

14. If $c \perp d$, then $b \perp \underline{\ ?\ }$.

Find $m\angle 1$.

15.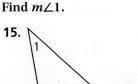

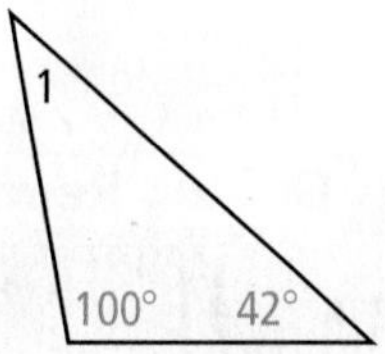

16.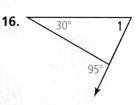

Find the value of x for which $a \parallel b$.

17.

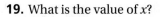
$(x + 66)°$ a
$(2x − 8)°$ b

18.

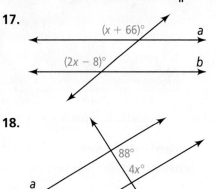
$88°$
$4x°$
a
b

19. What is the value of x?

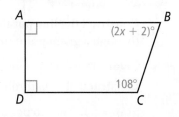

A B
$(2x + 2)°$
$108°$
D C

Do you UNDERSTAND?

20. Reasoning Can a pair of lines be both parallel and skew? Explain.

21. Open-Ended Give an example of parallel lines in the real world. Then describe how you could prove that the lines are parallel.

22. Reasoning Lines ℓ, r, and s are coplanar. Suppose ℓ is perpendicular to r and r is perpendicular to s. Is ℓ perpendicular to s? Explain.

Constructing Parallel and Perpendicular Lines

Ⓒ Content Standards
G.CO.12 Make formal geometric constructions with a variety of tools and methods . . . constructing perpendicular lines . . . and constructing a line parallel to a given line through a point not on the line.
Also G.CO.13

Objective To construct parallel and perpendicular lines

Getting Ready! ◀▶ ✕ ↻ ⌂

Draw a line m on a sheet of paper. Fold your paper so that line m falls on itself. Label your fold line n. Fold your paper again so that n falls on itself. Label your new fold line p. How are m and p related? How do you know?

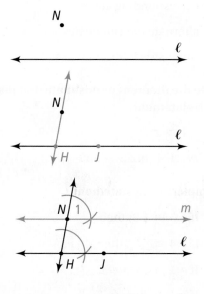

Check out the angles formed by the folded lines. How are they related?

Ⓒ MATHEMATICAL PRACTICES In the Solve It, you used paper-folding to construct lines.

Essential Understanding You can also use a straightedge and a compass to construct parallel and perpendicular lines.

In Lesson 3-5, you learned that through a point not on a line, there is a unique line parallel to the given line. Problem 1 shows the construction of this line.

Dynamic Activity
Constructing Parallel and Perpendicular Lines

Ⓒ Problem 1 Constructing Parallel Lines

Construct the line parallel to a given line and through a given point that is not on the line.

Given: line ℓ and point N not on ℓ

Construct: line m through N with $m \parallel \ell$

Step 1 Label two points H and J on ℓ. Draw $\overleftrightarrow{HN}$.

Think
What type of angles are $\angle 1$ and $\angle NHJ$? They are corresponding angles.

Step 2 At N, construct $\angle 1$ congruent to $\angle NHJ$. Label the new line m.

$m \parallel \ell$

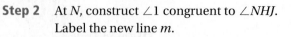

Ⓒ ✓ Got It? 1. **Reasoning** Why must lines ℓ and m be parallel?

Construct a quadrilateral with one pair of parallel sides of lengths a and b.

Given: segments of lengths a and b

Construct: quadrilateral $ABYZ$ with
$AZ = a$, $BY = b$, and $\overleftrightarrow{AZ} \parallel \overleftrightarrow{BY}$

Plan

How do you know which constructions to use?
Try sketching the final figure. This can help you visualize the construction steps you will need.

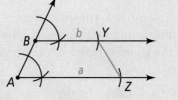

Write

Think

You need a pair of parallel sides, so construct parallel lines as you did in Problem 1. Start by drawing a ray with endpoint A. Then draw $\overrightarrow{AB}$ such that point B is not on the first ray.

Construct congruent corresponding angles to finish your parallel lines.

Now you need sides of lengths a and b. In Lesson 1-6, you learned how to construct congruent segments. Construct Y and Z so that $BY = b$ and $AZ = a$.

Draw $\overline{YZ}$.

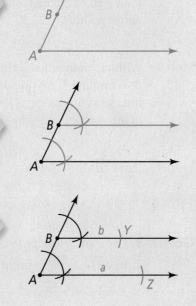

$ABYZ$ is a quadrilateral with parallel sides of lengths a and b.

© ✓ **Got It?** **2. a.** Draw a segment. Label its length m. Construct quadrilateral $ABCD$ with $\overleftrightarrow{AB} \parallel \overleftrightarrow{CD}$, so that $AB = m$ and $CD = 2m$.

       b. Reasoning Suppose you and a friend both use the steps in Problem 2 to construct $ABYZ$ independently. Will your quadrilaterals necessarily have the same angle measures and side lengths? Explain.

Problem 3   Perpendicular at a Point on a Line

Construct the perpendicular to a given line at a given point on the line.

Given: point P on line ℓ
Construct: $\overleftrightarrow{CP}$ with $\overleftrightarrow{CP} \perp \ell$

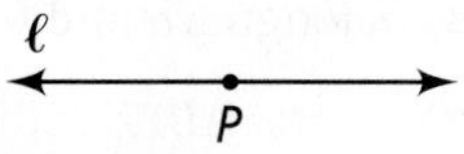

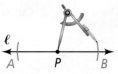

Step 1 Construct two points on ℓ that are equidistant from P. Label the points A and B.

Think
Why is it important to open your compass wider?
If you don't, you won't be able to draw intersecting arcs above point P.

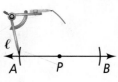

Step 2 Open the compass wider so the opening is greater than $\frac{1}{2}AB$. With the compass tip on A, draw an arc above point P.

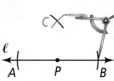

Step 3 Without changing the compass setting, place the compass point on point B. Draw an arc that intersects the arc from Step 2. Label the point of intersection C.

Step 4 Draw $\overleftrightarrow{CP}$.

$\overleftrightarrow{CP} \perp \ell$

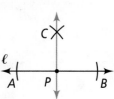

 Got It? **3.** Use a straightedge to draw $\overleftrightarrow{EF}$. Construct $\overleftrightarrow{FG}$ so that $\overleftrightarrow{FG} \perp \overleftrightarrow{EF}$ at point F.

You can also construct a perpendicular line from a point to a line. This perpendicular line is unique according to the Perpendicular Postulate. You will prove in Chapter 5 that the shortest path from any point to a line is along this unique perpendicular line.

take note

Postulate 3-3   Perpendicular Postulate

Through a point not on a line, there is one and only one line perpendicular to the given line.

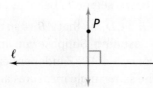

There is exactly one line through P perpendicular to ℓ.

Problem 4 **Perpendicular From a Point to a Line**

Plan

Can you use
similar steps for
this problem as in
Problem 3?
Yes. Mark two points on
ℓ that are equidistant
from the given point.
Then draw two
intersecting arcs.

Construct the perpendicular to a given line through a given point not on the line.

Given: line ℓ and point R not on ℓ

Construct: $\overleftrightarrow{RG}$ with $\overleftrightarrow{RG} \perp \ell$

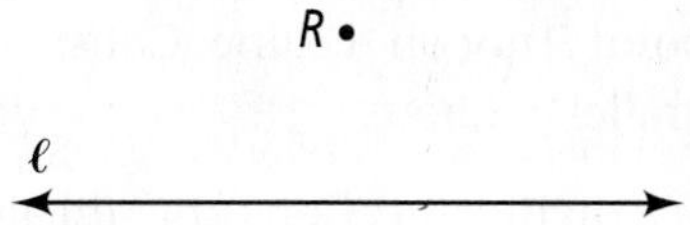

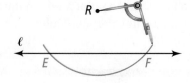

Step 1 Open your compass to a size greater than the distance from R to ℓ. With the compass on point R, draw an arc that intersects ℓ at two points. Label the points E and F.

Step 2 Place the compass point on E and make an arc.

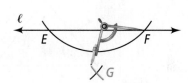

Step 3 Keep the same compass setting. With the compass tip on F, draw an arc that intersects the arc from Step 2. Label the point of intersection G.

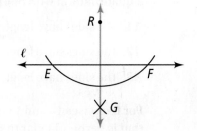

Step 4 Draw $\overleftrightarrow{RG}$.

$\overleftrightarrow{RG} \perp \ell$

Got It? **4.** Draw $\overleftrightarrow{CX}$ and a point Z not on $\overleftrightarrow{CX}$. Construct $\overleftrightarrow{ZB}$ so that $\overleftrightarrow{ZB} \perp \overleftrightarrow{CX}$.

Lesson Check

Do you know HOW?

1. Draw a line ℓ and a point P not on the line. Construct the line through P parallel to line ℓ.

2. Draw $\overleftrightarrow{QR}$ and a point S on the line. Construct the line perpendicular to $\overleftrightarrow{QR}$ at point S.

3. Draw a line w and a point X not on the line. Construct the line perpendicular to line w at point X.

Do you UNDERSTAND?

4. In Problem 3, is $\overline{AC}$ congruent to $\overline{BC}$? Explain.

5. Suppose you use a wider compass setting in Step 1 of Problem 4. Will you construct a different perpendicular line? Explain.

© 6. **Compare and Contrast** How are the constructions in Problems 3 and 4 similar? How are they different?

Practice and Problem-Solving Exercises

A Practice

For Exercises 7–10, draw a figure like the given one. Then construct the line through point J that is parallel to $\overleftrightarrow{AB}$.

See Problem 1.

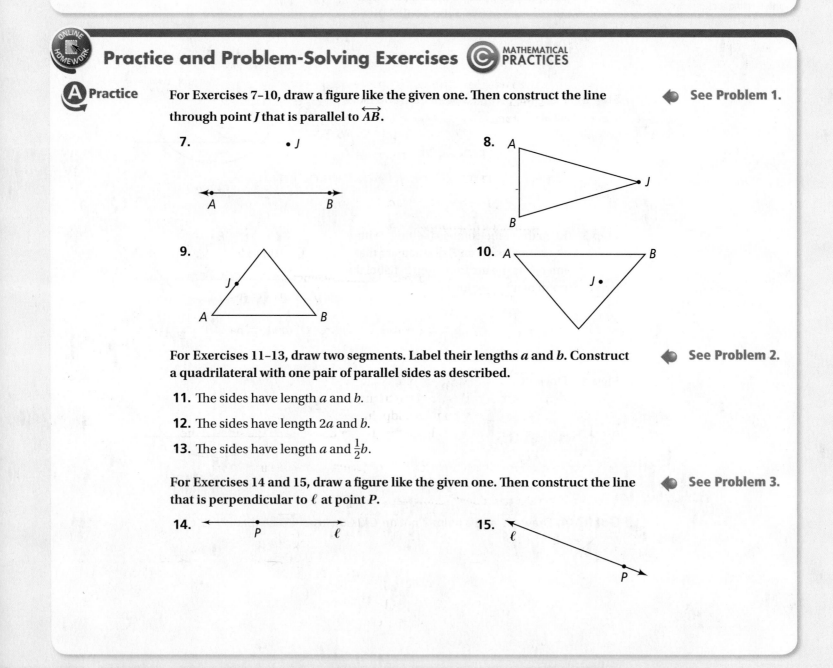

7.

8.

9.

10.

For Exercises 11–13, draw two segments. Label their lengths a and b. Construct a quadrilateral with one pair of parallel sides as described.

See Problem 2.

11. The sides have length a and b.

12. The sides have length $2a$ and b.

13. The sides have length a and $\frac{1}{2}b$.

For Exercises 14 and 15, draw a figure like the given one. Then construct the line that is perpendicular to ℓ at point P.

See Problem 3.

14.

15.

For Exercises 16–18, draw a figure like the given one. Then construct the line through point *P* that is perpendicular to $\overleftrightarrow{RS}$.

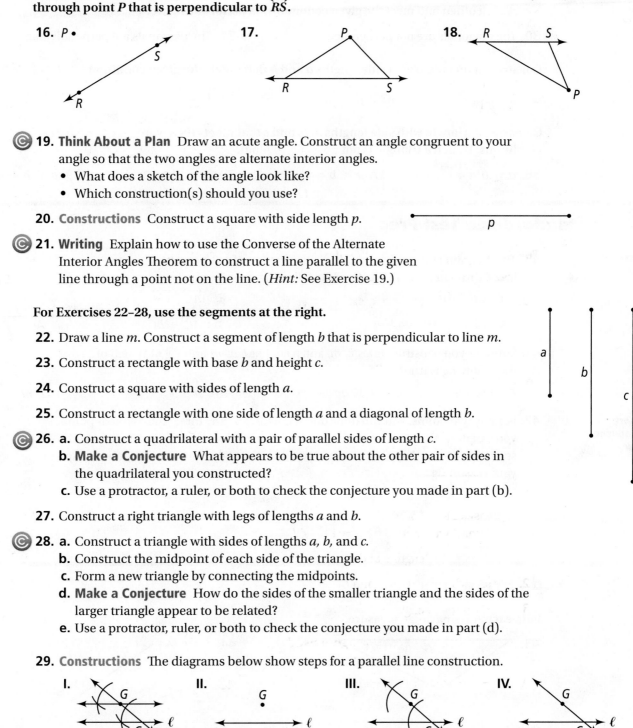

16. *P* •

S

R

17.

P

R   *S*

18.

R   *S*

P

B Apply

Ⓒ 19. Think About a Plan Draw an acute angle. Construct an angle congruent to your angle so that the two angles are alternate interior angles.
- What does a sketch of the angle look like?
- Which construction(s) should you use?

20. Constructions Construct a square with side length *p*.

•————————————————•
          p

Ⓒ 21. Writing Explain how to use the Converse of the Alternate Interior Angles Theorem to construct a line parallel to the given line through a point not on the line. (*Hint:* See Exercise 19.)

For Exercises 22–28, use the segments at the right.

a

b

c

22. Draw a line *m*. Construct a segment of length *b* that is perpendicular to line *m*.

23. Construct a rectangle with base *b* and height *c*.

24. Construct a square with sides of length *a*.

25. Construct a rectangle with one side of length *a* and a diagonal of length *b*.

Ⓒ 26. a. Construct a quadrilateral with a pair of parallel sides of length *c*.
  b. Make a Conjecture What appears to be true about the other pair of sides in the quadrilateral you constructed?
  c. Use a protractor, a ruler, or both to check the conjecture you made in part (b).

27. Construct a right triangle with legs of lengths *a* and *b*.

Ⓒ 28. a. Construct a triangle with sides of lengths *a, b,* and *c*.
  b. Construct the midpoint of each side of the triangle.
  c. Form a new triangle by connecting the midpoints.
  d. Make a Conjecture How do the sides of the smaller triangle and the sides of the larger triangle appear to be related?
  e. Use a protractor, ruler, or both to check the conjecture you made in part (d).

29. Constructions The diagrams below show steps for a parallel line construction.

I.
G
C
ℓ

II.
G
•
ℓ

III.
G
C
ℓ

IV.
G
C
ℓ

  a. List the construction steps in the correct order.
  b. For the steps that use a compass, describe the location(s) of the compass point.

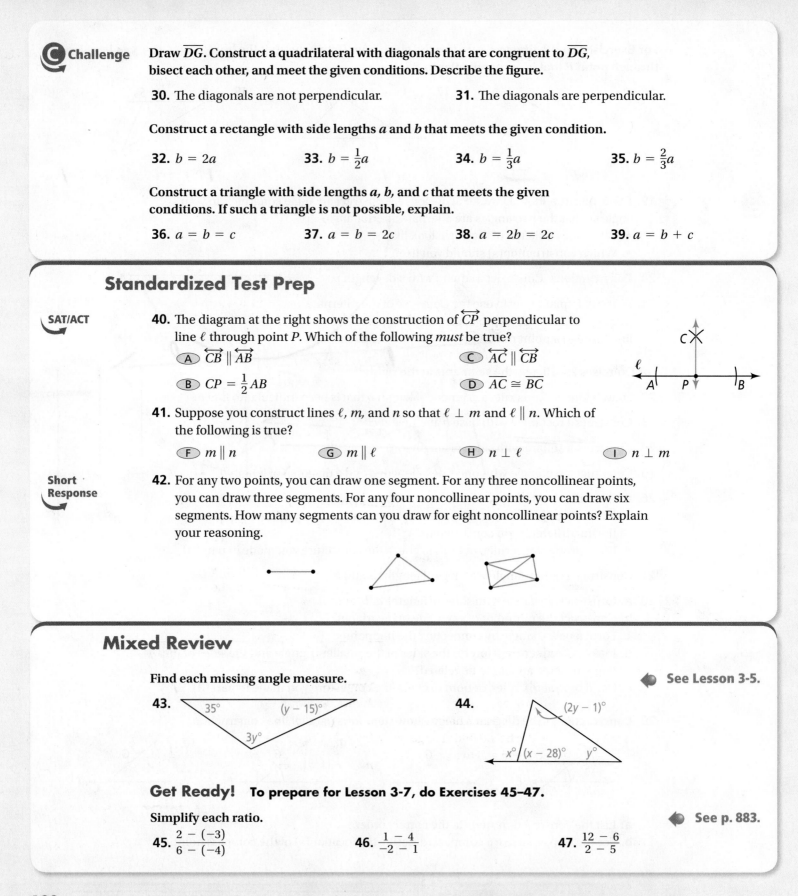

C **Challenge**   Draw $\overline{DG}$. Construct a quadrilateral with diagonals that are congruent to $\overline{DG}$, bisect each other, and meet the given conditions. Describe the figure.

30. The diagonals are not perpendicular.   **31.** The diagonals are perpendicular.

Construct a rectangle with side lengths *a* and *b* that meets the given condition.

32. $b = 2a$ **33.** $b = \frac{1}{2}a$ **34.** $b = \frac{1}{3}a$ **35.** $b = \frac{2}{3}a$

Construct a triangle with side lengths *a*, *b*, and *c* that meets the given conditions. If such a triangle is not possible, explain.

36. $a = b = c$ **37.** $a = b = 2c$ **38.** $a = 2b = 2c$ **39.** $a = b + c$

Standardized Test Prep

SAT/ACT

40. The diagram at the right shows the construction of $\overleftrightarrow{CP}$ perpendicular to line ℓ through point P. Which of the following *must* be true?

   A $\overleftrightarrow{CB} \parallel \overrightarrow{AB}$ **C** $\overleftrightarrow{AC} \parallel \overleftrightarrow{CB}$

   B $CP = \frac{1}{2}AB$ **D** $\overline{AC} \cong \overline{BC}$

41. Suppose you construct lines ℓ, m, and n so that $\ell \perp m$ and $\ell \parallel n$. Which of the following is true?

   F $m \parallel n$ **G** $m \parallel \ell$ **H** $n \perp \ell$ **I** $n \perp m$

Short Response

42. For any two points, you can draw one segment. For any three noncollinear points, you can draw three segments. For any four noncollinear points, you can draw six segments. How many segments can you draw for eight noncollinear points? Explain your reasoning.

Mixed Review

Find each missing angle measure.   ◀ **See Lesson 3-5.**

43. $35°$ $(y - 15)°$ $3y°$

44. $(2y - 1)°$ $x°$ $(x - 28)°$ $y°$

Get Ready!   **To prepare for Lesson 3-7, do Exercises 45–47.**

Simplify each ratio.   ◀ **See p. 883.**

45. $\dfrac{2 - (-3)}{6 - (-4)}$ **46.** $\dfrac{1 - 4}{-2 - 1}$ **47.** $\dfrac{12 - 6}{2 - 5}$

Equations of Lines in the Coordinate Plane

© **Content Standard**
Prepares for G.GPE.5 Prove the slope criteria for parallel and perpendicular lines . . .

Objective To graph and write linear equations

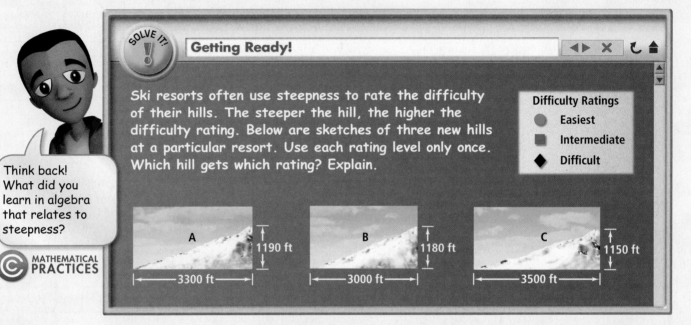

SOLVE IT!

Getting Ready!

Ski resorts often use steepness to rate the difficulty of their hills. The steeper the hill, the higher the difficulty rating. Below are sketches of three new hills at a particular resort. Use each rating level only once. Which hill gets which rating? Explain.

Difficulty Ratings
● Easiest
■ Intermediate
◆ Difficult

A ⊤1190 ft
|← 3300 ft →|

B ⊤1180 ft
|← 3000 ft →|

C ⊤1150 ft
|← 3500 ft →|

Think back! What did you learn in algebra that relates to steepness?

© **MATHEMATICAL PRACTICES**

Lesson Vocabulary
- slope
- slope-intercept form
- point-slope form

The Solve It involves using vertical and horizontal distances to determine steepness. The steepest hill has the greatest *slope*. In this lesson you will explore the concept of slope and how it relates to both the graph and the equation of a line.

Essential Understanding You can graph a line and write its equation when you know certain facts about the line, such as its slope and a point on the line.

take note

Key Concept Slope

Definition
The **slope** m of a line is the ratio of the vertical change (rise) to the horizontal change (run) between any two points.

Symbols
A line contains the points (x_1, y_1) and (x_2, y_2).

$$m = \frac{\text{rise}}{\text{run}} = \frac{y_2 - y_1}{x_2 - x_1}$$

Diagram

Problem 1 Finding Slopes of Lines

Plan

How do you know which numbers go where in a formula?
For slope, you can choose either point's y-coordinate as y_2. Just be sure to use the same point's x-coordinate as x_2.

A What is the slope of line b?

$$m = \frac{2 - (-2)}{-1 - 4}$$
$$= \frac{4}{-5}$$
$$= -\frac{4}{5}$$

B What is the slope of line d?

$$m = \frac{0 - (-2)}{4 - 4}$$
$$= \frac{2}{0} \quad \text{Undefined}$$

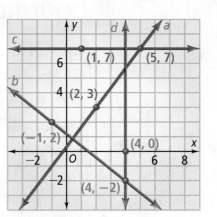

Got It? 1. Use the graph in Problem 1.

a. What is the slope of line a?

b. What is the slope of line c?

As you saw in Problem 1 and Got It 1 the slope of a line can be positive, negative, zero, or undefined. The sign of the slope tells you whether the line rises or falls to the right. A slope of zero tells you that the line is horizontal. An undefined slope tells you that the line is vertical.

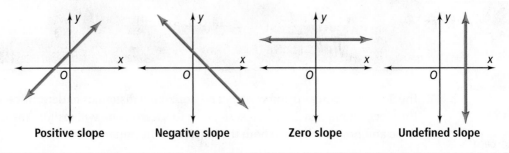

Positive slope **Negative slope** **Zero slope** **Undefined slope**

You can graph a line when you know its equation. The equation of a line has different forms. Two forms are shown below. Recall that the y-intercept of a line is the y-coordinate of the point where the line crosses the y-axis.

take note — Key Concept Forms of Linear Equations

Definition	Symbols
The **slope-intercept form** of an equation of a nonvertical line is $y = mx + b$, where m is the slope and b is the y-intercept.	$y = mx + b$ (m ↑ slope, b ↑ y-intercept)
The **point-slope form** of an equation of a nonvertical line is $y - y_1 = m(x - x_1)$, where m is the slope and (x_1, y_1) is a point on the line.	$y - y_1 = m(x - x_1)$ (y_1 ↑ y-coordinate, m ↑ slope, x_1 ↑ x-coordinate)

Plan

What do you do first?
Determine which form of linear equation you have. Then use the equation to identify the slope and a starting point.

A What is the graph of $y = \frac{2}{3}x + 1$?

The equation is in slope-intercept form, $y = mx + b$. The slope m is $\frac{2}{3}$ and the y-intercept b is 1.

Step 1 Graph a point at (0,1).

Step 2 Use the slope $\frac{2}{3}$. Go up 2 units and right 3 units. Graph a point.

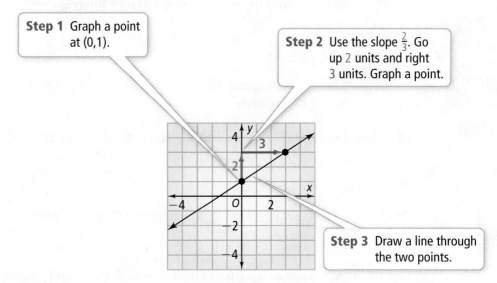

Step 3 Draw a line through the two points.

Think

What do you do when the slope is not a fraction?
A rise always needs a run. So, write the integer as a fraction with 1 as the denominator.

B What is the graph of $y - 3 = -2(x + 3)$?

The equation is in point-slope form, $y - y_1 = m(x - x_1)$. The slope m is -2 and a point (x_1, y_1) on the line is $(-3, 3)$.

Step 1 Graph a point at $(-3, 3)$.

Step 2 Use the slope -2. Go down 2 units and right 1 unit. Graph a point.

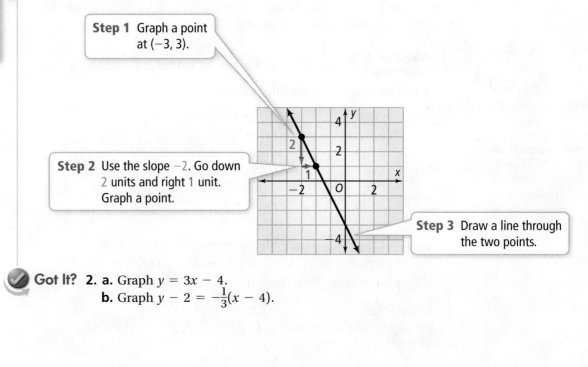

Step 3 Draw a line through the two points.

✓ **Got It?** **2. a.** Graph $y = 3x - 4$.
  b. Graph $y - 2 = -\frac{1}{3}(x - 4)$.

You can write an equation of a line when you know its slope and at least one point on the line.

 **Problem 3** **Writing Equations of Lines**

Plan

Which linear equation form should you use?
When you know the slope and the *y*-intercept, use slope-intercept form. When you know the slope and a point on the line, use point-slope form.

A **What is an equation of the line with slope 3 and *y*-intercept −5?**

$$y = mx + b$$

$m = 3$ $b = -5$

$y = 3x + (-5)$ Substitute 3 for *m* and −5 for *b*.
$y = 3x - 5$ Simplify.

B **What is an equation of the line through (−1, 5) with slope 2?**

$$y - y_1 = m(x - x_1)$$

$y_1 = 5$ $m = 2$ $x_1 = -1$

$y - 5 = 2[x - (-1)]$ Substitute (−1, 5) for (x_1, y_1) and 2 for *m*.
$y - 5 = 2(x + 1)$ Simplify.

Got It? **3. a.** What is an equation of the line with slope $-\frac{1}{2}$ and *y*-intercept 2?
  b. What is an equation of the line through (−1, 4) with slope −3?

Postulate 1-1 states that through any two points, there is exactly one line. So, you need only two points to write the equation of a line.

Problem 4 **Using Two Points to Write an Equation**

Plan

What is the first thing you need to know?
It doesn't matter yet what form of linear equation you plan to use. You'll need the slope for *both* slope-intercept form and point-slope form.

What is an equation of the line at the right?

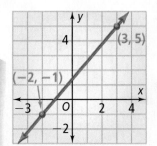

Think

Start by finding the slope *m* of the line through the given points.

Write

$$m = \frac{y_2 - y_1}{x_2 - x_1} = \frac{5 - (-1)}{3 - (-2)} = \frac{6}{5}$$

You have the slope and you know two points on the line. Use point-slope form.

$$y - y_1 = m(x - x_1)$$

Use either point for (x_1, y_1). For example, you can use (3, 5).

$$y - 5 = \frac{6}{5}(x - 3)$$

Got It? **4. a.** What is the equation of the line in Problem 4 if you use $(-2, -1)$ instead of $(3, 5)$ in the last step?

b. Rewrite the equations in Problem 4 and part (a) in slope-intercept form and compare them. What can you conclude?

You know that the slope of a horizontal line is 0 and the slope of a vertical line is undefined. Thus, horizontal and vertical lines have easily recognized equations.

Think

How is this different from writing other linear equations?
You don't need the slope. Just locate the point where the line crosses the x-axis (for vertical) or y-axis (for horizontal).

Problem 5 **Writing Equations of Horizontal and Vertical Lines**

What are the equations for the horizontal and vertical lines through $(2, 4)$?

Every point on the horizontal line through $(2, 4)$ has a y-coordinate of 4. The equation of the line is $y = 4$. It crosses the y-axis at $(0, 4)$.

Every point on the vertical line through $(2, 4)$ has an x-coordinate of 2. The equation of the line is $x = 2$. It crosses the x-axis at $(2, 0)$.

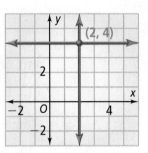

Got It? **5. a.** What are the equations for the horizontal and vertical lines through $(4, -3)$?

b. **Reasoning** Can you write the equation of a vertical line in slope-intercept form? Explain.

 Lesson Check

Do you know HOW?

For Exercises 1 and 2, find the slope of the line passing through the given points.

1. $(4, 5)$ and $(6, 15)$

2.

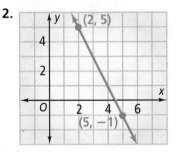

3. What is an equation of a line with slope 8 and y-intercept 10?

4. What is an equation of a line passing through $(3, 3)$ and $(4, 7)$?

Do you UNDERSTAND? **MATHEMATICAL PRACTICES**

5. **Vocabulary** Explain why you think *slope-intercept form* makes sense as a name for $y = mx + b$. Explain why you think *point-slope form* make sense as a name for $y - y_1 = m(x - x_1)$.

6. **Compare and Contrast** Graph $y = 2x + 5$ and $y = -\frac{1}{3}x + 5$. Describe how these lines are alike and how they are different.

7. **Error Analysis** A classmate found the slope of the line passing through $(8, -2)$ and $(8, 10)$, as shown at the right. Describe your classmate's error. Then find the correct slope of the line passing through the given points.

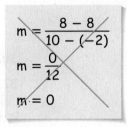

Practice and Problem-Solving Exercises

MATHEMATICAL PRACTICES

A Practice

Find the slope of the line passing through the given points.

◀ See Problem 1.

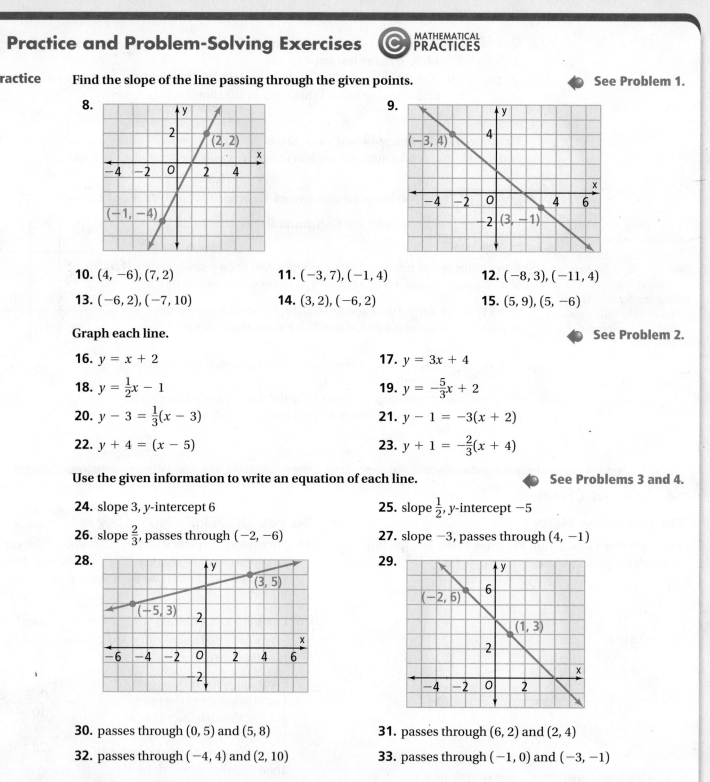

8.

9.

10. $(4, -6), (7, 2)$

11. $(-3, 7), (-1, 4)$

12. $(-8, 3), (-11, 4)$

13. $(-6, 2), (-7, 10)$

14. $(3, 2), (-6, 2)$

15. $(5, 9), (5, -6)$

Graph each line.

◀ See Problem 2.

16. $y = x + 2$

17. $y = 3x + 4$

18. $y = \frac{1}{2}x - 1$

19. $y = -\frac{5}{3}x + 2$

20. $y - 3 = \frac{1}{3}(x - 3)$

21. $y - 1 = -3(x + 2)$

22. $y + 4 = (x - 5)$

23. $y + 1 = -\frac{2}{3}(x + 4)$

Use the given information to write an equation of each line.

◀ See Problems 3 and 4.

24. slope 3, y-intercept 6

25. slope $\frac{1}{2}$, y-intercept -5

26. slope $\frac{2}{3}$, passes through $(-2, -6)$

27. slope -3, passes through $(4, -1)$

28.

29.

30. passes through $(0, 5)$ and $(5, 8)$

31. passes through $(6, 2)$ and $(2, 4)$

32. passes through $(-4, 4)$ and $(2, 10)$

33. passes through $(-1, 0)$ and $(-3, -1)$

Write the equation of the horizontal and vertical lines though the given point.

◀ See Problem 5.

34. $(4, 7)$

35. $(3, -2)$

36. $(0, -1)$

37. $(6, 4)$

Graph each line.

38. $x = 3$ **39.** $y = -2$ **40.** $x = 9$ **41.** $y = 4$

42. Open-Ended Write equations for three lines that contain the point (5, 6).

43. Think About a Plan You want to construct a "funbox" at a local skate park. The skate park's safety regulations allow for the ramp on the funbox to have a maximum slope of $\frac{4}{11}$. If you use the funbox plan at the right, can you build the ramp to meet the safety regulations? Explain.
- What information do you have that you can use to find the slope?
- How can you compare slopes?

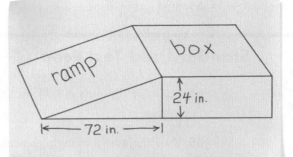

Write each equation in slope-intercept form.

44. $y - 5 = 2(x + 2)$ **45.** $y + 2 = -(x - 4)$ **46.** $-5x + y = 2$ **47.** $3x + 2y = 10$

STEM 48. Science The equation $P = -\frac{1}{33}d + 1$ represents the pressure P in atmospheres a scuba diver feels d feet below the surface of the water.
- **a.** What is the slope of the line?
- **b.** What does the slope represent in this situation?
- **c.** What is the y-intercept (P-intercept)?
- **d.** What does the y-intercept represent in this situation?

Graph each pair of lines. Then find their point of intersection.

49. $y = -4, x = 6$ **50.** $x = 0, y = 0$ **51.** $x = -1, y = 3$ **52.** $y = 5, x = 4$

STEM 53. Accessibility By law, the maximum slope of an access ramp in new construction is $\frac{1}{12}$. The plan for the new library shows a 3-ft height from the ground to the main entrance. The distance from the sidewalk to the building is 10 ft. If you assume the ramp does not have any turns, can you design a ramp that complies with the law? Explain.

54. a. What is the slope of the x-axis? Explain.
     b. Write an equation for the x-axis.

55. a. What is the slope of the y-axis? Explain.
     b. Write an equation for the y-axis.

56. Reasoning The x-intercept of a line is 2 and the y-intercept is 4. Use this information to write an equation for the line.

57. Coordinate Geometry The vertices of a triangle are $A(0, 0)$, $B(2, 5)$, and $C(4, 0)$.
- **a.** Write an equation for the line through A and B.
- **b.** Write an equation for the line through B and C.
- **c.** Compare the slopes and the y-intercepts of the two lines.

Do the three points lie on one line? Justify your answer.

58. $(5, 6), (3, 2), (6, 8)$ **59.** $(-2, -2), (4, -4), (0, 0)$ **60.** $(5, -4), (2, 3), (-1, 10)$

Find the value of a such that the graph of the equation has the given slope.

61. $y = \frac{2}{9}ax + 6; m = 2$ **62.** $y = -3ax - 4; m = \frac{1}{2}$ **63.** $y = -4ax - 10; m = -\frac{2}{3}$

Standardized Test Prep

64. $\overline{AB}$ has endpoints $A(8, k)$ and $B(7, -3)$. The slope of $\overline{AB}$ is 5. What is k?

    Ⓐ 1   Ⓑ 2   Ⓒ 5   Ⓓ 8

65. Two angles of a triangle measure 68 and 54. What is the measure of the third angle?

    Ⓕ 14   Ⓖ 58   Ⓗ 122   Ⓘ 180

66. Which of the following CANNOT be true?

    Ⓐ plane $ABCD \parallel$ plane $EFGH$

    Ⓑ Planes $ABCD$ and $CDHG$ intersect in $\overleftrightarrow{CD}$.

    Ⓒ $ABCD$ and ABC represent the same plane.

    Ⓓ plane $ADHE \parallel$ plane DCG

67. The length of a rectangle is $(x - 2)$ inches and the width is $5x$ inches. Which expression represents the perimeter of the rectangle in inches?

    Ⓕ $6x - 2$ Ⓖ $12x - 4$ Ⓗ $5x^2 - 10x$ Ⓘ $10x^2 - 20x$

68. One of the angles in a certain linear pair is acute. Your friend says the other angle must be obtuse. Is your friend's conjecture reasonable? Explain.

Mixed Review

For Exercises 69 and 70, construct the geometric figure.   ◀ See Lesson 3-6.

69. a rectangle with a length twice its width   **70.** a square

Name the property that justifies each statement.   ◀ See Lesson 2-5.

71. $4(2a - 3) = 8a - 12$ **72.** If $b + c = 7$ and $b = 2$, then $2 + c = 7$.

73. $\overline{RS} \cong \overline{RS}$ **74.** If $\angle 1 \cong \angle 4$, then $\angle 4 \cong \angle 1$.

Get Ready!   **To prepare for Lesson 3-8, do Exercises 75–77.**

Find the slope of the line passing through the given points.   ◀ See Lesson 3-7.

75. $(2, 5), (-2, 3)$ **76.** $(0, -5), (2, 0)$ **77.** $(1, 1), (2, -4)$

Slopes of Parallel and Perpendicular Lines

© Content Standard
G.GPE.5 Prove the slope criteria for parallel and perpendicular lines and use them to solve geometric problems.

Objective To relate slope to parallel and perpendicular lines

Think about what the slope of a line means in the context of this situation.

SOLVE IT!

Getting Ready!

You and a friend enjoy exercising together. One day, you are about to go running when your friend receives a phone call. You decide to start running and tell your friend to catch up after the call. The red line represents you and the blue line represents your friend. Will your friend catch up? Explain.

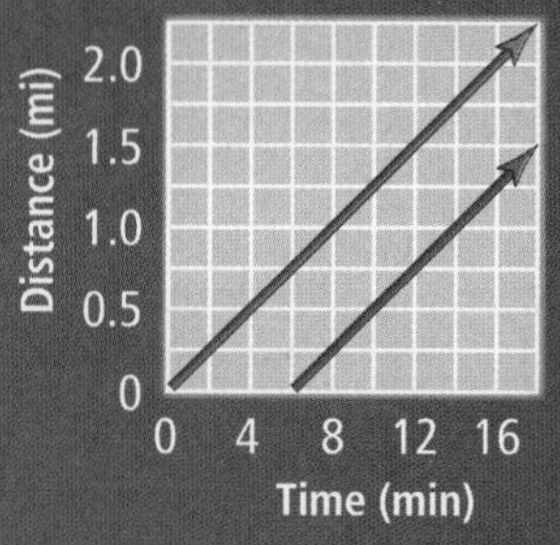

In the Solve It, slope represents the running rate, or speed. According to the graph, you and your friend run at the same speed, so the slopes of the lines are the same. In this lesson, you will learn how to use slopes to determine how two lines relate graphically to each other.

Essential Understanding You can determine whether two lines are parallel or perpendicular by comparing their slopes.

When two lines are parallel, their slopes are the same.

---

take note

Key Concept Slopes of Parallel Lines

- If two nonvertical lines are parallel, then their slopes are equal.
- If the slopes of two distinct nonvertical lines are equal, then the lines are parallel.
- Any two vertical lines or horizontal lines are parallel.

 **Problem 1** **Checking for Parallel Lines**

Are lines ℓ_1 and ℓ_2 parallel? Explain.

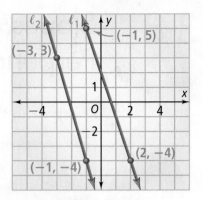

Step 1 Find the slope of each line.

$$\text{slope of } \ell_1 = \frac{5 - (-4)}{-1 - 2} = \frac{9}{-3} = -3$$

$$\text{slope of } \ell_2 = \frac{3 - (-4)}{-3 - (-1)} = \frac{7}{-2} = -\frac{7}{2}$$

Step 2 Compare the slopes.

Since $-3 \neq -\frac{7}{2}$, ℓ_1 and ℓ_2 are not parallel.

Got It? **1.** Line ℓ_3 contains $A(-13, 6)$ and $B(-1, 2)$. Line ℓ_4 contains $C(3, 6)$ and $D(6, 7)$. Are ℓ_3 and ℓ_4 parallel? Explain.

Plan

How does the given line help you?
Parallel lines have the same slope. Once you know the slope of the given line, you know the slope you need to write an equation.

 **Problem 2** **Writing Equations of Parallel Lines**

What is an equation of the line parallel to $y = -3x - 5$ that contains $(-1, 8)$?

Think

Identify the slope of the given line.

Write

$y = -3x - 5$

You now know the slope of the new line and that it passes through $(-1, 8)$. Use point-slope form to write the equation.

$y - y_1 = m(x - x_1)$

Substitute -3 for m and $(-1, 8)$ for (x_1, y_1) and simplify.

$y - 8 = -3(x - (-1))$
$y - 8 = -3(x + 1)$

Got It? **2.** What is an equation of the line parallel to $y = -x - 7$ that contains $(-5, 3)$?

When two lines are perpendicular, the product of their slopes is -1. Numbers with product -1 are opposite reciprocals. This proof will be presented in more detail in Chapter 7.

take note

Key Concept **Slopes of Perpendicular Lines**

- If two nonvertical lines are perpendicular, then the product of their slopes is -1.
- If the slopes of two lines have a product of -1, then the lines are perpendicular.
- Any horizontal line and vertical line are perpendicular.

Plan

Can you tell from the diagram whether the lines are perpendicular?
No. You can tell that the lines intersect, but not necessarily at right angles. So, you need to compare their slopes.

ⓒ Problem 3 **Checking for Perpendicular Lines**

Lines ℓ_1 and ℓ_2 are neither horizontal nor vertical. Are they perpendicular? Explain.

Step 1   Find the slope of each line.

$$m_1 = \text{slope of } \ell_1 = \frac{2-(-4)}{-4-0} = \frac{6}{-4} = -\frac{3}{2}$$

$$m_2 = \text{slope of } \ell_2 = \frac{3-(-3)}{4-(-5)} = \frac{6}{9} = \frac{2}{3}$$

Step 2   Find the product of the slopes.

$$m_1 \cdot m_2 = -\frac{3}{2} \cdot \frac{2}{3} = -1$$

Lines ℓ_1 and ℓ_2 are perpendicular because the product of their slopes is -1.

✓ Got It? 3. Line ℓ_3 contains $A(2, 7)$ and $B(3, -1)$. Line ℓ_4 contains $C(-2, 6)$ and $D(8, 7)$. Are ℓ_3 and ℓ_4 perpendicular? Explain.

Think

How is this similar to writing equations of parallel lines?
You follow the same process. The only difference here is that the slopes of perpendicular lines have product -1.

ⓒ Problem 4 **Writing Equations of Perpendicular Lines**

What is an equation of the line perpendicular to $y = \frac{1}{5}x + 2$ that contains $(15, -4)$?

Step 1   Identify the slope of the given line.

$$y = \frac{1}{5}x + 2$$
↑
slope

Step 2   Find the slope of the line perpendicular to the given line.

$m_1 \cdot m_2 = -1$ The product of the slopes of ⊥ lines is -1.

$\frac{1}{5} \cdot m_2 = -1$ Substitute $\frac{1}{5}$ for m_1.

$m_2 = -5$ Multiply each side by 5.

Step 3   Use point-slope form to write an equation of the new line.

$y - y_1 = m(x - x_1)$

$y - (-4) = -5(x - 15)$ Substitute -5 for m and $(15, -4)$ for (x_1, y_1).

$y + 4 = -5(x - 15)$ Simplify.

✓ Got It? 4. What is an equation of the line perpendicular to $y = -3x - 5$ that contains $(-3, 7)$?

Sports The baseball field below is on a coordinate grid with home plate at the origin. A batter hits a ground ball along the line shown. The player at (110, 70) runs along a path perpendicular to the path of the baseball. What is an equation of the line on which the player runs?

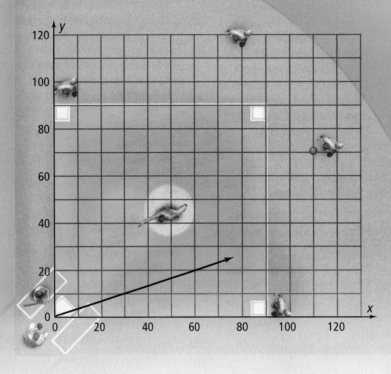

Step 1 Find the slope of the baseball's path.

$$m_1 = \frac{y_2 - y_1}{x_2 - x_1} = \frac{20 - 10}{60 - 30} = \frac{10}{30} = \frac{1}{3}$$

Points (30, 10) and (60, 20) are on the baseball's path.

Step 2 Find the slope of a line perpendicular to the baseball's path.

$m_1 \cdot m_2 = -1$ The product of the slopes of ⊥ lines is −1.

$\frac{1}{3} \cdot m_2 = -1$ Substitute $\frac{1}{3}$ for m_1.

$m_2 = -3$ Multiply each side by 3.

Think

Which linear equation form should you use?
You know the slope. The player is located at a point on the line. Use point-slope form.

Step 3 Write an equation of the line on which the player runs.
The slope is −3 and a point on the line is (110, 70).

$y - y_1 = m(x - x_1)$ Point-slope form

$y - 70 = -3(x - 110)$ Substitute −3 for m and (110, 70) for (x_1, y_1).

✓ **Got It?** **5.** Suppose a second player standing at (90, 40) misses the ball, turns around, and runs on a path parallel to the baseball's path. What is an equation of the line representing this player's path?

Lesson Check

Do you know HOW?

$\overleftrightarrow{AB}$ contains points A and B. $\overleftrightarrow{CD}$ contains points C and D. Are $\overleftrightarrow{AB}$ and $\overleftrightarrow{CD}$ *parallel*, *perpendicular*, or *neither*? Explain.

1. $A(-8, 3)$, $B(-4, 11)$, $C(-1, 3)$, $D(1, 2)$

2. $A(3, 5)$, $B(2, -1)$, $C(7, -2)$, $D(10, 16)$

3. $A(3, 1)$, $B(4, 1)$, $C(5, 9)$, $D(2, 6)$

4. What is an equation of the line perpendicular to $y = -4x + 1$ that contains $(2, -3)$?

Do you UNDERSTAND?

5. **Error Analysis** Your classmate tries to find an equation for a line parallel to $y = 3x - 5$ that contains $(-4, 2)$. What is your classmate's error?

> slope of given line = 3
> slope of parallel line = $\frac{1}{3}$
> $y - y_1 = m(x - x_1)$
> $y - 2 = \frac{1}{3}(x + 4)$

6. **Compare and Contrast** What are the differences between the equations of parallel lines and the equations of perpendicular lines? Explain.

Practice and Problem-Solving Exercises · MATHEMATICAL PRACTICES

A Practice For Exercises 7–10, are lines ℓ_1 and ℓ_2 parallel? Explain. ◀ See Problem 1.

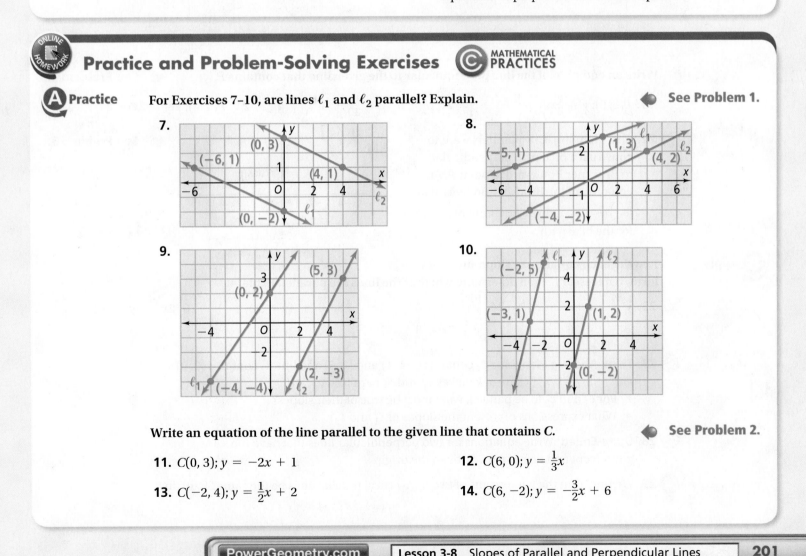

Write an equation of the line parallel to the given line that contains C. ◀ See Problem 2.

11. $C(0, 3)$; $y = -2x + 1$

12. $C(6, 0)$; $y = \frac{1}{3}x$

13. $C(-2, 4)$; $y = \frac{1}{2}x + 2$

14. $C(6, -2)$; $y = -\frac{3}{2}x + 6$

For Exercises 15–18, are lines ℓ_1 and ℓ_2 perpendicular? Explain. ⬅ **See Problem 3.**

15.

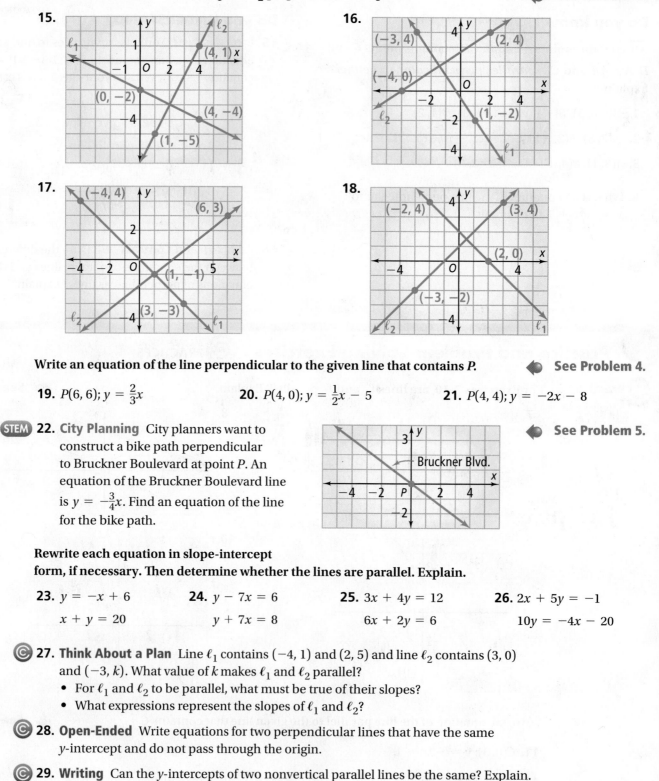

16.

17.

18.

Write an equation of the line perpendicular to the given line that contains P. ⬅ **See Problem 4.**

19. $P(6, 6)$; $y = \frac{2}{3}x$ **20.** $P(4, 0)$; $y = \frac{1}{2}x - 5$ **21.** $P(4, 4)$; $y = -2x - 8$

STEM **22. City Planning** City planners want to construct a bike path perpendicular to Bruckner Boulevard at point P. An equation of the Bruckner Boulevard line is $y = -\frac{3}{4}x$. Find an equation of the line for the bike path. ⬅ **See Problem 5.**

B Apply

Rewrite each equation in slope-intercept form, if necessary. Then determine whether the lines are parallel. Explain.

23. $y = -x + 6$ **24.** $y - 7x = 6$ **25.** $3x + 4y = 12$ **26.** $2x + 5y = -1$
     $x + y = 20$ $y + 7x = 8$ $6x + 2y = 6$ $10y = -4x - 20$

© **27. Think About a Plan** Line ℓ_1 contains $(-4, 1)$ and $(2, 5)$ and line ℓ_2 contains $(3, 0)$ and $(-3, k)$. What value of k makes ℓ_1 and ℓ_2 parallel?
- For ℓ_1 and ℓ_2 to be parallel, what must be true of their slopes?
- What expressions represent the slopes of ℓ_1 and ℓ_2?

© **28. Open-Ended** Write equations for two perpendicular lines that have the same y-intercept and do not pass through the origin.

© **29. Writing** Can the y-intercepts of two nonvertical parallel lines be the same? Explain.

Use slopes to determine whether the opposite sides of quadrilateral *ABCD* are parallel.

30. $A(0, 2)$, $B(3, 4)$, $C(2, 7)$, $D(-1, 5)$

31. $A(-3, 1)$, $B(1, -2)$, $C(0, -3)$, $D(-4, 0)$

32. $A(1, 1)$, $B(5, 3)$, $C(7, 1)$, $D(3, 0)$

33. $A(1, 0)$, $B(4, 0)$, $C(3, -3)$, $D(-1, -3)$

34. Reasoning Are opposite sides of hexagon *RSTUVW* at the right parallel? Justify your answer.

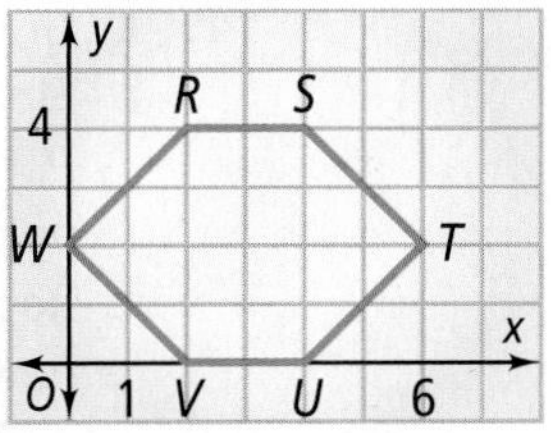

35. Which line is perpendicular to $3y + 2x = 12$?

Ⓐ $6x - 4y = 24$

Ⓒ $2x + 3y = 6$

Ⓑ $y + 3x = -2$

Ⓓ $y = -2x + 6$

Rewrite each equation in slope-intercept form, if necessary. Then determine whether the lines are perpendicular. Explain.

36. $y = -x - 7$

$y - x = 20$

37. $y = 3$

$x = -2$

38. $2x - 7y = -42$

$4y = -7x - 2$

Developing Proof **Explain why each theorem is true for three lines in the coordinate plane.**

39. Theorem 3-7: If two lines are parallel to the same line, then they are parallel to each other.

40. Theorem 3-8: In a plane, if two lines are perpendicular to the same line, then they are parallel to each other.

41. Rail Trail A community recently converted an old railroad corridor into a recreational trail. The graph at the right shows a map of the trail on a coordinate grid. They plan to construct a path to connect the trail to a parking lot. The new path will be perpendicular to the recreational trail.

a. Write an equation of the line representing the new path.

b. What are the coordinates of the point at which the path will meet the recreational trail?

c. If each grid space is 25 yd by 25 yd, how long is the path to the nearest yard?

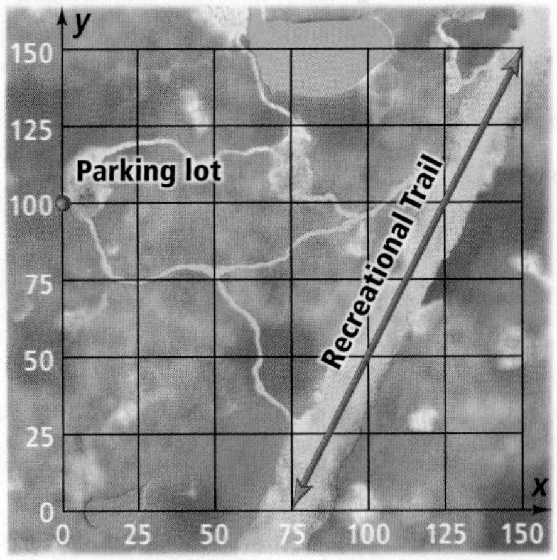

42. Reasoning Is a triangle with vertices $G(3, 2)$, $H(8, 5)$, and $K(0, 10)$ a right triangle? Justify your answer.

43. Graphing Calculator $\overleftrightarrow{AB}$ contains points $A(-3, 2)$ and $B(5, 1)$. $\overleftrightarrow{CD}$ contains points $C(2, 7)$ and $D(1, -1)$. Use your graphing calculator to find the slope of $\overleftrightarrow{AB}$. Enter the *x*-coordinates of *A* and *B* into the **L₁** list of your list editor. Enter the *y*-coordinates into the **L₂** list. In your **stat** **CALC** menu select **LinReg (*ax* + *b*)**. Press **enter** to find the slope *a*. Repeat to find the slope of $\overleftrightarrow{CD}$. Are $\overleftrightarrow{AB}$ and $\overleftrightarrow{CD}$ *parallel*, *perpendicular*, or *neither*?

Challenge

For Exercises 44 and 45, use the graph at the right.

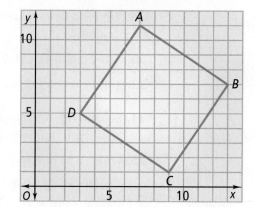

44. Show that the diagonals of the figure are congruent.

45. Show that the diagonals of the figure are perpendicular bisectors of each other.

46. a. Graph the points $P(2, 2)$, $Q(7, 4)$, and $R(3, 5)$.
   b. Find the coordinates of a point S that, along with points P, Q, and R, will form the vertices of a quadrilateral with opposite sides parallel. Graph the quadrilateral.
   c. Repeat part (b) to find a different point S. Graph the new quadrilateral.

47. Algebra A triangle has vertices $L(-5, 6)$, $M(-2, -3)$, and $N(4, 5)$. Write an equation for the line perpendicular to $\overline{LM}$ that contains point N.

Standardized Test Prep

GRIDDED RESPONSE

SAT/ACT

48. $\triangle ABC$ is right with right angle C. The slope of $\overline{AC}$ is -2. What is the slope of $\overline{BC}$?

49. In the diagram at the right, M is the midpoint of $\overline{AB}$. What is AB?

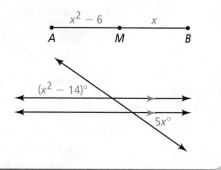

50. What is the distance between $(-4.5, 1.2)$ and $(3.5, -2.8)$ to the nearest tenth?

51. What is the value of x in the diagram at the right?

52. The perimeter of a square is 20 ft. What is the area of the square in square feet?

Mixed Review

Algebra Write an equation for the line containing the given points.

◀ **See Lesson 3-7.**

53. $A(0, 3)$, $B(6, 0)$ **54.** $C(-4, 2)$, $D(-1, 7)$ **55.** $E(3, -2)$, $F(-5, -8)$

Name the property that justifies each statement.

◀ **See Lesson 2-5.**

56. $\angle 4 \cong \angle 4$ **57.** If $m\angle B = 8$, then $2m\angle B = 16$.

58. $-3x + 6 = 3(-x + 2)$ **59.** If $\overline{RS} \cong \overline{MN}$, then $\overline{MN} \cong \overline{RS}$.

Get Ready! **To prepare for Lesson 4-1, do Exercises 60–62.**

Are $\angle 1$ and $\angle 2$ congruent? Explain.

◀ **See Lessons 1-5 and 2-6.**

60. **61.** **62.**

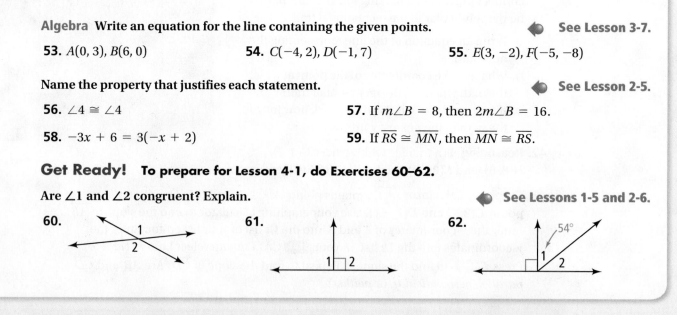

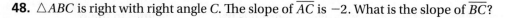

Pull It **All** Together © ASSESSMENT

To solve these problems, you will pull together many concepts and skills that you have studied about parallel lines.

BIG idea Reasoning and Proof

You can prove that lines are parallel if you know that certain pairs of angles formed by the lines and a transversal are congruent.

© Performance Task 1

You want to put tape on the ground to mark the lines for a volleyball court. What is the most efficient way to make sure that the opposite sides of the court are parallel? Support your answer with a diagram.

BIG idea Reasoning and Proof

You can prove that lines are parallel if you know that certain pairs of angles formed by the lines and a transversal are congruent.

BIG idea Measurement

You can find missing angle measures in triangles by using the fact that the sum of the measures of the angles of a triangle is 180.

© Performance Task 2

In the diagram below, $a \parallel b$. For lines p and q to be parallel, what is $m\angle 4$? Explain.

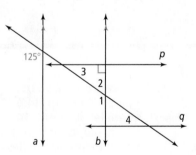

BIG idea Coordinate Geometry

You can write the equation of a line by using its slope and y-intercept.

© Performance Task 3

$\overleftrightarrow{AB}$ contains points $A(-6, -1)$ and $B(1, 4)$. $\overleftrightarrow{CD}$ contains point $D(7, 2)$. If $\angle ABC \cong \angle BCD$ and $m\angle ABC = 90$, what is an equation of $\overleftrightarrow{CD}$? Show your work.

Connecting BIG ideas and Answering the Essential Questions

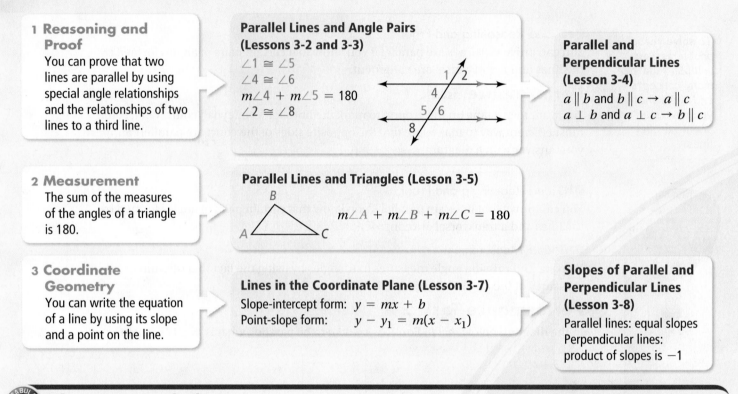

1 Reasoning and Proof
You can prove that two lines are parallel by using special angle relationships and the relationships of two lines to a third line.

Parallel Lines and Angle Pairs (Lessons 3-2 and 3-3)
$\angle 1 \cong \angle 5$
$\angle 4 \cong \angle 6$
$m\angle 4 + m\angle 5 = 180$
$\angle 2 \cong \angle 8$

Parallel and Perpendicular Lines (Lesson 3-4)
$a \parallel b$ and $b \parallel c \rightarrow a \parallel c$
$a \perp b$ and $a \perp c \rightarrow b \parallel c$

2 Measurement
The sum of the measures of the angles of a triangle is 180.

Parallel Lines and Triangles (Lesson 3-5)
$m\angle A + m\angle B + m\angle C = 180$

3 Coordinate Geometry
You can write the equation of a line by using its slope and a point on the line.

Lines in the Coordinate Plane (Lesson 3-7)
Slope-intercept form: $y = mx + b$
Point-slope form: $y - y_1 = m(x - x_1)$

Slopes of Parallel and Perpendicular Lines (Lesson 3-8)
Parallel lines: equal slopes
Perpendicular lines: product of slopes is -1

Chapter Vocabulary

- alternate exterior angles (p. 142)
- alternate interior angles (p. 142)
- auxiliary line (p. 172)
- corresponding angles (p. 142)
- exterior angle of a polygon (p. 173)
- flow proof (p. 158)
- parallel lines (p. 140)
- parallel planes (p. 140)
- point-slope form (p. 190)
- remote interior angles (p. 173)
- same-side interior angles (p. 142)
- skew lines (p. 140)
- slope (p. 189)
- slope-intercept form (p. 190)
- transversal (p. 141)

Choose the correct term to complete each sentence.

1. A(n) __?__ intersects two or more coplanar lines at distinct points.

2. The measure of a(n) __?__ of a triangle is equal to the sum of the measures of its two remote interior angles.

3. The linear equation $y - 3 = 4(x + 5)$ is in __?__ form.

4. When two coplanar lines are cut by a transversal, the angles formed between the two lines and on opposite sides of the transversal are __?__.

5. Noncoplanar lines that do not intersect are __?__.

6. The linear equation $y = 3x - 5$ is in __?__ form.

3-1 Lines and Angles

Quick Review

A **transversal** is a line that intersects two or more coplanar lines at distinct points.

∠1 and ∠3 are **corresponding angles.**

∠2 and ∠6 are **alternate interior angles.**

∠2 and ∠3 are **same-side interior angles.**

∠4 and ∠8 are **alternate exterior angles.**

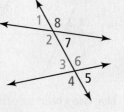

Example

Name two other pairs of corresponding angles in the diagram above.

∠5 and ∠7
∠2 and ∠4

Exercises

Identify all numbered angle pairs that form the given type of angle pair. Then name the two lines and transversal that form each pair.

7. alternate interior angles

8. same-side interior angles

9. corresponding angles

10. alternate exterior angles

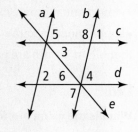

Classify the angle pair formed by ∠1 and ∠2.

11.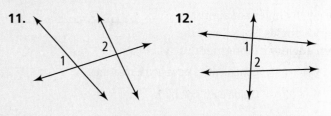
12.

3-2 Properties of Parallel Lines

Quick Review

If two parallel lines are cut by a transversal, then

- corresponding angles, alternate interior angles, and alternate exterior angles are congruent
- same-side interior angles are supplementary

Example

Which other angles measure 110?

∠6 (corresponding angles)

∠3 (alternate interior angles)

∠8 (vertical angles)

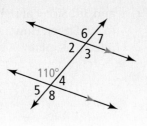

Exercises

Find *m*∠1 and *m*∠2. Justify your answers.

13.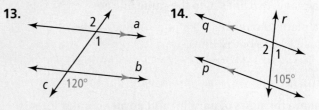
14.

15. Find the values of *x* and *y* in the diagram below.

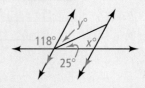

3-3 Proving Lines Parallel

Quick Review

If two lines and a transversal form
- congruent corresponding angles,
- congruent alternate interior angles,
- congruent alternate exterior angles, or
- supplementary same-side interior angles,

then the two lines are parallel.

Example

What is the value of x for which $\ell \parallel m$?

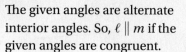

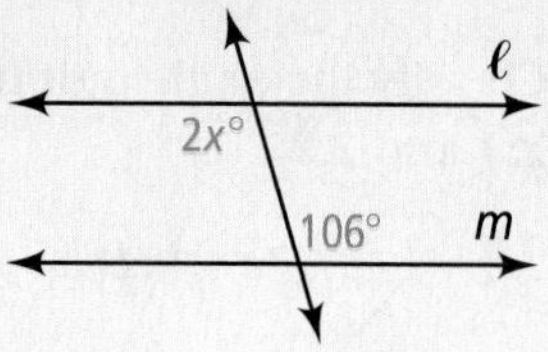

The given angles are alternate
interior angles. So, $\ell \parallel m$ if the
given angles are congruent.

$2x = 106$ Congruent ∡ have equal measures.

$x = 53$ Divide each side by 2.

Exercises

Find the value of x for which $\ell \parallel m$.

16.
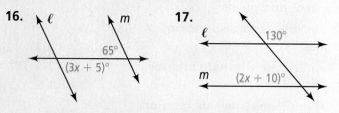
65°
$(3x + 5)°$

17.
ℓ 130°
m $(2x + 10)°$

Use the given information to decide which lines, if any, are parallel. Justify your conclusion.

18. $\angle 1 \cong \angle 9$

19. $m\angle 3 + m\angle 6 = 180$

20. $m\angle 2 + m\angle 3 = 180$

21. $\angle 5 \cong \angle 11$

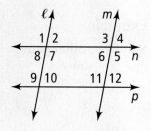

3-4 Parallel and Perpendicular Lines

Quick Review

- Two lines $\parallel$ to the same line are $\parallel$ to each other.
- In a plane, two lines $\perp$ to the same line are $\parallel$.
- In a plane, if one line is $\perp$ to one of two $\parallel$ lines, then it is $\perp$ to both $\parallel$ lines.

Example

What are the pairs of parallel and perpendicular lines in the diagram?

$\ell \parallel n$, $\ell \parallel m$, and $m \parallel n$.

$a \perp \ell$, $a \perp m$, and $a \perp n$.

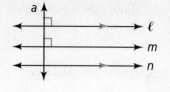

Exercises

Use the diagram at the right to complete each statement.

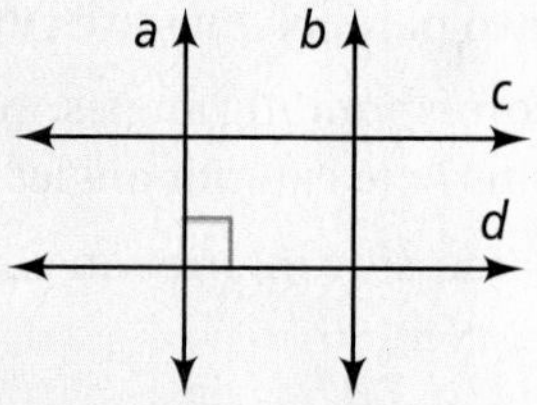

22. If $b \perp c$ and $b \perp d$, then
    $c \underline{\ ?\ } d$.

23. If $c \parallel d$, then $\underline{\ ?\ } \perp c$.

24. **Maps** Morris Avenue intersects both 1st Street
    and 3rd Street at right angles. 3rd Street is parallel
    to 5th Street. How are 1st Street and 5th Street
    related? Explain.

3-5 Parallel Lines and Triangles

Quick Review

The sum of the measures of the angles of a triangle is 180.

The measure of each **exterior angle** of a triangle equals the sum of the measures of its two **remote interior angles.**

Example

What are the values of x and y?

$x + 50 = 125$	Exterior Angle Theorem
$x = 75$	Simplify.

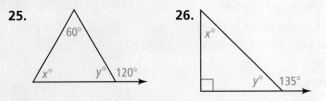

$x + y + 50 = 180$	Triangle Angle-Sum Theorem
$75 + y + 50 = 180$	Substitute 75 for x.
$y = 55$	Simplify.

Exercises

Find the values of the variables.

25.

26.

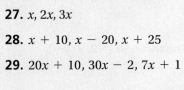

The measures of the three angles of a triangle are given. Find the value of x.

27. $x, 2x, 3x$

28. $x + 10, x - 20, x + 25$

29. $20x + 10, 30x - 2, 7x + 1$

3-6 Constructing Parallel and Perpendicular Lines

Quick Review

You can use a compass and a straightedge to construct

- a line parallel to a given line through a point not on the line
- a line perpendicular to a given line through a point on the line, or through a point not on the line

Example

Which step of the parallel lines construction guarantees the lines are parallel?

The parallel lines construction involves constructing a pair of congruent angles. Since the congruent angles are corresponding angles, the lines are parallel.

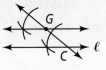

Exercises

30. Draw a line m and point Q not on m. Construct a line perpendicular to m through Q.

Use the segments below.

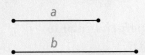

31. Construct a rectangle with side lengths a and b.

32. Construct a rectangle with side lengths a and $2b$.

33. Construct a quadrilateral with one pair of parallel opposite sides, each side of length $2a$.

3-7 Equations of Lines in the Coordinate Plane

Quick Review

Slope-intercept form is $y = mx + b$, where m is the slope and b is the y-intercept.

Point-slope form is $y - y_1 = m(x - x_1)$, where m is the slope and (x_1, y_1) is a point on the line.

Example

What is an equation of the line with slope -5 and y-intercept 6?

Use slope-intercept form: $y = -5x + 6$.

Example

What is an equation of the line through $(-2, 8)$ with slope 3?

Use point-slope form: $y - 8 = 3(x + 2)$.

Exercises

Find the slope of the line passing through the points.

34. $(6, -2), (1, 3)$ **35.** $(-7, 2), (-7, -5)$

36. Name the slope and y-intercept of $y = 2x - 1$. Then graph the line.

37. Name the slope of and a point on $y - 3 = -2(x + 5)$. Then graph the line.

Write an equation of the line.

38. slope $-\frac{1}{2}$, y-intercept 12

39. slope 3, passes through $(1, -9)$

40. passes through $(4, 2)$ and $(3, -2)$

3-8 Slopes of Parallel and Perpendicular Lines

Quick Review

Parallel lines have the same slopes.

The product of the slopes of two perpendicular lines is -1.

Example

What is an equation of the line perpendicular to $y = 2x - 5$ that contains $(1, -3)$?

Step 1 Identify the slope of $y = 2x - 5$.

The slope of the given line is 2.

Step 2 Find the slope of a line perpendicular to $y = 2x - 5$.

The slope is $-\frac{1}{2}$, because $2\left(-\frac{1}{2}\right) = -1$.

Step 3 Use point-slope form to write $y + 3 = -\frac{1}{2}(x - 1)$.

Exercises

Determine whether $\overleftrightarrow{AB}$ and $\overleftrightarrow{CD}$ are *parallel*, *perpendicular*, or *neither*.

41. $A(-1, -4), B(2, 11), C(1, 1), D(4, 10)$

42. $A(2, 8), B(-1, -2), C(3, 7), D(0, -3)$

43. $A(-3, 3), B(0, 2), C(1, 3), D(-2, -6)$

44. $A(-1, 3), B(4, 8), C(-6, 0), D(2, 8)$

45. Write an equation of the line parallel to $y = 8x - 1$ that contains $(-6, 2)$.

46. Write an equation of the line perpendicular to $y = \frac{1}{6}x + 4$ that contains $(3, -3)$.

Do you know HOW?

Find the measure of the third angle of a triangle given the measures of two angles.

1. 57 and 101

2. 72 and 72

3. x and 20

Find $m\angle 1$ and $m\angle 2$. Justify each answer.

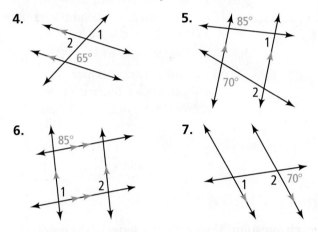

4.

5.

6.

7.

8. Draw a line m and a point T not on the line. Construct the line through T perpendicular to m.

9. Draw any $\angle ABC$. Then construct line m through A so that $m \parallel \overleftrightarrow{BC}$.

10. The measures of the angles of a triangle are $2x$, $x + 24$, and $x - 4$. Find the value of x. Then find the measures of the angles.

Determine whether the following are *parallel lines*, *skew lines*, or *neither*.

11. opposite sides of a rectangular picture frame

12. the center line of a soccer field and a sideline of the field

13. the path of an airplane flying north at 15,000 ft and the path of an airplane flying west at 10,000 ft

Use the given information to write an equation of each line.

14. slope -5, y-intercept -2

15. slope $\frac{1}{2}$, passes through $(4, -1)$

16. passes through $(1, 5)$ and $(3, 11)$

Algebra Find the value of x for which $\ell \parallel m$.

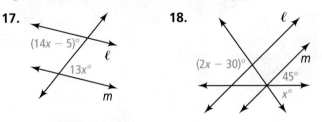

17.

18.

Graph each pair of lines. Tell whether they are *parallel*, *perpendicular*, or *neither*.

19. $y = 4x + 7$ and $y = -\frac{1}{4}x - 3$

20. $y = 3x - 4$ and $y = 3x + 1$

21. $y = x + 5$ and $y = -5x - 1$

Do you UNDERSTAND?

ⓒ **22. Developing Proof** Provide the reason for each step.

Given: $\ell \parallel m$, $\angle 2 \cong \angle 4$
Prove: $n \parallel p$

Statements	Reasons
1) $\ell \parallel m$	1) a. __?__
2) $\angle 1 \cong \angle 2$	2) b. __?__
3) $\angle 2 \cong \angle 4$	3) c. __?__
4) $\angle 1 \cong \angle 4$	4) d. __?__
5) $n \parallel p$	5) e. __?__

ⓒ **23. Reasoning** Suppose a line intersecting two planes A and B forms a right angle at exactly one point in each plane. What must be true about planes A and B? (*Hint:* Draw a picture.)

3 Cumulative Standards Review

ASSESSMENT

TIPS FOR SUCCESS

Some test questions ask you to analyze a diagram. Read the sample question at the right. Then follow the tips to answer it.

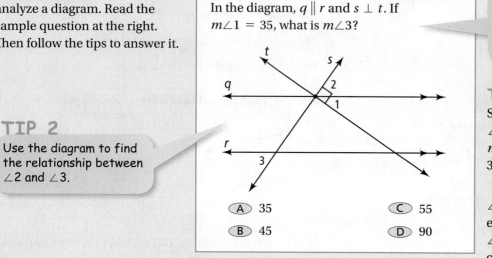

In the diagram, $q \parallel r$ and $s \perp t$. If $m\angle 1 = 35$, what is $m\angle 3$?

Ⓐ 35 Ⓒ 55

Ⓑ 45 Ⓓ 90

TIP 1

Use $m\angle 1$ and what you know about the measures of angles formed by perpendicular lines to find $m\angle 2$.

TIP 2

Use the diagram to find the relationship between $\angle 2$ and $\angle 3$.

Think It Through

Since $s \perp t$, the angle formed by $\angle 1$ and $\angle 2$ is a right angle. So $m\angle 1 + m\angle 2 = 90$. Substitute 35 for $m\angle 1$ and solve for $m\angle 2$.

$$m\angle 2 = 90 - 35 = 55$$

$\angle 2$ and $\angle 3$ are alternate exterior angles. Since $q \parallel r$, $\angle 2 \cong \angle 3$. So $m\angle 3 = 55$. The correct answer is C.

Vocabulary Builder

As you solve test items, you must understand the meanings of mathematical terms. Match each term with its mathematical meaning.

A. transversal

B. complementary angles

C. conditional

D. midpoint

E. supplementary angles

I. two angles whose measures have sum 90

II. a point that divides a segment into two congruent segments

III. an *if-then* statement

IV. a line that intersects two coplanar lines at two distinct points

V. two angles whose measures have sum 180

Multiple Choice

Read each question. Then write the letter of the correct answer on your paper.

1. Which expression describes the area of a square that has side lengths $7n^3$?

Ⓐ $14n^6$ Ⓒ $49n^6$

Ⓑ $14n^9$ Ⓓ $49n^9$

2. What is the area of $\triangle PQR$?

Ⓕ 10 units^2 Ⓗ 20 units^2

Ⓖ 15 units^2 Ⓘ 25 units^2

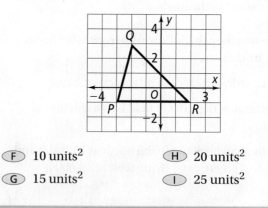

3. Which condition(s) will allow you to prove that $\ell \parallel m$?

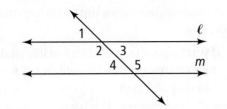

I. $\angle 1 \cong 4$

II. $\angle 2 \cong \angle 5$

III. $\angle 3 \cong \angle 4$

IV. $m\angle 2 + m\angle 4 = 180$

Ⓐ III only   Ⓒ II and III only

Ⓑ I and IV only   Ⓓ I, II, III, and IV

4. The length of your rectangular vegetable garden is 15 times its width. You used 160 ft of fencing to surround the garden. How much area do you have for planting?

Ⓕ 150 ft^2 Ⓗ 800 ft^2

Ⓖ 375 ft^2 Ⓘ 1600 ft^2

5. Which of the following angle relationships can you use to prove that two lines are parallel?

Ⓐ supplementary corresponding angles

Ⓑ congruent alternate interior angles

Ⓒ congruent vertical angles

Ⓓ congruent same-side interior angles

6. Which point lies farthest from the origin?

Ⓕ $(0, -7)$ Ⓗ $(-4, -3)$

Ⓖ $(-3, 8)$ Ⓘ $(5, 1)$

7. $\angle A$ and $\angle B$ are supplementary vertical angles. What is $m\angle B$?

Ⓐ 45   Ⓒ 135

Ⓑ 90   Ⓓ 180

8. Which types of angles can an obtuse triangle have?

I. a right angle   **II.** two acute angles

III. an obtuse angle   **IV.** two vertical angles

Ⓕ I and II   Ⓗ III and IV

Ⓖ II and III   Ⓘ I and IV

9. Given the diagram below, which expression could be used to find the sum of the angles in the triangle?

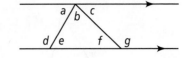

Ⓐ $f + g + c$

Ⓑ $a + b + c$

Ⓒ $a + b + e$

Ⓓ $d + e + g$

10. What is the value of x in the figure?

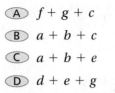

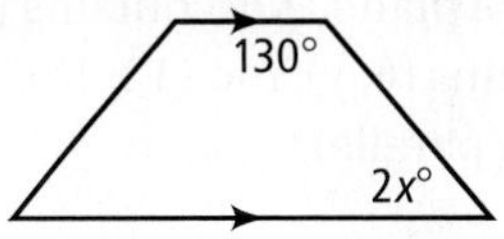

Ⓕ 20   Ⓗ 45

Ⓖ 25   Ⓘ 50

11. The net for a cylindrical container that holds a stack of DVDs is shown below. What is the total area of the net?

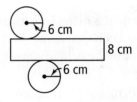

Ⓐ 226 cm^2 Ⓒ 528 cm^2

Ⓑ 302 cm^2 Ⓓ 582 cm^2

12. What are the coordinates of the midpoint of a segment with endpoints $(-1, 2)$ and $(5, 6)$?

Ⓕ $(2, 4)$ Ⓗ $(6, 4)$

Ⓖ $(4, 8)$ Ⓘ $(3, 4)$

13. What is the measure of any exterior angle of an equiangular triangle?

Ⓐ 30   Ⓒ 90

Ⓑ 60   Ⓓ 120

GRIDDED RESPONSE

14. What is the measure of ∠1?

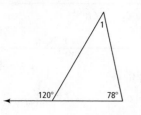

15. Two angles of an isosceles triangle have measures 54.5 and 71. What is the measure of the third angle?

16. In the coordinate plane, $\overleftrightarrow{AB}$ contains $(-2, -4)$ and $(6, 8)$. $\overleftrightarrow{CD}$ contains $(6, y)$ and $(12, 10)$. For what value of y are the lines parallel?

17. A circular wading pool has a diameter of 10 ft. What is the circumference of the wading pool in feet? Use 3.14 for π.

18. What is the measure of the complement of a 56° angle?

19. A new athletic field is being constructed, as shown below. The given coordinates are in terms of yards. What is the area of the field in square yards?

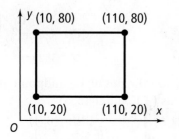

Short Response

20. Is your friend's argument for the following situation valid? Explain.

> **Given:** If you buy a one-year membership at the gym, then you get one month free. You got a free month at the gym.
>
> **Your friend's conclusion:** You bought a one-year membership.

21. Draw $\overline{MN}$. Then construct the perpendicular bisector of $\overline{MN}$.

22. What is the equation for a line that passes through the point $(3, -3)$ and is parallel to the line shown in the graph below?

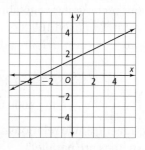

Extended Response

23. $\overline{CD}$ has endpoints $C(5, 7)$ and $D(10, -5)$. What are the coordinates of the midpoint of $\overline{CD}$? What is CD? Show your work.

24. Examples and nonexamples of *bleebles* are shown.

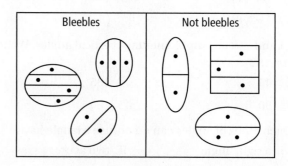

a. Is the figure at the right a *bleeble*? Explain your reasoning.
b. What is a definition for *bleeble*?

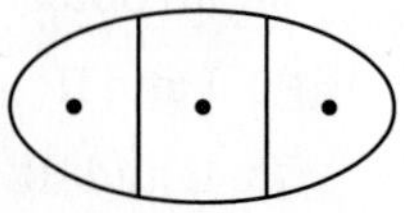

Get Ready!

Lesson 1-7 ◀◼ **The Distance Formula**

Find the side lengths of △ABC.

1. $A(3, 1), B(-1, 1), C(-1, -2)$

2. $A(-3, 2), B(-3, -6), C(8, 6)$

3. $A(-1, -2), B(6, 1), C(2, 5)$

Lesson 2-6 ◀◼ **Proving Angles Congruent**

Draw a conclusion based on the information given.

4. $\angle J$ is supplementary to $\angle K$;
$\angle L$ is supplementary to $\angle K$.

5. $\angle M$ is supplementary to $\angle N$;
$\angle M \cong \angle N$.

6. $\angle 1$ is complementary to $\angle 2$.

7. $\overrightarrow{FA} \perp \overrightarrow{FC}, \overrightarrow{FB} \perp \overrightarrow{FD}$

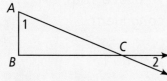

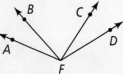

Lessons 3-2
and 3-5 ◀◼ **Parallel Lines and the Triangle Angle-Sum Theorem**

What can you conclude about the angles in each diagram?

8. **9.** **10.**

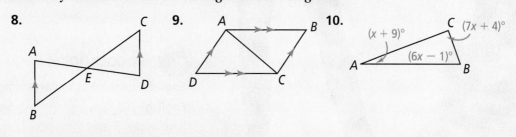

 Looking Ahead Vocabulary

11. The foundation of a building is the *base* of the building. How would you describe the *base of an isosceles triangle* in geometry?

12. The *legs* of a table support the tabletop and are equal in length. How might they be similar to the *legs of an isosceles triangle*?

13. A postal worker delivers each piece of mail to the mailbox that *corresponds* to the address on the envelope. What might the term *corresponding parts* of geometric figures mean?

Congruent Triangles

DOMAINS
- Congruence
- Mathematical Practice: Construct viable arguments
- Modeling with Geometry

Have you ever tried building a bridge out of dried spaghetti? The triangles in the bridge at the right are congruent—they have the same size and shape.

In this chapter, you'll learn how to prove that two triangles are congruent.

Vocabulary

English/Spanish Vocabulary Audio Online:

English	Spanish
base angles of an isosceles triangle, p. 250	ángulos de base de un triángulo isósceles
base of an isosceles triangle, p. 250	base de un triángulo isósceles
congruent polygons, p. 219	polígonos congruentes
corollary, p. 252	corolario
hypotenuse, p. 258	hipotenusa
legs of an isosceles triangle, p. 250	catetos de un triángulo isósceles
legs of a right triangle, p. 258	catetos de un triángulo rectángulo
vertex angle of an isosceles triangle, p. 250	ángulo en vértice de un triángulo isósceles

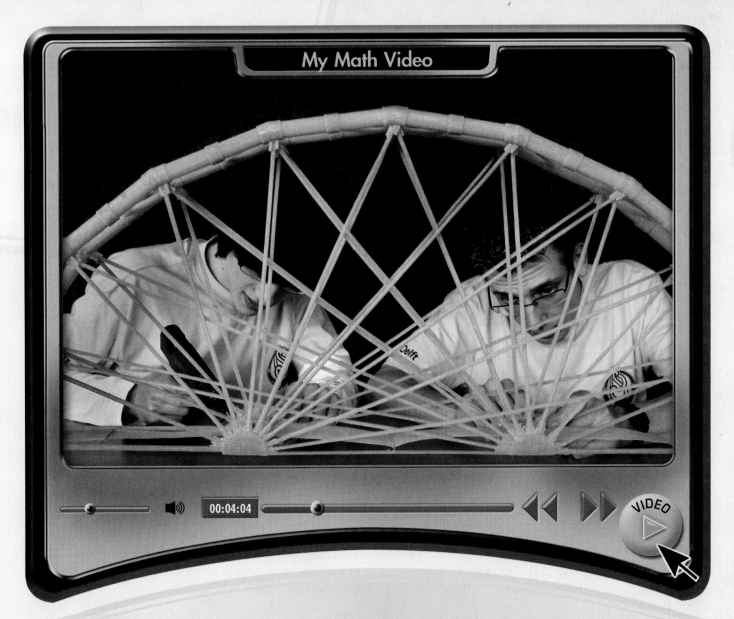

BIG ideas

1 Visualization
Essential Question How do you identify corresponding parts of congruent triangles?

2 Reasoning and Proof
Essential Question How do you show that two triangles are congruent?

3 Reasoning and Proof
Essential Question How can you tell whether a triangle is isosceles or equilateral?

Chapter Preview

4-1 Congruent Figures

© **Content Standard**
Prepares for G.SRT.5 Use congruence . . . criteria for triangles to solve problems and prove relationships in geometric figures.

Objective To recognize congruent figures and their corresponding parts

SOLVE IT!

Getting Ready!

You are working on a puzzle. You've almost finished, except for a few pieces of the sky. Place the remaining pieces in the puzzle. How did you figure out where to place the pieces?

Having trouble? How can tracing pieces 1, 2, and 3 help?

© MATHEMATICAL PRACTICES

Lesson Vocabulary
• congruent polygons

Congruent figures have the same size and shape. When two figures are congruent, you can slide, flip, or turn one so that it fits exactly on the other one, as shown below. In this lesson, you will learn how to determine if geometric figures are congruent.

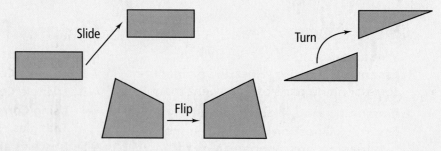

Slide

Flip

Turn

Essential Understanding You can determine whether two figures are congruent by comparing their corresponding parts.

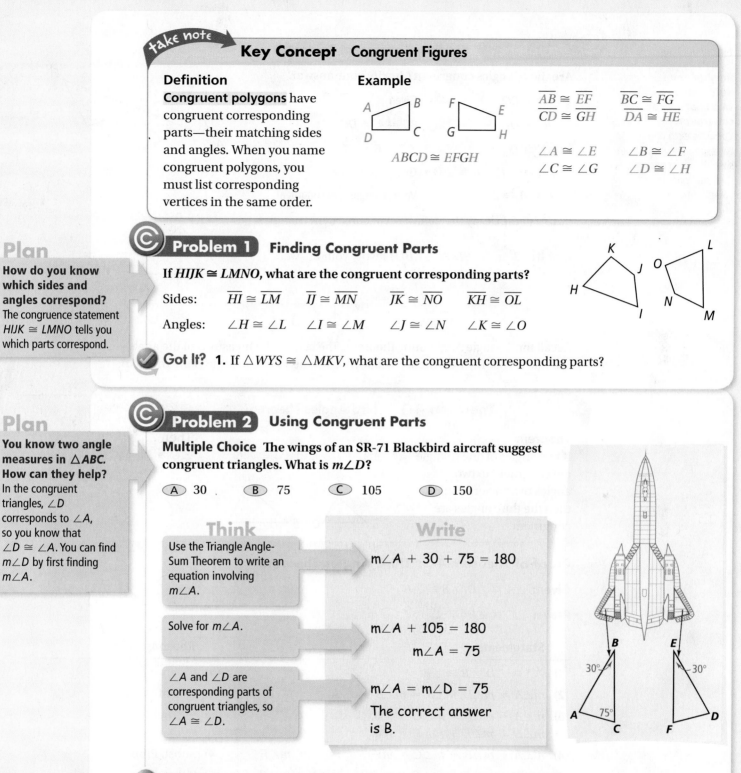

Key Concept   Congruent Figures

Definition
Congruent polygons have congruent corresponding parts—their matching sides and angles. When you name congruent polygons, you must list corresponding vertices in the same order.

Example

$ABCD \cong EFGH$

$\overline{AB} \cong \overline{EF}$ $\overline{BC} \cong \overline{FG}$
$\overline{CD} \cong \overline{GH}$ $\overline{DA} \cong \overline{HE}$

$\angle A \cong \angle E$ $\angle B \cong \angle F$
$\angle C \cong \angle G$ $\angle D \cong \angle H$

Plan

How do you know which sides and angles correspond?
The congruence statement $HIJK \cong LMNO$ tells you which parts correspond.

Problem 1   Finding Congruent Parts

If $HIJK \cong LMNO$, what are the congruent corresponding parts?

Sides: $\overline{HI} \cong \overline{LM}$ $\overline{IJ} \cong \overline{MN}$ $\overline{JK} \cong \overline{NO}$ $\overline{KH} \cong \overline{OL}$

Angles: $\angle H \cong \angle L$ $\angle I \cong \angle M$ $\angle J \cong \angle N$ $\angle K \cong \angle O$

Got It?   1. If $\triangle WYS \cong \triangle MKV$, what are the congruent corresponding parts?

Plan

You know two angle measures in $\triangle ABC$. How can they help?
In the congruent triangles, $\angle D$ corresponds to $\angle A$, so you know that $\angle D \cong \angle A$. You can find $m\angle D$ by first finding $m\angle A$.

Problem 2   Using Congruent Parts

Multiple Choice The wings of an SR-71 Blackbird aircraft suggest congruent triangles. What is $m\angle D$?

Ⓐ 30   Ⓑ 75   Ⓒ 105   Ⓓ 150

Think

Use the Triangle Angle-Sum Theorem to write an equation involving $m\angle A$.

Solve for $m\angle A$.

$\angle A$ and $\angle D$ are corresponding parts of congruent triangles, so $\angle A \cong \angle D$.

Write

$m\angle A + 30 + 75 = 180$

$m\angle A + 105 = 180$
$m\angle A = 75$

$m\angle A = m\angle D = 75$
The correct answer is B.

Got It?   2. Suppose that $\triangle WYS \cong \triangle MKV$. If $m\angle W = 62$ and $m\angle Y = 35$, what is $m\angle V$? Explain.

 **Problem 3** Finding Congruent Triangles

Are the triangles congruent? Justify your answer.

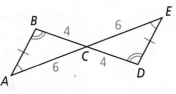

$\overline{AB} \cong \overline{ED}$	Given
$\overline{BC} \cong \overline{DC}$	$BC = 4 = DC$
$\overline{AC} \cong \overline{EC}$	$AC = 6 = EC$
$\angle A \cong \angle E, \angle B \cong \angle D$	Given
$\angle BCA \cong \angle DCE$	Vertical angles are congruent.

$\triangle ABC \cong \triangle EDC$ by the definition of congruent triangles.

Got It? **3.** Is $\triangle ABD \cong \triangle CBD$? Justify your answer.

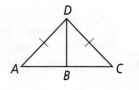

Recall the Triangle Angle-Sum Theorem: The sum of the measures of the angles in a triangle is 180. The next theorem follows from the Triangle Angle-Sum Theorem.

take note

Theorem 4-1 Third Angles Theorem

Theorem	**If . . .**	**Then . . .**
If two angles of one triangle are congruent to two angles of another triangle, then the third angles are congruent.	$\angle A \cong \angle D$ and $\angle B \cong \angle E$	$\angle C \cong \angle F$

Proof **Proof of Theorem 4-1: Third Angles Theorem**

Given: $\angle A \cong \angle D, \angle B \cong \angle E$

Prove: $\angle C \cong \angle F$

Statements	**Reasons**
1) $\angle A \cong \angle D, \angle B \cong \angle E$	**1)** Given
2) $m\angle A = m\angle D, m\angle B = m\angle E$	**2)** Def. of $\cong$ $\angle$s
3) $m\angle A + m\angle B + m\angle C = 180,$ $m\angle D + m\angle E + m\angle F = 180$	**3)** $\triangle$ Angle-Sum Thm.
4) $m\angle A + m\angle B + m\angle C = m\angle D + m\angle E + m\angle F$	**4)** Subst. Prop.
5) $m\angle D + m\angle E + m\angle C = m\angle D + m\angle E + m\angle F$	**5)** Subst. Prop.
6) $m\angle C = m\angle F$	**6)** Subtraction Prop. of $=$
7) $\angle C \cong \angle F$	**7)** Def. of $\cong$ $\angle$s

Plan

You know four pairs of congruent parts. What else do you need to prove the triangles congruent? You need a third pair of congruent sides and a third pair of congruent angles.

Proof

Problem 4 Proving Triangles Congruent

Given: $\overline{LM} \cong \overline{LO}$, $\overline{MN} \cong \overline{ON}$, $\angle M \cong \angle O$, $\angle MLN \cong \angle OLN$

Prove: $\triangle LMN \cong \triangle LON$

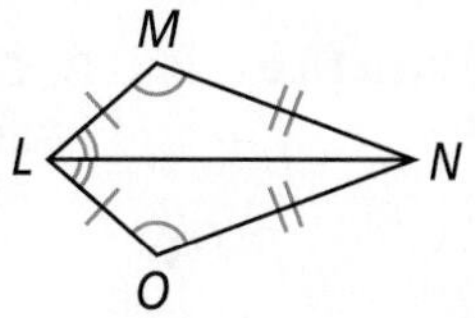

Statements	Reasons
1) $\overline{LM} \cong \overline{LO}$, $\overline{MN} \cong \overline{ON}$	**1)** Given
2) $\overline{LN} \cong \overline{LN}$	**2)** Reflexive Property of $\cong$
3) $\angle M \cong \angle O$, $\angle MLN \cong \angle OLN$	**3)** Given
4) $\angle MNL \cong \angle ONL$	**4)** Third Angles Theorem
5) $\triangle LMN \cong \triangle LON$	**5)** Definition of $\cong$ triangles

Got It? 4. Given: $\angle A \cong \angle D$, $\overline{AE} \cong \overline{DC}$, $\overline{EB} \cong \overline{CB}$, $\overline{BA} \cong \overline{BD}$

Prove: $\triangle AEB \cong \triangle DCB$

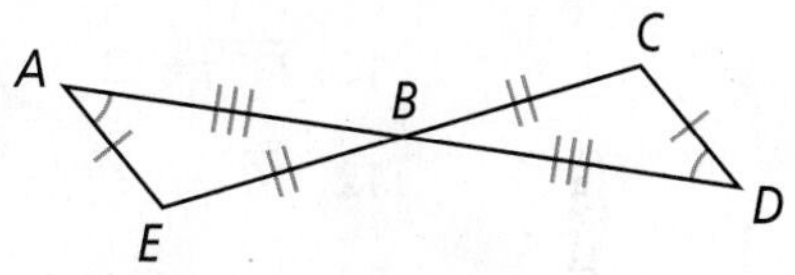

Lesson Check

Do you know HOW?

Complete the following statements.

1. Given: $\triangle QXR \cong \triangle NYC$

a. $\overline{QX} \cong$ __?__

b. $\angle Y \cong$ __?__

2. Given: $\triangle BAT \cong \triangle FOR$

a. $\overline{TA} \cong$ __?__

b. $\angle R \cong$ __?__

3. Given: $BAND \cong LUCK$

a. $\angle U \cong$ __?__

b. $\overline{DB} \cong$ __?__

c. $NDBA \cong$ __?__

4. In $\triangle MAP$ and $\triangle TIE$, $\angle A \cong \angle I$ and $\angle P \cong \angle E$.

a. What is the relationship between $\angle M$ and $\angle T$?

b. If $m\angle A = 52$ and $m\angle P = 36$, what is $m\angle T$?

Do you UNDERSTAND?

MATHEMATICAL PRACTICES

5. Open-Ended When do you think you might need to know that things are congruent in your everyday life?

6. If each angle in one triangle is congruent to its corresponding angle in another triangle, are the two triangles congruent? Explain.

7. Error Analysis Walter sketched the diagram below. He claims it shows that the two polygons are congruent. What information is missing to support his claim?

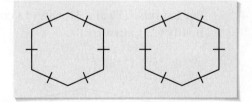

Practice and Problem-Solving Exercises

MATHEMATICAL PRACTICES

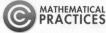

A Practice

8. **Construction** Builders use the king post truss (below left) for the top of a simple structure. In this truss, $\triangle ABC \cong \triangle ABD$. List the congruent corresponding parts.

See Problem 1.

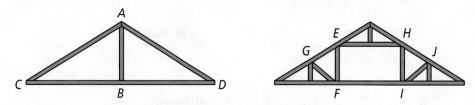

9. The attic frame truss (above right) provides open space in the center for storage. In this truss, $\triangle EFG \cong \triangle HIJ$. List the congruent corresponding parts.

$\triangle LMC \cong \triangle BJK$. Complete the congruence statements.

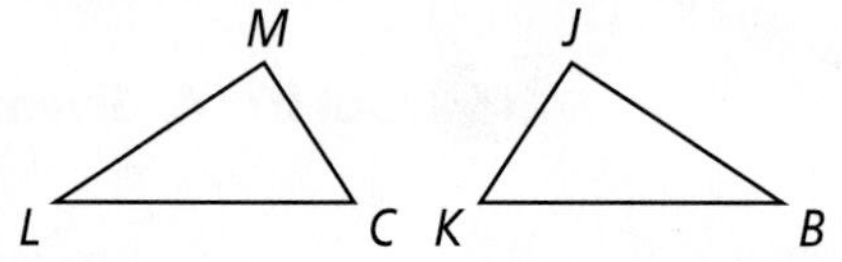

10. $\overline{LC} \cong$ __?__
11. $\overline{KJ} \cong$ __?__
12. $\overline{JB} \cong$ __?__
13. $\angle L \cong$ __?__
14. $\angle K \cong$ __?__
15. $\angle M \cong$ __?__
16. $\triangle CML \cong$ __?__
17. $\triangle KBJ \cong$ __?__
18. $\triangle MLC \cong$ __?__
19. $\triangle JKB \cong$ __?__

***POLY* ≅ *SIDE*. List each of the following.**

20. four pairs of congruent sides
21. four pairs of congruent angles

At an archeological site, the remains of two ancient step pyramids are congruent. If *ABCD* ≅ *EFGH*, find each of the following. (Diagrams are not to scale.)

See Problem 2.

22. AD
23. GH
24. $m\angle GHE$
25. $m\angle BAD$
26. EF
27. BC
28. $m\angle DCB$
29. $m\angle EFG$

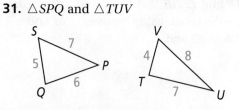

For Exercises 30 and 31, can you conclude that the triangles are congruent? Justify your answers.

See Problem 3.

30. $\triangle TRK$ and $\triangle TUK$

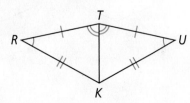

31. $\triangle SPQ$ and $\triangle TUV$

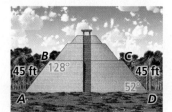

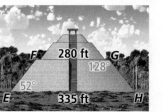

32. Given: $\overline{AB} \parallel \overline{DC}$, $\angle B \cong \angle D$,
Proof $\qquad \overline{AB} \cong \overline{DC}$, $\overline{BC} \cong \overline{AD}$

Prove: $\triangle ABC \cong \triangle CDA$

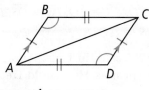

◀ **See Problem 4.**

B Apply

33. If $\triangle DEF \cong \triangle LMN$, which of the following must be a correct congruence statement?

Ⓐ $\overline{DE} \cong \overline{LN}$ Ⓒ $\angle N \cong \angle F$

Ⓑ $\overline{FE} \cong \overline{NL}$ Ⓓ $\angle M \cong \angle F$

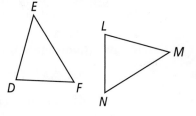

34. Reasoning Randall says he can use the information in the figure to prove $\triangle BCD \cong \triangle DAB$. Is he correct? Explain.

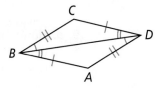

Algebra $\triangle ABC \cong \triangle DEF$. **Find the measures of the given angles or the lengths of the given sides.**

35. $m\angle A = x + 10$, $m\angle D = 2x$ **36.** $m\angle B = 3y$, $m\angle E = 6y - 12$

37. $BC = 3z + 2$, $EF = z + 6$ **38.** $AC = 7a + 5$, $DF = 5a + 9$

39. Think About a Plan $\triangle ABC \cong \triangle DBE$. Find the value of x.

- What does it mean for two triangles to be congruent?
- Which angle measures do you already know?
- How can you find the missing angle measure in a triangle?

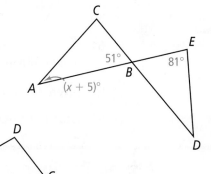

Algebra **Find the values of the variables.**

40.

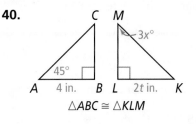

$\triangle ABC \cong \triangle KLM$

41.

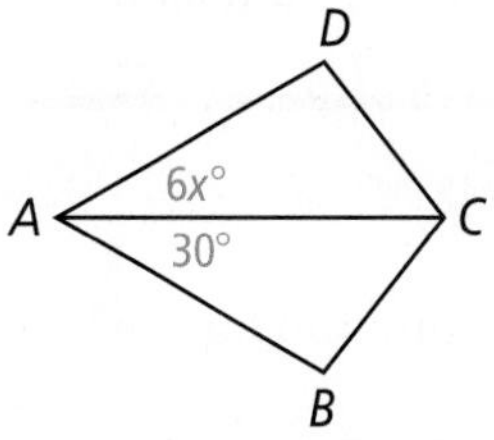

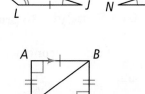

$\triangle ACD \cong \triangle ACB$

42. Complete in two different ways:
$\triangle JLM \cong \underline{}$.

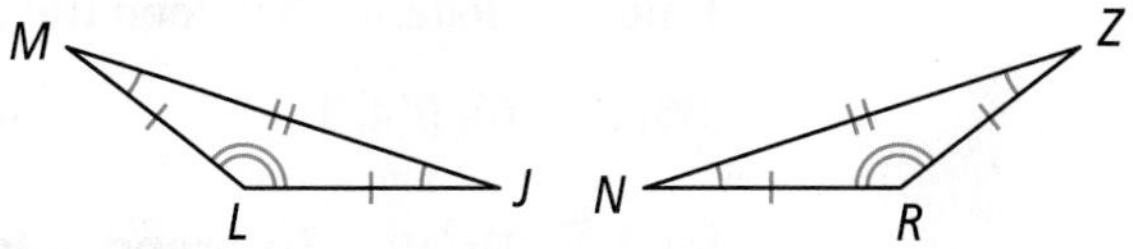

43. Open-Ended Write a congruence statement for two triangles. List the congruent sides and angles.

44. Given: $\overline{AB} \perp \overline{AD}$, $\overline{BC} \perp \overline{CD}$, $\overline{AB} \cong \overline{CD}$, $\overline{AD} \cong \overline{CB}$, $\overline{AB} \parallel \overline{CD}$
Proof **Prove:** $\triangle ABD \cong \triangle CDB$

45. Given: $\overline{PR} \parallel \overline{TQ}$, $\overline{PR} \cong \overline{TQ}$, $\overline{PS} \cong \overline{QS}$, $\overline{PQ}$ bisects $\overline{RT}$

Proof **Prove:** $\triangle PRS \cong \triangle QTS$

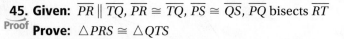

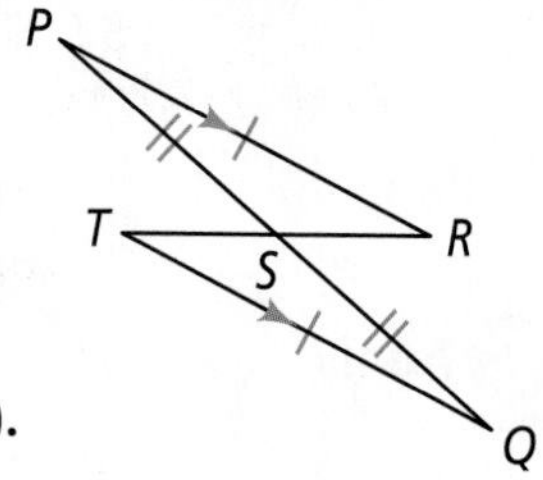

46. Writing The 225 cards in Tracy's sports card collection are rectangles of three different sizes. How could Tracy quickly sort the cards?

C Challenge **Coordinate Geometry** The vertices of $\triangle GHJ$ are $G(-2, -1)$, $H(-2, 3)$, and $J(1, 3)$.

47. $\triangle KLM \cong \triangle GHJ$. Find KL, LM, and KM.

48. If L and M have coordinates $L(3, -3)$ and $M(6, -3)$, how many pairs of coordinates are possible for K? Find one such pair.

49. a. How many quadrilaterals (convex and concave) with different shapes or sizes can you make on a three-by-three geoboard? Sketch them. One is shown at the right.
  b. How many quadrilaterals of each type are there?

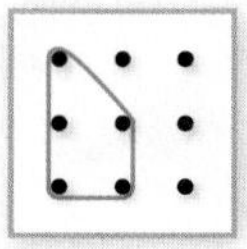

Standardized Test Prep

SAT/ACT

50. $\triangle HLN \cong \triangle GST$, $m\angle H = 66$, and $m\angle S = 42$. What is $m\angle T$?

51. The measure of one angle in a triangle is 80. The other two angles are congruent. What is the measure of each?

52. Given $A(2, -6)$ and $B(-3, 6)$, what is AB?

53. What is the number of feet in the perimeter of a square with side length 7 ft?

Mixed Review

Write an equation for the line perpendicular to the given line that contains P. ◆ **See Lesson 3-8.**

54. $P(2, 7)$; $y = \frac{3}{2}x - 2$ **55.** $P(1, 1)$; $y = 4x + 3$

Find the distance between the points. If necessary, round to the nearest tenth. ◆ **See Lesson 1-7.**

56. $A(0, 0)$, $B(4, 3)$ **57.** $X(11, 24)$, $Y(-7, 24)$ **58.** $E(1, -12)$, $F(1, -2)$

Get Ready! **To prepare for Lesson 4-2, do Exercises 59–61.**

What can you conclude from each diagram? ◆ **See Lessons 1-4 and 3-2.**

59.

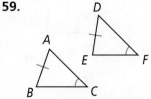

60.

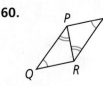

61.

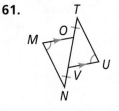

Building Congruent Triangles

© **Content Standard**
Prepares for G.SRT.5 Use congruence . . . criteria for triangles to solve problems and prove relationships in geometric figures.

Can you use shortcuts to find congruent triangles? Find out by building and comparing triangles.

Activity 1

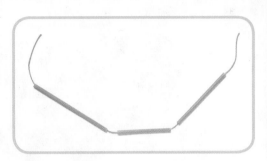

Step 1 Cut straws into three pieces of lengths 4 in., 5 in., and 6 in. Thread a string through the three pieces of straw. The straw pieces can be in any order.

Step 2 Bring the two ends of the string together to make a triangle. Tie the ends to hold your triangle in place.

Step 3 Compare your triangle with your classmates' triangles. Try to make your triangle fit exactly on top of the other triangles.

1. Is your triangle congruent to your classmates' triangles?

© **2. Make a Conjecture** What seems to be true about two triangles in which three sides of one are congruent to three sides of another?

3. As a class, choose three different lengths and repeat Steps 1–3. Are all the triangles congruent? Does this support your conjecture from Question 2?

Activity 2

Step 1 Use a straightedge to draw and label any △ABC on tracing paper.

Step 2 Use a ruler. Carefully measure $\overline{AB}$ and $\overline{AC}$. Use a protractor to measure the angle between them, ∠A.

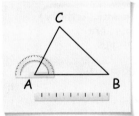

Step 3 Write the measurements on an index card and swap cards with a classmate. Draw a triangle using only your classmate's measurements.

Step 4 Compare your new triangle to your classmate's original △ABC. Try to make your classmate's △ABC fit exactly on top of your new triangle.

4. Is your new triangle congruent to your classmate's original △ABC?

© **5. Make a Conjecture** What seems to be true about two triangles when they have two congruent sides and a congruent angle between them?

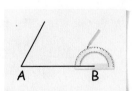

© **6. Make a Conjecture** At least how many triangle measurements must you know in order to guarantee that all triangles built with those measurements will be congruent?

Triangle Congruence by SSS and SAS

© Content Standard
G.SRT.5 Use congruence . . . criteria for triangles to solve problems and prove relationships in geometric figures.

Objective To prove two triangles congruent using the SSS and SAS Postulates

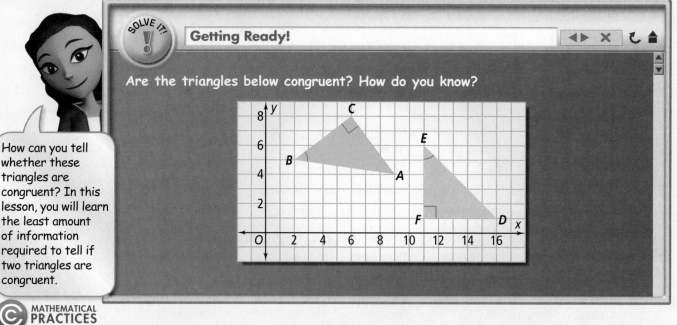

SOLVE IT!

Getting Ready!

Are the triangles below congruent? How do you know?

How can you tell whether these triangles are congruent? In this lesson, you will learn the least amount of information required to tell if two triangles are congruent.

© MATHEMATICAL PRACTICES

In the Solve It, you looked for relationships between corresponding sides and angles. In Lesson 4-1, you learned that if two triangles have three pairs of congruent corresponding angles and three pairs of congruent corresponding sides, then the triangles are congruent.

Dynamic Activity
Congruent
Triangles

If you know . . .

$\angle F \cong \angle J$ $\overline{FG} \cong \overline{JK}$

$\angle G \cong \angle K$ $\overline{GH} \cong \overline{KL}$

$\angle H \cong \angle L$ $\overline{FH} \cong \overline{JL}$

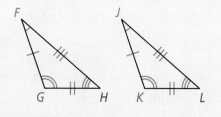

. . . then you know $\triangle FGH \cong \triangle JKL$.

However, this is more information about the corresponding parts than you need to prove triangles congruent.

Essential Understanding You can prove that two triangles are congruent without having to show that *all* corresponding parts are congruent. In this lesson, you will prove triangles congruent by using (1) three pairs of corresponding sides and (2) two pairs of corresponding sides and one pair of corresponding angles.

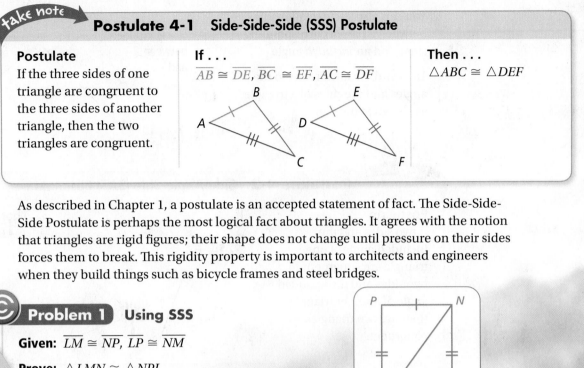

Postulate 4-1 Side-Side-Side (SSS) Postulate

take note

Postulate	**If . . .**	**Then . . .**
If the three sides of one triangle are congruent to the three sides of another triangle, then the two triangles are congruent.	$\overline{AB} \cong \overline{DE}$, $\overline{BC} \cong \overline{EF}$, $\overline{AC} \cong \overline{DF}$	$\triangle ABC \cong \triangle DEF$

As described in Chapter 1, a postulate is an accepted statement of fact. The Side-Side-Side Postulate is perhaps the most logical fact about triangles. It agrees with the notion that triangles are rigid figures; their shape does not change until pressure on their sides forces them to break. This rigidity property is important to architects and engineers when they build things such as bicycle frames and steel bridges.

Proof

Problem 1 Using SSS

Plan

You have two pairs of congruent sides. What else do you need?
You need a third pair of congruent corresponding sides. Notice that the triangles share a common side, $\overline{LN}$.

Given: $\overline{LM} \cong \overline{NP}$, $\overline{LP} \cong \overline{NM}$

Prove: $\triangle LMN \cong \triangle NPL$

$\overline{LM} \cong \overline{NP}$	$\overline{LN} \cong \overline{LN}$	$\overline{LP} \cong \overline{NM}$
Given	Reflexive Prop. of $\cong$	Given

$$\triangle LMN \cong \triangle NPL$$
SSS

✓ Got It? **1. Given:** $\overline{BC} \cong \overline{BF}$, $\overline{CD} \cong \overline{FD}$
Prove: $\triangle BCD \cong \triangle BFD$

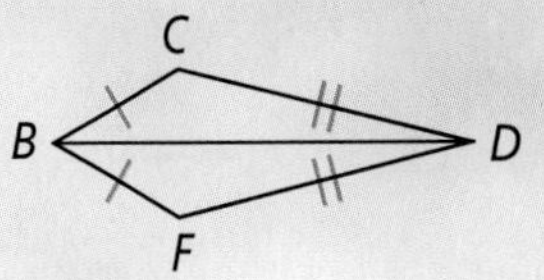

You can also show relationships between a pair of corresponding sides and an *included* angle.

The word *included* refers to the angles and the sides of a triangle as shown at the right.

∠A is included between $\overline{BA}$ and $\overline{AC}$.

$\overline{BC}$ is included between ∠B and ∠C.

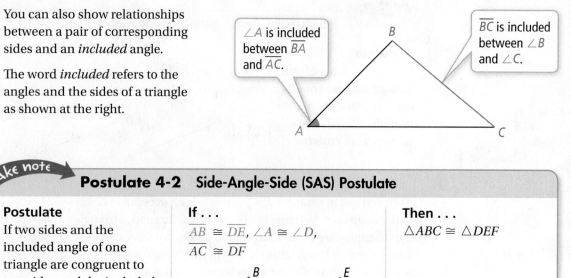

take note

Postulate 4-2   Side-Angle-Side (SAS) Postulate

Postulate	**If . . .**	**Then . . .**
If two sides and the included angle of one triangle are congruent to two sides and the included angle of another triangle, then the two triangles are congruent.	$\overline{AB} \cong \overline{DE}$, ∠A ≅ ∠D, $\overline{AC} \cong \overline{DF}$	△ABC ≅ △DEF

You likely have used the properties of the Side-Angle-Side Postulate before. For example, SAS can help you determine whether a box will fit through a doorway.

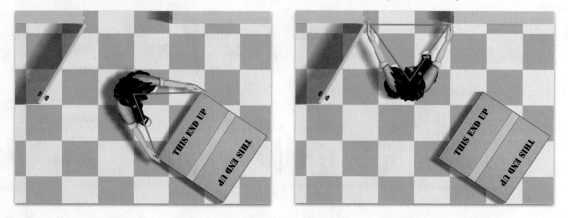

Suppose you keep your arms at a fixed angle as you move from the box to the doorway. The triangle you form with the box is congruent to the triangle you form with the doorway. The two triangles are congruent because two sides and the included angle of one triangle are congruent to the two sides and the included angle of the other triangle.

Plan

Do you need another pair of congruent sides?
Look at the diagram. The triangles share $\overline{DF}$. So you already have two pairs of congruent sides.

Ⓒ **Problem 2** Using SAS

What other information do you need to prove △DEF ≅ △FGD by SAS? Explain.

The diagram shows that $\overline{EF} \cong \overline{GD}$. Also, $\overline{DF} \cong \overline{DF}$ by the Reflexive Property of Congruence. To prove that △DEF ≅ △FGD by SAS, you must have congruent included angles. You need to know that $\angle EFD \cong \angle GDF$.

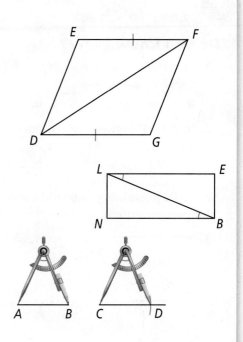

✓ **Got It?** **2.** What other information do you need to prove △LEB ≅ △BNL by SAS?

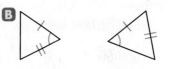

Recall that, in Lesson 1-6, you learned to construct segments using a compass open to a fixed angle. Now you can show that it works. Similar to the situation with the box and the doorway, the Side-Angle-Side Postulate tells you that the triangles outlined at the right are congruent. So, $\overline{AB} \cong \overline{CD}$.

Ⓒ **Problem 3** Identifying Congruent Triangles

Plan

What should you look for first, sides or angles?
Start with sides. If you have three pairs of congruent sides, use SSS. If you have two pairs of congruent sides, look for a pair of congruent included angles.

Would you use SSS or SAS to prove the triangles congruent? If there is not enough information to prove the triangles congruent by SSS or SAS, write *not enough information*. Explain your answer.

Ⓐ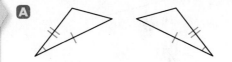

Use SAS because two pairs of corresponding sides and their included angles are congruent.

Ⓑ

There is not enough information; two pairs of corresponding sides are congruent, but one of the angles is not the included angle.

Ⓒ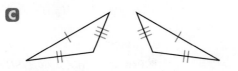

Use SSS because three pairs of corresponding sides are congruent.

Ⓓ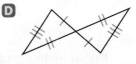

Use SSS or SAS because all three pairs of corresponding sides and a pair of included angles (the vertical angles) are congruent.

✓ **Got It?** **3.** Would you use SSS or SAS to prove the triangles at the right congruent? Explain.

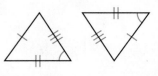

Lesson Check

Do you know HOW?

1. In △PEN, name the angle that is included between the given sides.
   a. $\overline{PE}$ and $\overline{EN}$ b. $\overline{NP}$ and $\overline{PE}$

2. In △HAT, between which sides is the given angle included?
   a. ∠H   b. ∠T

Name the postulate you would use to prove the triangles congruent.

3.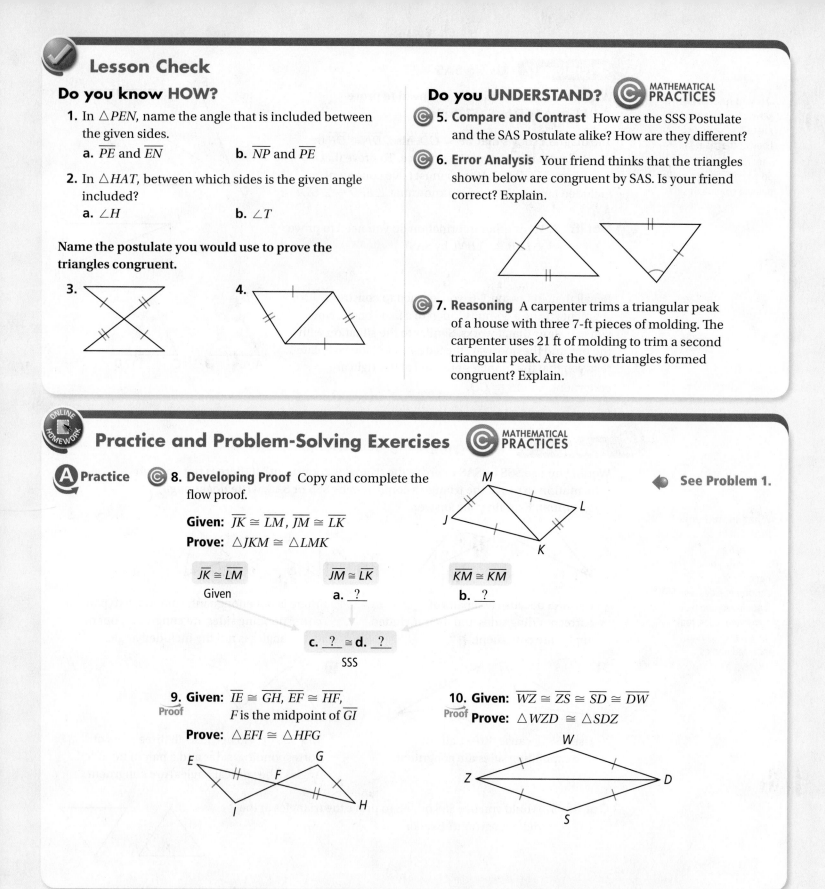

4.

Do you UNDERSTAND? © MATHEMATICAL PRACTICES

© 5. **Compare and Contrast** How are the SSS Postulate and the SAS Postulate alike? How are they different?

© 6. **Error Analysis** Your friend thinks that the triangles shown below are congruent by SAS. Is your friend correct? Explain.

© 7. **Reasoning** A carpenter trims a triangular peak of a house with three 7-ft pieces of molding. The carpenter uses 21 ft of molding to trim a second triangular peak. Are the two triangles formed congruent? Explain.

Practice and Problem-Solving Exercises © MATHEMATICAL PRACTICES

A Practice   © 8. **Developing Proof** Copy and complete the flow proof.

See Problem 1.

Given: $\overline{JK} \cong \overline{LM}$, $\overline{JM} \cong \overline{LK}$
Prove: △JKM ≅ △LMK

$\overline{JK} \cong \overline{LM}$
Given

$\overline{JM} \cong \overline{LK}$
a. ?

$\overline{KM} \cong \overline{KM}$
b. ?

c. ? ≅ d. ?
SSS

9. **Given:** $\overline{IE} \cong \overline{GH}$, $\overline{EF} \cong \overline{HF}$,
   F is the midpoint of $\overline{GI}$
   Prove: △EFI ≅ △HFG

10. **Given:** $\overline{WZ} \cong \overline{ZS} \cong \overline{SD} \cong \overline{DW}$
    Prove: △WZD ≅ △SDZ

What other information, if any, do you need to prove the two triangles congruent by SAS? Explain.

◀ See Problem 2.

11.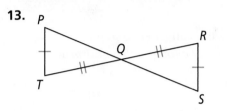

12.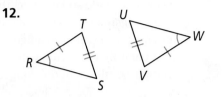

Would you use SSS or SAS to prove the triangles congruent? If there is not enough information to prove the triangles congruent by SSS or SAS, write *not enough information*. Explain your answer.

◀ See Problem 3.

13.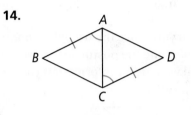

14.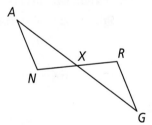

B Apply

© 15. **Think About a Plan** You and a friend are cutting triangles out of felt for an art project. You want all the triangles to be congruent. Your friend tells you that each triangle should have two 5-in. sides and a 40° angle. If you follow this rule, will all your felt triangles be congruent? Explain.
   • How can you use diagrams to help you?
   • Which postulate, SSS or SAS, are you likely to apply to the given situation?

16. **Given:** $\overline{BC} \cong \overline{DA}$, $\angle CBD \cong \angle ADB$
    Proof **Prove:** $\triangle BCD \cong \triangle DAB$

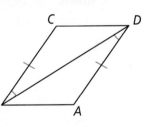

17. **Given:** X is the midpoint of $\overline{AG}$ and $\overline{NR}$.
    Proof **Prove:** $\triangle ANX \cong \triangle GRX$

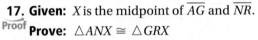

Use the Distance Formula to determine whether $\triangle ABC$ and $\triangle DEF$ are congruent. Justify your answer.

18. $A(1, 4)$, $B(5, 5)$, $C(2, 2)$;
    $D(-5, 1)$, $E(-1, 0)$, $F(-4, 3)$

19. $A(3, 8)$, $B(8, 12)$, $C(10, 5)$;
    $D(3, -1)$, $E(7, -7)$, $F(12, -2)$

20. $A(2, 9)$, $B(2, 4)$, $C(5, 4)$;
    $D(1, -3)$, $E(1, 2)$, $F(-2, 2)$

© 21. **Writing** List three real-life uses of congruent triangles. For each real-life use, describe why you think congruence is necessary.

22. Sierpinski's Triangle Sierpinski's triangle is a famous geometric pattern. To draw Sierpinski's triangle, start with a single triangle and connect the midpoints of the sides to draw a smaller triangle. If you repeat this pattern over and over, you will form a figure like the one shown. This particular figure started with an isosceles triangle. Are the triangles outlined in red congruent? Explain.

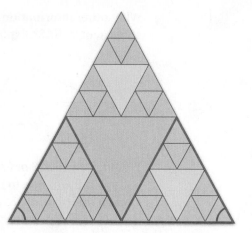

23. Constructions Use a straightedge to draw any triangle *JKL*. Then construct △*MNP* ≅ △*JKL* using the given postulate.
a. SSS
b. SAS

Can you prove the triangles congruent? If so, write the congruence statement and name the postulate you would use. If not, write *not enough information* and tell what other information you would need.

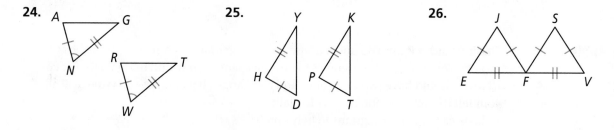

24.   **25.**   **26.**

ⓒ **27. Reasoning** Suppose $\overline{GH} \cong \overline{JK}$, $\overline{HI} \cong \overline{KL}$, and $\angle I \cong \angle L$. Is △*GHI* congruent to △*JKL*? Explain.

28. Given: $\overline{GK}$ bisects $\angle JGM$, $\overline{GJ} \cong \overline{GM}$
Proof **Prove:** △*GJK* ≅ △*GMK*

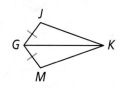

29. Given: $\overline{AE}$ and $\overline{BD}$ bisect each other.
Proof **Prove:** △*ACB* ≅ △*ECD*

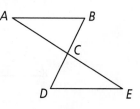

30. Given: $\overline{FG} \parallel \overline{KL}$, $\overline{FG} \cong \overline{KL}$
Proof **Prove:** △*FGK* ≅ △*KLF*

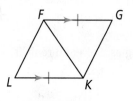

31. Given: $\overline{AB} \perp \overline{CM}$, $\overline{AB} \perp \overline{DB}$, $\overline{CM} \cong \overline{DB}$, *M* is the midpoint of $\overline{AB}$
Proof **Prove:** △*AMC* ≅ △*MBD*

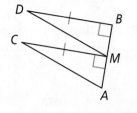

C Challenge

32. Given: $\overline{HK} \cong \overline{LG}$, $\overline{HF} \cong \overline{LJ}$, $\overline{FG} \cong \overline{JK}$
Proof **Prove:** $\triangle FGH \cong \triangle JKL$

33. Given: $\angle N \cong \angle L$, $\overline{MN} \cong \overline{OL}$, $\overline{NO} \cong \overline{LM}$
Proof **Prove:** $\overline{MN} \parallel \overline{OL}$

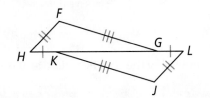

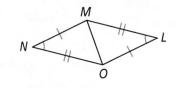

© 34. Reasoning Four sides of polygon $ABCD$ are congruent, respectively, to the four sides of polygon $EFGH$. Are $ABCD$ and $EFGH$ congruent? Is a quadrilateral a rigid figure? If not, what could you add to make it a rigid figure? Explain.

Standardized Test Prep

SAT/ACT

35. What additional information do you need to prove that $\triangle VWY \cong \triangle VWZ$ by SAS?

Ⓐ $\overline{YW} \cong \overline{ZW}$

Ⓒ $\angle Y \cong \angle Z$

Ⓑ $\angle WVY \cong \angle WVZ$

Ⓓ $\overline{VZ} \cong \overline{VY}$

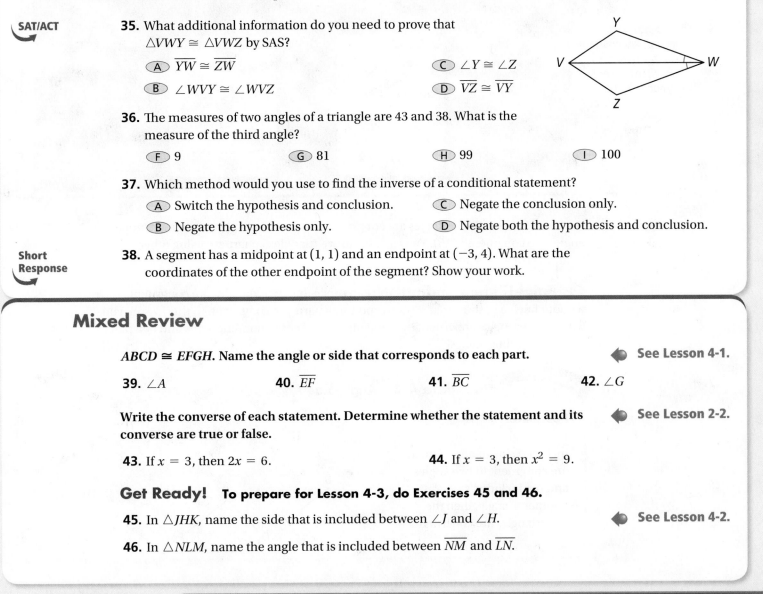

36. The measures of two angles of a triangle are 43 and 38. What is the measure of the third angle?

Ⓕ 9 Ⓖ 81 Ⓗ 99 Ⓘ 100

37. Which method would you use to find the inverse of a conditional statement?

Ⓐ Switch the hypothesis and conclusion.

Ⓒ Negate the conclusion only.

Ⓑ Negate the hypothesis only.

Ⓓ Negate both the hypothesis and conclusion.

Short Response

38. A segment has a midpoint at $(1, 1)$ and an endpoint at $(-3, 4)$. What are the coordinates of the other endpoint of the segment? Show your work.

Mixed Review

$ABCD \cong EFGH$. **Name the angle or side that corresponds to each part.**

See Lesson 4-1.

39. $\angle A$ **40.** $\overline{EF}$ **41.** $\overline{BC}$ **42.** $\angle G$

Write the converse of each statement. Determine whether the statement and its converse are true or false.

See Lesson 2-2.

43. If $x = 3$, then $2x = 6$. **44.** If $x = 3$, then $x^2 = 9$.

Get Ready! **To prepare for Lesson 4-3, do Exercises 45 and 46.**

45. In $\triangle JHK$, name the side that is included between $\angle J$ and $\angle H$.

See Lesson 4-2.

46. In $\triangle NLM$, name the angle that is included between $\overline{NM}$ and $\overline{LN}$.

Triangle Congruence by ASA and AAS

Content Standard
G.SRT.5 Use congruence . . . criteria for triangles to solve problems and prove relationships in geometric figures.

Objective   To prove two triangles congruent using the ASA Postulate and the AAS Theorem

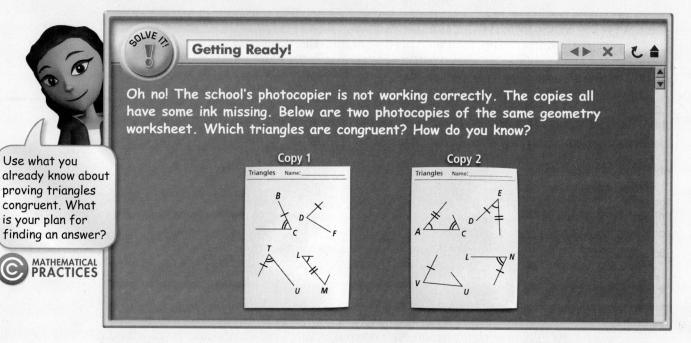

Getting Ready!

Oh no! The school's photocopier is not working correctly. The copies all have some ink missing. Below are two photocopies of the same geometry worksheet. Which triangles are congruent? How do you know?

Use what you already know about proving triangles congruent. What is your plan for finding an answer?

MATHEMATICAL PRACTICES

You already know that triangles are congruent if two pairs of sides and the included angles are congruent (SAS). You can also prove triangles congruent using other groupings of angles and sides.

Essential Understanding You can prove that two triangles are congruent without having to show that *all* corresponding parts are congruent. In this lesson, you will prove triangles congruent by using one pair of corresponding sides and two pairs of corresponding angles.

take note

Postulate 4-3   Angle-Side-Angle (ASA) Postulate

Postulate	**If . . .**	**Then . . .**
If two angles and the included side of one triangle are congruent to two angles and the included side of another triangle, then the two triangles are congruent.	$\angle A \cong \angle D$, $\overline{AC} \cong \overline{DF}$, $\angle C \cong \angle F$	$\triangle ABC \cong \triangle DEF$

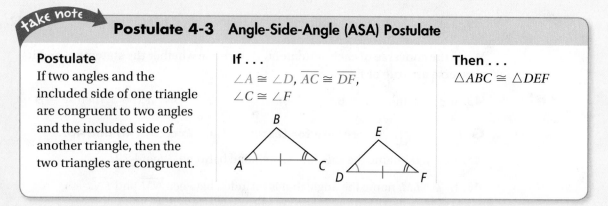

Which two triangles are congruent by ASA? Explain.

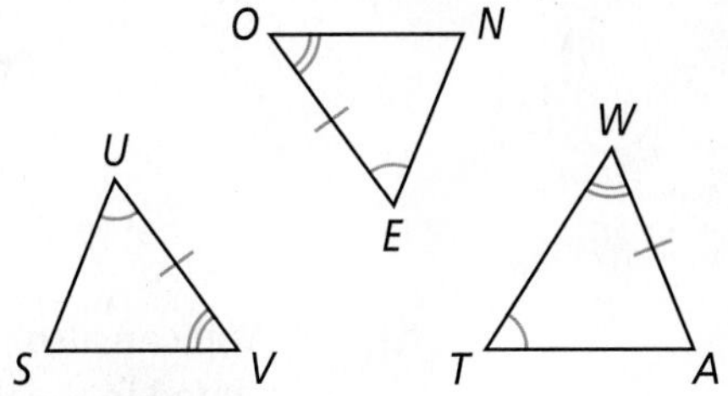

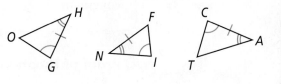

Know

From the diagram you know
- $\angle U \cong \angle E \cong \angle T$
- $\angle V \cong \angle O \cong \angle W$
- $\overline{UV} \cong \overline{EO} \cong \overline{AW}$

Need

To use ASA, you need two pairs of congruent angles and a pair of included congruent sides.

Plan

You already have pairs of congruent angles. So, identify the included side for each triangle and see whether it has a congruence marking.

In $\triangle SUV$, $\overline{UV}$ is included between $\angle U$ and $\angle V$ and has a congruence marking. In $\triangle NEO$, $\overline{EO}$ is included between $\angle E$ and $\angle O$ and has a congruence marking. In $\triangle ATW$, $\overline{TW}$ is included between $\angle T$ and $\angle W$ but does *not* have a congruence marking.

Since $\angle U \cong \angle E$, $\overline{UV} \cong \overline{EO}$, and $\angle V \cong \angle O$, $\triangle SUV \cong \triangle NEO$.

✅ **Got It?** **1.** Which two triangles are congruent by ASA? Explain.

Proof © **Problem 2** **Writing a Proof Using ASA**

Recreation Members of a teen organization are building a miniature golf course at your town's youth center. The design plan calls for the first hole to have two congruent triangular bumpers. Prove that the bumpers on the first hole, shown at the right, meet the conditions of the plan.

Given: $\overline{AB} \cong \overline{DE}$, $\angle A \cong \angle D$, $\angle B$ and $\angle E$ are right angles

Prove: $\triangle ABC \cong \triangle DEF$

Proof: $\angle B \cong \angle E$ because all right angles are congruent, and you are given that $\angle A \cong \angle D$. $\overline{AB}$ and $\overline{DE}$ are included sides between the two pairs of congruent angles. You are given that $\overline{AB} \cong \overline{DE}$. Thus, $\triangle ABC \cong \triangle DEF$ by ASA.

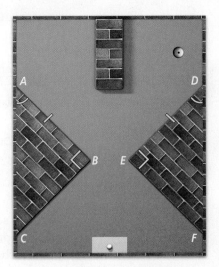

Plan

Can you use a plan similar to the plan in Problem 1?
Yes. Use the diagram to identify the included side for the marked angles in each triangle.

Got It? **2. Given:** $\angle CAB \cong \angle DAE$, $\overline{BA} \cong \overline{EA}$,
$\angle B$ and $\angle E$ are right angles
Prove: $\triangle ABC \cong \triangle AED$

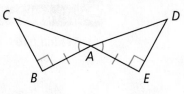

You can also prove triangles congruent by using two angles and a nonincluded side, as stated in the theorem below.

take note

Theorem 4-2 Angle-Angle-Side (AAS) Theorem

Theorem	**If . . .**	**Then . . .**
If two angles and a nonincluded side of one triangle are congruent to two angles and the corresponding nonincluded side of another triangle, then the triangles are congruent.	$\angle A \cong \angle D$, $\angle B \cong \angle E$, $\overline{AC} \cong \overline{DF}$ 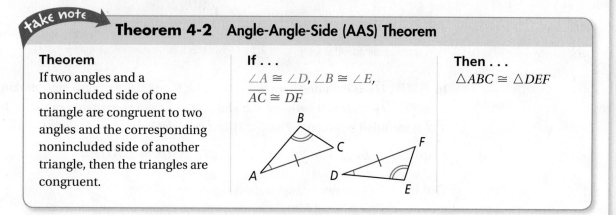	$\triangle ABC \cong \triangle DEF$

Proof **Proof of Theorem 4-2:   Angle-Angle-Side Theorem**

Given: $\angle A \cong \angle D$, $\angle B \cong \angle E$, $\overline{AC} \cong \overline{DF}$

Prove: $\triangle ABC \cong \triangle DEF$

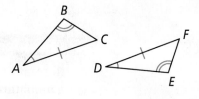

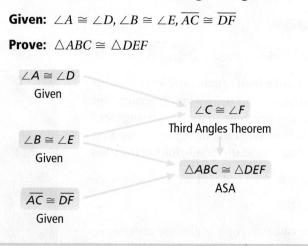

You have seen and used three methods of proof in this book—two-column, paragraph, and flow proof. Each method is equally as valid as the others. Unless told otherwise, you can choose any of the three methods to write a proof. Just be sure your proof always presents logical reasoning with justification.

Plan

How does information about parallel sides help?
You will need another pair of congruent angles to use AAS. Think back to what you learned in Chapter 3. $\overline{WR}$ is a transversal here.

Proof © **Problem 3** **Writing a Proof Using AAS**

Given: $\angle M \cong \angle K$, $\overline{WM} \parallel \overline{RK}$

Prove: $\triangle WMR \cong \triangle RKW$

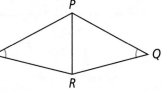

Statements	Reasons
1) $\angle M \cong \angle K$	1) Given
2) $\overline{WM} \parallel \overline{RK}$	2) Given
3) $\angle MWR \cong \angle KRW$	3) If lines are $\parallel$, then alternate interior $\angle$s are $\cong$.
4) $\overline{WR} \cong \overline{WR}$	4) Reflexive Property of Congruence
5) $\triangle WMR \cong \triangle RKW$	5) AAS

© ✓ **Got It?** **3. a. Given:** $\angle S \cong \angle Q$, $\overline{RP}$ bisects $\angle SRQ$

Prove: $\triangle SRP \cong \triangle QRP$

b. Reasoning In Problem 3, how could you prove that $\triangle WMR \cong \triangle RKW$ by ASA? Explain.

© **Problem 4** **Determining Whether Triangles Are Congruent**

Multiple Choice Use the diagram at the right. Which of the following statements best represents the answer and justification to the question, "Is $\triangle BIF \cong \triangle UTO$?"

A Yes, the triangles are congruent by ASA.

B No, $\overline{FB}$ and $\overline{OT}$ are not corresponding sides.

C Yes, the triangles are congruent by AAS.

D No, $\angle B$ and $\angle U$ are not corresponding angles.

Think

Can you eliminate any of the choices?
Yes. If $\triangle BIF \cong \triangle UTO$ then $\angle B$ and $\angle U$ would be corresponding angles. You can eliminate choice D.

The diagram shows that two pairs of angles and one pair of sides are congruent. The third pair of angles is congruent by the Third Angles Theorem. To prove these triangles congruent, you need to satisfy ASA or AAS.

ASA and AAS both fail because $\overline{FB}$ and $\overline{TO}$ are not included between the same pair of congruent corresponding angles, so they are not corresponding sides. The triangles are not necessarily congruent. The correct answer is B.

✓ **Got It?** **4.** Are $\triangle PAR$ and $\triangle SIR$ congruent? Explain.

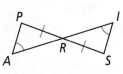

Lesson Check

Do you know HOW?

1. In △RST, which side is included between ∠R and ∠S?

2. In △NOM, $\overline{NO}$ is included between which angles?

Which postulate or theorem could you use to prove △ABC ≅ △DEF?

3.

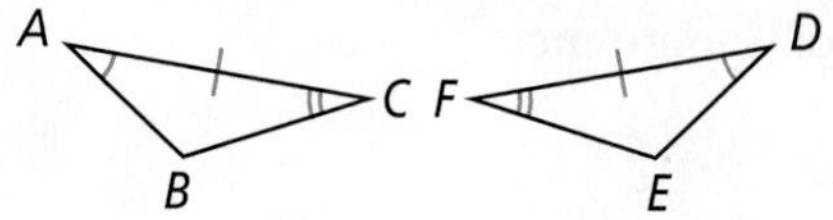

4.

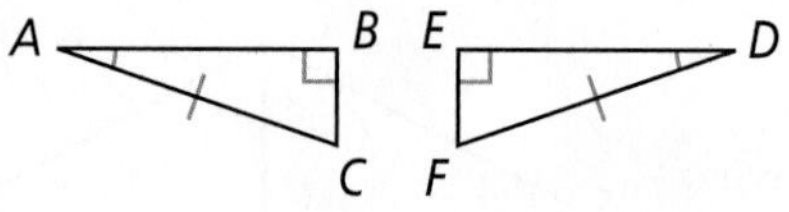

Do you UNDERSTAND? ⒸMATHEMATICAL PRACTICES

Ⓒ **5. Compare and Contrast** How are the ASA Postulate and the SAS Postulate alike? How are they different?

Ⓒ **6. Error Analysis** Your friend asks you for help on a geometry exercise. Below is your friend's paper. What error did your friend make? Explain.

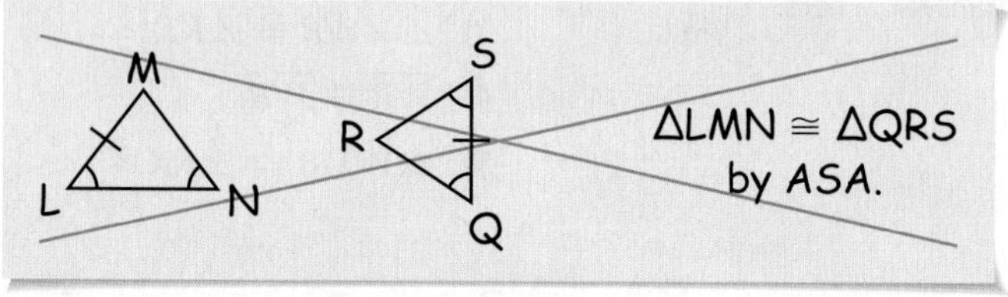

△LMN ≅ △QRS by ASA.

Ⓒ **7. Reasoning** Suppose ∠E ≅ ∠I and $\overline{FE} \cong \overline{GI}$. What else must you know in order to prove △FDE ≅ △GHI by ASA? By AAS?

Practice and Problem-Solving Exercises ⒸMATHEMATICAL PRACTICES

Ⓐ **Practice** **Name two triangles that are congruent by ASA.** ◀ See Problem 1.

8.

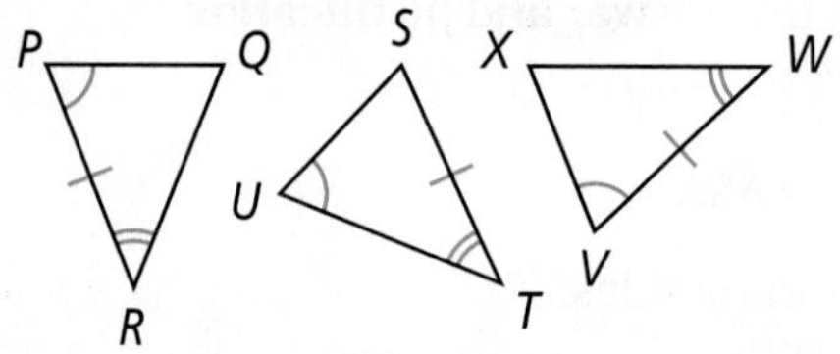

9.

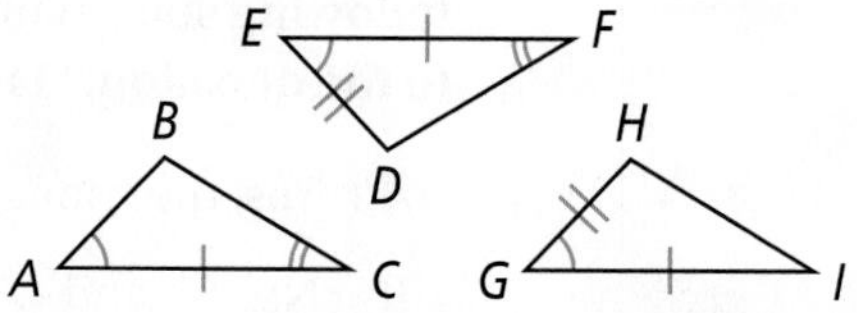

Ⓒ **10. Developing Proof** Complete the paragraph proof by filling in the blanks. ◀ See Problem 2.

Given: ∠LKM ≅ ∠JKM,
    ∠LMK ≅ ∠JMK

Prove: △LKM ≅ △JKM

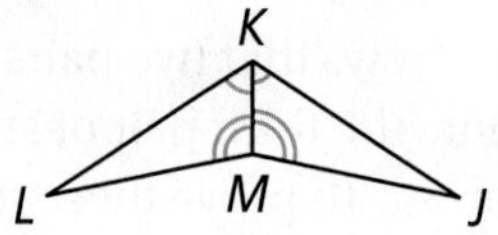

Proof: ∠LKM ≅ ∠JKM and ∠LMK ≅ ∠JMK are given. $\overline{KM} \cong \overline{KM}$ by the **a.** ? Property of Congruence. So, △LKM ≅ △JKM by **b.** ? .

11. Given: ∠BAC ≅ ∠DAC,
    $\overline{AC} \perp \overline{BD}$
Proof

Prove: △ABC ≅ △ADC

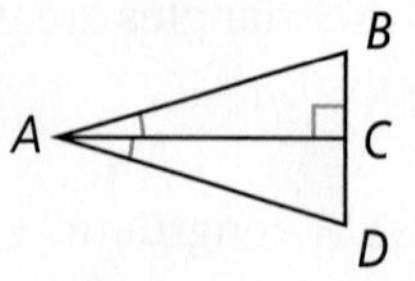

12. Given: $\overline{QR} \cong \overline{TS}$,
    $\overline{QR} \parallel \overline{TS}$
Proof

Prove: △QRT ≅ △TSQ

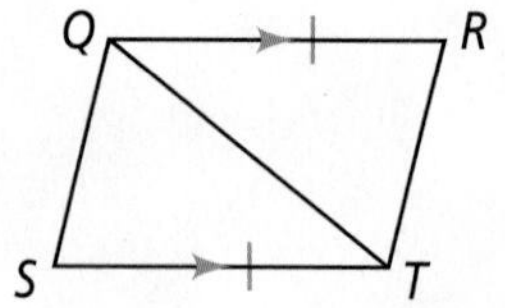

© 13. Developing Proof Complete the two-column proof by filling in the blanks. ◀ **See Problem 3.**

Given: ∠N ≅ ∠S,
line ℓ bisects $\overline{TR}$ at Q
Prove: △NQT ≅ △SQR

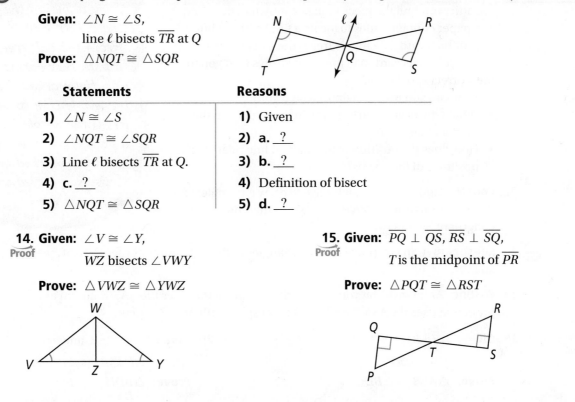

Statements	Reasons
1) ∠N ≅ ∠S	**1)** Given
2) ∠NQT ≅ ∠SQR	**2) a.** _?_
3) Line ℓ bisects $\overline{TR}$ at Q.	**3) b.** _?_
4) c. _?_	**4)** Definition of bisect
5) △NQT ≅ △SQR	**5) d.** _?_

14. Given: ∠V ≅ ∠Y,
$\overline{WZ}$ bisects ∠VWY
Prove: △VWZ ≅ △YWZ

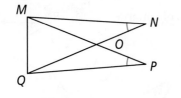

15. Given: $\overline{PQ} \perp \overline{QS}$, $\overline{RS} \perp \overline{SQ}$,
T is the midpoint of $\overline{PR}$
Prove: △PQT ≅ △RST

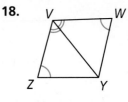

Determine whether the triangles must be congruent. If so, name the postulate or theorem that justifies your answer. If not, explain. ◀ **See Problem 4.**

16.

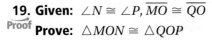

17.

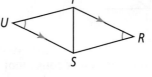

18.

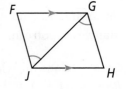

Ⓑ Apply

19. Given: ∠N ≅ ∠P, $\overline{MO} \cong \overline{QO}$
Prove: △MON ≅ △QOP

20. Given: ∠FJG ≅ ∠HGJ, $\overline{FG} \parallel \overline{JH}$
Prove: △FGJ ≅ △HJG

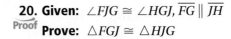

PowerGeometry.com   **Lesson 4-3** Triangle Congruence by ASA and AAS   **239**

21. Think About a Plan While helping your family clean out the attic, you find the piece of paper shown at the right. The paper contains clues to locate a time capsule buried in your backyard. The maple tree is due east of the oak tree in your backyard. Will the clues always lead you to the correct spot? Explain.

- How can you use a diagram to help you?
- What type of geometric figure do the paths and the marked line form?
- How does the position of the marked line relate to the positions of the angles?

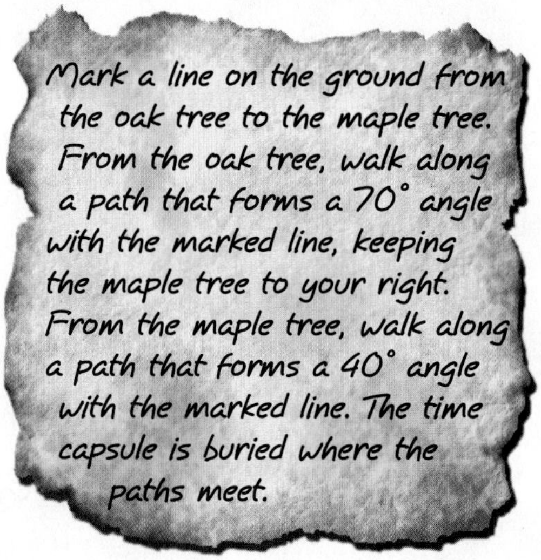

Mark a line on the ground from the oak tree to the maple tree. From the oak tree, walk along a path that forms a 70° angle with the marked line, keeping the maple tree to your right. From the maple tree, walk along a path that forms a 40° angle with the marked line. The time capsule is buried where the paths meet.

22. Constructions Use a straightedge to draw a triangle. Label it △JKL. Construct △MNP ≅ △JKL so that the triangles are congruent by ASA.

23. Reasoning Can you prove that the triangles at the right are congruent? Justify your answer.

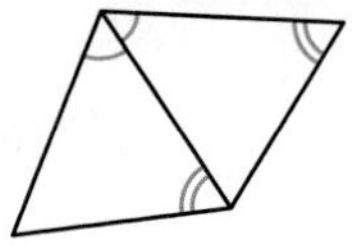

24. Writing Anita says that you can rewrite any proof that uses the AAS Theorem as a proof that uses the ASA Postulate. Do you agree with Anita? Explain.

25. Given: $\overrightarrow{AE} \parallel \overline{BD}, \overline{AE} \cong \overline{BD},$
$\angle E \cong \angle D$
Proof
Prove: △AEB ≅ △BDC

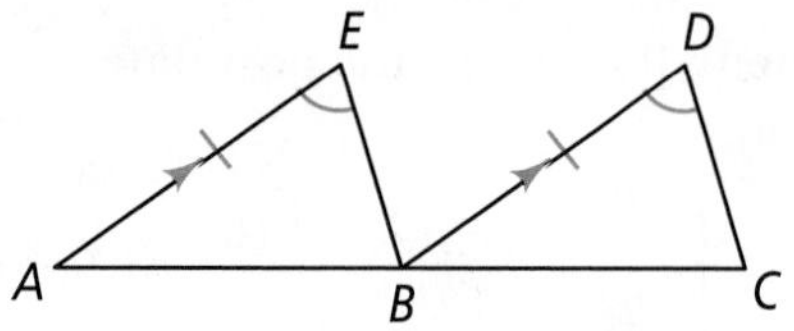

26. Given: $\angle 1 \cong \angle 2,$ and
$\overrightarrow{DH}$ bisects $\angle BDF$.
Proof
Prove: △BDH ≅ △FDH

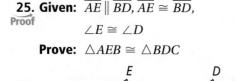

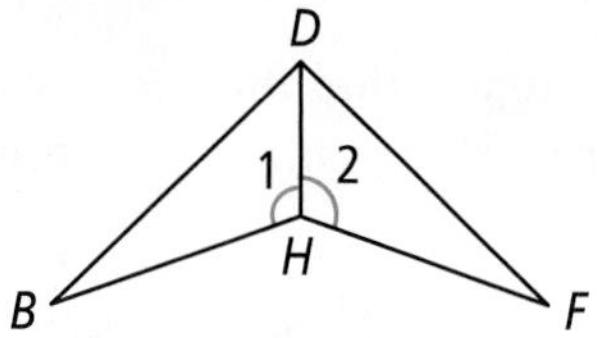

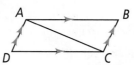

27. Draw a Diagram Draw two noncongruent triangles that have two pairs of congruent angles and one pair of congruent sides.

28. Given: $\overline{AB} \parallel \overline{DC}, \overline{AD} \parallel \overline{BC}$
Proof
Prove: △ABC ≅ △CDA

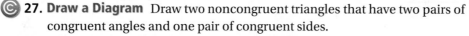

Challenge

29. Given $\overline{AD} \parallel \overline{BC}$ and $\overline{AB} \parallel \overline{DC}$, name as many pairs of congruent triangles as you can.

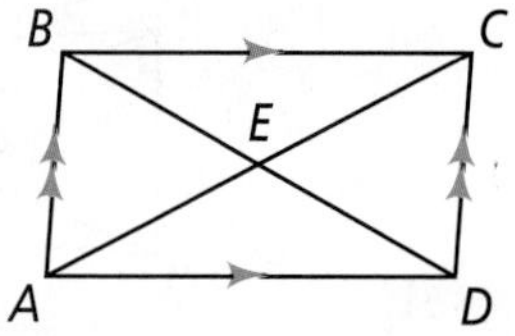

30. Constructions In △RST at the right, $RS = 5, RT = 9,$ and $m\angle T = 30$. Show that there is no SSA congruence rule by constructing △UVW with $UV = RS$, $UW = RT$, and $m\angle W = m\angle T$, but with △UVW ≇ △RST.

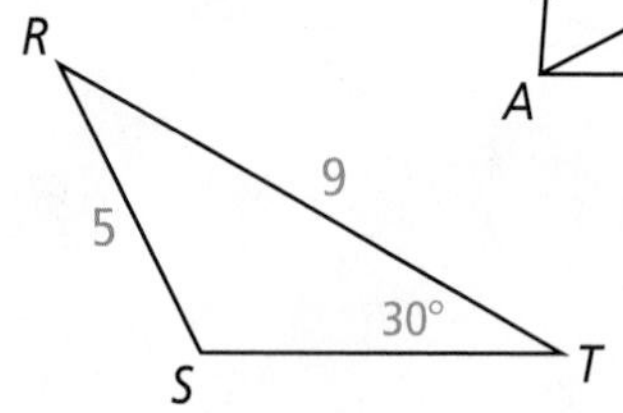

31. Probability Below are six statements about the triangles at the right.

$\angle A \cong \angle X$ $\angle B \cong \angle Y$ $\angle C \cong \angle Z$

$\overline{AB} \cong \overline{XY}$ $\overline{AC} \cong \overline{XZ}$ $\overline{BC} \cong \overline{YZ}$

There are 20 ways to choose a group of three statements from these six.
What is the probability that three statements chosen at random from the
six will guarantee that the triangles are congruent?

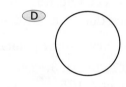

Standardized Test Prep

SAT/ACT

32. Suppose $\overline{RT} \cong \overline{ND}$ and $\angle R \cong \angle N$. What additional information do you need to
prove that $\triangle RTJ \cong \triangle NDF$ by ASA?

Ⓐ $\angle T \cong \angle D$ Ⓑ $\angle J \cong \angle F$ Ⓒ $\angle J \cong \angle D$ Ⓓ $\angle T \cong \angle F$

33. You plan to make a 2 ft-by-3 ft rectangular poster of class trip photos. Each photo is
a 4 in.-by-6 in. rectangle. If the photos do not overlap, what is the greatest number
of photos you can fit on your poster?

Ⓕ 4   Ⓖ 24   Ⓗ 32   Ⓘ 36

34. Which of the following figures is a concave polygon?

Ⓐ   Ⓑ   Ⓒ   Ⓓ

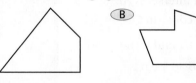

Short
Response

35. Write the converse of the true conditional statement below. Then determine
whether the converse is true or false.

If you are less than 18 years old, then you are too young to vote in the United States.

Mixed Review

Would you use SSS or SAS to prove the triangles congruent? Explain.   ◀ **See Lesson 4-2.**

36.   **37.**

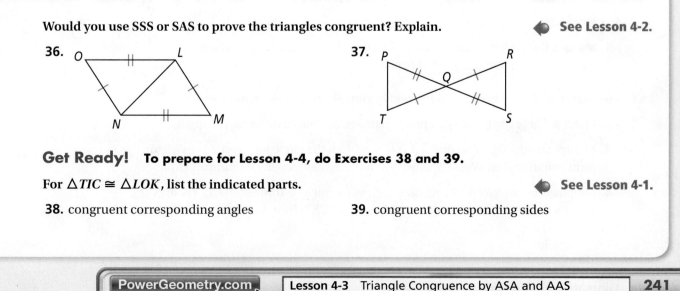

Get Ready!   **To prepare for Lesson 4-4, do Exercises 38 and 39.**

For $\triangle TIC \cong \triangle LOK$, list the indicated parts.   ◀ **See Lesson 4-1.**

38. congruent corresponding angles   **39.** congruent corresponding sides

Concept Byte

Use With Lesson 4-3

TECHNOLOGY

Exploring AAA and SSA

© **Content Standard**
Extends G.SRT.5 Use congruence . . . criteria for triangles to solve problems and prove relationships in geometric figures.

So far, you know four ways to conclude that two triangles are congruent—SSS, SAS, ASA, and AAS. It is good mathematics to wonder about the other two possibilities.

© MATHEMATICAL PRACTICES

Activity 1

Construct Use geometry software to construct $\overrightarrow{AB}$ and $\overrightarrow{AC}$. Construct $\overline{BC}$ to form $\triangle ABC$. Construct a line parallel to $\overline{BC}$ that intersects $\overrightarrow{AB}$ and $\overrightarrow{AC}$ at points D and E to form $\triangle ADE$.

Investigate Are the three angles of $\triangle ABC$ congruent to the three angles of $\triangle ADE$? Manipulate the figure to change the positions of $\overline{DE}$ and $\overline{BC}$. Do the corresponding angles of the triangles remain congruent? Are the two triangles congruent? Can the two triangles be congruent?

Activity 2

Construct Construct $\overrightarrow{AB}$. Draw a circle with center C that intersects $\overrightarrow{AB}$ at two points. Construct $\overline{AC}$. Construct point E on the circle and construct $\overline{CE}$.

Investigate Move point E around the circle until E is on $\overrightarrow{AB}$ and forms $\triangle ACE$. Then move E on the circle to the other point on $\overrightarrow{AB}$ to form another $\triangle ACE$.

Compare AC, CE, and $m\angle A$ in the two triangles. Are two sides and a nonincluded angle of one triangle congruent to two sides and a nonincluded angle of the other triangle? Are the triangles congruent? If you change the measure of $\angle A$ and the size of the circle, do you get the same results?

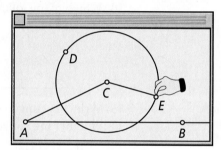

Exercises

© **1. Make a Conjecture** Based on your first investigation above, can you prove triangles congruent using AAA? Explain.

For Exercises 2–4, use what you learned in your second investigation above.

© **2. Make a Conjecture** Can you prove triangles congruent using SSA? Explain.

3. Manipulate the figure so that $\angle A$ is obtuse. Can the circle intersect $\overrightarrow{AB}$ twice to form two triangles? Would SSA work if the congruent angles are obtuse? Explain.

4. Suppose you are given $\overline{CE}$, $\overline{AC}$, and $\angle A$. What must be true about CE, AC, and $m\angle A$ so that you can construct exactly one $\triangle ACE$? (*Hint:* Consider cases.)

4 Mid-Chapter Quiz

MathXL® for School
Go to PowerGeometry.com

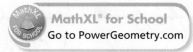

Do you know HOW?

1. $\triangle RST \cong \triangle JKL$. List the three pairs of congruent corresponding sides and the three pairs of congruent corresponding angles.

***LMNO* ≅ *PQRS*. Name the angle or side that corresponds to the given part.**

2. $\overline{RS}$

3. $\angle L$

4. $\angle Q$

5. $\overline{MN}$

State the postulate or theorem you can use to prove the triangles congruent. If you do not have enough information to prove the triangles congruent, write *not enough information*.

6.

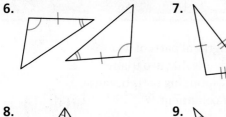

7.

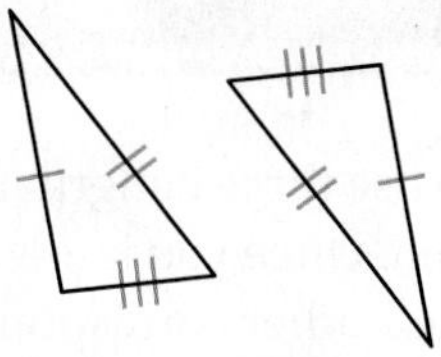

8.

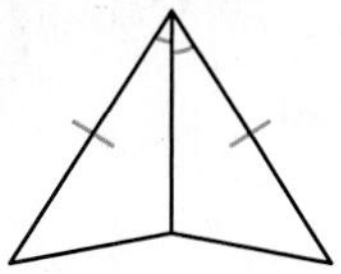

9.

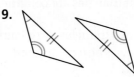

10.

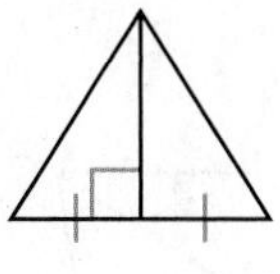

11.

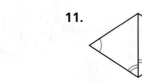

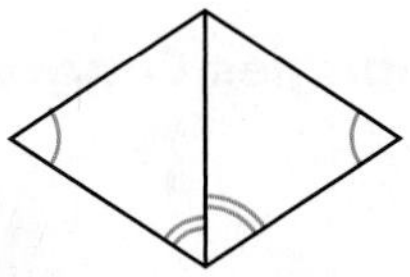

Use the diagram below. Tell why each statement is true.

12. $\angle H \cong \angle K$

13. $\angle HNL \cong \angle KNJ$

14. $\triangle HNL \cong \triangle KNJ$

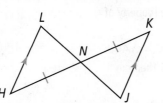

Determine what other information you need to prove the two triangles congruent. Then write the congruence statement and name the postulate or theorem you would use.

15.

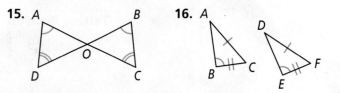

16.

17. Constructions Construct $\triangle LMN$ congruent to $\triangle PQR$ using SSS.

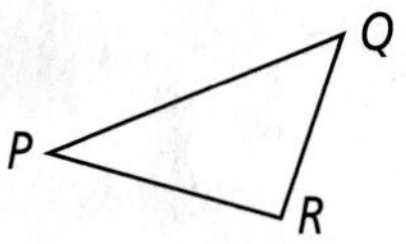

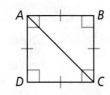

Do you UNDERSTAND?

18. Given: $\angle A \cong \angle D$, O is the midpoint of $\overline{AD}$

Prove: $\triangle AOB \cong \triangle DOC$

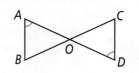

19. Given: $\overline{AB} \cong \overline{BC} \cong \overline{CD} \cong \overline{DA}$, $\angle A$, $\angle B$, $\angle C$, and $\angle D$ are right angles

Prove: $\triangle ABC \cong \triangle CDA$

20. Reasoning Three segments form a triangle. How many unique triangles can you construct using the same three segments? Using the same three angles? Explain.

21. Open-Ended Write a congruency postulate for quadrilaterals. Does your postulate always hold true? Explain.

Using Corresponding Parts of Congruent Triangles

Content Standards
G.SRT.5 Use congruence . . . criteria for triangles to solve problems and prove relationships in geometric figures.
Also G.CO.12

Objective To use triangle congruence and corresponding parts of congruent triangles to prove that parts of two triangles are congruent

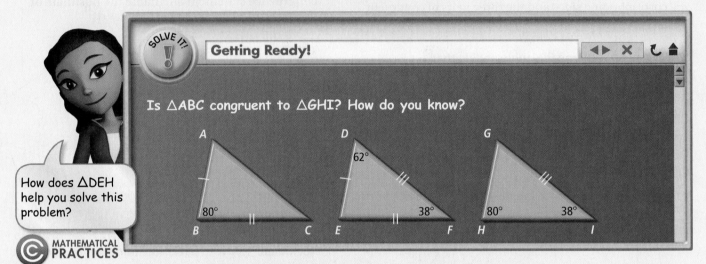

SOLVE IT!

Getting Ready!

Is △ABC congruent to △GHI? How do you know?

How does △DEH help you solve this problem?

MATHEMATICAL PRACTICES

With SSS, SAS, ASA, and AAS, you know how to use three congruent parts of two triangles to show that the triangles are congruent. Once you know that two triangles are congruent, you can make conclusions about their other corresponding parts because, by definition, corresponding parts of congruent triangles are congruent.

Essential Understanding If you know two triangles are congruent, then you know that every pair of their corresponding parts is also congruent.

Think

In the diagram, which congruent pair is not marked?
The third angles of both triangles are congruent. But there is no AAA congruence rule. So, find a congruent pair of sides.

Problem 1 Proving Parts of Triangles Congruent

Given: $\angle KBC \cong \angle ACB$, $\angle K \cong \angle A$

Prove: $\overline{KB} \cong \overline{AC}$

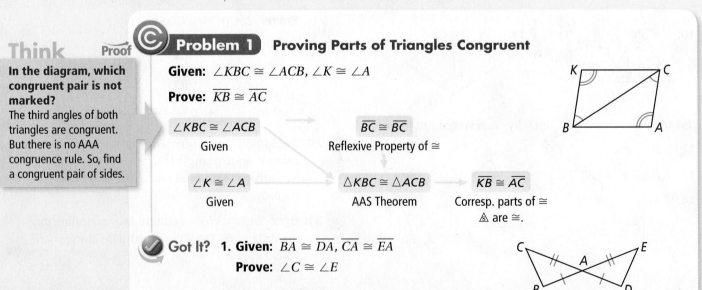

$\angle KBC \cong \angle ACB$ ⟶ $\overline{BC} \cong \overline{BC}$
Given ⟶ Reflexive Property of ≅

$\angle K \cong \angle A$ ⟶ $\triangle KBC \cong \triangle ACB$ ⟶ $\overline{KB} \cong \overline{AC}$
Given ⟶ AAS Theorem ⟶ Corresp. parts of ≅ △ are ≅.

Got It? **1. Given:** $\overline{BA} \cong \overline{DA}$, $\overline{CA} \cong \overline{EA}$
Prove: $\angle C \cong \angle E$

Measurement Thales, a Greek philosopher, is said to have developed a method to measure the distance to a ship at sea. He made a compass by nailing two sticks together. Standing on top of a tower, he would hold one stick vertical and tilt the other until he could see the ship S along the line of the tilted stick. With this compass setting, he would find a landmark L on the shore along the line of the tilted stick. How far would the ship be from the base of the tower?

Given: $\angle TRS$ and $\angle TRL$ are right angles, $\angle RTS \cong \angle RTL$

Prove: $\overline{RS} \cong \overline{RL}$

Plan

Which congruency rule can you use?
You have information about two pairs of angles. *Guess-and-check* AAS and ASA.

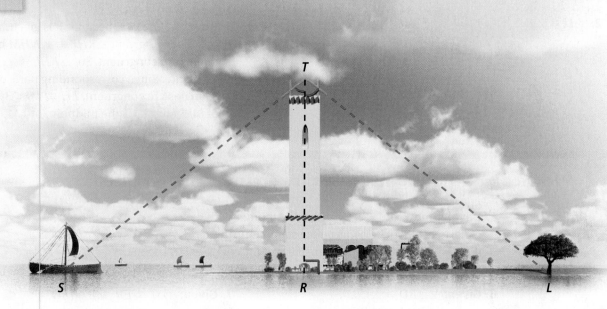

Statements	Reasons
1) $\angle RTS \cong \angle RTL$	1) Given
2) $\overline{TR} \cong \overline{TR}$	2) Reflexive Property of Congruence
3) $\angle TRS$ and $\angle TRL$ are right angles.	3) Given
4) $\angle TRS \cong \angle TRL$	4) All right angles are congruent.
5) $\triangle TRS \cong \triangle TRL$	5) ASA Postulate
6) $\overline{RS} \cong \overline{RL}$	6) Corresponding parts of $\cong \triangle$ are $\cong$.

The distance between the ship and the base of the tower would be the same as the distance between the base of the tower and the landmark.

© ✓ **Got It? 2. a. Given:** $\overline{AB} \cong \overline{AC}$, M is the midpoint of $\overline{BC}$
  Prove: $\angle AMB \cong \angle AMC$

  b. Reasoning If the landmark were not at sea level, would the method in Problem 2 work? Explain.

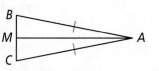

Lesson Check

Do you know HOW?

Name the postulate or theorem that you can use to show the triangles are congruent. Then explain why the statement is true.

1. $\overline{EA} \cong \overline{MA}$

2. $\angle U \cong \angle E$

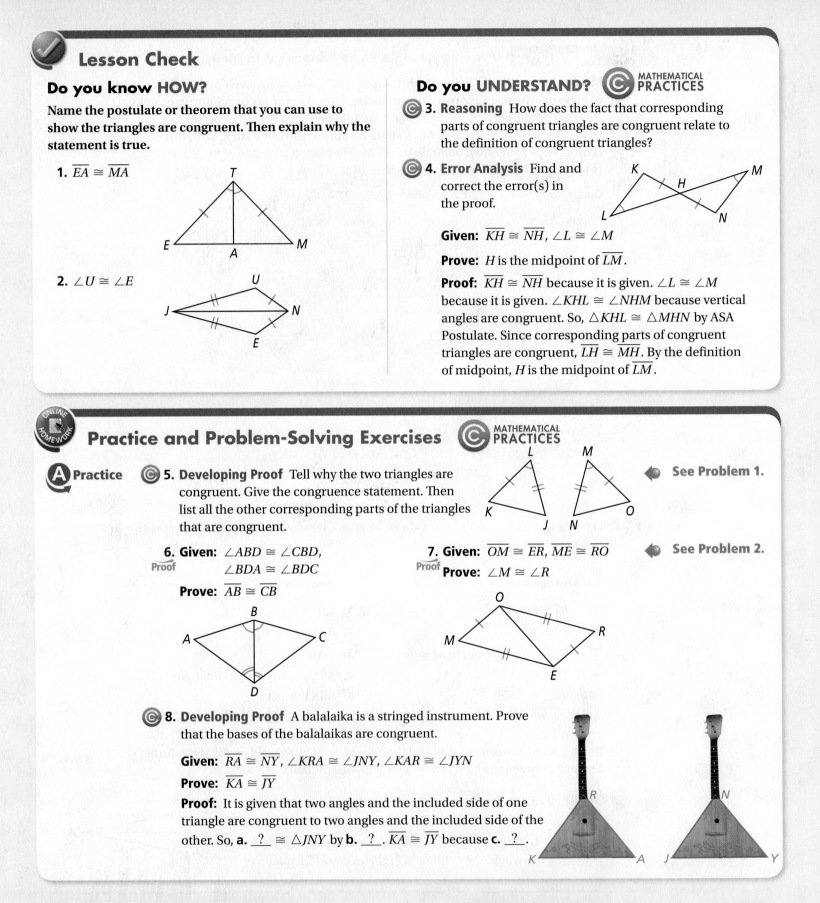

Do you UNDERSTAND? MATHEMATICAL PRACTICES

3. **Reasoning** How does the fact that corresponding parts of congruent triangles are congruent relate to the definition of congruent triangles?

4. **Error Analysis** Find and correct the error(s) in the proof.

Given: $\overline{KH} \cong \overline{NH}$, $\angle L \cong \angle M$

Prove: H is the midpoint of $\overline{LM}$.

Proof: $\overline{KH} \cong \overline{NH}$ because it is given. $\angle L \cong \angle M$ because it is given. $\angle KHL \cong \angle NHM$ because vertical angles are congruent. So, $\triangle KHL \cong \triangle MHN$ by ASA Postulate. Since corresponding parts of congruent triangles are congruent, $\overline{LH} \cong \overline{MH}$. By the definition of midpoint, H is the midpoint of $\overline{LM}$.

Practice and Problem-Solving Exercises MATHEMATICAL PRACTICES

A Practice

5. **Developing Proof** Tell why the two triangles are congruent. Give the congruence statement. Then list all the other corresponding parts of the triangles that are congruent.

◀ See Problem 1.

6. **Given:** $\angle ABD \cong \angle CBD$, $\angle BDA \cong \angle BDC$

   Prove: $\overline{AB} \cong \overline{CB}$

7. **Given:** $\overline{OM} \cong \overline{ER}$, $\overline{ME} \cong \overline{RO}$

   Prove: $\angle M \cong \angle R$

◀ See Problem 2.

8. **Developing Proof** A balalaika is a stringed instrument. Prove that the bases of the balalaikas are congruent.

   Given: $\overline{RA} \cong \overline{NY}$, $\angle KRA \cong \angle JNY$, $\angle KAR \cong \angle JYN$

   Prove: $\overline{KA} \cong \overline{JY}$

   Proof: It is given that two angles and the included side of one triangle are congruent to two angles and the included side of the other. So, **a.** __?__ $\cong \triangle JNY$ by **b.** __?__ . $\overline{KA} \cong \overline{JY}$ because **c.** __?__ .

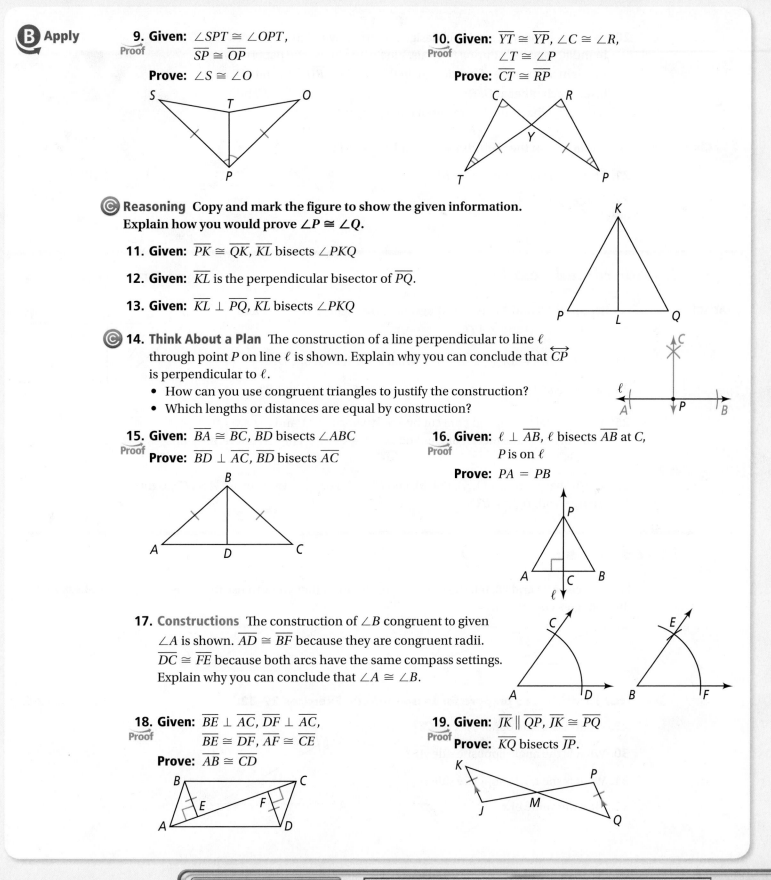

B Apply

9. Given: $\angle SPT \cong \angle OPT$, $\overline{SP} \cong \overline{OP}$
Proof
Prove: $\angle S \cong \angle O$

10. Given: $\overline{YT} \cong \overline{YP}$, $\angle C \cong \angle R$, $\angle T \cong \angle P$
Proof
Prove: $\overline{CT} \cong \overline{RP}$

C Reasoning Copy and mark the figure to show the given information. Explain how you would prove $\angle P \cong \angle Q$.

11. Given: $\overline{PK} \cong \overline{QK}$, $\overline{KL}$ bisects $\angle PKQ$

12. Given: $\overline{KL}$ is the perpendicular bisector of $\overline{PQ}$.

13. Given: $\overline{KL} \perp \overline{PQ}$, $\overline{KL}$ bisects $\angle PKQ$

C 14. Think About a Plan The construction of a line perpendicular to line ℓ through point P on line ℓ is shown. Explain why you can conclude that $\overleftrightarrow{CP}$ is perpendicular to ℓ.
- How can you use congruent triangles to justify the construction?
- Which lengths or distances are equal by construction?

15. Given: $\overline{BA} \cong \overline{BC}$, $\overline{BD}$ bisects $\angle ABC$
Proof
Prove: $\overline{BD} \perp \overline{AC}$, $\overline{BD}$ bisects $\overline{AC}$

16. Given: $\ell \perp \overline{AB}$, ℓ bisects $\overline{AB}$ at C, P is on ℓ
Proof
Prove: $PA = PB$

17. Constructions The construction of $\angle B$ congruent to given $\angle A$ is shown. $\overline{AD} \cong \overline{BF}$ because they are congruent radii. $\overline{DC} \cong \overline{FE}$ because both arcs have the same compass settings. Explain why you can conclude that $\angle A \cong \angle B$.

18. Given: $\overline{BE} \perp \overline{AC}$, $\overline{DF} \perp \overline{AC}$, $\overline{BE} \cong \overline{DF}$, $\overline{AF} \cong \overline{CE}$
Proof
Prove: $\overline{AB} \cong \overline{CD}$

19. Given: $\overline{JK} \parallel \overline{QP}$, $\overline{JK} \cong \overline{PQ}$
Proof
Prove: $\overline{KQ}$ bisects $\overline{JP}$.

20. Designs Rangoli is a colorful design pattern drawn outside houses in India, especially during festivals. Vina plans to use the pattern at the right as the base of her design. In this pattern, $\overline{RU}$, $\overline{SV}$, and $\overline{QT}$ bisect each other at O. $RS = 6$, $RU = 12$, $\overline{RU} \cong \overline{SV}$, $\overline{ST} \parallel \overline{RU}$, and $\overline{RS} \parallel \overline{QT}$. What is the perimeter of the hexagon?

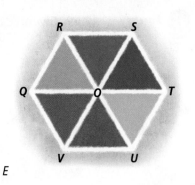

Challenge In the diagram at the right, $\overline{BA} \cong \overline{KA}$ and $\overline{BE} \cong \overline{KE}$.

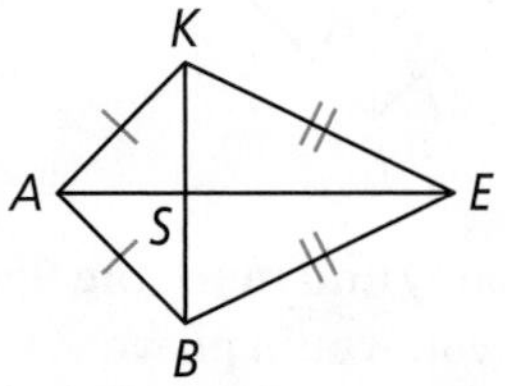

21. Prove: S is the midpoint of $\overline{BK}$.
Proof
22. Prove: $\overline{BK} \perp \overline{AE}$
Proof

Standardized Test Prep

GRIDDED RESPONSE

SAT/ACT   For Exercises 23 and 24, use the diagram at the right. $\overline{TM} \perp \overline{BD}$ and $\overline{TM}$ bisects $\angle BTD$ and $\angle ATC$.

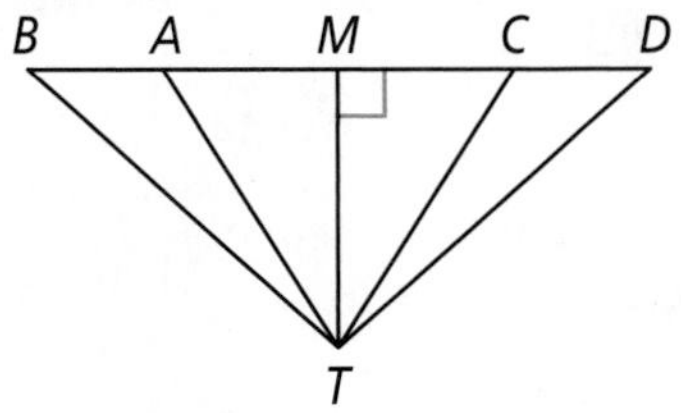

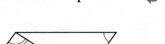

23. Suppose $BD = 17$ and $AM = 5$. What is the length of $\overline{CD}$?

24. Suppose $m\angle ATC = 64$, and $m\angle BTA = 16$. What is $m\angle B$?

25. Two parallel lines q and s are cut by a transversal t. $\angle 1$ and $\angle 2$ are a pair of alternate interior angles and $m\angle 2 = 38$. $\angle 1$ and $\angle 3$ are vertical angles. What is $m\angle 3$?

26. $\triangle ABC$ has vertices $A(1, 9)$, $B(4, 3)$, and $C(x, 6)$. For what value of x is $\triangle ABC$ a right triangle with right $\angle B$?

Mixed Review

For Exercises 27 and 28, tell the postulate or theorem that you can use to prove the triangles congruent.

◀ **See Lesson 4-3.**

27.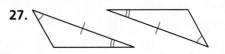

28.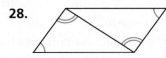

Get Ready!   To prepare for Lesson 4-5, do Exercises 29–32.

◀ **See Lesson 3-5.**

29. What is the side opposite $\angle ABC$?

30. What is the angle opposite side $\overline{AB}$?

31. What is the angle opposite side $\overline{BC}$?

32. Find the value of x.

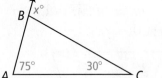

Paper-Folding Conjectures

© **Content Standard**

G.CO.12 Make formal geometric constructions with a variety of tools and methods (. . . paper folding . . .).

Isosceles triangles have two congruent sides. Folding one of the sides onto the other will suggest another important property of isosceles triangles.

Activity 1

Step 1 Construct an isosceles $\triangle ABC$ on tracing paper, with $\overline{AC} \cong \overline{BC}$.

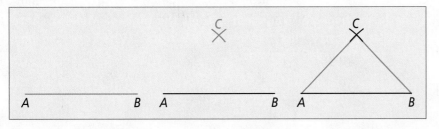

Step 2 Fold the paper so the two congruent sides fit exactly one on top of the other. Crease the paper. Label the intersection of the fold line and $\overline{AB}$ as point D.

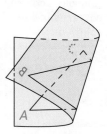

1. What do you notice about $\angle A$ and $\angle B$? Compare your results with others. Make a conjecture about the angles opposite the congruent sides in an isosceles triangle.

2. **a.** Study the fold line $\overline{CD}$ and the base $\overline{AB}$. What type of angles are $\angle CDA$ and $\angle CDB$? How do $\overline{AD}$ and $\overline{BD}$ seem to be related?

   b. Use your answers to part (a) to complete the conjecture:
   The fold line $\overline{CD}$ is the _?_ of the base $\overline{AB}$ of isosceles $\triangle ABC$.

Activity 2

In Activity 1, you made a conjecture about angles opposite the congruent sides of a triangle. You can also fold paper to study whether the converse is true.

Step 1 On tracing paper, draw acute angle F and one side $\overline{FG}$. Construct $\angle G$ as shown, so that $\angle G \cong \angle F$.

Step 2 Fold the paper so $\angle F$ and $\angle G$ fit exactly one on top of the other.

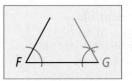

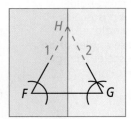

3. Why do sides 1 and 2 meet at point H on the fold line? Make a conjecture about sides $\overline{FH}$ and $\overline{GH}$ opposite congruent angles in a triangle.

4. Write your conjectures from Questions 1 and 3 as a biconditional.

Isosceles and Equilateral Triangles

Content Standards
G.CO.10 Prove theorems about triangles . . . base angles of isosceles triangles are congruent . . .
Also G.CO.13, G.SRT.5

Objective To use and apply properties of isosceles and equilateral triangles

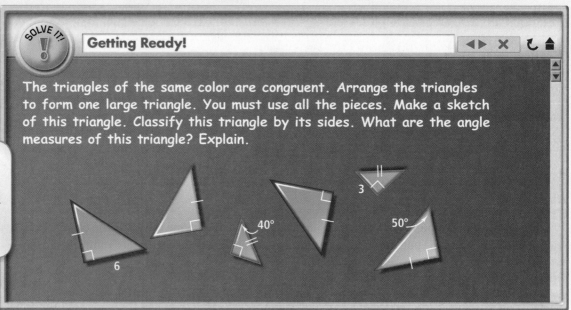

Solving puzzles is fun! Work with the pieces until they make a whole triangle. Look for patterns in your solution.

MATHEMATICAL PRACTICES

Getting Ready!

The triangles of the same color are congruent. Arrange the triangles to form one large triangle. You must use all the pieces. Make a sketch of this triangle. Classify this triangle by its sides. What are the angle measures of this triangle? Explain.

In the Solve It, you classified a triangle based on the lengths of its sides. You can also identify certain triangles based on information about their angles. In this lesson, you will learn how to use and apply properties of isosceles and equilateral triangles.

Dynamic Activity
Isosceles and Equilateral Triangles

Essential Understanding The angles and sides of isosceles and equilateral triangles have special relationships.

Isosceles triangles are common in the real world. You can frequently see them in structures such as bridges and buildings, as well as in art and design. The congruent sides of an isosceles triangle are its **legs.** The third side is the **base.** The two congruent legs form the **vertex angle.** The other two angles are the **base angles.**

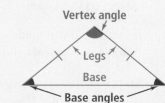

Lesson Vocabulary
• legs of an isosceles triangle
• base of an isosceles triangle
• vertex angle of an isosceles triangle
• base angles of an isosceles triangle
• corollary

Theorem 4-3 Isosceles Triangle Theorem

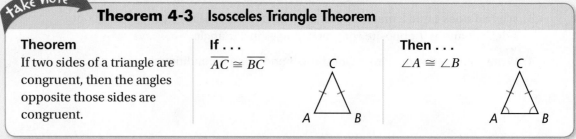

Theorem	**If . . .**	**Then . . .**
If two sides of a triangle are congruent, then the angles opposite those sides are congruent.	$\overline{AC} \cong \overline{BC}$	$\angle A \cong \angle B$

The proof of the Isosceles Triangle Theorem requires an auxiliary line.

Proof **Proof of Theorem 4-3: Isosceles Triangle Theorem**

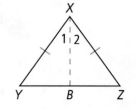

Begin with isosceles $\triangle XYZ$ with $\overline{XY} \cong \overline{XZ}$. Draw $\overline{XB}$, the bisector of the vertex angle $\angle YXZ$.

Given: $\overline{XY} \cong \overline{XZ}$, $\overline{XB}$ bisects $\angle YXZ$

Prove: $\angle Y \cong \angle Z$

Proof: $\overline{XY} \cong \overline{XZ}$ is given. By the definition of angle bisector, $\angle 1 \cong \angle 2$. By the Reflexive Property of Congruence, $\overline{XB} \cong \overline{XB}$. So by the SAS Postulate, $\triangle XYB \cong \triangle XZB$. $\angle Y \cong \angle Z$ since corresponding parts of congruent triangles are congruent.

take note

Theorem 4-4 Converse of the Isosceles Triangle Theorem		
Theorem If two angles of a triangle are congruent, then the sides opposite those angles are congruent.	**If . . .** $\angle A \cong \angle B$ 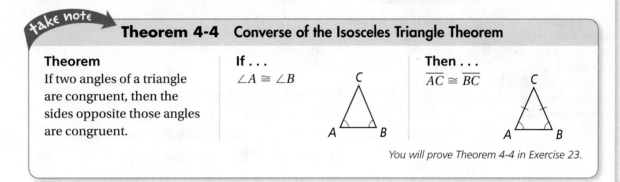	**Then . . .** $\overline{AC} \cong \overline{BC}$

You will prove Theorem 4-4 in Exercise 23.

© **Problem 1** **Using the Isosceles Triangle Theorems**

Think

What are you looking for in the diagram?
To use the Isosceles Triangle Theorems, you need a pair of congruent angles or a pair of congruent sides.

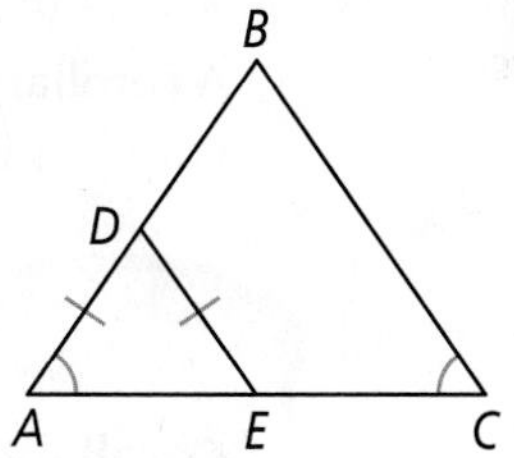

A Is $\overline{AB}$ congruent to $\overline{CB}$? Explain.

Yes. Since $\angle C \cong \angle A$, $\overline{AB} \cong \overline{CB}$ by the Converse of the Isosceles Triangle Theorem.

B Is $\angle A$ congruent to $\angle DEA$? Explain.

Yes. Since $\overline{AD} \cong \overline{ED}$, $\angle A \cong \angle DEA$ by the Isosceles Triangle Theorem.

© ✓ **Got It?** **1. a.** Is $\angle WVS$ congruent to $\angle S$? Is $\overline{TR}$ congruent to $\overline{TS}$? Explain.
b. Reasoning Can you conclude that $\triangle RUV$ is isosceles? Explain.

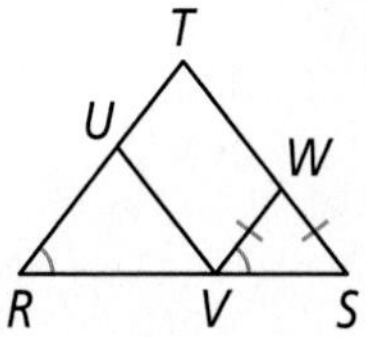

An isosceles triangle has a certain type of symmetry about a line through its vertex angle. The theorems in this lesson suggest this symmetry, which you will study in greater detail in Lesson 9-4.

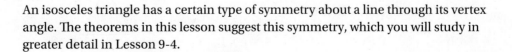

take note

Theorem 4-5

Theorem	**If . . .**	**Then . . .**
If a line bisects the vertex angle of an isosceles triangle, then the line is also the perpendicular bisector of the base.	$\overline{AC} \cong \overline{BC}$ and $\angle ACD \cong \angle BCD$	$\overline{CD} \perp \overline{AB}$ and $\overline{AD} \cong \overline{BD}$

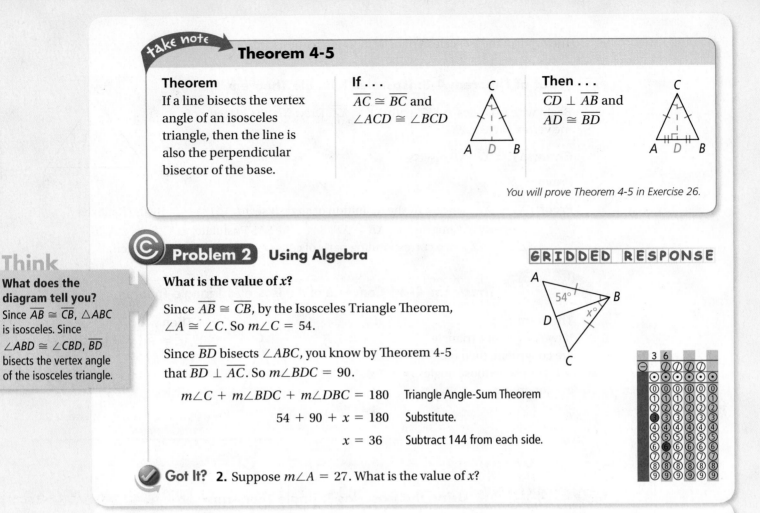

You will prove Theorem 4-5 in Exercise 26.

Think

What does the diagram tell you?
Since $\overline{AB} \cong \overline{CB}$, $\triangle ABC$ is isosceles. Since $\angle ABD \cong \angle CBD$, $\overline{BD}$ bisects the vertex angle of the isosceles triangle.

Problem 2 Using Algebra

What is the value of x?

GRIDDED RESPONSE

Since $\overline{AB} \cong \overline{CB}$, by the Isosceles Triangle Theorem, $\angle A \cong \angle C$. So $m\angle C = 54$.

Since $\overline{BD}$ bisects $\angle ABC$, you know by Theorem 4-5 that $\overline{BD} \perp \overline{AC}$. So $m\angle BDC = 90$.

$m\angle C + m\angle BDC + m\angle DBC = 180$	Triangle Angle-Sum Theorem
$54 + 90 + x = 180$	Substitute.
$x = 36$	Subtract 144 from each side.

✓ **Got It?** **2.** Suppose $m\angle A = 27$. What is the value of x?

A **corollary** is a theorem that can be proved easily using another theorem. Since a corollary is a theorem, you can use it as a reason in a proof.

take note

Corollary to Theorem 4-3

Corollary	**If . . .**	**Then . . .**
If a triangle is equilateral, then the triangle is equiangular.	$\overline{XY} \cong \overline{YZ} \cong \overline{ZX}$	$\angle X \cong \angle Y \cong \angle Z$

Corollary to Theorem 4-4

Corollary	**If . . .**	**Then . . .**
If a triangle is equiangular, then the triangle is equilateral.	$\angle X \cong \angle Y \cong \angle Z$	$\overline{XY} \cong \overline{YZ} \cong \overline{ZX}$

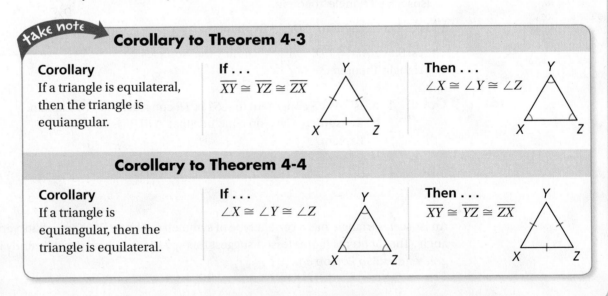

Problem 3 Finding Angle Measures

Design **What are the measures of $\angle A$, $\angle B$, and $\angle ADC$ in the photo at the right?**

Think	Write
The triangles are equilateral, so they are also equiangular. Find the measure of each angle of an equilateral triangle.	Let a = measure of one angle. $3a = 180$ $a = 60$
$\angle A$ and $\angle B$ are both angles in an equilateral triangle.	$m\angle A = m\angle B = 60$
Use the Angle Addition Postulate to find the measure of $\angle ADC$.	$m\angle ADC = m\angle ADE + m\angle CDE$
Both $\angle ADE$ and $\angle CDE$ are angles in an equilateral triangle. So $m\angle ADE = 60$ and $m\angle CDE = 60$. Substitute into the above equation and simplify.	$m\angle ADC = 60 + 60$ $m\angle ADC = 120$

Got It? **3.** Suppose the triangles in Problem 3 are isosceles triangles, where $\angle ADE$, $\angle DEC$, and $\angle ECB$ are vertex angles. If the vertex angles each have a measure of 58, what are $m\angle A$ and $m\angle BCD$?

Lesson Check

Do you know HOW?

1. What is $m\angle A$?

a.

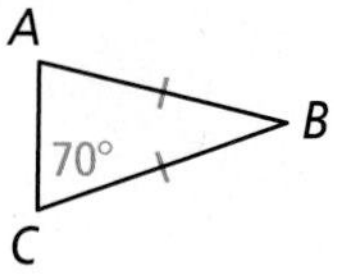

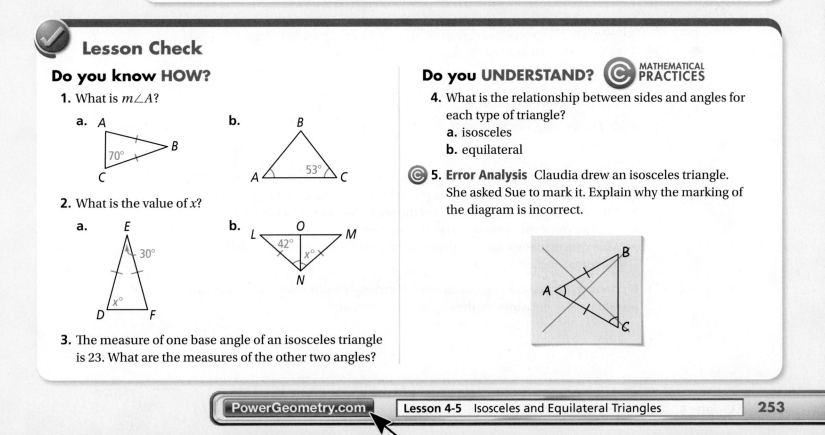

b.

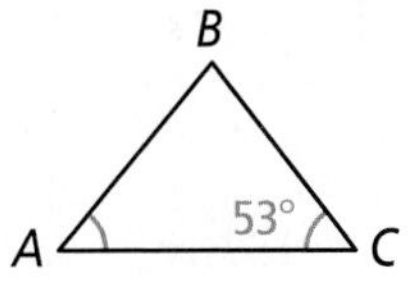

2. What is the value of x?

a.

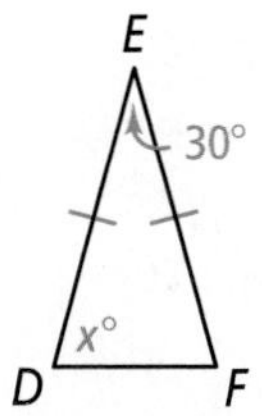

b.

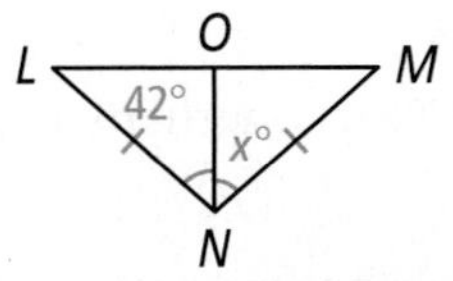

3. The measure of one base angle of an isosceles triangle is 23. What are the measures of the other two angles?

Do you UNDERSTAND? MATHEMATICAL PRACTICES

4. What is the relationship between sides and angles for each type of triangle?

a. isosceles

b. equilateral

5. Error Analysis Claudia drew an isosceles triangle. She asked Sue to mark it. Explain why the marking of the diagram is incorrect.

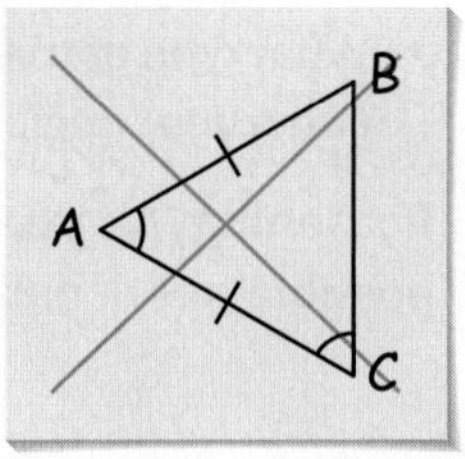

Practice · Complete each statement. Explain why it is true. See Problem 1.

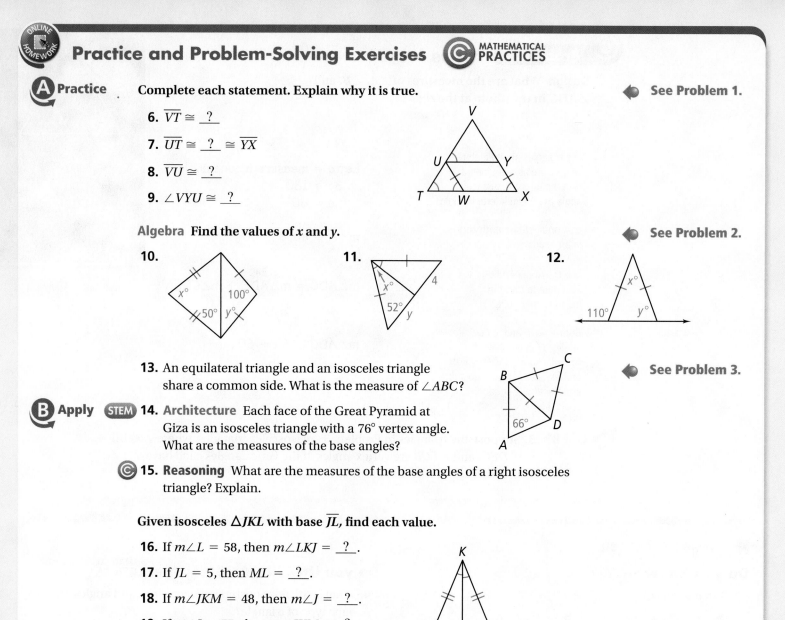

6. $\overline{VT} \cong$ __?__

7. $\overline{UT} \cong$ __?__ $\cong \overline{YX}$

8. $\overline{VU} \cong$ __?__

9. $\angle VYU \cong$ __?__

Algebra Find the values of x and y. See Problem 2.

10.

11.

12.

13. An equilateral triangle and an isosceles triangle share a common side. What is the measure of $\angle ABC$? See Problem 3.

Apply STEM **14. Architecture** Each face of the Great Pyramid at Giza is an isosceles triangle with a 76° vertex angle. What are the measures of the base angles?

15. Reasoning What are the measures of the base angles of a right isosceles triangle? Explain.

Given isosceles $\triangle JKL$ with base $\overline{JL}$, find each value.

16. If $m\angle L = 58$, then $m\angle LKJ =$ __?__ .

17. If $JL = 5$, then $ML =$ __?__ .

18. If $m\angle JKM = 48$, then $m\angle J =$ __?__ .

19. If $m\angle J = 55$, then $m\angle JKM =$ __?__ .

20. Think About a Plan A triangle has angle measures $x + 15$, $3x - 35$, and $4x$. What type of triangle is it? Be as specific as possible. Justify your answer.
- What do you know about the sum of the angle measures of a triangle?
- What do you need to know to classify a triangle?
- What type of triangle has no congruent angles? Two congruent angles? Three congruent angles?

21. Reasoning An exterior angle of an isosceles triangle has measure 100. Find two possible sets of measures for the angles of the triangle.

22. Developing Proof Here is another way to prove the Isosceles Triangle Theorem. Supply the missing information.

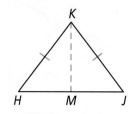

Begin with isosceles $\triangle HKJ$ with $\overline{KH} \cong \overline{KJ}$.
Draw **a.** __?__, a bisector of the base $\overline{HJ}$.

Given: $\overline{KH} \cong \overline{KJ}$, **b.** __?__ bisects $\overline{HJ}$
Prove: $\angle H \cong \angle J$

Statements	Reasons
1) $\overline{KM}$ bisects $\overline{HJ}$.	1) **c.** __?__
2) $\overline{HM} \cong \overline{JM}$	2) **d.** __?__
3) $\overline{KH} \cong \overline{KJ}$	3) Given
4) $\overline{KM} \cong \overline{KM}$	4) **e.** __?__
5) $\triangle KHM \cong \triangle KJM$	5) **f.** __?__
6) $\angle H \cong \angle J$	6) **g.** __?__

23. Supply the missing information in this statement of the Converse of the Isosceles Proof Triangle Theorem. Then write a proof.

Begin with $\triangle PRQ$ with $\angle P \cong \angle Q$.
Draw **a.** __?__, the bisector of $\angle PRQ$.
Given: $\angle P \cong \angle Q$, **b.** __?__ bisects $\angle PRQ$
Prove: $\overline{PR} \cong \overline{QR}$

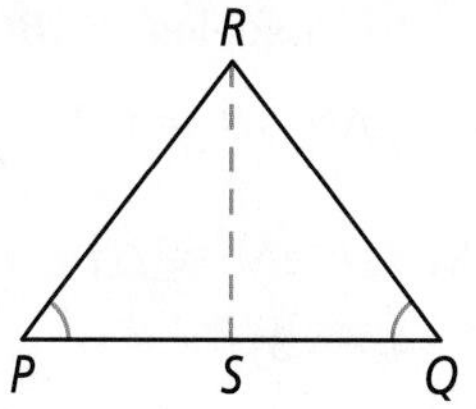

24. Writing Explain how the corollaries to the Isosceles Triangle Theorem and its converse follow from the theorems.

25. Given: $\overline{AE} \cong \overline{DE}$, $\overline{AB} \cong \overline{DC}$
Proof **Prove:** $\triangle ABE \cong \triangle DCE$

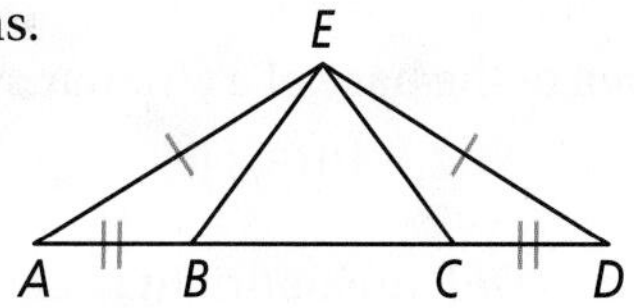

26. Prove Theorem 4-5. Use the diagram next to it on page 252.
Proof

STEM 27. a. Communications In the diagram at the right, what type of triangle is formed by the cables of the same height and the ground?
b. What are the two different base lengths of the triangles?
c. How is the tower related to each of the triangles?

28. Algebra The length of the base of an isosceles triangle is x. The length of a leg is $2x - 5$. The perimeter of the triangle is 20. Find x.

29. Constructions Construct equilateral triangle ABC. Justify your method.

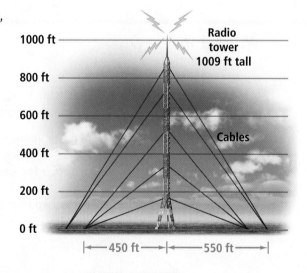

Radio tower 1009 ft tall

Cables

450 ft ⟶⟵ 550 ft

Algebra Find the values of *m* and *n*.

30.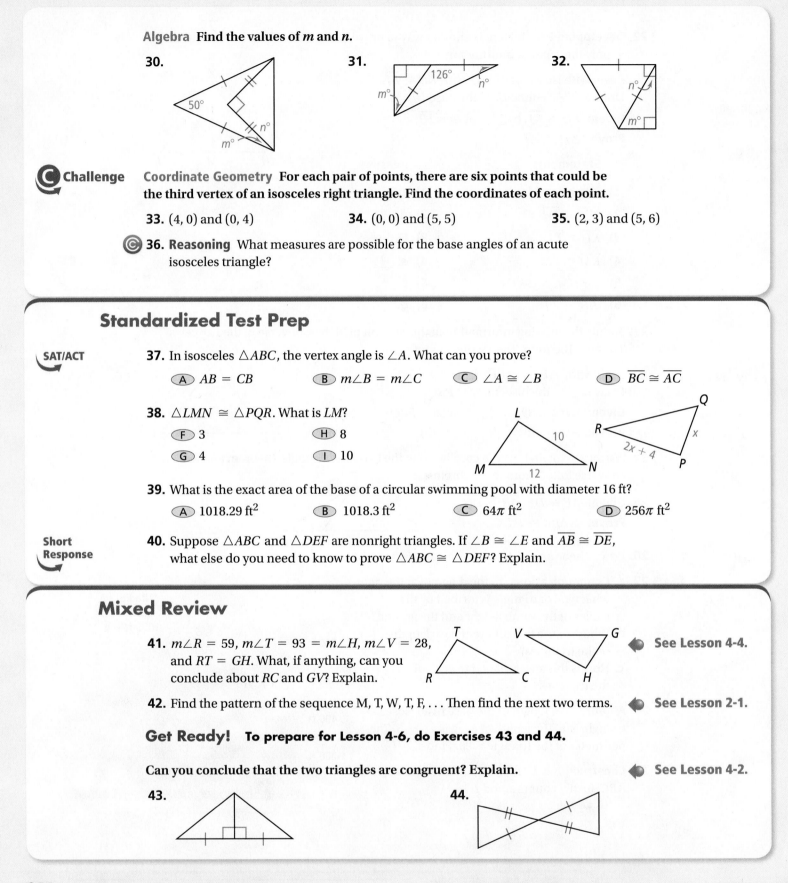

31.

32.

Challenge

Coordinate Geometry For each pair of points, there are six points that could be the third vertex of an isosceles right triangle. Find the coordinates of each point.

33. (4, 0) and (0, 4)

34. (0, 0) and (5, 5)

35. (2, 3) and (5, 6)

36. Reasoning What measures are possible for the base angles of an acute isosceles triangle?

Standardized Test Prep

SAT/ACT

37. In isosceles $\triangle ABC$, the vertex angle is $\angle A$. What can you prove?

Ⓐ $AB = CB$ Ⓑ $m\angle B = m\angle C$ Ⓒ $\angle A \cong \angle B$ Ⓓ $\overline{BC} \cong \overline{AC}$

38. $\triangle LMN \cong \triangle PQR$. What is LM?

Ⓕ 3   Ⓗ 8

Ⓖ 4   Ⓘ 10

39. What is the exact area of the base of a circular swimming pool with diameter 16 ft?

Ⓐ 1018.29 ft^2 Ⓑ 1018.3 ft^2 Ⓒ $64\pi \text{ ft}^2$ Ⓓ $256\pi \text{ ft}^2$

Short Response

40. Suppose $\triangle ABC$ and $\triangle DEF$ are nonright triangles. If $\angle B \cong \angle E$ and $\overline{AB} \cong \overline{DE}$, what else do you need to know to prove $\triangle ABC \cong \triangle DEF$? Explain.

Mixed Review

41. $m\angle R = 59$, $m\angle T = 93 = m\angle H$, $m\angle V = 28$, and $RT = GH$. What, if anything, can you conclude about RC and GV? Explain.

◀ **See Lesson 4-4.**

42. Find the pattern of the sequence M, T, W, T, F, . . . Then find the next two terms.

◀ **See Lesson 2-1.**

Get Ready! To prepare for Lesson 4-6, do Exercises 43 and 44.

Can you conclude that the two triangles are congruent? Explain.

◀ **See Lesson 4-2.**

43.

44.

Algebra Review

Systems of Linear Equations

Use With Lesson 4-6

You can solve a system of equations in two variables by using substitution.

Example 1

Algebra Solve the system.
$$y = 3x + 5$$
$$y = x + 1$$

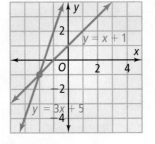

$$y = x + 1 \quad \text{Start with one equation.}$$
$$3x + 5 = x + 1 \quad \text{Substitute } 3x + 5 \text{ for } y.$$
$$2x = -4 \quad \text{Solve for } x.$$
$$x = -2$$

Substitute -2 for x in either equation and solve for y.

$$y = x + 1 = (-2) + 1 = -1$$

Since $x = -2$ and $y = 1$, the solution is $(-2, -1)$. This is the point of intersection of the two lines.

The graph of a linear system with *infinitely many solutions* is one line, and the graph of a linear system with *no solution* is two parallel lines.

Example 2

Algebra Solve the system.
$$x + y = 3$$
$$4x + 4y = 8$$

$$x + y = 3 \quad \text{Start with one equation.}$$
$$x = 3 - y \quad \text{Solve the equation for } x.$$
$$4(3 - y) + 4y = 8 \quad \text{Substitute } 3 - y \text{ for } x \text{ in the second equation.}$$
$$12 - 4y + 4y = 8 \quad \text{Solve for } y.$$
$$12 = 8 \quad \text{False!}$$

Since $12 = 8$ is a false statement, the system has no solution.

Exercises

Solve each system of equations.

1. $y = x - 4$
   $y = 3x + 2$

2. $2x - y = 8$
   $x + 2y = 9$

3. $3x + y = 4$
   $-6x - 2y = 12$

4. $2x - 3 = y + 3$
   $2x + y = -3$

5. $y = x + 1$
   $x = y - 1$

6. $x - y = 4$
   $3x - 3y = 6$

7. $y = -x + 2$
   $2y = 4 - 2x$

8. $y = 2x - 1$
   $y = 3x - 7$

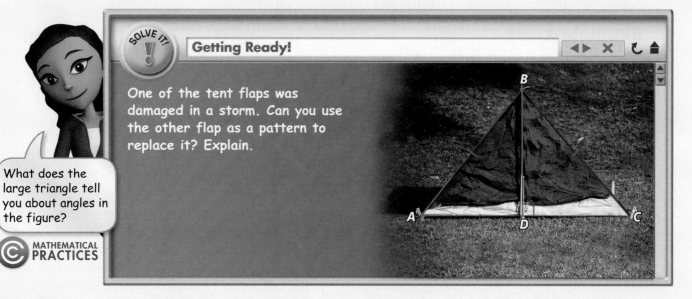

4-6 Congruence in Right Triangles

Content Standard
G.SRT.5 Use congruence . . . criteria to solve problems and prove relationships in geometric figures.

Objective To prove right triangles congruent using the Hypotenuse-Leg Theorem

Getting Ready!

One of the tent flaps was damaged in a storm. Can you use the other flap as a pattern to replace it? Explain.

What does the large triangle tell you about angles in the figure?

MATHEMATICAL PRACTICES

Dynamic Activity
Congruent Right Triangles

Lesson Vocabulary
• hypotenuse
• legs of a right triangle

In the diagram below, two sides and a nonincluded angle of one triangle are congruent to two sides and the nonincluded angle of another triangle.

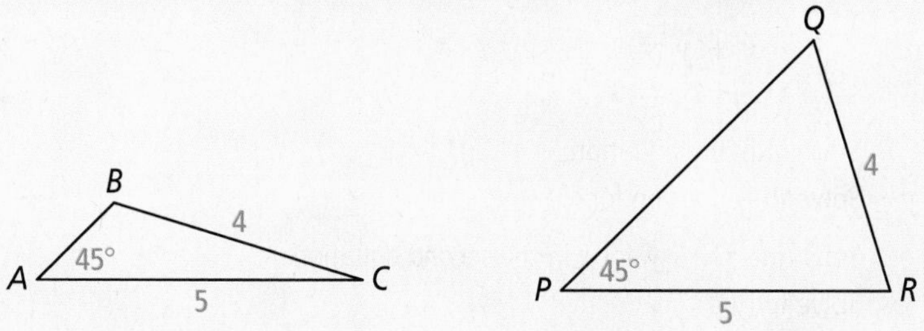

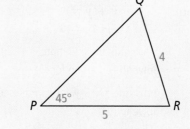

Notice that the triangles are not congruent. So, you can conclude that Side-Side-Angle is *not* a valid method for proving two triangles congruent. This method, however, works in the special case of right triangles, where the right angles are the nonincluded angles.

In a right triangle, the side opposite the right angle is called the **hypotenuse**. It is the longest side in the triangle. The other two sides are called **legs**.

The right angle always "points" to the hypotenuse.

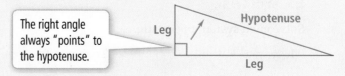

Essential Understanding You can prove that two triangles are congruent without having to show that *all* corresponding parts are congruent. In this lesson, you will prove right triangles congruent by using one pair of right angles, a pair of hypotenuses, and a pair of legs.

take note

Theorem 4-6   Hypotenuse-Leg (HL) Theorem

Theorem	If . . .	Then . . .
If the hypotenuse and a leg of one right triangle are congruent to the hypotenuse and a leg of another right triangle, then the triangles are congruent.	$\triangle PQR$ and $\triangle XYZ$ are right $\triangle$, $\overline{PR} \cong \overline{XZ}$, and $\overline{PQ} \cong \overline{XY}$ 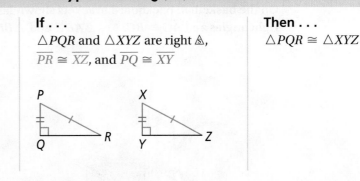	$\triangle PQR \cong \triangle XYZ$

To prove the HL Theorem you will need to draw auxiliary lines to make a third triangle.

Proof **Proof of Theorem 4-6: Hypotenuse-Leg Theorem**

Given: $\triangle PQR$ and $\triangle XYZ$ are right triangles, with right angles Q and Y. $\overline{PR} \cong \overline{XZ}$ and $\overline{PQ} \cong \overline{XY}$.

Prove: $\triangle PQR \cong \triangle XYZ$

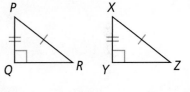

Proof: On $\triangle XYZ$, draw $\overrightarrow{ZY}$.

Mark point S so that $YS = QR$. Then, $\triangle PQR \cong \triangle XYS$ by SAS.

Since corresponding parts of congruent triangles are congruent, $\overline{PR} \cong \overline{XS}$. It is given that $\overline{PR} \cong \overline{XZ}$, so $\overline{XS} \cong \overline{XZ}$ by the Transitive Property of Congruence. By the Isosceles Triangle Theorem, $\angle S \cong \angle Z$, so $\triangle XYS \cong \triangle XYZ$ by AAS. Therefore, $\triangle PQR \cong \triangle XYZ$ by the Transitive Property of Congruence.

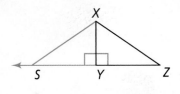

take note

Key Concept   Conditions for HL Theorem

To use the HL Theorem, the triangles must meet three conditions.

Conditions
- There are two right triangles.
- The triangles have congruent hypotenuses.
- There is one pair of congruent legs.

Problem 1 **Using the HL Theorem**

On the basketball backboard brackets shown below, ∠*ADC* and ∠*BDC* are right angles and $\overline{AC} \cong \overline{BC}$. Are △*ADC* and △*BDC* congruent? Explain.

Plan

How can you visualize the two right triangles?
Imagine cutting △*ABC* along $\overline{DC}$. On either side of the cut, you get triangles with the same leg $\overline{DC}$.

- You are given that ∠*ADC* and ∠*BDC* are right angles.
  So, △*ADC* and △*BDC* are right triangles.

- The hypotenuses of the two right triangles are $\overline{AC}$ and $\overline{BC}$.
  You are given that $\overline{AC} \cong \overline{BC}$.

- $\overline{DC}$ is a common leg of both △*ADC* and △*BDC*. $\overline{DC} \cong \overline{DC}$ by the Reflexive Property of Congruence.

Yes, △*ADC* ≅ △*BDC* by the HL Theorem.

Got It? **1. a. Given:** ∠*PRS* and ∠*RPQ* are
right angles, $\overline{SP} \cong \overline{QR}$

Prove: △*PRS* ≅ △*RPQ*

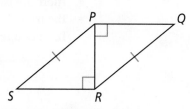

  b. Reasoning Your friend says, "Suppose you have two right triangles with congruent hypotenuses and one pair of congruent legs. It does not matter which leg in the first triangle is congruent to which leg in the second triangle. The triangles will be congruent." Is your friend correct? Explain.

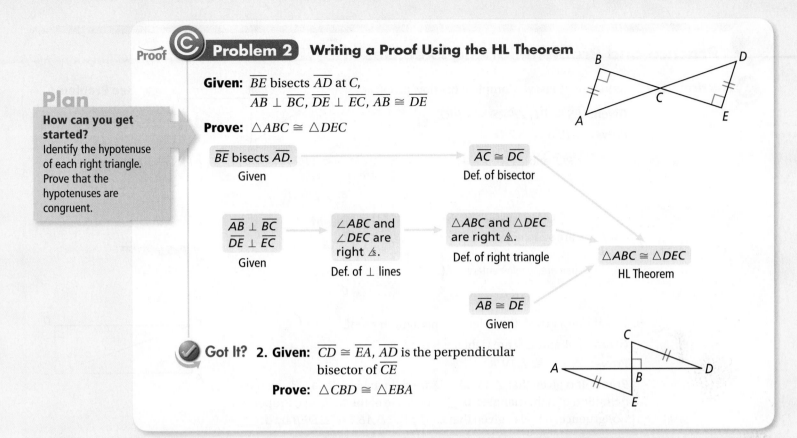

Proof © **Problem 2** **Writing a Proof Using the HL Theorem**

Given: $\overline{BE}$ bisects $\overline{AD}$ at C,
$\overline{AB} \perp \overline{BC}$, $\overline{DE} \perp \overline{EC}$, $\overline{AB} \cong \overline{DE}$

Prove: $\triangle ABC \cong \triangle DEC$

Plan

How can you get started?
Identify the hypotenuse of each right triangle. Prove that the hypotenuses are congruent.

| $\overline{BE}$ bisects $\overline{AD}$. Given | → | $\overline{AC} \cong \overline{DC}$ Def. of bisector |

| $\overline{AB} \perp \overline{BC}$ $\overline{DE} \perp \overline{EC}$ Given | → | $\angle ABC$ and $\angle DEC$ are right ∡. Def. of ⊥ lines | → | $\triangle ABC$ and $\triangle DEC$ are right ▲. Def. of right triangle | → | $\triangle ABC \cong \triangle DEC$ HL Theorem |

| $\overline{AB} \cong \overline{DE}$ Given |

✓ **Got It?** **2. Given:** $\overline{CD} \cong \overline{EA}$, $\overline{AD}$ is the perpendicular bisector of $\overline{CE}$

Prove: $\triangle CBD \cong \triangle EBA$

✓ **Lesson Check**

Do you know HOW?

Are the two triangles congruent? If so, write the congruence statement.

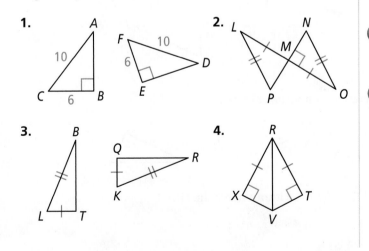

1.

2.

3.

4.

Do you UNDERSTAND? © MATHEMATICAL PRACTICES

© **5. Vocabulary** A right triangle has side lengths of 5 cm, 12 cm, and 13 cm. What is the length of the hypotenuse? How do you know?

© **6. Compare and Contrast** How do the HL Theorem and the SAS Postulate compare? How are they different? Explain.

© **7. Error Analysis** Your classmate says that there is not enough information to determine whether the two triangles below are congruent. Is your classmate correct? Explain.

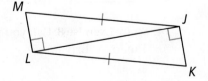

Practice and Problem-Solving Exercises

A Practice

8. Developing Proof Complete the flow proof.

See Problem 1.

Given: $\overline{PS} \cong \overline{PT}$, $\angle PRS \cong \angle PRT$

Prove: $\triangle PRS \cong \triangle PRT$

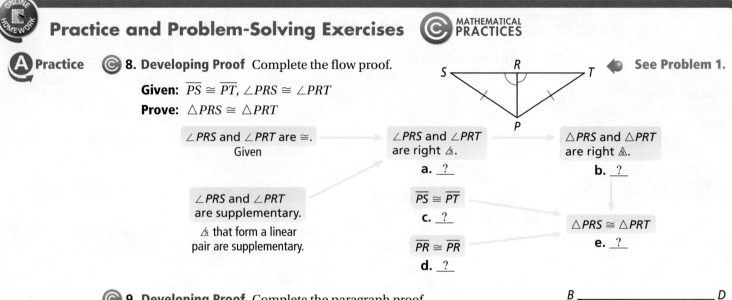

$\angle PRS$ and $\angle PRT$ are $\cong$.
Given

$\angle PRS$ and $\angle PRT$ are right $\triangle$.
a. ?

$\triangle PRS$ and $\triangle PRT$ are right $\triangle$.
b. ?

$\angle PRS$ and $\angle PRT$ are supplementary.
$\triangle$ that form a linear pair are supplementary.

$\overline{PS} \cong \overline{PT}$
c. ?

$\overline{PR} \cong \overline{PR}$
d. ?

$\triangle PRS \cong \triangle PRT$
e. ?

9. Developing Proof Complete the paragraph proof.

Given: $\angle A$ and $\angle D$ are right angles, $\overline{AB} \cong \overline{DE}$

Prove: $\triangle ABE \cong \triangle DEB$

Proof: It is given that $\angle A$ and $\angle D$ are right angles. So, **a.** ? by the definition of right triangles. **b.** ?, because of the Reflexive Property of Congruence. It is also given that **c.** ?. So, $\triangle ABE \cong \triangle DEB$ by **d.** ?.

10. Given: $\overline{HV} \perp \overline{GT}$, $\overline{GH} \cong \overline{TV}$,
   I is the midpoint of $\overline{HV}$

Prove: $\triangle IGH \cong \triangle ITV$

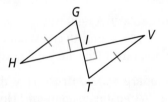

11. Given: $\overline{PM} \cong \overline{RJ}$,
   $\overline{PT} \perp \overline{TJ}$, $\overline{RM} \perp \overline{TJ}$,
   M is the midpoint of $\overline{TJ}$

Prove: $\triangle PTM \cong \triangle RMJ$

See Problem 2.

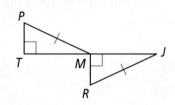

B Apply

Algebra For what values of x and y are the triangles congruent by HL?

12.

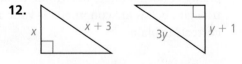

13.

14. Study Exercise 8. Can you prove that $\triangle PRS \cong \triangle PRT$ without using the HL Theorem? Explain.

15. Think About a Plan $\triangle ABC$ and $\triangle PQR$ are right triangular sections of a fire escape, as shown. Is each story of the building the same height? Explain.
- What can you tell from the diagram?
- How can you use congruent triangles here?

16. Writing "A HA!" exclaims your classmate. "There must be an HA Theorem, sort of like the HL Theorem!" Is your classmate correct? Explain.

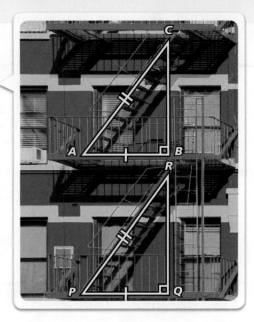

17. Given: $\overline{RS} \cong \overline{TU}, \overline{RS} \perp \overline{ST}, \overline{TU} \perp \overline{UV},$
Proof T is the midpoint of $\overline{RV}$
Prove: $\triangle RST \cong \triangle TUV$

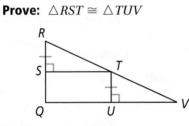

18. Given: $\triangle LNP$ is isosceles with base $\overline{NP}$,
Proof $\overline{MN} \perp \overline{NL}, \overline{QP} \perp \overline{PL}, \overline{ML} \cong \overline{QL}$
Prove: $\triangle MNL \cong \triangle QPL$

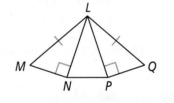

Constructions Copy the triangle and construct a triangle congruent to it using the given method.

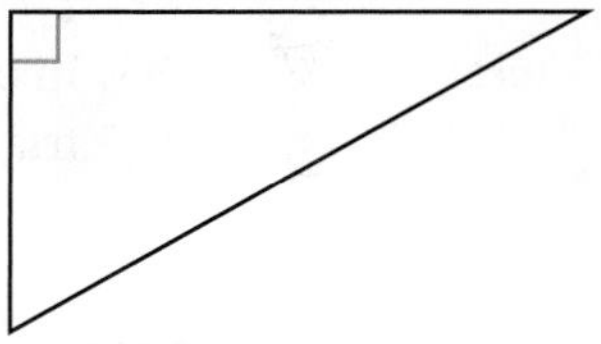

19. SAS

20. HL

21. ASA

22. SSS

23. Given: $\triangle GKE$ is isosceles with base $\overline{GE}$,
Proof $\angle L$ and $\angle D$ are right angles, and K is the midpoint of $\overline{LD}$.
Prove: $\overline{LG} \cong \overline{DE}$

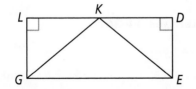

24. Given: $\overline{LO}$ bisects $\angle MLN$,
Proof $\overline{OM} \perp \overline{LM}, \overline{ON} \perp \overline{LN}$
Prove: $\triangle LMO \cong \triangle LNO$

25. Reasoning Are the triangles congruent? Explain.

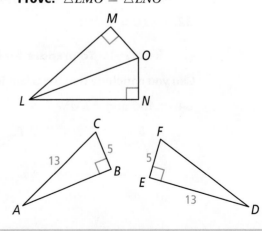

26. a. Coordinate Geometry Graph the points $A(-5, 6)$, $B(1, 3)$, $D(-8, 0)$, and $E(-2, -3)$. Draw $\overline{AB}$, $\overline{AE}$, $\overline{BD}$, and $\overline{DE}$. Label point C, the intersection of $\overline{AE}$ and $\overline{BD}$.

    b. Find the slopes of $\overline{AE}$ and $\overline{BD}$. How would you describe $\angle ACB$ and $\angle ECD$?

    c. Algebra Write equations for $\overleftrightarrow{AE}$ and $\overleftrightarrow{BD}$. What are the coordinates of C?

    d. Use the Distance Formula to find AB, BC, DC, and DE.

    e. Write a paragraph to prove that $\triangle ABC \cong \triangle EDC$.

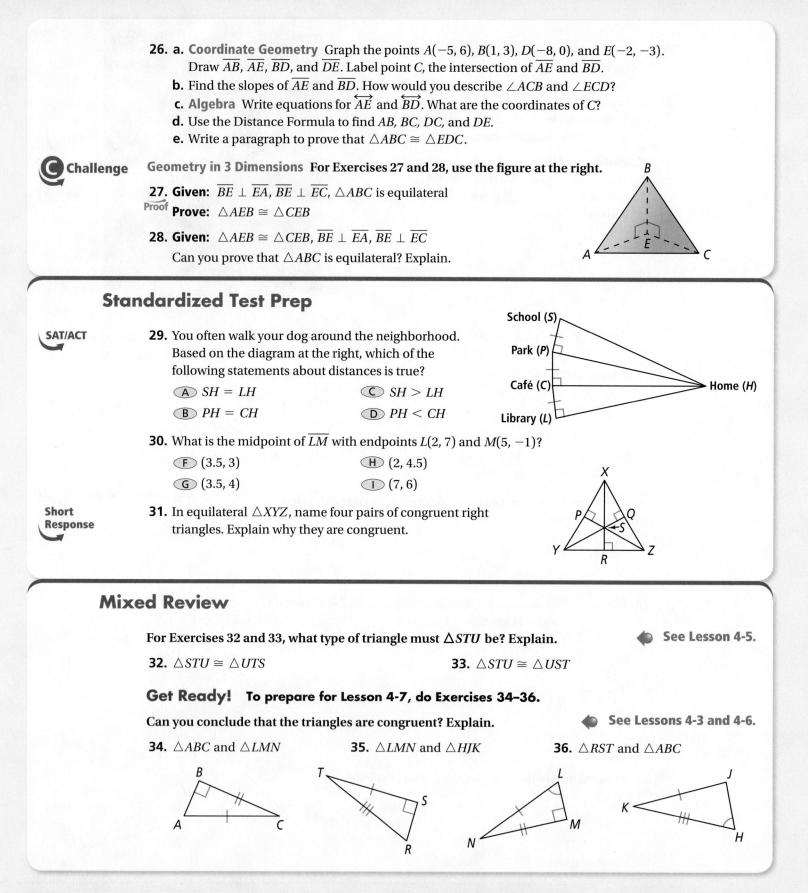

Challenge

Geometry in 3 Dimensions For Exercises 27 and 28, use the figure at the right.

27. Given: $\overline{BE} \perp \overline{EA}$, $\overline{BE} \perp \overline{EC}$, $\triangle ABC$ is equilateral

Proof **Prove:** $\triangle AEB \cong \triangle CEB$

28. Given: $\triangle AEB \cong \triangle CEB$, $\overline{BE} \perp \overline{EA}$, $\overline{BE} \perp \overline{EC}$

Can you prove that $\triangle ABC$ is equilateral? Explain.

Standardized Test Prep

SAT/ACT

29. You often walk your dog around the neighborhood. Based on the diagram at the right, which of the following statements about distances is true?

    Ⓐ $SH = LH$ Ⓒ $SH > LH$

    Ⓑ $PH = CH$ Ⓓ $PH < CH$

30. What is the midpoint of $\overline{LM}$ with endpoints $L(2, 7)$ and $M(5, -1)$?

    Ⓕ $(3.5, 3)$ Ⓗ $(2, 4.5)$

    Ⓖ $(3.5, 4)$ Ⓘ $(7, 6)$

Short Response

31. In equilateral $\triangle XYZ$, name four pairs of congruent right triangles. Explain why they are congruent.

Mixed Review

For Exercises 32 and 33, what type of triangle must $\triangle STU$ be? Explain.

◀ **See Lesson 4-5.**

32. $\triangle STU \cong \triangle UTS$ **33.** $\triangle STU \cong \triangle UST$

Get Ready! **To prepare for Lesson 4-7, do Exercises 34–36.**

Can you conclude that the triangles are congruent? Explain.

◀ **See Lessons 4-3 and 4-6.**

34. $\triangle ABC$ and $\triangle LMN$ **35.** $\triangle LMN$ and $\triangle HJK$ **36.** $\triangle RST$ and $\triangle ABC$

Congruence in Overlapping Triangles

© **Content Standard**
G.SRT.5 Use congruence . . . criteria to solve problems and prove relationships in geometric figures.

Objectives To identify congruent overlapping triangles
To prove two triangles congruent using other congruent triangles

Do all the triangles make you dizzy? Try to see each one. Then learn some tricks that may help you.

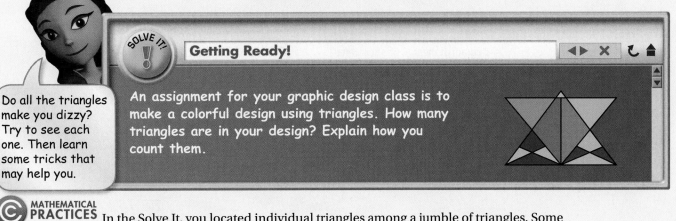

SOLVE IT!

Getting Ready!

An assignment for your graphic design class is to make a colorful design using triangles. How many triangles are in your design? Explain how you count them.

© **MATHEMATICAL PRACTICES**

In the Solve It, you located individual triangles among a jumble of triangles. Some triangle relationships are difficult to see because the triangles overlap.

Essential Understanding You can sometimes use the congruent corresponding parts of one pair of congruent triangles to prove another pair of triangles congruent. This often involves overlapping triangles.

Overlapping triangles may have a common side or angle. You can simplify your work with overlapping triangles by separating and redrawing the triangles.

© **Problem 1** **Identifying Common Parts**

Think

How can you see an individual triangle in order to redraw it?
Use your finger to trace along the lines connecting the three vertices. Then cover up any untraced lines.

What common angle do △ACD and △ECB share?

Separate and redraw △ACD and △ECB.

The common angle is ∠C.

✓ **Got It?** **1. a.** What is the common side in △ABD and △DCA?
b. What is the common side in △ABD and △BAC?

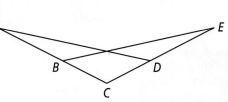

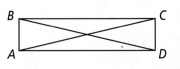

Given: $\angle ZXW \cong \angle YWX$, $\angle ZWX \cong \angle YXW$

Prove: $\overline{ZW} \cong \overline{YX}$

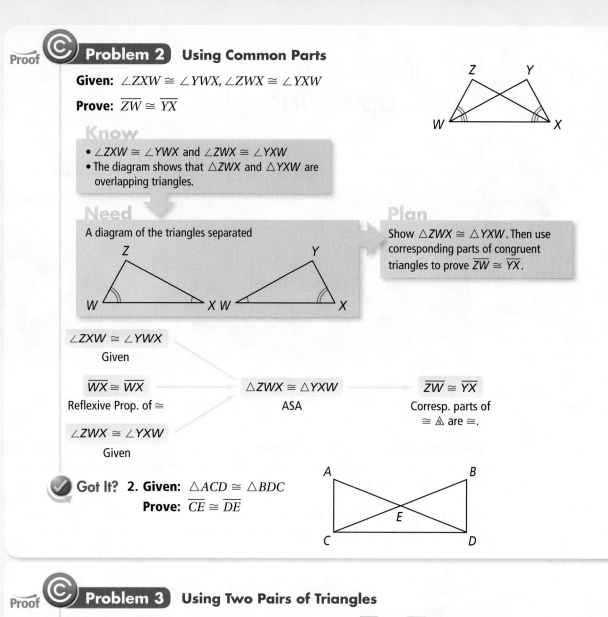

Know

- $\angle ZXW \cong \angle YWX$ and $\angle ZWX \cong \angle YXW$
- The diagram shows that $\triangle ZWX$ and $\triangle YXW$ are overlapping triangles.

Need

A diagram of the triangles separated

Plan

Show $\triangle ZWX \cong \triangle YXW$. Then use corresponding parts of congruent triangles to prove $\overline{ZW} \cong \overline{YX}$.

$\angle ZXW \cong \angle YWX$
Given

$\overline{WX} \cong \overline{WX}$
Reflexive Prop. of $\cong$

$\triangle ZWX \cong \triangle YXW$
ASA

$\overline{ZW} \cong \overline{YX}$
Corresp. parts of $\cong$ △ are $\cong$.

$\angle ZWX \cong \angle YXW$
Given

✓ **Got It?** **2. Given:** $\triangle ACD \cong \triangle BDC$

Prove: $\overline{CE} \cong \overline{DE}$

Plan

How do you choose another pair of triangles to help in your proof?
Look for triangles that share parts with $\triangle GED$ and $\triangle JEB$ and that you can prove congruent. In this case, first prove $\triangle AED \cong \triangle CEB$.

Given: In the origami design, E is the midpoint of $\overline{AC}$ and $\overline{DB}$.

Prove: $\triangle GED \cong \triangle JEB$

Proof: E is the midpoint of $\overline{AC}$ and $\overline{DB}$, so $\overline{AE} \cong \overline{CE}$ and $\overline{DE} \cong \overline{BE}$. $\angle AED \cong \angle CEB$ because vertical angles are congruent. Therefore, $\triangle AED \cong \triangle CEB$ by SAS. $\angle D \cong \angle B$ because corresponding parts of congruent triangles are congruent. $\angle GED \cong \angle JEB$ because vertical angles are congruent. Therefore, $\triangle GED \cong \triangle JEB$ by ASA.

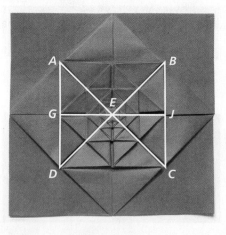

Got It? **3. Given:** $\overline{PS} \cong \overline{RS}$, $\angle PSQ \cong \angle RSQ$
Prove: $\triangle QPT \cong \triangle QRT$

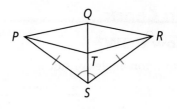

When several triangles overlap and you need to use one pair of congruent triangles to prove another pair congruent, you may find it helpful to draw a diagram of each pair of triangles.

Proof **Problem 4** **Separating Overlapping Triangles**

Given: $\overline{CA} \cong \overline{CE}$, $\overline{BA} \cong \overline{DE}$

Prove: $\overline{BX} \cong \overline{DX}$

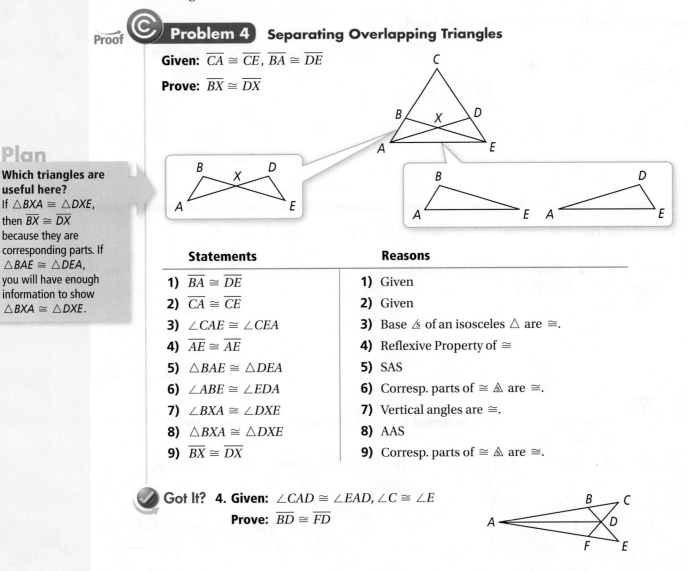

Plan

Which triangles are useful here?
If $\triangle BXA \cong \triangle DXE$, then $\overline{BX} \cong \overline{DX}$ because they are corresponding parts. If $\triangle BAE \cong \triangle DEA$, you will have enough information to show $\triangle BXA \cong \triangle DXE$.

Statements	Reasons
1) $\overline{BA} \cong \overline{DE}$	1) Given
2) $\overline{CA} \cong \overline{CE}$	2) Given
3) $\angle CAE \cong \angle CEA$	3) Base ⧊ of an isosceles △ are ≅.
4) $\overline{AE} \cong \overline{AE}$	4) Reflexive Property of ≅
5) $\triangle BAE \cong \triangle DEA$	5) SAS
6) $\angle ABE \cong \angle EDA$	6) Corresp. parts of ≅ ⧊ are ≅.
7) $\angle BXA \cong \angle DXE$	7) Vertical angles are ≅.
8) $\triangle BXA \cong \triangle DXE$	8) AAS
9) $\overline{BX} \cong \overline{DX}$	9) Corresp. parts of ≅ ⧊ are ≅.

 Got It? **4. Given:** $\angle CAD \cong \angle EAD$, $\angle C \cong \angle E$
Prove: $\overline{BD} \cong \overline{FD}$

Lesson Check

Do you know HOW?

Identify any common angles or sides.

1. △MKJ and △LJK   **2.** △DEH and △DFG

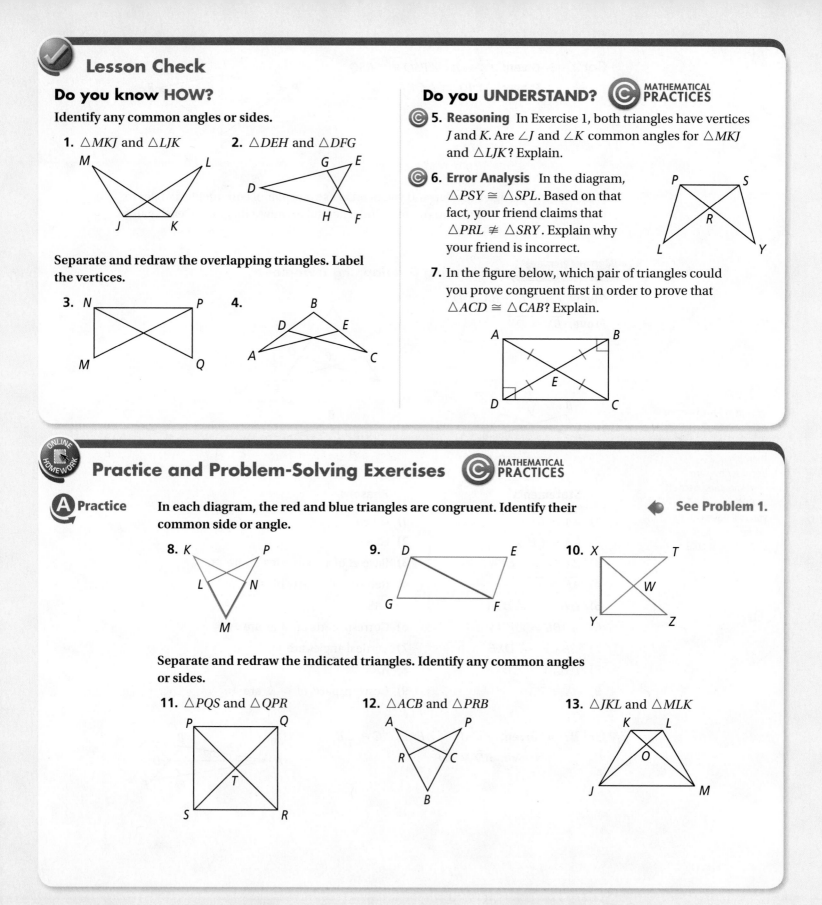

Separate and redraw the overlapping triangles. Label the vertices.

3.   **4.**

Do you UNDERSTAND?

MATHEMATICAL PRACTICES

5. Reasoning In Exercise 1, both triangles have vertices J and K. Are ∠J and ∠K common angles for △MKJ and △LJK? Explain.

6. Error Analysis In the diagram, △PSY ≅ △SPL. Based on that fact, your friend claims that △PRL ≇ △SRY. Explain why your friend is incorrect.

7. In the figure below, which pair of triangles could you prove congruent first in order to prove that △ACD ≅ △CAB? Explain.

Practice and Problem-Solving Exercises

MATHEMATICAL PRACTICES

A Practice   In each diagram, the red and blue triangles are congruent. Identify their common side or angle.

◀ See Problem 1.

8.   **9.**   **10.**

Separate and redraw the indicated triangles. Identify any common angles or sides.

11. △PQS and △QPR   **12.** △ACB and △PRB   **13.** △JKL and △MLK

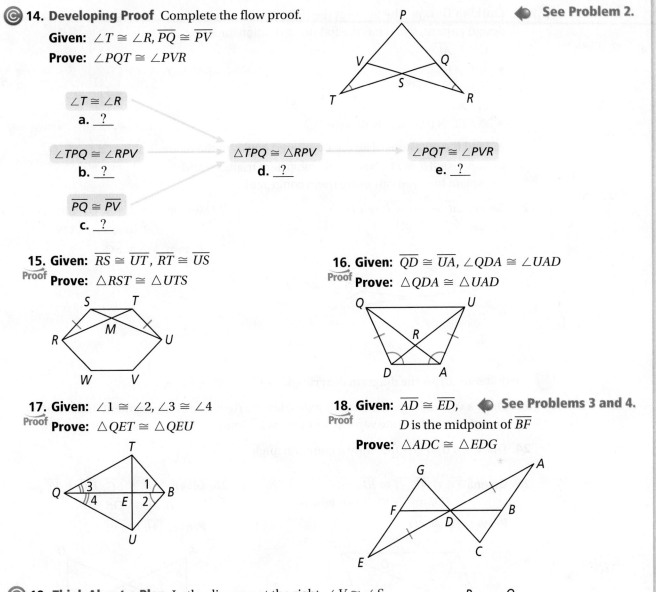

14. Developing Proof Complete the flow proof.

Given: $\angle T \cong \angle R$, $\overline{PQ} \cong \overline{PV}$

Prove: $\angle PQT \cong \angle PVR$

See Problem 2.

$\angle T \cong \angle R$
a. ?

$\angle TPQ \cong \angle RPV$
b. ?

$\overline{PQ} \cong \overline{PV}$
c. ?

$\triangle TPQ \cong \triangle RPV$
d. ?

$\angle PQT \cong \angle PVR$
e. ?

15. Given: $\overline{RS} \cong \overline{UT}$, $\overline{RT} \cong \overline{US}$
Proof Prove: $\triangle RST \cong \triangle UTS$

16. Given: $\overline{QD} \cong \overline{UA}$, $\angle QDA \cong \angle UAD$
Proof Prove: $\triangle QDA \cong \triangle UAD$

17. Given: $\angle 1 \cong \angle 2$, $\angle 3 \cong \angle 4$
Proof Prove: $\triangle QET \cong \triangle QEU$

18. Given: $\overline{AD} \cong \overline{ED}$, See Problems 3 and 4.
D is the midpoint of $\overline{BF}$
Proof Prove: $\triangle ADC \cong \triangle EDG$

B **Apply**

19. Think About a Plan In the diagram at the right, $\angle V \cong \angle S$, $\overline{VU} \cong \overline{ST}$, and $\overline{PS} \cong \overline{QV}$. Which two triangles are congruent by SAS? Explain.
- How can you use a new diagram to help you identify the triangles?
- What do you need to prove triangles congruent by SAS?

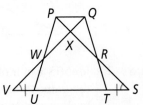

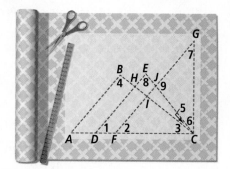

STEM **20. Clothing Design** The figure at the right is part of a clothing design pattern, and it has the following relationships.

- $\overline{GC} \perp \overline{AC}$
- $\overline{AB} \perp \overline{BC}$
- $\overline{AB} \parallel \overline{DE} \parallel \overline{FG}$
- $m\angle A = 50$
- $\triangle DEC$ is isosceles with base $\overline{DC}$.

a. Find the measures of all the numbered angles in the figure.

b. Suppose $\overline{AB} \cong \overline{FC}$. Name two congruent triangles and explain how you can prove them congruent.

21. Given: $\overline{AC} \cong \overline{EC}$, $\overline{CB} \cong \overline{CD}$
Proof **Prove:** $\angle A \cong \angle E$

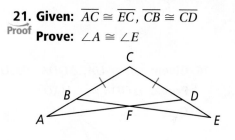

22. Given: $\overline{QT} \perp \overline{PR}$, $\overline{QT}$ bisects $\overline{PR}$,
Proof $\overline{QT}$ bisects $\angle VQS$
Prove: $\overline{VQ} \cong \overline{SQ}$

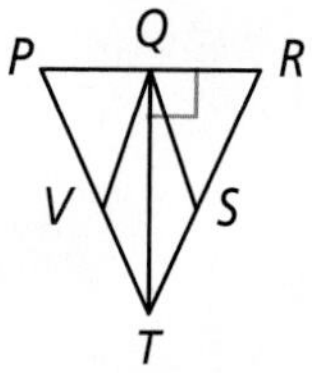

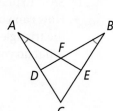

© **Open-Ended** **Draw the diagram described.**

23. Draw a vertical segment on your paper. On the right side of the segment draw two triangles that share the vertical segment as a common side.

24. Draw two triangles that have a common angle.

25. Given: $\overline{TE} \cong \overline{RI}$, $\overline{TI} \cong \overline{RE}$,
Proof $\angle TDI$ and $\angle ROE$ are right $\angle$s
Prove: $\overline{TD} \cong \overline{RO}$

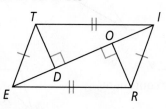

26. Given: $\overline{AB} \perp \overline{BC}$, $\overline{DC} \perp \overline{BC}$,
Proof $\overline{AC} \cong \overline{DB}$
Prove: $\overline{AE} \cong \overline{DE}$

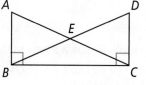

© **Challenge** **27.** Identify a pair of overlapping congruent triangles in the diagram. Then use the given information to write a proof to show that the triangles are congruent.

Given: $\overline{AC} \cong \overline{BC}$, $\angle A \cong \angle B$

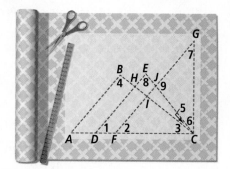

28. Reasoning Draw a quadrilateral $ABCD$ with $\overline{AB} \parallel \overline{DC}$, $\overline{AD} \parallel \overline{BC}$, and diagonals $\overline{AC}$

Proof and $\overline{DB}$ intersecting at E. Label your diagram to indicate the parallel sides.

    a. List all the pairs of congruent segments in your diagram.

    b. Writing Explain how you know that the segments you listed are congruent.

Standardized Test Prep

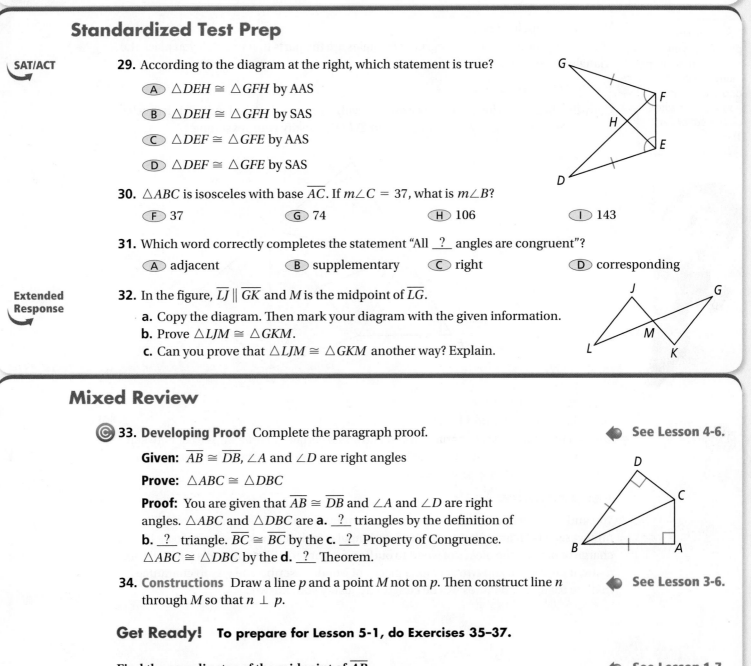

SAT/ACT

29. According to the diagram at the right, which statement is true?

    Ⓐ $\triangle DEH \cong \triangle GFH$ by AAS

    Ⓑ $\triangle DEH \cong \triangle GFH$ by SAS

    Ⓒ $\triangle DEF \cong \triangle GFE$ by AAS

    Ⓓ $\triangle DEF \cong \triangle GFE$ by SAS

30. $\triangle ABC$ is isosceles with base $\overline{AC}$. If $m\angle C = 37$, what is $m\angle B$?

    Ⓕ 37   Ⓖ 74   Ⓗ 106   Ⓘ 143

31. Which word correctly completes the statement "All __?__ angles are congruent"?

    Ⓐ adjacent   Ⓑ supplementary   Ⓒ right   Ⓓ corresponding

Extended Response

32. In the figure, $\overline{LJ} \parallel \overline{GK}$ and M is the midpoint of $\overline{LG}$.

    a. Copy the diagram. Then mark your diagram with the given information.

    b. Prove $\triangle LJM \cong \triangle GKM$.

    c. Can you prove that $\triangle LJM \cong \triangle GKM$ another way? Explain.

Mixed Review

33. Developing Proof Complete the paragraph proof.   ◀ **See Lesson 4-6.**

    Given: $\overline{AB} \cong \overline{DB}$, $\angle A$ and $\angle D$ are right angles

    Prove: $\triangle ABC \cong \triangle DBC$

    Proof: You are given that $\overline{AB} \cong \overline{DB}$ and $\angle A$ and $\angle D$ are right angles. $\triangle ABC$ and $\triangle DBC$ are **a.** __?__ triangles by the definition of **b.** __?__ triangle. $\overline{BC} \cong \overline{BC}$ by the **c.** __?__ Property of Congruence. $\triangle ABC \cong \triangle DBC$ by the **d.** __?__ Theorem.

34. Constructions Draw a line p and a point M not on p. Then construct line n ◀ **See Lesson 3-6.**
    through M so that $n \perp p$.

Get Ready! To prepare for Lesson 5-1, do Exercises 35–37.

Find the coordinates of the midpoint of $\overline{AB}$.   ◀ **See Lesson 1-7.**

35. $A(-2, 3)$, $B(4, 1)$ **36.** $A(0, 5)$, $B(3, 6)$ **37.** $A(7, 10)$, $B(-5, -8)$

Pull It All Together © ASSESSMENT

To solve these problems you will pull together many concepts and skills that you have studied about congruent triangles.

BIG idea Visualization

The corresponding parts of congruent triangles are the parts that match if you place the triangles on top of each other.

© Performance Task 1

Copy the diagram below. $\triangle GAB$ is isosceles with vertex angle A and $\triangle BCD$ is isosceles with vertex angle C. Is $\triangle BGH$ congruent to $\triangle BDH$? Justify your reasoning.

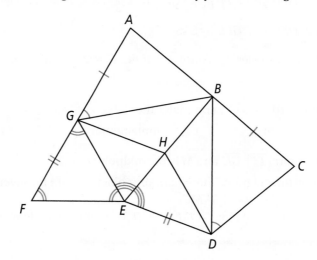

BIG idea Reasoning and Proof

You can prove triangles congruent if you know that certain relationships exist between corresponding parts. If you know that triangles are congruent, you know that all their corresponding parts are congruent.

© Performance Task 2

You and some neighbors are landscaping a community park. The organizer of the project selects an area for two congruent triangular rock gardens. You agree to be in charge of placing the pieces of wood to outline the gardens. The only tools you have are a saw, a protractor, and two very long pieces of wood. Describe one way to guarantee that the triangular outlines will be congruent. Justify your answer.

4 Chapter Review

Connecting **BIG** ideas and Answering the Essential Questions

1 Visualization
You can identify corresponding parts of congruent triangles by visualizing the figures placed on top of each other.

Congruent Figures (Lesson 4-1)

$\triangle ABC \cong \triangle DEF$

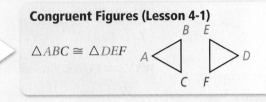

2 Reasoning and Proof
You can show two triangles are congruent by proving that certain relationships exist between three pairs of corresponding parts.

Proving Triangles Congruent (Lessons 4-2, 4-3, and 4-6)
Side-Side-Side (SSS), Side-Angle-Side (SAS), Angle-Side-Angle (ASA), Angle-Angle-Side (AAS), Hypotenuse-Leg (HL)

Using Corresponding Parts of Congruent Triangles (Lessons 4-4 and 4-7)

If $\triangle LMN \cong \triangle QRS$

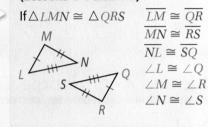

$\overline{LM} \cong \overline{QR}$
$\overline{MN} \cong \overline{RS}$
$\overline{NL} \cong \overline{SQ}$
$\angle L \cong \angle Q$
$\angle M \cong \angle R$
$\angle N \cong \angle S$

3 Reasoning and Proof
You can tell whether a triangle is isosceles or equilateral by looking at the number of congruent angles or sides.

Isosceles and Equilateral Triangles (Lesson 4-5)
• The base angles of an isosceles triangle are congruent.
• All equilateral triangles are equiangular.
• All equiangular triangles are equilateral.

Chapter Vocabulary

- base angles of an isosceles triangle (p. 250)
- base of an isosceles triangle (p. 250)
- congruent polygons (p. 219)
- corollary (p. 252)
- hypotenuse (p. 258)
- legs of an isosceles triangle (p. 250)
- legs of a right triangle (p. 258)
- vertex angle of an isosceles triangle (p. 250)

Choose the correct term to complete each sentence.

1. The two congruent sides of an isosceles triangle are the __?__.

2. The side opposite the right angle of a right triangle is the __?__.

3. A __?__ to a theorem is a statement that follows immediately from the theorem.

4. __?__ have congruent corresponding parts.

4-1 Congruent Figures

Quick Review

Congruent polygons have congruent corresponding parts. When you name congruent polygons, always list corresponding vertices in the same order.

Example

HIJK ≅ PQRS. **Write all possible congruence statements.**

The order of the parts in the congruence statement tells you which parts correspond.

Sides: $\overline{HI} \cong \overline{PQ}$, $\overline{IJ} \cong \overline{QR}$, $\overline{JK} \cong \overline{RS}$, $\overline{KH} \cong \overline{SP}$

Angles: $\angle H \cong \angle P$, $\angle I \cong \angle Q$, $\angle J \cong \angle R$, $\angle K \cong \angle S$

Exercises

RSTUV ≅ KLMNO. **Complete the congruence statements.**

5. $\overline{TS} \cong$ _?_

6. $\angle N \cong$ _?_

7. $\overline{LM} \cong$ _?_

8. *VUTSR ≅* _?_

WXYZ ≅ PQRS. **Find each measure or length.**

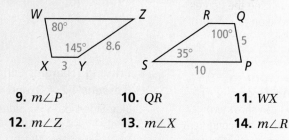

9. $m\angle P$

10. QR

11. WX

12. $m\angle Z$

13. $m\angle X$

14. $m\angle R$

4-2 and 4-3 Triangle Congruence by SSS, SAS, ASA, and AAS

Quick Review

You can prove triangles congruent with limited information about their congruent sides and angles.

Postulate or Theorem	You need
Side-Side-Side (SSS)	three sides
Side-Angle-Side (SAS)	two sides and an included angle
Angle-Side-Angle (ASA)	two angles and an included side
Angle-Angle-Side (AAS)	two angles and a nonincluded side

Example

What postulate would you use to prove the triangles congruent?

You know that three sides are congruent. Use SSS.

Exercises

15. In $\triangle HFD$, what angle is included between $\overline{DH}$ and $\overline{DF}$?

16. In $\triangle OMR$, what side is included between $\angle M$ and $\angle R$?

Which postulate or theorem, if any, could you use to prove the two triangles congruent? If there is not enough information to prove the triangles congruent, write *not enough information.*

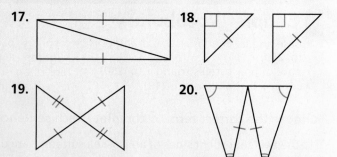

17.

18.

19.

20.

4-4 Using Corresponding Parts of Congruent Triangles

Quick Review

Once you know that triangles are congruent, you can make conclusions about corresponding sides and angles because, by definition, corresponding parts of congruent triangles are congruent. You can use congruent triangles in the proofs of many theorems.

Example

How can you use congruent triangles to prove $\angle Q \cong \angle D$?

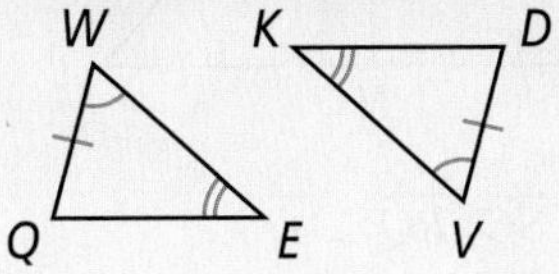

Since $\triangle QWE \cong \triangle DVK$ by AAS, you know that $\angle Q \cong \angle D$ because corresponding parts of congruent triangles are congruent.

Exercises

How can you use congruent triangles to prove the statement true?

21. $\overline{TV} \cong \overline{YW}$

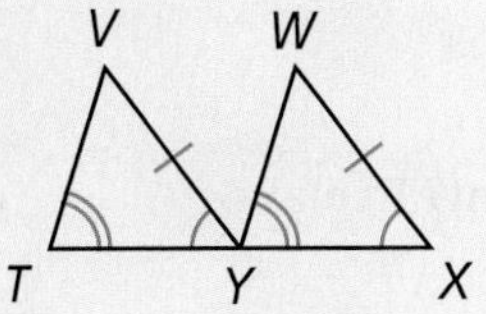

22. $\overline{BE} \cong \overline{DE}$

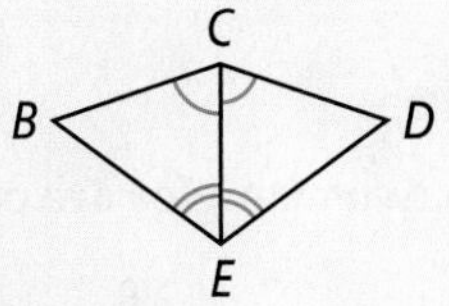

23. $\angle B \cong \angle D$

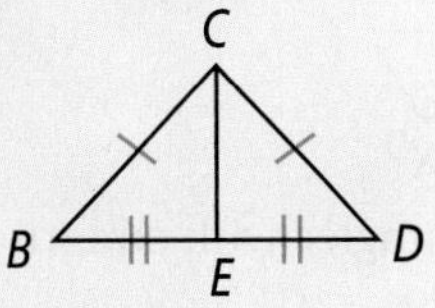

24. $\overline{KN} \cong \overline{ML}$

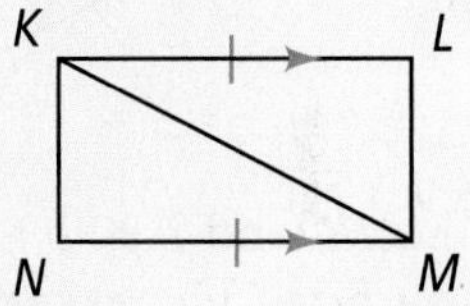

4-5 Isosceles and Equilateral Triangles

Quick Review

If two sides of a triangle are congruent, then the angles opposite those sides are also congruent by the **Isosceles Triangle Theorem.** If two angles of a triangle are congruent, then the sides opposite the angle are congruent by the **Converse of the Isosceles Triangle Theorem.**

Equilateral triangles are also equiangular.

Example

What is $m\angle G$?

Since $\overline{EF} \cong \overline{EG}$, $\angle F \cong \angle G$ by the Isosceles Triangle Theorem. So $m\angle G = 30$.

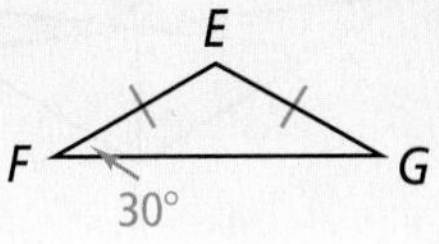

Exercises

Algebra Find the values of x and y.

25.

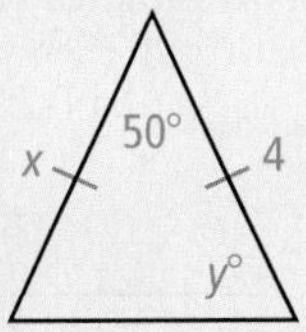

26.

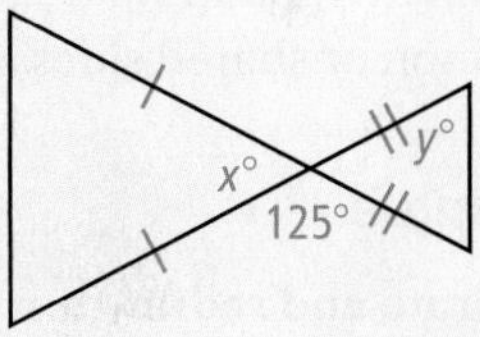

27.

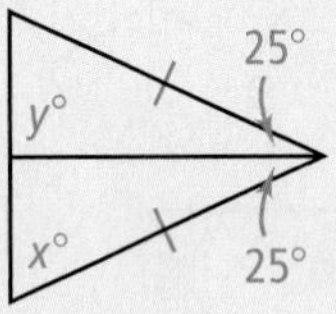

28.

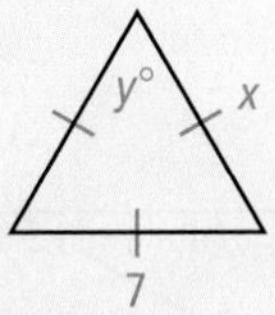

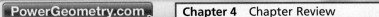

4-6 Congruence in Right Triangles

Quick Review

If the hypotenuse and a leg of one right triangle are congruent to the hypotenuse and a leg of another right triangle, then the triangles are congruent by the **Hypotenuse-Leg (HL) Theorem.**

Example

Which two triangles are congruent? Explain.

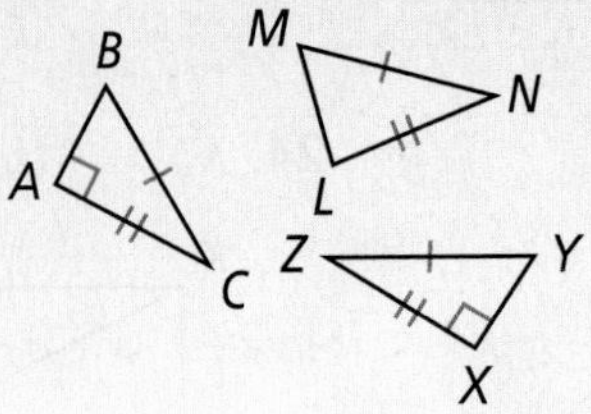

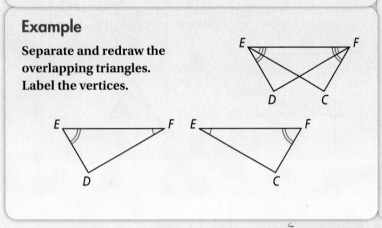

Since $\triangle ABC$ and $\triangle XYZ$ are right triangles with congruent legs, and $\overline{BC} \cong \overline{YZ}$, $\triangle ABC \cong \triangle XYZ$ by HL.

Exercises

Write a proof for each of the following.

29. Given: $\overline{LN} \perp \overline{KM}$, $\overline{KL} \cong \overline{ML}$

  Prove: $\triangle KLN \cong \triangle MLN$

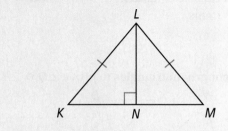

30. Given: $\overline{PS} \perp \overline{SQ}$, $\overline{RQ} \perp \overline{QS}$,
  $\overline{PQ} \cong \overline{RS}$

  Prove: $\triangle PSQ \cong \triangle RQS$

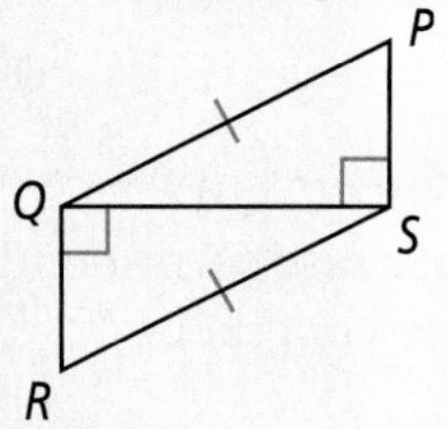

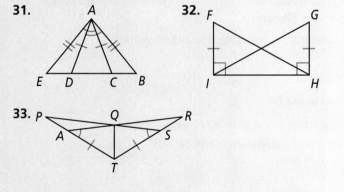

4-7 Congruence in Overlapping Triangles

Quick Review

To prove overlapping triangles congruent, you look for the common or shared sides and angles.

Example

Separate and redraw the overlapping triangles. Label the vertices.

Exercises

Name a pair of overlapping congruent triangles in each diagram. State whether the triangles are congruent by SSS, SAS, ASA, AAS, or HL.

31.

32.

33.

Do you know HOW?

Write a congruence statement for each pair of triangles.

1.

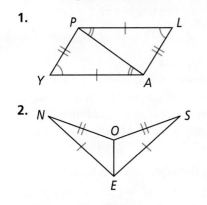

2. N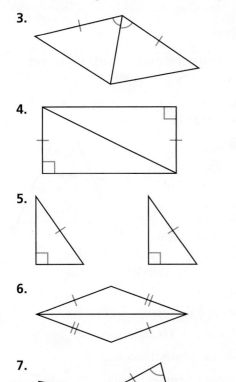

Which postulate or theorem, if any, could you use to prove the two triangles congruent? If not enough information is given, write *not enough information*.

3.

4.

5.

6.

7.

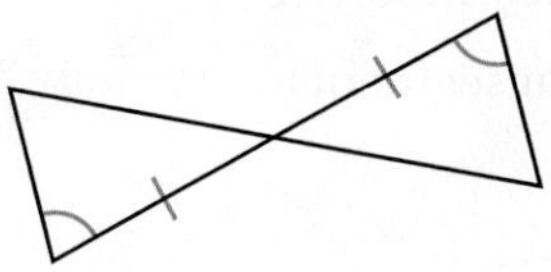

8. $\triangle CEO \cong \triangle HDF$. Name all of the pairs of corresponding congruent parts.

9. Algebra Find the value of x.

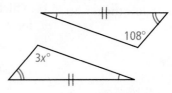

108°
$3x°$

Name a pair of overlapping congruent triangles in each diagram. State whether the triangles are congruent by SSS, SAS, ASA, AAS, or HL.

10. Given: $\overline{CE} \cong \overline{DF}$,
$\overline{CF} \cong \overline{DE}$

11. Given: $\overline{RT} \cong \overline{QT}$,
$\overline{AT} \cong \overline{ST}$

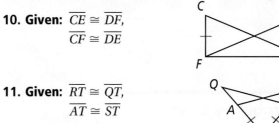

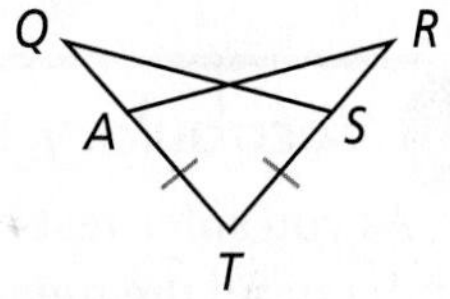

Do you UNDERSTAND?

12. Reasoning Isosceles $\triangle ABC$, with right $\angle B$, has a point D on $\overline{AC}$ such that $\overline{BD} \perp \overline{AC}$. What is the relationship between $\triangle ABD$ and $\triangle CBD$? Explain.

Write a proof for each of the following.

13. Given: $\overline{AT} \cong \overline{GS}$,
$\overline{AT} \parallel \overline{GS}$
Prove: $\triangle GAT \cong \triangle TSG$

14. Given: $\overline{LN}$ bisects $\angle OLM$
and $\angle ONM$.
Prove: $\overline{ON} \cong \overline{MN}$

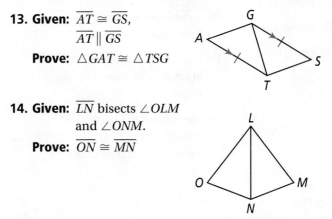

TIPS FOR SUCCESS

Some test questions ask you to compare geometric figures. Read the sample question at the right. Then follow the tips to answer it.

TIP 2

You can use SSS to show triangles are congruent. Find the corresponding side lengths of △DEF.

In the coordinate plane, the vertices of △ABC are $A(-2, 5)$, $B(2, 5)$, and $C(-2, 2)$. Which of the following are the side lengths for △DEF, such that △ABC ≅ △DEF?

- Ⓐ $DE = 3, DF = 4, EF = 4$
- Ⓑ $DE = 3, DF = 4, EF = 5$
- Ⓒ $DE = 4, DF = 3, EF = 5$
- Ⓓ $DE = 4, DF = 5, EF = 5$

TIP 1

Find the lengths of the sides of △ABC.

Think It Through

Vertices A and B have the same y-coordinate and vertices A and C have the same x-coordinate, so $AB = |-2 - 2| = |-4| = 4$ and $AC = |5 - 2| = |3| = 3$.

By the Distance Formula, $BC = \sqrt{(4)^2 + (3)^2} = \sqrt{25} = 5$.

So $DE = AB = 4$, $DF = AC = 3$, and $EF = BC = 5$. The correct answer is C.

Vocabulary Builder

As you solve test items, you must understand the meanings of mathematical terms. Match each term with its mathematical meaning.

A. slope

B. perpendicular lines

C. polygon

D. conjecture

E. congruent

I. lines that intersect to form right angles

II. having the same size and shape

III. a conclusion reached by inductive reasoning

IV. a closed plane figure with at least three sides that are segments

V. the ratio of the vertical change (rise) to the horizontal change (run)

Multiple Choice

Read each question. Then write the letter of the correct answer on your paper.

1. Given: $\overline{DE} \parallel \overline{CB}$,
 $\angle ADE \cong \angle AED$

Prove: $\overline{AC} \cong \overline{AB}$

Proof: Since $\overline{DE} \parallel \overline{CB}$, $\angle ACB \cong \angle ADE$ and $\angle AED \cong \angle ABC$ by the Corresponding Angles Theorem. Since $\angle ADE \cong \angle AED$, $\angle ACB \cong \angle ABC$ by the Transitive Property. Which theorem or definition proves that $\overline{AC} \cong \overline{AB}$?

- Ⓐ Isosceles Triangle Theorem
- Ⓑ Converse of Isosceles Triangle Theorem
- Ⓒ Alternate Interior Angles Theorem
- Ⓓ Definition of congruent segments

2. Which statement must be true for two polygons to be congruent?

   F All the corresponding sides should be congruent.

   G All the corresponding sides and angles should be congruent.

   H All the corresponding angles should be congruent.

   I All sides in each polygon should be congruent.

3. If $\triangle ABC \cong \triangle CDA$, which of the following must be true?

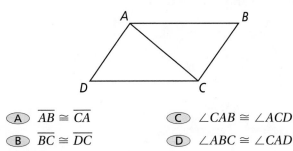

   A $\overline{AB} \cong \overline{CA}$ **C** $\angle CAB \cong \angle ACD$

   B $\overline{BC} \cong \overline{DC}$ **D** $\angle ABC \cong \angle CAD$

4. Given: $\angle 1 \cong \angle 2$, $\overline{AB} \cong \overline{AC}$

What additional information do you need to prove $\triangle ABD \cong \triangle ACE$ by AAS?

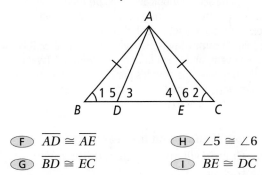

   F $\overline{AD} \cong \overline{AE}$ **H** $\angle 5 \cong \angle 6$

   G $\overline{BD} \cong \overline{EC}$ **I** $\overline{BE} \cong \overline{DC}$

5. Which of the following statements is true?

   A *Point, line,* and *plane* are undefined terms.

   B A theorem is an accepted statement of fact.

   C "Vertical angles are congruent" is a definition.

   D A postulate is a conjecture that is proven.

6. Which condition allows you to prove that $\ell \parallel m$?

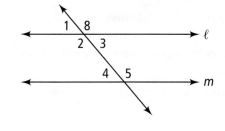

   F $\angle 1 \cong \angle 8$ **H** $\angle 3 \cong \angle 4$

   G $\angle 2 \cong \angle 8$ **I** $\angle 3 \cong \angle 5$

7. A line passes through $(3, -4)$ and has a slope of -5. How can you find the y-intercept of the graph?

   A Substitute -5 for b, -4 for x, and 3 for y in $y = mx + b$. Then solve for m, the y-intercept.

   B Substitute -5 for b, 3 for x, and -4 for y in $y = mx + b$. Then solve for m, the y-intercept.

   C Substitute -5 for m, -4 for x, and 3 for y in $y = mx + b$. Then solve for b, the y-intercept.

   D Substitute -5 for m, 3 for x, and -4 for y in $y = mx + b$. Then solve for b, the y-intercept.

8. Given: $\triangle RST \cong \triangle LMN$

Which reason could you use to prove that $\angle R \cong \angle L$?

   F SAS

   G SSS

   H ASA

   I Corresponding parts of congruent triangles are congruent.

9. Which equation represents the perpendicular bisector of the segment shown?

   A $y = x - 3$

   B $3x + 3y = 3$

   C $y = 3x$

   D $x + y = 3$

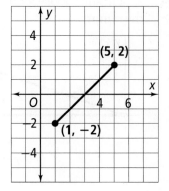

10. One bag of garden soil covers approximately 16 ft². If it takes 16π ft of fencing to enclose a circular garden, how many bags of soil do you need to cover the garden?

11. What is the value of *x* in the figure below?

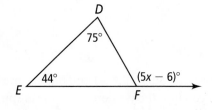

12. The length of a rectangle is seven more than three times its width. If the perimeter is 38 cm, what is the area, in square centimeters, of the rectangle?

13. What is the value of *x* in the figure below?

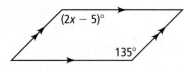

14. *ABCD* ≅ *WXYZ*. What is *WX*?

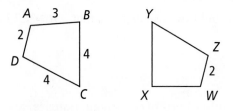

15. Amy is designing a ramp up to a 16-in.-high skateboarding platform, as shown on the graph below. If she wants the slope of the ramp to be $\frac{1}{3}$, what value should she choose for the *x*-coordinate at the top of the ramp?

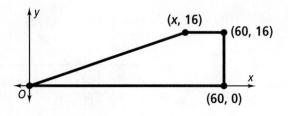

Short Response

16. Describe how the following everyday meanings of *acute* and *obtuse* help you to remember their mathematical meanings.

acute *adj.* Having a sharp point

obtuse *adj.* Not sharp or pointed; blunt

17. Write a proof for the following.

Given: $\overline{AE} \cong \overline{DE}$, $\overline{EB} \cong \overline{EC}$
Prove: $\triangle AEB \cong \triangle DEC$

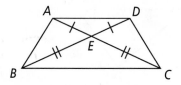

Extended Response

18. Read this excerpt from an online news article.

> Halley's Comet can be seen periodically at its *perihelion*, the shortest distance from the sun during its orbit. Mark Twain was born two weeks after the comet's perihelion. In his biography he said, "I came in with Halley's Comet in 1835. It is coming again next year, and I expect to go out with it." Twain died in 1910, the day after the comet's perihelion. The most recent sighting of Halley's Comet was in 1986. Its next appearance is expected in 2061.

a. Make a conjecture about the year in which Halley's Comet will appear after 2061. Explain your reasoning.

b. How confident are you about your conjecture? Explain.

19. The coordinates of the vertices of rectangle *LMNK* are *L*(−2, 5), *M*(2, 5), *N*(2, 3), and *K*(−2, 3). The coordinates of the vertices of rectangle *PQRS* are *P*(3, 0), *Q*(3, −3), *R*(1, −3), and *S*(1, 0). Are these two rectangles congruent? Explain why or why not. If not, how could you change the vertices of one of the rectangles to make them congruent?

Get Ready!

Lesson 1-6 ◆ **Basic Constructions**

Use a compass and straightedge for each construction.

1. Construct the perpendicular bisector of a segment.

2. Construct the bisector of an angle.

Lesson 1-7 ◆ **The Midpoint Formula and Distance Formula**

Find the coordinates of the midpoints of the sides of $\triangle ABC$. Then find the lengths of the three sides of the triangle.

3. $A(5, 1), B(-3, 3), C(1, -7)$

4. $A(-1, 2), B(9, 2), C(-1, 8)$

5. $A(-2, -3), B(2, -3), C(0, 3)$

Lesson 2-2 ◆ **Finding the Negation**

Write the negation of each statement.

6. The team won.   **7.** It is not too late.   **8.** $m\angle R > 60$

Lesson 3-7 ◆ **Slope**

Find the slope of the line passing through the given points.

9. $A(9, 6), B(8, 12)$ **10.** $C(3, -2), D(0, 6)$ **11.** $E(-3, 7), F(-3, 12)$

Looking Ahead Vocabulary

12. The *altitude* of an airplane is the height of the airplane above ground. What do you think an *altitude of a triangle* is?

13. The *distance* between your home and your school is the length of the shortest path connecting them. How might you define the *distance between a point and a line* in geometry?

14. In Chapter 1, you learned the definition of a *midpoint* of a segment. What do you think a *midsegment* of a triangle is?

15. If two parties are happening at the same time, they are *concurrent*. What would it mean for three lines to be *concurrent*?

Relationships Within Triangles

© DOMAINS
- Congruence
- Similarity, Right Triangles, and Trigonometry
- Mathematical Practice: Construct viable arguments

Look at that roof of triangles! It's interesting how the support beams in each triangle meet at one point.

In this chapter, you will learn how special lines and segments in triangles relate.

Vocabulary

English/Spanish Vocabulary Audio Online:

English	Spanish
altitude of a triangle, *p. 310*	altura de un triángulo
centroid, *p. 309*	centroide
circumcenter, *p. 301*	circuncentro
concurrent, *p. 301*	concurrente
equidistant, *p. 292*	equidistante
incenter, *p. 303*	incentro
indirect proof, *p. 317*	prueba indirecta
median, *p. 309*	mediana
midsegment of a triangle, *p. 285*	segmento medio de un triángulo
orthocenter, *p. 311*	ortocentro

My Math Video

00:04:04

VIDEO ▷

BIG ideas

1 **Coordinate Geometry**
Essential Question How do you use coordinate geometry to find relationships within triangles?

2 **Measurement**
Essential Question How do you solve problems that involve measurements of triangles?

3 **Reasoning and Proof**
Essential Question How do you write indirect proofs?

Chapter Preview

Concept Byte

Use With Lesson 5-1

TECHNOLOGY

Investigating Midsegments

© **Content Standard**
Prepares for G.CO.10 Prove theorems about triangles . . . the segment joining the midpoints of two sides of a triangle is parallel to the third side and half the length . . .

Activity

Step 1 Use geometry software to draw and label △*ABC*. Construct the midpoints *D* and *E* of $\overline{AB}$ and $\overline{AC}$. Connect the midpoints with a *midsegment*.

Step 2 Measure $\overline{DE}$ and $\overline{BC}$. Calculate $\frac{DE}{BC}$.

Step 3 Measure the slopes of $\overline{DE}$ and $\overline{BC}$.

Step 4 Manipulate the triangle and observe the lengths and slopes of $\overline{DE}$ and $\overline{BC}$.

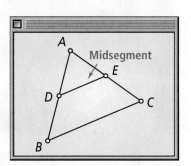

Exercises

© **1. Make a Conjecture** Make conjectures about the lengths and slopes of midsegments.

2. Construct the midpoint *F* of $\overline{BC}$. Then construct the other two midsegments of △*ABC*. Test whether these midsegments support your conjectures in Exercise 1.

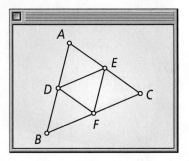

© **3.** △*ABC* and the three midsegments form four small triangles.
  a. Measure the sides of the four small triangles and list those that you find are congruent.
  b. Make a Conjecture Make a conjecture about the four small triangles formed by a triangle and its three midsegments.

For the remaining exercises, assume your conjectures in Exercises 1 and 3 are true.

4. What can you say about the areas of the four small triangles in the window above?

5. a. How does △*ABC* compare to each small triangle in area?
  b. How does △*ABC* compare to each small triangle in perimeter?

6. Construct the three midsegments of △*DEF*. Label this triangle △*GHI*.
  a. How does △*ABC* compare to △*GHI* in area?
  b. How does △*ABC* compare to △*GHI* in perimeter?
  c. Suppose you construct the midsegment triangle inside △*GHI*. How would △*ABC* compare to this third midsegment triangle in area and perimeter?

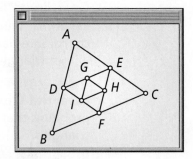

Midsegments of Triangles

© **Content Standards**
G.CO.10 Prove theorems about triangles . . . the segment joining the midpoints of two sides of a triangle is parallel to the third side and half the length . . .
Also G.CO.12 and G.SRT.5

Objective To use properties of midsegments to solve problems

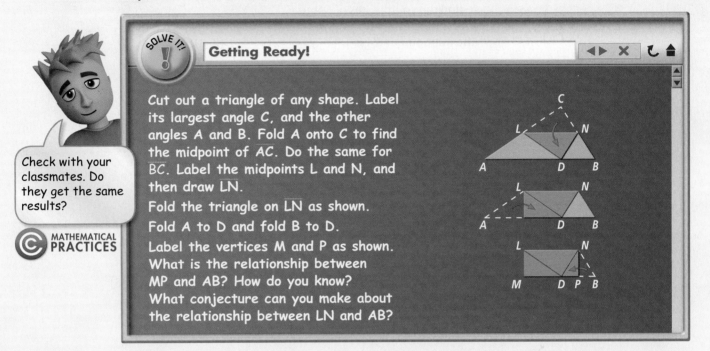

Check with your classmates. Do they get the same results?

© MATHEMATICAL PRACTICES

Getting Ready!

Cut out a triangle of any shape. Label its largest angle C, and the other angles A and B. Fold A onto C to find the midpoint of AC. Do the same for $\overline{BC}$. Label the midpoints L and N, and then draw LN.

Fold the triangle on $\overline{LN}$ as shown.

Fold A to D and fold B to D.

Label the vertices M and P as shown. What is the relationship between MP and AB? How do you know? What conjecture can you make about the relationship between LN and AB?

Lesson Vocabulary
• midsegment of a triangle

In the Solve It, $\overline{LN}$ is a midsegment of $\triangle ABC$. A **midsegment of a triangle** is a segment connecting the midpoints of two sides of the triangle.

Essential Understanding There are two special relationships between a midsegment of a triangle and the third side of the triangle.

take note

Theorem 5-1   Triangle Midsegment Theorem

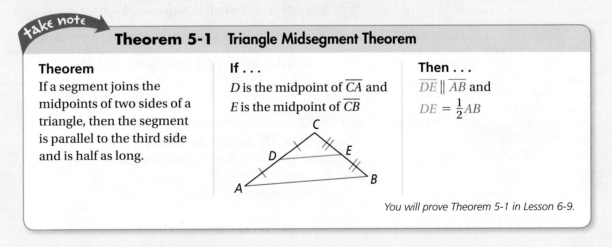

Theorem	**If . . .**	**Then . . .**
If a segment joins the midpoints of two sides of a triangle, then the segment is parallel to the third side and is half as long.	D is the midpoint of $\overline{CA}$ and E is the midpoint of $\overline{CB}$	$\overline{DE} \parallel \overline{AB}$ and $DE = \frac{1}{2}AB$

You will prove Theorem 5-1 in Lesson 6-9.

Here's Why It Works You can verify that the Triangle Midsegment Theorem works for a particular triangle. Use the following steps to show that $\overline{DE} \parallel \overline{AB}$ and that $DE = \frac{1}{2}AB$ for a triangle with vertices at $A(4, 6)$, $B(6, 0)$, and $C(0, 0)$, where D and E are the midpoints of $\overline{CA}$ and $\overline{CB}$.

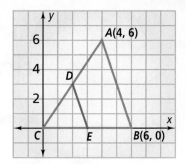

Step 1 Use the Midpoint Formula, $M = \left(\frac{x_1 + x_2}{2}, \frac{y_1 + y_2}{2}\right)$, to find the coordinates of D and E.

The midpoint of $\overline{CA}$ is $D\left(\frac{0 + 4}{2}, \frac{0 + 6}{2}\right) = D(2, 3)$.

The midpoint of $\overline{CB}$ is $E\left(\frac{0 + 6}{2}, \frac{0 + 0}{2}\right) = E(3, 0)$.

Step 2 To show that the midsegment $\overline{DE}$ is parallel to the side $\overline{AB}$, find the slope, $m = \frac{y_2 - y_1}{x_2 - x_1}$, of each segment.

slope of $\overline{DE} = \frac{0 - 3}{3 - 2}$ slope of $AB = \frac{0 - 6}{6 - 4}$

$= \frac{-3}{1}$ $= \frac{-6}{2}$

$= -3$ $= -3$

Step 3 To show $DE = \frac{1}{2}AB$, use the Distance Formula, $d = \sqrt{(x_2 - x_1)^2 + (y_2 - y_1)^2}$ to find DE and AB.

$DE = \sqrt{(3 - 2)^2 + (0 - 3)^2}$ $AB = \sqrt{(6 - 4)^2 + (0 - 6)^2}$

$= \sqrt{1 + 9}$ $= \sqrt{4 + 36}$

$= \sqrt{10}$ $= \sqrt{40}$

                             $= 2\sqrt{10}$

Since $\sqrt{10} = \frac{1}{2}(2\sqrt{10})$, you know that $DE = \frac{1}{2}AB$.

© **Problem 1**   **Identifying Parallel Segments**

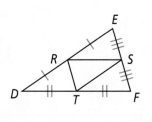

Think box on left

Think

How do you identify a midsegment?
Look for indications that the endpoints of a segment are the midpoints of a side of the triangle.

What are the three pairs of parallel segments in $\triangle DEF$?

$\overline{RS}$, $\overline{ST}$, and $\overline{TR}$ are the midsegments of $\triangle DEF$. By the Triangle Midsegment Theorem, $\overline{RS} \parallel \overline{DF}$, $\overline{ST} \parallel \overline{ED}$, and $\overline{TR} \parallel \overline{FE}$.

✓ **Got It?**   **1. a.** In $\triangle XYZ$, A is the midpoint of $\overline{XY}$, B is the midpoint of $\overline{YZ}$, and C is the midpoint of $\overline{ZX}$. What are the three pairs of parallel segments?

     © **b. Reasoning** What is $m\angle VUO$ in the figure at the right? Explain your reasoning.

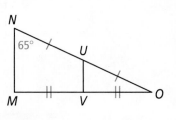

Problem 2 **Finding Lengths**

In $\triangle QRS$, T, U, and B are midpoints. What are the lengths of $\overline{TU}$, $\overline{UB}$, and $\overline{QR}$?

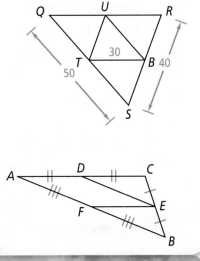

Use the relationship

length of a midsegment $= \frac{1}{2}$ (length of the third side)
to write an equation about the length of each midsegment.

$$TU = \frac{1}{2}SR \qquad\qquad UB = \frac{1}{2}QS \qquad\qquad TB = \frac{1}{2}QR$$
$$\;\;\;\;\; = \frac{1}{2}(40) \qquad\qquad\quad = \frac{1}{2}(50) \qquad\qquad 30 = \frac{1}{2}QR$$
$$\;\;\;\;\; = 20 \qquad\qquad\qquad\;\; = 25 \qquad\qquad\;\; 60 = QR$$

Plan

Which relationship stated in the Triangle Midsegment Theorem should you use?
You are asked to find lengths, so use the relationship that refers to the lengths of a midsegment and the third side.

✓ **Got It? 2.** In the figure at the right, $AD = 6$ and $DE = 7.5$. What are the lengths of $\overline{DC}$, $\overline{AC}$, $\overline{EF}$, and $\overline{AB}$?

You can use the Triangle Midsegment Theorem to find lengths of segments that might be difficult to measure directly.

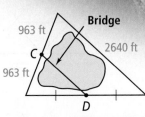

 Problem 3 **Using a Midsegment of a Triangle** (STEM)

Environmental Science A geologist wants to determine the distance, AB, across a sinkhole. Choosing a point E outside the sinkhole, she finds the distances AE and BE. She locates the midpoints C and D of $\overline{AE}$ and $\overline{BE}$ and then measures $\overline{CD}$. What is the distance across the sinkhole?

CD is a midsegment of $\triangle AEB$.

$$CD = \frac{1}{2}AB \qquad \triangle \text{ Midsegment Thm.}$$
$$46 = \frac{1}{2}AB \qquad \text{Substitute 46 for } CD.$$
$$92 = AB \qquad \text{Multiply each side by 2.}$$

The distance across the sinkhole is 92 ft.

Think

Why does the geologist find the length of $\overline{CD}$?
$\overline{CD}$ is a midsegment of $\triangle AEB$, so the geologist can use its length to find AB, the distance across the sinkhole.

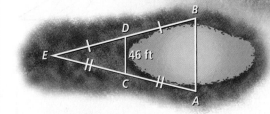

✓ **Got It? 3.** $\overline{CD}$ is a bridge being built over a lake, as shown in the figure at the right. What is the length of the bridge?

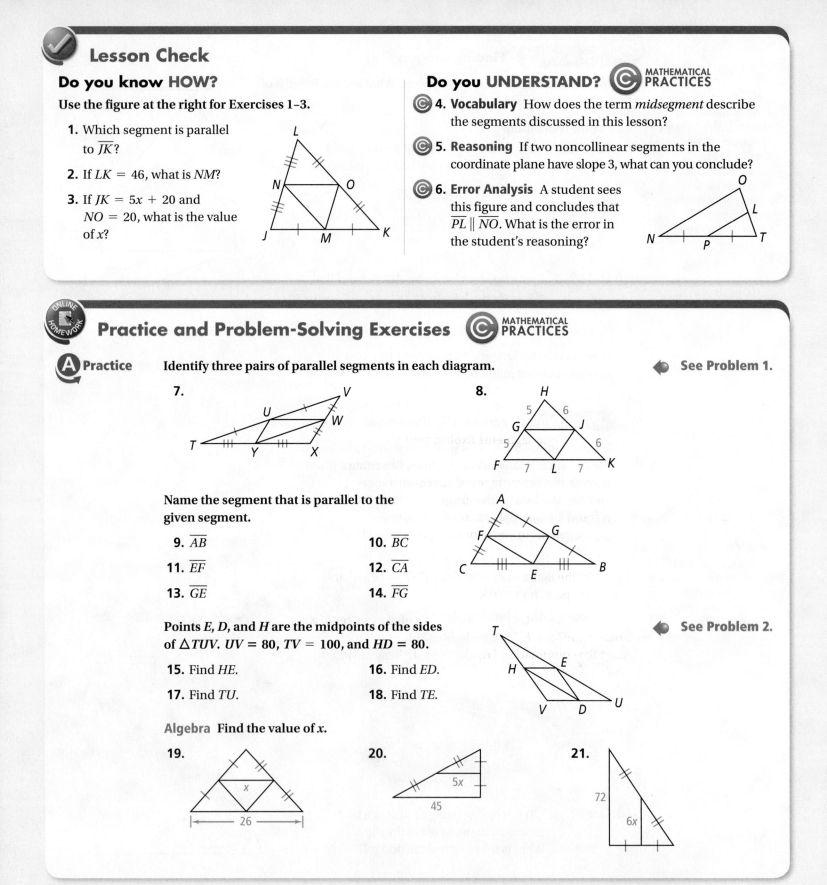

Lesson Check

Do you know HOW?

Use the figure at the right for Exercises 1–3.

1. Which segment is parallel to $\overline{JK}$?

2. If $LK = 46$, what is NM?

3. If $JK = 5x + 20$ and $NO = 20$, what is the value of x?

Do you UNDERSTAND? MATHEMATICAL PRACTICES

4. **Vocabulary** How does the term *midsegment* describe the segments discussed in this lesson?

5. **Reasoning** If two noncollinear segments in the coordinate plane have slope 3, what can you conclude?

6. **Error Analysis** A student sees this figure and concludes that $\overline{PL} \parallel \overline{NO}$. What is the error in the student's reasoning?

Practice and Problem-Solving Exercises MATHEMATICAL PRACTICES

A Practice

Identify three pairs of parallel segments in each diagram.

See Problem 1.

7.

8.

Name the segment that is parallel to the given segment.

9. $\overline{AB}$

10. $\overline{BC}$

11. $\overline{EF}$

12. $\overline{CA}$

13. $\overline{GE}$

14. $\overline{FG}$

Points E, D, and H are the midpoints of the sides of $\triangle TUV$. $UV = 80$, $TV = 100$, and $HD = 80$.

See Problem 2.

15. Find HE.

16. Find ED.

17. Find TU.

18. Find TE.

Algebra Find the value of x.

19.

20.

21.

Algebra Find the value of *x*.

22.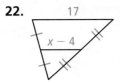
17
x − 4

23.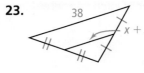
38
x + 2

24.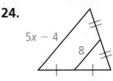
5*x* − 4
8

25. Surveying A surveyor needs to measure the distance *PQ* across the lake. Beginning at point *S*, she locates the midpoints of $\overline{SQ}$ and $\overline{SP}$ at *M* and *N*. She then measures $\overline{NM}$. What is *PQ*?

See Problem 3.

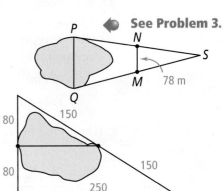

B Apply

26. Kayaking You want to paddle your kayak across a lake. To determine how far you must paddle, you pace out a triangle, counting the number of strides, as shown.
   a. If your strides average 3.5 ft, what is the length of the longest side of the triangle?
   b. What distance must you paddle across the lake?

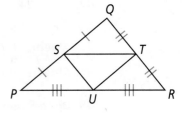

27. Architecture The triangular face of the Rock and Roll Hall of Fame in Cleveland, Ohio, is isosceles. The length of the base is 229 ft 6 in. Each leg is divided into four congruents parts by the red segments. What is the length of the white segment? Explain your reasoning.

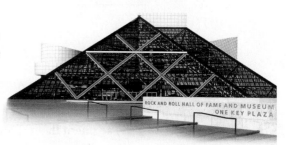

ROCK AND ROLL HALL OF FAME AND MUSEUM
ONE KEY PLAZA

28. Think About a Plan Draw △*ABC*. Construct another triangle so that the three sides of △*ABC* are the midsegments of the new triangle.
   • Can you visualize or sketch the final figure?
   • Which segments in your final construction will be parallel?

29. Writing In the figure at the right, *m*∠*QST* = 40. What is *m*∠*QPR*? Explain how you know.

30. Coordinate Geometry The coordinates of the vertices of a triangle are *E*(1, 2), *F*(5, 6), and *G*(3, −2).
   a. Find the coordinates of *H*, the midpoint of $\overline{EG}$, and *J*, the midpoint of $\overline{FG}$.
   b. Show that $\overline{HJ} \parallel \overline{EF}$.
   c. Show that $HJ = \frac{1}{2}EF$.

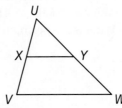

X **is the midpoint of** $\overline{UV}$. *Y* **is the midpoint of** $\overline{UW}$.

31. If *m*∠*UXY* = 60, find *m*∠*V*.

32. If *m*∠*W* = 45, find *m*∠*UYX*.

33. If *XY* = 50, find *VW*.

34. If *VW* = 110, find *XY*.

$\overline{IJ}$ is a midsegment of $\triangle FGH$. $IJ = 7$, $FH = 10$, and $GH = 13$. Find the perimeter of each triangle.

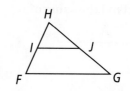

35. $\triangle IJH$

36. $\triangle FGH$

37. Kite Design You design a kite to look like the one at the right. Its diagonals measure 64 cm and 90 cm. You plan to use ribbon, represented by the purple rectangle, to connect the midpoints of its sides. How much ribbon do you need?

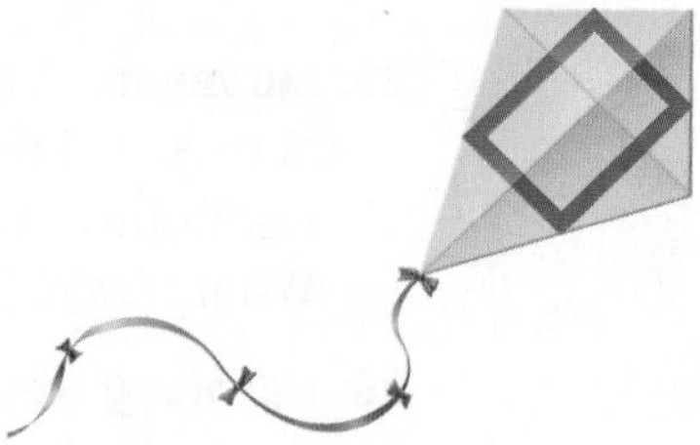

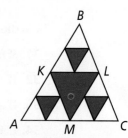

   Ⓐ 77 cm   Ⓒ 154 cm

   Ⓑ 122 cm   Ⓓ 308 cm

Algebra Find the value of each variable.

38.

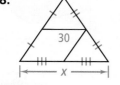

39.

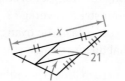

40.

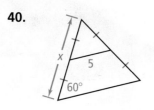

41.

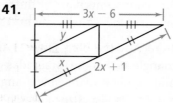

Use the figure at the right for Exercises 42–44.

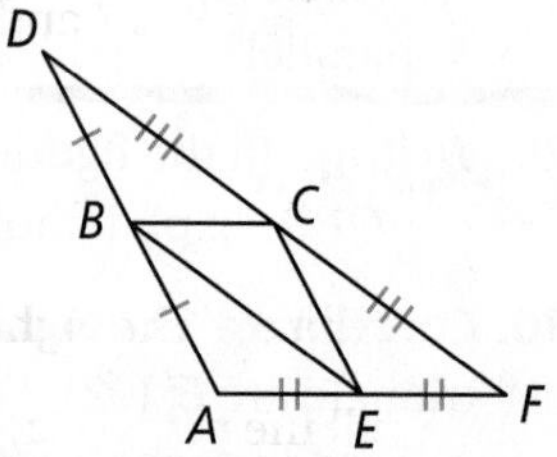

42. $DF = 24$, $BC = 6$, and $DB = 8$. Find the perimeter of $\triangle ADF$.

43. Algebra If $BE = 2x + 6$ and $DF = 5x + 9$, find DF.

44. Algebra If $EC = 3x - 1$ and $AD = 5x + 7$, find EC.

Ⓒ **45. Open-Ended** Explain how you could use the Triangle Midsegment Theorem as the basis for this construction: Draw $\overline{CD}$. Draw point A not on $\overline{CD}$. Construct $\overline{AB}$ so that $\overline{AB} \parallel \overline{CD}$ and $AB = \frac{1}{2}CD$.

Ⓒ **Challenge** Ⓒ **46. Reasoning** In the diagram at the right, K, L, and M are the midpoints of the sides of $\triangle ABC$. The vertices of the three small purple triangles are the midpoints of the sides of $\triangle KBL$, $\triangle AKM$, and $\triangle MLC$. The perimeter of $\triangle ABC$ is 24 cm. What is the perimeter of the shaded region?

47. Coordinate Geometry In △GHJ, K(2, 3) is the midpoint of $\overline{GH}$, L(4, 1) is the midpoint of $\overline{HJ}$, and M(6, 2) is the midpoint of $\overline{GJ}$. Find the coordinates of G, H, and J.

48. Complete the Prove statement and then write a proof.

Proof **Given:** In △VYZ, S, T, and U are midpoints.

Prove: △YST ≅ △TUZ ≅ △SVU ≅ ___?___

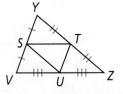

Standardized Test Prep

GRIDDED RESPONSE

SAT/ACT

Use the figure at the right for Exercises 49 and 50. Your home is at point H. Your friend lives at point F, the midpoint of Elm Street. Elm Street intersects Beech Street and Maple Street at their midpoints.

49. Your friend walks to school by going east on Elm and then turning right on Maple. How far in miles does she walk?

50. You walk your dog along this route: Walk from home to Elm along Maple. Walk west on Elm to Beech, south on Beech to the library, and east on Oak to school. Then walk back home along Maple. How far in miles do you walk?

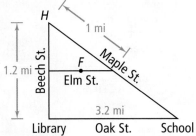

For Exercises 51 and 52, △ABC is a triangle in which $m\angle A = 30$ **and** $m\angle B = 70$. **P, Q, and R are the midpoints of** $\overline{AB}$, $\overline{BC}$, **and** $\overline{CA}$, **respectively.**

51. What is the measure, in degrees, of ∠RPQ?

52. If $QP = 2x + 17$ and $CA = x + 97$, what is CA?

Mixed Review

Use the figure at the right for Exercises 53 and 54.

53. List all the pairs of congruent triangles that you can find in the figure.

54. Given: $\overline{FD} \cong \overline{FE}$, $\overline{BF} \cong \overline{CF}$, $\angle 1 \cong \angle 2$

Prove: $\overline{AB} \cong \overline{AC}$

◀ See Lesson 4-7.

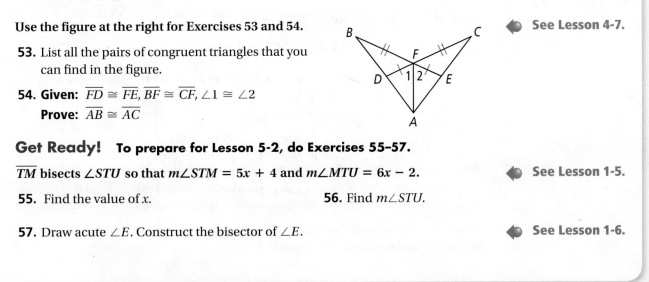

Get Ready! To prepare for Lesson 5-2, do Exercises 55–57.

$\overline{TM}$ bisects ∠STU so that $m\angle STM = 5x + 4$ and $m\angle MTU = 6x - 2$.

◀ See Lesson 1-5.

55. Find the value of x.   **56.** Find m∠STU.

57. Draw acute ∠E. Construct the bisector of ∠E.

◀ See Lesson 1-6.

© **Content Standards**

G.CO.9 Prove theorems about lines and angles . . . points on a perpendicular bisector of a line segment are exactly those equidistant from the segment's endpoints.

G.SRT.5 Use congruence . . . criteria to solve problems and prove relationships in geometric figures.

Objective To use properties of perpendicular bisectors and angle bisectors

Confused? Try drawing a diagram to "straighten" yourself out.

© MATHEMATICAL
PRACTICES

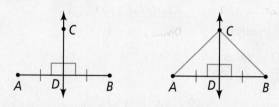

Getting Ready!

You hang a bulletin board over your desk using string. The bulletin board is crooked. When you straighten the bulletin board, what type of triangle does the string form with the top of the board? How do you know? Visualize the vertical line along the wall that passes through the nail. What relationships exist between this line and the top edge of the straightened bulletin board? Explain.

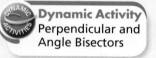

Dynamic Activity
Perpendicular and Angle Bisectors

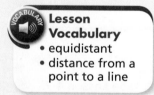

Lesson Vocabulary
• equidistant
• distance from a point to a line

In the Solve It, you thought about the relationships that must exist in order for a bulletin board to hang straight. You will explore these relationships in this lesson.

Essential Understanding There is a special relationship between the points on the perpendicular bisector of a segment and the endpoints of the segment.

In the diagram below on the left, $\overleftrightarrow{CD}$ is the perpendicular bisector of $\overline{AB}$. $\overleftrightarrow{CD}$ is perpendicular to $\overline{AB}$ at its midpoint. In the diagram on the right, $\overline{CA}$ and $\overline{CB}$ are drawn to complete $\triangle CAD$ and $\triangle CBD$.

You should recognize from your work in Chapter 4 that $\triangle CAD \cong \triangle CBD$. So you can conclude that $\overline{CA} \cong \overline{CB}$, or that $CA = CB$. A point is **equidistant** from two objects if it is the same distance from the objects. So point C is equidistant from points A and B.

This suggests a proof of Theorem 5-2, the Perpendicular Bisector Theorem. Its converse is also true and is stated as Theorem 5-3.

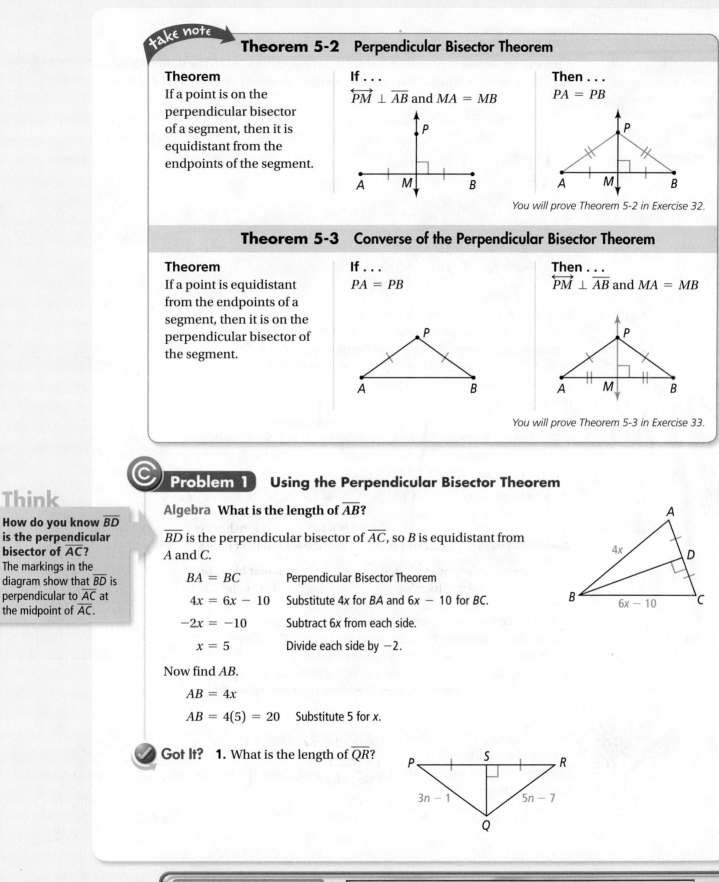

take note

Theorem 5-2   Perpendicular Bisector Theorem

Theorem

If a point is on the perpendicular bisector of a segment, then it is equidistant from the endpoints of the segment.

If . . .

$\overleftrightarrow{PM} \perp \overline{AB}$ and $MA = MB$

Then . . .

$PA = PB$

You will prove Theorem 5-2 in Exercise 32.

Theorem 5-3   Converse of the Perpendicular Bisector Theorem

Theorem

If a point is equidistant from the endpoints of a segment, then it is on the perpendicular bisector of the segment.

If . . .

$PA = PB$

Then . . .

$\overleftrightarrow{PM} \perp \overline{AB}$ and $MA = MB$

You will prove Theorem 5-3 in Exercise 33.

ⓒ Problem 1   Using the Perpendicular Bisector Theorem

Think

How do you know $\overline{BD}$ is the perpendicular bisector of $\overline{AC}$?
The markings in the diagram show that $\overline{BD}$ is perpendicular to $\overline{AC}$ at the midpoint of $\overline{AC}$.

Algebra   **What is the length of $\overline{AB}$?**

$\overline{BD}$ is the perpendicular bisector of $\overline{AC}$, so B is equidistant from A and C.

$BA = BC$	Perpendicular Bisector Theorem
$4x = 6x - 10$	Substitute $4x$ for BA and $6x - 10$ for BC.
$-2x = -10$	Subtract $6x$ from each side.
$x = 5$	Divide each side by -2.

Now find AB.

$AB = 4x$

$AB = 4(5) = 20$ Substitute 5 for x.

✓ **Got It?**   **1.** What is the length of $\overline{QR}$?

Plan

How do you find points that are equidistant from two given points?

By the Converse of the Perpendicular Bisector Theorem, points equidistant from two given points are on the perpendicular bisector of the segment that joins the two points.

Ⓒ **Problem 2** **Using a Perpendicular Bisector**

A park director wants to build a T-shirt stand equidistant from the Rollin' Coaster and the Spaceship Shoot. What are the possible locations of the stand? Explain.

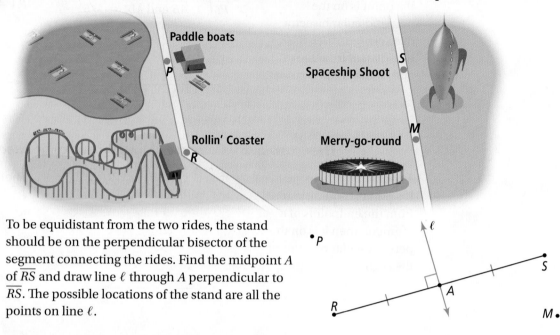

To be equidistant from the two rides, the stand should be on the perpendicular bisector of the segment connecting the rides. Find the midpoint A of $\overline{RS}$ and draw line ℓ through A perpendicular to $\overline{RS}$. The possible locations of the stand are all the points on line ℓ.

Ⓒ ✅ **Got It?** **2. a.** Suppose the director wants the T-shirt stand to be equidistant from the paddle boats and the Spaceship Shoot. What are the possible locations?

b. Reasoning Can you place the T-shirt stand so that it is equidistant from the paddle boats, the Spaceship Shoot, and the Rollin' Coaster? Explain.

Essential Understanding There is a special relationship between the points on the bisector of an angle and the sides of the angle.

The **distance from a point to a line** is the length of the perpendicular segment from the point to the line. This distance is also the length of the shortest segment from the point to the line. You will prove this in Lesson 5-6. In the figure at the right, the distances from A to ℓ and from B to ℓ are represented by the red segments.

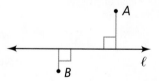

In the diagram, $\overrightarrow{AD}$ is the bisector of $\angle CAB$. If you measure the lengths of the perpendicular segments from D to the two sides of the angle, you will find that the lengths are equal. Point D is equidistant from the sides of the angle.

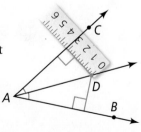

Theorem 5-4 Angle Bisector Theorem

Theorem
If a point is on the bisector of an angle, then the point is equidistant from the sides of the angle.

If . . .
$\overrightarrow{QS}$ bisects $\angle PQR$, $\overline{SP} \perp \overrightarrow{QP}$, and $\overline{SR} \perp \overrightarrow{QR}$

Then . . .
$SP = SR$

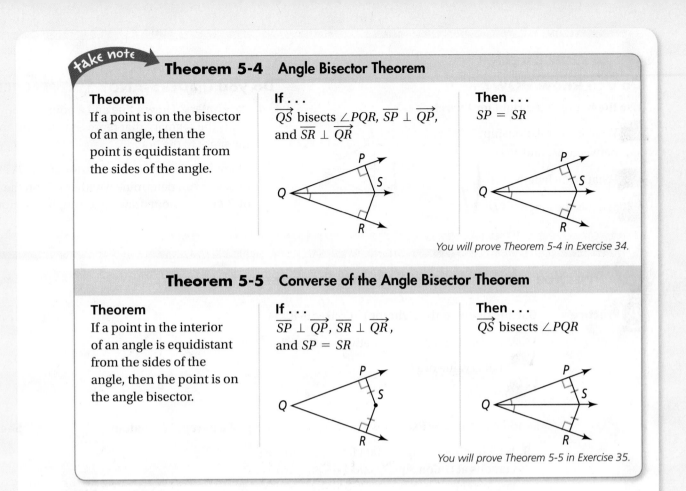

You will prove Theorem 5-4 in Exercise 34.

Theorem 5-5 Converse of the Angle Bisector Theorem

Theorem
If a point in the interior of an angle is equidistant from the sides of the angle, then the point is on the angle bisector.

If . . .
$\overline{SP} \perp \overrightarrow{QP}$, $\overline{SR} \perp \overrightarrow{QR}$, and $SP = SR$

Then . . .
$\overrightarrow{QS}$ bisects $\angle PQR$

You will prove Theorem 5-5 in Exercise 35.

Problem 3 Using the Angle Bisector Theorem

Algebra What is the length of $\overline{RM}$?

Know	Need	Plan
$\overrightarrow{NR}$ bisects $\angle LNQ$. $\overline{RM} \perp \overline{NL}$ and $\overline{RP} \perp \overline{NQ}$.	The length of $\overline{RM}$	Use the Angle Bisector Theorem to write an equation you can solve for x.

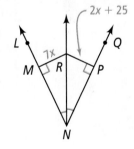

$RM = RP$ Angle Bisector Theorem

$7x = 2x + 25$ Substitute.

$5x = 25$ Subtract $2x$ from each side.

$x = 5$ Divide each side by 5.

Now find RM.

$RM = 7x$

$= 7(5) = 35$ Substitute 5 for x.

Think

How can you use the expression given for *RP* to check your answer?
Substitute 5 for x in the expression $2x + 25$ and verify that the result is 35.

✓ **Got It?** **3.** What is the length of $\overline{FB}$?

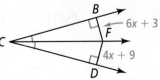

Lesson Check

Do you know HOW?

Use the figure at the right for Exercises 1–3.

1. What is the relationship between $\overline{AC}$ and $\overline{BD}$?

2. What is the length of $\overline{AB}$?

3. What is the length of $\overline{DC}$?

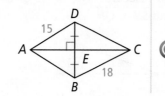

Do you UNDERSTAND?

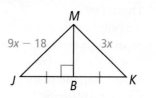MATHEMATICAL PRACTICES

4. **Vocabulary** Draw a line and a point not on the line. Draw the segment that represents the distance from the point to the line.

5. **Writing** Point P is in the interior of $\angle LOX$. Describe how you can determine whether P is on the bisector of $\angle LOX$ without drawing the angle bisector.

Practice and Problem-Solving Exercises

MATHEMATICAL PRACTICES

A Practice

Use the figure at the right for Exercises 6–8.

6. What is the relationship between $\overline{MB}$ and $\overline{JK}$?

7. What is value of x?

8. Find JM.

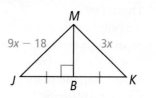

See Problem 1.

Reading Maps For Exercises 9 and 10, use the map of a part of Manhattan.

9. Which school is equidistant from the subway stations at Union Square and 14th Street? How do you know?

10. Is St. Vincent's Hospital equidistant from Village Kids Nursery School and Legacy School? How do you know?

11. **Writing** On a piece of paper, mark a point H for home and a point S for school. Describe how to find the set of points equidistant from H and S.

See Problem 2.

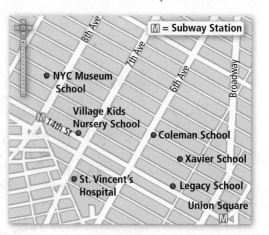

Use the figure at the right for Exercises 12–15.

12. According to the diagram, how far is L from $\overrightarrow{HK}$? From $\overrightarrow{HF}$?

13. How is $\overrightarrow{HL}$ related to $\angle KHF$? Explain.

14. Find the value of y.

15. Find $m\angle KHL$ and $m\angle FHL$.

See Problem 3.

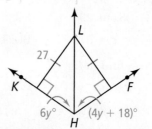

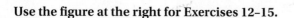

16. Algebra Find x, JK, and JM.

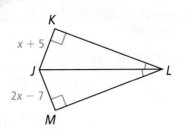

17. Algebra Find y, ST, and TU.

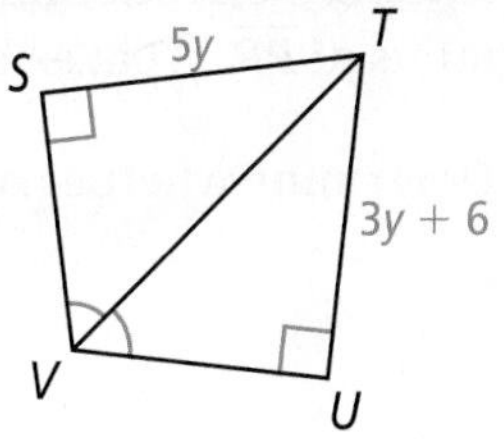

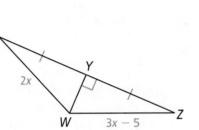

B **Apply**

Algebra **Use the figure at the right for Exercises 18–22.**

18. Find the value of x.

19. Find TW.

20. Find WZ.

21. What kind of triangle is $\triangle TWZ$? Explain.

22. If R is on the perpendicular bisector of $\overline{TZ}$, then R is ___?___ from T and Z, or ___?___ = ___?___.

© **23. Think About a Plan** In the diagram at the right, the soccer goalie will prepare for a shot from the player at point P by moving out to a point on $\overline{XY}$. To have the best chance of stopping the ball, should the goalie stand at the point on $\overline{XY}$ that lies on the perpendicular bisector of $\overline{GL}$ or at the point on $\overline{XY}$ that lies on the bisector of $\angle GPL$? Explain your reasoning.
  • How can you draw a diagram to help?
  • Would the goalie want to be the same distance from G and L or from $\overline{PG}$ and $\overline{PL}$?

© **24. a. Constructions** Draw $\angle CDE$. Construct the angle bisector of the angle.
  b. Reasoning Use the converse of the angle bisector theorem to justify your construction.

© **25. a. Constructions** Draw $\overline{QR}$. Construct the perpendicular bisector of $\overline{QR}$ to construct $\triangle PQR$.
  b. Reasoning Use the perpendicular bisector theorem to justify that your construction is an isosceles triangle.

26. Write Theorems 5-2 and 5-3 as a single biconditional statement.

27. Write Theorems 5-4 and 5-5 as a single biconditional statement.

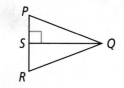

28. Error Analysis To prove that △PQR is isosceles, a student began by stating that since Q is on the segment perpendicular to $\overline{PR}$, Q is equidistant from the endpoints of $\overline{PR}$. What is the error in the student's reasoning?

Writing Determine whether A must be on the bisector of ∠TXR. Explain.

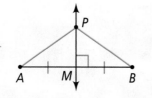

29.

30.
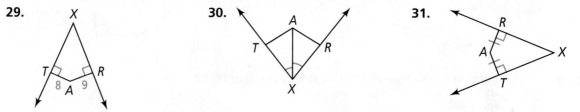

31.
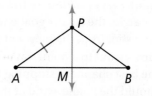

32. Prove the Perpendicular
Proof Bisector Theorem.

    Given: $\overleftrightarrow{PM} \perp \overline{AB}$, $\overleftrightarrow{PM}$ bisects $\overline{AB}$
    Prove: $AP = BP$

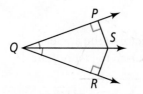

33. Prove the Converse of the
Proof Perpendicular Bisector Theorem.

    Given: $PA = PB$ with $\overleftrightarrow{PM} \perp \overline{AB}$ at M.
    Prove: P is on the perpendicular bisector of $\overline{AB}$.

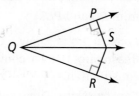

34. Prove the Angle
Proof Bisector Theorem.

    Given: $\overrightarrow{QS}$ bisects ∠PQR,
    $\overline{SP} \perp \overrightarrow{QP}, \overline{SR} \perp \overrightarrow{QR}$
    Prove: $SP = SR$

35. Prove the Converse of the
Proof Angle Bisector Theorem.

    Given: $\overline{SP} \perp \overrightarrow{QP}, \overline{SR} \perp \overrightarrow{QR}$,
    $SP = SR$
    Prove: $\overrightarrow{QS}$ bisects ∠PQR.

36. Coordinate Geometry Use points $A(6, 8)$, $O(0, 0)$, and $B(10, 0)$.
    a. Write equations of lines ℓ and m such that $\ell \perp \overleftrightarrow{OA}$ at A and $m \perp \overleftrightarrow{OB}$ at B.
    b. Find the intersection C of lines ℓ and m.
    c. Show that $CA = CB$.
    d. Explain why C is on the bisector of ∠AOB.

37. *A*, *B*, and *C* are three noncollinear points. Describe and sketch a line in plane *ABC* such that points *A*, *B*, and *C* are equidistant from the line. Justify your response.

38. Reasoning *M* is the intersection of the perpendicular bisectors of two sides of △*ABC*. Line ℓ is perpendicular to plane *ABC* at *M*. Explain why a point *E* on ℓ is equidistant from *A*, *B*, and *C*. (*Hint:* See page 48, Exercise 33. Explain why △*EAM* ≅ △*EBM* ≅ △*ECM*.)

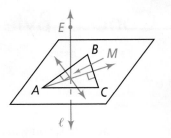

Standardized Test Prep

39. For *A*(1, 3) and *B*(1, 9), which point lies on the perpendicular bisector of $\overline{AB}$?

   (A) (3, 3)   (B) (1, 5)   (C) (6, 6)   (D) (3, 12)

40. What is the converse of the following conditional statement?

If a triangle is isosceles, then it has two congruent angles.

   (F) If a triangle is isosceles, then it has two congruent sides.

   (G) If a triangle has congruent sides, then it is equilateral.

   (H) If a triangle has two congruent angles, then it is isosceles.

   (I) If a triangle is not isosceles, then it does not have two congruent angles.

41. Which figure represents the statement $\overline{BD}$ bisects ∠*ABC*?

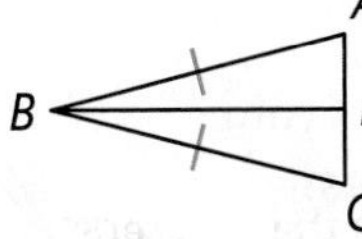

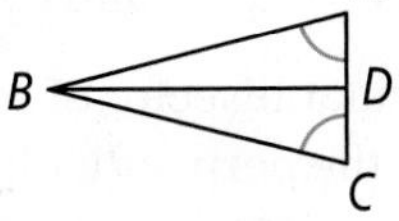

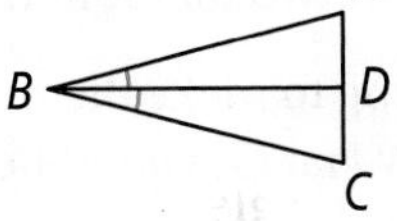

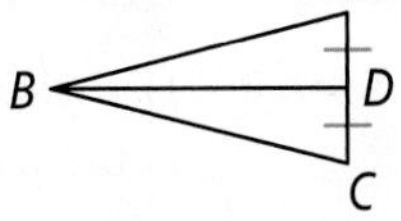

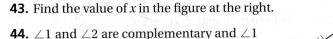

42. The line *y* = 7 is the perpendicular bisector of the segment with endpoints *A*(2, 10) and *B*(2, *k*). What is the value of *k*? Explain your reasoning.

Mixed Review

43. Find the value of *x* in the figure at the right.   ◖ **See Lesson 5-1.**

44. ∠1 and ∠2 are complementary and ∠1 and ∠3 are supplementary. If *m*∠2 = 30, what is *m*∠3?   ◖ **See Lesson 1-5.**

Get Ready!   **To prepare for Lesson 5-3, do Exercises 45–47.**

45. What is the slope of a line that is perpendicular to the line *y* = −3*x* + 4?   ◖ **See Lesson 3-8.**

46. Line ℓ is a horizontal line. What is the slope of a line perpendicular to ℓ?

47. Describe the line *x* = 5.

Concept Byte

Use With Lesson 5-3

ACTIVITY

Paper Folding Bisectors

© **Content Standard**
Prepares for G.C.3 Construct the inscribed and circumscribed circles of a triangle . . .

In Activity 1, you will use paper folding to investigate the bisectors of the angles of a triangle.

Activity 1

Step 1 Draw and cut out three different triangles: one acute, one right, and one obtuse.

Step 2 Use paper folding to make the angle bisectors of each angle of your acute triangle. What do you notice about the angle bisectors?

Step 3 Repeat Step 2 with your right triangle and your obtuse triangle. Does your discovery from Step 2 still hold true?

Folding an angle bisector

In Activity 2, you will use paper folding to investigate the perpendicular bisectors of the sides of a triangle.

Activity 2

Step 1 Draw and cut out two different triangles: one acute and one right.

Step 2 Use paper folding to make the perpendicular bisector of each side of your acute triangle. What do you notice about the perpendicular bisectors?

Step 3 Repeat Step 1 with your right triangle. Does your discovery from Step 2 still hold true?

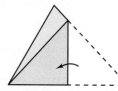

Folding a perpendicular bisector

Exercises

© **1. Make a Conjecture** Make a conjecture about the bisectors of the angles of a triangle.

© **2. Make a Conjecture** Make a conjecture about the perpendicular bisectors of the sides of a triangle.

© **3. Extend** Draw and cut out an obtuse triangle. Fold the perpendicular bisectors.
   a. How do the results for your obtuse triangle compare to the results for your acute and right triangles from Activity 2?
   b. Based on your answer to part (a), how would you revise your conjecture in Exercise 2?

© **4. Extend** For what type of triangle would the three perpendicular bisectors and the three angle bisectors intersect at the same point?

Bisectors in Triangles

©Content Standard
G.C.3 Construct the inscribed and circumscribed circles of a triangle . . .

Objective To identify properties of perpendicular bisectors and angle bisectors

Can you conjecture any other properties that the perpendicular lines might have?

© MATHEMATICAL PRACTICES

SOLVE IT!

Getting Ready!

Construct a circle and label its center C. Choose any three points on the circle and connect them to form a triangle. Draw three lines from C such that each line is perpendicular to one side of the triangle. What conjecture can you make about the two segments into which each side of the triangle is divided? Justify your reasoning.

In the Solve It, the three lines you drew intersect at one point, the center of the circle. When three or more lines intersect at one point, they are **concurrent**. The point at which they intersect is the **point of concurrency.**

Essential Understanding For any triangle, certain sets of lines are always concurrent. Two of these sets of lines are the perpendicular bisectors of the triangle's three sides and the bisectors of the triangle's three angles.

Lesson Vocabulary

- concurrent
- point of concurrency
- circumcenter of a triangle
- circumscribed about
- incenter of a triangle
- inscribed in

Theorem 5-6 Concurrency of Perpendicular Bisectors Theorem

Theorem
The perpendicular bisectors of the sides of a triangle are concurrent at a point equidistant from the vertices.

Diagram

Symbols
Perpendicular bisectors $\overline{PX}$, $\overline{PY}$, and $\overline{PZ}$ are concurrent at P.

$$PA = PB = PC$$

The point of concurrency of the perpendicular bisectors of a triangle is called the **circumcenter of the triangle.**

Since the circumcenter is equidistant from the vertices, you can use the circumcenter as the center of the circle that contains each vertex of the triangle. You say the circle is **circumscribed about** the triangle.

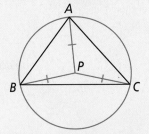

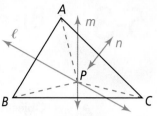

Proof of Theorem 5-6

Given: Lines ℓ, m, and n are the perpendicular bisectors of the sides of $\triangle ABC$. P is the intersection of lines ℓ and m.

Prove: Line n contains point P, and $PA = PB = PC$.

Proof: A point on the perpendicular bisector of a segment is equidistant from the endpoints of the segment. Point P is on ℓ, which is the perpendicular bisector of $\overline{AB}$, so $PA = PB$. Using the same reasoning, since P is on m, and m is the perpendicular bisector of $\overline{BC}$, $PB = PC$. Thus, $PA = PC$ by the Transitive Property. Since $PA = PC$, P is equidistant from the endpoints of $\overline{AC}$. Then, by the converse of the Perpendicular Bisector Theorem, P is on line n, the perpendicular bisector of $\overline{AC}$.

The circumcenter of a triangle can be inside, on, or outside a triangle.

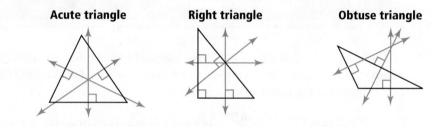

Acute triangle   **Right triangle**   **Obtuse triangle**

ⓒ **Problem 1** Finding the Circumcenter of a Triangle

What are the coordinates of the circumcenter of the triangle with vertices $P(0, 6)$, $O(0, 0)$, and $S(4, 0)$?

Find the intersection point of two of the triangle's perpendicular bisectors. Here, it is easiest to find the perpendicular bisectors of $\overline{PO}$ and $\overline{OS}$.

Think

Does the location of the circumcenter make sense?
Yes, $\triangle POS$ is a right triangle, so its circumcenter should lie on its hypotenuse.

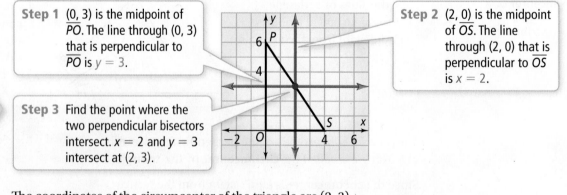

Step 1 $(0, 3)$ is the midpoint of $\overline{PO}$. The line through $(0, 3)$ that is perpendicular to $\overline{PO}$ is $y = 3$.

Step 2 $(2, 0)$ is the midpoint of $\overline{OS}$. The line through $(2, 0)$ that is perpendicular to $\overline{OS}$ is $x = 2$.

Step 3 Find the point where the two perpendicular bisectors intersect. $x = 2$ and $y = 3$ intersect at $(2, 3)$.

The coordinates of the circumcenter of the triangle are $(2, 3)$.

Got It? **1.** What are the coordinates of the circumcenter of the triangle with vertices $A(2, 7)$, $B(10, 7)$, and $C(10, 3)$?

Think

How do you find a point equidistant from three points?
As long as the three points are noncollinear, they are vertices of a triangle. Find the circumcenter of the triangle.

Problem 2 Using a Circumcenter

A town planner wants to locate a new fire station equidistant from the elementary, middle, and high schools. Where should he locate the station?

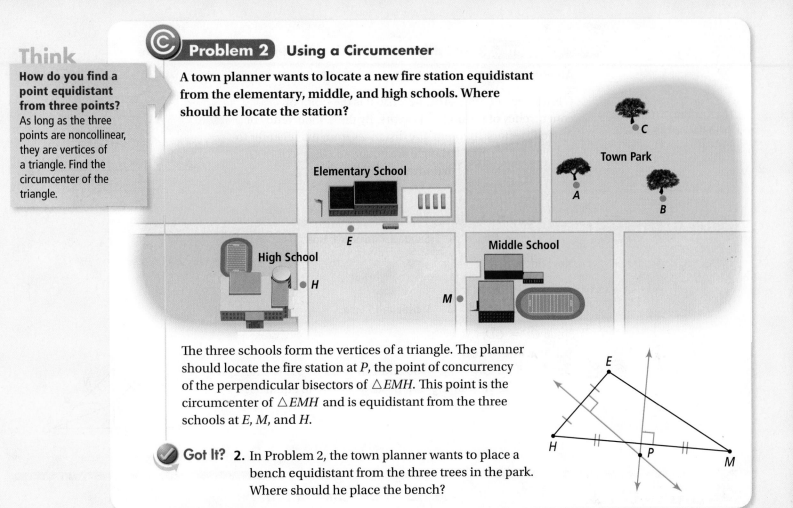

The three schools form the vertices of a triangle. The planner should locate the fire station at *P*, the point of concurrency of the perpendicular bisectors of △*EMH*. This point is the circumcenter of △*EMH* and is equidistant from the three schools at *E*, *M*, and *H*.

Got It? 2. In Problem 2, the town planner wants to place a bench equidistant from the three trees in the park. Where should he place the bench?

take note

Theorem 5-7   Concurrency of Angle Bisectors Theorem

Theorem	Diagram	Symbols
The bisectors of the angles of a triangle are concurrent at a point equidistant from the sides of the triangle.	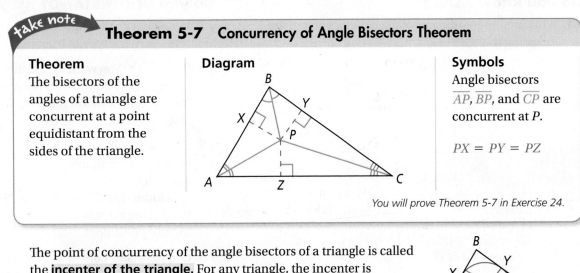	Angle bisectors $\overline{AP}$, $\overline{BP}$, and $\overline{CP}$ are concurrent at *P*. $PX = PY = PZ$

You will prove Theorem 5-7 in Exercise 24.

The point of concurrency of the angle bisectors of a triangle is called the **incenter of the triangle.** For any triangle, the incenter is always inside the triangle. In the diagram, points *X*, *Y*, and *Z* are equidistant from *P*, the incenter of △*ABC*. *P* is the center of the circle that is **inscribed in** the triangle.

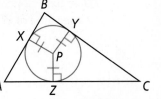

Problem 3 **Identifying and Using the Incenter of a Triangle**

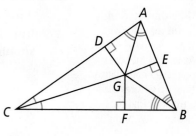

Think

What is the distance from a point to a line?

The distance from a point to a line is the length of the perpendicular segment that joins the point to the line.

Algebra $GE = 2x - 7$ and $GF = x + 4$. What is GD?

G is the incenter of $\triangle ABC$ because it is the point of concurrency of the angle bisectors. By the Concurrency of Angle Bisectors Theorem, the distances from the incenter to the three sides of the triangle are equal, so $GE = GF = GD$. Use this relationship to find x.

$$2x - 7 = x + 4 \qquad GE = GF$$

$$2x = x + 11 \qquad \text{Add 7 to each side.}$$

$$x = 11 \qquad \text{Subtract } x \text{ from each side.}$$

Now find GF.

$$GF = x + 4$$

$$= 11 + 4 = 15 \qquad \text{Substitute 11 for } x.$$

Since $GF = GD$, $GD = 15$.

Got It? **3. a.** $QN = 5x + 36$ and $QM = 2x + 51$. What is QO?

b. Reasoning Is it possible for QP to equal 50? Explain.

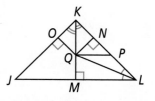

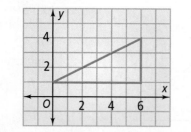

Lesson Check

Do you know HOW?

1. What are the coordinates of the circumcenter of the following triangle?

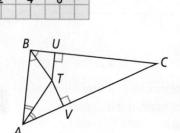

2. In the figure at the right, $TV = 3x - 12$ and $TU = 5x - 24$. What is the value of x?

Do you UNDERSTAND?   **MATHEMATICAL PRACTICES**

3. **Vocabulary** A triangle's circumcenter is outside the triangle. What type of triangle is it?

4. **Reasoning** You want to find the circumcenter of a triangle. Why do you only need to find the intersection of two of the triangle's perpendicular bisectors, instead of all three?

5. **Error Analysis** Your friend sees the triangle at the right and concludes that $CT = CP$. What is the error in your friend's reasoning?

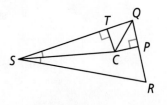

6. **Compare and Contrast** How are the circumcenter and incenter of a triangle alike? How are they different?

Practice and Problem-Solving Exercises

 MATHEMATICAL PRACTICES

A Practice

Coordinate Geometry Find the coordinates of the circumcenter of each triangle.

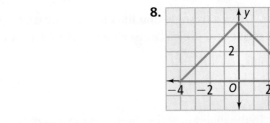

◀ **See Problem 1.**

Coordinate Geometry Find the coordinates of the circumcenter of △*ABC*.

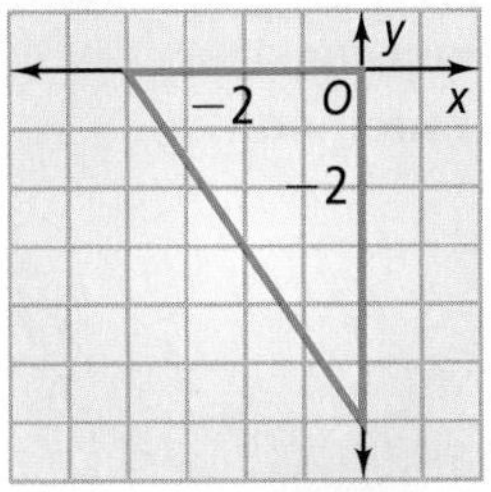

9. $A(0, 0)$
$B(3, 0)$
$C(3, 2)$

10. $A(0, 0)$
$B(4, 0)$
$C(4, -3)$

11. $A(-4, 5)$
$B(-2, 5)$
$C(-2, -2)$

12. $A(-1, -2)$
$B(-5, -2)$
$C(-1, -7)$

13. $A(1, 4)$
$B(1, 2)$
$C(6, 2)$

14. City Planning Copy the diagram of the beach. Show where town officials should place a recycling barrel so that it is equidistant from the lifeguard chair, the snack bar, and the volleyball court. Explain.

◀ **See Problem 2.**

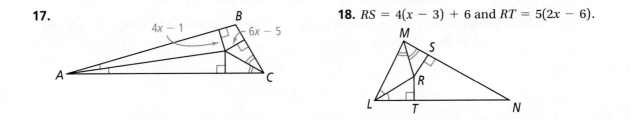

Name the point of concurrency of the angle bisectors.

◀ **See Problem 3.**

15.

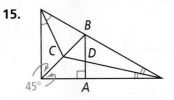

16.

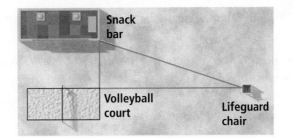

Find the value of *x*.

17.

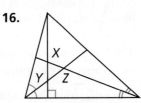

18. $RS = 4(x - 3) + 6$ and $RT = 5(2x - 6)$.

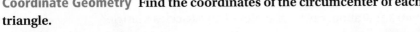

19. Think About a Plan In the figure at the right, *P* is the incenter of isosceles △*RST*. What type of triangle is △*RPT*? Explain.
- What segments determine the incenter of a triangle?
- What do you know about the base angles of an isosceles triangle?

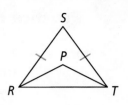

Constructions Draw a triangle that fits the given description. Then construct the inscribed circle and the circumscribed circle. Describe your method.

20. right triangle, △*DEF*

21. obtuse triangle, △*STU*

22. Algebra In the diagram at the right, *G* is the incenter of △*DEF*, $m\angle DEF = 60$, and $m\angle EFD = 2 \cdot m\angle EDF$. What are $m\angle DGE$, $m\angle DGF$, and $m\angle EGF$?

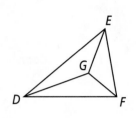

23. Writing Ivars found an old piece of paper inside an antique book. It read,

From the spot I buried Olaf's treasure, equal sets of paces did I measure; each of three directions in a line, there to plant a seedling Norway pine. I could not return for failing health; now the hounds of Haiti guard my wealth. —Karl

After searching Caribbean islands for five years, Ivars found an island with three tall Norway pines. How might Ivars find where Karl buried Olaf's treasure?

24. Use the diagram at the right to prove the Concurrrency of Angle
Proof Bisectors Theorem.

Given: Rays ℓ, *m*, and *n* are bisectors of the angles of △*ABC*. *X* is the intersection of rays ℓ and *m*, $\overline{XD} \perp \overline{AC}$, $\overline{XE} \perp \overline{AB}$, and $\overline{XF} \perp \overline{BC}$.

Prove: Ray *n* contains point *X*, and $XD = XE = XF$.

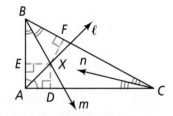

25. Noise Control You are trying to talk to a friend on the phone in a busy bus station. The buses are so loud that you can hardly hear. Referring to the figure at the right, should you stand at *P* or *C* to be as far as possible from all the buses? Explain.

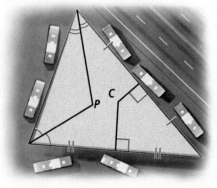

Reasoning Determine whether each statement is *true* or *false*. If the statement is false, give a counterexample.

26. The incenter of a triangle is equidistant from all three vertices.

27. The incenter of a triangle always lies inside the triangle.

28. You can circumscribe a circle about any three points in a plane.

29. If point *C* is the circumcenter of △*PQR* and the circumcenter of △*PQS*, then *R* and *S* must be the same point.

Challenge

30. Reasoning Explain why the circumcenter of a right triangle is on one of the triangle's sides.

Determine whether each statement is *always*, *sometimes*, or *never* true. Explain.

31. It is possible to find a point equidistant from three parallel lines in a plane.

32. The circles inscribed in and circumscribed about an isosceles triangle have the same center.

Standardized Test Prep

SAT/ACT

33. Which of the following statements is *false*?

Ⓐ The bisectors of the angles of a triangle are concurrent.

Ⓑ The midsegments of a triangle are concurrent.

Ⓒ The perpendicular bisectors of the sides of a triangle are concurrent.

Ⓓ Four lines intersecting in one point are concurrent.

34. What type of triangle is △*PUT*?

Ⓕ right isosceles

Ⓖ acute isosceles

Ⓗ obtuse scalene

Ⓘ acute scalene

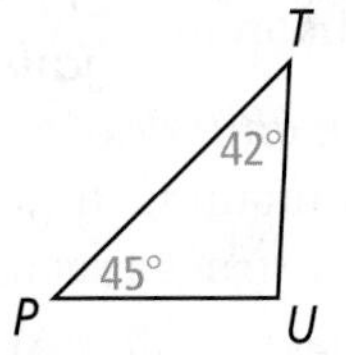

35. Which statement is logically equivalent to the following statement?

If a triangle is right isosceles, then it has exactly two acute angles.

Ⓐ If a triangle is right isosceles, then it has one right angle.

Ⓑ If a triangle has exactly two acute angles, then it is right isosceles.

Ⓒ If a triangle does not have exactly two acute angles, then it is not right isosceles.

Ⓓ If a triangle is not right isosceles, then it does not have a right angle.

Short Response

36. Refer to the figure at the right above. Explain in two different ways why $\overline{MV}$ is the angle bisector of ∠*KVR*.

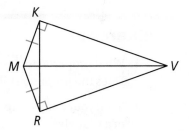

Mixed Review

Use the figure at the right for Exercises 37 and 38.

37. Find the value of *x*.

38. Find the length of $\overline{AD}$.

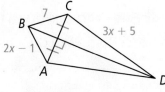

◀ **See Lesson 5-2.**

Get Ready! **To prepare for Lesson 5-4, do Exercises 39 and 40.**

Find the coordinates of the midpoint of $\overline{AB}$ with the given endpoints.

◀ **See Lesson 1-7.**

39. *A*(3, 0), *B*(3, 16)

40. *A*(6, 8), *B*(4, −1)

Special Segments in Triangles

© **Content Standard**
Prepares for G.CO.9 Prove theorems about triangles . . . the medians of a triangle meet at a point.

You already know about two sets of lines that are concurrent for any triangle. In the following activity, you will use geometry software to confirm what you know about the concurrency of a triangle's perpendicular bisectors and angle bisectors. Then you will explore two more sets of special segments in triangles.

© MATHEMATICAL PRACTICES

Activity

Use geometry software.

- Construct a triangle and the three perpendicular bisectors of its sides. Use your result to confirm Theorem 5-6, the Concurrency of Perpendicular Bisectors Theorem.

- Construct a triangle and its three angle bisectors. Use your result to confirm Theorem 5-7, the Concurrency of Angle Bisectors Theorem.

- An *altitude* of a triangle is the perpendicular segment from a vertex to the line containing the opposite side. Construct a triangle. Through a vertex of the triangle construct a segment that is perpendicular to the line containing the side opposite that vertex. Next construct the altitudes from the other two vertices.

- A *median* of a triangle is the segment joining the midpoint of a side and the opposite vertex. Construct a triangle. Construct the midpoint of one side. Draw the median. Then construct the other two medians.

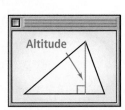

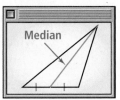

Exercises

1. What property do the lines containing altitudes and the medians seem to have? Does the property still hold as you manipulate the triangles?

2. State your conjectures about the lines containing altitudes and about the medians of a triangle.

3. Copy the table. Think about acute, right, and obtuse triangles. Use *inside*, *on*, or *outside* to describe the location of each point of concurrency.

	Perpendicular Bisectors	Angle Bisectors	Lines Containing the Altitudes	Medians
Acute Triangle	■	■	■	■
Right Triangle	■	■	■	■
Obtuse Triangle	■	■	■	■

© 4. **Extend** What observations, if any, can you make about these special segments for isosceles triangles? For equilateral triangles?

Medians and Altitudes

Content Standards
G.CO.10 Prove theorems about triangles . . . the medians of a triangle meet at a point.
Also G.SRT.5

Objective To identify properties of medians and altitudes of a triangle

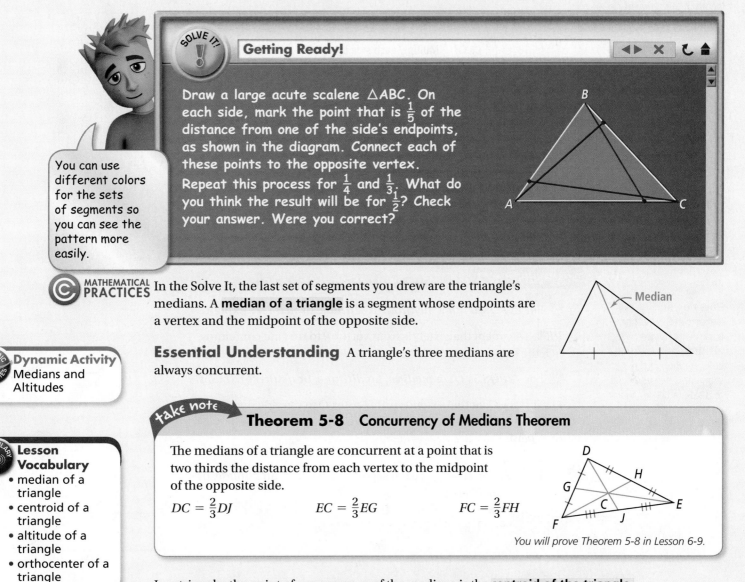

SOLVE IT!

Getting Ready!

Draw a large acute scalene △ABC. On each side, mark the point that is $\frac{1}{5}$ of the distance from one of the side's endpoints, as shown in the diagram. Connect each of these points to the opposite vertex. Repeat this process for $\frac{1}{4}$ and $\frac{1}{3}$. What do you think the result will be for $\frac{1}{2}$? Check your answer. Were you correct?

You can use different colors for the sets of segments so you can see the pattern more easily.

MATHEMATICAL PRACTICES In the Solve It, the last set of segments you drew are the triangle's medians. A **median of a triangle** is a segment whose endpoints are a vertex and the midpoint of the opposite side.

Median

Dynamic Activity
Medians and Altitudes

Essential Understanding A triangle's three medians are always concurrent.

Lesson Vocabulary
• median of a triangle
• centroid of a triangle
• altitude of a triangle
• orthocenter of a triangle

take note

Theorem 5-8 Concurrency of Medians Theorem

The medians of a triangle are concurrent at a point that is two thirds the distance from each vertex to the midpoint of the opposite side.

$$DC = \frac{2}{3}DJ \qquad\qquad EC = \frac{2}{3}EG \qquad\qquad FC = \frac{2}{3}FH$$

You will prove Theorem 5-8 in Lesson 6-9.

In a triangle, the point of concurrency of the medians is the **centroid of the triangle.** The point is also called the *center of gravity* of a triangle because it is the point where a triangular shape will balance. For any triangle, the centroid is always inside the triangle.

Problem 1 **Finding the Length of a Median**   **GRIDDED RESPONSE**

In the diagram at the right, $XA = 8$. What is the length of $\overline{XB}$?

A is the centroid of $\triangle XYZ$ because it is the point of concurrency of the triangle's medians.

$XA = \frac{2}{3}XB$ Concurrency of Medians Theorem

$8 = \frac{2}{3}XB$ Substitute 8 for XA.

$\left(\frac{3}{2}\right)8 = \left(\frac{3}{2}\right)\frac{2}{3}XB$ Multiply each side by $\frac{3}{2}$.

$12 = XB$ Simplify.

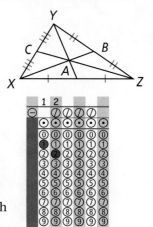

Got It? **1. a.** In the diagram for Problem 1, $ZA = 9$. What is the length of $\overline{ZC}$?

b. Reasoning What is the ratio of ZA to AC? Explain.

An **altitude of a triangle** is the perpendicular segment from a vertex of the triangle to the line containing the opposite side. An altitude of a triangle can be inside or outside the triangle, or it can be a side of the triangle.

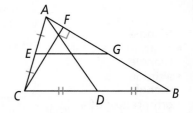

Problem 2 **Identifying Medians and Altitudes**

A **For $\triangle PQS$, is $\overline{PR}$ a _median_, an _altitude_, or _neither_? Explain.**

$\overline{PR}$ is a segment that extends from vertex P to the line containing $\overrightarrow{SQ}$, the side opposite P. $\overline{PR} \perp \overrightarrow{QR}$, so $\overline{PR}$ is an altitude of $\triangle PQS$.

B **For $\triangle PQS$, is $\overline{QT}$ a _median_, an _altitude_, or _neither_? Explain.**

$\overline{QT}$ is a segment that extends from vertex Q to the side opposite Q. Since $\overline{PT} \cong \overline{TS}$, T is the midpoint of $\overline{PS}$. So $\overline{QT}$ is a median of $\triangle PQS$.

Got It? **2.** For $\triangle ABC$, is each segment a _median_, an _altitude_, or _neither_? Explain.

a. $\overline{AD}$ **b.** $\overline{EG}$ **c.** $\overline{CF}$

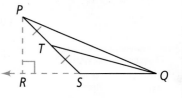

 **take note**

Theorem 5-9 Concurrency of Altitudes Theorem

The lines that contain the altitudes of a triangle are concurrent.

You will prove Theorem 5-9 in Lesson 6-9.

Plan

How do you use the centroid?
Write an equation relating the length of the whole median to the length of the segment from the vertex to the centroid.

Plan

How do you determine whether a segment is an altitude or a median?
Look at whether the segment is perpendicular to a side (altitude) and/or bisects a side (median).

The lines that contain the altitudes of a triangle are concurrent at the **orthocenter of the triangle.** The orthocenter of a triangle can be inside, on, or outside the triangle.

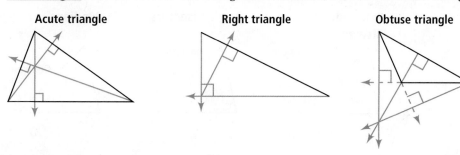

Acute triangle   Right triangle   Obtuse triangle

© **Problem 3**   **Finding the Orthocenter**

$\triangle ABC$ has vertices $A(1, 3)$, $B(2, 7)$, and $C(6, 3)$. **What are the coordinates of the orthocenter of $\triangle ABC$?**

Know → **Need** → **Plan**

Know	Need	Plan
The coordinates of the three vertices	The intersection point of the triangle's altitudes	Write the equations of the lines that contain two of the altitudes. Then solve the system of equations.

Think

Which two altitudes should you choose?
It does not matter, but the altitude to $\overline{AC}$ is a vertical line, so its equation will be easy to find.

Step 1   Find the equation of the line containing the altitude to $\overline{AC}$. Since $\overline{AC}$ is horizontal, the line containing the altitude to $\overline{AC}$ is vertical. The line passes through the vertex $B(2, 7)$. The equation of the line is $x = 2$.

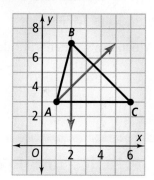

Step 2   Find the equation of the line containing the altitude to $\overline{BC}$. The slope of the line containing $\overline{BC}$ is $\frac{3 - 7}{6 - 2} = -1$. Since the product of the slopes of two perpendicular lines is -1, the line containing the altitude to $\overline{BC}$ has slope 1.

The line passes through the vertex $A(1, 3)$. The equation of the line is $y - 3 = 1(x - 1)$, which simplifies to $y = x + 2$.

Step 3   Find the orthocenter by solving this system of equations: $x = 2$
                                                                      $y = x + 2$

$y = 2 + 2$ Substitute 2 for x in the second equation.

$y = 4$ Simplify.

The coordinates of the orthocenter are $(2, 4)$.

✓ Got It? 3. $\triangle DEF$ has vertices $D(1, 2)$, $E(1, 6)$, and $F(4, 2)$. What are the coordinates of the orthocenter of $\triangle DEF$?

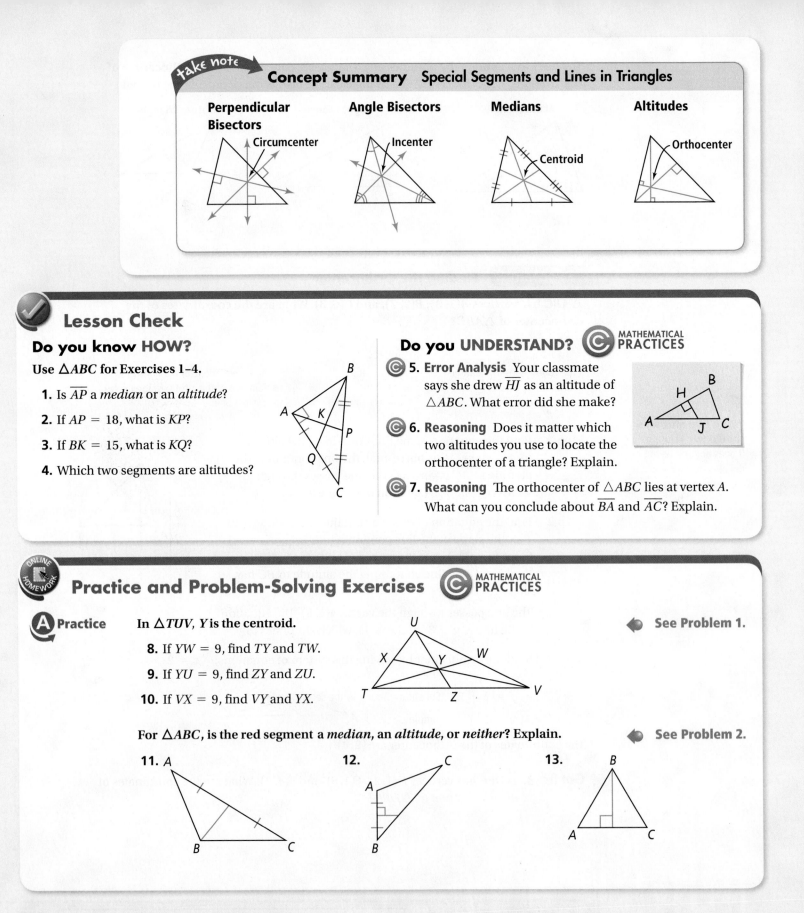

Concept Summary Special Segments and Lines in Triangles

Perpendicular Bisectors	Angle Bisectors	Medians	Altitudes
Circumcenter	Incenter	Centroid	Orthocenter

Lesson Check

Do you know HOW?

Use △ABC for Exercises 1–4.

1. Is $\overline{AP}$ a *median* or an *altitude*?

2. If $AP = 18$, what is KP?

3. If $BK = 15$, what is KQ?

4. Which two segments are altitudes?

Do you UNDERSTAND? MATHEMATICAL PRACTICES

5. Error Analysis Your classmate says she drew $\overline{HJ}$ as an altitude of △ABC. What error did she make?

6. Reasoning Does it matter which two altitudes you use to locate the orthocenter of a triangle? Explain.

7. Reasoning The orthocenter of △ABC lies at vertex A. What can you conclude about $\overline{BA}$ and $\overline{AC}$? Explain.

Practice and Problem-Solving Exercises MATHEMATICAL PRACTICES

Practice

In △TUV, Y is the centroid.

8. If $YW = 9$, find TY and TW.

9. If $YU = 9$, find ZY and ZU.

10. If $VX = 9$, find VY and YX.

See Problem 1.

For △ABC, is the red segment a *median*, an *altitude*, or *neither*? Explain.

See Problem 2.

11.

12.

13.

Coordinate Geometry Find the coordinates of the orthocenter of △ABC.

◀ See Problem 3.

14. A(0, 0)
B(4, 0)
C(4, 2)

15. A(2, 6)
B(8, 6)
C(6, 2)

16. A(0, −2)
B(4, −2)
C(−2, −8)

B **Apply**

Name the centroid.

17.

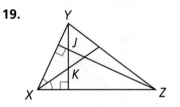

18.

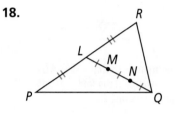

Name the orthocenter of △XYZ.

19.

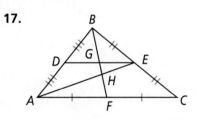

20.

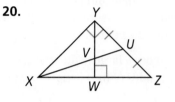

© 21. Think About a Plan In the diagram at the right, $\overline{QS}$ and $\overline{PT}$ are altitudes and $m\angle R = 55$. What is $m\angle POQ$?
- What does it mean for a segment to be an altitude?
- What do you know about the sum of the angle measures in a triangle?
- How do you sketch overlapping triangles separately?

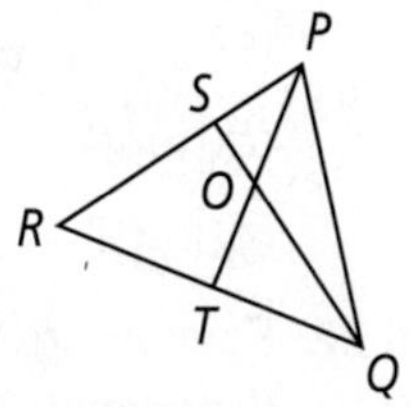

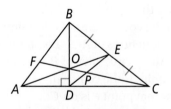

Constructions Draw a triangle that fits the given description. Then construct the centroid and the orthocenter.

22. acute scalene triangle, △LMN

23. obtuse isosceles triangle, △RST

In Exercises 24–27, name each segment.

24. a median in △ABC

25. an altitude in △ABC

26. a median in △BDC

27. an altitude in △AOC

© 28. Reasoning A centroid separates a median into two segments. What is the ratio of the length of the shorter segment to the length of the longer segment?

Paper Folding The figures below show how to construct altitudes and medians by paper folding. Refer to them for Exercises 29 and 30.

Folding an Altitude   **Folding a Median**

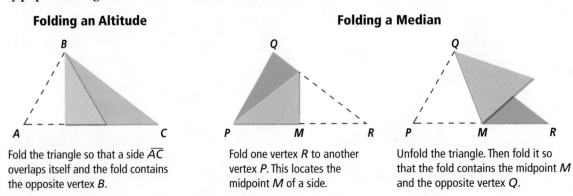

Fold the triangle so that a side $\overline{AC}$ overlaps itself and the fold contains the opposite vertex B.

Fold one vertex R to another vertex P. This locates the midpoint M of a side.

Unfold the triangle. Then fold it so that the fold contains the midpoint M and the opposite vertex Q.

29. Cut out a large triangle. Fold the paper carefully to construct the three medians of the triangle and demonstrate the Concurrency of Medians Theorem. Use a ruler to measure the length of each median and the distance of each vertex from the centroid.

30. Cut out a large acute triangle. Fold the paper carefully to construct the three altitudes of the triangle and demonstrate the Concurrency of Altitudes Theorem.

31. In the figure at the right, C is the centroid of $\triangle DEF$.
If $GF = 12x^2 + 6y$, which expression represents CF?

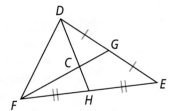

   Ⓐ $6x^2 + 3y$ Ⓒ $8x^2 + 4y$

   Ⓑ $4x^2 + 2y$ Ⓓ $8x^2 + 3y$

Ⓒ **32. Reasoning** What type of triangle has its orthocenter on the exterior of the triangle? Draw a sketch to support your answer.

Ⓒ **33. Writing** Explain why the median to the base of an isosceles triangle is also an altitude.

34. Coordinate Geometry $\triangle ABC$ has vertices $A(0, 0)$, $B(2, 6)$, and $C(8, 0)$. Complete the following steps to verify the Concurrency of Medians Theorem for $\triangle ABC$.

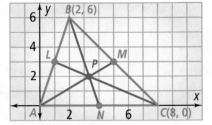

   a. Find the coordinates of midpoints L, M, and N.
   b. Find equations of $\overleftrightarrow{AM}$, $\overleftrightarrow{BN}$, and $\overleftrightarrow{CL}$.
   c. Find the coordinates of P, the intersection of $\overleftrightarrow{AM}$ and $\overleftrightarrow{BN}$. This point is the centroid.
   d. Show that point P is on $\overleftrightarrow{CL}$.
   e. Use the Distance Formula to show that point P is two-thirds of the distance from each vertex to the midpoint of the opposite side.

Ⓒ **Challenge**

35. Constructions A, B, and O are three noncollinear points. Construct point C such that O is the orthocenter of $\triangle ABC$. Describe your method.

Ⓒ **36. Reasoning** In an isosceles triangle, show that the circumcenter, incenter, centroid, and orthocenter can be four different points, but all four must be collinear.

A, *B*, *C*, and *D* are points of concurrency for the triangle. Determine whether each point is a *circumcenter*, *incenter*, *centroid*, or *orthocenter*. Explain.

37.

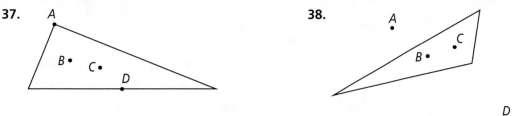

38.

39. History In 1765, Leonhard Euler proved that, for any triangle, three of the four points of concurrency are collinear. The line that contains these three points is known as Euler's Line. Use Exercises 37 and 38 to determine which point of concurrency does not necessarily lie on Euler's Line.

Standardized Test Prep

SAT/ACT

For Exercises 40 and 41, use the figure at the right.

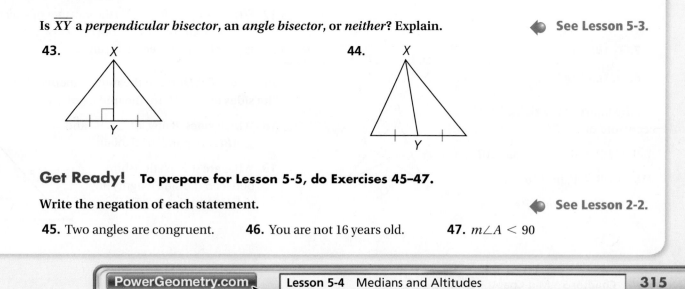

40. If *CR* = 24, what is *KR*?

    Ⓐ 6   Ⓒ 12

    Ⓑ 8   Ⓓ 16

41. If *TR* = 12 what is *CP*?

    Ⓕ 16   Ⓗ 24

    Ⓖ 18   Ⓘ 36

Extended Response

42. The orthocenter of a triangle lies outside the triangle. Where are its circumcenter, incenter, and centroid located in relation to the triangle? Draw and label diagrams to support your answers.

Mixed Review

Is $\overline{XY}$ a *perpendicular bisector*, an *angle bisector*, or *neither*? Explain.

◀ **See Lesson 5-3.**

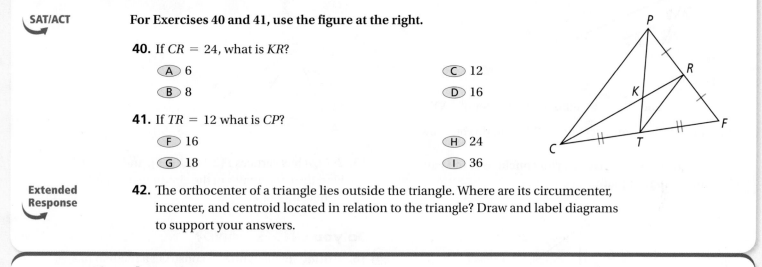

43.

44.

Get Ready! **To prepare for Lesson 5-5, do Exercises 45–47.**

Write the negation of each statement.

◀ **See Lesson 2-2.**

45. Two angles are congruent.   **46.** You are not 16 years old.   **47.** $m\angle A < 90$

5 Mid-Chapter Quiz

MathXL® for School
Go to PowerGeometry.com

Do you know HOW?

Algebra **Find the value of *x*.**

1.

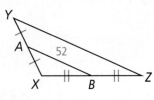

2.

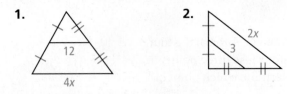

Use the figure below for Exercises 3–5.

3. Find YZ.

4. $AX = 26$ and $BZ = 36$. Find the perimeter of $\triangle XYZ$.

5. Which angle is congruent to $\angle XBA$? How do you know?

For the figure below, what can you conclude about each of the following? Explain.

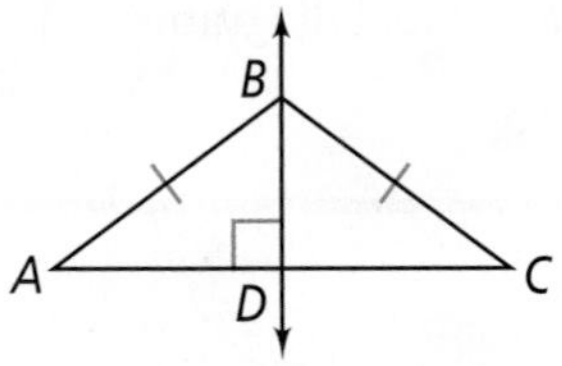

6. $\angle CDB$

7. $\triangle ABD$ and $\triangle CBD$

8. $\overline{AD}$ and $\overline{DC}$

In the figure at the right, *P* is the centroid of $\triangle ABC$.

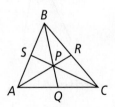

9. If $PR = 6$, find AP and AR.

10. If $PB = 6$, find QP and QB.

11. If $SC = 6$, find CP and PS.

For $\triangle ABC$, is the red line a *perpendicular bisector*, an *angle bisector*, a *median*, an *altitude*, or *none of these*? Explain.

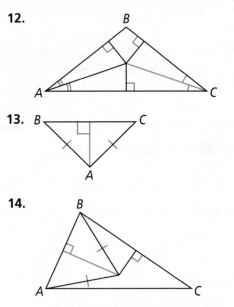

12.

13.

14.

15. $\triangle PQR$ has vertices $P(2, 5)$, $Q(8, 5)$, and $R(8, 1)$. Find the coordinates of the circumcenter and the orthocenter of $\triangle PQR$.

Do you UNDERSTAND?

16. Writing Explain how to construct a median of a triangle and an altitude of a triangle.

17. Error Analysis Point O is the incenter of scalene $\triangle XYZ$. Your friend says that $m\angle YXO = m\angle YZO$. Is your friend correct? Explain.

The sides of $\triangle DEF$ are the midsegments of $\triangle ABC$. The sides of $\triangle GHI$ are the midsegments of $\triangle DEF$.

18. Which sides, if any, of $\triangle GHI$ and $\triangle ABC$ are parallel? Explain.

19. What are the relationships between the side lengths of $\triangle GHI$ and $\triangle ABC$? Explain.

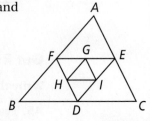

Indirect Proof

© **Content Standard**
Extends G.CO.10 Prove theorems about triangles . . .

Objective To use indirect reasoning to write proofs

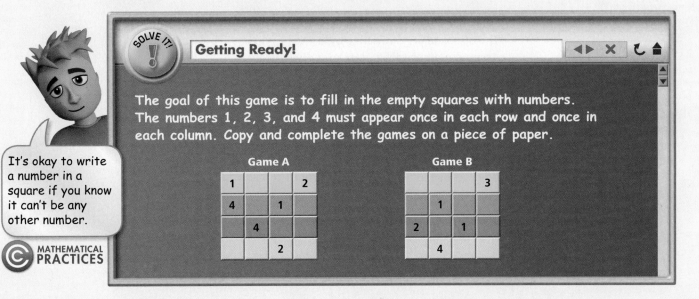

SOLVE IT!

Getting Ready!

The goal of this game is to fill in the empty squares with numbers. The numbers 1, 2, 3, and 4 must appear once in each row and once in each column. Copy and complete the games on a piece of paper.

Game A

Game B

It's okay to write a number in a square if you know it can't be any other number.

© MATHEMATICAL PRACTICES

Lesson Vocabulary
• indirect reasoning
• indirect proof

In the Solve It, you can conclude that a square must contain a certain number if you can eliminate the other three numbers as possibilities. This type of reasoning is called indirect reasoning. In **indirect reasoning,** all possibilities are considered and then all but one are proved false. The remaining possibility must be true.

Essential Understanding You can use indirect reasoning as another method of proof.

A proof involving indirect reasoning is an **indirect proof.** Often in an indirect proof, a statement and its negation are the only possibilities. When you see that one of these possibilities leads to a conclusion that contradicts a fact you know to be true, you can eliminate that possibility. For this reason, indirect proof is sometimes called *proof by contradiction.*

take note

Key Concept Writing an Indirect Proof

Step 1 State as a temporary assumption the opposite (negation) of what you want to prove.

Step 2 Show that this temporary assumption leads to a contradiction.

Step 3 Conclude that the temporary assumption must be false and that what you want to prove must be true.

In the first step of an indirect proof you assume as true the opposite of what you want to prove.

Problem 1 Writing the First Step of an Indirect Proof

Suppose you want to write an indirect proof of each statement. As the first step of the proof, what would you assume?

A An integer n is divisible by 5.

The opposite of "is divisible by" is "is not divisible by."
Assume temporarily that n is not divisible by 5.

B You do not have soccer practice today.

The opposite of "do not have" is "do have."
Assume temporarily that you do have soccer practice today.

Think

How do you find the opposite of a statement?
Write the negation of the statement. This often involves adding or removing the word *not*.

Got It? **1.** Suppose you want to write an indirect proof of each statement. As the first step of the proof, what would you assume?
  a. $\triangle BOX$ is not acute.
  b. At least one pair of shoes you bought cost more than $25.

To write an indirect proof, you have to be able to identify a contradiction.

Problem 2 Identifying Contradictions

Which two statements contradict each other?

I. $\overline{FG} \parallel \overline{KL}$ **II.** $\overline{FG} \cong \overline{KL}$ **III.** $\overline{FG} \perp \overline{KL}$

Segments can be parallel and congruent. Statements I and II do not contradict each other.

Segments can be congruent and perpendicular. Statements II and III do not contradict each other.

Parallel segments do not intersect, so they cannot be perpendicular. Statements I and III contradict each other.

Think

How do you know that two statements contradict each other?
A statement contradicts another statement if it is impossible for both to be true at the same time.

Got It? **2. a.** Which two statements contradict each other?
  I. $\triangle XYZ$ is acute.
  II. $\triangle XYZ$ is scalene.
  III. $\triangle XYZ$ is equiangular.
  b. Reasoning Statements I and II below contradict each other. Statement III is the negation of Statement I. Are Statements II and III equivalent? Explain your reasoning.
  I. $\triangle ABC$ is scalene.
  II. $\triangle ABC$ is equilateral.
  III. $\triangle ABC$ is not scalene.

Given: $\triangle ABC$ is scalene.

Prove: $\angle A$, $\angle B$, and $\angle C$ all have different measures.

Think

Assume temporarily the opposite of what you want to prove.

Show that this assumption leads to a contradiction.

Conclude that the temporary assumption must be false and that what you want to prove must be true.

Write

Assume temporarily that two angles of $\triangle ABC$ have the same measure. Assume that $m\angle A = m\angle B$.

By the Converse of the Isosceles Triangle Theorem, the sides opposite $\angle A$ and $\angle B$ are congruent. This contradicts the given information that $\triangle ABC$ is scalene.

The assumption that two angles of $\triangle ABC$ have the same measure must be false. Therefore, $\angle A$, $\angle B$, and $\angle C$ all have different measures.

Got It? **3. Given:** $7(x + y) = 70$ and $x \neq 4$.

Prove: $y \neq 6$

Lesson Check

Do you know HOW?

1. Suppose you want to write an indirect proof of the following statement. As the first step of the proof, what would you assume?

Quadrilateral $ABCD$ has four right angles.

2. Write a statement that contradicts the following statement. Draw a diagram to support your answer.

Lines a and b are parallel.

Do you UNDERSTAND?

3. Error Analysis A classmate began an indirect proof as shown below. Explain and correct your classmate's error.

> Given: $\triangle ABC$
> Prove: $\angle A$ is obtuse.
> Assume temporarily that $\angle A$ is acute.

Practice and Problem-Solving Exercises 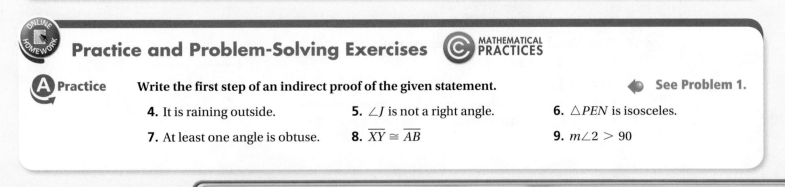 MATHEMATICAL PRACTICES

Ⓐ Practice **Write the first step of an indirect proof of the given statement.** ◀ **See Problem 1.**

4. It is raining outside.

5. $\angle J$ is not a right angle.

6. $\triangle PEN$ is isosceles.

7. At least one angle is obtuse.

8. $\overline{XY} \cong \overline{AB}$

9. $m\angle 2 > 90$

Identify the two statements that contradict each other. See Problem 2.

10. **I.** $\triangle PQR$ is equilateral.
II. $\triangle PQR$ is a right triangle.
III. $\triangle PQR$ is isosceles.

11. **I.** $\ell \parallel m$
II. ℓ and m do not intersect.
III. ℓ and m are skew.

12. **I.** Each of the two items that Val bought costs more than $10.
II. Val spent $34 for the two items.
III. Neither of the two items that Val bought costs more than $15.

13. **I.** In right $\triangle ABC$, $m\angle A = 60$.
II. In right $\triangle ABC$, $\angle A \cong \angle C$.
III. In right $\triangle ABC$, $m\angle B = 90$.

14. Developing Proof Fill in the blanks to prove the following statement. See Problem 3.

If the Yoga Club and Go Green Club together have fewer than 20 members and the Go Green Club has 10 members, then the Yoga Club has fewer than 10 members.

Given: The total membership of the Yoga Club and the Go Green Club is fewer than 20. The Go Green Club has 10 members.

Prove: The Yoga Club has fewer than 10 members.

Proof: Assume temporarily that the Yoga Club has 10 or more members. This means that together the two clubs have **a.** _?_ members. This contradicts the given information that **b.** _?_ . The temporary assumption is false. Therefore, it is true that **c.** _?_ .

15. Developing Proof Fill in the blanks to prove the following statement.

In a given triangle, $\triangle LMN$, there is at most one right angle.

Given: $\triangle LMN$

Prove: $\triangle LMN$ has at most one right angle.

Proof: Assume temporarily that $\triangle LMN$ has more than one **a.** _?_ . That is, assume that both $\angle M$ and $\angle N$ are **b.** _?_ . If $\angle M$ and $\angle N$ are both right angles, then $m\angle M = m\angle N = $ **c.** _?_ . By the Triangle Angle-Sum Theorem, $m\angle L + m\angle M + m\angle N = $ **d.** _?_ . Use substitution to write the equation $m\angle L + $ **e.** _?_ $ + $ **f.** _?_ $ = 180$. When you solve for $m\angle L$, you find that $m\angle L = $ **g.** _?_ . This means that there is no $\triangle LMN$, which contradicts the given statement. So the temporary assumption that $\triangle LMN$ has **h.** _?_ must be false. Therefore, $\triangle LMN$ has **i.** _?_ .

B Apply

16. History Use indirect reasoning to eliminate all but one of the following answers. In what year was George Washington born?

Ⓐ 1492 Ⓑ 1732 Ⓒ 1902 Ⓓ 2002

17. Think About a Plan Write an indirect proof.
Proof **Given:** $\angle 1 \not\cong \angle 2$

Prove: $\ell \not\parallel p$

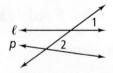

- What assumption should be the first step of your proof?
- In the figure, what type of angle pair do $\angle 1$ and $\angle 2$ form?

Write the first step of an indirect proof of the given statement.

18. If a number n ends in 5, then it is not divisible by 2.

19. If point X is on the perpendicular bisector of $\overline{AB}$, then $\overline{XB} \cong \overline{XA}$.

20. If a transversal intersects two parallel lines, then alternate exterior angles are congruent.

Ⓒ 21. Reasoning Identify the two statements that contradict each other.

  I. The orthocenter of $\triangle JRK$ is on the triangle.

  II. The centroid of $\triangle JRK$ is inside the triangle.

  III. $\triangle JRK$ is an obtuse triangle.

Write an indirect proof.

22. Use the figure at the right.

Proof **Given:** $\triangle ABC$ with $BC > AC$

  Prove: $\angle A \not\cong \angle B$

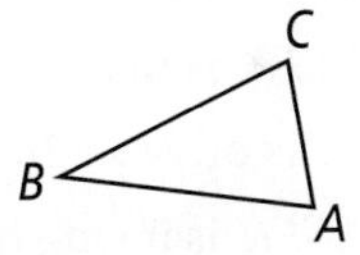

23. Given: $\triangle XYZ$ is isosceles.

Proof **Prove:** Neither base angle is a right angle.

Ⓒ Writing **For Exercises 24 and 25, write a convincing argument that uses indirect reasoning.**

STEM 24. Chemistry Ice is forming on the sidewalk in front of Toni's house. Show that the temperature of the sidewalk surface must be 32°F or lower.

25. Show that an obtuse triangle cannot contain a right angle.

Ⓒ 26. Error Analysis Your friend wants to prove indirectly that $\triangle ABC$ is equilateral. For a first step, he writes, "Assume temporarily that $\triangle ABC$ is scalene." What is wrong with your friend's statement? How can he correct himself?

27. Literature In Arthur Conan Doyle's story "The Sign of the Four," Sherlock Holmes talks to his friend Watson about how a culprit enters a room that has only four entrances: a door, a window, a chimney, and a hole in the roof.

"You will not apply my precept," he said, shaking his head. "How often have I said to you that when you have eliminated the impossible, whatever remains, however improbable, must be the truth? We know that he did not come through the door, the window, or the chimney. We also know that he could not have been concealed in the room, as there is no concealment possible. Whence, then, did he come?"

How did the culprit enter the room? Explain.

Ⓒ 28. Open-Ended Describe a real-life situation in which you used an indirect argument to convince someone of your point of view. Outline your argument.

Use the figure at the right for Exercises 29 and 30.

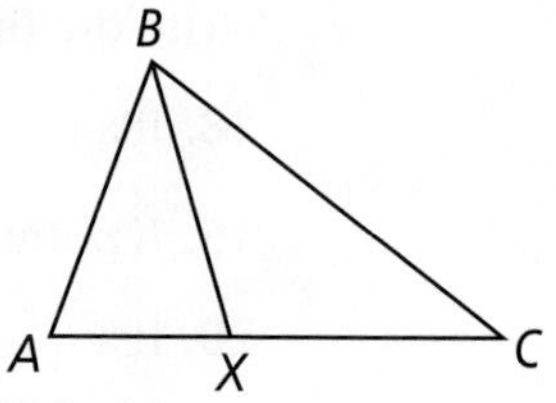

29. Given: $\triangle ABC$ is scalene, $m\angle ABX = 36$, $m\angle CBX = 36$
Proof **Prove:** $\overline{XB}$ is not perpendicular to $\overline{AC}$.

30. Given: $\triangle ABC$ is scalene, $m\angle ABX = 36$, $m\angle CBX = 36$
Proof **Prove:** $\overline{AX} \not\cong \overline{XC}$

Standardized Test Prep

31. What temporary assumption is the first step of the following indirect proof?

Given: The sides of $\triangle SFK$ measure 3 cm, 4 cm, and 5 cm.

Prove: The orthocenter of $\triangle SFK$ is on the triangle.

Ⓐ The incenter of $\triangle SFK$ is inside the triangle.

Ⓑ The orthocenter of $\triangle SFK$ is inside the triangle.

Ⓒ The centroid of $\triangle SFK$ is outside the triangle.

Ⓓ The orthocenter of $\triangle SFK$ is inside or outside the triangle.

32. $\triangle LMN \cong \triangle OPQ$, $m\angle L = 39$, and $m\angle P = 61$. What is $m\angle Q$?

Ⓕ 39   Ⓖ 80   Ⓗ 100   Ⓘ 141

33. In the diagram, what are the coordinates of the circumcenter of $\triangle KMF$?

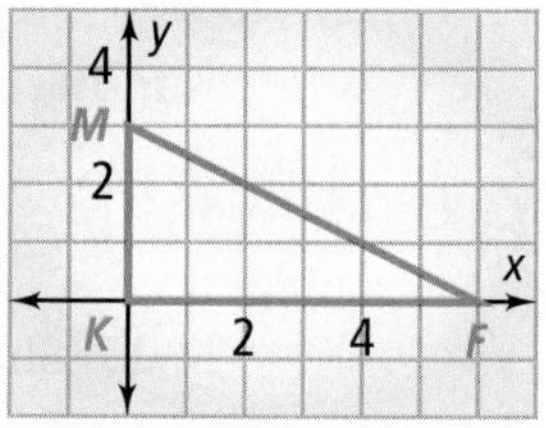

Ⓐ (1.5, 1.5)   Ⓒ (3, 1.5)

Ⓑ (6, 0)   Ⓓ (0, 0)

34. For what types of triangles are the centroid, circumcenter, incenter, and orthocenter all inside the triangle? Explain.

Mixed Review

35. The distances from the centroid of a triangle to its vertices are 16 cm, 17 cm, and 18 cm. What is the length of the shortest median?

◀ **See Lesson 5-4.**

36. The orthocenter of isosceles $\triangle ABC$ lies outside the triangle and $m\angle A = 30$. What are the measures of the two other angles?

37. You think, "If I leave home at 7:10, I'll catch the 7:25 bus. If I catch the 7:25 bus, I'll get to school before class starts. I am leaving and it's 7:10, so I'll get to school on time." Which law of deductive reasoning are you using?

◀ **See Lesson 2-4.**

Get Ready! **To prepare for Lesson 5-6, do Exercises 38–40.**

Graph $\triangle ABC$. List the sides in order from shortest to longest.

◀ **See Lesson 1-7.**

38. $A(5, 0)$, $B(0, 8)$, $C(0, 0)$ **39.** $A(2, 4)$, $B(-5, 1)$, $C(0, 0)$ **40.** $A(3, 0)$, $B(4, 3)$, $C(8, 0)$

Algebra Review

Use With Lesson 5-6

Solving Inequalities

Content Standard

Reviews A-CED.1 Create equations and inequalities in one variable and use them to solve problems.

The solutions of an inequality are all the numbers that make the inequality true. The following chart reviews the Properties of Inequality.

take note

Key Concept **Properties of Inequality**

	For all real numbers a, b, c, and d:
Addition Property	If $a > b$ and $c \geq d$, then $a + c > b + d$.
Multiplication Property	If $a > b$ and $c > 0$, then $ac > bc$. If $a > b$ and $c < 0$, then $ac < bc$.
Transitive Property	If $a > b$ and $b > c$, then $a > c$.

You can use the Addition and Multiplication Properties of Inequality to solve inequalities.

Example

Algebra **Solve $-6x + 7 > 25$.**

$-6x + 7 > 25$

$-6x > 18$ Subtract 7 from each side.

$\frac{-6x}{-6} < \frac{18}{-6}$ Divide each side by -6. Remember to reverse the inequality symbol.

$x < -3$ Simplify.

Exercises

Algebra **Solve each inequality.**

1. $7x - 13 \leq -20$

2. $3x + 8 > 16$

3. $-2x - 5 < 16$

4. $8y + 2 \geq 14$

5. $a + 1 \leq 91$

6. $-x - 2 > 17$

7. $-4z - 10 < -12$

8. $9x - 8 \geq 82$

9. $6n + 3 \leq -18$

10. $c + 13 > 34$

11. $3x - 5x + 2 < 12$

12. $2(y - 5) > -24$

13. $-3(4x - 1) \geq 15$

14. $-n - 27 \leq 92$

15. $8x - 4 + x > -76$

16. $8y - 4y + 11 \leq -33$

17. $x + 78 \geq -284$

18. $4(5a + 3) < -8$

Inequalities in One Triangle

Content Standard
Extends G.CO.10 Prove theorems about triangles . . .

Objective To use inequalities involving angles and sides of triangles

SOLVE IT!

Getting Ready!

◄► ✗ ⤺ ⬛

For a neighborhood improvement project, you volunteer to help build a new sandbox at the town playground. You have two boards that will make up two sides of the triangular sandbox. One is 5 ft long and the other is 8 ft long. Boards come in the lengths shown. Which boards can you use for the third side of the sandbox? Explain.

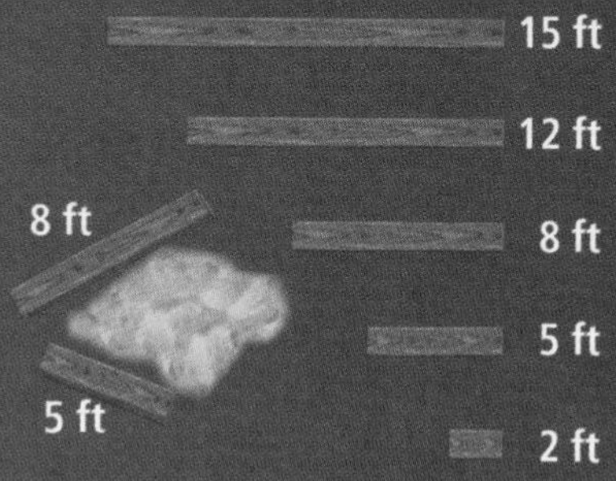

15 ft

12 ft

8 ft

8 ft

5 ft

5 ft

2 ft

If you are having trouble, try making a model of the situation to see which boards you can use to make a triangle.

In the Solve It, you explored triangles formed by various lengths of board. You may have noticed that changing the angle formed by two sides of the sandbox changes the length of the third side.

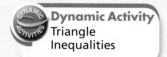

Dynamic Activity
Triangle Inequalities

Essential Understanding The angles and sides of a triangle have special relationships that involve inequalities.

take note

Property   Comparison Property of Inequality

If $a = b + c$ and $c > 0$, then $a > b$.

Proof **Proof of the Comparison Property of Inequality**

Given: $a = b + c, c > 0$

Prove: $a > b$

Statements	Reasons
1) $c > 0$	1) Given
2) $b + c > b + 0$	2) Addition Property of Inequality
3) $b + c > b$	3) Identity Property of Addition
4) $a = b + c$	4) Given
5) $a > b$	5) Substitution

The Comparison Property of Inequality allows you to prove the following corollary to the Triangle Exterior Angle Theorem (Theorem 3-12).

take note

Corollary Corollary to the Triangle Exterior Angle Theorem

Corollary	**If . . .**	**Then . . .**
The measure of an exterior angle of a triangle is greater than the measure of each of its remote interior angles.	$\angle 1$ is an exterior angle 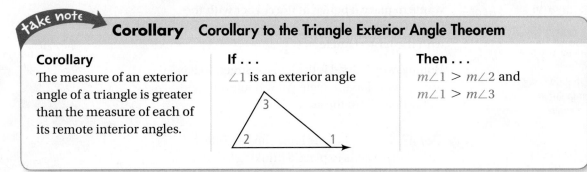	$m\angle 1 > m\angle 2$ and $m\angle 1 > m\angle 3$

Proof **Proof of the Corollary**

Given: $\angle 1$ is an exterior angle of the triangle.

Prove: $m\angle 1 > m\angle 2$ and $m\angle 1 > m\angle 3$.

Proof: By the Triangle Exterior Angle Theorem, $m\angle 1 = m\angle 2 + m\angle 3$. Since $m\angle 2 > 0$ and $m\angle 3 > 0$, you can apply the Comparison Property of Inequality and conclude that $m\angle 1 > m\angle 2$ and $m\angle 1 > m\angle 3$.

© **Problem 1** **Applying the Corollary**

Think

How do you identify an exterior angle?
An exterior angle must be formed by the extension of a side of the triangle. Here, $\angle 1$ is an exterior angle of $\triangle ABD$, but $\angle 2$ is not.

Use the figure at the right. Why is $m\angle 2 > m\angle 3$?

In $\triangle ACD$, $\overline{CB} \cong \overline{CD}$, so by the Isosceles Triangle Theorem, $m\angle 1 = m\angle 2$. $\angle 1$ is an exterior angle of $\triangle ABD$, so by the Corollary to the Triangle Exterior Angle Theorem, $m\angle 1 > m\angle 3$. Then $m\angle 2 > m\angle 3$ by substitution.

✓ **Got It?** **1.** Why is $m\angle 5 > m\angle C$?

You can use the corollary to Theorem 3-12 to prove the following theorem.

take note

Theorem 5-10

Theorem	**If . . .**	**Then . . .**
If two sides of a triangle are not congruent, then the larger angle lies opposite the longer side.	$XZ > XY$ 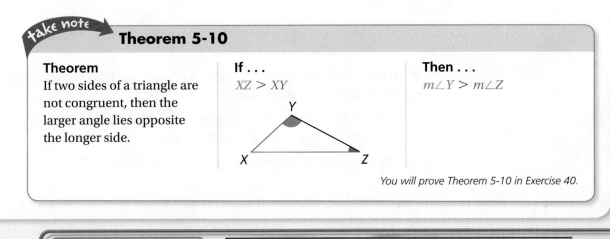	$m\angle Y > m\angle Z$

You will prove Theorem 5-10 in Exercise 40.

ⓒ **Problem 2** Using Theorem 5-10

A town park is triangular. A landscape architect wants to place a bench at the corner with the largest angle. Which two streets form the corner with the largest angle?

Hollingsworth Road is the longest street, so it is opposite the largest angle. MLK Boulevard and Valley Road form the largest angle.

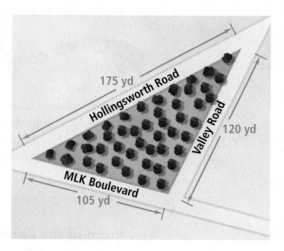

✓ **Got It?** 2. Suppose the landscape architect wants to place a drinking fountain at the corner with the second largest angle. Which two streets form the corner with the second-largest angle?

Theorem 5-11 below is the converse of Theorem 5-10. The proof of Theorem 5-11 relies on indirect reasoning.

 take note

Theorem 5-11

Theorem	**If . . .**	**Then . . .**
If two angles of a triangle are not congruent, then the longer side lies opposite the larger angle.	$m\angle A > m\angle B$ 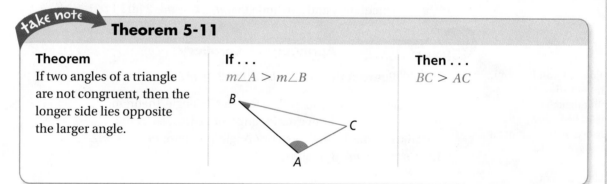	$BC > AC$

Proof Indirect Proof of Theorem 5-11

Given: $m\angle A > m\angle B$

Prove: $BC > AC$

Step 1 Assume temporarily that $BC \not> AC$. That is, assume temporarily that either $BC < AC$ or $BC = AC$.

Step 2 If $BC < AC$, then $m\angle A < m\angle B$ (Theorem 5-10). This contradicts the given fact that $m\angle A > m\angle B$. Therefore, $BC < AC$ must be false.

If $BC = AC$, then $m\angle A = m\angle B$ (Isosceles Triangle Theorem). This also contradicts $m\angle A > m\angle B$. Therefore, $BC = AC$ must be false.

Step 3 The temporary assumption $BC \not> AC$ is false, so $BC > AC$.

How do you use
the angle measures
to order the
side lengths?
List the angle measures
in order from smallest to
largest. Then replace the
measure of each angle
with the length of the
side opposite.

© **Problem 3** Using Theorem 5-11

Multiple Choice Which choice shows the sides of $\triangle TUV$ in order from shortest to longest?

Ⓐ $\overline{TV}, \overline{UV}, \overline{UT}$ Ⓒ $\overline{UV}, \overline{UT}, \overline{TV}$

Ⓑ $\overline{UT}, \overline{UV}, \overline{TV}$ Ⓓ $\overline{TV}, \overline{UT}, \overline{UV}$

By the Triangle Angle-Sum Theorem, $m\angle T = 60.\ 58 < 60 < 62$, so $m\angle U < m\angle T < m\angle V$. By Theorem 5-11, $TV < UV < UT$. Choice A is correct.

✓ **Got It?** **3. Reasoning** In the figure at the right, $m\angle S = 24$ and $m\angle O = 130$. Which side of $\triangle SOX$ is the shortest side? Explain your reasoning.

For three segments to form a triangle, their lengths must be related in a certain way. Notice that only one of the sets of segments below can form a triangle. The sum of the smallest two lengths must be greater than the greatest length.

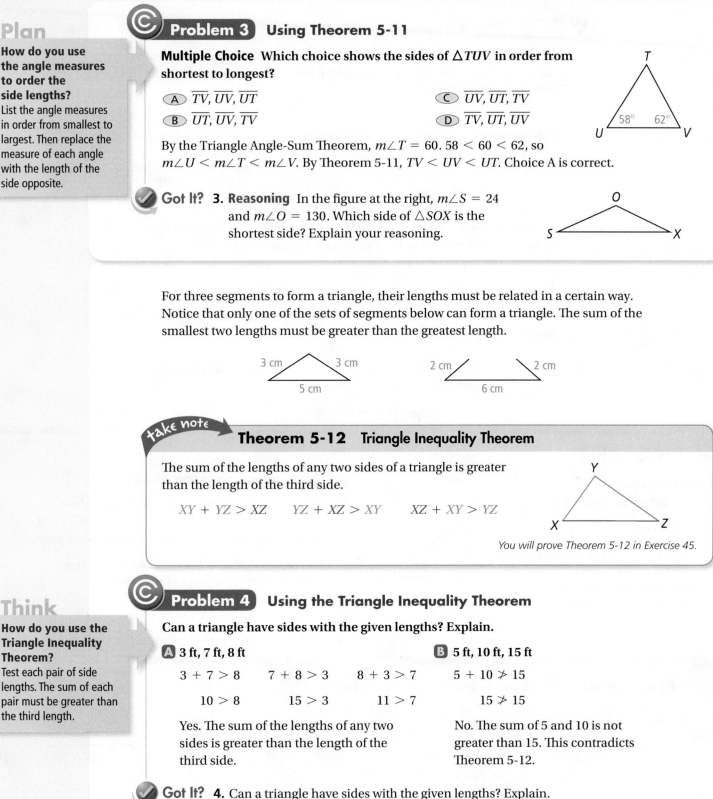

3 cm   3 cm

5 cm

2 cm   2 cm

6 cm

take note → **Theorem 5-12** Triangle Inequality Theorem

The sum of the lengths of any two sides of a triangle is greater than the length of the third side.

$XY + YZ > XZ$ $YZ + XZ > XY$ $XZ + XY > YZ$

You will prove Theorem 5-12 in Exercise 45.

How do you use the
Triangle Inequality
Theorem?
Test each pair of side
lengths. The sum of each
pair must be greater than
the third length.

© **Problem 4** Using the Triangle Inequality Theorem

Can a triangle have sides with the given lengths? Explain.

Ⓐ **3 ft, 7 ft, 8 ft**

$3 + 7 > 8$ $7 + 8 > 3$ $8 + 3 > 7$

$10 > 8$ $15 > 3$ $11 > 7$

Yes. The sum of the lengths of any two sides is greater than the length of the third side.

Ⓑ **5 ft, 10 ft, 15 ft**

$5 + 10 \not> 15$

$15 \not> 15$

No. The sum of 5 and 10 is not greater than 15. This contradicts Theorem 5-12.

✓ **Got It?** **4.** Can a triangle have sides with the given lengths? Explain.

a. 2 m, 6 m, and 9 m   **b.** 4 yd, 6 yd, and 9 yd

Problem 5   Finding Possible Side Lengths

Algebra In the Solve It, you explored the possible dimensions of a triangular sandbox. Two of the sides are 5 ft and 8 ft long. What is the range of possible lengths for the third side?

Know	Need	Plan
The lengths of two sides of the triangle are 5 ft and 8 ft.	The range of possible lengths of the third side	Use the Triangle Inequality Theorem to write three inequalities. Use the solutions of the inequalities to determine the greatest and least possible lengths.

Let x represent the length of the third side. Use the Triangle Inequality Theorem to write three inequalities. Then solve each inequality for x.

$$x + 5 > 8 \qquad\qquad x + 8 > 5 \qquad\qquad 5 + 8 > x$$
$$x > 3 \qquad\qquad\quad x > -3 \qquad\qquad\quad x < 13$$

Numbers that satisfy $x > 3$ and $x > -3$ must be greater than 3. So, the third side must be greater than 3 ft and less than 13 ft.

Got It?   **5.** A triangle has side lengths of 4 in. and 7 in. What is the range of possible lengths for the third side?

Lesson Check

Do you know HOW?

Use △ABC for Exercises 1 and 2.

1. Which side is the longest?

2. Which angle is the smallest?

3. Can a triangle have sides of lengths 4, 5, and 10? Explain.

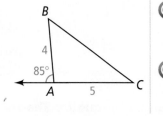

Do you UNDERSTAND?   MATHEMATICAL PRACTICES

4. Error Analysis A friend tells you that she drew a triangle with perimeter 16 and one side of length 8. How do you know she made an error in her drawing?

5. Reasoning Is it possible to draw a right triangle with an exterior angle measuring 88? Explain your reasoning.

Practice and Problem-Solving Exercises   MATHEMATICAL PRACTICES

A Practice   Explain why $m\angle 1 > m\angle 2$.   See Problem 1.

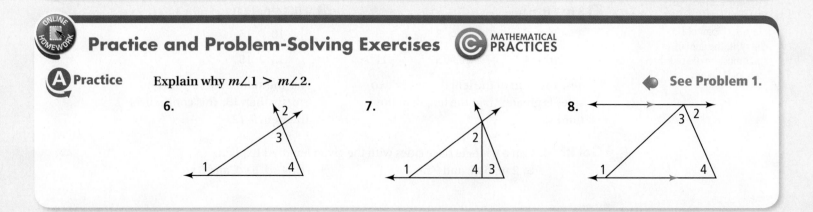

6.   **7.**   **8.**

For Exercises 9–14, list the angles of each triangle in order from smallest to largest. 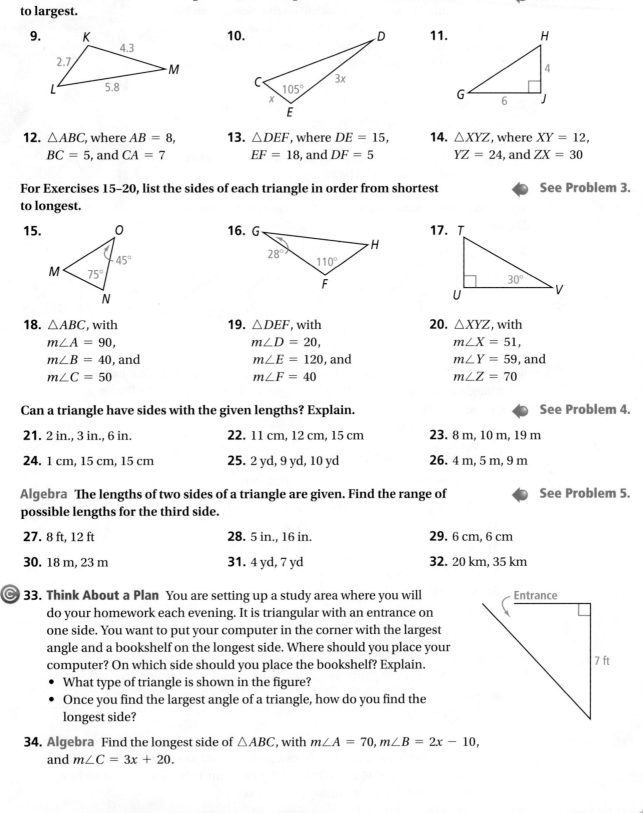 See Problem 2.

9.

10.

11.

12. △ABC, where AB = 8, BC = 5, and CA = 7

13. △DEF, where DE = 15, EF = 18, and DF = 5

14. △XYZ, where XY = 12, YZ = 24, and ZX = 30

For Exercises 15–20, list the sides of each triangle in order from shortest to longest. See Problem 3.

15.

16.

17.

18. △ABC, with
m∠A = 90,
m∠B = 40, and
m∠C = 50

19. △DEF, with
m∠D = 20,
m∠E = 120, and
m∠F = 40

20. △XYZ, with
m∠X = 51,
m∠Y = 59, and
m∠Z = 70

Can a triangle have sides with the given lengths? Explain. See Problem 4.

21. 2 in., 3 in., 6 in.

22. 11 cm, 12 cm, 15 cm

23. 8 m, 10 m, 19 m

24. 1 cm, 15 cm, 15 cm

25. 2 yd, 9 yd, 10 yd

26. 4 m, 5 m, 9 m

Algebra The lengths of two sides of a triangle are given. Find the range of possible lengths for the third side. See Problem 5.

27. 8 ft, 12 ft

28. 5 in., 16 in.

29. 6 cm, 6 cm

30. 18 m, 23 m

31. 4 yd, 7 yd

32. 20 km, 35 km

B Apply **©** 33. **Think About a Plan** You are setting up a study area where you will do your homework each evening. It is triangular with an entrance on one side. You want to put your computer in the corner with the largest angle and a bookshelf on the longest side. Where should you place your computer? On which side should you place the bookshelf? Explain.
- What type of triangle is shown in the figure?
- Once you find the largest angle of a triangle, how do you find the longest side?

34. **Algebra** Find the longest side of △ABC, with $m\angle A = 70$, $m\angle B = 2x - 10$, and $m\angle C = 3x + 20$.

35. Writing You and a friend compete in a scavenger hunt at a museum. The two of you walk from the Picasso exhibit to the Native American gallery along the dashed red line. When he sees that another team is ahead of you, your friend says, "They must have cut through the courtyard." Explain what your friend means.

Native American Gallery

Picasso Exhibit

36. Error Analysis Your family drives across Kansas on Interstate 70. A sign reads, "Wichita 90 mi, Topeka 110 mi." Your little brother says, "I didn't know that it was only 20 miles from Wichita to Topeka." Explain why the distance between the two cities does not have to be 20 mi.

Reasoning Determine which segment is shortest in each diagram.

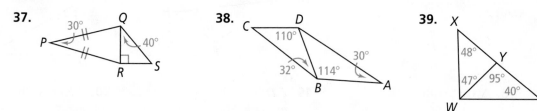

37.

30° Q
P 40°
R S

38.

C 110° D
32° 114° 30°
B A

39.

X
48° Y
47° 95°
W 40° Z

40. Developing Proof Fill in the blanks for a proof of Theorem 5-10: If two sides of a triangle are not congruent, then the larger angle lies opposite the longer side.

Given: △TOY, with YO > YT

Prove: a. ? **> b.** ?

Mark P on $\overline{YO}$ so that $\overline{YP} \cong \overline{YT}$. Draw $\overline{TP}$.

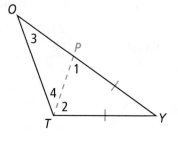

Statements	Reasons
1) $\overline{YP} \cong \overline{YT}$	1) Ruler Postulate
2) $m\angle 1 = m\angle 2$	2) **c.** ?
3) $m\angle OTY = m\angle 4 + m\angle 2$	3) **d.** ?
4) $m\angle OTY > m\angle 2$	4) **e.** ?
5) $m\angle OTY > m\angle 1$	5) **f.** ?
6) $m\angle 1 > m\angle 3$	6) **g.** ?
7) $m\angle OTY > m\angle 3$	7) **h.** ?

41. Prove this corollary to Theorem 5-11: The perpendicular segment from
Proof a point to a line is the shortest segment from the point to the line.

Given: $\overline{PT} \perp \overline{TA}$

Prove: PA > PT

P
T A

C Challenge

42. Probability A student has two straws. One is 6 cm long and the other is 9 cm long. She picks a third straw at random from a group of four straws whose lengths are 3 cm, 5 cm, 11 cm, and 15 cm. What is the probability that the straw she picks will allow her to form a triangle? Justify your answer.

For Exercises 43 and 44, x and y are integers such that 1 < x < 5 and 2 < y < 9.

43. The sides of a triangle are 5 cm, x cm, and y cm. List all possible (x, y) pairs.

44. Probability What is the probability that you can draw an isosceles triangle that has sides 5 cm, x cm, and y cm, with x and y chosen at random?

45. Prove the Triangle Inequality Theorem: The sum of the lengths of any two sides of
Proof a triangle is greater than the length of the third side.

    Given: $\triangle ABC$

    Prove: $AC + CB > AB$

    (*Hint:* On $\overrightarrow{BC}$, mark a point D not on $\overline{BC}$, so that $DC = AC$.
    Draw $\overline{DA}$ and use Theorem 5-11 with $\triangle ABD$.)

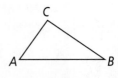

Standardized Test Prep

SAT/ACT

46. The figure shows the walkways connecting four dormitories on a college campus. What is the greatest possible whole-number length, in yards, for the walkway between South dorm and East dorm?

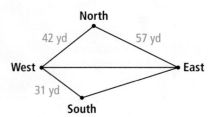

47. What is the length of a segment with endpoints $A(-13, -16)$ and $B(-9, -13)$?

48. How many sides does a convex quadrilateral have?

49. $\angle 1$ and $\angle 2$ are corresponding angles formed by two parallel lines and a transversal. If $m\angle 1 = 33x + 2$ and $m\angle 2 = 68$, what is the value of x?

Mixed Review

Write the first step of an indirect proof of the given statement.   ◀ **See Lesson 5-5.**

50. The side is at least 2 ft long.   **51.** $\triangle PQR$ has two congruent angles.

52. You know that $\overline{AB} \cong \overline{XY}$, $\overline{BC} \cong \overline{YZ}$, and $\overline{CA} \cong \overline{ZX}$. By what theorem or   ◀ **See Lesson 4-2.**
postulate can you conclude that $\triangle ABC \cong \triangle XYZ$?

Get Ready!   **To prepare for Lesson 5-7, do Exercises 53–55.**

Use the figure at the right.   ◀ **See Lesson 3-5.**

53. What is $m\angle P$?

54. What is $m\angle D$?

55. Is it possible for AW to equal OG?

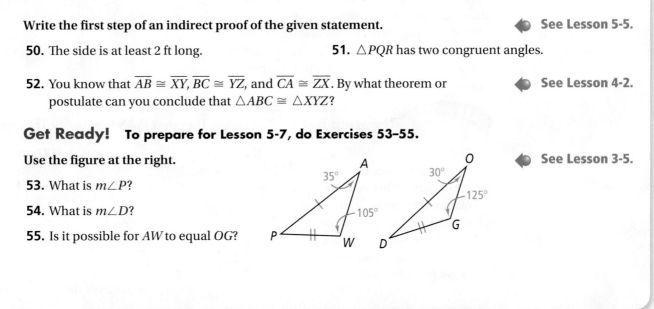

5-7 Inequalities in Two Triangles

Content Standard
Extends G.CO.10 Prove theorems about triangles. . . .

Objective To apply inequalities in two triangles

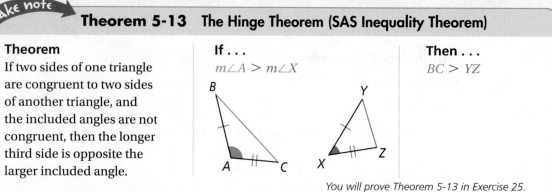

Getting Ready!

Think of a clock or watch that has an hour hand and a minute hand. As minutes pass, the distance between the tip of the hour hand and the tip of the minute hand changes. This distance is x in the figure at the right. What is the order of the times below from least to greatest length of x? How do you know?

1:00, 3:00, 5:00, 8:30, 1:30, 12:20

Try to make the problem simpler by finding a way to find the distance x without measuring directly.

In the Solve It, the hands of the clock and the segment labeled x form a triangle. As the time changes, the shape of the triangle changes, but the lengths of two of its sides do not change.

Essential Understanding In triangles that have two pairs of congruent sides, there is a relationship between the included angles and the third pair of sides.

When you close a door, the angle between the door and the frame (at the hinge) gets smaller. The relationship between the measure of the hinge angle and the length of the opposite side is the basis for the SAS Inequality Theorem, also known as the Hinge Theorem.

take note

Theorem 5-13 The Hinge Theorem (SAS Inequality Theorem)

Theorem	**If . . .**	**Then . . .**
If two sides of one triangle are congruent to two sides of another triangle, and the included angles are not congruent, then the longer third side is opposite the larger included angle.	$m\angle A > m\angle X$	$BC > YZ$

You will prove Theorem 5-13 in Exercise 25.

Problem 1 Using the Hinge Theorem

Multiple Choice Which of the following statements must be true?

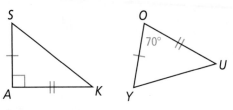

Ⓐ $AS < YU$ Ⓒ $SK < YU$

Ⓑ $SK > YU$ Ⓓ $AK = YU$

Plan

How do you apply the Hinge Theorem?
After you identify the angles included between the pairs of congruent sides, locate the sides opposite those angles.

$\overline{SA} \cong \overline{YO}$ and $\overline{AK} \cong \overline{OU}$, so the triangles have two pairs of congruent sides. The included angles, $\angle A$ and $\angle O$, are not congruent. Since $m\angle A > m\angle O$, $SK > YU$ by the Hinge Theorem. The correct answer is B.

Got It? **1. a.** What inequality relates LN and OQ in the figure at the right?

b. Reasoning In $\triangle ABC$, $AB = 3$, $BC = 4$, and $CA = 6$. In $\triangle PQR$, $PQ = 3$, $QR = 5$, and $RP = 6$. How can you use indirect reasoning to explain why $m\angle P > m\angle A$?

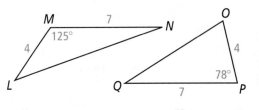

Problem 2 Applying the Hinge Theorem

Swing Ride The diagram below shows the position of a swing at two different times. As the speed of the swing ride increases, the angle between the chain and $\overline{AB}$ increases. Is the rider farther from point A at Time 1 or Time 2? Explain how the Hinge Theorem justifies your answer.

Think

For $\triangle ABC$, which side lengths are the same at Time 1 and Time 2?
The lengths of the chain and $\overline{AB}$ do not change. So, AB and BC are the same at Time 1 and Time 2.

The rider is farther from point A at Time 2. The lengths of $\overline{AB}$ and $\overline{BC}$ stay the same throughout the ride. Since the angle formed at Time 2 ($\angle 2$) is greater than the angle formed at Time 1 ($\angle 1$), you can use the Hinge Theorem to conclude that $\overline{AC}$ at Time 2 is longer than $\overline{AC}$ at Time 1.

Time 1   Time 2

Got It? **2.** The diagram below shows a pair of scissors in two different positions. In which position is the distance between the tips of the two blades greater? Use the Hinge Theorem to justify your answer.

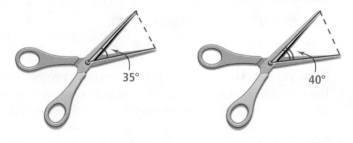

35°   40°

The Converse of the Hinge Theorem is also true. The proof of the converse is an indirect proof.

> **Theorem 5-14** **Converse of the Hinge Theorem (SSS Inequality)**
>
> **Theorem**
> If two sides of one triangle are congruent to two sides of another triangle, and the third sides are not congruent, then the larger included angle is opposite the longer third side.
>
> **If . . .**
> $BC > YZ$
>
>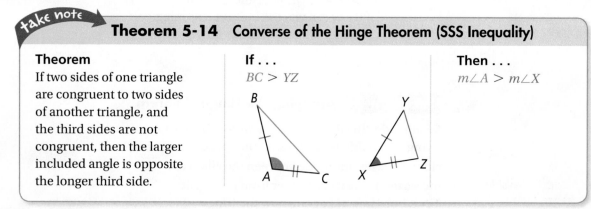
>
> **Then . . .**
> $m\angle A > m\angle X$

Proof **Indirect Proof of the Converse of the Hinge Theorem (SSS Inequality)**

Given: $\overline{AB} \cong \overline{XY}$, $\overline{AC} \cong \overline{XZ}$,
$BC > YZ$

Prove: $m\angle A > m\angle X$

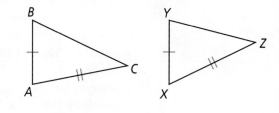

Step 1 Assume temporarily that $m\angle A \not> m\angle X$. This means either $m\angle A < m\angle X$ or $m\angle A = m\angle X$.

Step 2 If $m\angle A < m\angle X$, then $BC < YZ$ by the Hinge Theorem. This contradicts the given information that $BC > YZ$. Therefore, the assumption that $m\angle A < m\angle X$ must be false.

If $m\angle A = m\angle X$, then $\triangle ABC \cong \triangle XYZ$ by SAS. If the two triangles are congruent, then $BC = YZ$ because corresponding parts of congruent triangles are congruent. This contradicts the given information that $BC > YZ$. Therefore, the assumption that $m\angle A = m\angle X$ must be false.

Step 3 The temporary assumption that $m\angle A \not> m\angle X$ is false. Therefore, $m\angle A > m\angle X$.

Using the Converse of the Hinge Theorem

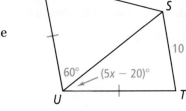

Algebra What is the range of possible values for x?

Step 1 Find an upper limit for the value of x. $\overline{UT} \cong \overline{UR}$ and $\overline{US} \cong \overline{US}$, so $\triangle TUS$ and $\triangle RUS$ have two pairs of congruent sides. $RS > TS$, so you can use the Converse of the Hinge Theorem to write an inequality.

$m\angle RUS > m\angle TUS$	Converse of the Hinge Theorem
$60 > 5x - 20$	Substitute.
$80 > 5x$	Add 20 to each side.
$16 > x$	Divide each side by 5.

Step 2 Find a lower limit for the value of x.

$m\angle TUS > 0$	The measure of an angle of a triangle is greater than 0.
$5x - 20 > 0$	Substitute.
$5x > 20$	Add 20 to each side.
$x > 4$	Divide each side by 5.

Rewrite $16 > x$ and $x > 4$ as $4 < x < 16$.

Plan

How do you put upper and lower limits on the value of x?
Use the largest possible value of $m\angle TUS$ as the upper limit for $5x - 20$ and the smallest possible value of $m\angle TUS$ as the lower limit for $5x - 20$.

✅ **Got It?** **3.** What is the range of possible values for x in the figure at the right?

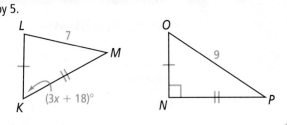

Proof **Problem 4** **Proving Relationships in Triangles**

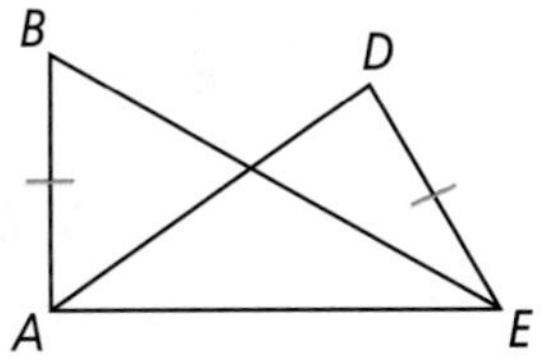

Given: $BA = DE,\ BE > DA$

Prove: $m\angle BAE > m\angle BEA$

Statement	Reasons
1) $BA = DE$	**1)** Given
2) $AE = AE$	**2)** Reflexive Property of Equality
3) $BE > DA$	**3)** Given
4) $m\angle BAE > m\angle DEA$	**4)** Converse of the Hinge Theorem
5) $m\angle DEA = m\angle DEB + m\angle BEA$	**5)** Angle Addition Postulate
6) $m\angle DEA > m\angle BEA$	**6)** Comparison Property of Inequality
7) $m\angle BAE > m\angle BEA$	**7)** Transitive Property of Inequality

Think

How do you know $m\angle BAE > m\angle BEA$?
Use the Transitive Property of Inequality on the inequalities in Statements 4 and 6.

✅ **Got It?** **4. Given:** $m\angle MON = 80$, O is the midpoint of $\overline{LN}$

  Prove: $LM > MN$

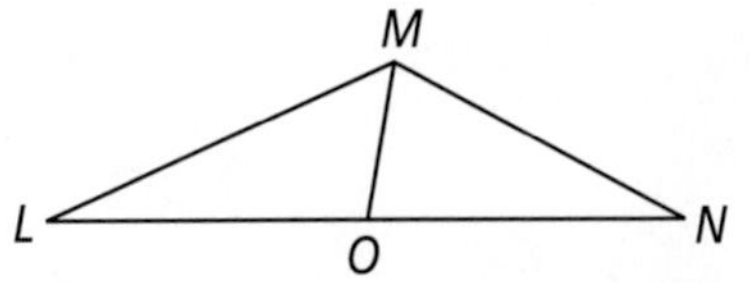

Lesson Check

Do you know HOW?

Write an inequality relating the given side lengths or angle measures.

1. *FD* and *BC*

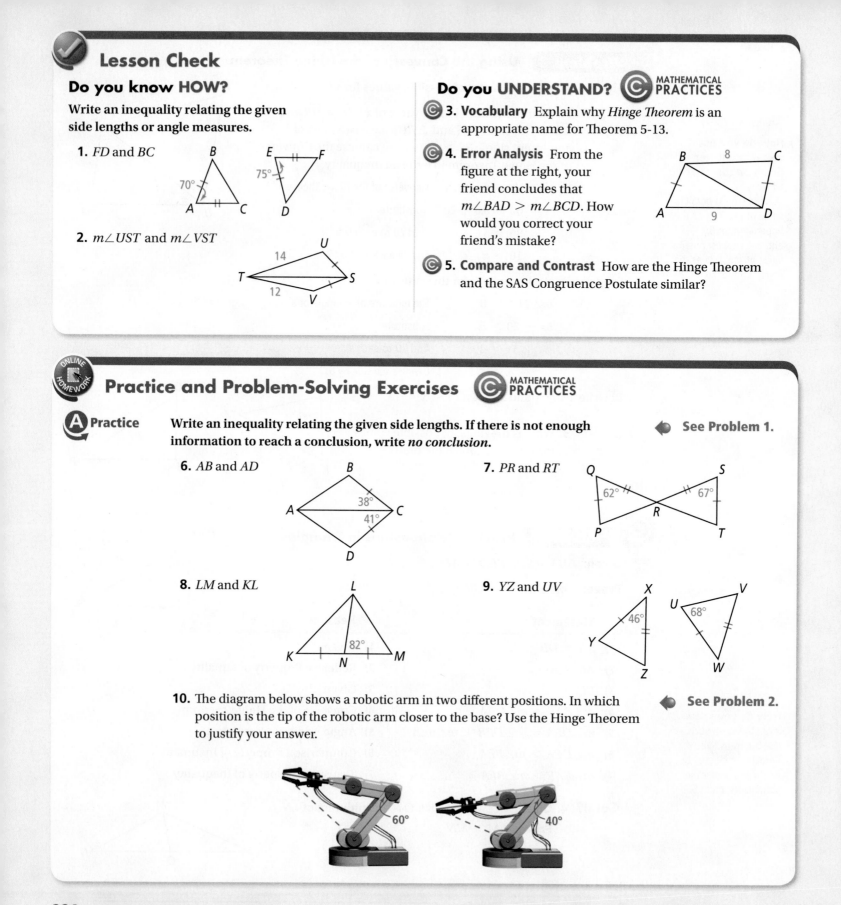

2. *m∠UST* and *m∠VST*

Do you UNDERSTAND? MATHEMATICAL PRACTICES

3. **Vocabulary** Explain why *Hinge Theorem* is an appropriate name for Theorem 5-13.

4. **Error Analysis** From the figure at the right, your friend concludes that *m∠BAD* > *m∠BCD*. How would you correct your friend's mistake?

5. **Compare and Contrast** How are the Hinge Theorem and the SAS Congruence Postulate similar?

Practice and Problem-Solving Exercises MATHEMATICAL PRACTICES

A Practice

Write an inequality relating the given side lengths. If there is not enough information to reach a conclusion, write *no conclusion*.

See Problem 1.

6. *AB* and *AD*

7. *PR* and *RT*

8. *LM* and *KL*

9. *YZ* and *UV*

10. The diagram below shows a robotic arm in two different positions. In which position is the tip of the robotic arm closer to the base? Use the Hinge Theorem to justify your answer.

See Problem 2.

Algebra Find the range of possible values for each variable.

See Problem 3.

11.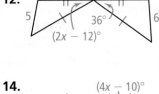

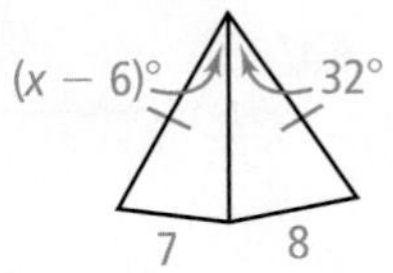

$(x - 6)°$ $32°$

7 8

12.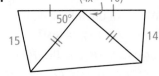

5 $36°$ 6

$(2x - 12)°$

13.

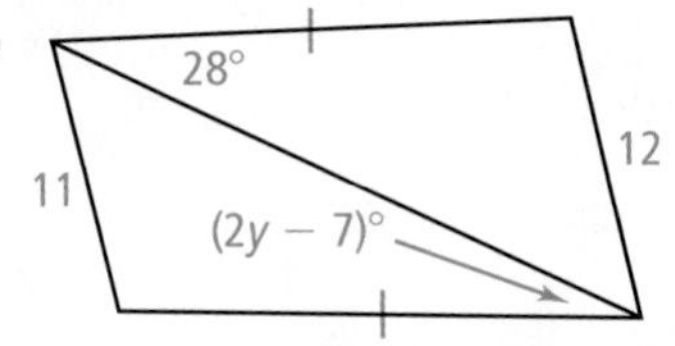

28°

11 12

$(2y - 7)°$

14.

$(4x - 10)°$

50°

15 14

15. Developing Proof Complete the following proof.

See Problem 4.

Given: C is the midpoint of $\overline{BD}$,
$m\angle EAC = m\angle AEC$,
$m\angle BCA > m\angle DCE$

Prove: $AB > ED$

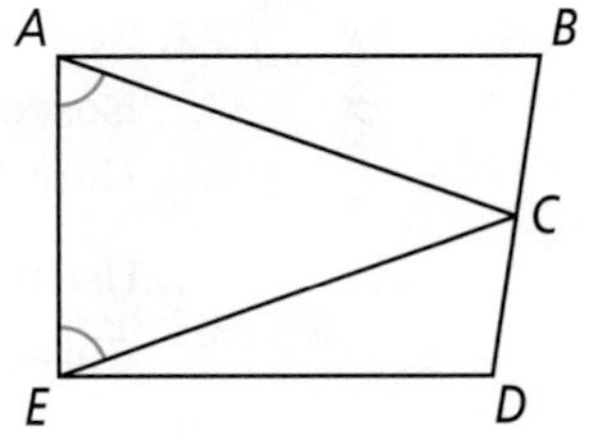

A B

C

E D

	Statements		Reasons
1)	$m\angle EAC = m\angle AEC$	1)	Given
2)	$AC = EC$	2)	**a.** ?
3)	C is the midpoint of $\overline{BD}$.	3)	**b.** ?
4)	$\overline{BC} \cong \overline{CD}$	4)	**c.** ?
5)	**d.** ?	5)	$\cong$ segments have $=$ length.
6)	$m\angle BCA > m\angle DCE$	6)	**e.** ?
7)	$AB > ED$	7)	**f.** ?

Ⓑ Apply Copy and complete with $>$ or $<$. Explain your reasoning.

16. $PT \blacksquare QR$

17. $m\angle QTR \blacksquare m\angle RTS$

18. $PT \blacksquare RS$

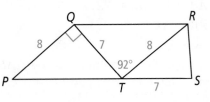

Q R

8 7 8

92°

P T 7 S

19. a. Error Analysis Your classmate draws the figure at the right. Explain why the figure cannot have the labeled dimensions.

 b. Open-Ended Describe a way you could change the dimensions to make the figure possible.

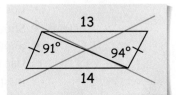

13

91° 94°

14

20. Think About a Plan Ship A and Ship B leave from the same point in the ocean. Ship A travels 150 mi due west, turns 65° toward north, and then travels another 100 mi. Ship B travels 150 mi due east, turns 70° toward south, and then travels another 100 mi. Which ship is farther from the starting point? Explain.

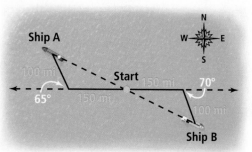
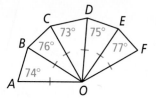

- How can you use the given angle measures?
- How does the Hinge Theorem help you to solve this problem?

21. Which of the following lists the segment lengths in order from least to greatest?

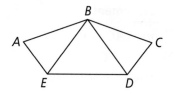

Ⓐ CD, AB, DE, BC, EF

Ⓑ EF, DE, AB, BC, CD

Ⓒ BC, DE, EF, AB, CD

Ⓓ EF, BC, DE, AB, CD

22. Reasoning The legs of a right isosceles triangle are congruent to the legs of an isosceles triangle with an 80° vertex angle. Which triangle has a greater perimeter? How do you know?

23. Use the figure at the right.

Proof

Given: △ABE is isosceles with vertex ∠B,
△ABE ≅ △CBD,
m∠EBD > m∠ABE

Prove: ED > AE

Challenge

24. Coordinate Geometry △ABC has vertices A(0, 7), B(−1, −2), C(2, −1), and O(0, 0). Show that m∠AOB > m∠AOC.

25. Use the plan below to complete a proof of the Hinge Theorem: If two sides of one triangle are congruent to two sides of another triangle and the included angles are not congruent, then the longer third side is opposite the larger included angle.

Proof

Given: $\overline{AB} \cong \overline{XY}$, $\overline{BC} \cong \overline{YZ}$, m∠B > m∠Y

Prove: AC > XZ

Plan for proof:

- Copy △ABC. Locate point D outside △ABC so that m∠CBD = m∠ZYX and BD = YX. Show that △DBC ≅ △XYZ.
- Locate point F on $\overline{AC}$, so that $\overline{BF}$ bisects ∠ABD.
- Show that △ABF ≅ △DBF and that $\overline{AF} \cong \overline{DF}$.
- Show that AC = FC + DF.
- Use the Triangle Inequality Theorem to write an inequality that relates DC to the lengths of the other sides of △FCD.
- Relate DC and XZ.

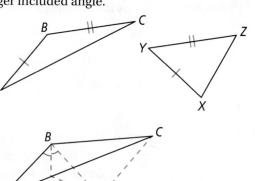

Standardized Test Prep

SAT/ACT

26. Which is a possible value for x in the figure at the right?

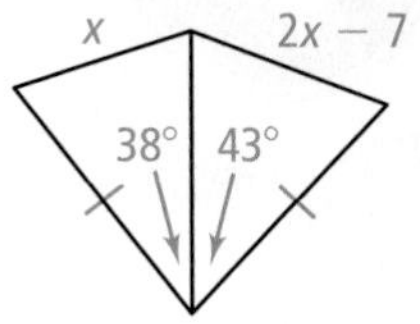

(A) 2   (C) 6

(B) 4   (D) 8

27. A quilter cuts out and sews together two triangles, as shown in the figure. She tries to make $\triangle ABD$ and $\triangle BDC$ equilateral triangles, but the angle measures are as shown. Assuming that the side $\overline{BD}$ is actually the same length in both triangles, which segment is the longest side in the figure?

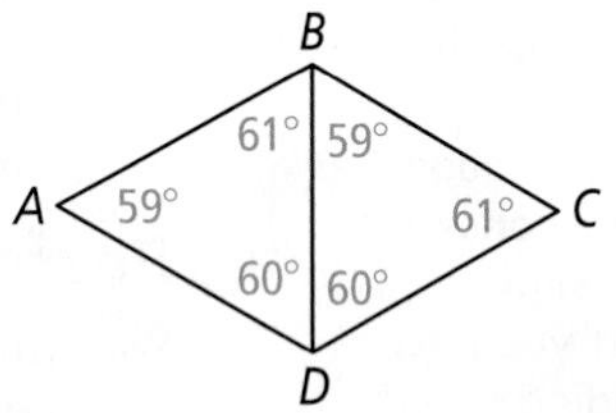

(F) $\overline{AB}$ (H) $\overline{CD}$

(G) $\overline{BC}$ (I) $\overline{AD}$

28. What is the equation of the perpendicular bisector of $\overline{MN}$ with endpoints $M(8, 0)$ and $N(0, 0)$?

(A) $y = x - 8$ (C) $x = 4$

(B) $x = 8$ (D) $y = 4$

29. The orthocenter of a triangle lies outside the triangle. Which of the following statements cannot be true?

(F) The triangle has a 120° angle.   (H) The incenter is inside the triangle.

(G) The triangle is isosceles.   (I) The triangle is acute.

Short Response

30. Use indirect reasoning to write a convincing argument that a triangle has at most one obtuse angle.

Mixed Review

List the angles of each triangle in order from smallest to largest.

See Lesson 5-6.

31.

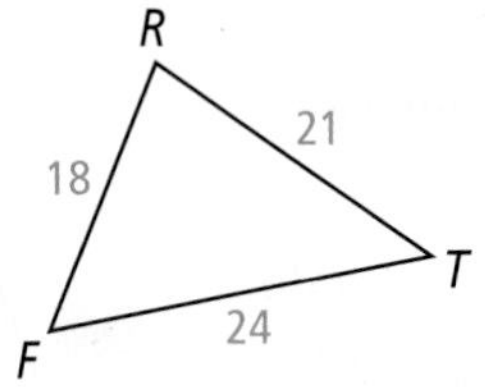

32.

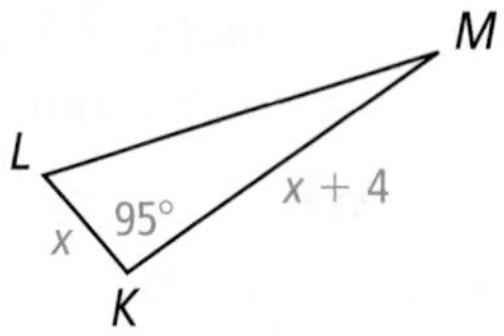

Algebra The lengths of two sides of a triangle are given. Find the range of possible lengths for the third side.

33. 15 cm, 19 cm   **34.** 6 ft, 11 ft   **35.** 3 in., 3 in.

36. In $\triangle GHI$, which side is included between $\angle G$ and $\angle H$?

See Lesson 4-2.

Get Ready!   **To prepare for Lesson 6-1, do Exercises 37–39.**

Find the slope of the line through each pair of points.

See Lesson 3-7.

37. $X(0, 6)$, $Y(4, 9)$ **38.** $R(3, 8)$, $S(6, 0)$ **39.** $A(4, 3)$, $B(2, 1)$

Pull It **All Together** © ASSESSMENT

To solve these problems you will pull together many concepts and skills that you have studied about relationships within triangles.

BIG idea Coordinate Geometry

You can use the Midpoint Formula, the slope formula, and the relationship between perpendicular lines to find points of concurrency.

© Performance Task 1

Your math teacher manages a campground during summer vacation. He loves math so much that he has mapped the campground on a coordinate grid. The campsites have the following coordinates: Brighton Bluff at $B(2, 2)$, Ponaganset Peak at $P(4, 10)$, and Harmony Hill at $H(12, 2)$. He wants to build showers that are equidistant from all three campsites. Find the coordinates of the point where the showers should be placed.

BIG idea Reasoning and Proof

You can use proven theorems to explore relationships among sides, angles, and special lines and segments in triangles.

© Performance Task 2

a. Draw $\triangle ABC$ with obtuse $\angle C$ and construct its orthocenter O. Then find the orthocenters of $\triangle ABO$, $\triangle ACO$, and $\triangle BCO$. What conjecture can you make about these orthocenters? Explain why you will get this result.

b. Will your conjecture be true for any acute or right $\triangle ABC$? Explain your reasoning.

BIG idea Reasoning and Proof

You can use indirect reasoning to prove relationships within triangles.

Performance Task 3

In $\triangle ABC$, $AB \neq BC$. Show that there does not exist a point P on altitude $\overline{BD}$ that is equidistant from A and C.

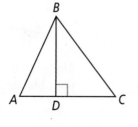

5 Chapter Review

Connecting BIG ideas and Answering the Essential Questions

1 Coordinate Geometry
Use parallel and perpendicular lines, and the slope, midpoint, and distance formulas to find intersection points and unknown lengths.

Midsegments of Triangles (Lesson 5-1)
If $\overline{DE}$ is a midsegment, then $\overline{AC} \parallel \overline{DE}$ and $DE = \frac{1}{2}AC$.

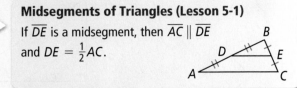

2 Measurement
Use theorems about perpendicular bisectors, angle bisectors, medians, and altitudes to find points of concurrency, angle measures, and segment lengths.

Concurrent Lines and Segments in Triangles (Lessons 5-2, 5-3, and 5-4)

Concurrent Lines and Segments	Intersection
• perpendicular bisectors	• circumcenter
• angle bisectors	• incenter
• medians	• centroid
• lines containing altitudes	• orthocenter

3 Reasoning and Proof
You can write an indirect proof by showing that a temporary assumption is false.

Indirect Proof (Lesson 5-5)
1) Assume temporarily the opposite of what you want to prove.
2) Show that this temporary assumption leads to a contradiction.
3) Conclude that what you want to prove is true.

Inequalities in Triangles (Lessons 5-6 and 5-7)
Use indirect reasoning to prove that the longer of two sides of a triangle lies opposite the larger angle, and to prove the Converse of the Hinge Theorem.

Chapter Vocabulary

- altitude of a triangle (p. 310)
- centroid of a triangle (p. 309)
- circumcenter of a triangle (p. 301)
- circumscribed about (p. 301)
- concurrent (p. 301)
- distance from a point to a line (p. 294)
- equidistant (p. 292)
- incenter of a triangle (p. 303)
- indirect proof (p. 317)
- indirect reasoning (p. 317)
- inscribed in (p. 303)
- median of a triangle (p. 309)
- midsegment of a triangle (p. 285)
- orthocenter of a triangle (p. 311)
- point of concurrency (p. 301)

Choose the correct vocabulary term to complete each sentence.

1. A *(centroid, median)* of a triangle is a segment from a vertex of the triangle to the midpoint of the side opposite the vertex.

2. The length of the perpendicular segment from a point to a line is the *(midsegment, distance from a point to the line)*.

3. The *(circumcenter, incenter)* of a triangle is the point of concurrency of the angle bisectors of the triangle.

5-1 Midsegments of Triangles

Quick Review

A **midsegment of a triangle** is a segment that connects the midpoints of two sides. A midsegment is parallel to the third side and is half as long.

Example

Algebra Find the value of x.

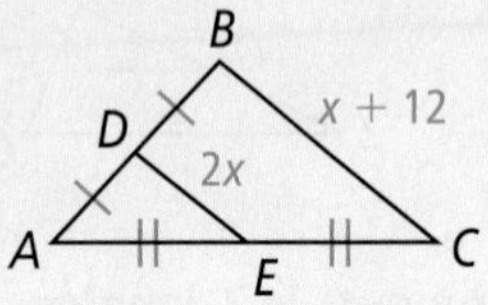

$\overline{DE}$ is a midsegment because D and E are midpoints.

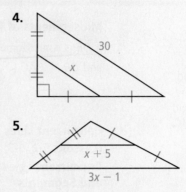

$DE = \frac{1}{2}BC$ △ Midsegment Theorem

$2x = \frac{1}{2}(x + 12)$ Substitute.

$4x = x + 12$ Simplify.

$3x = 12$ Subtract x from each side.

$x = 4$ Divide each side by 3.

Exercises

Algebra Find the value of x.

4.

5.

6. $\triangle ABC$ has vertices $A(0, 0)$, $B(2, 2)$, and $C(5, -1)$. Find the coordinates of L, the midpoint of $\overline{AC}$, and M, the midpoint of $\overline{BC}$. Verify that $\overline{LM} \parallel \overline{AB}$ and $LM = \frac{1}{2}AB$.

5-2 Perpendicular and Angle Bisectors

Quick Review

The **Perpendicular Bisector Theorem** together with its converse states that P is equidistant from A and B if and only if P is on the perpendicular bisector of $\overline{AB}$.

The **distance from a point to a line** is the length of the perpendicular segment from the point to the line.

The **Angle Bisector Theorem** together with its converse states that P is equidistant from the sides of an angle if and only if P is on the angle bisector.

Example

In the figure, QP = 4 and AB = 8. Find QR and CB.

Q is on the bisector of $\angle ABC$, so $QR = QP = 4$.

B is on the perpendicular bisector of $\overline{AC}$, so $CB = AB = 8$.

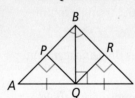

Exercises

7. Writing Describe how to find all the points on a baseball field that are equidistant from second base and third base.

In the figure, $m\angle DBE = 50$. Find each of the following.

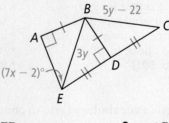

8. $m\angle BED$ **9.** $m\angle BEA$

10. x **11.** y

12. BE **13.** BC

5-3 Bisectors in Triangles

Quick Review

When three or more lines intersect in one point, they are **concurrent**.

- The point of concurrency of the perpendicular bisectors of a triangle is the **circumcenter of the triangle.**
- The point of concurrency of the angle bisectors of a triangle is the **incenter of the triangle.**

Example

Identify the incenter of the triangle.

The incenter of a triangle is the point of concurrency of the angle bisectors. $\overline{MR}$ and $\overline{LQ}$ are angle bisectors that intersect at Z. So, Z is the incenter.

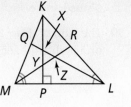

Exercises

Find the coordinates of the circumcenter of $\triangle DEF$.

14. $D(6, 0), E(0, 6), F(-6, 0)$

15. $D(0, 0), E(6, 0), F(0, 4)$

16. $D(5, -1), E(-1, 3), F(3, -1)$

17. $D(2, 3), E(8, 3), F(8, -1)$

P is the incenter of $\triangle XYZ$. Find the indicated angle measure.

18. $m\angle PXY$

19. $m\angle XYZ$

20. $m\angle PZX$

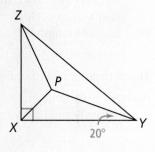

5-4 Medians and Altitudes

Quick Review

A **median of a triangle** is a segment from a vertex to the midpoint of the opposite side. An **altitude of a triangle** is a perpendicular segment from a vertex to the line containing the opposite side.

- The point of concurrency of the medians of a triangle is the **centroid of the triangle.** The centroid is two thirds the distance from each vertex to the midpoint of the opposite side.
- The point of concurrency of the altitudes of a triangle is the **orthocenter of the triangle.**

Example

If $PB = 6$, what is SB?

S is the centroid because $\overline{AQ}$ and $\overline{CR}$ are medians. So, $SB = \frac{2}{3}PB = \frac{2}{3}(6) = 4.$

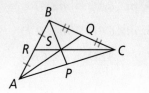

Exercises

Determine whether $\overline{AB}$ is a *median*, an *altitude*, or *neither*. Explain.

21.

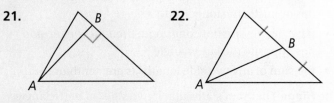

22.

23. $\triangle PQR$ has medians $\overline{QM}$ and $\overline{PN}$ that intersect at Z. If $ZM = 4$, find QZ and QM.

$\triangle ABC$ has vertices $A(2, 3), B(-4, -3),$ and $C(2, -3)$. Find the coordinates of each point of concurrency.

24. centroid

25. orthocenter

5-5 Indirect Proof

Quick Review

In an **indirect proof,** you first assume temporarily the opposite of what you want to prove. Then you show that this temporary assumption leads to a contradiction.

Example

Which two statements contradict each other?

I. The perimeter of $\triangle ABC$ is 14.

II. $\triangle ABC$ is isosceles.

III. The side lengths of $\triangle ABC$ are 3, 5, and 6.

An isosceles triangle can have a perimeter of 14.

The perimeter of a triangle with side lengths 3, 5, and 6 is 14.

An isosceles triangle must have two sides of equal length. Statements II and III contradict each other.

Exercises

Write a convincing argument that uses indirect reasoning.

26. The product of two numbers is even. Show that at least one of the numbers must be even.

27. Two lines in the same plane are not parallel. Show that a third line in the plane must intersect at least one of the two lines.

28. Show that a triangle can have at most one obtuse angle.

29. Show that an equilateral triangle cannot have an obtuse angle.

30. The sum of three integers is greater than 9. Show that one of the integers must be greater than 3.

5-6 and 5-7 Inequalities in Triangles

Quick Review

For any triangle,

- the measure of an exterior angle is greater than the measure of each of its remote interior angles
- if two sides are not congruent, then the larger angle lies opposite the longer side
- if two angles are not congruent, then the longer side lies opposite the larger angle
- the sum of any two side lengths is greater than the third

The **Hinge Theorem** states that if two sides of one triangle are congruent to two sides of another triangle, and the included angles are not congruent, then the longer third side is opposite the larger included angle.

Example

Which is greater, BC or AD?

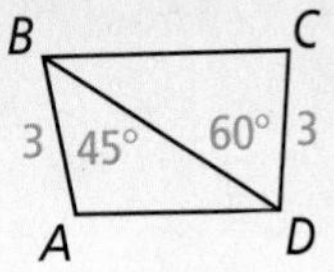

$\overline{BA} \cong \overline{CD}$ and $\overline{BD} \cong \overline{DB}$, so $\triangle ABD$ and $\triangle CDB$ have two pairs of congruent corresponding sides. Since $60 > 45$, you know $BC > AD$ by the Hinge Theorem.

Exercises

31. In $\triangle RST$, $m\angle R = 70$ and $m\angle S = 80$. List the sides of $\triangle RST$ in order from shortest to longest.

Is it possible for a triangle to have sides with the given lengths? Explain.

32. 5 in., 8 in., 15 in.

33. 10 cm, 12 cm, 20 cm

34. The lengths of two sides of a triangle are 12 ft and 13 ft. Find the range of possible lengths for the third side.

Use the figure below. Complete each statement with >, <, or =.

35. $m\angle BAD$ ■ $m\angle ABD$

36. $m\angle CBD$ ■ $m\angle BCD$

37. $m\angle ABD$ ■ $m\angle CBD$

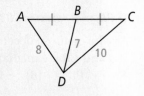

Do you know HOW?

Find the coordinates of the circumcenter of $\triangle ABC$.

1. $A(3, -1)$, $B(-2, -1)$, $C(3, -8)$

2. $A(0, 5)$, $B(-4, 5)$, $C(-4, -3)$

Find the coordinates of the orthocenter of $\triangle ABC$.

3. $A(-1, -1)$, $B(-1, 5)$, $C(-4, -1)$

4. $A(0, 0)$, $B(5, 0)$, $C(5, 3)$

Identify the two statements that contradict each other.

5. **I.** $\triangle PQR$ is a right triangle.

   II. $\triangle PQR$ is an obtuse triangle.

   III. $\triangle PQR$ is scalene.

6. **I.** $\angle DOS \cong \angle CAT$

   II. $\angle DOS$ and $\angle CAT$ are vertical.

   III. $\angle DOS$ and $\angle CAT$ are adjacent.

7. If $AB = 9$, $BC = 4\frac{1}{2}$, and $AC = 12$, list the angles of $\triangle ABC$ from smallest to largest.

8. Point P is inside $\triangle ABC$ and equidistant from all three sides. If $m\angle ABC = 60$, what is $m\angle PBC$?

List the sides from shortest to longest.

9.

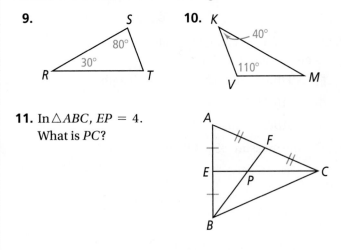

10.

11. In $\triangle ABC$, $EP = 4$. What is PC?

12. Which is greater, AD or DC? Explain.

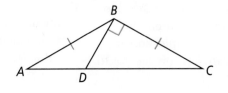

Algebra Find the value of x.

13.

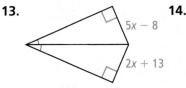

14.

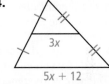

Do you UNDERSTAND?

15. What can you conclude from the diagram at the right? Justify your answer.

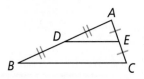

16. Reasoning In $\triangle ABC$, $BC > BA$. Draw $\triangle ABC$ and the median $\overline{BD}$. Use the Converse of the Hinge Theorem to explain why $\angle BDC$ is obtuse.

17. Given: $\overleftrightarrow{PQ}$ is the perpendicular bisector of $\overline{AB}$. $\overleftrightarrow{QT}$ is the perpendicular bisector of $\overline{AC}$.

   Prove: $QC = QB$

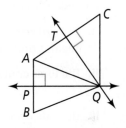

18. Writing Use indirect reasoning to explain why the following statement is true: If an isosceles triangle is obtuse, then the obtuse angle is the vertex angle.

19. In the figure, $WK = KR$. What can you conclude about point A? Explain.

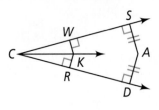

ⓉⒾⓅⓈ ⒻⓄⓇ ⓈⓊⒸⒸⒺⓈⓈ

Some test questions ask you to find missing measurements. Read the sample question at the right. Then follow the tips to answer it.

The diagram below shows the walkways in a triangular park. What is the distance from point S to point Z using the walkways $\overline{ST}$ and $\overline{TZ}$?

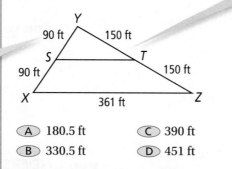

Ⓐ 180.5 ft   Ⓒ 390 ft
Ⓑ 330.5 ft   Ⓓ 451 ft

TIP 1

You need to find the length of $\overline{ST}$ and add it to 150 ft, the length of $\overline{TZ}$.

Think It Through

$\overline{ST}$ is a midsegment of $\triangle XYZ$. By the Triangle Midsegment Theorem, you know $ST = \frac{1}{2}(XZ) = \frac{1}{2}(361)$, or 180.5 ft. The distance from point S to point Z using walkways $\overline{ST}$ and $\overline{TZ}$ is $ST + TZ = 180.5 + 150$, or 330.5 ft. The correct answer is B.

TIP 2

From the dimensions given in the diagram, you can conclude that S and T are midpoints of sides of $\triangle XYZ$. So you can use the Triangle Midsegment Theorem to find ST.

Vocabulary Builder

As you solve test items, you must understand the meanings of mathematical terms. Choose the correct term to complete each sentence.

I. The (*inverse, converse*) of a conditional statement negates both the hypothesis and the conclusion.

II. The lines containing the altitudes of a triangle are concurrent at the (*orthocenter, centroid*) of the triangle.

III. The side opposite the vertex angle of an isosceles triangle is the (*hypotenuse, base*).

IV. The linear equation $y = mx + b$ is in (*slope-intercept form, point-slope form*).

Multiple Choice

Read each question. Then write the letter of the correct answer on your paper.

1. One side of a triangle has length 6 in. and another side has length 3 in. Which is the greatest possible value for the length of the third side?

Ⓐ 3 in.   Ⓒ 8 in.
Ⓑ 6 in.   Ⓓ 9 in.

2. $\triangle ABC$ is an equilateral triangle. Which is NOT a true statement about $\triangle ABC$?

Ⓕ All three sides have the same length.

Ⓖ $\triangle ABC$ is isosceles.

Ⓗ $\triangle ABC$ is equiangular.

Ⓘ The measure of $\angle A$ is 50.

3. In the figure below, $\triangle ABC$ has vertices $A(-2, -4)$, $B(-3, 2)$, and $C(3, 0)$. D is the midpoint of $\overline{AB}$, E is the midpoint of $\overline{BC}$, and F is the midpoint of $\overline{AC}$. What are the coordinates of the vertices of $\triangle DEF$?

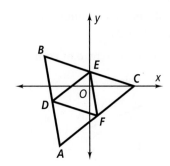

  (A) $D(-2.5, -1)$, $E(0, 1)$, $F(0.5, -2)$

  (B) $D(-2, -1)$, $E(0, 0.5)$, $F(0.5, -1)$

  (C) $D(-2, -1.5)$, $E(0, 1.5)$, $F(0.8, -2)$

  (D) $D(-2.2, -0.8)$, $E(0, 1.5)$, $F(0.5, -2)$

4. A square and a rectangle have the same area. The square has side length 8 in. The length of the rectangle is four times its width. What is the length of the rectangle?

  (F) 4 in.   (H) 32 in.

  (G) 16 in.   (I) 64 in.

5. In the figure below, $\angle A \cong \angle DBE$ and $\overline{AB} \cong \overline{BE}$. What additional information do you need in order to prove a pair of triangles congruent by AAS?

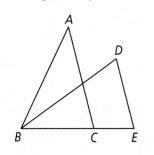

  (A) $\angle ABC \cong \angle BED$

  (B) C is the midpoint of $\overline{BE}$.

  (C) $\angle ACB \cong \angle BDE$

  (D) $\angle ABC \cong \angle BDE$

6. What is the value of y in the figure below?

  (F) $\dfrac{3}{7}$

  (G) 1

  (H) $\dfrac{7}{3}$

  (I) 3

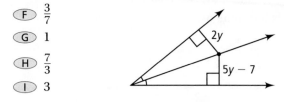

7. Which statement is the inverse of the following statement?

  If $\overline{PQ}$ is a midsegment of $\triangle ABC$, then $\overline{PQ}$ is parallel to a side of $\triangle ABC$.

  (A) If $\overline{PQ}$ is a midsegment of $\triangle ABC$, then $\overline{PQ}$ is not parallel to a side of $\triangle ABC$.

  (B) If $\overline{PQ}$ is not a midsegment of $\triangle ABC$, then $\overline{PQ}$ is not parallel to a side of $\triangle ABC$.

  (C) If $\overline{PQ}$ is not parallel to a side of $\triangle ABC$, then $\overline{PQ}$ is not a midsegment of $\triangle ABC$.

  (D) If $\overline{PQ}$ is parallel to a side of $\triangle ABC$, then $\overline{PQ}$ is a midsegment of $\triangle ABC$.

8. $\overline{AB}$ has endpoints $A(0, -4)$ and $B(8, -2)$. What is the slope-intercept form of the equation of the perpendicular bisector of $\overline{AB}$?

  (F) $y = -4x + 13$ (H) $y + 3 = -4(x - 4)$

  (G) $y = 4x - 19$ (I) $y + 3 = 4(x - 4)$

9. How can you prove that the two triangles at the right are congruent?

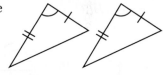

  (A) ASA   (C) SAS

  (B) SSS   (D) HL

10. Which statement contradicts the statement $\triangle ABC \cong \triangle JMK$ by SAS?

  (F) $\angle A$ and $\angle J$ are vertical angles.

  (G) $\overline{BC}$ is the hypotenuse of $\triangle ABC$.

  (H) $\triangle ABC$ is isosceles and $\triangle JMK$ is scalene.

  (I) $\overline{AB}$ is not congruent to $\overline{MK}$.

GRIDDED RESPONSE

11. What is the value of x?

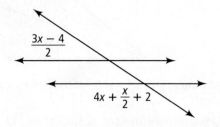

$$\frac{3x-4}{2}$$

$$4x + \frac{x}{2} + 2$$

12. $A(-2, -2)$ and $B(5, 7)$ are the endpoints of $\overline{AB}$. Point C lies on $\overline{AB}$ and is $\frac{1}{3}$ of the way from B to A. What is the y-coordinate of Point C?

13. $\triangle ABC$ is a right triangle with area 14 in.² $\overline{BM}$ and $\overline{CN}$ are medians and $BN = 2$ in. What is the area of $\triangle CNM$ in square inches?

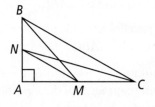

14. The measure of one angle of an isosceles triangle is 70. What is the measure of the largest angle?

15. $\overline{DE}$ is a midsegment of $\triangle ABC$. In millimeters, what is DE?

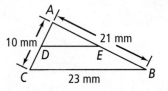

16. What is the area in square units of a rectangle with vertices $(-2, 5)$, $(3, 5)$, $(3, -1)$, and $(-2, -1)$?

17. In the figure below, the area of $\triangle ADB$ is 4.5 cm². In centimeters, what is DF?

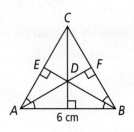

Short Response

18. If two lines are parallel, then they do not intersect. If two lines do not intersect, then they do not have any points in common.

   a. Suppose $\ell \parallel m$. What conclusion can you make about lines ℓ and m?

   b. Explain how you arrived at your conclusion.

19. Copy the triangle below.

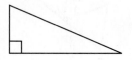

   a. Construct a congruent triangle.

   b. Describe your method.

Extended Response

20. Write an indirect proof.

   Given: $\triangle ABC$ is obtuse.

   Prove: $\triangle ABC$ is not a right triangle.

21. The towns of Westfield, Bayville, and Oxboro need a cell phone tower. The strength of the signal from the tower should be the same for each of the three towns. The map below shows the location of each town, with each grid square representing 1 square mile.

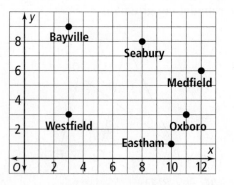

   a. At what coordinates should the new cell phone tower be located? Explain.

   b. What is the distance from the new tower to Westfield? To Bayville? To Oxboro?

   c. Can any of the other towns shown on the map benefit from the new cell phone tower? Explain.

Get Ready!

Lesson 3-2

Properties of Parallel Lines

Algebra Use properties of parallel lines to find the value of *x*.

1.

$(x + 9)°$ $(2x - 21)°$

2.
$(3x - 14)°$
$(2x - 16)°$

3.

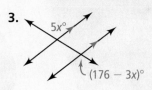

$5x°$
$(176 - 3x)°$

Lesson 3-3

Proving Lines Parallel

Algebra Determine whether $\overline{AB}$ is parallel to $\overline{CD}$.

4.
A
B
C
D

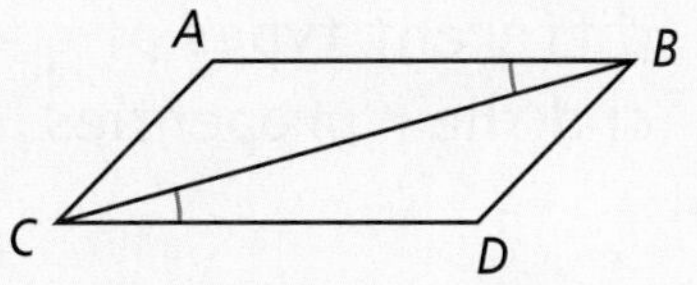

5.
B
D $3x°$
$4x°$ F
C $2x°$
A $(3x + 18)°$

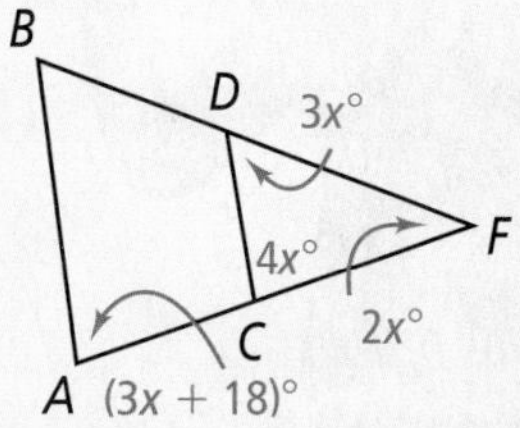

6.
$(2x + 11)°$ $(3x - 9)°$ B
A
$(6x + 9)°$
C   D

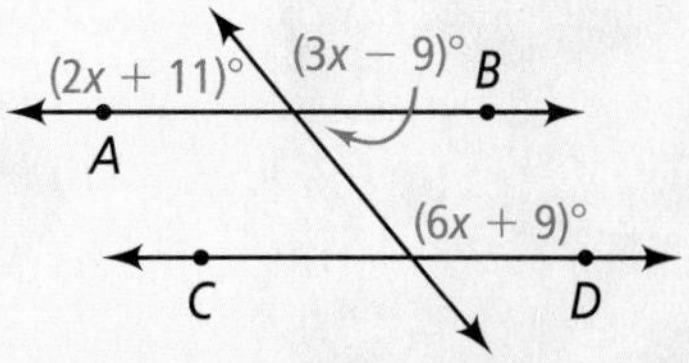

Lesson 3-8

Using Slope to Determine Parallel and Perpendicular Lines

Algebra Determine whether each pair of lines is *parallel, perpendicular,* or *neither*.

7. $y = -2x; y = -2x + 4$ **8.** $y = -\frac{3}{5}x + 1; y = \frac{5}{3}x - 3$ **9.** $2x - 3y = 1; 3x - 2y = 8$

Lessons 4-2 and 4-3

Proving Triangles Congruent

Determine the postulate or theorem that makes each pair of triangles congruent.

10.
A
B
D
C

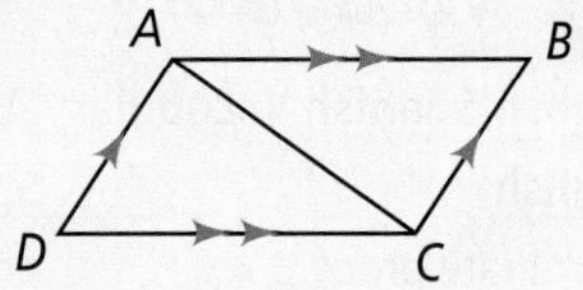

11.

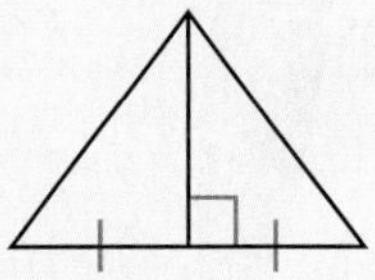

12.

Looking Ahead Vocabulary

13. You know the meaning of *equilateral*. What do you think an *equiangular* polygon is?

14. Think about what a *kite* looks like. What characteristics might a *kite* in geometry have?

15. When a team wins two *consecutive* gold medals, it means they have won two gold medals in a row. What do you think two *consecutive* angles in a quadrilateral means?

Polygons and Quadrilaterals

© **DOMAINS**
- Congruence
- Similarity, Right Triangles, and Trigonometry
- Expressing Geometric Properties with Equations

That is one cool building! The side of the building has the shape of a quadrilateral that is not a rectangle.

In this chapter, you'll learn about different types of quadrilaterals and their properties.

Vocabulary

English/Spanish Vocabulary Audio Online:

English	Spanish
coordinate proof, p. 408	prueba de coordenadas
equiangular polygon, p. 354	polígono equiángulo
equilateral polygon, p. 354	polígono equilátero
isosceles trapezoid, p. 389	trapecio isósceles
kite, p. 392	cometa
midsegment of a trapezoid, p. 391	segmento medio de un trapecio
parallelogram, p. 359	paralelogramo
rectangle, p. 375	rectángulo
regular polygon, p. 354	polígono regular
rhombus, p. 375	rombo
trapezoid, p. 389	trapecio

My Math Video

00:04:04

VIDEO

BIG ideas

1 **Measurement**
Essential Question How can you find the sum of the measures of polygon angles?

2 **Reasoning and Proof**
Essential Question How can you classify quadrilaterals?

3 **Coordinate Geometry**
Essential Question How can you use coordinate geometry to prove general relationships?

Chapter Preview

Exterior Angles of Polygons

© **Content Standard**
Prepares for G.SRT.5 Use congruence . . . criteria to solve problems and prove relationships in geometric figures.

Activity

© **Construct** Use geometry software. Construct a polygon similar to the one at the right. Extend each side as shown. Mark a point on each ray so that you can measure the exterior angles.

© **Investigate** Use your figure to explore properties of a polygon.

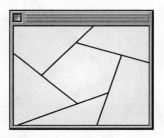

- Measure each exterior angle.

- Calculate the sum of the measures of the exterior angles.

- Manipulate the polygon. Observe the sum of the measures of the exterior angles of the new polygon.

Exercises

1. Write a conjecture about the sum of the measures of the exterior angles (one at each vertex) of a convex polygon. Test your conjecture with another polygon.

© **2. Extend** The figures below show a polygon that is decreasing in size until it finally becomes a point. Describe how you could use this to justify your conjecture in Exercise 1.

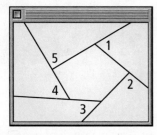
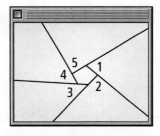
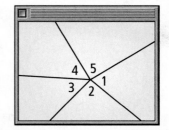

3. The figure at the right shows a square that has been copied several times. Notice that you can use the square to completely cover, or tile, a plane, without gaps or overlaps.

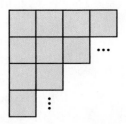

   a. Using geometry software, make several copies of other regular polygons with 3, 5, 6, and 8 sides. Regular polygons have sides of equal length and angles of equal measure.

   b. Which of the polygons you made can tile a plane?

   c. Measure *one* exterior angle of each polygon (including the square).

   d. Write a conjecture about the relationship between the measure of an exterior angle and your ability to tile a plane with the polygon. Test your conjecture with another regular polygon.

6-1 The Polygon Angle-Sum Theorems

Content Standard
G.SRT.5 Use congruence . . . criteria to solve problems and prove relationships in geometric figures.

Objectives To find the sum of the measures of the interior angles of a polygon
To find the sum of the measures of the exterior angles of a polygon

SOLVE IT!

Getting Ready!

◀▶ ✕ ↻ 🏠

Sketch a convex pentagon, hexagon, and heptagon. For each figure, draw all the diagonals you can from one vertex. What conjecture can you make about the relationship between the number of sides of a polygon and the number of triangles formed by the diagonals from one vertex?

If you can find a pattern, you won't have to draw all those diagonals.

MATHEMATICAL PRACTICES The Solve It is related to a formula for the sum of the interior angle measures of a polygon. (In this textbook, a polygon is convex unless otherwise stated.)

Essential Understanding The sum of the interior angle measures of a polygon depends on the number of sides the polygon has.

By dividing a polygon with n sides into $(n - 2)$ triangles, you can show that the sum of the interior angle measures of any polygon is a multiple of 180.

Dynamic Activity
Polygon Angle-Sum Theorem

Lesson Vocabulary
• equilateral polygon
• equiangular polygon
• regular polygon

take note

Theorem 6-1 Polygon Angle-Sum Theorem

The sum of the measures of the interior angles of an n-gon is $(n - 2)180$.

Problem 1 Finding a Polygon Angle Sum

Think
How many sides does a heptagon have?
A heptagon has 7 sides.

What is the sum of the interior angle measures of a heptagon?

$$\begin{aligned} \text{Sum} &= (n - 2)180 && \text{Polygon Angle-Sum Theorem} \\ &= (7 - 2)180 && \text{Substitute 7 for } n. \\ &= 5 \cdot 180 && \text{Simplify.} \\ &= 900 \end{aligned}$$

The sum of the interior angle measures of a heptagon is 900.

Got It? **1. a.** What is the sum of the interior angle measures of a 17-gon?
b. Reasoning The sum of the interior angle measures of a polygon is 1980. How can you find the number of sides in the polygon?

An **equilateral polygon** is a polygon with all sides congruent.

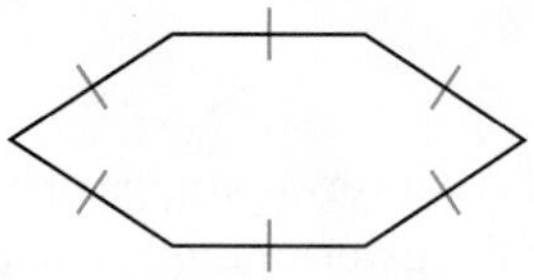

An **equiangular polygon** is a polygon with all angles congruent.

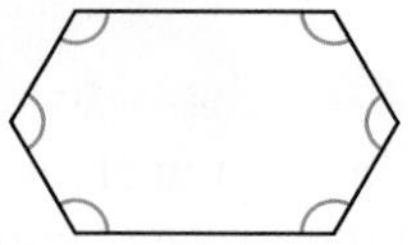

A **regular polygon** is a polygon that is both equilateral and equiangular.

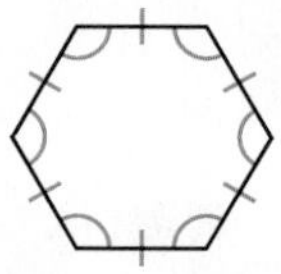

Corollary to the Polygon Angle-Sum Theorem

The measure of each interior angle of a regular n-gon is $\dfrac{(n-2)180}{n}$.

You will prove the Corollary to the Polygon Angle-Sum Theorem in Exercise 43.

Problem 2   Using the Polygon Angle-Sum Theorem   STEM

Biology The common housefly, *Musca domestica*, has eyes that consist of approximately 4000 facets. Each facet is a regular hexagon. What is the measure of each interior angle in one hexagonal facet?

Think

How does the word *regular* help you answer the question?
The word *regular* tells you that each angle has the same measure.

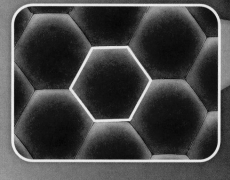

Measure of an angle $= \dfrac{(n-2)180}{n}$ Corollary to the Polygon Angle-Sum Theorem

$= \dfrac{(6-2)180}{6}$ Substitute 6 for n.

$= \dfrac{4 \cdot 180}{6}$ Simplify.

$= 120$

The measure of each interior angle in one hexagonal facet is 120.

Got It? 2. What is the measure of each interior angle in a regular nonagon?

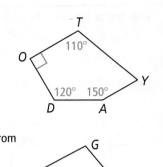

Think

How does the diagram help you?
You know the number of sides and four of the five angle measures.

 Problem 3 Using the Polygon Angle-Sum Theorem

What is $m\angle Y$ in pentagon *TODAY*?

Use the Polygon Angle-Sum Theorem for $n = 5$.

$m\angle T + m\angle O + m\angle D + m\angle A + m\angle Y = (5 - 2)180$

$\quad\quad 110 + 90 + 120 + 150 + m\angle Y = 3 \cdot 180$ Substitute.

$\quad\quad\quad\quad\quad\quad\quad\quad 470 + m\angle Y = 540$ Simplify.

$\quad\quad\quad\quad\quad\quad\quad\quad\quad\quad\quad\quad m\angle Y = 70$ Subtract 470 from each side.

Got It? 3. What is $m\angle G$ in quadrilateral *EFGH*?

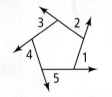

You can draw exterior angles at any vertex of a polygon. The figures below show that the sum of the measures of the exterior angles, one at each vertex, is 360.

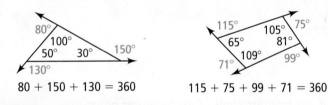

$80 + 150 + 130 = 360$ $115 + 75 + 99 + 71 = 360$

take note

Theorem 6-2 Polygon Exterior Angle-Sum Theorem

The sum of the measures of the exterior angles of a polygon, one at each vertex, is 360.

For the pentagon, $m\angle 1 + m\angle 2 + m\angle 3 + m\angle 4 + m\angle 5 = 360$.

You will prove Theorem 6-2 in Exercise 39.

Think

What kind of angle is ∠1?
Looking at the diagram, you know that ∠1 is an exterior angle.

Problem 4 Finding an Exterior Angle Measure

What is $m\angle 1$ in the regular octagon at the right?

By the Polygon Exterior Angle-Sum Theorem, the sum of the exterior angle measures is 360. Since the octagon is regular, the interior angles are congruent. So their supplements, the exterior angles, are also congruent.

$m\angle 1 = \dfrac{360}{8}$ Divide 360 by 8, the number of sides in an octagon.

$\quad\quad = 45$ Simplify.

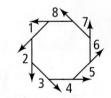

Got It? 4. What is the measure of an exterior angle of a regular nonagon?

Lesson Check

Do you know HOW?

1. What is the sum of the interior angle measures of an 11-gon?

2. What is the sum of the measures of the exterior angles of a 15-gon?

3. Find the measures of an interior angle and an exterior angle of a regular decagon.

Do you UNDERSTAND? Ⓒ MATHEMATICAL PRACTICES

Ⓒ **4. Vocabulary** Can you draw an equiangular polygon that is not equilateral? Explain.

Ⓒ **5. Reasoning** Which angles are the exterior angles for ∠1? What do you know about their measures? Explain.

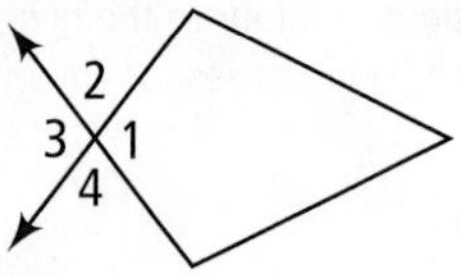

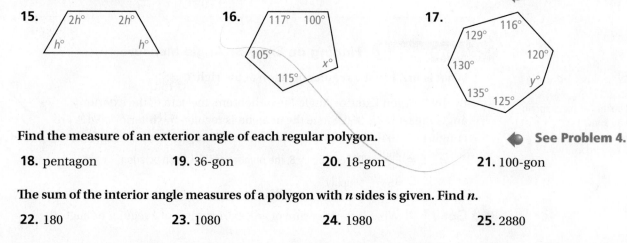

Ⓒ **6. Error Analysis** Your friend says that she measured an interior angle of a regular polygon as 130. Explain why this result is impossible.

Practice and Problem-Solving Exercises Ⓒ MATHEMATICAL PRACTICES

Ⓐ **Practice**   Find the sum of the interior angle measures of each polygon.   ◀ See Problem 1.

7.
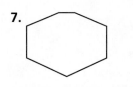

8. 35-gon

9. 14-gon

10. 20-gon

11. 1002-gon

Find the measure of one interior angle in each regular polygon.   ◀ See Problem 2.

12.

13.

14.

Algebra Find the missing angle measures.   ◀ See Problem 3.

15.

16.

17.

Find the measure of an exterior angle of each regular polygon.   ◀ See Problem 4.

18. pentagon

19. 36-gon

20. 18-gon

21. 100-gon

Ⓑ **Apply**   The sum of the interior angle measures of a polygon with n sides is given. Find n.

22. 180

23. 1080

24. 1980

25. 2880

26. Open-Ended Sketch an equilateral polygon that is not equiangular.

27. Stage Design A theater-in-the-round allows for a play to have an audience on all sides. The diagram at the right shows a platform constructed for a theater-in-the-round stage. What type of regular polygon is the largest platform? Find the measure of each numbered angle.

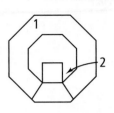

28. Think About a Plan A triangle has two congruent interior angles and an exterior angle that measures 100. Find two possible sets of interior angle measures for the triangle.
- How can a diagram help you?
- What is the sum of the angle measures in a triangle?

Algebra Find the value of each variable.

29.   **30.**   **31.**

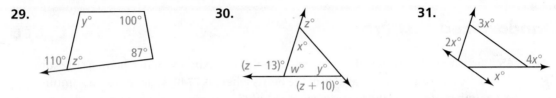

The measure of an exterior angle of a regular polygon is given. Find the measure of an interior angle. Then find the number of sides.

32. 72   **33.** 36   **34.** 18   **35.** 30   **36.** x

Packaging The gift package at the right contains fruit and cheese. The fruit is in a container that has the shape of a regular octagon. The fruit container fits in a square box. A triangular cheese wedge fills each corner of the box.

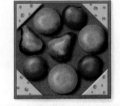

37. Find the measure of each interior angle of a cheese wedge.

38. Reasoning Show how to rearrange the four pieces of cheese to make a regular polygon. What is the measure of each interior angle of the polygon?

39. Algebra A polygon has n sides. An interior angle of the polygon and an adjacent exterior angle form a straight angle.
- **a.** What is the sum of the measures of the n straight angles?
- **b.** What is the sum of the measures of the n interior angles?
- **c.** Using your answers above, what is the sum of the measures of the n exterior angles?
- **d.** What theorem do the steps above prove?

40. Reasoning Your friend says she has another way to find the sum of the interior angle measures of a polygon. She picks a point inside the polygon, draws a segment to each vertex, and counts the number of triangles. She multiplies the total by 180, and then subtracts 360 from the product. Does her method work? Explain.

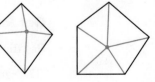

41. Algebra The measure of an interior angle of a regular polygon is three times the measure of an exterior angle of the same polygon. What is the name of the polygon?

42. **Probability** Find the probability that the measure of an interior angle of a regular n-gon is a positive integer when n is an integer and $3 \le n \le 12$.

43. **a.** In the Corollary to the Polygon Angle-Sum Theorem, explain why the measure of an interior angle of a regular n-gon is given by the formulas $\frac{180(n-2)}{n}$ and $180 - \frac{360}{n}$.

   b. Use the second formula to explain what happens to the measures of the interior angles of regular n-gons as n becomes a large number. Explain also what happens to the polygons.

44. *ABCDEFGHJK* is a regular decagon. A ray bisects $\angle C$, and another ray bisects $\angle D$. The two rays intersect in the decagon's interior. Find the measure of the acute angles formed by the intersecting rays.

Standardized Test Prep

GRIDDED RESPONSE

SAT/ACT

45. The car at each vertex of a Ferris wheel holds a maximum of five people. The sum of the interior angle measures of the Ferris wheel is 7740. What is the maximum number of people the Ferris wheel can hold?

46. A rectangle and a square have equal areas. The rectangle has length 9 cm and width 4 cm. What is the perimeter of the square, in centimeters?

47. The Public Garden is located between two parallel streets: Maple Street and Oak Street. The garden faces Maple Street and is bordered by rows of shrubs that intersect Oak Street at point B. What is $m\angle ABC$, the angle formed by the shrubs?

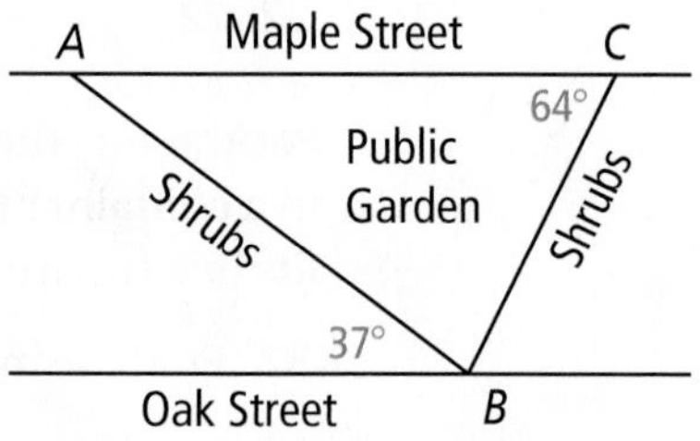

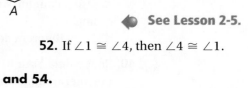

48. $\triangle ABC \cong \triangle DEF$. If $m\angle A = 3x + 4$, $m\angle C = 2x$, and $m\angle E = 4x + 5$, what is $m\angle B$?

Mixed Review

49. If $\overline{AB} \cong \overline{CB}$ and $m\angle ABD < m\angle CBD$, which is longer, $\overline{AD}$ or $\overline{CD}$? Explain.

◀ See Lesson 5-7.

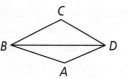

Name the property that justifies each statement.

◀ See Lesson 2-5.

50. $4(2a - 3) = 8a - 12$ 51. $\overline{RS} \cong \overline{RS}$ 52. If $\angle 1 \cong \angle 4$, then $\angle 4 \cong \angle 1$.

Get Ready! **To prepare for Lesson 6-2, do Exercises 53 and 54.**

Use the figure below.

◀ See Lessons 4-1 and 4-3.

53. Name the postulate or theorem that justifies $\triangle EFG \cong \triangle GHE$.

54. Complete each statement.

    a. $\angle FEG \cong$ ■   **b.** $\angle EFG \cong$ ■   **c.** $\angle FGE \cong$ ■

    d. $\overline{EF} \cong$ ■   **e.** $\overline{FG} \cong$ ■   **f.** $\overline{GE} \cong$ ■

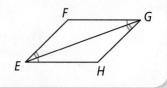

Properties of Parallelograms

Content Standards
G.CO.11 Prove theorems about parallelogram. Theorems include: opposite sides are congruent, opposite angles are congruent, the diagonals of a parallelogram bisect each other . . .
Also G.SRT.5

Objectives To use relationships among sides and angles of parallelograms
To use relationships among diagonals of parallelograms

Don't settle for your answer too soon! Be sure to find all pairs of congruent triangles.

MATHEMATICAL PRACTICES

SOLVE IT!

Getting Ready!

Use the information given in the diagram. Which triangles are congruent? How do you know?

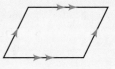

Lesson Vocabulary
• parallelogram
• opposite sides
• opposite angles
• consecutive angles

A **parallelogram** is a quadrilateral with both pairs of opposite sides parallel. In the Solve It, you made some conjectures about the characteristics of a parallelogram. In this lesson, you will verify whether your conjectures are correct.

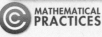

Essential Understanding Parallelograms have special properties regarding their sides, angles, and diagonals.

In a quadrilateral, **opposite sides** do not share a vertex and **opposite angles** do not share a side.

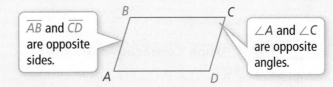

$\overline{AB}$ and $\overline{CD}$ are opposite sides.

$\angle A$ and $\angle C$ are opposite angles.

You can abbreviate *parallelogram* with the symbol $\square$ and *parallelograms* with the symbol $\varoslash$. You can use what you know about parallel lines and transversals to prove some theorems about parallelograms.

take note

Theorem 6-3

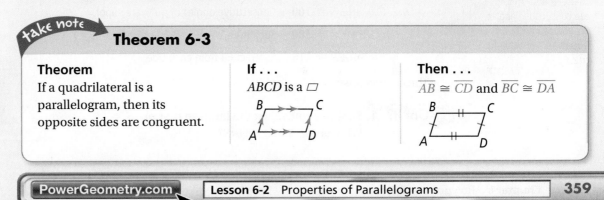

Theorem	**If . . .**	**Then . . .**
If a quadrilateral is a parallelogram, then its opposite sides are congruent.	$ABCD$ is a $\square$	$\overline{AB} \cong \overline{CD}$ and $\overline{BC} \cong \overline{DA}$

Proof of Theorem 6-3

Given: ▱ABCD

Prove: $\overline{AB} \cong \overline{CD}$ and $\overline{BC} \cong \overline{DA}$

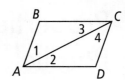

Statements	Reasons
1) $ABCD$ is a parallelogram.	1) Given
2) $\overline{AB} \parallel \overline{CD}$ and $\overline{BC} \parallel \overline{DA}$	2) Definition of parallelogram
3) $\angle 1 \cong \angle 4$ and $\angle 3 \cong \angle 2$	3) If lines are $\parallel$, then alt. int. $\angle$s are $\cong$.
4) $\overline{AC} \cong \overline{AC}$	4) Reflexive Property of $\cong$
5) $\triangle ABC \cong \triangle CDA$	5) ASA
6) $\overline{AB} \cong \overline{CD}$ and $\overline{BC} \cong \overline{DA}$	6) Corresp. parts of $\cong$ $\triangle$s are $\cong$.

Angles of a polygon that share a side are **consecutive angles.** In the diagram, $\angle A$ and $\angle B$ are consecutive angles because they share side $\overline{AB}$.

∠B and ∠C are also consecutive angles.

The theorem below uses the fact that consecutive angles of a parallelogram are same-side interior angles of parallel lines.

take note

Theorem 6-4

Theorem	If . . .	Then . . .
If a quadrilateral is a parallelogram, then its consecutive angles are supplementary.	$ABCD$ is a ▱	$m\angle A + m\angle B = 180$ $m\angle B + m\angle C = 180$ $m\angle C + m\angle D = 180$ $m\angle D + m\angle A = 180$

You will prove Theorem 6-4 in Exercise 32.

© Problem 1 Using Consecutive Angles

Plan

What information from the diagram helps you get started?
From the diagram, you know $m\angle PSR$ and that $\angle P$ and $\angle PSR$ are consecutive angles. So, you can *write an equation* and solve for $m\angle P$.

Multiple Choice What is $m\angle P$ in ▱$PQRS$?

Ⓐ 26   Ⓒ 116

Ⓑ 64   Ⓓ 126

$$m\angle P + m\angle S = 180 \quad \text{Consecutive angles of a ▱ are suppl.}$$
$$m\angle P + 64 = 180 \quad \text{Substitute.}$$
$$m\angle P = 116 \quad \text{Subtract 64 from each side.}$$

The correct answer is C.

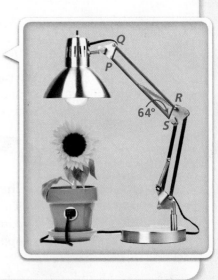

Got It? **1.** Suppose you adjust the lamp so that $m\angle S = 86$. What is $m\angle R$ in ▱$PQRS$?

Theorem 6-5

Theorem	**If . . .**	**Then . . .**
If a quadrilateral is a parallelogram, then its opposite angles are congruent.	$ABCD$ is a ▱. 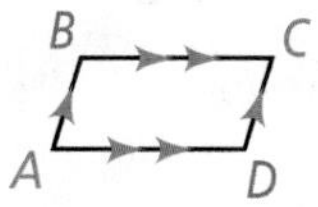 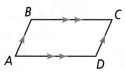	$\angle A \cong \angle C$ and $\angle B \cong \angle D$ 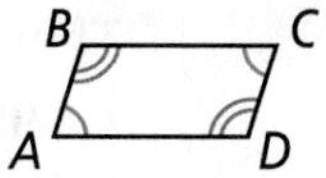

A proof of Theorem 6-5 in Problem 2 uses the consecutive angles of a parallelogram and the fact that supplements of the same angle are congruent.

Proof **Problem 2** **Using Properties of Parallelograms in a Proof**

Given: ▱$ABCD$

Prove: $\angle A \cong \angle C$ and $\angle B \cong \angle D$

$ABCD$ is a ▱.
Given

$\angle A$ and $\angle B$ are consecutive ∠s.	$\angle B$ and $\angle C$ are consecutive ∠s.	$\angle C$ and $\angle D$ are consecutive ∠s.
Def. of consecutive ∠s	Def. of consecutive ∠s	Def. of consecutive ∠s
$\angle A$ and $\angle B$ are supplementary.	$\angle B$ and $\angle C$ are supplementary.	$\angle C$ and $\angle D$ are supplementary.
Consecutive ∠s are supplementary.	Consecutive ∠s are supplementary.	Consecutive ∠s are supplementary.

$\angle A \cong \angle C$
Supplements of the same ∠ are ≅.

$\angle B \cong \angle D$
Supplements of the same ∠ are ≅.

Think

Why is a flow proof useful here?
A flow proof allows you to see how the pairing of two statements leads to a conclusion.

✓ **Got It?** **2.** Use the diagram at the right.

Given: ▱$ABCD$, $\overline{AK} \cong \overline{MK}$

Prove: $\angle BCD \cong \angle CMD$

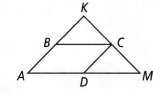

The diagonals of parallelograms have a special property.

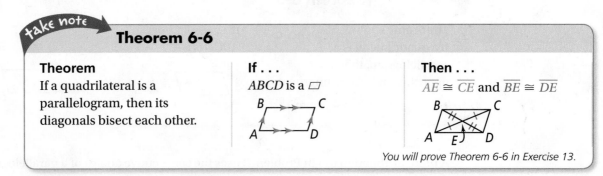

take note

Theorem 6-6

Theorem	**If . . .**	**Then . . .**
If a quadrilateral is a parallelogram, then its diagonals bisect each other.	$ABCD$ is a $\square$	$\overline{AE} \cong \overline{CE}$ and $\overline{BE} \cong \overline{DE}$

You will prove Theorem 6-6 in Exercise 13.

You can use Theorem 6-6 to find unknown lengths in parallelograms.

Ⓒ **Problem 3** **Using Algebra to Find Lengths**

Solve a system of linear equations to find the values of x and y in $\square KLMN$. What are KM and LN?

Think

The diagonals of a parallelogram bisect each other.

Set up a system of linear equations by substituting the algebraic expressions for each segment length.

Substitute $(y + 2)$ for x in equation ①. Then solve for y.

Substitute 14 for y in equation ②. Then solve for x.

Use the values of x and y to find KM and LN.

Write

$\overline{KP} \cong \overline{MP}$
$\overline{LP} \cong \overline{NP}$

① $y + 10 = 2x - 8$
② $x = y + 2$

$y + 10 = 2(y + 2) - 8$
$y + 10 = 2y + 4 - 8$
$y + 10 = 2y - 4$
$\quad\;\; 10 = y - 4$
$\quad\;\; 14 = y$

$x = 14 + 2$
$\;\; = 16$

$KM = 2(KP)$	$LN = 2(LP)$
$= 2(y + 10)$	$= 2(x)$
$= 2(14 + 10)$	$= 2(16)$
$= 48$	$= 32$

Ⓒ ✓ **Got It? 3. a.** Find the values of x and y in $\square PQRS$ at the right. What are PR and SQ?

b. Reasoning In Problem 3, does it matter which variable you solve for first? Explain.

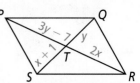

You will use parallelograms to prove the following theorem.

take note

Theorem 6-7

Theorem	**If . . .**	**Then . . .**
If three (or more) parallel lines cut off congruent segments on one transversal, then they cut off congruent segments on every transversal.	$\overleftrightarrow{AB} \parallel \overleftrightarrow{CD} \parallel \overleftrightarrow{EF}$ and $\overline{AC} \cong \overline{CE}$	$\overline{BD} \cong \overline{DF}$

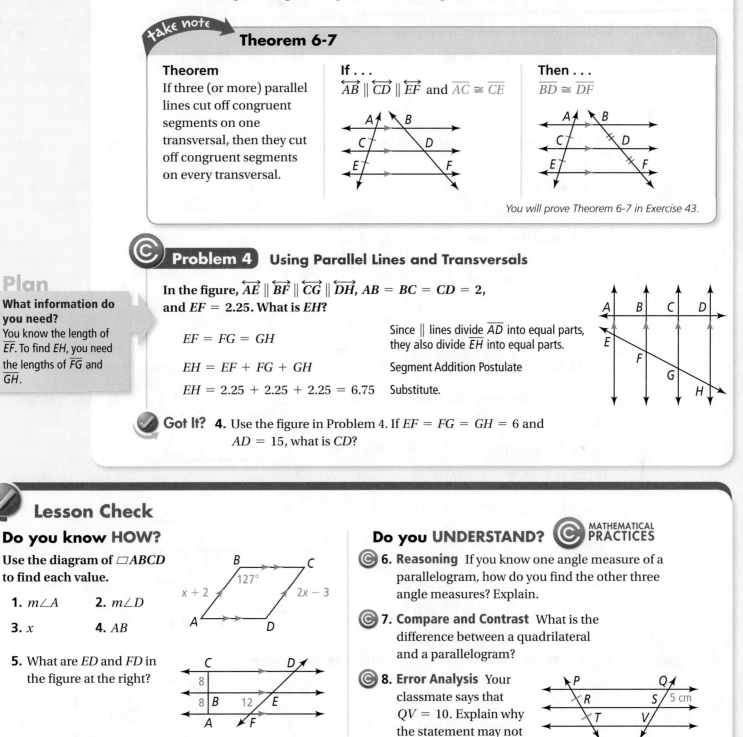

You will prove Theorem 6-7 in Exercise 43.

Problem 4 Using Parallel Lines and Transversals

Plan

What information do you need?
You know the length of $\overline{EF}$. To find EH, you need the lengths of $\overline{FG}$ and $\overline{GH}$.

In the figure, $\overleftrightarrow{AE} \parallel \overleftrightarrow{BF} \parallel \overleftrightarrow{CG} \parallel \overleftrightarrow{DH}$, $AB = BC = CD = 2$, and $EF = 2.25$. What is EH?

$EF = FG = GH$ — Since $\parallel$ lines divide $\overline{AD}$ into equal parts, they also divide $\overline{EH}$ into equal parts.

$EH = EF + FG + GH$ — Segment Addition Postulate

$EH = 2.25 + 2.25 + 2.25 = 6.75$ — Substitute.

Got It? **4.** Use the figure in Problem 4. If $EF = FG = GH = 6$ and $AD = 15$, what is CD?

Lesson Check

Do you know HOW?

Use the diagram of $\square ABCD$ to find each value.

1. $m\angle A$ **2.** $m\angle D$

3. x **4.** AB

5. What are ED and FD in the figure at the right?

Do you UNDERSTAND?   MATHEMATICAL PRACTICES

6. Reasoning If you know one angle measure of a parallelogram, how do you find the other three angle measures? Explain.

7. Compare and Contrast What is the difference between a quadrilateral and a parallelogram?

8. Error Analysis Your classmate says that $QV = 10$. Explain why the statement may not be correct.

A Practice

Algebra Find the value of *x* in each parallelogram.

See Problem 1.

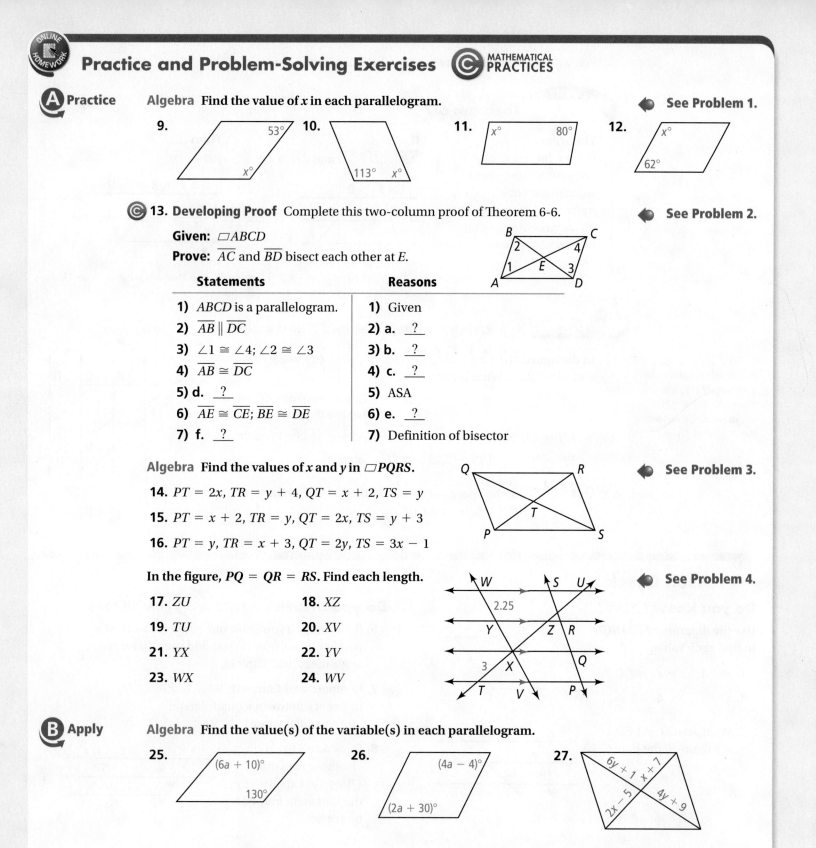

9.   53°   *x*°

10.   113°   *x*°

11.   *x*°   80°

12.   *x*°   62°

13. Developing Proof Complete this two-column proof of Theorem 6-6.

See Problem 2.

Given: ▱*ABCD*

Prove: $\overline{AC}$ and $\overline{BD}$ bisect each other at *E*.

Statements	Reasons
1) *ABCD* is a parallelogram.	**1)** Given
2) $\overline{AB} \parallel \overline{DC}$	**2) a.** ?
3) $\angle 1 \cong \angle 4; \angle 2 \cong \angle 3$	**3) b.** ?
4) $\overline{AB} \cong \overline{DC}$	**4) c.** ?
5) d. ?	**5)** ASA
6) $\overline{AE} \cong \overline{CE}; \overline{BE} \cong \overline{DE}$	**6) e.** ?
7) f. ?	**7)** Definition of bisector

Algebra Find the values of *x* and *y* in ▱*PQRS*.

See Problem 3.

14. $PT = 2x$, $TR = y + 4$, $QT = x + 2$, $TS = y$

15. $PT = x + 2$, $TR = y$, $QT = 2x$, $TS = y + 3$

16. $PT = y$, $TR = x + 3$, $QT = 2y$, $TS = 3x - 1$

In the figure, $PQ = QR = RS$. Find each length.

See Problem 4.

17. *ZU*   **18.** *XZ*

19. *TU*   **20.** *XV*

21. *YX*   **22.** *YV*

23. *WX*   **24.** *WV*

B Apply

Algebra Find the value(s) of the variable(s) in each parallelogram.

25.   (6*a* + 10)°   130°

26.   (4*a* − 4)°   (2*a* + 30)°

27.   6*y* + 1   *x* + 7   2*x* − 5   4*y* + 9

© 28. Think About a Plan What are the values of x and y in the parallelogram? 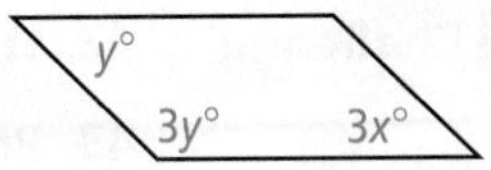
  - How are the angles related?
  - Which variable should you solve for first?

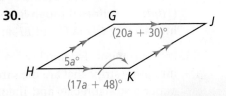

Algebra Find the value of a. Then find each side length or angle measure.

29.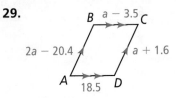

30.

31. Studio Lighting A pantograph is an expandable device shown at the right. Pantographs are used in the television industry in positioning lighting and other equipment. In the photo, points D, E, F, and G are the vertices of a parallelogram. $\square DEFG$ is one of many parallelograms that change shape as the pantograph extends and retracts.
a. If $DE = 2.5$ ft, what is FG? **b.** If $m\angle E = 129$, what is $m\angle G$?
c. What happens to $m\angle D$ as $m\angle E$ increases or decreases? Explain.

32. Prove Theorem 6-4.
Proof
   Given: $\square ABCD$
   Prove: $\angle A$ is supplementary to $\angle B$.
      $\angle A$ is supplementary to $\angle D$.

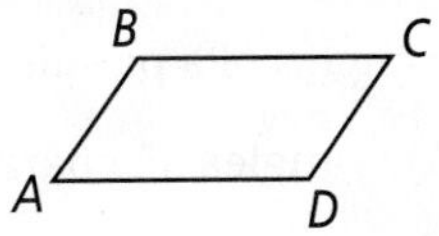

Use the diagram at the right for each proof.
Proof
33. Given: $\square LENS$ and $\square NGTH$
   Prove: $\angle L \cong \angle T$

34. Given: $\square LENS$ and $\square NGTH$
   Prove: $\overline{LS} \parallel \overline{GT}$

35. Given: $\square LENS$ and $\square NGTH$
   Prove: $\angle E$ is supplementary to $\angle T$.

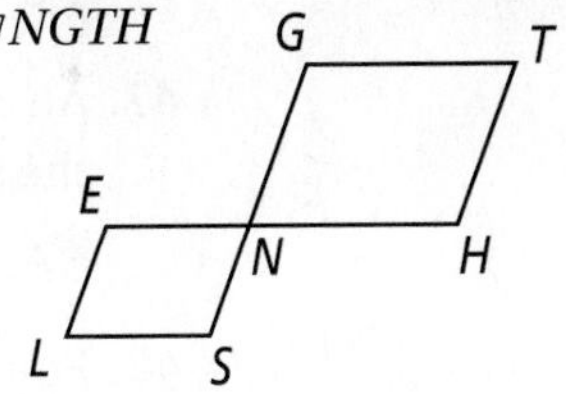

Use the diagram at the right for each proof.
Proof
36. Given: $\square RSTW$ and $\square XYTZ$
   Prove: $\angle R \cong \angle X$

37. Given: $\square RSTW$ and $\square XYTZ$
   Prove: $\overline{XY} \parallel \overline{RS}$

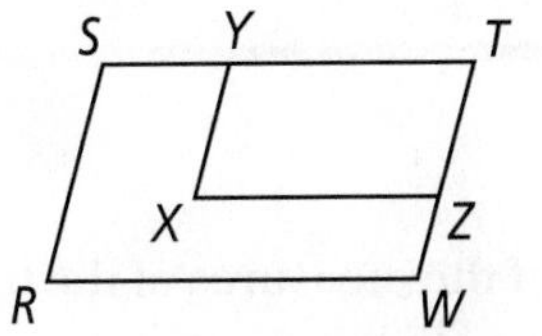

Find the measures of the numbered angles for each parallelogram.

38. **39.** **40.**

41. Algebra The perimeter of $\square ABCD$ is 92 cm. AD is 7 cm more than twice AB. Find the lengths of all four sides of $\square ABCD$.

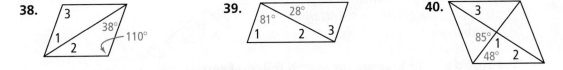

Challenge

42. Writing Is there an SSSS congruence theorem for parallelograms? Explain.

43. Prove Theorem 6-7. Use the diagram at the right.

Proof   **Given:** $\overleftrightarrow{AB} \parallel \overleftrightarrow{CD} \parallel \overleftrightarrow{EF}$, $\overline{AC} \cong \overline{CE}$

       Prove: $\overline{BD} \cong \overline{DF}$

       (*Hint:* Draw lines through B and D parallel to $\overleftrightarrow{AE}$ and intersecting $\overleftrightarrow{CD}$ at G and $\overleftrightarrow{EF}$ at H.)

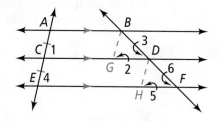

44. Measurement Explain how to separate a blank card into three strips that are the same height by using lined paper, a straightedge, and Theorem 6-7.

Standardized Test Prep

SAT/ACT

45. $PQRS$ is a parallelogram with $m\angle Q = 4x$ and $m\angle R = x + 10$. Which statement explains why you can use the equation $4x + (x + 10) = 180$ to solve for x?

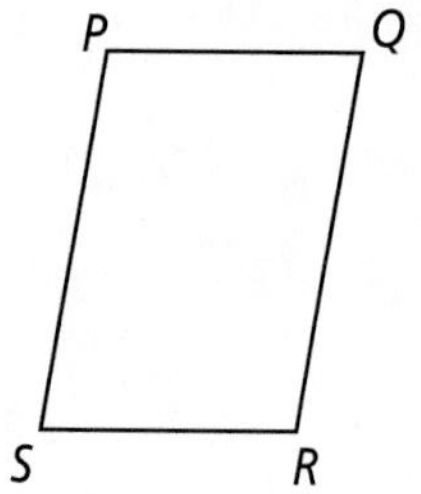

   Ⓐ The measures of the interior angles of a quadrilateral have a sum of 360.

   Ⓑ Opposite sides of a parallelogram are congruent.

   Ⓒ Opposite angles of a parallelogram are congruent.

   Ⓓ Consecutive angles of a parallelogram are supplementary.

46. In the figure of $DEFG$ at the right, $\overline{DE} \parallel \overline{GF}$. Which statement must be true?

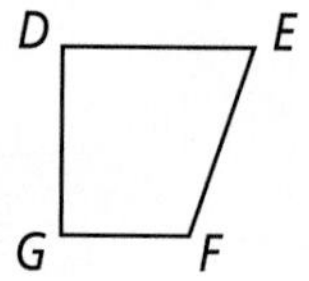

   Ⓕ $m\angle D + m\angle E = 180$ Ⓗ $\overline{DE} \cong \overline{GF}$

   Ⓖ $m\angle D + m\angle G = 180$ Ⓘ $\overline{DG} \cong \overline{EF}$

47. An obtuse triangle has side lengths of 5 cm, 9 cm, and 12 cm. What is the length of the side opposite the obtuse angle?

   Ⓐ 5 cm   Ⓑ 9 cm   Ⓒ 12 cm   Ⓓ not enough information

Short Response

48. Find the measure of one exterior angle of a regular hexagon. Explain your method.

Mixed Review

Find the sum of the measures of the interior angles of each polygon.   ◀ See Lesson 6-1.

49. decagon   **50.** 16-gon   **51.** 25-gon   **52.** 40-gon

53. What additional information do you need to prove $\triangle ADC \cong \triangle ABC$ by the HL Theorem?   ◀ See Lesson 4-6.

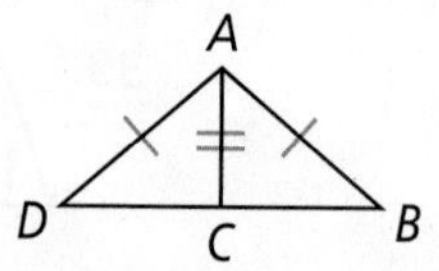

Get Ready!   **To prepare for Lesson 6-3, do Exercise 54.**

54. Two consecutive angles in a parallelogram have measures $x + 5$ and $4x - 10$. Find the measure of the smaller angle.   ◀ See Lesson 6-2.

Proving That a Quadrilateral Is a Parallelogram

© **Content Standards**
G.CO.11 Prove theorems about parallelograms . . . the diagonals of a parallelogram bisect each other and its converse . . .
Also G.SRT.5

Objective   To determine whether a quadrilateral is a parallelogram

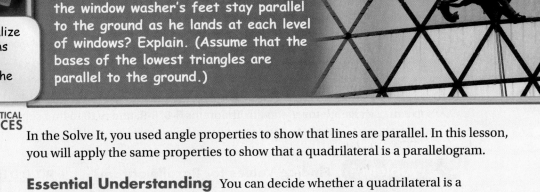

SOLVE IT!

Getting Ready!

Each section of glass in the exterior of a building in Macau, China, forms an equilateral triangle. Do you think the window washer's feet stay parallel to the ground as he lands at each level of windows? Explain. (Assume that the bases of the lowest triangles are parallel to the ground.)

Can you visualize parallelograms composed of triangles in the pattern?

© **MATHEMATICAL PRACTICES**

Dynamic Activity
Parallelogram Conditions

In the Solve It, you used angle properties to show that lines are parallel. In this lesson, you will apply the same properties to show that a quadrilateral is a parallelogram.

Essential Understanding   You can decide whether a quadrilateral is a parallelogram if its sides, angles, and diagonals have certain properties.

In Lesson 6-2, you learned theorems about the properties of parallelograms. In this lesson, you will learn the converses of those theorems. That is, if a quadrilateral has certain properties, then it must be a parallelogram. Theorem 6-8 is the converse of Theorem 6-3.

take note

Theorem 6-8

Theorem	**If . . .**	**Then . . .**
If both pairs of opposite sides of a quadrilateral are congruent, then the quadrilateral is a parallelogram.	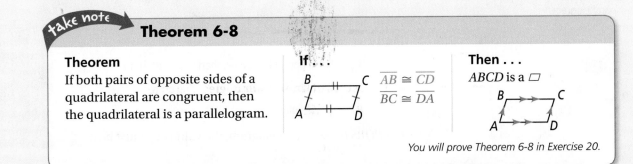 $\overline{AB} \cong \overline{CD}$ $\overline{BC} \cong \overline{DA}$	$ABCD$ is a $\square$

You will prove Theorem 6-8 in Exercise 20.

Theorems 6-9 and 6-10 are the converses of Theorems 6-4 and 6-5, respectively. They use angle relationships to conclude that a quadrilateral is a parallelogram.

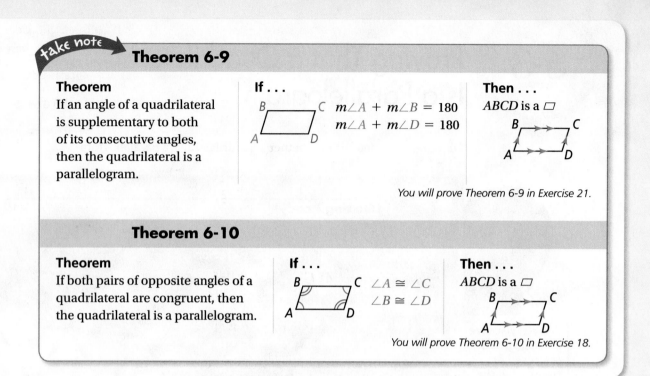

Theorem 6-9

Theorem
If an angle of a quadrilateral is supplementary to both of its consecutive angles, then the quadrilateral is a parallelogram.

If . . .
$m\angle A + m\angle B = 180$
$m\angle A + m\angle D = 180$

Then . . .
$ABCD$ is a $\square$

You will prove Theorem 6-9 in Exercise 21.

Theorem 6-10

Theorem
If both pairs of opposite angles of a quadrilateral are congruent, then the quadrilateral is a parallelogram.

If . . .
$\angle A \cong \angle C$
$\angle B \cong \angle D$

Then . . .
$ABCD$ is a $\square$

You will prove Theorem 6-10 in Exercise 18.

You can use algebra together with Theorems 6-8, 6-9, and 6-10 to find segment lengths and angle measures that assume that a quadrilateral is a parallelogram.

Plan

Which theorem should you use?
The diagram gives you information about sides. Use Theorem 6-8 because it uses sides to conclude that a quadrilateral is a parallelogram.

Ⓒ Problem 1 **Finding Values for Parallelograms**   **GRIDDED RESPONSE**

For what value of y must $PQRS$ be a parallelogram?

Step 1 Find x.

$$3x - 5 = 2x + 1$$ If opp. sides are $\cong$, then the quad. is a $\square$.

$$x - 5 = 1$$ Subtract $2x$ from each side.

$$x = 6$$ Add 5 to each side.

Step 2 Find y.

$$y = x + 2$$ If opp. sides are $\cong$, then the quad. is a $\square$.

$$= 6 + 2$$ Substitute 6 for x.

$$= 8$$ Simplify.

For $PQRS$ to be a parallelogram, the value of y must be 8.

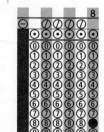

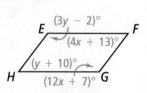

✔ Got It? **1.** Use the diagram at the right. For what values of x and y must $EFGH$ be a parallelogram?

You know that the converses of Theorems 6-3, 6-4, and 6-5 are true. Using what you have learned, you can show that the converse of Theorem 6-6 is also true.

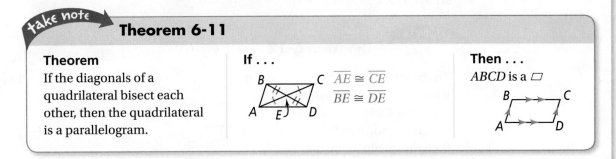

take note

Theorem 6-11

Theorem	**If . . .**	**Then . . .**
If the diagonals of a quadrilateral bisect each other, then the quadrilateral is a parallelogram.	$\overline{AE} \cong \overline{CE}$ $\overline{BE} \cong \overline{DE}$	$ABCD$ is a ▱

Proof Proof of Theorem 6-11

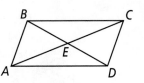

Given: $\overline{AC}$ and $\overline{BD}$ bisect each other at E.

Prove: $ABCD$ is a parallelogram.

$\overline{AC}$ and $\overline{BD}$ bisect each other at E.
Given

$\angle AEB \cong \angle CED$	$\overline{AE} \cong \overline{CE}$ $\overline{BE} \cong \overline{DE}$	$\angle BEC \cong \angle DEA$
Vertical △ are ≅.	Def. of segment bisector	Vertical △ are ≅.

$\triangle AEB \cong \triangle CED$
SAS

$\triangle BEC \cong \triangle DEA$
SAS

$\angle BAE \cong \angle DCE$
Corresp. parts of ≅ △ are ≅.

$\angle ECB \cong \angle EAD$
Corresp. parts of ≅ △ are ≅.

$\overline{AB} \parallel \overline{CD}$
If alternate interior △ ≅, then lines are ∥.

$\overline{BC} \parallel \overline{AD}$
If alternate interior △ ≅, then lines are ∥.

$ABCD$ is a parallelogram.
Def. of parallelogram

Theorem 6-12 suggests that if you keep two objects of the same length parallel, such as cross-country skis, then the quadrilateral formed by connecting their endpoints is always a parallelogram.

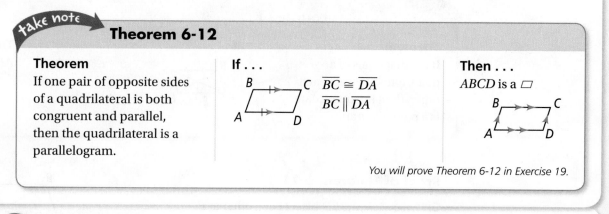

take note

Theorem 6-12

Theorem
If one pair of opposite sides of a quadrilateral is both congruent and parallel, then the quadrilateral is a parallelogram.

If . . .
$\overline{BC} \cong \overline{DA}$
$\overline{BC} \parallel \overline{DA}$

Then . . .
$ABCD$ is a $\square$

You will prove Theorem 6-12 in Exercise 19.

Problem 2 Deciding Whether a Quadrilateral Is a Parallelogram

Think

How do you decide if you have enough information?
If you can satisfy every condition of a theorem about parallelograms, then you have enough information.

Can you prove that the quadrilateral is a parallelogram based on the given information? Explain.

A Given: $AB = 5$, $CD = 5$,
$m\angle A = 50$, $m\angle D = 130$

Prove: $ABCD$ is a parallelogram.

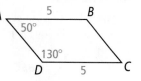

Yes. Same-side interior angles A and D are supplementary, so $\overline{AB} \parallel \overline{CD}$. Since $\overline{AB} \cong \overline{CD}$, $ABCD$ is a parallelogram by Theorem 6-12.

B Given: $\overline{HI} \cong \overline{HK}$, $\overline{JI} \cong \overline{JK}$

Prove: $HIJK$ is a parallelogram.

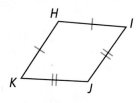

No. By Theorem 6-8, you need to show that both pairs of *opposite* sides are congruent, not consecutive sides.

Got It? **2.** Can you prove that the quadrilateral is a parallelogram based on the given information? Explain.

a. Given: $\overline{EF} \cong \overline{GD}$, $\overline{DE} \parallel \overline{FG}$
Prove: $DEFG$ is a parallelogram.

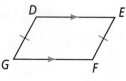

b. Given: $\angle ALN \cong \angle DNL$, $\angle ANL \cong \angle DLN$
Prove: $LAND$ is a parallelogram.

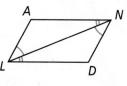

Think

As the arms of the lift move, what changes and what stays the same?
The angles the arms form with the ground and the platform change, but the lengths of the arms and the platform stay the same.

 **Problem 3** **Identifying Parallelograms**

Vehicle Lifts A truck sits on the platform of a vehicle lift. Two moving arms raise the platform until a mechanic can fit underneath. Why will the truck always remain parallel to the ground as it is lifted? Explain.

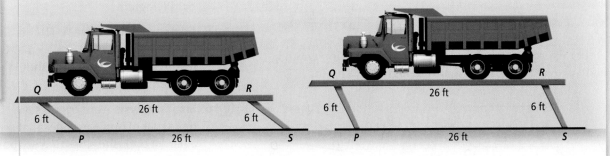

The angles of *PQRS* change as platform $\overline{QR}$ rises, but its side lengths remain the same. Both pairs of opposite sides are congruent, so *PQRS* is a parallelogram by Theorem 6-8. By the definition of a parallelogram, $\overline{PS} \parallel \overline{QR}$. Since the base of the lift $\overline{PS}$ lies along the ground, platform $\overline{QR}$, and therefore the truck, will always be parallel to the ground.

Got It? **3. Reasoning** What is the maximum height that the vehicle lift can elevate the truck? Explain.

 Concept Summary **Proving That a Quadrilateral Is a Parallelogram**

Method	Source	Diagram
Prove that both pairs of opposite sides are parallel.	Definition of parallelogram	
Prove that both pairs of opposite sides are congruent.	Theorem 6-8	
Prove that an angle is supplementary to both of its consecutive angles.	Theorem 6-9	75° 75° 105°
Prove that both pairs of opposite angles are congruent.	Theorem 6-10	
Prove that the diagonals bisect each other.	Theorem 6-11	
Prove that one pair of opposite sides is congruent and parallel.	Theorem 6-12	

Lesson Check

Do you know HOW?

1. For what value of y must $LMNP$ be a parallelogram?

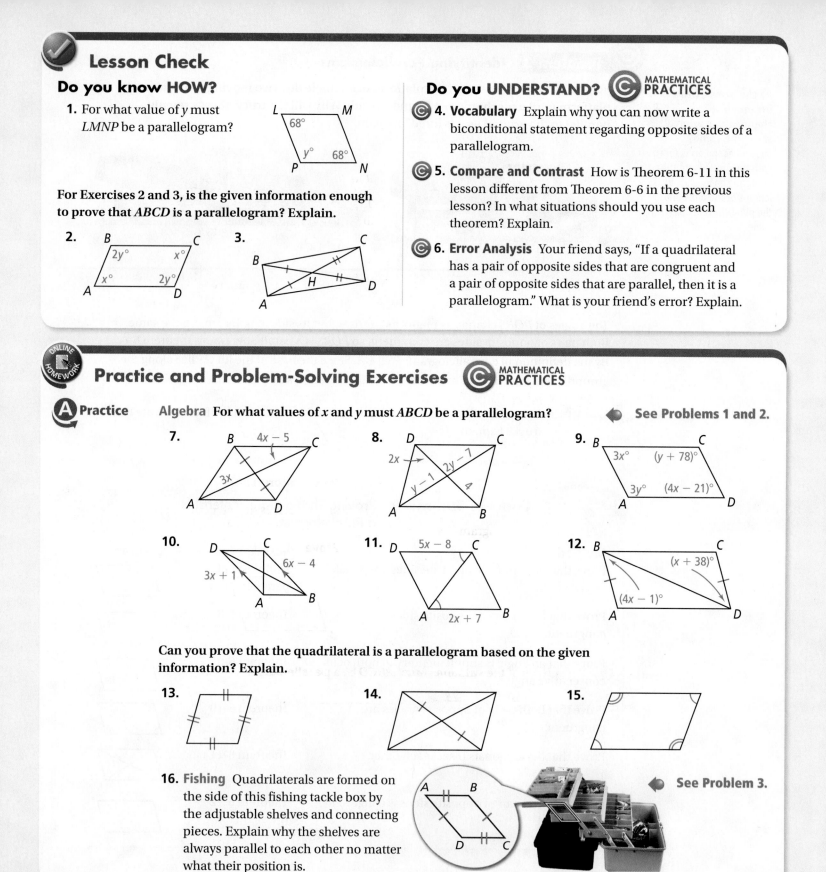

For Exercises 2 and 3, is the given information enough to prove that $ABCD$ is a parallelogram? Explain.

2.

3.

Do you UNDERSTAND?

MATHEMATICAL PRACTICES

4. **Vocabulary** Explain why you can now write a biconditional statement regarding opposite sides of a parallelogram.

5. **Compare and Contrast** How is Theorem 6-11 in this lesson different from Theorem 6-6 in the previous lesson? In what situations should you use each theorem? Explain.

6. **Error Analysis** Your friend says, "If a quadrilateral has a pair of opposite sides that are congruent and a pair of opposite sides that are parallel, then it is a parallelogram." What is your friend's error? Explain.

Practice and Problem-Solving Exercises

MATHEMATICAL PRACTICES

A **Practice**   **Algebra** For what values of x and y must $ABCD$ be a parallelogram?   **See Problems 1 and 2.**

7.

8.

9.

10.

11.

12.

Can you prove that the quadrilateral is a parallelogram based on the given information? Explain.

13.

14.

15.

16. **Fishing** Quadrilaterals are formed on the side of this fishing tackle box by the adjustable shelves and connecting pieces. Explain why the shelves are always parallel to each other no matter what their position is.   **See Problem 3.**

17. Writing Combine each of Theorems 6-3, 6-4, 6-5, and 6-6 with its converse from this lesson into biconditional statements.

18. Developing Proof Complete this two-column proof of Theorem 6-10.

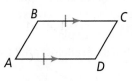

Given: $\angle A \cong \angle C$, $\angle B \cong \angle D$
Prove: $ABCD$ is a parallelogram.

Statements	Reasons
1) $x + y + x + y = 360$	**1)** The sum of the measures of the angles of a quadrilateral is 360.
2) $2(x + y) = 360$	**2) a.** __?__
3) $x + y = 180$	**3) b.** __?__
4) $\angle A$ and $\angle B$ are supplementary. $\angle A$ and $\angle D$ are supplementary.	**4)** Definition of supplementary
5) c. __?__ ∥ __?__ , __?__ ∥ __?__	**5) d.** __?__
6) $ABCD$ is a parallelogram.	**6) e.** __?__

19. Think About a Plan Prove Theorem 6-12.

Proof

Given: $\overline{BC} \parallel \overline{DA}$, $\overline{BC} \cong \overline{DA}$
Prove: $ABCD$ is a parallelogram.

• How can drawing diagonals help you?
• How can you use triangles in this proof?

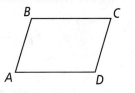

20. Prove Theorem 6-8.

Proof

Given: $\overline{AB} \cong \overline{CD}$, $\overline{BC} \cong \overline{DA}$
Prove: $ABCD$ is a parallelogram.

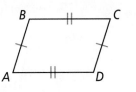

21. Prove Theorem 6-9.

Proof

Given: $\angle A$ is supplementary to $\angle B$.
$\angle A$ is supplementary to $\angle D$.
Prove: $ABCD$ is a parallelogram.

Algebra For what values of the variables must $ABCD$ be a parallelogram?

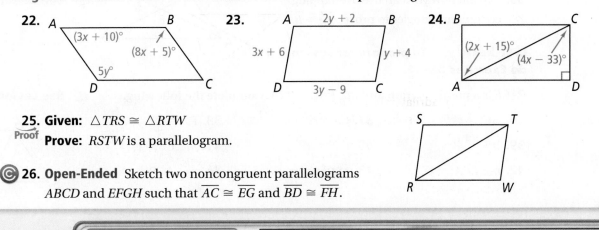

22.

23.

24.

25. Given: $\triangle TRS \cong \triangle RTW$

Proof **Prove:** $RSTW$ is a parallelogram.

26. Open-Ended Sketch two noncongruent parallelograms $ABCD$ and $EFGH$ such that $\overline{AC} \cong \overline{EG}$ and $\overline{BD} \cong \overline{FH}$.

27. Construction In the figure at the right, point D is constructed by
Proof drawing two arcs. One has center C and radius AB. The other has center
B and radius AC. Prove that $\overline{AM}$ is a median of $\triangle ABC$.

28. Probability If two opposite angles of a quadrilateral measure 120
and the measures of the other angles are multiples of 10, what is the
probability that the quadrilateral is a parallelogram?

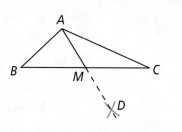

Standardized Test Prep

SAT/ACT

29. From which set of information can you conclude that $RSTW$ is a parallelogram?

   Ⓐ $\overline{RS} \parallel \overline{WT}, \overline{RS} \cong \overline{ST}$ Ⓒ $\overline{RS} \cong \overline{ST}, \overline{RW} \cong \overline{WT}$

   Ⓑ $\overline{RS} \parallel \overline{WT}, \overline{ST} \cong \overline{RW}$ Ⓓ $\overline{RZ} \cong \overline{TZ}, \overline{SZ} \cong \overline{WZ}$

Short Response

30. Write a proof using the diagram.

   Given: $\triangle NRJ \cong \triangle CPT, \overline{JN} \parallel \overline{CT}$

   Prove: $JNTC$ is a parallelogram.

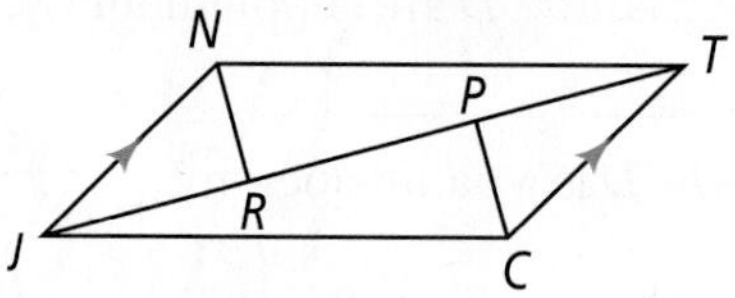

Extended Response

31. Use the figure at the right.
   a. Write an equation and solve for x.
   b. Is $\overline{AF} \parallel \overline{DE}$? Explain.
   c. Is $BDEF$ a parallelogram? Explain.

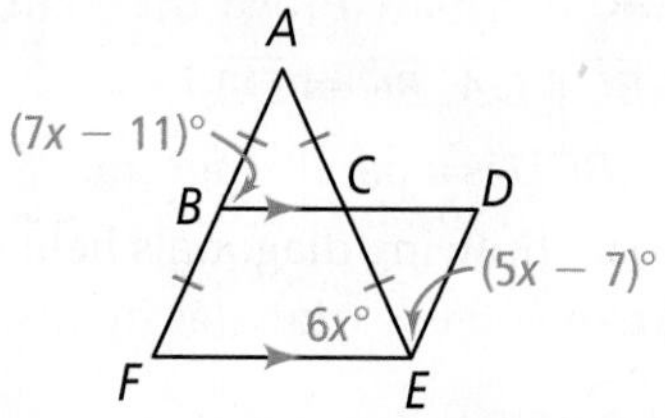

Mixed Review

Algebra **Find the value of each variable in each parallelogram.**   ◀ **See Lesson 6-2.**

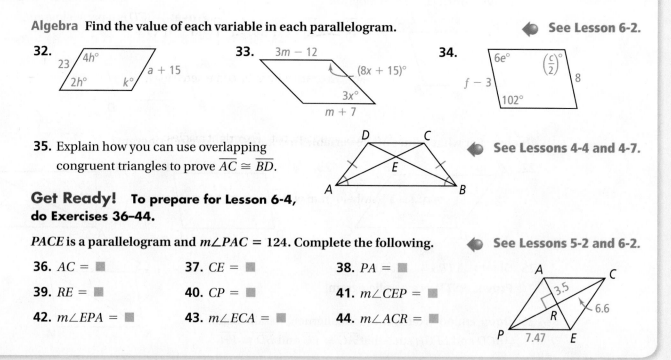

32.

33.

34.

35. Explain how you can use overlapping
congruent triangles to prove $\overline{AC} \cong \overline{BD}$.   ◀ **See Lessons 4-4 and 4-7.**

Get Ready!   **To prepare for Lesson 6-4,
do Exercises 36–44.**

$PACE$ is a parallelogram and $m\angle PAC = 124$. Complete the following.   ◀ **See Lessons 5-2 and 6-2.**

36. $AC = \blacksquare$ **37.** $CE = \blacksquare$ **38.** $PA = \blacksquare$

39. $RE = \blacksquare$ **40.** $CP = \blacksquare$ **41.** $m\angle CEP = \blacksquare$

42. $m\angle EPA = \blacksquare$ **43.** $m\angle ECA = \blacksquare$ **44.** $m\angle ACR = \blacksquare$

Properties of Rhombuses, Rectangles, and Squares

© **Content Standards**
G.CO.11 Prove theorems about parallelograms . . . rectangles are parallelograms with congruent diagonals.
Also G.SRT.5

Objectives To define and classify special types of parallelograms
To use properties of diagonals of rhombuses and rectangles

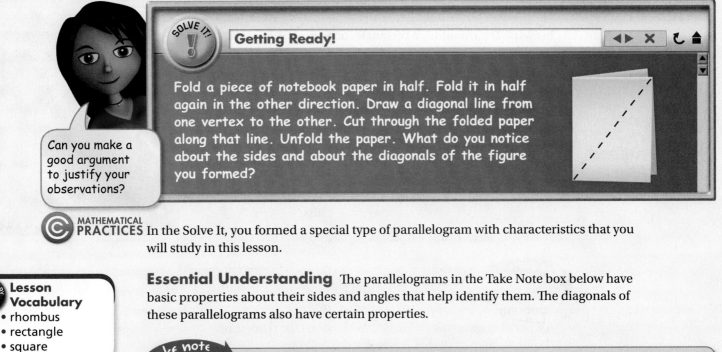

Can you make a good argument to justify your observations?

Getting Ready!

Fold a piece of notebook paper in half. Fold it in half again in the other direction. Draw a diagonal line from one vertex to the other. Cut through the folded paper along that line. Unfold the paper. What do you notice about the sides and about the diagonals of the figure you formed?

© MATHEMATICAL **PRACTICES** In the Solve It, you formed a special type of parallelogram with characteristics that you will study in this lesson.

Lesson Vocabulary
• rhombus
• rectangle
• square

Essential Understanding The parallelograms in the Take Note box below have basic properties about their sides and angles that help identify them. The diagonals of these parallelograms also have certain properties.

take note

Key Concept Special Parallelograms

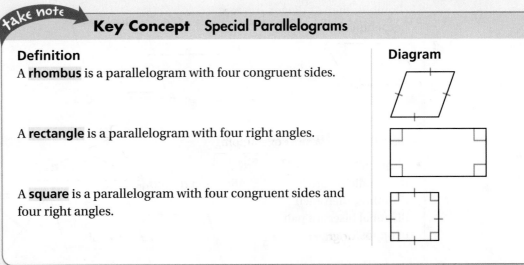

Definition

A **rhombus** is a parallelogram with four congruent sides.

A **rectangle** is a parallelogram with four right angles.

A **square** is a parallelogram with four congruent sides and four right angles.

Diagram

The Venn diagram at the right shows the relationships among special parallelograms.

Special Parallelograms

Rhombuses Squares Rectangles

Think

How do you decide whether *ABCD* is a rhombus, rectangle, or square?
Use the definitions of *rhombus*, *rectangle*, and *square* along with the markings on the figure.

Problem 1 **Classifying Special Parallelograms**

Is ▱*ABCD* a rhombus, a rectangle, or a square? Explain.

▱*ABCD* is a rectangle. Opposite angles of a parallelogram are congruent so $m\angle D$ is 90. By the Same-Side Interior Angles Theorem, $m\angle A = 90$ and $m\angle C = 90$. Since ▱*ABCD* has four right angles, it is a rectangle. You cannot conclude that *ABCD* is a square because you do not know its side lengths.

Got It? **1.** Is ▱*EFGH* a rhombus, a rectangle, or a square? Explain.

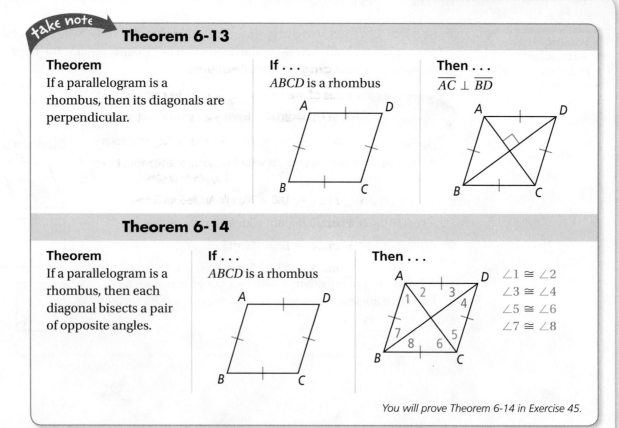

Theorem 6-13

Theorem	**If . . .**	**Then . . .**
If a parallelogram is a rhombus, then its diagonals are perpendicular.	*ABCD* is a rhombus	$\overline{AC} \perp \overline{BD}$

Theorem 6-14

Theorem	**If . . .**	**Then . . .**
If a parallelogram is a rhombus, then each diagonal bisects a pair of opposite angles.	*ABCD* is a rhombus	$\angle 1 \cong \angle 2$ $\angle 3 \cong \angle 4$ $\angle 5 \cong \angle 6$ $\angle 7 \cong \angle 8$

You will prove Theorem 6-14 in Exercise 45.

Proof **Proof of Theorem 6-13**

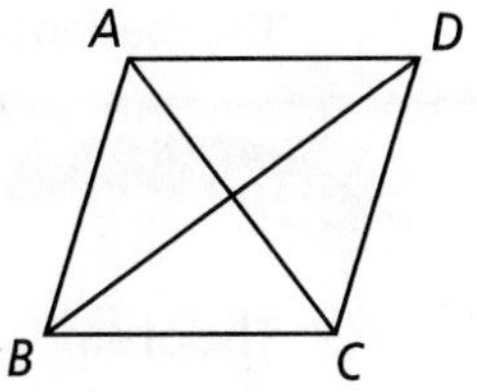

Given: $ABCD$ is a rhombus.

Prove: The diagonals of $ABCD$ are perpendicular.

Statements	Reasons
1) A and C are equidistant from B and D; B and D are equidistant from A and C.	**1)** All sides of a rhombus are $\cong$.
2) A and C are on the perpendicular bisector of $\overline{BD}$; B and D are on the perpendicular bisector of $\overline{AC}$.	**2)** Converse of the Perpendicular Bisector Theorem
3) $\overline{AC} \perp \overline{BD}$	**3)** Through two points, there is one unique line perpendicular to a given line.

You can use Theorems 6-13 and 6-14 to find angle measures in a rhombus.

© **Problem 2** **Finding Angle Measures**

Think

How are the numbered angles formed?
The angles are formed by diagonals. Use what you know about the diagonals of a rhombus to find the angle measures.

What are the measures of the numbered angles in rhombus $ABCD$?

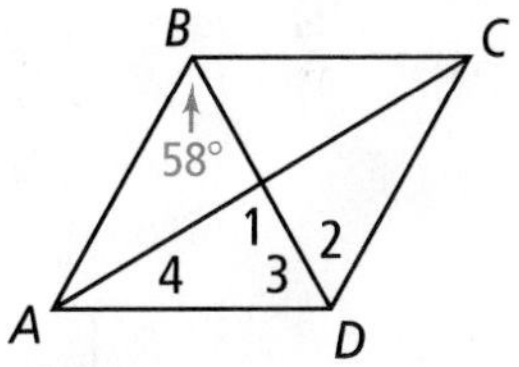

$m\angle 1 = 90$	The diagonals of a rhombus are $\perp$.
$m\angle 2 = 58$	Alternate Interior Angles Theorem
$m\angle 3 = 58$	Each diagonal of a rhombus bisects a pair of opposite angles.
$m\angle 1 + m\angle 3 + m\angle 4 = 180$	Triangle Angle-Sum Theorem
$90 + 58 + m\angle 4 = 180$	Substitute.
$148 + m\angle 4 = 180$	Simplify.
$m\angle 4 = 32$	Subtract 148 from each side.

✔ **Got It?** **2.** What are the measures of the numbered angles in rhombus $PQRS$?

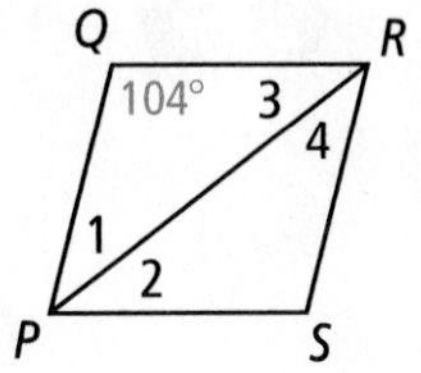

The diagonals of a rectangle also have a special property.

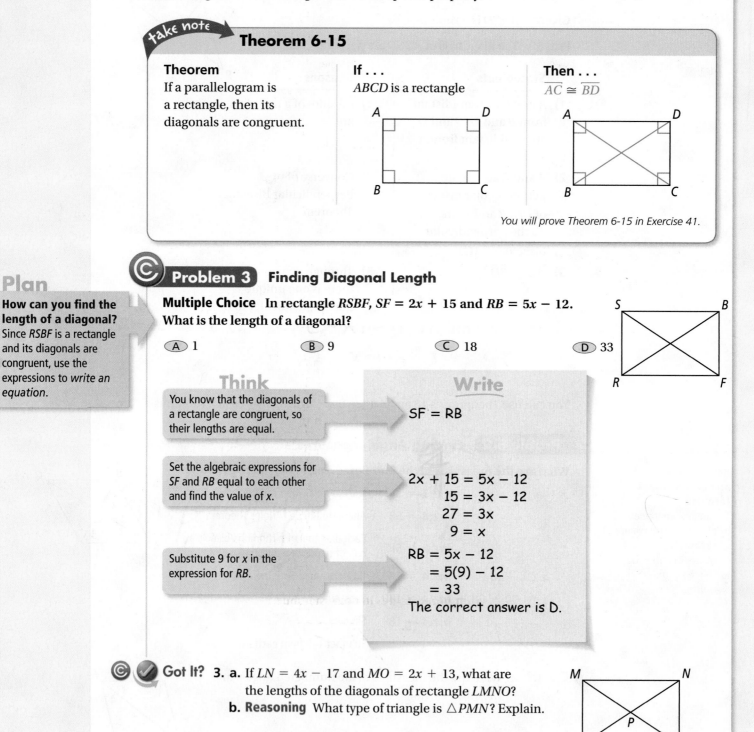

take note

Theorem 6-15

Theorem	**If . . .**	**Then . . .**
If a parallelogram is a rectangle, then its diagonals are congruent.	$ABCD$ is a rectangle	$\overline{AC} \cong \overline{BD}$

You will prove Theorem 6-15 in Exercise 41.

Plan

How can you find the length of a diagonal?
Since *RSBF* is a rectangle and its diagonals are congruent, use the expressions to *write an equation*.

Problem 3 Finding Diagonal Length

Multiple Choice In rectangle *RSBF*, $SF = 2x + 15$ and $RB = 5x - 12$. What is the length of a diagonal?

   Ⓐ 1   Ⓑ 9   Ⓒ 18   Ⓓ 33

Think

You know that the diagonals of a rectangle are congruent, so their lengths are equal.

Set the algebraic expressions for *SF* and *RB* equal to each other and find the value of *x*.

Substitute 9 for *x* in the expression for *RB*.

Write

$SF = RB$

$2x + 15 = 5x - 12$
$15 = 3x - 12$
$27 = 3x$
$9 = x$

$RB = 5x - 12$
$= 5(9) - 12$
$= 33$

The correct answer is D.

Got It? **3. a.** If $LN = 4x - 17$ and $MO = 2x + 13$, what are the lengths of the diagonals of rectangle *LMNO*?

b. Reasoning What type of triangle is $\triangle PMN$? Explain.

Lesson Check

Do you know HOW?

Is each parallelogram a rhombus, rectangle, or square? Explain.

1.
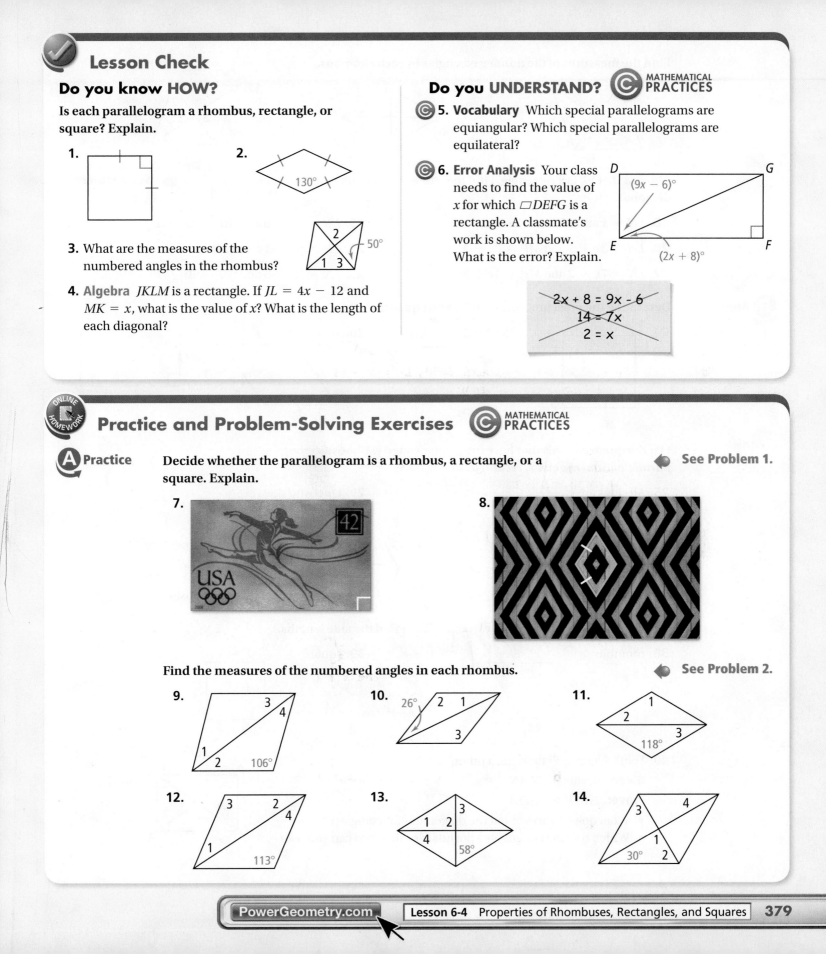

2.

3. What are the measures of the numbered angles in the rhombus?

4. **Algebra** $JKLM$ is a rectangle. If $JL = 4x - 12$ and $MK = x$, what is the value of x? What is the length of each diagonal?

Do you UNDERSTAND? MATHEMATICAL PRACTICES

5. **Vocabulary** Which special parallelograms are equiangular? Which special parallelograms are equilateral?

6. **Error Analysis** Your class needs to find the value of x for which ▱$DEFG$ is a rectangle. A classmate's work is shown below. What is the error? Explain.

$$2x + 8 = 9x - 6$$
$$14 = 7x$$
$$2 = x$$

Practice and Problem-Solving Exercises MATHEMATICAL PRACTICES

A Practice

Decide whether the parallelogram is a rhombus, a rectangle, or a square. Explain.

See Problem 1.

7.

8.

Find the measures of the numbered angles in each rhombus.

See Problem 2.

9.   106°

10.   26°

11.   118°

12.   113°

13.   58°

14.   30°

Find the measures of the numbered angles in each rhombus.

15.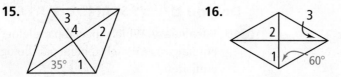

16.

17.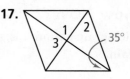

Algebra *LMNP* is a rectangle. Find the value of *x* and the length of each diagonal.

◀ **See Problem 3.**

18. $LN = x$ and $MP = 2x - 4$

19. $LN = 5x - 8$ and $MP = 2x + 1$

20. $LN = 3x + 1$ and $MP = 8x - 4$

21. $LN = 9x - 14$ and $MP = 7x + 4$

22. $LN = 7x - 2$ and $MP = 4x + 3$

23. $LN = 3x + 5$ and $MP = 9x - 10$

Ⓑ Apply

Determine the most precise name for each quadrilateral.

24.   **25.**   **26.**   **27.**

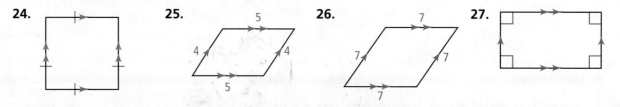

List the quadrilaterals that have the given property. Choose among *parallelogram, rhombus, rectangle,* **and** *square.*

28. All sides are ≅.

29. Opposite sides are ≅.

30. Opposite sides are ‖.

31. Opposite ∠ are ≅.

32. All ∠ are right ∠.

33. Consecutive ∠ are supplementary.

34. Diagonals bisect each other.

35. Diagonals are ≅.

36. Diagonals are ⊥.

37. Each diagonal bisects opposite ∠.

Algebra Find the values of the variables. Then find the side lengths.

38. rhombus

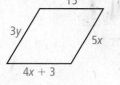

39. square

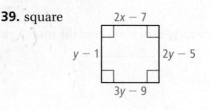

Ⓒ 40. Think About a Plan Write a proof.

Proof **Given:** Rectangle *PLAN*

Prove: $\triangle LTP \cong \triangle NTA$

• What do you know about the diagonals of rectangles?

• Which triangle congruence postulate or theorem can you use?

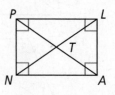

41. **Developing Proof** Complete the flow proof of Theorem 6-15.

Given: $ABCD$ is a rectangle.

Prove: $\overline{AC} \cong \overline{BD}$

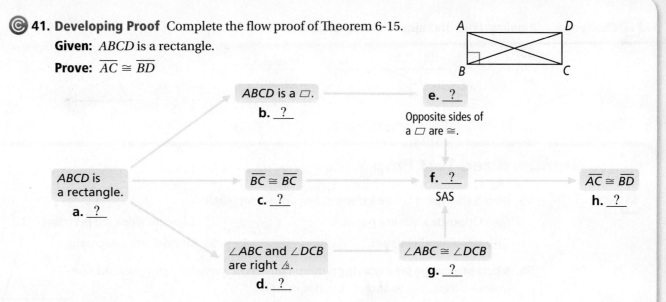

42. $RZ = 2x + 5$, $SW = 5x - 20$

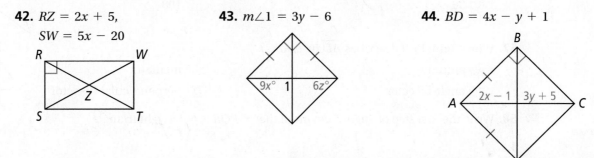

43. $m\angle 1 = 3y - 6$

44. $BD = 4x - y + 1$

Algebra Find the value(s) of the variable(s) for each parallelogram.

45. Prove Theorem 6-14.

Proof **Given:** $ABCD$ is a rhombus.

Prove: $\overline{AC}$ bisects $\angle BAD$ and $\angle BCD$.

46. **Writing** Summarize the properties of squares that follow from a square being (a) a parallelogram, (b) a rhombus, and (c) a rectangle.

47. **Algebra** Find the angle measures and the side lengths of the rhombus at the right.

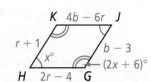

48. **Open-Ended** On graph paper, draw a parallelogram that is neither a rectangle nor a rhombus.

Algebra $ABCD$ is a rectangle. Find the length of each diagonal.

49. $AC = 2(x - 3)$ and $BD = x + 5$

50. $AC = 2(5a + 1)$ and $BD = 2(a + 1)$

51. $AC = \dfrac{3y}{5}$ and $BD = 3y - 4$

52. $AC = \dfrac{3c}{9}$ and $BD = 4 - c$

Algebra Find the value of *x* in the rhombus.

53.

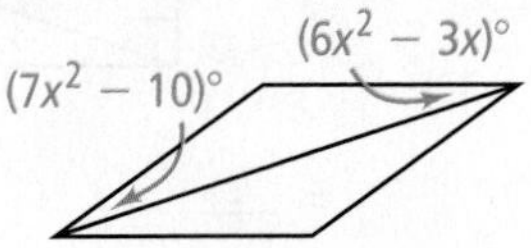

$(7x^2 - 10)°$ $(6x^2 - 3x)°$

54.

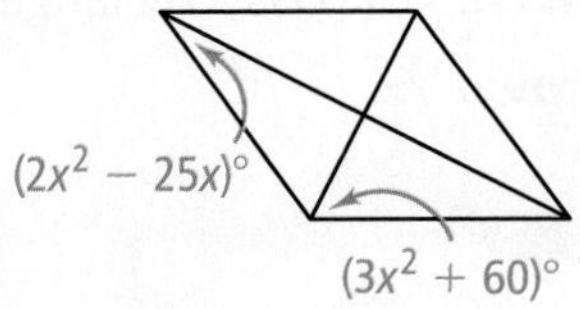

$(2x^2 - 25x)°$ $(3x^2 + 60)°$

Standardized Test Prep

SAT/ACT

55. Which statement is true for some, but not all, rectangles?

Ⓐ Opposite sides are parallel.

Ⓒ Adjacent sides are perpendicular.

Ⓑ It is a parallelogram.

Ⓓ All sides are congruent.

56. A part of a design for a quilting pattern consists of a regular pentagon and five isosceles triangles, as shown. What is $m\angle 1$?

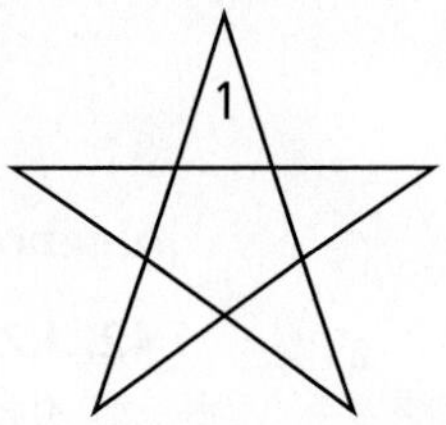

Ⓕ 18

Ⓗ 72

Ⓖ 36

Ⓘ 108

57. Which term best describes $\overline{AD}$ in $\triangle ABC$?

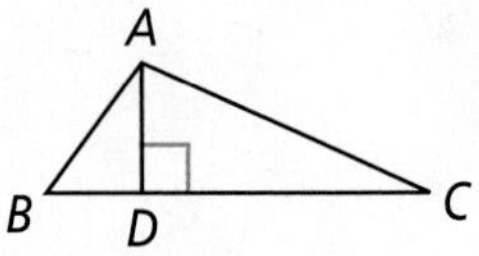

Ⓐ altitude

Ⓒ median

Ⓑ angle bisector

Ⓓ perpendicular bisector

Short Response

58. Write the first step of an indirect proof that $\triangle PQR$ is not a right triangle.

Mixed Review

Can you conclude that the quadrilateral is a parallelogram? Explain.

◀ See Lesson 6-3.

59.

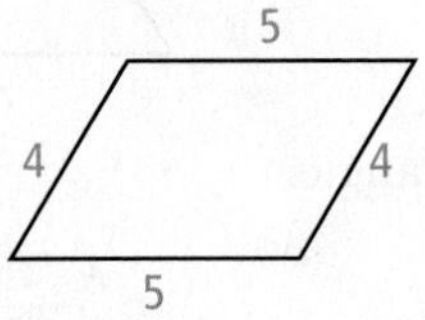

5

4   4

5

60.

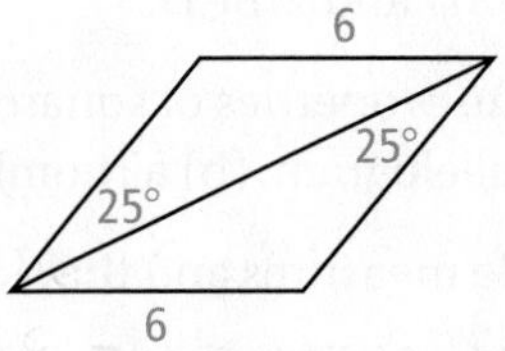

6

25°

25°

6

61.

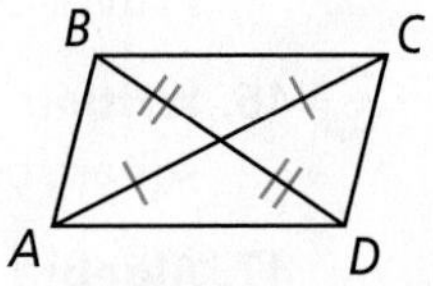

B   C

A   D

In $\triangle PQR$, points *S*, *T*, and *U* are midpoints. Complete each statement.

◀ See Lesson 5-1.

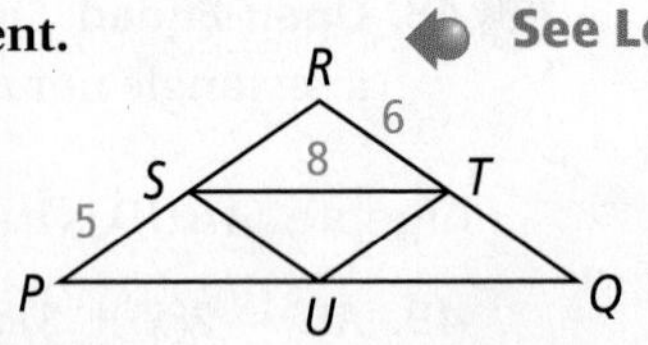

62. $TQ = \underline{\ ?\ }$

63. $PQ = \underline{\ ?\ }$

64. $TU = \underline{\ ?\ }$

65. $\overline{SU} \parallel \underline{\ ?\ }$

66. $\overline{TU} \parallel \underline{\ ?\ }$

67. $\overline{PQ} \parallel \underline{\ ?\ }$

R

6

S   8   T

5

P   U   Q

Get Ready! To prepare for Lesson 6-5, do Exercises 68 and 69.

68. Draw a rhombus that is not a square.

◀ See Lesson 6-4.

69. Draw a rectangle that is not a square.

Conditions for Rhombuses, Rectangles, and Squares

© **Content Standards**
G.CO.11 Prove theorems about parallelograms . . . rectangles are parallelograms with congruent diagonals.
Also G.SRT.5

Objective To determine whether a parallelogram is a rhombus or rectangle

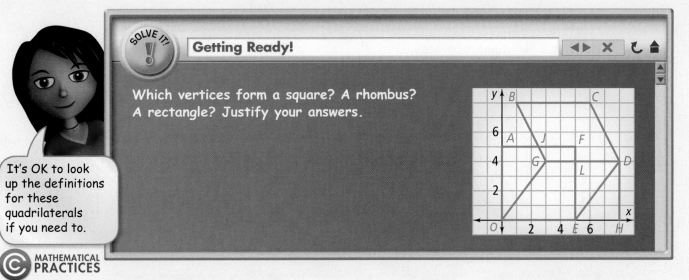

It's OK to look up the definitions for these quadrilaterals if you need to.

MATHEMATICAL PRACTICES

Essential Understanding You can determine whether a parallelogram is a rhombus or a rectangle based on the properties of its diagonals.

Dynamic Activity
Special Parallelograms

take note

Theorem 6-16

Theorem	**If . . .**	**Then . . .**
If the diagonals of a parallelogram are perpendicular, then the parallelogram is a rhombus.	$ABCD$ is a $\square$ and $\overline{AC} \perp \overline{BD}$	$ABCD$ is a rhombus

Proof of Theorem 6-16

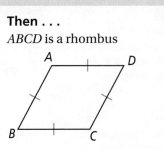

Given: $ABCD$ is a parallelogram, $\overline{AC} \perp \overline{BD}$

Prove: $ABCD$ is a rhombus.

Since $ABCD$ is a parallelogram, $\overline{AC}$ and $\overline{BD}$ bisect each other, so $\overline{BE} \cong \overline{DE}$. Since $\overline{AC} \perp \overline{BD}$, $\angle AED$ and $\angle AEB$ are congruent right angles. By the Reflexive Property of Congruence, $\overline{AE} \cong \overline{AE}$. So $\triangle AEB \cong \triangle AED$ by SAS. Corresponding parts of congruent triangles are congruent, so $\overline{AB} \cong \overline{AD}$. Since opposite sides of a parallelogram are congruent, $\overline{AB} \cong \overline{DC} \cong \overline{BC} \cong \overline{AD}$. By definition, $ABCD$ is a rhombus.

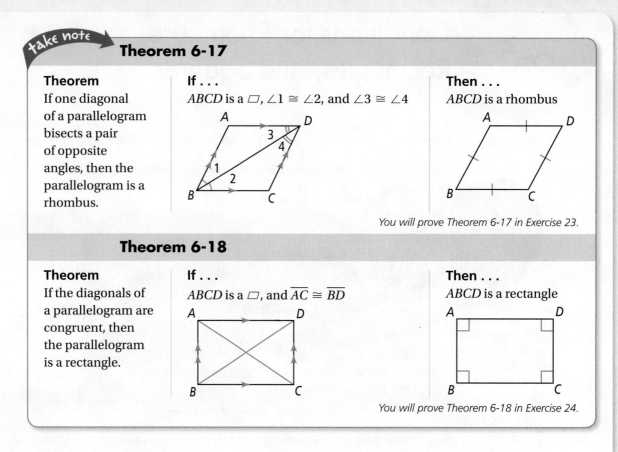

Theorem 6-17

Theorem	If . . .	Then . . .
If one diagonal of a parallelogram bisects a pair of opposite angles, then the parallelogram is a rhombus.	$ABCD$ is a $\square$, $\angle 1 \cong \angle 2$, and $\angle 3 \cong \angle 4$	$ABCD$ is a rhombus

You will prove Theorem 6-17 in Exercise 23.

Theorem 6-18

Theorem	If . . .	Then . . .
If the diagonals of a parallelogram are congruent, then the parallelogram is a rectangle.	$ABCD$ is a $\square$, and $\overline{AC} \cong \overline{BD}$	$ABCD$ is a rectangle

You will prove Theorem 6-18 in Exercise 24.

You can use Theorems 6-16, 6-17, and 6-18 to classify parallelograms. Notice that if a parallelogram is both a rectangle and a rhombus, then it is a square.

Problem 1 Identifying Special Parallelograms

Think

How can you determine whether a figure is a special parallelogram?
See if you can satisfy every condition of a definition or theorem about rhombuses, rectangles, or squares.

Can you conclude that the parallelogram is a rhombus, a rectangle, or a square? Explain.

A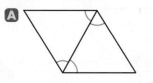

B

Yes. A diagonal bisects two angles. By Theorem 6-17, this parallelogram is a rhombus.

Yes. The diagonals are congruent, so by Theorem 6-18, this parallelogram is a rectangle. The diagonals are perpendicular, so by Theorem 6-16, it is a rhombus. Therefore, this parallelogram is a square.

Got It? 1. **a.** A parallelogram has angle measures of 20, 160, 20, and 160. Can you conclude that it is a rhombus, a rectangle, or a square? Explain.

b. Reasoning Suppose the diagonals of a quadrilateral bisect each other. Can you conclude that it is a rhombus, a rectangle, or a square? Explain.

Problem 2 | Using Properties of Special Parallelograms

Algebra For what value of *x* is ▱*ABCD* a rhombus?

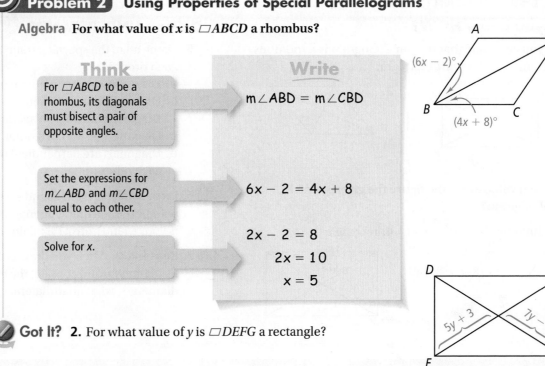

$(6x - 2)°$

$(4x + 8)°$

Think

For ▱*ABCD* to be a rhombus, its diagonals must bisect a pair of opposite angles.

Set the expressions for *m*∠*ABD* and *m*∠*CBD* equal to each other.

Solve for *x*.

Write

$m∠ABD = m∠CBD$

$6x - 2 = 4x + 8$

$2x - 2 = 8$

$2x = 10$

$x = 5$

Got It? **2.** For what value of *y* is ▱*DEFG* a rectangle?

$5y + 3$ $7y - 5$

Problem 3 | Using Properties of Parallelograms

Community Service Builders use properties of diagonals to "square up" rectangular shapes like building frames and playing-field boundaries. Suppose you are on the volunteer building team at the right. You are helping to lay out a rectangular patio for a youth center. How can you use properties of diagonals to locate the four corners?

You can use two theorems.

* Theorem 6-11: If the diagonals of a quadrilateral bisect each other, then the quadrilateral is a parallelogram.

* Theorem 6-18: If the diagonals of a parallelogram are congruent, then the parallelogram is a rectangle.

Step 1 Cut two pieces of rope that will be the diagonals of the foundation rectangle. Cut them the same length because of Theorem 6-18.

Step 2 Join the two pieces of rope at their midpoints because of Theorem 6-11.

Step 3 Pull the ropes straight and taut. The ends of the ropes will be the corners of a rectangle.

Think

Is there only one rectangle that can be formed by pulling the ropes taut?

No, you can change the shape of the rectangle. Have two of the people move closer together. Then the other two people move until the ropes are taut again.

Got It? **3.** Can you adapt this method slightly to stake off a square play area? Explain.

Lesson Check

Do you know HOW?

Can you conclude that the parallelogram is a rhombus, a rectangle, or a square? Explain.

1.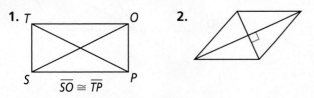
$\overline{SO} \cong \overline{TP}$

2.

For what value of *x* is the figure the given special parallelogram?

3. rhombus

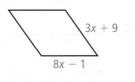
3x + 9
8x − 1

4. rectangle

3x − 5   x + 1

Do you UNDERSTAND? ⓒ MATHEMATICAL PRACTICES

5. Name all of the special parallelograms that have each property.
 a. Diagonals are perpendicular.
 b. Diagonals are congruent.
 c. Diagonals are angle bisectors.
 d. Diagonals bisect each other.
 e. Diagonals are perpendicular bisectors of each other.

ⓒ **6. Error Analysis** Your friend says, "A parallelogram with perpendicular diagonals is a rectangle." What is your friend's error? Explain.

ⓒ **7. Reasoning** When you draw a circle and two of its diameters and connect the endpoints of the diameters, what quadrilateral do you get? Explain.

Practice and Problem-Solving Exercises ⓒ MATHEMATICAL PRACTICES

Ⓐ Practice Can you conclude that the parallelogram is a rhombus, a rectangle, or a square? Explain.

◀ See Problem 1.

8.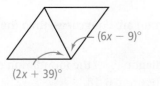

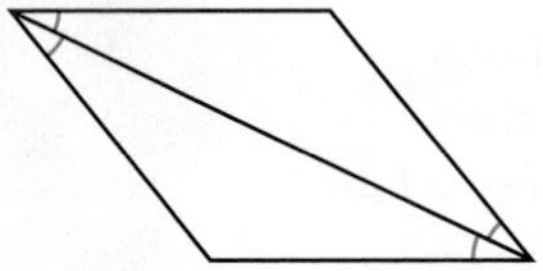

9.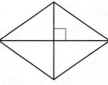

10.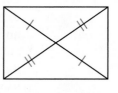

For what value of *x* is the figure the given special parallelogram?

◀ See Problem 2.

11. rhombus

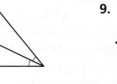
(6x − 9)°
(2x + 39)°

12. rectangle

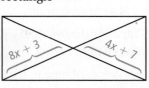
8x + 3   4x + 7

13. rectangle

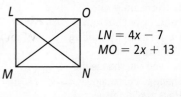
LN = 4x − 7
MO = 2x + 13

14. Carpentry A carpenter is building a bookcase. How can the carpenter use a tape measure to check that the bookshelf is rectangular? Justify your answer and name any theorems used.

◀ See Problem 3.

B Apply **STEM** **15. Hardware** You can use a simple device called a turnbuckle to "square up" structures that are parallelograms. For the gate pictured at the right, you tighten or loosen the turnbuckle on the diagonal cable so that the rectangular frame will keep the shape of a parallelogram when it sags. What are two ways you can make sure that the turnbuckle works? Explain.

C 16. Reasoning Suppose the diagonals of a parallelogram are both perpendicular and congruent. What type of special quadrilateral is it? Explain your reasoning.

Algebra **For what value of x is the figure the given special parallelogram?**

17. rectangle

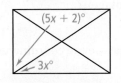
$(5x + 2)°$
$3x°$

18. rhombus

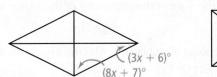
$(3x + 6)°$
$(8x + 7)°$

19. rectangle

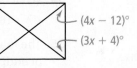
$(4x - 12)°$
$(3x + 4)°$

C Open-Ended **Given two segments with lengths a and b ($a \neq b$), what special parallelograms meet the given conditions? Show each sketch.**

20. Both diagonals have length a.

21. The two diagonals have lengths a and b.

22. One diagonal has length a, and one side of the quadrilateral has length b.

23. Prove Theorem 6-17.
Proof **Given:** $ABCD$ is a parallelogram.
$\overrightarrow{AC}$ bisects $\angle BAD$ and $\angle BCD$.
Prove: $ABCD$ is a rhombus.

24. Prove Theorem 6-18.
Proof **Given:** $\square ABCD$, $\overline{AC} \cong \overline{BD}$
Prove: $ABCD$ is a rectangle.

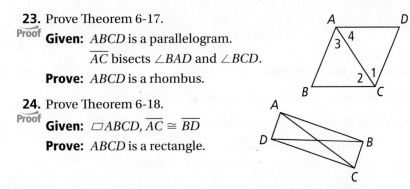

C Think About a Plan **Explain how to construct each figure given its diagonals.**
- What do you know about the diagonals of each figure?
- How can you apply constructions to what you know about the diagonals?

25. parallelogram

26. rectangle

27. rhombus

C Challenge **Determine whether the quadrilateral can be a parallelogram. Explain.**

28. The diagonals are congruent, but the quadrilateral has no right angles.

29. Each diagonal is 3 cm long and two opposite sides are 2 cm long.

30. Two opposite angles are right angles, but the quadrilateral is not a rectangle.

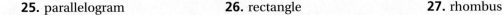

31. In Theorem 6-17, replace "a pair of opposite angles" with "one angle." Write a
Proof paragraph that proves this new statement to be true, or give a counterexample to
prove it to be false.

Standardized Test Prep

SAT/ACT

32. Each diagonal of a quadrilateral bisects a pair of opposite angles of the
quadrilateral. What is the most precise name for the quadrilateral?

     (A) parallelogram   (B) rhombus   (C) rectangle   (D) not enough information

33. Given a triangle with side lengths 7 and 11, which value could NOT be the length of
the third side of the triangle?

     (F) 13   (G) 7   (H) 5   (I) 2

34. What is the sum of the measures of the exterior angles in a pentagon?

     (A) 180   (B) 360   (C) 540   (D) 108

Short
Response

35. The midpoint of $\overline{PQ}$ is $(-1, 4)$. One endpoint is $P(-7, 10)$. What are the coordinates
of endpoint Q? Explain your work.

Mixed Review

Find the measures of the numbered angles in each rhombus.

See Lesson 6-4.

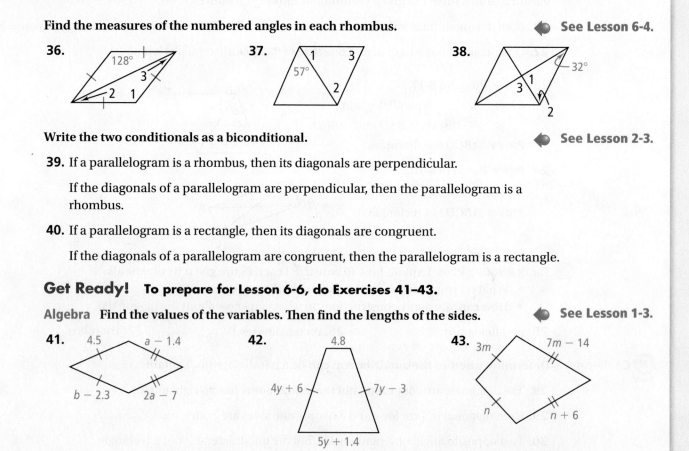

36.

37.

38.

Write the two conditionals as a biconditional.

See Lesson 2-3.

39. If a parallelogram is a rhombus, then its diagonals are perpendicular.

If the diagonals of a parallelogram are perpendicular, then the parallelogram is a
rhombus.

40. If a parallelogram is a rectangle, then its diagonals are congruent.

If the diagonals of a parallelogram are congruent, then the parallelogram is a rectangle.

Get Ready! To prepare for Lesson 6-6, do Exercises 41–43.

Algebra Find the values of the variables. Then find the lengths of the sides.

See Lesson 1-3.

41.

42.

43.

Trapezoids and Kites

© **Content Standard**
G.SRT.5 Use congruence . . . criteria to solve problems and prove relationships in geometric figures.

Objective   To verify and use properties of trapezoids and kites

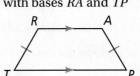

Getting Ready!

Two isosceles triangles form the figure at the right. Each white segment is a midsegment of a triangle. What can you determine about the angles in the orange region? In the green region? Explain.

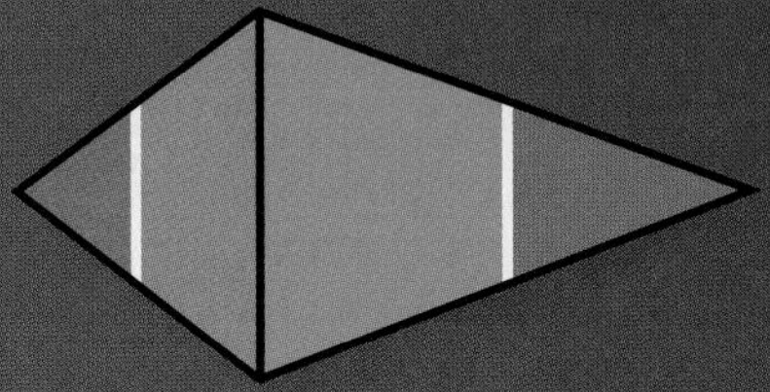

Make a sketch and number the angles to help make sense of the problem.

MATHEMATICAL PRACTICES   In the Solve It, the orange and green regions are trapezoids. The entire figure is a kite. In this lesson, you will learn about these special quadrilaterals that are not parallelograms.

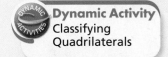

Dynamic Activity
Classifying Quadrilaterals

Lesson Vocabulary
• trapezoid
• base
• leg
• base angle
• isosceles trapezoid
• midsegment of a trapezoid
• kite

Essential Understanding   The angles, sides, and diagonals of a trapezoid have certain properties.

A **trapezoid** is a quadrilateral with exactly one pair of parallel sides. The parallel sides of a trapezoid are called **bases**. The nonparallel sides are called **legs**. The two angles that share a base of a trapezoid are called **base angles**. A trapezoid has two pairs of base angles.

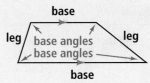

An **isosceles trapezoid** is a trapezoid with legs that are congruent. *ABCD* at the right is an isosceles trapezoid. The angles of an isosceles trapezoid have some unique properties.

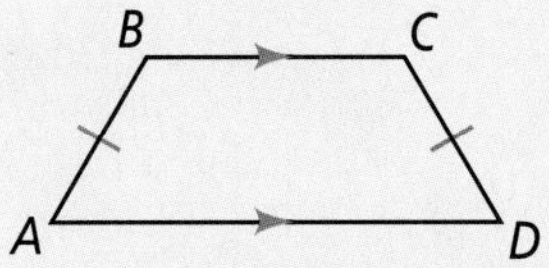

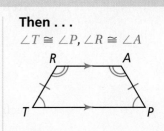

take note

Theorem 6-19

Theorem	**If . . .**	**Then . . .**
If a quadrilateral is an isosceles trapezoid, then each pair of base angles is congruent.	*TRAP* is an isosceles trapezoid with bases $\overline{RA}$ and $\overline{TP}$	$\angle T \cong \angle P$, $\angle R \cong \angle A$

You will prove Theorem 6-19 in Exercise 45.

Problem 1 Finding Angle Measures in Trapezoids

CDEF is an isosceles trapezoid and $m\angle C = 65$. What are $m\angle D$, $m\angle E$, and $m\angle F$?

$m\angle C + m\angle D = 180$	Two angles that form same-side interior angles along one leg are supplementary.
$65 + m\angle D = 180$	Substitute.
$m\angle D = 115$	Subtract 65 from each side.

Since each pair of base angles of an isosceles trapezoid is congruent, $m\angle C = m\angle F = 65$ and $m\angle D = m\angle E = 115$.

Got It? **1. a.** In the diagram, *PQRS* is an isosceles trapezoid and $m\angle R = 106$. What are $m\angle P$, $m\angle Q$, and $m\angle S$?

   b. Reasoning In Problem 1, if *CDEF* were not an isosceles trapezoid, would $\angle C$ and $\angle D$ still be supplementary? Explain.

Problem 2 Finding Angle Measures in Isosceles Trapezoids

Paper Fans The second ring of the paper fan shown at the right consists of 20 congruent isosceles trapezoids that appear to form circles. What are the measures of the base angles of these trapezoids?

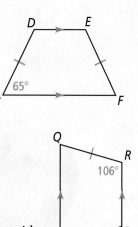

Step 1 Find the measure of each angle at the center of the fan. This is the measure of the vertex angle of an isosceles triangle.

$$m\angle 1 = \frac{360}{20} = 18$$

Step 2 Find the measure of each acute base angle of an isosceles triangle.

$18 + x + x = 180$	Triangle Angle-Sum Theorem
$18 + 2x = 180$	Combine like terms.
$2x = 162$	Subtract 18 from each side.
$x = 81$	Divide each side by 2.

Step 3 Find the measure of each obtuse base angle of the isosceles trapezoid.

$81 + y = 180$	Two angles that form same-side interior angles along one leg are supplementary.
$y = 99$	Subtract 81 from each side.

Each acute base angle measures 81. Each obtuse base angle measures 99.

Got It? **2.** A fan like the one in Problem 2 has 15 angles meeting at the center. What are the measures of the base angles of the trapezoids in its second ring?

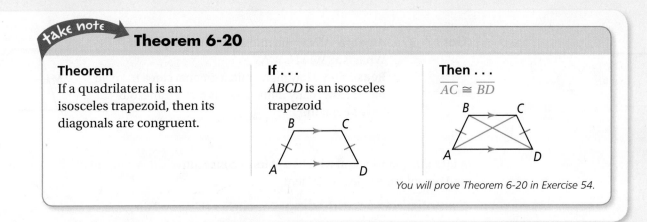

Theorem 6-20

Theorem
If a quadrilateral is an isosceles trapezoid, then its diagonals are congruent.

If . . .
ABCD is an isosceles trapezoid

Then . . .
$\overline{AC} \cong \overline{BD}$

You will prove Theorem 6-20 in Exercise 54.

In Lesson 5-1, you learned about midsegments of triangles. Trapezoids also have midsegments. The **midsegment of a trapezoid** is the segment that joins the midpoints of its legs. The midsegment has two unique properties.

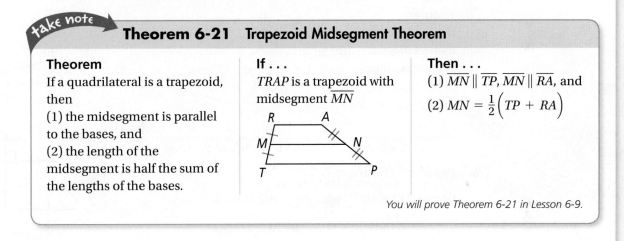

Theorem 6-21 Trapezoid Midsegment Theorem

Theorem
If a quadrilateral is a trapezoid, then
(1) the midsegment is parallel to the bases, and
(2) the length of the midsegment is half the sum of the lengths of the bases.

If . . .
TRAP is a trapezoid with midsegment $\overline{MN}$

Then . . .
(1) $\overline{MN} \parallel \overline{TP}, \overline{MN} \parallel \overline{RA}$, and
(2) $MN = \frac{1}{2}\left(TP + RA\right)$

You will prove Theorem 6-21 in Lesson 6-9.

ⓒ Problem 3 Using the Midsegment of a Trapezoid

Algebra $\overline{QR}$ is the midsegment of trapezoid *LMNP*.
What is *x*?

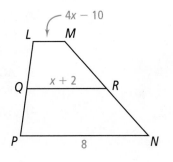

$QR = \frac{1}{2}(LM + PN)$ Trapezoid Midsegment Theorem

$x + 2 = \frac{1}{2}[(4x - 10) + 8]$ Substitute.

$x + 2 = \frac{1}{2}(4x - 2)$ Simplify.

$x + 2 = 2x - 1$ Distributive Property

$3 = x$ Subtract *x* and add 1 to each side.

Think
How can you check your answer?
Find *LM* and *QR*. Then see if *QR* equals half of the sum of the base lengths.

Got It? **3. a. Algebra** $\overline{MN}$ is the midsegment of trapezoid *PQRS*.
What is *x*? What is *MN*?

b. Reasoning How many midsegments can a triangle have? How many midsegments can a trapezoid have? Explain.

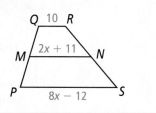

A **kite** is a quadrilateral with two pairs of consecutive sides congruent and no opposite sides congruent.

Essential Understanding The angles, sides, and diagonals of a kite have certain properties.

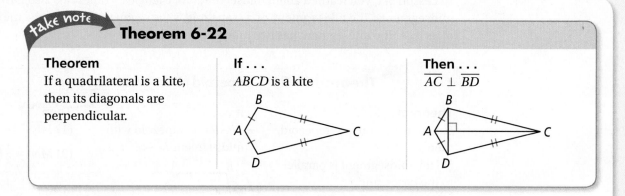

take note

Theorem 6-22

Theorem	**If . . .**	**Then . . .**
If a quadrilateral is a kite, then its diagonals are perpendicular.	*ABCD* is a kite	$\overline{AC} \perp \overline{BD}$

Proof **Proof of Theorem 6-22**

Given: Kite *ABCD* with $\overline{AB} \cong \overline{AD}$ and $\overline{CB} \cong \overline{CD}$

Prove: $\overline{AC} \perp \overline{BD}$

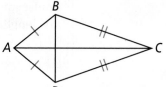

Statements	Reasons
1) Kite *ABCD* with $\overline{AB} \cong \overline{AD}$ and $\overline{CB} \cong \overline{CD}$	**1)** Given
2) *A* and *C* lie on the perpendicular bisector of $\overline{BD}$.	**2)** Converse of Perpendicular Bisector Theorem
3) $\overline{AC}$ is the perpendicular bisector of $\overline{BD}$.	**3)** Two points determine a line.
4) $\overline{AC} \perp \overline{BD}$	**4)** Definition of perpendicular bisector

© **Problem 4**   **Finding Angle Measures in Kites**

Quadrilateral *DEFG* is a kite. What are $m\angle 1$, $m\angle 2$, and $m\angle 3$?

$m\angle 1 = 90$	Diagonals of a kite are $\perp$.
$90 + m\angle 2 + 52 = 180$	Triangle Angle-Sum Theorem
$142 + m\angle 2 = 180$	Simplify.
$m\angle 2 = 38$	Subtract 142 from each side.

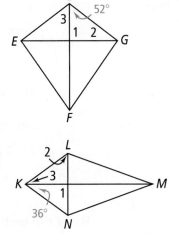

Think

How are the triangles congruent by SSS?
$\overline{DE} \cong \overline{DG}$ and $\overline{FE} \cong \overline{FG}$ because a kite has congruent consecutive sides. $\overline{DF} \cong \overline{DF}$ by the Reflexive Property of Congruence.

$\triangle DEF \cong \triangle DGF$ by SSS. Since corresponding parts of congruent triangles are congruent, $m\angle 3 = m\angle GDF = 52$.

✓ **Got It? 4.** Quadrilateral *KLMN* is a kite. What are $m\angle 1$, $m\angle 2$, and $m\angle 3$?

take note

Concept Summary   Relationships Among Quadrilaterals

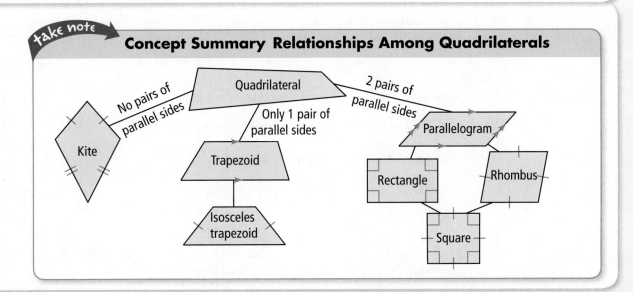

✓ **Lesson Check**

Do you know HOW?

What are the measures of the numbered angles?

1.

2.

3. What is the length of the midsegment of a trapezoid with bases of length 14 and 26?

Do you UNDERSTAND?   © MATHEMATICAL PRACTICES

© **4. Vocabulary** Is a kite a parallelogram? Explain.

© **5. Compare and Contrast** How is a kite similar to a rhombus? How is it different? Explain.

© **6. Error Analysis** Since a parallelogram has two pairs of parallel sides, it certainly has one pair of parallel sides. Therefore, a parallelogram must also be a trapezoid. What is the error in this reasoning? Explain.

Practice and Problem-Solving Exercises

MATHEMATICAL PRACTICES

A Practice   Find the measures of the numbered angles in each isosceles trapezoid.   ◀ See Problems 1 and 2.

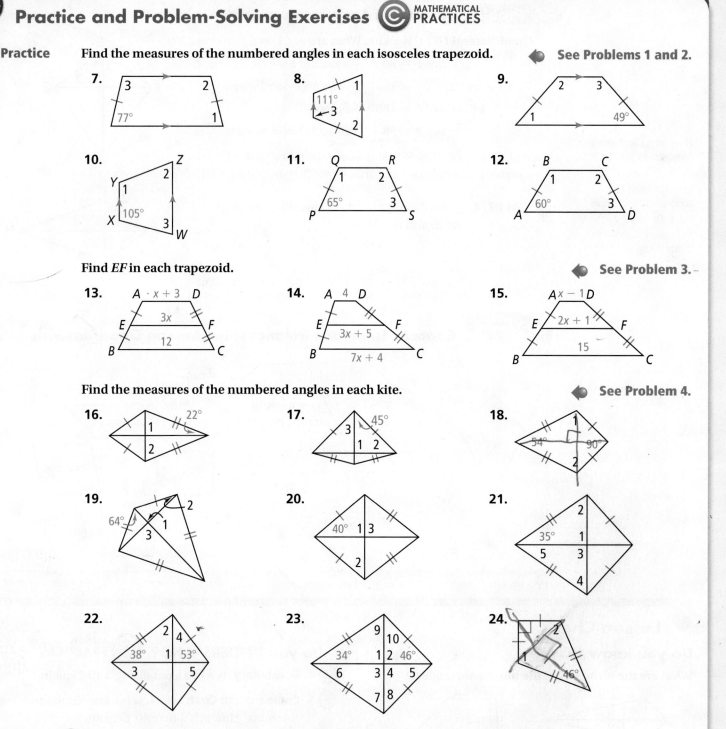

7. 3   2
77°   1

8. 1
111°
3
2

9. 2   3
1   49°

10. Z
Y 2
1
X 105°
3   W

11. Q   R
1   2
65°   3
P   S

12. B   C
1   2
60°   3
A   D

Find *EF* in each trapezoid.   ◀ See Problem 3.

13. A · x + 3 D
3x
E   F
12
B   C

14. A 4 D
E   F
3x + 5
B   7x + 4   C

15. A x − 1 D
2x + 1
E   F
15
B   C

Find the measures of the numbered angles in each kite.   ◀ See Problem 4.

16. 22°
1
2

17. 45°
3
1 2

18. 1
54°   90°
2

19. 2
64° 1
3

20. 40° 1 3
2

21. 2
35° 1
5   3
4

22. 2 4
38° 1 53°
3   5

23. 9 10
34° 1 2 46°
6   3 4 5
7 8

24. 2
1   46°

B Apply   **25. Open-Ended** Sketch two noncongruent kites such that the diagonals of one are congruent to the diagonals of the other.

26. Think About a Plan The perimeter of a kite is 66 cm. The length of one of its sides is 3 cm less than twice the length of another. Find the length of each side of the kite.
- Can you draw a diagram?
- How can you write algebraic expressions for the lengths of the sides?

27. Reasoning If *KLMN* is an isosceles trapezoid, is it possible for $\overline{KM}$ to bisect $\angle LMN$ and $\angle LKN$? Explain.

Algebra Find the value of the variable in each isosceles trapezoid.

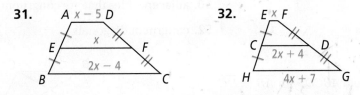

28.

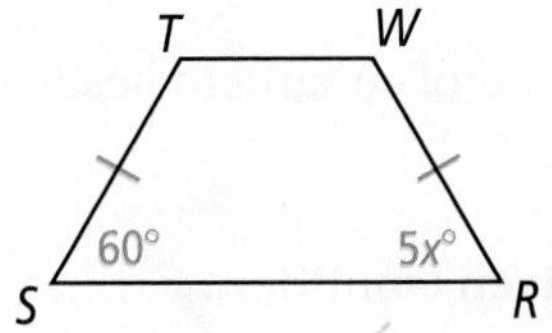

29. *B* $60°$ $(3x + 15)°$ *C*

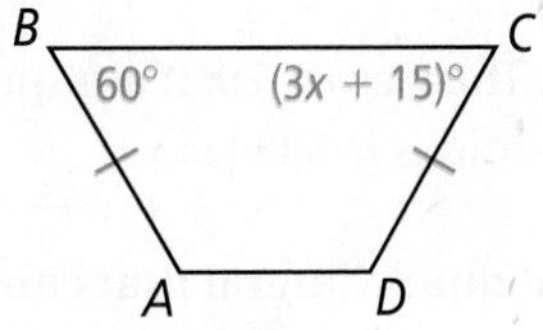

30. *Q* *R*
$QS = x + 5$
$RP = 3x + 3$

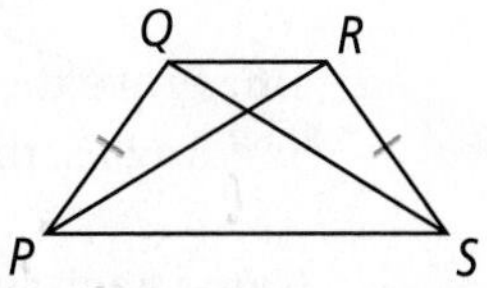

Algebra Find the lengths of the segments with variable expressions.

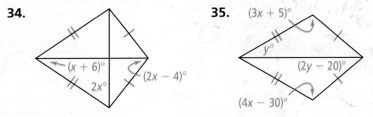

31. *A* $x - 5$ *D*
x
$2x - 4$

32. *E* x *F*
$2x + 4$
$4x + 7$

33. *H* $x - 3$ *G*
x
$2x - 2$

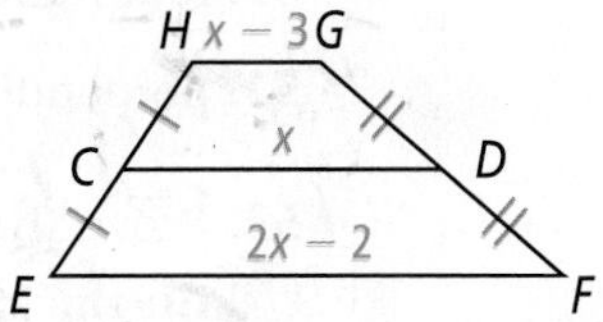

Algebra Find the value(s) of the variable(s) in each kite.

34. $(x + 6)°$
$2x°$
$(2x - 4)°$

35. $(3x + 5)°$
$y°$
$(2y - 20)°$
$(4x - 30)°$

36. $y°$
$6x°$
$\dfrac{3x°}{2}$

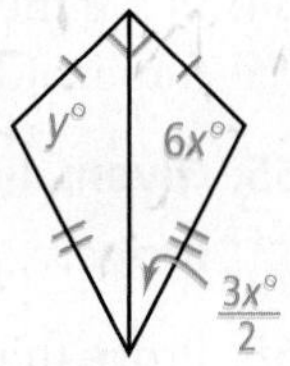

STEM Bridge Design The beams of the bridge at the right form quadrilateral *ABCD*. $\triangle AED \cong \triangle CDE \cong \triangle BEC$ and $m\angle DCB = 120$.

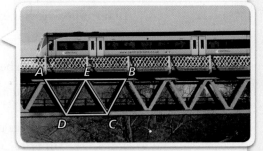

37. Classify the quadrilateral. Explain your reasoning.

38. Find the measures of the other interior angles of the quadrilateral.

Reasoning Can two angles of a kite be as follows? Explain.

39. opposite and acute

40. consecutive and obtuse

41. opposite and supplementary

42. consecutive and supplementary

43. opposite and complementary

44. consecutive and complementary

45. Developing Proof The plan suggests a proof of Theorem 6-19. Write a proof that follows the plan.

Given: Isosceles trapezoid $ABCD$ with $\overline{AB} \cong \overline{DC}$

Prove: $\angle B \cong \angle C$ and $\angle BAD \cong \angle D$

Plan: Begin by drawing $\overline{AE} \parallel \overline{DC}$ to form parallelogram $AECD$ so that $\overline{AE} \cong \overline{DC} \cong \overline{AB}$. $\angle B \cong \angle C$ because $\angle B \cong \angle 1$ and $\angle 1 \cong \angle C$. Also, $\angle BAD \cong \angle D$ because they are supplements of the congruent angles, $\angle B$ and $\angle C$.

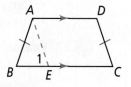

46. Prove the converse of Theorem 6-19: If a trapezoid has a pair of congruent base
Proof angles, then the trapezoid is isosceles.

Name each type of special quadrilateral that can meet the given condition. Make sketches to support your answers.

47. exactly one pair of congruent sides

48. two pairs of parallel sides

49. four right angles

50. adjacent sides that are congruent

51. perpendicular diagonals

52. congruent diagonals

53. Prove Theorem 6-20.
Proof

Given: Isosceles trapezoid $ABCD$ with $\overline{AB} \cong \overline{DC}$

Prove: $\overline{AC} \cong \overline{DB}$

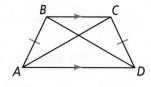

54. Prove the converse of Theorem 6-20: If the diagonals of a
Proof trapezoid are congruent, then the trapezoid is isosceles.

55. Given: Isosceles trapezoid $TRAP$ with $\overline{TR} \cong \overline{PA}$
Proof **Prove:** $\angle RTA \cong \angle APR$

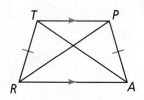

56. Prove that the angles formed by the noncongruent sides of a
Proof kite are congruent. (*Hint:* Draw a diagonal of the kite.)

Determine whether each statement is *true* or *false*. Justify your response.

57. All squares are rectangles.

58. A trapezoid is a parallelogram.

59. A rhombus can be a kite.

60. Some parallelograms are squares.

61. Every quadrilateral is a parallelogram.

62. All rhombuses are squares.

Challenge

63. Given: Isosceles trapezoid $TRAP$ with $\overline{TR} \cong \overline{PA}$;
Proof $\overline{BI}$ is the perpendicular bisector of $\overline{RA}$, intersecting $\overline{RA}$ at B and $\overline{TP}$ at I.

Prove: $\overline{BI}$ is the perpendicular bisector of $\overline{TP}$.

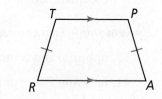

For a trapezoid, consider the segment joining the midpoints of the two given segments. How are its length and the lengths of the two parallel sides of the trapezoid related? Justify your answer.

64. the two nonparallel sides

65. the diagonals

66. $\overleftrightarrow{BN}$ is the perpendicular bisector of $\overline{AC}$ at N. Describe the set of points, D, for which $ABCD$ is a kite.

Standardized Test Prep

SAT/ACT

67. Which statement is never true?

    Ⓐ Square $ABCD$ is a rhombus.

    Ⓑ Trapezoid $GHJK$ is a parallelogram.

    Ⓒ Parallelogram $PQRS$ is a square.

    Ⓓ Square $WXYZ$ is a parallelogram.

68. A quadrilateral has four congruent sides. Which name best describes the figure?

    Ⓕ trapezoid   Ⓖ parallelogram   Ⓗ rhombus   Ⓘ kite

69. How would you classify triangle LMN?

    Ⓐ acute isosceles

    Ⓑ right isosceles

    Ⓒ obtuse scalene

    Ⓓ acute scalene

Extended Response

70. Given $\overline{DE}$ is congruent to $\overline{FG}$ and $\overline{EF}$ is congruent to $\overline{GD}$, prove $\angle E \cong \angle G$.

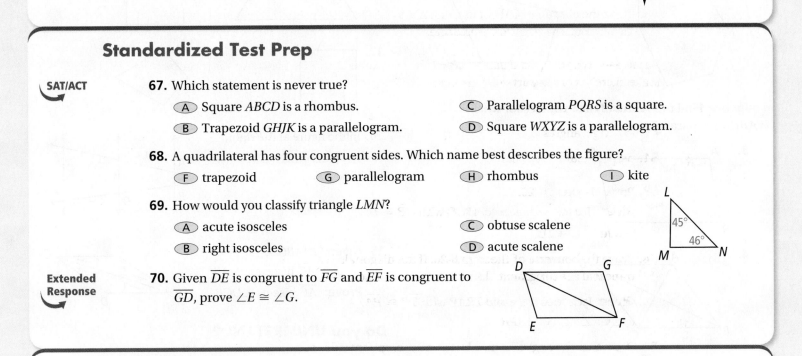

Mixed Review

Find the value of x for which the figure is the given special parallelogram.

◀ **See Lesson 6-5.**

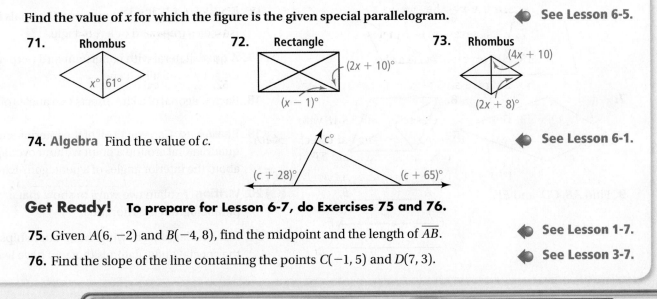

71. Rhombus

72. Rectangle

73. Rhombus

74. Algebra Find the value of c.

◀ **See Lesson 6-1.**

Get Ready! To prepare for Lesson 6-7, do Exercises 75 and 76.

75. Given $A(6, -2)$ and $B(-4, 8)$, find the midpoint and the length of $\overline{AB}$.

◀ **See Lesson 1-7.**

76. Find the slope of the line containing the points $C(-1, 5)$ and $D(7, 3)$.

◀ **See Lesson 3-7.**

Do you know HOW?

Find the value of each variable.

1.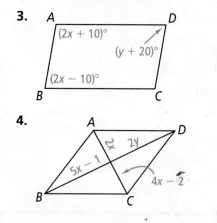

58° $x°$ $w°$
$(2w - 25)°$
$(x + 6)°$
125° 100°

2.
50°
89°
$a°$ $m°$ 64°

Algebra Find the values of the variables for which *ABCD* is a parallelogram.

3.
A D
$(2x + 10)°$
$(y + 20)°$
$(2x - 10)°$
B C

4.
A D
$2x$ $2y$
$5x - 1$
$4x - 2$
B C

Algebra Classify the quadrilateral. Then find the value(s) of the variable(s).

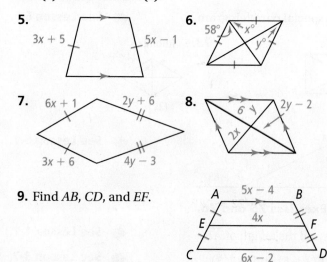

5.
$3x + 5$ $5x - 1$

6.
58° $x°$
$y°$

7.
$6x + 1$ $2y + 6$
$3x + 6$ $4y - 3$

8.
6 y $2y - 2$
$2x$

9. Find *AB*, *CD*, and *EF*.
A $5x - 4$ B
E $4x$ F
C $6x - 2$ D

Classify the quadrilateral as precisely as possible. Explain your reasoning.

10.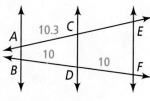

11.

12.

13.

14. In the figure at the right, $\overleftrightarrow{AB} \parallel \overleftrightarrow{CD} \parallel \overleftrightarrow{EF}$. Find *AE*.

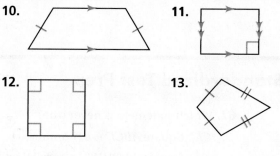

A 10.3 C E
B 10 D 10 F

15. Given: □*ABCD*,
$\overline{AC}$ bisects ∠*DAB*
Prove: $\overline{AC}$ bisects ∠*DCB*.

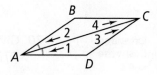

B C
2 4
1 3
A D

Do you UNDERSTAND?

Decide whether the statement is *true* or *false*. If true, explain why. If false, show a counterexample.

16. A quadrilateral with congruent diagonals is either an isosceles trapezoid or a rectangle.

17. A quadrilateral with congruent and perpendicular diagonals must be a kite.

18. Each diagonal of a kite bisects two angles of the kite.

Ⓒ **19. Reasoning** Can you fit all of the interior angles of a quadrilateral around a point without overlap? What about the interior angles of a pentagon? Explain.

Ⓒ **20. Writing** Explain two ways to show that a parallelogram is a rhombus.

21. Draw a diagram showing the relationships among the special quadrilaterals that you have learned.

Algebra Review

Use With Lesson 6-7

Simplifying Radicals

A radical expression is in simplest form when all of the following are true.

- The radicand has no perfect square factors other than 1.
- The radicand does not contain a fraction.
- A denominator does not contain a radical expression.

radical symbol ↓ $\sqrt{a}$ ← radicand

Example 1

Simplify the expressions $\sqrt{2} \cdot \sqrt{8}$ and $\sqrt{294} \div \sqrt{3}$.

$\sqrt{2} \cdot \sqrt{8} = \sqrt{2 \cdot 8}$ Write both numbers under one radical.

$= \sqrt{16}$ Simplify the expression under the radical.

$= 4$ Factor out perfect squares and simplify.

$$\sqrt{294} \div \sqrt{3} = \sqrt{\frac{294}{3}} = \sqrt{98} = \sqrt{49 \cdot 2} = 7\sqrt{2}$$

Example 2

Write $\sqrt{\frac{4}{3}}$ in simplest form.

$\sqrt{\frac{4}{3}} = \frac{\sqrt{4}}{\sqrt{3}}$ Rewrite the single radical as a quotient.

$= \frac{2}{\sqrt{3}}$ Simplify the numerator.

$= \frac{2}{\sqrt{3}} \cdot \frac{\sqrt{3}}{\sqrt{3}}$ Multiply by $\frac{\sqrt{3}}{\sqrt{3}}$ (a form of 1) to remove the radical from the denominator. This is called *rationalizing the denominator*.

$= \frac{2\sqrt{3}}{3}$

Exercises

Simplify each expression.

1. $\sqrt{5} \cdot \sqrt{10}$ **2.** $\sqrt{243}$ **3.** $\sqrt{128} \div \sqrt{2}$ **4.** $\sqrt{\frac{125}{4}}$ **5.** $\sqrt{6} \cdot \sqrt{8}$

6. $\frac{\sqrt{36}}{\sqrt{3}}$ **7.** $\frac{\sqrt{144}}{\sqrt{2}}$ **8.** $\sqrt{3} \cdot \sqrt{12}$ **9.** $\sqrt{72} \div \sqrt{2}$ **10.** $\sqrt{169}$

11. $28 \div \sqrt{8}$ **12.** $\sqrt{300} \div \sqrt{5}$ **13.** $\sqrt{12} \cdot \sqrt{2}$ **14.** $\frac{\sqrt{6} \cdot \sqrt{3}}{\sqrt{9}}$ **15.** $\frac{\sqrt{3} \cdot \sqrt{15}}{\sqrt{2}}$

Polygons in the Coordinate Plane

© **Content Standard**
G.GPE.7 Use coordinates to compute perimeters of polygons . . .

Objective To classify polygons in the coordinate plane

Apply what you learned "B-4" about classifying polygons.

© **MATHEMATICAL PRACTICES**

Getting Ready!

You and a friend are playing a board game. Players place rubber bands on their own square grid to form different shapes. The object of the game is to guess the vertices of your opponent's shape. How would you place pieces on the grid shown to complete a right isosceles triangle? Sketch the triangle and justify the placement of each piece.

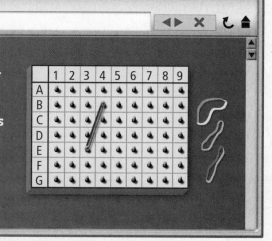

In the Solve It, you formed a polygon on a grid. In this lesson, you will classify polygons in the coordinate plane.

Essential Understanding You can classify figures in the coordinate plane using the formulas for slope, distance, and midpoint.

The chart below reviews these formulas and tells when to use them.

take note

Key Concept   Formulas and the Coordinate Plane

Formula	When to Use It
Distance Formula $d = \sqrt{(x_2 - x_1)^2 + (y_2 - y_1)^2}$	To determine whether • sides are congruent • diagonals are congruent
Midpoint Formula $M = \left(\dfrac{x_1 + x_2}{2}, \dfrac{y_1 + y_2}{2}\right)$	To determine • the coordinates of the midpoint of a side • whether diagonals bisect each other
Slope Formula $m = \dfrac{y_2 - y_1}{x_2 - x_1}$	To determine whether • opposite sides are parallel • diagonals are perpendicular • sides are perpendicular

Problem 1 **Classifying a Triangle**

Is $\triangle ABC$ scalene, isosceles, or equilateral?

The vertices of the triangle are $A(0, 1)$, $B(4, 4)$, and $C(7, 0)$.

Find the lengths of the sides using the Distance Formula.

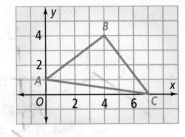

$$AB = \sqrt{(4 - 0)^2 + (4 - 1)^2}$$

$$= \sqrt{16 + 9}$$ Simplify within parentheses.
Then simplify the powers.

$$= \sqrt{25}$$ Simplify the radicand.

$$= 5$$ Simplify.

$$BC = \sqrt{(7 - 4)^2 + (0 - 4)^2} \qquad CA = \sqrt{(0 - 7)^2 + (1 - 0)^2}$$

$$= \sqrt{9 + 16} \qquad\qquad\qquad = \sqrt{49 + 1}$$

$$= \sqrt{25} \qquad\qquad\qquad\quad = \sqrt{50}$$

$$= 5 \qquad\qquad\qquad\qquad = 5\sqrt{2}$$

Since $AB = BC = 5$, $\triangle ABC$ is isosceles.

Got It? **1.** $\triangle DEF$ has vertices $D(0, 0)$, $E(1, 4)$, and $F(5, 2)$. Is $\triangle DEF$ *scalene, isosceles,* or *equilateral*?

Problem 2 **Classifying a Parallelogram**

Is $\square ABCD$ a rhombus? Explain.

Step 1 Use the Slope Formula to find the slopes of the diagonals.

$$\text{slope of } \overline{AC} = \frac{5 - 0}{4 - (-2)} = \frac{5}{6}$$

$$\text{slope of } \overline{BD} = \frac{1 - 4}{2 - 0} = -\frac{3}{2}$$

Step 2 Find the product of the slopes.

$$\frac{5}{6} \cdot \left(-\frac{3}{2}\right) = -\frac{15}{12}$$

Since the product of the slopes is not -1, the diagonals are not perpendicular. So $ABCD$ is not a rhombus.

Got It? **2.** $\square MNPQ$ has vertices $M(0, 1)$, $N(-1, 4)$, $P(2, 5)$, and $Q(3, 2)$.

   a. Is $\square MNPQ$ a rectangle? Explain.

   b. Is $\square MNPQ$ a square? Explain.

   c. Reasoning Is the triangle in Problem 1 a right triangle? Explain.

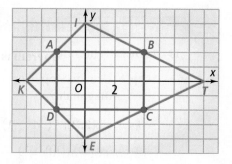

© **Problem 3** **Classifying a Quadrilateral**

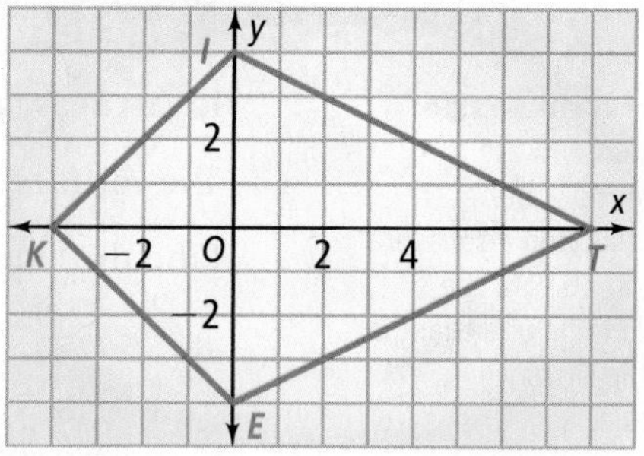

A kite is shown at the right. What is the most precise classification of the quadrilateral formed by connecting the midpoints of the sides of the kite?

Know

$K(-4, 0)$, $I(0, 4)$, $T(8, 0)$, and $E(0, -4)$

Need

The midpoints of the sides of the kite

Plan

Use the Midpoint Formula to find the coordinates of the vertices of the inner quadrilateral. *Draw a diagram* to see it. Then classify the figure.

Step 1 Find the midpoint of each side of the kite.

$A = $ midpoint of $\overline{KI} = \left(\dfrac{-4 + 0}{2}, \dfrac{0 + 4}{2}\right) = (-2, 2)$

$B = $ midpoint of $\overline{IT} = \left(\dfrac{0 + 8}{2}, \dfrac{4 + 0}{2}\right) = (4, 2)$

$C = $ midpoint of $\overline{TE} = \left(\dfrac{8 + 0}{2}, \dfrac{0 + (-4)}{2}\right) = (4, -2)$

$D = $ midpoint of $\overline{EK} = \left(\dfrac{0 + (-4)}{2}, \dfrac{-4 + 0}{2}\right) = (-2, -2)$

Step 2 Draw a diagram of $ABCD$.

Step 3 Classify $ABCD$.

$AB = |4 - (-2)| = 6$ Use the definition of distance $BC = |-2 - 2| = 4$
$CD = |-2 - 4| = 6$ on a number line. $DA = |2 - (-2)| = 4$

Since opposite sides are congruent, $ABCD$ is a parallelogram.

Since $\overline{AB}$ and $\overline{CD}$ are both horizontal, and $\overline{BC}$ and $\overline{DA}$ are both vertical, the segments form right angles. So, $ABCD$ is a rectangle.

✔ **Got It?** **3.** An isosceles trapezoid has vertices $A(0, 0)$, $B(2, 4)$, $C(6, 4)$, and $D(8, 0)$. What special quadrilateral is formed by connecting the midpoints of the sides of $ABCD$?

Lesson Check

Do you know HOW?

1. △TRI has vertices $T(-3, 4)$, $R(3, 4)$, and $I(0, 0)$. Is △TRI *scalene, isosceles,* or *equilateral*?

2. Is *QRST* below a rectangle? Explain.

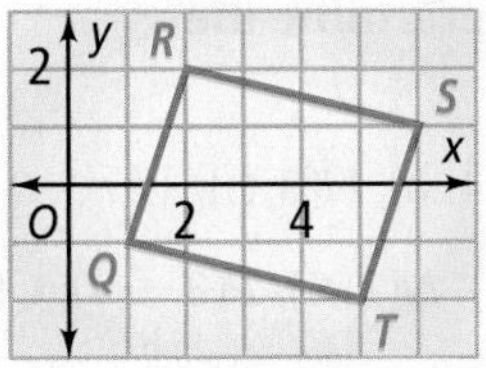

Do you UNDERSTAND?

MATHEMATICAL PRACTICES

3. **Writing** Describe how you would determine whether the lengths of the medians from base angles D and F are congruent.

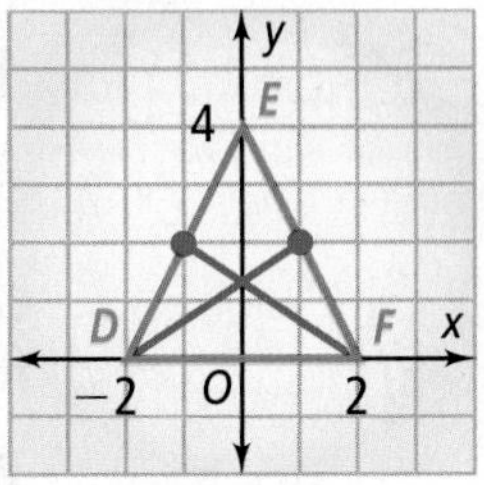

4. **Error Analysis** A student says that the quadrilateral with vertices $D(1, 2)$, $E(0, 7)$, $F(5, 6)$, and $G(7, 0)$ is a rhombus because its diagonals are perpendicular. What is the student's error?

Practice and Problem-Solving Exercises

MATHEMATICAL PRACTICES

A Practice

Determine whether △ABC is *scalene, isosceles,* or *equilateral.* Explain.

See Problem 1.

5.

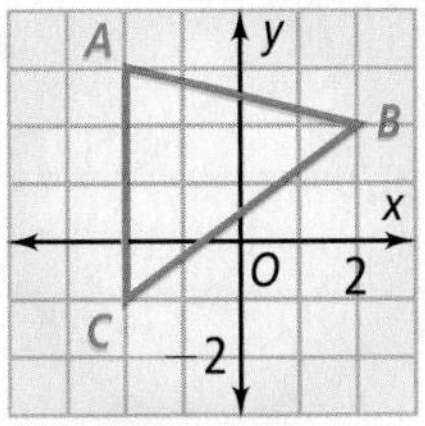

6.

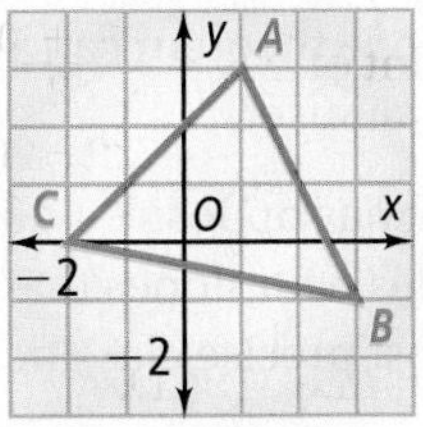

7.

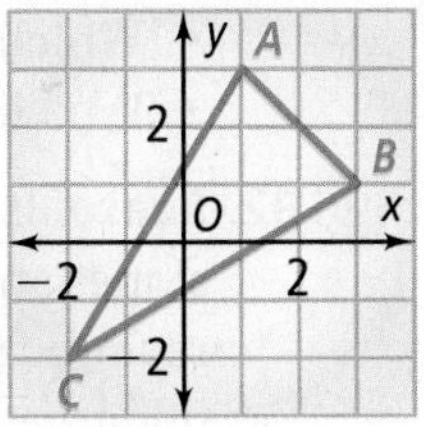

Determine whether the parallelogram is a *rhombus, rectangle, square,* or *none.* Explain.

See Problem 2.

8. $P(-1, 2)$, $O(0, 0)$, $S(4, 0)$, $T(3, 2)$

9. $L(1, 2)$, $M(3, 3)$, $N(5, 2)$, $P(3, 1)$

10. $R(-2, -3)$, $S(4, 0)$, $T(3, 2)$, $V(-3, -1)$

11. $G(0, 0)$, $H(6, 0)$, $I(9, 1)$, $J(3, 1)$

12. $W(-3, 0)$, $I(0, 3)$, $N(3, 0)$, $D(0, -3)$

13. $S(1, 3)$, $P(4, 4)$, $A(3, 1)$, $T(0, 0)$

What is the most precise classification of the quadrilateral formed by connecting in order the midpoints of each figure below?

See Problem 3.

14. parallelogram *PART*

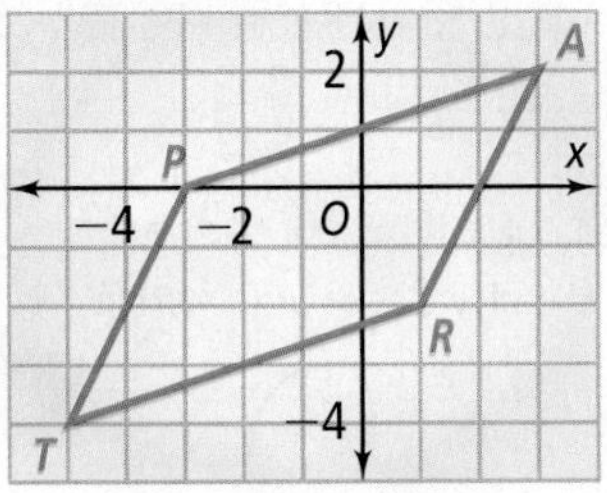

15. rectangle *EFGH*

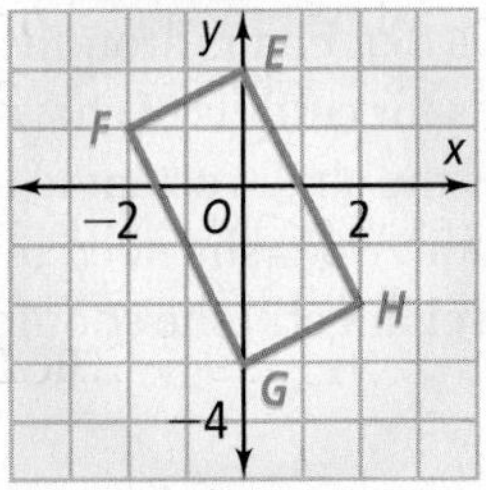

16. isosceles trapezoid *JKLM*

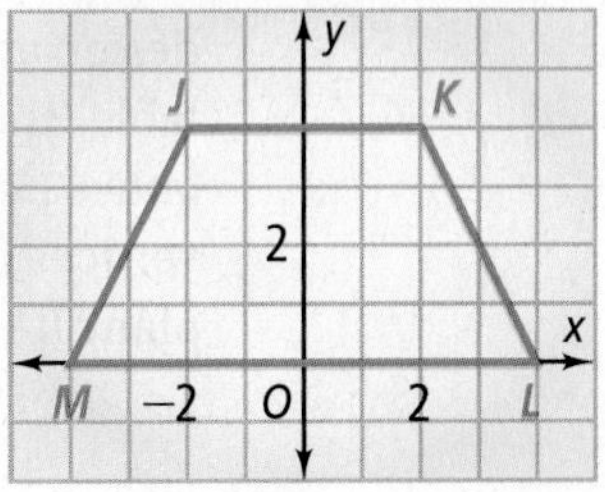

Graph and label each triangle with the given vertices. Determine whether each triangle is *scalene, isosceles,* or *equilateral*. Then tell whether each triangle is a right triangle.

17. $T(1, 1)$, $R(3, 8)$, $I(6, 4)$

18. $J(-5, 0)$, $K(5, 8)$, $L(4, -1)$

19. $A(3, 2)$, $B(-10, 4)$, $C(-5, -8)$

20. $H(1, -2)$, $B(-1, 4)$, $F(5, 6)$

Graph and label each quadrilateral with the given vertices. Then determine the most precise name for each quadrilateral.

21. $P(-5, 0)$, $Q(-3, 2)$, $R(3, 2)$, $S(5, 0)$

22. $S(0, 0)$, $T(4, 0)$, $U(3, 2)$, $V(-1, 2)$

23. $F(0, 0)$, $G(5, 5)$, $H(8, 4)$, $I(7, 1)$

24. $M(-14, 4)$, $N(1, 6)$, $P(3, -9)$, $Q(-12, -11)$

25. $A(3, 5)$, $B(7, 6)$, $C(6, 2)$, $D(2, 1)$

26. $N(-6, 4)$, $P(-3, 1)$, $Q(0, 2)$, $R(-3, 5)$

27. $J(2, 1)$, $K(5, 4)$, $L(8, 1)$, $M(2, -3)$

28. $H(-2, -3)$, $I(4, 0)$, $J(3, 2)$, $K(-3, -1)$

29. $W(-1, 1)$, $X(0, 2)$, $Y(1, 1)$, $Z(0, -2)$

30. $D(-3, 1)$, $E(-7, -3)$, $F(6, -3)$, $G(2, 1)$

© 31. Think About a Plan Are the triangles at the right congruent? How do you know?
- Which triangle congruence theorem can you use?
- Which formula should you use?

© 32. Reasoning A quadrilateral has opposite sides with equal slopes and consecutive sides with slopes that are negative reciprocals. What is the most precise classification of the quadrilateral? Explain.

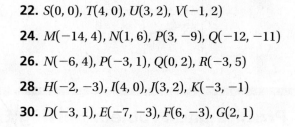

Determine the most precise name for each quadrilateral. Then find its area.

33. $A(0, 2)$, $B(4, 2)$, $C(-3, -4)$, $D(-7, -4)$

34. $J(1, -3)$, $K(3, 1)$, $L(7, -1)$, $M(5, -5)$

35. $\overline{DE}$ is a midsegment of $\triangle ABC$ at the right. Show that the Triangle Midsegment Theorem holds true for $\triangle ABC$.

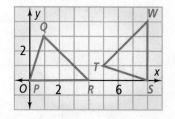

© 36. a. Writing Describe two ways you can show whether a quadrilateral in the coordinate plane is a square.
   b. Reasoning Which method is more efficient? Explain.

37. Interior Design Interior designers often use grids to plan the placement of furniture in a room. The design at the right shows four chairs around a coffee table. The designer plans for cutouts of chairs on lattice points. She wants the chairs oriented at the vertices of a parallelogram. Does she need to fix her plan? If so, describe the change(s) she should make.

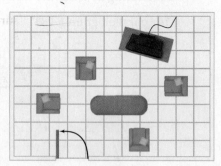

38. Use the diagram at the right.

a. What is the most precise classification of *ABCD*?

b. What is the most precise classification of *EFGH*?

c. Are *ABCD* and *EFGH* congruent? Explain.

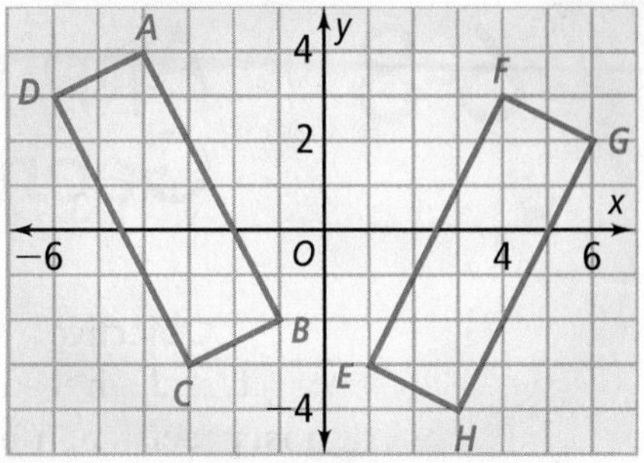

Challenge **39. Coordinate Geometry** The diagonals of quadrilateral *EFGH* intersect at $D(-1, 4)$. *EFGH* has vertices at $E(2, 7)$ and $F(-3, 5)$. What must be the coordinates of *G* and *H* to ensure that *EFGH* is a parallelogram?

The endpoints of $\overline{AB}$ are $A(-3, 5)$ and $B(9, 15)$. Find the coordinates of the points that divide $\overline{AB}$ into the given number of congruent segments.

40. 4 **41.** 6 **42.** 10 **43.** 50 **44.** *n*

Standardized Test Prep

SAT/ACT **45.** $K(-3, 0)$, $I(0, 2)$, and $T(3, 0)$ are three vertices of a kite. Which point could be the fourth vertex?

Ⓐ $E(0, 5)$ Ⓑ $E(0, 0)$ Ⓒ $E(0, -2)$ Ⓓ $E(0, -10)$

46. In the diagram, lines ℓ and m are parallel. What is the value of *x*?

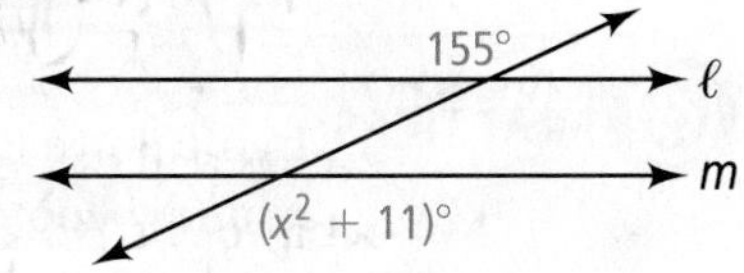

Ⓕ 5 Ⓗ 13

Ⓖ 12 Ⓘ 25

47. In the diagram, which segment is shortest?

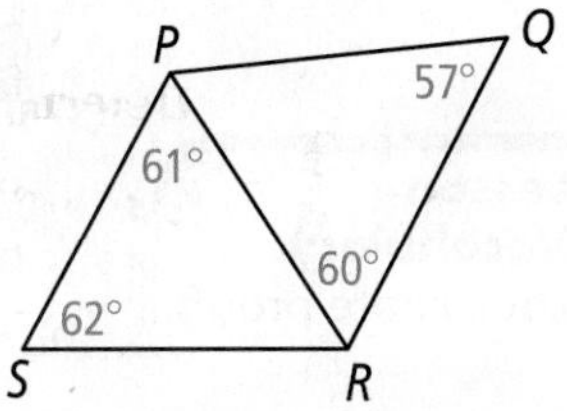

Ⓐ $\overline{PS}$ Ⓒ $\overline{PQ}$

Ⓑ $\overline{PR}$ Ⓓ $\overline{QR}$

Short Response **48.** $A(-3, 1)$, $B(-1, -2)$, and $C(2, 1)$ are three vertices of quadrilateral *ABCD*. Could *ABCD* be a rectangle? Explain.

Mixed Review

49. Algebra Find the measure of each angle and the value of *x* in the isosceles trapezoid.

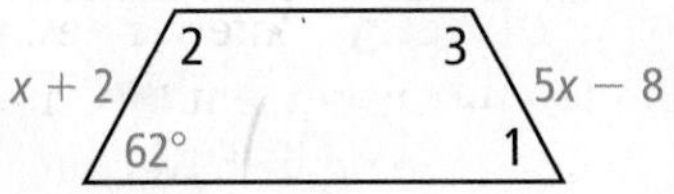

◀ **See Lesson 6-6.**

Find the circumcenter of $\triangle ABC$. ◀ **See Lesson 5-3.**

50. $A(1, 1)$, $B(5, 3)$, $C(5, 1)$ **51.** $A(-5, 0)$, $B(-1, -8)$, $C(-1, 0)$

Get Ready! To prepare for Lesson 6-8, do Exercises 52–54.

Find the slope of $\overline{XY}$. ◀ **See Lesson 3-7.**

52. $X(0, a)$, $Y(-a, 2a)$ **53.** $X(-a, b)$, $Y(a, b)$ **54.** $X(a, 0)$, $Y(c + d, b)$

Applying Coordinate Geometry

© Content Standard
Prepares for G.GPE.4 Use coordinates to prove simple geometric theorems algebraically.

Objective To name coordinates of special figures by using their properties

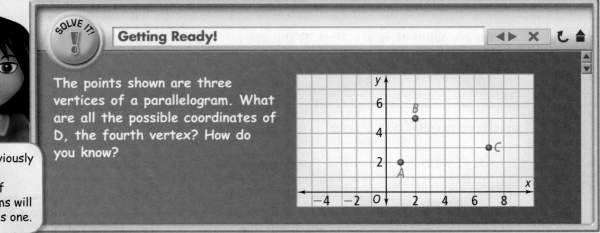

SOLVE IT!

Getting Ready!

The points shown are three vertices of a parallelogram. What are all the possible coordinates of D, the fourth vertex? How do you know?

Knowing previously established properties of parallelograms will help with this one.

© MATHEMATICAL PRACTICES In the Solve It, you found coordinates of a point and named it using numbers for the *x*- and *y*-coordinates. In this lesson, you will learn to use variables for the coordinates.

Essential Understanding You can use variables to name the coordinates of a figure. This allows you to show that relationships are true for a general case.

Lesson Vocabulary
• coordinate proof

In Chapter 5, you learned about the segment joining the midpoints of two sides of a triangle. Here are three possible ways to place a triangle and its midsegment.

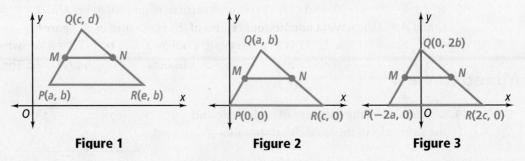

Figure 1　　　　**Figure 2**　　　　**Figure 3**

Figure 1 does not use the axes, so it requires more variables. Figures 2 and 3 have good placement. In Figure 2, the midpoint coordinates are $M\left(\frac{a}{2}, \frac{b}{2}\right)$ and $N\left(\frac{a+c}{2}, \frac{b}{2}\right)$. In Figure 3, the coordinates are $M(-a, b)$ and $N(c, b)$. You can see that Figure 3 is the easiest to work with.

To summarize, to place a figure in the coordinate plane, it is usually helpful to place at least one side on an axis or to center the figure at the origin. For the coordinates, try to anticipate what you will need to do in the problem. Then multiply the coordinates by the appropriate number to make your work easier.

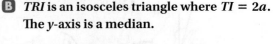

Problem 1 Naming Coordinates

Plan

How do you start the problem?
Look at the position of the figure. Use the given information to determine how far each vertex is from the x- and y-axes.

What are the coordinates of the vertices of each figure?

A *SQRE* is a square where $SQ = 2a$. The axes bisect each side.

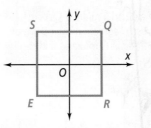

Since *SQRE* is a square centered at the origin and $SQ = 2a$, S and Q are each a units from each axis. The same is true for the other vertices.

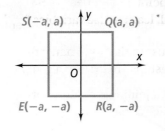

B *TRI* is an isosceles triangle where $TI = 2a$. The y-axis is a median.

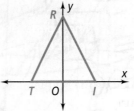

The y-axis is a median, so it bisects $\overline{TI}$. $TI = 2a$, so T and I are both a units from the y-axis. The height of *TRI* does not depend on a, so use a different variable for R.

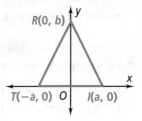

 Got It? 1. What are the coordinates of the vertices of each figure?

a. *RECT* is a rectangle with height a and length $2b$. The y-axis bisects $\overline{EC}$ and $\overline{RT}$.

b. *KITE* is a kite where $IE = 2a$, $KO = b$, and $OT = c$. The x-axis bisects $\overline{IE}$.

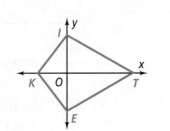

Problem 2 Using Variable Coordinates

The diagram shows a general parallelogram with a vertex at the origin and one side along the *x*-axis. What are the coordinates of *D*, the point of intersection of the diagonals of □*ABCO*? How do you know?

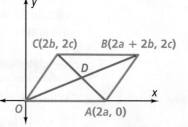

Know

- The coordinates of the vertices of □*ABCO*
- $\overline{OB}$ bisects $\overline{AC}$ and $\overline{AC}$ bisects $\overline{OB}$

Need

The coordinates of *D*

Plan

Since the diagonals of a parallelogram bisect each other, the midpoint of each segment is their point of intersection. Use the Midpoint Formula to find the midpoint of one diagonal.

Use the Midpoint Formula to find the midpoint of $\overline{AC}$.

$$D = \text{midpoint of } \overline{AC} = \left(\frac{2a + 2b}{2}, \frac{0 + 2c}{2}\right) = (a + b, c)$$

The coordinates of the point of intersection of the diagonals of □*ABCO* are $(a + b, c)$.

Got It? **2. a. Reasoning** In Problem 2, explain why the *x*-coordinate of *B* is the sum of 2*a* and 2*b*.

b. The diagram below shows a trapezoid with the base centered at the origin. Is the trapezoid isosceles? Explain.

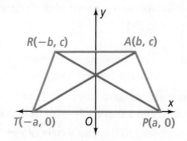

You can use coordinate geometry and algebra to prove theorems in geometry. This kind of proof is called a **coordinate proof**. Sometimes it is easier to show that a theorem is true by using a coordinate proof rather than a standard deductive proof. It is useful to write a plan for a coordinate proof. Problem 3 shows you how.

Problem 3 **Planning a Coordinate Proof**

Plan a coordinate proof of the Trapezoid Midsegment Theorem (Theorem 6-21).

(1) The midsegment of a trapezoid is parallel to the bases.

(2) The length of the midsegment of a trapezoid is half the sum of the lengths of the bases.

Step 1 Draw and label a figure.

Midpoints will be involved, so use multiples of 2 to name coordinates.

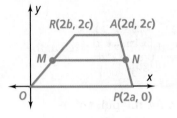

Step 2 Write the *Given* and *Prove* statements.

Use the information on the diagram to write the statements.

Given: $\overline{MN}$ is the midsegment of trapezoid *ORAP*.

Prove: $\overline{MN} \parallel \overline{OP}$, $\overline{MN} \parallel \overline{RA}$, $MN = \frac{1}{2}(OP + RA)$

Step 3 Determine the formulas you will need. Then write the plan.

- First, use the Midpoint Formula to find the coordinates of *M* and *N*.

- Then, use the Slope Formula to determine whether the slopes of $\overline{MN}$, $\overline{OP}$, and $\overline{RA}$ are equal. If they are, $\overline{MN}$, $\overline{OP}$, and $\overline{RA}$ are parallel.

- Finally, use the Distance Formula to find and compare the lengths of $\overline{MN}$, $\overline{OP}$, and $\overline{RA}$.

Got It? **3.** Plan a coordinate proof of the Triangle Midsegment Theorem (Theorem 5-1).

Lesson Check

Do you know HOW?

Use the diagram at the right.

1. In $\square KLMO$, $OM = 2a$. What are the coordinates of *K* and *M*?

2. What are the slopes of the diagonals of *KLMN*?

3. What are the coordinates of the point of intersection of $\overline{KM}$ and $\overline{OL}$?

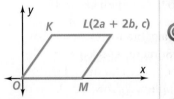

Do you UNDERSTAND? MATHEMATICAL PRACTICES

4. Reasoning How do variable coordinates generalize figures in the coordinate plane?

5. Reasoning A vertex of a quadrilateral has coordinates (a, b). The *x*-coordinates of the other three vertices are *a* or $-a$, and the *y*-coordinates are *b* or $-b$. What kind of quadrilateral is the figure?

6. Error Analysis A classmate says the endpoints of the midsegment of the trapezoid in Problem 3 are $\left(\frac{b}{2}, \frac{c}{2}\right)$ and $\left(\frac{d+a}{2}, \frac{c}{2}\right)$. What is your classmate's error? Explain.

Practice and Problem-Solving Exercises

MATHEMATICAL PRACTICES

A Practice   **Algebra** What are the coordinates of the vertices of each figure?   See Problem 1.

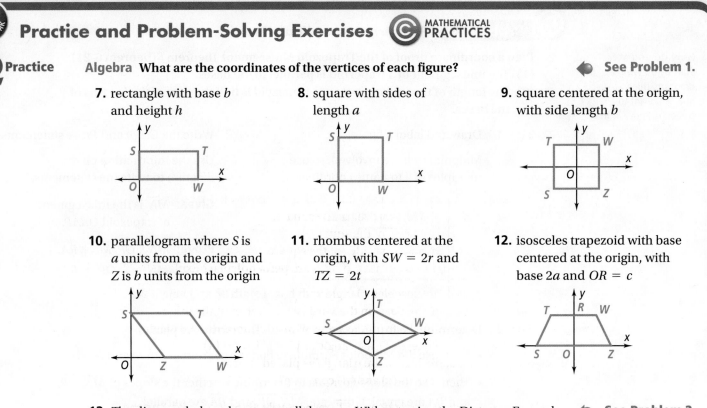

7. rectangle with base b and height h

8. square with sides of length a

9. square centered at the origin, with side length b

10. parallelogram where S is a units from the origin and Z is b units from the origin

11. rhombus centered at the origin, with $SW = 2r$ and $TZ = 2t$

12. isosceles trapezoid with base centered at the origin, with base $2a$ and $OR = c$

13. The diagram below shows a parallelogram. Without using the Distance Formula, determine whether the parallelogram is a rhombus. How do you know?   See Problem 2.

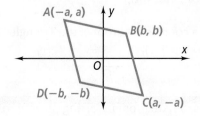

14. Plan a coordinate proof to show that the midpoints of the sides of an isosceles trapezoid form a rhombus.   See Problem 3.
  a. Name the coordinates of isosceles trapezoid *TRAP* at the right, with bottom base length $4a$, top base length $4b$, and $EG = 2c$. The y-axis bisects the bases.
  b. Write the *Given* and *Prove* statements.
  c. How will you find the coordinates of the midpoints of each side?
  d. How will you determine whether *DEFG* is a rhombus?

B Apply   **15. Open-Ended** Place a general quadrilateral in the coordinate plane.

16. Reasoning A rectangle *LMNP* is centered at the origin with $M(r, -s)$. What are the coordinates of P?

Give the coordinates for point P without using any new variables.

17. isosceles trapezoid

18. trapezoid with a right $\angle$

19. kite

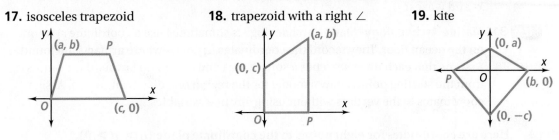

20. a. Draw a square whose diagonals of length $2b$ lie on the x- and y-axes.
 b. Give the coordinates of the vertices of the square.
 c. Compute the length of a side of the square.
 d. Find the slopes of two adjacent sides of the square.
 e. Writing Do the slopes show that the sides are perpendicular? Explain.

21. Make two drawings of an isosceles triangle with base length $2b$ and height $2c$.
 a. In one drawing, place the base on the x-axis with a vertex at the origin.
 b. In the second, place the base on the x-axis with its midpoint at the origin.
 c. Find the lengths of the legs of the triangle as placed in part (a).
 d. Find the lengths of the legs of the triangle as placed in part (b).
 e. How do the results of parts (c) and (d) compare?

22. W and Z are the midpoints of $\overline{OR}$ and $\overline{ST}$, respectively. In parts (a)–(c), find the coordinates of W and Z.

 a.   **b.**   **c.**

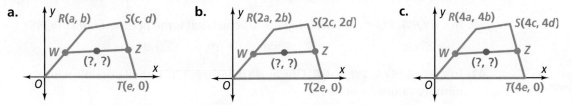

 d. You are to plan a coordinate proof involving the midpoint of $\overline{WZ}$. Which of the figures (a)–(c) would you prefer to use? Explain.

Plan the coordinate proof of each statement.

23. Think About a Plan The opposite sides of a parallelogram are congruent (Theorem 6-3).
 • How will you place the parallelogram in a coordinate plane?
 • What formulas will you need to use?

24. The diagonals of a rectangle bisect each other.

25. The consecutive sides of a square are perpendicular.

Classify each quadrilateral as precisely as possible.

26. $A(b, 2c)$, $B(4b, 3c)$, $C(5b, c)$, $D(2b, 0)$

27. $O(0, 0)$, $P(t, 2s)$, $Q(3t, 2s)$, $R(4t, 0)$

28. $E(a, b)$, $F(2a, 2b)$, $G(3a, b)$, $H(2a, -b)$

29. $O(0, 0)$, $L(-e, f)$, $M(f - e, f + e)$, $N(f, e)$

30. What property of a rhombus makes it convenient to place its diagonals on the x- and y-axes?

STEM **31. Marine Archaeology** Marine archaeologists sometimes use a coordinate system on the ocean floor. They record the coordinates of points where artifacts are found. Assume that each diver searches a square area and can go no farther than b units from the starting point. Draw a model for the region one diver can search. Assign coordinates to the vertices without using any new variables.

C Challenge Here are coordinates for eight points in the coordinate plane $(q > p > 0)$. $A(0, 0), B(p, 0), C(q, 0), D(p + q, 0), E(0, q), F(p, q), G(q, q), H(p + q, q)$. Which four points, if any, are the vertices for each type of figure?

32. parallelogram

33. rhombus

34. rectangle

35. square

36. trapezoid

37. isosceles trapezoid

Standardized Test Prep

SAT/ACT

38. Which number of right angles is NOT possible for a quadrilateral to have?

Ⓐ exactly one   Ⓑ exactly two   Ⓒ exactly three   Ⓓ exactly four

39. The vertices of a rhombus are located at $(a, 0)$, $(0, b)$, $(-a, 0)$, and $(0, -b)$, where $a > 0$ and $b > 0$. What is the midpoint of the side that is in Quadrant II?

Ⓕ $\left(\frac{a}{2}, \frac{b}{2}\right)$ Ⓖ $\left(-\frac{a}{2}, \frac{b}{2}\right)$ Ⓗ $\left(-\frac{a}{2}, -\frac{b}{2}\right)$ Ⓘ $\left(\frac{a}{2}, -\frac{b}{2}\right)$

40. In $\square PQRS$, $PQ = 35$ cm and $QR = 12$ cm. What is the perimeter of $\square PQRS$?

Ⓐ 23 cm   Ⓑ 47 cm   Ⓒ 94 cm   Ⓓ 420 cm

Short Response

41. In $\triangle PQR$, $PQ > PR > QR$. One angle measures 170. List all possible whole number values for $m\angle P$.

Mixed Review

42. Let $X(-2, 3)$, $Y(5, 5)$, and $Z(4, 10)$. Is $\triangle XYZ$ a right triangle? Explain.   ◀ See Lesson 6-7.

Write (a) the inverse and (b) the contrapositive of each statement.   ◀ See Lesson 2-2.

43. If $x = 51$, then $2x = 102$.

44. If $a = 5$, then $a^2 = 25$.

45. If $b < -4$, then b is negative.

46. If $c > 0$, then c is positive.

47. If the sum of the interior angle measures of a polygon is not 360, then the polygon is not a quadrilateral.

Get Ready! **To prepare for Lesson 6-9, do Exercises 48 and 49.**   ◀ See Lesson 3-7.

48. Find the equation for the line that contains the origin and $(4, 5)$.

49. Find the equation for the line that contains (p, q) and has slope $\frac{a}{b}$.

Quadrilaterals in Quadrilaterals

© Content Standard
G.CO.12 Make formal geometric constructions with a variety of tool and methods . . .

Activity

MATHEMATICAL PRACTICES

© Construct

- Use geometry software to construct a quadrilateral *ABCD*.
- Construct the midpoint of each side of *ABCD*.
- Construct segments joining the midpoints, in order, to form quadrilateral *EFGH*.

© Investigate

- Measure the lengths of the sides of *EFGH* and their slopes.
- Measure the angles of *EFGH*.

What kind of quadrilateral does *EFGH* appear to be?

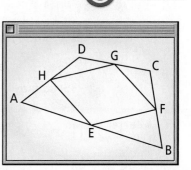

Exercises

1. Manipulate quadrilateral *ABCD*.
   a. Make a conjecture about the quadrilateral with vertices that are the midpoints of the sides of a quadrilateral.
   b. Does your conjecture hold when *ABCD* is concave?
   c. Can you manipulate *ABCD* so that your conjecture doesn't hold?

© 2. **Extend** Draw the diagonals of *ABCD*.
   a. Describe *EFGH* when the diagonals are perpendicular.
   b. Describe *EFGH* when the diagonals are congruent.
   c. Describe *EFGH* when the diagonals are both perpendicular and congruent.

3. Construct the midpoints of *EFGH* and use them to construct quadrilateral *IJKL*. Construct the midpoints of *IJKL* and use them to construct quadrilateral *MNOP*. For *MNOP* and *EFGH*, compare the ratios of the lengths of the sides, perimeters, and areas. How are the sides of *MNOP* and *EFGH* related?

© 4. **Writing** In Exercise 1, you made a conjecture as to the type of quadrilateral *EFGH* appears to be. Prove your conjecture. Include in your proof the Triangle Midsegment Theorem, "If a segment joins the midpoints of two sides of a triangle, then the segment is parallel to the third side and half its length."

5. Describe the quadrilateral formed by joining the midpoints, in order, of the sides of each of the following. Justify each response.
   a. parallelogram   **b.** rectangle   **c.** rhombus   **d.** square
   e. trapezoid   **f.** isosceles trapezoid   **g.** kite

Proofs Using Coordinate Geometry

© Content Standard
G.GPE.4 Use coordinates to prove simple geometric theorems algebraically.

Objective To prove theorems using figures in the coordinate plane

SOLVE IT!

Getting Ready!   ◀▶ ✕ ↻ ▲

The coordinates of three vertices of a rectangle are (−2a, 0), (2a, 0), and (2a, 2b). A diagonal joins one of these points with the fourth vertex. What are the coordinates of the midpoint of the diagonal? Justify your answer.

Better draw a diagram!

© MATHEMATICAL PRACTICES In the Solve It, the coordinates of the point include variables. In this lesson, you will use coordinates with variables to write a coordinate proof.

Essential Understanding You can prove geometric relationships using variable coordinates for figures in the coordinate plane.

Proof © **Problem 1** **Writing a Coordinate Proof**

Plan

What formulas do you need?
You need to find the distance to a midpoint, so use the midpoint and distance formulas.

Use coordinate geometry to prove that the midpoint of the hypotenuse of a right triangle is equidistant from the three vertices.

Given: $\triangle OEF$ is a right triangle.

   M is the midpoint of $\overline{EF}$.

Prove: $EM = FM = OM$

Coordinate Proof:

By the Midpoint Formula, $M = \left(\dfrac{2a + 0}{2}, \dfrac{0 + 2b}{2}\right) = (a, b)$.

By the Distance Formula,

$$OM = \sqrt{a^2 + b^2}$$

$$FM = \sqrt{(2a - a)^2 + (0 - b)^2} \qquad EM = \sqrt{(0 - a)^2 + (2b - b)^2}$$
$$= \sqrt{a^2 + b^2} \qquad\qquad\qquad = \sqrt{a^2 + b^2}$$

Since $EM = FM = OM$, the midpoint of the hypotenuse is equidistant from the vertices of the right triangle.

© ✓ Got It? **1. Reasoning** What is the advantage of using coordinates $O(0, 0)$, $E(0, 2b)$, and $F(2a, 0)$ rather than $O(0, 0)$, $E(0, b)$, and $F(a, 0)$?

In Lesson 6-8, you wrote a plan for the proof of the Trapezoid Midsegment Theorem. Now you will write the full coordinate proof.

Plan

Proof

Problem 2 Writing a Coordinate Proof

Refer to the plan from Lesson 6-8. Find the coordinates of M and N. Determine whether $\overline{MN}$ is parallel to $\overline{OP}$ and $\overline{RA}$. Then find and compare the lengths of $\overline{MN}$, $\overline{OP}$, and $\overline{RA}$.

Write a coordinate proof of the Trapezoid Midsegment Theorem.

Given: $\overline{MN}$ is the midsegment of trapezoid $ORAP$.

Prove: $\overline{MN} \parallel \overline{OP}$, $\overline{MN} \parallel \overline{RA}$, $MN = \frac{1}{2}(OP + RA)$

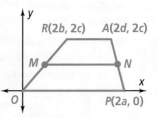

Coordinate Proof:

Use the Midpoint Formula to find the coordinates of M and N.

$$M = \left(\frac{2b + 0}{2}, \frac{2c + 0}{2}\right) = (b, c)$$
$$N = \left(\frac{2a + 2d}{2}, \frac{0 + 2c}{2}\right) = (a + d, c)$$

Use the Slope Formula to determine whether $\overline{MN}$ is parallel to $\overline{OP}$ and $\overline{RA}$.

$$\text{slope of } \overline{MN} = \frac{c - c}{(a + d) - b} = 0$$
$$\text{slope of } \overline{RA} = \frac{2c - 2c}{2d - 2b} = 0$$
$$\text{slope of } \overline{OP} = \frac{0 - 0}{2a - 0} = 0$$

The three slopes are equal, so $\overline{MN} \parallel \overline{OP}$ and $\overline{MN} \parallel \overline{RA}$.

Use the Distance Formula to find and compare MN, OP, and RA.

$$MN = \sqrt{[(a + d) - b]^2 + (c - c)^2} = a + d - b$$
$$OP = \sqrt{(2a - 0)^2 + (0 - 0)^2} = 2a$$
$$RA = \sqrt{(2d - 2b)^2 + (2c - 2c)^2} = 2d - 2b$$

$$MN \overset{?}{=} \tfrac{1}{2}(OP + RA) \qquad \text{Check that } MN = \tfrac{1}{2}(OP + RA) \text{ is true.}$$
$$a + d - b \overset{?}{=} \tfrac{1}{2}[2a + (2d - 2b)] \quad \text{Substitute.}$$
$$a + d - b = a + d - b \checkmark \qquad \text{Simplify.}$$

So, (1) the midsegment of a trapezoid is parallel to its bases, and
   (2) the length of the midsegment of a trapezoid is half the sum of the lengths of the bases.

✓ **Got It?** **2.** Write a coordinate proof of the Triangle Midsegment Theorem (Theorem 5-1).

Lesson Check

Do you know HOW?

1. Use coordinate geometry to prove that the diagonals of a rectangle are congruent.
   a. Place rectangle *PQRS* in the coordinate plane with *P* at $(0, 0)$.
   b. What are the coordinates of *Q*, *R*, and *S*?
   c. Write the *Given* and *Prove* statements.
   d. Write a coordinate proof.

Do you UNDERSTAND? MATHEMATICAL PRACTICES

2. **Reasoning** Describe a good strategy for placing the vertices of a rhombus for a coordinate proof.

3. **Error Analysis** Your classmate places a trapezoid on the coordinate plane. What is the error?

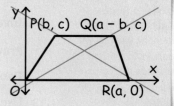

Practice and Problem-Solving Exercises MATHEMATICAL PRACTICES

A Practice **Developing Proof** Complete the following coordinate proofs. ◀ See Problems 1 and 2.

4. The diagonals of an isosceles trapezoid are congruent.
   Given: Trapezoid *EFGH* with $\overline{EF} \cong \overline{GH}$
   Prove: $\overline{EG} \cong \overline{FH}$
   a. Find *EG*.
   b. Find *FH*.
   c. Explain why $\overline{EG} \cong \overline{FH}$.

   $F(-b, c)$ y $G(b, c)$
   $E(-a, 0)$ O x $H(a, 0)$

5. The medians drawn to the congruent sides of an isosceles triangle are congruent.
   Given: $\triangle PQR$ with $\overline{PQ} \cong \overline{RQ}$, *M* is the midpoint of $\overline{PQ}$, *N* is the midpoint of $\overline{RQ}$
   Prove: $\overline{PN} \cong \overline{RM}$
   a. What are the coordinates of *M* and *N*?
   b. What are *PN* and *RM*?
   c. Explain why $\overline{PN} \cong \overline{RM}$.

   y $Q(0, 2b)$
   M N
   $P(-2a, 0)$ O $R(2a, 0)$ x

B Apply Tell whether you can reach each type of conclusion below using coordinate methods. Give a reason for each answer.

6. $\overline{AB} \cong \overline{CD}$

7. $\overline{AB} \parallel \overline{CD}$

8. $\overline{AB} \perp \overline{CD}$

9. $\overline{AB}$ bisects $\overline{CD}$.

10. $\overline{AB}$ bisects $\angle CAD$.

11. $\angle A \cong \angle B$

12. $\angle A$ is a right angle.

13. $AB + BC = AC$

14. $\triangle ABC$ is isosceles.

15. Quadrilateral *ABCD* is a rhombus.

16. $\overline{AB}$ and $\overline{CD}$ bisect each other.

17. $\angle A$ is the supplement of $\angle B$.

18. $\overline{AB}$, $\overline{CD}$, and $\overline{EF}$ are concurrent.

19. Flag Design The flag design at the right is made by connecting **Proof** the midpoints of the sides of a rectangle. Use coordinate geometry to prove that the quadrilateral formed is a rhombus.

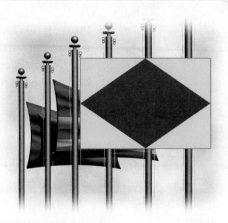

20. Open-Ended Give an example of a statement that you think is easier to prove with a coordinate geometry proof than with a proof method that does not require coordinate geometry. Explain your choice.

Use coordinate geometry to prove each statement.

Proof
21. Think About a Plan If a parallelogram is a rhombus, its diagonals are perpendicular (Theorem 6-13).
  - How will you place the rhombus in a coordinate plane?
  - What formulas will you need to use?

22. The altitude to the base of an isosceles triangle bisects the base.

23. If the midpoints of a trapezoid are joined to form a quadrilateral, then the quadrilateral is a parallelogram.

24. One diagonal of a kite divides the kite into two congruent triangles.

25. You learned in Theorem 5-8 that the centroid of a triangle is two thirds the distance **Proof** from each vertex to the midpoint of the opposite side. Complete the steps to prove this theorem.
  a. Find the coordinates of points L, M, and N, the midpoints of the sides of $\triangle ABC$.
  b. Find equations of $\overleftrightarrow{AM}$, $\overleftrightarrow{BN}$, and $\overleftrightarrow{CL}$.
  c. Find the coordinates of point P, the intersection of $\overleftrightarrow{AM}$ and $\overleftrightarrow{BN}$.
  d. Show that point P is on $\overleftrightarrow{CL}$.
  e. Use the Distance Formula to show that point P is two thirds the distance from each vertex to the midpoint of the opposite side.

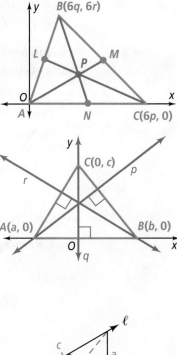

26. Complete the steps to prove Theorem 5-9. You are given $\triangle ABC$ **Proof** with altitudes p, q, and r. Show that p, q, and r intersect at a point (called the orthocenter of the triangle).
  a. The slope of $\overline{BC}$ is $\frac{c}{-b}$. What is the slope of line p?
  b. Show that the equation of line p is $y = \frac{b}{c}(x - a)$.
  c. What is the equation of line q?
  d. Show that lines p and q intersect at $\left(0, \frac{-ab}{c}\right)$.
  e. The slope of $\overline{AC}$ is $\frac{c}{-a}$. What is the slope of line r?
  f. Show that the equation of line r is $y = \frac{a}{c}(x - b)$.
  g. Show that lines r and q intersect at $\left(0, \frac{-ab}{c}\right)$.
  h. What are the coordinates of the orthocenter of $\triangle ABC$?

Challenge

27. Multiple Representations Use the diagram at the right.
  a. Explain using area why $\frac{1}{2}ad = \frac{1}{2}bc$ and therefore $ad = bc$.
  b. Find two ratios for the slope of ℓ. Use these two ratios to show that $ad = bc$.

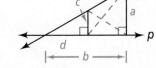

28. Prove: If two lines are perpendicular, the product of their slopes is −1.

Proof **a.** Two nonvertical lines, ℓ_1 and ℓ_2, intersect as shown at the right. Find the coordinates of C.

b. Choose coordinates for D and B. (*Hint:* Find the relationship between $\angle 1$, $\angle 2$, and $\angle 3$. Then use congruent triangles.)

c. Complete the proof that the product of slopes is −1.

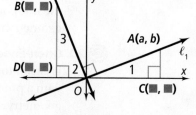

Standardized Test Prep

GRIDDED RESPONSE

SAT/ACT

29. The endpoint of a segment is $(27, -3)$. The midpoint is $(3, 4)$. What is the length of the segment?

30. In the diagram of $\triangle POR$ at the right, $PO = 16$ and $OR = 12$. What is OM?

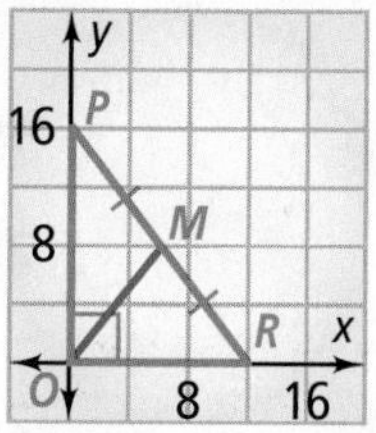

31. $\square FGHI$ has sides with lengths $FG = 2x + 5$, $GH = x + 7$, $HI = 3x - 2$, and $FI = 2x$. What is the length of the longer sides of $\square FGHI$?

32. In $\triangle ABC$, $m\angle A = 55$. If $m\angle C$ is twice $m\angle A$, what is $m\angle B$?

Mixed Review

33. A rectangle $ABCD$ is centered at the origin with $A(-a, b)$. Without using any new variables, what are the coordinates of point C?

◀ See Lesson 6-8.

Explain how you can use SSS, SAS, ASA, or AAS with corresponding parts of congruent triangles to prove each statement true.

◀ See Lessons 4-2, 4-3, and 4-4.

34. $\overline{AB} \cong \overline{CB}$

35. $\angle 1 \cong \angle 2$

36. $\angle K = \angle M$

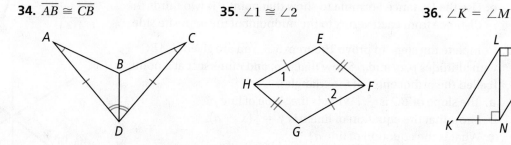

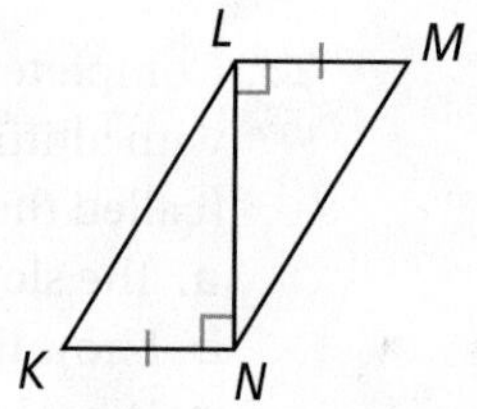

Get Ready! To prepare for Lesson 7-1, do Exercises 37–40.

Algebra Solve. Round to the nearest tenth if necessary.

◀ See p. 827.

37. $x^2 = 144$

38. $r^2 - 3 = 61$

39. $y^2 + 10 = 35$

40. $7^2 + k^2 = 18^2$

Pull It **All Together** © ASSESSMENT

> To solve these problems you will pull together many concepts and skills that you have learned about polygons.

BIG idea Measurement and Reasoning and Proof

You can find the interior angle measures in a regular polygon and then use what you know about triangles and special quadrilaterals to analyze complex figures.

© **Performance Task 1**

ABCDEF is a regular hexagon. What is the most precise classification of quadrilateral *GBHE*? How do you know? What are the interior angle measures of *GBHE*?

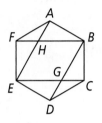

© **Performance Task 2**

JKLM is a parallelogram. If you extend each side by a distance *x*, what kind of quadrilateral is *PQRS*? How do you know?

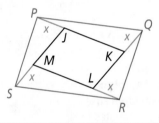

BIG idea Coordinate Geometry and Reasoning and Proof

You can use coordinates and certain relationships between pairs of corresponding parts to prove triangles congruent in the coordinate plane.

© **Performance Task 3**

What are two methods for proving the two triangles congruent? Use one of your methods along with coordinate geometry to prove that the two triangles are congruent.

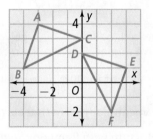

Connecting **BIG** ideas and Answering the Essential Questions

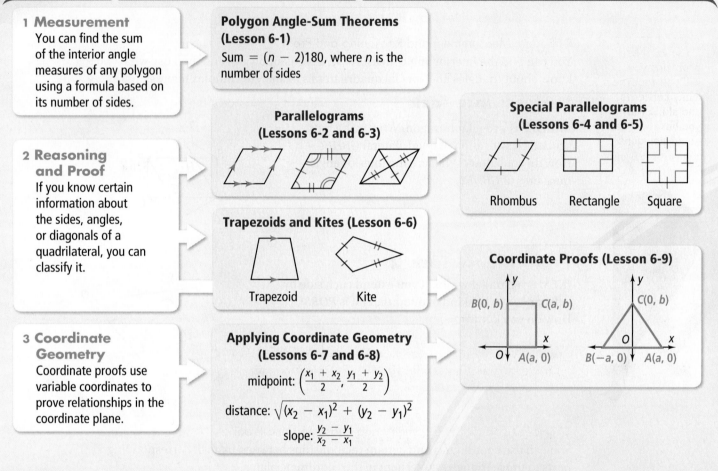

1 Measurement
You can find the sum of the interior angle measures of any polygon using a formula based on its number of sides.

Polygon Angle-Sum Theorems (Lesson 6-1)
Sum $= (n - 2)180$, where n is the number of sides

2 Reasoning and Proof
If you know certain information about the sides, angles, or diagonals of a quadrilateral, you can classify it.

Parallelograms (Lessons 6-2 and 6-3)

Special Parallelograms (Lessons 6-4 and 6-5)

Rhombus   Rectangle   Square

Trapezoids and Kites (Lesson 6-6)

Trapezoid   Kite

3 Coordinate Geometry
Coordinate proofs use variable coordinates to prove relationships in the coordinate plane.

Applying Coordinate Geometry (Lessons 6-7 and 6-8)
midpoint: $\left(\dfrac{x_1 + x_2}{2}, \dfrac{y_1 + y_2}{2} \right)$

distance: $\sqrt{(x_2 - x_1)^2 + (y_2 - y_1)^2}$

slope: $\dfrac{y_2 - y_1}{x_2 - x_1}$

Coordinate Proofs (Lesson 6-9)

$B(0, b)$ $C(a, b)$
O $A(a, 0)$

$C(0, b)$
$B(-a, 0)$ O $A(a, 0)$

Chapter Vocabulary

- base, base angle, and leg of a trapezoid (p. 389)
- consecutive angles (p. 360)
- coordinate proof (p. 408)
- equiangular, equilateral polygon (p. 354)
- isosceles trapezoid (p. 389)
- kite (p. 392)
- midsegment of a trapezoid (p. 391)
- opposite angles (p. 359)
- opposite sides (p. 359)
- parallelogram (p. 359)
- rectangle (p. 375)
- regular polygon (p. 354)
- rhombus (p. 375)
- square (p. 375)
- trapezoid (p. 389)

Choose the vocabulary term that correctly completes the sentence.

1. A parallelogram with four congruent sides is a(n) __?__ .

2. A polygon with all angles congruent is a(n) __?__ .

3. Angles of a polygon that share a side are __?__ .

4. A(n) __?__ is a quadrilateral with exactly one pair of parallel sides.

6-1 The Polygon Angle-Sum Theorems

Quick Review

The sum of the measures of the interior angles of an n-gon is $(n - 2)180$. The measure of one interior angle of a regular n-gon is $\frac{(n - 2)180}{n}$. The sum of the measures of the exterior angles of a polygon, one at each vertex, is 360.

Example

Find the measure of an interior angle of a regular 20-gon.

$$\text{Measure} = \frac{(n - 2)\,180}{n} \qquad \text{Corollary to the Polygon Angle-Sum Theorem}$$

$$= \frac{(20 - 2)180}{20} \qquad \text{Substitute.}$$

$$= \frac{18 \cdot 180}{20} \qquad \text{Simplify.}$$

$$= 162$$

The measure of an interior angle is 162.

Exercises

Find the measure of an interior angle and an exterior angle of each regular polygon.

5. hexagon   **6.** 16-gon   **7.** pentagon

8. What is the sum of the exterior angles for each polygon in Exercises 5–7?

Find the measure of the missing angle.

9.
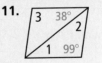

10.

6-2 Properties of Parallelograms

Quick Review

Opposite sides and **opposite angles** of a **parallelogram** are congruent. **Consecutive angles** in a parallelogram are supplementary. The diagonals of a parallelogram bisect each other. If three (or more) parallel lines cut off congruent segments on one transversal, then they cut off congruent segments on every transversal.

Example

Find the measures of the numbered angles in the parallelogram.

Since consecutive angles are supplementary, $m\angle 1 = 180 - 56$, or 124. Since opposite angles are congruent, $m\angle 2 = 56$ and $m\angle 3 = 124$.

Exercises

Find the measures of the numbered angles for each parallelogram.

11.   **12.**

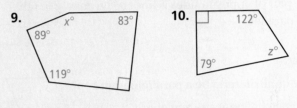

13.   **14.**

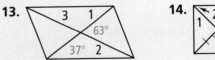

Find the values of x and y in $\square ABCD$.

15. $AB = 2y,\ BC = y + 3,\ CD = 5x - 1,\ DA = 2x + 4$

16. $AB = 2y + 1,\ BC = y + 1,\ CD = 7x - 3,\ DA = 3x$

6-3 Proving That a Quadrilateral Is a Parallelogram

Quick Review

A quadrilateral is a parallelogram if any one of the following is true.

- Both pairs of opposite sides are parallel.
- Both pairs of opposite sides are congruent.
- Consecutive angles are supplementary.
- Both pairs of opposite angles are congruent.
- The diagonals bisect each other.
- One pair of opposite sides is both congruent and parallel.

Example

Must the quadrilateral be a parallelogram?

Yes, both pairs of opposite angles are congruent.

Exercises

Determine whether the quadrilateral must be a parallelogram.

17. 18.

Algebra Find the values of the variables for which *ABCD* must be a parallelogram.

19. 20.

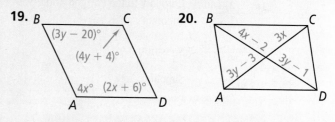

6-4 Properties of Rhombuses, Rectangles, and Squares

Quick Review

A **rhombus** is a parallelogram with four congruent sides.

A **rectangle** is a parallelogram with four right angles.

A **square** is a parallelogram with four congruent sides and four right angles.

The diagonals of a rhombus are perpendicular. Each diagonal bisects a pair of opposite angles.

The diagonals of a rectangle are congruent.

Example

What are the measures of the numbered angles in the rhombus?

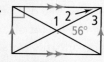

$m\angle 1 = 60$ Each diagonal of a rhombus bisects a pair of opposite angles.

$m\angle 2 = 90$ The diagonals of a rhombus are $\perp$.

$60 + m\angle 2 + m\angle 3 = 180$ Triangle Angle-Sum Thm.

$60 + 90 + m\angle 3 = 180$ Substitute.

$m\angle 3 = 30$ Simplify.

Exercises

Find the measures of the numbered angles in each special parallelogram.

21. 22.

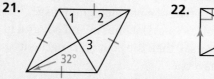

Determine whether each statement is *always*, *sometimes*, or *never* true.

23. A rhombus is a square.

24. A square is a rectangle.

25. A rhombus is a rectangle.

26. The diagonals of a parallelogram are perpendicular.

27. The diagonals of a parallelogram are congruent.

28. Opposite angles of a parallelogram are congruent.

6-5 Conditions for Rhombuses, Rectangles, and Squares

Quick Review

If one diagonal of a parallelogram bisects two angles of the parallelogram, then the parallelogram is a rhombus. If the diagonals of a parallelogram are perpendicular, then the parallelogram is a rhombus. If the diagonals of a parallelogram are congruent, then the parallelogram is a rectangle.

Example

Can you conclude that the parallelogram is a rhombus, rectangle, or square? Explain.

Yes, the diagonals are perpendicular, so the parallelogram is a rhombus.

Exercises

Can you conclude that the parallelogram is a rhombus, rectangle, or square? Explain.

29. 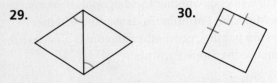   **30.**

For what value of x is the figure the given parallelogram? Justify your answer.

31. Rhombus   **32.** Rectangle

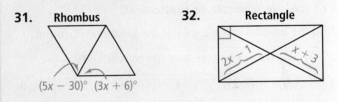

31. $(5x - 30)°$ $(3x + 6)°$

32. $2x - 1$ $x + 3$

6-6 Trapezoids and Kites

Quick Review

The parallel sides of a **trapezoid** are its **bases** and the nonparallel sides are its **legs.** Two angles that share a base of a trapezoid are **base angles** of the trapezoid. The **midsegment of a trapezoid** joins the midpoints of its legs.

The base angles of an isosceles trapezoid are congruent. The diagonals of an isosceles trapezoid are congruent.

The diagonals of a kite are perpendicular.

Example

ABCD is an isosceles trapezoid. What is $m\angle C$?

Since $\overline{BC} \parallel \overline{AD}$, $\angle C$ and $\angle D$ are same-side interior angles.

$m\angle C + m\angle D = 180$ — Same-side interior angles are supplementary.

$m\angle C + 60 = 180$ — Substitute.

$m\angle C = 120$ — Subtract 60 from each side.

Exercises

Find the measures of the numbered angles in each isosceles trapezoid.

33. 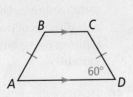   **34.**

Find the measures of the numbered angles in each kite.

35. 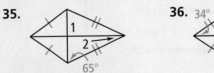   **36.** $34°$ $38°$

37. Algebra A trapezoid has base lengths of $(6x - 1)$ units and 3 units. Its midsegment has a length of $(5x - 3)$ units. What is the value of x?

6-7 Polygons in the Coordinate Plane

Quick Review

To determine whether sides or diagonals are congruent, use the Distance Formula. To determine the coordinate of the midpoint of a side, or whether the diagonals bisect each other, use the Midpoint Formula. To determine whether opposite sides are parallel, or whether diagonals or sides are perpendicular, use the Slope Formula.

Example

$\triangle XYZ$ has vertices $X(1, 0)$, $Y(-2, -4)$, and $Z(4, -4)$. Is $\triangle XYZ$ *scalene*, *isosceles*, or *equilateral*?

To find the lengths of the legs, use the Distance Formula.

$XY = \sqrt{(-2 - 1)^2 + (-4 - 0)^2} = \sqrt{9 + 16} = 5$

$YZ = \sqrt{(4 - (-2))^2 + (-4 - (-4))^2} = \sqrt{36 + 0} = 6$

$XZ = \sqrt{(4 - 1)^2 + (-4 - 0)^2} = \sqrt{9 + 16} = 5$

Two side lengths are equal, so $\triangle XYZ$ is isosceles.

Exercises

Determine whether $\triangle ABC$ is *scalene*, *isosceles*, or *equilateral*.

38. 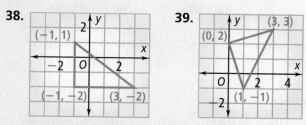   **39.**

What is the most precise classification of the quadrilateral?

40. $G(2, 5)$, $R(5, 8)$, $A(-2, 12)$, $D(-5, 9)$

41. $F(-13, 7)$, $I(1, 12)$, $N(15, 7)$, $E(1, -5)$

42. $Q(4, 5)$, $U(12, 14)$, $A(20, 5)$, $D(12, -4)$

43. $W(-11, 4)$, $H(-9, 10)$, $A(2, 10)$, $T(4, 4)$

6-8 and 6-9 Coordinate Geometry and Coordinate Proofs

Quick Review

When placing a figure in the coordinate plane, it is usually helpful to place at least one side on an axis. Use variables when naming the coordinates of a figure in order to show that relationships are true for a general case.

Example

Rectangle PQRS has length a and width $4b$. The x-axis bisects $\overline{PS}$ and $\overline{QR}$. What are the coordinates of the vertices?

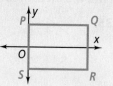

Since the width of $PQRS$ is $4b$ and the x-axis bisects $\overline{PS}$ and $\overline{QR}$, all the vertices are $2b$ units from the x-axis. $\overline{PS}$ is on the y-axis, so $P = (0, 2b)$ and $S = (0, -2b)$. The length of $PQRS$ is a, so $Q = (a, 2b)$ and $R = (a, -2b)$.

Exercises

44. In rhombus *FLPS*, the axes form the diagonals. If $SL = 2a$ and $FP = 4b$, what are the coordinates of the vertices?

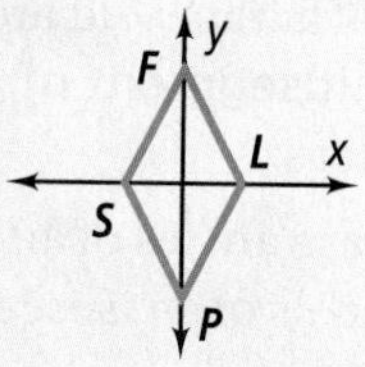

45. The figure at the right is a parallelogram. Give the coordinates of point P without using any new variables.

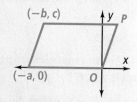

46. Use coordinate geometry to prove that the quadrilateral formed by connecting the midpoints of a kite is a rectangle.

Do you know HOW?

1. What is the sum of the interior angle measures of a polygon with 15 sides?

2. What is the measure of an exterior angle of a 25-gon?

Graph each quadrilateral ABCD. Then determine the most precise name for it.

3. $A(1, 2)$, $B(11, 2)$, $C(7, 5)$, $D(4, 5)$

4. $A(3, -2)$, $B(5, 4)$, $C(3, 6)$, $D(1, 4)$

5. $A(1, -4)$, $B(1, 1)$, $C(-2, 2)$, $D(-2, -3)$

Algebra Find the values of the variables for each quadrilateral.

6.   **7.**

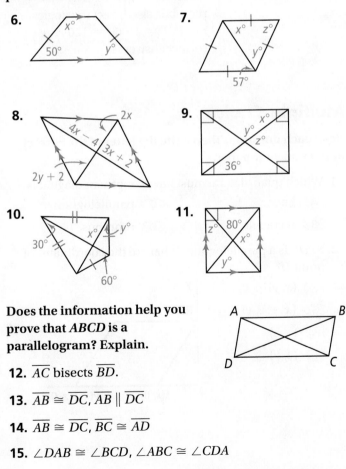

8.   **9.**

10.   **11.**

Does the information help you prove that ABCD is a parallelogram? Explain.

12. $\overline{AC}$ bisects $\overline{BD}$.

13. $\overline{AB} \cong \overline{DC}$, $\overline{AB} \parallel \overline{DC}$

14. $\overline{AB} \cong \overline{DC}$, $\overline{BC} \cong \overline{AD}$

15. $\angle DAB \cong \angle BCD$, $\angle ABC \cong \angle CDA$

16. Algebra Determine the values of the variables for which ABCD is a parallelogram.

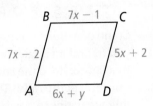

Give the coordinates for points S and T without using any new variables. Then find the midpoint and the slope of $\overline{ST}$.

17. rectangle

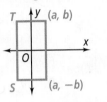

18. parallelogram

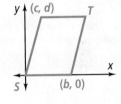

19. Prove that the diagonals of square ABCD are congruent.

20. Sketch two noncongruent parallelograms ABCD and EFGH such that $\overline{AC} \cong \overline{BD} \cong \overline{EG} \cong \overline{FH}$.

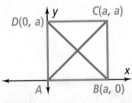

Do you UNDERSTAND?

21. Open-Ended Write the coordinates of four points that determine each figure with the given conditions. One vertex is at the origin and one side is 3 units long.

   a. square   **b.** parallelogram

   c. rectangle   **d.** trapezoid

22. Writing Explain why a square cannot be a kite.

23. Error Analysis Your classmate says, "If the diagonals of a quadrilateral intersect to form four congruent triangles, then the quadrilateral is always a parallelogram." How would you correct this statement? Explain your answer.

24. Reasoning PQRS has vertices $P(0, 0)$, $Q(4, 2)$, and $S(4, -2)$. Its diagonals intersect at $H(4, 0)$. What are the possible coordinates of R for PQRS to be a kite? Explain.

TIPS FOR SUCCESS

Some questions on tests ask you to find an angle measure in a figure. Read the sample question at the right. Then follow the tips to answer it.

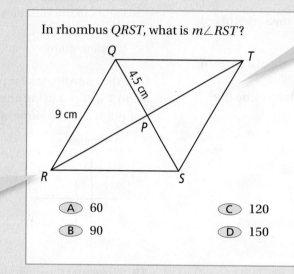

In rhombus $QRST$, what is $m\angle RST$?

A. 60
B. 90
C. 120
D. 150

TIP 1

The diagonal $\overline{QS}$ forms $\triangle QRS$.

TIP 2

Use the properties of a rhombus to find RS and QS.

Think It Through

$RS = 9$ cm because a rhombus has four congruent sides. The diagonals of a rhombus bisect each other, so $QS = 2 \cdot 4.5 = 9$ cm.

Since $QR = RS = QS$, $\triangle QRS$ is equilateral, and thus equiangular. So $m\angle RSQ = 60$. Since the diagonals of a rhombus also bisect the angles, $m\angle RST = 2 \cdot 60 = 120$. The correct answer is C.

Vocabulary Builder

As you solve test items, you must understand the meanings of mathematical terms. Choose the correct term to complete each sentence.

A. Two lines that intersect to form right angles are (parallel, perpendicular) lines.

B. A quadrilateral with two pairs of adjacent sides congruent and no opposite sides congruent is called a (rhombus, kite).

C. The (circumcenter, incenter) of a triangle is the point of concurrency of the perpendicular bisectors of the sides of the triangle.

D. Two angles are (complementary, supplementary) angles if the sum of their measures is 90.

E. In a plane, two lines that never intersect are (parallel, skew) lines.

Multiple Choice

Read each question. Then write the letter of the correct answer on your paper.

1. Which quadrilateral must have congruent diagonals?
A. kite
B. rectangle
C. parallelogram
D. rhombus

2. $STUV$ is a parallelogram. What are the coordinates of point U?
F. (x, y)
G. $(x + z, y)$
H. (y, z)
I. (z, y)

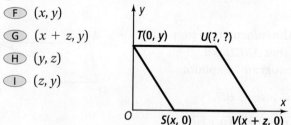

3. Which list could represent the lengths of the sides of a triangle?

(A) 7 cm, 10 cm, 25 cm

(B) 4 in., 6 in., 10 in.

(C) 1 ft, 2 ft, 4 ft

(D) 3 m, 5 m, 7 m

4. Which quadrilateral CANNOT contain four right angles?

(F) square

(H) trapezoid

(G) rhombus

(I) rectangle

5. What is the circumcenter of $\triangle ABC$ with vertices $A(-7, 0)$, $B(-3, 8)$, and $C(-3, 0)$?

(A) $(-7, -3)$

(C) $(-4, 3)$

(B) $(-5, 4)$

(D) $(-3, 4)$

6. $ABCD$ is a rhombus. To prove that the diagonals of a rhombus are perpendicular, which pair of angles below must you prove congruent by using corresponding parts of congruent triangles?

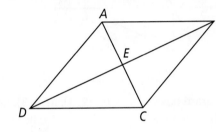

(F) $\angle AEB$ and $\angle DEC$

(G) $\angle AEB$ and $\angle AED$

(H) $\angle BEC$ and $\angle AED$

(I) $\angle DAB$ and $\angle ABC$

7. $FGHJ$ is a quadrilateral. If at least one pair of opposite angles in quadrilateral $FGHJ$ is congruent, which statement is false?

(A) Quadrilateral $FGHJ$ is a trapezoid.

(B) Quadrilateral $FGHJ$ is a rhombus.

(C) Quadrilateral $FGHJ$ is a kite.

(D) Quadrilateral $FGHJ$ is a parallelogram.

8. For which value of x are lines g and h parallel?

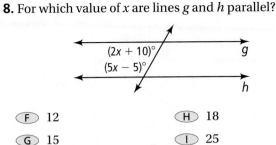

(F) 12

(H) 18

(G) 15

(I) 25

9. In $\triangle GHJ$, $\overline{GH} \cong \overline{HJ}$. Using the indirect proof method, you attempt to prove that $\angle G$ and $\angle J$ are right angles. Which theorem will contradict this claim?

(A) Triangle Angle-Sum Theorem

(B) Side-Angle-Side Theorem

(C) Converse of the Isosceles Triangle Theorem

(D) Angle-Angle-Side Theorem

10. Which angles could an obtuse triangle have?

(F) two right angles

(H) two obtuse angles

(G) two acute angles

(I) two vertical angles

11. What values of x and y make the quadrilateral below a parallelogram?

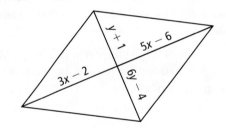

(A) $x = 2, y = 1$

(C) $x = 1, y = 2$

(B) $x = 3, y = 5$

(D) $x = 2, y = \frac{9}{7}$

12. Which is the most valid conclusion based on the statements below?

If a triangle is equilateral, then it is isosceles. $\triangle ABC$ is not equilateral.

(F) $\triangle ABC$ is not isosceles.

(G) $\triangle ABC$ is isosceles.

(H) $\triangle ABC$ may or may not be isosceles.

(I) $\triangle ABC$ is equilateral.

13. What is $m\angle 1$ in the figure below?

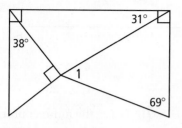

14. $\angle ABE$ and $\angle CBD$ are vertical angles and both are complementary with $\angle FGH$. If $m\angle ABE = (3x - 1)$, and $m\angle FGH = 4x$, what is $m\angle CBD$?

15. What is the value of x in the kite below?

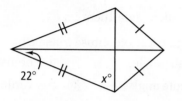

16. A 3-ft-wide walkway is placed around an animal exhibit at the zoo. The exhibit is rectangular in shape and has length 15 ft and width 8 ft. What is the area, in square feet, of the walkway around the exhibit?

17. The outer walls of the Pentagon in Arlington, Virginia, are formed by two regular pentagons, as shown below. What is the value of x?

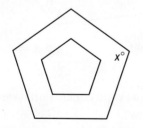

Short Response

18. What are the possible values for n to make ABC a valid triangle? Show your work.

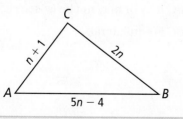

19. The pattern of a soccer ball contains regular hexagons and regular pentagons. The figure at the right shows what a section of the pattern would look like on a flat surface. Use the fact that there are 360° in a circle to explain why there are gaps between the hexagons.

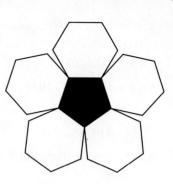

Extended Response

20. Prove that $GHIJ$ is a rhombus.

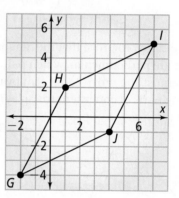

21. A parallelogram has vertices $L(-2, 5)$, $M(3, 3)$, and $N(1, 0)$. What are possible coordinates for its fourth vertex? Explain.

22. Jim is clearing out trees and planting grass to enlarge his backyard. He plans to double both the width and the length of his current rectangular backyard, as shown below.

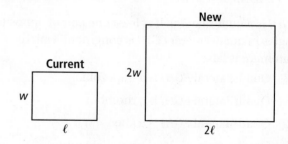

a. How will the area of Jim's new backyard compare to the area of his current backyard?

b. Will it take Jim twice as long to mow his new backyard than his old backyard? Explain. (Assume Jim mows at the same rate for both yards.)

Get Ready!

Lesson 3-2 ◀ **Properties of Parallel Lines**

**Use the diagram at the right. Find the measure of each angle.
Justify your answer.**

1. ∠1 **2.** ∠2 **3.** ∠3 **4.** ∠4

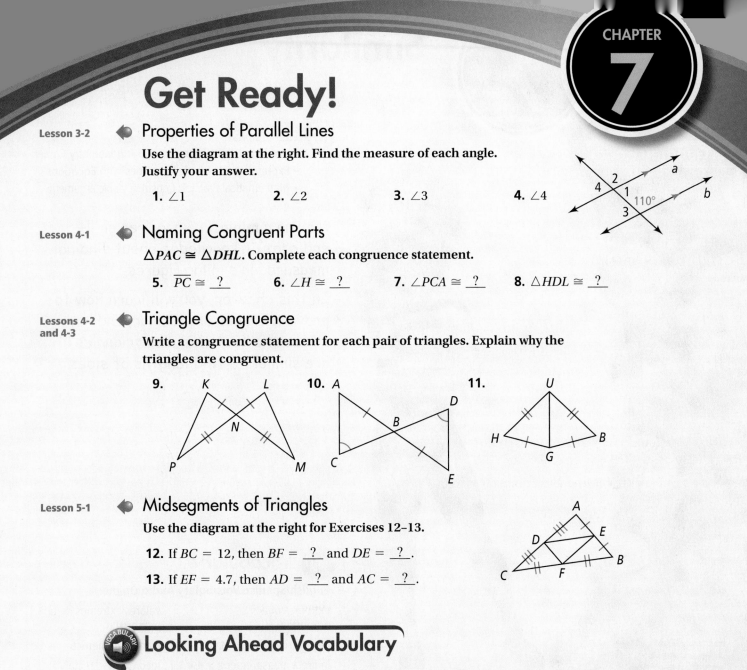

Lesson 4-1 ◀ **Naming Congruent Parts**

△*PAC* ≅ △*DHL*. Complete each congruence statement.

5. $\overline{PC} \cong$? **6.** ∠*H* ≅ ? **7.** ∠*PCA* ≅ ? **8.** △*HDL* ≅ ?

**Lessons 4-2
and 4-3** ◀ **Triangle Congruence**

**Write a congruence statement for each pair of triangles. Explain why the
triangles are congruent.**

9. **10.** **11.**

Lesson 5-1 ◀ **Midsegments of Triangles**

Use the diagram at the right for Exercises 12–13.

12. If *BC* = 12, then *BF* = ? and *DE* = ? .

13. If *EF* = 4.7, then *AD* = ? and *AC* = ? .

🔊 Looking Ahead Vocabulary

14. An artist sketches a person. She is careful to draw the different parts of the person's
body in *proportion*. What does *proportion* mean in this situation?

15. Siblings often look *similar* to each other. How might two geometric figures be
similar?

16. A road map has a *scale* on it that tells you how many miles are equivalent to
a distance of 1 inch on the map. How would you use the scale to estimate the
distance between two cities on the map?

Similarity

© **DOMAINS**
• Similarity, Right Triangles, and Trigonometry
• Expressing Geometric Properties with Equations
• Mathematical Practice: Construct viable arguments

Making a miniature requires skill— and a lot of knowledge about finding measures in similar figures.

In this chapter, you will learn how to prove triangles are similar, and how to use the fact that two triangles are similar to find lengths of sides.

Vocabulary

English/Spanish Vocabulary Audio Online:

English	Spanish
extremes of a proportion, *p. 434*	valores extremos de una proporción
geometric mean, *p. 462*	media geométrica
indirect measurement, *p. 454*	medición indirecta
means of a proportion, *p. 434*	valores medios de una proporción
proportion, *p. 434*	proporción
ratio, *p. 432*	razón
scale drawing, *p. 443*	dibujo a escala
scale factor, *p. 440*	factor de escala
similar, *p. 440*	semejante
similar polygons, *p. 440*	polígonos semejantes

My Math Video

00:04:04

VIDEO ▷

BIG ideas

1 Similarity
Essential Question How do you use proportions to find side lengths in similar polygons?

2 Reasoning and Proof
Essential Question How do you show two triangles are similar?

3 Visualization
Essential Question How do you identify corresponding parts of similar triangles?

Chapter Preview

7-1 Ratios and Proportions
7-2 Similar Polygons
7-3 Proving Triangles Similar
7-4 Similarity in Right Triangles
7-5 Proportions in Triangles

Ratios and Proportions

© Content Standard
Prepares for G.SRT.5 Use . . . similarity criteria for triangles to solve problems and to prove relationships in geometric figures.

Objective To write ratios and solve proportions

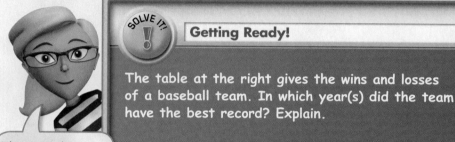

SOLVE IT!

Getting Ready!

The table at the right gives the wins and losses of a baseball team. In which year(s) did the team have the best record? Explain.

Year	Wins	Losses
1890	60	24
1930	110	44
1970	110	52
2010	108	54

The year the team had the most wins is not necessarily the year in which it had the best record.

© MATHEMATICAL PRACTICES

In the Solve It, you compared two quantities for four years.

Essential Understanding You can write a *ratio* to compare two quantities.

Lesson Vocabulary
• ratio
• extended ratio
• proportion
• extremes
• means
• Cross Products Property

A **ratio** is a comparison of two quantities by division. You can write the ratio of two numbers a and b, where $b \neq 0$, in three ways: $\frac{a}{b}$, $a : b$, and a to b. You usually express a and b in the same unit and write the ratio in simplest form.

© Problem 1 Writing a Ratio

Bonsai Trees The bonsai bald cypress tree is a small version of a full-size tree. A Florida bald cypress tree called the Senator stands 118 ft tall. What is the ratio of the height of the bonsai to the height of the Senator?

Think

How can you write the heights using the same unit?
You can convert the height of the Senator to inches or the height of the bonsai tree to feet.

Express both heights in the same unit. To convert 118 ft to inches, multiply by the conversion factor $\frac{12 \text{ in.}}{1 \text{ ft}}$.

$$118 \text{ ft} = \frac{118 \text{ ft}}{1} \cdot \frac{12 \text{ in.}}{1 \text{ ft}} = (118 \cdot 12) \text{ in.} = 1416 \text{ in.}$$

Write the ratio as a fraction in simplest form.

height of bonsai → $\frac{15 \text{ in.}}{118 \text{ ft}} = \frac{15 \text{ in.}}{1416 \text{ in.}} = \frac{(3 \cdot 5) \text{ in.}}{(3 \cdot 472) \text{ in.}} = \frac{5}{472}$
height of Senator →

15 in.

The ratio of the height of the bonsai to the height of the Senator is $\frac{5}{472}$ or 5 : 472.

Got It? **1.** A bonsai tree is 18 in. wide and stands 2 ft tall. What is the ratio of the width of the bonsai to its height?

Problem 2 Dividing a Quantity Into a Given Ratio

Fundraising Members of the school band are buying pots of tulips and pots of daffodils to sell at their fundraiser. They plan to buy 120 pots of flowers. The ratio $\frac{\text{number of tulip pots}}{\text{number of daffodil pots}}$ will be $\frac{2}{3}$. How many pots of each type of flower should they buy?

Plan

How do you write expressions for the numbers of pots?
Multiply the numerator 2 and the denominator 3 by the factor x. $\frac{2x}{3x} = \frac{2}{3}$.

Think

If the ratio $\frac{\text{number of tulip pots}}{\text{number of daffodil pots}}$ is $\frac{2}{3}$, it must be in the form $\frac{2x}{3x}$.

The total number of flower pots is 120. Use this fact to write an equation. Then solve for x.

Substitute 24 for x in the expressions for the numbers of pots.

Write the answer in words.

Write

Let $2x$ = the number of tulip pots.

Let $3x$ = the number of daffodil pots.

$2x + 3x = 120$
$5x = 120$
$x = 24$

$2x = 2(24) = 48$

$3x = 3(24) = 72$

The band members should buy 48 tulip pots and 72 daffodil pots.

Got It? **2.** The measures of two supplementary angles are in the ratio 1 : 4. What are the measures of the angles?

An **extended ratio** compares three (or more) numbers. In the extended ratio $a : b : c$, the ratio of the first two numbers is $a : b$, the ratio of the last two numbers is $b : c$, and the ratio of the first and last numbers is $a : c$.

Problem 3 Using an Extended Ratio

The lengths of the sides of a triangle are in the extended ratio 3 : 5 : 6. The perimeter of the triangle is 98 in. What is the length of the longest side?

Sketch the triangle. Use the extended ratio to label the sides with expressions for their lengths.

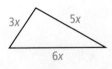

$3x + 5x + 6x = 98$ The perimeter is 98 in.

$14x = 98$ Simplify.

$x = 7$ Divide each side by 14.

Think

How do you use the solution of the equation to answer the question?
Substitute the value for x in the expression for the length of the longest side.

The expression that represents the length of the longest side is $6x$. $6(7) = 42$, so the length of the longest side is 42 in.

Got It? **3.** The lengths of the sides of a triangle are in the extended ratio 4 : 7 : 9. The perimeter is 60 cm. What are the lengths of the sides?

Essential Understanding If two ratios are equivalent, you can write an equation stating that the ratios are equal. If the equation contains a variable, you can solve the equation to find the value of the variable.

An equation that states that two ratios are equal is called a **proportion.** The first and last numbers in a proportion are the **extremes.** The middle two numbers are the **means.**

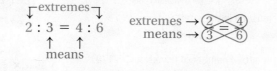

$$2 : 3 = 4 : 6$$

extremes →
means →

take note

Key Concept   Cross Products Property

Words	**Symbols**	**Example**
In a proportion, the product of the extremes equals the product of the means.	If $\frac{a}{b} = \frac{c}{d}$, where $b \neq 0$ and $d \neq 0$, then $ad = bc$.	$\frac{2}{3} = \frac{4}{6}$ $2 \cdot 6 = 3 \cdot 4$ $12 = 12$

Here's Why It Works Begin with $\frac{a}{b} = \frac{c}{d}$, where $b \neq 0$ and $d \neq 0$.

$$bd \cdot \frac{a}{b} = \frac{c}{d} \cdot bd \qquad \text{Multiply each side of the proportion by } bd.$$

$$\frac{\cancel{b}d}{1} \cdot \frac{a}{\cancel{b}} = \frac{c}{\cancel{d}} \cdot \frac{b\cancel{d}}{1} \qquad \text{Divide the common factors.}$$

$$ad = bc \qquad \text{Simplify.}$$

Problem 4   Solving a Proportion

Algebra What is the solution of each proportion?

A
$$\frac{6}{x} = \frac{5}{4}$$
$$6(4) = 5x \qquad \text{Cross Products Property}$$
$$24 = 5x \qquad \text{Simplify.}$$
$$x = \frac{24}{5} \qquad \text{Solve for the variable.}$$

The solution is $\frac{24}{5}$ or 4.8.

B
$$\frac{y+4}{9} = \frac{y}{3}$$
$$3(y + 4) = 9y$$
$$3y + 12 = 9y$$
$$12 = 6y$$
$$y = 2$$

The solution is 2.

Think

Does the solution check?

$$\frac{6}{\frac{24}{5}} \overset{?}{=} \frac{5}{4}$$

$$6 \cdot 4 \overset{?}{=} \frac{24}{5} \cdot 5$$

$$24 = 24 \ \checkmark$$

Got It? **4.** What is the solution of each proportion?

a. $\dfrac{9}{2} = \dfrac{a}{14}$

b. $\dfrac{15}{m+1} = \dfrac{3}{m}$

Using the Properties of Equality, you can rewrite proportions in equivalent forms.

take note

Key Concept Properties of Proportions

a, b, c, and d do not equal zero.

Property

(1) $\frac{a}{b} = \frac{c}{d}$ is equivalent to $\frac{b}{a} = \frac{d}{c}$.

(2) $\frac{a}{b} = \frac{c}{d}$ is equivalent to $\frac{a}{c} = \frac{b}{d}$.

(3) $\frac{a}{b} = \frac{c}{d}$ is equivalent to $\frac{a+b}{b} = \frac{c+d}{d}$.

How to Apply It

Write the reciprocal of each ratio.

$\left(\frac{2}{3} = \frac{4}{6}\right)$ becomes $\frac{3}{2} = \frac{6}{4}$.

Switch the means.

$\frac{2}{3} \diagup\!\!\!\!\diagdown \frac{4}{6}$ becomes $\frac{2}{4} = \frac{3}{6}$.

In each ratio, add the denominator to the numerator.

$\frac{2}{3} = \frac{4}{6}$ becomes $\frac{2+3}{3} = \frac{4+6}{6}$.

© **Problem 5** **Writing Equivalent Proportions**

In the diagram, $\frac{x}{6} = \frac{y}{7}$. What ratio completes the equivalent proportion $\frac{x}{y} = \frac{\blacksquare}{\blacksquare}$? Justify your answer.

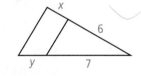

Plan

How do you decide which property of proportions applies?
Look at how the positions of the known parts of the incomplete proportion relate to their positions in the original proportion.

Method 1

$\frac{x}{6} = \frac{y}{7}$

$\frac{x}{y} = \frac{6}{7}$ Property of Proportions (2)

Method 2

$\frac{x}{6} = \frac{y}{7}$

$7x = 6y$ Cross Products Property

$\frac{7x}{7y} = \frac{6y}{7y}$ To solve for $\frac{x}{y}$, divide each side by $7y$.

$\frac{x}{y} = \frac{6}{7}$ Simplify.

The ratio that completes the proportion is $\frac{6}{7}$.

© ✓ **Got It?** **5.** For parts (a) and (b), use the proportion $\frac{x}{6} = \frac{y}{7}$. What ratio completes the equivalent proportion? Justify your answer.

a. $\frac{6}{x} = \frac{\blacksquare}{\blacksquare}$

b. $\frac{\blacksquare}{\blacksquare} = \frac{y+7}{7}$

c. Reasoning Explain why $\frac{6}{x-6} = \frac{7}{y-7}$ is an equivalent proportion to $\frac{x}{6} = \frac{y}{7}$.

Lesson Check

Do you know HOW?

1. To the nearest millimeter, a cell phone is 84 mm long and 46 mm wide. What is the ratio of the width to the length?

2. Two angle measures are in the ratio 5 : 9. Write expressions for the two angle measures in terms of the variable x.

3. What is the solution of the proportion $\frac{20}{z} = \frac{5}{3}$?

4. For $\frac{a}{7} = \frac{13}{b}$ complete each equivalent proportion.

    a. $\frac{a}{\blacksquare} = \frac{7}{\blacksquare}$ **b.** $\frac{a-7}{7} = \frac{\blacksquare}{\blacksquare}$ **c.** $\frac{7}{a} = \frac{\blacksquare}{\blacksquare}$

Do you UNDERSTAND?

5. **Vocabulary** What is the difference between a ratio and a proportion?

6. **Open-Ended** The lengths of the sides of a triangle are in the extended ratio 3 : 6 : 7. What are two possible sets of side lengths, in inches, for the triangle?

7. **Error Analysis** What is the error in the solution of the proportion shown at the right?

    $$\frac{7}{3} = \frac{4}{x}$$
    $$28 = 3x$$
    $$\frac{28}{3} = x$$

8. What is a proportion that has means 6 and 18 and extremes 9 and 12?

Practice and Problem-Solving Exercises

A Practice   **Write the ratio of the first measurement to the second measurement.**   ◄ See Problem 1.

9. length of a tennis racket: 2 ft 4 in.
   length of a table tennis paddle: 10 in.

10. height of a table tennis net: 6 in.
    height of a tennis net: 3 ft

11. diameter of a table tennis ball: 40 mm
    diameter of a tennis ball: 6.8 cm

12. length of a tennis court: 26 yd
    length of a table tennis table: 9 ft

13. **Baseball** A baseball team played 154 regular season games. The ratio of the number of games they won to the number of games they lost was $\frac{5}{2}$. How many games did they win? How many games did they lose?   ◄ See Problem 2.

14. The measures of two supplementary angles are in the ratio 5 : 7. What is the measure of the larger angle?

15. The lengths of the sides of a triangle are in the extended ratio 6 : 7 : 9. The perimeter of the triangle is 88 cm. What are the lengths of the sides?   ◄ See Problem 3.

16. The measures of the angles of a triangle are in the extended ratio 4 : 3 : 2. What is the measure of the largest angle?

Algebra Solve each proportion.   ◄ See Problem 4.

17. $\frac{1}{3} = \frac{x}{12}$ 18. $\frac{9}{5} = \frac{3}{x}$ 19. $\frac{4}{x} = \frac{5}{9}$ 20. $\frac{y}{10} = \frac{15}{25}$ 21. $\frac{9}{24} = \frac{12}{n}$

22. $\frac{11}{14} = \frac{b}{21}$ 23. $\frac{3}{5} = \frac{6}{x+3}$ 24. $\frac{y+7}{9} = \frac{8}{5}$ 25. $\frac{5}{x-3} = \frac{10}{x}$ 26. $\frac{n+4}{8} = \frac{n}{4}$

In the diagram, $\frac{a}{b} = \frac{3}{4}$. Complete each statement. Justify your answer.

See Problem 5.

27. $\frac{b}{a} = \frac{\blacksquare}{\blacksquare}$ **28.** $4a = \blacksquare$ **29.** $\frac{\blacksquare}{\blacksquare} = \frac{b}{4}$

30. $\frac{\blacksquare}{\blacksquare} = \frac{7}{4}$ **31.** $\frac{a + b}{b} = \frac{\blacksquare}{\blacksquare}$ **32.** $\frac{b}{\blacksquare} = \frac{4}{\blacksquare}$

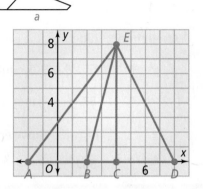

B Apply

Coordinate Geometry Use the graph. Write each ratio in simplest form.

33. $\frac{AC}{BD}$ **34.** $\frac{AE}{EC}$ **35.** slope of $\overline{EB}$ **36.** slope of $\overline{ED}$

37. Think About a Plan The area of a rectangle is 150 in.^{2}. The ratio of the length to the width is $3 : 2$. Find the length and the width.
- What is the formula for the area of a rectangle?
- How can you use the given ratio to write expressions for the length and width?

Art To draw a face, you can sketch the head as an oval and then lightly draw horizontal lines to help locate the eyes, nose, and mouth. You can use the extended ratios shown in the diagrams to help you place the lines for an adult's face or for a baby's face.

38. If $AE = 72$ cm in the diagram, find AB, BC, CD, and DE.

39. You draw a baby's head as an oval that is 21 in. from top to bottom.
  a. How far from the top should you place the line for the eyes?
  b. Suppose you decide to make the head an adult's head. How far up should you move the line for the eyes?

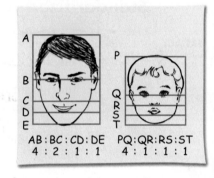

AB : BC : CD : DE   PQ : QR : RS : ST
  4 : 2 : 1 : 1   4 : 1 : 1 : 1

Algebra Solve each proportion.

40. $\frac{1}{7y - 5} = \frac{2}{9y}$ **41.** $\frac{4a + 1}{7} = \frac{2a}{3}$ **42.** $\frac{5}{x + 2} = \frac{3}{x + 1}$ **43.** $\frac{2b - 1}{4} = \frac{b - 2}{12}$

44. The ratio of the length to the width of a rectangle is $9 : 4$. The width of the rectangle is 52 mm. Write and solve a proportion to find the length.

45. Open-Ended Draw a quadrilateral that satisfies this condition: The measures of the consecutive angles are in the extended ratio $4 : 5 : 4 : 7$.

46. Reasoning The means of a proportion are 4 and 15. List all possible pairs of positive integers that could be the extremes of the proportion.

47. Writing Describe how to use the Cross Products Property to determine whether $\frac{10}{26} = \frac{16}{42}$ is a true proportion.

48. Reasoning Explain how to use two different properties of proportions to change the proportion $\frac{3}{4} = \frac{12}{16}$ into the proportion $\frac{12}{3} = \frac{16}{4}$.

Complete each statement. Justify your answer.

49. If $4m = 9n$, then $\frac{m}{n} = \frac{\blacksquare}{\blacksquare}$.

50. If $\frac{30}{t} = \frac{18}{r}$, then $\frac{t}{r} = \frac{\blacksquare}{\blacksquare}$.

51. If $\frac{a+5}{5} = \frac{b+2}{2}$, then $\frac{a}{5} = \frac{\blacksquare}{\blacksquare}$.

52. If $\frac{a}{b} = \frac{c}{d}$, then $\frac{a+b}{c+d} = \frac{\blacksquare}{\blacksquare}$.

53. If $\frac{a}{b} = \frac{c}{d}$, then $\frac{a+c}{b+d} = \frac{\blacksquare}{\blacksquare}$.

54. If $\frac{a}{b} = \frac{c}{d}$, then $\frac{a+2b}{b} = \frac{\blacksquare}{\blacksquare}$.

 Challenge **Algebra** Use properties of equality to justify each property of proportions.

55. $\frac{a}{b} = \frac{c}{d}$ is equivalent to $\frac{b}{a} = \frac{d}{c}$.

56. $\frac{a}{b} = \frac{c}{d}$ is equivalent to $\frac{a}{c} = \frac{b}{d}$.

57. $\frac{a}{b} = \frac{c}{d}$ is equivalent to $\frac{a+b}{b} = \frac{c+d}{d}$.

Algebra Solve each proportion for the variable(s).

58. $\frac{x-3}{3} = \frac{2}{x+2}$

59. $\frac{3-4x}{1+5x} = \frac{1}{2+3x}$

60. $\frac{x}{6} = \frac{x+10}{18} = \frac{4x}{y}$

Standardized Test Prep

GRIDDED RESPONSE

SAT/ACT

61. In the diagram at the right, $\overline{BD}$ is the perpendicular bisector of $\overline{AC}$. What is the length of $\overline{EA}$?

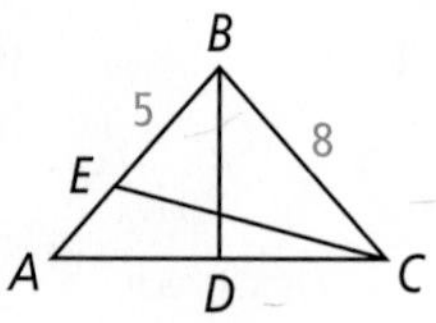

62. The measures of the angles of a triangle are in the extended ratio $6 : 3 : 1$. What is the measure of the largest angle?

63. In the diagram at the right, $\triangle ABC$ is an isosceles triangle with base $\overline{AC}$ and $m\angle ABC = 40$. What is the value of x?

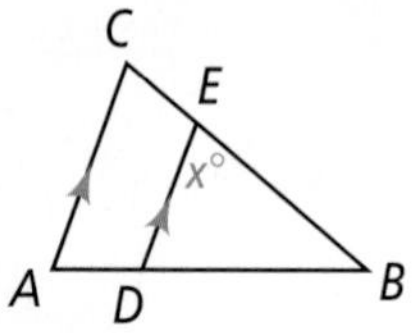

64. In $\triangle PQR$, $PQ = 10$ in. and $QR = 12$ in. RP must be less than x inches. What is the value of x?

65. What is the length of the line segment joining $P(-3, -2)$ and $Q(9, 3)$?

Mixed Review

66. Write a coordinate proof of this statement: The diagonals of a square are perpendicular.

◀ **See Lesson 6-9.**

In each exercise, identify two statements that contradict each other.

◀ **See Lessons 1-5 and 5-5.**

67. I. $\triangle PQR$ is isosceles.
II. $\triangle PQR$ is an obtuse triangle.
III. $\triangle PQR$ is scalene.

68. I. $\angle 1 \cong \angle 2$
II. $\angle 1$ and $\angle 2$ are complementary.
III. $m\angle 1 + m\angle 2 = 180$

Get Ready! **To prepare for Lesson 7-2, do Exercise 69.**

69. $\triangle ABC \cong \triangle HIJ$. Name three pairs of congruent angles and three pairs of congruent sides.

◀ **See Lesson 4-1.**

Solving Quadratic Equations

© **Content Standard**
Reviews A-CED.1 Create equations and inequalities in one variable and use them to solve problems.

Equations in the form $ax^2 + bx + c = 0$, where $a \neq 0$, are quadratic equations in standard form. You can solve some quadratic equations in standard form by factoring and using the Zero-Product Property:

If $ab = 0$, then $a = 0$ or $b = 0$.

You can solve all quadratic equations in standard form using the quadratic formula:

If $ax^2 + bx + c = 0$, where $a \neq 0$, then $x = \dfrac{-b \pm \sqrt{b^2 - 4ac}}{2a}$.

Example

Algebra Solve for x. For irrational solutions, give both the exact answer and the answer rounded to the nearest hundredth.

A $7x^2 + 6x - 1 = 0$ The equation is in standard form.

  $(7x - 1)(x + 1) = 0$ Factor.

  $7x - 1 = 0$ or $x + 1 = 0$ Use the Zero-Product Property.

   $x = \dfrac{1}{7}$ or $x = -1$ Solve for x.

B $-3x^2 - 5x + 1 = 0$ The equation is in standard form.

  $a = -3, b = -5, c = 1$ Identify a, b, and c.

  $x = \dfrac{-(-5) \pm \sqrt{(-5)^2 - 4(-3)(1)}}{2(-3)}$ Substitute in the quadratic formula.

  $x = \dfrac{5 \pm \sqrt{37}}{-6}$ Simplify.

  $x = -\dfrac{5 + \sqrt{37}}{6}$ or $x = -\dfrac{5 - \sqrt{37}}{6}$ Write the two solutions separately.

  $x \approx -1.85$ or $x \approx 0.18$ Use a calculator and round to the nearest hundredth.

Exercises

Algebra Solve for x. For irrational solutions, give both the exact answer and the answer rounded to the nearest hundredth.

1. $x^2 + 5x - 14 = 0$ **2.** $4x^2 - 13x + 3 = 0$ **3.** $2x^2 + 7x + 3 = 0$

4. $5x^2 + 2x - 2 = 0$ **5.** $2x^2 - 10x + 11 = 0$ **6.** $8x^2 - 2x - 3 = 0$

7. $2x^2 + 3x - 20 = 0$ **8.** $x^2 - x - 210 = 0$ **9.** $x^2 - 4x = 0$

10. $x^2 - 25 = 0$ **11.** $6x^2 + 10x = 5$ **12.** $1 = 2x^2 - 6x$

Similar Polygons

Content Standard
G.SRT.5 Use . . . similarity criteria for triangles to solve problems and to prove relationships in geometric figures.

Objective   To identify and apply similar polygons

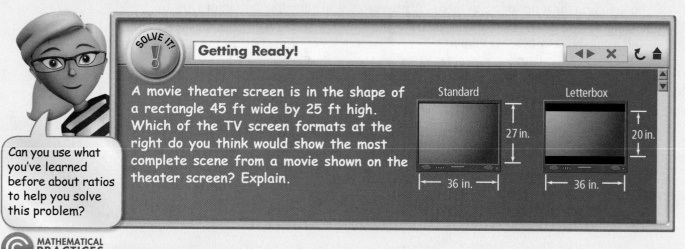

SOLVE IT!

Getting Ready!

A movie theater screen is in the shape of a rectangle 45 ft wide by 25 ft high. Which of the TV screen formats at the right do you think would show the most complete scene from a movie shown on the theater screen? Explain.

Standard
27 in.
36 in.

Letterbox
20 in.
36 in.

Can you use what you've learned before about ratios to help you solve this problem?

MATHEMATICAL PRACTICES

Similar figures have the same shape but not necessarily the same size. You can abbreviate *is similar to* with the symbol ~.

Dynamic Activity
Similar Polygons

Essential Understanding You can use ratios and proportions to decide whether two polygons are similar and to find unknown side lengths of similar figures.

Lesson Vocabulary
• similar figures
• similar polygons
• extended proportion
• scale factor
• scale drawing
• scale

take note

Key Concept   Similar Polygons

Define	**Diagram**	**Symbols**
Two polygons are **similar polygons** if corresponding angles are congruent and if the lengths of corresponding sides are proportional.	$ABCD \sim GHIJ$ 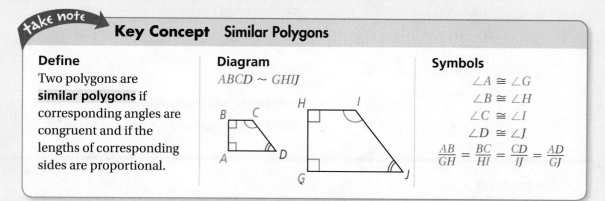	$\angle A \cong \angle G$ $\angle B \cong \angle H$ $\angle C \cong \angle I$ $\angle D \cong \angle J$ $\dfrac{AB}{GH} = \dfrac{BC}{HI} = \dfrac{CD}{IJ} = \dfrac{AD}{GJ}$

You write a similarity statement with corresponding vertices in order, just as you write a congruence statement. When three or more ratios are equal, you can write an **extended proportion.** The proportion $\dfrac{AB}{GH} = \dfrac{BC}{HI} = \dfrac{CD}{IJ} = \dfrac{AD}{GJ}$ is an extended proportion.

A **scale factor** is the ratio of corresponding linear measurements of two similar figures. The ratio of the lengths of corresponding sides $\overline{BC}$ and $\overline{YZ}$, or more simply stated, the ratio of corresponding sides, is $\dfrac{BC}{YZ} = \dfrac{20}{8} = \dfrac{5}{2}$. So the scale factor of $\triangle ABC$ to $\triangle XYZ$ is $\dfrac{5}{2}$ or $5:2$.

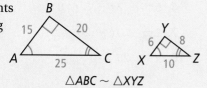

$\triangle ABC \sim \triangle XYZ$

Problem 1 **Understanding Similarity**

$\triangle MNP \sim \triangle SRT$

A **What are the pairs of congruent angles?**

$\angle M \cong \angle S$, $\angle N \cong \angle R$, and $\angle P \cong \angle T$

B **What is the extended proportion for the ratios of corresponding sides?**

$$\frac{MN}{SR} = \frac{NP}{RT} = \frac{MP}{ST}$$

Think

How can you use the similarity statement to write ratios of corresponding sides?
Use the order of the sides in the similarity statement. $\overline{MN}$ corresponds to $\overline{SR}$, so $\frac{MN}{SR}$ is a ratio of corresponding sides.

Got It? **1.** $DEFG \sim HJKL$.

  a. What are the pairs of congruent angles?

  b. What is the extended proportion for the ratios of the lengths of corresponding sides?

Problem 2 **Determining Similarity**

Are the polygons similar? If they are, write a similarity statement and give the scale factor.

A *JKLM* and *TUVW*

 Step 1 Identify pairs of congruent angles.

   $\angle J \cong \angle T$, $\angle K \cong \angle U$, $\angle L \cong \angle V$, and $\angle M \cong \angle W$

 Step 2 Compare the ratios of corresponding sides.

$$\frac{JK}{TU} = \frac{12}{6} = \frac{2}{1} \qquad \frac{KL}{UV} = \frac{24}{16} = \frac{3}{2}$$

$$\frac{LM}{VW} = \frac{24}{14} = \frac{12}{7} \qquad \frac{JM}{TW} = \frac{6}{6} = \frac{1}{1}$$

Corresponding sides are not proportional, so the polygons are not similar.

Think

How do you identify corresponding sides?
The included side between a pair of angles of one polygon corresponds to the included side between the corresponding pair of congruent angles of another polygon.

B $\triangle ABC$ and $\triangle EFD$

 Step 1 Identify pairs of congruent angles.

   $\angle A \cong \angle D$, $\angle B \cong \angle E$, and $\angle C \cong \angle F$

 Step 2 Compare the ratios of corresponding sides.

$$\frac{AB}{DE} = \frac{12}{15} = \frac{4}{5} \qquad \frac{BC}{EF} = \frac{16}{20} = \frac{4}{5} \qquad \frac{AC}{DF} = \frac{8}{10} = \frac{4}{5}$$

Yes; $\triangle ABC \sim \triangle DEF$ and the scale factor is $\frac{4}{5}$ or $4 : 5$.

Got It? **2.** Are the polygons similar? If they are, write a similarity statement and give the scale factor.

  a.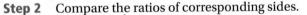

  b.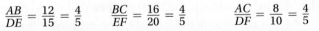

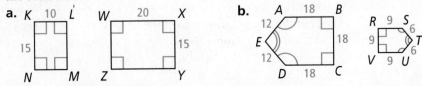

 **Problem 3** **Using Similar Polygons**

Algebra $ABCD \sim EFGD$. **What is the value of** x?

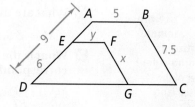

   Ⓐ 4.5   Ⓒ 7.2

   Ⓑ 5   Ⓓ 11.25

$\dfrac{FG}{BC} = \dfrac{ED}{AD}$ Corresponding sides of similar polygons are proportional.

$\dfrac{x}{7.5} = \dfrac{6}{9}$ Substitute.

$9x = 45$ Cross Products Property

$x = 5$ Divide each side by 9.

The value of x is 5. The correct answer is B.

✓ **Got It? 3.** Use the diagram in Problem 3. What is the value of y?

 **Problem 4** **Using Similarity**

Design Your class is making a rectangular poster for a rally. The poster's design is 6 in. high by 10 in. wide. The space allowed for the poster is 4 ft high by 8 ft wide. What are the dimensions of the largest poster that will fit in the space?

Step 1   Determine whether the height or width will fill the space first.

     Height: 4 ft = 48 in.   Width: 8 ft = 96 in.

                    48 in. ÷ 6 in. = 8   96 in. ÷ 10 in. = 9.6

     The design can be enlarged at most 8 times.

Step 2   The greatest height is 48 in., so find the width.

$\dfrac{6}{48} = \dfrac{10}{x}$ Corresponding sides of similar polygons are proportional.

$6x = 480$ Cross Products Property

$x = 80$ Divide each side by 6.

The largest poster is 48 in. by 80 in. or 4 ft by $6\frac{2}{3}$ ft.

✓ **Got It? 4.** Use the same poster design in Problem 4. What are the dimensions of the largest complete poster that will fit in a space 3 ft high by 4 ft wide?

In a **scale drawing,** all lengths are proportional to their corresponding actual lengths. The **scale** is the ratio that compares each length in the scale drawing to the actual length. The lengths used in a scale can be in different units. For example, a scale might be written as 1 cm to 50 km, 1 in. = 100 mi, or 1 in. : 10 ft.

You can use proportions to find the actual dimensions represented in a scale drawing.

Ⓒ Problem 5 **Using a Scale Drawing** (STEM)

Design The diagram shows a scale drawing of the Golden Gate Bridge in San Francisco. The distance between the two towers is the main span. What is the actual length of the main span of the bridge?

Scale: 1 cm = 200 m

The length of the main span in the scale drawing is 6.4 cm. Let s represent the main span of the bridge. Use the scale to set up a proportion.

$$\frac{1}{200} = \frac{6.4}{s} \qquad \frac{\text{length in drawing (cm)}}{\text{actual length (m)}}$$

$$s = 1280 \qquad \text{Cross Products Property}$$

The actual length of the main span of the bridge is 1280 m.

Think

Why is it helpful to use a scale in different units?
1 cm : 200 m in the same units would be 1 cm : 20,000 cm. When solving the problem, $\frac{1}{200}$ is easier to work with than $\frac{1}{20,000}$.

✓ Got It? **5. a.** Use the scale drawing in Problem 5. What is the actual height of the towers above the roadway?

    Ⓒ **b. Reasoning** The Space Needle in Seattle is 605 ft tall. A classmate wants to make a scale drawing of the Space Needle on an $8\frac{1}{2}$ in.–by-11 in. sheet of paper. He decides to use the scale 1 in. = 50 ft. Is this a reasonable scale? Explain.

Lesson Check

Do you know HOW?

JDRT ~ WHYX. Complete each statement.

1. $\angle D \cong$ ___?___

2. $\dfrac{RT}{YX} = \dfrac{\blacksquare}{WX}$

3. Are the polygons similar? If they are, write a similarity statement and give the scale factor.

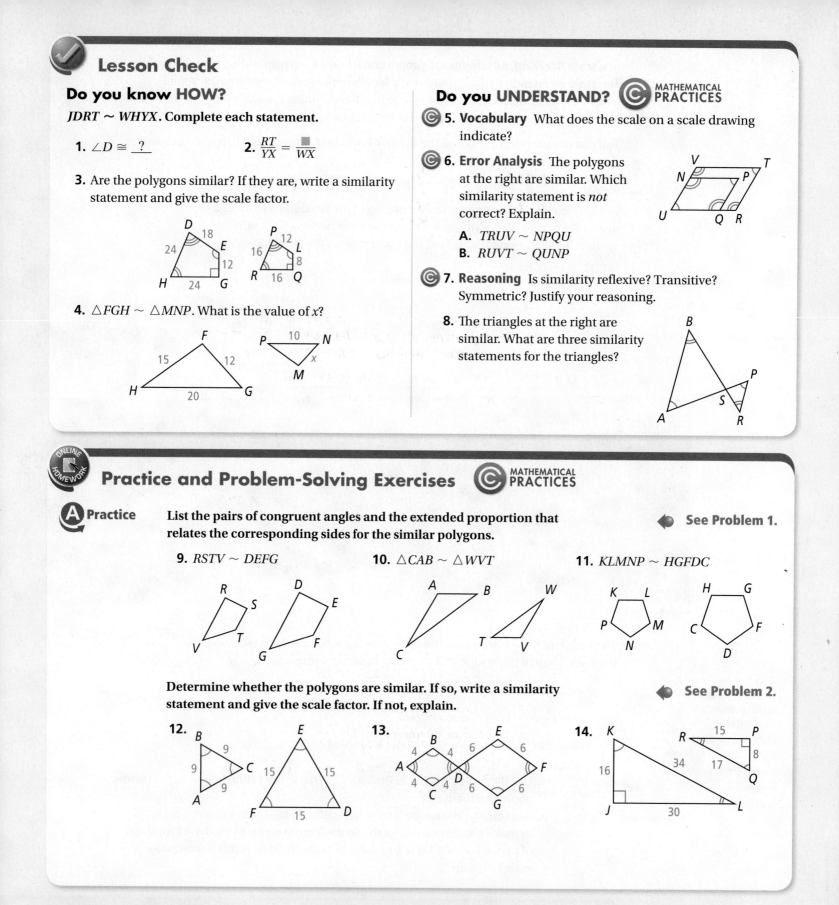

4. $\triangle FGH \sim \triangle MNP$. What is the value of *x*?

Do you UNDERSTAND?

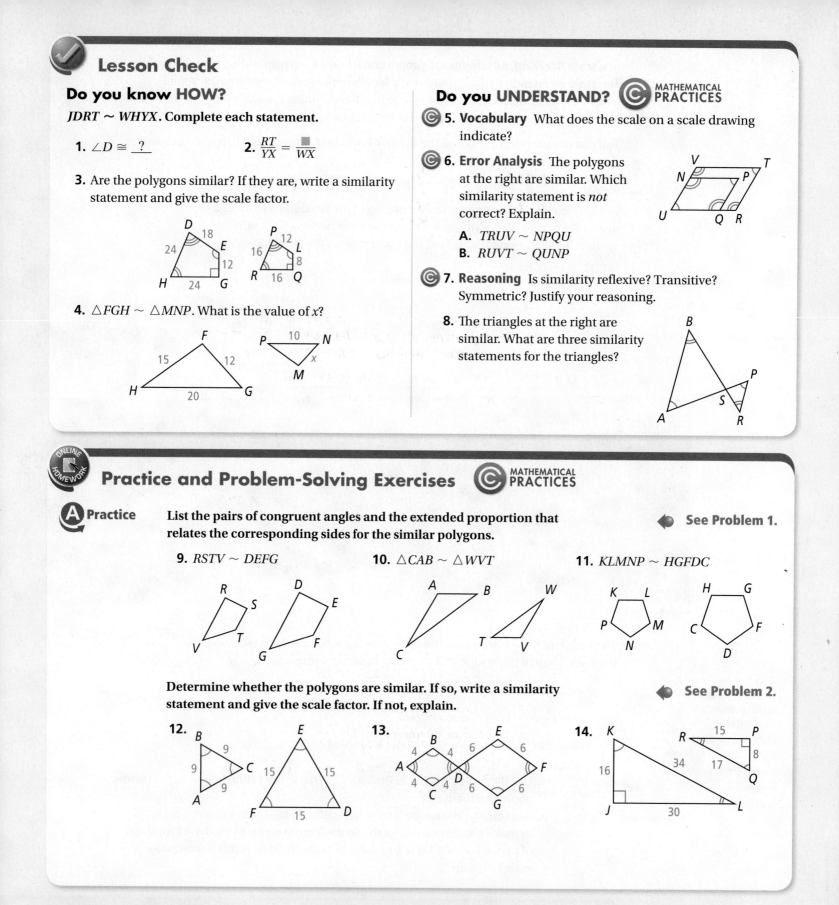
MATHEMATICAL PRACTICES

5. Vocabulary What does the scale on a scale drawing indicate?

6. Error Analysis The polygons at the right are similar. Which similarity statement is *not* correct? Explain.

  A. *TRUV ~ NPQU*

  B. *RUVT ~ QUNP*

7. Reasoning Is similarity reflexive? Transitive? Symmetric? Justify your reasoning.

8. The triangles at the right are similar. What are three similarity statements for the triangles?

Practice and Problem-Solving Exercises

MATHEMATICAL PRACTICES

A Practice

List the pairs of congruent angles and the extended proportion that relates the corresponding sides for the similar polygons.

◀ See Problem 1.

9. *RSTV ~ DEFG*

10. $\triangle CAB \sim \triangle WVT$

11. *KLMNP ~ HGFDC*

Determine whether the polygons are similar. If so, write a similarity statement and give the scale factor. If not, explain.

◀ See Problem 2.

12.

13.

14.

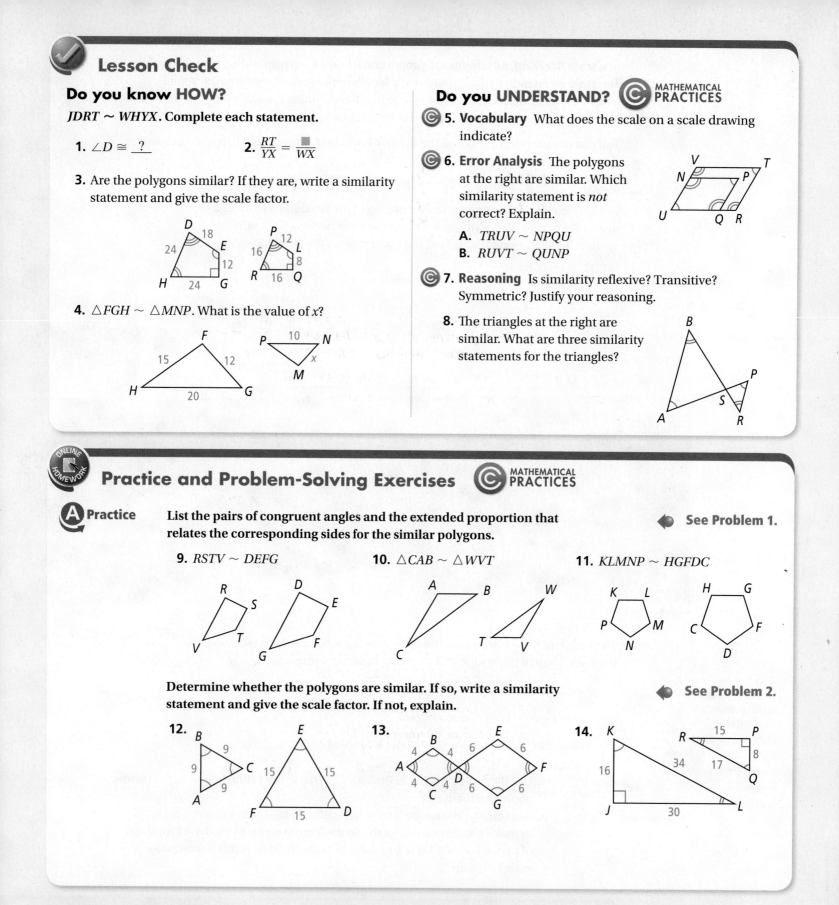

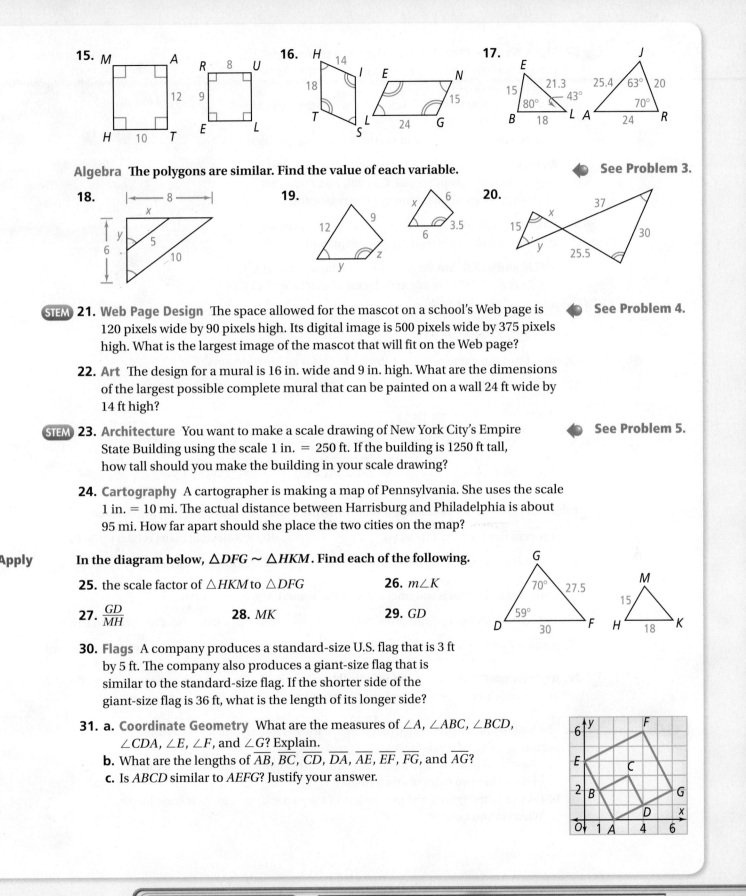

15. M   A   R   8   U
12   9
H   10   T   E   L

16. H   14   I   E   N
18   15
T   S   L   24   G

17. E   21.3   43°   J   25.4   63°   20
15   80°   70°
B   18   L   A   24   R

Algebra The polygons are similar. Find the value of each variable.

◀ **See Problem 3.**

18.
|← 8 →|
x
y   5
6   10

19.
x   6
12   9   3.5
y   z   6
y

20.
37
15   x
y   25.5   30

STEM **21. Web Page Design** The space allowed for the mascot on a school's Web page is 120 pixels wide by 90 pixels high. Its digital image is 500 pixels wide by 375 pixels high. What is the largest image of the mascot that will fit on the Web page?

◀ **See Problem 4.**

22. Art The design for a mural is 16 in. wide and 9 in. high. What are the dimensions of the largest possible complete mural that can be painted on a wall 24 ft wide by 14 ft high?

STEM **23. Architecture** You want to make a scale drawing of New York City's Empire State Building using the scale 1 in. = 250 ft. If the building is 1250 ft tall, how tall should you make the building in your scale drawing?

◀ **See Problem 5.**

24. Cartography A cartographer is making a map of Pennsylvania. She uses the scale 1 in. = 10 mi. The actual distance between Harrisburg and Philadelphia is about 95 mi. How far apart should she place the two cities on the map?

B **Apply**

In the diagram below, $\triangle DFG \sim \triangle HKM$. Find each of the following.

25. the scale factor of $\triangle HKM$ to $\triangle DFG$

26. $m\angle K$

27. $\dfrac{GD}{MH}$

28. MK

29. GD

G
70°   27.5   M
59°   15
D   30   F   H   18   K

30. Flags A company produces a standard-size U.S. flag that is 3 ft by 5 ft. The company also produces a giant-size flag that is similar to the standard-size flag. If the shorter side of the giant-size flag is 36 ft, what is the length of its longer side?

31. a. Coordinate Geometry What are the measures of $\angle A$, $\angle ABC$, $\angle BCD$, $\angle CDA$, $\angle E$, $\angle F$, and $\angle G$? Explain.
b. What are the lengths of $\overline{AB}$, $\overline{BC}$, $\overline{CD}$, $\overline{DA}$, $\overline{AE}$, $\overline{EF}$, $\overline{FG}$, and $\overline{AG}$?
c. Is $ABCD$ similar to $AEFG$? Justify your answer.

32. Think About a Plan The Davis family is planning to drive from San Antonio to Houston. About how far will they have to drive?
- How can you find the distance between the two cities on the map?
- What proportion can you set up to solve the problem?

33. Reasoning Two polygons have corresponding side lengths that are proportional. Can you conclude that the polygons are similar? Justify your reasoning.

34. Writing Explain why two congruent figures must also be similar. Include scale factor in your explanation.

35. △*JLK* and △*RTS* are similar. The scale factor of △*JLK* to △*RTS* is 3 : 1. What is the scale factor of △*RTS* to △*JLK*?

36. Open-Ended Draw and label two different similar quadrilaterals. Write a similarity statement for each and give the scale factor.

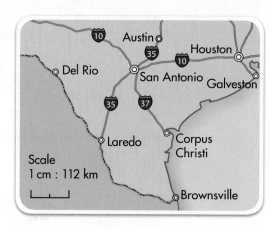

Scale
1 cm : 112 km

Algebra Find the value of *x*. Give the scale factor of the polygons.

37. △*WLJ* ~ △*QBV*

38. *GKNM* ~ *VRPT*

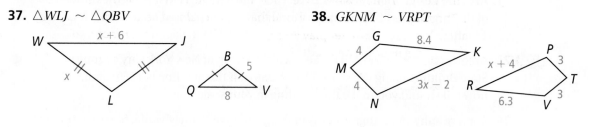

Sports Choose a scale and make a scale drawing of each rectangular playing surface.

39. A soccer field is 110 yd by 60 yd.

40. A volleyball court is 60 ft by 30 ft.

41. A tennis court is 78 ft by 36 ft.

42. A football field is 360 ft by 160 ft.

Determine whether each statement is *always*, *sometimes*, or *never* true.

43. Any two regular pentagons are similar.

44. A hexagon and a triangle are similar.

45. A square and a rhombus are similar.

46. Two similar rectangles are congruent.

47. Architecture The scale drawing at the right is part of a floor plan for a home. The scale is 1 cm = 10 ft. What are the actual dimensions of the family room?

Challenge

48. The lengths of the sides of a triangle are in the extended ratio 2 : 3 : 4. The perimeter of the triangle is 54 in.
a. The length of the shortest side of a similar triangle is 16 in. What are the lengths of the other two sides of this triangle?
b. Compare the ratio of the perimeters of the two triangles to their scale factor. What do you notice?

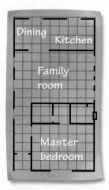

49. In rectangle $BCEG$, $BC : CE = 2 : 3$. In rectangle $LJAW$, $LJ : JA = 2 : 3$. Show that $BCEG \sim LJAW$.

50. Prove the following statement: If $\triangle ABC \sim \triangle DEF$ and $\triangle DEF \sim \triangle GHK$, then $\triangle ABC \sim \triangle GHK$.

Standardized Test Prep

SAT/ACT

51. $PQRS \sim JKLM$ with a scale factor of $4 : 3$. $QR = 8$ cm. What is the value of KL?

 Ⓐ 6 cm   Ⓑ 8 cm   Ⓒ $10\frac{2}{3}$ cm   Ⓓ 24 cm

52. Which of the following is NOT a property of an isosceles trapezoid?

 Ⓕ The base angles are congruent.   Ⓗ The diagonals are perpendicular.

 Ⓖ The legs are congruent.   Ⓘ The diagonals are congruent.

53. In the diagram at the right, what is $m\angle 1$?

 Ⓐ 45   Ⓑ 75   Ⓒ 125   Ⓓ 135

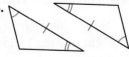

Short
Response

54. A high school community-action club plans to build a circular play area in a city park. The club members need to buy materials to enclose the area and sand to fill the area. For a 9-ft-diameter play area, what will be the circumference and area rounded to the nearest hundredth?

Mixed Review

If $\frac{x}{7} = \frac{y}{9}$, complete each statement using the properties of proportions.   ◀ See Lesson 7-1.

55. $9x = \blacksquare$ **56.** $\frac{x}{y} = \frac{\blacksquare}{\blacksquare}$ **57.** $\frac{x+7}{7} = \frac{\blacksquare}{\blacksquare}$

Use the diagram for Exercises 58–61.   ◀ See Lesson 4-5.

58. Name the isosceles triangles in the figure.

59. $\overline{CD} \cong \underline{\ ?\ } \cong \underline{\ ?\ }$

60. $AE = \underline{\ ?\ }$ **61.** $m\angle A = \underline{\ ?\ }$

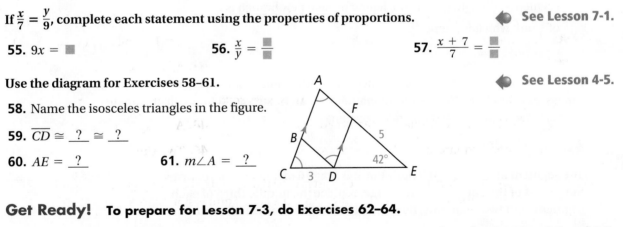

Get Ready! **To prepare for Lesson 7-3, do Exercises 62–64.**

How can you prove that the triangles are congruent?   ◀ See Lessons 4-2 and 4-3.

62.   **63.**   **64.**

Fractals

© **Content Standard**
Extends G.SRT.5 Use . . . similarity criteria for triangles to solve problems and to prove relationships in geometric figures.

Fractals are objects that have three important properties:

- You can form fractals by repeating steps. This process is called *iteration*.

- They require infinitely many iterations. In practice, you can continue until the objects become too small to draw. Even then the steps could continue in your mind.

- At each stage, a portion of the object is a reduced copy of the entire object at the previous stage. This property is called *self-similarity*.

Example 1

The segment at the right of length 1 unit is Stage 0 of a fractal tree. Draw Stage 1 and Stage 2 of the tree. For each stage, draw two branches from the top third of each segment.

- To draw Stage 1, find the point that is $\frac{1}{3}$ unit from the top of the segment. From this point, draw two segments of length $\frac{1}{3}$ unit.

- To draw Stage 2, find the point that is $\frac{1}{9}$ unit from the top of each branch of Stage 1. From each of these points, draw two segments of length $\frac{1}{9}$ unit. The length of each new branch is $\frac{1}{3}$ of $\frac{1}{3}$ unit which is $\frac{1}{9}$ unit.

Stage 0

1 unit

Stage 1   Stage 2

Amazingly, some fractals are used to describe natural formations such as mountain ranges and clouds. In 1904, Swedish mathematician Helge von Koch made the Koch Curve, a fractal that is used to model coastlines.

Example 2

The segment at the right of length 1 unit is Stage 0 of a Koch Curve. Draw Stages 1–4 of the curve. For each stage, replace the middle third of each segment with two segments, both equal in length to the middle third.

- For Stage 1, replace the middle third with two segments that are each $\frac{1}{3}$ unit long.

- For Stage 2, replace the middle third of each segment of Stage 1 with two segments that are each $\frac{1}{9}$ unit long.

- Continue with a third and fourth iteration.

Stage 0 _____ 1 unit

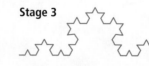

Stage 1   Stage 2

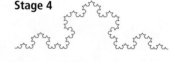

Stage 3   Stage 4

Example 3

The equilateral triangle at the right is Stage 0 of a Koch Snowflake. Draw Stage 1 of the snowflake by first drawing an equilateral triangle on the middle third of each side. Then erase the middle third of each side of the original triangle.

- To draw an equilateral triangle on the middle third of a side, find the two points that are $\frac{1}{3}$ unit from an endpoint of the side. From each point, draw a segment of length $\frac{1}{3}$ unit. Each segment must make a 60° angle with the side of the original triangle.

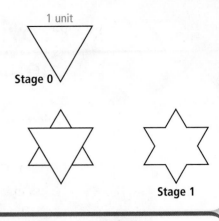

1 unit

Stage 0

Stage 1

Exercises

1. Draw Stage 3 of the fractal tree in Example 1.

Use the Koch Curve in Example 2 for Exercises 2–4.

2. Complete the table to find the length of the Koch Curve at each stage.

3. Examine the results of Exercise 2 and look for a pattern. Use this pattern to predict the length of the Koch Curve at Stage 3 and at Stage 4.

4. Suppose you are able to complete a Koch Curve to Stage n.
  a. Write an expression for the length of the curve.
  b. What happens to the length of the curve as n increases?

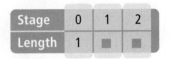

Stage	0	1	2
Length	1	■	■

5. Draw Stage 2 of the Koch Snowflake in Example 3.

Stage 3 of the Koch Snowflake is shown at the right. Use it and the earlier stages to answer Exercises 6–8.

6. At each stage, is the snowflake equilateral?

7. a. Complete the table to find the perimeter at each stage.

Stage	Number of Sides	Length of a Side	Perimeter
0	3	1	3
1	■	$\frac{1}{3}$	■
2	48	■	■
3	■	■	■

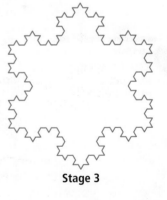

Stage 3

b. Predict the perimeter at Stage 4.
c. Will there be a stage at which the perimeter is greater than 100 units? Explain.

8. Exercises 4 and 7 suggest that there is no bound on the perimeter of the Koch Snowflake. Is this true about the area of the Koch Snowflake? Explain.

Proving Triangles Similar

© Content Standards
G.SRT.5 Use . . . similarity criteria for triangles to solve problems and to prove relationships in geometric figures.
G.GPE.5 Prove the slope criteria for parallel and perpendicular lines and use them to solve geometric problems.

Objectives To use the AA ~ Postulate and the SAS ~ and SSS ~ Theorems
To use similarity to find indirect measurements

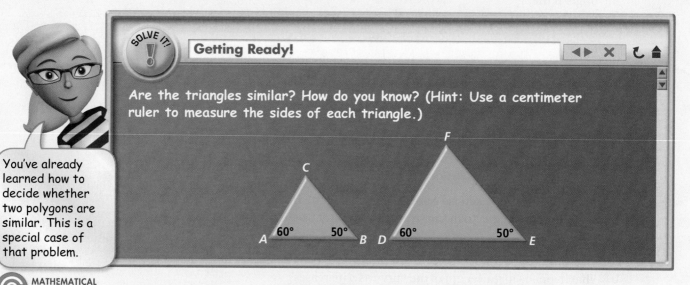

You've already learned how to decide whether two polygons are similar. This is a special case of that problem.

Getting Ready!

Are the triangles similar? How do you know? (Hint: Use a centimeter ruler to measure the sides of each triangle.)

© MATHEMATICAL PRACTICES
In the Solve It, you determined whether the two triangles are similar. That is, you needed information about all three pairs of angles and all three pairs of sides. In this lesson, you'll learn an easier way to determine whether two triangles are similar.

Lesson Vocabulary
• indirect measurement

Essential Understanding You can show that two triangles are similar when you know the relationships between only two or three pairs of corresponding parts.

take note

Postulate 7-1	Angle-Angle Similarity (AA ~) Postulate	
Postulate If two angles of one triangle are congruent to two angles of another triangle, then the triangles are similar.	**If . . .** $\angle S \cong \angle M$ and $\angle R \cong \angle L$ 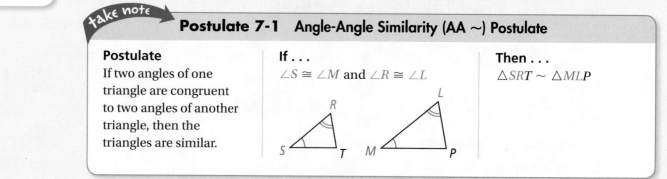	**Then . . .** $\triangle SRT \sim \triangle MLP$

Problem 1 Using the AA ~ Postulate

What do you need to show that the triangles are similar?
To use the AA ~ Postulate, you need to prove that two pairs of angles are congruent.

Are the two triangles similar? How do you know?

A △RSW and △VSB

∠R ≅ ∠V because both angles measure 45°.
∠RSW ≅ ∠VSB because vertical angles are congruent.
So, △RSW ~ △VSB by the AA ~ Postulate.

B △JKL and △PQR

∠L ≅ ∠R because both angles measure 70°.
By the Triangle Angle-Sum Theorem,
m∠K = 180 − 30 − 70 = 80 and
m∠P = 180 − 85 − 70 = 25. Only one pair of
angles is congruent. So, △JKL and △PQR are *not* similar.

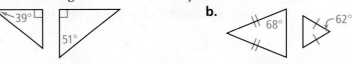

Got It? **1.** Are the two triangles similar? How do you know?

a. **b.**

Here are two other ways to determine whether two triangles are similar.

take note

Theorem 7-1 Side-Angle-Side Similarity (SAS ~) Theorem

Theorem
If an angle of one triangle is congruent to an angle of a second triangle, and the sides that include the two angles are proportional, then the triangles are similar.

If . . .
$\frac{AB}{QR} = \frac{AC}{QS}$ and ∠A ≅ ∠Q

Then . . .
△ABC ~ △QRS

You will prove Theorem 7-1 in Exercise 35.

Theorem 7-2 Side-Side-Side Similarity (SSS ~) Theorem

Theorem
If the corresponding sides of two triangles are proportional, then the triangles are similar.

If . . .
$\frac{AB}{QR} = \frac{AC}{QS} = \frac{BC}{RS}$

Then . . .
△ABC ~ △QRS

You will prove Theorem 7-2 in Exercise 36.

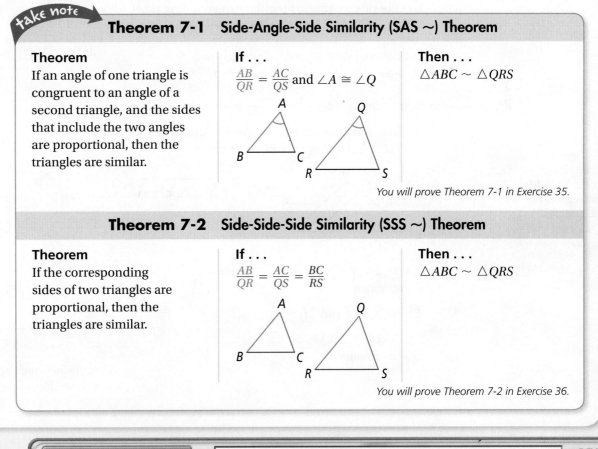

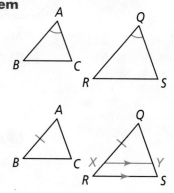

Proof **Proof of Theorem 7-1: Side-Angle-Side Similarity Theorem**

Given: $\dfrac{AB}{QR} = \dfrac{AC}{QS}$, $\angle A \cong \angle Q$

Prove: $\triangle ABC \sim \triangle QRS$

Plan for Proof: Choose X on $\overline{RQ}$ so that $QX = AB$. Draw $\overleftrightarrow{XY} \parallel \overline{RS}$. Show that $\triangle QXY \sim \triangle QRS$ by the AA $\sim$ Postulate. Then use the proportion $\dfrac{QX}{QR} = \dfrac{QY}{QS}$ and the given proportion $\dfrac{AB}{QR} = \dfrac{AC}{QS}$ to show that $AC = QY$. Then prove that $\triangle ABC \cong \triangle QXY$. Finally, prove that $\triangle ABC \sim \triangle QRS$ by the AA $\sim$ Postulate.

 Problem 2 **Verifying Triangle Similarity**

Are the triangles similar? If so, write a similarity statement for the triangles.

A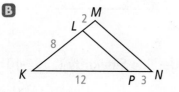

Use the side lengths to identify corresponding sides. Then set up ratios for each pair of corresponding sides.

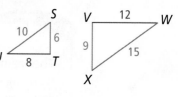

Shortest sides	$\dfrac{ST}{XV} = \dfrac{6}{9} = \dfrac{2}{3}$
Longest sides	$\dfrac{US}{WX} = \dfrac{10}{15} = \dfrac{2}{3}$
Remaining sides	$\dfrac{TU}{VW} = \dfrac{8}{12} = \dfrac{2}{3}$

All three ratios are equal, so corresponding sides are proportional. $\triangle STU \sim \triangle XVW$ by the SSS $\sim$ Theorem.

Plan

How can you make it easier to identify corresponding sides and angles? Sketch and label two separate triangles.

B

$\angle K \cong \angle K$ by the Reflexive Property of Congruence.

$\dfrac{KL}{KM} = \dfrac{8}{10} = \dfrac{4}{5}$ and $\dfrac{KP}{KN} = \dfrac{12}{15} = \dfrac{4}{5}$.

So, $\triangle KLP \sim \triangle KMN$ by the SAS $\sim$ Theorem.

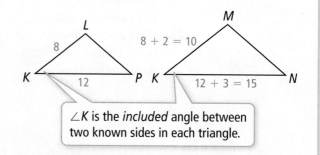

$\angle K$ is the *included* angle between two known sides in each triangle.

Got It? 2. Are the triangles similar? If so, write a similarity statement for the triangles and explain how you know the triangles are similar.

a.

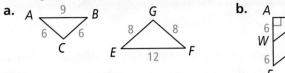

b.

Problem 3 Proving Triangles Similar

Proof

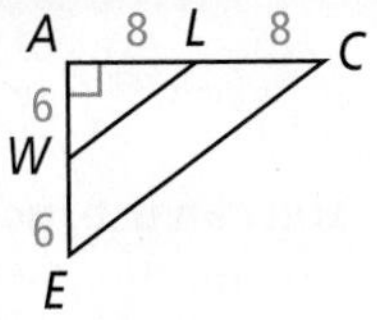

Given: $\overline{FG} \cong \overline{GH}$,
$\overline{JK} \cong \overline{KL}$,
$\angle F \cong \angle J$

Prove: $\triangle FGH \sim \triangle JKL$

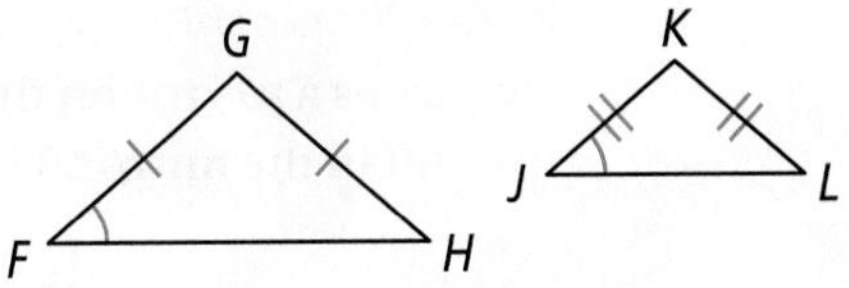

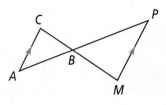

Know

The triangles are isosceles, so the base angles are congruent.

Need

You need to show that the triangles are similar.

Plan

Find two pairs of corresponding congruent angles and use the AA $\sim$ Postulate to prove the triangles are similar.

Statements	Reasons
1) $\overline{FG} \cong \overline{GH}$, $\overline{JK} \cong \overline{KL}$	**1)** Given
2) $\triangle FGH$ is isosceles. $\triangle JKL$ is isosceles.	**2)** Def. of an isosceles $\triangle$
3) $\angle F \cong \angle H$, $\angle J \cong \angle L$	**3)** Base $\measuredangle$ of an isosceles $\triangle$ are $\cong$.
4) $\angle F \cong \angle J$	**4)** Given
5) $\angle H \cong \angle J$	**5)** Transitive Property of $\cong$
6) $\angle H \cong \angle L$	**6)** Transitive Property of $\cong$
7) $\triangle FGH \sim \triangle JKL$	**7)** AA $\sim$ Postulate

Got It? 3. a. Given: $\overline{MP} \parallel \overline{AC}$

Prove: $\triangle ABC \sim \triangle PBM$

b. Reasoning For the figure at the right, suppose you are given only that $\frac{CA}{PM} = \frac{CB}{MB}$. Could you prove that the triangles are similar? Explain.

Essential Understanding Sometimes you can use similar triangles to find lengths that cannot be measured easily using a ruler or other measuring device.

You can use **indirect measurement** to find lengths that are difficult to measure directly. One method of indirect measurement uses the fact that light reflects off a mirror at the same angle at which it hits the mirror.

ⓒ Problem 4 Finding Lengths in Similar Triangles

Rock Climbing Before rock climbing, Darius wants to know how high he will climb. He places a mirror on the ground and walks backward until he can see the top of the cliff in the mirror. What is the height of the cliff?

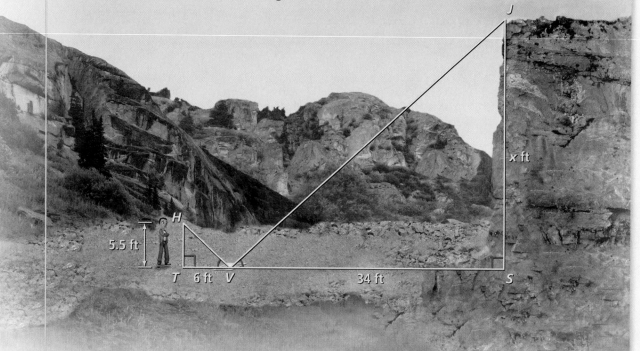

Plan

Before solving for x, verify that the triangles are similar. △HTV ~ △JSV by the AA ~ Postulate because ∠T ≅ ∠S and ∠HVT ≅ ∠JVS.

$\triangle HTV \sim \triangle JSV$	AA ~ Postulate
$\dfrac{HT}{JS} = \dfrac{TV}{SV}$	Corresponding sides of ~ triangles are proportional.
$\dfrac{5.5}{x} = \dfrac{6}{34}$	Substitute.
$187 = 6x$	Cross Products Property
$31.2 \approx x$	Solve for x.

The cliff is about 31 ft high.

ⓒ ✓ Got It? **4. Reasoning** Why is it important that the ground be flat to use the method of indirect measurement illustrated in Problem 4? Explain.

Lesson Check

Do you know HOW?

Are the triangles similar? If yes, write a similarity statement and explain how you know they are similar.

1.
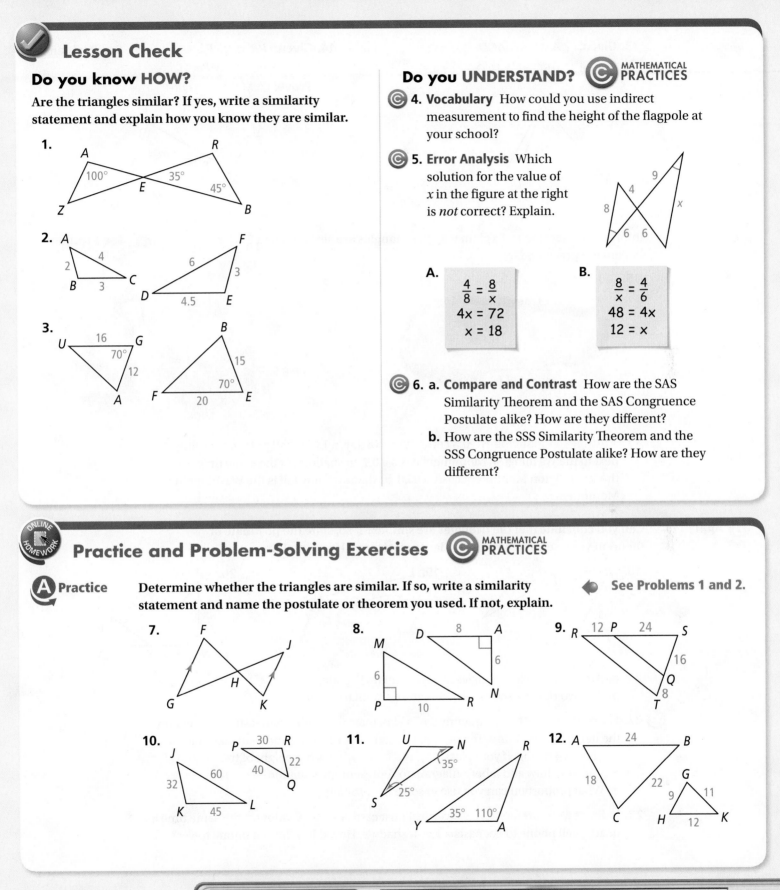

2.

3.

Do you UNDERSTAND? MATHEMATICAL PRACTICES

4. Vocabulary How could you use indirect measurement to find the height of the flagpole at your school?

5. Error Analysis Which solution for the value of x in the figure at the right is *not* correct? Explain.

A.
$$\frac{4}{8} = \frac{8}{x}$$
$$4x = 72$$
$$x = 18$$

B.
$$\frac{8}{x} = \frac{4}{6}$$
$$48 = 4x$$
$$12 = x$$

6. a. Compare and Contrast How are the SAS Similarity Theorem and the SAS Congruence Postulate alike? How are they different?

b. How are the SSS Similarity Theorem and the SSS Congruence Postulate alike? How are they different?

Practice and Problem-Solving Exercises MATHEMATICAL PRACTICES

Practice Determine whether the triangles are similar. If so, write a similarity statement and name the postulate or theorem you used. If not, explain.

See Problems 1 and 2.

7.

8.

9.

10.

11.

12.

13. Given: $\angle ABC \cong \angle ACD$

Proof **Prove:** $\triangle ABC \sim \triangle ACD$

14. Given: $PR = 2NP,$

Proof $PQ = 2MP$

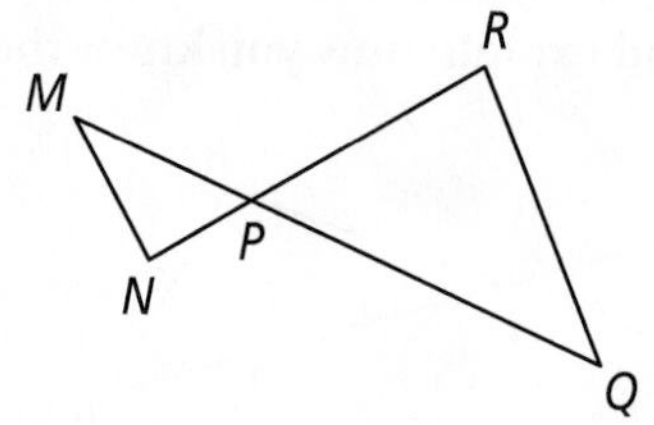

See Problem 3.

Prove: $\triangle MNP \sim \triangle QRP$

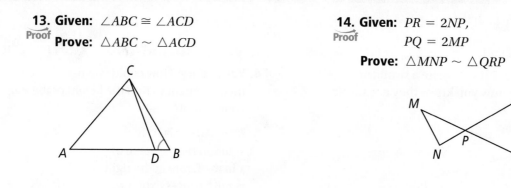

Indirect Measurement Explain why the triangles are similar. Then find the distance represented by *x*.

See Problem 4.

15.

16.

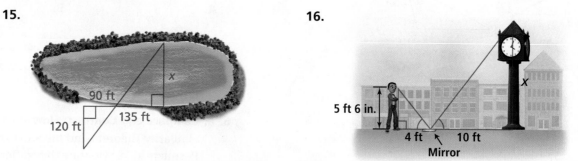

17. Washington Monument At a certain time of day, a 1.8-m-tall person standing next to the Washington Monument casts a 0.7-m shadow. At the same time, the Washington Monument casts a 65.8-m shadow. How tall is the Washington Monument?

B Apply

Can you conclude that the triangles are similar? If so, state the postulate or theorem you used and write a similarity statement. If not, explain.

18.   **19.**   **20.**

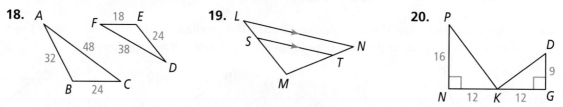

21. a. Are two isosceles triangles always similar? Explain.
b. Are two right isosceles triangles always similar? Explain.

© 22. Think About a Plan On a sunny day, a classmate uses indirect measurement to find the height of a building. The building's shadow is 12 ft long and your classmate's shadow is 4 ft long. If your classmate is 5 ft tall, what is the height of the building?
• Can you draw and label a diagram to represent the situation?
• What proportion can you use to solve the problem?

23. Indirect Measurement A 2-ft vertical post casts a 16-in. shadow at the same time a nearby cell phone tower casts a 120-ft shadow. How tall is the cell phone tower?

Algebra For each pair of similar triangles, find the value of x.

24.

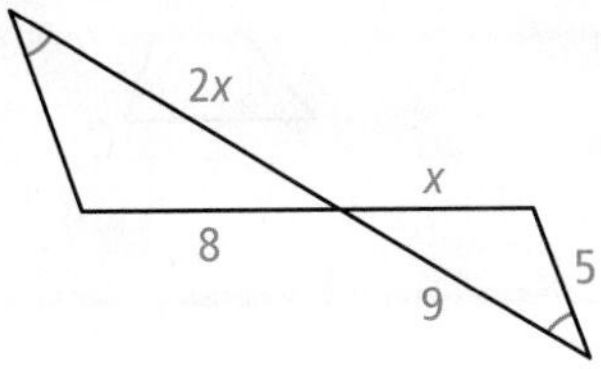

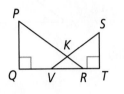

25.

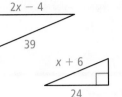

26.

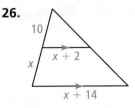

27. Given: $\overline{PQ} \perp \overline{QT}, \overline{ST} \perp \overline{TQ}, \dfrac{PQ}{ST} = \dfrac{QR}{TV}$
Proof **Prove:** $\triangle VKR$ is isosceles.

28. Given: $\overline{AB} \parallel \overline{CD}, \overline{BC} \parallel \overline{DG}$
Proof **Prove:** $AB \cdot CG = CD \cdot AC$

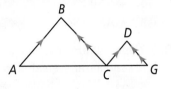

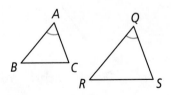

© 29. Reasoning Does any line that intersects two sides of a triangle and is parallel to the third side of the triangle form two similar triangles? Justify your reasoning.

30. Constructions Draw any $\triangle ABC$ with $m\angle C = 30$. Use a straightedge and compass to construct $\triangle LKJ$ so that $\triangle LKJ \sim \triangle ABC$.

© 31. Reasoning In the diagram at the right, $\triangle PMN \sim \triangle SRW$. $\overline{MQ}$ and $\overline{RT}$ are altitudes. The scale factor of $\triangle PMN$ to $\triangle SRW$ is $4:3$. What is the ratio of $\overline{MQ}$ to $\overline{RT}$? Explain how you know.

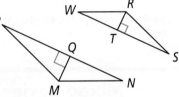

32. Coordinate Geometry $\triangle ABC$ has vertices $A(0, 0)$, $B(2, 4)$, and
Proof $C(4, 2)$. $\triangle RST$ has vertices $R(0, 3)$, $S(-1, 5)$, and $T(-2, 4)$. Prove that $\triangle ABC \sim \triangle RST$. (*Hint:* Graph $\triangle ABC$ and $\triangle RST$ in the coordinate plane.)

33. Write a proof of the following: Any two nonvertical parallel
Proof lines have equal slopes.

Given: Nonvertical lines ℓ_1 and ℓ_2, $\ell_1 \parallel \ell_2$,
$\overline{EF}$ and $\overline{BC}$ are $\perp$ to the x-axis
Prove: $\dfrac{BC}{AC} = \dfrac{EF}{DF}$

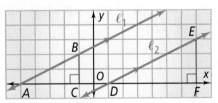

34. Use the diagram in Exercise 33. Prove: Any two nonvertical lines with equal slopes
Proof are parallel.

© Challenge

35. Prove the Side-Angle-Side Similarity Theorem (Theorem 7-1).
Proof **Given:** $\dfrac{AB}{QR} = \dfrac{AC}{QS}$, $\angle A \cong \angle Q$
Prove: $\triangle ABC \sim \triangle QRS$

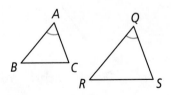

36. Prove the Side-Side-Side Similarity Theorem (Theorem 7-2).

Proof **Given:** $\dfrac{AB}{QR} = \dfrac{AC}{QS} = \dfrac{BC}{RS}$

Prove: $\triangle ABC \sim \triangle QRS$

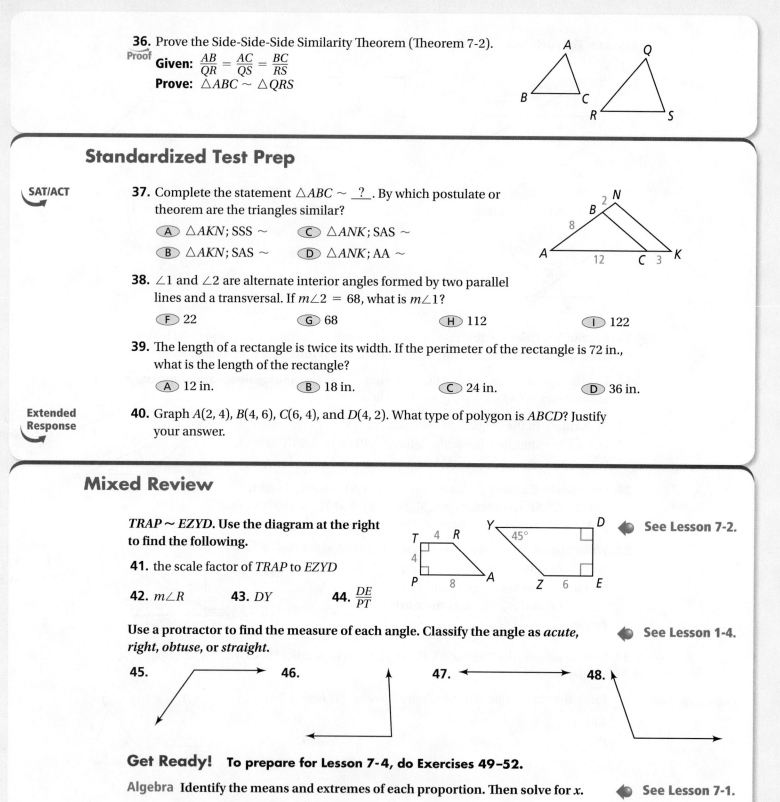

Standardized Test Prep

SAT/ACT

37. Complete the statement $\triangle ABC \sim \underline{\;?\;}$. By which postulate or theorem are the triangles similar?

Ⓐ $\triangle AKN$; SSS $\sim$

Ⓒ $\triangle ANK$; SAS $\sim$

Ⓑ $\triangle AKN$; SAS $\sim$

Ⓓ $\triangle ANK$; AA $\sim$

38. $\angle 1$ and $\angle 2$ are alternate interior angles formed by two parallel lines and a transversal. If $m\angle 2 = 68$, what is $m\angle 1$?

Ⓕ 22 Ⓖ 68 Ⓗ 112 Ⓘ 122

39. The length of a rectangle is twice its width. If the perimeter of the rectangle is 72 in., what is the length of the rectangle?

Ⓐ 12 in. Ⓑ 18 in. Ⓒ 24 in. Ⓓ 36 in.

Extended Response

40. Graph $A(2, 4)$, $B(4, 6)$, $C(6, 4)$, and $D(4, 2)$. What type of polygon is $ABCD$? Justify your answer.

Mixed Review

$TRAP \sim EZYD$. Use the diagram at the right to find the following.

See Lesson 7-2.

41. the scale factor of $TRAP$ to $EZYD$

42. $m\angle R$ **43.** DY **44.** $\dfrac{DE}{PT}$

Use a protractor to find the measure of each angle. Classify the angle as *acute, right, obtuse,* or *straight.*

See Lesson 1-4.

45.   **46.**   **47.**   **48.**

Get Ready! To prepare for Lesson 7-4, do Exercises 49–52.

Algebra Identify the means and extremes of each proportion. Then solve for x.

See Lesson 7-1.

49. $\dfrac{x}{8} = \dfrac{18}{24}$ **50.** $\dfrac{12}{m} = \dfrac{18}{20}$ **51.** $\dfrac{15}{x + 2} = \dfrac{9}{x}$ **52.** $\dfrac{x - 3}{x + 4} = \dfrac{5}{9}$

Do you know HOW?

1. A bookcase is 4 ft tall. A model of the bookcase is 6 in. tall. What is the ratio of the height of the model bookcase to the height of the real bookcase?

2. If $\frac{a}{b} = \frac{9}{10}$, complete this statement: $\frac{a}{9} = \frac{\blacksquare}{\blacksquare}$.

3. Are the two polygons shown below similar? If so, give the similarity ratio of the first polygon to the second. If not, explain.

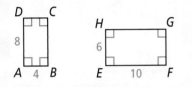

Solve each proportion.

4. $\frac{y}{6} = \frac{18}{54}$

5. $\frac{5}{7} = \frac{x-2}{4}$

6. On the scale drawing of a floor plan 2 in. = 5 ft. A room is 7 in. long on the scale drawing. Find the actual length of the room.

$\triangle ABC \sim \triangle DBF$. **Complete each statement.**

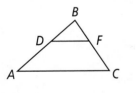

7. $m\angle A = m\angle \underline{\ ?\ }$

8. $\frac{AB}{DB} = \frac{BC}{\blacksquare}$

9. A postcard is 6 in. by 4 in. A printing shop will enlarge it so that the longer side is any length up to 3 ft. Find the dimensions of the biggest enlargement.

10. **Algebra** Find the value of x.

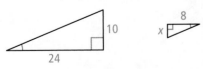

Are the triangles similar? If so, write a similarity statement and name the postulate or theorem you used. If not, explain.

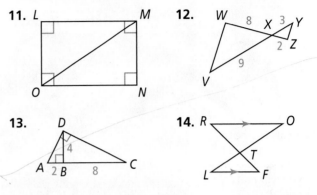

11.

12.

13.

14.

Algebra Explain why the triangles are similar. Then find the value of x.

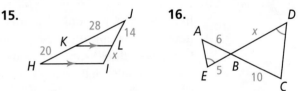

15.

16.

17. In a garden, a birdbath 2 ft 6 in. tall casts an 18-in. shadow at the same time an oak tree casts a 90-ft shadow. How tall is the oak tree?

Do you UNDERSTAND?

18. **Writing** You find an old scale drawing of your home, but the scale has faded and you cannot read it. How can you find the scale of the drawing?

19. **Reasoning** The sides of one triangle are twice as long as the corresponding sides of a second triangle. What is the relationship between the angles?

20. **Error Analysis** Your classmate says that since all congruent polygons are similar, all similar polygons must be congruent. Is he right? Explain.

Similarity in Right Triangles

© Content Standards
G.SRT.5 Use . . . similarity criteria for triangles to solve problems and to prove relationships in geometric figures.
G.GPE.5 Prove the slope criteria for parallel and perpendicular lines and use them to solve geometric problems.

Objective   To find and use relationships in similar right triangles

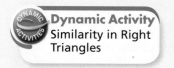

SOLVE IT!

Getting Ready!

Draw a diagonal of a rectangular piece of paper to form two right triangles. In one triangle, draw the altitude from the right angle to the hypotenuse. Number the angles as shown. Cut out the three triangles. How can you match the angles of the triangles to show that all three triangles are similar? Explain how you know the matching angles are congruent.

Analyze the situation first. Think about how you will match angles.

MATHEMATICAL PRACTICES

In the Solve It, you looked at three similar right triangles. In this lesson, you will learn new ways to think about the proportions that come from these similar triangles. You began with three separate, nonoverlapping triangles in the Solve It. Now you will see the two smaller right triangles fitting side-by-side to form the largest right triangle.

Dynamic Activity
Similarity in Right Triangles

Essential Understanding   When you draw the *altitude to the hypotenuse* of a right triangle, you form three pairs of similar right triangles.

Lesson Vocabulary
• geometric mean

take note

Theorem 7-3

Theorem
The altitude to the hypotenuse of a right triangle divides the triangle into two triangles that are similar to the original triangle and to each other.

If . . .
$\triangle ABC$ is a right triangle with right $\angle ACB$, and $\overline{CD}$ is the altitude to the hypotenuse

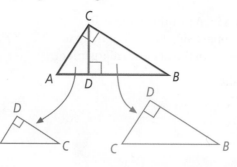

Then . . .
$\triangle ABC \sim \triangle ACD$
$\triangle ABC \sim \triangle CBD$
$\triangle ACD \sim \triangle CBD$

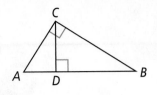

Proof Proof of Theorem 7-3

Given: Right $\triangle ABC$ with right $\angle ACB$ and altitude $\overline{CD}$

Prove: $\triangle ACD \sim \triangle ABC$, $\triangle CBD \sim \triangle ABC$, $\triangle ACD \sim \triangle CBD$

Statements	Reasons
1) $\angle ACB$ is a right angle.	1) Given
2) $\overline{CD}$ is an altitude.	2) Given
3) $\overline{CD} \perp \overline{AB}$	3) Definition of altitude
4) $\angle ADC$ and $\angle CDB$ are right angles.	4) Definition of $\perp$
5) $\angle ADC \cong \angle ACB$, $\angle CDB \cong \angle ACB$	5) All right $\angle$ are $\cong$.
6) $\angle A \cong \angle A$, $\angle B \cong \angle B$	6) Reflexive Property of $\cong$
7) $\triangle ACD \sim \triangle ABC$, $\triangle CBD \sim \triangle ABC$	7) AA $\sim$ Postulate
8) $\angle ACD \cong \angle B$	8) Corresponding $\angle$ of $\sim$ $\triangle$ are $\cong$.
9) $\angle ADC \cong \angle CDB$	9) All right $\angle$ are $\cong$.
10) $\triangle ACD \sim \triangle CBD$	10) AA $\sim$ Postulate

---

© Problem 1 Identifying Similar Triangles

Plan

What will help you see the corresponding vertices?
Sketch the triangles separately in the same orientation.

What similarity statement can you write relating the three triangles in the diagram?

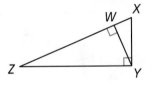

$\overline{YW}$ is the altitude to the hypotenuse of right $\triangle XYZ$, so you can use Theorem 7-3. There are three similar triangles.

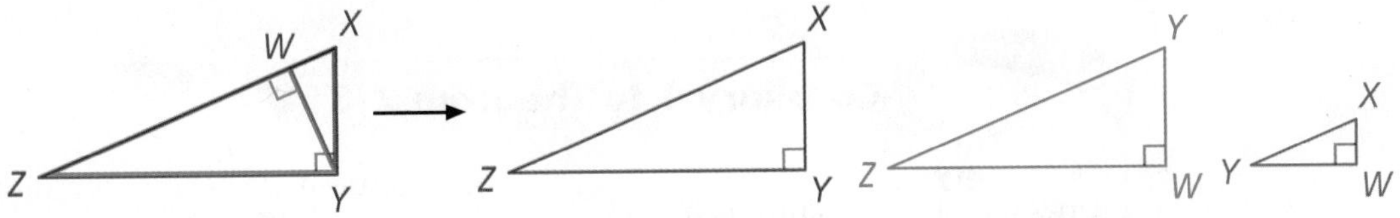

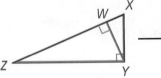

$\triangle XYZ \sim \triangle YWZ \sim \triangle XWY$

© **Got It?** **1. a.** What similarity statement can you write relating the three triangles in the diagram?

b. Reasoning From the similarity statement in part (a), write two different proportions using the ratio $\frac{SR}{SP}$.

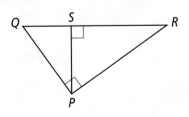

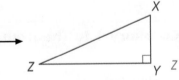

Proportions in which the means are equal occur frequently in geometry. For any two positive numbers a and b, the **geometric mean** of a and b is the positive number x such that $\frac{a}{x} = \frac{x}{b}$.

Finding the Geometric Mean

Multiple Choice What is the geometric mean of 6 and 15?

Ⓐ 90   Ⓑ $3\sqrt{10}$ Ⓒ $9\sqrt{10}$ Ⓓ 30

Think

How do you use the definition of geometric mean?
Set up a proportion with x in both means positions. The numbers 6 and 15 go into the extremes positions.

$\frac{6}{x} = \frac{x}{15}$	Definition of geometric mean
$x^2 = 90$	Cross Products Property
$x = \sqrt{90}$	Take the positive square root of each side.
$x = 3\sqrt{10}$	Write in simplest radical form.

The geometric mean of 6 and 15 is $3\sqrt{10}$. The correct answer is B.

Got It? **2.** What is the geometric mean of 4 and 18?

In Got It 1 part (b), you used a pair of similar triangles to write a proportion with a geometric mean.

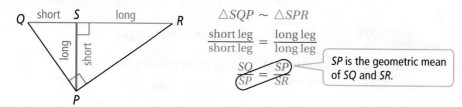

$\triangle SQP \sim \triangle SPR$

$\dfrac{\text{short leg}}{\text{short leg}} = \dfrac{\text{long leg}}{\text{long leg}}$

$\dfrac{SQ}{SP} = \dfrac{SP}{SR}$

SP is the geometric mean of SQ and SR.

This illustrates the first of two important corollaries of Theorem 7-3.

take note

Corollary 1 to Theorem 7-3

Corollary
The length of the altitude to the hypotenuse of a right triangle is the geometric mean of the lengths of the segments of the hypotenuse.

If . . .

Then . . .
$\dfrac{AD}{CD} = \dfrac{CD}{DB}$

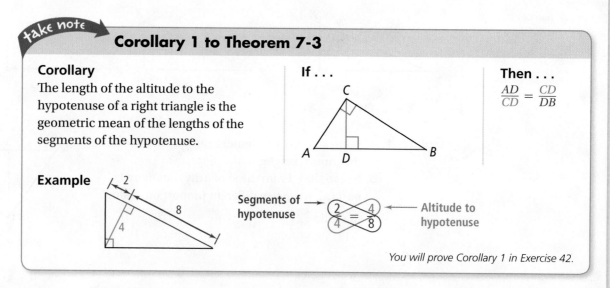

Example

Segments of hypotenuse → $\dfrac{2}{4} = \dfrac{4}{8}$ ← Altitude to hypotenuse

You will prove Corollary 1 in Exercise 42.

Corollary 2 to Theorem 7-3

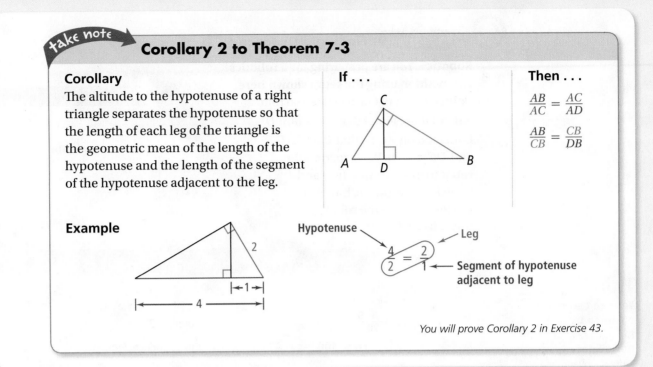

Corollary

The altitude to the hypotenuse of a right triangle separates the hypotenuse so that the length of each leg of the triangle is the geometric mean of the length of the hypotenuse and the length of the segment of the hypotenuse adjacent to the leg.

If . . .

Then . . .

$$\frac{AB}{AC} = \frac{AC}{AD}$$

$$\frac{AB}{CB} = \frac{CB}{DB}$$

Example

Hypotenuse → $\frac{4}{2} = \frac{2}{1}$ ← Leg

← Segment of hypotenuse adjacent to leg

You will prove Corollary 2 in Exercise 43.

The corollaries to Theorem 7-3 give you ways to write proportions using lengths in right triangles without thinking through the similar triangles. To help remember these corollaries, consider the diagram and these properties.

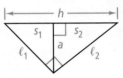

Corollary 1

$$\frac{s_1}{a} = \frac{a}{s_2}$$

Corollary 2

$$\frac{h}{\ell_1} = \frac{\ell_1}{s_1}, \ \frac{h}{\ell_2} = \frac{\ell_2}{s_2}$$

Problem 3 Using the Corollaries

Algebra What are the values of x and y?

Plan

How do you decide which corollary to use?
If you are using or finding an altitude, use Corollary 1. If you are using or finding a leg or hypotenuse, use Corollary 2.

Use Corollary 2.	$\frac{4+12}{x} = \frac{x}{4}$	Write a proportion.	$\frac{4}{y} = \frac{y}{12}$	Use Corollary 1.
	$x^2 = 64$	Cross Products Property	$y^2 = 48$	
	$x = \sqrt{64}$	Take the positive square root.	$y = \sqrt{48}$	
	$x = 8$	Simplify.	$y = 4\sqrt{3}$	

Got It? 3. What are the values of x and y?

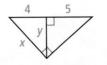

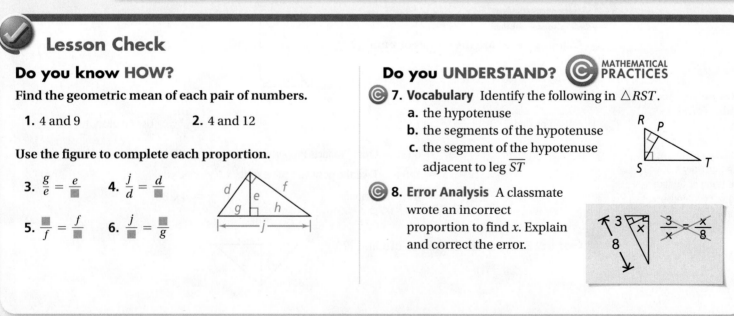

Problem 4 **Finding a Distance** STEM

Robotics You are preparing for a robotics competition using the setup shown here. Points A, B, and C are located so that $AB = 20$ in., and $\overline{AB} \perp \overline{BC}$. Point D is located on $\overline{AC}$ so that $\overline{BD} \perp \overline{AC}$ and $DC = 9$ in. You program the robot to move from A to D and to pick up the plastic bottle at D. How far does the robot travel from A to D?

B

20 in.

C   9 in.   *D*   *x*   *A*

Think

You can't solve this equation by taking the square root. What do you do?

Write the quadratic equation in the standard form $ax^2 + bx + c = 0$. Then solve by factoring or use the quadratic formula.

$\dfrac{x + 9}{20} = \dfrac{20}{x}$	Corollary 2
$x^2 + 9x = 400$	Cross Products Property
$x^2 + 9x - 400 = 0$	Subtract 400 from each side.
$(x - 16)(x + 25) = 0$	Factor.
$x - 16 = 0$ or $(x + 25) = 0$	Zero-Product Property
$x = 16$ or $x = -25$	Solve for x.

Only the positive solution makes sense in this situation. The robot travels 16 in.

Got It? **4.** From point D, the robot must turn right and move to point B to put the bottle in the recycling bin. How far does the robot travel from D to B?

Lesson Check

Do you know HOW?

Find the geometric mean of each pair of numbers.

1. 4 and 9   **2.** 4 and 12

Use the figure to complete each proportion.

3. $\dfrac{g}{e} = \dfrac{e}{\blacksquare}$ **4.** $\dfrac{j}{d} = \dfrac{d}{\blacksquare}$

5. $\dfrac{\blacksquare}{f} = \dfrac{f}{\blacksquare}$ **6.** $\dfrac{j}{\blacksquare} = \dfrac{\blacksquare}{g}$

Do you UNDERSTAND? MATHEMATICAL PRACTICES

7. **Vocabulary** Identify the following in $\triangle RST$.
   a. the hypotenuse
   b. the segments of the hypotenuse
   c. the segment of the hypotenuse adjacent to leg $\overline{ST}$

8. **Error Analysis** A classmate wrote an incorrect proportion to find x. Explain and correct the error.

$\dfrac{3}{x} = \dfrac{x}{8}$

Practice and Problem-Solving Exercises

MATHEMATICAL PRACTICES

A Practice   Write a similarity statement relating the three triangles in each diagram. **See Problem 1.**

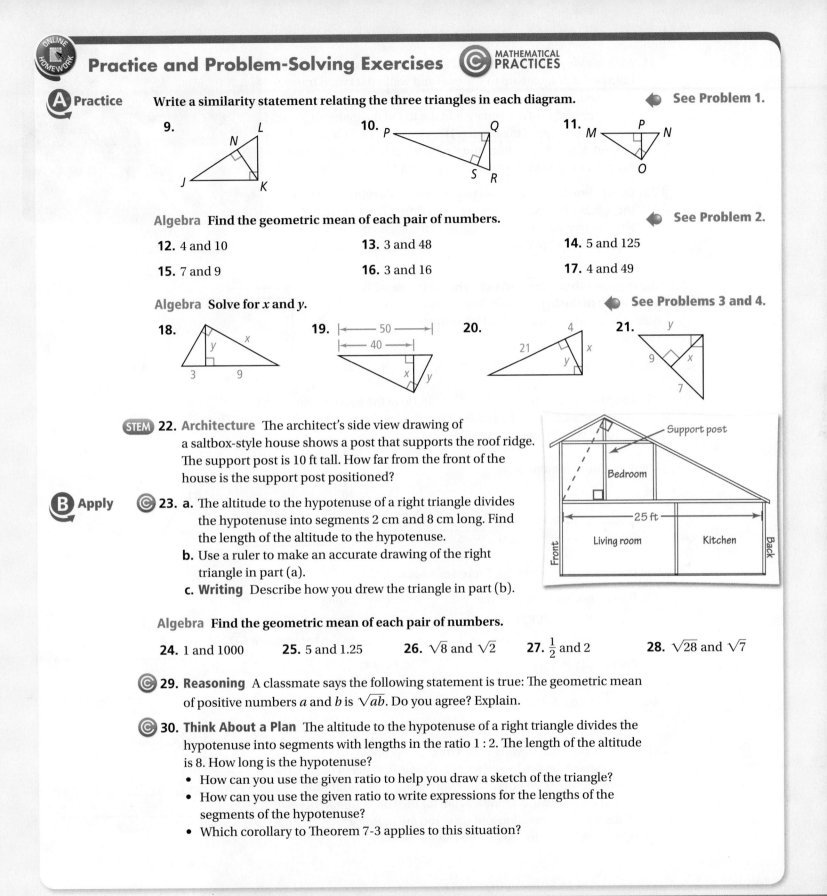

9.

10.

11.

Algebra Find the geometric mean of each pair of numbers. **See Problem 2.**

12. 4 and 10
13. 3 and 48
14. 5 and 125

15. 7 and 9
16. 3 and 16
17. 4 and 49

Algebra Solve for x and y. **See Problems 3 and 4.**

18.

19.

20.

21.

STEM **22. Architecture** The architect's side view drawing of a saltbox-style house shows a post that supports the roof ridge. The support post is 10 ft tall. How far from the front of the house is the support post positioned?

B Apply   **©** **23. a.** The altitude to the hypotenuse of a right triangle divides the hypotenuse into segments 2 cm and 8 cm long. Find the length of the altitude to the hypotenuse.
  b. Use a ruler to make an accurate drawing of the right triangle in part (a).
  c. Writing Describe how you drew the triangle in part (b).

Algebra Find the geometric mean of each pair of numbers.

24. 1 and 1000
25. 5 and 1.25
26. $\sqrt{8}$ and $\sqrt{2}$
27. $\frac{1}{2}$ and 2
28. $\sqrt{28}$ and $\sqrt{7}$

© **29. Reasoning** A classmate says the following statement is true: The geometric mean of positive numbers a and b is $\sqrt{ab}$. Do you agree? Explain.

© **30. Think About a Plan** The altitude to the hypotenuse of a right triangle divides the hypotenuse into segments with lengths in the ratio 1 : 2. The length of the altitude is 8. How long is the hypotenuse?
  • How can you use the given ratio to help you draw a sketch of the triangle?
  • How can you use the given ratio to write expressions for the lengths of the segments of the hypotenuse?
  • Which corollary to Theorem 7-3 applies to this situation?

31. Archaeology To estimate the height of a stone figure, Anya holds a small square up to her eyes and walks backward from the figure. She stops when the bottom of the figure aligns with the bottom edge of the square and the top of the figure aligns with the top edge of the square. Her eye level is 1.84 m from the ground. She is 3.50 m from the figure. What is the height of the figure to the nearest hundredth of a meter?

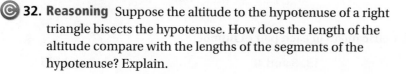

Ⓒ **32. Reasoning** Suppose the altitude to the hypotenuse of a right triangle bisects the hypotenuse. How does the length of the altitude compare with the lengths of the segments of the hypotenuse? Explain.

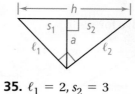

The diagram shows the parts of a right triangle with an altitude to the hypotenuse. For the two given measures, find the other four. Use simplest radical form.

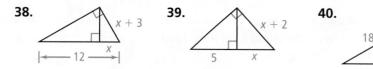

33. $h = 2, s_1 = 1$ **34.** $a = 6, s_1 = 6$ **35.** $\ell_1 = 2, s_2 = 3$ **36.** $s_1 = 3, \ell_2 = 6\sqrt{3}$

37. Coordinate Geometry $\overline{CD}$ is the altitude to the hypotenuse of right $\triangle ABC$. The coordinates of A, D, and B are $(4, 2)$, $(4, 6)$, and $(4, 15)$, respectively. Find all possible coordinates of point C.

Algebra Find the value of x.

38.
39.
40. 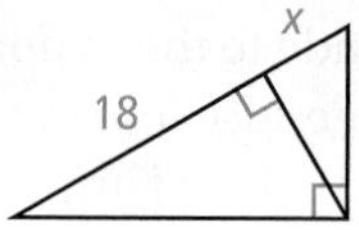
41.

Use the figure at the right for Exercises 42–43.

42. Prove Corollary 1 to Theorem 7-3.
Proof **Given:** Right $\triangle ABC$ with altitude to the hypotenuse $\overline{CD}$
Prove: $\dfrac{AD}{CD} = \dfrac{CD}{DB}$

43. Prove Corollary 2 to Theorem 7-3.
Proof **Given:** Right $\triangle ABC$ with altitude to the hypotenuse $\overline{CD}$
Prove: $\dfrac{AB}{AC} = \dfrac{AC}{AD}, \dfrac{AB}{BC} = \dfrac{BC}{DB}$

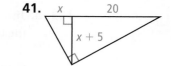

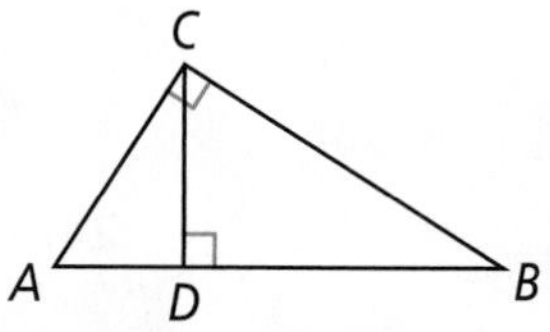

44. Given: Right $\triangle ABC$ with altitude $\overline{CD}$ to the hypotenuse $\overline{AB}$
Proof **Prove:** The product of the slopes of perpendicular lines is -1.

Ⓒ **Challenge** **45. a.** Consider the following conjecture: The product of the lengths of the two legs of a right triangle is equal to the product of the lengths of the hypotenuse and the altitude to the hypotenuse. Draw a figure for the conjecture. Write the *Given* information and what you are to *Prove*.
Ⓒ **b. Reasoning** Is the conjecture true? Explain.

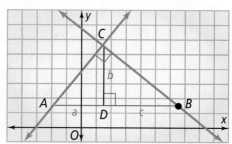

46. a. In the diagram, $c = x + y$. Use Corollary 2 to Theorem 7-3 to write two more equations involving a, b, c, x, and y.

   b. The equations in part (a) form a system of three equations in five variables. Reduce the system to one equation in three variables by eliminating x and y.

   c. State in words what the one resulting equation tells you.

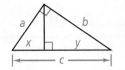

47. Given: In right $\triangle ABC$, $\overline{BD} \perp \overline{AC}$, and $\overline{DE} \perp \overline{BC}$.

Proof **Prove:** $\dfrac{AD}{DC} = \dfrac{BE}{EC}$

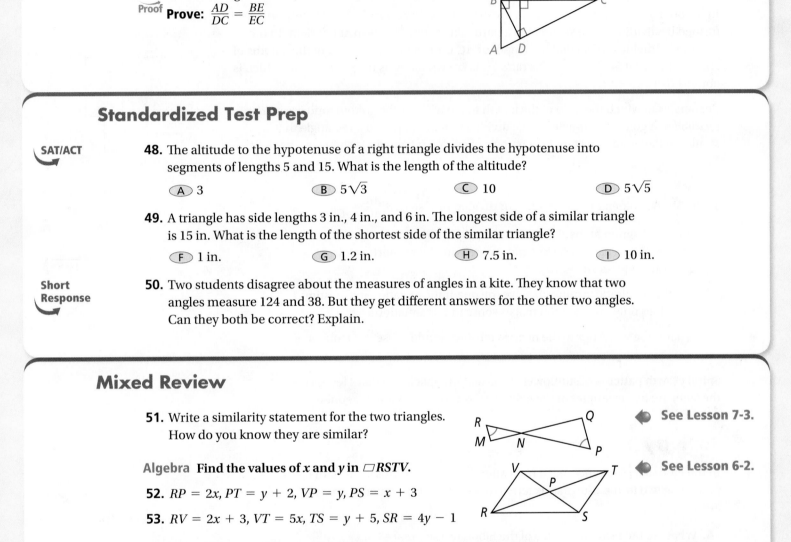

Standardized Test Prep

48. The altitude to the hypotenuse of a right triangle divides the hypotenuse into segments of lengths 5 and 15. What is the length of the altitude?

   Ⓐ 3   Ⓑ $5\sqrt{3}$ Ⓒ 10   Ⓓ $5\sqrt{5}$

49. A triangle has side lengths 3 in., 4 in., and 6 in. The longest side of a similar triangle is 15 in. What is the length of the shortest side of the similar triangle?

   Ⓕ 1 in.   Ⓖ 1.2 in.   Ⓗ 7.5 in.   Ⓘ 10 in.

50. Two students disagree about the measures of angles in a kite. They know that two angles measure 124 and 38. But they get different answers for the other two angles. Can they both be correct? Explain.

Mixed Review

51. Write a similarity statement for the two triangles. How do you know they are similar?

   ◀ See Lesson 7-3.

Algebra Find the values of x and y in $\square RSTV$.

   ◀ See Lesson 6-2.

52. $RP = 2x$, $PT = y + 2$, $VP = y$, $PS = x + 3$

53. $RV = 2x + 3$, $VT = 5x$, $TS = y + 5$, $SR = 4y - 1$

Get Ready!   **To prepare for Lesson 7-5, do Exercises 54–56.**

The two triangles in each diagram are similar. Find the value of x in each.

   ◀ See Lesson 7-2.

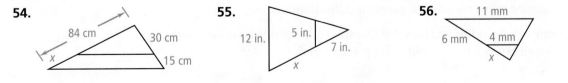

54. 84 cm, 30 cm, 15 cm, x

55. 12 in., 5 in., 7 in., x

56. 11 mm, 6 mm, 4 mm, x

Concept Byte

Use With Lesson 7-4

ACTIVITY

The Golden Ratio

© **Content Standard**
Extends G.SRT.5 Use . . . similarity criteria for triangles to solve problems and to prove relationships in geometric figures.

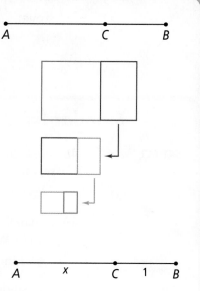

In his book *Elements*, Euclid defined the *extreme and mean ratio* using a proportion formed by dividing a line segment at a particular point, as shown at the right. In the diagram, C divides $\overline{AB}$ so that the length of $\overline{AC}$ is the geometric mean of the lengths of $\overline{AB}$ and $\overline{CB}$. That is, $\frac{AB}{AC} = \frac{AC}{CB}$. The ratio $\frac{AC}{CB}$ is known today as the *golden ratio*, which is about $1.618 : 1$.

Rectangles in which the ratio of the length to the width is the golden ratio are *golden rectangles*. A golden rectangle can be divided into a square and a rectangle that is similar to the original rectangle. A pattern of golden rectangles is shown at the right.

Activity 1

To derive the golden ratio, consider $\overline{AB}$ divided by C so that $\frac{AB}{AC} = \frac{AC}{CB}$.

1. Use the diagram at the right to write a proportion that relates the lengths of the segments. How can you rewrite the proportion as a quadratic equation?

2. Use the quadratic formula to solve the quadratic equation in Question 1. Why does only one solution makes sense in this situation?

3. What is the value of x to the nearest ten-thousandth? Use a calculator.

Spiral growth patterns of sunflower seeds and the spacing of plant leaves on the stem are two examples of the golden ratio and the Fibonacci sequence in nature.

Activity 2

In the Fibonacci sequence, each term after the first two terms is the sum of the preceding two terms. The first six terms of the Fibonacci sequence are 1, 1, 2, 3, 5, and 8.

4. What are the next nine terms of the Fibonacci sequence?

5. Starting with the second term, the ratios of each term to the previous term for the first six terms are $\frac{1}{1} = 1$, $\frac{2}{1} = 2$, $\frac{3}{2} = 1.5$, $\frac{5}{3} = 1.666\ldots$, and $\frac{8}{5} = 1.6$. What are the next nine ratios rounded to the nearest thousandth?

6. Compare the ratios you found in Question 5. What do you notice? How is the Fibonacci sequence related to the golden ratio?

Exercises

7. The golden rectangle is considered to be pleasing to the human eye. Of the following rectangles, which do you prefer? Is it a golden rectangle?

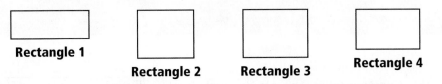

Rectangle 1

Rectangle 2 **Rectangle 3**

Rectangle 4

8. A drone is a male honeybee. Drones have only one parent, a queen. Workers and queens are female honeybees. Females have two parents, a drone and a queen. Part of the family tree showing the ancestors of a drone is shown below, where D represents a drone and Q represents a queen.

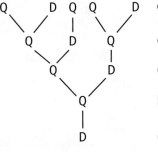

Q   D Q Q   D   **Great-Great-Grandparents**

Q   D   Q   **Great-Grandparents**

Q   D   **Grandparents**

Q   **Parent**

D   **Child**

   a. Continue the family tree for three more generations of ancestors.
   b. Count the number of honeybees in each generation. What pattern do you notice?

9. What is the relationship between the flowers and the Fibonacci sequence?

10. In $\triangle ABC$, point D divides the hypotenuse into the golden ratio. That is, $AD : DB$ is about $1.618 : 1$. $\overline{CD}$ is an altitude. Using the value 1.618 for AD and the value 1 for DB, solve for x. What do you notice?

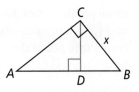

Concept Byte

Use With Lesson 7-5

TECHNOLOGY

Exploring Proportions in Triangles

© **Content Standard**
G.CO.12 Make formal geometric constructions with a variety of tools and methods . . .

Activity 1

Use geometry software to draw $\triangle ABC$. Construct point D on $\overline{AB}$. Next, construct a line through D parallel to $\overline{AC}$. Then construct the intersection E of the parallel line with $\overline{BC}$.

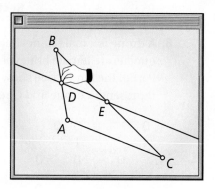

1. Measure $\overline{BD}$, $\overline{DA}$, $\overline{BE}$, and $\overline{EC}$. Calculate the ratios $\frac{BD}{DA}$ and $\frac{BE}{EC}$.

2. Manipulate $\triangle ABC$ and observe $\frac{BD}{DA}$ and $\frac{BE}{EC}$. What do you notice?

3. Make a conjecture about the four segments formed by a line parallel to one side of a triangle intersecting the other two sides.

Activity 2

Use geometry software to construct $\triangle ADE$ with vertices $A(3, 3)$, $D(-1, 0)$, and $E(5, 1)$.

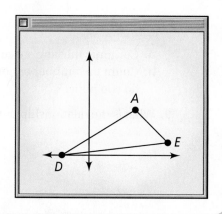

4. Measure AD, AE, and DE.

5. Suppose you draw $\overline{AD}$ so that it partitions $\triangle ADE$ and $AB = \frac{2}{3} AD$ and $CB \parallel DE$. Describe how you could approximate the coordinates of points A and D.

6. Now use the geometry software to draw $\overline{AD}$ and manipulate the segment to most closely find the points B and C. What are the coordinates of points B and C?

Exercises

7. Construct $\overleftrightarrow{AB} \parallel \overleftrightarrow{CD} \parallel \overleftrightarrow{EF}$. Then construct two transversals that intersect all three parallel lines. Measure $\overline{AC}$, $\overline{CE}$, $\overline{BD}$, and $\overline{DF}$. Calculate the ratios $\frac{AC}{CE}$ and $\frac{BD}{DF}$. Manipulate the locations of A and B and observe $\frac{AC}{CE}$ and $\frac{BD}{DF}$. Make a conjecture about the segments of the transversals formed by the three parallel lines intersecting two transversals.

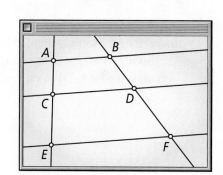

8. Suppose four or more parallel lines intersect two transversals. Make a conjecture about the segments of the transversals.

Proportions in Triangles

© **Content Standard**
G.SRT.4 Prove theorems about triangles . . .a line parallel to one side of a triangle divides the other two proportionally . . .

Objective To use the Side-Splitter Theorem and the Triangle-Angle-Bisector Theorem

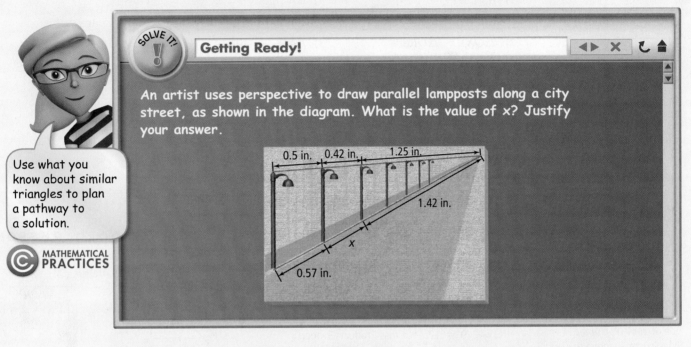

SOLVE IT!

Getting Ready!

An artist uses perspective to draw parallel lampposts along a city street, as shown in the diagram. What is the value of *x*? Justify your answer.

0.5 in. 0.42 in. 1.25 in.

1.42 in.

x

0.57 in.

Use what you know about similar triangles to plan a pathway to a solution.

© **MATHEMATICAL PRACTICES**

The Solve It involves parallel lines cut by two transversals that intersect. In this lesson, you will learn how to use proportions to find lengths of segments formed by parallel lines that intersect two or more transversals.

Essential Understanding When two or more parallel lines intersect other lines, proportional segments are formed.

take note

Theorem 7-4   Side-Splitter Theorem

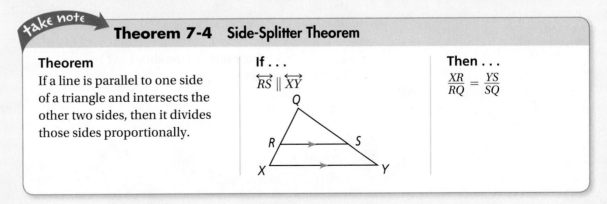

Theorem	**If . . .**	**Then . . .**
If a line is parallel to one side of a triangle and intersects the other two sides, then it divides those sides proportionally.	$\overleftrightarrow{RS} \parallel \overleftrightarrow{XY}$	$\dfrac{XR}{RQ} = \dfrac{YS}{SQ}$

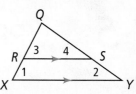

Proof **Proof of Theorem 7-4: Side-Splitter Theorem**

Given: $\triangle QXY$ with $\overleftrightarrow{RS} \parallel \overleftrightarrow{XY}$

Prove: $\dfrac{XR}{RQ} = \dfrac{YS}{SQ}$

Statements	Reasons
1) $\overleftrightarrow{RS} \parallel \overleftrightarrow{XY}$	1) Given
2) $\angle 1 \cong \angle 3, \angle 2 \cong \angle 4$	2) If lines are $\parallel$, then corresponding $\angle$s are $\cong$.
3) $\triangle QXY \sim \triangle QRS$	3) AA $\sim$ Postulate
4) $\dfrac{XQ}{RQ} = \dfrac{YQ}{SQ}$	4) Corresponding sides of $\sim \triangle$s are proportional.
5) $XQ = XR + RQ,$ $YQ = YS + SQ$	5) Segment Addition Postulate
6) $\dfrac{XR + RQ}{RQ} = \dfrac{YS + SQ}{SQ}$	6) Substitution Property
7) $\dfrac{XR}{RQ} = \dfrac{YS}{SQ}$	7) Property of Proportions (3)

Plan

How can you use the parallel lines in the diagram?
$\overline{KL}$ is parallel to one side of $\triangle MNP$. Use the Side-Splitter Theorem to set up a proportion.

Ⓒ **Problem 1** **Using the Side-Splitter Theorem**

GRIDDED RESPONSE

What is the value of x in the diagram at the right?

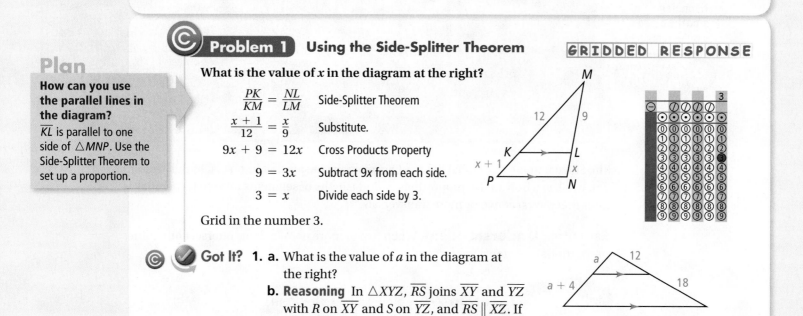

$$\frac{PK}{KM} = \frac{NL}{LM} \qquad \text{Side-Splitter Theorem}$$

$$\frac{x + 1}{12} = \frac{x}{9} \qquad \text{Substitute.}$$

$$9x + 9 = 12x \qquad \text{Cross Products Property}$$

$$9 = 3x \qquad \text{Subtract } 9x \text{ from each side.}$$

$$3 = x \qquad \text{Divide each side by 3.}$$

Grid in the number 3.

Ⓒ ✓ **Got It?** **1. a.** What is the value of a in the diagram at the right?

b. **Reasoning** In $\triangle XYZ$, $\overline{RS}$ joins $\overline{XY}$ and $\overline{YZ}$ with R on $\overline{XY}$ and S on $\overline{YZ}$, and $\overline{RS} \parallel \overline{XZ}$. If $\dfrac{YR}{RX} = \dfrac{YS}{SZ} = 1$, what must be true about RS? Justify your reasoning.

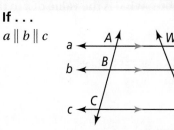

Corollary Corollary to the Side-Splitter Theorem

Corollary
If three parallel lines intersect two transversals, then the segments intercepted on the transversals are proportional.

If . . .
$a \parallel b \parallel c$

Then . . .
$\dfrac{AB}{BC} = \dfrac{WX}{XY}$

You will prove the Corollary to Theorem 7-4 in Exercise 46.

Plan

What information does the diagram give you?
The lines separating the campsites are parallel. Think of the river and the edge of the road as transversals. Then the boundaries along the road and river for each campsite are proportional.

Problem 2 Finding a Length

Camping Three campsites are shown in the diagram. What is the length of Site A along the river?

Let x be the length of Site A along the river.

$\dfrac{x}{8} = \dfrac{9}{7.2}$ Corollary to the Side-Splitter Theorem

$7.2x = 72$ Cross Products Property

$x = 10$ Divide each side by 7.2.

The length of Site A along the river is 10 yd.

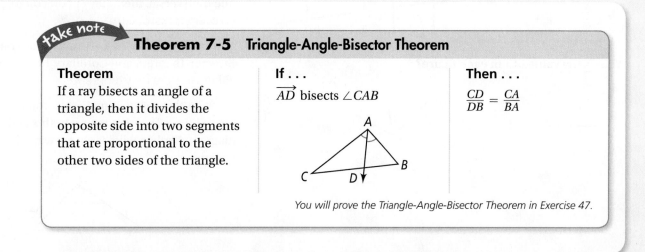

Got It? **2.** What is the length of Site C along the road?

Essential Understanding The bisector of an angle of a triangle divides the opposite side into two segments with lengths proportional to the sides of the triangle that form the angle.

Theorem 7-5 Triangle-Angle-Bisector Theorem

Theorem
If a ray bisects an angle of a triangle, then it divides the opposite side into two segments that are proportional to the other two sides of the triangle.

If . . .
$\overrightarrow{AD}$ bisects $\angle CAB$

Then . . .
$\dfrac{CD}{DB} = \dfrac{CA}{BA}$

You will prove the Triangle-Angle-Bisector Theorem in Exercise 47.

Problem 3 Using the Triangle-Angle-Bisector Theorem

Algebra What is the value of x in the diagram at the right?

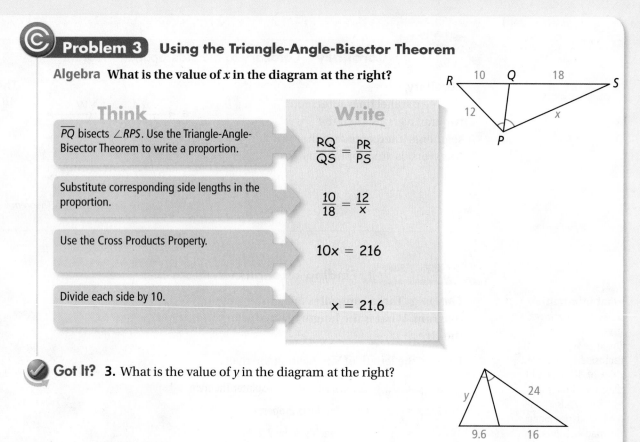

Think	Write
$\overline{PQ}$ bisects $\angle RPS$. Use the Triangle-Angle-Bisector Theorem to write a proportion.	$\dfrac{RQ}{QS} = \dfrac{PR}{PS}$
Substitute corresponding side lengths in the proportion.	$\dfrac{10}{18} = \dfrac{12}{x}$
Use the Cross Products Property.	$10x = 216$
Divide each side by 10.	$x = 21.6$

Got It? 3. What is the value of y in the diagram at the right?

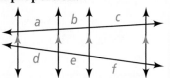

Lesson Check

Do you know HOW?

Use the figure to complete each proportion.

1. $\dfrac{a}{b} = \dfrac{\blacksquare}{e}$ **2.** $\dfrac{b}{\blacksquare} = \dfrac{e}{f}$

3. $\dfrac{a}{b+c} = \dfrac{\blacksquare}{e+f}$

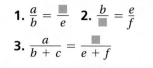

What is the value of x in each figure?

4.

5.

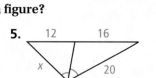

Do you UNDERSTAND? MATHEMATICAL PRACTICES

6. Compare and Contrast How is the Corollary to the Side-Splitter Theorem related to Theorem 6-7: If three (or more) parallel lines cut off congruent segments on one transversal, then they cut off congruent segments on every transversal?

7. Compare and Contrast How are the Triangle-Angle-Bisector Theorem and Corollary 1 to Theorem 7-3 alike? How are they different?

8. Error Analysis A classmate says you can use the Side-Splitter Theorem to find both x and y in the diagram. Explain what is wrong with your classmate's statement.

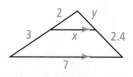

Practice and Problem-Solving Exercises

See Problem 1.

A Practice

Algebra Solve for *x*.

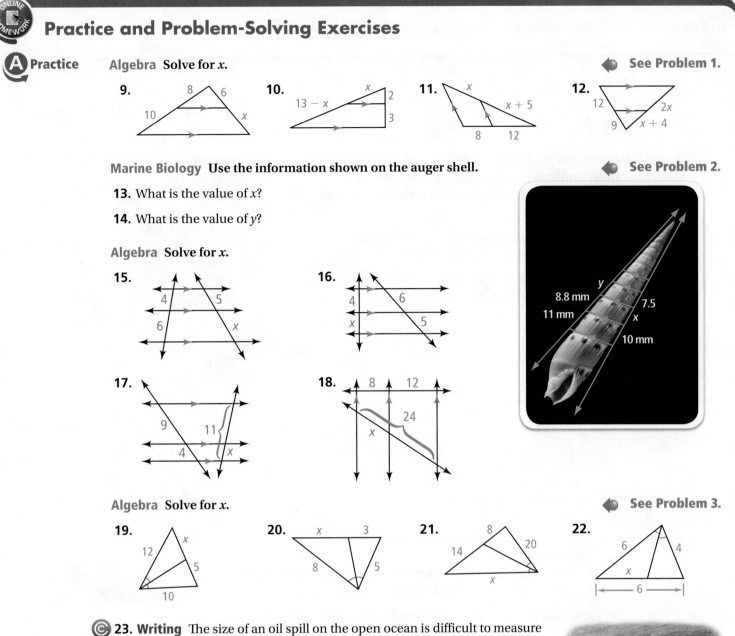

9.

10. $13 - x$, x, 2, 3

11. x, $x + 5$, 8, 12

12. 12, 9, $2x$, $x + 4$

Marine Biology Use the information shown on the auger shell.

See Problem 2.

13. What is the value of *x*?

14. What is the value of *y*?

Algebra Solve for *x*.

15. 4, 5, 6, *x*

16. 4, 6, *x*, 5

8.8 mm
11 mm
y
7.5
x
10 mm

17. 9, 11, 4, *x*

18. 8, 12, 24, *x*

Algebra Solve for *x*.

See Problem 3.

19. *x*, 12, 5, 10

20. *x*, 3, 8, 5

21. 8, 14, 20, *x*

22. 6, 4, *x*, 6

C 23. Writing The size of an oil spill on the open ocean is difficult to measure directly. Use the figure at the right to describe how you could find the length of the oil spill indirectly. What measurements and calculations would you use?

24. The lengths of the sides of a triangle are 5 cm, 12 cm, and 13 cm. Find the lengths, to the nearest tenth, of the segments into which the bisector of each angle divides the opposite side.

B **Apply**

Use the figure at the right to complete each proportion. Justify your answer.

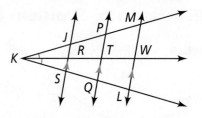

25. $\dfrac{RS}{\blacksquare} = \dfrac{JR}{KJ}$

26. $\dfrac{KJ}{JP} = \dfrac{KS}{\blacksquare}$

27. $\dfrac{QL}{PM} = \dfrac{SQ}{\blacksquare}$

28. $\dfrac{PT}{\blacksquare} = \dfrac{TQ}{KQ}$

29. $\dfrac{KL}{LW} = \dfrac{\blacksquare}{MW}$

30. $\dfrac{\blacksquare}{KP} = \dfrac{LQ}{KQ}$

STEM **Urban Design** In Washington, D.C., E. Capitol Street, Independence Avenue, C Street, and D Street are parallel streets that intersect Kentucky Avenue and 12th Street.

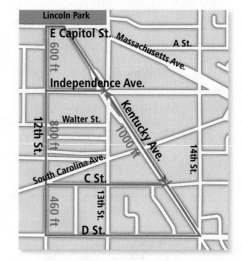

31. How long (to the nearest foot) is Kentucky Avenue between C Street and D Street?

32. How long (to the nearest foot) is Kentucky Avenue between E. Capitol Street and Independence Avenue?

Algebra Solve for x.

33.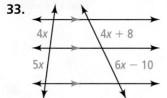

34.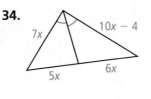

35.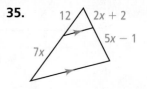

© **36. Think About a Plan** The perimeter of the triangular lot at the right is 50 m. The surveyor's tape bisects an angle. Find the lengths x and y.
  • How can you use the perimeter to write an equation in x and y?
  • What other relationship do you know between x and y?

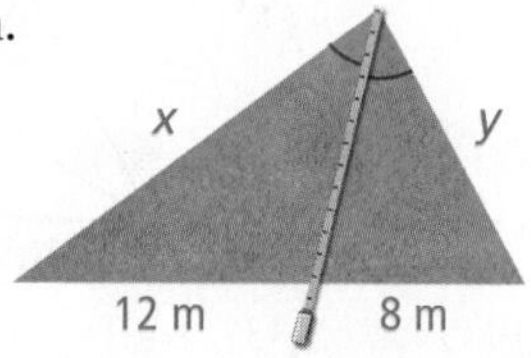

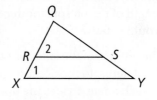

37. Prove the Converse of the Side-Splitter Theorem: If a line divides two
Proof sides of a triangle proportionally, then it is parallel to the third side.

Given: $\dfrac{XR}{RQ} = \dfrac{YS}{SQ}$

Prove: $\overline{RS} \parallel \overline{XY}$

Determine whether the red segments are parallel. Explain each answer. You can use the theorem proved in Exercise 37.

38.
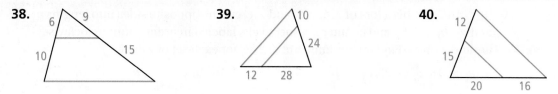

39.

40.

41. An angle bisector of a triangle divides the opposite side of the triangle into segments 5 cm and 3 cm long. A second side of the triangle is 7.5 cm long. Find all possible lengths for the third side of the triangle.

42. Open-Ended In a triangle, the bisector of an angle divides the opposite side into two segments with lengths 6 cm and 9 cm. How long could the other two sides of the triangle be? (*Hint:* Make sure the three sides satisfy the Triangle Inequality Theorem.)

43. Reasoning In $\triangle ABC$, the bisector of $\angle C$ bisects the opposite side. What type of triangle is $\triangle ABC$? Explain your reasoning.

Algebra Solve for *x*.

44.

45.

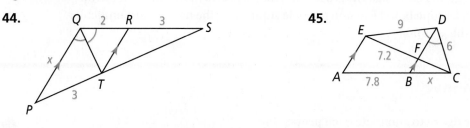

46. Prove the Corollary to the Side-Splitter Theorem. In the diagram from page 473, draw the auxiliary line $\overleftrightarrow{CW}$ and label its intersection with line *b* as point *P*.

Given: $a \parallel b \parallel c$

Prove: $\dfrac{AB}{BC} = \dfrac{WX}{XY}$

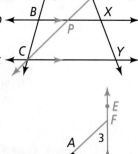

47. Prove the Triangle-Angle-Bisector Theorem. In the diagram from page 473, draw the auxiliary line $\overleftrightarrow{BE}$ so that $\overleftrightarrow{BE} \parallel \overline{DA}$. Extend $\overline{CA}$ to meet $\overleftrightarrow{BE}$ at point *F*.

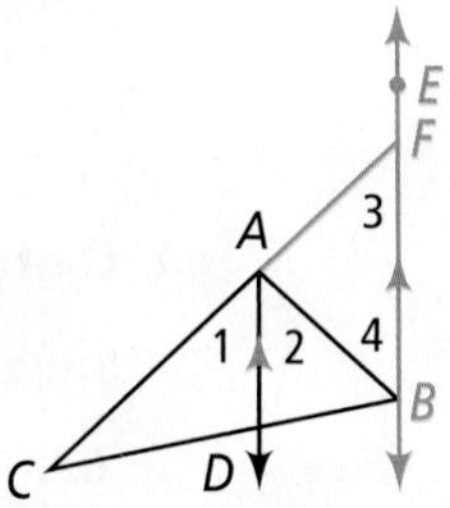

Given: $\overleftrightarrow{AD}$ bisects $\angle CAB$.

Prove: $\dfrac{CD}{DB} = \dfrac{CA}{BA}$

Challenge

48. Use the definition in part (a) to prove the statements in parts (b) and (c).
  a. Write a definition for a midsegment of a parallelogram.
  b. A parallelogram midsegment is parallel to two sides of the parallelogram.
  c. A parallelogram midsegment bisects the diagonals of a parallelogram.

49. State the converse of the Triangle-Angle-Bisector Theorem. Give a convincing argument that the converse is true or a counterexample to prove that it is false.

50. In $\triangle ABC$, the bisectors of $\angle A$, $\angle B$, and $\angle C$ cut the opposite sides into lengths a_1 and a_2, b_1 and b_2, and c_1 and c_2, respectively, labeled in order counterclockwise around $\triangle ABC$. Find the perimeter of $\triangle ABC$ for each set of values.

    a. $b_1 = 16$, $b_2 = 20$, $c_1 = 18$ **b.** $a_1 = \frac{5}{3}$, $a_2 = \frac{10}{3}$, $b_1 = \frac{15}{4}$

Standardized Test Prep

51. What is the value of x in the figure at the right?

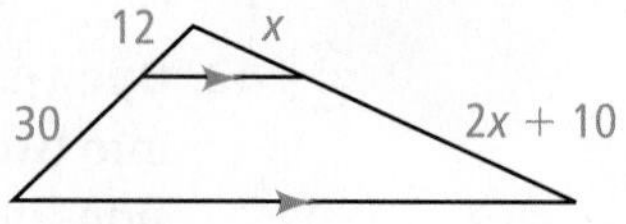

52. Suppose $\triangle VLQ \sim \triangle PSX$. If $m\angle V = 48$ and $m\angle L = 80$, what is $m\angle X$?

53. In the diagram at the right, $\overline{PR} \cong \overline{QR}$. For what value of x is $\overline{TS}$ parallel to $\overline{QP}$?

54. Leah is playing basketball on an outdoor basketball court. The 10-ft pole supporting the basketball net casts a 15-ft shadow. At the same time, the length of Leah's shadow is 8 ft 3 in. What is Leah's height in inches? You can assume both Leah and the pole supporting the net are perpendicular to the ground.

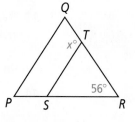

Mixed Review

Use the figure to complete each proportion.

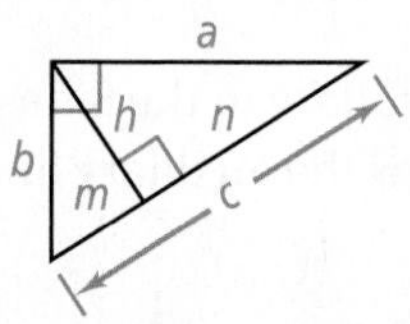

55. $\dfrac{n}{h} = \dfrac{h}{\blacksquare}$ **56.** $\dfrac{\blacksquare}{b} = \dfrac{b}{c}$

57. $\dfrac{n}{a} = \dfrac{a}{\blacksquare}$ **58.** $\dfrac{m}{h} = \dfrac{\blacksquare}{n}$

➦ **See Lesson 7-4.**

Find the center of the circle that you can circumscribe about each $\triangle ABC$.

➦ **See Lesson 5-3.**

59. $A(0, 0)$
    $B(6, 0)$
    $C(0, -6)$

60. $A(2, 5)$
    $B(-2, 5)$
    $C(-2, -1)$

61. $A(-2, 0)$
    $B(5, 5)$
    $C(-2, 5)$

Get Ready! **To prepare for Lesson 8-1, do Exercises 62–64.**

Square the lengths of the sides of each triangle.

➦ **See p. 829.**

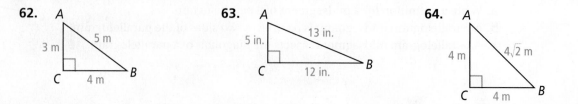

62.   **63.**   **64.**

Pull It All Together © ASSESSMENT

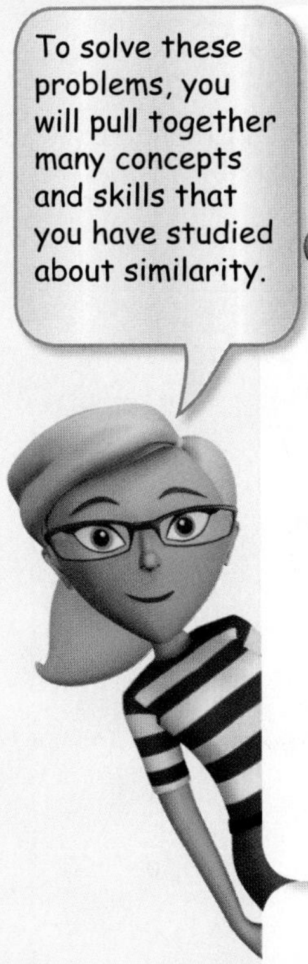

To solve these problems, you will pull together many concepts and skills that you have studied about similarity.

BIG idea Visualization, Reasoning and Proof, and Similarity

You can show that two triangles are similar when certain relationships exist between two or three pairs of corresponding parts. If you know two triangles are similar, then you know their corresponding sides are proportional.

© Performance Task 1

In the diagram below, $\overline{AC} \parallel \overline{DF} \parallel \overline{BH}$ and $\overline{CB} \parallel \overline{FE}$.

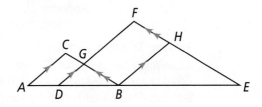

a. Find four similar triangles. Explain how you know that they are all similar.

b. Using the similar triangles you found in part (a), complete the following extended proportion:

$$\frac{AB}{AC} = \frac{DE}{\blacksquare} = \frac{\blacksquare}{DG} = \frac{\blacksquare}{\blacksquare}$$

BIG idea Similarity

Lines with special relationships to the sides and angles of a triangle determine proportional segments. When you know the lengths of some of the segments, you can use a proportion to find an unknown length.

© Performance Task 2

You are making the kite shown at the right from five pairs of congruent panels. In parts (a)–(d) below, use the given information to find the side lengths of the kite's panels. Show your work.

$ABCD$ is a kite.
$EB = 15$ in., $BC = 25$ in.
The extended ratio $XY : YZ : ZC$ is $3 : 1 : 4$.
$\overline{EX} \perp \overline{BC}, \overline{EX} \parallel \overline{YF} \parallel \overline{GZ}$

a. $\triangle BEX$ **b.** $XEFY$

c. $YFGZ$ **d.** $\triangle ZGC$

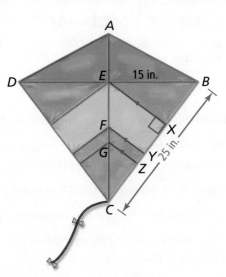

Connecting **BIG** ideas and Answering the Essential Questions

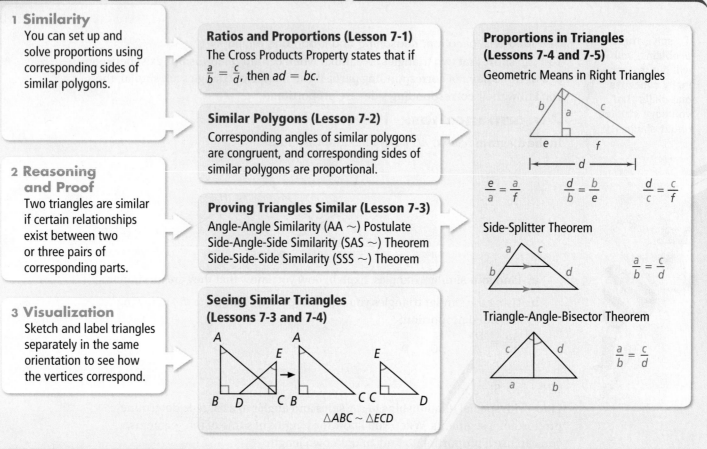

1 Similarity
You can set up and solve proportions using corresponding sides of similar polygons.

Ratios and Proportions (Lesson 7-1)
The Cross Products Property states that if $\frac{a}{b} = \frac{c}{d}$, then $ad = bc$.

Similar Polygons (Lesson 7-2)
Corresponding angles of similar polygons are congruent, and corresponding sides of similar polygons are proportional.

2 Reasoning and Proof
Two triangles are similar if certain relationships exist between two or three pairs of corresponding parts.

Proving Triangles Similar (Lesson 7-3)
Angle-Angle Similarity (AA ~) Postulate
Side-Angle-Side Similarity (SAS ~) Theorem
Side-Side-Side Similarity (SSS ~) Theorem

3 Visualization
Sketch and label triangles separately in the same orientation to see how the vertices correspond.

Seeing Similar Triangles (Lessons 7-3 and 7-4)

$\triangle ABC \sim \triangle ECD$

Proportions in Triangles (Lessons 7-4 and 7-5)

Geometric Means in Right Triangles

$\frac{e}{a} = \frac{a}{f}$ $\frac{d}{b} = \frac{b}{e}$ $\frac{d}{c} = \frac{c}{f}$

Side-Splitter Theorem

$\frac{a}{b} = \frac{c}{d}$

Triangle-Angle-Bisector Theorem

$\frac{a}{b} = \frac{c}{d}$

Chapter Vocabulary

- extended proportion (p. 440)
- extended ratio (p. 433)
- extremes (p. 434)
- geometric mean (p. 462)
- indirect measurement (p. 454)
- means (p. 434)
- proportion (p. 434)
- ratio (p. 432)
- scale drawing (p. 443)
- scale factor (p. 440)
- similar figures (p. 440)
- similar polygons (p. 440)

Choose the correct term to complete each sentence.

1. Two polygons are ? if their corresponding angles are congruent and corresponding sides are proportional.

2. A(n) ? is a statement that two ratios are equal.

3. The ratio of the lengths of corresponding sides of two similar polygons is the ? .

4. The Cross Products Property states that the product of the ? is equal to the product of the ? .

7-1 Ratios and Proportions

Quick Review

A **ratio** is a comparison of two quantities by division. A **proportion** is a statement that two ratios are equal. The **Cross Products Property** states that if $\frac{a}{b} = \frac{c}{d}$, where $b \neq 0$ and $d \neq 0$, then $ad = bc$.

Example

What is the solution of $\frac{x}{x+3} = \frac{4}{6}$?

$6x = 4(x + 3)$	Cross Products Property
$6x = 4x + 12$	Distributive Property
$2x = 12$	Subtract $4x$ from each side.
$x = 6$	Divide each side by 2.

Exercises

5. A high school has 16 math teachers for 1856 math students. What is the ratio of math teachers to math students?

6. The measures of two complementary angles are in the ratio $2 : 3$. What is the measure of the smaller angle?

Algebra Solve each proportion.

7. $\frac{x}{7} = \frac{18}{21}$

8. $\frac{6}{11} = \frac{15}{2x}$

9. $\frac{x}{3} = \frac{x+4}{5}$

10. $\frac{8}{x+9} = \frac{2}{x-3}$

7-2 and 7-3 Similar Polygons and Proving Triangles Similar

Quick Review

Similar polygons have congruent corresponding angles and proportional corresponding sides. You can prove triangles similar with limited information about congruent corresponding angles and proportional corresponding sides.

Postulate or Theorem	What You Need
Angle-Angle (AA $\sim$)	two pairs of $\cong$ angles
Side-Angle-Side (SAS $\sim$)	two pairs of proportional sides and the included angles $\cong$
Side-Side-Side (SSS $\sim$)	three pairs of proportional sides

Example

Is $\triangle ABC$ similar to $\triangle RQP$? How do you know?

You know that $\angle A \cong \angle R$.

$\frac{AB}{RQ} = \frac{AC}{RP} = \frac{2}{1}$, so the triangles are similar by the SAS $\sim$ Theorem.

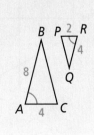

Exercises

The polygons are similar. Write a similarity statement and give the scale factor.

11. **12.**

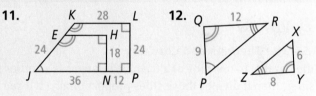

13. **City Planning** The length of a rectangular playground in a scale drawing is 12 in. If the scale is 1 in. = 10 ft, what is the actual length?

14. **Indirect Measurement** A 3-ft vertical post casts a 24-in. shadow at the same time a pine tree casts a 30-ft shadow. How tall is the pine tree?

Are the triangles similar? How do you know?

15. **16.**

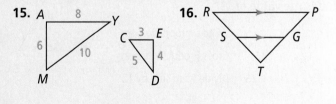

7-4 Similarity in Right Triangles

Quick Review

$\overline{CD}$ is the altitude to the hypotenuse of right $\triangle ABC$.

- $\triangle ABC \sim \triangle ACD$,
  $\triangle ABC \sim \triangle CBD$, and
  $\triangle ACD \sim \triangle CBD$

- $\dfrac{AD}{CD} = \dfrac{CD}{DB}$, $\dfrac{AB}{AC} = \dfrac{AC}{AD}$, and $\dfrac{AB}{CB} = \dfrac{CB}{DB}$

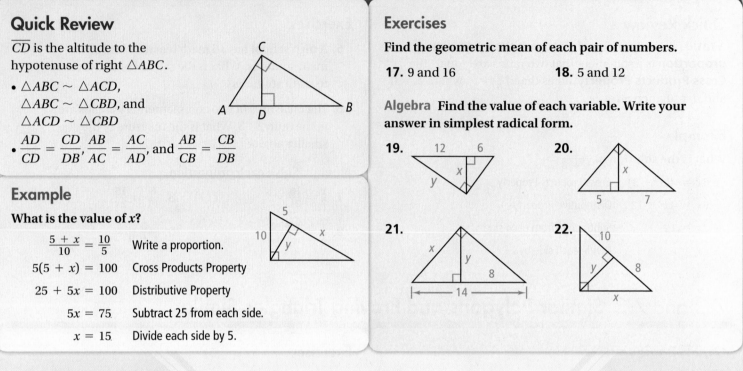

Example

What is the value of x?

$\dfrac{5 + x}{10} = \dfrac{10}{5}$ Write a proportion.

$5(5 + x) = 100$ Cross Products Property

$25 + 5x = 100$ Distributive Property

$5x = 75$ Subtract 25 from each side.

$x = 15$ Divide each side by 5.

Exercises

Find the geometric mean of each pair of numbers.

17. 9 and 16

18. 5 and 12

Algebra Find the value of each variable. Write your answer in simplest radical form.

19.

20.

21.

22.

7-5 Proportions in Triangles

Quick Review

Side-Splitter Theorem and Corollary
If a line parallel to one side of a triangle intersects the other two sides, then it divides those sides proportionally. If three parallel lines intersect two transversals, then the segments intercepted on the transversals are proportional.

Triangle-Angle-Bisector Theorem
If a ray bisects an angle of a triangle, then it divides the opposite side into two segments that are proportional to the other two sides of the triangle.

Example

What is the value of x?

$\dfrac{12}{15} = \dfrac{9}{x}$ Write a proportion.

$12x = 135$ Cross Products Property

$x = 11.25$ Divide each side by 12.

Exercises

Algebra Find the value of x.

23.

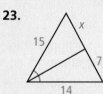

24.

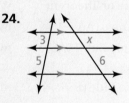

25.

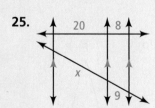

26.

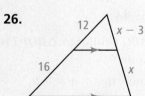

27.

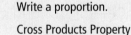

28.

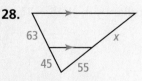

Do you know HOW?

Algebra Solve each proportion.

1. $\frac{x}{3} = \frac{8}{12}$

2. $\frac{4}{x+2} = \frac{16}{9}$

3. Are the polygons below similar? If they are, write a similarity statement and give the scale factor.

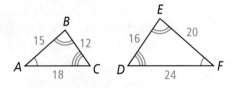

Algebra The figures in each pair are similar. Find the value of each variable.

4.

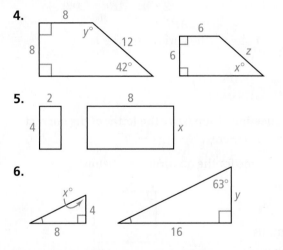

5.

6.

Determine whether the triangles are similar. If so, write a similarity statement and name the postulate or the theorem you used. If not, explain.

7.

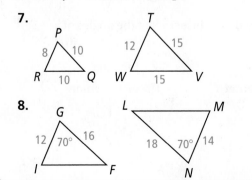

8.

9. Indirect Measurement A meter stick perpendicular to the ground casts a 1.5-m shadow. At the same time, a telephone pole casts a shadow that is 9 m. How tall is the telephone pole?

10. Photography A photographic negative is 3 cm by 2 cm. A similar print from the negative is 9 cm long on its shorter side. What is the length of the longer side?

11. What is the geometric mean of 10 and 15?

Algebra Find the value of x.

12.

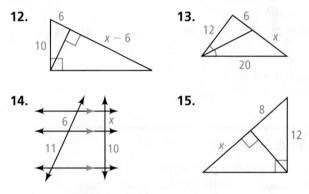

13.

14.

15.

Do you UNDERSTAND?

16. Reasoning In the diagram, $\overline{MN} \parallel \overline{LK}$. Is $\frac{JM}{ML}$ equal to $\frac{MN}{LK}$? Explain.

17. Reasoning $\square ABCD \sim \square PQRS$. $\overline{DB}$ is a diagonal of $\square ABCD$ and $\overline{SQ}$ is a diagonal of $\square PQRS$. Is $\triangle BCD$ similar to $\triangle QRS$? Justify your reasoning.

Determine whether each statement is *always*, *sometimes*, or *never* true.

18. A parallelogram is similar to a trapezoid.

19. Two rectangles are similar.

20. If the vertex angles of two isosceles triangles are congruent, then the triangles are similar.

TIPS FOR SUCCESS

Some test questions ask you to find the measure of an interior or exterior angle of a polygon. Read the sample question at the right. Then follow the tips to answer it.

In the figure below, *ABCDE* is a regular pentagon. What is the measure, in degrees, of $\angle ABE$?

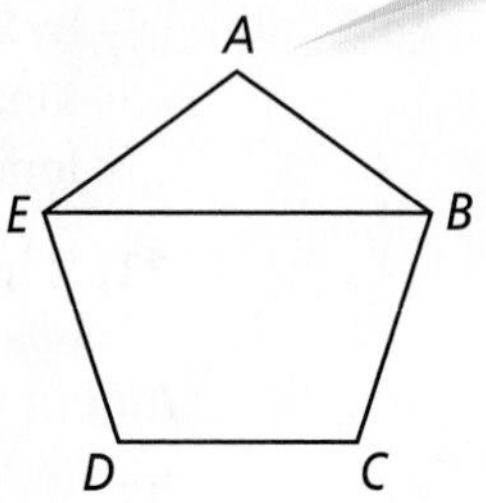

A

E   B

D   C

Ⓐ 36   Ⓒ 72

Ⓑ 54   Ⓓ 108

TIP 2

Find $m\angle A$ and then use it to find $m\angle ABE$.

TIP 1

List what you know about $\triangle ABE$.

• $\overline{AE} \cong \overline{AB}$ because the pentagon is regular.

• $\angle ABE \cong \angle AEB$ because $\triangle ABE$ is isosceles.

• $m\angle A + m\angle ABE + m\angle AEB = 180$

Think It Through

By the Polygon Angle-Sum Theorem, the sum of the interior angle measures of *ABCDE* is $(5 - 2)180 = 3(180) = 540$. So $m\angle A = \frac{540}{5} = 108$. Then $108 + m\angle ABE + m\angle AEB = 180$. Since $\angle ABE \cong \angle AEB$, $108 + 2 \cdot m\angle ABE = 180$. So $m\angle ABE = \frac{180 - 108}{2} = \frac{72}{2} = 36$. The correct answer is A.

Vocabulary Builder

As you solve test items, you must understand the meanings of mathematical terms. Match each term with its mathematical meaning.

A. corollary

B. geometric mean

C. midsegment

D. scale

I. the ratio of a length in a scale drawing to the actual length

II. a segment connecting the midpoints of two sides of a triangle

III. a statement that follows immediately from a theorem

IV. for positive numbers *a* and *b*, the positive number *x* such that $\frac{a}{x} = \frac{x}{b}$

Multiple Choice

Read each question. Then write the letter of the correct answer on your paper.

1. What is a name for the quadrilateral below?

I. square

II. rectangle

III. rhombus

IV. parallelogram

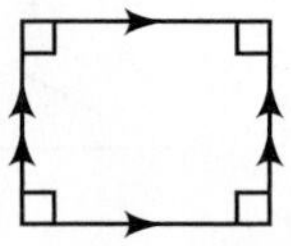

Ⓐ I only   Ⓒ II and IV

Ⓑ IV only   Ⓓ I, II, and IV

2. In which point do the bisectors of the angles of a triangle meet?

Ⓕ centroid   Ⓗ incenter

Ⓖ circumcenter   Ⓘ orthocenter

3. Which quadrilateral does NOT always have perpendicular diagonals?

   (A) square   (C) kite

   (B) rhombus   (D) isosceles trapezoid

4. Which of the following facts would be sufficient to prove $\triangle ACE \sim \triangle BCD$?

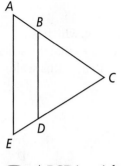

   (F) $\triangle BCD$ is a right triangle.

   (G) $\overline{AB} \cong \overline{ED}$

   (H) $m\angle A = m\angle E$

   (I) $\overline{AE} \parallel \overline{BD}$

5. What is the midpoint of the segment whose endpoints are $M(6, -11)$ and $N(-18, 7)$?

   (A) $(-6, -2)$ (C) $(-12, 9)$

   (B) $(6, 2)$ (D) $(12, -9)$

6. Use the figure below. By which theorem or postulate does $x = 3$?

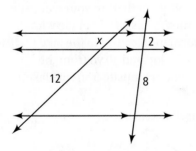

   (F) SAS Postulate

   (G) If three parallel lines intersect two transversals, then the segments intercepted on the transversals are proportional.

   (H) Opposite sides of a parallelogram are congruent.

   (I) If two lines are parallel to the same line, then they are parallel to each other.

7. Which angle is congruent to $\angle DCB$?

   (A) $\angle B$ (C) $\angle A$

   (B) $\angle CDB$ (D) $\angle ACD$

8. In the figure below, $\overline{EF}$ is a midsegment of $\triangle ABC$ and $ADGC$ is a rectangle. What is the area of $\triangle EDA$?

   (F) 49 cm^2

   (G) 98 cm^2

   (H) 294 cm^2

   (I) 588 cm^2

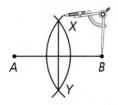

9. Andrew is looking at a map that uses the scale 1 in. = 5 mi. On the map, the distance from Westville to Allentown is 9 in. Which proportion CANNOT be used to find the actual distance?

   (A) $\dfrac{1 \text{ in.}}{5 \text{ mi}} = \dfrac{9 \text{ in.}}{d}$ (C) $\dfrac{d}{9 \text{ in.}} = \dfrac{5 \text{ mi}}{1 \text{ in.}}$

   (B) $\dfrac{5 \text{ mi}}{d} = \dfrac{9 \text{ in.}}{1 \text{ in.}}$ (D) $\dfrac{5 \text{ mi}}{1 \text{ in.}} = \dfrac{d}{9 \text{ in.}}$

10. What type of construction is shown below?

   (F) angle bisector

   (G) perpendicular bisector

   (H) congruent angles

   (I) congruent triangles

11. A student is sketching an 11-sided regular polygon. What is the sum of the measures of the polygon's first five angles to the nearest degree?

   (A) 147   (C) 736

   (B) 720   (D) 1620

12. Triangle *ABC* is similar to triangle *HIJ*. Find the area of rectangle *HIJK*.

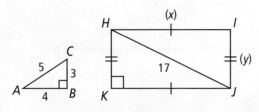

13. What is the value of *x* in the figure below?

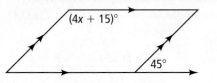

14. In hexagon *ABCDEF*, ∠*A* and ∠*B* are right angles. If ∠*C* ≅ ∠*D* ≅ ∠*E* ≅ ∠*F*, what is the measure of ∠*F* in degrees?

15. A scale drawing of a swimming pool and deck is shown below. Use the scale 1 in. = 2 m. What is the area of the deck in square meters?

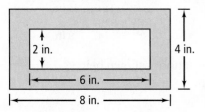

16. In parallelogram *ABCD* below, *DB* is 15. What is *DE*?

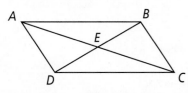

17. The measure of the vertex angle of an isosceles triangle is 112. What is the measure of a base angle?

18. What is the value of *x* in the figure below?

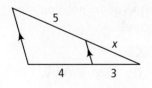

Short Response

19. In rectangle *ABCD*, *AC* = 5(*x* − 2) and *BD* = 3(*x* + 2). What is the value of *x*? Justify your answer.

20. Draw line *m* with point *A* on it. Construct a line perpendicular to *m* at *A*. What steps did you take to perform the construction?

21. Petra visited the Empire State Building, which is approximately 1454 ft tall. She estimates that the scale of the model she bought is 1 in. = 12 ft. Is this scale reasonable? Explain.

Extended Response

22. In the diagram, *AB* = *FE*, *BC* = *ED*, and *AE* = *FB*.

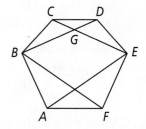

   a. Is there enough information to prove △*BCG* ≅ △*EDG*? Explain.

   b. What one additional piece of information would allow you to prove △*BCD* ≅ △*EDC*? Explain.

   c. What can you conclude from the diagram that would help you prove △*BAF* ≅ △*EFA*?

   d. In part (c), is △*BAF* ≅ △*EFA* by SAS or SSS? Explain.

23. At a campground, the 50-yd path from your campsite to the information center forms a right angle with the path from the information center to the lake. The information center is located 30 yd from the bathhouse. How far is your campsite from the lake? Show your work.

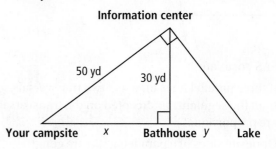

Get Ready!

Lesson 7-1 ◀ **Solving Proportions**

Algebra Solve for *x*. If necessary, round answers to the nearest thousandth.

1. $0.2734 = \dfrac{x}{17}$ **2.** $0.5858 = \dfrac{24}{x}$ **3.** $0.8572 = \dfrac{5271}{x}$ **4.** $0.5 = \dfrac{x}{3x + 5}$

Lesson 7-3 ◀ **Proving Triangles Similar**

Name the postulate or theorem that proves each pair of triangles similar.

5. $\overline{CD} \parallel \overline{AB}$ **6.** **7.** $\overline{JK} \perp \overline{ML}$

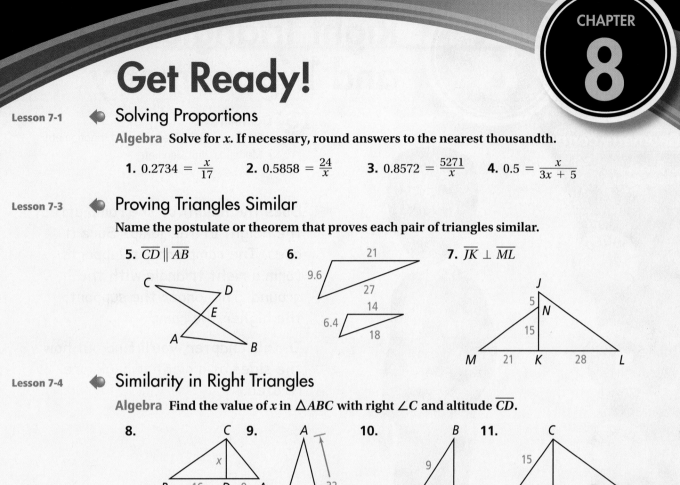

Lesson 7-4 ◀ **Similarity in Right Triangles**

Algebra Find the value of *x* in $\triangle ABC$ with right $\angle C$ and altitude $\overline{CD}$.

8. **9.** **10.** **11.**

Looking Ahead Vocabulary

12. People often describe the height of a mountain as its *elevation*. How might you describe an *angle of elevation* in geometry?

13. You see the prefix *tri-* in many words, such as *triad, triathlon, trilogy,* and *trimester*. What does the prefix indicate in these words? What geometric figure do you think is associated with the phrase *trigonometric ratio*? Explain.

14. You can use the Pythagorean Theorem to derive the *Law of Cosines*. What do you think you can find by using the *Law of Cosines*?

CHAPTER 8

Right Triangles and Trigonometry

© **DOMAINS**
- Similarity, Right Triangles, and Trigonometry
- Modeling with Geometry

Does the height of the ramp affect the height of the jump? Sure it does. The ramp and its supports form a right triangle with the ground. The longer the support, the higher the ramp.

In this chapter, you'll find out how the sides of a right triangle are related.

 Vocabulary

English/Spanish Vocabulary Audio Online:

English	Spanish
angle of depression, *p. 516*	ángulo de depresión
angle of elevation, *p. 516*	ángulo de elevación
cosine, *p. 507*	coseno
Law of Cosines, *p. 526*	Ley de cosenos
Law of Sines, *p. 522*	Ley de senos
Pythagorean triple, *p. 492*	tripleta de Pitágoras
sine, *p. 507*	seno
tangent, *p. 507*	tangente

My Math Video

00:04:04

BIG ideas

1 **Measurement**

Essential Question How do you find a side length or angle measure in a right triangle?

2 **Similarity**

Essential Question How do trigonometric ratios relate to similar right triangles?

Chapter Preview

The Pythagorean Theorem

© **Content Standard**
Prepares for G.SRT.4 Prove theorems about triangles . . . the Pythagorean Theorem . . .

You will learn the Pythagorean Theorem in Lesson 8-1. The activity below will help you understand why the theorem is true.

Activity

Step 1 Using graph paper, draw any rectangle and label the width a and the length b.

Step 2 Cut four rectangles with width a and length b from the graph paper. Then cut each rectangle on its diagonal, c, forming eight congruent triangles.

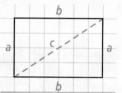

Step 3 Cut three squares from colored paper, one with sides of length a, one with sides of length b, and one with sides of length c.

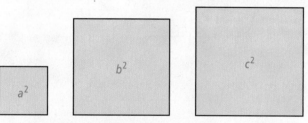

Step 4 Separate the 11 pieces into two groups.

Group 1: four triangles and the two smaller squares
Group 2: four triangles and the largest square

Step 5 Arrange the pieces of each group to form a square.

1. a. How do the areas of the two squares you formed in Step 5 compare?
  b. Write an algebraic expression for the area of each of these squares.
  c. What can you conclude about the areas of the three squares you cut from colored paper? Explain.
  d. Repeat the activity using a new rectangle and different a and b values. What do you notice?

2. a. Express your conclusion as an algebraic equation.
  b. Use a ruler with any rectangle to find actual measures for a, b, and c. Do these measures confirm your equation in part (a)?

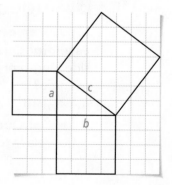

3. Explain how the diagram at the right represents your equation in Question 2.

4. Does your equation work for nonright triangles? Explore and explain.

The Pythagorean Theorem and Its Converse

© **Content Standards**
G.SRT.8 Use . . . the Pythagorean Theorem to solve right triangles in applied problems.
Also G.SRT.4

Objective To use the Pythagorean Theorem and its converse

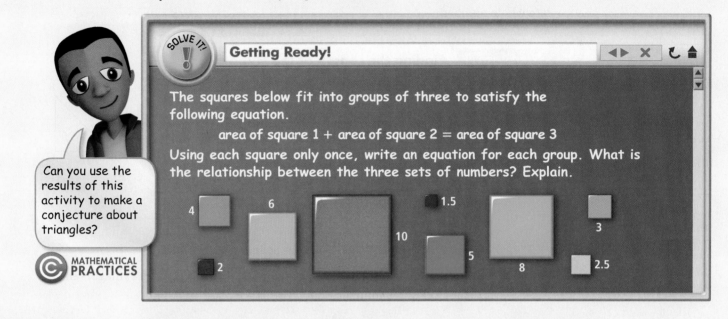

SOLVE IT!

Getting Ready!

The squares below fit into groups of three to satisfy the following equation.

area of square 1 + area of square 2 = area of square 3

Using each square only once, write an equation for each group. What is the relationship between the three sets of numbers? Explain.

4   6   1.5   3

10   2   5   8   2.5

Can you use the results of this activity to make a conjecture about triangles?

© **MATHEMATICAL PRACTICES**

Dynamic Activity
Pythagorean Theorem

The equations in the Solve It demonstrate an important relationship in right triangles called the Pythagorean Theorem. This theorem is named for Pythagoras, a Greek mathematician who lived in the 500s B.C. We now know that the Babylonians, Egyptians, and Chinese were aware of this relationship before its discovery by Pythagoras. There are many proofs of the Pythagorean Theorem. You will see one proof in this lesson and others later in the book.

Lesson Vocabulary
• Pythagorean triple

Essential Understanding If you know the lengths of any two sides of a right triangle, you can find the length of the third side by using the Pythagorean Theorem.

take note

Theorem 8-1 Pythagorean Theorem

Theorem	**If . . .**	**Then . . .**
If a triangle is a right triangle, then the sum of the squares of the lengths of the legs is equal to the square of the length of the hypotenuse.	$\triangle ABC$ is a right triangle 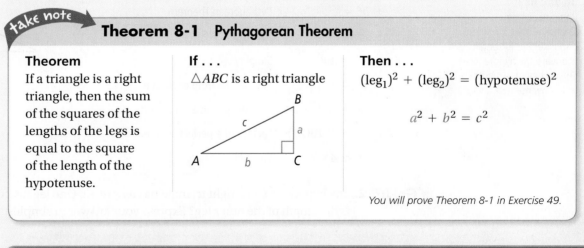	$(\text{leg}_1)^2 + (\text{leg}_2)^2 = (\text{hypotenuse})^2$ $$a^2 + b^2 = c^2$$

You will prove Theorem 8-1 in Exercise 49.

A **Pythagorean triple** is a set of nonzero whole numbers a, b, and c that satisfy the equation $a^2 + b^2 = c^2$. Below are some common Pythagorean triples.

| 3, 4, 5 | 5, 12, 13 | 8, 15, 17 | 7, 24, 25 |

If you multiply each number in a Pythagorean triple by the same whole number, the three numbers that result also form a Pythagorean triple. For example, the Pythagorean triples 6, 8, 10, and 9, 12, 15 each result from multiplying the numbers in the triple 3, 4, 5 by a whole number.

Ⓒ Problem 1   Finding the Length of the Hypotenuse

What is the length of the hypotenuse of △ABC? Do the side lengths of △ABC form a Pythagorean triple? Explain.

$(leg_1)^2 + (leg_2)^2 = (hypotenuse)^2$ Pythagorean Theorem

$a^2 + b^2 = c^2$

$21^2 + 20^2 = c^2$ Substitute 21 for a and 20 for b.

$441 + 400 = c^2$ Simplify.

$841 = c^2$

$c = 29$ Take the positive square root.

Think

Is the answer reasonable?
Yes. The hypotenuse is the longest side of a right triangle. The value for c, 29, is greater than 20 and 21.

The length of the hypotenuse is 29. The side lengths 20, 21, and 29 form a Pythagorean triple because they are whole numbers that satisfy $a^2 + b^2 = c^2$.

✓ **Got It?** **1. a.** The legs of a right triangle have lengths 10 and 24. What is the length of the hypotenuse?

 b. Do the side lengths in part (a) form a Pythagorean triple? Explain.

Ⓒ Problem 2   Finding the Length of a Leg

Algebra What is the value of x? Express your answer in simplest radical form.

$a^2 + b^2 = c^2$ Pythagorean Theorem

$8^2 + x^2 = 20^2$ Substitute.

$64 + x^2 = 400$ Simplify.

$x^2 = 336$ Subtract 64 from each side.

$x = \sqrt{336}$ Take the positive square root.

$x = \sqrt{16(21)}$ Factor out a perfect square.

$x = 4\sqrt{21}$ Simplify.

Plan

Which side lengths do you have?
Remember from Chapter 4 that the side opposite the 90° angle is always the hypotenuse. So you have the lengths of the hypotenuse and one leg.

✓ **Got It?** **2.** The hypotenuse of a right triangle has length 12. One leg has length 6. What is the length of the other leg? Express your answer in simplest radical form.

Problem 3 Finding Distance

Dog Agility **Dog agility courses often contain a seesaw obstacle, as shown below. To the nearest inch, how far above the ground are the dog's paws when the seesaw is parallel to the ground?**

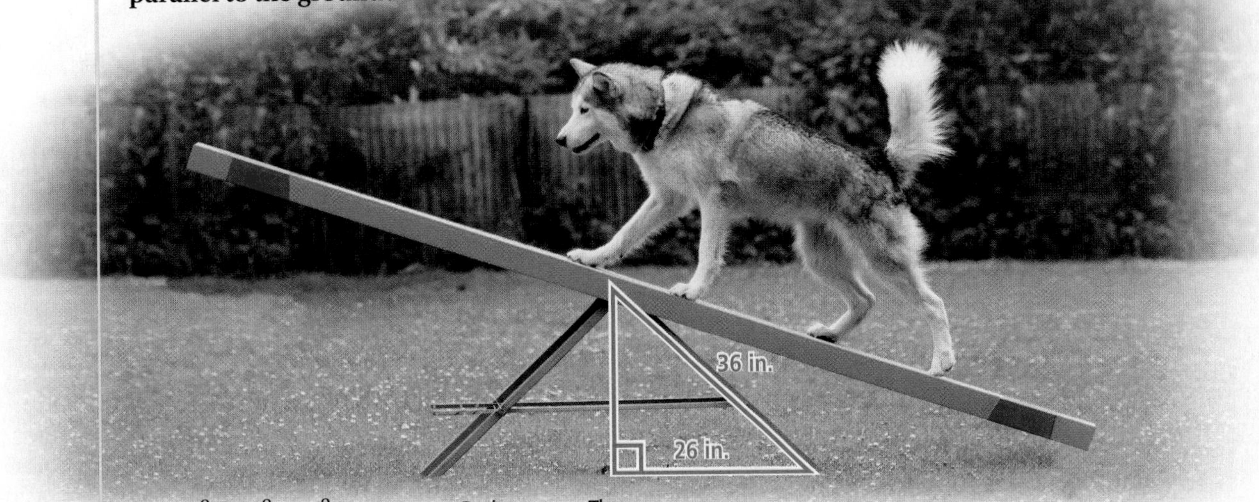

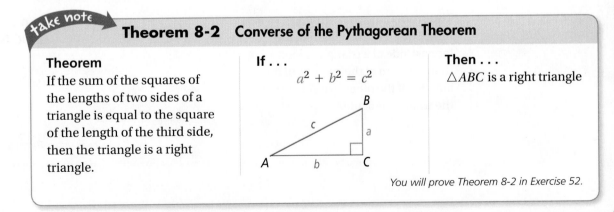

36 in.
26 in.

$a^2 + b^2 = c^2$	Pythagorean Theorem
$26^2 + b^2 = 36^2$	Substitute.
$676 + b^2 = 1296$	Simplify.
$b^2 = 620$	Subtract 676 from each side.
$b \approx 24.8997992$	Use a calculator to take the positive square root.

The dog's paws are 25 in. above the ground.

Think

How do you know when to use a calculator?
This is a real-world situation. Real-world distances are not usually expressed in radical form.

Got It? 3. The size of a computer monitor is the length of its diagonal. You want to buy a 19-in. monitor that has a height of 11 in. What is the width of the monitor? Round to the nearest tenth of an inch.

You can use the Converse of the Pythagorean Theorem to determine whether a triangle is a right triangle.

take note Theorem 8-2 Converse of the Pythagorean Theorem

Theorem
If the sum of the squares of the lengths of two sides of a triangle is equal to the square of the length of the third side, then the triangle is a right triangle.

If . . .
$a^2 + b^2 = c^2$

B
c
a
A
b
C

Then . . .
$\triangle ABC$ is a right triangle

You will prove Theorem 8-2 in Exercise 52.

Problem 4 **Identifying a Right Triangle**

Plan

How do you know where each of the side lengths goes in the equation?
Work backward. If the triangle is a right triangle, then the hypotenuse is the longest side. So use the greatest number for *c*.

A triangle has side lengths 85, 84, and 13. Is the triangle a right triangle? Explain.

$$a^2 + b^2 \stackrel{?}{=} c^2 \qquad \text{Pythagorean Theorem}$$
$$13^2 + 84^2 \stackrel{?}{=} 85^2 \qquad \text{Substitute 13 for } a, \text{ 84 for } b, \text{ and 85 for } c.$$
$$169 + 7056 \stackrel{?}{=} 7225 \qquad \text{Simplify.}$$
$$7225 = 7225 \checkmark$$

Yes, the triangle is a right triangle because $13^2 + 84^2 = 85^2$.

Got It? **4. a.** A triangle has side lengths 16, 48, and 50. Is the triangle a right triangle? Explain.

  b. Reasoning Once you know which length represents the hypotenuse, does it matter which length you substitute for *a* and which length you substitute for *b*? Explain.

The theorems below allow you to determine whether a triangle is acute or obtuse. These theorems relate to the Hinge Theorem, which states that the longer side is opposite the larger angle and the shorter side is opposite the smaller angle.

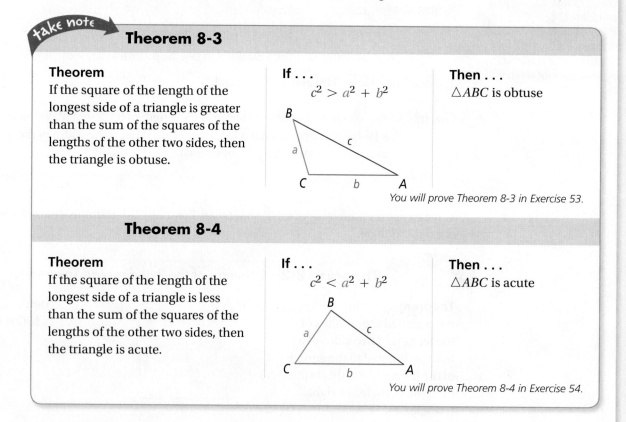

take note

Theorem 8-3

Theorem
If the square of the length of the longest side of a triangle is greater than the sum of the squares of the lengths of the other two sides, then the triangle is obtuse.

If . . .
$$c^2 > a^2 + b^2$$

Then . . .
$\triangle ABC$ is obtuse

You will prove Theorem 8-3 in Exercise 53.

Theorem 8-4

Theorem
If the square of the length of the longest side of a triangle is less than the sum of the squares of the lengths of the other two sides, then the triangle is acute.

If . . .
$$c^2 < a^2 + b^2$$

Then . . .
$\triangle ABC$ is acute

You will prove Theorem 8-4 in Exercise 54.

Problem 5 Classifying a Triangle

A triangle has side lengths 6, 11, and 14. Is it *acute*, *obtuse*, or *right*?

Plan

What information do you need?
You need to know how the square of the longest side compares to the sum of the squares of the other two sides.

$c^2 \blacksquare a^2 + b^2$ Compare c^2 to $a^2 + b^2$.

$14^2 \blacksquare 6^2 + 11^2$ Substitute the greatest value for c.

$196 \blacksquare 36 + 121$ Simplify.

$196 > 157$

Since $c^2 > a^2 + b^2$, the triangle is obtuse.

Got It? **5.** Is a triangle with side lengths 7, 8, and 9 *acute*, *obtuse*, or *right*?

Lesson Check

Do you know HOW?

What is the value of x in simplest radical form?

1.

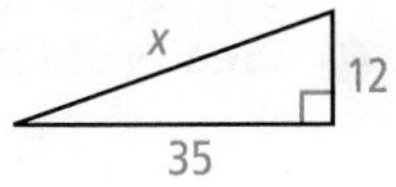

2.

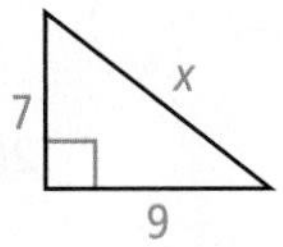

3.

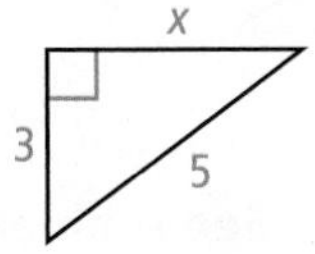

4.

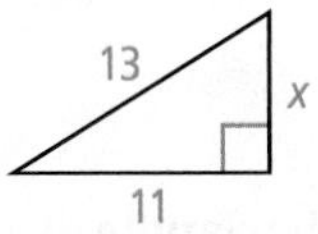

Do you UNDERSTAND? MATHEMATICAL PRACTICES

5. Vocabulary Describe the conditions that a set of three numbers must meet in order to form a Pythagorean triple.

6. Error Analysis A triangle has side lengths 16, 34, and 30. Your friend says it is not a right triangle. Look at your friend's work and describe the error.

$16^2 + 34^2 \overset{?}{=} 30^2$
$256 + 1156 \overset{?}{=} 900$
$1412 \neq 900$

Practice and Problem-Solving Exercises MATHEMATICAL PRACTICES

A Practice   **Algebra** Find the value of x.   ◀ See Problem 1.

7.

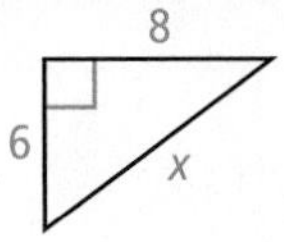

8.

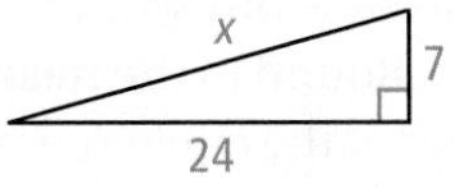

9.

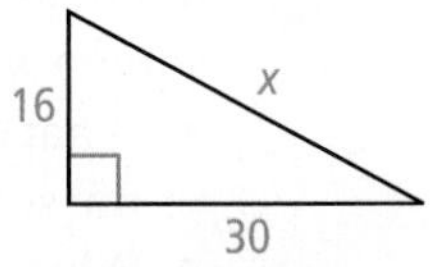

10.

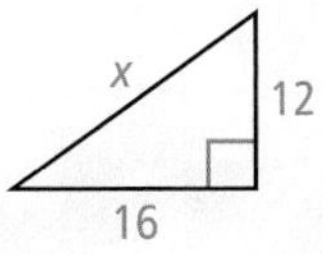

11.

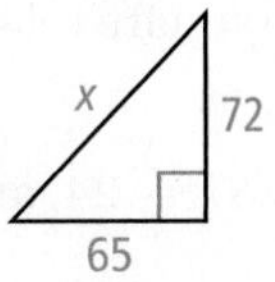

12.

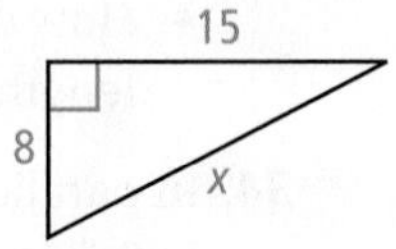

Does each set of numbers form a Pythagorean triple? Explain.

13. 4, 5, 6   **14.** 10, 24, 26   **15.** 15, 20, 25

Algebra Find the value of *x*. Express your answer in simplest radical form. ◀ **See Problem 2.**

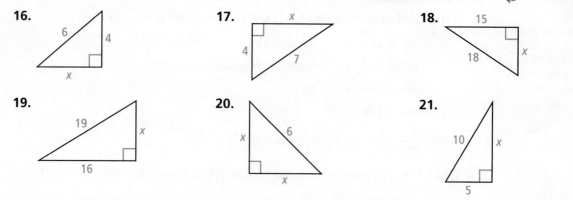

16.

17.

18.

19.

20.

21.

22. Home Maintenance A painter leans a 15-ft ladder against a house. The base of the ladder is 5 ft from the house. To the nearest tenth of a foot, how high on the house does the ladder reach? ◀ **See Problem 3.**

23. A walkway forms one diagonal of a square playground. The walkway is 24 m long. To the nearest meter, how long is a side of the playground?

Is each triangle a right triangle? Explain. ◀ **See Problem 4.**

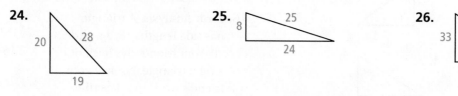

24.

25.

26.

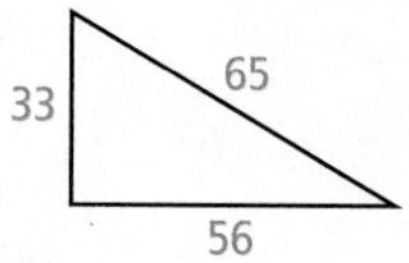

The lengths of the sides of a triangle are given. Classify each triangle as *acute*, *right*, or *obtuse*. ◀ **See Problem 5.**

27. 4, 5, 6

28. 0.3, 0.4, 0.6

29. 11, 12, 15

30. $\sqrt{3}$, 2, 3

31. 30, 40, 50

32. $\sqrt{11}$, $\sqrt{7}$, 4

Ⓑ Apply

Ⓒ 33. Think About a Plan You want to embroider a square design. You have an embroidery hoop with a 6-in. diameter. Find the largest value of *x* so that the entire square will fit in the hoop. Round to the nearest tenth.
- What does the diameter of the circle represent in the square?
- What do you know about the sides of a square?
- How do the side lengths of the square relate to the length of the diameter?

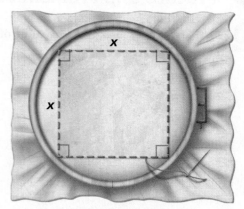

34. In parallelogram *RSTW*, *RS* = 7, *ST* = 24, and *RT* = 25. Is *RSTW* a rectangle? Explain.

35. Coordinate Geometry You can use the Pythagorean Theorem to prove the
Proof Distance Formula. Let points $P(x_1, y_1)$ and $Q(x_2, y_2)$ be the endpoints of the
hypotenuse of a right triangle.

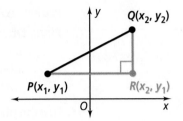

a. Write an algebraic expression to complete each of the
   following: $PR = \underline{\ ?\ }$ and $QR = \underline{\ ?\ }$.

b. By the Pythagorean Theorem, $PQ^2 = PR^2 + QR^2$. Rewrite
   this statement by substituting the algebraic expressions you
   found for PR and QR in part (a).

c. Complete the proof by taking the square root of each side of
   the equation that you wrote in part (b).

Algebra Find the value of x. If your answer is not an integer, express it in
simplest radical form.

36.
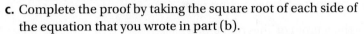

37.

38.

For each pair of numbers, find a third whole number such that
the three numbers form a Pythagorean triple.

39. 20, 21   **40.** 14, 48   **41.** 13, 85   **42.** 12, 37

© Open-Ended Find integers j and k such that (a) the two given integers and j
represent the side lengths of an acute triangle and (b) the two given integers
and k represent the side lengths of an obtuse triangle.

43. 4, 5   **44.** 2, 4   **45.** 6, 9   **46.** 5, 10   **47.** 6, 7   **48.** 9, 12

49. Prove the Pythagorean Theorem.
Proof

   Given: $\triangle ABC$ is a right triangle.

   Prove: $a^2 + b^2 = c^2$

   (*Hint:* Begin with proportions suggested by Theorem 7-3 or
   its corollaries.)

STEM **50. Astronomy** The Hubble Space Telescope orbits 600 km above Earth's
   surface. Earth's radius is about 6370 km. Use the Pythagorean Theorem
   to find the distance x from the telescope to Earth's horizon. Round your
   answer to the nearest ten kilometers. (Diagram is not to scale.)

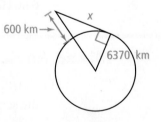

51. Prove that if the slopes of two lines have product -1, then the lines are
   perpendicular. Use parts (a)–(c) to write a coordinate proof.

   a. First, argue that neither line can be horizontal nor vertical.

   b. Then, tell why the lines must intersect. (*Hint:* Use indirect reasoning.)

   c. Place the lines in the coordinate plane. Choose a point on ℓ_1 and find a related
      point on ℓ_2. Complete the proof.

52. Use the plan and write a proof of Theorem 8-2 (Converse of the
Proof Pythagorean Theorem).

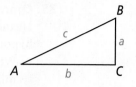

> **Given:** $\triangle ABC$ with sides of length a, b, and c, where $a^2 + b^2 = c^2$
>
> **Prove:** $\triangle ABC$ is a right triangle.
>
> **Plan:** Draw a right triangle (not $\triangle ABC$) with legs of lengths a and b. Label the hypotenuse x. By the Pythagorean Theorem, $a^2 + b^2 = x^2$. Use substitution to compare the lengths of the sides of your triangle and $\triangle ABC$. Then prove the triangles congruent.

53. Use the plan and write a proof of Theorem 8-3.
Proof

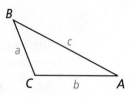

> **Given:** $\triangle ABC$ with sides of length a, b, and c, where $c^2 > a^2 + b^2$
>
> **Prove:** $\triangle ABC$ is an obtuse triangle.
>
> **Plan:** Draw a right triangle (not $\triangle ABC$) with legs of lengths a and b. Label the hypotenuse x. By the Pythagorean Theorem, $a^2 + b^2 = x^2$. Use substitution to compare lengths c and x. Then use the Converse of the Hinge Theorem to compare $\angle C$ to the right angle.

54. Prove Theorem 8-4.
Proof

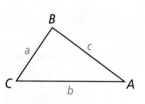

> **Given:** $\triangle ABC$ with sides of length a, b, and c, where $c^2 < a^2 + b^2$
>
> **Prove:** $\triangle ABC$ is an acute triangle.

Standardized Test Prep

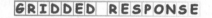

GRIDDED RESPONSE

55. A 16-ft ladder leans against a building, as shown. To the nearest foot, how far is the base of the ladder from the building?

Ladder

15.5 ft

56. What is the measure of the complement of a 67° angle?

57. The measure of the vertex angle of an isosceles triangle is 58. What is the measure of one of the base angles?

58. The length of rectangle $ABCD$ is 4 in. The length of similar rectangle $DEFG$ is 6 in. How many times greater than the area of $ABCD$ is the area of $DEFG$?

Mixed Review

59. $\triangle ABC$ has side lengths $AB = 8$, $BC = 9$, and $AC = 10$. Find the lengths of the segments formed on $\overline{BC}$ by the bisector of $\angle A$.

See Lesson 7-5.

Get Ready! **To prepare for Lesson 8-2, do Exercises 60–62.**

Simplify each expression.

See Review, p. 399.

60. $\sqrt{9} \div \sqrt{3}$ **61.** $30 \div \sqrt{2}$ **62.** $\dfrac{16}{\sqrt{3}}$

Special Right Triangles

© Content Standard
G.SRT.8 Use . . . the Pythagorean Theorem to solve right triangles in applied problems.

Objective   To use the properties of 45°-45°-90° and 30°-60°-90° triangles

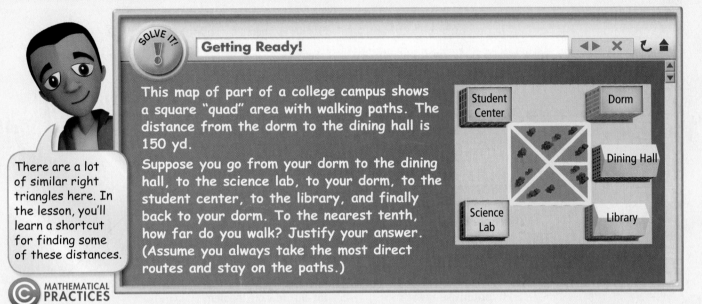

> There are a lot of similar right triangles here. In the lesson, you'll learn a shortcut for finding some of these distances.

© MATHEMATICAL PRACTICES

SOLVE IT!

Getting Ready!

This map of part of a college campus shows a square "quad" area with walking paths. The distance from the dorm to the dining hall is 150 yd.

Suppose you go from your dorm to the dining hall, to the science lab, to your dorm, to the student center, to the library, and finally back to your dorm. To the nearest tenth, how far do you walk? Justify your answer. (Assume you always take the most direct routes and stay on the paths.)

The Solve It involves triangles with angles 45°, 45°, and 90°.

Essential Understanding   Certain right triangles have properties that allow you to use shortcuts to determine side lengths without using the Pythagorean Theorem.

The acute angles of a right isosceles triangle are both 45° angles. Another name for an isosceles right triangle is a 45°-45°-90° triangle. If each leg has length x and the hypotenuse has length y, you can solve for y in terms of x.

$x^2 + x^2 = y^2$ Use the Pythagorean Theorem.

$2x^2 = y^2$ Simplify.

$x\sqrt{2} = y$ Take the positive square root of each side.

You have just proved the following theorem.

take note

Theorem 8-5   45°-45°-90° Triangle Theorem

In a 45°-45°-90° triangle, both legs are congruent and the length of the hypotenuse is $\sqrt{2}$ times the length of a leg.

$$\text{hypotenuse} = \sqrt{2} \cdot \text{leg}$$

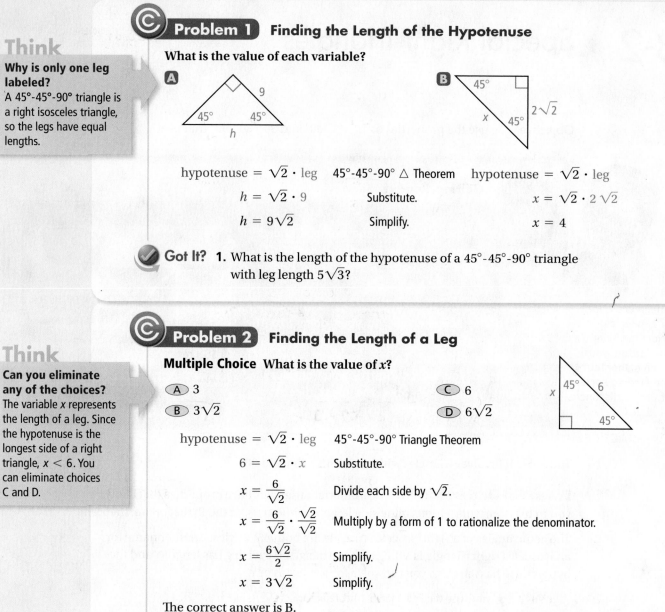

Problem 1 Finding the Length of the Hypotenuse

What is the value of each variable?

Ⓐ

Ⓑ

hypotenuse $= \sqrt{2} \cdot$ leg	45°-45°-90° △ Theorem	hypotenuse $= \sqrt{2} \cdot$ leg
$h = \sqrt{2} \cdot 9$	Substitute.	$x = \sqrt{2} \cdot 2\sqrt{2}$
$h = 9\sqrt{2}$	Simplify.	$x = 4$

Got It? 1. What is the length of the hypotenuse of a 45°-45°-90° triangle with leg length $5\sqrt{3}$?

Problem 2 Finding the Length of a Leg

Multiple Choice What is the value of x?

Ⓐ 3

Ⓑ $3\sqrt{2}$

Ⓒ 6

Ⓓ $6\sqrt{2}$

hypotenuse $= \sqrt{2} \cdot$ leg	45°-45°-90° Triangle Theorem
$6 = \sqrt{2} \cdot x$	Substitute.
$x = \dfrac{6}{\sqrt{2}}$	Divide each side by $\sqrt{2}$.
$x = \dfrac{6}{\sqrt{2}} \cdot \dfrac{\sqrt{2}}{\sqrt{2}}$	Multiply by a form of 1 to rationalize the denominator.
$x = \dfrac{6\sqrt{2}}{2}$	Simplify.
$x = 3\sqrt{2}$	Simplify.

The correct answer is B.

Got It? 2. a. The length of the hypotenuse of a 45°-45°-90° triangle is 10. What is the length of one leg?

b. **Reasoning** In Problem 2, why can you multiply $\dfrac{6}{\sqrt{2}}$ by $\dfrac{\sqrt{2}}{\sqrt{2}}$?

When you apply the 45°-45°-90° Triangle Theorem to a real-life example, you can use a calculator to evaluate square roots.

Problem 3 **Finding Distance**

Softball A high school softball diamond is a square. The distance from base to base is 60 ft. To the nearest foot, how far does a catcher throw the ball from home plate to second base?

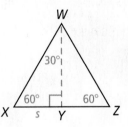

The distance d is the length of the hypotenuse of a 45°-45°-90° triangle.

$d = 60\sqrt{2}$ hypotenuse $= \sqrt{2} \cdot$ leg

$d \approx 84.85281374$ Use a calculator.

The catcher throws the ball about 85 ft from home plate to second base.

Think

How do you know that d is a hypotenuse?
The diagonal d is part of two right triangles. The hypotenuse of a right triangle is always opposite the 90° angle. So d must be a hypotenuse.

Got It? **3.** You plan to build a path along one diagonal of a 100 ft-by-100 ft square garden. To the nearest foot, how long will the path be?

Another type of special right triangle is a 30°-60°-90° triangle.

take note

Theorem 8-6 **30°-60°-90° Triangle Theorem**

In a 30°-60°-90° triangle, the length of the hypotenuse is twice the length of the shorter leg. The length of the longer leg is $\sqrt{3}$ times the length of the shorter leg.

$$\text{hypotenuse} = 2 \cdot \text{shorter leg}$$
$$\text{longer leg} = \sqrt{3} \cdot \text{shorter leg}$$

Proof **Proof of Theorem 8-6: 30°-60°-90° Triangle Theorem**

For equilateral $\triangle WXZ$, altitude $\overline{WY}$ bisects $\angle W$ and is the perpendicular bisector of $\overline{XZ}$. So, $\overline{WY}$ divides $\triangle WXZ$ into two congruent 30°-60°-90° triangles.

Thus, $XY = \frac{1}{2}XZ = \frac{1}{2}XW$, or $XW = 2XY = 2s$.

$XY^2 + YW^2 = XW^2$ Use the Pythagorean Theorem.

$s^2 + YW^2 = (2s)^2$ Substitute s for XY and $2s$ for XW.

$YW^2 = 4s^2 - s^2$ Subtract s^2 from each side.

$YW^2 = 3s^2$ Simplify.

$YW = s\sqrt{3}$ Take the positive square root of each side.

You can also use the 30°-60°-90° Triangle Theorem to find side lengths.

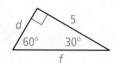

Problem 4 Using the Length of One Side

Algebra **What is the value of *d* in simplest radical form?**

Think	Write
In a 30°-60°-90° triangle, the leg opposite the 60° angle is the longer leg. So *d* represents the length of the shorter leg. Write an equation relating the legs.	longer leg = $\sqrt{3}$ · shorter leg $5 = d\sqrt{3}$
Divide each side by $\sqrt{3}$ to solve for *d*.	$d = \dfrac{5}{\sqrt{3}}$
The value of *d* is not in simplest radical form because there is a radical in the denominator. Multiply *d* by a form of 1.	$\dfrac{5}{\sqrt{3}} \cdot \dfrac{\sqrt{3}}{\sqrt{3}} = \dfrac{5\sqrt{3}}{3}$ So $d = \dfrac{5\sqrt{3}}{3}$.

✓ **Got It?** **4.** In Problem 4, what is the value of *f* in simplest radical form?

Problem 5 Applying the 30°-60°-90° Triangle Theorem

Plan

How does knowing the shape of the pendants help?
Since the triangle is equilateral, you know that an altitude divides the triangle into two congruent 30°-60°-90° triangles.

Jewelry Making **An artisan makes pendants in the shape of equilateral triangles. The height of each pendant is 18 mm. What is the length *s* of each side of a pendant to the nearest tenth of a millimeter?**

The hypotenuse of each 30°-60°-90° triangle is *s*. The shorter leg is $\frac{1}{2}s$.

$$18 = \sqrt{3}\left(\tfrac{1}{2}s\right) \qquad \text{longer leg} = \sqrt{3} \cdot \text{shorter leg}$$

$$18 = \frac{\sqrt{3}}{2}s \qquad \text{Simplify.}$$

$$\frac{2}{\sqrt{3}} \cdot 18 = s \qquad \text{Multiply each side by } \frac{2}{\sqrt{3}}.$$

$$s \approx 20.78460969 \qquad \text{Use a calculator.}$$

Each side of a pendant is about 20.8 mm long.

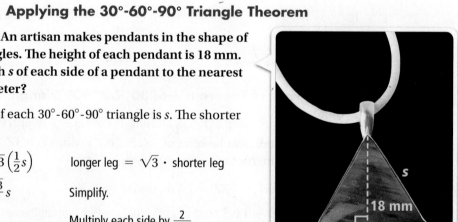

✓ **Got It?** **5.** Suppose the sides of a pendant are 18 mm long. What is the height of the pendant to the nearest tenth of a millimeter?

Lesson Check

Do you know HOW?

What is the value of x? If your answer is not an integer, express it in simplest radical form.

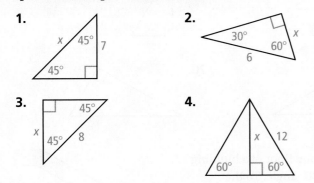

1.

2.

3.

4.

Do you UNDERSTAND?

5. Error Analysis Sandra drew the triangle below. Rika said that the labeled lengths are not possible. With which student do you agree? Explain.

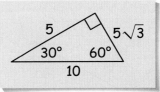

6. Reasoning A test question asks you to find two side lengths of a 45°-45°-90° triangle. You know that the length of one leg is 6, but you forgot the special formula for 45°-45°-90° triangles. Explain how you can still determine the other side lengths. What are the other side lengths?

Practice and Problem-Solving Exercises

MATHEMATICAL PRACTICES

A Practice Find the value of each variable. If your answer is not an integer, express it in simplest radical form.

See Problems 1 and 2.

7.

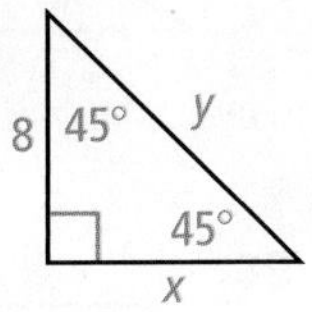

8.

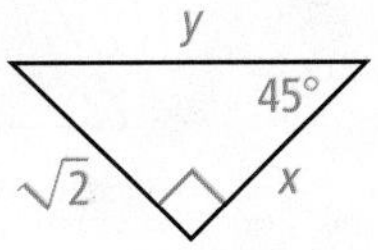

9.

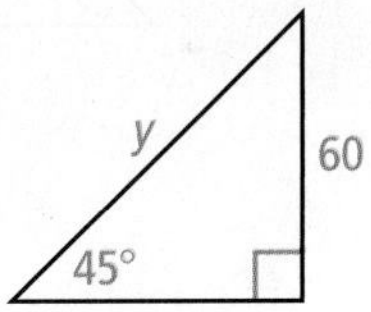

10.

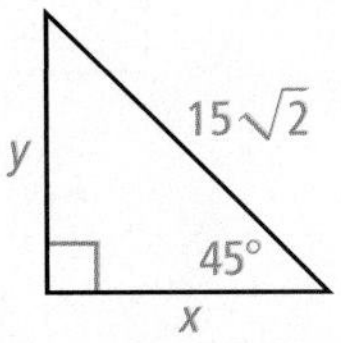

11.

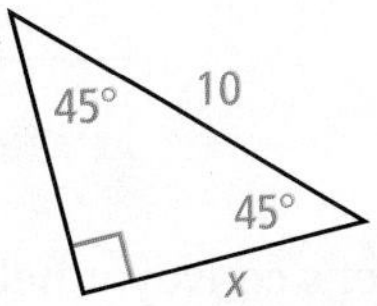

12.

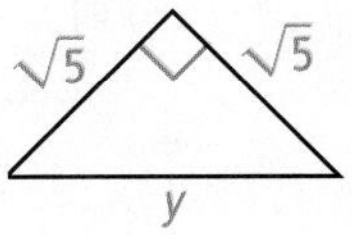

13. Dinnerware Design What is the side length of the smallest square plate on which a 20-cm chopstick can fit along a diagonal without any overhang? Round your answer to the nearest tenth of a centimeter.

See Problem 3.

14. Aviation The four blades of a helicopter meet at right angles and are all the same length. The distance between the tips of two adjacent blades is 36 ft. How long is each blade? Round your answer to the nearest tenth of a foot.

Algebra Find the value of each variable. If your answer is not an integer, express it in simplest radical form.

See Problems 4 and 5.

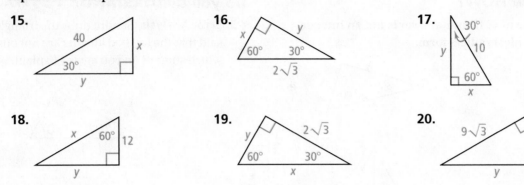

15.

16.

17.

18.

19.

20.

STEM **21. Architecture** An escalator lifts people to the second floor of a building, 25 ft above the first floor. The escalator rises at a 30° angle. To the nearest foot, how far does a person travel from the bottom to the top of the escalator?

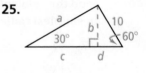

STEM **22. City Planning** Jefferson Park sits on one square city block 300 ft on each side. Sidewalks across the park join opposite corners. To the nearest foot, how long is each diagonal sidewalk?

B Apply

Algebra Find the value of each variable. If your answer is not an integer, express it in simplest radical form.

23.

24.

25.

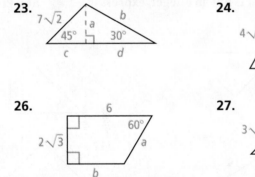

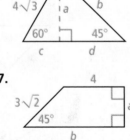

26.

27.

28.

© 29. Think About a Plan A farmer's conveyor belt carries bales of hay from the ground to the barn loft. The conveyor belt moves at 100 ft/min. How many seconds does it take for a bale of hay to go from the ground to the barn loft?

• Which part of a right triangle does the conveyor belt represent?
• You know the speed. What other information do you need to find time?
• How are minutes and seconds related?

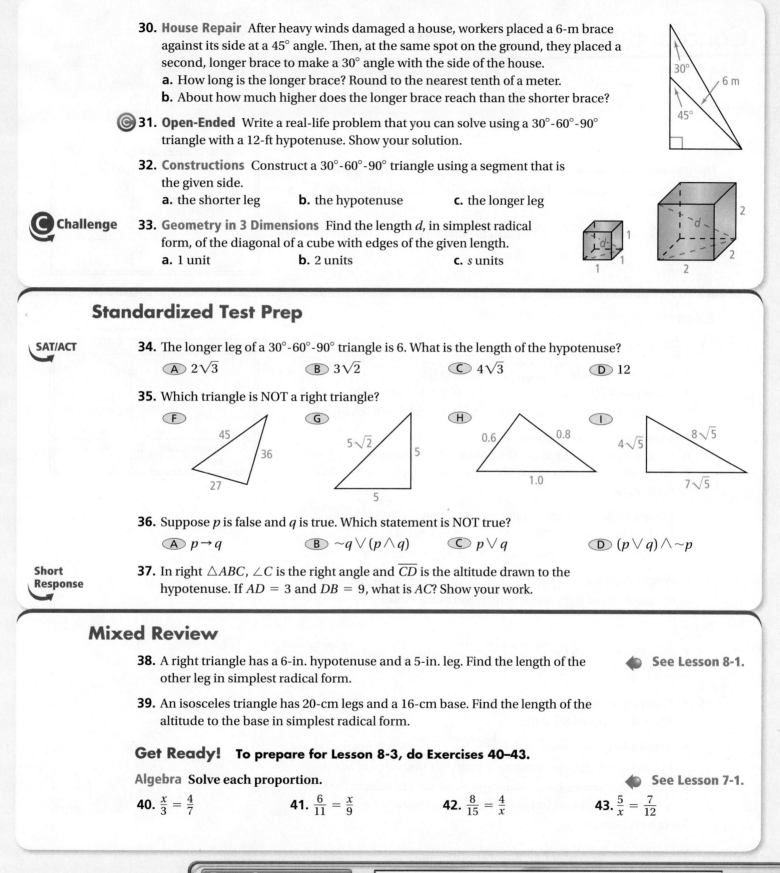

30. House Repair After heavy winds damaged a house, workers placed a 6-m brace against its side at a 45° angle. Then, at the same spot on the ground, they placed a second, longer brace to make a 30° angle with the side of the house.
 a. How long is the longer brace? Round to the nearest tenth of a meter.
 b. About how much higher does the longer brace reach than the shorter brace?

Ⓒ **31. Open-Ended** Write a real-life problem that you can solve using a 30°-60°-90° triangle with a 12-ft hypotenuse. Show your solution.

32. Constructions Construct a 30°-60°-90° triangle using a segment that is the given side.
 a. the shorter leg   **b.** the hypotenuse   **c.** the longer leg

Ⓒ **Challenge** **33. Geometry in 3 Dimensions** Find the length d, in simplest radical form, of the diagonal of a cube with edges of the given length.
 a. 1 unit   **b.** 2 units   **c.** s units

Standardized Test Prep

SAT/ACT

34. The longer leg of a 30°-60°-90° triangle is 6. What is the length of the hypotenuse?
 Ⓐ $2\sqrt{3}$ Ⓑ $3\sqrt{2}$ Ⓒ $4\sqrt{3}$ Ⓓ 12

35. Which triangle is NOT a right triangle?
 Ⓕ 45 36 27
 Ⓖ $5\sqrt{2}$ 5 5
 Ⓗ 0.6 0.8 1.0
 Ⓘ $4\sqrt{5}$ $8\sqrt{5}$ $7\sqrt{5}$

36. Suppose p is false and q is true. Which statement is NOT true?
 Ⓐ $p \rightarrow q$ Ⓑ $\sim q \vee (p \wedge q)$ Ⓒ $p \vee q$ Ⓓ $(p \vee q) \wedge \sim p$

Short Response

37. In right $\triangle ABC$, $\angle C$ is the right angle and $\overline{CD}$ is the altitude drawn to the hypotenuse. If $AD = 3$ and $DB = 9$, what is AC? Show your work.

Mixed Review

38. A right triangle has a 6-in. hypotenuse and a 5-in. leg. Find the length of the other leg in simplest radical form.

◆ See Lesson 8-1.

39. An isosceles triangle has 20-cm legs and a 16-cm base. Find the length of the altitude to the base in simplest radical form.

Get Ready! **To prepare for Lesson 8-3, do Exercises 40–43.**

Algebra Solve each proportion.

◆ See Lesson 7-1.

40. $\frac{x}{3} = \frac{4}{7}$ **41.** $\frac{6}{11} = \frac{x}{9}$ **42.** $\frac{8}{15} = \frac{4}{x}$ **43.** $\frac{5}{x} = \frac{7}{12}$

Exploring Trigonometric Ratios

© **Content Standard**
G.SRT.6 Understand that by similarity, side ratios in right triangles are properties of the angles in the triangle, leading to definitions of trigonometric ratios for acute angles.

Construct

Use geometry software to construct $\overrightarrow{AB}$ and $\overrightarrow{AC}$ so that $\angle A$ is acute. Through a point D on $\overrightarrow{AB}$, construct a line perpendicular to $\overrightarrow{AB}$ that intersects $\overrightarrow{AC}$ in point E.

Moving point D changes the size of $\triangle ADE$. Moving point C changes the size of $\angle A$.

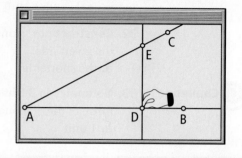

Exercises

1. • Measure $\angle A$ and find the lengths of the sides of $\triangle ADE$.
   • Calculate the ratio $\frac{\text{leg opposite } \angle A}{\text{hypotenuse}}$, which is $\frac{ED}{AE}$.
   • Move point D to change the size of $\triangle ADE$ without changing $m\angle A$. What do you observe about the ratio as the size of $\triangle ADE$ changes?

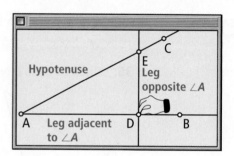

2. • Move point C to change $m\angle A$.
   a. What do you observe about the ratio as $m\angle A$ changes?
   b. What value does the ratio approach as $m\angle A$ approaches 0? As $m\angle A$ approaches 90?

3. • Make a table that shows values for $m\angle A$ and the ratio $\frac{\text{leg opposite } \angle A}{\text{hypotenuse}}$. In your table, include 10, 20, 30, . . . , 80 for $m\angle A$.
   • Compare your table with a table of trigonometric ratios.

   Do your values for $\frac{\text{leg opposite } \angle A}{\text{hypotenuse}}$ match the values in one of the columns of the table? What is the name of this ratio in the table?

Extend

4. Repeat Exercises 1–3 for $\frac{\text{leg adjacent to } \angle A}{\text{hypotenuse}}$, which is $\frac{AD}{AE}$, and $\frac{\text{leg opposite} \angle A}{\text{leg adjacent to} \angle A}$, which is $\frac{ED}{AD}$.

5. • Choose a measure for $\angle A$ and determine the ratio $r = \frac{\text{leg opposite } \angle A}{\text{hypotenuse}}$. Record $m\angle A$ and this ratio.
   • Manipulate the triangle so that $\frac{\text{leg adjacent to } \angle A}{\text{hypotenuse}}$ has the same value r. Record this $m\angle A$ and compare it with your first value of $m\angle A$.
   • Repeat this procedure several times. Look for a pattern in the two measures of $\angle A$ that you found for different values of r.
   Make a conjecture.

Trigonometry

© **Content Standards**

G.SRT.8 Use trigonometric ratios and the Pythagorean Theorem to solve right triangles in applied problems.

G.MG.1 Use geometric shape, their measures, and their properties to describe objects.

Also, G.SRT.7

Objective   To use the sine, cosine, and tangent ratios to determine side lengths and angle measures in right triangles

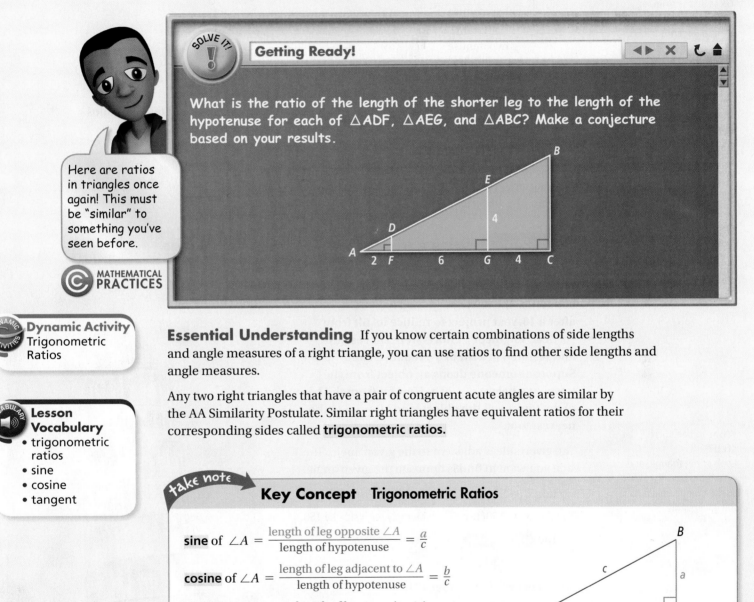

SOLVE IT!

Getting Ready!

What is the ratio of the length of the shorter leg to the length of the hypotenuse for each of △ADF, △AEG, and △ABC? Make a conjecture based on your results.

Here are ratios in triangles once again! This must be "similar" to something you've seen before.

© **MATHEMATICAL PRACTICES**

Dynamic Activity
Trigonometric Ratios

Lesson Vocabulary
• trigonometric ratios
• sine
• cosine
• tangent

Essential Understanding   If you know certain combinations of side lengths and angle measures of a right triangle, you can use ratios to find other side lengths and angle measures.

Any two right triangles that have a pair of congruent acute angles are similar by the AA Similarity Postulate. Similar right triangles have equivalent ratios for their corresponding sides called **trigonometric ratios.**

take note

Key Concept   Trigonometric Ratios

sine of $\angle A = \dfrac{\text{length of leg opposite } \angle A}{\text{length of hypotenuse}} = \dfrac{a}{c}$

cosine of $\angle A = \dfrac{\text{length of leg adjacent to } \angle A}{\text{length of hypotenuse}} = \dfrac{b}{c}$

tangent of $\angle A = \dfrac{\text{length of leg opposite } \angle A}{\text{length of leg adjacent to } \angle A} = \dfrac{a}{b}$

You can abbreviate the ratios as

$$\sin A = \frac{\text{opposite}}{\text{hypotenuse}}, \cos A = \frac{\text{adjacent}}{\text{hypotenuse}}, \text{ and } \tan A = \frac{\text{opposite}}{\text{adjacent}}.$$

Problem 1 Writing Trigonometric Ratios

Think

How do the sides relate to ∠T?
$\overline{GR}$ is across from, or *opposite*, ∠T. $\overline{TR}$ is next to, or *adjacent* to, ∠T. $\overline{TG}$ is the *hypotenuse* because it is opposite the 90° angle.

What are the sine, cosine, and tangent ratios for ∠T?

$$\sin T = \frac{\text{opposite}}{\text{hypotenuse}} = \frac{8}{17}$$

$$\cos T = \frac{\text{adjacent}}{\text{hypotenuse}} = \frac{15}{17}$$

$$\tan T = \frac{\text{opposite}}{\text{adjacent}} = \frac{8}{15}$$

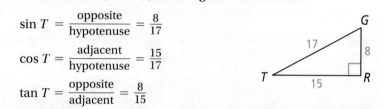

Got It? **1.** Use the triangle in Problem 1. What are the sine, cosine, and tangent ratios for ∠G?

In Chapter 7, you used similar triangles to measure distances indirectly. You can also use trigonometry for indirect measurement.

Problem 2 Using a Trigonometric Ratio to Find Distance

Plan

What is the first step?
Look at the triangle and determine how the sides of the triangle relate to the given angle.

Landmarks In 1990, the Leaning Tower of Pisa was closed to the public due to safety concerns. The tower reopened in 2001 after a 10-year project to reduce its tilt from vertical. Engineers' efforts were successful and resulted in a tilt of 5°, reduced from 5.5°. Suppose someone drops an object from the tower at a height of 150 ft. How far from the base of the tower will the object land? Round to the nearest foot.

The given side is adjacent to the given angle. The side you want to find is opposite the given angle.

$$\tan 5° = \frac{x}{150} \qquad \text{Use the tangent ratio.}$$

$$x = 150(\tan 5°) \qquad \text{Multiply each side by 150.}$$

$$150 \boxed{\text{tan}} \ 5 \boxed{\text{enter}} \qquad \text{Use a calculator.}$$

$$x \approx 13.12329953$$

The object will land about 13 ft from the base of the tower.

5°

150 ft

Got It? **2.** For parts (a)–(c), find the value of w to the nearest tenth.

a.

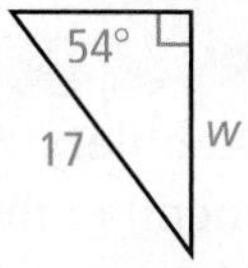

54°

17   w

b.

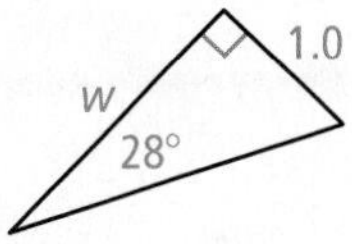

w

1.0

28°

c.

w

33°

4.5

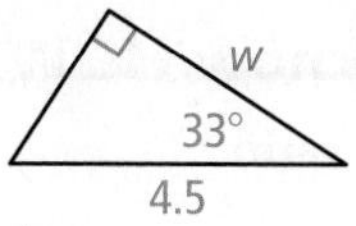

d. A section of Filbert Street in San Francisco rises at an angle of about 17°. If you walk 150 ft up this section, what is your vertical rise? Round to the nearest foot.

If you know the sine, cosine, or tangent ratio for an angle, you can use an inverse $(\sin^{-1}, \cos^{-1}, \text{or } \tan^{-1})$ to find the measure of the angle.

Problem 3 **Using Inverses**

What is $m\angle X$ to the nearest degree?

A

H

6   10

B   X

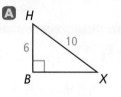

B

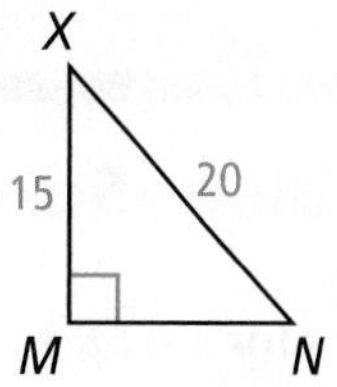

X

15   20

M   N

You know the lengths of the hypotenuse and the side opposite $\angle X$.

Use the sine ratio.

$\sin X = \frac{6}{10}$ Write the ratio.

$m\angle X = \sin^{-1}\left(\frac{6}{10}\right)$ Use the inverse.

[sin⁻¹] 6 [÷] 10 [enter]   Use a calculator.

$m\angle X \approx 36.86989765$

≈ 37

You know the lengths of the hypotenuse and the side adjacent to $\angle X$.

Use the cosine ratio.

$\cos X = \frac{15}{20}$

$m\angle X = \cos^{-1}\left(\frac{15}{20}\right)$

[cos⁻¹] 15 [÷] 20 [enter]

$m\angle X \approx 41.40962211$

≈ 41

Think

When should you use an inverse?
Use an inverse when you know two side lengths of a right triangle and you want to find the measure of one of the acute angles.

Got It? **3. a.** Use the figure at the right. What is $m\angle Y$ to the nearest degree?

b. **Reasoning** Suppose you know the lengths of all three sides of a right triangle. Does it matter which trigonometric ratio you use to find the measure of any of the three angles? Explain.

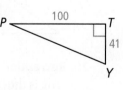

P   100   T

41

Y

Lesson Check

Do you know HOW?

Write each ratio.

1. sin A
2. cos A
3. tan A
4. sin B
5. cos B
6. tan B

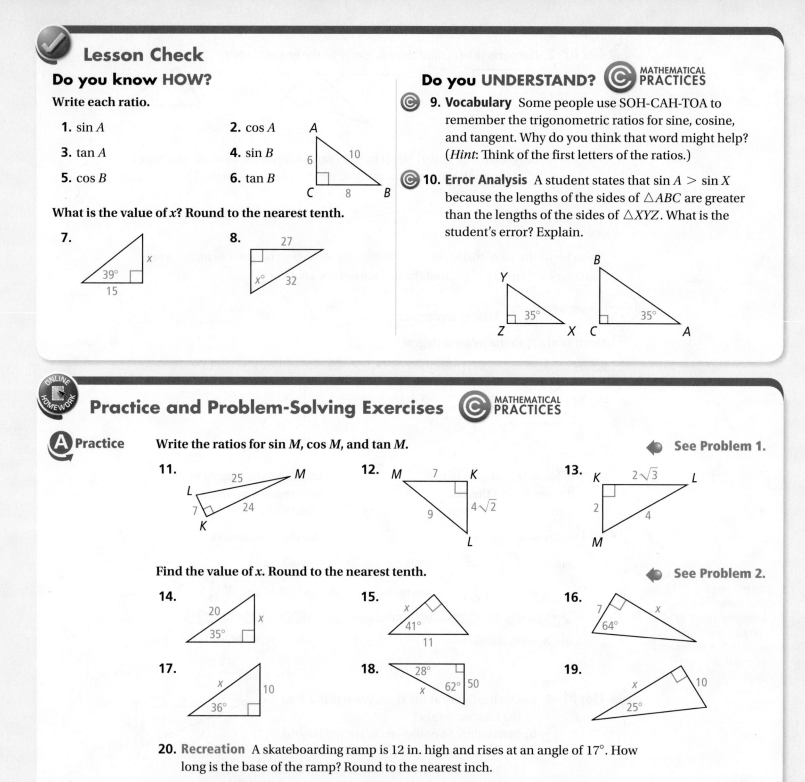

What is the value of x? Round to the nearest tenth.

7.

8.

Do you UNDERSTAND? ⓒ MATHEMATICAL PRACTICES

ⓒ **9. Vocabulary** Some people use SOH-CAH-TOA to remember the trigonometric ratios for sine, cosine, and tangent. Why do you think that word might help? (*Hint*: Think of the first letters of the ratios.)

ⓒ **10. Error Analysis** A student states that sin A > sin X because the lengths of the sides of △ABC are greater than the lengths of the sides of △XYZ. What is the student's error? Explain.

Practice and Problem-Solving Exercises ⓒ MATHEMATICAL PRACTICES

Ⓐ **Practice**   Write the ratios for sin M, cos M, and tan M.   ◀ See Problem 1.

11.

12.

13.

Find the value of x. Round to the nearest tenth.   ◀ See Problem 2.

14.

15.

16.

17.

18.

19.

20. Recreation A skateboarding ramp is 12 in. high and rises at an angle of 17°. How long is the base of the ramp? Round to the nearest inch.

21. Public Transportation An escalator in the subway station has a vertical rise of 195 ft 9.5 in., and rises at an angle of 10.4°. How long is the escalator? Round to the nearest foot.

Find the value of *x*. Round to the nearest degree.

See Problem 3.

22.
14
5
x°

23.
x°
5 8

24.
13
x°
9

25.
3.0 *x*° 5.8

26.
x°
17
41

27.
0.34
x° 0.15

B Apply

28. The lengths of the diagonals of a rhombus are 2 in. and 5 in. Find the measures of the angles of the rhombus to the nearest degree.

© 29. Think About a Plan Carlos plans to build a grain bin with a radius of 15 ft. The recommended slant of the roof is 25°. He wants the roof to overhang the edge of the bin by 1 ft. What should the length *x* be? Give your answer in feet and inches.

- What is the position of the side of length *x* in relation to the given angle?
- What information do you need to find a side length of a right triangle?
- Which trigonometric ratio could you use?

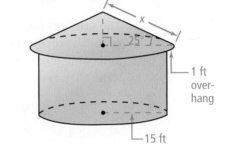

x
25°
1 ft over-hang
15 ft

An *identity* is an equation that is true for all the allowed values of the variable. Use what you know about trigonometric ratios to show that each equation is an identity.

30. $\tan X = \dfrac{\sin X}{\cos X}$

31. $\sin X = \cos X \cdot \tan X$

32. $\cos X = \dfrac{\sin X}{\tan X}$

Find the values of *w* and then *x*. Round lengths to the nearest tenth and angle measures to the nearest degree.

33.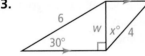
6
w *x*° 4
30°

34.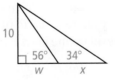
10
56° 34°
w *x*

35.
102 *w* 102
42°
x

STEM 36. Pyramids All but two of the pyramids built by the ancient Egyptians have faces inclined at 52° angles. Suppose an archaeologist discovers the ruins of a pyramid. Most of the pyramid has eroded, but the archaeologist is able to determine that the length of a side of the square base is 82 m. How tall was the pyramid, assuming its faces were inclined at 52°? Round your answer to the nearest meter.

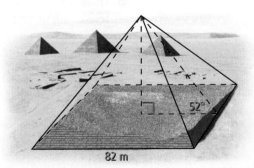

x
52°
82 m

37. a. In △ABC at the right, how does sin A compare to cos B? Is this true for the acute angles of other right triangles?

b. Reading Math The word cosine is derived from the words *complement's sine*. Which angle in △ABC is the complement of ∠A? Of ∠B?

c. Explain why the derivation of the word cosine makes sense.

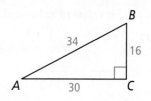

38. For right △ABC with right ∠C, prove each of the following.

Proof **a.** sin A < 1

b. cos A < 1

39. a. Writing Explain why tan 60° = √3. Include a diagram with your explanation.

b. Make a Conjecture How are the sine and cosine of a 60° angle related? Explain.

The sine, cosine, and tangent ratios each have a reciprocal ratio. The reciprocal ratios are cosecant (csc), secant (sec), and cotangent (cot). Use △ABC and the definitions below to write each ratio.

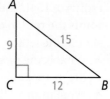

$$\csc X = \frac{1}{\sin X} \qquad \sec X = \frac{1}{\cos X} \qquad \cot X = \frac{1}{\tan X}$$

40. csc A

41. sec A

42. cot A

43. csc B

44. sec B

45. cot B

46. Graphing Calculator Use the `table` feature of your graphing calculator to study sin X as X gets close to (but not equal to) 90. In the `y=` screen, enter Y1 = sin X.

a. Use the `tblset` feature so that X starts at 80 and changes by 1. Access the `table`. From the table, what is sin X for X = 89?

b. Perform a "numerical zoom-in." Use the `tblset` feature, so that X starts with 89 and changes by 0.1. What is sin X for X = 89.9?

c. Continue to zoom-in numerically on values close to 90. What is the greatest value you can get for sin X on your calculator? How close is X to 90? Does your result contradict what you are asked to prove in Exercise 38a?

d. Use right triangles to explain the behavior of sin X found above.

47. a. Reasoning Does tan A + tan B = tan (A + B) when A + B < 90? Explain.

b. Does tan A − tan B = tan (A − B) when A − B > 0? Use part (a) and indirect reasoning to explain.

Challenge Verify that each equation is an identity by showing that each expression on the left simplifies to 1.

48. $(\sin A)^2 + (\cos A)^2 = 1$

49. $(\sin B)^2 + (\cos B)^2 = 1$

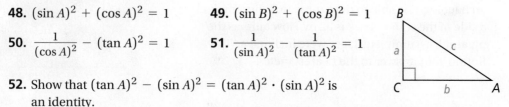

50. $\dfrac{1}{(\cos A)^2} - (\tan A)^2 = 1$

51. $\dfrac{1}{(\sin A)^2} - \dfrac{1}{(\tan A)^2} = 1$

52. Show that $(\tan A)^2 - (\sin A)^2 = (\tan A)^2 \cdot (\sin A)^2$ is an identity.

STEM **53. Astronomy** The Polish astronomer Nicolaus Copernicus devised a method for determining the sizes of the orbits of planets farther from the sun than Earth. His method involved noting the number of days between the times that a planet was in the positions labeled A and B in the diagram. Using this time and the number of days in each planet's year, he calculated c and d.

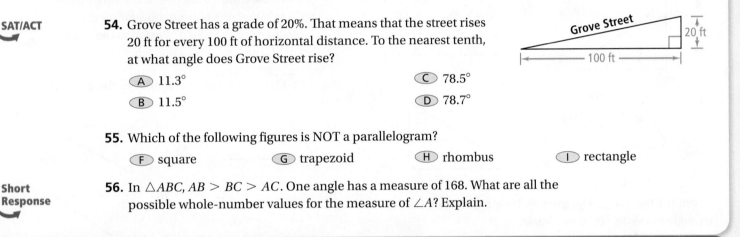

Outer planet's orbit
Earth's orbit
Sun
A A
1 AU
$c°$ $d°$
B
B
Not to scale

    a. For Mars, $c = 55.2$ and $d = 103.8$. How far is Mars from the sun in astronomical units (AU)? One astronomical unit is defined as the average distance from Earth to the center of the sun, about 93 million miles.

    b. For Jupiter, $c = 21.9$ and $d = 100.8$. How far is Jupiter from the sun in astronomical units?

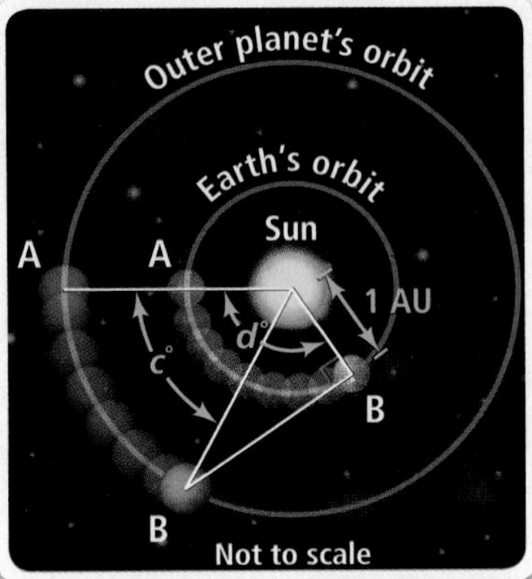

Standardized Test Prep

SAT/ACT

54. Grove Street has a grade of 20%. That means that the street rises 20 ft for every 100 ft of horizontal distance. To the nearest tenth, at what angle does Grove Street rise?

Grove Street
20 ft
100 ft

    Ⓐ 11.3°   Ⓒ 78.5°

    Ⓑ 11.5°   Ⓓ 78.7°

55. Which of the following figures is NOT a parallelogram?

    Ⓕ square   Ⓖ trapezoid   Ⓗ rhombus   Ⓘ rectangle

Short Response

56. In $\triangle ABC$, $AB > BC > AC$. One angle has a measure of 168. What are all the possible whole-number values for the measure of $\angle A$? Explain.

Mixed Review

57. The length of the hypotenuse of a 30°-60°-90° triangle is 8. What are the lengths of the legs?

◀ See Lesson 8-2.

58. A diagonal of a square is 10 units. Find the length of a side of the square. Express your answer in simplest radical form.

Get Ready! **To prepare for Lesson 8-4, do Exercises 59–62.**

Use rectangle *ABCD* to complete each statement.

◀ See Lessons 3-2 and 6-4.

59. $\angle 1 \cong$ _?_

60. $\angle 5 \cong$ _?_

61. $\angle 3 \cong$ _?_

62. $m\angle 1 + m\angle 5 =$ _?_

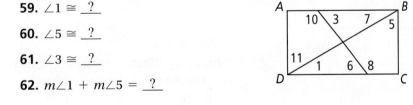

A   B
10   3   7   5
11
1   6   8
D   C

8 Mid-Chapter Quiz

MathXL® for School
Go to PowerGeometry.com

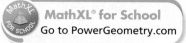

Do you know HOW?

Algebra **Find the value of each variable. Express your answer in simplest radical form.**

1.

2.

3.

4.

5.

6.

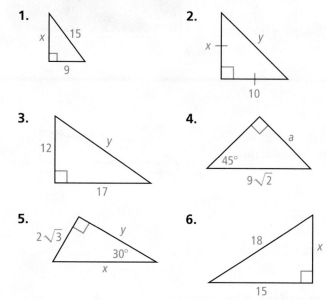

Given the following triangle side lengths, identify the triangle as *acute*, *right*, or *obtuse*.

7. 7, 8, 9

8. 15, 36, 39

9. 10, 12, 16

10. A square has a 40-cm diagonal. How long is each side of the square? Round to the nearest tenth of a centimeter.

Write the sine, cosine, and tangent ratios for $\angle A$ and $\angle B$.

11.

12.

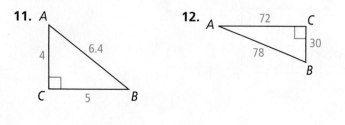

Does each set of numbers form a Pythagorean triple? Explain.

13. 32, 60, 68

14. 1, 2, 3

15. 2.5, 6, 6.5

16. Landscaping A landscaper uses a 13-ft wire to brace a tree. The wire is attached to a protective collar around the trunk of the tree. If the wire makes a 60° angle with the ground, how far up the tree is the protective collar located? Round to the nearest tenth of a foot.

Algebra **Find the value of x. Round to the nearest tenth.**

17.

18.

19.

20.

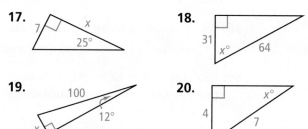

Do you UNDERSTAND?

21. Compare and Contrast What are the similarities between the methods you use to determine whether a triangle is acute, obtuse, or right? What are the differences?

22. In the figure below, which angle has the greater sine value? The greater cosine value? Explain.

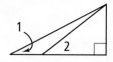

23. Reasoning What angle has a tangent of 1? Explain. (Do not use a calculator or a table.)

Concept Byte

Use With Lesson 8-4

ACTIVITY

Measuring From Afar

Content Standard
G.SRT.8 Use trigonometric ratios . . . to solve right triangles in applied problems.

In this activity, you will make a tool to find heights. The tool requires you to use right triangles and trigonometric ratios.

Activity

The device shown below is an inclinometer. Make your own inclinometer using a protractor, a piece of string, and a washer.

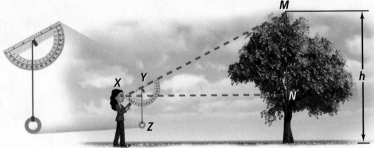

Not to scale

1. The string on the inclinometer shows $m\angle XYZ$. Explain how you can find $m\angle X$ if you know $m\angle XYZ$.

2. You can calculate an approximate height of the tree by using a trigonometric ratio involving $\angle X$.

a. Which side length of $\triangle XMN$ could you most easily measure? Explain.

b. What trigonometric ratio would you use? Explain.

c. Show how to find the height of the tree.

Try using your inclinometer to find the heights of tall objects at your school.

Exercises

3. You are on a steep hillside directly across from the top of a tree. Explain how you could use an inclinometer, a trigonometric ratio, and the distance from the hilltop to the base of the tree to find the approximate height of the tree.

4. Suppose you could climb up only to point P. You know the distance, d, from P to the base of the tree. Explain how you could use an inclinometer and trigonometric ratios to find the height of the tree from P.

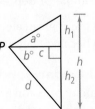

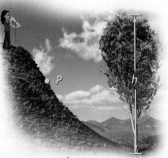

Not to scale

5. Use the diagram from Exercise 4. Show that you can find the height of the tree using the following formula.

$$h = d(\cos b^\circ)(\tan a^\circ + \tan b^\circ)$$

Angles of Elevation and Depression

© **Content Standard**
G.SRT.8 Use trigonometric ratios . . . to solve right triangles in applied problems.

Objective   To use angles of elevation and depression to solve problems

Did you know you could use geometry in theater? You can find math anywhere you look . . . up or down.

© MATHEMATICAL PRACTICES

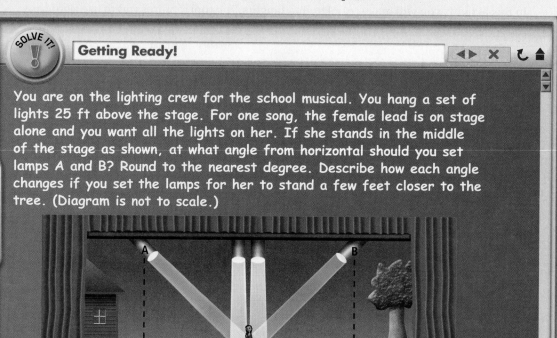

SOLVE IT!

Getting Ready!

You are on the lighting crew for the school musical. You hang a set of lights 25 ft above the stage. For one song, the female lead is on stage alone and you want all the lights on her. If she stands in the middle of the stage as shown, at what angle from horizontal should you set lamps A and B? Round to the nearest degree. Describe how each angle changes if you set the lamps for her to stand a few feet closer to the tree. (Diagram is not to scale.)

A   B

|← 10 ft →|← 10 ft →|

Lesson Vocabulary
• angle of elevation
• angle of depression

The angles in the Solve It are formed below the horizontal black pipe. Angles formed above and below a horizontal line have specific names.

Suppose a person on the ground sees a hang glider at a 38° angle above a horizontal line. This angle is the **angle of elevation.**

At the same time, a person in the hang glider sees the person on the ground at a 38° angle below a horizontal line. This angle is the **angle of depression.**

Notice that the angle of elevation is congruent to the angle of depression because they are alternate interior angles.

Essential Understanding   You can use the angles of elevation and depression as the acute angles of right triangles formed by a horizontal distance and a vertical height.

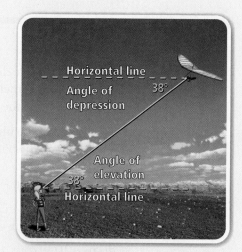

Horizontal line
Angle of depression   38°
Angle of elevation
38°
Horizontal line

How can you tell
if it is an angle
of elevation or
depression?
Place your finger on the
vertex of the angle. Trace
along the nonhorizontal
side of the angle. See
if your finger is above
(elevation) or below
(depression) the vertex.

© **Problem 1** **Identifying Angles of Elevation and Depression**

What is a description of the angle as it
relates to the situation shown?

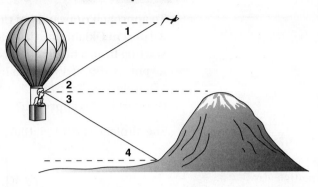

A ∠1

∠1 is the angle of depression from
the bird to the person in the hot-air
balloon.

B ∠4

∠4 is the angle of elevation from
the base of the mountain to the
person in the hot-air balloon.

✓ **Got It?** **1.** Use the diagram in Problem 1. What is a description of the angle as it relates to
the situation shown?

a. ∠2 **b.** ∠3

© **Problem 2** **Using the Angle of Elevation**

Wind Farm Suppose you stand 53 ft from a wind
farm turbine. Your angle of elevation to the hub of
the turbine is 56.5°. Your eye level is 5.5 ft above
the ground. Approximately how tall is the
turbine from the ground to its hub?

$$\tan 56.5° = \frac{x}{53}$$ Use the tangent ratio.

$$x = 53(\tan 56.5°)$$ Solve for x.

53 **tan** 56.5 **enter**   Use a calculator.

80.07426526

So $x \approx 80$, which is the height from your
eye level to the hub of the turbine.
To find the total height of the
turbine, add the height from
the ground to your eyes.
Since $80 + 5.5 = 85.5$,
the wind turbine is
about 85.5 ft tall from
the ground to its hub.

Why does your eye
level matter here?
Your normal line of sight
is a horizontal line. The
angle of elevation starts
from this eye level, not
from the ground.

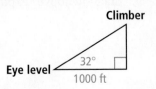

✓ **Got It?** **2.** You sight a rock climber on a cliff at a 32° angle of
elevation. Your eye level is 6 ft above the ground and
you are 1000 ft from the base of the cliff. What is the
approximate height of the rock climber from the ground?

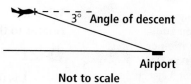
To approach runway 17 of the Ponca City Municipal Airport in Oklahoma, the pilot must begin a 3° descent starting from a height of 2714 ft above sea level. The airport is 1007 ft above sea level. To the nearest tenth of a mile, how far from the runway is the airplane at the start of this approach?

The airplane is $2714 - 1007$, or 1707 ft, above the level of the airport.

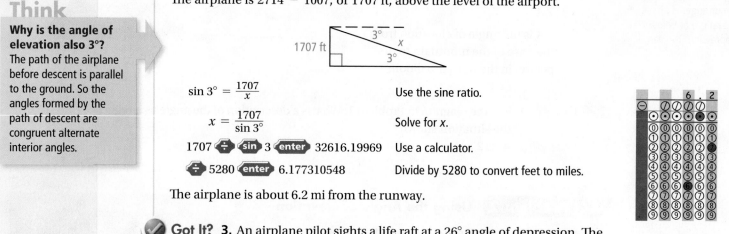

$$\sin 3° = \frac{1707}{x}$$ Use the sine ratio.

$$x = \frac{1707}{\sin 3°}$$ Solve for x.

1707 ÷ sin 3 enter 32616.19969   Use a calculator.

÷ 5280 enter 6.177310548   Divide by 5280 to convert feet to miles.

The airplane is about 6.2 mi from the runway.

Got It? 3. An airplane pilot sights a life raft at a 26° angle of depression. The airplane's altitude is 3 km. What is the airplane's horizontal distance d from the raft?

Lesson Check

Do you know HOW?

What is a description of each angle as it relates to the diagram?

1. $\angle 1$

2. $\angle 2$

3. $\angle 3$

4. $\angle 4$

5. $\angle 5$

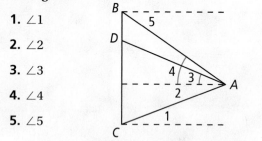

6. What are two pairs of congruent angles in the diagram above? Explain why they are congruent.

Do you UNDERSTAND? MATHEMATICAL PRACTICES

7. Vobabulary How is an angle of elevation formed?

8. Error Analysis A homework question says that the angle of depression from the bottom of a house window to a ball on the ground is 20°. Below is your friend's sketch of the situation. Describe your friend's error.

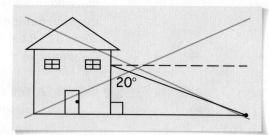

A Practice Describe each angle as it relates to the situation in the diagram.

See Problem 1.

9. ∠1   **10.** ∠2   **11.** ∠3   **12.** ∠4

13. ∠5   **14.** ∠6   **15.** ∠7   **16.** ∠8

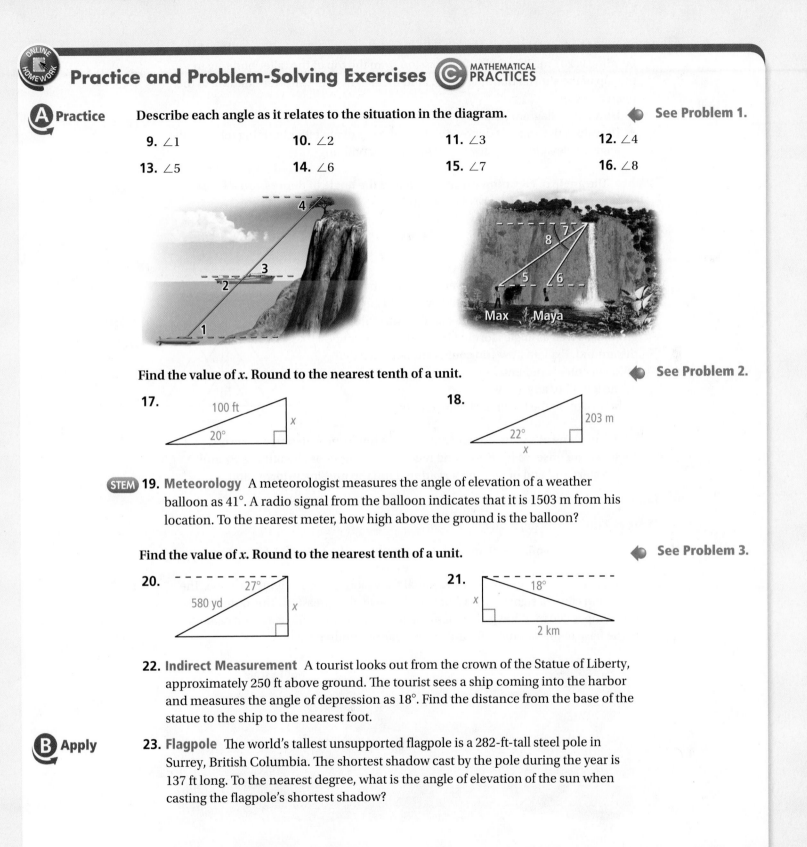

Find the value of x. Round to the nearest tenth of a unit.

See Problem 2.

17. 100 ft, 20°, x

18. 203 m, 22°, x

STEM 19. Meteorology A meteorologist measures the angle of elevation of a weather balloon as 41°. A radio signal from the balloon indicates that it is 1503 m from his location. To the nearest meter, how high above the ground is the balloon?

Find the value of x. Round to the nearest tenth of a unit.

See Problem 3.

20. 27°, 580 yd, x

21. 18°, x, 2 km

22. Indirect Measurement A tourist looks out from the crown of the Statue of Liberty, approximately 250 ft above ground. The tourist sees a ship coming into the harbor and measures the angle of depression as 18°. Find the distance from the base of the statue to the ship to the nearest foot.

B Apply

23. Flagpole The world's tallest unsupported flagpole is a 282-ft-tall steel pole in Surrey, British Columbia. The shortest shadow cast by the pole during the year is 137 ft long. To the nearest degree, what is the angle of elevation of the sun when casting the flagpole's shortest shadow?

24. Think About a Plan Two office buildings are 51 m apart. The height of the taller building is 207 m. The angle of depression from the top of the taller building to the top of the shorter building is 15°. Find the height of the shorter building to the nearest meter.

- How can a diagram help you?
- How does the angle of depression from the top of the taller building relate to the angle of elevation from the top of the shorter building?

Algebra The angle of elevation *e* from *A* to *B* and the angle of depression *d* from *B* to *A* are given. Find the measure of each angle.

25. $e: (7x - 5)°$, $d: 4(x + 7)°$

26. $e: (3x + 1)°$, $d: 2(x + 8)°$

27. $e: (x + 21)°$, $d: 3(x + 3)°$

28. $e: 5(x - 2)°$, $d: (x + 14)°$

29. Writing A communications tower is located on a plot of flat land. The tower is supported by several guy wires. Assume that you are able to measure distances along the ground, as well as angles formed by the guy wires and the ground. Explain how you could estimate each of the following measurements.

a. the length of any guy wire

b. how high on the tower each wire is attached

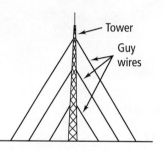

Tower

Guy wires

Flying An airplane at a constant altitude *a* flies a horizontal distance *d* toward you at velocity *v*. You observe for time *t* and measure its angles of elevation $\angle E_1$ and $\angle E_2$ at the start and end of your observation. Find the missing information.

30. $a = \blacksquare$ mi, $v = 5$ mi/min, $t = 1$ min, $m\angle E_1 = 45$, $m\angle E_2 = 90$

31. $a = 2$ mi, $v = \blacksquare$ mi/min, $t = 15$ s, $m\angle E_1 = 40$, $m\angle E_2 = 50$

32. $a = 4$ mi, $d = 3$ mi, $v = 6$ mi/min, $t = \blacksquare$ min, $m\angle E_1 = 50$, $m\angle E_2 = \blacksquare$

33. Aerial Television A blimp provides aerial television views of a football game. The television camera sights the stadium at a 7° angle of depression. The altitude of the blimp is 400 m. What is the line-of-sight distance from the television camera to the base of the stadium? Round to the nearest hundred meters.

Not to scale

7°

400 m

Challenge

34. **Firefighting** A firefighter on the ground sees fire break through a window near the top of the building. The angle of elevation to the windowsill is 28°. The angle of elevation to the top of the building is 42°. The firefighter is 75 ft from the building and her eyes are 5 ft above the ground. What roof-to-windowsill distance can she report by radio to firefighters on the roof?

35. **Geography** For locations in the United States, the relationship between the latitude ℓ and the greatest angle of elevation a of the sun at noon on the first day of summer is $a = 90° - \ell + 23.5°$. Find the latitude of your town. Then determine the greatest angle of elevation of the sun for your town on the first day of summer.

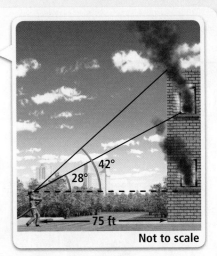

Not to scale

Standardized Test Prep

SAT/ACT

36. A 107-ft-tall building casts a shadow of 90 ft. To the nearest whole degree, what is the angle of elevation of the sun?

    Ⓐ 33°   Ⓑ 40°   Ⓒ 50°   Ⓓ 57°

37. Which assumption should you make to prove indirectly that the sum of the measures of the angles of a parallelogram is 360?

    Ⓕ The sum of the measures of the angles of a parallelogram is 360.

    Ⓖ The sum of the measures of the angles of a parallelogram is not 360.

    Ⓗ The sum of the measures of consecutive angles of a parallelogram is 180.

    Ⓘ The sum of the measures of the angles of a parallelogram is 180.

Extended Response

38. A parallelogram has four congruent sides.
   a. Name the types of parallelograms that have this property.
   b. What is the most precise name for the figure, based only on the given description? Explain.
   c. Draw a diagram to show the categorization of parallelograms.

Mixed Review

Find the value of x. Round to the nearest tenth of a unit.   ◀ **See Lesson 8-3.**

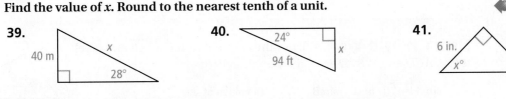

39. 40 m, x, 28°   40. 24°, x, 94 ft   41. 6 in., 6 in., $x°$

Get Ready! To prepare for Lesson 8-5, do Exercises 42–44.

Find the distance between each pair of points.   ◀ **See Lesson 1-7.**

42. $(0, 0)$ and $(8, 2)$ 43. $(-15, -2)$ and $(0, 0)$ 44. $(-2, 12)$ and $(0, 0)$

Law of Sines

© Content Standards
G.SRT.11 Understand and apply the Law of Sines . . . to find unknown measurements in right and non-right triangles . . .
Also G.SRT.10

Objectives To apply the Law of Sines

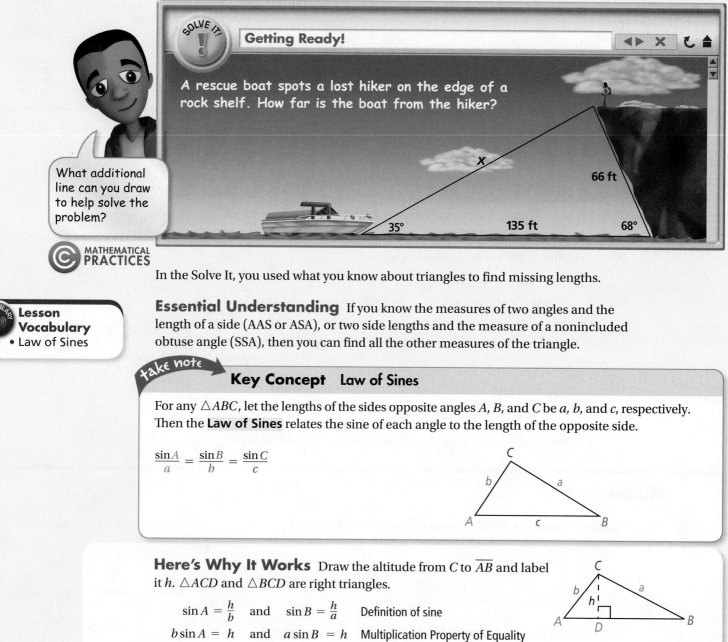

Getting Ready!

A rescue boat spots a lost hiker on the edge of a rock shelf. How far is the boat from the hiker?

x

66 ft

35°   135 ft   68°

What additional line can you draw to help solve the problem?

MATHEMATICAL PRACTICES

Lesson Vocabulary
• Law of Sines

In the Solve It, you used what you know about triangles to find missing lengths.

Essential Understanding If you know the measures of two angles and the length of a side (AAS or ASA), or two side lengths and the measure of a nonincluded obtuse angle (SSA), then you can find all the other measures of the triangle.

take note

Key Concept   Law of Sines

For any $\triangle ABC$, let the lengths of the sides opposite angles A, B, and C be a, b, and c, respectively. Then the **Law of Sines** relates the sine of each angle to the length of the opposite side.

$$\frac{\sin A}{a} = \frac{\sin B}{b} = \frac{\sin C}{c}$$

Here's Why It Works Draw the altitude from C to $\overline{AB}$ and label it h. $\triangle ACD$ and $\triangle BCD$ are right triangles.

$$\sin A = \frac{h}{b} \quad \text{and} \quad \sin B = \frac{h}{a} \qquad \text{Definition of sine}$$

$$b \sin A = h \quad \text{and} \quad a \sin B = h \qquad \text{Multiplication Property of Equality}$$

$$b \sin A = a \sin B \qquad \text{Transitive Property of Equality}$$

$$\frac{\sin A}{a} = \frac{\sin B}{b} \qquad \text{Division Property of Equality}$$

Problem 1 Using the Law of Sines (AAS)

In $\triangle ABC$, $m\angle A = 48$, $m\angle B = 93$, and $AC = 15$. To the nearest tenth, what is the length of $\overline{BC}$?

Plan

How will drawing a diagram help you solve the problem?
Drawing a diagram will help you to visualize the problem. Carefully draw a diagram and label it with all of the given information.

Think

Draw and label $\triangle ABC$. You are given two angle measures and the length of a nonincluded side (AAS).

Use the Law of Sines to write an equation.

Solve for BC.

Use a calculator to find BC.

Write

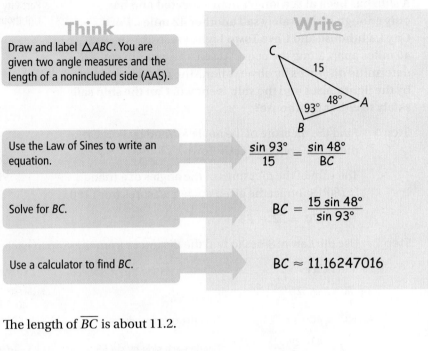

$$\frac{\sin 93°}{15} = \frac{\sin 48°}{BC}$$

$$BC = \frac{15 \sin 48°}{\sin 93°}$$

$$BC \approx 11.16247016$$

The length of $\overline{BC}$ is about 11.2.

Got It? **1.** In $\triangle ABC$ above, what is AB to the nearest tenth?

You can also use the Law of Sines to find missing angle measures.

Problem 2 Using the Law of Sines (SSA)

In $\triangle RST$, $RT = 11$, $ST = 18$, and $m\angle R = 120$. To the nearest tenth, what is $m\angle S$?

Think

What unknown angle should you use?
Use $\angle S$ because it is opposite a known side length.

Step 1 Draw and label a diagram.

Step 2 Use the Law of Sines to set up an equation.

$$\frac{\sin 120°}{18} = \frac{\sin S}{11}$$

Step 3 Find $m\angle S$.

$$\sin S = \frac{11 \sin 120°}{18}$$ Solve for $\sin S$.

$$m\angle S = \sin^{-1}\left(\frac{11 \sin 120°}{18}\right) \approx 31.95396690$$ Use the inverse.

$m\angle S$ is about 32.0.

Got It? **2.** In $\triangle KLM$, $LM = 9$, $KM = 14$, and $m\angle L = 105$. To the nearest tenth, what is $m\angle K$?

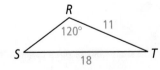

You can apply the Law of Sines to real-world problems involving triangles.

Problem 3 Using the Law of Sines to Solve a Problem

A ship has been at sea longer than expected and has only enough fuel to safely sail another 42 miles. Port City Lighthouse and Cove Town Lighthouse are located 40 miles apart along the coast. At sea, the captain cannot determine distances by observation. The triangle formed by the lighthouses and the ship is shown. Can the ship sail safely to either lighthouse?

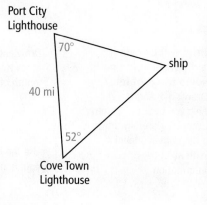

Think

What other information do you need to know to use the Law of Sines?
You can use the Law of Sines to find the distances from the ship to each lighthouse if you can find the angle with the ship as its vertex.

Step 1 Find the measure of the angle formed by Port City, the cruise ship, and Cove Town.

The sum of the measures of the angles of a triangle is 180°. Subtract the given angle measures from 180.

$$180 - 70 - 52 = 58$$

Step 2 Use the Law of Sines to find the distances from the ship to each lighthouse.

Port City		**Cove Town**
$\dfrac{\sin 58°}{40} = \dfrac{\sin 52°}{d}$	Law of Sines	$\dfrac{\sin 58°}{40} = \dfrac{\sin 70°}{d}$
$d \cdot \sin 58° = 40 \cdot \sin 52°$	Find the cross products.	$d \cdot \sin 58° = 40 \cdot \sin 70°$
$d = \dfrac{40 \cdot \sin 52°}{\sin 58°}$	Divide each side by sin 58°.	$d = \dfrac{40 \cdot \sin 70°}{\sin 58°}$
$d \approx 37.16821049$	Use a calculator.	$d \approx 44.32260977$

The distance from the ship to Port City Lighthouse is about 37.2 miles, and the distance to Cove Town Lighthouse is about 44.3 miles.

The ship can sail safely to Port City Lighthouse but not to Cove Town Lighthouse.

Got It? 3. The right-fielder fields a softball between first base and second base as shown in the figure. If the right-fielder throws the ball to second base, how far does she throw the ball?

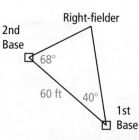

Lesson Check

Do you know HOW?

1. In $\triangle ABC$, $AB = 7$, $BC = 10$, and $m\angle A = 80$. To the nearest tenth, what is $m\angle C$?

2. What is x?

3. What is y?

Do you UNDERSTAND? MATHEMATICAL PRACTICES

4. Reasoning If you know the three side lengths of a triangle, can you use the Law of Sines to find the missing angle measures? Explain.

5. Error Analysis In $\triangle PQR$, $PQ = 4$ cm, $QR = 3$ cm, and $m\angle R = 75$. Your friend uses the Law of Sines to write $\dfrac{\sin 75°}{3} = \dfrac{\sin P}{4}$ to find $m\angle P$. Explain the error.

A) Practice   Use the information given to solve.   See Problems 1 and 2.

6. In $\triangle ABC$, $m\angle A = 70$, $m\angle C = 62$, and $BC = 7.3$. To the nearest tenth, what is AB?

7. In $\triangle XYZ$, $m\angle Y = 80$, $XY = 14$, and $XZ = 17$. To the nearest tenth, what is $m\angle Z$?

Use the Law of Sines to find the values of x and y. Round to the nearest tenth.

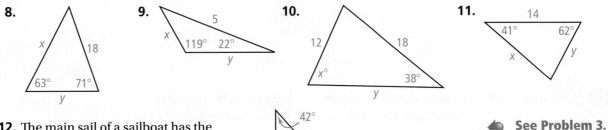

8.

9. 5 / x / 119° 22° / y

10. 12 / 18 / $x°$ 38° / y

11. 14 / 41° 62° / x / y

(8.) x / 18 / 63° 71° / y

12. The main sail of a sailboat has the dimensions shown in the figure at the right. To the nearest tenth of a foot, what is the height of the main sail?   See Problem 3.

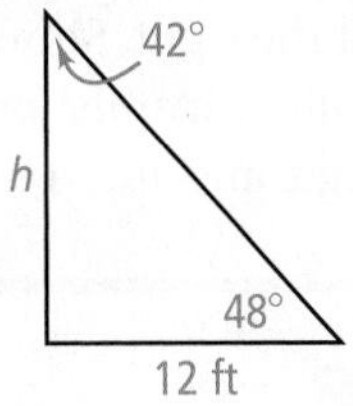

B) Apply

13. A portion of a city map is shown in the figure at the right. If you walk along Maple Street between 2nd Street and Elm Grove Lane, how far do you walk? Round your answer to the nearest tenth of a yard.

14. Navigation The Bermuda Triangle is a historically famous region of the Atlantic Ocean. The vertices of the triangle are formed by Miami, FL; Bermuda; and San Juan, Puerto Rico. The approximate dimensions of the Bermuda Triangle are shown in the figure at the right. Explain how you would find the distance from Bermuda to Miami. What is this distance to the nearest mile?

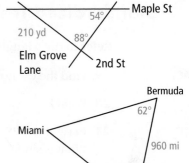

15. Think About a Plan An airplane took off from an airport and started flying toward its destination 210 miles due east. After flying 80 miles east, it encountered a storm and altered its course by turning left 22°. When it was past the storm, it turned right 30° and flew in a straight line until it reached its destination. How far was the plane from its destination when it made the 30° turn?
• Would drawing a diagram help you visualize the problem situation?
• What are you being asked to find?
• What measures do you know?

16. Zipline A zipline is constructed over a ravine as shown in the diagram at the right. What is the horizontal distance from the bottom of the ladder to the platform where the zipline ends? Round your answer to the nearest tenth of a foot.

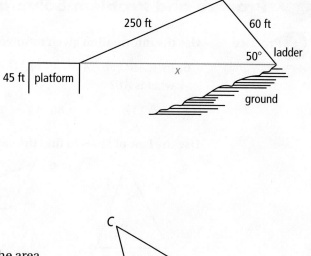

250 ft
60 ft
50° ladder
x
45 ft | platform
ground

17. If $m\angle DEG = m\angle D + m\angle G + 43$, what is $m\angle EFG$?

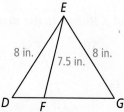

E
8 in.
7.5 in.
8 in.
D   F   G

 Challenge

18. You can use the formula Area $= \frac{1}{2}bc \sin A$ to find the area of the triangle shown at the right. Show how this formula becomes the more familiar formula for the area of a triangle if $\triangle ABC$ is a right triangle and $m\angle A = 90$.

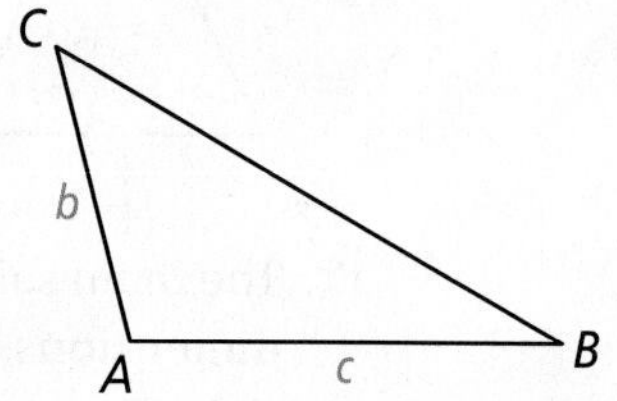

C
b
A   c   B

Standardized Test Prep

GRIDDED RESPONSE

Refer to the triangle at the right to solve Exercises 19 and 20.

SAT/ACT

19. Find the value of x to the nearest hundredth of an inch.

20. What is the value of y to the nearest hundredth of an inch?

21. In right $\triangle TRS$, $m\angle R = 90$, $TR = 3$, $RS = 4$, and $TS = 5$. What is $\sin S$?

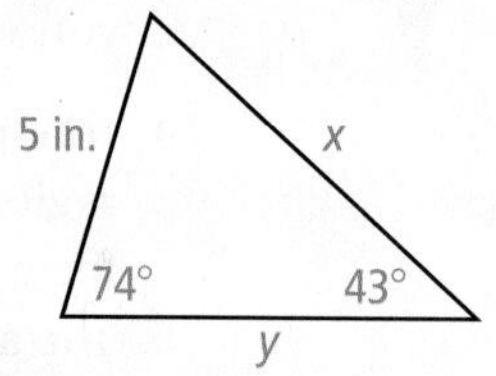

5 in.   x
74°   43°
        y

Mixed Review

22. Indirect Measurement A hot-air balloon pilot sights the landing field from a height of 2000 ft. The angle of depression is 24°. To the nearest foot, what is the ground distance from the hot-air balloon to the landing field?

 See Lesson 8-4.

23. What is the value of x?

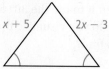

$x + 5$ $2x - 3$

See Lesson 4-5.

Get Ready! **To Prepare for Lesson 8-6, do Exercises 24–27.**

Evaluate each expression. Round your answers to the nearest hundredth.

See Lesson 8-3.

24. $\cos 58°$

25. $\cos^{-1} 0.5875$

26. $\dfrac{25 \cos 62°}{18}$

27. $\cos^{-1}\left(\dfrac{5}{8}\right)$

Law of Cosines

© **Content Standards**

G.SRT.11 Understand and apply the . . . Law of Cosines . . .

Also G.SRT.10

Objective   To apply the Law of Cosines

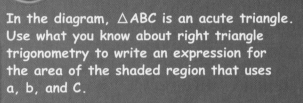

Getting Ready!

In the diagram, △ABC is an acute triangle. Use what you know about right triangle trigonometry to write an expression for the area of the shaded region that uses a, b, and C.

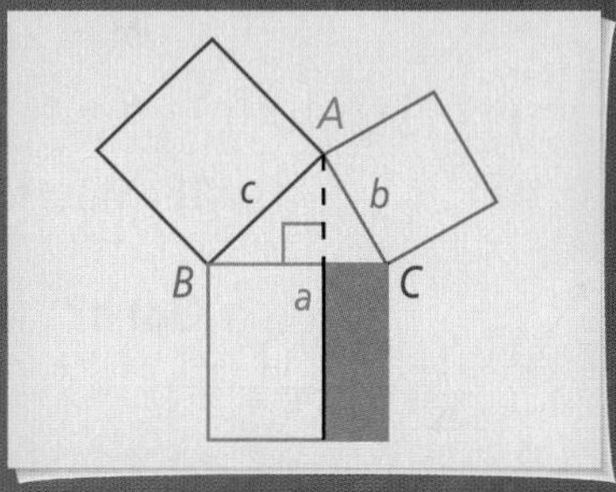

Think about how the areas of the squares are related to the side lengths of the triangle. What does this remind you of?

© MATHEMATICAL PRACTICES   In the Solve It, you used right triangle trigonometry to write an expression to describe a side length. You can also find relationships between the angle measures and the side lengths of nonright triangles.

Essential Understanding   If you know the measures of two side lengths and the measure of the included angle (SAS), or all three side lengths (SSS), then you can find all the other measures of the triangle.

take note

Key Concept   Law of Cosines

Lesson Vocabulary
• Law of Cosines

For any △ABC, the **Law of Cosines** relates the cosine of each angle to the side lengths of the triangle.

$$a^2 = b^2 + c^2 - 2bc \cos A$$
$$b^2 = a^2 + c^2 - 2ac \cos B$$
$$c^2 = a^2 + b^2 - 2ab \cos C$$

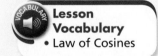

Here's Why It Works   Note that $b^2 = x^2 + h^2$ and $x = b \cos A$. Use the Pythagorean Theorem with △BCD and simplify.

$a^2 = (c - x)^2 + h^2$	Pythagorean Theorem
$a^2 = c^2 - 2cx + x^2 + h^2$	Simplify.
$a^2 = c^2 - 2cb\cos A + b^2$	Substitute b^2 for $x^2 + h^2$ and $b \cos A$ for x.
$a^2 = b^2 + c^2 - 2bc\cos A$	Commutative Property

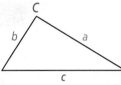

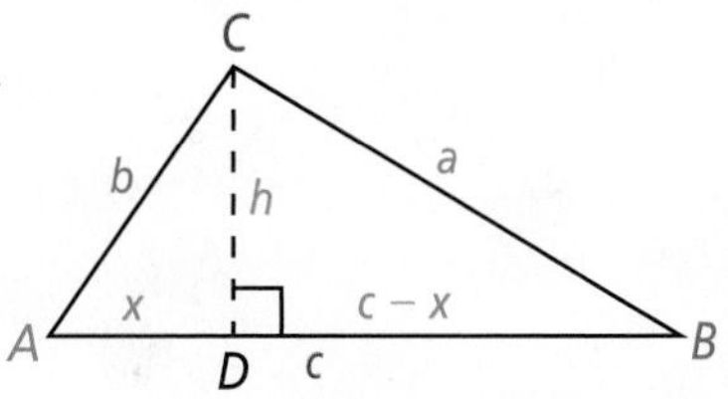

Problem 1 Using the Law of Cosines (SAS)

Find *b* to the nearest tenth.

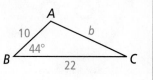

Know	Need	Plan
$\overline{AB}$ is opposite $\angle C$, so $AB = c = 10$. $\overline{BC}$ is opposite $\angle A$, so $BC = a = 22$. $m\angle B = 44$	the length of $\overline{AC}$	Because you know $m\angle B$ and need *b*, substitute the angle measure and the two side lengths into $b^2 = a^2 + c^2 - 2ac \cos B$ and solve for *b*.

$$b^2 = a^2 + c^2 - 2ac \cos B \qquad \text{Law of Cosines}$$

$$b^2 = 22^2 + 10^2 - 2(22)(10) \cos 44° \qquad \text{Substitute.}$$

$$b \approx 16.35513644 \qquad \text{Use a calculator.}$$

The value of *b* is about 16.4.

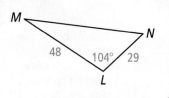

Got It? 1. Find *MN* to the nearest tenth.

Problem 2 Using the Law of Cosines (SSS)

In $\triangle TUV$, $TU = 4.4$, $UV = 7.1$, and $TV = 6.7$. Find $m\angle V$ to the nearest tenth of a degree.

Plan

How can you use what you know to find $m\angle V$?
You know the three side lengths (SSS) so you can use the Law of Cosines to find $m\angle V$.

Step 1 Draw and label a diagram.

Step 2 Use the Law of Cosines to set up an equation.

$$TU^2 = UV^2 + TV^2 - 2(UV)(TV) \cos V$$

$$4.4^2 = 7.1^2 + 6.7^2 - 2(7.1)(6.7) \cos V$$

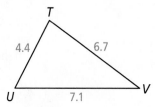

Step 3 Solve for $m\angle V$.

$19.36 = 50.41 + 44.89 - 95.14 \cos V$	Simplify.
$\dfrac{-75.94}{-95.14} = \cos V$	Solve for cos *V*.
$V = \cos^{-1}\left(\dfrac{-75.94}{-95.14}\right)$	Solve for $m\angle V$.
$m\angle V \approx 37.04219062$	Use a caclulator.

The measure of $\angle V$ is about 37.0.

Got It? 2. In $\triangle TUV$ above, find $m\angle T$ to the nearest tenth degree.

You can use the Law of Cosines to solve real world problems involving triangles.

 **Problem 3** **Using the Law of Cosines to Solve a Problem**

Think

What do you need to find before you can use the law of Cosines?
You need to find the measure of the angle opposite d in the triangle before you can apply the Law of Cosines.

An air traffic controller is tracking a plane 2.1 kilometers due south of the radar tower. A second plane is located 3.5 kilometers from the tower at a heading of N 75° E (75° east of north). To the nearest tenth of a kilometer, how far apart are the two planes?

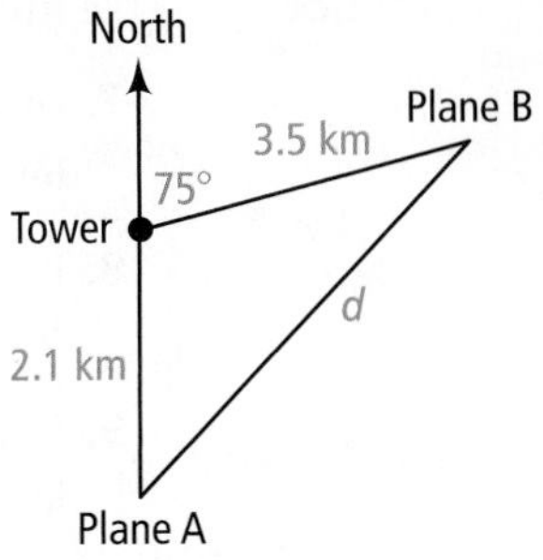

The north-south line in the figure represents a straight angle. Let the angle opposite d be $\angle D$. Use supplementary angles to find $m\angle D$. Use supplementary angles to find the measure of the angle opposite d.

North

Plane B

3.5 km

75°

Tower

2.1 km

d

Plane A

$$m\angle D = 180 - 75 = 105 \qquad \text{Supplementary angles}$$

Use the Law of Cosines to solve for d.

$$d^2 = a^2 + b^2 - 2ab \cos D \qquad \text{Law of Cosines}$$

$$d^2 = 3.5^2 + 2.1^2 - 2(3.5)(2.1) \cos 105° \qquad \text{Substitute.}$$

$$d \approx 4.52378602 \qquad \text{Use a calculator.}$$

The distance between the two planes is about 4.5 kilometers.

Got It? **3.** You and a friend hike 1.4 miles due west from a campsite. At the same time, two other friends hike 1.9 miles at a heading of S 11° W (11° west of south) from the campsite. To the nearest tenth of a mile, how far apart are the two groups?

Lesson Check

Do you know HOW?

1. In $\triangle ABC$, $AB = 7$, $BC = 10$, and $m\angle B = 80$. To the nearest tenth, what is b?

2. In $\triangle QRS$, $QR = 31.9$, $RS = 25.2$, and $QS = 37.6$. To the nearest tenth, what is $m\angle R$?

3. In $\triangle LMN$, $LN = 7$, $MN = 10$, and $m\angle N = 48$. To the nearest tenth, what is the area of $\triangle LMN$?

4. What are $m\angle X$, $m\angle Y$, and $m\angle Z$?

X

4 m   6 m

Y   7 m   Z

Do you UNDERSTAND?

MATHEMATICAL PRACTICES

5. Error Analysis In $\triangle ABC$, $AC = 15$ ft, $BC = 12$ ft, and $m\angle C = 32$. A student solved for c for $a = 12$ ft, $b = 15$ ft, and $m\angle C = 32$. What was the error?

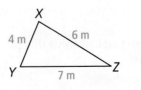

$C = 12^2 + 15^2 - 2(12)(15)\cos 32°$
$C = 369 - 360 \cos 32°$
$C = 63.7$

6. Reasoning Explain how you would find the measure of the largest angle of a triangle if given the measures of the three side lengths.

Practice and Problem-Solving Exercises

A Practice

Use the information given to solve.

See Problems 1 and 2.

7. In $\triangle QRS$, $m\angle R = 38$, $QR = 11$, and $RS = 16$. To the nearest tenth, what is the length of $\overline{QS}$?

8. In $\triangle WXY$, $WX = 20.4$, $XY = 16.4$, and $WY = 25.3$. To the nearest tenth, what is $m\angle W$?

9. In $\triangle JKL$, $JK = 2.6$, $KL = 6.4$, and $m\angle K = 10.5$. To the nearest tenth, what is the length of $\overline{JL}$?

10. In $\triangle DEF$, $DE = 13$, $EF = 24$, and $FD = 27$. To the nearest tenth, what is $m\angle E$?

Use the Law of Cosines to find the values of x and y. Round to the nearest tenth.

11. **12.** **13.** **14.**

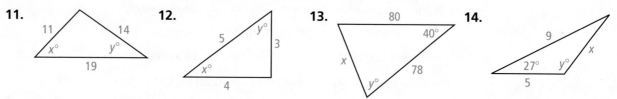

Use the Law of Cosines to solve each problem.

See Problem 3.

15. Baseball After fielding a ground ball, a pitcher is located 110 feet from first base and 57 feet from home plate as shown in the figure at the right. To the nearest tenth, what is the measure of the angle with its vertex at the pitcher?

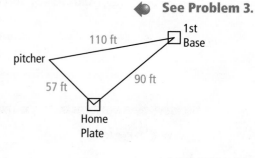

16. Zipline One side of a ravine is 14 ft long. The other side is 12 ft long. A 20 ft zipline runs from the top of one side of the ravine to the other. To the nearest tenth, at what angle do the sides of the ravine meet?

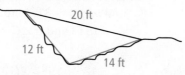

B Apply

© **17. Think About a Plan** A walking path around the outside of a garden is shaped like a triangle. Two sides of the path that measure 32 ft and 39 ft form a 76° angle. If you walk around the entire path one time, how far have you walked? Write your answer to the nearest foot.
  • What information do you need to find before you can solve this problem?
  • How can you find the information you need?
  • Can drawing a diagram help you solve this problem?

18. **Airplane** A commuter plane flies from City A to City B, a distance of 90 mi due north. Due to bad weather, the plane is redirected at take-off to a heading N 60° W (60° west of north). After flying 57 mi, the plane is directed to turn northeast and fly directly toward City B. To the nearest tenth, how many miles did the plane fly on the last leg of the trip?

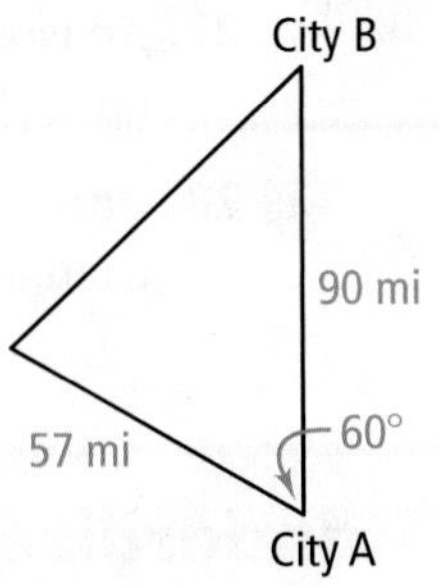

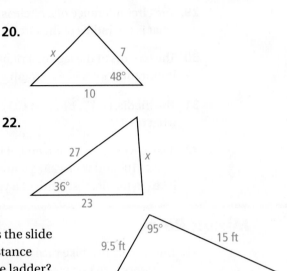

For each triangle shown below, determine whether you would use the Law of Sines or Law of Cosines to find the value of *x*. Then find the value of *x* to the nearest tenth.

19.

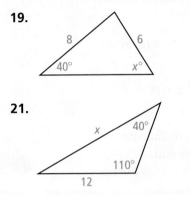

20.

21.

22.

23. A 15-ft water slide has a 9.5-ft ladder which meets the slide at a 95° angle. To the nearest tenth, what is the distance between the end of the slide and the bottom of the ladder?

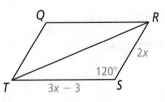

24. **Flags** The dimensions of a triangular flag are 18 ft by 25 ft by 27 ft. To the nearest tenth, what is the measure of the angle formed by the two shorter sides?

 **Challenge**

25. Parallelogram *QRST* has a perimeter of 62 mm. To the nearest tenth, what is the length of *TR*?

26. **Surveying** A surveyor measures the distance to the base of a monument to be 12.4 meters at an angle of elevation of 11°. At an angle of elevation of 26°, the distance to the top of the monument is 13.3 meters. What is the height of the monument to the nearest tenth?

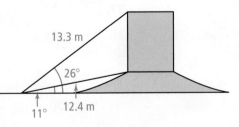

27. An isosceles triangle *XYZ* has a base of 12 in. and a height of 8 in. To the nearest tenth, what are the measures of the angles?

Ⓒ **28. Open-Ended** Describe a situation in which you are given three measures of a triangle but are unable to solve the triangle for the other three measures.

Standardized Test Prep

GRIDDED RESPONSE

SAT/ACT

29. The circumference of a circle is 24 mm. To the nearest tenth of a millimeter, what is the radius of the circle? Use 3.14 for π.

30. The lengths of the legs of a right triangle are 6 and 11. To the nearest hundredth, what is the length of the hypotenuse?

31. The medians $\overline{AE}$, $\overline{BF}$, and $\overline{CD}$ of $\triangle ABC$ intersect at *G*. If $GD = 4$, what is *CD*?

32. A surveyor starts at one end of the lake and walks 192 yd. Then he turns 110° and walks 237 yd until he arrives at the other end of the lake. To the nearest tenth of a yard, what is the distance across the lake?

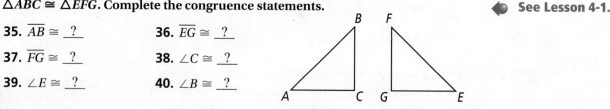

110°

237 yd   192 yd

Mixed Review

33. The first leg of a bike race is 6 km due east. For the second leg of the race, the riders turn northwest and ride 9 km. The final leg of the race runs at a 40° angle to the second leg and brings the racers back to the starting point. To the nearest tenth, at what angle measure does the third leg of the race meet the first leg?

◀ See Lesson 8-4.

34. Is the quadrilateral with vertices $A(-1, -5)$, $B(6, -5)$, $C(9, 3)$, and $D(2, 3)$ a parallelogram? Explain.

◀ See Lesson 6-7.

Get Ready! **To prepare for Lesson 9-1, do Exercises 35–40.**

$\triangle ABC \cong \triangle EFG$. Complete the congruence statements.

◀ See Lesson 4-1.

35. $\overline{AB} \cong$?

36. $\overline{EG} \cong$?

37. $\overline{FG} \cong$?

38. $\angle C \cong$?

39. $\angle E \cong$?

40. $\angle B \cong$?

Pull It **All Together** © ASSESSMENT

To solve these problems, you will pull together many concepts and skills that you have studied about right triangles and trigonometry.

BIG idea Measurement

You can use the Pythagorean Theorem or trigonometric ratios to find side lengths or angle measures of a right triangle.

© **Performance Task 1**

The diagram below shows equilateral $\triangle ABC$ sharing a side with square *ACDE*. The square has side lengths of 4. What is *BE*? Justify your answer.

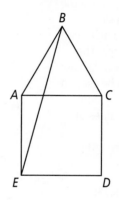

BIG idea Similarity

You can use trigonometric ratios to find the lengths of the sides of right triangles.

© **Performance Task 2**

A construction crew wants to hoist a heavy beam so that it is standing up straight. They tie a rope to the beam, secure the base, and pull the rope through a pulley to raise one end of the beam from the ground. When the beam makes an angle of 40° with the ground, the top of the beam is 8 ft above the ground.

The construction site has some telephone wires crossing it. The workers are concerned that the beam may hit the wires. When the beam makes an angle of 60° with the ground, the wires are 2 ft above the top of the beam. Will the beam clear the wires on its way to standing up straight? Explain.

Connecting **BIG** ideas and Answering the Essential Questions

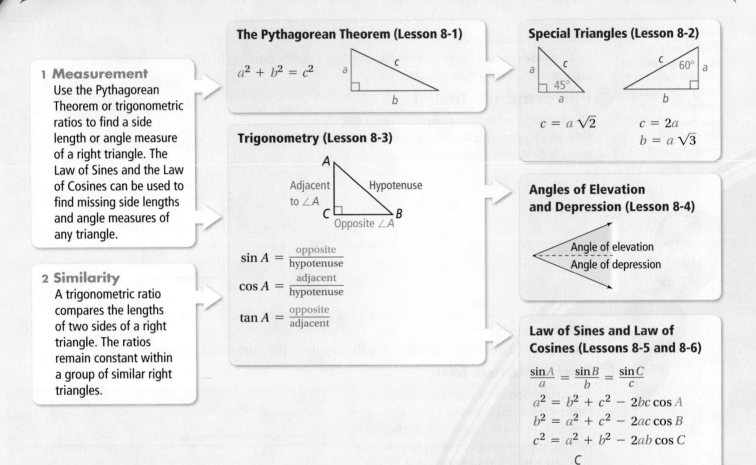

1 Measurement
Use the Pythagorean Theorem or trigonometric ratios to find a side length or angle measure of a right triangle. The Law of Sines and the Law of Cosines can be used to find missing side lengths and angle measures of any triangle.

2 Similarity
A trigonometric ratio compares the lengths of two sides of a right triangle. The ratios remain constant within a group of similar right triangles.

The Pythagorean Theorem (Lesson 8-1)

$$a^2 + b^2 = c^2$$

Special Triangles (Lesson 8-2)

$$c = a\sqrt{2}$$

$$c = 2a$$
$$b = a\sqrt{3}$$

Trigonometry (Lesson 8-3)

$$\sin A = \frac{\text{opposite}}{\text{hypotenuse}}$$

$$\cos A = \frac{\text{adjacent}}{\text{hypotenuse}}$$

$$\tan A = \frac{\text{opposite}}{\text{adjacent}}$$

Angles of Elevation and Depression (Lesson 8-4)

Angle of elevation

Angle of depression

Law of Sines and Law of Cosines (Lessons 8-5 and 8-6)

$$\frac{\sin A}{a} = \frac{\sin B}{b} = \frac{\sin C}{c}$$

$$a^2 = b^2 + c^2 - 2bc \cos A$$
$$b^2 = a^2 + c^2 - 2ac \cos B$$
$$c^2 = a^2 + b^2 - 2ab \cos C$$

Chapter Vocabulary

- angle of depression (p. 516)
- angle of elevation (p. 516)
- cosine (p. 507)
- Law of Cosines (p. 526)
- Law of Sines (p. 522)
- Pythagorean triple (p. 492)
- sine (p. 507)
- tangent (p. 507)
- trigonometric ratios (p. 507)

Choose the correct term to complete each sentence.

1. __?__ are equivalent ratios for the corresponding sides of two triangles.

2. A(n) __?__ is formed by a horizontal line and the line of sight above that line.

3. A set of three nonzero whole numbers that satisfy $a^2 + b^2 = c^2$ form a(n) __?__.

8-1 The Pythagorean Theorem and Its Converse

Quick Review

The **Pythagorean Theorem** holds true for any right triangle.

$$(\text{leg}_1)^2 + (\text{leg}_2)^2 = (\text{hypotenuse})^2$$
$$a^2 + b^2 = c^2$$

The Converse of the Pythagorean Theorem states that if $a^2 + b^2 = c^2$, where c is the greatest side length of a triangle, then the triangle is a right triangle.

Example

What is the value of x?

$a^2 + b^2 = c^2$	Pythagorean Theorem
$x^2 + 12^2 = 20^2$	Substitute.
$x^2 = 256$	Simplify.
$x = 16$	Take the square root.

Exercises

Find the value of x. If your answer is not an integer, express it in simplest radical form.

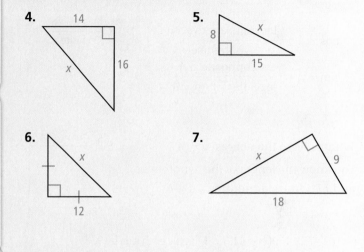

4. 14, x, 16

5. 8, x, 15

6. x, 12

7. x, 9, 18

8-2 Special Right Triangles

Quick Review

$45°$-$45°$-$90°$ Triangle

$$\text{hypotenuse} = \sqrt{2} \cdot \text{leg}$$

$30°$-$60°$-$90°$ Triangle

$$\text{hypotenuse} = 2 \cdot \text{shorter leg}$$
$$\text{longer leg} = \sqrt{3} \cdot \text{shorter leg}$$

Example

What is the value of x?

The triangle is a $30°$-$60°$-$90°$ triangle, and x represents the length of the longer leg.

$$\text{longer leg} = \sqrt{3} \cdot \text{shorter leg}$$
$$x = 20\sqrt{3}$$

Exercises

Find the value of each variable. If your answer is not an integer, express it in simplest radical form.

8. 7, y, x, $45°$

9. $45°$, 10, x

10. 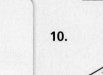 y, 6, $30°$, x

11. 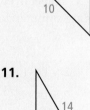 14, y, $60°$, x

12. A square garden has sides 50 ft long. You stretch a hose from one corner of the garden to another corner along the garden's diagonal. To the nearest tenth, how long is the hose?

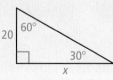

8-3 and 8-4 Trigonometry and Angles of Elevation and Depression

Quick Review

In right $\triangle ABC$, C is the right angle.

$$\sin \angle A = \frac{\text{leg opposite } \angle A}{\text{hypotenuse}}$$

$$\cos \angle A = \frac{\text{leg adjacent to } \angle A}{\text{hypotenuse}}$$

$$\tan \angle A = \frac{\text{leg opposite } \angle A}{\text{leg adjacent to } \angle A}$$

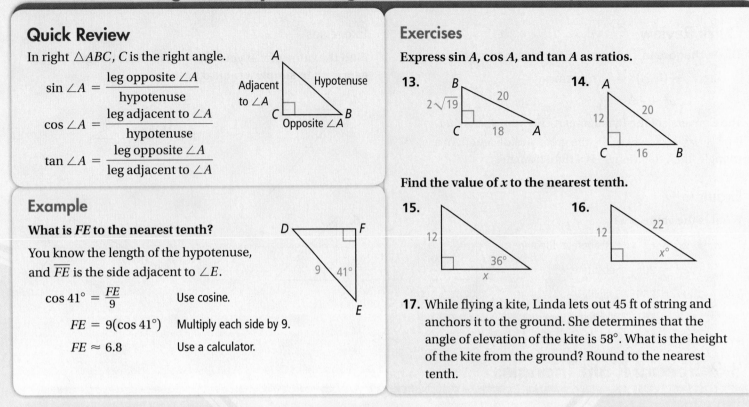

Example

What is FE to the nearest tenth?

You know the length of the hypotenuse, and $\overline{FE}$ is the side adjacent to $\angle E$.

$$\cos 41° = \frac{FE}{9} \qquad \text{Use cosine.}$$

$$FE = 9(\cos 41°) \qquad \text{Multiply each side by 9.}$$

$$FE \approx 6.8 \qquad \text{Use a calculator.}$$

Exercises

Express sin A, cos A, and tan A as ratios.

13.

14.

Find the value of x to the nearest tenth.

15.

16.

17. While flying a kite, Linda lets out 45 ft of string and anchors it to the ground. She determines that the angle of elevation of the kite is 58°. What is the height of the kite from the ground? Round to the nearest tenth.

8-5 and 8-6 Law of Sines and Law of Cosines

Quick Review

In $\triangle ABC$, a, b, and c are the lengths of the sides opposite $\angle A$, $\angle B$, and $\angle C$, respectively. The Law of Sines and the Law of Cosines are summarized below.

$$\frac{\sin A}{a} = \frac{\sin B}{b} = \frac{\sin C}{c}$$

$$a^2 = b^2 + c^2 - 2bc \cos A$$

$$b^2 = a^2 + c^2 - 2ac \cos B$$

$$c^2 = a^2 + b^2 - 2ab \cos C$$

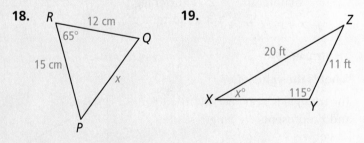

Example

What is GH?

Use the Law of Sines to find GH.

$$\frac{\sin 46°}{GH} = \frac{\sin 80°}{14.1}$$

$$GH \sin 80° = 14.1 \sin 46°$$

$$0.9848\,GH = 10.1427$$

$$GH \approx 10.3$$

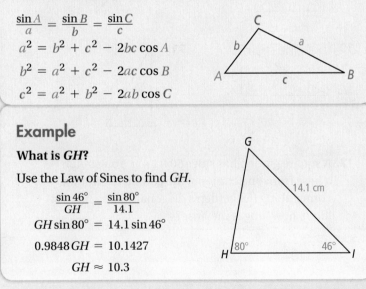

Exercises

Find the value of x to the nearest tenth.

18.

19.

20. In $\triangle DEF$, sides d, e, and f are opposite $\angle D$, $\angle E$, and $\angle F$ respectively. The side lengths are $d = 25$ in., $e = 18$ in., and $f = 20$ in. Find the $m\angle D$ to the nearest tenth.

21. In $\triangle LMN$, sides ℓ, m, and n are opposite $\angle L$, $\angle M$, and $\angle N$ respectively. You know that $m = 3$ cm, $n = 8$ cm, and $m\angle L = 72°$. Find the $m\angle N$ to the nearest tenth.

Do you know HOW?

Algebra Find the value of each variable. Express your answer in simplest radical form.

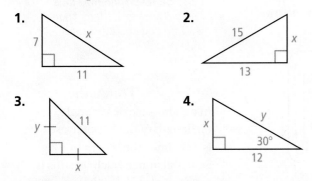

1.

2.

3.

4.

Given the following triangle side lengths, identify the triangle as *acute*, *right*, or *obtuse*.

5. 9 cm, 10, cm, 12, cm

6. 8 m, 15 m, 17 m

7. 5 in., 6 in., 10 in.

Express sin *B*, cos *B*, and tan *B* as ratios.

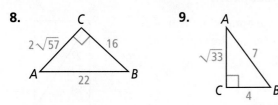

8.

9.

Find each missing value to the nearest tenth.

10. $\tan \blacksquare° = 1.11$

11. $\sin 34° = \dfrac{5}{\blacksquare}$

12. $\cos \blacksquare° = \dfrac{12}{15}$

13. A woman stands 15 ft from a statue. She looks up at an angle of 60° to see the top of the statue. Her eye level is 5 ft above the ground. How tall is the statue to the nearest foot?

Find the value of *x*. Round lengths to the nearest tenth and angle measures to the nearest degree.

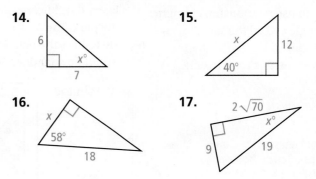

14.

15.

16.

17.

18. Find the $m\angle A$ to the nearest tenth.

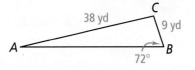

19. Find *TU* to the nearest tenth.

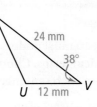

20. In $\triangle KLP$, $k = 13$ mi, $\ell = 10$ mi, and $p = 8$ mi. Find $m\angle K$ to the nearest tenth.

21. In $\triangle ABC$, $a = 8$, $b = 10$, and $m\angle B = 120$. Find the $m\angle C$ to the nearest tenth.

Do you UNDERSTAND?

22. Writing Explain why $\sin x° = \cos(90 - x)°$. Include a diagram with your explanation.

23. Reasoning Suppose that you know all three angle measures of a triangle. Can you use Law of Sines or Law of Cosines to find the side lengths? Explain.

24. Reasoning If you know the measures of both acute angles of a right triangle, can you determine the lengths of the sides? Explain.

ⓣⓘⓟⓢ ⓕⓞⓡ ⓢⓤⓒⓒⓔⓢⓢ

Some test questions require you to use relationships in right triangles. Read the sample question at the right. Then follow the tips to answer it.

Every day, Michael goes for a run through the park. The park is shaped like a square. He follows the path shown by the arrows. About how far does Michael run?

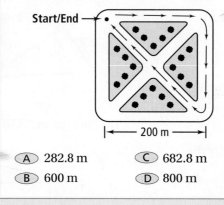

Start/End →

← 200 m →

Ⓐ 282.8 m   Ⓒ 682.8 m
Ⓑ 600 m   Ⓓ 800 m

TIP 1

Michael's path is an isosceles right triangle with each leg about 200 m long. This means that he runs more than 600 m. You can eliminate choices A and B.

TIP 2

The length of the hypotenuse must be less than 400 m, so the path must be less than 800 m. You can eliminate choice D.

Think It Through

In a right isosceles triangle, the length of the hypotenuse is $\sqrt{2}$ times the length of a leg. So the distance Michael runs is $200 + 200 + \sqrt{2}(200) \approx 682.8$. The correct answer is C.

Vocabulary Builder

As you solve test items, you must understand the meanings of mathematical terms. Choose the correct term to complete each sentence.

I. In a right triangle, the (*sine*, *cosine*) of an acute angle is the ratio of the length of the side opposite the angle to the length of the hypotenuse.

II. Polygons that have congruent corresponding angles and corresponding sides that are proportional are (*similar*, *congruent*) polygons.

III. Angles of a polygon that share a side are (*adjacent*, *consecutive*) angles.

IV. A (*proportion*, *ratio*) is a comparison of two numbers using division.

Multiple Choice

Read each question. Then write the letter of the correct answer on your paper.

1. What is the approximate area of the rectangle at the right?

Ⓐ 102 cm²

Ⓑ 75 cm²

Ⓒ 63 cm²

Ⓓ 45 cm²

x 32°

8 cm

2. △ABC has $AB = 7$, $BC = 24$, and $CA = 24$. Which statement is true?

Ⓕ △ABC is an equilateral triangle.

Ⓖ △ABC is an isosceles triangle.

Ⓗ ∠C is the largest angle.

Ⓘ ∠B is the smallest angle.

3. From the top of a 45-ft-tall building, the angle of depression to the edge of a parking lot is 48°. About how many feet is the base of the building from the edge of the parking lot?

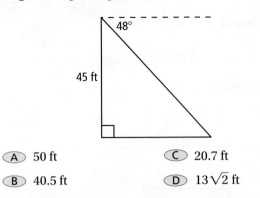

Ⓐ 50 ft

Ⓒ 20.7 ft

Ⓑ 40.5 ft

Ⓓ $13\sqrt{2}$ ft

4. What is the converse of the following statement?

If you study in front of the television, then you do not score well on exams.

Ⓕ If you do not study in front of the television, then you score well on exams.

Ⓖ If you score well on exams, then you do not study in front of the television.

Ⓗ If you do not score well on exams, then you study in front of the television.

Ⓘ If you study in front of the television, then you score well on exams.

5. What are the values of x and of y in the parallelogram below?

Ⓐ $x = 3, y = 5$

Ⓑ $x = 6, y = 10$

Ⓒ $x = 6, y = 14$

Ⓓ $x = 2, y = 2$

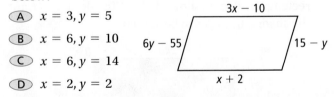

6. In $\triangle HTQ$, if $m\angle H = 72$ and $m\angle Q = 55$, what is the correct order of the lengths of the sides from least to greatest?

Ⓕ TQ, HQ, HT

Ⓖ TQ, HT, HQ

Ⓗ HQ, HT, TQ

Ⓘ HQ, TQ, HT

7. If $m\angle 2 = m\angle 3$, which statement must be true?

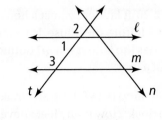

Ⓐ $\ell \parallel m$

Ⓒ $t \perp \ell$

Ⓑ $t \perp m$

Ⓓ $m\angle 1 = m\angle 2$

8. A bike messenger has just been asked to make an additional stop. Now, instead of biking straight from the law office to the court, she is going to stop at City Hall in between. Approximately how many additional miles will she bike?

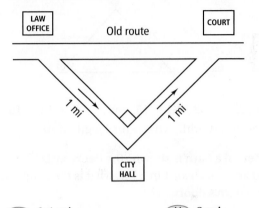

Ⓕ 1.4 mi

Ⓗ 2 mi

Ⓖ 0.6 mi

Ⓘ 3.4 mi

9. In $\triangle XYZ$, $XY = 12$, $YZ = 10$, and $ZX = 8$. To the nearest tenth of a degree, what is the $m\angle Z$?

Ⓐ 12.8

Ⓒ 41.4

Ⓑ 55.8

Ⓓ 82.8

10. What is the value of y?

Ⓕ 16

Ⓖ $8\sqrt{2}$

Ⓗ 8

Ⓘ $8\sqrt{3}$

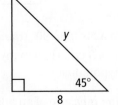

11. A roofer leans a 20-ft ladder against a house. The base of the ladder is 7 ft from the house. How high, in feet, on the house does the ladder reach? Round to the nearest tenth of a foot.

12. A ship's loading ramp is 15 ft long and makes an angle of 18° with the dock. How many feet above the dock, to the nearest tenth of a foot, is the ship's deck?

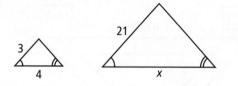

13. What is the value of *x* in the figure below?

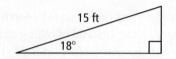

14. In $\triangle ABC$, $AC = 32$, $BC = 15$, and $m\angle C = 124°$. To the nearest hundredth, what is the length of $\overline{AB}$?

15. A front view of a barn is shown. The doorway is a square. Using a scale of 1 in. : 7 ft, what is the height, in feet, of the barn's doorway?

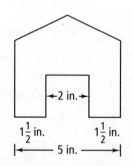

16. In a 30°-60°-90° triangle, the longer leg measures $56\sqrt{3}$ cm. How many centimeters long is the hypotenuse?

17. The measure of an angle is 12 more than 5 times its complement. What is the measure of the angle?

Short Response

18. In the trapezoid at the right, $BE = 2x - 8$, $DE = x - 4$, and $AC = x + 2$.

    a. Write and solve an equation for *x*.

    b. Find the length of each diagonal.

19. Use the diagram below.

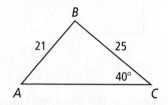

    a. Explain whether you would use Law of Sines or Law of Cosines to find $m\angle A$.

    b. To the nearest tenth, what is $m\angle A$?

Extended Response

20. A clothing store window designer is preparing a new window display. Using the lower left-hand corner of the window as the origin, she marks points at (1, 7), (4, 3), (9, 3), (12, 7), and (6.5, 11). The designer uses tape to connect the points to form a polygon. Is the polygon an equilateral pentagon? Justify your answer.

21. A youth organization is designing a 25 in.-by-40 in. rectangular flag, as shown below. The designers want the shaded triangles to be similar.

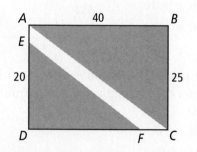

    a. What should *DF* be in order to make $\triangle EDF \sim \triangle CBA$? Explain your reasoning.

    b. In square inches, how much dark fabric do the designers need? How much light fabric?

    c. Are the hypotenuses of the two triangles parallel? How do you know? (*Hint:* Use corresponding angles.)

Get Ready!

CHAPTER 9

Lesson 4-1 ◆ **Congruent Figures**

The triangles in each exercise are congruent. For each pair, complete the
congruence statement △*ABC* ≅ ___?___ .

1.

2.

3.

4.

Lesson 6-1 ◆ **Regular Polygons**

Determine the measure of an angle of the given regular polygon. Draw a figure
to help explain your reasoning.

5. pentagon **6.** octagon **7.** decagon **8.** 18-gon

Lessons 6-2,
6-4, 6-5,
and 6-6 ◆ **Quadrilaterals**

Determine whether a diagonal of the given quadrilateral *always, sometimes,* or
never produces congruent triangles.

9. rectangle **10.** isosceles trapezoid **11.** kite **12.** parallelogram

Lesson 7-2 ◆ **Scale Drawing**

The scale of a blueprint is 1 in. = 20 ft.

13. The length of a wall in the blueprint is 2.5 in. What is the length of the actual wall?
Explain your reasoning.

14. What is the blueprint length of an entrance that is 5 ft wide? Explain your reasoning.

Looking Ahead Vocabulary

15. Think about your *reflection* in a mirror. If you raise your right hand, which hand
appears to be raised in your reflection? If you are standing 2 ft from the mirror,
how far away from you does your reflection appear to be?

16. The minute hand of a clock *rotates* as the minutes go by. What part of the minute
hand stays fixed as the hand rotates?

17. The pupils in your eyes *dilate* in the dark. What do you think it means to *dilate* a
geometric figure?

Transformations

 DOMAINS
- Congruence
- Similarity, Right Triangles, and Trigonometry

Have you ever watched a video in slow motion? The screen at the right shows the position of a butterfly over time. In geometry, a change in the position, size, or shape of a figure is called a transformation.

In this chapter, you'll learn about transformations and apply them to the real world.

🔊 Vocabulary

English/Spanish Vocabulary Audio Online:

English	Spanish
congruence transformation, *p. 580*	transformación de congruencia
dilation, *p. 587*	dilatación
image, *p. 545*	imagen
isometry, *p. 570*	isometría
preimage, *p. 545*	preimagen
reflection, *p. 554*	reflexión
rigid motion, *p. 545*	movimiento rígido
rotation, *p. 561*	rotación
similarity transformation, *p. 596*	transformación de semejanza
translation, *p. 547*	traslación

My Math Video

00:04:04

VIDEO ▷

BIG ideas

1 Transformations

Essential Questions How can you change a figure's position without changing its size and shape? How can you change a figure's size without changing its shape?

2 Coordinate Geometry

Essential Question How can you represent a transformation in the coordinate plane?

3 Visualization

Essential Question How do you recognize congruence and similarity in figures?

Chapter Preview

9-1 **Translations**
9-2 **Reflections**
9-3 **Rotations**
9-4 **Compositions of Isometries**
9-5 **Congruence Transformations**
9-6 **Dilations**
9-7 **Similarity Transformations**

Tracing Paper Transformations

© **Content Standard**
G.CO.2 Represent transformations in the plane . . .

In this activity, you will use tracing paper to perform translations, rotations, and reflections.

Activity

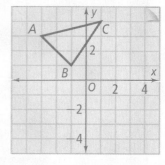

Step 1 Copy △ABC and the x- and y-axis on graph paper. Trace the copy of △ABC on tracing paper.

Step 2 Translate △ABC up 4 units and to the right 2 units by sliding the tracing paper. Draw the new triangle on the graph paper and label it △A′B′C′ so that the original vertices A, B, and C correspond to the vertices A′, B′, and C′ of the new triangle. What are the coordinates of the vertices of △A′B′C′? What is the same about the triangles? What is different?

Step 3 Align your tracing of △ABC with the original and then trace the positive x-axis and the origin.

Step 4 Rotate △ABC 90° about the origin by keeping the origin in place and aligning the traced axis with the positive y-axis. You can use the point of your pencil to hold the origin in place as you rotate the triangle. Draw the image of △ABC after the rotation on the graph paper and label it △A″B″C″. Compare the coordinates of the vertices of △ABC with the coordinates of the vertices of △A″B″C″. Describe the pattern.

Step 5 Flip your tracing of △ABC over and align the origin and the traced positive x-axis to reflect △ABC across the x-axis. Draw and label the reflected triangle △A‴B‴C‴ on the graph paper. What do you notice about the orientations of the triangles?

Exercises

Use tracing paper. Find the images of each triangle for a translation 3 units left and 5 units down, a 90° rotation about the origin, and a reflection across the x-axis.

1.

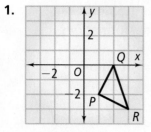

2.

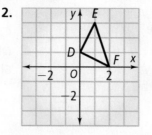

3.

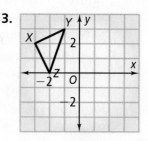

Translations

© Content Standards

G.CO.2 Represent transformations in the plane . . . describe transformations as functions that take points in the plane as inputs and give other points as outputs . . .

Also G.CO.4, G.CO.6

Objectives To identify isometries
To find translation images of figures

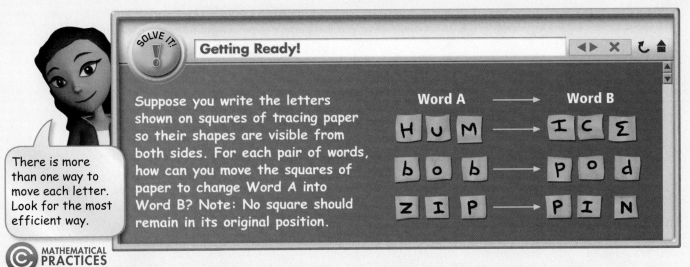

SOLVE IT!

Getting Ready!

Suppose you write the letters shown on squares of tracing paper so their shapes are visible from both sides. For each pair of words, how can you move the squares of paper to change Word A into Word B? Note: No square should remain in its original position.

Word A → Word B

HUM → ICE

bob → Pod

ZIP → PIN

There is more than one way to move each letter. Look for the most efficient way.

© MATHEMATICAL PRACTICES

In the Solve It, you described changes in positions of letters. In this lesson, you will learn some of the mathematical language used to describe changes in positions of geometric figures.

Essential Understanding You can change the position of a geometric figure so that the angle measures and the distance between any two points of a figure stay the same.

A **transformation** of a geometric figure is a function, or *mapping* that results in a change in the position, shape, or size of the figure. When you play dominoes, you often move the dominoes by flipping them, sliding them, or turning them. Each move is a type of transformation. The diagrams below illustrate some basic transformations that you will study.

Lesson Vocabulary
• transformation
• preimage
• image
• rigid motion
• translation
• composition of transformations

The domino flips.

The domino slides.

The domino turns.

In a transformation, the original figure is the **preimage.** The resulting figure is the **image.** Some transformations, like those shown by the dominoes, preserve distance and angle measures. To preserve distance means that the distance between any two points of the image is the same as the distance between the corresponding points of the preimage. To preserve angles means that the angles of the image have the same angle measure as the corresponding angles of the preimage. A transformation that preserves distance and angle measures is called a **rigid motion.**

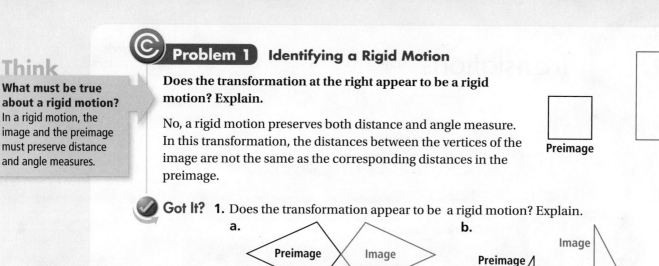

Think

What must be true about a rigid motion?
In a rigid motion, the image and the preimage must preserve distance and angle measures.

Problem 1 Identifying a Rigid Motion

Does the transformation at the right appear to be a rigid motion? Explain.

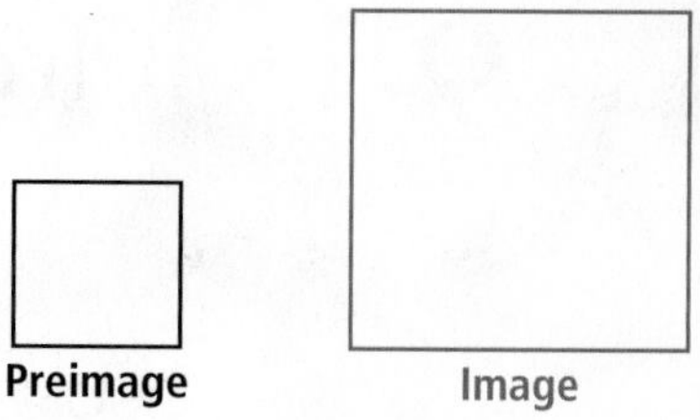

No, a rigid motion preserves both distance and angle measure. In this transformation, the distances between the vertices of the image are not the same as the corresponding distances in the preimage.

Got It? 1. Does the transformation appear to be a rigid motion? Explain.

a.

b.

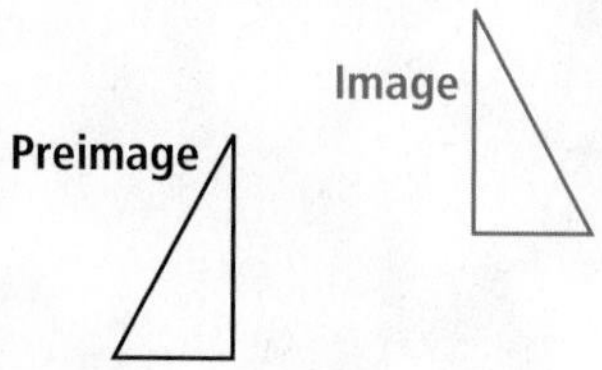

A transformation maps every point of a figure onto its image and may be described with arrow notation (→). Prime notation (′) is sometimes used to identify image points. In the diagram below, K' is the image of K.

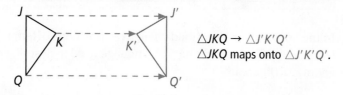

$\triangle JKQ \rightarrow \triangle J'K'Q'$
$\triangle JKQ$ maps onto $\triangle J'K'Q'$.

Notice that you list corresponding points of the preimage and image in the same order, as you do for corresponding points of congruent or similar figures.

Problem 2 Naming Images and Corresponding Parts

Plan

How do you identify corresponding points?
Corresponding points have the same position in the names of the preimage and image. You can use the statement $EFGH \rightarrow E'F'G'H'$.

In the diagram, $EFGH \rightarrow E'F'G'H'$.

A What are the images of $\angle F$ and $\angle H$?

   $\angle F'$ is the image of $\angle F$. $\angle H'$ is the image of $\angle H$.

B What are the pairs of corresponding sides?

   $\overline{EF}$ and $\overline{E'F'}$ $\overline{FG}$ and $\overline{F'G'}$

   $\overline{EH}$ and $\overline{E'H'}$ $\overline{GH}$ and $\overline{G'H'}$

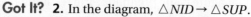

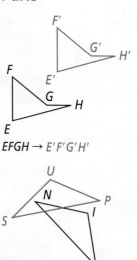

$EFGH \rightarrow E'F'G'H'$

Got It? 2. In the diagram, $\triangle NID \rightarrow \triangle SUP$.
   a. What are the images of $\angle I$ and point D?
   b. What are the pairs of corresponding sides?

Key Concept Translation

A **translation** is a transformation that maps all points of a figure the same distance in the same direction.

You write the translation that maps $\triangle ABC$ onto $\triangle A'B'C'$ as $T(\triangle ABC) = \triangle A'B'C'$. A translation is a rigid motion with the following properties.

If $T(\triangle ABC) = \triangle A'B'C'$, then
- $AA' = BB' = CC'$
- $AB = A'B', BC = B'C', AC = A'C'$
- $m\angle A = m\angle A', m\angle B = m\angle B', m\angle C = m\angle C'$

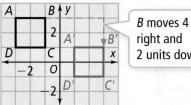

The diagram at the right shows a translation in the coordinate plane. Each point of $ABCD$ is translated 4 units right and 2 units down. So each (x, y) pair in $ABCD$ is mapped to $(x + 4, y - 2)$. You can use the function notation $T_{<4, -2>}(ABCD) = A'B'C'D'$ to describe this translation, where 4 represents the translation of each point of the figure along the x-axis and -2 represents the translation along the y-axis.

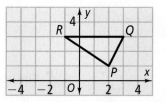

B moves 4 units right and 2 units down.

© **Problem 3** **Finding the Image of a Translation**

What are the vertices of $T_{<-2, -5>}(\triangle PQR)$? Graph the image of $\triangle PQR$.

Identify the coordinates of each vertex. Use the translation rule to find the coordinates of each vertex of the image.

$T_{<-2, -5>}(P) = (2 - 2, 1 - 5)$, or $P'(0, -4)$.

$T_{<-2, -5>}(Q) = (3 - 2, 3 - 5)$, or $Q'(1, -2)$.

$T_{<-2, -5>}(R) = (-1 - 2, 3 - 5)$, or $R'(-3, -2)$.

To graph the image of $\triangle PQR$, first graph P', Q', and R'. Then draw $\overline{P'Q'}$, $\overline{Q'R'}$, and $\overline{R'P'}$.

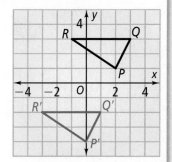

Think

What does the rule tell you about the direction each point moves?
-2 means that each point moves 2 units left. -5 means that each point moves 5 units down.

© ✓ **Got It? 3. a.** What are the vertices of $T_{<1, -4>}(\triangle ABC)$? Copy $\triangle ABC$ and graph its image.

   b. Reasoning Draw $\overline{AA'}$, $\overline{BB'}$, and $\overline{CC'}$. What relationships exist among these three segments? How do you know?

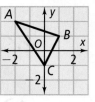

Problem 4 — Writing a Rule to Describe a Translation

What is a rule that describes the translation that maps _PQRS_ onto _P′Q′R′S′_?

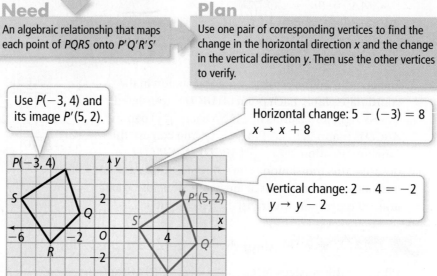

Know
The coordinates of the vertices of both figures

Need
An algebraic relationship that maps each point of _PQRS_ onto _P′Q′R′S′_

Plan
Use one pair of corresponding vertices to find the change in the horizontal direction x and the change in the vertical direction y. Then use the other vertices to verify.

Use $P(-3, 4)$ and its image $P'(5, 2)$.

Horizontal change: $5 - (-3) = 8$
$x \rightarrow x + 8$

Vertical change: $2 - 4 = -2$
$y \rightarrow y - 2$

Think

How do you know which pair of corresponding vertices to use?
A translation moves all points the same distance and the same direction. You can use any pair of corresponding vertices.

The translation maps each (x, y) to $(x + 8, y - 2)$. The translation rule is $T_{<8, -2>}(PQRS)$.

Got It? **4.** The translation image of $\triangle LMN$ is $\triangle L'M'N'$ with $L'(1, -2)$, $M'(3, -4)$, and $N'(6, -2)$. What is a rule that describes the translation?

A **composition of transformations** is a combination of two or more transformations. In a composition, you perform each transformation on the image of the preceding transformation.

In the diagram at the right, the field hockey ball can move from Player 3 to Player 5 by a direct pass. This translation is represented by the blue arrow. The ball can also be passed from Player 3 to Player 9, and then from Player 9 to Player 5. The two red arrows represent this composition of translations.

In general, the composition of any two translations is another translation.

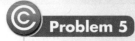 **Problem 5** **Composing Translations**

Chess The diagram at the right shows two moves of the black bishop in a chess game. Where is the bishop in relation to its original position?

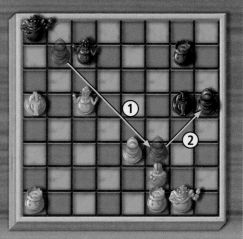

Think

How can you define the bishop's original position?
You can think of the chessboard as a coordinate plane with the bishop's original position at the origin.

Use $(0, 0)$ to represent the bishop's original position. Write translation rules to represent each move.

$T_{<4, -4>}(x, y) = (x + 4, y - 4)$ The bishop moves 4 squares right and 4 squares down.

$T_{<2, 2>}(x, y) = (x + 2, y + 2)$ The bishop moves 2 squares right and 2 squares up.

The bishop's current position is the composition of the two translations.

First, $T_{<4, -4>}(0, 0) = (0 + 4, 0 - 4)$, or $(4, -4)$.

Then, $T_{<2, 2>}(4, -4) = (4 + 2, -4 + 2)$, or $(6, -2)$.

The bishop is 6 squares right and 2 squares down from its original position.

Got It? **5.** The bishop next moves 3 squares left and 3 squares down. Where is the bishop in relation to its original position?

Lesson Check

Do you know HOW?

1. If $\triangle JPT \rightarrow \triangle J'P'T'$, what are the images of P and $\overline{TJ}$?

2. Copy the graph at the right. Graph $T_{<-3, -4>}(NILE)$.

3. Point $H(x, y)$ moves 12 units left and 4 units up. What is a rule that describes this translation?

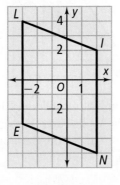

Do you UNDERSTAND? © MATHEMATICAL PRACTICES

© **4. Vocabulary** What is true about a transformation that is not a rigid motion? Include a sketch of an example.

© **5. Error Analysis** Your friend says the transformation $\triangle ABC \rightarrow \triangle PQR$ is a translation. Explain and correct her error.

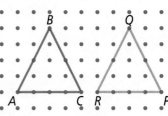

© **6. Reasoning** Write the translation $T_{<1, -3>}(x, y)$ as a composition of a horizontal translation and a vertical translation.

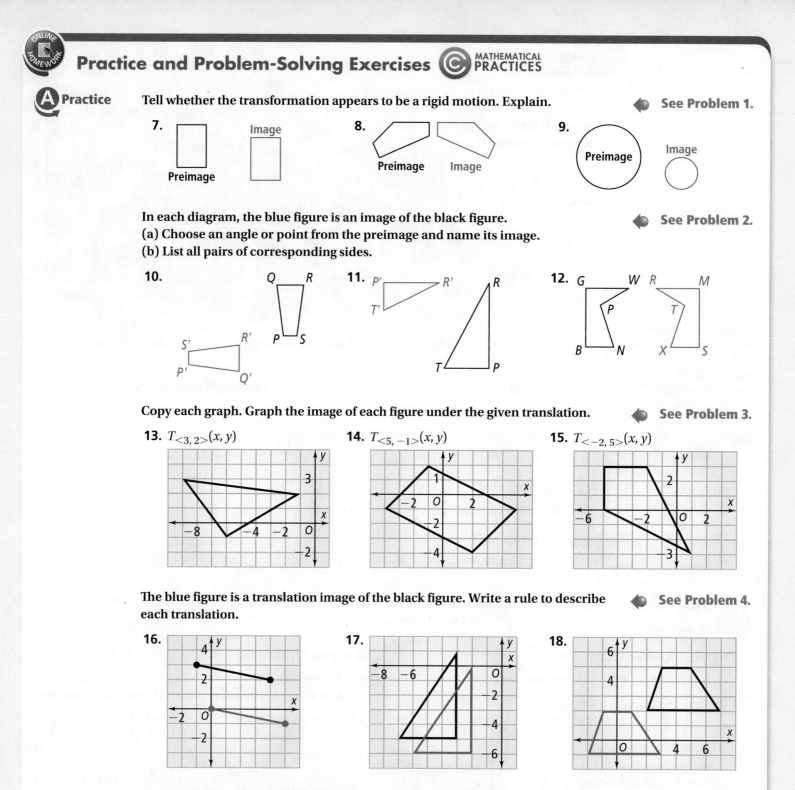

Practice and Problem-Solving Exercises ⓒ MATHEMATICAL PRACTICES

A Practice Tell whether the transformation appears to be a rigid motion. Explain.   ◀ See Problem 1.

7.

Preimage   Image

8.

Preimage   Image

9.

Preimage   Image

In each diagram, the blue figure is an image of the black figure.   ◀ See Problem 2.
(a) Choose an angle or point from the preimage and name its image.
(b) List all pairs of corresponding sides.

10.

Q   R
P   S
S'   R'
P'   Q'

11.

P'   R'
T'
R
T   P

12.

G   W   R   M
P   T
B   N   X   S

Copy each graph. Graph the image of each figure under the given translation.   ◀ See Problem 3.

13. $T_{<3, 2>}(x, y)$

14. $T_{<5, -1>}(x, y)$

15. $T_{<-2, 5>}(x, y)$

The blue figure is a translation image of the black figure. Write a rule to describe   ◀ See Problem 4.
each translation.

16.

17.

18.

19. Travel You are visiting San Francisco. From your hotel near Union Square, you   ◀ See Problem 5.
walk 4 blocks east and 4 blocks north to the Wells Fargo History Museum. Then
you walk 5 blocks west and 3 blocks north to the Cable Car Barn Museum. Where
is the Cable Car Barn Museum in relation to your hotel?

20. Travel Your friend and her parents are visiting colleges. They leave their home in Enid, Oklahoma, and drive to Tulsa, which is 107 mi east and 18 mi south of Enid. From Tulsa, they go to Norman, 83 mi west and 63 mi south of Tulsa. Where is Norman in relation to Enid?

B Apply

21. In the diagram at the right, the orange figure is a translation image of the red figure. Write a rule that describes the translation.

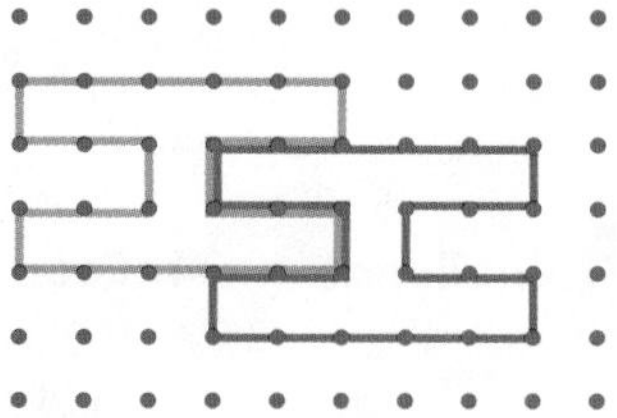

22. Think About a Plan $\triangle MUG$ has coordinates $M(2, -4)$, $U(6, 6)$, and $G(7, 2)$. A translation maps point M to $M'(-3, 6)$. What are the coordinates of U' and G' for this translation?
- How can you use a graph to help you visualize the problem?
- How can you find a rule that describes the translation?

23. Coordinate Geometry $PLAT$ has vertices $P(-2, 0)$, $L(-1, 1)$, $A(0, 1)$, and $T(-1, 0)$. The translation $T_{<2, -3>}(PLAT) = P'L'A'T'$. Show that $\overline{PP'}$, $\overline{LL'}$, $\overline{AA'}$, and $\overline{TT'}$ are all parallel.

Geometry in 3 Dimensions
Follow the sample at the right. Use each figure, graph paper, and the given translation to draw a three-dimensional figure.

SAMPLE Use the rectangle and the translation $T_{<3, 1>}(x, y)$ to draw a box.

Step 1   Step 2

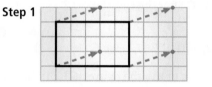

24. $T_{<2, -1>}(x, y)$

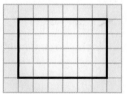

25. $T_{<-2, 2>}(x, y)$

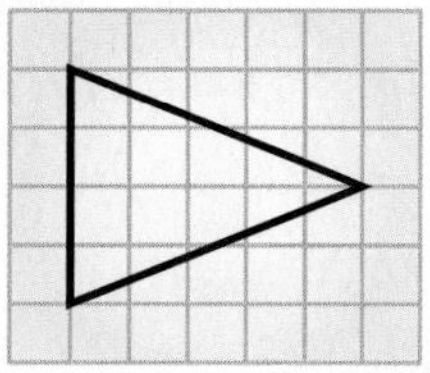

26. $T_{<-3, -5>}(x, y)$

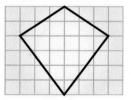

27. Open-Ended You are a graphic designer for a company that manufactures wrapping paper. Make a design for wrapping paper that involves translations.

28. Reasoning If $T_{<5, 7>}(\triangle MNO) = \triangle M'N'O'$, what translation rule maps $\triangle M'N'O'$ onto $\triangle MNO$?

29. Landscaping The diagram at the right shows the site plan for a backyard storage shed. Local law, however, requires the shed to sit at least 15 ft from property lines. Describe how to move the shed to comply with the law.

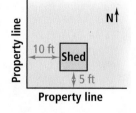

STEM **30. Computer Animation** You write a computer animation program to help young children learn the alphabet. The program draws a letter, erases the letter, and makes it reappear in a new location two times. The program uses the following composition of translations to move the letter.

$$T_{<5, 7>}(x, y) \text{ followed by } T_{<-9, -2>}(x, y)$$

Suppose the program makes the letter W by connecting the points $(1, 2)$, $(2, 0)$, $(3, 2)$, $(4, 0)$ and $(5, 2)$. What points does the program connect to make the last W?

31. Use the graph at the right. Write three different translation rules for which the image of $\triangle JKL$ has a vertex at the origin.

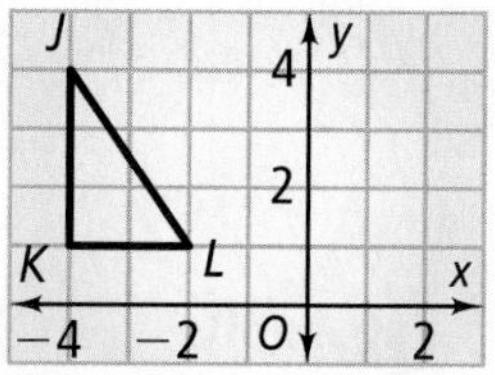

Find a translation that has the same effect as each composition of translations.

32. $T_{<2,\,5>}(x, y)$ followed by $T_{<-4,\,9>}(x, y)$

33. $T_{<12,\,0.5>}(x, y)$ followed by $T_{<1,\,-3>}(x, y)$

© Challenge **34. Coordinate Geometry** $\triangle ABC$ has vertices $A(-2, 5)$, $B(-4, -1)$, and $C(2, -3)$. If $T_{<4,\,2>}(\triangle ABC) = \triangle A'B'C'$, show that the images of the midpoints of the sides of $\triangle ABC$ are the midpoints of the sides of $\triangle A'B'C'$.

© 35. Writing Explain how to use translations to draw a parallelogram.

Standardized Test Prep

SAT/ACT

36. $\triangle ABC$ has vertices $A(-5, 2)$, $B(0, -4)$, and $C(3, 3)$. What are the vertices of the image of $\triangle ABC$ after the translation $T_{<7,\,-5>}(\triangle ABC)$?

Ⓐ $A'(2, -3)$, $B'(7, -9)$, $C'(10, -2)$ Ⓒ $A'(-12, 7)$, $B'(-7, 1)$, $C'(-4, 8)$

Ⓑ $A'(-12, -3)$, $B'(-7, -9)$, $C'(-4, -2)$ Ⓓ $A'(2, -3)$, $B'(10, -2)$, $C'(7, -9)$

37. What is the value of x in the figure at the right?

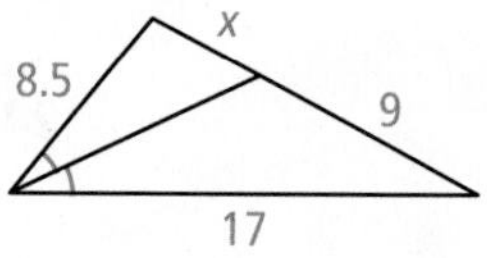

Ⓕ 4.5   Ⓗ 18

Ⓖ 16   Ⓘ 18.5

38. In $\triangle PQR$, $PQ = 4.5$, $QR = 4.4$, and $RP = 4.6$. Which statement is true?

Ⓐ $m\angle P + m\angle Q < m\angle R$ Ⓒ $\angle R$ is the largest angle.

Ⓑ $\angle Q$ is the largest angle.   Ⓓ $m\angle R < m\angle P$

Short Response

39. $\square ABCD$ has vertices $A(0, -3)$, $B(-4, -2)$, and $D(-1, 1)$.
a. What are the coordinates of C?   **b.** Is $\square ABCD$ a rhombus? Explain.

Mixed Review

40. Navigation An airplane landed at a point 100 km east and 420 km south from where it took off. If the airplane flew in a straight line from where it took off to where it lands, how far did it fly?

◀ **See Lesson 8-1.**

41. Given: $\overline{BC} \cong \overline{EF}$, $\overline{BC} \parallel \overline{EF}$, $\overline{AD} \cong \overline{DC} \cong \overline{CF}$

Prove: $\overline{AB} \cong \overline{DE}$

◀ **See Lesson 4-7.**

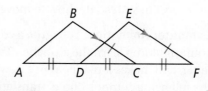

Get Ready! **To prepare for Lesson 9-2, do Exercises 42–44.**

Write an equation for the line through A perpendicular to the given line.

◀ **See Lesson 3-8.**

42. $A(1, -2)$; $x = -2$ **43.** $A(-1, -1)$; $y = 1$ **44.** $A(-1, 2)$; $y = x$

Paper Folding and Reflections

© **Content Standard**
G.CO.5 Given a geometric figure and a rotation, reflection, or translation, draw the transformed figure . . .

In Activity 1, you will see how a figure and its *reflection* image are related. In Activity 2, you will use these relationships to construct a reflection image.

Activity 1

Step 1 Use a piece of tracing paper and a straightedge. Using less than half the page, draw a large, scalene triangle. Label its vertices *A, B,* and *C*.

Step 2 Fold the paper so that your triangle is covered. Trace △*ABC* using a straightedge.

Step 3 Unfold the paper. Label the traced points corresponding to *A, B,* and *C* as *A′, B′,* and *C′*, respectively. △*A′B′C′* is a reflection image of △*ABC*. The fold is the reflection line.

1. Use a ruler to draw $\overline{AA'}$. Measure the perpendicular distances from *A* to the fold and from *A′* to the fold. What do you notice?

2. Measure the angles formed by the fold and $\overline{AA'}$. What are the angle measures?

3. Repeat Exercises 1 and 2 for *B* and *B′* and for *C* and *C′*. Then, make a conjecture: How is the reflection line related to the segment joining a point and its image?

Activity 2

Step 1 On regular paper, draw a simple shape or design made of segments. Use less than half the page. Draw a reflection line near your figure.

Step 2 Use a compass and straightedge to construct a perpendicular to the reflection line through one point of your drawing.

4. Explain how you can use a compass and the perpendicular you drew to find the reflection image of the point you chose.

5. Connect the reflection images for several points of your shape and complete the image. Check the accuracy of the reflection image by folding the paper along the reflection line and holding it up to a light source.

9-2 Reflections

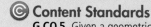
Content Standards
G.CO.5 Given a geometric figure and a rotation, reflection, or translation, draw the transformed figure . . . . Specify a sequence of transformations that will carry a given figure onto another.
Also G.CO.2, G.CO.4, G.CO.6

Objective To find reflection images of figures

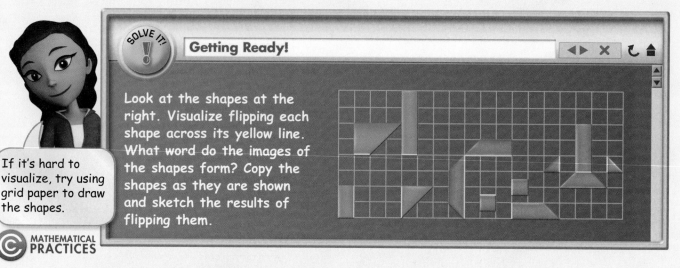

SOLVE IT!

Getting Ready!

Look at the shapes at the right. Visualize flipping each shape across its yellow line. What word do the images of the shapes form? Copy the shapes as they are shown and sketch the results of flipping them.

If it's hard to visualize, try using grid paper to draw the shapes.

MATHEMATICAL PRACTICES

In the Solve It, you reflected shapes across lines. Notice that when you reflect a figure, the shapes have *opposite orientations*. Two figures have opposite orientations if the corresponding vertices of the preimage and image read in opposite directions.

Lesson Vocabulary
• reflection
• line of reflection

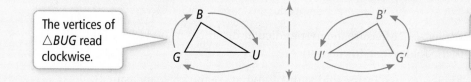

The vertices of △*BUG* read clockwise.

The vertices of △*B'U'G'* read counterclockwise.

Essential Understanding When you reflect a figure across a line, each point of the figure maps to another point the same distance from the line but on the other side. The orientation of the figure reverses.

take note

Key Concept Reflection Across a Line

A **reflection** across a line *m*, called the **line of reflection,** is a transformation with the following properties:
• If a point *A* is on line *m*, then the image of *A* is itself (that is, *A'* = *A*).
• If a point *B* is not on line *m*, then *m* is the perpendicular bisector of $\overline{BB'}$.

You write the reflecion across *m* that takes *P* to *P'* as $R_m(P) = P'$.

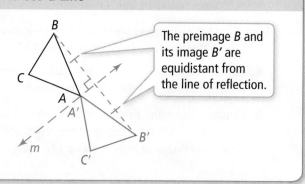

The preimage *B* and its image *B'* are equidistant from the line of reflection.

Dynamic Activity
Reflections

You can use the equation of a line of reflection in the function notation. For example, $R_{y=x}$ describes the reflection across the line $y = x$.

Problem 1 Reflecting a Point Across a Line

Multiple Choice Point P has coordinates $(3, 4)$. What are the coordinates of $R_{y=1}(P)$?

A $(3, -4)$ B $(0, 4)$ C $(3, -2)$ D $(-3, -2)$

Graph point P and the line of reflection $y = 1$. P and its reflection image across the line must be equidistant from the line of reflection.

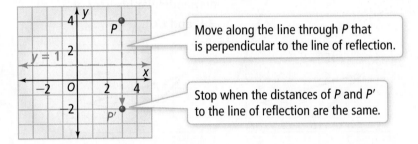

Move along the line through P that is perpendicular to the line of reflection.

Stop when the distances of P and P' to the line of reflection are the same.

Think

How does a graph help you visualize the problem?
A graph shows that $y = 1$ is a horizontal line, so the line through P that is perpendicular to the line of reflection is a vertical line.

P is 3 units above the line $y = 1$, so P' must be 3 units below the line $y = 1$. The line $y = 1$ is the perpendicular bisector of $\overline{PP'}$ if P' is $(3, -2)$. The correct answer is C.

Got It? **1.** $R_{x=1}(P) = P'$. What are the coordinates of P'?

You can also use the notation R_m to describe reflections of figures. The diagram below shows $R_m(\triangle ABC)$, and function notation is used to describe some of the properties of reflections.

take note

Property   Properties of Reflections

- Reflections preserve distance.
  If $R_m(A) = A'$, and $R_m(B) = B'$, then $AB = A'B'$.
- Reflections preserve angle measure.
  If $R_m(\angle ABC) = \angle A'B'C'$, then $m\angle ABC = m\angle A'B'C'$.
- Reflections map each point of the preimage to one and only one corresponding point of its image.
  $R_m(A) = A'$ if and only if $R_m(A') = A$.

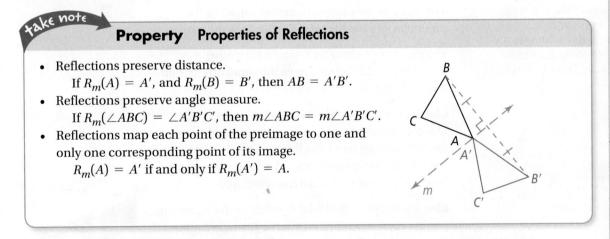

Observe that the above properties mean that reflections are rigid motions, which you learned about in Lesson 9-1.

Problem 2 **Graphing a Reflection Image**

Coordinate Geometry Graph points $A(-3, 4)$, $B(0, 1)$, and $C(4, 2)$. Graph and label $R_{y\text{-axis}}(\triangle ABC)$.

Step 1
Graph $\triangle ABC$. Show the y-axis as the dashed line of reflection.

Step 2
Find A', B', and C'. B' is in the same position as B because B is on the line of reflection. Locate A' and C' so that the y-axis is the perpendicular bisector of $\overline{AA'}$ and $\overline{CC'}$.

Step 3
Draw $\triangle A'B'C'$.

Think

$\triangle ABC$ intersects the line of reflection. How will the image relate to the line of reflection?
The image will also intersect the line of reflection.

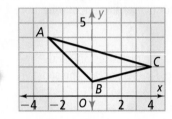 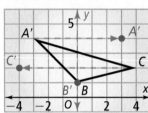 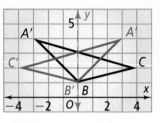

✅ **Got It?** **2.** Graph $\triangle ABC$ from Problem 2. Graph and label $R_{x\text{-axis}}(\triangle ABC)$.

Problem 3 **Writing a Reflection Rule**

Each triangle in the diagram is a reflection of another triangle across one of the given lines. How can you describe Triangle 2 by using a reflection rule?

Triangle 2 is the image of a reflection, so find the preimage and the line of reflection to write a rule.

The preimage cannot be Triangle 3 because Triangle 2 and Triangle 3 have the same orientation and reflections reverse orientation.

Check Triangles 1 and 4 by drawing line segments that connect the corresponding vertices of Triangle 2. Because neither line k nor line m is the perpendicular bisector of the segment drawn from Triangle 1 to Triangle 2, Triangle 1 is not the preimage.

Line k is the perpendicular bisector of the segments joining corresponding vertices of Triangle 2 and Triangle 4. So, Triangle 2 $= R_k(\text{Triangle 4})$.

Plan

If Triangle 2 is the image of a reflection, what do you know about the preimage?
The preimage has opposite orientation, and lies on the opposite side of the line of reflection.

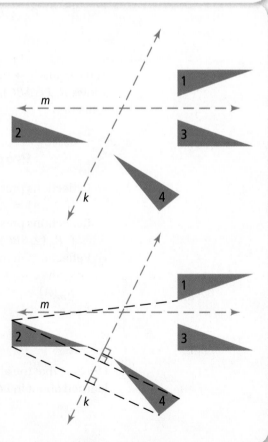

✅ **Got It?** **3.** How can you use a reflection rule to describe Triangle 1? Explain.

You can use the properties of reflections to prove statements about figures.

Problem 4 Using Properties of Reflections

Plan

What do you have to know about $\triangle GHJ$ to show that it is an isosceles triangle? Isosceles triangles have at least two congruent sides.

In the diagram, $R_t(G) = G$, $R_t(H) = J$, and $R_t(D) = D$. Use the properties of reflections to describe how you know that $\triangle GHJ$ is an isosceles triangle.

Since $R_t(G) = G$, $R_t(H) = J$, and reflections preserve distance, $R_t(\overline{GH}) = \overline{GJ}$. So, $GH = GJ$ and, by definition, $\triangle GHJ$ is an isosceles triangle.

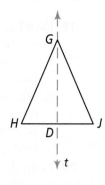

Got It? 4. Can you use properties of reflections to prove that $\triangle GHJ$ is equilateral? Explain.

Lesson Check

Do you know HOW?

Use the graph of $\triangle FGH$.

1. What are the coordinates of $R_{y\text{-axis}}(H)$?
2. What are the coordinates of $R_{x=3}(G)$?
3. Graph and label $R_{y=4}(\triangle FGH)$.

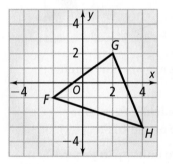

Do you UNDERSTAND? MATHEMATICAL PRACTICES

4. **Vocabulary** What is the relationship between a line of reflection and a segment joining corresponding points of the preimage and image?
5. **Error Analysis** A classmate sketched $R_s(A) = A'$ as shown in the diagram.

   a. Explain your classmate's error.
   b. Copy point A and line s and show the correct location of A'.
6. What are the coordinates of a point $P(x, y)$ reflected across the y-axis? Across the x-axis? Use reflection notation to write your answer.

Practice and Problem-Solving Exercises MATHEMATICAL PRACTICES

A Practice

Find the coordinates of each image.

See Problem 1.

7. $R_{x=1}(Q)$
8. $R_{y=-1}(P)$
9. $R_{y\text{-axis}}(S)$
10. $R_{y=0.5}(T)$
11. $R_{x=-3}(U)$
12. $R_{x\text{-axis}}(V)$

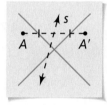

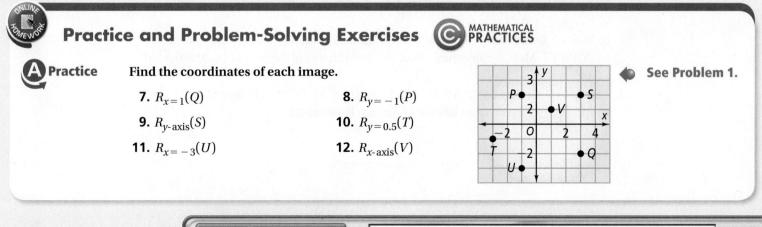

Coordinate Geometry Given points $J(1, 4)$, $A(3, 5)$, and $G(2, 1)$, graph $\triangle JAG$ and its reflection image as indicated.

 See Problem 2.

13. $R_{x\text{-axis}}$ **14.** $R_{y\text{-axis}}$ **15.** $R_{y=2}$ **16.** $R_{y=5}$ **17.** $R_{x=-1}$ **18.** $R_{x=2}$

19. Each figure in the diagram at the right is a reflection of another figure across one of the reflection lines.

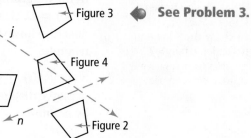

See Problem 3.

 a. Write a reflection rule to describe Figure 3. Justify your answer.
 b. Write a reflection rule to describe Figure 2. Justify your answer.
 c. Write a reflection rule to describe Figure 4. Justify your answer.

20. In the diagram at the right, *LMNP* is a rectangle with $LM = 2MN$.

See Problem 4.

 a. Copy the diagram. Then sketch $R_{\overline{LM}}(LMNP)$.
 b. What figure results from the reflection? Use properties of reflections to justify your solution.

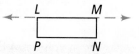

Ⓑ **Apply**   Copy each figure and line ℓ. Draw each figure's reflection image across line ℓ.

21.

22.

Ⓒ **23. Coordinate Geometry** The following steps explain how to reflect point *A* across the line $y = x$.

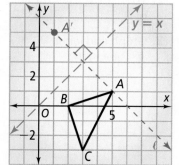

 Step 1 Draw line ℓ through $A(5, 1)$ perpendicular to the line $y = x$. The slope of $y = x$ is 1, so the slope of line ℓ is $1 \cdot (-1)$, or -1.

 Step 2 From *A*, move two units left and two units up to $y = x$. Then move two more units left and two more units up to find the location of A' on line ℓ. The coordinates of A' are $(1, 5)$.

 a. Copy the diagram. Then draw the lines through *B* and *C* that are perpendicular to the line $y = x$. What is the slope of each line?
 b. $R_{y=x}(B) = B'$ and $R_{y=x}(C) = C'$. What are the coordinates of B' and C'?
 c. Graph $\triangle A'B'C'$.
 d. Make a Conjecture Compare the coordinates of the vertices of $\triangle ABC$ and $\triangle A'B'C'$. Make a conjecture about the coordinates of the point $P(a, b)$ reflected across the line $y = x$.

24. Coordinate Geometry $\triangle ABC$ has vertices $A(-3, 5)$, $B(-2, -1)$, and $C(0, 3)$. Graph $R_{y=-x}(\triangle ABC)$ and label it. (*Hint:* See Exercise 23.)

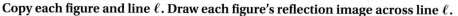
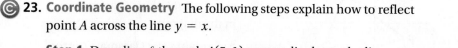

25. Recreation When you play pool, you can use the fact that the ball bounces off the side of the pool table at the same angle at which it hits the side. Suppose you want to put the ball at point B into the pocket at point P by bouncing it off side $\overline{RS}$. Off what point on $\overline{RS}$ should the ball bounce? Draw a diagram and explain your reasoning.

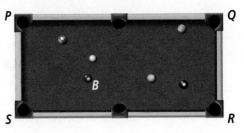

ⓒ **26. Think About a Plan** The coordinates of the vertices of $\triangle FGH$ are $F(2, -1)$, $G(-2, -2)$, and $H(-4, 3)$. Graph $\triangle FGH$ and $R_{y=x-3}(\triangle FGH)$.
- What is the relationship between the line $y = x - 3$ and $\overline{FF'}$, $\overline{GG'}$, and $\overline{HH'}$?
- How can you use slope to find the image of each vertex?

27. In the diagram $R(ABCDE) = A'B'C'D'E'$.
- **a.** What are the midpoints of $\overline{AA'}$ and $\overline{DD'}$?
- **b.** What is the equation of the line of reflection?
- **c.** Write a rule that describes this reflection.

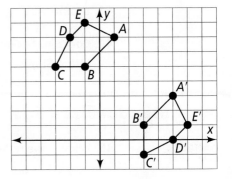

Copy each pair of figures. Then draw the line of reflection you can use to map one figure onto the other.

28.

29.

30. History The work of artist and scientist Leonardo da Vinci (1452–1519) has an unusual characteristic. His handwriting is a mirror image of normal handwriting.
- **a.** Write the mirror image of the sentence, "Leonardo da Vinci was left-handed." Use a mirror to check how well you did.
- **b.** Explain why the fact about da Vinci in part (a) might have made mirror writing seem natural to him.

ⓒ **31. Open-Ended** Give three examples from everyday life of objects or situations that show or use reflections.

Find the image of $O(0, 0)$ after two reflections, first across line ℓ_1 and then across line ℓ_2.

32. $\ell_1: y = 3$, ℓ_2: x-axis   **33.** $\ell_1: x = -2$, ℓ_2: y-axis   **34.** ℓ_1: x-axis, ℓ_2: y-axis

ⓒ **35. Reasoning** When you reflect a figure across a line, does every point on the preimage move the same distance? Explain.

36. Security Recall that when a ray of light hits a mirror, it bounces off the mirror at the same angle at which it hits the mirror. You are installing a security camera. At what point on the mirrored wall should you aim the camera at C in order to view the door at D? Draw a diagram and explain your reasoning.

Mirrored wall

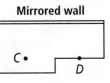

37. Use the diagram at the right. Find the coordinates of each image point.

  a. $R_{y=x}(A) = A'$
  b. $R_{y=-x}(A') = A''$
  c. $R_{y=x}(A'') = A'''$
  d. $R_{y=-x}(A''') = A''''$
  e. How are A and A'''' related?

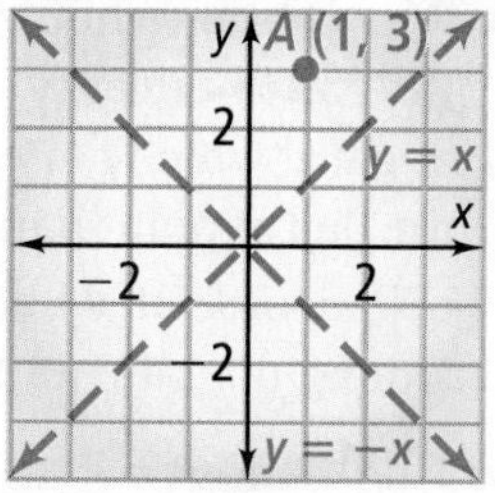

Challenge

Reasoning Can you form the given type of quadrilateral by drawing a triangle and then reflecting one or more times? Explain.

38. parallelogram

39. isosceles trapezoid

40. kite

41. rhombus

42. rectangle

43. square

44. Coordinate Geometry Show that $R_{y=x}(A) = B$ for point $A(a, b)$ and $B(b, a)$. (*Hint:* Show that $y = x$ is the perpendicular bisector of $\overline{AB}$.)

Standardized Test Prep

SAT/ACT

45. What is the reflection image of (a, b) across the line $y = -6$?

  Ⓐ $(a - 6, b)$ Ⓑ $(a, b - 6)$ Ⓒ $(-12 - a, b)$ Ⓓ $(a, -12 - b)$

46. The diagonals of a quadrilateral are perpendicular and bisect each other. What is the most precise name for the quadrilateral?

  Ⓕ rectangle   Ⓖ parallelogram   Ⓗ rhombus   Ⓘ kite

47. $\overrightarrow{AD}$ bisects $\angle A$ of $\triangle CAB$, with point D on $\overline{CB}$. Which of the following is true?

  Ⓐ $\dfrac{CD}{DB} = \dfrac{CA}{BA}$ Ⓑ $\dfrac{CD}{DB} = \dfrac{BA}{CA}$ Ⓒ $\dfrac{CB}{CD} = \dfrac{CA}{BA}$ Ⓓ $\dfrac{CA}{CB} = \dfrac{AB}{DB}$

Extended Response

48. Write an indirect proof of the following statement: The hypotenuse of a right triangle is the longest side of the right triangle.

Mixed Review

For the given points, $T(\overline{AB}) = \overline{A'B'}$. Write a rule to describe each translation.   ◀ **See Lesson 9-1.**

49. $A(-1, 5), B(2, 0), A'(3, 3), B'(6, -2)$

50. $A(-9, -4), B(-7, 1), A'(-4, -3), B'(-2, 2)$

51. Maps A map of Alberta, Canada, uses the scale 1 cm = 25 km. On the map, the distance from Calgary to Edmonton is 11.1 cm. How far apart are the two cities?   ◀ **See Lesson 7-2.**

Get Ready!   **To prepare for Lesson 9-3, do Exercises 52–57.**

Use a protractor to draw an angle with the given measure.   ◀ **See p. 884.**

52. 120°   **53.** 90°   **54.** 72°   **55.** 60°   **56.** 45°   **57.** 36°

Rotations

© **Content Standards**
G.CO.4 Develop definitions of rotations . . . in terms of angles, circles, perpendicular lines, parallel lines, and line segments.
Also G.CO.2, G.CO.6

Objective To draw and identify rotation images of figures

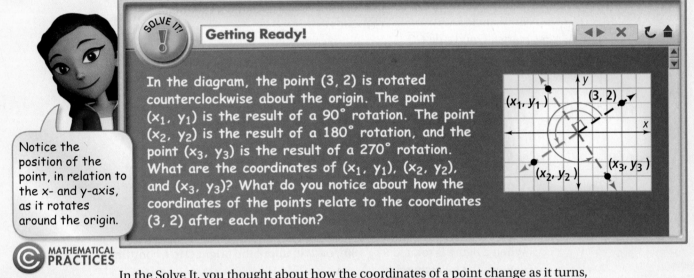

SOLVE IT!

Getting Ready!

In the diagram, the point (3, 2) is rotated counterclockwise about the origin. The point (x_1, y_1) is the result of a 90° rotation. The point (x_2, y_2) is the result of a 180° rotation, and the point (x_3, y_3) is the result of a 270° rotation. What are the coordinates of (x_1, y_1), (x_2, y_2), and (x_3, y_3)? What do you notice about how the coordinates of the points relate to the coordinates (3, 2) after each rotation?

Notice the position of the point, in relation to the x- and y-axis, as it rotates around the origin.

© **MATHEMATICAL PRACTICES**

Dynamic Activity
Rotations, Reflections, and Translations

Lesson Vocabulary
• rotation
• center of rotation
• angle of rotation

In the Solve It, you thought about how the coordinates of a point change as it turns, or *rotates*, about the origin on a coordinate grid. In this lesson, you will learn how to recognize and construct rotations of geometric figures.

Essential Understanding Rotations preserve distance, angle measures, and orientation of figures.

take note

Key Concept   Rotation About a Point

A **rotation** of $x°$ about a point Q, called the **center of rotation,** is a transformation with these two properties:
- The image of Q is itself (that is, $Q' = Q$).
- For any other point V, $QV' = QV$ and $m\angle VQV' = x$.

The number of degrees a figure rotates is the **angle of rotation.**

A rotation about a point is a rigid motion. You write the $x°$ rotation of $\triangle UVW$ about point Q as $r_{(x°, Q)}(\triangle UVW) = \triangle U'V'W'$.

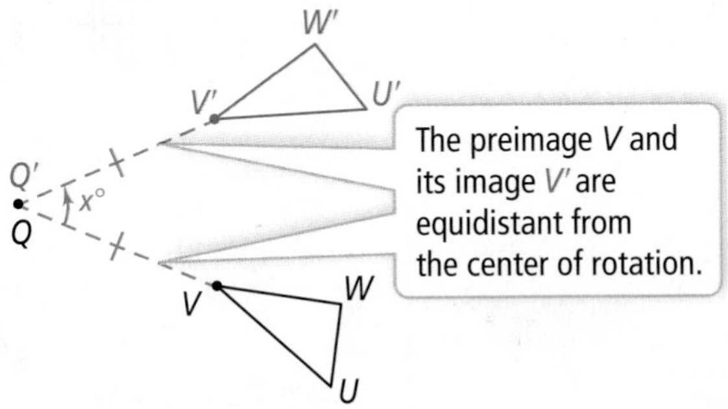

The preimage V and its image V' are equidistant from the center of rotation.

Unless stated otherwise, rotations in this book are counterclockwise.

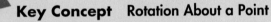

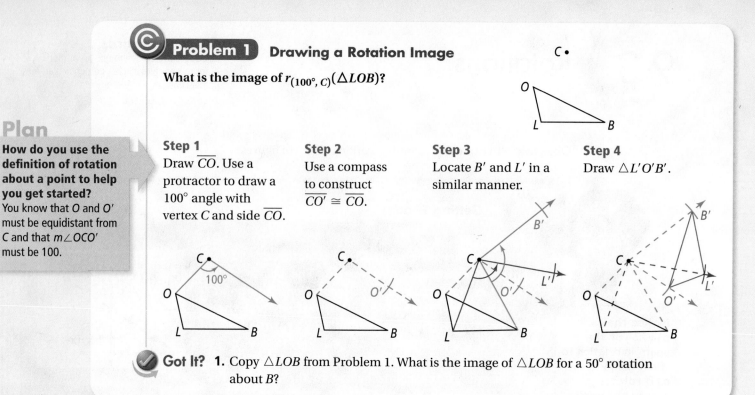

Problem 1   Drawing a Rotation Image

C •

What is the image of $r_{(100°, C)}(\triangle LOB)$?

Plan

How do you use the definition of rotation about a point to help you get started?
You know that O and O' must be equidistant from C and that $m\angle OCO'$ must be 100.

Step 1
Draw $\overline{CO}$. Use a protractor to draw a 100° angle with vertex C and side $\overline{CO}$.

Step 2
Use a compass to construct $\overline{CO'} \cong \overline{CO}$.

Step 3
Locate B' and L' in a similar manner.

Step 4
Draw $\triangle L'O'B'$.

Got It?   **1.** Copy $\triangle LOB$ from Problem 1. What is the image of $\triangle LOB$ for a 50° rotation about B?

When a figure is rotated 90°, 180°, or 270° about the origin O in a coordinate plane, you can use the following rules.

take note

Key Concept   Rotation in the Coordinate Plane

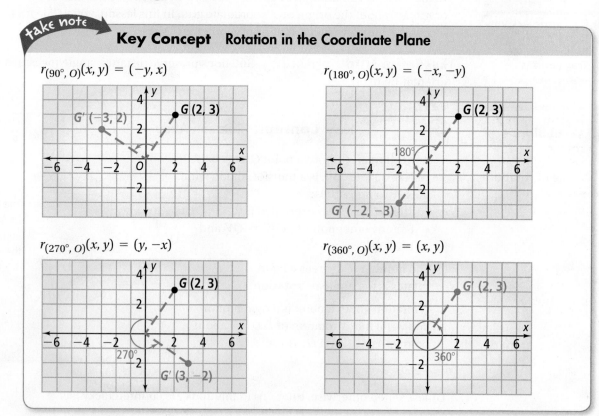

$r_{(90°, O)}(x, y) = (-y, x)$

$r_{(180°, O)}(x, y) = (-x, -y)$

$r_{(270°, O)}(x, y) = (y, -x)$

$r_{(360°, O)}(x, y) = (x, y)$

Problem 2 Drawing Rotations in a Coordinate Plane

Plan

How do you know where to draw the vertices on the coordinate plane?
Use the rules for rotating a point and apply them to each vertex of the figure. Then graph the points and connect them to draw the image.

PQRS has vertices $P(1, 1)$, $Q(3, 3)$, $R(4, 1)$, and $S(3, 0)$.
What is the graph of $r_{(90°, O)}(PQRS)$.

First, graph the images of each vertex.

$P' = r_{(90°, O)}(1, 1) = (-1, 1)$

$Q' = r_{(90°, O)}(3, 3) = (-3, 3)$

$R' = r_{(90°, O)}(4, 1) = (-1, 4)$

$S' = r_{(90°, O)}(3, 0) = (0, 3)$

Next, connect the vertices to graph $P'Q'R'S'$.

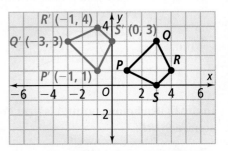

✔ **Got It? 2.** Graph $r_{(270°, O)}(FGHI)$.

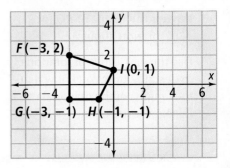

You can use the properties of rotations to solve problems.

 Problem 3 Using Properties of Rotations

Think

What do you know about rotations that can help you show that opposite sides of the parallelogram are equal?
You know that rotations are rigid motions, so if you show that the opposite sides can be mapped to each other, then the side lengths must be equal.

In the diagram, *WXYZ* is a parallelogram, and *T* is the midpoint of the diagonals. How can you use the properties of rotations to show that the lengths of the opposite sides of the parallelogram are equal?

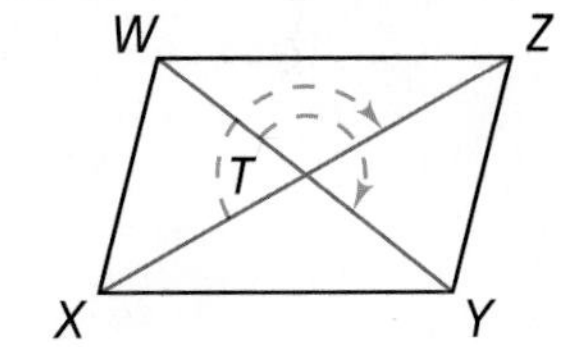

Because *T* is the midpoint of the diagonals, $XT = ZT$ and $WT = YT$. Since *W* and *Y* are equidistant from *T*, and the measure of $\angle WTY = 180$, you know that $r_{(180°, T)}(W) = Y$. Similarly, $r_{(180°, T)}(X) = Z$.

You can rotate every point on $\overline{WX}$ in this same way, so $r_{(180°, T)}(\overline{WX}) = \overline{YZ}$. Likewise, you can map $\overline{WZ}$ to $\overline{YX}$ with $r_{(180°, T)}(\overline{WZ}) = \overline{YX}$.

Because rotations are rigid motions and preserve distance, $WX = YZ$ and $WZ = YX$.

✔ **Got It? 3.** Can you use the properties of rotations to prove that *WXYZ* is a rhombus? Explain.

Lesson Check

Do you know HOW?

1. Copy the figure and point P. Draw $r_{(70°, P)}(\triangle ABC)$.

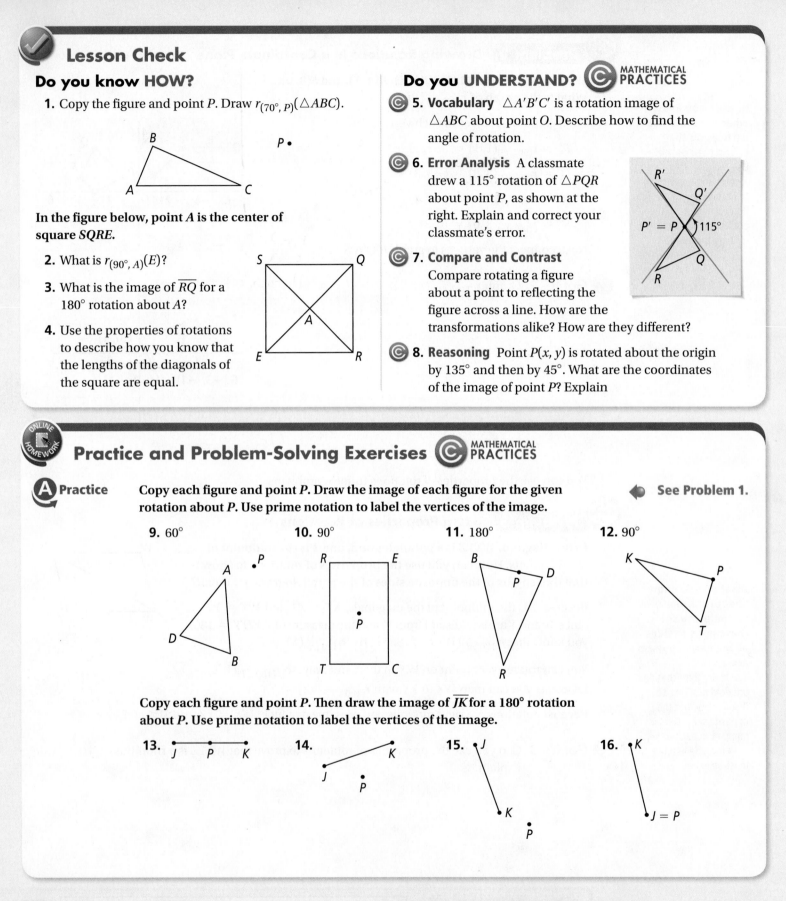

In the figure below, point A is the center of square $SQRE$.

2. What is $r_{(90°, A)}(E)$?

3. What is the image of $\overline{RQ}$ for a 180° rotation about A?

4. Use the properties of rotations to describe how you know that the lengths of the diagonals of the square are equal.

Do you UNDERSTAND? MATHEMATICAL PRACTICES

5. **Vocabulary** $\triangle A'B'C'$ is a rotation image of $\triangle ABC$ about point O. Describe how to find the angle of rotation.

6. **Error Analysis** A classmate drew a 115° rotation of $\triangle PQR$ about point P, as shown at the right. Explain and correct your classmate's error.

7. **Compare and Contrast** Compare rotating a figure about a point to reflecting the figure across a line. How are the transformations alike? How are they different?

8. **Reasoning** Point $P(x, y)$ is rotated about the origin by 135° and then by 45°. What are the coordinates of the image of point P? Explain

Practice and Problem-Solving Exercises MATHEMATICAL PRACTICES

A Practice Copy each figure and point P. Draw the image of each figure for the given rotation about P. Use prime notation to label the vertices of the image.

See Problem 1.

9. 60°   10. 90°   11. 180°   12. 90°

Copy each figure and point P. Then draw the image of $\overline{JK}$ for a 180° rotation about P. Use prime notation to label the vertices of the image.

13.   14.   15.   16.

For Exercises 17–19, use the graph at the right.

See Problem 2.

17. Graph $r_{(90°, O)}(FGHJ)$.

18. Graph $r_{(180°, O)}(FGHJ)$.

19. Graph $r_{(270°, O)}(FGHJ)$.

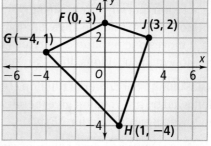

20. The coordinates of $\triangle PRS$ are $P(-3, 2)$, $R(2, 5)$, and $S(0, 0)$. What are the coordinates of the vertices of $r_{(270°, O)}(\triangle PRS)$?

21. $V'W'X'Y'$ has vertices $V'(-3, 2)$, $W'(5, 1)$, $X'(0, 4)$, and $Y'(-2, 0)$. If $r_{(90°, O)}(VWXY) = V'W'X'Y'$, what are the coordinates of $VWXY$?

22. **Ferris Wheel** A Ferris wheel is drawn on a coordinate plane so that the first car is located at the point $(30, 0)$. What are the coordinates of the first car after a rotation of $270°$ about the origin?

For Exercises 23–25, use the diagram at the right. *TQNV* is a rectangle. *M* is the midpoint of the diagonals.

See Problem 3.

23. Use the properties of rotations to show that the measures of both pairs of opposite sides are equal in length.

Ⓒ 24. **Reasoning** Can you use the properties of rotations to show that the measures of the lengths of the diagonals are equal?

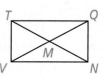

Ⓒ 25. **Reasoning** Can you use properties of rotations to conclude that the diagonals of *TQNV* bisect the angles of *TQNV*? Explain.

Ⓑ Apply

26. In the diagram at the right, $\overline{M'N'}$ is the rotation image of $\overline{MN}$ about point E. Name all pairs of angles and all pairs of segments that have equal measures in the diagram.

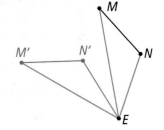

27. **Language Arts** Symbols are used in dictionaries to help users pronounce words correctly. The symbol ə is called a *schwa*. It is used in dictionaries to represent neutral vowel sounds such as *a* in *ago*, *i* in *sanity*, and *u* in *focus*. What transformation maps a ə to a lowercase e?

Find the angle of rotation about *C* that maps the black figure to the blue figure.

28.

29.

30.

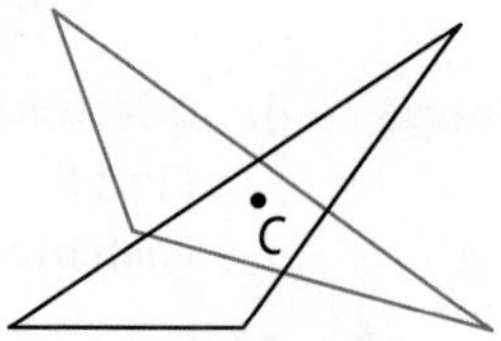

31. **Think About a Plan** The Millenium Wheel, also known as the London Eye, contains 32 observation cars. Determine the angle of rotation that will bring Car 3 to the position of Car 18.
   - How do you find the angle of rotation that a car travels when it moves one position counterclockwise?
   - How many positions does Car 3 move?

Car 3

Car 18

32. **Reasoning** For center of rotation P, does an $x°$ rotation followed by a $y°$ rotation give the same image as a $y°$ rotation followed by an $x°$ rotation? Explain.

33. **Writing** Describe how a series of rotations can have the same effect as a 360° rotation about a point X.

34. **Coordinate Geometry** Graph $A(5, 2)$. Graph B, the image of A for a 90° rotation about the origin O. Graph C, the image of A for a 180° rotation about O. Graph D, the image of A for a 270° rotation about O. What type of quadrilateral is $ABCD$? Explain.

Point O is the center of the regular nonagon shown at the right.

35. Find the angle of rotation that maps F to H.

36. **Open-Ended** Describe a rotation that maps H to C.

37. **Error Analysis** Your friend says that $\overline{AB}$ is the image of $\overline{ED}$ for a 120° rotation about O. What is wrong with your friend's statement?

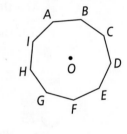

In the figure at the right, the large triangle, the quadrilateral, and the hexagon are regular. Find the image of each point or segment for the given rotation or composition of rotations. (*Hint:* Adjacent green segments form 30° angles.)

38. $r_{(120°, O)}(B)$

39. $r_{(270°, O)}(L)$

40. $r_{(300°, O)}(\overline{IB})$

41. $r_{(60°, O)}(E)$

42. $r_{(180°, O)}(\overline{JK})$

43. $r_{(240°, O)}(G)$

44. $r_{(120°, H)}(F)$

45. $r_{(270°, L)}(M)$

46. $r_{(180°, O)}(I)$

47. $r_{(270°, O)}(M)$

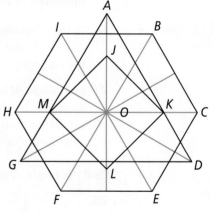

 Challenge

48. **Coordinate Geometry** Draw $\triangle LMN$ with vertices $L(2, -1)$, $M(6, -2)$, and $N(4, 2)$. Find the coordinates of the vertices after a 90° rotation about the origin and about each of the points L, M, and N.

49. **Reasoning** If you are given a figure and a rotation image of the figure, how can you find the center and angle of rotation?

Standardized Test Prep

50. What is the image of $(1, -6)$ for a 90° counterclockwise rotation about the origin?

Ⓐ $(6, 1)$　　　Ⓑ $(-1, 6)$　　　Ⓒ $(-6, -1)$　　　Ⓓ $(-1, -6)$

51. The costume crew for your school musical makes aprons like the one shown. If blue ribbon costs $1.50 per foot, what is the cost of ribbon for six aprons?

Ⓕ $15.75　　　Ⓗ $42.00

Ⓖ $31.50　　　Ⓘ $63.00

52. In $\triangle ABC$, $m\angle A + m\angle B = 84$. Which statement must be true?

Ⓐ $BC > AC$　　　Ⓑ $AC > BC$　　　Ⓒ $AB > BC$　　　Ⓓ $BC > AB$

53. Use the following statement: If two lines are parallel, then the lines do not intersect.
  a. What are the converse, inverse, and contrapositive of the statement?
  b. What is the truth value of each statement you wrote in part (a)? If a statement is false, give a counterexample.

Mixed Review

$\triangle BIG$ has vertices $B(-4, 2)$, $I(0, -3)$, and $G(1, 0)$. Graph $\triangle BIG$ and its reflection image across the given line.

◀ See Lesson 9-2.

54. the y-axis　　　**55.** the x-axis　　　**56.** $x = 4$

Find the value of x. Round answers to the nearest tenth.

◀ See Lessons 8-2 and 8-3.

57.

58.

Get Ready!　To prepare for Lesson 9-4, do Exercises 59–61.　◀ See Lessons 9-2 and 9-3.

59. What are the coordinates of the image of point $A(-2, 3)$ after two 90° rotations about the origin?

60. What are the coordinates of the image of point $T(3, 0)$ after a reflection across the y-axis followed by a 180° rotation about the origin?

61. The image of point H after a 90° rotation about the origin followed by a reflection across the x-axis is $K(3, 2)$. What are the coordinates of H?

© **Content Standard**
G.CO.3 Given a rectangle, parallelogram, trapezoid, or regular polygon, describe the rotations and reflections that carry it onto itself.

You can use what you know about reflections and rotations to identify types of **symmetry**. A figure has symmetry if there is a rigid motion that maps the figure onto itself.

A figure has **line symmetry**, or **reflectional symmetry**, if there is a reflection for which the figure is its own image. The line of reflection is called the **line of symmetry**.

A figure has a **rotational symmetry**, if its image, after a rotation of less than 360°, is exactly the same as the original figure. A figure has **point symmetry** if a 180° rotation about a center of rotation maps the figure onto itself.

Activity 1

1. Use a straightedge to copy the rhombus at the right.
   a. How many lines of reflection, or lines of symmetry, does the rhombus have?
   b. Draw all of the lines of symmetry.

2. Do all parallelograms have reflectional symmetry? Explain your reasoning.

3. The isosceles trapezoid at the right has only 1 pair of parallel sides. How many lines of symmetry does the trapezoid have?

4. Do all isosceles trapezoids have reflectional symmetry? Do all trapezoids have reflectional symmetry? Explain.

Activity 2

5. Use a straightedge to copy the regular hexagon at the right.
   a. How many lines of symmetry does a regular hexagon have?
   b. Draw all of the lines of symmetry.

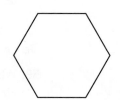

6. What are the center and angle(s) of the rotations that map the regular hexagon onto itself?

7. Do all regular polygons have rotational symmetry? Explain your reasoning.

8. Do all regular polygons have point symmetry? Explain.

Activity 3

Copy and cut out the shapes below. Shade $\frac{1}{2}$ of each square to represent the orange sections. Arrange the shapes to make a design that has both reflectional symmetry and rotational symmetry.

9. Draw the design you made.

10. How many lines of symmetry does your design have? Sketch each line of symmetry.

11. Why are the colors of the tiles important to the symmetry?

12. Does your design have more than one of angle of rotation that maps it onto itself? If so, what are they?

13. Can you change the center of rotation and still map the figure onto itself? Explain.

Exercises

Tell what type(s) of symmetry each figure has. Sketch the figure and the line(s) of symmetry, and give the angle(s) of rotation when appropriate.

14.

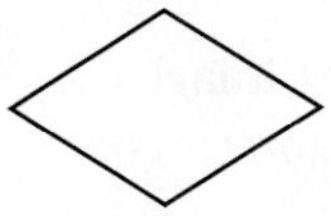

15.

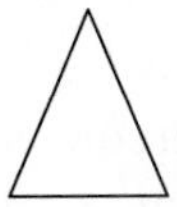

16.

17. Vocabulary If a figure has point symmetry, must it also have rotational symmetry? Explain.

18. Writing A quadrilateral with vertices (1, 5) and (–2, –3) has point symmetry about the origin.
   a. Show that the quadrilateral is a parallelogram.
   b. How can you use point symmetry to find the other vertices?

19. Error Analysis Your friend thinks that the regular pentagon in the diagram has 10 lines of symmetry. Explain and correct your friend's error.

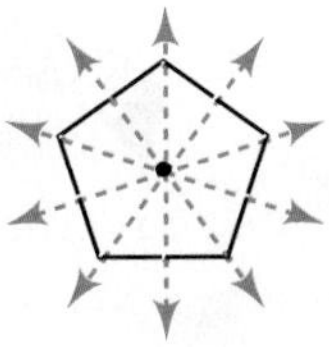

9-4 Compositions of Isometries

Content Standards
G.CO.5 . . . Specify a sequence of transformation that will carry a given figure onto another.
G.CO.6 Use geometric descriptions of rigid motions to transform figures and to predict the effect of a given rigid motion on a given figure . . .

Objectives To find compositions of isometries, including glide reflections
To classify isometries

SOLVE IT!

Getting Ready!

The blue E is a horizontal translation of the red E. How can you use two reflections, one after the other, to move the red E to the position of the blue E? Copy the figure exactly as shown and draw in the two lines of reflection. Explain how you found the lines.

Can you find more than one way? Which way is the most efficient?

MATHEMATICAL PRACTICES

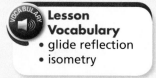

Lesson Vocabulary
• glide reflection
• isometry

In the Solve It, you looked for a way to use two reflections to produce the same image as a given horizontal translation. In this lesson, you will learn that any rigid motion can be expressed as a composition of reflections.

The term *isometry* means same distance. An **isometry** is a transformation that preserves distance, or length. So, translations, reflections, and rotations are isometries.

Essential Understanding You can express all isometries as compositions of reflections.

Expressing isometries as compositions of reflections depends on the following theorem.

take note

Theorem 9-1

The composition of two or more isometries is an isometry.

There are only four kinds of isometries.

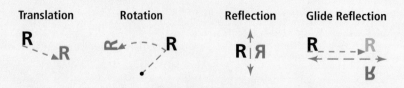

Translation   Rotation   Reflection   Glide Reflection

You will learn about *glide reflections* later in the lesson.

In Lesson 9-1, you learned that a composition of transformations is a combination of two or more transformations, one performed after the other.

take note

Theorem 9-2   Reflections Across Parallel Lines

A composition of reflections across two parallel lines is a translation.
You can write this composition as
$(R_m \circ R_\ell)(\triangle ABC) = \triangle A''B''C''$
or $R_m(R_\ell(\triangle ABC)) = \triangle A''B''C''$.

$\overline{AA''}$, $\overline{BB''}$, and $\overline{CC''}$ are all perpendicular to lines ℓ and m.

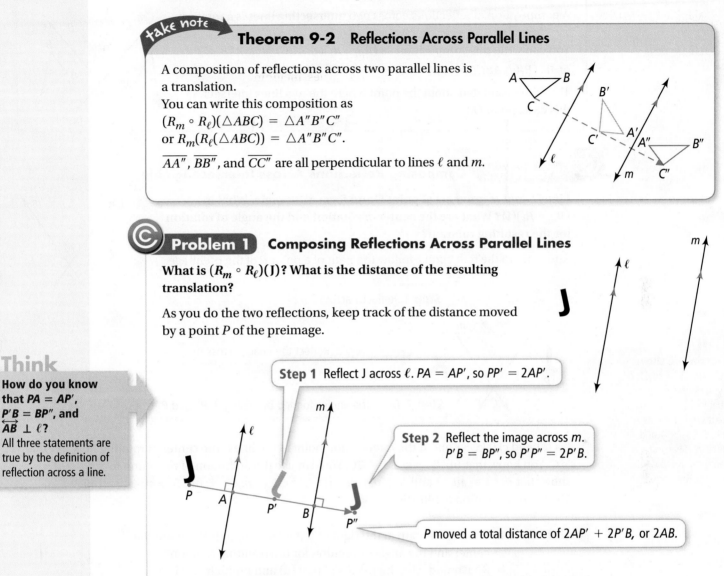

© **Problem 1**   **Composing Reflections Across Parallel Lines**

What is $(R_m \circ R_\ell)(J)$? What is the distance of the resulting translation?

As you do the two reflections, keep track of the distance moved by a point P of the preimage.

Think

How do you know that $PA = AP'$, $P'B = BP''$, and $\overleftrightarrow{AB} \perp \ell$?
All three statements are true by the definition of reflection across a line.

Step 1 Reflect J across ℓ. $PA = AP'$, so $PP' = 2AP'$.

Step 2 Reflect the image across m. $P'B = BP''$, so $P'P'' = 2P'B$.

P moved a total distance of $2AP' + 2P'B$, or $2AB$.

The red arrow shows the translation. The total distance P moved is $2 \cdot AB$. Because $\overleftrightarrow{AB} \perp \ell$, AB is the distance between ℓ and m. The distance of the translation is twice the distance between ℓ and m.

© ✓ **Got It?**   **1. a.** Draw parallel lines ℓ and m as in Problem 1. Draw J between ℓ and m. What is the image of $(R_m \circ R_\ell)(J)$? What is the distance of the resulting translation?

   b. Reasoning Use the results of part (a) and Problem 1. Make a conjecture about the distance of any translation that is the result of a composition of reflections across two parallel lines.

Theorem 9-3   Reflections Across Intersecting Lines

A composition of reflections across two intersecting lines is a rotation.

You can write this composition as $(R_m \circ R_\ell)(\triangle ABC) = \triangle A''B''C''$
or $R_m(R_\ell(\triangle ABC)) = \triangle A''B''C''$.

The figure is rotated about the point where the two lines intersect. In
this case, point Q.

Problem 2   Composing Reflections Across Intersecting Lines

Lines ℓ and m intersect at point C and form a $70°$ angle. What is
$(R_m \circ R_\ell)(J)$? What are the center of rotation and the angle of rotation
for the resulting rotation?

After you do the reflections, follow the path of a point P of the preimage.

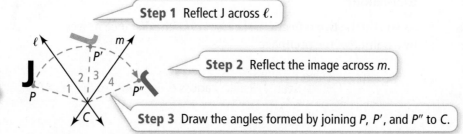

Step 1 Reflect J across ℓ.

Step 2 Reflect the image across m.

Step 3 Draw the angles formed by joining P, P', and P'' to C.

Think

**How do you show
that $m\angle 1 = m\angle 2$?**
If you draw $\overline{PP'}$ and
label its intersection
point with line ℓ as A,
then $PA = P'A$ and
$PP' \perp \ell$. So, by the
Converse of the Angle
Bisector Theorem,
$m\angle 1 = m\angle 2$.

J is rotated clockwise about the intersection point of the lines. The center of rotation
is C. You know that $m\angle 2 + m\angle 3 = 70$. You can use the definition of reflection to
show that $m\angle 1 = m\angle 2$ and $m\angle 3 = m\angle 4$. So, $m\angle 1 + m\angle 2 + m\angle 3 + m\angle 4 = 140$.
The angle of rotation is $140°$ clockwise.

Got It? 2. a. Use the diagram at the right. What is $(R_b \circ R_a)(J)$? What are the
center and the angle of rotation for the resulting rotation?

   b. Reasoning Use the results of part (a) and Problem 2. Make
   a conjecture about the center of rotation and the angle of
   rotation for any rotation that is the result of any composition of
   reflections across two intersecting lines.

Any composition of isometries can be represented by either a reflection,
translation, rotation, or glide reflection. A **glide reflection** is the composition
of a translation (a glide) and a reflection across a line parallel to the direction
of translation. You can map a left paw print onto a right paw print with a
glide reflection.

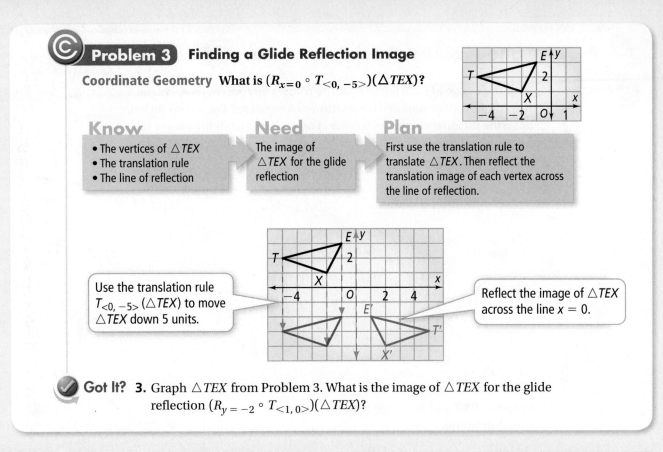

Problem 3 Finding a Glide Reflection Image

Coordinate Geometry What is $(R_{x=0} \circ T_{<0, -5>})(\triangle TEX)$?

Know
- The vertices of $\triangle TEX$
- The translation rule
- The line of reflection

Need
The image of $\triangle TEX$ for the glide reflection

Plan
First use the translation rule to translate $\triangle TEX$. Then reflect the translation image of each vertex across the line of reflection.

Use the translation rule $T_{<0, -5>}(\triangle TEX)$ to move $\triangle TEX$ down 5 units.

Reflect the image of $\triangle TEX$ across the line $x = 0$.

Got It? **3.** Graph $\triangle TEX$ from Problem 3. What is the image of $\triangle TEX$ for the glide reflection $(R_{y = -2} \circ T_{<1, 0>})(\triangle TEX)$?

Lesson Check

Do you know HOW?

Copy the diagrams below. Sketch the image of Z reflected across line a, then across line b.

1.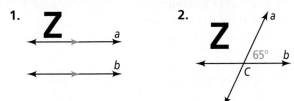

2.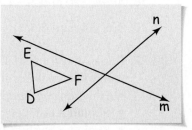

3. $\triangle PQR$ has vertices $P(0, 5)$, $Q(5, 3)$, and $R(3, 1)$. What are the vertices of the image of $\triangle PQR$ for the glide reflection $(R_{y = -2} \circ T_{<3, -1>})(\triangle PQR)$?

Do you UNDERSTAND? MATHEMATICAL PRACTICES

4. Vocabulary In a glide reflection, what is the relationship between the direction of the translation and the line of reflection?

5. Error Analysis You reflect $\triangle DEF$ first across line m and then across line n. Your friend says you can get the same result by reflecting $\triangle DEF$ first across line n and then across line m. Explain your friend's error.

Practice and Problem-Solving Exercises © MATHEMATICAL PRACTICES

A) Practice

Find the image of each letter after the transformation $R_m \circ R_\ell$. Is the resulting transformation a translation or a rotation? For a translation, describe the direction and distance. For a rotation, tell the center of rotation and the angle of rotation.

◀ **See Problems 1 and 2.**

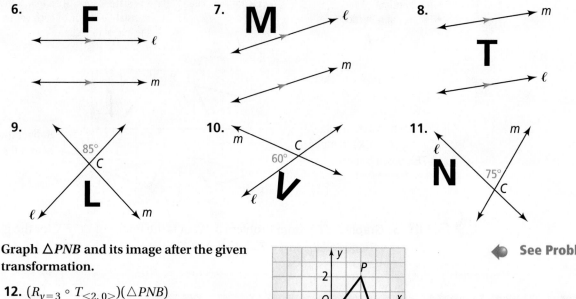

6. F ℓ m

7. M ℓ m

8. m T ℓ

9. 85° C L ℓ m

10. m 60° C V ℓ

11. m ℓ N 75° C

Graph $\triangle PNB$ and its image after the given transformation.

◀ **See Problem 3.**

12. $(R_{y=3} \circ T_{<2,\, 0>})(\triangle PNB)$

13. $(R_{x=0} \circ T_{<0,\, -3>})(\triangle PNB)$

14. $(R_{y=0} \circ T_{<2,\, 2>})(\triangle PNB)$

15. $(R_{y=x} \circ T_{<-1,\, 1>})(\triangle PNB)$

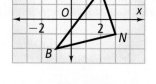

B) Apply

Use the given points and lines. Graph $\overline{AB}$ and its image $\overline{A''B''}$ after a reflection first across ℓ_1 and then across ℓ_2. Is the resulting transformation a translation or a rotation? For a translation, describe the direction and distance. For a rotation, tell the center of rotation and the angle of rotation.

16. $A(1, 5)$ and $B(2, 1)$; ℓ_1: $x = 3$; ℓ_2: $x = 7$

17. $A(2, 4)$ and $B(3, 1)$; ℓ_1: x-axis; ℓ_2: y-axis

18. $A(-4, -3)$ and $B(-4, 0)$; ℓ_1: $y = x$; ℓ_2: $y = -x$

19. $A(2, -5)$ and $B(-1, -3)$; ℓ_1: $y = 0$; ℓ_2: $y = 2$

20. $A(6, -4)$ and $B(5, 0)$; ℓ_1: $x = 6$; ℓ_2: $x = 4$

21. $A(-1, 0)$ and $B(0, -2)$; ℓ_1: $y = -1$; ℓ_2: $y = 1$

© 22. **Think About a Plan** Let A' be the point $(1, 5)$. If $(R_{y=1} \circ T_{<3,\, 0>})(A) = A'$, then what are the coordinates of A?
   - How can you *work backwards* to find the coordinates of A?
   - Should A be to left or to the right of A'?
   - Should A be above or below A'?

Describe the isometry that maps the black figure onto the blue figure.

23.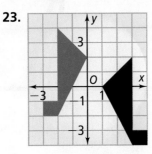

24.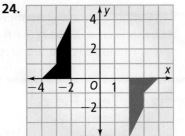

25. Which transformation maps the black triangle onto the blue triangle?

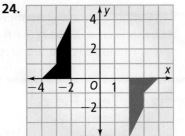

Ⓐ $R_{x=2} \circ T_{<0, -3>}$

Ⓑ $r_{(180°, O)}$

Ⓒ $R_{y=-\frac{1}{2}}$

Ⓓ $r_{(180°, O)} \circ R_{x\text{-axis}}$

Ⓒ **26. Writing** Reflections and glide reflections are *odd isometries*, while translations and rotations are *even isometries*. Use what you have learned in this lesson to explain why these categories make sense.

Ⓒ **27. Open-Ended** Draw $\triangle ABC$. Describe a reflection, a translation, a rotation, and a glide reflection. Then draw the image of $\triangle ABC$ for each transformation.

Ⓒ **28. Reasoning** The definition states that a glide reflection is the composition of a translation and a reflection. Explain why these can occur in either order.

Identify each mapping as a translation, reflection, rotation, or glide reflection. Write the rule for each translation, reflection, rotation, or glide reflection. For glide reflections, write the rule as a composition of a translation and a reflection.

29. $\triangle ABC \rightarrow \triangle EDC$

30. $\triangle EDC \rightarrow \triangle PQM$

31. $\triangle MNJ \rightarrow \triangle EDC$

32. $\triangle HIF \rightarrow \triangle HGF$

33. $\triangle PQM \rightarrow \triangle JLM$

34. $\triangle MNP \rightarrow \triangle EDC$

35. $\triangle JLM \rightarrow \triangle MNJ$

36. $\triangle PQM \rightarrow \triangle KJN$

37. $\triangle KJN \rightarrow \triangle ABC$

38. $\triangle HGF \rightarrow \triangle KJN$

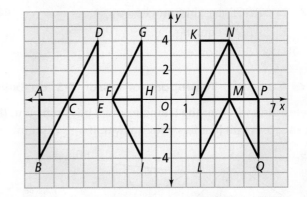

39. Describe a glide reflection that maps the black R to the blue R.

R

© 40. Reasoning Does an $x°$ rotation about a point P followed by a reflection across a line ℓ give the same image as a reflection across ℓ followed by an $x°$ rotation about P? Explain.

Standardized Test Prep

41. What is $(R_{x=0} \circ T_{<-12, -6>})(11, -5)$?

Ⓐ $(1, -11)$ Ⓑ $(-1, 11)$ Ⓒ $(1, 11)$ Ⓓ $(-1, -11)$

42. $ABCD$ is a rectangular window divided into 12 panes of glass. E, F, G, and H are midpoints of $\overline{AB}$, $\overline{BC}$, $\overline{CD}$, and $\overline{AD}$, respectively. Which statement must be true?

Ⓕ The quadrilateral panes are squares.

Ⓖ The quadrilateral panes are rhombuses.

Ⓗ The triangular panes are all congruent.

Ⓘ The triangular panes are right triangles.

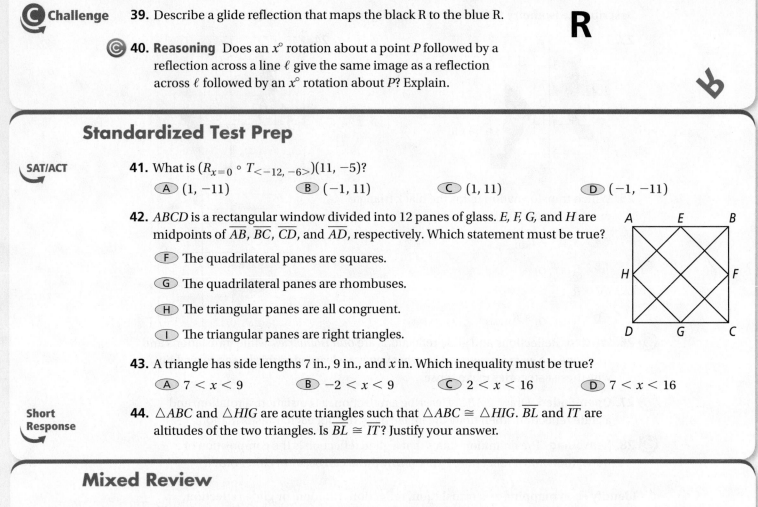

43. A triangle has side lengths 7 in., 9 in., and x in. Which inequality must be true?

Ⓐ $7 < x < 9$ Ⓑ $-2 < x < 9$ Ⓒ $2 < x < 16$ Ⓓ $7 < x < 16$

44. $\triangle ABC$ and $\triangle HIG$ are acute triangles such that $\triangle ABC \cong \triangle HIG$. $\overline{BL}$ and $\overline{IT}$ are altitudes of the two triangles. Is $\overline{BL} \cong \overline{IT}$? Justify your answer.

Mixed Review

Coordinate Geometry Find the image of $\triangle ABC$ for the rotation described.   ◀ See Lesson 9-3.

45. $A(0, 4)$, $B(0, 0)$, $C(-3, -1)$; $r_{(90°, O)}(\triangle ABC)$ **46.** $A(4, 2)$, $B(2, 8)$, $C(8, 0)$; $r_{(180°, O)}(\triangle ABC)$

Identify the two statements that contradict each other.   ◀ See Lesson 5-5.

47. I. $\triangle ABC$ is a right triangle.
   II. $\triangle ABC$ is equiangular.
   III. $\triangle ABC$ is isosceles.

48. I. In right $\triangle ABC$, $m\angle B = 90$.
   II. In right $\triangle ABC$, $m\angle A = 80$.
   III. In right $\triangle ABC$, $m\angle C = 90$.

Get Ready!   **To prepare for Lesson 9-5, do Exercises 49–51.**

Determine whether the triangles must be congruent. If so, name the postulate or theorem that justifies your answer. If not, explain.   ◀ See Lesson 4-2.

49.   **50.**   **51.**

Do you know HOW?

Tell whether the transformation appears to be an isometry. Explain.

1.

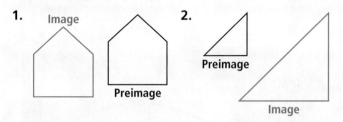

Image

Preimage

2.

Preimage

Image

3. What rule describes the translation 5 units left and 10 units up?

4. Describe the translation $T_{<-3, 5>}(x, y)$ in words.

5. Find a translation that has the same effect as the composition $T_{<-3, 2>} \circ T_{<7, -2>}$.

Find each reflection image.

6. $R_{x\text{-axis}}(5, -3)$ **7.** $R_{y\text{-axis}}(5, -3)$

8. $\triangle WXY$ has vertices $W(-4, 1)$, $X(2, -7)$, and $Y(0, -3)$. What are the vertices of $T_{<-3, 5>}(\triangle WXY)$?

9. $\triangle ABC$ has vertices $A(-1, 4)$, $B(2, 0)$, and $C(4, 3)$. Graph the reflection image of $\triangle ABC$ across the line $x = -1$.

Copy each figure and point A. Draw the image of each figure for the given angle of rotation about A. Use prime notation to label the vertices of the image.

10. 40°

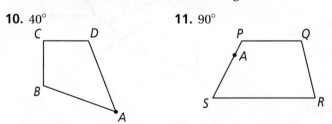

C   D

B

A

11. 90°

P   Q

A

S   R

Graph $\triangle JKL$ and its image after each composition of transformations.

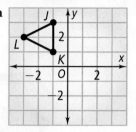

12. $(T_{<0, -4>} \circ R_{y\text{-axis}})(\triangle JKL)$

13. $(R_{x\text{-axis}} \circ r_{(180°, O)})(\triangle JKL)$

14. $(r_{(90°, O)} \circ T_{<-1, -3>})(\triangle JKL)$

Do you UNDERSTAND?

15. Reasoning The point $(5, -9)$ is the image under the translation $T_{<-3, 2>}(x, y) = (5, -9)$. What is its preimage?

16. Reasoning The point $T(5, -1)$ is reflected first across the x-axis and then across the y-axis. What is the distance between the image and the preimage? Explain.

17. Coordinate Geometry The point $L(a, b)$ is in Quadrant I. What are the coordinates of the image of L after the composition of transformations $R_{y\text{-axis}} \circ r_{(90°, O)}$?

The two letters in each pair can be mapped onto each other. Does one figure appear to be a translation image, a reflection image, or a rotation image of the other?

18.

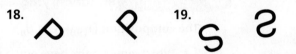

19.

20. Error Analysis Your friend draws the diagram at the right to show the reflection of $\square PQRS$ across the x-axis. Explain and correct your friend's error.

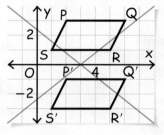

Congruence Transformations

Content Standards
G.CO.7 Use the definition of congruence in terms of rigid motions to show that two triangles are congruent . . .
Also G.CO.6, G.CO.8

Objective To identify congruence transformations
To prove triangle congruence using isometries

Getting Ready!

Suppose that you want to create two identical wings for a model airplane. You draw one wing on a large sheet of tracing paper, fold it along the dashed line, and then trace the first wing. How do you know that the two wings are identical?

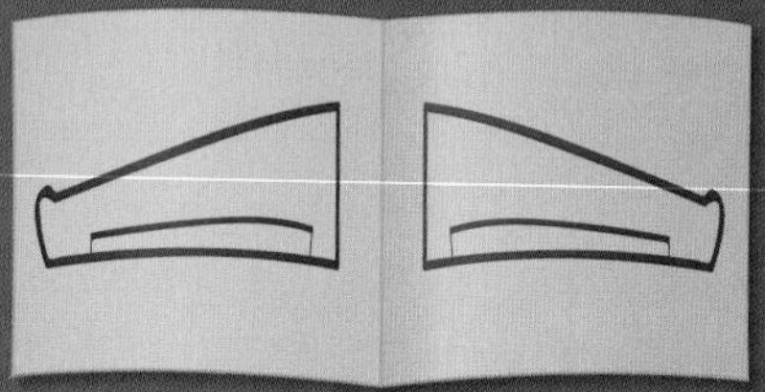

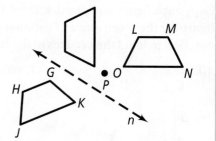

Are there other methods you could use to create two identical wings?

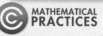
MATHEMATICAL
PRACTICES

In the Solve It, you may have used the properties of rigid motions to describe why the wings are identical.

Lesson Vocabulary
• congruent
• congruence transformation

Essential Understanding You can use compositions of rigid motions to understand congruence.

Plan

How can you use the properties of isometries to find equal angle measures and equal side lengths?
Isometries preserve angle measure and distance, so identify corresponding angles and corresponding side lengths.

Problem 1 Identifying Equal Measures

The composition $(r_{(90°, P)} \circ R_n)(LMNO) = GHJK$ is shown at the right.

A Which angle pairs have equal measures?

Because compositions of isometries preserve angle measure, corresponding angles have equal measures.

$m\angle L = m\angle G$, $m\angle M = m\angle H$, $m\angle N = m\angle J$, and $m\angle O = m\angle K$.

B Which sides have equal lengths?

By definition, isometries preserve distance. So, corresponding side lengths have equal measures.

$LM = GH$, $MN = HJ$, $NO = JK$, and $LO = GK$.

Got It? 1. The composition $(R_t \circ T_{<2, 3>})(\triangle ABC) = \triangle XYZ$. List all of the pairs of angles and sides with equal measures.

In Problem 1 you saw that compositions of rigid motions preserve corresponding side lengths and angle measures. This suggests another way to define congruence.

take note

Key Concept Congruent Figures

Two figures are **congruent** if and only if there is a sequence of one or more rigid motions that maps one figure onto the other.

© **Problem 2** Identifying Congruent Figures

Which pairs of figures in the grid are congruent? For each pair, what is a sequence of rigid motions that maps one figure to the other?

Figures are congruent if and only if there is a sequence of rigid motions that maps one figure to the other. So, to find congruent figures, look for sequences of translations, rotations, and reflections that map one figure to another.

Because $r_{(180°,\, O)}(\triangle DEF) = \triangle LMN$, the triangles are congruent.

Because $(T_{<-1,\, 5>} \circ R_{y\text{-axis}})(ABCJ) = WXYZ$, the trapezoids are congruent.

Because $T_{<-2,\, 9>}(\overline{HG}) = \overline{PQ}$, the line segments are congruent.

Think

Does one rigid motion count as a sequence?
Yes. It is a sequence of length 1.

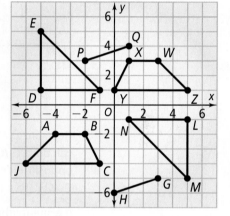

✓ **Got It?** **2.** Which pairs of figures in the grid are congruent? For each pair, what is a sequence of rigid motions that map one figure to the other?

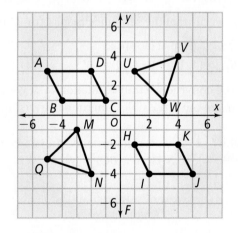

Because compositions of rigid motions take figures to congruent figures, they are also called **congruence transformations**.

Problem 3 Identifying Congruence Transformations

In the diagram at the right, $\triangle JQV \cong \triangle EWT$. What is a congruence transformation that maps $\triangle JQV$ onto $\triangle EWT$?

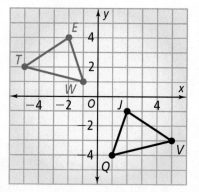

Know

The coordinates of the vertices of the triangles

Need

A sequence of rigid motions that maps $\triangle JQV$ onto $\triangle EWT$

Plan

Identify the corresponding parts and find a congruence transformation that maps the preimage to the image. Then use the vertices to verify the congruence transformation.

Because $\triangle EWT$ lies above $\triangle JQV$ on the plane, a translation can map $\triangle JQV$ up on the plane. Also, notice that $\triangle EWT$ is on the opposite side of the y-axis and has the opposite orientation of $\triangle JQV$. This suggests that the triangle is reflected across the y-axis.

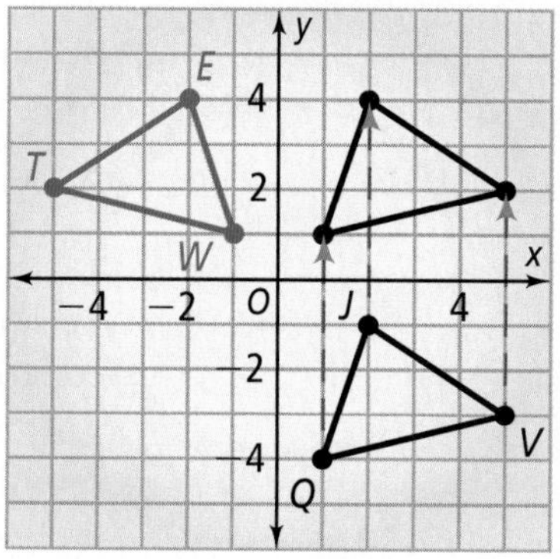

It appears that a translation of $\triangle JQV$ up 5 units, followed by a reflection across the y-axis maps $\triangle JQV$ to $\triangle EWT$. Verify by using the coordinates of the vertices.

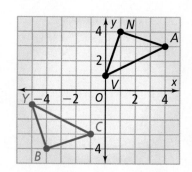

$$T_{<5,\,0>}(x, y) = (x + 5, y)$$
$$T_{<5,\,0>}(J) = (2, 4)$$
$$R_{y\text{-axis}}(2, 4) = (-2, 4) = E$$

Next, verify that the sequence maps Q to W and V to T.

$$T_{<5,\,0>}(Q) = (1, 1)$$
$$R_{y\text{-axis}}(1, 1) = (-1, 1) = W$$

$$T_{<5,\,0>}(V) = (5, 2)$$
$$R_{y\text{-axis}}(5, 2) = (-5, 2) = T$$

So, the congruence transformation $R_{y\text{-axis}} \circ T_{<5,\,0>}$ maps $\triangle JQV$ onto $\triangle EWT$. Note that there are other possible congruence transformations that map $\triangle JQV$ onto $\triangle EWT$.

Got It? 3. What is a congruence transformation that maps $\triangle NAV$ to $\triangle BCY$?

In Chapter 4, you studied triangle congruence postulates and theorems. You can use congruence transformations to justify criteria for determining triangle congruence.

 **Proof** **Problem 4** **Verifying the SAS Postulate**

Think

How do you show that the two triangles are congruent?
Find a congruence transformation that maps one onto the other.

Given: $\angle J \cong \angle P$, $\overline{PA} \cong \overline{JO}$, $\overline{FP} \cong \overline{SJ}$
Prove: $\triangle JOS \cong \triangle PAF$

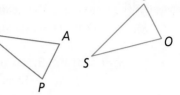

Step 1 Translate $\triangle PAF$ so that points A and O coincide.

Step 2 Because $\overline{PA} \cong \overline{JO}$, you can rotate $\triangle PAF$ about point A so that $\overline{PA}$ and $\overline{JO}$ coincide.

Step 3 Reflect $\triangle PAF$ across $\overline{PA}$. Because reflections preserve angle measure and distance, and because $\angle J \cong \angle P$ and $\overline{FP} \cong \overline{SJ}$, you know that the reflection maps $\angle P$ to $\angle J$ and $\overline{FP}$ to $\overline{SJ}$. Since points S and F coincide, $\triangle PAF$ coincides with $\triangle JOS$.

There is a congruence transformation that maps $\triangle PAF$ onto $\triangle JOS$, so $\triangle PAF \cong \triangle JOS$.

Got It? **4.** Verify the SSS postulate.
  Given: $\overline{TD} \cong \overline{EN}$, $\overline{YT} \cong \overline{SE}$, $\overline{YD} \cong \overline{SN}$
  Prove: $\triangle YDT \cong \triangle SNE$

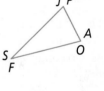

In Problem 4, you used the transformational approach to prove triangle congruence. Because this approach is more general, you can use what you know about congruence transformations to determine whether any two figures are congruent.

Problem 5 Determining Congruence

Is Figure A congruent to Figure B? Explain how you know.

Figure A can be mapped to Figure B by a sequence of reflections or a simple translation. So, Figure A is congruent to Figure B because there is a congruence transformation that maps one to the other.

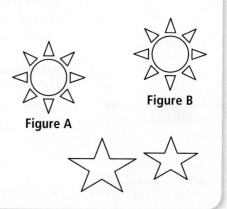

Figure B

Figure A

Got It? **5.** Are the figures shown at the right congruent? Explain.

Lesson Check

Do you know HOW?

Use the graph for Exercises 1 and 2.

1. Identify a pair of congruent figures and write a congruence statement.

2. What is a congruence transformation that relates two congruent figures?

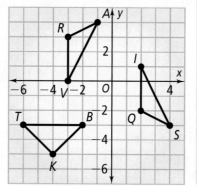

Do you UNDERSTAND? MATHEMATICAL PRACTICES

3. How can the definition of congruence in terms of rigid motions be more useful than a definition of congruence that relies on corresponding angles and sides?

4. **Reasoning** Is a composition of a rotation followed by a glide reflection a congruence transformation? Explain.

5. **Open Ended** What is an example of a board game in which a game piece is moved by using a congruence transformation?

Practice and Problem-Solving Exercises MATHEMATICAL PRACTICES

A Practice

For each coordinate grid, identify a pair of congruent figures. Then determine a congruence transformation that maps the preimage to the congruent image.

See Problem 1 and 2.

6.
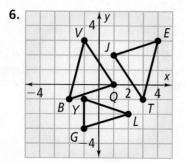

7.
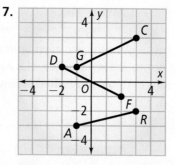

8.
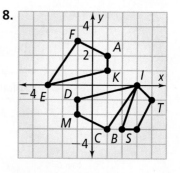

In Exercises 9–11, find a congruence transformation that maps
△*LMN* to △*RST*.

See Problem 3.

9.

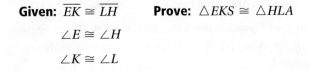

10.

11.

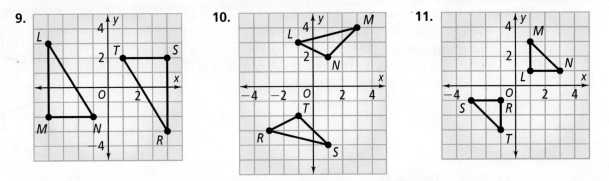

12. Verify the ASA Postulate for triangle congruence by using congruence
Proof transformations.

See Problem 4.

    Given: $\overline{EK} \cong \overline{LH}$ **Prove:** △*EKS* ≅ △*HLA*

            ∠*E* ≅ ∠*H*

            ∠*K* ≅ ∠*L*

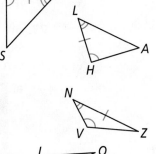

13. Verify the AAS Postulate for triangle congruence by using congruence
Proof transformations.

    Given: ∠*I* ≅ ∠*V*   **Prove:** △*NVZ* ≅ △*CIQ*

            ∠*C* ≅ ∠*N*

            $\overline{QC} \cong \overline{NZ}$

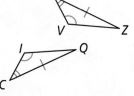

In Exercises 14–16, determine whether the figures are congruent. If so, describe
a congruence transformation that maps one to the other. If not, explain.

See Problem 5.

14.   **15.**   **16.**

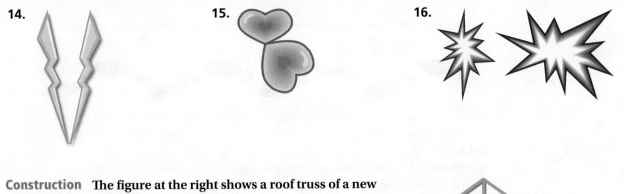

B **Apply**

Construction The figure at the right shows a roof truss of a new
building. Identify an isometry or composition of isometries
to justify each of the following statements.

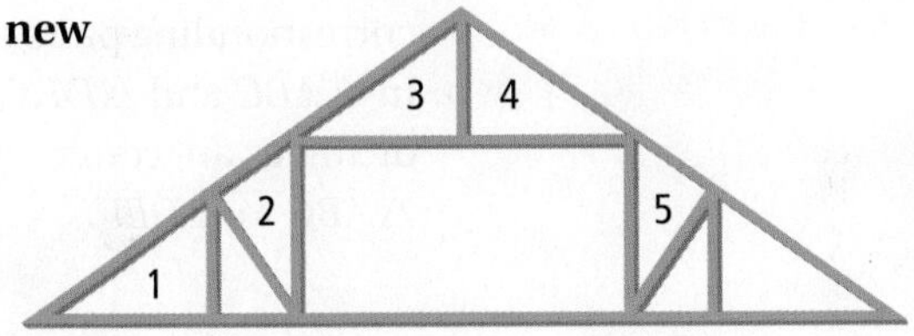

    17. Triangle 1 is congruent to triangle 3.

    18. Triangle 1 is congruent to triangle 4.

    19. Triangle 2 is congruent to triangle 5.

20. Vocabulary If two figures are _____, then there is an isometry that maps one figure onto the other.

 21. Think About a Plan The figure at the right shows two congruent, isosceles triangles. What are four different isometries that map the top triangle onto the bottom triangle?

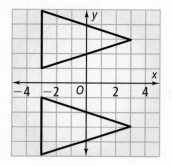

- How can you use the three basic rigid motions to map the top triangle onto the bottom triangle?
- What other isometries can you use?

22. Graphic Design Most companies have a logo that is used on company letterhead and signs. A graphic designer sketches the logo at the right. What congruence transformations might she have used to draw this logo?

23. Art Artists frequently use congruence transformations in their work. The artworks shown below are called *tessellations*. What types of congruence transformations can you identify in the tessellations?

a.

b.

24. In the footprints shown below, what congruence transformations can you use to extend the footsteps?

25. Prove the statements in parts (a) and (b) to show congruence in terms of
Proof transformations is equivalent to the criteria of for triangle congruence you learned in Chapter 4.

  a. If there is a congruence transformation that maps $\triangle ABC$ to $\triangle DEF$ then corresponding pairs of sides and corresponding pairs of angles are congruent.

  b. In $\triangle ABC$ and $\triangle DEF$, if corresponding pairs of sides and corresponding pairs of angles are congruent, then there is a congruence transformation that maps $\triangle ABC$ to $\triangle DEF$.

26. Baking Cookie makers often use a cookie press so that the cookies all look the same. The baker fills a cookie sheet for baking in the pattern shown. What types of congruence transformations are being used to set each cookie on the sheet?

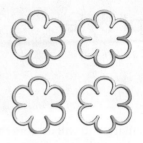

27. Use congruence transformations to prove the Isosceles Triangle Theorem.
Proof

  Given: $\overline{FG} \cong \overline{FH}$

  Prove: $\angle G \cong \angle H$

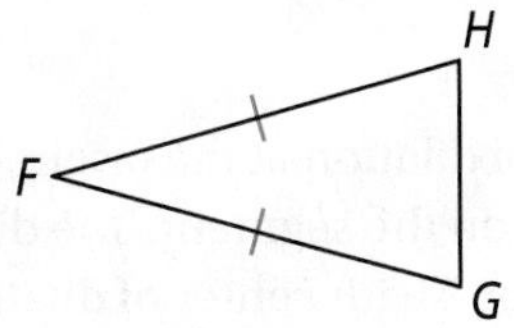

Challenge

28. Reasoning You project an image for viewing in a large classroom. Is the projection of the image an example of a congruence transformation? Explain your reasoning.

Standardized Test Prep

GRIDDED RESPONSE

SAT/ACT

29. To the nearest hundredth, what is the value of x in the diagram at the right?

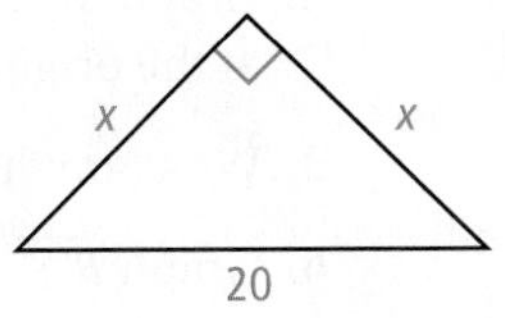

30. In $\triangle FGH$ and $\triangle XYZ$, $\angle G$ and $\angle Y$ are right angles. $\overline{FH} \cong \overline{XZ}$ and $\overline{GH} \cong \overline{YZ}$. If $GH = 7$ ft and $XY = 9$ ft, what is the area of $\triangle FGH$ in square inches?

31. $\triangle ACB$ is isosceles with base $\overline{AB}$. Point D is on $\overline{AB}$ and $\overline{CD}$ is the bisector of $\angle C$. If $CD = 5$ in. and $DB = 4$ in., what is BC to the nearest tenth of an inch?

32. Two angle measures of $\triangle JKL$ are 30 and 60. The shortest side measures 10 cm. What is the length, in centimeters, of the longest side of the triangle?

Mixed Review

33. A triangle has vertices $A(3, 2)$, $B(4, 1)$, and $C(4, 3)$. Find the coordinates of the images of A, B, and C for a glide reflection with translation $(x, y) \rightarrow (x, y + 1)$ and reflection line $x = 0$.

See Lesson 9-4.

The lengths of two sides of a triangle are given. What are the possible lengths for the third side?

See Lesson 5-6.

34. 16 in., 26 in.   **35.** 19.5 ft, 20.5 ft   **36.** 9 m, 9 m   **37.** $4\frac{1}{2}$ yd, 8 yd

Get Ready! **To prepare for Lessons 9-6, do Exercises 38–40.**

Determine the scale drawing dimensions of each room using a scale of $\frac{1}{4}$ in. = 1 ft.

See Lesson 7-2.

38. kitchen: 12 ft by 16 ft   **39.** bedroom: 8 ft by 10 ft   **40.** laundry room: 6 ft by 9 ft

Exploring Dilations

© **Content Standards**

G.SRT.1b The dilation of a line segment is longer or shorter in the ratio given by the scale factor.

Also G.SRT.1a

In this activity, you will explore the properties of dilations. A dilation is defined by a center of dilation and a scale factor.

Activity 1

To dilate a segment by a scale factor n with center of dilation at the origin, you measure the distance from the origin to each point on the segment. The diagram at the right shows the dilation of $\overline{GH}$ by the scale factor 3 with center of dilation at the origin. To locate the dilation image of $\overline{GH}$, draw rays from the origin through points G and H. Then, measure the distance from the origin to G. Next, find the point along the same ray that is 3 times that distance. Label the point G'. Now dilate the endpoint H similarly. Draw $\overline{G'H'}$.

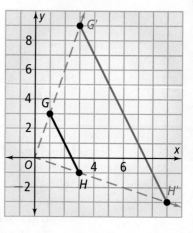

1. Graph $\overline{RS}$ with $R(1, 4)$ and $S(2, -1)$. What is the length of $\overline{RS}$?

2. Graph the dilations of the endpoint of $\overline{RS}$ by scale factor 2 and center of dilation at the origin. Label the dilated endpoints R' and S'.

3. What are the coordinates of R' and S'?

4. Graph $\overline{R'S'}$.

5. What is $R'S'$?

6. How do the lengths of $\overline{RS}$ and $\overline{R'S'}$ compare?

7. Graph the dilation of $\overline{RS}$ by scale factor $\frac{1}{2}$ with center of dilation at the origin. Label the dilation $\overline{R''S''}$.

8. What is $R''S''$?

9. How do the lengths of $\overline{R'S'}$ and $\overline{R''S''}$ compare?

10. What can you conjecture about the length of a line segment that has been dilated by scale factor n?

Activity 2

11. Graph $L'M'N'P'$, the dilation of $LMNP$ with scale factor 3 and center of dilation at the origin. What are the coordinates of L', M', N', and P'?

12. How are the coordinates of the vertices of $LMNP$ and $L'M'N'P'$ related?

13. Compare the shape, size, and orientation of the preimage $LMNP$ with the image $L'M'N'P'$. What conjecture can you make about the properties of dilations?

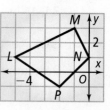

Dilations

© Content Standards

G.SRT.1a A dilation takes a line not passing through the center of the dilation to a parallel line, and leaves a line passing through the center unchanged.

Also G.SRT.1b, G.CO.2, G.SRT.2

Objective To understand dilation images of figures

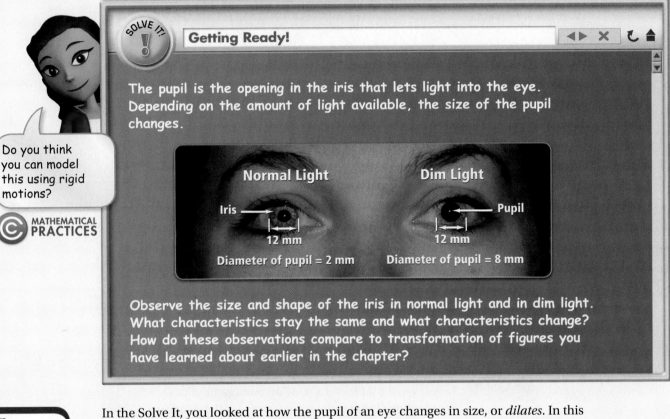

SOLVE IT!

Getting Ready!

The pupil is the opening in the iris that lets light into the eye. Depending on the amount of light available, the size of the pupil changes.

Normal Light — Iris — 12 mm — Diameter of pupil = 2 mm

Dim Light — Pupil — 12 mm — Diameter of pupil = 8 mm

Observe the size and shape of the iris in normal light and in dim light. What characteristics stay the same and what characteristics change? How do these observations compare to transformation of figures you have learned about earlier in the chapter?

Do you think you can model this using rigid motions?

© MATHEMATICAL PRACTICES

In the Solve It, you looked at how the pupil of an eye changes in size, or *dilates*. In this lesson, you will learn how to dilate geometric figures.

Essential Understanding You can use a scale factor to make a larger or smaller copy of a figure that is also similar to the original figure.

Lesson Vocabulary
- dilation
- center of dilation
- scale factor of a dilation
- enlargement
- reduction

take note

Key Concept Dilation

A **dilation** with **center of dilation** C and **scale factor** n, $n > 0$, can be written as $D_{(n, C)}$. A dilation is a transformation with the following properties.
- The image of C is itself (that is, $C' = C$).
- For any other point R, R' is on $\overrightarrow{CR}$ and $CR' = n \cdot CR$, or $n = \frac{CR'}{CR}$.
- Dilations preserve angle measure.

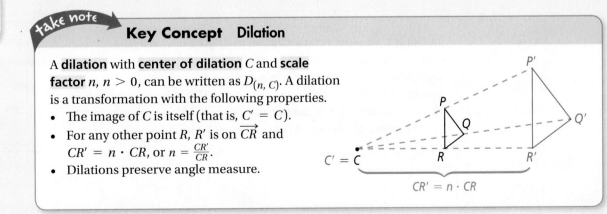

$CR' = n \cdot CR$

The scale factor n of a dilation is the ratio of a length of the image to the corresponding length in the preimage, with the image length always in the numerator. For the figure shown on page 587, $n = \frac{CR'}{CR} = \frac{R'P'}{RP} = \frac{P'Q'}{PQ} = \frac{Q'R'}{QR}$.

A dilation is an **enlargement** if the scale factor n is greater than 1. The dilation is a **reduction** if the scale factor n is between 0 and 1.

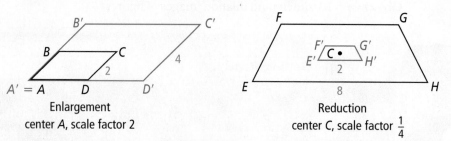

Enlargement
center A, scale factor 2

Reduction
center C, scale factor $\frac{1}{4}$

Problem 1 Finding a Scale Factor

Multiple Choice Is $D_{(n, X)}(\triangle XTR) = \triangle X'T'R'$ an enlargement or a reduction? What is the scale factor n of the dilation?

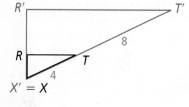

Ⓐ enlargement; $n = 2$ Ⓒ reduction; $n = \frac{1}{3}$

Ⓑ enlargement; $n = 3$ Ⓓ reduction; $n = 3$

Think

Why is the scale factor not $\frac{4}{12}$, or $\frac{1}{3}$?

The scale factor of a dilation always has the image length (or the distance between a point on the image and the center of dilation) in the numerator.

The image is larger than the preimage, so the dilation is an enlargement.

Use the ratio of the lengths of corresponding sides to find the scale factor.

$$n = \frac{X'T'}{XT} = \frac{4 + 8}{4} = \frac{12}{4} = 3$$

$\triangle X'T'R'$ is an enlargement of $\triangle XTR$, with a scale factor of 3. The correct answer is B.

Got It? **1.** Is $D_{(n, O)}(JKLM) = J'K'L'M'$ an enlargement or a reduction? What is the scale factor n of the dilation?

In Got It 1, you looked at a dilation of a figure drawn in the coordinate plane. In this book, all dilations of figures in the coordinate plane have the origin as the center of dilation. So you can find the dilation image of a point $P(x, y)$ by multiplying the coordinates of P by the scale factor n. A dilation of scale factor n with center of dilation at the origin can be written as

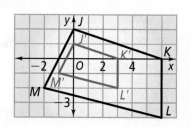

$$D_n(x, y) = (nx, ny)$$

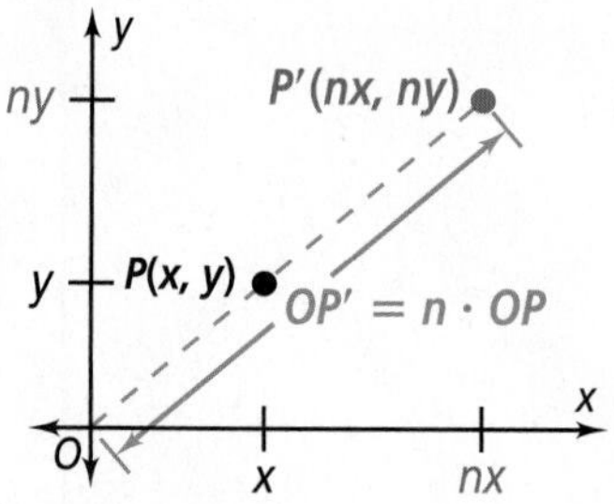

Think

Will the vertices of the triangle move closer to (0, 0) or farther from (0, 0)?
The scale factor is 2, so the dilation is an enlargement. The vertices will move farther from (0, 0).

Problem 2 Finding a Dilation Image

What are the coordinates of the vertices of $D_2(\triangle PZG)$? Graph the image of $\triangle PZG$.

Identify the coordinates of each vertex. The center of dilation is the origin and the scale factor is 2, so use the dilation rule $D_2(x, y) = (2x, 2y)$.

$D_2(P) = (2 \cdot 2, 2 \cdot (-1))$, or $P'(4, -2)$.

$D_2(Z) = (2 \cdot (-2), 2 \cdot 1)$, or $Z'(-4, 2)$.

$D_2(G) = (2 \cdot 0, 2 \cdot (-2))$, or $G'(0, -4)$.

To graph the image of $\triangle PZG$, graph P', Z', and G'. Then draw $\triangle P'Z'G'$.

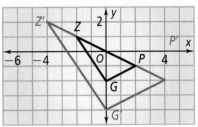

Got It? **2. a.** What are the coordinates of the vertices of $D_{\frac{1}{2}}(\triangle PZG)$?

b. Reasoning How are $\overline{PZ}$ and $\overline{P'Z'}$ related? How are $\overline{PG}$ and $\overline{P'G'}$, and $\overline{GZ}$ and $\overline{G'Z'}$ related? Use these relationships to make a conjecture about the effects of dilations on lines.

Dilations and scale factors help you understand real-world enlargements and reductions, such as images seen through a microscope or on a computer screen.

Think

What does a scale factor of 7 tell you?
A scale factor of 7 tells you that the ratio of the image length to the actual length is 7, or
$\frac{\text{image length}}{\text{actual length}} = 7$.

Problem 3 Using a Scale Factor to Find a Length

Biology A magnifying glass shows you an image of an object that is 7 times the object's actual size. So the scale factor of the enlargement is 7. The photo shows an apple seed under this magnifying glass. What is the actual length of the apple seed?

$1.75 = 7 \cdot p$ image length = scale factor · actual length

$0.25 = p$ Divide each side by 7.

The actual length of the apple seed is 0.25 in.

Got It? **3.** The height of a document on your computer screen is 20.4 cm. When you change the zoom setting on your screen from 100% to 25%, the new image of your document is a dilation of the previous image with scale factor 0.25. What is the height of the new image?

1.75 in.

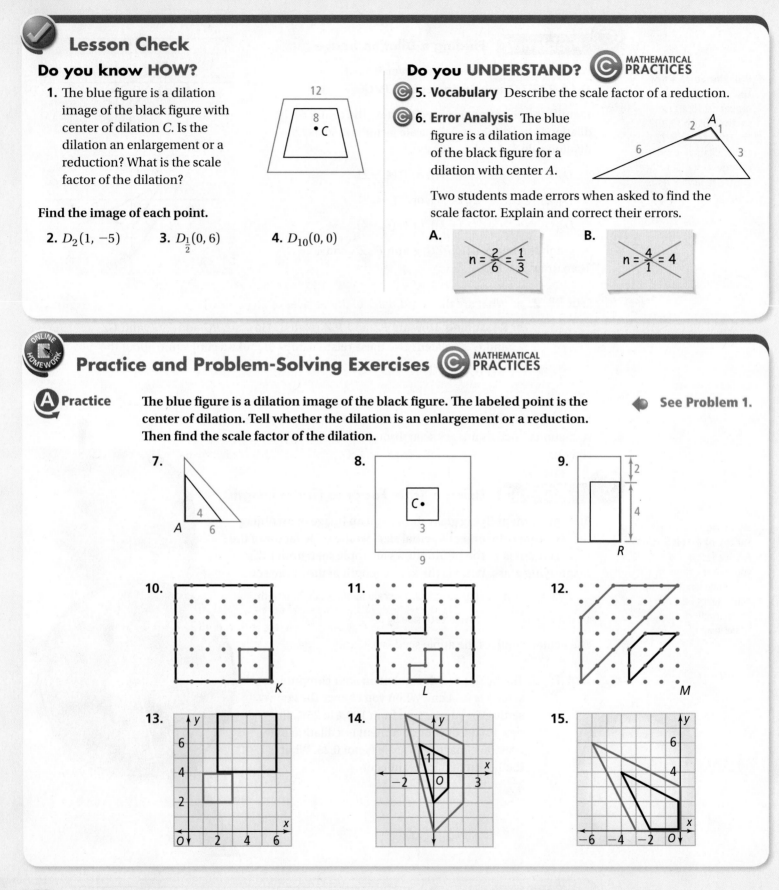

Lesson Check

Do you know HOW?

1. The blue figure is a dilation image of the black figure with center of dilation C. Is the dilation an enlargement or a reduction? What is the scale factor of the dilation?

12

8
•C

Find the image of each point.

2. $D_2(1, -5)$ 3. $D_{\frac{1}{2}}(0, 6)$ 4. $D_{10}(0, 0)$

Do you UNDERSTAND? MATHEMATICAL PRACTICES

5. **Vocabulary** Describe the scale factor of a reduction.

6. **Error Analysis** The blue figure is a dilation image of the black figure for a dilation with center A.

2   A
  1
6   3

Two students made errors when asked to find the scale factor. Explain and correct their errors.

A.

$n = \dfrac{2}{6} = \dfrac{1}{3}$

B.

$n = \dfrac{4}{1} = 4$

Practice and Problem-Solving Exercises MATHEMATICAL PRACTICES

A Practice

The blue figure is a dilation image of the black figure. The labeled point is the center of dilation. Tell whether the dilation is an enlargement or a reduction. Then find the scale factor of the dilation.

◀ See Problem 1.

7.
4
A   6

8.
C•
3
9

9.
2
4
R

10.
K

11.
L

12.
M

13.
(graph with y-axis and x-axis, marks at 2, 4, 6)

14.
(graph with y-axis and x-axis, marks at -2, O, 3, 1)

15.
(graph with y-axis and x-axis, marks at -6, -4, -2, O, 4, 6)

Find the images of the vertices of △*PQR* for each dilation. Graph the image.

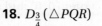 **See Problem 2.**

16. $D_3 (\triangle PQR)$ **17.** $D_{10} (\triangle PQR)$ **18.** $D_{\frac{3}{4}} (\triangle PQR)$

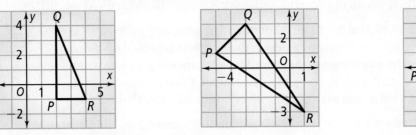

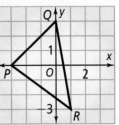

Magnification You look at each object described in Exercises 19–22 under a magnifying glass. Find the actual dimension of each object.

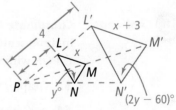 **See Problem 3.**

19. The image of a button is 5 times the button's actual size and has a diameter of 6 cm.

20. The image of a pinhead is 8 times the pinhead's actual size and has a width of 1.36 cm.

21. The image of an ant is 7 times the ant's actual size and has a length of 1.4 cm.

22. The image of a capital letter N is 6 times the letter's actual size and has a height of 1.68 cm.

B Apply

Find the image of each point for the given scale factor.

23. $L(-3, 0); D_5 (L)$ **24.** $N(-4, 7); D_{0.2} (N)$ **25.** $A(-6, 2); D_{1.5} (A)$

26. $F(3, -2); D_{\frac{1}{3}} (F)$ **27.** $B\left(\frac{5}{4}, -\frac{3}{2}\right); D_{\frac{1}{10}} (B)$ **28.** $Q\left(6, \frac{\sqrt{3}}{2}\right); D_{\sqrt{6}} (Q)$

Use the graph at the right. Find the vertices of the image of *QRTW* for a dilation with center (0, 0) and the given scale factor.

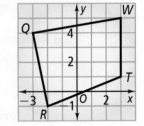

29. $\frac{1}{4}$ **30.** 0.6   **31.** 0.9   **32.** 10   **33.** 100

© 34. Compare and Contrast Compare the definition of scale factor of a dilation to the definition of scale factor of two similar polygons. How are they alike? How are they different?

© 35. Think About a Plan The diagram at the right shows △*LMN* and its image △*L'M'N'* for a dilation with center *P*. Find the values of *x* and *y*. Explain your reasoning.
- What is the relationship between △*LMN* and △*L'M'N'*?
- What is the scale factor of the dilation?
- Which variable can you find using the scale factor?

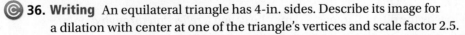

© 36. Writing An equilateral triangle has 4-in. sides. Describe its image for a dilation with center at one of the triangle's vertices and scale factor 2.5.

Coordinate Geometry Graph $MNPQ$ and its image $M'N'P'Q'$ for a dilation with center $(0, 0)$ and the given scale factor.

37. $M(1, 3)$, $N(-3, 3)$, $P(-5, -3)$, $Q(-1, -3)$; 3   **38.** $M(2, 6)$, $N(-4, 10)$, $P(-4, -8)$, $Q(-2, -12)$; $\frac{1}{4}$

© **39. Open-Ended** Use the dilation command in geometry software or drawing software to create a design that involves repeated dilations, such as the one shown at the right. The software will prompt you to specify a center of dilation and a scale factor. Print your design and color it. Feel free to use other transformations along with dilations.

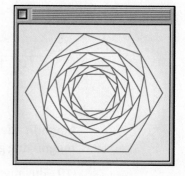

40. Copy Reduction Your picture of your family crest is 4.5 in. wide. You need a reduced copy for the front page of the family newsletter. The copy must fit in a space 1.8 in. wide. What scale factor should you use on the copy machine to adjust the size of your picture of the crest?

A dilation maps $\triangle HIJ$ onto $\triangle H'I'J'$. Find the missing values.

41. $HI = 8$ in. $H'I' = 16$ in.   **42.** $HI = 7$ cm $H'I' = 5.25$ cm   **43.** $HI = \blacksquare$ ft $H'I' = 8$ ft

$IJ = 5$ in. $I'J' = \blacksquare$ in. $IJ = 7$ cm $I'J' = \blacksquare$ cm $IJ = 30$ ft $I'J' = \blacksquare$ ft

$HJ = 6$ in. $H'J' = \blacksquare$ in. $HJ = \blacksquare$ cm $H'J' = 9$ cm $HJ = 24$ ft $H'J' = 6$ ft

Copy $\triangle TBA$ and point O for each of Exercises 44–47. Draw the dilation image $\triangle T'B'A'$.

44. $D_{(2,\ O)}(\triangle TBA)$ **45.** $D_{(3,\ B)}(\triangle TBA)$

46. $D_{\left(\frac{1}{3},\ T\right)}(\triangle TBA)$ **47.** $D_{\left(\frac{1}{2},\ O\right)}(\triangle TBA)$

© **48. Reasoning** You are given $\overline{AB}$ and its dilation image $\overline{A'B'}$ with A, B, A', and B' noncollinear. Explain how to find the center of dilation and scale factor.

© **Reasoning** Write *true* or *false* for Exercises 49–52. Explain your answers.

49. A dilation is an isometry.

50. A dilation with a scale factor greater than 1 is a reduction.

51. For a dilation, corresponding angles of the image and preimage are congruent.

52. A dilation image cannot have any points in common with its preimage.

© **Challenge**   **Coordinate Geometry** In the coordinate plane, you can extend dilations to include scale factors that are negative numbers. For Exercises 53 and 54, use $\triangle PQR$ with vertices $P(1, 2)$, $Q(3, 4)$, and $R(4, 1)$.

53. Graph $D_{-3}(\triangle PQR)$.

54. a. Graph $D_{-1}(\triangle PQR)$.
     b. Explain why the dilation in part (a) may be called a *reflection through a point*. Extend your explanation to a new definition of point symmetry.

55. Shadows A flashlight projects an image of rectangle *ABCD* on a wall so that each vertex of *ABCD* is 3 ft away from the corresponding vertex of *A'B'C'D'*. The length of $\overline{AB}$ is 3 in. The length of $\overline{A'B'}$ is 1 ft. How far from each vertex of *ABCD* is the light?

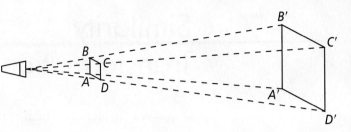

Standardized Test Prep

SAT/ACT

56. A dilation maps $\triangle CDE$ onto $\triangle C'D'E'$. If $CD = 7.5$ ft, $CE = 15$ ft, $D'E' = 3.25$ ft, and $C'D' = 2.5$ ft, what is *DE*?

    Ⓐ 1.08 ft   Ⓑ 5 ft   Ⓒ 9.75 ft   Ⓓ 19 ft

57. You want to prove indirectly that the diagonals of a rectangle are congruent. As the first step of your proof, what should you assume?

    Ⓕ A quadrilateral is not a rectangle.

    Ⓖ The diagonals of a rectangle are not congruent.

    Ⓗ A quadrilateral has no diagonals.

    Ⓘ The diagonals of a rectangle are congruent.

58. Which word can describe a kite?

    Ⓐ equilateral   Ⓑ equiangular   Ⓒ convex   Ⓓ scalene

Short Response

59. Use the figure at the right to answer the questions below. 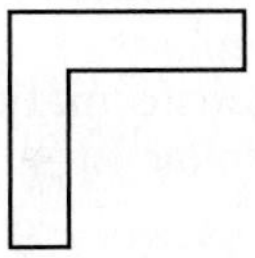
    a. Does the figure have rotational symmetry? If so, identify the angle of rotation.
    b. Does the figure have reflectional symmetry? If so, how many lines of symmetry does it have?

Mixed Review

60. $\triangle JKL$ has vertices $J(23, 2)$, $K(4, 1)$, and $L(1, 23)$. What are the coordinates of J', K', and L' if $(R_{x\text{-axis}} \circ T_{<2, -3>})(\triangle JKL) = \triangle J'K'L'$?

    ◀ See Lesson 9-5.

Get Ready! **To prepare for Lesson 9-7, do Exercises 61–63.**

Algebra *TRSU* ~ *NMYZ*. **Find the value of each variable.**

    ◀ See Lesson 7-2.

61. *a*   **62.** *b*   **63.** *c*

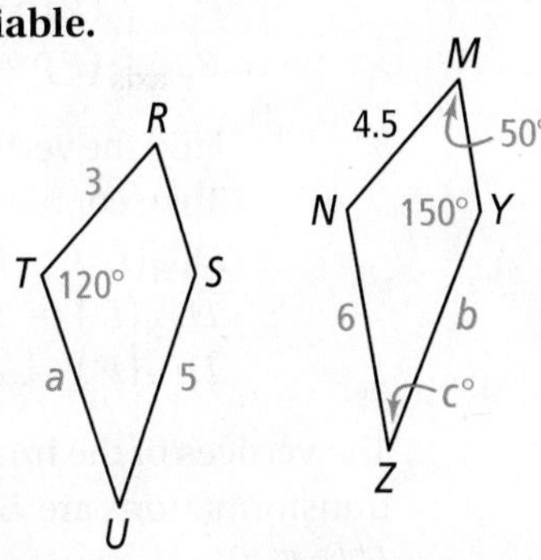

Similarity Transformations

© **Content Standards**
G.SRT.2 Given two figures, use the definition of similarity in terms of similarity transformations to decide if they are similar . . .
Also G.SRT.3

Objectives To identify similarity transformations and verify properties of similarity

SOLVE IT!

Getting Ready!

Your friend says that she performed a composition of transformations to map △ABC to △A'B'C'. Describe the composition of transformations.

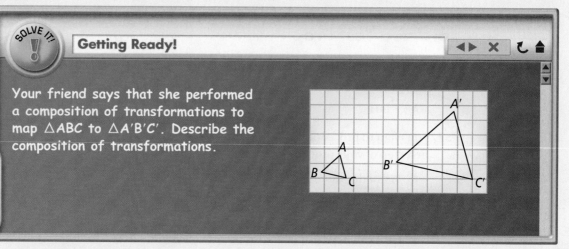

Is there more than one composition of transformations possible to map △ABC to △A'B'C'?

© **MATHEMATICAL PRACTICES**

In the Solve It, you used a composition of a rigid motion and a dilation to describe the mapping from △ABC to △A'B'C'.

Essential Understanding You can use compositions of rigid motions and dilations to help you understand the properties of similarity.

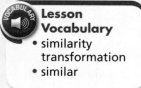

Lesson Vocabulary
• similarity transformation
• similar

© **Problem 1**   **Drawing Transformations**

△DEF has vertices D(2, 0), E(1, 4), and F(4, 2). What is the image of △DEF when you apply the composition $D_{1.5} \circ R_{y\text{-axis}}$?

Step 1 Find the vertices of $R_{y\text{-axis}}(\triangle DEF)$. Then connect the vertices to draw the image.
$R_{y\text{-axis}}(D) = D'(-2, 0)$
$R_{y\text{-axis}}(E) = E'(-1, 4)$
$R_{y\text{-axis}}(F) = F'(-4, 2)$

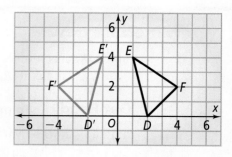

Step 2 Find the vertices of the dilation of △D'E'F'. Then connect the vertices to draw the image.
$D_{1.5}(D') = D''(-3, 0)$
$D_{1.5}(E') = E''(-1.5, 6)$
$D_{1.5}(F') = F''(-6, 3)$

The vertices of the image after the composition of transformations are $D''(-3, 0)$, $E''(-1.5, 6)$, and $F''(-6, 3)$.

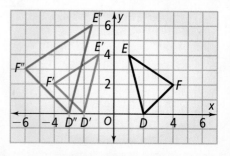

Got It? 1. Reasoning $\triangle LMN$ has vertices $L(-4, 2)$, $M(-3, -3)$, and $N(-1, 1)$. Suppose the triangle is translated 4 units right and 2 units up and then dilated by a scale factor of 0.5 with center of dilation at the origin. Sketch the resulting image of the composition of transformations.

Problem 2 Describing Transformations

What is a composition of rigid motions and a dilation that maps $\triangle RST$ to $\triangle PYZ$?

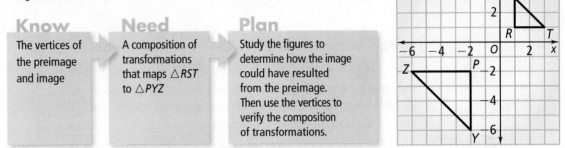

Know	Need	Plan
The vertices of the preimage and image	A composition of transformations that maps $\triangle RST$ to $\triangle PYZ$	Study the figures to determine how the image could have resulted from the preimage. Then use the vertices to verify the composition of transformations.

It appears that $\triangle RST$ was rotated and then enlarged to create $\triangle PYZ$. To verify the composition of transformations, begin by rotating the triangle 180° about the origin.

$r_{(180°, O)}(R) = R'(-1, -1)$ Use the rule $r_{(180°, O)}(x, y) = (-x, -y)$.

$r_{(180°, O)}(S) = S'(-1, -3)$

$r_{(180°, O)}(T) = T'(-3, -1)$

$\triangle PYZ$ appears to be about twice as large as $\triangle RST$. Scale the vertices of the intermediate image $R'S'T'$ to verify the composition.

$D_2(-1, -1) = P(-2, -2)$ Use the rule $D_2(x, y) = (2x, 2y)$.

$D_2(-1, -3) = Y(-2, -6)$

$D_2(-3, -1) = Z(-6, -2)$

The vertices of the dilation of $\triangle R'S'T'$ match the vertices of $\triangle PYZ$.

A rotation of 180° about the origin followed by a dilation with scale factor 2 maps $\triangle RST$ to $\triangle PYZ$.

 Got It? 2. What is a composition of rigid motions and a dilation that maps trapezoid $ABCD$ to trapezoid $MNHP$?

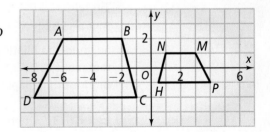

Notice that the figures in Problems 1 and 2 appear to have the same shape but different sizes. Compositions of rigid motions and dilations map preimages to similar images. For this reason, they are called **similarity transformations**. Similarity transformations give you another way to think about similarity.

take note

Key Concept Similar Figures

Two figures are **similar** if and only if there is a similarity transformation that maps one figure onto the other.

Here's Why It Works Consider the composition of a rigid motion and a dilation shown at the right.

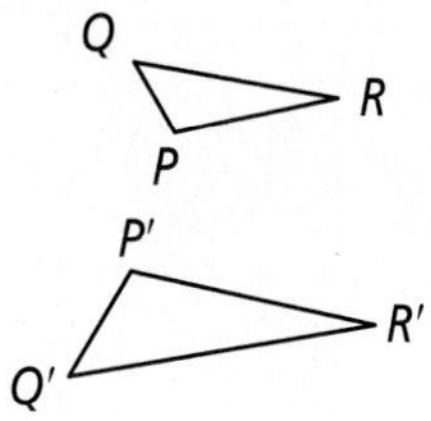

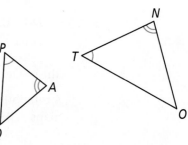

Because rigid motions and dilations preserve angle measure, $m\angle P = m\angle P'$, $m\angle Q = m\angle Q'$, and $m\angle R = m\angle R'$. So, corresponding angles are congruent.

Because there is a dilation, there is some scale factor k such that:

$$PQ = kP'Q' \quad QR = kQ'R' \quad\quad PR = kP'R'$$

$$k = \frac{PQ}{P'Q'} \quad\quad k = \frac{QR}{Q'R'} \quad\quad k = \frac{PR}{P'R'}$$

So, $\dfrac{PQ}{P'Q'} = \dfrac{QR}{Q'R'} = \dfrac{PR}{P'R'}$.

© Problem 3 Finding Similarity Transformations

Is there a similarity transformation that maps $\triangle PAQ$ to $\triangle TNO$? If so, identify the similarity transformation and write a similarity statement. If not, explain.

Think

Does it matter what the center of dilation is?
No. All that matters is that $k \cdot PA = TN$.

Although $PA \neq TN$, there is a scale factor k such that $k \cdot PA = TN$. Dilate $\triangle PAQ$ using this scale factor. Then $\overline{P'A'} \cong \overline{TN}$. Since dilations preserve angle measure, you also know that $\angle P' \cong \angle T$ and $\angle A' \cong \angle N$. Therefore, $\triangle P'A'Q' \cong \triangle TNO$ by ASA. This means that there is a sequence of rigid motions that maps $\triangle P'A'Q'$ onto $\triangle TNO$.

So, there is a dilation that maps $\triangle PAQ$ to $\triangle P'A'Q'$, and a sequence of rigid motions that maps $\triangle P'A'Q'$ to $\triangle TNO$. Therefore, there is a composition of a dilation and rigid motions that maps $\triangle PAQ$ onto $\triangle TNO$.

Got It? 3. Is there a similarity transformation that maps $\triangle JKL$ to $\triangle RST$? If so, identify the similarity transformation and write a similarity statement. If not, explain.

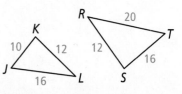

Similarity transformations provide a powerful general approach to similarity. In Problem 3, you used similarity transformations to verify the AA Postulate for triangle similarity. Another advantage to the transformational approach to similarity is that you can apply it figures other than polygons.

 Problem 4 **Determining Similarity**

Plan

How can you determine whether two figures are similar if you have no information about side lengths or angle measures?
Any two plane figures are similar if you can find a similarity transformation that maps one onto the other.

A new company is using a computer program to design its logo. Are the two figures used in the logo so far similar?

If you can find a similarity transformation between two figures, then you know they are similar. The smaller lightning bolt can be translated so that the tips coincide. Then it can be enlarged by some scale factor so that the two bolts overlap.

The figures are similar because there is a similarity transformation that maps one figure onto the other. The transformation is a translation followed by a dilation.

Got It? **4.** Are the figures at the right similar? Explain.

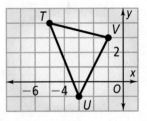

Lesson Check

Do you know HOW?

Use the diagram below for Exercises 1 and 2.

1. What is a similarity transformation that maps $\triangle RST$ to $\triangle JKL$?

2. What are the coordinates of $(D_{\frac{1}{4}} \circ r_{(180°, O)})(\triangle RST)$?

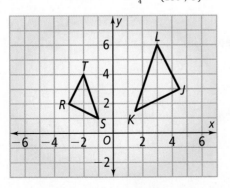

Do you UNDERSTAND? MATHEMATICAL PRACTICES

3. Vobabulary Describe how the word *dilation* is used in areas outside of mathematics. How do these applications relate the mathematical definition?

4. Open Ended For $\triangle TUV$ at the right, give the vertices of a similar triangle after a similarity transformation that uses at least 1 rigid motion.

Practice and Problem-Solving Exercises  MATHEMATICAL PRACTICES

A Practice   △*MAT* has vertices *M*(6, −2), *A*(4, −5), and *T*(1, −2). For each of the following, sketch the image of the composition of transformations.

⬅ **See Problem 1.**

5. reflection across the *x*-axis followed by a dilation by a scale factor of 0.5

6. rotation of 180° about the origin followed by a dilation by a scale factor of 1.5

7. translation 6 units up followed by a reflection across the *y*-axis and then a dilation by a scale factor of 2

For each graph, describe the composition of transformations that maps △*FGH* to △*QRS*.

⬅ **See Problem 2.**

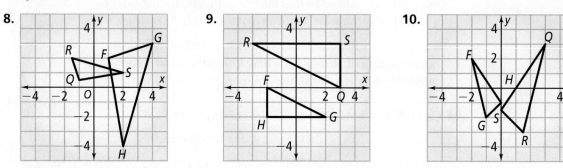

8.   **9.**   **10.**

For each pair of figures, determine if there is a similarity transformation that maps one figure onto the other. If so, identify the similarity transformation and write a similarity statement. If not, explain.

⬅ **See Problem 3.**

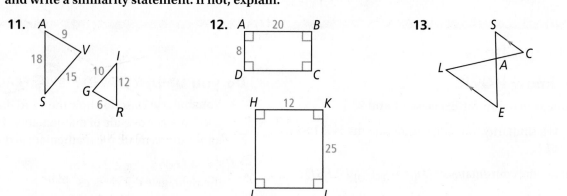

11.   **12.**   **13.**

Determine whether or not each pair of figures below are similar. Explain your reasoning.

⬅ **See Problem 4.**

14.    **15.**

16. Writing Your teacher uses geometry software program to plot $\triangle ABC$ with vertices $A(2, 1)$, $B(6, 1)$, and $C(6, 4)$. Then he used a similarity transformation to plot $\triangle DEF$ with vertices $D(-4, -2)$, $E(-12, -2)$, and $F(-12, -8)$. The corresponding angles of the two triangles are congruent. How can the Distance Formula be used to verify that the ratios of the corresponding sides are proportional? Verify that the figures are similar.

17. Think About a Plan Suppose that $\triangle JKL$ is formed by connecting the midpoints of $\triangle ABC$. Is $\triangle AJL$ similar to $\triangle ABC$? Explain.
- How are the side lengths of $\triangle AJL$ related to the side lengths of $\triangle ABC$?
- Can you find a similarity transformation that maps $\triangle AJL$ to $\triangle ABC$? Explain.

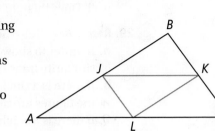

18. Writing What properties are preserved by rigid motions but not by similarity transformations?

Determine whether each statement is *always*, *sometimes*, or *never* true.

19. There is a similarity transformation between two rectangles.

20. There is a similarity transformation between two squares.

21. There is a similarity transformation between two circles.

22. There is a similarity transformation between a right triangle and an equilateral triangle.

23. Indirect Measurement A surveyor wants to use similar triangles to determine the distance across a lake as shown at the right.
- **a.** Are the two triangles in the figure similar? Justify your reasoning.
- **b.** What is the distance d across the lake?

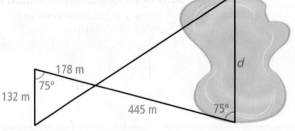

24. Photography A 4-inch by 6-inch rectangular photo is enlarged to fit an 8-inch by 10-inch frame. Are the two photographs similar? Explain.

25. Reasoning Is a rigid motion an example of a similarity transformation? Explain your reasoning and give an example.

26. Art A printing company enlarges a banner for a graduation party by a scale factor of 8.
- **a.** What are the dimensions of the larger banner?
- **b.** How can the printing company be sure that the enlarged banner is similar to the original?

13 in.

3 in.

Challenge

27. If △ABC has vertices given by A(u, v), B(w, x), and C(y, z), and △NOP has vertices given by N(5u, −4v), O(5w, −4x), and P(5y, −4z), is there a similarity transformation that maps △ABC to △NOP? Explain.

28. Overhead Projector When Mrs. Sheldon places a transparency on the screen of the overhead projector, the projector shows an enlargement of the transparency on the wall. Does this situation represent a similarity transformation? Explain.

29. Reasoning Tell whether each statement below is *true* or *false*.
   a. In order to show that two figures are similar, it is sufficient to show that there is a similarity transformation that maps one figure to the other.
   b. If there is a similarity transformation that maps one figure to another figure, then the figures are similar.
   c. If there is a similarity transformation that maps one figure to another figure, then the figures are congruent.

Standardized Test Prep

GRIDDED RESPONSE

SAT/ACT

30. △STU has vertices S(1, 2), T(0, 5), and U(−8, 0). What is the x-coordinate of S after a 270° rotation about the origin?

31. The diagonals of rectangle PQRS intersect at O. PO = 2x − 5 and OR = 7 − x. What is the length of $\overline{QS}$?

32. The length of the hypotenuse of a 45°-45°-90° triangle is 55 in. What is the length of one of its legs to the nearest tenth of an inch?

33. You place a sprinkler so that it is equidistant from three rose bushes at points A, B, and C. How many feet is the sprinkler from A?

B
3 yd
4 yd
A
C

Mixed Review

34. Which capital letters of the alphabet are rotation images of themselves? Draw each letter and give an angle of rotation (< 360°).

See Lesson 9-3.

35. Three vertices of an isosceles trapezoid are (−2, 1), (1, 4), and (4, 4). Find all possible coordinates for the fourth vertex.

See Lesson 6-7.

Get Ready! **To prepare for Lesson 10-1, do Exercises 34–37.**

Find the area of each figure.

See Lesson 1-8.

36. a square with 5-cm sides

37. a rectangle with base 4 in. and height 7 in.

38. a 4.6 m-by-2.5 m rectangle

39. a rectangle with length 3 ft and width $\frac{1}{2}$ ft

Pull It All Together © ASSESSMENT

To solve these problems you will pull together many concepts and skills that you have studied about transformations.

BIG idea Visualization

You can use visualization to find the image of a figure for a transformation.

© Performance Task 1

Each figure below is part of a capital letter in the English alphabet. To find the whole letter, combine the figure with its image for the appropriate rotation or reflection. What letter corresponds to each figure? What transformation produces each letter?

BIG idea Transformations

You can use transformations to describe a change in the position of a point.

© Performance Task 2

The arcs in the photo at the right appear to be paths of stars rotating about the North Star. To produce this effect, the photographer set a camera on a tripod and left the shutter open for an extended time. If the photographer left the shutter open for a full 24 hours, each arc would be a complete circle.

You can model a star's "rotation" in the coordinate plane. Place the North Star at the origin. Let $P(1, 0)$ be the position of the star at the moment the camera's shutter opens. Suppose the shutter is left open for 2 h 40 min, with the arc ending at P'.

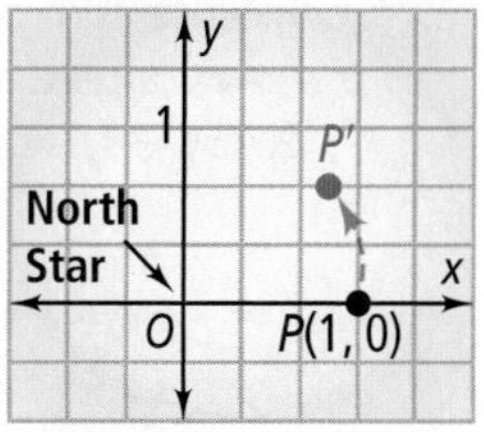

a. What angle of rotation maps P onto P'?
b. What are the x- and y-coordinates of P' to the nearest thousandth?
c. What translation rule maps P onto P'?

BIG idea Coordinate Geometry

You can use coordinate geometry to determine whether composing transformations is commutative.

© Performance Task 3

Copy the graph at the right. On the same set of axes, graph $(T_{<-4,3>} \circ R_{x\text{-axis}})(MNOP)$ and $(R_{x\text{-axis}} \circ T_{<-4,3>})(MNOP)$.

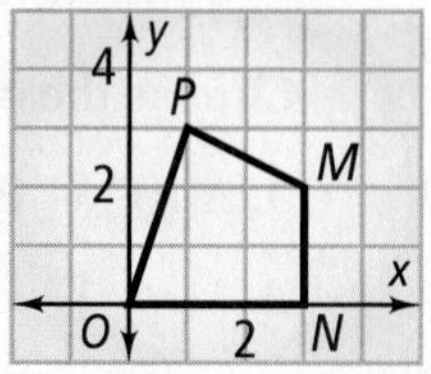

a. Does order matter for this composition of transformations? Explain.
b. Make a conjecture about whether composition of reflections and translations is commutative. Explain your reasoning.

9 Chapter Review

Connecting BIG ideas and Answering the Essential Questions

1 Transformations

When you translate, reflect, or rotate a geometric figure, its size and shape stay the same. When you dilate a geometric figure, the figure is enlarged or reduced.

2 Coordinate Geometry

You can show a transformation in the coordinate plane by graphing a figure and its image.

3 Visualization

If two figures are congruent, then you can visualize a congruence transformation that maps one figure to the other. If you can visualize a composition of rigid motions and dilations that map one figure to another, then the figures are similar.

Transformations (Lessons 9-1, 9-2, 9-3, and 9-6)

The black triangle is the preimage of each transformation.

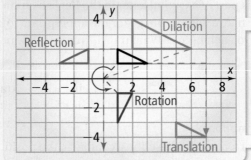

Congruence Transformations (Lesson 9-5)

Triangles are congruent if and only if there is a sequence of rigid motions that maps one triangle to the other.
Because $\triangle A''B''C''$ is the image of $\triangle ABC$ after a reflection and a translation, $\triangle ABC \cong \triangle A''B''C''$.

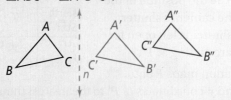

Composing Transformations (Lesson 9-4)

A glide reflection moves the black triangle down 3 units and then reflects it across the line $x = -2$.

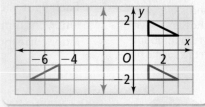

Similarity Transformations (Lesson 9-7)

Figures are similar if and only if there is a sequence of rigid motions and dilations that maps one figure onto the other.
Because $\triangle L''M''N''$ is the image of $\triangle LMN$ after a reflection and a dilation, $\triangle LMN$ is similar to $\triangle L''M''N''$.

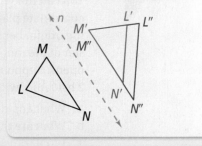

Chapter Vocabulary

- composition of transformations (p. 548)
- congruence transformation (p. 580)
- dilation (p. 587)
- glide reflection (p. 572)
- image (p. 545)
- isometry (p. 570)
- preimage (p. 545)
- reflection (p. 554)
- rigid motion (p. 545)
- rotation (p. 561)
- similarity transformation (p. 596)
- transformation (p. 545)
- translation (p. 547)

Choose the correct term to complete each sentence.

1. A(n) __?__ is a change in the position, shape, or size of a figure.

2. A(n) __?__ is a composition of rigid motions and dilations.

3. In a(n) __?__, all points of a figure move the same distance in the same direction.

4. A(n) __?__ is the result of a transformation.

9-1 Translations

Quick Review

A **transformation** of a geometric figure is a change in its position, shape, or size.

A **translation** is a rigid motion that maps all points of a figure the same distance in the same direction.

In a **composition of transformations,** each transformation is performed on the image of the preceding transformation.

Example

What are the coordinates of $T_{<-2,3>}(5, -9)$?

Add -2 to the x-coordinate, and 3 to the y-coordinate.

$A(5, -9) \rightarrow (5 - 2, -9 + 3)$, or $A'(3, -6)$.

Exercises

5. a. A transformation maps *ZOWE* onto *LFMA*. Does the transformation appear to be a rigid motion? Explain.

   b. What is the image of $\overline{ZE}$? What is the preimage of M?

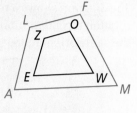

6. $\triangle RST$ has vertices $R(0, -4)$, $S(-2, -1)$, and $T(-6, 1)$. Graph $T_{<-4, 7>}(\triangle RST)$.

7. Write a rule to describe a translation 5 units left and 10 units up.

8. Find a single translation that has the same effect as the following composition of translations.

$T_{<-4, 7>}$ followed by $T_{<3, 0>}$

9-2 Reflections

Quick Review

The diagram shows a **reflection** across line r. A reflection is rigid motion that preserves distance and angle measure. The image and preimage of a reflection have opposite orientations.

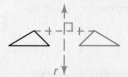

Example

Use points $P(1, 0)$, $Q(3, -2)$, and $R(4, 0)$. What is $R_{y\text{-axis}}(\triangle PQR)$?

Graph $\triangle PQR$. Find P', Q', and R' such that the y-axis is the perpendicular bisector of $\overline{PP'}$, $\overline{QQ'}$, and $\overline{RR'}$. Draw $\triangle P'Q'R'$.

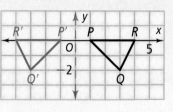

Exercises

Given points $A(6, 4)$, $B(-2, 1)$, and $C(5, 0)$, graph $\triangle ABC$ and each reflection image.

9. $R_{x\text{-axis}}(\triangle ABC)$

10. $R_{x=4}(\triangle ABC)$

11. $R_{y=x}(\triangle ABC)$

12. Copy the diagram. Then draw $R_{y\text{-axis}}(BGHT)$. Label the vertices of the image by using prime notation.

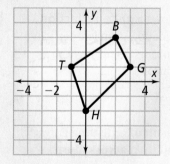

9-3 Rotations

Quick Review

The diagram shows a **rotation** of $x°$ about point R. A rotation is rigid motion in which a figure and its image have the same orientation.

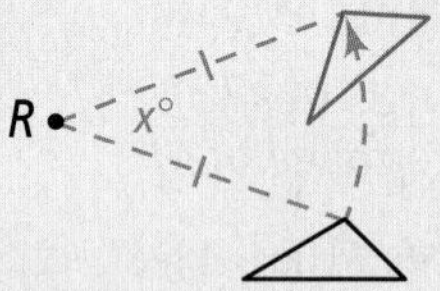

Example

$GHIJ$ has vertices $G(0, -3)$, $H(4, 1)$, $I(-1, 2)$, and $J(-5, -2)$. What are the vertices of $r_{(90°, O)}(GHIJ)$?

Use the rule $r_{(90°, O)}(x, y) = (-y, x)$.

$r_{(90°, O)}(G) = (3, 0)$

$r_{(90°, O)}(H) = (-1, 4)$

$r_{(90°, O)}(I) = (-2, -1)$

$r_{(90°, O)}(J) = (2, -5)$

Exercises

13. Copy the diagram below. Then draw $r_{(90°, P)}(\triangle ZXY)$. Label the vertices of the image by using prime notation.

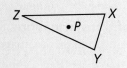

14. What are the coordinates of $r_{(180°, O)}(-4, 1)$?

15. $WXYZ$ is a quadrilateral with vertices $W(3, -1)$, $X(5, 2)$, $Y(0, 8)$, and $Z(2, -1)$. Graph $WXYZ$ and $r_{(270°, O)}(WXYZ)$.

9-4 Compositions of Isometries

Quick Review

An isometry is a transformation that preserves distance. All of the rigid motions, translations, reflections, and rotations, are isometries. A composition of isometries is also an isometry. All rigid motions can be expressed as a composition of reflections.

The diagram shows a **glide reflection** of N. A glide reflection is an isometry in which a figure and its image have opposite orientations.

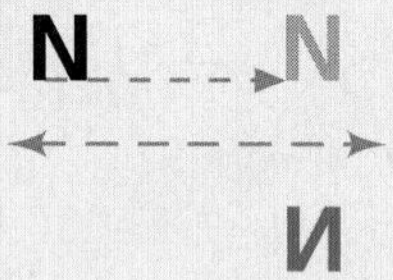

Example

Describe the result of reflecting P first across line ℓ and then across line m.

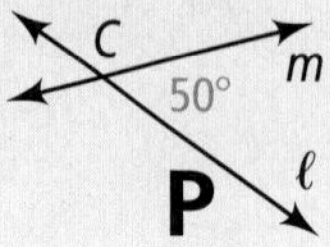

A composition of two reflections across intersecting lines is a rotation. The angle of rotation is twice the measure of the acute angle formed by the intersecting lines. P is rotated $100°$ about C.

Exercises

16. Sketch and describe the result of reflecting E first across line ℓ and then across line m.

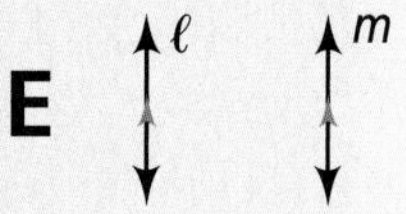

Each figure is an isometry image of the figure at the right. Tell whether their orientations are the same or opposite. Then classify the isometry.

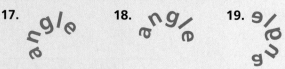

17. **18.** **19.**

20. $\triangle TAM$ has vertices $T(0, 5)$, $A(4, 1)$, and $M(3, 6)$. Find the image of $R_{y=-2} \circ T_{(-4, 0)}(\triangle TAM)$.

9-5 Congruence Transformations

Quick Review

Two figures are congruent if and only if there is a sequence of rigid motions that maps one figure onto the other.

Example

$R_{y\text{-axis}}(TGMB) = KWAV$.
What are all of the congruent angles and all of the congruent sides?

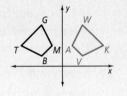

A reflection is a congruence transformation, so $TGMB \cong KWAV$, and corresponding angles and corresponding sides are congruent.

$\angle T \cong \angle K$, $\angle G \cong \angle W$, $\angle M \cong \angle A$, and $\angle B \cong \angle V$
$TG = KW$, $GM = WA$, $MB = AV$ and $TB = KV$

Exercises

21. In the diagram at the right, $\triangle LMN \cong \triangle XYZ$. Identify a congruence transformation that maps $\triangle LMN$ onto $\triangle XYZ$.

22. Fonts Graphic designers use some fonts because they have pleasing proportions or are easy to read from far away. The letters p and d below are used on a sign using a special font. Are the letters congruent? If so, describe a congruence transformation that maps one onto the other. If not, explain why not.

9-6 Dilations

Quick Review

The diagram shows a **dilation** with center C and scale factor n. The preimage and image are similar.

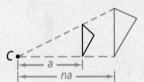

In the coordinate plane, if the origin is the center of a dilation with scale factor n, then $P(x, y) \rightarrow P'(nx, ny)$.

Example

The blue figure is a dilation image of the black figure. The center of dilation is A. Is the dilation an enlargement or a reduction? What is the scale factor?

The image is smaller than the preimage, so the dilation is a reduction. The scale factor is $\frac{\text{image length}}{\text{original length}} = \frac{2}{2+4} = \frac{2}{6}$, or $\frac{1}{3}$.

Exercises

23. The blue figure is a dilation image of the black figure. The center of dilation is O. Tell whether the dilation is an enlargement or a reduction. Then find the scale factor.

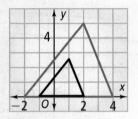

Graph the polygon with the given vertices. Then graph its image for a dilation with center (0, 0) and the given scale factor.

24. $M(-3, 4)$, $A(-6, -1)$, $T(0, 0)$, $H(3, 2)$; scale factor 5

25. $F(-4, 0)$, $U(5, 0)$, $N(-2, -5)$; scale factor $\frac{1}{2}$

26. A dilation maps $\triangle LMN$ onto $\triangle L'M'N'$. $LM = 36$ ft, $LN = 26$ ft, $MN = 45$ ft, and $L'M' = 9$ ft. Find $L'N'$ and $M'N'$.

Quick Review

Two figures are similar if and only if there is a similarity transformation that maps one figure onto the other.

When a figure is transformed by a composition of rigid motions and dilations, the corresponding angles of the image and preimage are congruent, and the ratios of corresponding sides are proportional.

Example

Is △JKL similar to △DCX? If so, write a similarity transformation rule. If not, explain why not.

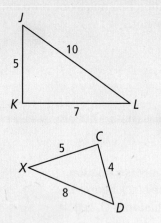

△JKL can be rotated and then translated so that J and D coincide and $\overline{JK}$ and $\overline{CD}$ are collinear. Then if △JKL is dilated by scale factor $\frac{4}{5}$, then △JKL will coincide with △DCX. So, the △JKL is similar to △DCX, and the similarity transformation is a rotation, followed by a translation, followed by a dilation of scale factor $\frac{4}{5}$.

Exercises

27. □GHJK has vertices $G(-3, -1)$, $H(-3, 2)$, $J(4, 2)$, and $K(4, -1)$. Draw □GHJK and its image when you apply the composition $D_2 \circ R_{x\text{-axis}}$.

28. **Writing** Suppose that you have an 8 in. by 12 in. photo of your friends and a 2 in. by 6 in. copy of the same picture. Are the two photos similar figures? How do you know?

29. **Reasoning** A model airplane has an overall length that is $\frac{1}{20}$ the actual plane's length, and an overall height that is $\frac{1}{18}$ the actual plane's height. Are the model airplane and the actual airplane similar figures? Explain.

30. Determine whether the figures below are similar. If so, write the similarity transformation rule. If not, explain.

Do you know HOW?

For Exercises 1–7, find the coordinates of the vertices of the image of *ABCD* for each transformation.

1. $R_{x=-4}(ABCD)$

2. $T_{<-6,\,8>}(ABCD)$

3. $r_{(90°,\,O)}(ABCD)$

4. $D_{\frac{2}{3}}(ABCD)$

5. $(R_{x=0} \circ T_{<0,\,5>})(ABCD)$

6. $R_{y=x}(ABCD)$

7. $D_3(ABCD)$

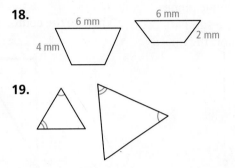

8. Write the translation rule that maps $P(-4, 2)$ to $P'(-1, -1)$.

What type of transformation has the same effect as each composition of transformations?

9. $R_{x=6} \circ T_{<0,\,-5>}$

10. $T_{<-3,\,2>} \circ T_{<8,\,-4>}$

11. $R_{x=4} \circ R_{x=-2}$

12. $R_{y=x} \circ R_{y=-x}$

What type(s) of symmetry does each figure have?

13.   14.   15.

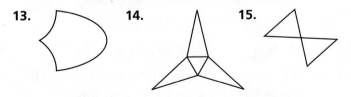

For Exercises 16 and 17, find the coordinates of the vertices of $\triangle XYZ$ with vertices $X(3, 4)$, $Y(2, 1)$, and $Z(-2, 2)$ for each similarity transformation.

16. $(r_{(90°,\,O)} \circ R_{y=1})(\triangle XYZ)$

17. $(r_{(180°,\,O)} \circ T_{<2,\,-1>})(\triangle XYZ)$

Determine whether the figures are similar. If so, identify a similarity transformation that maps one to the other. If not, explain.

18.

19.

Identify the type of isometry that maps the black figure to the blue figure.

20.   21.

Do you UNDERSTAND?

22. **Vocabulary** Is a dilation an isometry? Explain.

23. **Writing** Line *m* intersects $\overline{UH}$ at *N*, and $UN = NH$. Must *H* be the reflection image of *U* across line *m*? Explain your reasoning.

24. **Coordinate Geometry** A dilation with center $(0, 0)$ and scale factor 2.5 maps (a, b) to $(10, -25)$. What are the values of *a* and *b*?

25. **Error Analysis** A classmate says that a certain regular polygon has 50° rotational symmetry. Explain your classmate's error.

26. **Reasoning** Choose points *A*, *B*, and *C* in the first quadrant. Find the coordinates of A', B', and C' by multiplying the coordinates of *A*, *B*, and *C* by -2. What composition of transformations maps $\triangle ABC$ onto $\triangle A'B'C'$? Explain.

TIPS FOR SUCCESS

Some problems ask you to perform a transformation on a figure in the coordinate plane. Read the sample question at the right. Then follow the tips to answer it.

$\triangle G'H'K'$ is the image of $\triangle GHK$ for a dilation with center $(0, 0)$ and $H'K' = 8$. What are the coordinates of H'?

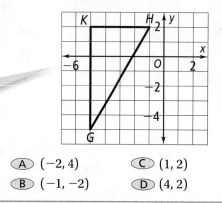

TIP 1

To calculate the scale factor of a dilation, you need to know the lengths of a pair of corresponding sides. You are given $H'K'$. You can use the graph to find HK.

TIP 2

Identify the coordinates of G, H, and K from the graph.

Think It Through

The scale factor is
$$\frac{H'K'}{HK} = \frac{8}{4} = 2.$$

To find the coordinates of H', multiply the coordinates of H by the scale factor. H' is at $(2 \cdot (-1), 2 \cdot 2)$, or $(-2, 4)$. The correct answer is A.

Ⓐ $(-2, 4)$ Ⓒ $(1, 2)$
Ⓑ $(-1, -2)$ Ⓓ $(4, 2)$

Vocabulary Builder

As you solve problems, you must understand the meanings of mathematical terms. Match each term with its mathematical meaning.

A. centroid

B. circumcenter

C. isometry

D. similarity transformation

E. proportion

F. transformation

I. a transformation that preserves distance

II. the point of concurrency of the medians in a triangle

III. an equation that states that two ratios are equal

IV. a mapping that may result in a change in the position, shape, or size of a figure

V. a composition of rigid motions and dilations

VI. the point of concurrency of the perpendicular bisectors of a triangle

Multiple Choice

Read each question. Then write the letter of the correct answer on your paper.

1. Which quadrilateral must have congruent diagonals?

Ⓐ kite   Ⓒ parallelogram
Ⓑ rectangle   Ⓓ rhombus

2. In a 30°-60°-90° triangle, the shortest leg measures 13 in. What is the measure of the longer leg?

Ⓕ 13 in.   Ⓗ $13\sqrt{3}$ in.
Ⓖ $13\sqrt{2}$ in.   Ⓘ 26 in.

3. The vertices of $\square ABCD$ are $A(1, 7)$, $B(0, 0)$, $C(7, -1)$, and $D(8, 6)$. What is the perimeter of $\square ABCD$?

Ⓐ 50   Ⓒ $\sqrt{200}$
Ⓑ 100   Ⓓ $20\sqrt{2}$

4. Mica and Joy are standing at corner *A* of the rectangular field shown below.

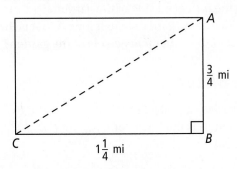

Mica walks diagonally across the field, from corner *A* to corner *C*. Joy walks from corner *A* to corner *B*, and then to corner *C*. To the nearest hundredth of a mile, how much farther did Joy walk than Mica?

- Ⓕ 0.54 mi
- Ⓖ 1 mi
- Ⓗ 1.46 mi
- Ⓘ 2 mi

5. What is the area of the square floor tile?

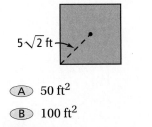

- Ⓐ 50 ft²
- Ⓑ 100 ft²
- Ⓒ 100√2 ft²
- Ⓓ 150 ft²

6. What type of symmetry does the figure have?

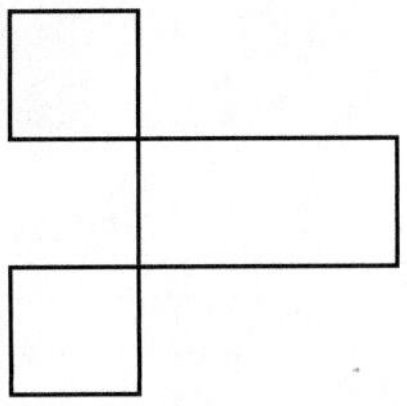

- Ⓕ 60° rotational symmetry
- Ⓖ 90° rotational symmetry
- Ⓗ line symmetry
- Ⓘ point symmetry

7. Which conditions allow you to conclude that a quadrilateral is a parallelogram?

- Ⓐ one pair of sides congruent, the other pair of sides parallel
- Ⓑ perpendicular, congruent diagonals
- Ⓒ diagonals that bisect each other
- Ⓓ one diagonal bisects opposite angles

8. If you are given a line and a point not on the line, what is the first step to construct the line parallel to the given line through the point?

- Ⓕ Construct an angle from a point on the line to the given point.
- Ⓖ Draw a straight line through the given point.
- Ⓗ Draw a ray from the given point that does not intersect the line.
- Ⓘ Label a point on the given line, and draw a line through that point and the given point

9. In a right triangle, which point lies on the hypotenuse?

- Ⓐ incenter
- Ⓒ centroid
- Ⓑ orthocenter
- Ⓓ circumcenter

10. In △*LMN*, *P* is the centroid and *LE* = 24. What is *PE*?

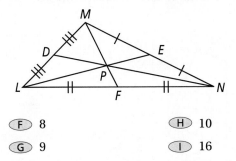

- Ⓕ 8
- Ⓖ 9
- Ⓗ 10
- Ⓘ 16

11. What is the sum of the angle measures of a 32-gon?

- Ⓐ 3200°
- Ⓑ 3800°
- Ⓒ 5400°
- Ⓓ 5580°

12. The diagonals of rectangle *PQRS* intersect at *H*. What is the length of $\overline{QS}$?

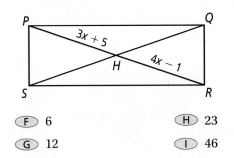

- Ⓕ 6
- Ⓖ 12
- Ⓗ 23
- Ⓘ 46

13. What is the value of x for which $p \parallel q$?

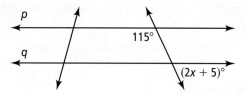

14. What is the measure of $\angle H$?

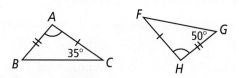

15. What is the area of the square, in square units, in the figure below?

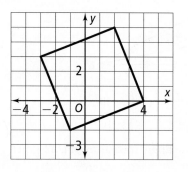

16. In $\square PQRS$, what is the value of x?

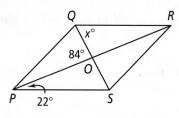

17. For what value of x are the two triangles similar?

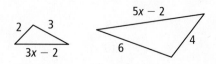

18. Your friend is 5 ft 6 in. tall. When your friend's shadow is 6 ft long, the shadow of a nearby sculpture is 30 ft long. What is the height, in feet, of the sculpture?

19. Lucia makes a triangular garden in one corner of her fenced rectangular backyard. She has 25 ft of edging to use along the unfenced side of the garden. One of the fenced sides of the garden is 15 ft long. What is the length, in feet, of the other fenced side of the garden?

Short Response

20. $\triangle DEB$ has vertices $D(3, 7)$, $E(1, 4)$, and $B(-1, 5)$. In which quadrant(s) is the image of $r_{(270°, O)}(\triangle DEB)$? Draw a diagram.

21. What is the area of an isosceles right triangle whose hypotenuse is $7\sqrt{2}$? Show your work.

22. In $\triangle BGT$, $m\angle B = 48$, $m\angle G = 52$, and $GT = 6$ mm. What is BT? Write your answer to the nearest hundredth of a millimeter.

23. In $\triangle ABC$ below, $\overline{AB} \cong \overline{CB}$ and $\overline{BD} \perp \overline{AC}$. Prove that $\triangle ABD \cong \triangle CBD$.

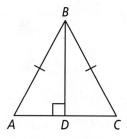

Extended Response

24. Is $\triangle ABC$ a right triangle? Justify your answer.

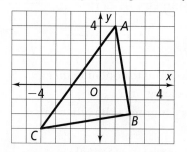

25. $LMNO$ has vertices $L(-4, 0)$, $M(-2, 3)$, $N(1, 1)$, and $O(-1, -2)$. $RSTV$ has vertices $R(1, 1)$, $S(3, -2)$, $T(6, 0)$, and $V(4, 3)$. Graph the two quadrilaterals. Is $LMNO \cong RSTV$? If so, write the rule for the congruence transformation that maps $LMNO$ to $RSTV$. If not, explain why not.

Get Ready!

Skills
Handbook,
p. 888

Squaring Numbers and Finding Square Roots

Simplify.

1. 3^2 **2.** 8^2 **3.** 12^2 **4.** 15^2

5. $\sqrt{16}$ **6.** $\sqrt{64}$ **7.** $\sqrt{100}$ **8.** $\sqrt{169}$

Solve each quadratic equation.

9. $x^2 = 64$ **10.** $b^2 - 225 = 0$ **11.** $a^2 = 144$

Review,
p. 399

Simplifying Radicals

Simplify. Leave your answer in simplest radical form.

12. $\sqrt{8}$ **13.** $\sqrt{27}$ **14.** $\sqrt{75}$ **15.** $4\sqrt{72}$

Lesson 1-8

Area

16. A garden that is 6 ft by 8 ft has a walkway that is 3 ft wide around it. What is the ratio of the area of the garden to the area of the garden and walkway? Write your answer in simplest form.

17. A rectangular rose garden is 8 m by 10 m. One bag of fertilizer can cover 16 m². How many bags of fertilizer will be needed to cover the entire garden?

Lessons 6-3
and 6-5

Classifying Quadrilaterals

Classify each quadrilateral as specifically as possible.

18.   **19.**   **20.**

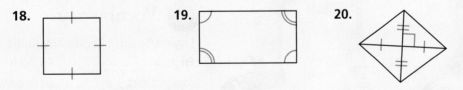

Looking Ahead Vocabulary

21. A *semi*annual school fundraiser is an event that occurs every half year. What might a *semi*circle look like in geometry?

22. A *major* skill is an important skill. How would you describe a *major* arc in geometry?

23. Two buildings are *adjacent* if they are next to each other. What do you think *adjacent* arcs on a geometric figure could be?

CHAPTER
10

Area

Ⓒ **DOMAINS**
- Circles
- Expressing Geometric Properties with Equations
- Modeling with Geometry

The mural at the right must have required a lot of paint. Before you paint, you can estimate the amount to buy based on the area of the wall.

You'll learn and use area formulas in this chapter.

Vocabulary

English/Spanish Vocabulary Audio Online:

English	Spanish
adjacent arcs, *p. 650*	arcos adyacentes
apothem, *p. 629*	apotema
arc length, *p. 653*	longitud de un arco
central angle, *p. 649*	ángulo central
concentric circles, *p. 651*	círculos concéntricos
congruent arcs, *p. 653*	arcos congruentes
diameter, *p. 649*	diámetro
major arc, *p. 649*	arco mayor
minor arc, *p. 649*	arco menor
radius, *pp. 629, 649*	radio
sector of a circle, *p. 661*	sector de un círculo
segment of a circle, *p. 662*	segmento de un círculo

00:04:04   VIDEO

BIG ideas

1 **Measurement**

Essential Question How do you find the area of a polygon or find the circumference and area of a circle?

2 **Similarity**

Essential Question How do perimeters and areas of similar polygons compare?

Chapter Preview

Transforming to Find Area

© **Content Standard**
Prepares for G.GMD.3 Use volume formulas for cylinders, pyramids, cones, and spheres to solve problems.

You can use transformations to find formulas for the areas of polygons. In these activities, you will cut polygons into pieces and use the pieces to form different polygons.

Activity 1

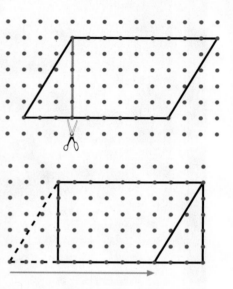

Step 1 Count and record the number of units in the base and the height of the parallelogram at the right.

Step 2 Copy the parallelogram onto grid paper.

Step 3 Cut out the parallelogram. Then cut it into two pieces as shown.

Step 4 Translate the triangle to the right through a distance equal to the base of the parallelogram.

The translation results in a rectangle. Since their pieces are congruent, the parallelogram and rectangle have the same area.

1. How many units are in the base of the rectangle? The height of the rectangle?

2. How do the base and height of the rectangle compare to the base and height of the parallelogram?

3. Write the formula for the area of the rectangle. Explain how you can use this formula to find the area of a parallelogram.

Activity 2

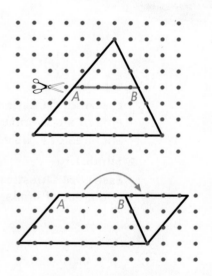

Step 1 Count and record the number of units in the base and the height of the triangle at the right.

Step 2 Copy the triangle onto grid paper. Mark the midpoints A and B and draw midsegment $\overline{AB}$.

Step 3 Cut out the triangle. Then cut it along $\overline{AB}$.

Step 4 Rotate the small triangle 180° about the point B.

The bottom part of the triangle and the image of the top part form a parallelogram.

4. How many units are in the base of the parallelogram? The height of the parallelogram?

5. How do the base and height of the parallelogram compare to the base and height of the original triangle? Write an expression for the height of the parallelogram in terms of the height h of the triangle.

6. Write your formula for the area of a parallelogram from Activity 1. Substitute the expression you wrote for the height of the parallelogram into this formula. You now have a formula for the area of a triangle.

Activity 3

Step 1 Count and record the bases and height of the trapezoid at the right.

Step 2 Copy the trapezoid. Mark the midpoints M and N, and draw midsegment $\overline{MN}$.

Step 3 Cut out the trapezoid. Then cut it along $\overline{MN}$.

Step 4 Transform the trapezoid into a parallelogram.

7. What transformation did you apply to form a parallelogram?

8. What is an expression for the base of the parallelogram in terms of the two bases, b_1 and b_2, of the trapezoid?

9. If h represents the height of the trapezoid, what is an expression in terms of h for the height of the parallelogram?

10. Substitute your expressions from Questions 8 and 9 into your area formula for a parallelogram. What is the formula for the area of a trapezoid?

Exercises

11. In Activity 2, can a different rotation of the small triangle form a parallelogram? If so, does using that rotation change your results? Explain.

12. Make another copy of the Activity 2 triangle. Find a rotation of the entire triangle so that the preimage and image together form a parallelogram. How can you use the parallelogram and your formula for the area of a parallelogram to find the formula for the area of a triangle?

13. a. In the trapezoid at the right, a cut is shown from the midpoint of one leg to a vertex. What transformation can you apply to the top piece to form a triangle from the trapezoid?
  b. Use your formula for the area of a triangle to find a formula for the area of a trapezoid.

14. Count and record the lengths of the diagonals, d_1 and d_2, of the kite at the right. Copy and cut out the kite. Reflect half of the kite across the line of symmetry d_1 by folding the kite along d_1. Use your formula for the area of a triangle to find a formula for the area of a kite.

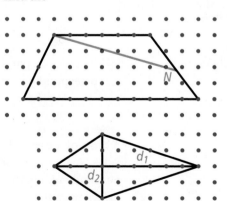

© **Content Standards**
G.MG.1 Use geometric shapes, their measures, and their properties to describe objects.
G.GPE.7 Use coordinates to compute perimeters of polygons and areas of triangles and rectangles.

Objective To find the area of parallelograms and triangles

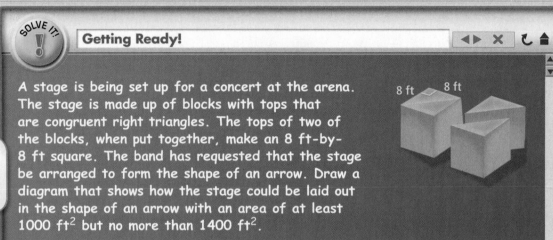

SOLVE IT!

Getting Ready!

A stage is being set up for a concert at the arena. The stage is made up of blocks with tops that are congruent right triangles. The tops of two of the blocks, when put together, make an 8 ft-by-8 ft square. The band has requested that the stage be arranged to form the shape of an arrow. Draw a diagram that shows how the stage could be laid out in the shape of an arrow with an area of at least 1000 ft² but no more than 1400 ft².

You can combine triangles to make just about any shape!

© MATHEMATICAL PRACTICES

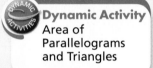

Dynamic Activity
Area of Parallelograms and Triangles

Essential Understanding You can find the area of a parallelogram or a triangle when you know the length of its base and its height.

A parallelogram with the same base and height as a rectangle has the same area as the rectangle.

Lesson Vocabulary
• base of a parallelogram
• altitude of a parallelogram
• height of a parallelogram
• base of a triangle
• height of a triangle

take note

Theorem 10-1   Area of a Rectangle

The area of a rectangle is the product of its base and height.
$$A = bh$$

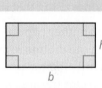

Theorem 10-2   Area of a Parallelogram

The area of a parallelogram is the product of a base and the corresponding height.
$$A = bh$$

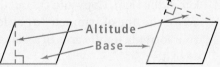

A **base of a parallelogram** can be any one of its sides. The corresponding **altitude** is a segment perpendicular to the line containing that base, drawn from the side opposite the base. The **height** is the length of an altitude.

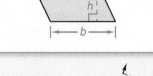

Think

Why aren't the sides of the parallelogram considered altitudes?
Altitudes must be perpendicular to the bases. Unless the parallelogram is also a rectangle, the sides are not perpendicular to the bases.

© **Problem 1** **Finding the Area of a Parallelogram**

What is the area of each parallelogram?

Ⓐ

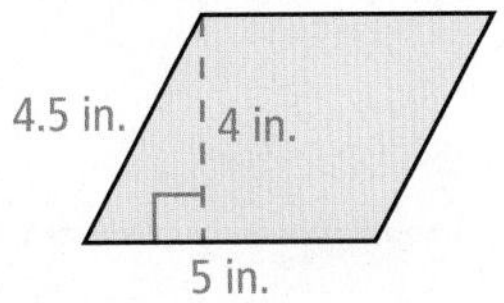

4.5 in. 4 in.

5 in.

Ⓑ

4.6 cm 3.5 cm

2 cm

You are given each height. Choose the corresponding side to use as the base.

$A = bh$

$= 5(4) = 20$ Substitute for *b* and *h*.

The area is 20 in.²

$A = bh$

$= 2(3.5) = 7$

The area is 7 cm².

✅ **Got It?** **1.** What is the area of a parallelogram with base length 12 m and height 9 m?

© **Problem 2** **Finding a Missing Dimension**

Think

What does $\overline{CF}$ represent?
$\overline{CF}$ is an altitude of the parallelogram when $\overline{AD}$ and $\overline{BC}$ are used as bases.

For ☐ABCD, what is DE to the nearest tenth?

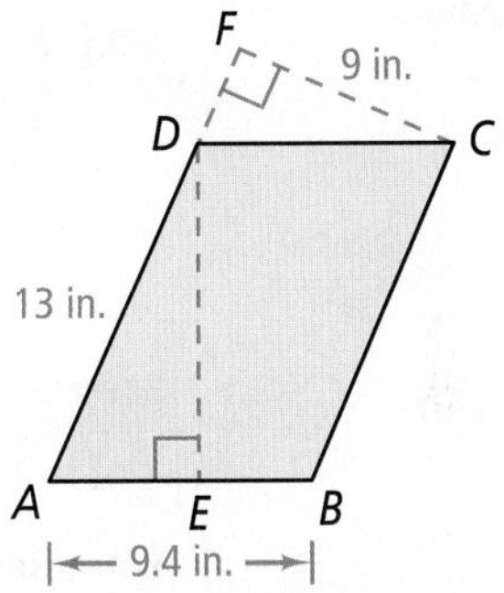

First, find the area of ☐ABCD. Then use the area formula a second time to find DE.

$A = bh$

$= 13(9) = 117$ Use base *AD* and height *CF*.

The area of ☐ABCD is 117 in.²

$A = bh$

$117 = 9.4(DE)$ Use base *AB* and height *DE*.

$DE = \frac{117}{9.4} \approx 12.4$

DE is about 12.4 in.

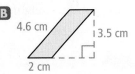

✅ **Got It?** **2.** A parallelogram has sides 15 cm and 18 cm. The height corresponding to a 15-cm base is 9 cm. What is the height corresponding to an 18-cm base?

You can rotate a triangle about the midpoint of a side to form a parallelogram.

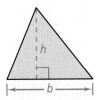

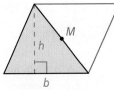

The area of the triangle is half the area of the parallelogram.

Theorem 10-3   Area of a Triangle

The area of a triangle is half the product of a base and the corresponding height.

$$A = \tfrac{1}{2}bh$$

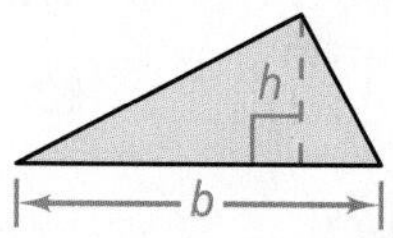

A **base of a triangle** can be any of its sides. The corresponding **height** is the length of the altitude to the line containing that base.

Problem 3   Finding the Area of a Triangle

Plan

Why do you need to convert the base and the height into inches?
You must convert them both because you can only multiply measurements with like units.

Sailing You want to make a triangular sail like the one at the right. How many square feet of material do you need?

Step 1 Convert the dimensions of the sail to inches.

$(12 \text{ ft} \cdot \tfrac{12 \text{ in.}}{1 \text{ ft}}) + 2 \text{ in.} = 146 \text{ in.}$ Use a conversion factor.

$(13 \text{ ft} \cdot \tfrac{12 \text{ in.}}{1 \text{ ft}}) + 4 \text{ in.} = 160 \text{ in.}$

Step 2 Find the area of the triangle.

$A = \tfrac{1}{2}bh$

$ = \tfrac{1}{2}(160)(146)$ Substitute 160 for b and 146 for h.

$ = 11{,}680$ Simplify.

Step 3 Convert 11,680 in.^{2} to square feet.

$11{,}680 \text{ in.}^2 \cdot \dfrac{1 \text{ ft}}{12 \text{ in.}} \cdot \dfrac{1 \text{ ft}}{12 \text{ in.}} = 81\tfrac{1}{9} \text{ ft}^2$

You need $81\tfrac{1}{9}$ ft^2 of material.

12 ft 2 in.

13 ft 4 in.

Got It? 3. What is the area of the triangle?

5 in.

1 ft 1 in.

1 ft

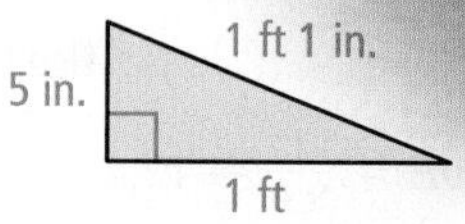

Problem 4   Finding the Area of an Irregular Figure

Plan

How do you know the length of the base of the triangle?
The lower part of the figure is a square. The base length of the triangle is the same as the base length of the square.

What is the area of the figure at the right?

Find the area of each part of the figure.

triangle area $= \tfrac{1}{2}bh = \tfrac{1}{2}(6)8 = 24$ in.^{2}

square area $= bh = 6(6) = 36$ in.^{2}

area of the figure $= 24$ in.$^2 + 36$ in.$^2 = 60$ in.^{2}

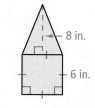

8 in.

6 in.

Got It? 4. Reasoning Suppose the base lengths of the square and triangle in the figure above are doubled to 12 in., but the height of each polygon remains the same. How is the area of the figure affected?

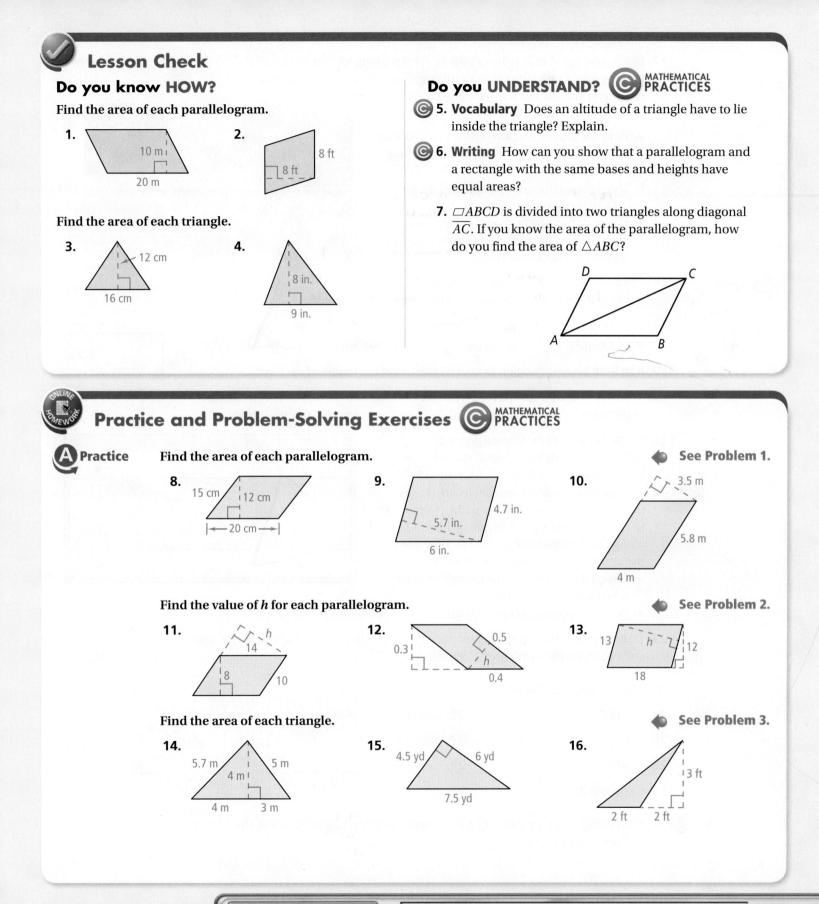

Lesson Check

Do you know HOW?

Find the area of each parallelogram.

1.
10 m
20 m

2.
8 ft
8 ft

Find the area of each triangle.

3.
12 cm
16 cm

4.
8 in.
9 in.

Do you UNDERSTAND? ⓒ MATHEMATICAL PRACTICES

ⓒ **5. Vocabulary** Does an altitude of a triangle have to lie inside the triangle? Explain.

ⓒ **6. Writing** How can you show that a parallelogram and a rectangle with the same bases and heights have equal areas?

7. □*ABCD* is divided into two triangles along diagonal $\overline{AC}$. If you know the area of the parallelogram, how do you find the area of △*ABC*?

D   C

A   B

Practice and Problem-Solving Exercises ⓒ MATHEMATICAL PRACTICES

A Practice   Find the area of each parallelogram.   ◀ See Problem 1.

8.
15 cm   12 cm
20 cm

9.
5.7 in.   4.7 in.
6 in.

10.
3.5 m
5.8 m
4 m

Find the value of *h* for each parallelogram.   ◀ See Problem 2.

11.
h
14
8   10

12.
0.5
0.3
h
0.4

13.
13   h   12
18

Find the area of each triangle.   ◀ See Problem 3.

14.
5.7 m   5 m
4 m
4 m   3 m

15.
4.5 yd   6 yd
7.5 yd

16.
3 ft
2 ft   2 ft

17. Urban Design A bakery has a 50 ft-by-31 ft parking lot. The four parking spaces are congruent parallelograms, the driving region is a rectangle, and the two areas for flowers are congruent triangles.

◀ See Problem 4.

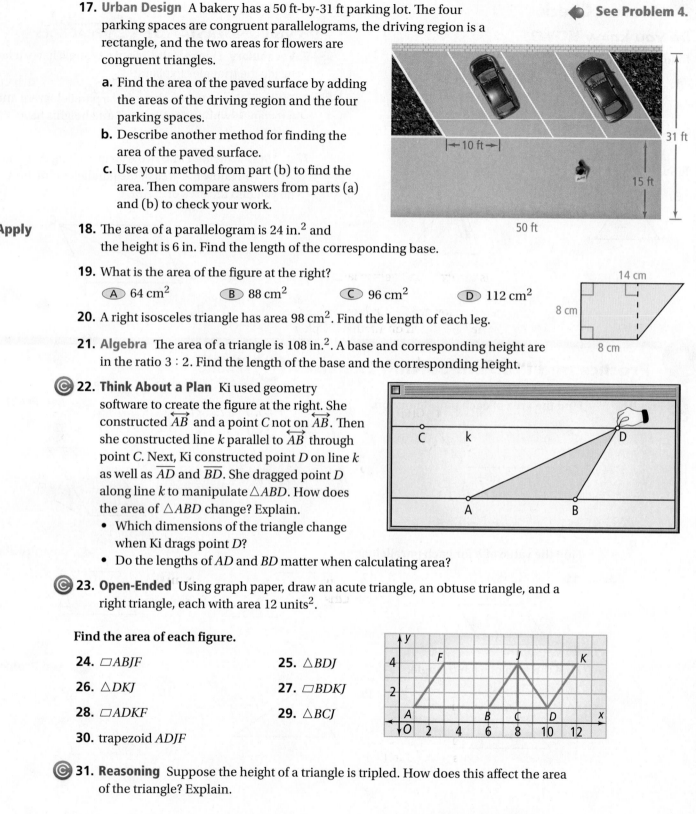

a. Find the area of the paved surface by adding the areas of the driving region and the four parking spaces.

b. Describe another method for finding the area of the paved surface.

c. Use your method from part (b) to find the area. Then compare answers from parts (a) and (b) to check your work.

 Apply

18. The area of a parallelogram is 24 in.^{2} and the height is 6 in. Find the length of the corresponding base.

19. What is the area of the figure at the right?

　Ⓐ 64 cm^2　　　Ⓑ 88 cm^2　　　Ⓒ 96 cm^2　　　Ⓓ 112 cm^2

20. A right isosceles triangle has area 98 cm^2. Find the length of each leg.

21. Algebra The area of a triangle is 108 in.^{2}. A base and corresponding height are in the ratio 3 : 2. Find the length of the base and the corresponding height.

22. Think About a Plan Ki used geometry software to create the figure at the right. She constructed $\overleftrightarrow{AB}$ and a point C not on $\overleftrightarrow{AB}$. Then she constructed line k parallel to $\overleftrightarrow{AB}$ through point C. Next, Ki constructed point D on line k as well as $\overline{AD}$ and $\overline{BD}$. She dragged point D along line k to manipulate $\triangle ABD$. How does the area of $\triangle ABD$ change? Explain.

- Which dimensions of the triangle change when Ki drags point D?
- Do the lengths of AD and BD matter when calculating area?

23. Open-Ended Using graph paper, draw an acute triangle, an obtuse triangle, and a right triangle, each with area 12 units^2.

Find the area of each figure.

24. ▱$ABJF$

25. △BDJ

26. △DKJ

27. ▱$BDKJ$

28. ▱$ADKF$

29. △BCJ

30. trapezoid $ADJF$

31. Reasoning Suppose the height of a triangle is tripled. How does this affect the area of the triangle? Explain.

For Exercises 32–35, (a) graph the lines and (b) find the area of the triangle enclosed by the lines.

32. $y = x, x = 0, y = 7$

33. $y = x + 2, y = 2, x = 6$

34. $y = -\frac{1}{2}x + 3, y = 0, x = -2$

35. $y = \frac{3}{4}x - 2, y = -2, x = 4$

Ⓒ 36. Probability Your friend drew these three figures on a grid. A fly lands at random at a point on the grid.

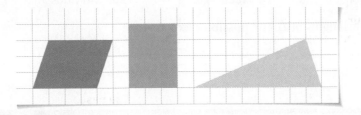

    a. Writing Is the fly more likely to land on one of the figures or on the blank grid? Explain.

    b. Suppose you know the fly lands on one of the figures. Is the fly more likely to land on one figure than on another? Explain.

Coordinate Geometry Find the area of a polygon with the given vertices.

37. $A(3, 9), B(8, 9), C(2, -3), D(-3, -3)$

38. $E(1, 1), F(4, 5), G(11, 5), H(8, 1)$

39. $D(0, 0), E(2, 4), F(6, 4), G(6, 0)$

40. $K(-7, -2), L(-7, 6), M(1, 6), N(7, -2)$

Find the area of each figure.

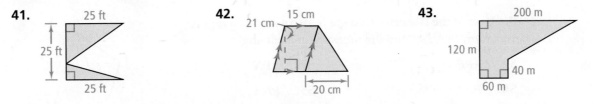

41. 25 ft, 25 ft, 25 ft   **42.** 15 cm, 21 cm, 20 cm   **43.** 200 m, 120 m, 40 m, 60 m

Ⓒ Challenge

History The Greek mathematician Heron is most famous for this formula for the area of a triangle in terms of the lengths of its sides a, b, and c.

$$A = \sqrt{s(s - a)(s - b)(s - c)}, \text{ where } s = \frac{1}{2}(a + b + c)$$

Use Heron's Formula and a calculator to find the area of each triangle. Round your answer to the nearest whole number.

44. $a = 8$ in., $b = 9$ in., $c = 10$ in.

45. $a = 15$ m, $b = 17$ m, $c = 21$ m

46. a. Use Heron's Formula to find the area of this triangle.
    b. Verify your answer to part (a) by using the formula $A = \frac{1}{2}bh$.

15 in.   9 in.   12 in.

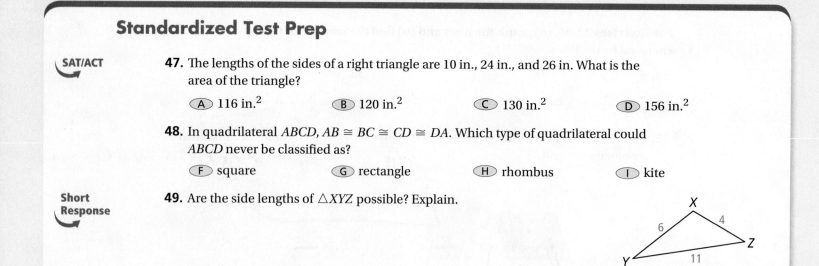

SAT/ACT

47. The lengths of the sides of a right triangle are 10 in., 24 in., and 26 in. What is the area of the triangle?

    Ⓐ 116 in.^{2} Ⓑ 120 in.^{2} Ⓒ 130 in.^{2} Ⓓ 156 in.^{2}

48. In quadrilateral $ABCD$, $AB \cong BC \cong CD \cong DA$. Which type of quadrilateral could $ABCD$ never be classified as?

    Ⓕ square   Ⓖ rectangle   Ⓗ rhombus   Ⓘ kite

Short Response

49. Are the side lengths of $\triangle XYZ$ possible? Explain.

Mixed Review

For each pair of figures, determine whether the figures are similar. If so, describe the similarity transformation.

◀ **See Lesson 9-7.**

50.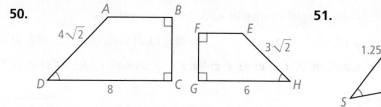

51.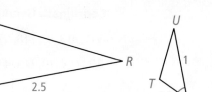

The base of the isosceles triangle is also a side of a regular pentagon *PENTA*. Find the measure of each angle.

◀ **See Lesson 4-5.**

52. $\angle APE$ **53.** $\angle APN$

54. $\angle PAN$ **55.** $\angle PNA$

56. $\angle EPN$ **57.** $\angle ANT$

Get Ready!   To prepare for Lesson 10-2, do Exercises 58–62.

Write the formula for the area of each type of figure.

◀ **See Lesson 10-1.**

58. a rectangle   **59.** a triangle

Find the area of each trapezoid by using the formulas for area of a rectangle and area of a triangle.

60.   **61.**   **62.**

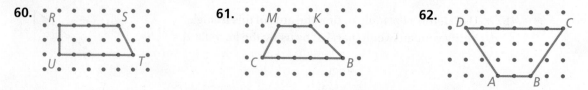

Areas of Trapezoids, Rhombuses, and Kites

© **Content Standard**
G.MG.1 Use geometric shapes, their measures, and their properties to describe objects.

Objective To find the area of a trapezoid, rhombus, or kite

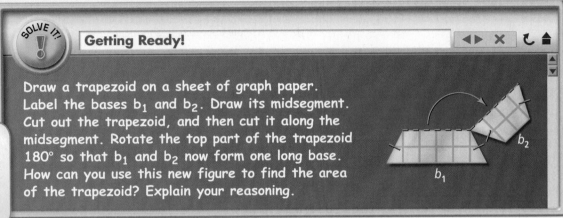

SOLVE IT!

Getting Ready!

Draw a trapezoid on a sheet of graph paper. Label the bases b_1 and b_2. Draw its midsegment. Cut out the trapezoid, and then cut it along the midsegment. Rotate the top part of the trapezoid 180° so that b_1 and b_2 now form one long base. How can you use this new figure to find the area of the trapezoid? Explain your reasoning.

Rearranging figures into familiar shapes is an example of the Solve a Simpler Problem strategy.

© **MATHEMATICAL PRACTICES** **Essential Understanding** You can find the area of a trapezoid when you know its height and the lengths of its bases.

The **height of a trapezoid** is the perpendicular distance between the bases.

Lesson Vocabulary
• height of a trapezoid

take note

Theorem 10-4   Area of a Trapezoid

The area of a trapezoid is half the product of the height and the sum of the bases.

$$A = \frac{1}{2}h(b_1 + b_2)$$

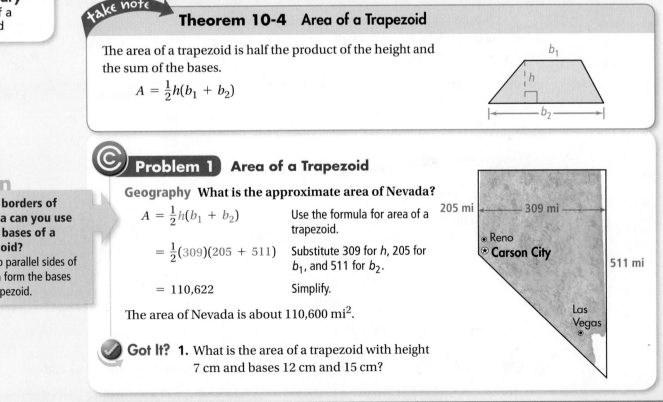

© **Problem 1**   **Area of a Trapezoid**

Plan

Which borders of Nevada can you use as the bases of a trapezoid?
The two parallel sides of Nevada form the bases of a trapezoid.

Geography What is the approximate area of Nevada?

$A = \frac{1}{2}h(b_1 + b_2)$ Use the formula for area of a trapezoid.

$= \frac{1}{2}(309)(205 + 511)$ Substitute 309 for h, 205 for b_1, and 511 for b_2.

$= 110,622$ Simplify.

The area of Nevada is about 110,600 mi^2.

✓ **Got It?** **1.** What is the area of a trapezoid with height 7 cm and bases 12 cm and 15 cm?

Problem 2 Finding Area Using a Right Triangle

What is the area of trapezoid _PQRS_?

You can draw an altitude that divides the trapezoid into a rectangle and a 30°-60°-90° triangle. Since the opposite sides of a rectangle are congruent, the longer base of the trapezoid is divided into segments of lengths 2 m and 5 m.

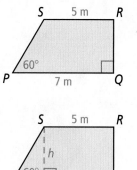

$h = 2\sqrt{3}$	longer leg = shorter leg $\cdot \sqrt{3}$
$A = \frac{1}{2}h(b_1 + b_2)$	Use the trapezoid area formula.
$= \frac{1}{2}(2\sqrt{3})(7 + 5)$	Substitute $2\sqrt{3}$ for h, 7 for b_1, and 5 for b_2.
$= 12\sqrt{3}$	Simplify.

The area of trapezoid _PQRS_ is $12\sqrt{3}$ m².

Got It? **2. Reasoning** In Problem 2, suppose h decreases so that $m\angle P = 45$ while angles R and Q and the bases stay the same. What is the area of trapezoid _PQRS_?

Essential Understanding You can find the area of a rhombus or a kite when you know the lengths of its diagonals.

Theorem 10-5 Area of a Rhombus or a Kite

The area of a rhombus or a kite is half the product of the lengths of its diagonals.

$$A = \frac{1}{2}d_1d_2$$

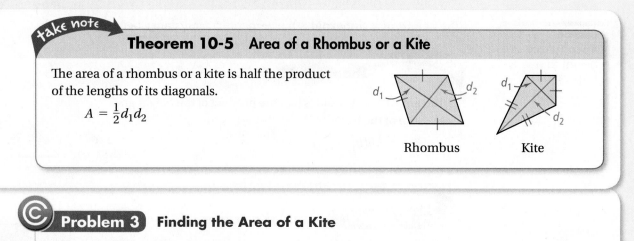

Rhombus   Kite

Problem 3 Finding the Area of a Kite

What is the area of kite _KLMN_?

Find the lengths of the two diagonals:
$KM = 2 + 5 = 7$ m and $LN = 3 + 3 = 6$ m.

$A = \frac{1}{2}d_1d_2$	Use the formula for area of a kite.
$= \frac{1}{2}(7)(6)$	Substitute 7 for d_1 and 6 for d_2.
$= 21$	Simplify.

The area of kite _KLMN_ is 21 m².

Got It? **3.** What is the area of a kite with diagonals that are 12 in. and 9 in. long?

 **Problem 4** **Finding the Area of a Rhombus**

Car Pooling The High Occupancy Vehicle (HOV) lane is marked by a series of "diamonds," or rhombuses painted on the pavement. What is the area of the HOV lane diamond shown at the right?

Think

How can you find the length of $\overline{AB}$?
$\overline{AB}$ is a leg of right $\triangle ABC$. You can use the Pythagorean Theorem, $a^2 + b^2 = c^2$, to find its length.

$\triangle ABC$ is a right triangle. Using the Pythagorean Theorem, $AB = \sqrt{6.5^2 - 2.5^2} = 6$. Since the diagonals of a rhombus bisect each other, the diagonals of the HOV lane diamond are 5 ft and 12 ft.

$A = \frac{1}{2}d_1 d_2$ Use the formula for area of a rhombus.

$= \frac{1}{2}(5)(12)$ Substitute 5 for d_1 and 12 for d_2.

$= 30$ Simplify.

The area of the HOV lane diamond is 30 ft^2.

Got It? **4.** A rhombus has sides 10 cm long. If the longer diagonal is 16 cm, what is the area of the rhombus?

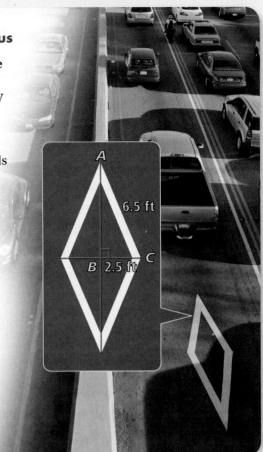

Lesson Check

Do you know HOW?

Find the area of each figure.

Do you UNDERSTAND?
MATHEMATICAL PRACTICES

7. Vocabulary Can a trapezoid and a parallelogram with the same base and height have the same area? Explain.

8. Reasoning Do you need to know all the side lengths to find the area of a trapezoid?

9. Reasoning Can you find the area of a rhombus if you only know the lengths of its sides? Explain.

10. Reasoning Do you need to know the lengths of the sides to find the area of a kite? Explain.

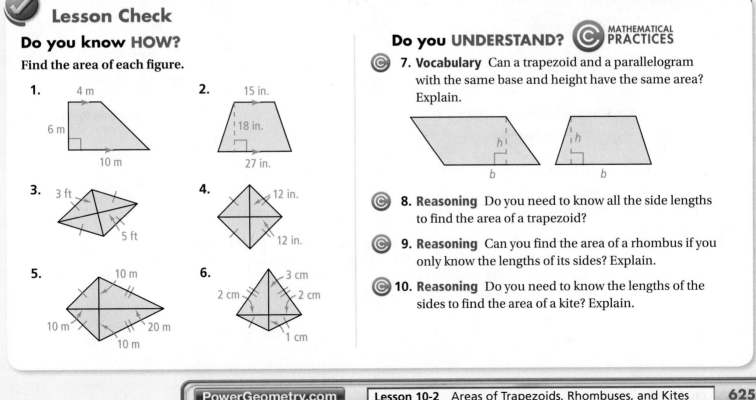

A Practice   Find the area of each trapezoid   ◀ **See Problem 1.**

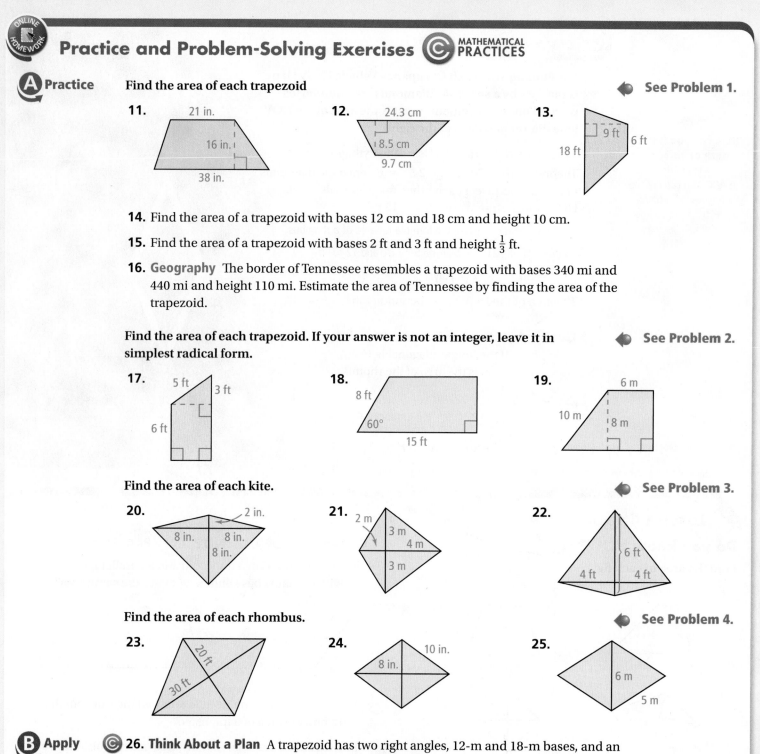

11. 21 in. / 16 in. / 38 in.

12. 24.3 cm / 8.5 cm / 9.7 cm

13. 9 ft / 6 ft / 18 ft

14. Find the area of a trapezoid with bases 12 cm and 18 cm and height 10 cm.

15. Find the area of a trapezoid with bases 2 ft and 3 ft and height $\frac{1}{3}$ ft.

16. Geography The border of Tennessee resembles a trapezoid with bases 340 mi and 440 mi and height 110 mi. Estimate the area of Tennessee by finding the area of the trapezoid.

Find the area of each trapezoid. If your answer is not an integer, leave it in simplest radical form.   ◀ **See Problem 2.**

17. 5 ft / 3 ft / 6 ft

18. 8 ft / 60° / 15 ft

19. 6 m / 10 m / 8 m

Find the area of each kite.   ◀ **See Problem 3.**

20. 2 in. / 8 in. / 8 in. / 8 in.

21. 2 m / 3 m / 4 m / 3 m

22. 6 ft / 4 ft / 4 ft

Find the area of each rhombus.   ◀ **See Problem 4.**

23. 20 ft / 30 ft

24. 10 in. / 8 in.

25. 6 m / 5 m

B Apply   © **26. Think About a Plan** A trapezoid has two right angles, 12-m and 18-m bases, and an 8-m height. Sketch the trapezoid and find its perimeter and area.
   • Are the right angles consecutive or opposite angles?
   • How does knowing the height help you find the perimeter?

27. Metallurgy The end of a gold bar has the shape of a trapezoid with the measurements shown. Find the area of the end.

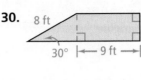
6.9 cm
4.4 cm
9.2 cm

Ⓒ **28. Open-Ended** Draw a kite. Measure the lengths of its diagonals. Find its area.

Find the area of each trapezoid to the nearest tenth.

29.
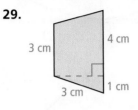
4 cm
3 cm
3 cm
1 cm

30.
8 ft
30°
9 ft

31.
1.7 m 45°
2.1 m
0.9 m

Coordinate Geometry Find the area of quadrilateral *QRST*.

32.
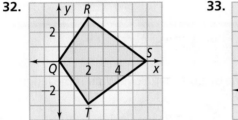

33.
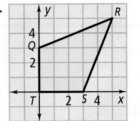

34.
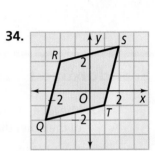

35. What is the area of the kite at the right?

Ⓐ 90 m^2

Ⓒ 135 m^2

Ⓑ 108 m^2

Ⓓ 216 m^2

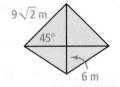
$9\sqrt{2}$ m
45°
6 m

36. a. Coordinate Geometry Graph the lines $x = 0$, $x = 6$, $y = 0$, and $y = x + 4$.
b. What type of quadrilateral do the lines form?
c. Find the area of the quadrilateral.

Find the area of each rhombus. Leave your answer in simplest radical form.

37.
45°
3 cm

38.
60°
4 m

39.
30°
8 in.

Ⓒ **40. Visualization** The kite has diagonals d_1 and d_2 congruent to the sides of the rectangle. Explain why the area of the kite is $\frac{1}{2}d_1d_2$.

Ⓒ **41.** Draw a trapezoid. Label its bases b_1 and b_2 and its height h. Then draw a diagonal of the trapezoid.
a. Write equations for the area of each of the two triangles formed.
b. Writing Explain how you can justify the trapezoid area formula using the areas of the two triangles.

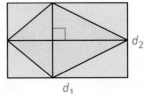
d_2
d_1

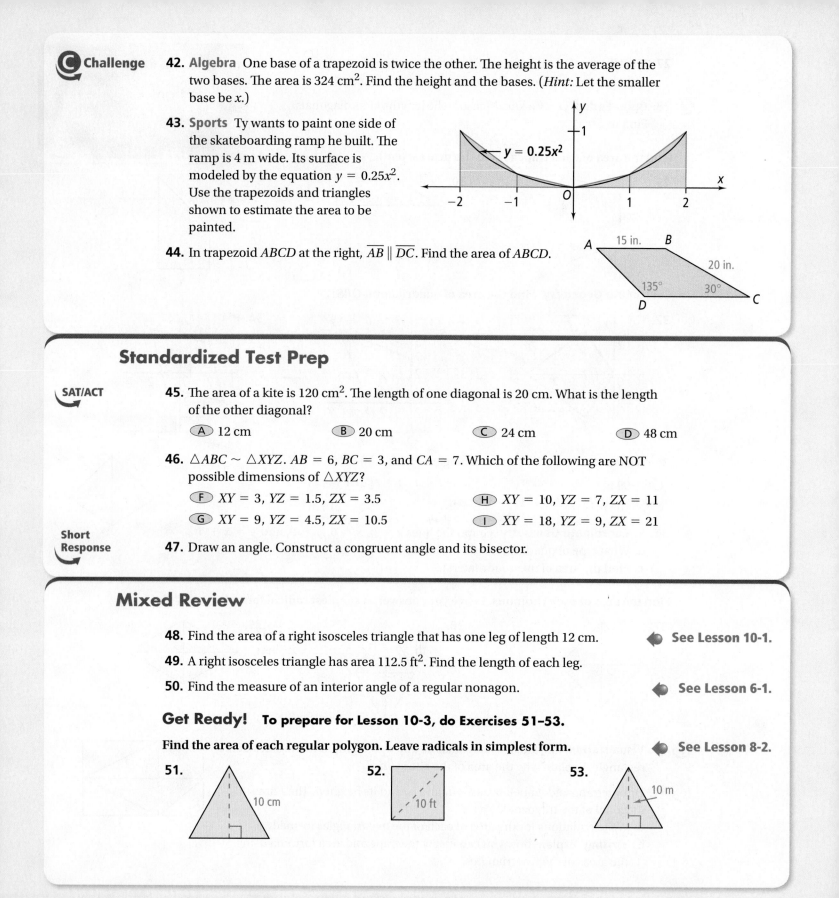

C Challenge

42. Algebra One base of a trapezoid is twice the other. The height is the average of the two bases. The area is 324 cm². Find the height and the bases. (*Hint:* Let the smaller base be *x*.)

43. Sports Ty wants to paint one side of the skateboarding ramp he built. The ramp is 4 m wide. Its surface is modeled by the equation $y = 0.25x^2$. Use the trapezoids and triangles shown to estimate the area to be painted.

$y = 0.25x^2$

44. In trapezoid *ABCD* at the right, $\overline{AB} \parallel \overline{DC}$. Find the area of *ABCD*.

15 in.
20 in.
135°   30°

Standardized Test Prep

45. The area of a kite is 120 cm². The length of one diagonal is 20 cm. What is the length of the other diagonal?

Ⓐ 12 cm   Ⓑ 20 cm   Ⓒ 24 cm   Ⓓ 48 cm

46. $\triangle ABC \sim \triangle XYZ$. $AB = 6$, $BC = 3$, and $CA = 7$. Which of the following are NOT possible dimensions of $\triangle XYZ$?

Ⓕ $XY = 3$, $YZ = 1.5$, $ZX = 3.5$ Ⓗ $XY = 10$, $YZ = 7$, $ZX = 11$

Ⓖ $XY = 9$, $YZ = 4.5$, $ZX = 10.5$ Ⓘ $XY = 18$, $YZ = 9$, $ZX = 21$

47. Draw an angle. Construct a congruent angle and its bisector.

Mixed Review

48. Find the area of a right isosceles triangle that has one leg of length 12 cm.   ◀ **See Lesson 10-1.**

49. A right isosceles triangle has area 112.5 ft². Find the length of each leg.

50. Find the measure of an interior angle of a regular nonagon.   ◀ **See Lesson 6-1.**

Get Ready! **To prepare for Lesson 10-3, do Exercises 51–53.**

Find the area of each regular polygon. Leave radicals in simplest form.   ◀ **See Lesson 8-2.**

51.

10 cm

52.

10 ft

53.

10 m

Areas of Regular Polygons

© Content Standards
G.MG.1 Use geometric shapes, their measures, and their properties to describe objects.
Also G.CO.13

Objective To find the area of a regular polygon

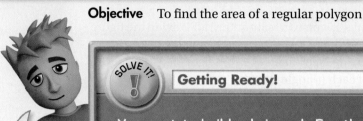

Getting Ready!

You want to build a koi pond. For the border, you plan to use 3-ft-long pieces of wood. You have 12 pieces that you can connect together at any angle, including a straight angle. If you want to maximize the area of the pond, in what shape should you arrange the pieces? Explain your reasoning.

Solve a simpler problem. Try using fewer sides to see what happens.

© MATHEMATICAL PRACTICES

The Solve It involves the area of a polygon.

Lesson Vocabulary
- radius of a regular polygon
- apothem

Essential Understanding The area of a regular polygon is related to the distance from the center to a side.

You can circumscribe a circle about any regular polygon. The center of a regular polygon is the center of the circumscribed circle. The **radius of a regular polygon** is the distance from the center to a vertex. The **apothem** is the perpendicular distance from the center to a side.

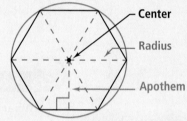

Center
Radius
Apothem

Problem 1 Finding Angle Measures

Think

How do you know the radii make isosceles triangles?
Since the pentagon is a regular polygon, the radii are congruent. So, the triangle made by two adjacent radii and a side of the polygon is an isosceles triangle.

The figure at the right is a regular pentagon with radii and an apothem drawn. What is the measure of each numbered angle?

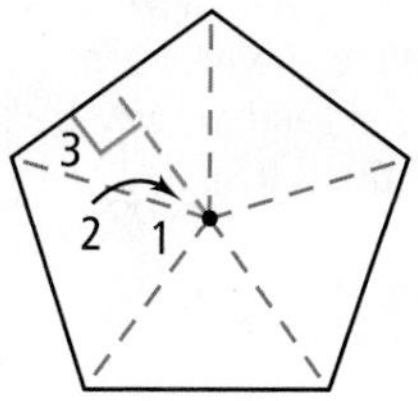

$m\angle 1 = \frac{360}{5} = 72$ Divide 360 by the number of sides.

$m\angle 2 = \frac{1}{2}m\angle 1$ The apothem bisects the vertex angle of the isosceles triangle formed by the radii.

$= \frac{1}{2}(72) = 36$

$90 + 36 + m\angle 3 = 180$ The sum of the measures of the angles of a triangle is 180.

$m\angle 3 = 54$

$m\angle 1 = 72, m\angle 2 = 36,$ and $m\angle 3 = 54.$

Got It? **1.** At the right, a portion of a regular octagon has radii and an apothem drawn. What is the measure of each numbered angle?

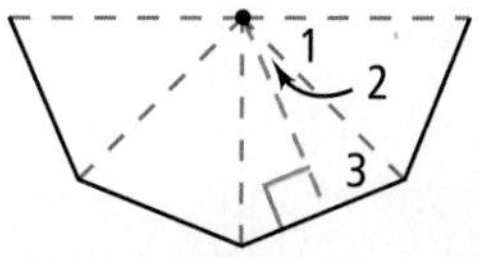

Postulate 10-1

If two figures are congruent, then their areas are equal.

Suppose you have a regular *n*-gon with side *s*. The radii divide the figure into *n* congruent isosceles triangles. By Postulate 10-1, the areas of the isosceles triangles are equal. Each triangle has a height of *a* and a base of length *s*, so the area of each triangle is $\frac{1}{2}as$.

Since there are *n* congruent triangles, the area of the *n*-gon is $A = n \cdot \frac{1}{2}as$. The perimeter *p* of the *n*-gon is the number of sides *n* times the length of a side *s*, or *ns*. By substitution, the area can be expressed as $A = \frac{1}{2}ap$.

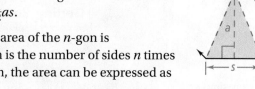

Theorem 10-6   Area of a Regular Polygon

The area of a regular polygon is half the product of the apothem and the perimeter.

$$A = \frac{1}{2}ap$$

Problem 2   **Finding the Area of a Regular Polygon**

Plan

What do you know about the regular decagon?
A decagon has 10 sides, so *n* = 10. From the diagram, you know that the apothem *a* is 12.3 in., and the side length *s* is 8 in.

What is the area of the regular decagon at the right?

Step 1   Find the perimeter of the regular decagon.

$p = ns$ Use the formula for the perimeter of an *n*-gon.

$\quad = 10(8)$ Substitute 10 for *n* and 8 for *s*.

$\quad = 80$ in.

Step 2   Find the area of the regular decagon.

$A = \frac{1}{2}ap$ Use the formula for the area of a regular polygon.

$\quad = \frac{1}{2}(12.3)(80)$ Substitute 12.3 for *a* and 80 for *p*.

$\quad = 492$

The regular decagon has an area of 492 in.^{2}.

Got It? **2. a.** What is the area of a regular pentagon with an 8-cm apothem and 11.6-cm sides?

b. Reasoning If the side of a regular polygon is reduced to half its length, how does the perimeter of the polygon change? Explain.

Problem 3 Using Special Triangles to Find Area **STEM**

Zoology A honeycomb is made up of regular hexagonal cells. The length of a side of a cell is 3 mm. What is the area of a cell?

Know
You know the length of a side, which you can use to find the perimeter.

Need
The apothem

Plan
Draw a diagram to help find the apothem. Then use the area formula for a regular polygon.

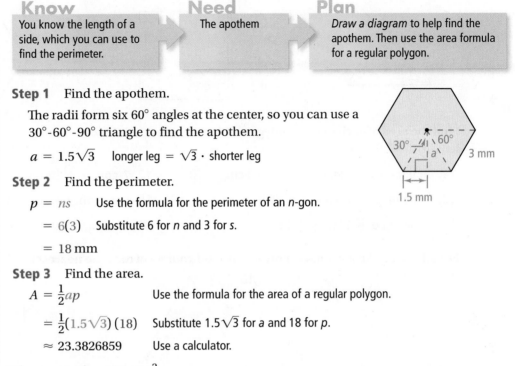

Step 1 Find the apothem.

The radii form six 60° angles at the center, so you can use a 30°-60°-90° triangle to find the apothem.

$a = 1.5\sqrt{3}$ longer leg $= \sqrt{3} \cdot$ shorter leg

Step 2 Find the perimeter.

$p = ns$ Use the formula for the perimeter of an *n*-gon.

$\quad = 6(3)$ Substitute 6 for *n* and 3 for *s*.

$\quad = 18$ mm

Step 3 Find the area.

$A = \frac{1}{2}ap$ Use the formula for the area of a regular polygon.

$\quad = \frac{1}{2}(1.5\sqrt{3})(18)$ Substitute $1.5\sqrt{3}$ for *a* and 18 for *p*.

$\quad \approx 23.3826859$ Use a calculator.

The area is about 23 mm^2.

Got It? **3.** The side of a regular hexagon is 16 ft. What is the area of the hexagon? Round your answer to the nearest square foot.

Lesson Check

Do you know HOW?

What is the area of each regular polygon? Round your answer to the nearest tenth.

1.

5 in.

2.

3 ft

3.

2 m

4.

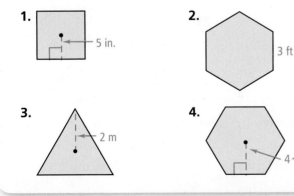

$4\sqrt{3}$

Do you UNDERSTAND? **MATHEMATICAL PRACTICES**

5. Vocabulary What is the difference between a radius and an apothem?

6. What is the relationship between the side length and the apothem in each figure?
   a. a square
   b. a regular hexagon
   c. an equilateral triangle

7. Error Analysis Your friend says you can use special triangles to find the apothem of any regular polygon. What is your friend's error? Explain.

Practice and Problem-Solving Exercises

MATHEMATICAL PRACTICES

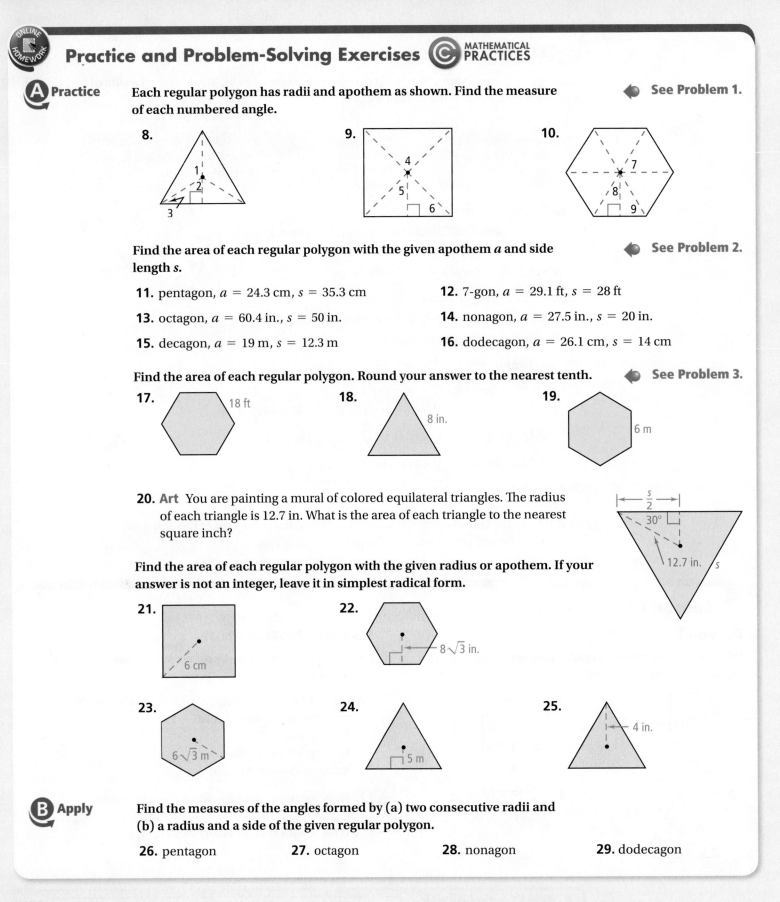

A Practice

Each regular polygon has radii and apothem as shown. Find the measure of each numbered angle.

See Problem 1.

8.

9.

10.

Find the area of each regular polygon with the given apothem *a* and side length *s*.

See Problem 2.

11. pentagon, $a = 24.3$ cm, $s = 35.3$ cm

12. 7-gon, $a = 29.1$ ft, $s = 28$ ft

13. octagon, $a = 60.4$ in., $s = 50$ in.

14. nonagon, $a = 27.5$ in., $s = 20$ in.

15. decagon, $a = 19$ m, $s = 12.3$ m

16. dodecagon, $a = 26.1$ cm, $s = 14$ cm

Find the area of each regular polygon. Round your answer to the nearest tenth.

See Problem 3.

17.

18 ft

18.

8 in.

19.

6 m

20. Art You are painting a mural of colored equilateral triangles. The radius of each triangle is 12.7 in. What is the area of each triangle to the nearest square inch?

Find the area of each regular polygon with the given radius or apothem. If your answer is not an integer, leave it in simplest radical form.

21.

6 cm

22.

$8\sqrt{3}$ in.

23.

$6\sqrt{3}$ m

24.

5 m

25.

4 in.

B Apply

Find the measures of the angles formed by (a) two consecutive radii and (b) a radius and a side of the given regular polygon.

26. pentagon

27. octagon

28. nonagon

29. dodecagon

STEM 30. Satellites One of the smallest space satellites ever developed has the shape of a pyramid. Each of the four faces of the pyramid is an equilateral triangle with sides about 13 cm long. What is the area of one equilateral triangular face of the satellite? Round your answer to the nearest whole number.

31. Think About a Plan The gazebo in the photo is built in the shape of a regular octagon. Each side is 8 ft long, and the enclosed area is 310.4 ft². What is the length of the apothem?
   • How can you *draw a diagram* to help you solve the problem?
   • How can you use the area of a regular polygon formula?

32. A regular hexagon has perimeter 120 m. Find its area.

33. The area of a regular polygon is 36 in.². Find the length of a side if the polygon has the given number of sides. Round your answer to the nearest tenth.
   a. 3   **b.** 4   **c.** 6
   d. Estimation Suppose the polygon is a pentagon. What would you expect the length of a side to be? Explain.

34. A portion of a regular decagon has radii and an apothem drawn. Find the measure of each numbered angle.

35. Writing Explain why the radius of a regular polygon is greater than the apothem.

36. Constructions Use a compass to construct a circle.
   a. Construct two perpendicular diameters of the circle.
   b. Construct diameters that bisect each of the four right angles.
   c. Connect the consecutive points where the diameters intersect the circle. What regular polygon have you constructed?
   d. Reasoning How can a circle help you construct a regular hexagon?

Find the perimeter and area of each regular polygon. Round to the nearest tenth, as necessary.

37. a square with vertices at $(-1, 0)$, $(2, 3)$, $(5, 0)$ and $(2, -3)$

38. an equilateral triangle with two vertices at $(-4, 1)$ and $(4, 7)$

39. a hexagon with two adjacent vertices at $(-2, 1)$ and $(1, 2)$

40. To find the area of an equilateral triangle, you can use the formula $A = \frac{1}{2}bh$ or $A = \frac{1}{2}ap$. A third way to find the area of an equilateral triangle is to use the formula $A = \frac{1}{4}s^2\sqrt{3}$. Verify the formula $A = \frac{1}{4}s^2\sqrt{3}$ in two ways as follows:
   a. Find the area of Figure 1 using the formula $A = \frac{1}{2}bh$.
   b. Find the area of Figure 2 using the formula $A = \frac{1}{2}ap$.

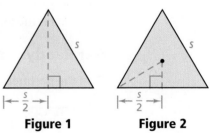

Figure 1   **Figure 2**

41. For Problem 1 on page 629, write a proof that the apothem
Proof bisects the vertex angle of an isosceles triangle formed by two radii.

 **Challenge**

42. Prove that the bisectors of the angles of a regular polygon are concurrent and that
Proof they are, in fact, radii of the polygon. (*Hint:* For regular *n*-gon *ABCDE* . . ., let *P* be
the intersection of the bisectors of $\angle ABC$ and $\angle BCD$. Show that $\overrightarrow{DP}$ must be the
bisector of $\angle CDE$.)

43. Coordinate Geometry A regular octagon with center at the
origin and radius 4 is graphed in the coordinate plane.
  a. Since V_2 lies on the line $y = x$, its *x*- and *y*-coordinates are
  equal. Use the Distance Formula to find the coordinates of V_2
  to the nearest tenth.
  b. Use the coordinates of V_2 and the formula $A = \frac{1}{2}bh$ to find
  the area of $\triangle V_1OV_2$ to the nearest tenth.
  c. Use your answer to part (b) to find the area of the octagon to
  the nearest whole number.

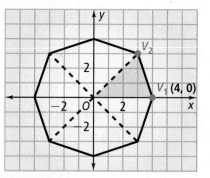

Standardized Test Prep

SAT/ACT

44. What is the area of a regular pentagon with an apothem of 25.1 mm and perimeter
of 182 mm?

  (A) 913.6 mm^2 (B) 2284.1 mm^2 (C) 3654.6 mm^2 (D) 4568.2 mm^2

45. What is the most precise name for a regular polygon with four right angles?

  (F) square   (G) parallelogram   (H) trapezoid   (I) rectangle

46. $\triangle ABC$ has coordinates $A(-2, 4)$, $B(3, 1)$, and $C(0, -2)$. If you reflect $\triangle ABC$ across
the *x*-axis, what are the coordinates of the vertices of the image $\triangle A'B'C'$?

  (A) $A'(2, 4)$, $B'(-3, 1)$, $C'(0, -2)$ (C) $A'(4, -2)$, $B'(1, 3)$, $C'(-2, 0)$
  (B) $A'(-2, -4)$, $B'(3, -1)$, $C'(0, 2)$ (D) $A'(4, 2)$, $B'(1, -3)$, $C'(-2, 0)$

**Short
Response**

47. An equilateral triangle on a coordinate grid has vertices at $(0, 0)$ and $(4, 0)$. What are
the possible locations of the third vertex?

Mixed Review

48. What is the area of a kite with diagonals 8 m and 11.5 m?   ◀ **See Lesson 10-2.**

49. The area of a trapezoid is 42 m^2. The trapezoid has a height of 7 m and one base of 4 m.
What is the length of the other base?

Get Ready!   **To prepare for Lesson 10-4, do Exercises 50–52.**   ◀ **See Lesson 1-8.**

Find the perimeter and area of each figure.

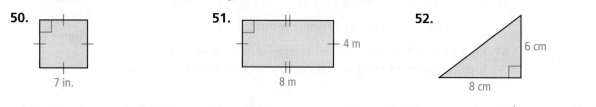

50.   **51.**   **52.**

7 in.   8 m   4 m   8 cm   6 cm

Perimeters and Areas of Similar Figures

Content Standard
Prepares for **G.GMD.3** Use volume formulas for cylinders, pyramids, cones, and spheres to solve problems.

Objective   To find the perimeters and areas of similar polygons

Getting Ready!

◄▷ ✕ ↻ ⌂

On a piece of grid paper, draw a 3 unit-by-4 unit rectangle. Then draw three different rectangles, each similar to the original rectangle. Label them I, II, and III. Use your drawings to complete a chart like this.

You already know that if you double the length and width of a rectangle, its area quadruples.

MATHEMATICAL PRACTICES

Rectangle	Perimeter	Area
Original	■	■
I	■	■
II	■	■
III	■	■

Use the information from the first chart to complete a chart like this.

Rectangle	Scale Factor	Ratio of Perimeters	Ratio of Areas
I to Original	■	■	■
II to Original	■	■	■
III to Original	■	■	■

How do the ratios of perimeters and the ratios of areas compare with the scale factors?

Dynamic Activity
Perimeters and Areas of Similar Figures

In the Solve It, you compared the areas of similar figures.

Essential Understanding You can use ratios to compare the perimeters and areas of similar figures.

take note

Theorem 10-7   Perimeters and Areas of Similar Figures

If the scale factor of two similar figures is $\frac{a}{b}$, then

(1) the ratio of their perimeters is $\frac{a}{b}$ and

(2) the ratio of their areas is $\frac{a^2}{b^2}$.

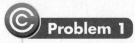

Problem 1   Finding Ratios in Similar Figures

The trapezoids at the right are similar. The ratio of the lengths of corresponding sides is $\frac{6}{9}$, or $\frac{2}{3}$.

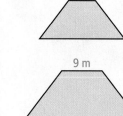

6 m

9 m

A **What is the ratio (smaller to larger) of the perimeters?**

The ratio of the perimeters is the same as the ratio of corresponding sides, which is $\frac{2}{3}$.

B **What is the ratio (smaller to larger) of the areas?**

The ratio of the areas is the square of the ratio of corresponding sides, which is $\frac{2^2}{3^2}$, or $\frac{4}{9}$.

Got It? **1.** Two similar polygons have corresponding sides in the ratio 5 : 7.
   a. What is the ratio (larger to smaller) of their perimeters?
   b. What is the ratio (larger to smaller) of their areas?

When you know the area of one of two similar polygons, you can use a proportion to find the area of the other polygon.

Problem 2   Finding Areas Using Similar Figures

Multiple Choice The area of the smaller regular pentagon is about 27.5 cm^2. What is the best approximation for the area of the larger regular pentagon?

  Ⓐ 11 cm^2 Ⓑ 69 cm^2 Ⓒ 172 cm^2 Ⓓ 275 cm^2

4 cm   10 cm

Regular pentagons are similar because all angles measure 108 and all sides in each pentagon are congruent. Here the ratio of corresponding side lengths is $\frac{4}{10}$, or $\frac{2}{5}$. The ratio of the areas is $\frac{2^2}{5^2}$, or $\frac{4}{25}$.

$$\frac{4}{25} = \frac{27.5}{A}$$ Write a proportion using the ratio of the areas.

$$4A = 687.5$$ Cross Products Property

$$A = \frac{687.5}{4}$$ Divide each side by 4.

$$A = 171.875$$ Simplify.

The area of the larger pentagon is about 172 cm^2. The correct answer is C.

Got It? **2.** The scale factor of two similar parallelograms is $\frac{3}{4}$. The area of the larger parallelogram is 96 in.^{2}. What is the area of the smaller parallelogram?

Think

Do you need to know the shapes of the two plots of land?
No. As long as the plots are similar, you can compare their areas using their scale factor.

Agriculture During the summer, a group of high school students cultivated a plot of city land and harvested 13 bushels of vegetables that they donated to a food pantry. Next summer, the city will let them use a larger, similar plot of land. In the new plot, each dimension is 2.5 times the corresponding dimension of the original plot. How many bushels can the students expect to harvest next year?

The ratio of the dimensions is $2.5 : 1$. So, the ratio of the areas is $(2.5)^2 : 1^2$, or $6.25 : 1$. With 6.25 times as much land next year, the students can expect to harvest $6.25(13)$, or about 81, bushels.

Got It? 3. a. The scale factor of the dimensions of two similar pieces of window glass is $3 : 5$. The smaller piece costs $2.50. How much should the larger piece cost?

b. Reasoning In Problem 3, why is it important that *each* dimension is 2.5 times the corresponding dimension of the original plot? Explain.

When you know the ratio of the areas of two similar figures, you can work backward to find the ratio of their perimeters.

Problem 4 Finding Perimeter Ratios

The triangles at the right are similar. What is the scale factor? What is the ratio of their perimeters?

Know

The areas of the two similar triangles

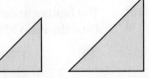

Area = 50 cm²   Area = 98 cm²

Need

The scale factor

Plan

Write a proportion using the ratios of the areas.

$\dfrac{a^2}{b^2} = \dfrac{50}{98}$ Use $a^2 : b^2$ for the ratio of the areas.

$\dfrac{a^2}{b^2} = \dfrac{25}{49}$ Simplify.

$\dfrac{a}{b} = \dfrac{5}{7}$ Take the positive square root of each side.

The ratio of the perimeters equals the scale factor $5 : 7$.

Got It? 4. The areas of two similar rectangles are 1875 ft² and 135 ft². What is the ratio of their perimeters?

Lesson Check

Do you know HOW?

The figures in each pair are similar. What is the ratio of the perimeters and the ratio of the areas?

1.

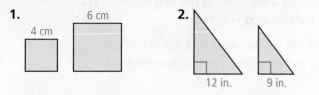

2.

3. In Exercise 2, if the area of the smaller triangle is about 39 ft^2, what is the area of the larger triangle to the nearest tenth?

4. The areas of two similar rhombuses are 48 m^2 and 128 m^2. What is the ratio of their perimeters?

Do you UNDERSTAND?

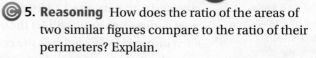

5. Reasoning How does the ratio of the areas of two similar figures compare to the ratio of their perimeters? Explain.

6. Reasoning The area of one rectangle is twice the area of another. What is the ratio of their perimeters? How do you know?

7. Error Analysis Your friend says that since the ratio of the perimeters of two polygons is $\frac{1}{2}$, the area of the smaller polygon must be one half the area of the larger polygon. What is wrong with this statement? Explain.

8. Compare and Contrast How is the relationship between the areas of two congruent figures different from the relationship between the areas of two similar figures?

Practice and Problem-Solving Exercises

A Practice The figures in each pair are similar. Compare the first figure to the second. Give the ratio of the perimeters and the ratio of the areas.

See Problem 1.

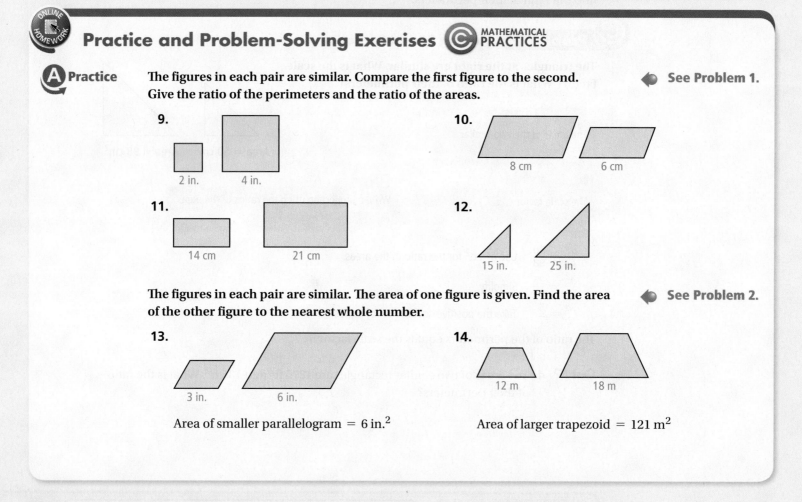

9.

10.

11.

12.

The figures in each pair are similar. The area of one figure is given. Find the area of the other figure to the nearest whole number.

See Problem 2.

13.

Area of smaller parallelogram = 6 in.^{2}

14.

Area of larger trapezoid = 121 m^2

15.

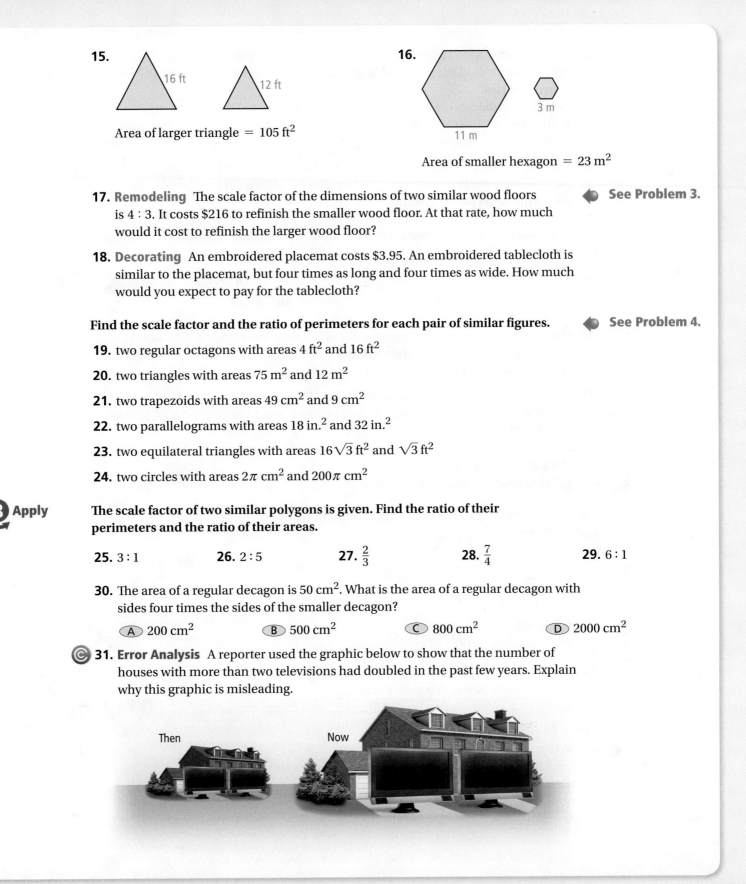

Area of larger triangle $= 105 \text{ ft}^2$

16.

Area of smaller hexagon $= 23 \text{ m}^2$

17. Remodeling The scale factor of the dimensions of two similar wood floors is 4 : 3. It costs \$216 to refinish the smaller wood floor. At that rate, how much would it cost to refinish the larger wood floor?

◀ See Problem 3.

18. Decorating An embroidered placemat costs \$3.95. An embroidered tablecloth is similar to the placemat, but four times as long and four times as wide. How much would you expect to pay for the tablecloth?

Find the scale factor and the ratio of perimeters for each pair of similar figures.

◀ See Problem 4.

19. two regular octagons with areas 4 ft^2 and 16 ft^2

20. two triangles with areas 75 m^2 and 12 m^2

21. two trapezoids with areas 49 cm^2 and 9 cm^2

22. two parallelograms with areas 18 in.^2 and 32 in.^2

23. two equilateral triangles with areas $16\sqrt{3} \text{ ft}^2$ and $\sqrt{3} \text{ ft}^2$

24. two circles with areas $2\pi \text{ cm}^2$ and $200\pi \text{ cm}^2$

B Apply

The scale factor of two similar polygons is given. Find the ratio of their perimeters and the ratio of their areas.

25. $3 : 1$ **26.** $2 : 5$ **27.** $\dfrac{2}{3}$ **28.** $\dfrac{7}{4}$ **29.** $6 : 1$

30. The area of a regular decagon is 50 cm^2. What is the area of a regular decagon with sides four times the sides of the smaller decagon?

Ⓐ 200 cm^2 Ⓑ 500 cm^2 Ⓒ 800 cm^2 Ⓓ 2000 cm^2

© **31. Error Analysis** A reporter used the graphic below to show that the number of houses with more than two televisions had doubled in the past few years. Explain why this graphic is misleading.

Then   Now

32. **Think About a Plan** Two similar rectangles have areas 27 in.^{2} and 48 in.^{2}. The length of one side of the larger rectangle is 16 in. What are the dimensions of both rectangles?
   - How does the ratio of the similar rectangles compare to their scale factor?
   - How can you use the dimensions of the larger rectangle to find the dimensions of the smaller rectangle?

33. The longer sides of a parallelogram are 5 m. The longer sides of a similar parallelogram are 15 m. The area of the smaller parallelogram is 28 m^2. What is the area of the larger parallelogram?

Algebra Find the values of x and y when the smaller triangle shown here has the given area.

34. 3 cm^2

35. 6 cm^2

36. 12 cm^2

37. 16 cm^2

38. 24 cm^2

39. 48 cm^2

STEM 40. **Medicine** For some medical imaging, the scale of the image is 3 : 1. That means that if an image is 3 cm long, the corresponding length on the person's body is 1 cm. Find the actual area of a lesion if its image has area 2.7 cm^2.

41. In $\triangle RST$, $RS = 20$ m, $ST = 25$ m, and $RT = 40$ m.
   a. **Open-Ended** Choose a convenient scale. Then use a ruler and compass to draw $\triangle R'S'T' \sim \triangle RST$.
   b. **Constructions** Construct an altitude of $\triangle R'S'T'$ and measure its length. Find the area of $\triangle R'S'T'$.
   c. **Estimation** Estimate the area of $\triangle RST$.

Compare the blue figure to the red figure. Find the ratios of (a) their perimeters and (b) their areas.

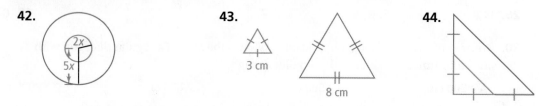

42.

43.

44.

45. a. Find the area of a regular hexagon with sides 2 cm long. Leave your answer in simplest radical form.
   b. Use your answer to part (a) and Theorem 10-7 to find the areas of the regular hexagons shown at the right.

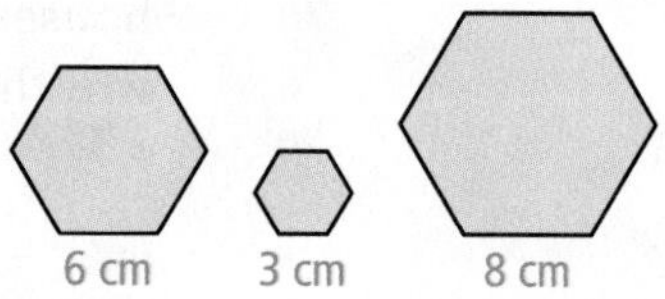

46. **Writing** The enrollment at an elementary school is going to increase from 200 students to 395 students. A parents' group is planning to increase the 100 ft-by-200 ft playground area to a larger area that is 200 ft by 400 ft. What would you tell the parents' group when they ask your opinion about whether the new playground will be large enough?

STEM **47. a. Surveying** A surveyor measured one side and two angles of a field, as shown in the diagram. Use a ruler and a protractor to draw a similar triangle.

   b. Measure the sides and altitude of your triangle and find its perimeter and area.

   © **c. Estimation** Estimate the perimeter and area of the field.

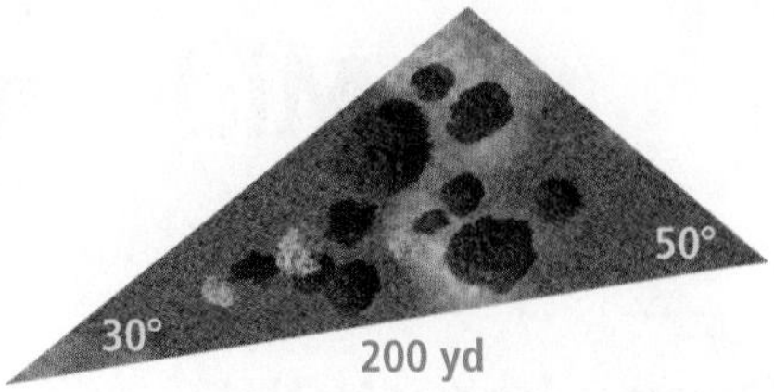

© **Challenge** **Reasoning** Complete each statement with *always, sometimes,* or *never.* **Justify your answers.**

48. Two similar rectangles with the same perimeter are __?__ congruent.

49. Two rectangles with the same area are __?__ similar.

50. Two rectangles with the same area and different perimeters are __?__ similar.

51. Similar figures __?__ have the same area.

Standardized Test Prep

GRIDDED RESPONSE

SAT/ACT

52. Two regular hexagons have sides in the ratio 3 : 5. The area of the smaller hexagon is 81 m^2. In square meters, what is the area of the larger hexagon?

53. What is the value of x in the diagram at the right?

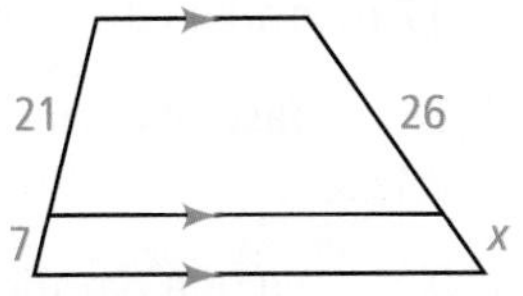

54. A trapezoid has base lengths of 9 in. and 4 in. and a height of 3 in. What is the area of the trapezoid in square inches?

55. In quadrilateral $ABCD$, $m\angle A = 62$, $m\angle B = 101$, and $m\angle C = 42$. What is $m\angle D$?

Mixed Review

Find the area of each regular polygon.

◀ See Lesson 10-3.

56. a square with a 5-cm radius

57. a pentagon with apothem 13.8 and side length 20

58. an octagon with apothem 12 and side length 10

59. An angle bisector divides the opposite side of a triangle into segments 4 cm and 6 cm long. A second side of the triangle is 8 cm long. What are all possible lengths for the third side of the triangle?

◀ See Lesson 7-5.

Get Ready! **To prepare for Lesson 10-5, do Exercises 60–62.**

Find the area of each regular polygon.

◀ See Lesson 10-3.

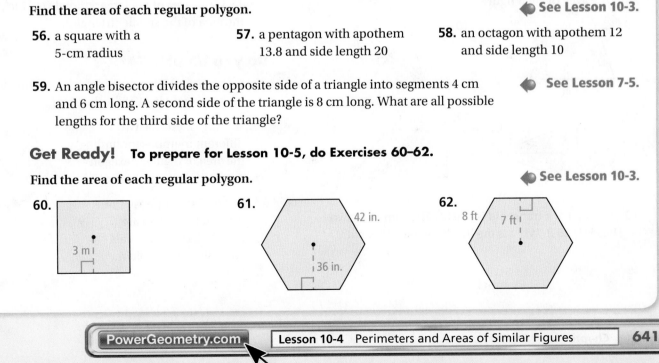

60.

3 m

61.

42 in.

36 in.

62.

8 ft   7 ft

Do you know HOW?

Find the area of each figure.

1.
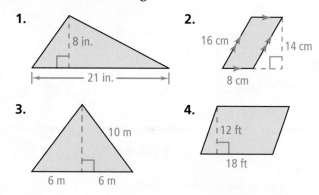
8 in.
21 in.

2.
16 cm   14 cm
8 cm

3.
10 m
6 m   6 m

4.
12 ft
18 ft

5. What is the area of a parallelogram with a base of 17 in. and a corresponding height of 12 in.?

6. If the base of a triangle is 10 cm, and its area is 35 cm², what is the height of the triangle?

7. The area of a parallelogram is 36 in.², and its height is 3 in. How long is the corresponding base?

8. An equilateral triangle has a perimeter of 60 m and a height of 17.3 m. What is its area?

Find the area of each figure.

9.
12 in.
9 in.
18 in.

10.
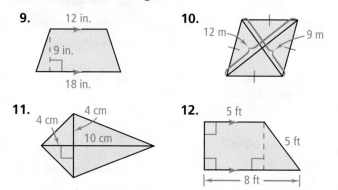
12 m   9 m

11.
4 cm   4 cm
10 cm

12.
5 ft
5 ft
8 ft

13. The area of a trapezoid is 100 ft². The sum of its two bases is 25 ft. What is the height of the trapezoid?

Find the area of each regular polygon. Round your answer to the nearest tenth.

14.
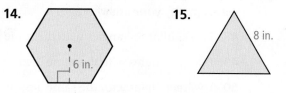
6 in.

15.
8 in.

16. A regular octagon has sides 15 cm long. The apothem is 18.5 cm long. What is the area of the octagon?

17. The radius of a regular hexagon is 5 ft. What is the area of the hexagon to the nearest square foot?

The scale factor of $\triangle ABC$ to $\triangle DEF$ is 3 : 5. Fill in the missing information.

18. The perimeter of $\triangle ABC$ is 36 in.
The perimeter of $\triangle DEF$ is __?__.

19. The area of $\triangle ABC$ is __?__.
The area of $\triangle DEF$ is 125 in.².

20. The areas of two similar triangles are 1.44 and 1.00. Find their scale factor.

21. The ratio of the perimeters of two similar triangles is 1 : 3. The area of the larger triangle is 27 ft². What is the area of the smaller triangle?

Do you UNDERSTAND?

22. Open-Ended Draw a rhombus. Measure the lengths of the diagonals. What is the area?

23. Writing Describe two different methods for finding the area of regular hexagon *ABCDEF*. What is the area?

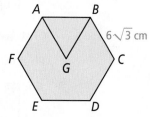
A   B
6√3 cm
F   C
G
E   D

24. Reasoning Suppose the diagonals of a kite are doubled. How does this affect its area? Explain.

Trigonometry and Area

© Content Standard
G.SRT.9 Derive the formula $A = \frac{1}{2}ab\sin(C)$ for the area of a triangle by drawing an auxiliary line from a vertex perpendicular to the opposite side.

Objective   To find areas of regular polygons and triangles using trigonometry

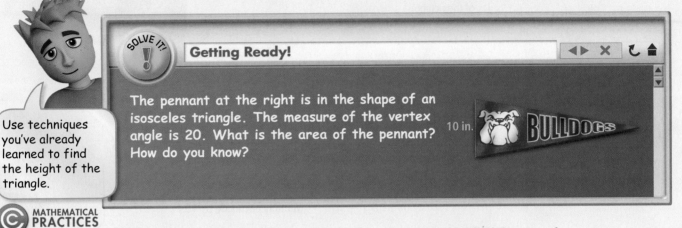

SOLVE IT!

Getting Ready!

The pennant at the right is in the shape of an isosceles triangle. The measure of the vertex angle is 20. What is the area of the pennant? How do you know?

10 in.

Use techniques you've already learned to find the height of the triangle.

MATHEMATICAL PRACTICES

In this lesson you will use isosceles triangles and trigonometry to find the area of a regular polygon.

Essential Understanding   You can use trigonometry to find the area of a regular polygon when you know the length of a side, radius, or apothem.

© Problem 1   **Finding Area**

Think

What is the apothem in the diagram?
The apothem is the altitude of the isosceles triangle. The apothem bisects the central angle and the side of the polygon.

What is the area of a regular nonagon with 10-cm sides?

Draw a regular nonagon with center C. Draw $\overline{CP}$ and $\overline{CR}$ to form isosceles $\triangle PCR$. The measure of central $\angle PCR$ is $\frac{360}{9}$, or 40. The perimeter is $9 \cdot 10$, or 90 cm. Draw the apothem $\overline{CS}$.

$m\angle PCS = \frac{1}{2}m\angle PCR = 20$ and $PS = \frac{1}{2}PR = 5$ cm.

Let a represent CS. Find a and substitute into the area formula.

$\tan 20° = \dfrac{5}{a}$ Use the tangent ratio.

$a = \dfrac{5}{\tan 20°}$ Solve for a.

$A = \frac{1}{2}ap$

$= \frac{1}{2} \cdot \dfrac{5}{\tan 20°} \cdot 90$ Substitute $\frac{5}{\tan 20°}$ for a and 90 for p.

≈ 618.1824194 Use a calculator.

The area of the regular nonagon is about 618 cm².

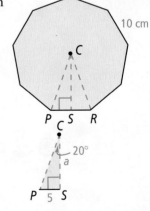

10 cm

$\bullet C$

P S R

C

$20°$

a

P 5 S

✓ Got It?   **1.** What is the area of a regular pentagon with 4-in. sides? Round your answer to the nearest square inch.

Road Signs A stop sign is a regular octagon. The standard size has a 16.2-in. radius. What is the area of the stop sign to the nearest square inch?

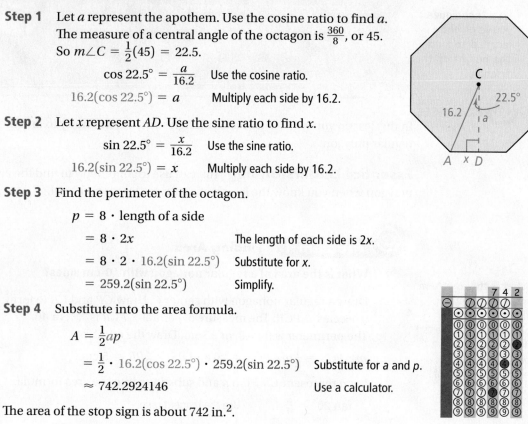

Know

The radius and the number of sides of the octagon

Need

The apothem and the length of a side

Plan

Use trigonometric ratios to find the apothem and the length of a side

Step 1   Let a represent the apothem. Use the cosine ratio to find a.
The measure of a central angle of the octagon is $\frac{360}{8}$, or 45.
So $m\angle C = \frac{1}{2}(45) = 22.5$.

$$\cos 22.5° = \frac{a}{16.2} \qquad \text{Use the cosine ratio.}$$

$$16.2(\cos 22.5°) = a \qquad \text{Multiply each side by 16.2.}$$

Step 2   Let x represent AD. Use the sine ratio to find x.

$$\sin 22.5° = \frac{x}{16.2} \qquad \text{Use the sine ratio.}$$

$$16.2(\sin 22.5°) = x \qquad \text{Multiply each side by 16.2.}$$

Step 3   Find the perimeter of the octagon.

$$p = 8 \cdot \text{length of a side}$$

$$= 8 \cdot 2x \qquad\qquad\qquad \text{The length of each side is } 2x.$$

$$= 8 \cdot 2 \cdot 16.2(\sin 22.5°) \qquad \text{Substitute for } x.$$

$$= 259.2(\sin 22.5°) \qquad\qquad \text{Simplify.}$$

Step 4   Substitute into the area formula.

$$A = \frac{1}{2}ap$$

$$= \frac{1}{2} \cdot 16.2(\cos 22.5°) \cdot 259.2(\sin 22.5°) \qquad \text{Substitute for } a \text{ and } p.$$

$$\approx 742.2924146 \qquad\qquad\qquad\qquad\qquad \text{Use a calculator.}$$

The area of the stop sign is about 742 in.^{2}.

Got It?   **2. a.** A tabletop has the shape of a regular decagon with a radius of 9.5 in. What is the area of the tabletop to the nearest square inch?

     b. Reasoning Suppose the radius of a regular polygon is doubled. How does the area of the polygon change? Explain.

Essential Understanding You can use trigonometry to find the area of a triangle when you know the length of two sides and the included angle.

Suppose you want to find the area of $\triangle ABC$, but you know only $m\angle A$ and the length b and c. To use the formula $A = \frac{1}{2}bh$, you need to know the height. You can find the height by using the sine ratio.

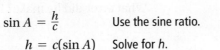

$\sin A = \frac{h}{c}$ Use the sine ratio.

$h = c(\sin A)$ Solve for h.

Now substitute for h in the formula $Area = \frac{1}{2}bh$.

$Area = \frac{1}{2}bc(\sin A)$

This completes the proof of the following theorem for the case in which $\angle A$ is acute.

take note

Theorem 10-8   Area of a Triangle Given SAS

The area of a triangle is half the product of the lengths of two sides and the sine of the included angle.

Area of $\triangle ABC = \frac{1}{2}bc(\sin A)$

Plan

Which formula should you use?
The diagram gives the lengths of two sides and the measure of the included angle. Use the formula for the area of a triangle given SAS.

© Problem 3   Finding Area

What is the area of the triangle?

Area $= \frac{1}{2} \cdot$ side length $\cdot$ side length $\cdot$ sine of included angle

$= \frac{1}{2} \cdot 12 \cdot 21 \cdot \sin 48°$ Substitute.

≈ 93.63624801 Use a calculator.

The area of the triangle is about 94 cm^2.

Got It? 3. What is the area of the triangle? Round your answer to the nearest square inch.

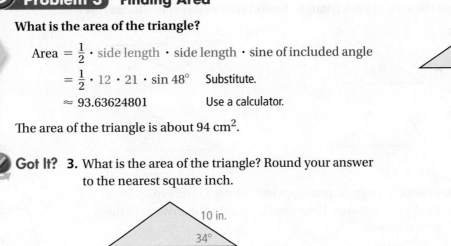

Lesson Check

Do you know HOW?

What is the area of each regular polygon? Round your answers to the nearest tenth.

1.

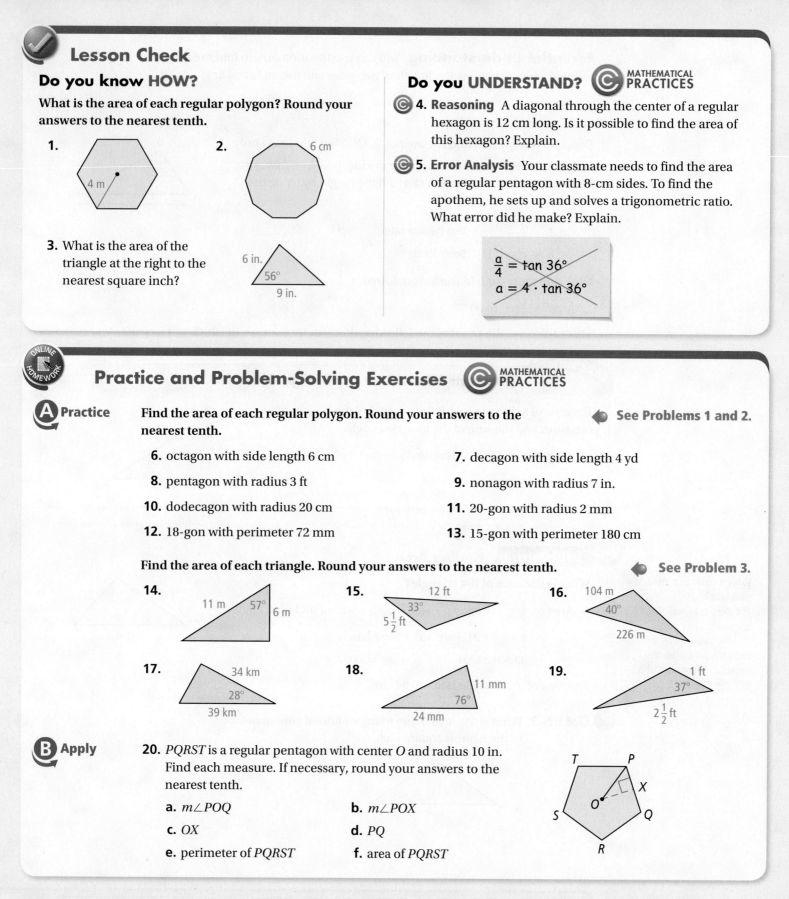

4 m

2.

6 cm

3. What is the area of the triangle at the right to the nearest square inch?

6 in.

56°

9 in.

Do you UNDERSTAND? MATHEMATICAL PRACTICES

Ⓒ 4. Reasoning A diagonal through the center of a regular hexagon is 12 cm long. Is it possible to find the area of this hexagon? Explain.

Ⓒ 5. Error Analysis Your classmate needs to find the area of a regular pentagon with 8-cm sides. To find the apothem, he sets up and solves a trigonometric ratio. What error did he make? Explain.

$$\frac{a}{4} = \tan 36°$$

$$a = 4 \cdot \tan 36°$$

Practice and Problem-Solving Exercises MATHEMATICAL PRACTICES

Ⓐ Practice

Find the area of each regular polygon. Round your answers to the nearest tenth.

◀ See Problems 1 and 2.

6. octagon with side length 6 cm

7. decagon with side length 4 yd

8. pentagon with radius 3 ft

9. nonagon with radius 7 in.

10. dodecagon with radius 20 cm

11. 20-gon with radius 2 mm

12. 18-gon with perimeter 72 mm

13. 15-gon with perimeter 180 cm

Find the area of each triangle. Round your answers to the nearest tenth.

◀ See Problem 3.

14.

11 m 57° 6 m

15.

12 ft

33°

$5\frac{1}{2}$ ft

16.

104 m

40°

226 m

17.

34 km

28°

39 km

18.

11 mm

76°

24 mm

19.

1 ft

37°

$2\frac{1}{2}$ ft

Ⓑ Apply

20. *PQRST* is a regular pentagon with center *O* and radius 10 in. Find each measure. If necessary, round your answers to the nearest tenth.

a. $m\angle POQ$

b. $m\angle POX$

c. *OX*

d. *PQ*

e. perimeter of *PQRST*

f. area of *PQRST*

T P

X

O

S Q

R

© **21. Writing** Describe three ways to find the area of a regular hexagon if you know only the length of a side.

© **22. Think About a Plan** The surveyed lengths of two adjacent sides of a triangular plot of land are 80 yd and 150 yd. The angle between the sides is 67°. What is the area of the parcel of land to the nearest square yard?
- Can you *draw a diagram* to represent the situation?
- Which formula for the area of a triangle should you use?

Find the perimeter and area of each regular polygon to the nearest tenth.

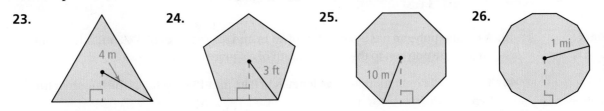

23.   24.   25.   26.

   4 m   3 ft   10 m   1 mi

27. Architecture The Pentagon in Arlington, Virginia, is one of the world's largest office buildings. It is a regular pentagon, and the length of each of its sides is 921 ft. What is the area of land that the Pentagon covers to the nearest thousand square feet?

28. What is the area of the triangle shown at the right?

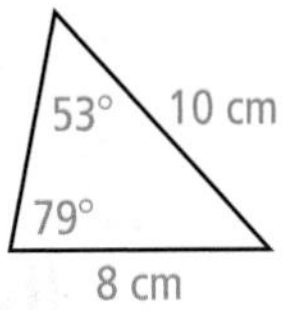

29. The central angle of a regular polygon is 10°. The perimeter of the polygon is 108 cm. What is the area of the polygon?

30. Replacement glass for energy-efficient windows costs $5/ft². About how much will you pay for replacement glass for a regular hexagonal window with a radius of 2 ft?

   Ⓐ $10.39   Ⓑ $27.78   Ⓒ $45.98   Ⓓ $51.96

53° 10 cm
79°
8 cm

Regular polygons A and B are similar. Compare their areas.

31. The apothem of Pentagon A equals the radius of Pentagon B.

32. The length of a side of Hexagon A equals the radius of Hexagon B.

33. The radius of Octagon A equals the apothem of Octagon B.

34. The perimeter of Decagon A equals the length of a side of Decagon B.

The polygons are regular polygons. Find the area of the shaded region.

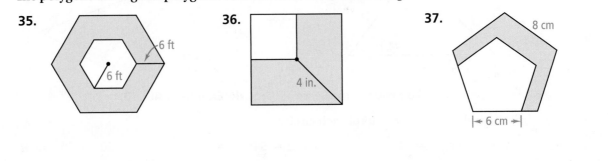

35.   36.   37.

6 ft   8 cm
6 ft   4 in.   6 cm

Challenge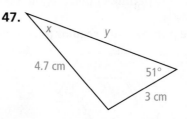

38. Segments are drawn between the midpoints of consecutive sides of a regular pentagon to form another regular pentagon. Find, to the nearest hundredth, the ratio of the area of the smaller pentagon to the area of the larger pentagon.

STEM 39. Surveying A surveyor wants to mark off a triangular parcel with an area of 1 acre (1 acre = 43,560 ft²). One side of the triangle extends 300 ft along a straight road. A second side extends at an angle of 65° from one end of the first side. What is the length of the second side to the nearest foot?

Standardized Test Prep   **GRIDDED RESPONSE**

SAT/ACT

40. A regular polygon has a perimeter of 54 m and an apothem of $3\sqrt{3}$ m. What is the area of the polygon to the nearest tenth of a square meter?

41. The legs of a right triangle have lengths of 8 in. and 15 in. What is the length of the hypotenuse in inches?

42. $\triangle PEN \cong \triangle LIV$. If $m\angle P = 36$ and $m\angle N = 82$, what is $m\angle I$?

43. The perimeter of a parallelogram is 23.6 ft. If its length and width are doubled, what is the perimeter of the parallelogram in feet?

44. The altitude to the hypotenuse of a right triangle divides the hypotenuse into segments of lengths 8 and 10. What is the length of the shorter leg of the triangle?

Mixed Review

45. Two regular octagons are shown.
  a. What is the scale factor of the smaller octagon to the larger octagon?
  b. The area of the larger octagon is 391.1 in.². What is the area of the smaller octagon to the nearest tenth of a square inch?

9 in.   ◀ **See Lesson 10-4.**
6 in.

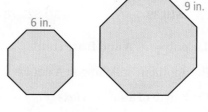

Use the Law of Sines to find the values of x and y. Round to the nearest tenth.   ◀ **See Lesson 8-5.**

46.

x   14 m
44°   53°
y

47.

x   y
4.7 cm
51°
3 cm

Get Ready!   **To prepare for Lesson 10-6, do Exercises 48–50.**

Find the diameter or radius of each circle.   ◀ **See Lesson 1-8.**

48. $r = 7$ cm, $d = $ ■   **49.** $d = 5$ in., $r = $ ■   **50.** $r = 1.6$ m, $d = $ ■

Circles and Arcs

© **Content Standards**
G.CO.1 Know precise definitions of . . . circle . . .
G.C.1 Prove that all circles are similar.
Also G.C.2, G.C.5

Objectives To find the measures of central angles and arcs
To find the circumference and arc length

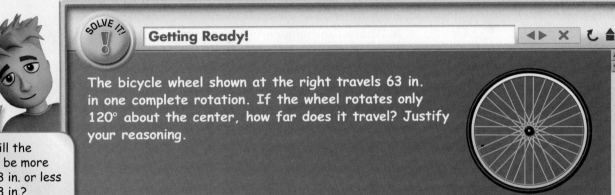

Hm. Will the answer be more than 63 in. or less than 63 in.?

© **MATHEMATICAL PRACTICES**

In a plane, a **circle** is the set of all points equidistant from a given point called the **center.** You name a circle by its center. Circle P ($\odot P$) is shown below.

A **diameter** is a segment that contains the center of a circle and has both endpoints on the circle. A **radius** is a segment that has one endpoint at the center and the other endpoint on the circle. **Congruent circles** have congruent radii. A **central angle** is an angle whose vertex is the center of the circle.

Lesson Vocabulary
- circle
- center
- diameter
- radius
- congruent circles
- central angle
- semicircle
- minor arc
- major arc
- adjacent arcs
- circumference
- pi
- concentric circles
- arc length

P is the center of the circle.

$\overline{AB}$ is a diameter.

$\angle APC$ is a central angle.

$\overline{PC}$ is a radius.

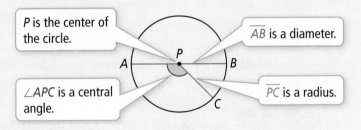

Essential Understanding You can find the length of part of a circle's circumference by relating it to an angle in the circle.

An arc is a part of a circle. One type of arc, a **semicircle,** is half of a circle. A **minor arc** is smaller than a semicircle. A **major arc** is larger than a semicircle. You name a minor arc by its endpoints and a major arc or a semicircle by its endpoints and another point on the arc.

$\overset{\frown}{STR}$ is a major arc.

$\overset{\frown}{RS}$ is a minor arc.

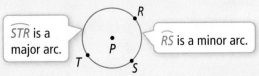

© **Problem 1** **Naming Arcs**

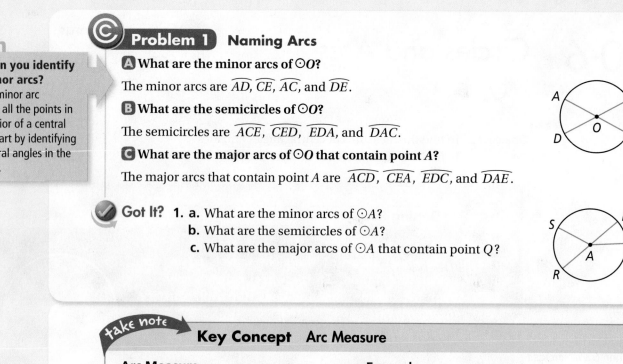

A What are the minor arcs of ⊙*O*?

The minor arcs are $\overset{\frown}{AD}$, $\overset{\frown}{CE}$, $\overset{\frown}{AC}$, and $\overset{\frown}{DE}$.

B What are the semicircles of ⊙*O*?

The semicircles are $\overset{\frown}{ACE}$, $\overset{\frown}{CED}$, $\overset{\frown}{EDA}$, and $\overset{\frown}{DAC}$.

C What are the major arcs of ⊙*O* that contain point *A*?

The major arcs that contain point *A* are $\overset{\frown}{ACD}$, $\overset{\frown}{CEA}$, $\overset{\frown}{EDC}$, and $\overset{\frown}{DAE}$.

Got It? **1. a.** What are the minor arcs of ⊙*A*?

b. What are the semicircles of ⊙*A*?

c. What are the major arcs of ⊙*A* that contain point *Q*?

take note **Key Concept** **Arc Measure**

Arc Measure
The measure of a minor arc is equal to the measure of its corresponding central angle.

The measure of a major arc is the measure of the related minor arc subtracted from 360.

The measure of a semicircle is 180.

Example

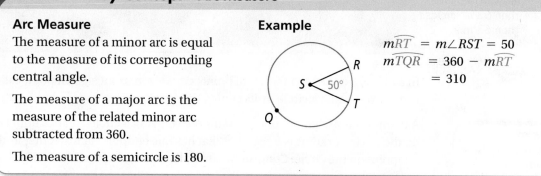

$m\overset{\frown}{RT} = m\angle RST = 50$
$m\overset{\frown}{TQR} = 360 - m\overset{\frown}{RT}$
$\qquad\quad = 310$

Adjacent arcs are arcs of the same circle that have exactly one point in common. You can add the measures of adjacent arcs just as you can add the measures of adjacent angles.

take note **Postulate 10-2** **Arc Addition Postulate**

The measure of the arc formed by two adjacent arcs is the sum of the measures of the two arcs.

$$m\overset{\frown}{ABC} = m\overset{\frown}{AB} + m\overset{\frown}{BC}$$

Think

How can you find $m\widehat{BD}$?
$\widehat{BD}$ is formed by adjacent arcs $\widehat{BC}$ and $\widehat{CD}$. Use the Arc Addition Postulate.

Problem 2 Finding the Measures of Arcs

What is the measure of each arc in $\odot O$?

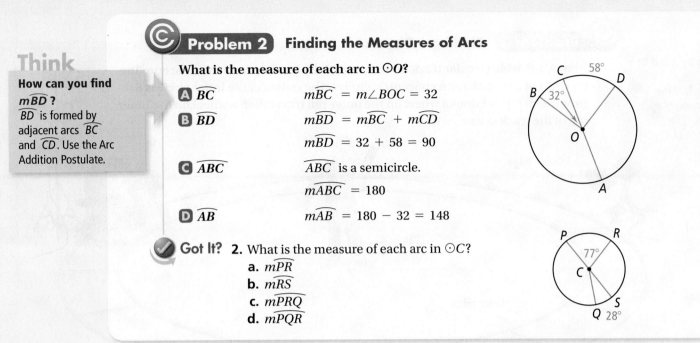

A $\widehat{BC}$ $m\widehat{BC} = m\angle BOC = 32$

B $\widehat{BD}$ $m\widehat{BD} = m\widehat{BC} + m\widehat{CD}$

                 $m\widehat{BD} = 32 + 58 = 90$

C $\widehat{ABC}$ $\widehat{ABC}$ is a semicircle.

                 $m\widehat{ABC} = 180$

D $\widehat{AB}$ $m\widehat{AB} = 180 - 32 = 148$

Got It? **2.** What is the measure of each arc in $\odot C$?

     a. $m\widehat{PR}$
     b. $m\widehat{RS}$
     c. $m\widehat{PRQ}$
     d. $m\widehat{PQR}$

The **circumference** of a circle is the distance around the circle. The number **pi** (π) is the ratio of the circumference of a circle to its diameter.

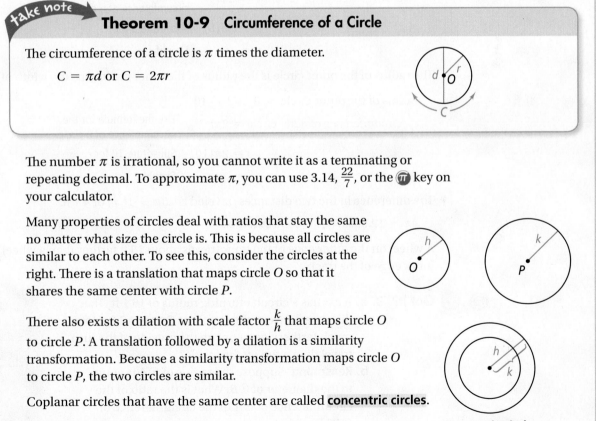

take note

Theorem 10-9   Circumference of a Circle

The circumference of a circle is π times the diameter.

     $C = \pi d$ or $C = 2\pi r$

The number π is irrational, so you cannot write it as a terminating or repeating decimal. To approximate π, you can use 3.14, $\frac{22}{7}$, or the π key on your calculator.

Many properties of circles deal with ratios that stay the same no matter what size the circle is. This is because all circles are similar to each other. To see this, consider the circles at the right. There is a translation that maps circle O so that it shares the same center with circle P.

There also exists a dilation with scale factor $\frac{k}{h}$ that maps circle O to circle P. A translation followed by a dilation is a similarity transformation. Because a similarity transformation maps circle O to circle P, the two circles are similar.

Coplanar circles that have the same center are called **concentric circles**.

Concentric circles

Plan

What do you need to find?
You need to find the distance around the track, which is the circumference of a circle.

Problem 3 Finding a Distance

Film A 2-ft-wide circular track for a camera dolly is set up for a movie scene. The two rails of the track form concentric circles. The radius of the inner circle is 8 ft. How much farther does a wheel on the outer rail travel than a wheel on the inner rail of the track in one turn?

Outer edge

8 ft

← 2 ft →

Inner edge

$$\text{circumference of inner circle} = 2\pi r \quad \text{Use the formula for the circumference of a circle.}$$

$$= 2\pi(8) \quad \text{Substitute 8 for } r.$$

$$= 16\pi \quad \text{Simplify.}$$

The radius of the outer circle is the radius of the inner circle plus the width of the track.

$$\text{radius of the outer circle} = 8 + 2 = 10$$

$$\text{circumference of outer circle} = 2\pi r \quad \text{Use the formula for the circumference of a circle.}$$

$$= 2\pi(10) \quad \text{Substitute 10 for } r.$$

$$= 20\pi \quad \text{Simplify.}$$

The difference in the two distances traveled is $20\pi - 16\pi$, or 4π ft.

$$4\pi \approx 12.56637061 \quad \text{Use a calculator.}$$

A wheel on the outer edge of the track travels about 13 ft farther than a wheel on the inner edge of the track.

Got It? **3. a.** A car has a circular turning radius of 16.1 ft. The distance between the two front tires is 4.7 ft. How much farther does a tire on the outside of the turn travel than a tire on the inside?

b. Reasoning Suppose the radius of $\odot A$ is equal to the diameter of $\odot B$. What is the ratio of the circumference of $\odot A$ to the circumference of $\odot B$? Explain.

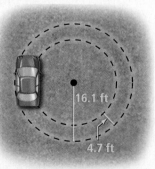

16.1 ft

4.7 ft

The measure of an arc is in degrees, while the **arc length** is a fraction of the circumference.

Consider the arcs shown at the right. Since the circles are concentric, there is a dilation that maps C_1 to C_2. The same dilation maps the slice of the small circle to the slice of the large circle. Since corresponding lengths of similar figures are proportional,

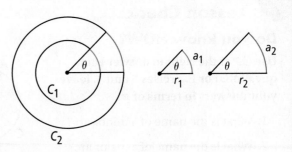

$$\frac{r_1}{r_2} = \frac{a_1}{a_2}$$

$$r_1 a_2 = r_2 a_1$$

$$a_1 = r_1 \frac{a_2}{r_2}$$

This means that the arc length a_1 is equal to the radius r_1 times some number. So for a given central angle, the length of the arc it intercepts depends only on the radius.

An arc of 60° represents $\frac{60}{360}$, or $\frac{1}{6}$, of the circle. So its arc length is $\frac{1}{6}$ of the circumference. This observation suggests the following theorem.

take note

Theorem 10-10   Arc Length

The length of an arc of a circle is the product of the ratio $\frac{\text{measure of the arc}}{360}$ and the circumference of the circle.

$$\text{length of } \widehat{AB} = \frac{m\widehat{AB}}{360} \cdot 2\pi r$$

$$= \frac{m\widehat{AB}}{360} \cdot \pi d$$

© **Problem 4**   **Finding Arc Length**

What is the length of each arc shown in red? Leave your answer in terms of π.

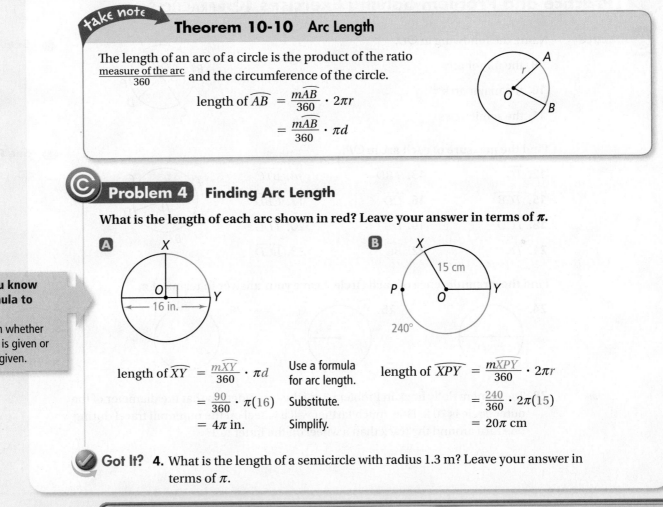

A

$$\text{length of } \widehat{XY} = \frac{m\widehat{XY}}{360} \cdot \pi d \qquad \text{Use a formula for arc length.}$$

$$= \frac{90}{360} \cdot \pi(16) \qquad \text{Substitute.}$$

$$= 4\pi \text{ in.} \qquad \text{Simplify.}$$

B

$$\text{length of } \widehat{XPY} = \frac{m\widehat{XPY}}{360} \cdot 2\pi r$$

$$= \frac{240}{360} \cdot 2\pi(15)$$

$$= 20\pi \text{ cm}$$

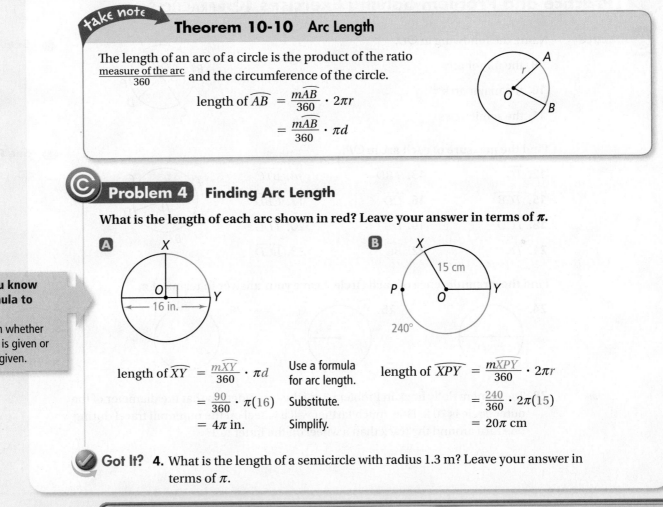

Think

How do you know which formula to use?
It depends on whether the diameter is given or the radius is given.

✓ **Got It?**   **4.** What is the length of a semicircle with radius 1.3 m? Leave your answer in terms of π.

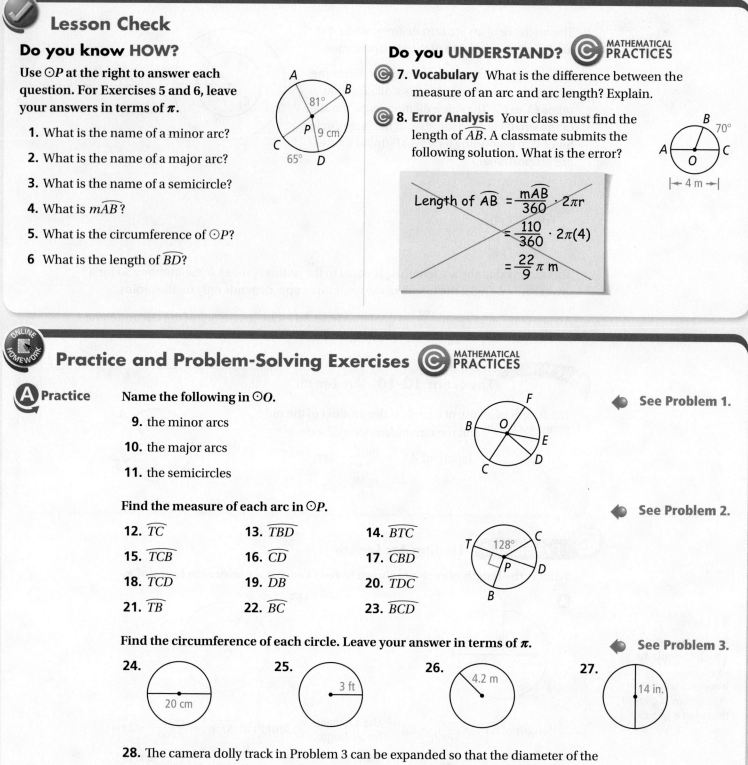

Lesson Check

Do you know HOW?

Use ⊙P at the right to answer each question. For Exercises 5 and 6, leave your answers in terms of π.

1. What is the name of a minor arc?

2. What is the name of a major arc?

3. What is the name of a semicircle?

4. What is $m\widehat{AB}$?

5. What is the circumference of ⊙P?

6 What is the length of $\widehat{BD}$?

Do you UNDERSTAND? MATHEMATICAL PRACTICES

7. Vocabulary What is the difference between the measure of an arc and arc length? Explain.

8. Error Analysis Your class must find the length of $\widehat{AB}$. A classmate submits the following solution. What is the error?

$$\text{Length of } \widehat{AB} = \frac{m\widehat{AB}}{360} \cdot 2\pi r$$
$$= \frac{110}{360} \cdot 2\pi(4)$$
$$= \frac{22}{9} \pi \text{ m}$$

Practice and Problem-Solving Exercises MATHEMATICAL PRACTICES

Practice

Name the following in ⊙O.

See Problem 1.

9. the minor arcs

10. the major arcs

11. the semicircles

Find the measure of each arc in ⊙P.

See Problem 2.

12. $\widehat{TC}$	**13.** $\widehat{TBD}$	**14.** $\widehat{BTC}$
15. $\widehat{TCB}$	**16.** $\widehat{CD}$	**17.** $\widehat{CBD}$
18. $\widehat{TCD}$	**19.** $\widehat{DB}$	**20.** $\widehat{TDC}$
21. $\widehat{TB}$	**22.** $\widehat{BC}$	**23.** $\widehat{BCD}$

Find the circumference of each circle. Leave your answer in terms of π.

See Problem 3.

24. 20 cm

25. 3 ft

26. 4.2 m

27. 14 in.

28. The camera dolly track in Problem 3 can be expanded so that the diameter of the outer circle is 70 ft. How much farther will a wheel on the outer rail travel during one turn around the track than a wheel on the inner rail?

29. The wheel of a compact car has a 25-in. diameter. The wheel of a pickup truck has a 31-in. diameter. To the nearest inch, how much farther does the pickup truck wheel travel in one revolution than the compact car wheel?

Find the length of each arc shown in red. Leave your answer in terms of π.

See Problem 4.

30.

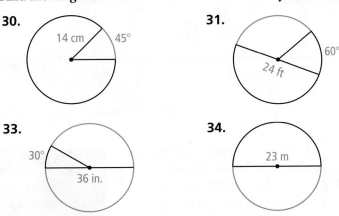

14 cm 45°

31.

60°
24 ft

32.

18 m

33.

30°
36 in.

34.

23 m

35.

9 m   25°

Ⓑ **Apply**

Ⓒ **36. Think About a Plan** Nina designed a semicircular arch made of wrought iron for the top of a mall entrance. The nine segments between the two concentric semicircles are each 3 ft long. What is the total length of wrought iron used to make this structure? Round your answer to the nearest foot.
- What do you know from the diagram?
- What formula should you use to find the amount of wrought iron used in the semicircular arches?

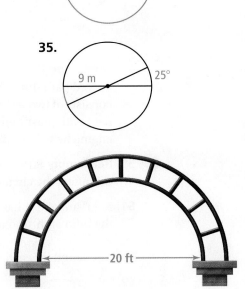

—20 ft—

Find each indicated measure for ⊙*O*.

37. $m\angle EOF$

38. $m\widehat{EJH}$

39. $m\widehat{FH}$

40. $m\angle FOG$

41. $m\widehat{JEG}$

42. $m\widehat{HFJ}$

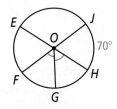

E   J
O   70°
F   H
G

43. Pets A hamster wheel has a 7-in. diameter. How many feet will a hamster travel in 100 revolutions of the wheel?

STEM **44. Traffic** Five streets come together at a traffic circle, as shown at the right. The diameter of the circle traveled by a car is 200 ft. If traffic travels counterclockwise, what is the approximate distance from East St. to Neponset St.?

Ⓐ 227 ft

Ⓒ 454 ft

Ⓑ 244 ft

Ⓓ 488 ft

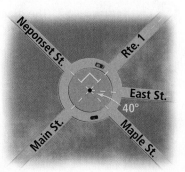

Ⓒ **45. Writing** Describe two ways to find the arc length of a major arc if you are given the measure of the corresponding minor arc and the radius of the circle.

46. Time Hands of a clock suggest an angle whose measure is continually changing. How many degrees does a minute hand move through during each time interval?

a. 1 min   **b.** 5 min   **c.** 20 min

Algebra Find the value of each variable.

47.

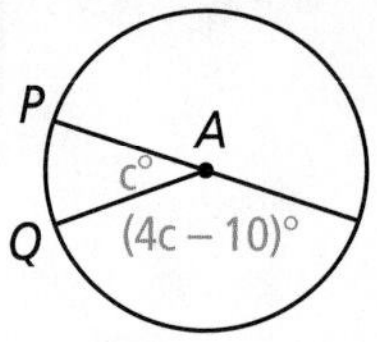

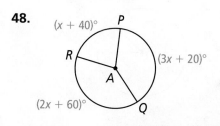

48.

49. Landscape Design A landscape architect is constructing a curved path through a rectangular yard. The curved path consists of two 90° arcs. He plans to edge the two sides of the path with plastic edging. What is the total length of plastic edging he will need? Round your answer to the nearest meter.

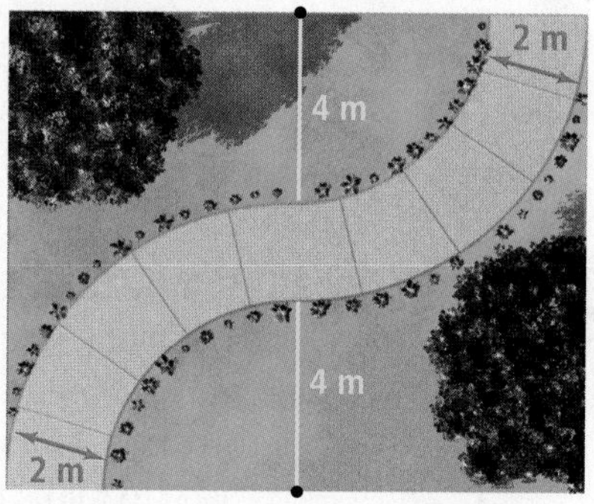

Ⓒ **50. Reasoning** Suppose the radius of a circle is doubled. How does this affect the circumference of the circle? Explain.

51. A 60° arc of ⊙*A* has the same length as a 45° arc of ⊙*B*. What is the ratio of the radius of ⊙*A* to the radius of ⊙*B*?

Find the length of each arc shown in red. Leave your answer in terms of π.

52.   **53.**   **54.**

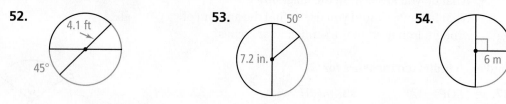

55. Coordinate Geometry Find the length of a semicircle with endpoints $(1, 3)$ and $(4, 7)$. Round your answer to the nearest tenth.

56. In ⊙*O*, the length of $\overset{\frown}{AB}$ is 6π cm and $m\overset{\frown}{AB}$ is 120. What is the diameter of ⊙*O*?

Ⓒ **Challenge**   **57.** The diagram below shows two concentric circles. $\overline{AR} \cong \overline{RW}$. Show that the length of $\overset{\frown}{ST}$ is equal to the length of $\overset{\frown}{QR}$.

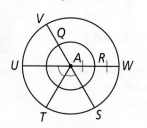

58. Given: ⊙*P* with $\overline{AB} \parallel \overline{PC}$
Proof **Prove:** $m\overset{\frown}{BC} = m\overset{\frown}{CD}$

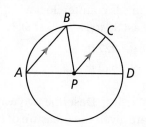

59. Sports An athletic field is a 100 yd-by-40 yd rectangle, with a semicircle at each of the short sides. A running track 10 yd wide surrounds the field. If the track is divided into eight lanes of equal width, what is the distance around the track along the inside edge of each lane?

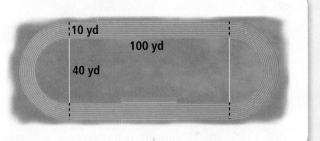

Standardized Test Prep

SAT/ACT

60. The radius of a circle is 12 cm. What is the length of a 60° arc?

- Ⓐ 3π cm
- Ⓑ 4π cm
- Ⓒ 5π cm
- Ⓓ 6π cm

61. What is the image of P for a 135° clockwise rotation about the center of the regular octagon?

- Ⓕ S
- Ⓖ T
- Ⓗ U
- Ⓘ R

62. Which of the following are the sides of a right triangle?

- Ⓐ 6, 8, 12
- Ⓑ 8, 15, 17
- Ⓒ 9, 11, 23
- Ⓓ 5, 12, 15

Extended Response

63. Quadrilateral $ABCD$ has vertices $A(1, 1)$, $B(4, 1)$, $C(4, 6)$, and $D(1, 6)$. Quadrilateral $RSTV$ has vertices $R(-3, 4)$, $S(-3, -2)$, $T(-13 -2)$, and $V(-13, 4)$. Show that $ABCD$ and $RSTV$ are similar rectangles.

Mixed Review

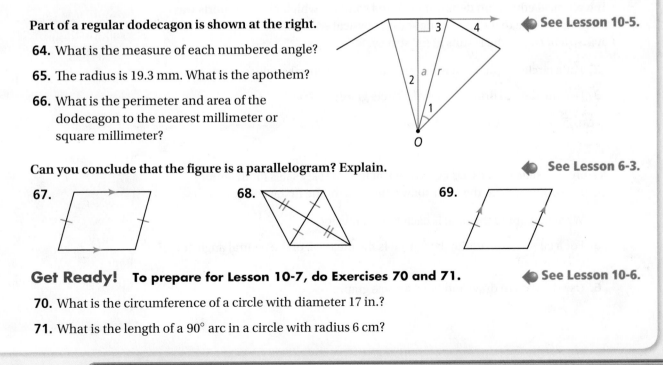

Part of a regular dodecagon is shown at the right.　　See Lesson 10-5.

64. What is the measure of each numbered angle?

65. The radius is 19.3 mm. What is the apothem?

66. What is the perimeter and area of the dodecagon to the nearest millimeter or square millimeter?

Can you conclude that the figure is a parallelogram? Explain.　　See Lesson 6-3.

67.　　　　**68.**　　　　**69.**

Get Ready!　To prepare for Lesson 10-7, do Exercises 70 and 71.　　See Lesson 10-6.

70. What is the circumference of a circle with diameter 17 in.?

71. What is the length of a 90° arc in a circle with radius 6 cm?

Concept Byte

Use With Lesson 10-6

ACTIVITY

Circle Graphs

© **Content Standard**

G.C.2 Identify and describe relationships among inscribed angles, radii, and chords.

Circle graphs show data as percents or fractions of a whole. The total of the data must be 100% or 1. The measure of the central angle for a particular category of the data is proportional to the percent or fraction of the total that the category represents. The measures of the central angles in a circle graph have a total of 360. To find a central angle for a category of data, you multiply the percent or fraction that the category represents by 360.

Activity 1

The circle graph at the right shows the results of a time study in which participants recorded how they spent their time over a 24-h period.

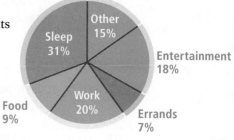

1. What is the measure of the central angle that represents the time spent on each activity? Round to the nearest tenth where necessary.

   a. Sleep   **b.** Food   **c.** Work

   d. Errands   **e.** Entertainment **f.** Other

To make a circle graph, use a compass to make the circle and use a protractor to measure the central angles.

Activity 2

The physical education department asked students which of three sports was their first choice to include in the spring physical education classes. The table at the right shows the results of the survey.

Sport	First Choice
Volleyball	30%
Basketball	25%
Tennis	45%

2. For a circle graph, what is the measure of the central angle for each sport?

3. Use the data to draw and label a circle graph.

Activity 3

A store that sells music CDs keeps track of their weekly sales for inventory purposes. The table at the right shows the sales for the first week in March.

4. What percent of the total is each type of music?

5. For a circle graph of the data, what is the measure of the central angle for each type of music?

6. Use the data to draw and label a circle graph.

Music Genre	Sales (dollars)
Rock	3150
Country	1800
Rap	2250
Classical	1350
Other	450

Concept Byte

Use With Lesson 10-7

ACTIVITY

Exploring the Area of a Circle

Content Standard

G.GMD.1 Give an informal argument for the formulas for the circumference of a circle, area of a circle, volume of a cylinder, pyramid, and cone.

You can use transformations to find the formula for the area of a circle.

Activity

Step 1 Use a compass to draw a large circle. Fold the circle horizontally and then vertically. Cut the circle into four wedges on the fold lines.

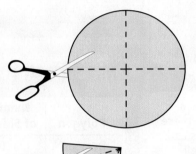

Step 2 Fold each wedge into quarters. Cut each wedge on the fold lines. You will have 16 wedges.

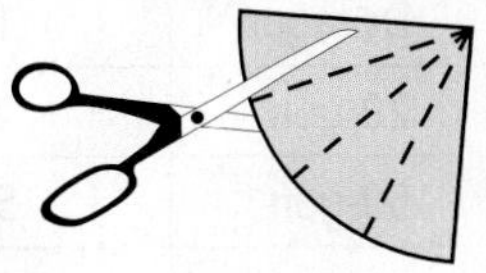

Step 3 Tape the wedges to a piece of paper to form the figure shown at the right. The figure resembles a parallelogram.

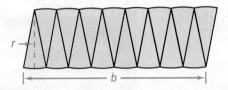

1. How does the area of the parallelogram compare with the area of the circle?

2. The base of the parallelogram is formed by arcs of the circle. Explain how the length b relates to the circumference C of the circle.

3. Explain how the length b relates to the radius r of the circle.

4. Write an expression for the area of the parallelogram in terms of r to write a formula for the area of a circle.

Exercises

Repeat Steps 1 and 2 from the activity. Tape the wedges to a piece of paper to form another figure that resembles a parallelogram, as shown at the right.

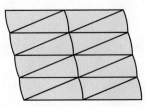

5. What are the base and height of the figure in terms of r?

6. Write an expression for the area of the figure to write a formula for the area of a circle. Is this expression the same as the one you wrote in the activity?

10-7 Areas of Circles and Sectors

© **Content Standard**
G.C.5 Derive using similarity the fact that the length of the arc intercepted by an angle is proportional to the radius, and define the radian measure of the angle as the constant of proportionality; derive the formula for the area of a sector.

Objective To find the areas of circles, sectors, and segments of circles

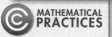

Getting Ready! ◀▶ ✕ ↻ ⌂

Each of the regular polygons in the table has radius 1. Use a calculator to complete the table for the perimeter and area of each polygon. Write out the first five decimal places.

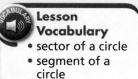

Try to find a pattern in these perimeters and areas to tell you what the circumference and area of a circle should be.

© MATHEMATICAL PRACTICES

Polygon	Number of Sides, n	Length of Side, s	Apothem, a	Perimeter $(P = ns)$	Area $(A = \frac{1}{2}ap)$
Decagon	10	2(sin 18°)	cos 18°	6.18033 . . .	2.93892 . . .
20-gon	20	2(sin 9°)	cos 9°	■	■
50-gon	50	2(sin 3.6°)	cos 3.6°	■	■
100-gon	100	2(sin 1.8°)	cos 1.8°	■	■
1000-gon	1000	2(sin 0.18°)	cos 0.18°	■	■

Look at the results in your table. Notice the perimeter and area of an n-gon as n gets very large. Now consider a circle with radius 1. What are the circumference and area of the circle? Explain your reasoning.

In the Solve It, you explored the area of a circle.

Essential Understanding You can find the area of a circle when you know its radius. You can use the area of a circle to find the area of part of a circle formed by two radii and the arc the radii form when they intersect with the circle.

Lesson Vocabulary
- sector of a circle
- segment of a circle

take note

Theorem 10-11 Area of a Circle

The area of a circle is the product of π and the square of the radius.

$$A = \pi r^2$$

Problem 1 Finding the Area of a Circle

Think

What do you need in order to use the area formula?
You need the radius. The diameter is given, so you can find the radius by dividing the diameter by 2.

Sports What is the area of the circular region on the wrestling mat?

Since the diameter of the region is 32 ft, the radius is $\frac{32}{2}$, or 16 ft.

$A = \pi r^2$ — Use the area formula.

$= \pi(16)^2$ — Substitute 16 for r.

$= 256\pi$ — Simplify.

≈ 804.2477193 — Use a calculator.

The area of the wrestling region is about 804 ft².

Got It? **1. a.** What is the area of a circular wrestling region with a 42-ft diameter?

b. Reasoning If the radius of a circle is halved, how does its area change? Explain.

A **sector of a circle** is a region bounded by an arc of the circle and the two radii to the arc's endpoints. You name a sector using one arc endpoint, the center of the circle, and the other arc endpoint.

The area of a sector is a fractional part of the area of a circle. The area of a sector formed by a 60° arc is $\frac{60}{360}$, or $\frac{1}{6}$, of the area of the circle.

Sector *RPS*

Theorem 10-12 Area of a Sector of a Circle

The area of a sector of a circle is the product of the ratio $\frac{\text{measure of the arc}}{360}$ and the area of the circle.

$$\text{Area of sector } AOB = \frac{m\widehat{AB}}{360} \cdot \pi r^2$$

Problem 2 Finding the Area of a Sector of a Circle

Think

What fraction of a circle's area is the area of a sector formed by a 72° arc?
The area of a sector formed by a 72° arc is $\frac{72}{360}$, or $\frac{1}{5}$, of the area of the circle.

What is the area of sector *GPH*? Leave your answer in terms of π.

area of sector $GPH = \frac{m\widehat{GH}}{360} \cdot \pi r^2$

$= \frac{72}{360} \cdot \pi(15)^2$ — Substitute 72 for $m\widehat{GH}$ and 15 for r.

$= 45\pi$ — Simplify.

The area of sector *GPH* is 45π cm².

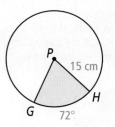

Got It? **2.** A circle has a radius of 4 in. What is the area of a sector bounded by a 45° minor arc? Leave your answer in terms of π.

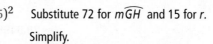

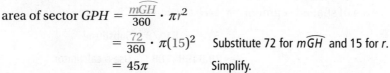

A part of a circle bounded by an arc and the segment joining its endpoints is a **segment of a circle.**

To find the area of a segment for a minor arc, draw radii to form a sector. The area of the segment equals the area of the sector minus the area of the triangle formed.

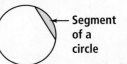
Segment of a circle

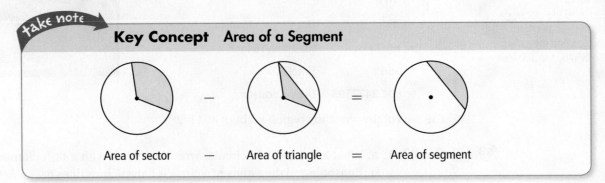

take note

Key Concept   Area of a Segment

Area of sector   −   Area of triangle   =   Area of segment

© **Problem 3**   **Finding the Area of a Segment of a Circle**

What is the area of the shaded segment shown at the right? Round your answer to the nearest tenth.

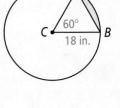

Know
- The radius and $m\widehat{AB}$
- $\overline{CA} \cong \overline{CB}$ and $m\angle ACB$

Need
The area of sector ACB and the area of $\triangle ACB$

Plan
Subtract the area of $\triangle ACB$ from the area of sector ACB.

$$\text{area of sector } ACB = \frac{m\widehat{AB}}{360} \cdot \pi r^2 \qquad \text{Use the formula for area of a sector.}$$

$$= \frac{60}{360} \cdot \pi(18)^2 \qquad \text{Substitute 60 for } m\widehat{AB} \text{ and 18 for } r.$$

$$= 54\pi \qquad \text{Simplify.}$$

$\triangle ACB$ is equilateral. The altitude forms a 30°-60°-90° triangle.

Think

What kind of triangle is $\triangle ACB$?
Since $\overline{CA} \cong \overline{CB}$, the base angles of $\triangle ACB$ are congruent. By the Triangle-Angle-Sum Theorem, $m\angle A = m\angle B = 60$. So, $\triangle ACB$ is equiangular, and therefore equilateral.

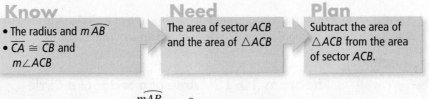

$$\text{area of } \triangle ACB = \frac{1}{2}bh \qquad \text{Use the formula for area of a triangle.}$$

$$= \frac{1}{2}(18)(9\sqrt{3}) \qquad \text{Substitute 18 for } b \text{ and } 9\sqrt{3} \text{ for } h.$$

$$= 81\sqrt{3} \qquad \text{Simplify.}$$

$$\text{area of shaded segment} = \text{area of sector } ACB - \text{area of } \triangle ACB$$

$$= 54\pi - 81\sqrt{3} \qquad \text{Substitute.}$$

$$\approx 29.34988788 \qquad \text{Use a calculator.}$$

The area of the shaded segment is about 29.3 in.^{2}.

✓ **Got It? 3.** What is the area of the shaded segment shown at the right? Round your answer to the nearest tenth.

Lesson Check

Do you know HOW?

1. What is the area of a circle with diameter 16 in.? Leave your answer in terms of π.

Find the area of the shaded region of the circle. Leave your answer in terms of π.

2.
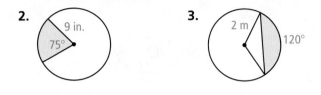
9 in.
75°

3.
2 m
120°

Do you UNDERSTAND? MATHEMATICAL PRACTICES

4. Vocabulary What is the difference between a sector of a circle and a segment of a circle?

5. Reasoning Suppose a sector of $\odot P$ has the same area as a sector of $\odot O$. Can you conclude that $\odot P$ and $\odot O$ have the same area? Explain.

6. Error Analysis Your class must find the area of a sector of a circle determined by a 150° arc. The radius of the circle is 6 cm. What is your classmate's error? Explain.

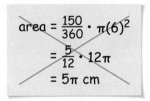

$$\text{area} = \frac{150}{360} \cdot \pi(6)^2$$
$$= \frac{5}{12} \cdot 12\pi$$
$$= 5\pi \text{ cm}$$

Practice and Problem-Solving Exercises MATHEMATICAL PRACTICES

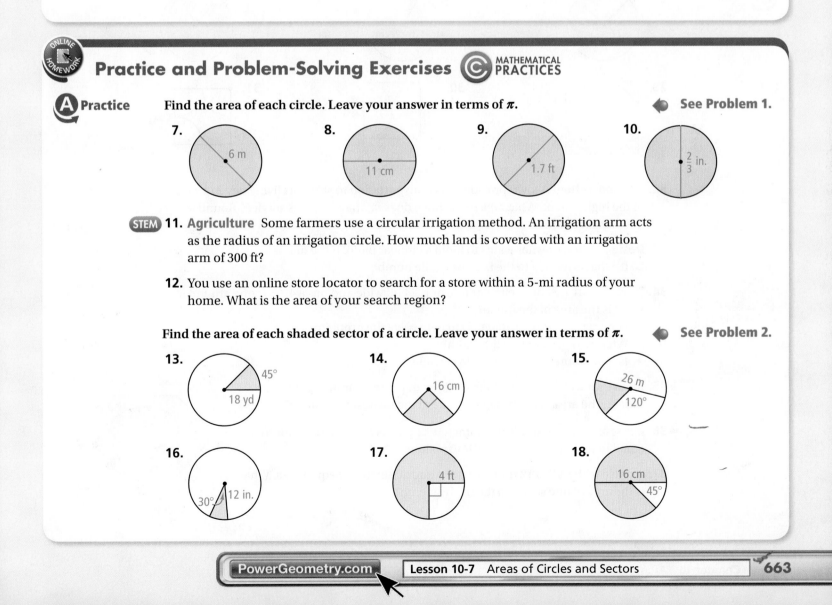

A Practice   **Find the area of each circle. Leave your answer in terms of π.**   ◀ **See Problem 1.**

7.
6 m

8.
11 cm

9.
1.7 ft

10.
$\frac{2}{3}$ in.

STEM **11. Agriculture** Some farmers use a circular irrigation method. An irrigation arm acts as the radius of an irrigation circle. How much land is covered with an irrigation arm of 300 ft?

12. You use an online store locator to search for a store within a 5-mi radius of your home. What is the area of your search region?

Find the area of each shaded sector of a circle. Leave your answer in terms of π.   ◀ **See Problem 2.**

13.
45°
18 yd

14.
16 cm

15.
26 m
120°

16.
30° 12 in.

17.
4 ft

18.
16 cm
45°

Find the area of sector *TOP* in ⊙*O* using the given information. Leave your answer in terms of *π*.

19. $r = 5$ m, $m\widehat{TP} = 90$

20. $r = 6$ ft, $m\widehat{TP} = 15$

21. $d = 16$ in., $m\widehat{PT} = 135$

22. $d = 15$ cm, $m\widehat{POT} = 180$

Find the area of each shaded segment. Round your answer to the nearest tenth.   ◀ **See Problem 3.**

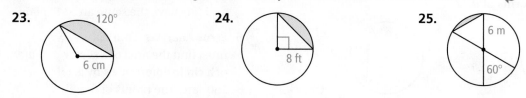

23. 120° / 6 cm

24. 8 ft

25. 6 m / 60°

Find the area of the shaded region. Leave your answer in terms of *π* and in simplest radical form.

Ⓑ Apply

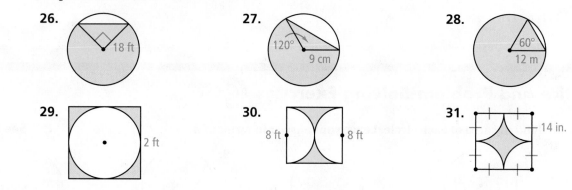

26. 18 ft

27. 120° / 9 cm

28. 60° / 12 m

29. 2 ft

30. 8 ft ⎯ 8 ft

31. 14 in.

32. Transportation A town provides bus transportation to students living beyond 2 mi of the high school. What area of the town does *not* have the bus service? Round to the nearest tenth.

33. Design A homeowner wants to build a circular patio. If the diameter of the patio is 20 ft, what is its area to the nearest whole number?

Ⓒ **34. Think About a Plan** A circular mirror is 24 in. wide and has a 4-in. frame around it. What is the area of the frame?
- How can you *draw a diagram* to help solve the problem?
- What part of a circle is the width?
- Is there more than one area to consider?

STEM 35. Industrial Design Refer to the diagram of the regular hexagonal nut. What is the area of the hexagonal face to the nearest millimeter?

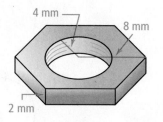

4 mm / 8 mm / 2 mm

Ⓒ **36. Reasoning** $\overline{AB}$ and $\overline{CD}$ are diameters of ⊙*O*. Is the area of sector *AOC* equal to the area of sector *BOD*? Explain.

37. A circle with radius 12 mm is divided into 20 sectors of equal area. What is the area of one sector to the nearest tenth?

38. The circumference of a circle is 26π in. What is its area? Leave your answer in terms of π.

39. In a circle, a 90° sector has area 36π in.² What is the radius of the circle?

© 40. Open-Ended Draw a circle and a sector so that the area of the sector is 16π cm². Give the radius of the circle and the measure of the sector's arc.

© 41. A method for finding the area of a segment determined by a minor arc is described in this lesson.

   a. Writing Describe two ways to find the area of a segment determined by a major arc.

   b. If $m\overarc{AB} = 90$ in a circle of radius 10 in., find the areas of the two segments determined by $\overarc{AB}$.

Find the area of the shaded segment to the nearest tenth.

42.
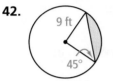
9 ft
45°

43.
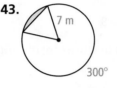
7 m
300°

44.
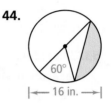
60°
|← 16 in. →|

 Challenge **Find the area of the shaded region. Leave your answer in terms of π.**

45.
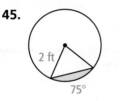
2 ft
75°

46.
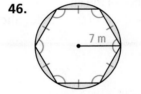
7 m

47.
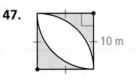
10 m

48. Recreation An 8 ft-by-10 ft floating dock is anchored in the middle of a pond. The bow of a canoe is tied to a corner of the dock with a 10-ft rope, as shown in the picture below.

   a. Sketch a diagram of the region in which the bow of the canoe can travel.

   b. What is the area of that region? Round your answer to the nearest square foot.

49. $\odot O$ at the right is inscribed in square $ABCD$ and circumscribed about square $PQRS$. Which is smaller, the blue region or the yellow region? Explain.

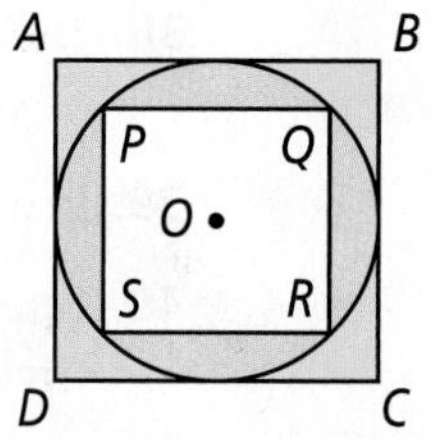

50. Circles T and U each have radius 10 and $TU = 10$. Find the area of the region that is contained inside both circles. (*Hint:* Think about where T and U must lie in a diagram of $\odot T$ and $\odot U$.)

Standardized Test Prep

SAT/ACT

51. A circular tabletop has a diameter of 6 ft. What is its area?

Ⓐ 6π ft^2 Ⓑ 9π ft^2 Ⓒ 12π ft^2 Ⓓ 36π ft^2

52. What is the value of x in the diagram at the right?

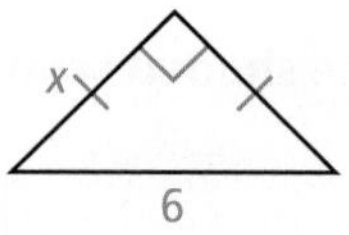

Ⓕ 3 Ⓗ $6\sqrt{2}$ Ⓖ $3\sqrt{2}$ Ⓘ 9

53. The radius of $\odot P$ is 3 cm and the measure of central $\angle APB$ is 100. What is the measure of $\overset{\frown}{AB}$?

Ⓐ 50 Ⓑ 100 Ⓒ 260 Ⓓ 300

Extended Response

54. A circle has area 81π yd^2.
  a. What is the circumference of the circle?
  b. What is the length of a 45° arc of this circle? Show all your work.

Mixed Review

Find the length of $\overset{\frown}{AB}$ in each circle. Leave your answers in terms of π. ◀ **See Lesson 10-6.**

55.

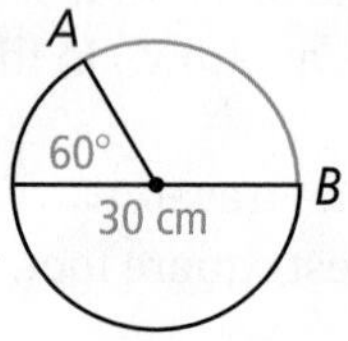

56.

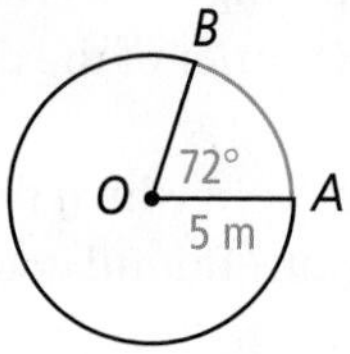

57.

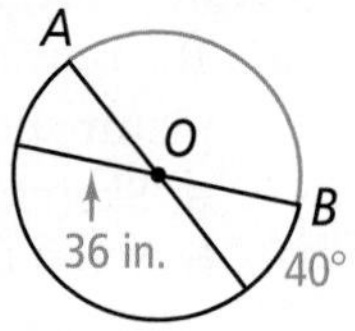

58. Three sides of a trapezoid are congruent. The fourth side is 4 in. longer than each of the other three. The perimeter is 49 in. What is the length of each side? ◀ **See Lesson 6-6.**

Get Ready! **To prepare for Lesson 10-8, do Exercises 59–60.**

59. $\odot A$ has radius 4 cm and $\odot B$ has radius 6 cm. What is the ratio of the area of $\odot A$ to the area of $\odot B$? Write your answer in simplest form. ◀ **See Lesson 10-7.**

60. A large square has side length 24 in. A small square has side length 16 in. What is the ratio of the area of the small square to the large square? Write your answer in simplest form.

Inscribed and Circumscribed Figures

© Content Standard
Extends G.GPE.7 Use coordinates to compute perimeters of polygons and areas of triangles and rectangles, e.g., using the distance formula.

In this Activity, you will compare the circumference and area of a circle with the perimeter and area of regular polygons inscribed in and circumscribed about the circle.

Activity 1

Write your answers as decimals rounded to the nearest tenth.

1. The square is inscribed in a circle.
   a. What is the length of a side of the square?
   b. What is the perimeter and area of the square?
   c. What are the circumference and the area of the circle? Use 3.14 for π.
   d. What is the ratio of the perimeter of the square to the circumference of the circle? What is the ratio of the area of the square to the area of the circle?

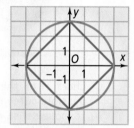

2. The regular octagon is inscribed in a circle.
   a. What is the length of a side of the octagon?
   b. What is the perimeter and area of the octagon?
   c. The radius of the circle is the same as in Exercise 1. What is the ratio of the perimeter of the octagon to the circumference of the circle? What is the ratio of the area of the octagon to the area of the circle?

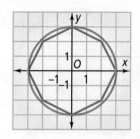

© 3. **Make a Conjecture** What will happen to the ratios as you increase the number of sides of the regular polygon?

Activity 2

Write your answers as decimals rounded to the nearest tenth, as necessary.

4. The square circumscribes the circle.
   a. What is the perimeter and area of the square?
   b. The radius of the circle is the same as in Exercise 1. What is the ratio of the circumference of the circle to the perimeter of the square? What is the ratio of the area of the circle to the area of the square?

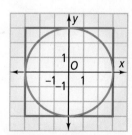

5. The octagon circumscribes the circle.
   a. What is the perimeter and area of the octagon?
   b. The radius of the circle is the same as in Exercise 1. What is the ratio of the circumference of the circle to the perimeter of the octagon? What is the ratio of the area of the circle to the area of the octagon?

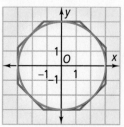

© 6. **Make a Conjecture** What will happen to the ratios as you increase the number of sides of the regular polygon?

Geometric Probability

© **Content Standard**
Prepares for S.CP.1 Describe events as subsets of a sample space (the set of outcomes) using characteristics (or categories) of the outcomes, or as unions, intersections, or complements of other events ("or," "and," "not").

Objective   To use segment and area models to find the probabilities of events

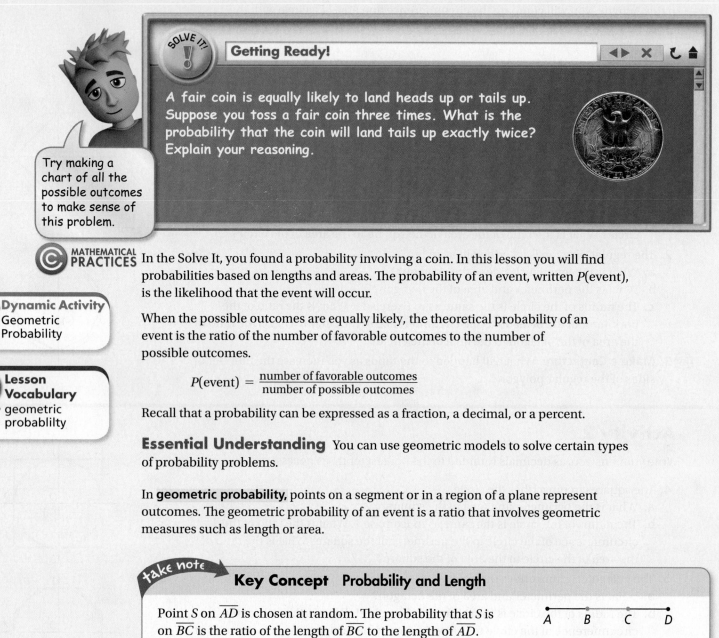

SOLVE IT!

Getting Ready!

A fair coin is equally likely to land heads up or tails up. Suppose you toss a fair coin three times. What is the probability that the coin will land tails up exactly twice? Explain your reasoning.

Try making a chart of all the possible outcomes to make sense of this problem.

© **MATHEMATICAL PRACTICES**

Dynamic Activity
Geometric Probability

Lesson Vocabulary
• geometric probablilty

In the Solve It, you found a probability involving a coin. In this lesson you will find probabilities based on lengths and areas. The probability of an event, written P(event), is the likelihood that the event will occur.

When the possible outcomes are equally likely, the theoretical probability of an event is the ratio of the number of favorable outcomes to the number of possible outcomes.

$$P(\text{event}) = \frac{\text{number of favorable outcomes}}{\text{number of possible outcomes}}$$

Recall that a probability can be expressed as a fraction, a decimal, or a percent.

Essential Understanding   You can use geometric models to solve certain types of probability problems.

In **geometric probability,** points on a segment or in a region of a plane represent outcomes. The geometric probability of an event is a ratio that involves geometric measures such as length or area.

take note

Key Concept   Probability and Length

Point S on $\overline{AD}$ is chosen at random. The probability that S is on $\overline{BC}$ is the ratio of the length of $\overline{BC}$ to the length of $\overline{AD}$.

$$P(S \text{ on } \overline{BC}) = \frac{BC}{AD}$$

Problem 1 Using Segments to Find Probability

Think

How can you find the length of each segment?
You can use the Ruler Postulate to find the length of each segment.

Point K on $\overline{ST}$ is chosen at random. What is the probability that K lies on $\overline{QR}$?

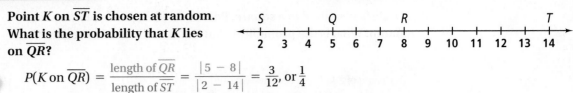

$$P(K \text{ on } \overline{QR}) = \frac{\text{length of } \overline{QR}}{\text{length of } \overline{ST}} = \frac{|5 - 8|}{|2 - 14|} = \frac{3}{12}, \text{ or } \frac{1}{4}$$

The probability that K is on $\overline{QR}$ is $\frac{1}{4}$, or 25%.

✓ **Got It?** 1. Use the diagram in Problem 1. Point H on $\overline{ST}$ is selected at random. What is the probability that H lies on $\overline{SR}$?

Problem 2 Using Segments to Find Probability

Plan

How can you draw a diagram to model the situation?
Draw a line segment with endpoints 0 and 25 to represent the length of time between trains. Each point on the segment represents an arrival time.

Transportation A commuter train runs every 25 min. If a commuter arrives at the station at a random time, what is the probability that the commuter will have to wait at least 10 min for the train?

Assume that a stop takes very little time. Draw a line segment to model the situation. The length of the entire segment represents the amount of time between trains. A commuter will have to wait at least 10 min for the train if the commuter arrives at any time between 0 and 15 min.

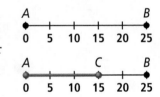

$$P(\text{waiting at least 10 min}) = \frac{\text{length of favorable segment}}{\text{length of entire segment}} = \frac{15}{25}, \text{ or } \frac{3}{5}$$

The probability that a commuter will have to wait at least 10 min for the train is $\frac{3}{5}$, or 60%.

✓ **Got It?** 2. What is the probability that a commuter will have to wait no more than 5 min for the train?

When the points of a region represent equally likely outcomes, you can find probabilities by comparing areas.

take note

Key Concept Probability and Area

Point S in region R is chosen at random. The probability that S is in region N is the ratio of the area of region N to the area of region R.

$$P(S \text{ in region } N) = \frac{\text{area of region } N}{\text{area of region } R}$$

Problem 3　Using Area to Find Probability

A circle is inscribed in a square. Point Q in the square is chosen at random. What is the probability that Q lies in the shaded region?

6 cm

Know

The length of a side of the square, which is also the length of the diameter of the inscribed circle

Need

The areas of the square and the shaded region

Plan

Subtract the area of the circle from the area of the square to find the area of the shaded region. Then use it to find the probability.

area of shaded region = area of square − area of circle

$$= 6^2 - \pi(3)^2$$

$$= 36 - 9\pi$$

$$P(Q \text{ lies in shaded region}) = \frac{\text{area of shaded region}}{\text{area of square}}$$

$$= \frac{36 - 9\pi}{36} \approx 0.215$$

The probability that Q lies in the shaded region is about 0.215, or 21.5%.

Got It?　3. A triangle is inscribed in a square. Point T in the square is selected at random. What is the probability that T lies in the shaded region?

5 in.

Problem 4　Using Area to Find Probability

Archery An archery target has 5 colored scoring zones formed by concentric circles. The target's diameter is 122 cm. The radius of the yellow zone is 12.2 cm. The width of each of the other zones is also 12.2 cm. If an arrow hits the target at a random point, what is the probability that it hits the red zone?

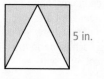

Plan

How can you find the area of the red zone?
The red zone lies between two concentric circles. To find the area of the red zone, subtract the areas of the two concentric circles.

The red zone is the region between a circle with radius $12.2 + 12.2$, or 24.4 cm and the yellow circle with radius 12.2 cm. The target is a circle with radius $\frac{122}{2}$, or 61 cm.

$$P(\text{arrow hits red zone}) = \frac{\text{area of red zone}}{\text{area of entire target}}$$

$$= \frac{\pi(24.4)^2 - \pi(12.2)^2}{\pi(61)^2} = 0.12$$

The probability of an arrow hitting a point in the red zone is 0.12, or 12%.

Got It?　4. a. What is the probability that an arrow hits the yellow zone?

　　b. Reasoning If an arrow hits the target at a random point, is it more likely to hit the black zone or the red zone? Explain.

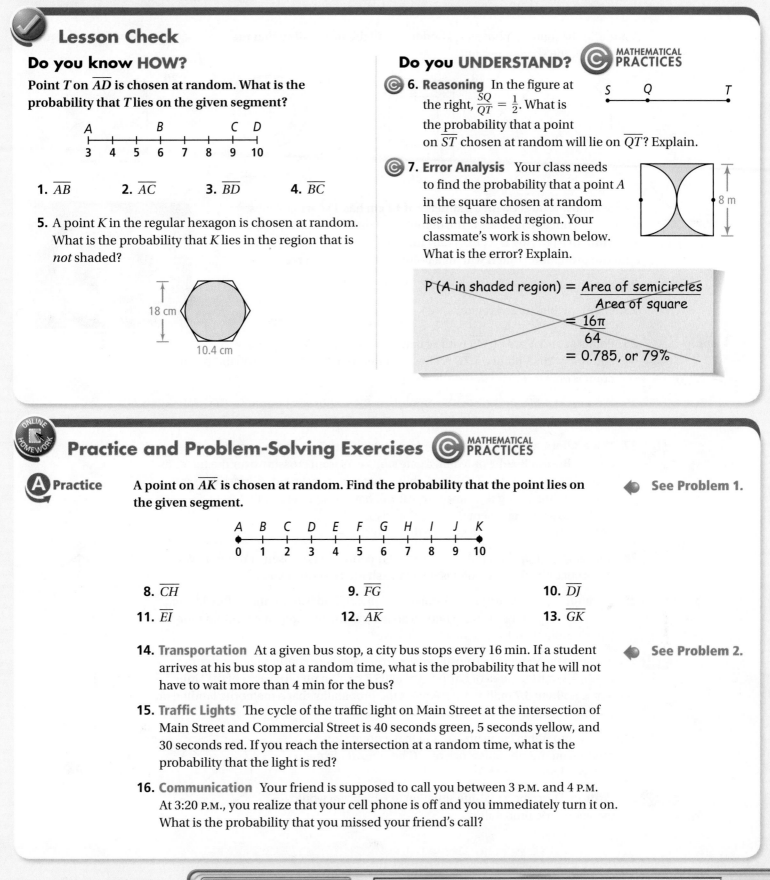

Lesson Check

Do you know HOW?

Point T on $\overline{AD}$ is chosen at random. What is the probability that T lies on the given segment?

A   B   C   D
3   4   5   6   7   8   9   10

1. $\overline{AB}$ **2.** $\overline{AC}$ **3.** $\overline{BD}$ **4.** $\overline{BC}$

5. A point K in the regular hexagon is chosen at random. What is the probability that K lies in the region that is *not* shaded?

18 cm

10.4 cm

Do you UNDERSTAND? MATHEMATICAL PRACTICES

6. Reasoning In the figure at the right, $\frac{SQ}{QT} = \frac{1}{2}$. What is the probability that a point on $\overline{ST}$ chosen at random will lie on $\overline{QT}$? Explain.

S   Q   T

7. Error Analysis Your class needs to find the probability that a point A in the square chosen at random lies in the shaded region. Your classmate's work is shown below. What is the error? Explain.

8 m

P (A in shaded region) = $\dfrac{\text{Area of semicircles}}{\text{Area of square}}$

$= \dfrac{16\pi}{64}$

$= 0.785$, or 79%

Practice and Problem-Solving Exercises MATHEMATICAL PRACTICES

A Practice   A point on $\overline{AK}$ is chosen at random. Find the probability that the point lies on the given segment.

See Problem 1.

A   B   C   D   E   F   G   H   I   J   K
0   1   2   3   4   5   6   7   8   9   10

8. $\overline{CH}$ **9.** $\overline{FG}$ **10.** $\overline{DJ}$

11. $\overline{EI}$ **12.** $\overline{AK}$ **13.** $\overline{GK}$

14. Transportation At a given bus stop, a city bus stops every 16 min. If a student arrives at his bus stop at a random time, what is the probability that he will not have to wait more than 4 min for the bus?

See Problem 2.

15. Traffic Lights The cycle of the traffic light on Main Street at the intersection of Main Street and Commercial Street is 40 seconds green, 5 seconds yellow, and 30 seconds red. If you reach the intersection at a random time, what is the probability that the light is red?

16. Communication Your friend is supposed to call you between 3 P.M. and 4 P.M. At 3:20 P.M., you realize that your cell phone is off and you immediately turn it on. What is the probability that you missed your friend's call?

A point in the figure is chosen at random. Find the probability that the point lies in the shaded region.

See Problems 3 and 4.

17.

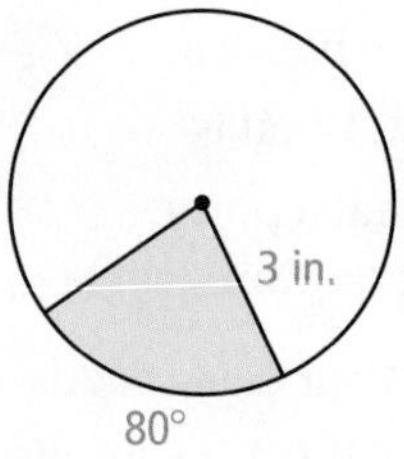

18.

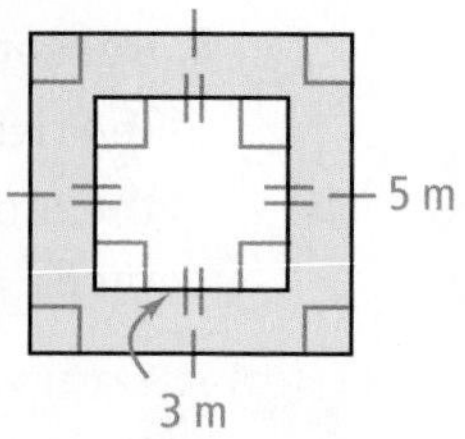

19.

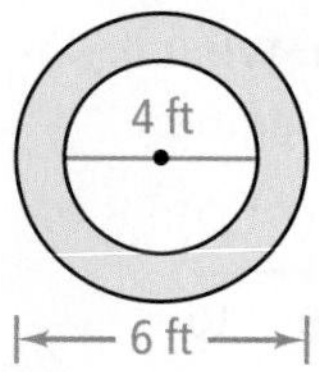

20.

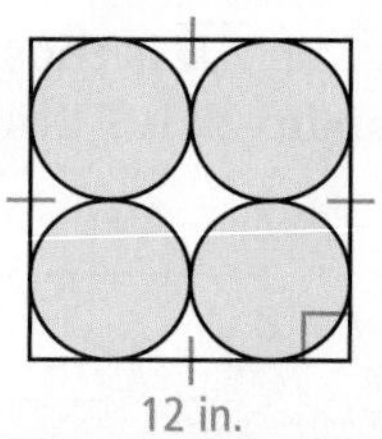

3 in.

80°

5 m

3 m

4 ft

6 ft

12 in.

Target Game A target with a diameter of 14 cm has 4 scoring zones formed by concentric circles. The diameter of the center circle is 2 cm. The width of each ring is 2 cm. A dart hits the target at a random point. Find the probability that it will hit a point in the indicated region.

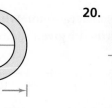

21. the center region

22. the blue region

23. either the blue or red region

24. any region

B **Apply**

25. Points M and N are on $\overline{ZB}$ with M between Z and N. $ZM = 5$, $NB = 9$, and $ZB = 20$. A point on $\overline{ZB}$ is chosen at random. What is the probability that the point is on $\overline{MN}$?

26. $\overline{BZ}$ contains $\overline{MN}$ and $BZ = 20$. A point on $\overline{BZ}$ is chosen at random. The probability that the point is also on $\overline{MN}$ is 0.3, or 30%. Find MN.

Ⓒ **27. Think About a Plan** Every 20 min from 4:00 P.M. to 7:00 P.M., a commuter train crosses Boston Road. For 3 min, a gate stops cars from crossing over the tracks as the train goes by. What is the probability that a motorist randomly arriving at the train crossing during this time interval will have to stop for a train?
   • How can you represent the situation visually?
   • What ratio can you use to solve the problem?

Ⓒ **28. Reasoning** Suppose a point in the regular pentagon is chosen at random. What is the probability that the point is *not* in the shaded region? Explain.

29. Commuting A bus arrives at a stop every 16 min and waits 3 min before leaving. What is the probability that a person arriving at the bus stop at a random time has to wait more than 10 min for a bus to leave?

STEM **30. Astronomy** Meteorites (mostly dust-particle size) are continually bombarding Earth. The surface area of Earth is about 65.7 million mi². The area of the United States is about 3.7 million mi². What is the probability that a meteorite landing on Earth will land in the United States?

Ⓒ **31. Reasoning** What is the probability that a point chosen at random on the circumference of $\odot C$ lies on $\overarc{AB}$? Explain how you know.

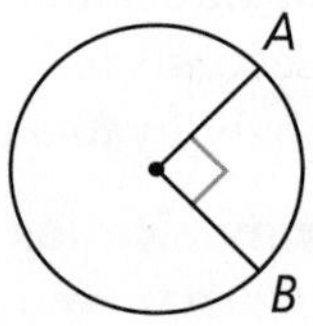

Ⓒ **32. Writing** Describe a real-life situation in which you would use geometric probability.

Algebra Find the probability that coordinate x of a point chosen at random on $\overline{AK}$ satisfies the inequality.

A B C D E F G H I J K
0 1 2 3 4 5 6 7 8 9 10

33. $2 \le x \le 8$ **34.** $2x \le 8$ **35.** $5 \le 11 - 6x$

36. $\frac{1}{2}x - 5 > 0$ **37.** $2 \le 4x \le 3$ **38.** $-7 \le 1 - 2x \le 1$

39. One type of dartboard is a square of radius 10 in. You throw a dart and hit the target. What is the probability that the dart lies within $\sqrt{10}$ in. of the square's center?

Ⓒ **40. Games** To win a prize at a carnival game, you must toss a quarter, so that it lands entirely within a circle as shown at the right. Assume that the center of a tossed quarter is equally likely to land at any point within the 8-in. square.
  a. What is the probability that the quarter lands entirely in the circle in one toss?
  b. Reasoning On average, how many coins must you toss to win a prize? Explain.

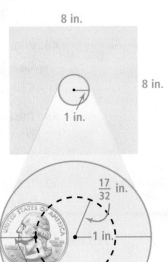

8 in.

8 in.

1 in.

41. Traffic Patterns The traffic lights at Fourth and State Streets repeat themselves in 1-min cycles. A motorist will face a red light 60% of the time. Use this information to estimate how long the Fourth Street light is red during each 1-min cycle.

42. You have a 4-in. straw and a 6-in. straw. You want to cut the 6-in. straw into two pieces so that the three pieces form a triangle.
  a. If you cut the straw to get two 3-in. pieces, can you form a triangle?
  b. If the two pieces are 1 in. and 5 in., can you form a triangle?
  c. If you cut the straw at a random point, what is the probability that you can form a triangle?

$\frac{17}{32}$ in.

1 in.

$\frac{15}{32}$ in.

43. Target Game Assume that a dart you throw will land on the 12 in.-by-12 in. square dartboard and is equally likely to land at any point on the board. The diameter of the center circle is 2 in., and the width of each ring is 1 in.
  a. What is the probability of hitting either the blue or yellow region?
  b. What is the probability the dart will *not* hit the gray region?

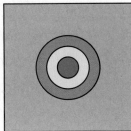

Ⓒ **Challenge**

44. Graphing Calculator A circular dartboard has radius 1 m and a yellow circle in the center. Assume you hit the target at a random point. For what radius of the yellow center region would P(hitting yellow) equal each of the following? Use the table feature of a calculator to generate all six answers. Round to the nearest centimeter.
  a. 0.2 **b.** 0.4 **c.** 0.5
  d. 0.6 **e.** 0.8 **f.** 1.0

45. You and your friend agree to meet for lunch between 12 P.M. and 1 P.M. Each of you agrees to wait 15 min for the other before giving up and eating lunch alone. If you arrive at 12:20, what is the probability you and your friend will eat lunch together?

Standardized Test Prep

SAT/ACT

46. A dart hits the square dartboard shown. What is the probability that it lands in the shaded region?

    Ⓐ 21%   Ⓒ 50%

    Ⓑ 25%   Ⓓ 79%

4 m

47. A dilation maps $\triangle JKL$ onto $\triangle J'K'L'$ with a scale factor of 1.2. If $J'L' = 54$ cm, what is JL?

    Ⓕ 43.2 cm   Ⓖ 45 cm   Ⓗ 54 cm   Ⓘ 64.8 cm

48. What is the value of x in the figure at the right?

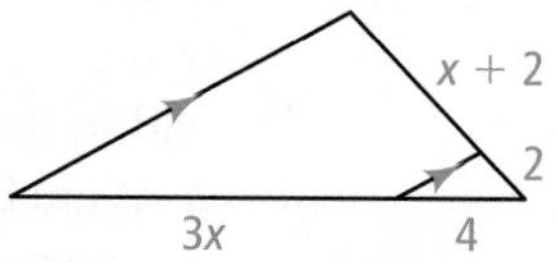

    Ⓐ $\frac{1}{3}$ Ⓒ 3

    Ⓑ 2   Ⓓ 4

$x + 2$

2

$3x$ 4

49. The radius of $\odot P$ is 4.5 mm. The measure of central $\angle APB$ is 160. What is the length of $\overset{\frown}{AB}$?

    Ⓕ 2π mm   Ⓖ 4π mm   Ⓗ 5π mm   Ⓘ 9π mm

Short Response

50. The perimeter of a square is 24 ft. What is the length of the square's diagonal?

Mixed Review

51. A circle has circumference 20π ft. What is its area in terms of π?   ◀ **See Lesson 10-7.**

52. A circle has radius 12 cm. What is the area of a sector of the circle with a 30° central angle in terms of π?

For Exercises 53–55, find the coordinates of the image, $\triangle A'''B'''C'''$, that result from the composition of transformations described.   ◀ **See Lesson 9-4.**

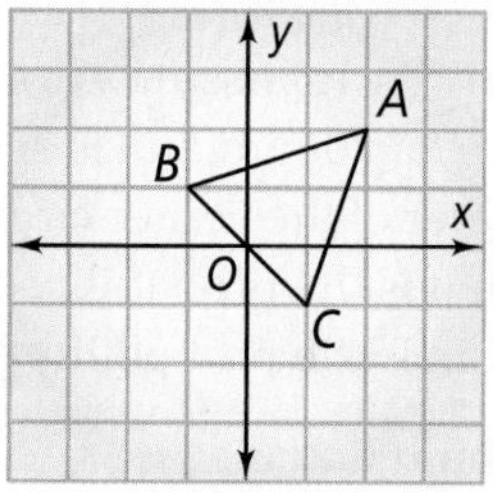

53. $R_{y\text{-axis}} \circ R_{x\text{-axis}}(\triangle ABC)$

54. $T_{<-3,4>} \circ r_{(90°, \, O)}(\triangle ABC)$

55. $R_{y=1} \circ T_{<2,-1>}(\triangle ABC)$

Get Ready!   To prepare for Lesson 11-1, do Exercises 56–62.

For each exercise, make a copy of the cube below. Shade the plane that contains the indicated points.   ◀ **See Lessons 1-1 and 1-2.**

56. A, B, and C **57.** B, F, and G

58. E, F, and H **59.** A, D, and G

60. F, D, and G **61.** A, C, and G

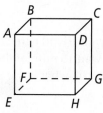

62. Draw a net for the cube at the right.

10 *Pull It* All Together ASSESSMENT

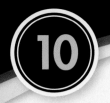

To solve these problems you will pull together many concepts and skills that you have studied about area and perimeter of polygons and circles.

BIG idea Measurement
You can use formulas to find areas of polygons.

Performance Task 1

A real estate company sells plots of land. The plot shown at the right costs \$84,120. What is the price per square foot of the land? Explain how you found your answer.

125 ft
260 ft
189 ft
127.5°
150 ft
265 ft

BIG idea Measurement
An arc length is a fractional part of the circumference of a circle. The area of a sector is a fractional part of the area of a circle.

Performance Task 2

The stained glass circle-head window has a 2-in. wide frame. The grills divide the semicircular glass pane into four congruent regions.

a. What is the area of the blue region? Show your work.

b. What is the outside perimeter of the window frame? Justify your reasoning.

6 in.
28 in.

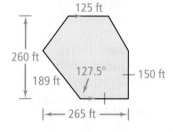

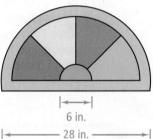

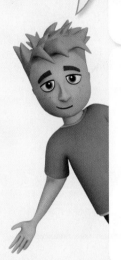

BIG idea Similarity
The ratio of the perimeters and the ratio of the areas of two similar figures are related to the ratio of the corresponding measures.

Performance Task 3

Regular hexagon $ABCDEF$ has vertices at $A(4, 4\sqrt{3})$, $B(8, 4\sqrt{3})$, $C(10, 2\sqrt{3})$, $D(8, 0)$, $E(4, 0)$ and $F(2, 2\sqrt{3})$. Suppose the sides of the hexagon are reduced by 40% to produce a similar regular hexagon. What are the perimeter and area of the smaller hexagon rounded to the nearest tenth? Explain how you found your answer.

Connecting **BIG** ideas and Answering the Essential Questions

1 Measurement
You can find the area of a polygon, or the circumference or area of a circle, by first determining which formula to use. Then you can substitute the needed measures into the formula.

2 Similarity
The perimeters of similar polygons are proportional to the ratio of corresponding measures. The areas are proportional to the squares of corresponding measures.

**Areas of Polygons
(Lessons 10-1, 10-2, and 10-3)**

Parallelogram	$A = bh$
Triangle	$A = \frac{1}{2}bh$
Trapezoid	$A = \frac{1}{2}h(b_1 + b_2)$
Rhombus or kite	$A = \frac{1}{2}d_1d_2$
Regular polygon	$A = \frac{1}{2}ap$

Circles and Arcs (Lesson 10-6)

$C = \pi d$ or $C = 2\pi r$

$m\widehat{ABC} = m\widehat{AB} + m\widehat{BC}$

length of $\widehat{AB} = \dfrac{m\widehat{AB}}{360} \cdot 2\pi r$

Perimeter and Area (Lesson 10-4)
If the scale factor of two similar figures is $\frac{a}{b}$, then

(1) the ratio of their perimeters is $\frac{a}{b}$ and

(2) the ratio of their areas is $\frac{a^2}{b^2}$.

**Area of a Triangle Given SAS
(Lesson 10-5)**
Area of $\triangle ABC = \frac{1}{2}bc(\sin A)$

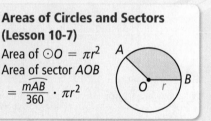

**Areas of Circles and Sectors
(Lesson 10-7)**
Area of $\odot O = \pi r^2$
Area of sector AOB
$= \dfrac{m\widehat{AB}}{360} \cdot \pi r^2$

Chapter Vocabulary

- adjacent arcs (p. 650)
- altitude (p. 616)
- apothem (p. 629)
- arc length (p. 653)
- base (pp. 616, 618)
- central angle (p. 649)
- circle (p. 649)

- circumference (p. 651)
- concentric circles (p. 651)
- congruent circles (p. 649)
- diameter (p. 649)
- geometric probability (p. 668)
- height (pp. 616, 618, 623)

- major arc (p. 649)
- minor arc (p. 649)
- radius (pp. 629, 649)
- sector of a circle (p. 661)
- segment of a circle (p. 662)
- semicircle (p. 649)

Choose the correct term to complete each sentence.

1. You can use any side as the ? of a triangle.

2. A(n) ? is a region bounded by an arc and the two radii to the arc's endpoints.

3. The distance from the center to a vertex is the ? of a regular polygon.

4. Two arcs of a circle with exactly one point in common are ? .

10-1 Areas of Parallelograms and Triangles

Quick Review

You can find the area of a rectangle, a parallelogram, or a triangle if you know the **base** b and the **height** h.

The area of a rectangle or parallelogram is $A = bh$.

The area of a triangle is $A = \frac{1}{2}bh$.

Example

What is area of the parallelogram?

$A = bh$ Use the area formula.

$= (12)(8) = 96$ Substitute and simplify.

The area of the parallelogram is 96 cm².

Exercises

Find the area of each figure.

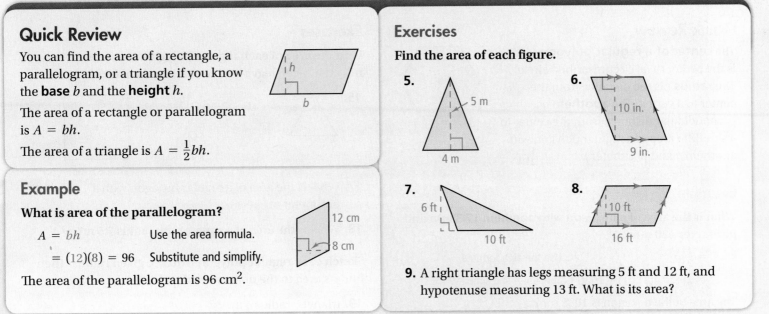

5.

6.
10 in.
9 in.

7.
6 ft
10 ft

8.
10 ft
16 ft

5 m
4 m

9. A right triangle has legs measuring 5 ft and 12 ft, and hypotenuse measuring 13 ft. What is its area?

12 cm
8 cm

10-2 Areas of Trapezoids, Rhombuses, and Kites

Quick Review

The **height of a trapezoid** h is the perpendicular distance between the bases, b_1 and b_2.

The area of a trapezoid is $A = \frac{1}{2}h(b_1 + b_2)$.

The area of a rhombus or a kite is $A = \frac{1}{2}d_1 d_2$, where d_1 and d_2 are the lengths of its diagonals.

Example

What is the area of the trapezoid?

$A = \frac{1}{2}h(b_1 + b_2)$ Use the area formula.

$= \frac{1}{2}(8)(7 + 3)$ Substitute.

$= 40$ Simplify.

The area of the trapezoid is 40 cm².

3 cm
8 cm
7 cm

Exercises

Find the area of each figure. If necessary, leave your answer in simplest radical form.

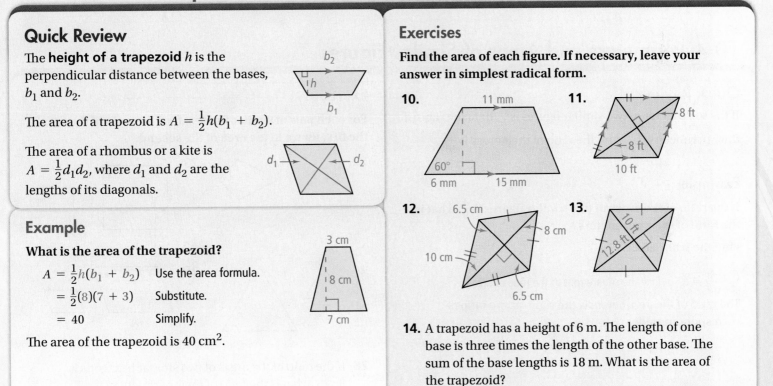

10.
11 mm
60°
6 mm   15 mm

11.
8 ft
8 ft
10 ft

12.
6.5 cm
10 cm
8 cm
6.5 cm

13.
10 ft
12.8 ft

14. A trapezoid has a height of 6 m. The length of one base is three times the length of the other base. The sum of the base lengths is 18 m. What is the area of the trapezoid?

10-3 Areas of Regular Polygons

Quick Review

The **center of a regular polygon** C is the center of its circumscribed circle. The **radius** r is the distance from the center to a vertex. The **apothem** a is the perpendicular distance from the center to a side. The area of a regular polygon with apothem a and perimeter p is $A = \frac{1}{2}ap$.

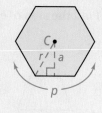

Example

What is the area of a hexagon with apothem 17.3 mm and perimeter 120 mm?

$\quad A = \frac{1}{2}ap$ Use the area formula.

$\quad\quad = \frac{1}{2}(17.3)(120) = 1038$ Substitute and simplify.

The area of the hexagon is 1038 mm².

Exercises

Find the area of each regular polygon. If your answer is not an integer, leave it in simplest radical form.

15.
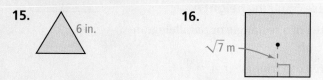
6 in.

16.
$\sqrt{7}$ m

17. What is the area of a regular hexagon with a perimeter of 240 cm?

18. What is the area of a square with radius 7.5 m?

Sketch each regular polygon with the given radius. Then find its area to the nearest tenth.

19. triangle; radius 4 in.

20. square; radius 8 mm

21. hexagon; radius 7 cm

10-4 Perimeters and Areas of Similar Figures

Quick Review

If the scale factor of two similar figures is $\frac{a}{b}$, then the ratio of their perimeters is $\frac{a}{b}$, and the ratio of their areas is $\frac{a^2}{b^2}$.

Example

If the ratio of the areas of two similar figures is $\frac{4}{9}$, what is the ratio of their perimeters?

Find the scale factor.

$\quad \frac{\sqrt{4}}{\sqrt{9}} = \frac{2}{3}$ Take the square root of the ratio of areas.

The ratio of the perimeters is the same as the ratio of corresponding sides, $\frac{2}{3}$.

Exercises

For each pair of similar figures, find the ratio of the area of the first figure to the area of the second.

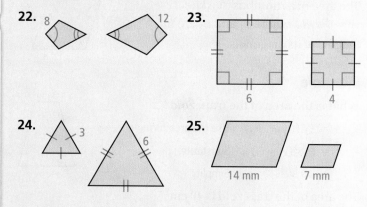

22. 8   12

23. 6   4

24. 3   6

25. 14 mm   7 mm

26. If the ratio of the areas of two similar hexagons is 8 : 25, what is the ratio of their perimeters?

10-5 Trigonometry and Area

Quick Review

You can use trigonometry to find the areas of regular polygons. You can also use trigonometry to find the area of a triangle when you know the lengths of two sides and the measure of the included angle.

Area of a $\triangle$
$= \frac{1}{2} \cdot$ side length $\cdot$ side length $\cdot$ sine of included angle

Example

What is the area of $\triangle XYZ$?

Area $= \frac{1}{2} \cdot XY \cdot XZ \cdot \sin X$

$ = \frac{1}{2} \cdot 15 \cdot 13 \cdot \sin 65°$

$ \approx 88.36500924$

The area of $\triangle XYZ$ is approximately 88 ft².

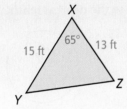

Exercises

Find the area of each polygon. Round your answers to the nearest tenth.

27. regular decagon with radius 5 ft

28. regular pentagon with apothem 8 cm

29. regular hexagon with apothem 6 in.

30. regular quadrilateral with radius 2 m

31. regular octagon with apothem 10 ft

32. regular heptagon with radius 3 ft

33.

34.

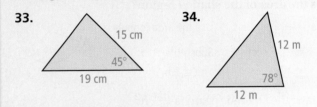

10-6 Circles and Arcs

Quick Review

A **circle** is the set of all points in a plane equidistant from a point called the **center.**

The **circumference** of a circle is $C = \pi d$ or $C = 2\pi r$.

Arc length is a fraction of a circle's circumference. The length of $\widehat{AB} = \frac{m\widehat{AB}}{360} \cdot 2\pi r$.

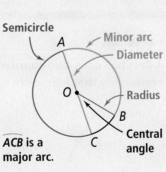

$\widehat{ACB}$ is a major arc.

Example

A circle has a radius of 5 cm. What is the length of an arc measuring 80°?

length of $\widehat{AB} = \frac{m\widehat{AB}}{360} \cdot 2\pi r$ Use the arc length formula.

$ = \frac{80}{360} \cdot 2\pi(5)$ Substitute.

$ = \frac{20}{9}\pi$ Simplify.

The length of the arc is $\frac{20}{9}\pi$ cm.

Exercises

Find each measure.

35. $m\angle APD$

36. $m\widehat{AC}$

37. $m\,\widehat{ABD}$

38. $m\angle CPA$

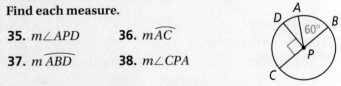

Find the length of each arc shown in red. Leave your answer in terms of π.

39.

40.

41.

42.

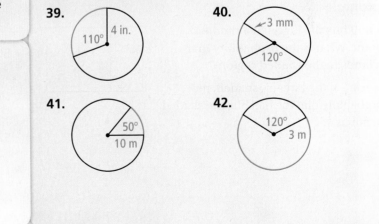

10-7 Areas of Circles and Sectors

Quick Review

The area of a circle is $A = \pi r^2$.

A **sector of a circle** is a region bounded by two radii and their intercepted arc. The area of sector $APB = \frac{m\widehat{AB}}{360} \cdot \pi r^2$.

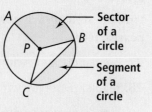

A **segment of a circle** is the part bounded by an arc and the segment joining its endpoints.

Example

What is the area of the shaded region?

$$\text{Area} = \frac{m\widehat{AB}}{360} \cdot \pi r^2 \quad \text{Use the area formula.}$$

$$= \frac{120}{360} \cdot \pi(4)^2 \quad \text{Substitute.}$$

$$= \frac{16\pi}{3} \quad \text{Simplify.}$$

The area of the shaded region is $\frac{16\pi}{3}$ ft^2.

Exercises

What is the area of each circle? Leave your answer in terms of π.

43.

44.

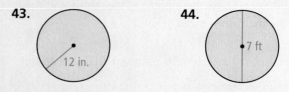

Find the area of each shaded region. Round your answer to the nearest tenth.

45.

46.

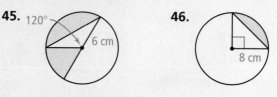

47. A circle has a radius of 20 cm. What is the area of the smaller segment of the circle formed by a 60° arc? Round to the nearest tenth.

10-8 Geometric Probability

Quick Review

Geometric probability uses geometric figures to represent occurrences of events. You can use a segment model or an area model. Compare the part that represents favorable outcomes to the whole, which represents all outcomes.

Example

A ball hits the target at a random point. What is the probability that it lands in the shaded region?

Since $\frac{1}{3}$ of the target is shaded, the probability that the ball hits the shaded region is $\frac{1}{3}$.

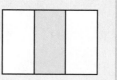

Exercises

A dart hits each dartboard at a random point. Find the probability that it lands in the shaded region.

48.

49.

50.

51.

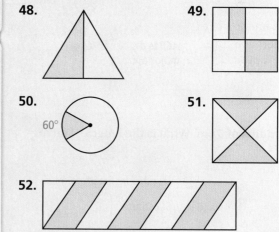

52.

Do you know HOW?

Find the area of each figure. Round to the nearest tenth.

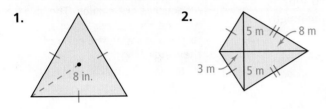

1.

2. 5 m / 8 m / 3 m / 5 m

Find the area of each regular polygon. Round to the nearest tenth.

3. 10 mm

4. 18 yd

For each pair of similar figures, find the ratio of the area of the first figure to the area of the second.

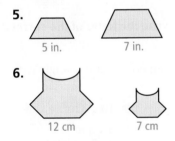

5. 5 in.   7 in.

6. 12 cm   7 cm

Find the area of each polygon to the nearest tenth.

7. 9 ft / 30° / 6 ft

8. 6 m / 60° / 10 m

Find each measure for ⊙P.

9. $m\angle BPC$ 10. $m\,\overarc{AB}$

11. $m\,\overarc{ADC}$ 12. $m\,\overarc{ADB}$

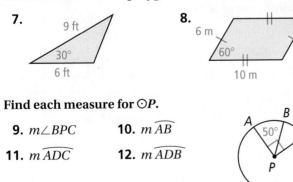

A / 50° / B / C / P / D

Find the length of each arc shown in red. Leave your answer in terms of π.

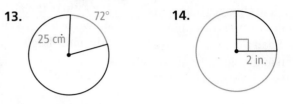

13. 72° / 25 cm

14. 2 in.

Find the area of each shaded region to the nearest hundredth.

15. 6 cm / 80°

16. 6 m / 6 m / 3 m

17. **Probability** Fly A lands on the edge of the ruler at a random point. Fly B lands on the surface of the target at a random point. Which fly is more likely to land in a yellow region? Explain.

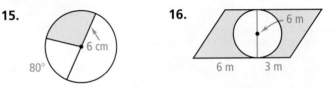

8 cm / 4 cm   8 cm / 4 cm

Do you UNDERSTAND?

18. **Reasoning** A right triangle has height 7 cm and base 4 cm. Find its area using the formulas $A = \frac{1}{2}bh$ and $A = \frac{1}{2}ab(\sin C)$. Are the results the same? Explain.

19. **Mental Math** Garden A has the shape of a quarter circle with radius 28 ft. Garden B has the shape of a circle with radius 14 ft. What is the ratio of the area of Garden A to the area of Garden B? Explain how you can find the answer without calculating the areas.

20. **Reasoning** Can a regular polygon have an apothem and a radius of the same length? Explain.

21. **Open-Ended** Use a compass to draw a circle. Shade a sector of the circle and find its area.

Some test questions require you to relate changes in lengths to changes in areas of similar figures. Read the sample question at the right. Then follow the tips to answer it.

TIP 1

The larger square must have a side length greater than that of the smaller square. You can eliminate any answer choices that are less than or equal to 6 in.

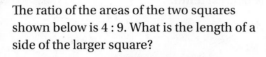

The ratio of the areas of the two squares shown below is 4 : 9. What is the length of a side of the larger square?

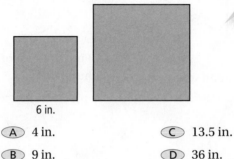

6 in.

(A) 4 in.   (C) 13.5 in.

(B) 9 in.   (D) 36 in.

TIP 2

All squares are similar. The ratio of their areas is equal to the square of the ratio of their side lengths.

Think It Through

The ratio of the areas is 4 : 9.
The ratio of the side lengths is $\sqrt{4} : \sqrt{9}$, or 2 : 3.

Use a proportion to find the length s of the larger square.

$$\frac{2}{3} = \frac{6}{s}$$

$$s = 9$$

The correct answer is B.

Vocabulary Builder

As you solve test items, you must understand the meanings of mathematical terms. Match each term with its mathematical meaning.

A. perimeter

B. segment of a circle

C. sector of a circle

D. area

I. the part of a circle bounded by an arc and the segment joining its endpoints

II. the region of a circle bounded by two radii and their intercepted arc

III. the number of square units enclosed by a figure

IV. the distance around a figure

Multiple Choice

Read each question. Then write the letter of the correct answer on your paper.

1. What is the exact area of an equilateral triangle with sides of length 10 m?

(A) $25\sqrt{3} \, m^2$ (C) $10\sqrt{3} \, m^2$

(B) $25 \, m^2$ (D) $5\sqrt{3} \, m^2$

2. If $\triangle CAT$ is rotated 90° around vertex C, what are the coordinates of A'?

(F) $(1, 5)$

(G) $(3, 3)$

(H) $(-1, 3)$

(I) $(3, 5)$

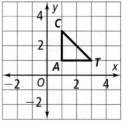

3. The vertices of $\triangle ABC$ have coordinates $A(-2, 3)$, $B(3, 4)$, and $C(1, -2)$. The vertices of $\triangle PQR$ have coordinates $P(-2, -3)$, $Q(3, -4)$, and $R(1, 2)$. Which transformation can you use to justify that $\triangle ABC$ is congruent to $\triangle PQR$?

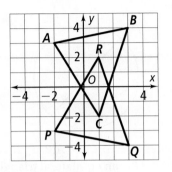

- (A) a reflection across the y-axis
- (B) a reflection across the x-axis
- (C) a rotation $180°$ clockwise around the origin
- (D) a translation 5 units left and 6 units down

4. If a truck's tire has a radius of 2 ft, what is the circumference of the tire in feet?

- (F) 8π
- (H) 4π
- (G) 6π
- (I) 2π

5. Your neighbor has a square garden, as shown below. He wants to install a sprinkler in the center of the garden so that the water sprays only as far as the corners of the garden. What is the approximate radius at which your neighbor should set the sprinkler?

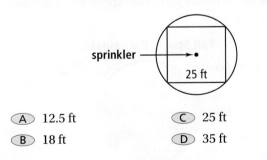

- (A) 12.5 ft
- (C) 25 ft
- (B) 18 ft
- (D) 35 ft

6. Which of the following could be the side lengths of a right triangle?

- (F) 4.1, 6.2, 7.3
- (H) 3.2, 5.4, 6.2
- (G) 40, 60, 72
- (I) 33, 56, 65

7. Every triangle in the figure at the right is an equilateral triangle. What is the total area of the shaded triangles?

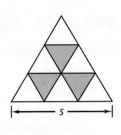

- (A) $\dfrac{s^2 \sqrt{3}}{36}$
- (C) $\dfrac{s^2 \sqrt{3}}{9}$
- (B) $\dfrac{s^2 \sqrt{3}}{12}$
- (D) $\dfrac{s^2 \sqrt{3}}{6}$

8. The shaded part of the circle below represents the portion of a garden where a landscaper planted rose bushes. What is the approximate area of the part of the garden used for rose bushes?

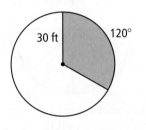

- (F) 300 ft^2
- (H) 3769 ft^2
- (G) 942 ft^2
- (I) 8482 ft^2

9. The vertices of $\triangle ART$ are $A(1, -2)$, $R(5, -1)$, and $T(4, -4)$. If $T_{<5, -2>}(\triangle ART) = A'R'T'$, what are the coordinates of the image triangle $A'R'T'$?

- (A) $A'(5, -1)$, $R'(4, -4)$, $T'(1, -2)$
- (B) $A'(6, -4)$, $R'(10, -3)$, $T'(9, -6)$
- (C) $A'(3, 3)$, $R'(7, 4)$, $T'(6, 1)$
- (D) $A'(-1, 3)$, $R'(3, 4)$, $T'(2, 1)$

10. In the figure below, what is the length of $\overline{AD}$?

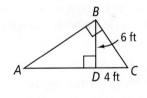

- (F) $2\sqrt{13}$ ft
- (H) 10 ft
- (G) 9 ft
- (I) $3\sqrt{13}$ ft

11. A triangle has a perimeter of 81 in. If you divide the length of each side by 3, what is the perimeter, in inches, of the new triangle?

12. A jewelry maker designed a pendant like the one shown below. It is a regular octagon set in a circle. Opposite vertices are connected by line segments. What is the measure of angle *P* in degrees?

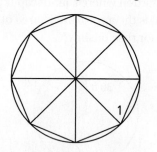

13. The diagonal of a rectangular patio makes a 70° angle with a side of the patio that is 60 ft long. What is the area, to the nearest square foot, of the patio?

14. What is the area, in square feet, of the unshaded part of the rectangle below?

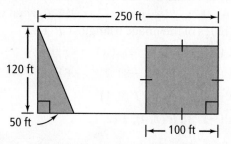

15. You are making a scale model of a building. The front of the actual building is 60 ft wide and 100 ft tall. The front of your model is 3 ft wide and 5 ft tall. What is the scale factor of the reduction?

16. The clock has a diameter of 18 in. At 8 o'clock, what is the area of the sector formed by the two hands of the clock? Round to the nearest tenth.

17. A square has an area of 225 cm². If you double the length of each side, what is the area, in square centimeters, of the new square?

Short Response

18. Use a compass and straightedge to copy the diagram of the circle below. Label the center, radius, and diameter. Then label a central angle, an arc, a sector, and a segment.

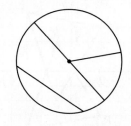

19. The coordinates of $\triangle ABC$ are $A(2, 3)$, $B(10, 9)$, and $C(10, -3)$. Is $\triangle ABC$ an equilateral triangle? Explain.

20. Draw a right triangle. Then construct a second triangle congruent to the first. Show your steps.

21. One diagonal of a parallelogram has endpoints at $P(-2, 5)$ and $R(1, -1)$. The other diagonal has endpoints at $Q(1, 5)$ and $S(-2, -1)$. What type of parallelogram is *PQRS*? Explain.

Extended Response

22. The coordinates of the vertices of isosceles trapezoid *ABCD* are $A(0, 0)$, $B(6, 8)$, $C(10, 8)$, and $D(16, 0)$. The coordinates of the vertices of isosceles trapezoid *AFGH* are $A(0, 0)$, $F(3, 4)$, $G(5, 4)$, and $H(8, 0)$. Are the two trapezoids similar? Justify your answer.

23. The circle graph below shows fall sports participation at one school.

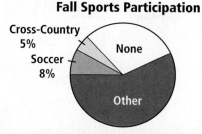

a. How many degrees greater is the measure of the angle that represents soccer than the measure of the angle that represents cross-country? Show your work.

b. If 48 students play soccer, how many run cross-country? Explain your reasoning.

Get Ready!

Lesson 8-1

The Pythagorean Theorem

Algebra Solve for *a, b,* or *c* in right $\triangle ABC$, where *a* and *b* are the lengths of the legs and *c* is the length of the hypotenuse.

1. $a = 8, b = 15, c = \blacksquare$ **2.** $a = \blacksquare, b = 4, c = 12$ **3.** $a = 2\sqrt{3}, b = 2\sqrt{6}, c = \blacksquare$

Lesson 8-2

Special Right Triangles

4. Find the length of the shorter leg of a 30°-60°-90° triangle with hypotenuse $8\sqrt{5}$.

5. Find the length of the diagonal of a square whose perimeter is 24.

6. Find the height of an equilateral triangle with sides of length 8.

Lessons 10-1, 10-2, and 10-3

Area

Find the area of each figure. Leave your answer in simplest radical form.

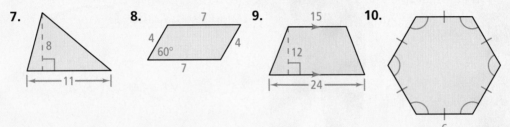

7. 8

├── 11 ──┤

8. 7

4   60°

4

7

9. 15

12

├── 24 ──┤

10. 6

Lesson 10-4

Perimeters and Areas of Similar Figures

11. Two similar triangles have corresponding sides in a ratio of 3 : 5. Find the perimeter of the smaller triangle if the larger triangle has perimeter 40.

12. Two regular hexagons have areas of 8 and 25. Find the ratio of corresponding sides.

Looking Ahead Vocabulary

13. In Chapter 10, you learned about the *altitude* of a triangle. What do you think is the *altitude* of a three-dimensional figure?

14. You can turn a rock over in your hands to examine its *surface*. What do you think the *surface area* of a three-dimensional figure is?

15. What does an Egyptian *pyramid* look like?

CHAPTER 11

Surface Area and Volume

PowerGeometry.com

Your place to get all things digital

VIDEO — Download videos connecting math to your world.

VOCABULARY — Math definitions in English and Spanish

SOLVE IT! — The online Solve It will get you in gear for each lesson.

DYNAMIC ACTIVITIES — Interactive! Vary numbers, graphs, and figures to explore math concepts.

ONLINE PROBLEMS — Online access to stepped-out problems aligned to Common Core

ONLINE HOMEWORK — Get and view your assignments online.

MathXL FOR SCHOOL — Extra practice and review online

DOMAINS

- Geometric Measurement and Dimension
- Modeling with Geometry

Doesn't that look like fun, riding on a lake in a sphere? How big would the sphere need to be to hold you?

In this chapter, you will learn about finding the surface areas and volumes of three-dimensional figures.

Vocabulary

English/Spanish Vocabulary Audio Online:

English	Spanish
cone, *p. 711*	cono
cross section, *p. 690*	sección de corte
cylinder, *p. 701*	cilindro
face, *p. 688*	cara
polyhedron, *p. 688*	poliedro
prism, *p. 699*	prisma
pyramid, *p. 708*	pirámide
similar solids, *p. 742*	cuerpos geométricos semejantes
sphere, *p. 733*	esfera
surface area, *pp. 700, 702, 709, 711*	área total
volume, *p. 717*	volumen

My Math Video

`00:04:04`

BIGideas

1 Visualization

Essential Question How can you determine the intersection of a solid and a plane?

2 Measurement

Essential Question How do you find the surface area and volume of a solid?

3 Similarity

Essential Question How do the surface areas and volumes of similar solids compare?

Chapter Preview

Space Figures and Cross Sections

© **Content Standard**
G.GMD.4 Identify the shapes of two-dimensional cross-sections of three-dimensional objects, and identify three-dimensional objects generated by rotations of two-dimensional objects.

Objectives To recognize polyhedra and their parts
To visualize cross sections of space figures

Getting Ready!

If you can shift your perspective by reflecting or rotating to get another net, then those nets are the same.

The tissue box at the right is a rectangular solid. Let x = the number of corners, y = the number of flat surfaces, and z = the number of folded creases. What is an equation that relates the quantities x, y, and z for a rectangular solid? Will your equation hold true for a cube? A solid with a triangular top and bottom? Explain.

© MATHEMATICAL PRACTICES

In the Solve It, you used two-dimensional nets to represent a three-dimensional object.

Lesson Vocabulary
• polyhedron
• face
• edge
• vertex
• cross section

A **polyhedron** is a space figure, or three-dimensional figure, whose surfaces are polygons. Each polygon is a **face** of the polyhedron. An **edge** is a segment that is formed by the intersection of two faces. A **vertex** is a point where three or more edges intersect.

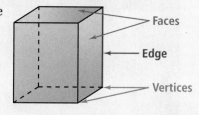

Faces

Edge

Vertices

Essential Understanding You can analyze a three-dimensional figure by using the relationships among its vertices, edges, and faces.

Plan

Can you see the solid?
A dashed line indicates an edge that is hidden from view. This figure has one four-sided face and four triangular faces.

© **Problem 1** **Identifying Vertices, Edges, and Faces**

How many vertices, edges, and faces are in the polyhedron at the right? List them.

There are five vertices: D, E, F, G, and H.

There are eight edges: $\overline{DE}$, $\overline{EF}$, $\overline{FG}$, $\overline{GD}$, $\overline{DH}$, $\overline{EH}$, $\overline{FH}$, and $\overline{GH}$.

There are five faces: $\triangle DEH$, $\triangle EFH$, $\triangle FGH$, $\triangle GDH$, and quadrilateral $DEFG$.

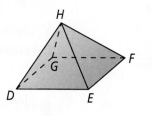

© ✓ **Got It?** **1. a.** How many vertices, edges, and faces are in the polyhedron at the right? List them.
b. Reasoning Is $\overline{TV}$ an edge? Explain why or why not.

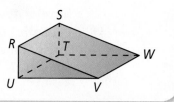

Leonhard Euler, a Swiss mathematician, discovered a relationship among the numbers of faces, vertices, and edges of any polyhedron. The result is known as Euler's Formula.

take note

Key Concept Euler's Formula

The sum of the number of faces (*F*) and vertices (*V*) of a polyhedron is two more than the number of its edges (*E*).

$$F + V = E + 2$$

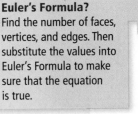

Plan

How do you verify Euler's Formula?
Find the number of faces, vertices, and edges. Then substitute the values into Euler's Formula to make sure that the equation is true.

Problem 2 Using Euler's Formula

How many vertices, edges, and faces does the polyhedron at the right have? Use your results to verify Euler's Formula.

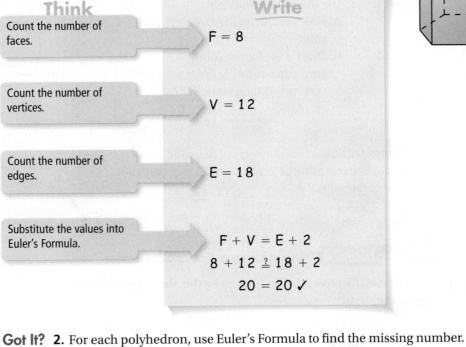

Think

Count the number of faces.

Count the number of vertices.

Count the number of edges.

Substitute the values into Euler's Formula.

Write

$F = 8$

$V = 12$

$E = 18$

$F + V = E + 2$

$8 + 12 \stackrel{?}{=} 18 + 2$

$20 = 20 \checkmark$

Got It? **2.** For each polyhedron, use Euler's Formula to find the missing number.

a.

faces: ■
edges: 30
vertices: 20

b.

faces: 20
edges: ■
vertices: 12

In two dimensions, Euler's Formula reduces to $F + V = E + 1$, where F is the number of regions formed by V vertices linked by E segments.

Plan

What do you use for the variables?

In 3-D	In 2-D
F: Faces	→ Regions
V: Vertices	→ Vertices
E: Edges	→ Segments

Problem 3 **Verifying Euler's Formula in Two Dimensions**

How can you verify Euler's Formula for a net for the solid in Problem 2?

Draw a net for the solid.

Number of regions: $F = 8$

Number of vertices: $V = 22$

Number of segments: $E = 29$

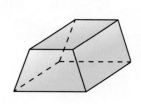

$F + V = E + 1$ Euler's Formula for two dimensions

$8 + 22 = 29 + 1$ Substitute.

$30 = 30$ ✓

Got It? **3.** Use the solid at the right.
  a. How can you verify Euler's Formula $F + V = E + 2$ for the solid?
  b. Draw a net for the solid.
  c. How can you verify Euler's Formula $F + V = E + 1$ for your two-dimensional net?

A **cross section** is the intersection of a solid and a plane. You can think of a cross section as a very thin slice of the solid.

This cross section is a triangle.

Think

How can you see the cross section?
Mentally rotate the solid so that the plane is parallel to your face.

Problem 4 **Describing a Cross Section**

What is the cross section formed by the plane and the solid at the right?

The cross section is a rectangle.

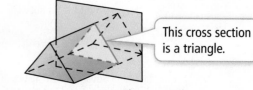

Got It? **4.** For the solid at the right, what is the cross section formed by each of the following planes?
  a. a horizontal plane
  b. a vertical plane that divides the solid in half

To draw a cross section, you can sometimes use the idea from Postulate 1-3 that the intersection of two planes is exactly one line.

 Problem 5 **Drawing a Cross Section**

Visualization Draw a cross section formed by a vertical plane intersecting the front and right faces of the cube. What shape is the cross section?

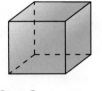

Think

How can you see parallel segments?
Focus on the plane intersecting the front and right faces. The plane and both faces are vertical, so the intersections are vertical parallel lines.

Step 1
Visualize a vertical plane intersecting the vertical faces in parallel segments.

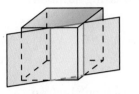

Step 2
Draw the parallel segments.

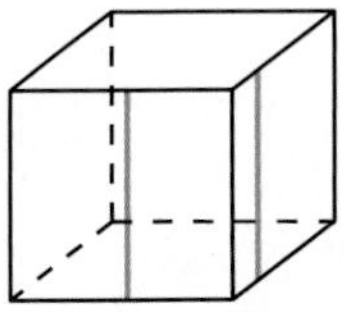

Step 3
Join their endpoints. Shade the cross section.

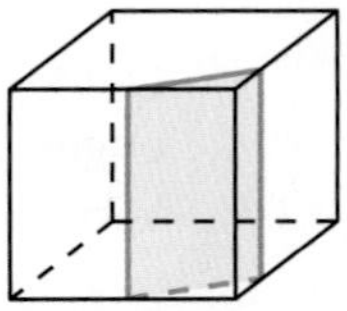

The cross section is a rectangle.

Got It? **5.** Draw the cross section formed by a horizontal plane intersecting the left and right faces of the cube. What shape is the cross section?

Lesson Check

Do you know HOW?

1. How many faces, edges, and vertices are in the solid? List them.

2. What is a net for the solid in Exercise 1? Verify Euler's Formula for the net.

3. What is the cross section formed by the cube and the plane containing the diagonals of a pair of opposite faces?

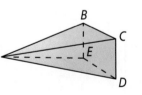

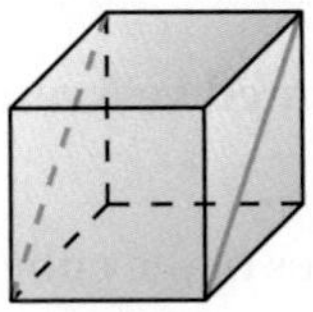

Do you UNDERSTAND?  **MATHEMATICAL PRACTICES**

4. Vocabulary Suppose you build a polyhedron from two octagons and eight squares. Without using Euler's Formula, how many edges does the solid have? Explain.

5. Error Analysis Your math class is drawing polyhedrons. Which figure does not belong in the diagram below? Explain.

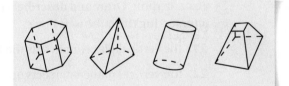

Practice and Problem-Solving Exercises

MATHEMATICAL PRACTICES

A Practice For each polyhedron, how many vertices, edges, and faces are there? List them. ◀ See Problem 1.

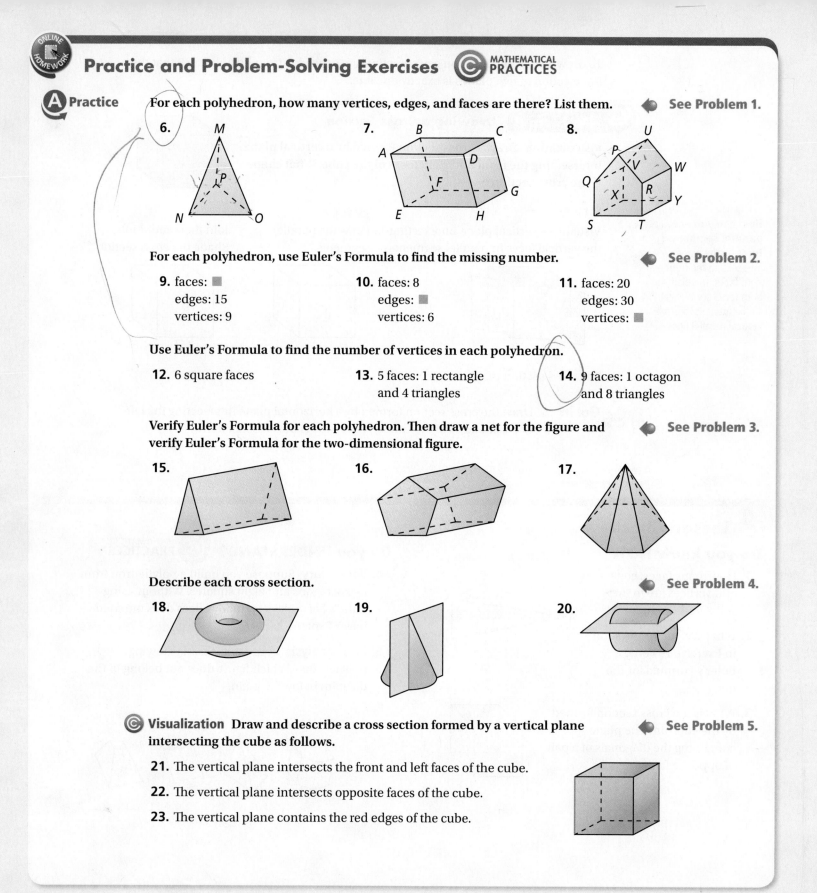

6.

7.

8.

For each polyhedron, use Euler's Formula to find the missing number. ◀ See Problem 2.

9. faces: ■
   edges: 15
   vertices: 9

10. faces: 8
   edges: ■
   vertices: 6

11. faces: 20
   edges: 30
   vertices: ■

Use Euler's Formula to find the number of vertices in each polyhedron.

12. 6 square faces

13. 5 faces: 1 rectangle and 4 triangles

14. 9 faces: 1 octagon and 8 triangles

Verify Euler's Formula for each polyhedron. Then draw a net for the figure and verify Euler's Formula for the two-dimensional figure. ◀ See Problem 3.

15.

16.

17.

Describe each cross section. ◀ See Problem 4.

18.

19.

20.

© Visualization Draw and describe a cross section formed by a vertical plane intersecting the cube as follows. ◀ See Problem 5.

21. The vertical plane intersects the front and left faces of the cube.

22. The vertical plane intersects opposite faces of the cube.

23. The vertical plane contains the red edges of the cube.

Apply

© **24. a. Open-Ended** Sketch a polyhedron whose faces are all rectangles. Label the lengths of its edges.
   b. Use graph paper to draw two different nets for the polyhedron.

25. For the figure shown at the right, sketch each of following.
   a. a horizontal cross section
   b. a vertical cross section that contains the vertical line of symmetry

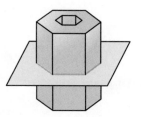

© **26. Reasoning** Can you find a cross section of a cube that forms a triangle? Explain.

© **27. Reasoning** Suppose the number of faces in a certain polyhedron is equal to the number of vertices. Can the polyhedron have nine edges? Explain.

© **Visualization** Draw and describe a cross section formed by a plane intersecting the cube as follows.

28. The plane is tilted and intersects the left and right faces of the cube.

29. The plane contains the red edges of the cube.

30. The plane cuts off a corner of the cube.

© **Visualization** A plane region that revolves completely about a line sweeps out a **solid of revolution.** Use the sample to help you describe the *solid of revolution* you get by revolving each region about line ℓ.

   Sample: Revolve the rectangular region about the line ℓ. You get a cylinder as the solid of revolution.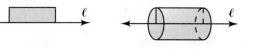

31.    **32.**    **33.**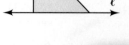

© **34. Think About a Plan** Some balls are made from panels that suggest polygons. A soccer ball suggests a polyhedron with 20 regular hexagons and 12 regular pentagons. How many vertices does this polyhedron have?
   • How can you determine the number of edges in a solid if you know the types of polygons that form the faces?
   • What relationship can you use to find the number of vertices?

Euler's Formula $F + V = E + 1$ applies to any two-dimensional network where F is the number of regions formed by V vertices linked by E edges (or paths). Verify Euler's Formula for each network shown.

35.   **36.**   **37.**

38. Platonic Solids There are five regular polyhedrons. They are called *regular* because all their faces are congruent regular polygons, and the same number of faces meet at each vertex. They are also called *Platonic solids* after the Greek philosopher Plato, who first described them in his work *Timaeus* (about 350 B.C.).

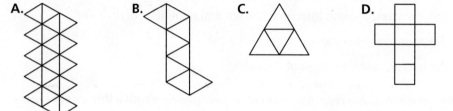

Tetrahedron   Octahedron   Icosahedron

Hexahedron   Dodecahedron

a. Match each net below with a Platonic solid.

A.   B.   C.   D.   E.

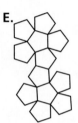

b. The first two Platonic solids have more familiar names. What are they?

c. Verify that Euler's Formula is true for the first three Platonic solids.

39. A cube has a net with area 216 in.^{2}. How long is an edge of the cube?

© **40. Writing** Cross sections are used in medical training and research. Research and write a paragraph on how magnetic resonance imaging (MRI) is used to study cross sections of the brain.

© **Challenge** **41. Open-Ended** Draw a single solid that has the following three cross sections.

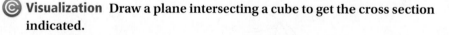

Horizontal   Vertical

© **Visualization** Draw a plane intersecting a cube to get the cross section indicated.

42. scalene triangle   **43.** isosceles triangle   **44.** equilateral triangle

45. trapezoid   **46.** isosceles trapezoid   **47.** parallelogram

48. rhombus   **49.** pentagon   **50.** hexagon

Standardized Test Prep

SAT/ACT

51. A polyhedron has four vertices and six edges. How many faces does it have?

    Ⓐ 2   Ⓑ 4   Ⓒ 5   Ⓓ 10

52. Suppose the circumcenter of $\triangle ABC$ lies on one of its sides. What type of triangle must $\triangle ABC$ be?

    Ⓕ scalene   Ⓖ isosceles   Ⓗ equilateral   Ⓘ right

53. What is the area of a regular hexagon whose perimeter is 36 in.?

    Ⓐ $18\sqrt{3}$ in.^{2} Ⓑ $27\sqrt{3}$ in.^{2} Ⓒ $36\sqrt{3}$ in.^{2} Ⓓ $54\sqrt{3}$ in.^{2}

54. What is the best description of the polygon at the right?

    Ⓕ concave decagon   Ⓗ regular pentagon

    Ⓖ convex decagon   Ⓘ regular decagon

Short Response

55. The coordinates of three vertices of a parallelogram are $A(2, 1)$, $B(1, -2)$ and $C(4, -1)$. What are the coordinates of the fourth vertex D? Explain.

Mixed Review

56. Probability Shuttle buses to an airport terminal leave every 20 min from a remote parking lot. Draw a geometric model and find the probability that a traveler who arrives at a random time will have to wait at least 8 min for the bus to leave the parking lot.

◀ **See Lesson 10-8.**

57. Games A dartboard is a circle with a 12-in. radius. What is the probability that you throw a dart that lands within 6 in. of the center of the dartboard?

Find the value of x to the nearest tenth.

◀ **See Lesson 8-3.**

58.
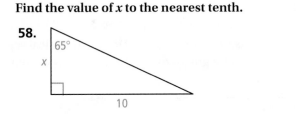

59.
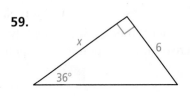

Get Ready! To prepare for Lesson 11-2, do Exercises 60–62.

Find the area of each net.

◀ **See Lessons 1-8 and 10-3.**

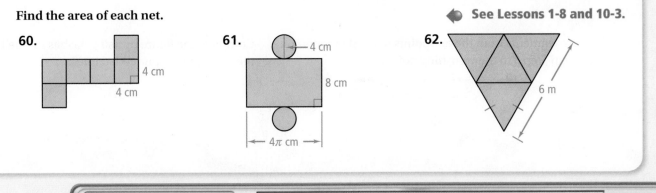

60. 4 cm, 4 cm   **61.** 4 cm, 8 cm, 4π cm   **62.** 6 m

Concept Byte

Use With Lesson 11-1

EXTENSION

Perspective Drawing

© **Content Standard**
Extends G.GMD.4 Identify the shapes of two-dimensional cross-sections of three-dimensional objects, and identify three-dimensional objects generated by rotations of two-dimensional objects.

You can draw a three-dimensional space figure using a two-dimensional *perspective drawing*. Suppose two lines are parallel in three dimensions, but extend away from the viewer. You draw them—and create perspective—so that they meet at a *vanishing point* on a *horizon line*.

Example 1

Draw a cube in one-point perspective.

Step 1

Draw a square. Then draw a horizon line and a vanishing point on the line.

Step 2

Lightly draw segments from the vertices of the square to the vanishing point.

Step 3

Draw a square for the back of the cube. Each vertex should lie on a segment you drew in Step 2.

Step 4

Complete the figure by using dashes for the hidden edges of the cube. Erase unneeded lines.

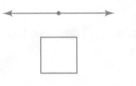 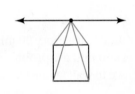 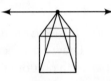

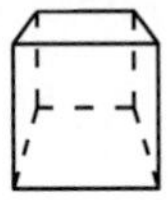

Example 2

Draw a box in two-point perspective.

Step 1

Draw a vertical segment. Then draw a horizon line and two vanishing points on the line.

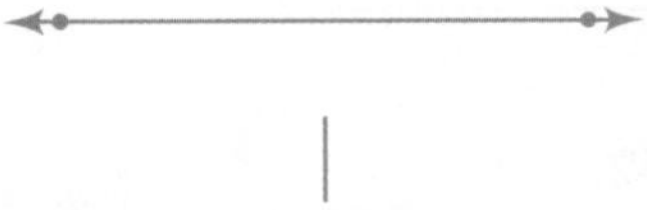

Step 2

Lightly draw segments from the endpoints of the vertical segment to each vanishing point.

Step 3

Draw two vertical segments between the segments of Step 2.

 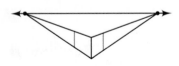

Step 4

Draw segments from the endpoints of the segments you drew in Step 3 to the vanishing points.

Step 5

Complete the figure by using dashes for the hidden edges of the figure. Erase unneeded lines.

In one-point perspective, the front of the cube is parallel to the drawing surface. A two-point perspective drawing generally looks like a corner view. For either type of drawing, you should be able to envision each vanishing point.

Exercises

Is each object drawn in one-point or two-point perspective?

1.

2.

3.

4.

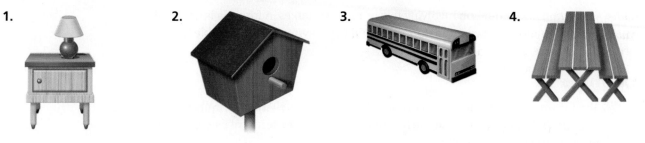

Draw each object in one-point perspective and then in two-point perspective.

5. a shoe box

6. a building in your town that sits on a street corner

Draw each container using one-point perspective. Show a base at the front.

7. a triangular carton

8. a hexagonal box

Copy each figure and locate the vanishing point(s).

9.

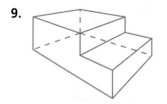

10.

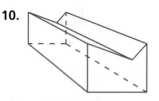

Optical Illusions What is the optical illusion? Explain the role of perspective in each illusion.

11.

12.

ⓒ **13. Open-Ended** You can draw block letters in either one-point or two-point perspective. Write your initials in block letters using one-point perspective and two-point perspective.

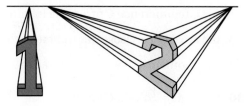

Algebra Review

Literal Equations

© **Content Standard**
Reviews A-CED.4 Identify the shapes of two-dimensional cross-sections of three-dimensional objects, and identify three-dimensional objects generated by rotations of two-dimensional objects.

A *literal equation* is an equation involving two or more variables. A formula is a special type of literal equation. You can transform a formula by solving for one variable in terms of the others.

Example

A The formula for the volume of a cylinder is $V = \pi r^2 h$. Find a formula for the height in terms of the radius and volume.

$V = \pi r^2 h$ Use the formula for the volume of a cylinder.

$\dfrac{V}{\pi r^2} = \dfrac{\pi r^2 h}{\pi r^2}$ Divide each side by πr^2, with $r \neq 0$.

$\dfrac{V}{\pi r^2} = h$ Simplify.

The formula for the height is $h = \dfrac{V}{\pi r^2}$.

B Find a formula for the area of a square in terms of its perimeter.

$P = 4s$ Use the formula for the perimeter of a square.

$\dfrac{P}{4} = s$ Solve for s in terms of P.

$A = s^2$ Use the formula for area.

$= \left(\dfrac{P}{4}\right)^2$ Substitute $\dfrac{P}{4}$ for s.

$= \dfrac{P^2}{16}$ Simplify.

The formula for the area is $A = \dfrac{P^2}{16}$.

Exercises

Algebra Solve each equation for the variable in red.

1. $C = 2\pi r$

2. $A = \frac{1}{2}bh$

3. $A = \pi r^2$

Algebra Solve for the variable in red. Then solve for the variable in blue.

4. $P = 2w + 2\ell$

5. $\tan A = \frac{y}{x}$

6. $A = \frac{1}{2}(b_1 + b_2)h$

Find a formula as described below.

7. the circumference C of a circle in terms of its area A

8. the area A of an isosceles right triangle in terms of the hypotenuse h

9. the apothem a of a regular hexagon in terms of the area A of the hexagon

10. Solve $A = \frac{1}{2}ab \sin C$ for $m\angle C$.

Surface Areas of Prisms and Cylinders

Content Standard
G.MG.1 Use geometric shapes, their measures, and their properties to describe objects.

Objective To find the surface area of a prism and a cylinder

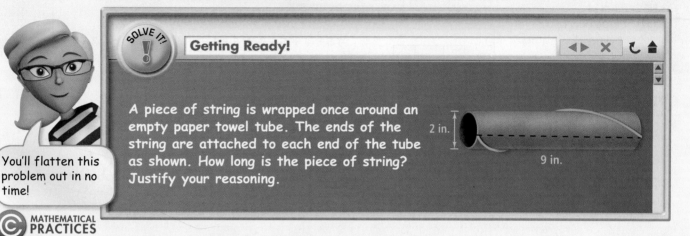

SOLVE IT!

Getting Ready!

A piece of string is wrapped once around an empty paper towel tube. The ends of the string are attached to each end of the tube as shown. How long is the piece of string? Justify your reasoning.

2 in.

9 in.

You'll flatten this problem out in no time!

MATHEMATICAL PRACTICES

Lesson Vocabulary

• prism (base, lateral face, altitude, height, lateral area, surface area)
• right prism
• oblique prism
• cylinder (base, altitude, height, lateral area, surface area)
• right cylinder
• oblique cylinder

In the Solve It, you investigated the structure of a tube. In this lesson, you will learn properties of three-dimensional figures by investigating their surfaces.

Essential Understanding To find the surface area of a three-dimensional figure, find the sum of the areas of all the surfaces of the figure.

A **prism** is a polyhedron with two congruent, parallel faces, called **bases.** The other faces are **lateral faces.** You can name a prism using the shape of its bases.

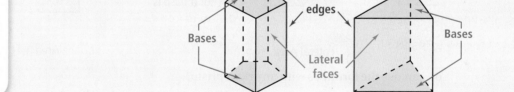

Lateral edges

Bases

Lateral faces

Bases

Pentagonal prism

Triangular prism

An **altitude** of a prism is a perpendicular segment that joins the planes of the bases. The **height** h of a prism is the length of an altitude. A prism may either be right or oblique.

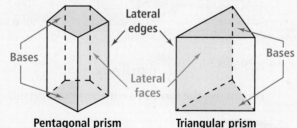

h

h

h h

Right prisms

Oblique prisms

In a **right prism,** the lateral faces are rectangles and a lateral edge is an altitude. In an **oblique prism,** some or all of the lateral faces are nonrectangular. In this book, you may assume that a prism is a right prism unless stated or pictured otherwise.

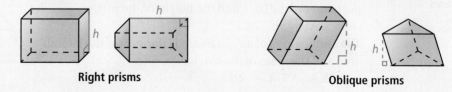

The **lateral area** (L.A.) of a prism is the sum of the areas of the lateral faces. The **surface area** (S.A.) is the sum of the lateral area and the area of the two bases.

Problem 1 **Using a Net to Find Surface Area of a Prism**

What is the surface area of the prism at the right? Use a net.

Draw a net for the prism. Then calculate the surface area.

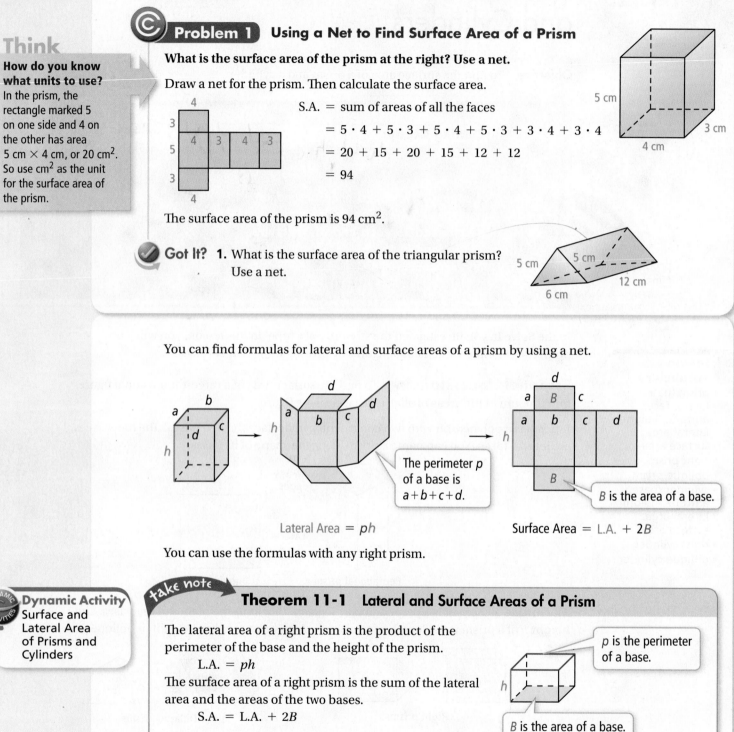

$$S.A. = \text{sum of areas of all the faces}$$
$$= 5 \cdot 4 + 5 \cdot 3 + 5 \cdot 4 + 5 \cdot 3 + 3 \cdot 4 + 3 \cdot 4$$
$$= 20 + 15 + 20 + 15 + 12 + 12$$
$$= 94$$

The surface area of the prism is 94 cm².

Got It? **1.** What is the surface area of the triangular prism? Use a net.

Think

How do you know what units to use?
In the prism, the rectangle marked 5 on one side and 4 on the other has area 5 cm × 4 cm, or 20 cm². So use cm² as the unit for the surface area of the prism.

You can find formulas for lateral and surface areas of a prism by using a net.

The perimeter p of a base is $a+b+c+d$.

B is the area of a base.

Lateral Area = ph

Surface Area = L.A. + 2B

You can use the formulas with any right prism.

Dynamic Activity
Surface and Lateral Area of Prisms and Cylinders

take note

Theorem 11-1 Lateral and Surface Areas of a Prism

The lateral area of a right prism is the product of the perimeter of the base and the height of the prism.

L.A. = ph

The surface area of a right prism is the sum of the lateral area and the areas of the two bases.

S.A. = L.A. + 2B

p is the perimeter of a base.

B is the area of a base.

Plan

What do you need to find first?
You first need to find the missing side length of a triangular base so that you can find the perimeter of a base.

What is the surface area of the prism at the right?

Step 1 Find the perimeter of a base.

The perimeter of the base is the sum of the side lengths of the triangle. Since the base is a right triangle, the hypotenuse is $\sqrt{3^2 + 4^2}$ cm, or 5 cm, by the Pythagorean Theorem.

$$p = 3 + 4 + 5 = 12$$

Step 2 Find the lateral area of the prism.

$$\text{L.A.} = ph \qquad \text{Use the formula for lateral area.}$$
$$= 12 \cdot 6 \qquad \text{Substitute 12 for } p \text{ and 6 for } h.$$
$$= 72 \qquad \text{Simplify.}$$

Think

Which height do you need?
For problems involving solids, make it a habit to note which height the formula requires. In Step 3, you need the height of the triangle, not the height of the prism.

Step 3 Find the area of a base.

$$B = \tfrac{1}{2}bh \qquad \text{Use the formula for the area of a triangle.}$$
$$= \tfrac{1}{2}(3 \cdot 4) \qquad \text{Substitute 3 for } b \text{ and 4 for } h.$$
$$= 6$$

Step 4 Find the surface area of the prism.

$$\text{S.A.} = \text{L.A.} + 2B \qquad \text{Use the formula for surface area.}$$
$$= 72 + 2(6) \qquad \text{Substitute 72 for L.A. and 6 for } B.$$
$$= 84 \qquad \text{Simplify.}$$

The surface area of the prism is 84 cm².

✓ **Got It?** **2. a.** What is the lateral area of the prism at the right?
  b. What is the area of a base in simplest radical form?
  c. What is the surface area of the prism rounded to a whole number?

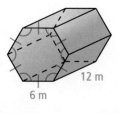

A **cylinder** is a solid that has two congruent parallel **bases** that are circles. An **altitude** of a cylinder is a perpendicular segment that joins the planes of the bases. The **height** h of a cylinder is the length of an altitude.

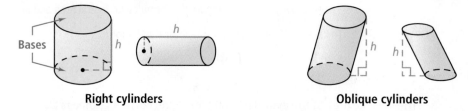

Right cylinders   **Oblique cylinders**

In a **right cylinder,** the segment joining the centers of the bases is an altitude. In an **oblique cylinder,** the segment joining the centers is not perpendicular to the planes containing the bases. In this book, you may assume that a cylinder is a right cylinder unless stated or pictured otherwise.

To find the area of the curved surface of a cylinder, visualize "unrolling" it. The area of the resulting rectangle is the **lateral area** of the cylinder. The **surface area** of a cylinder is the sum of the lateral area and the areas of the two circular bases. You can find formulas for these areas by looking at a net for a cylinder.

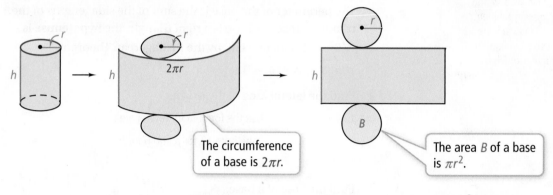

The circumference of a base is $2\pi r$.

The area B of a base is πr^2.

Lateral Area $= 2\pi rh$

Surface Area $=$ L.A. $+ 2\pi r^2$

---

take note

Theorem 11-2 Lateral and Surface Areas of a Cylinder

The lateral area of a right cylinder is the product of the circumference of the base and the height of the cylinder.

  L.A. $= 2\pi r \cdot h$, or L.A. $= \pi dh$

The surface area of a right cylinder is the sum of the lateral area and the areas of the two bases.

  S.A. $=$ L.A. $+ 2B$, or S.A. $= 2\pi rh + 2\pi r^2$

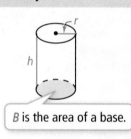

B is the area of a base.

Ⓒ **Problem 3** Finding Surface Area of a Cylinder

Multiple Choice The radius of the base of a cylinder is 4 in. and its height is 6 in. What is the surface area of the cylinder in terms of π?

Ⓐ 32π in.^{2} Ⓑ 42π in.^{2} Ⓒ 80π in.^{2} Ⓓ 120π in.^{2}

Think

How is finding the surface area of a cylinder like finding the surface area of a prism?
For both, you need to find the L.A. and add it to twice the area of a base.

$$\text{S.A.} = \text{L.A.} + 2B \qquad \text{Use the formula for surface area of a cylinder.}$$

$$= 2\pi rh + 2(\pi r^2) \qquad \text{Substitute the formulas for lateral area and area of a circle.}$$

$$= 2\pi(4)(6) + 2(\pi 4^2) \qquad \text{Substitute 4 for } r \text{ and 6 for } h.$$

$$= 48\pi + 32\pi \qquad \text{Simplify.}$$

$$= 80\pi$$

The surface area of the cylinder is 80π in.^{2}. The correct choice is C.

✓ **Got It?** **3.** A cylinder has a height of 9 cm and a radius of 10 cm. What is the surface area of the cylinder in terms of π?

Problem 4 **Finding Lateral Area of a Cylinder**

Interior Design You are using the cylindrical stencil roller below to paint patterns on your floor. What area does the roller cover in one full turn?

6 in.   2.5 in.

The area covered is the lateral area of a cylinder with height 6 in. and diameter 2.5 in.

L.A. $= \pi dh$ Use the formula for lateral area of a cylinder.

     $= \pi(2.5)(6)$ Substitute 2.5 for d and 6 for h.

     $= 15\pi \approx 47.1$ Simplify.

In one full turn, the stencil roller covers about 47.1 in.^{2}.

Got It? **4. a.** A smaller stencil roller has a height of 1.5 in. and the same diameter as the roller in Problem 4. What area does the smaller roller cover in one turn? Round your answer to the nearest tenth.

     b. Reasoning What is the ratio of the smaller roller's height to the larger roller's height? What is the ratio of the areas the rollers can cover in one turn (smaller to larger)?

 Lesson Check

Do you know HOW?

What is the surface area of each prism?

1.
5 in.
4 in.
5 in.

2. 7 ft   7 ft
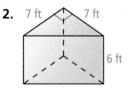
6 ft

What is the surface area of each cylinder?

3. 3 cm
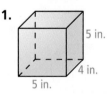
5 cm

4. 12 m
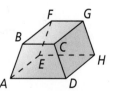
10 m

Do you UNDERSTAND? MATHEMATICAL PRACTICES

5. Vocabulary Name the lateral faces and the bases of the prism at the right.

F   G
B   C
E   H
A   D

6. Error Analysis Your friend drew a net of a cylinder. What is your friend's error? Explain.

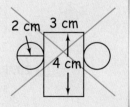
2 cm   3 cm
4 cm

Practice and Problem-Solving Exercises

MATHEMATICAL PRACTICES

A **Practice** Use a net to find the surface area of each prism.

See Problem 1.

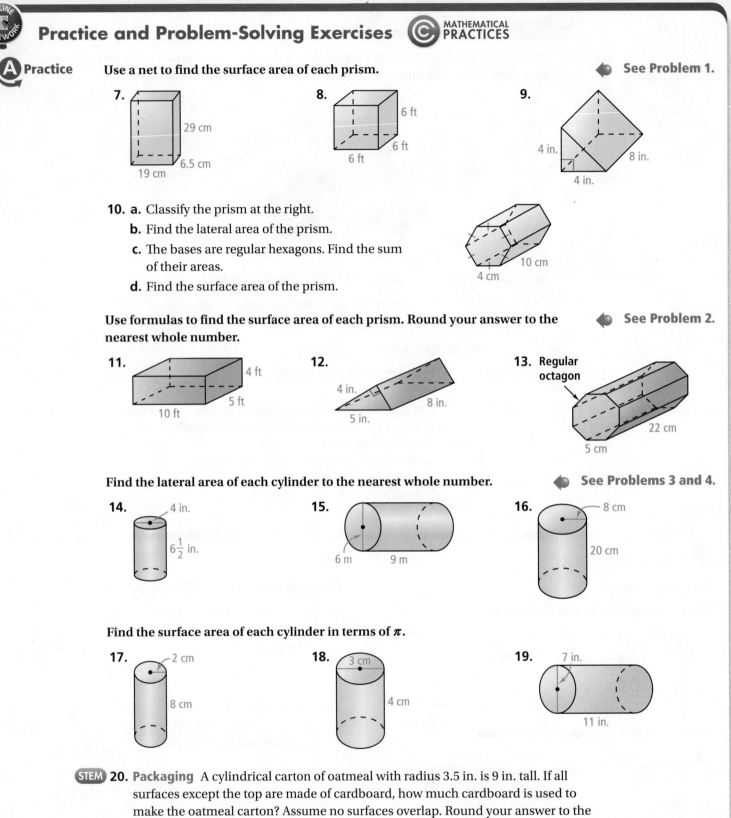

7. 29 cm, 6.5 cm, 19 cm

8. 6 ft, 6 ft, 6 ft

9. 4 in., 8 in., 4 in.

10. a. Classify the prism at the right.

   b. Find the lateral area of the prism.

   c. The bases are regular hexagons. Find the sum of their areas.

   d. Find the surface area of the prism.

   10 cm, 4 cm

Use formulas to find the surface area of each prism. Round your answer to the nearest whole number.

See Problem 2.

11. 4 ft, 5 ft, 10 ft

12. 4 in., 8 in., 5 in.

13. Regular octagon, 22 cm, 5 cm

Find the lateral area of each cylinder to the nearest whole number.

See Problems 3 and 4.

14. 4 in., $6\frac{1}{2}$ in.

15. 6 m, 9 m

16. 8 cm, 20 cm

Find the surface area of each cylinder in terms of π.

17. 2 cm, 8 cm

18. 3 cm, 4 cm

19. 7 in., 11 in.

STEM 20. Packaging A cylindrical carton of oatmeal with radius 3.5 in. is 9 in. tall. If all surfaces except the top are made of cardboard, how much cardboard is used to make the oatmeal carton? Assume no surfaces overlap. Round your answer to the nearest square inch.

21. A triangular prism has base edges 4 cm, 5 cm, and 6 cm long. Its lateral area is 300 cm². What is the height of the prism?

22. **Writing** Explain how a cylinder and a prism are alike and how they are different.

23. **Pencils** A hexagonal pencil is a regular hexagonal prism, as shown at the right. A base edge of the pencil has a length of 4 mm. The pencil (without eraser) has a height of 170 mm. What is the area of the surface of the pencil that gets painted?

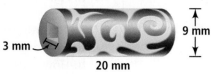

24. **Open-Ended** Draw a net for a rectangular prism with a surface area of 220 cm².

25. Consider a box with dimensions 3, 4, and 5.
    a. Find its surface area.
    b. Double each dimension and then find the new surface area.
    c. Find the ratio of the new surface area to the original surface area.
    d. Repeat parts (a)–(c) for a box with dimensions 6, 9, and 11.
    e. **Make a Conjecture** How does doubling the dimensions of a rectangular prism affect the surface area?

26. **Think About a Plan** A cylindrical bead has a square hole, as shown at the right. Find the area of the surface of the entire bead.
    • Can you visualize the bead as a combination of familiar figures?
    • How do you find the area of the surface of the inner part of the bead?

27. **Estimation** Estimate the surface area of a cube with edges 4.95 cm long.

28. **Error Analysis** Your class is drawing right triangular prisms. Your friend's paper is below. What is your friend's error?

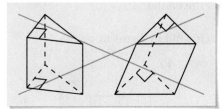

29. **Packaging** A cylindrical can of cocoa has the dimensions shown at the right. What is the approximate surface area available for the label? Round to the nearest square inch.

30. **Pest Control** A flour moth trap has the shape of a triangular prism that is open on both ends. An environmentally safe chemical draws the moth inside the prism, which is lined with an adhesive. What is the area of the trap that is lined with the adhesive?

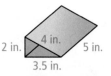

31. Suppose that a cylinder has a radius of r units, and that the height of the cylinder is also r units. The lateral area of the cylinder is 98π square units.
   a. Algebra Find the value of r.
   b. Find the surface area of the cylinder.

32. Geometry in 3 Dimensions Use the diagram at the right.
   a. Find the three coordinates of each vertex A, B, C, and D of the rectangular prism.
   b. Find AB.
   c. Find BC.
   d. Find CD.
   e. Find the surface area of the prism.

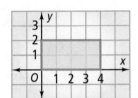

Ⓒ **Visualization** **Suppose you revolve the plane region completely about the given line to sweep out a solid of revolution. Describe the solid and find its surface area in terms of π.**

33. the y-axis

34. the x-axis

35. the line $y = 2$

36. the line $x = 4$

Ⓒ **37. Reasoning** Suppose you double the radius of a right cylinder.
   a. How does that affect the lateral area?
   b. How does that affect the surface area?
   c. Use the formula for surface area of a right cylinder to explain why the surface area in part (b) was not doubled.

38. Packaging Some cylinders have wrappers with a spiral seam. Peeled off, the wrapper has the shape of a parallelogram. The wrapper for a biscuit container has base 7.5 in. and height 6 in.
   a. Find the radius and height of the container.
   b. Find the surface area of the container.

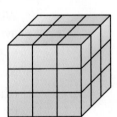

Ⓒ **Challenge**   **What is the surface area of each solid in terms of π?**

39.

40.

41.

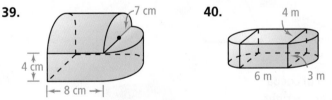
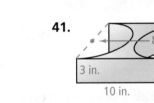

42. Each edge of the large cube at the right is 12 inches long. The cube is painted on the outside, and then cut into 27 smaller cubes. Answer these questions about the 27 cubes.
   a. How many are painted on 4, 3, 2, 1, and 0 faces?
   b. What is the total surface area that is unpainted?

43. Algebra The sum of the height and radius of a cylinder is 9 m. The surface area of the cylinder is 54π m². Find the height and the radius.

SAT/ACT

44. The height of a cylinder is twice the radius of the base. The surface area of the cylinder is 56π ft^2. What is the diameter of the base to the nearest tenth of a foot?

45. Two sides of a triangle measure 11 ft and 23 ft. What is the smallest possible whole number length, in feet, for the third side?

46. A polyhedron has one hexagonal face and six triangular faces. How many vertices does the polyhedron have?

47. The shortest shadow cast by a tree is 8 m long. The height of the tree is 20 m. To the nearest degree, what is the angle of elevation of the sun when the shortest shadow is cast?

Mixed Review

Sketch each space figure and then draw a net for it. Label the net with its dimensions.

◀ See Lessons 1-1 and 11-1.

48. a rectangular prism with height 5 cm and a base 3 cm by 4 cm

49. a cylinder with a 72π-in. circumference and a 22-in. height

Find the area of each part of the circle to the nearest tenth.

◀ See Lesson 10-7.

50. sector QOP

51. the segment of the circle bounded by $\overline{QP}$ and $\overparen{QP}$

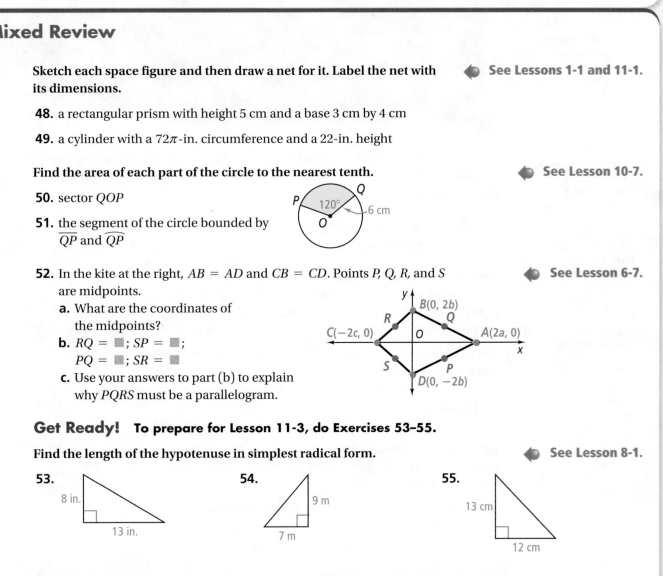

52. In the kite at the right, $AB = AD$ and $CB = CD$. Points P, Q, R, and S are midpoints.

◀ See Lesson 6-7.

  a. What are the coordinates of the midpoints?

  b. $RQ = \blacksquare$; $SP = \blacksquare$; $PQ = \blacksquare$; $SR = \blacksquare$

  c. Use your answers to part (b) to explain why $PQRS$ must be a parallelogram.

Get Ready! **To prepare for Lesson 11-3, do Exercises 53–55.**

Find the length of the hypotenuse in simplest radical form.

◀ See Lesson 8-1.

53.

8 in.

13 in.

54.

9 m

7 m

55.

13 cm

12 cm

Surface Areas of Pyramids and Cones

© Content Standard
G.MG.1 Use geometric shapes, their measures, and their properties to describe objects.

Objective To find the surface area of a pyramid and a cone

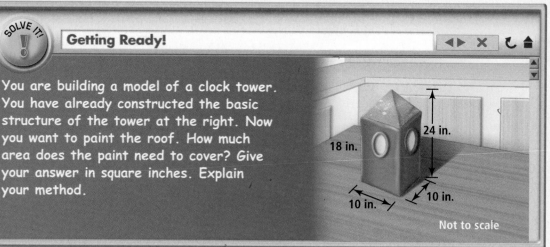

SOLVE IT!

Getting Ready!

You are building a model of a clock tower. You have already constructed the basic structure of the tower at the right. Now you want to paint the roof. How much area does the paint need to cover? Give your answer in square inches. Explain your method.

24 in.

18 in.

10 in.   10 in.

Not to scale

Think about what dimensions you need and how to get them from what you already know.

© MATHEMATICAL PRACTICES

The Solve It involves the triangular faces of a roof and the three-dimensional figures they form. In this lesson, you will learn to name such figures and to use formulas to find their areas.

Essential Understanding To find the surface area of a three-dimensional figure, find the sum of the areas of all the surfaces of the figure.

A **pyramid** is a polyhedron in which one face (the **base**) can be any polygon and the other faces (the **lateral faces**) are triangles that meet at a common vertex (called the **vertex** of the pyramid).

You name a pyramid by the shape of its base. The **altitude** of a pyramid is the perpendicular segment from the vertex to the plane of the base. The length of the altitude is the **height** h of the pyramid.

A **regular pyramid** is a pyramid whose base is a regular polygon and whose lateral faces are congruent isosceles triangles. The **slant height** ℓ is the length of the altitude of a lateral face of the pyramid.

In this book, you can assume that a pyramid is regular unless stated otherwise.

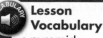
Lesson Vocabulary
- pyramid (base, lateral face, vertex, altitude, height, slant height, lateral area, surface area)
- regular pyramid
- cone (base, altitude, vertex, height, slant height, lateral area, surface area)
- right cone

Hexagonal pyramid
Vertex
Lateral edge
Lateral face
Altitude
Base
Base edge

Square pyramid
Height
Slant height
h ℓ

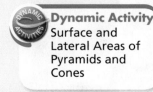
The **lateral area** of a pyramid is the sum of the areas of the congruent lateral faces. You can find a formula for the lateral area of a pyramid by looking at its net.

$A = \frac{1}{2}s\ell$

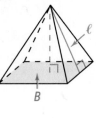

$$L.A. = 4\left(\tfrac{1}{2}s\ell\right) \quad \text{The area of each lateral face is } \tfrac{1}{2}s\ell.$$
$$= \tfrac{1}{2}(4s)\ell \quad \begin{array}{l}\text{Commutative and Associative} \\ \text{Properties of Multiplication}\end{array}$$
$$= \tfrac{1}{2}p\ell \quad \text{The perimeter } p \text{ of the base is } 4s.$$

To find the **surface area** of a pyramid, add the area of its base to its lateral area.

take note

Theorem 11-3   Lateral and Surface Areas of a Pyramid

The lateral area of a regular pyramid is half the product of the perimeter p of the base and the slant height ℓ of the pyramid.

$$L.A. = \tfrac{1}{2}p\ell$$

The surface area of a regular pyramid is the sum of the lateral area and the area B of the base.

$$S.A. = L.A. + B$$

© **Problem 1**   Finding the Surface Area of a Pyramid

Think

What is B?
B is the area of the base, which is a hexagon. You are given the apothem of the hexagon and the length of a side. Use them to find the area of the base.

What is the surface area of the hexagonal pyramid?

$$S.A. = L.A. + B \qquad \text{Use the formula for surface area.}$$
$$= \tfrac{1}{2}p\ell + \tfrac{1}{2}ap \qquad \text{Substitute the formulas for L.A. and } B.$$
$$= \tfrac{1}{2}(36)(9) + \tfrac{1}{2}\left(3\sqrt{3}\right)(36) \qquad \text{Substitute.}$$
$$\approx 255.5307436 \qquad \text{Use a calculator.}$$

The surface area of the pyramid is about 256 in.^{2}.

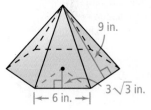
9 in.
$3\sqrt{3}$ in.
6 in.

© ✓ **Got It?**   **1. a.** A square pyramid has base edges of 5 m and a slant height of 3 m. What is the surface area of the pyramid?

   b. Reasoning Suppose the slant height of a pyramid is doubled. How does this affect the lateral area of the pyramid? Explain.

When the slant height of a pyramid is not given, you must calculate it before you can find the lateral area or surface area.

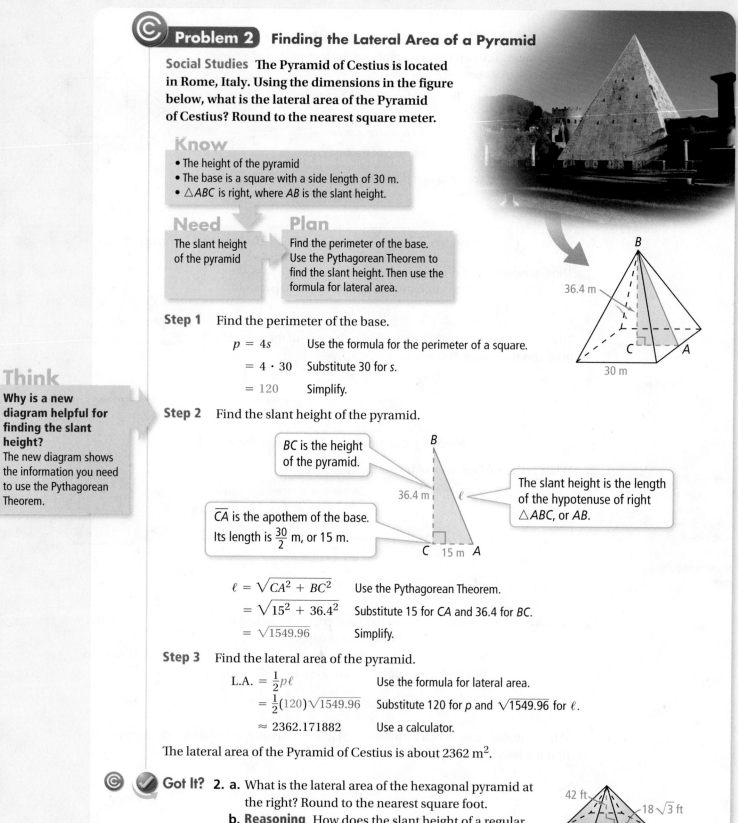

Problem 2 Finding the Lateral Area of a Pyramid

Social Studies The Pyramid of Cestius is located in Rome, Italy. Using the dimensions in the figure below, what is the lateral area of the Pyramid of Cestius? Round to the nearest square meter.

Know

- The height of the pyramid
- The base is a square with a side length of 30 m.
- $\triangle ABC$ is right, where AB is the slant height.

Need

The slant height of the pyramid

Plan

Find the perimeter of the base. Use the Pythagorean Theorem to find the slant height. Then use the formula for lateral area.

Step 1 Find the perimeter of the base.

$$p = 4s \qquad \text{Use the formula for the perimeter of a square.}$$
$$= 4 \cdot 30 \qquad \text{Substitute 30 for } s.$$
$$= 120 \qquad \text{Simplify.}$$

Think

Why is a new diagram helpful for finding the slant height?
The new diagram shows the information you need to use the Pythagorean Theorem.

Step 2 Find the slant height of the pyramid.

BC is the height of the pyramid.

The slant height is the length of the hypotenuse of right $\triangle ABC$, or *AB*.

$\overline{CA}$ is the apothem of the base. Its length is $\frac{30}{2}$ m, or 15 m.

$$\ell = \sqrt{CA^2 + BC^2} \qquad \text{Use the Pythagorean Theorem.}$$
$$= \sqrt{15^2 + 36.4^2} \qquad \text{Substitute 15 for } CA \text{ and 36.4 for } BC.$$
$$= \sqrt{1549.96} \qquad \text{Simplify.}$$

Step 3 Find the lateral area of the pyramid.

$$\text{L.A.} = \frac{1}{2}p\ell \qquad \text{Use the formula for lateral area.}$$
$$= \frac{1}{2}(120)\sqrt{1549.96} \qquad \text{Substitute 120 for } p \text{ and } \sqrt{1549.96} \text{ for } \ell.$$
$$\approx 2362.171882 \qquad \text{Use a calculator.}$$

The lateral area of the Pyramid of Cestius is about 2362 m^2.

Got It? **2. a.** What is the lateral area of the hexagonal pyramid at the right? Round to the nearest square foot.

      b. Reasoning How does the slant height of a regular pyramid relate to its height? Explain.

Like a pyramid, a **cone** is a solid that has one base and a vertex that is not in the same plane as the base. However, the **base** of a cone is a circle. In a **right cone,** the **altitude** is a perpendicular segment from the **vertex** to the center of the base. The **height** h is the length of the altitude. The **slant height** ℓ is the distance from the vertex to a point on the edge of the base. In this book, you can assume that a cone is a right cone unless stated or pictured otherwise.

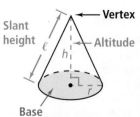

The **lateral area** is half the circumference of the base times the slant height. The formulas for the lateral area and **surface area** of a cone are similar to those for a pyramid.

Theorem 11-4   Lateral and Surface Areas of a Cone

The lateral area of a right cone is half the product of the circumference of the base and the slant height of the cone.

$$\text{L.A.} = \tfrac{1}{2} \cdot 2\pi r \cdot \ell, \text{ or L.A.} = \pi r \ell$$

The surface area of a cone is the sum of the lateral area and the area of the base.

$$\text{S.A.} = \text{L.A.} + B$$

 **Problem 3**   Finding the Surface Area of a Cone

Think

How is this different from finding the surface area of a pyramid?
For a pyramid, you need to find the perimeter of the base. For a cone, you need to find the circumference.

What is the surface area of the cone in terms of π?

$\text{S.A.} = \text{L.A.} + B$	Use the formula for surface area.
$= \pi r \ell + \pi r^2$	Substitute the formulas for L.A. and B.
$= \pi(15)(25) + \pi(15)^2$	Substitute 15 for r and 25 for ℓ.
$= 375\pi + 225\pi$	Simplify.
$= 600\pi$	

The surface area of the cone is 600π cm^2.

25 cm

15 cm

Got It?   3. The radius of the base of a cone is 16 m. Its slant height is 28 m. What is the surface area in terms of π?

By cutting a cone and laying it out flat, you can see how the formula for the lateral area of a cone $(\text{L.A.} = \tfrac{1}{2} \cdot C_{\text{base}} \cdot \ell)$ resembles that for the area of a triangle $(A = \tfrac{1}{2}bh)$.

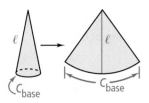

Think

What is the problem asking you to find?
The problem is asking you to find the area that the filter paper covers. This is the lateral area of a cone.

© **Problem 4** **Finding the Lateral Area of a Cone** STEM

Chemistry In a chemistry lab experiment, you use the conical filter funnel shown at the right. How much filter paper do you need to line the funnel?

The top part of the funnel has the shape of a cone with a diameter of 80 mm and a height of 45 mm.

L.A. = $\pi r \ell$	Use the formula for lateral area of a cone.
$= \pi r \left(\sqrt{r^2 + h^2} \right)$	To find the slant height, use the Pythagorean Theorem.
$= \pi (40) \left(\sqrt{40^2 + 45^2} \right)$	Substitute $\frac{1}{2} \cdot 80$, or 40, for r and 45 for h.
≈ 7565.957013	Use a calculator.

You need about 7566 mm² of filter paper to line the funnel.

© ✓ **Got It?** **4. a.** What is the lateral area of a traffic cone with radius 10 in. and height 28 in.? Round to the nearest whole number.

b. Reasoning Suppose the radius of a cone is halved, but the slant height remains the same. How does this affect the lateral area of the cone? Explain.

✓ **Lesson Check**

Do you know HOW?

Use the diagram of the square pyramid at the right.

1. What is the lateral area of the pyramid?

2. What is the surface area of the pyramid?

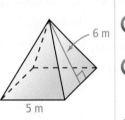

Use the diagram of the cone at the right.

3. What is the lateral area of the cone?

4. What is the surface area of the cone?

Do you UNDERSTAND? © **MATHEMATICAL PRACTICES**

© **5. Vocabulary** How do the height and the slant height of a pyramid differ?

© **6. Compare and Contrast** How are the formulas for the surface area of a prism and the surface area of a pyramid alike? How are they different?

© **7. Vocabulary** How many lateral faces does a pyramid have if its base is pentagonal? Hexagonal? n-sided?

© **8. Error Analysis** A cone has height 7 and radius 3. Your classmate calculates its lateral area. What is your classmate's error? Explain.

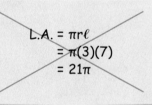

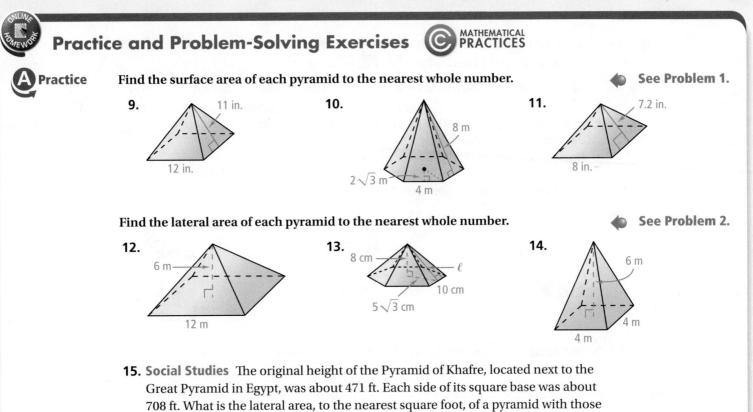

Practice and Problem-Solving Exercises

MATHEMATICAL PRACTICES

A **Practice**

Find the surface area of each pyramid to the nearest whole number.

◀ See Problem 1.

9. 11 in. 12 in.

10. 8 m $2\sqrt{3}$ m 4 m

11. 7.2 in. 8 in.

Find the lateral area of each pyramid to the nearest whole number.

◀ See Problem 2.

12. 6 m 12 m

13. 8 cm ℓ 10 cm $5\sqrt{3}$ cm

14. 6 m 4 m 4 m

15. Social Studies The original height of the Pyramid of Khafre, located next to the Great Pyramid in Egypt, was about 471 ft. Each side of its square base was about 708 ft. What is the lateral area, to the nearest square foot, of a pyramid with those dimensions?

Find the lateral area of each cone to the nearest whole number.

◀ See Problems 3 and 4.

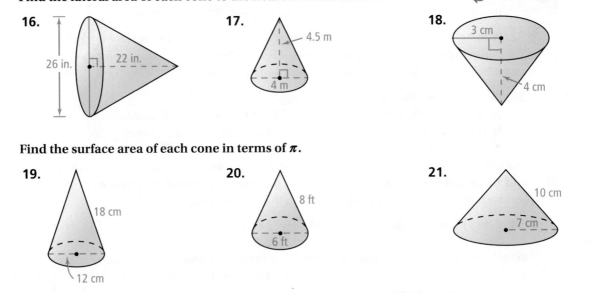

16. 26 in. 22 in.

17. 4.5 m 4 m

18. 3 cm 4 cm

Find the surface area of each cone in terms of π.

19. 18 cm 12 cm

20. 8 ft 6 ft

21. 10 cm 7 cm

B **Apply**

22. Reasoning Suppose you could climb to the top of the Great Pyramid. Which route would be shorter, a route along a lateral edge or a route along the slant height of a side? Which of these routes is steeper? Explain your answers.

23. The lateral area of a cone is 4.8π in.^{2}. The radius is 1.2 in. Find the slant height.

© **24. Writing** Explain why the altitude $\overline{PT}$ in the pyramid at the right must be shorter than all of the lateral edges $\overline{PA}$, $\overline{PB}$, $\overline{PC}$, and $\overline{PD}$.

© **25. Think About a Plan** The lateral area of a pyramid with a square base is 240 ft². Its base edges are 12 ft long. Find the height of the pyramid.
- What additional information do you know about the pyramid based on the given information?
- How can a diagram help you identify what you need to find?

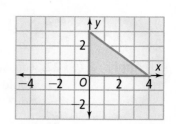

Find the surface area to the nearest whole number.

26. 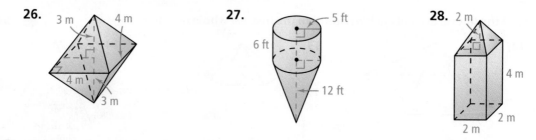 3 m   4 m   4 m   3 m

27. 5 ft   6 ft   12 ft

28. 2 m   4 m   2 m   2 m

© **29. Open-Ended** Draw a square pyramid with a lateral area of 48 cm². Label its dimensions. Then find its surface area.

STEM **30. Architecture** The roof of a tower in a castle is shaped like a cone. The height of the roof is 30 ft and the radius of the base is 15 ft. What is the lateral area of the roof? Round your answer to the nearest tenth.

© **31. Reasoning** The figure at the right shows two glass cones inside a cylinder. Which has a greater surface area, the two cones or the cylinder? Explain.

6 in.
8 in.

© **32. Writing** You can use the formula S.A. $= (\ell + r)r\pi$ to find the surface area of a cone. Explain why this formula works. Also, explain why you may prefer to use this formula when finding surface area with a calculator.

33. Find a formula for each of the following.
a. the slant height of a cone in terms of the surface area and radius
b. the radius of a cone in terms of the surface area and slant height

The length of a side (s) of the base, slant height (ℓ), height (h), lateral area (L.A.), and surface area (S.A.) are measurements of a square pyramid. Given two of the measurements, find the other three to the nearest tenth.

34. $s = 3$ in., S.A. $= 39$ in.²   **35.** $h = 8$ m, $\ell = 10$ m   **36.** L.A. $= 118$ cm², S.A. $= 182$ cm²

© **Visualization** Suppose you revolve the plane region completely about the given line to sweep out a solid of revolution. Describe the solid. Then find its surface area in terms of π.

37. the y-axis   **38.** the x-axis

C **Challenge**   **39.** the line $x = 4$ **40.** the line $y = 3$

41. A sector has been cut out of the disk at the right. The radii of the part that remains are taped together, without overlapping, to form the cone. The cone has a lateral area of 64π cm^2. Find the measure of the central angle of the cutout sector.

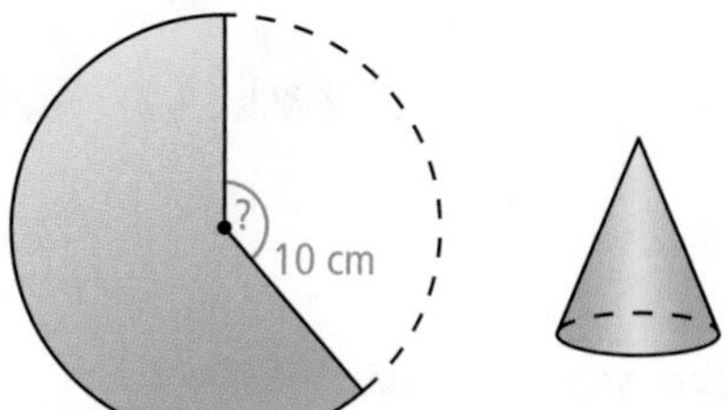

Each given figure fits inside a 10-cm cube. The figure's base is in one face of the cube and is as large as possible. The figure's vertex is in the opposite face of the cube. Draw a sketch and find the lateral and surface areas of the figure.

42. a square pyramid

43. a cone

SAT/ACT

Standardized Test Prep

44. To the nearest whole number, what is the surface area of a cone with diameter 27 m and slant height 19 m?

   Ⓐ 1378 m^2 Ⓑ 1951 m^2 Ⓒ 2757 m^2 Ⓓ 3902 m^2

45. What is the hypotenuse of a right isosceles triangle with leg $2\sqrt{6}$?

   Ⓕ $4\sqrt{6}$ Ⓖ $2\sqrt{3}$ Ⓗ $4\sqrt{3}$ Ⓘ $4\sqrt{2}$

46. Two angles in a triangle have measures 54 and 61. What is the measure of the smallest exterior angle?

   Ⓐ 119   Ⓑ 115   Ⓒ 112   Ⓓ 126

Short Response

47. A diagonal divides a parallelogram into two isosceles triangles. Can the parallelogram be a rhombus? Explain.

Extended Response

48. $ABCD$ has vertices at $A(3, 4)$, $B(7, 5)$, $C(6, 1)$, and $D(-2, -4)$. A dilation with center $(0, 0)$ maps A to $A'\left(\frac{15}{2}, 10\right)$. What are the coordinates of B', C', and D'? Show your work.

Mixed Review

49. How much cardboard do you need to make a closed box 4 ft by 5 ft by 2 ft?   ◀ **See Lesson 11-2.**

50. How much posterboard do you need to make a cylinder, open at each end, with height 9 in. and diameter $4\frac{1}{2}$ in.? Round your answer to the nearest square inch.

51. A kite with area 195 in.^{2} has a 15-in. diagonal. How long is the other diagonal?   ◀ **See Lesson 10-2.**

Get Ready!   **To prepare for Lesson 11-4, do Exercises 52 and 53.**

Find the area of each figure. If necessary, round to the nearest tenth.   ◀ **See Lessons 1-8 and 10-7.**

52. a square with side length 2 cm

53. a circle with diameter 15 in.

Do you know HOW?

Draw a net for each figure. Label the net with its dimensions.

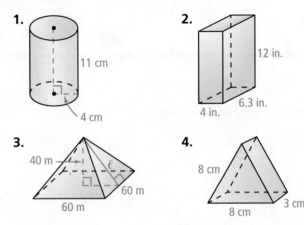

1. 11 cm, 4 cm

2. 12 in., 6.3 in., 4 in.

3. 40 m, ℓ, 60 m, 60 m

4. 8 cm, 8 cm, 3 cm

Find the surface area of each solid. Round to the nearest square unit.

5. cylinder in Exercise 1

6. prism in Exercise 2

7. pyramid in Exercise 3

8. cone in Exercise 4

For Exercises 9–11, use Euler's Formula. Show your work.

9. A polyhedron has 10 vertices and 15 edges. How many faces does it have?

10. A polyhedron has 2 hexagonal faces and 12 triangular faces. How many vertices does it have?

11. How many vertices does the net of a pentagonal pyramid have?

Draw a cube. Shade a cross section of the cube that forms each shape.

12. a rectangle   13. a trapezoid   14. a triangle

15. A square prism with base edges 2 in. has surface area 32 in.². What is its height?

Find the surface area of each figure to the nearest whole number.

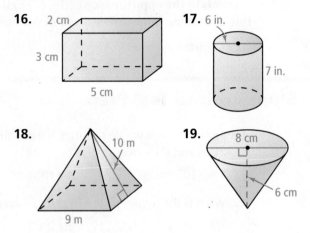

16. 2 cm, 3 cm, 5 cm

17. 6 in., 7 in.

18. 10 m, 9 m

19. 8 cm, 6 cm

Do you UNDERSTAND?

20. **Open-Ended** Draw a net for a regular hexagonal prism.

21. **Compare and Contrast** How are a right prism and a regular pyramid alike? How are they different?

22. **Algebra** The height of a cylinder is twice the radius of the base. A cube has an edge length equal to the radius of the cylinder. Find the ratio between the surface area of the cylinder and the surface area of the cube. Leave your answer in terms of π.

23. **Error Analysis** Your class learned that the number of regions in the net for a solid is the same as the number of faces in that solid. Your friend concludes that a solid and its net must also have the same number of edges. What is your friend's error? Explain.

24. **Visualization** The dimensions of a rectangular prism are 3 in. by 5 in. by 10 in. Is it possible to intersect this prism with a plane so that the resulting cross section is a square? If so, draw and describe your cross section. Indicate the length of the edge of the square.

Volumes of Prisms and Cylinders

© Content Standards
G.GMD.1 Give an informal argument for the formulas for . . . volume of a cylinder . . . Use . . . Cavalieri's principle . . .
G.GMD.3 Use volume formulas for cylinders . . .
Also G.GMD.2 and G.MG.1

Objective To find the volume of a prism and the volume of a cylinder

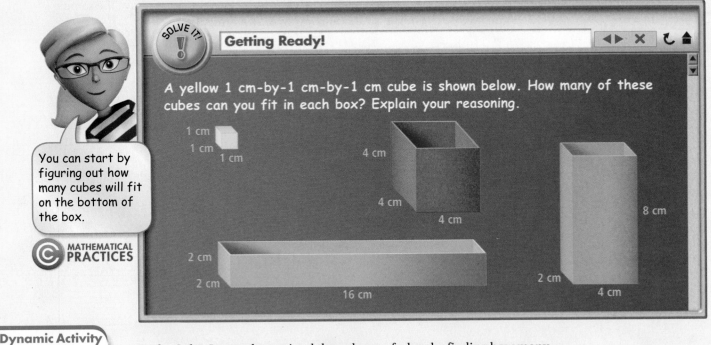

SOLVE IT!

Getting Ready!

A yellow 1 cm-by-1 cm-by-1 cm cube is shown below. How many of these cubes can you fit in each box? Explain your reasoning.

1 cm
1 cm
1 cm

4 cm
4 cm
4 cm

8 cm
2 cm
4 cm

2 cm
2 cm
16 cm

You can start by figuring out how many cubes will fit on the bottom of the box.

© MATHEMATICAL PRACTICES

Dynamic Activity
Volumes of Prisms and Cylinders

Lesson Vocabulary
• volume
• composite space figure

In the Solve It, you determined the volume of a box by finding how many 1 cm-by-1 cm-by-1 cm cubes the box holds.

Volume is the space that a figure occupies. It is measured in cubic units such as cubic inches (in.^{3}), cubic feet (ft^3), or cubic centimeters (cm^3). The volume V of a cube is the cube of the length of its edge e, or $V = e^3$.

Essential Understanding You can find the volume of a prism or a cylinder when you know its height and the area of its base.

Both stacks of paper below contain the same number of sheets.

The first stack forms an oblique prism. The second forms a right prism. The stacks have the same height. The area of every cross section parallel to a base is the area of one sheet of paper. The stacks have the same volume. These stacks illustrate the following principle.

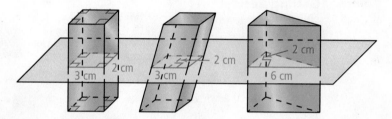

Theorem 11-5 Cavalieri's Principle

If two space figures have the same height and the same cross-sectional area at every level, then they have the same volume.

The area of each shaded cross section below is 6 cm². Since the prisms have the same height, their volumes must be the same by Cavalieri's Principle.

You can find the volume of a right prism by multiplying the area of the base by the height. Cavalieri's Principle lets you extend this idea to any prism.

Theorem 11-6 Volume of a Prism

The volume of a prism is the product of the area of the base and the height of the prism.

$$V = Bh$$

Problem 1   Finding the Volume of a Rectangular Prism

Plan

What do you need to use the formula?
You need to find B, the area of the base. The prism has a rectangular base, so the area of the base is length × width.

What is the volume of the rectangular prism at the right?

$V = Bh$	Use the formula for the volume of a prism.
$= 480 \cdot 10$	The area of the base B is 24 · 20, or 480 cm², and the height is 10 cm.
$= 4800$	Simplify.

The volume of the rectangular prism is 4800 cm³.

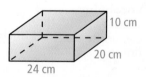

Got It? **1. a.** What is the volume of the rectangular prism at the right?
   b. Reasoning Suppose the prism at the right is turned so that the base is 4 ft by 5 ft and the height is 3 ft. Does the volume change? Explain.

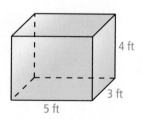

Problem 2 **Finding the Volume of a Triangular Prism**

Multiple Choice What is the approximate volume of the triangular prism?

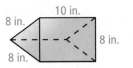

(A) 188 in.³   (C) 295 in.³

(B) 277 in.³   (D) 554 in.³

Step 1 Find the area of the base of the prism.

Each base of the triangular prism is an equilateral triangle, as shown at the right. An altitude of the triangle divides it into two 30°-60°-90° triangles. The height of the triangle is $\sqrt{3}$ · shorter leg, or $4\sqrt{3}$.

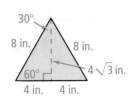

$$B = \tfrac{1}{2}bh \qquad \text{Use the formula for the area of a triangle.}$$
$$= \tfrac{1}{2}(8)(4\sqrt{3}) \qquad \text{Substitute 8 for } b \text{ and } 4\sqrt{3} \text{ for } h.$$
$$= 16\sqrt{3} \qquad \text{Simplify.}$$

Think

Which height do you use in the formula?
Remember that the h in the formula for volume represents the height of the entire prism, not the height of the triangular base.

Step 2 Find the volume of the prism.

$$V = Bh \qquad \text{Use the formula for the volume of a prism.}$$
$$= 16\sqrt{3} \cdot 10 \qquad \text{Substitute } 16\sqrt{3} \text{ for } B \text{ and 10 for } h.$$
$$= 160\sqrt{3} \qquad \text{Simplify.}$$
$$\approx 277.1281292 \qquad \text{Use a calculator.}$$

The volume of the triangular prism is about 277 in.³. The correct answer is B.

Got It? **2. a.** What is the volume of the triangular prism at the right?

b. Reasoning Suppose the height of a prism is doubled. How does this affect the volume of the prism? Explain.

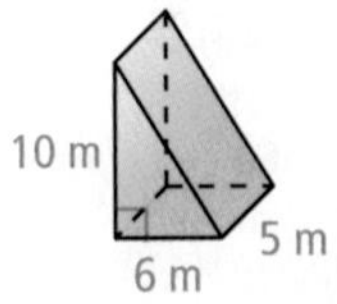

To find the volume of a cylinder, you use the same formula $V = Bh$ that you use to find the volume of a prism. Now, however, B is the area of the circle, so you use the formula $B = \pi r^2$ to find its value.

take note

Theorem 11-7 **Volume of a Cylinder**

The volume of a cylinder is the product of the area of the base and the height of the cylinder.

$$V = Bh, \text{ or } V = \pi r^2 h$$

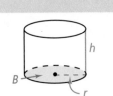

Problem 3 Finding the Volume of a Cylinder

Plan

What do you know from the diagram?
You know that the radius r is 3 cm and the height h is 8 cm.

What is the volume of the cylinder in terms of π?

$$V = \pi r^2 h \qquad \text{Use the formula for the volume of a cylinder.}$$
$$= \pi(3)^2(8) \qquad \text{Substitute 3 for } r \text{ and 8 for } h.$$
$$= \pi(72) \qquad \text{Simplify.}$$

The volume of the cylinder is 72π cm^3.

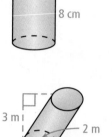

Got It? 3. a. What is the volume of the cylinder at the right in terms of π?
    b. Reasoning Suppose the radius of a cylinder is halved. How does this affect the volume of the cylinder? Explain.

A **composite space figure** is a three-dimensional figure that is the combination of two or more simpler figures. You can find the volume of a composite space figure by adding the volumes of the figures that are combined.

Problem 4 Finding Volume of a Composite Figure

What is the approximate volume of the bullnose aquarium to the nearest cubic inch?

Plan

How can you find the volume by *solving a simpler problem*?
The aquarium is the combination of a rectangular prism and half of a cylinder. Find the volume of each figure.

Think

The length of the prism is the total length minus the radius of the cylinder. The radius of the cylinder is half the width of the prism.

Find the volume of the prism and the half cylinder.

Add the two volumes together.

Write

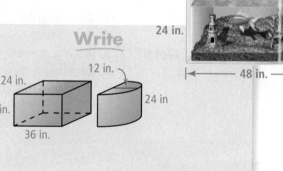

$$V_1 = Bh \qquad\qquad V_2 = \tfrac{1}{2}\pi r^2 h$$
$$= (24 \cdot 36)(24) \qquad = \tfrac{1}{2}\pi(12)^2(24)$$
$$= 20{,}736 \qquad\qquad \approx 5429$$

$$20{,}736 + 5429 = 26{,}165$$
The approximate volume of the aquarium is 26,165 in.^{3}.

Got It? 4. What is the approximate volume of the lunch box shown at the right? Round to the nearest cubic inch.

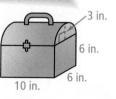

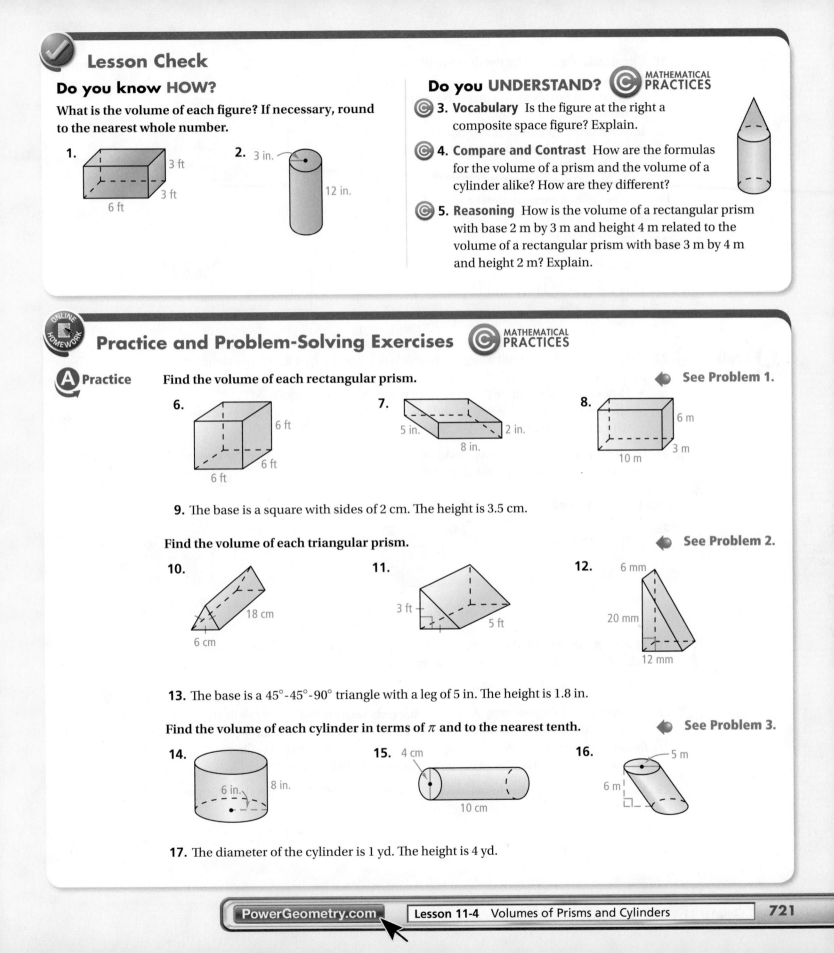

Lesson Check

Do you know HOW?

What is the volume of each figure? If necessary, round to the nearest whole number.

1.   3 ft / 3 ft / 6 ft

2.   3 in. / 12 in.

Do you UNDERSTAND? MATHEMATICAL PRACTICES

3. Vocabulary Is the figure at the right a composite space figure? Explain.

4. Compare and Contrast How are the formulas for the volume of a prism and the volume of a cylinder alike? How are they different?

5. Reasoning How is the volume of a rectangular prism with base 2 m by 3 m and height 4 m related to the volume of a rectangular prism with base 3 m by 4 m and height 2 m? Explain.

Practice and Problem-Solving Exercises MATHEMATICAL PRACTICES

A Practice   Find the volume of each rectangular prism.   **See Problem 1.**

6.   6 ft / 6 ft / 6 ft

7.   5 in. / 2 in. / 8 in.

8.   6 m / 3 m / 10 m

9. The base is a square with sides of 2 cm. The height is 3.5 cm.

Find the volume of each triangular prism.   **See Problem 2.**

10.   18 cm / 6 cm

11.   3 ft / 5 ft

12.   6 mm / 20 mm / 12 mm

13. The base is a 45°-45°-90° triangle with a leg of 5 in. The height is 1.8 in.

Find the volume of each cylinder in terms of π and to the nearest tenth.   **See Problem 3.**

14.   6 in. / 8 in.

15.   4 cm / 10 cm

16.   5 m / 6 m

17. The diameter of the cylinder is 1 yd. The height is 4 yd.

18. Composite Figures Use the diagram of the backpack at the right.
See Problem 4.
   a. What two figures approximate the shape of the backpack?
   b. What is the volume of the backpack in terms of π?
   c. What is the volume of the backpack to the nearest cubic inch?

17 in.

4 in.

12 in.

Find the volume of each composite space figure to the nearest whole number.

19.

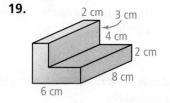

2 cm 3 cm

4 cm

2 cm

8 cm

6 cm

20.

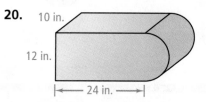

10 in.

12 in.

24 in.

B Apply

© 21. Think About a Plan A full waterbed mattress is 7 ft by 4 ft by 1 ft. If water weighs 62.4 lb/ft³, what is the weight of the water in the mattress to the nearest pound?
   • How can you determine the amount of water the mattress can hold?
   • The weight of the water is in pounds per cubic feet. How can you get an answer with a unit of pounds?

© 22. Open-Ended Give the dimensions of two rectangular prisms that have volumes of 80 cm³ each but also have different surface areas.

Find the height of each figure with the given volume.

23.

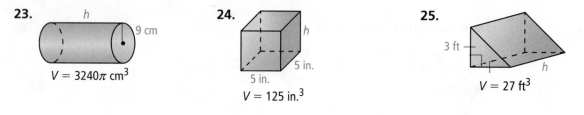

h

9 cm

$V = 3240\pi$ cm³

24.

h

5 in.

5 in.

$V = 125$ in.³

25.

3 ft

h

$V = 27$ ft³

26. Sports A can of tennis balls has a diameter of 3 in. and a height of 8 in. Find the volume of the can to the nearest cubic inch.

27. What is the volume of the oblique prism shown at the right?

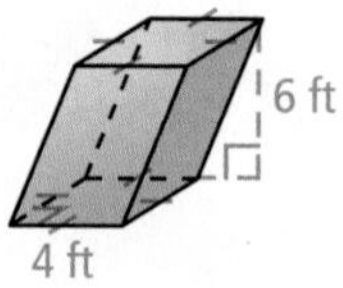

6 ft

4 ft

STEM 28. Environmental Engineering A scientist suggests keeping indoor air relatively clean as follows: For a room with a ceiling 8 ft high, provide two or three pots of flowers for every 100 ft² of floor space. If your classroom has an 8-ft ceiling and measures 35 ft by 40 ft, how many pots of flowers should it have?

© 29. Reasoning Suppose the dimensions of a prism are tripled. How does this affect its volume? Explain.

30. Swimming Pool The approximate dimensions of an Olympic-size swimming pool are 164 ft by 82 ft by 6.6 ft.
   a. Find the volume of the pool to the nearest cubic foot.
   b. If 1 ft^3 ≈ 7.48 gal, about how many gallons does the pool hold?

Ⓒ **31. Writing** The figures at the right can be covered by equal numbers of straws that are the same length. Describe how Cavalieri's Principle could be adapted to compare the areas of these figures.

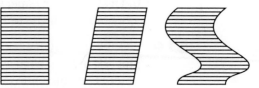

32. Algebra The volume of a cylinder is 600π cm^3. The radius of a base of the cylinder is 5 cm. What is the height of the cylinder?

33. Coordinate Geometry Find the volume of the rectangular prism at the right.

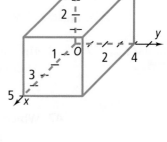

34. Algebra The volume of a cylinder is 135π cm^3. The height of the cylinder is 15 cm. What is the radius of a base of the cylinder?

35. Landscaping To landscape her 70 ft-by-60 ft rectangular backyard, your aunt is planning first to put down a 4-in. layer of topsoil. She can buy bags of topsoil at $2.50 per 3-ft^3 bag, with free delivery. Or, she can buy bulk topsoil for $22.00/yd^3, plus a $20 delivery fee. Which option is less expensive? Explain.

36. The closed box at the right is shaped like a regular pentagonal prism. The exterior of the box has base edge 10 cm and height 14 cm. The interior has base edge 7 cm and height 11 cm. Find each measurement.
   a. the outside surface area   **b.** the inside surface area
   c. the volume of the material needed to make the box

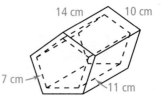

A cylinder has been cut out of each solid. Find the volume of the remaining solid. Round your answer to the nearest tenth.

37.

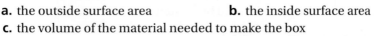

38.

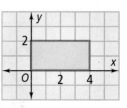

Ⓒ **Visualization** Suppose you revolve the plane region completely about the given line to sweep out a solid of revolution. Describe the solid and find its volume in terms of π.

39. the x-axis

40. the y-axis

41. the line $y = 2$

42. the line $x = 5$

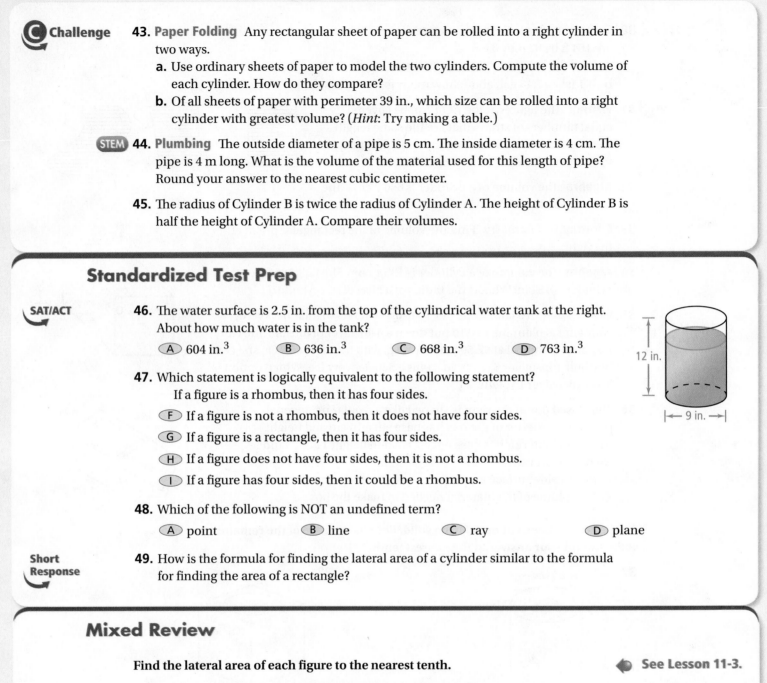

C Challenge

43. Paper Folding Any rectangular sheet of paper can be rolled into a right cylinder in two ways.

    a. Use ordinary sheets of paper to model the two cylinders. Compute the volume of each cylinder. How do they compare?

    b. Of all sheets of paper with perimeter 39 in., which size can be rolled into a right cylinder with greatest volume? (*Hint:* Try making a table.)

STEM **44. Plumbing** The outside diameter of a pipe is 5 cm. The inside diameter is 4 cm. The pipe is 4 m long. What is the volume of the material used for this length of pipe? Round your answer to the nearest cubic centimeter.

45. The radius of Cylinder B is twice the radius of Cylinder A. The height of Cylinder B is half the height of Cylinder A. Compare their volumes.

Standardized Test Prep

SAT/ACT

46. The water surface is 2.5 in. from the top of the cylindrical water tank at the right. About how much water is in the tank?

    Ⓐ 604 in.³   Ⓑ 636 in.³   Ⓒ 668 in.³   Ⓓ 763 in.³

47. Which statement is logically equivalent to the following statement?

    If a figure is a rhombus, then it has four sides.

    Ⓕ If a figure is not a rhombus, then it does not have four sides.

    Ⓖ If a figure is a rectangle, then it has four sides.

    Ⓗ If a figure does not have four sides, then it is not a rhombus.

    Ⓘ If a figure has four sides, then it could be a rhombus.

48. Which of the following is NOT an undefined term?

    Ⓐ point   Ⓑ line   Ⓒ ray   Ⓓ plane

Short Response

49. How is the formula for finding the lateral area of a cylinder similar to the formula for finding the area of a rectangle?

Mixed Review

Find the lateral area of each figure to the nearest tenth.   ◀ **See Lesson 11-3.**

50. a right circular cone with height 12 mm and radius 5 mm

51. a regular hexagonal pyramid with base edges 9.2 ft long and slant height 17 ft

Get Ready! **To prepare for Lesson 11-5, do Exercises 52 and 53.**

52. Find *h* in the figure at the right.   ◀ **See Lesson 8-1.**

53. A right triangle has hypotenuse 300 ft and leg 180 ft. What is the length of the other leg?

Concept Byte
Use With Lesson 11-5

ACTIVITY

Finding Volume

Content Standard
Prepares for G.GMD.1 Give an informal argument for the formulas for the circumference of a circle, area of a circle, volume of a cylinder, pyramid, and cone.

You know how to find the volumes of a prism and of a cylinder. Use the following activities to explore finding the volumes of a pyramid and of a cone.

Activity 1

Step 1 Draw the nets shown at the right on heavy paper.

Step 2 Cut out the nets and tape them together to make a cube and a square pyramid. Each model will have one open face.

1. How do the areas of the bases of the cube and the pyramid compare?

2. How do the heights of the cube and the pyramid compare?

3. Fill the pyramid with rice or other material. Then pour the rice from the pyramid into the cube. How many pyramids full of rice does the cube hold?

4. The volume of the pyramid is what fractional part of the volume of the cube?

5. Make a Conjecture What do you think is the formula for the volume of a pyramid? Explain.

Activity 2

Step 1 Draw the nets shown at the right on heavy paper.

Step 2 Cut out the nets and tape them together to make a cylinder and a cone. Each model will have one open face.

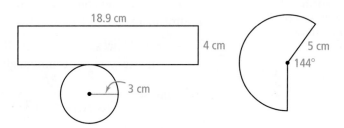

6. How do the areas of the bases of the cylinder and of the cone compare?

7. How do the heights of the cylinder and of the cone compare?

8. Fill the cone with rice or other material. Then pour the rice from the cone into the cylinder. How many cones full of rice does the cylinder hold?

9. What fractional part of the volume of the cylinder is the volume of the cone?

10. Make a Conjecture What do you think is the formula for the volume of a cone? Explain.

Volumes of Pyramids and Cones

© Content Standards
G.GMD.3 Use volume formulas for . . . pyramids, cones . . . to solve problems.
G.MG.1 Use geometric shapes, their measures, and their properties to describe objects.

Objective   To find the volume of a pyramid and of a cone

Make a table and look for a pattern.

MATHEMATICAL PRACTICES

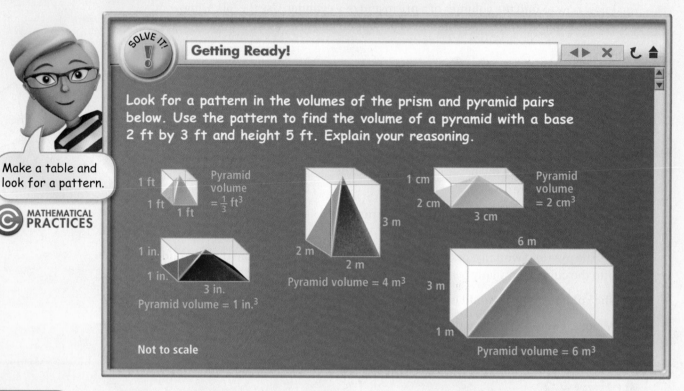

SOLVE IT!

Getting Ready!

Look for a pattern in the volumes of the prism and pyramid pairs below. Use the pattern to find the volume of a pyramid with a base 2 ft by 3 ft and height 5 ft. Explain your reasoning.

1 ft / 1 ft / 1 ft — Pyramid volume = $\frac{1}{3}$ ft^3

1 cm / 2 cm / 3 cm — Pyramid volume = 2 cm^3

1 in. / 1 in. / 3 in. — Pyramid volume = 1 in.^{3}

3 m / 2 m / 2 m — Pyramid volume = 4 m^3

6 m / 3 m / 1 m — Pyramid volume = 6 m^3

Not to scale

Dynamic Activity
Volumes of Pyramids and Cones

In the Solve It, you analyzed the relationship between the volume of a prism and the volume of an embedded pyramid.

Essential Understanding The volume of a pyramid is related to the volume of a prism with the same base and height.

take note

Theorem 11-8   Volume of a Pyramid

The volume of a pyramid is one third the product of the area of the base and the height of the pyramid.

$$V = \frac{1}{3}Bh$$

Because of Cavalieri's Principle, the volume formula is true for all pyramids. The height h of an oblique pyramid is the length of the perpendicular segment from its vertex to the plane of the base.

Oblique pyramid

Problem 1 Finding Volume of a Pyramid STEM

Architecture **The entrance to the Louvre Museum in Paris, France, is a square pyramid with a height of 21.64 m. What is the approximate volume of the Louvre Pyramid?**

35.4 m

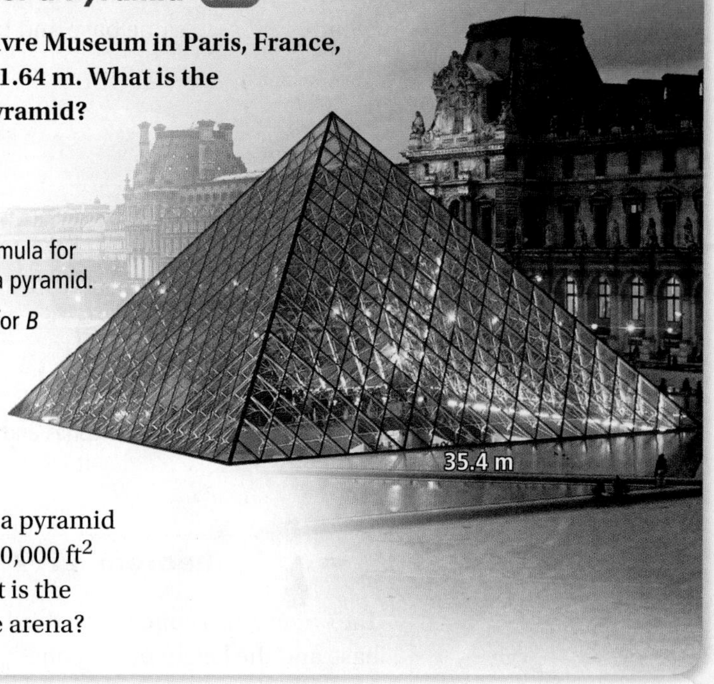

The area of the base of the pyramid is $35.4 \text{ m} \cdot 35.4 \text{ m}$, or 1253.16 m^2.

> **Think**
> **How is this similar to finding the volume of a prism?**
> In both cases, you need the area of the base and the height.

$V = \frac{1}{3}Bh$	Use the formula for volume of a pyramid.
$= \frac{1}{3}(1253.16)(21.64)$	Substitute for B and h.
$= 9039.4608$	Simplify.

The volume is about 9039 m^3.

Got It? **1.** A sports arena shaped like a pyramid has a base area of about $300{,}000 \text{ ft}^2$ and a height of 321 ft. What is the approximate volume of the arena?

Problem 2 Finding the Volume of a Pyramid GRIDDED RESPONSE

What is the volume in cubic feet of a square pyramid with base edges 40 ft and slant height 25 ft?

> **Think**
> **How do you use the slant height?**
> The slant height is the length of the hypotenuse of the right triangle. Use the slant height to find the height of the pyramid.

h
25 ft
40 ft

h
25 ft
20 ft

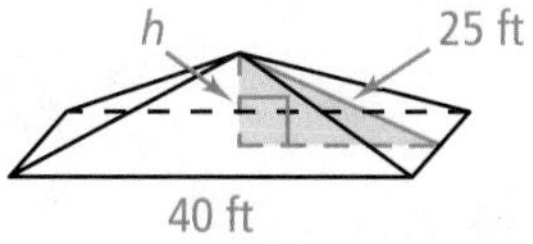

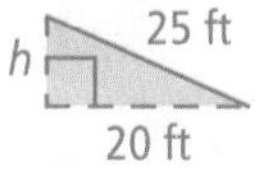

Step 1 Find the height of the pyramid.

$c^2 = a^2 + b^2$	Use the Pythagorean Theorem.
$25^2 = h^2 + 20^2$	Substitute 25 for c, h for a, and $\frac{40}{2}$, or 20, for b.
$625 = h^2 + 400$	Simplify.
$h^2 = 225$	Solve for h^2.
$h = 15$	Take the positive square root of both sides.

Step 2 Find the volume of the pyramid.

$V = \frac{1}{3}Bh$	Use the formula for volume of a pyramid.
$= \frac{1}{3}(40 \cdot 40)(15)$	Substitute $40 \cdot 40$ for B and 15 for h.
$= 8000$	Simplify.

8 0 0 0

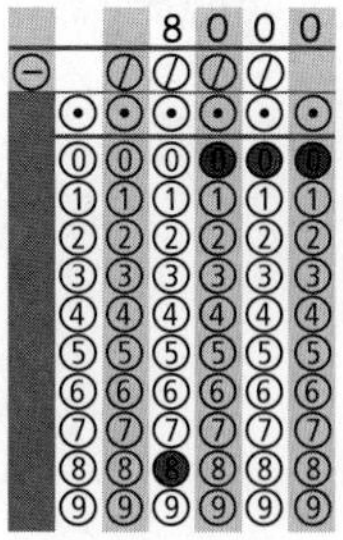

The volume of the pyramid is 8000 ft^3.

Got It? **2.** What is the volume of a square pyramid with base edges 24 m and slant height 13 m?

Essential Understanding The volume of a cone is related to the volume of a cylinder with the same base and height.

The cones and the cylinder have the same base and height.
It takes three cones full of rice to fill the cylinder.

take note

Theorem 11-9   Volume of a Cone

The volume of a cone is one third the product of the area of the base and the height of the cone.

$$V = \frac{1}{3}Bh, \text{ or } V = \frac{1}{3}\pi r^2 h$$

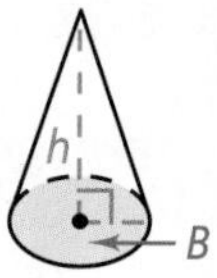

A cone-shaped structure can be particularly strong, as downward forces at the vertex are distributed to all points in its circular base.

© **Problem 3**   **Finding the Volume of a Cone** **STEM**

Traditional Architecture The covering on a tepee rests on poles that come together like concurrent lines. The resulting structure approximates a cone. If the tepee pictured is 12 ft high with a base diameter of 14 ft, what is its approximate volume?

Think

How is this similar to finding the volume of a cylinder?
In both cases, you need to find the base area of a circle.

$V = \frac{1}{3}\pi r^2 h$ Use the formula for the volume of a cone.

$= \frac{1}{3}\pi(7)^2(12)$ Substitute $\frac{14}{2}$, or 7, for r and 12 for h.

≈ 615.7521601 Use a calculator.

The volume of the tepee is approximately 616 ft³.

© **Got It?** **3. a.** The height and radius of a child's tepee are half those of the tepee in Problem 3. What is the volume of the child's tepee to the nearest cubic foot?

b. Reasoning What is the relationship between the volume of the original tepee and the child's tepee?

This volume formula applies to all cones, including oblique cones.

 Problem 4 Finding the Volume of an Oblique Cone

What is the volume of the oblique cone at the right? Give your answer in terms of π and also rounded to the nearest cubic foot.

$$V = \frac{1}{3}\pi r^2 h \qquad \text{Use the formula for volume of a cone.}$$

$$= \frac{1}{3}\pi(15)^2(25) \qquad \text{Substitute 15 for } r \text{ and 25 for } h.$$

$$= 1875\pi \qquad \text{Simplify.}$$

$$\approx 5890.486225 \qquad \text{Use a calculator.}$$

The volume of the cone is 1875π ft^3, or about 5890 ft^3.

Think

What is the height of the oblique cone?
The height is the length of the perpendicular segment from the vertex of the cone to the base, which is 25 ft. In an oblique cone, the segment does not intersect the center of the base.

Got It? **4. a.** What is the volume of the oblique cone at the right in terms of π and rounded to the nearest cubic meter?

   b. Reasoning How does the volume of an oblique cone compare to the volume of a right cone with the same diameter and height? Explain.

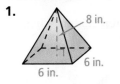

Lesson Check

Do you know HOW?

What is the volume of each figure? If necessary, round to the nearest tenth.

1.
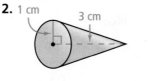

2.
1 cm, 3 cm

Do you UNDERSTAND? MATHEMATICAL PRACTICES

3. Compare and Contrast How are the formulas for the volume of a pyramid and the volume of a cone alike? How are they different?

4. Error Analysis A square pyramid has base edges 13 ft and height 10 ft. A cone has diameter 13 ft and height 10 ft. Your friend claims the figures have the same volume because the volume formulas for a pyramid and a cone are the same: $V = \frac{1}{3}Bh$. What is her error?

Practice and Problem-Solving Exercises   MATHEMATICAL PRACTICES

A Practice Find the volume of each square pyramid.   ◆ See Problem 1.

5. base edges 10 cm, height 6 cm

6. base edges 18 in., height 12 in.

7. base edges 5 m, height 6 m

8. Buildings The Transamerica Pyramid Building in San Francisco is 853 ft tall with a square base that is 149 ft on each side. To the nearest thousand cubic feet, what is the volume of the Transamerica Pyramid?

Find the volume of each square pyramid. Round to the nearest tenth if necessary.

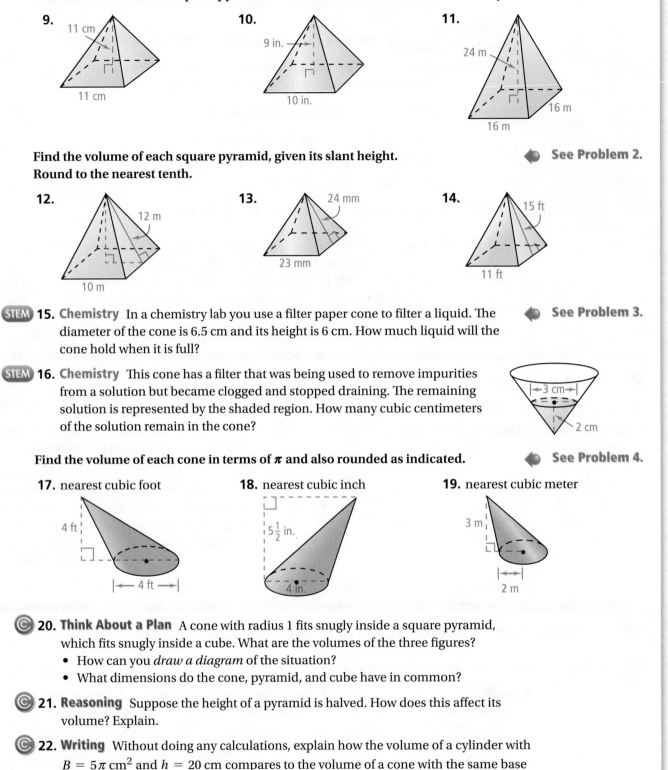

9.

11 cm

11 cm

10.

9 in.

10 in.

11.

24 m

16 m

16 m

**Find the volume of each square pyramid, given its slant height.
Round to the nearest tenth.**

See Problem 2.

12.

12 m

10 m

13.

24 mm

23 mm

14.

15 ft

11 ft

STEM **15. Chemistry** In a chemistry lab you use a filter paper cone to filter a liquid. The diameter of the cone is 6.5 cm and its height is 6 cm. How much liquid will the cone hold when it is full?

See Problem 3.

STEM **16. Chemistry** This cone has a filter that was being used to remove impurities from a solution but became clogged and stopped draining. The remaining solution is represented by the shaded region. How many cubic centimeters of the solution remain in the cone?

3 cm

2 cm

Find the volume of each cone in terms of π and also rounded as indicated.

See Problem 4.

17. nearest cubic foot

4 ft

4 ft

18. nearest cubic inch

$5\frac{1}{2}$ in.

4 in.

19. nearest cubic meter

3 m

2 m

B **Apply**

20. Think About a Plan A cone with radius 1 fits snugly inside a square pyramid, which fits snugly inside a cube. What are the volumes of the three figures?
- How can you *draw a diagram* of the situation?
- What dimensions do the cone, pyramid, and cube have in common?

21. Reasoning Suppose the height of a pyramid is halved. How does this affect its volume? Explain.

22. Writing Without doing any calculations, explain how the volume of a cylinder with $B = 5\pi$ cm^2 and $h = 20$ cm compares to the volume of a cone with the same base area and height.

Find the volume to the nearest whole number.

23.

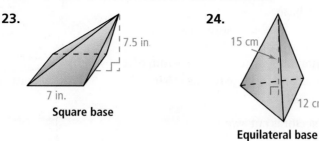

7.5 in.

7 in.

Square base

24.

15 cm

12 cm

Equilateral base

25.

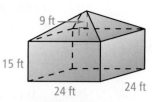

9 ft

15 ft

24 ft   24 ft

Square base

© **26. Writing** The two cylinders pictured at the right are congruent. How does the volume of the larger cone compare to the total volume of the two smaller cones? Explain.

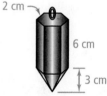

STEM **27. Architecture** The Pyramid of Peace is an opera house in Astana, Kazakhstan. The height of the pyramid is approximately 62 m and one side of its square base is approximately 62 m.
a. What is its volume to the nearest thousand cubic meters?
b. How tall would a prism-shaped building with the same square base as the Pyramid of Peace have to be to have the same volume as the pyramid?

STEM **28. Hardware** Builders use a plumb bob to find a vertical line. The plumb bob shown at the right combines a regular hexagonal prism with a pyramid. Find its volume to the nearest cubic centimeter.

2 cm

6 cm

3 cm

© **29. Reasoning** A cone with radius 3 ft and height 10 ft has a volume of 30π ft³. What is the volume of the cone formed when the following happens to the original cone?
a. The radius is doubled.   **b.** The height is doubled.
c. The radius and the height are both doubled.

Algebra Find the value of the variable in each figure. Leave answers in simplest radical form. The diagrams are not to scale.

30.

6

x

x

x

Volume $=18\sqrt{3}$

31.

x

7

Volume $= 21\pi$

32.

4

r

Volume $= 24\pi$

© **Visualization** Suppose you revolve the plane region completely about the given line to sweep out a solid of revolution. Describe the solid. Then find its volume in terms of π.

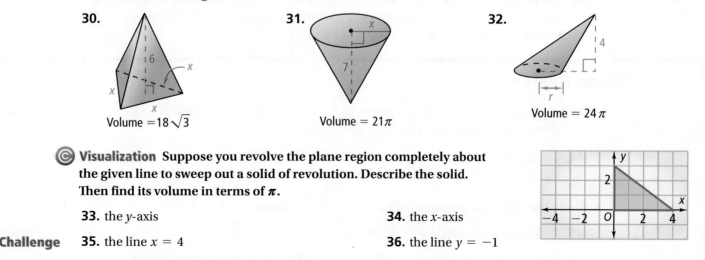

33. the y-axis   **34.** the x-axis

C **Challenge**   **35.** the line $x = 4$ **36.** the line $y = -1$

37. A *frustum* of a cone is the part that remains when the top of the cone is cut off by a plane parallel to the base.

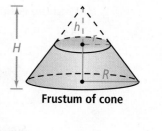

Frustum of cone

a. Explain how to use the formula for the volume of a cone to find the volume of a frustum of a cone.

b. **Containers** A popcorn container 9 in. tall is the frustum of a cone. Its small radius is 4.5 in. and its large radius is 6 in. What is its volume?

38. A disk has radius 10 m. A 90° sector is cut away, and a cone is formed.

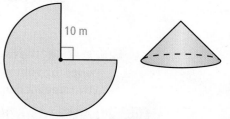

10 m

a. What is the circumference of the base of the cone?

b. What is the area of the base of the cone?

c. What is the volume of the cone? (*Hint:* Use the slant height and the radius of the base to find the height.)

Standardized Test Prep

SAT/ACT

39. A cone has diameter 8 in. and height 14 in. A rectangular prism is 6 in. by 4 in. by 10 in. A square pyramid has base edge 8 in. and height 12 in. What are the volumes of the three figures in order from least to greatest?

Ⓐ cone, prism, pyramid   Ⓒ pyramid, cone, prism

Ⓑ prism, cone, pyramid   Ⓓ prism, pyramid, cone

40. One row of a truth table lists *p* as true and *q* as false. Which of the following statements is true?

Ⓕ $p \rightarrow q$ Ⓖ $p \vee q$ Ⓗ $p \wedge q$ Ⓘ $\sim p$

41. If a polyhedron has 8 vertices and 12 edges, how many faces does it have?

Ⓐ 4   Ⓑ 6   Ⓒ 12   Ⓓ 24

Extended Response

42. The point of concurrency of the three altitudes of a triangle lies outside the triangle. Where are its circumcenter, incenter, and centroid located in relation to the triangle? Draw diagrams to support your answers.

Mixed Review

43. A triangular prism has height 30 cm. Its base is a right triangle with legs 10 cm and 24 cm. What is the volume of the prism?

◀ See Lesson 11-4.

44. Given $\triangle JAC$ and $\triangle KIN$, you know $\overline{JA} \cong \overline{KI}$, $\overline{AC} \cong \overline{IN}$, and $m\angle A > m\angle I$. What can you conclude about JC and KN?

◀ See Lesson 5-7.

Get Ready! To prepare for Lesson 11-6, do Exercises 45 and 46.

45. Find the area of a circle with diameter 3 in. to the nearest tenth.

◀ See Lesson 1-8.

46. Find the circumference of a circle with radius 2 cm to the nearest centimeter.

11-6

Surface Areas and Volumes of Spheres

@ Content Standards
G.GMD.3 Use volume formulas for . . . spheres to solve problems.
G.MG.1 Use geometric shapes, their measures, and their properties to describe objects.

Objective To find the surface area and volume of a sphere

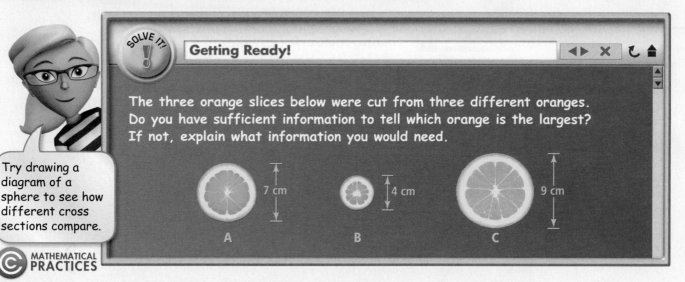

SOLVE IT!

Getting Ready!

The three orange slices below were cut from three different oranges. Do you have sufficient information to tell which orange is the largest? If not, explain what information you would need.

Try drawing a diagram of a sphere to see how different cross sections compare.

@ MATHEMATICAL PRACTICES

A   7 cm   B   4 cm   C   9 cm

Lesson Vocabulary
- sphere
- center of a sphere
- radius of a sphere
- diameter of a sphere
- circumference of a sphere
- great circle
- hemisphere

In the Solve It, you considered the sizes of objects with circular cross sections.

A **sphere** is the set of all points in space equidistant from a given point called the **center**. A **radius** is a segment that has one endpoint at the center and the other endpoint on the sphere. A **diameter** is a segment passing through the center with endpoints on the sphere.

r is the length of the radius of the sphere.

Essential Understanding You can find the surface area and the volume of a sphere when you know its radius.

When a plane and a sphere intersect in more than one point, the intersection is a circle. If the center of the circle is also the center of the sphere, it is called a **great circle**.

The circumference of a great circle is the **circumference** of the sphere.

A great circle divides a sphere into two **hemispheres**.

A baseball can model a sphere. To approximate its surface area, you can take apart its covering. Each of the two pieces suggests a pair of circles with radius r, which is approximately the radius of the ball. The area of the four circles, $4\pi r^2$, suggests the surface area of the ball.

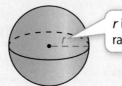

The area of each circle is πr^2.

Theorem 11-10 Surface Area of a Sphere

The surface area of a sphere is four times the product of π and the square of the radius of the sphere.

$$\text{S.A.} = 4\pi r^2$$

Problem 1 Finding the Surface Area of a Sphere

What is the surface area of the sphere in terms of π?

The diameter is 10 m, so the radius is $\frac{10}{2}$ m, or 5 m.

$$
\begin{aligned}
\text{S.A.} &= 4\pi r^2 && \text{Use the formula for surface area of a sphere.}\\
&= 4\pi(5)^2 && \text{Substitute 5 for } r.\\
&= 100\pi && \text{Simplify.}
\end{aligned}
$$

The surface area is 100π m^2.

Plan

What are you given?
In sphere problems, make it a habit to note whether you are given the radius or the diameter. In this case, you are given the diameter.

Got It? 1. What is the surface area of a sphere with a diameter of 14 in.? Give your answer in terms of π and rounded to the nearest square inch.

You can use spheres to approximate the surface areas of real-world objects.

Problem 2 Finding Surface Area

Geography Earth's equator is about 24,902 mi long. What is the approximate surface area of Earth? Round to the nearest thousand square miles.

Plan

How can you use the length of Earth's equator?
Earth's equator is a great circle that divides Earth into two hemispheres. Its length is Earth's circumference. Use it to find Earth's radius.

Step 1 Find the radius of Earth.

$$
\begin{aligned}
C &= 2\pi r && \text{Use the formula for circumference.}\\
24{,}902 &= 2\pi r && \text{Substitute 24,902 for } C.\\
\frac{24{,}902}{2\pi} &= r && \text{Divide each side by } 2\pi.\\
r &\approx 3963.276393 && \text{Use a calculator.}
\end{aligned}
$$

Step 2 Use the radius to find the surface area of Earth.

$$
\begin{aligned}
\text{S.A.} &= 4\pi r^2 && \text{Use the formula for surface area.}\\
&= 4\pi \text{ ANS } \boxed{x^2}\ \boxed{\text{enter}} && \text{Use a calculator. ANS uses the value of } r \text{ from Step 1.}\\
&\approx 197387017.5
\end{aligned}
$$

The surface area of Earth is about 197,387,000 mi^2.

Got It? 2. What is the surface area of a melon with circumference 18 in.? Round your answer to the nearest ten square inches.

In the previous lesson, you learned that the volume of a cone is $\frac{1}{3}\pi r^3$. You can use this with Cavalieri's Principle to find the formula for the volume of a sphere.

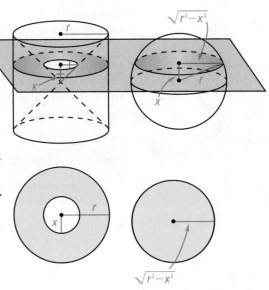

Both figures at the right have a parallel plane x units above their centers that form circular cross sections.

The area of the cross section of the cylinder minus the area of the cross section of the cone is the same as the area of the cross section of the sphere. Every horizontal plane will cut the figures into cross sections of equal area. By Cavalieri's Principle, the volume of the sphere = the volume of the cylinder = the volume of two cones.

$$\begin{aligned} \text{Volume of a sphere} &= \pi r^2(2r) - 2(\tfrac{1}{3}\pi r^3) \\ &= 2\pi r^3 - \tfrac{2}{3}\pi r^3 \\ &= \tfrac{4}{3}\pi r^3 \end{aligned}$$

take note

Theorem 11-11   Volume of a Sphere

The volume of a sphere is four thirds the product of π and the cube of the radius of the sphere.

$$V = \tfrac{4}{3}\pi r^3$$

Problem 3   Finding the Volume of a Sphere

What is the volume of the sphere in terms of π?

$$\begin{aligned} V &= \tfrac{4}{3}\pi r^3 \qquad \text{Use the formula for volume of a sphere.} \\ &= \tfrac{4}{3}\pi (6)^3 \qquad \text{Substitute.} \\ &= 288\pi \end{aligned}$$

The volume of the sphere is 288π m^3.

Think

What are the units of the answer?
You are cubing the radius, which is in meters (m), so your answer should be in cubic meters (m^3).

Got It? 3. a. A sphere has a diameter of 60 in. What is its volume to the nearest cubic inch?

 b. Reasoning Suppose the radius of a sphere is halved. How does this affect the volume of the sphere? Explain.

Notice that you only need to know the radius of a sphere to find its volume and surface area. This means that if you know the volume of a sphere, you can find its surface area.

 **Problem 4** **Using Volume to Find Surface Area**

The volume of a sphere is 5000 m³. What is its surface area to the nearest square meter?

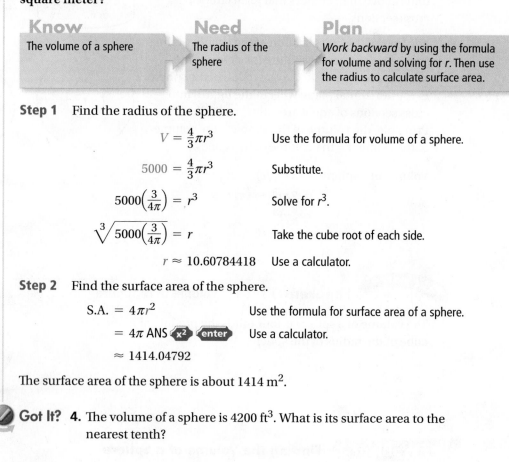

Know	Need	Plan
The volume of a sphere	The radius of the sphere	*Work backward* by using the formula for volume and solving for *r*. Then use the radius to calculate surface area.

Step 1 Find the radius of the sphere.

$$V = \frac{4}{3}\pi r^3$$ Use the formula for volume of a sphere.

$$5000 = \frac{4}{3}\pi r^3$$ Substitute.

$$5000\left(\frac{3}{4\pi}\right) = r^3$$ Solve for r^3.

$$\sqrt[3]{5000\left(\frac{3}{4\pi}\right)} = r$$ Take the cube root of each side.

$$r \approx 10.60784418$$ Use a calculator.

Step 2 Find the surface area of the sphere.

$$\text{S.A.} = 4\pi r^2$$ Use the formula for surface area of a sphere.

$$= 4\pi \text{ ANS } \boxed{x^2} \boxed{\text{enter}}$$ Use a calculator.

$$\approx 1414.04792$$

The surface area of the sphere is about 1414 m².

Got It? 4. The volume of a sphere is 4200 ft³. What is its surface area to the nearest tenth?

Lesson Check

Do you know HOW?

The diameter of a sphere is 12 ft.

1. What is its surface area in terms of π?

2. What is its volume to the nearest tenth?

3. The volume of a sphere is 80π cm³. What is its surface area to the nearest whole number?

Do you UNDERSTAND? MATHEMATICAL PRACTICES

4. Vocabulary What is the ratio of the area of a great circle to the surface area of the sphere?

5. Error Analysis Your classmate claims that if you double the radius of a sphere, its surface area and volume will quadruple. What is your classmate's error? Explain.

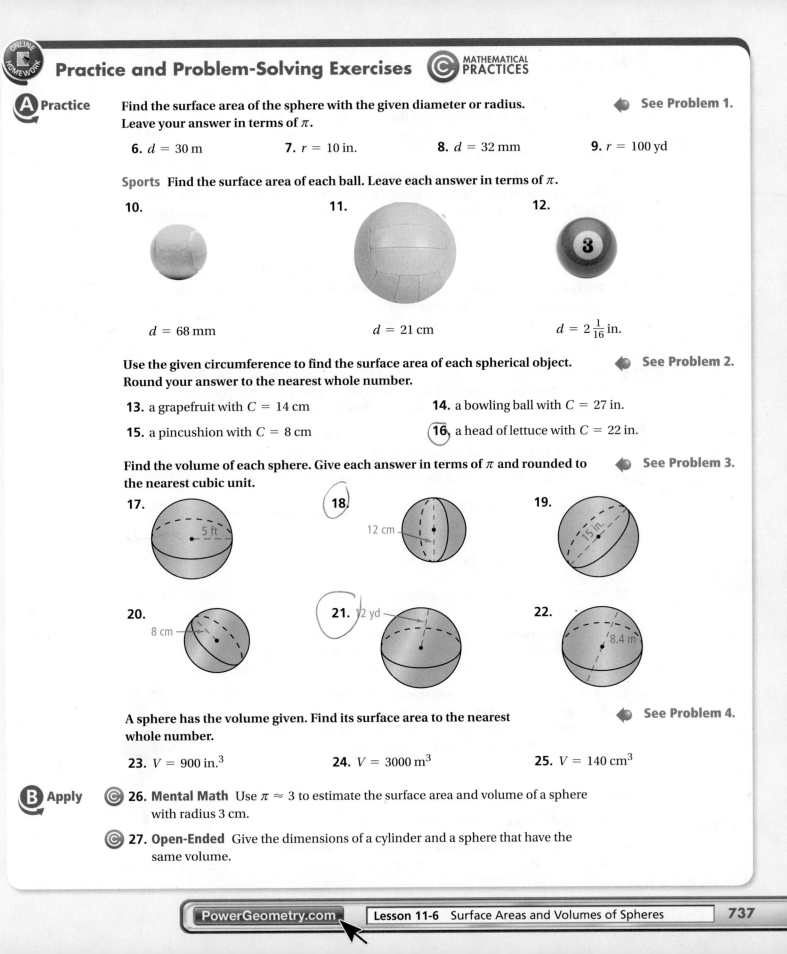

Practice and Problem-Solving Exercises · MATHEMATICAL PRACTICES

A Practice Find the surface area of the sphere with the given diameter or radius. Leave your answer in terms of π. ◄ **See Problem 1.**

6. $d = 30$ m   **7.** $r = 10$ in.   **8.** $d = 32$ mm   **9.** $r = 100$ yd

Sports Find the surface area of each ball. Leave each answer in terms of π.

10.

$d = 68$ mm

11.

$d = 21$ cm

12.

$d = 2\frac{1}{16}$ in.

Use the given circumference to find the surface area of each spherical object. Round your answer to the nearest whole number. ◄ **See Problem 2.**

13. a grapefruit with $C = 14$ cm

14. a bowling ball with $C = 27$ in.

15. a pincushion with $C = 8$ cm

16. a head of lettuce with $C = 22$ in.

Find the volume of each sphere. Give each answer in terms of π and rounded to the nearest cubic unit. ◄ **See Problem 3.**

17.

5 ft

18.

12 cm

19.

15 in.

20.

8 cm

21. 12 yd

22.

8.4 m

A sphere has the volume given. Find its surface area to the nearest whole number. ◄ **See Problem 4.**

23. $V = 900$ in.³   **24.** $V = 3000$ m³   **25.** $V = 140$ cm³

B Apply **26. Mental Math** Use $\pi \approx 3$ to estimate the surface area and volume of a sphere with radius 3 cm.

27. Open-Ended Give the dimensions of a cylinder and a sphere that have the same volume.

28. Visualization The region enclosed by the semicircle at the right is revolved completely about the x-axis.

    a. Describe the solid of revolution that is formed.

    b. Find its volume in terms of π.

    c. Find its surface area in terms of π.

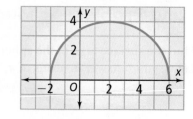

29. Think About a Plan A cylindrical tank with diameter 20 in. is half filled with water. How much will the water level in the tank rise if you place a metallic ball with radius 4 in. in the tank? Give your answer to the nearest tenth.

- What causes the water level in the tank to rise?
- Which volume formulas should you use?

30. The sphere at the right fits snugly inside a cube with 6-in. edges. What is the approximate volume of the space between the sphere and cube?

  Ⓐ 28.3 in.^{3} Ⓒ 102.9 in.^{3}

  Ⓑ 76.5 in.^{3} Ⓓ 113.1 in.^{3}

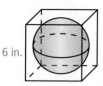

6 in.

STEM 31. Meteorology On September 3, 1970, a hailstone with diameter 5.6 in. fell at Coffeyville, Kansas. It weighed about 0.018 lb/in.^{3} compared to the normal 0.033 lb/in.^{3} for ice. About how heavy was this Kansas hailstone?

32. Reasoning Which is greater, the total volume of three spheres, each of which has diameter 3 in., or the volume of one sphere that has diameter 8 in.?

33. Reasoning How many great circles does a sphere have? Explain.

Find the volume in terms of π of each sphere with the given surface area.

34. $4\pi \text{ m}^2$ **35.** $36\pi \text{ in.}^2$ **36.** $9\pi \text{ ft}^2$ **37.** $100\pi \text{ mm}^2$

38. $25\pi \text{ yd}^2$ **39.** $144\pi \text{ cm}^2$ **40.** $49\pi \text{ m}^2$ **41.** $225\pi \text{ mi}^2$

42. Recreation A spherical balloon has a 14-in. diameter when it is fully inflated. Half of the air is let out of the balloon. Assume that the balloon remains a sphere.

    a. Find the volume of the fully inflated balloon in terms of π.

    b. Find the volume of the half-inflated balloon in terms of π.

    c. What is the diameter of the half-inflated balloon to the nearest inch?

43. Sports Equipment The diameter of a golf ball is 1.68 in.

    a. Approximate the surface area of the golf ball.

    b. Reasoning Do you think that the value you found in part (a) is greater than or less than the actual surface area of the golf ball? Explain.

Geometry in 3 Dimensions A sphere has center $(0, 0, 0)$ and radius 5.

44. Name the coordinates of six points on the sphere.

45. Tell whether each of the following points is *inside, outside,* or *on the sphere.*

    a. $A(0, -3, 4)$ **b.** $B(1, -1, -1)$ **c.** $C(4, -6, -10)$

46. Food An ice cream vendor presses a sphere of frozen yogurt into a cone, as shown at the right. If the yogurt melts into the cone, will the cone overflow? Explain.

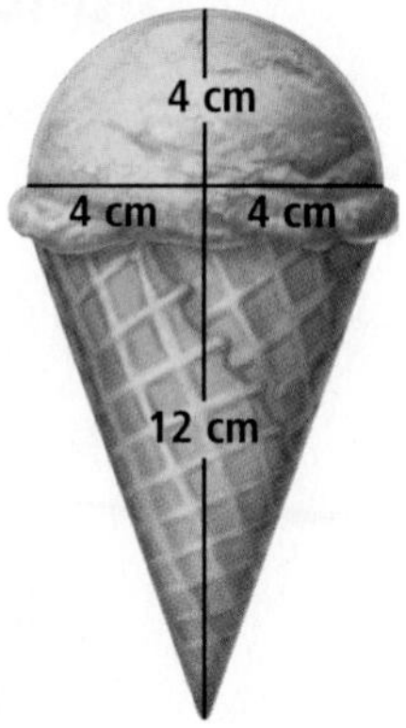

47. The surface area of a sphere is 5541.77 ft^2. What is its volume to the nearest tenth?

48. Geography The circumference of Earth at the equator is approximately 40,075 km. About 71% of Earth is covered by oceans and other bodies of water. To the nearest thousand square kilometers, how much of Earth's surface is land?

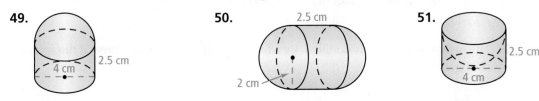

Find the surface area and volume of each figure.

49. 2.5 cm / 4 cm

50. 2.5 cm / 2 cm

51. 2.5 cm / 4 cm

STEM **52. Astronomy** The diameter of the Earth is about 7926 mi. The diameter of the moon is about 27% of the diameter of Earth. What percent of the volume of Earth is the volume of the moon? Round your answer to the nearest whole percent.

STEM **53. Science** The density of steel is about 0.28 lb/in.^{3}. Could you lift a solid steel ball with radius 4 in.? With radius 6 in.? Explain.

54. A cube with edges 6 in. long fits snugly inside a sphere as shown at the right. The diagonal of the cube is the diameter of the sphere.
a. Find the length of the diagonal and the radius of the sphere. Leave your answer in simplest radical form.
b. What is the volume of the space between the sphere and the cube to the nearest tenth?

Challenge **Find the radius of a sphere with the given property.**

55. The number of square meters of surface area equals the number of cubic meters of volume.

56. The ratio of surface area in square meters to volume in cubic meters is 1 : 5.

57. Suppose a cube and a sphere have the same volume.
a. Which has the greater surface area? Explain.
b. Writing Explain why spheres are rarely used for packaging.

58. A plane intersects a sphere to form a circular cross section. The radius of the sphere is 17 cm and the plane comes to within 8 cm of the center. Draw a sketch and find the area of the cross section, to the nearest whole number.

59. History At the right, the sphere fits snugly inside the cylinder. Archimedes (about 287–212 B.C.) requested that such a figure be put on his gravestone along with the ratio of their volumes, a finding that he regarded as his greatest discovery. What is that ratio?

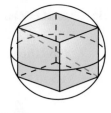

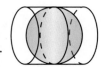

Standardized Test Prep

SAT/ACT

60. What is the diameter of a sphere whose surface area is 100π m^2?

     (A) 5 m   (B) 10 m   (C) 5π m   (D) 25π m

61. Which of the following statements contradict each other?

    I. Opposite sides of $\square ABCD$ are parallel.

    II. Diagonals of $\square ABCD$ are perpendicular.

    III. $\square ABCD$ is not a rhombus.

     (F) I and II   (G) II and III   (H) I and III   (I) none

62. What is the reflection image of (3, 7) across the line $y = 4$?

     (A) (3, 3)   (B) (−7, 3)   (C) (3, 1)   (D) (3, −7)

63. The radius of a sphere is doubled. By what factor does the surface area of the sphere change?

     (F) $\frac{1}{4}$ (G) $\frac{1}{2}$ (H) 2   (I) 4

Short Response

64. Find the values of x and y. Show your work.

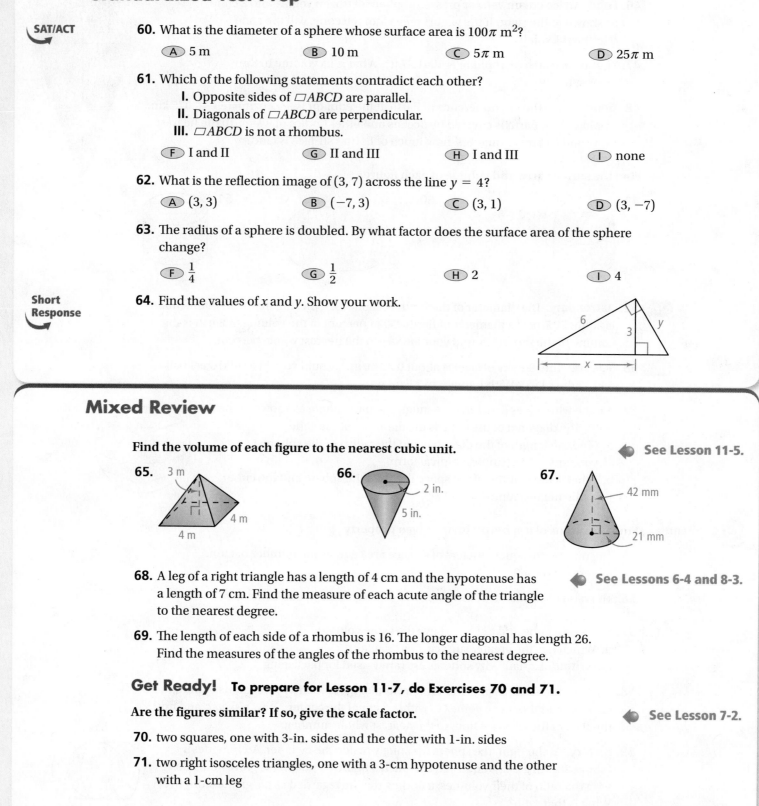

Mixed Review

Find the volume of each figure to the nearest cubic unit.

                                 ◀ **See Lesson 11-5.**

65. 3 m

    4 m

  4 m

66. 2 in.

  5 in.

67. 42 mm

  21 mm

68. A leg of a right triangle has a length of 4 cm and the hypotenuse has a length of 7 cm. Find the measure of each acute angle of the triangle to the nearest degree.

                                   ◀ **See Lessons 6-4 and 8-3.**

69. The length of each side of a rhombus is 16. The longer diagonal has length 26. Find the measures of the angles of the rhombus to the nearest degree.

Get Ready!   **To prepare for Lesson 11-7, do Exercises 70 and 71.**

Are the figures similar? If so, give the scale factor.

                                   ◀ **See Lesson 7-2.**

70. two squares, one with 3-in. sides and the other with 1-in. sides

71. two right isosceles triangles, one with a 3-cm hypotenuse and the other with a 1-cm leg

Exploring Similar Solids

© Content Standard
Prepares for G.MG.2 Apply concepts of density based on area and volume in modeling situations . . .

To explore surface areas and volumes of similar rectangular prisms, you can set up a spreadsheet like the one below. You choose the numbers for length, width, height, and scale factor. The computer uses formulas to calculate all the other numbers.

© MATHEMATICAL PRACTICES

Activity

	A	B	C	D	E	F	G	H	I	
1								Ratio of	Ratio of	
2					Surface		Scale	Surface	Volumes	
3		Length	Width	Height	Area	Volume	Factor (II : I)	Areas (II : I)	(II : I)	
4	Rectangular Prism I	6	4	23	508	552	2	4	8	
5	Similar Prism II	12	8	46	2032	4416				

In cell E4, enter the formula =2*(B4*C4+B4*D4+C4*D4). This will calculate the sum of the areas of the six faces of Prism I. In cell F4, enter the formula =B4*C4*D4. This will calculate the volume of Prism I.

In cells B5, C5, and D5 enter the formulas =G4*B4, =G4*C4, and =G4*D4, respectively. These will calculate the dimensions of similar Prism II. Copy the formulas from E4 and F4 into E5 and F5 to calculate the surface area and volume of Prism II.

In cell H4 enter the formula =E5/E4 and in cell I4 enter the formula =F5/F4. These will calculate the ratios of the surface areas and volumes.

© Investigate In row 4, enter numbers for the length, width, height, and scale factor. Change the numbers to explore how the ratio of the surface areas and the ratio of the volumes are related to the scale factor.

Exercises

State a relationship that seems to be true about the scale factor and the given ratio.

1. the ratio of volumes

2. the ratio of surface areas

Set up spreadsheets that allow you to investigate the following ratios. State a conclusion from each investigation.

3. the volumes of similar cylinders

4. the lateral areas of similar cylinders

5. the surface areas of similar cylinders

6. the volumes of similar square pyramids

11-7 Areas and Volumes of Similar Solids

© Content Standards
G.MG.1 Use geometric shapes, their measures, and their properties to describe objects.
G.MG.2 Apply concepts of density based on area and volume in modeling situations . . .

Objective To compare and find the areas and volumes of similar solids

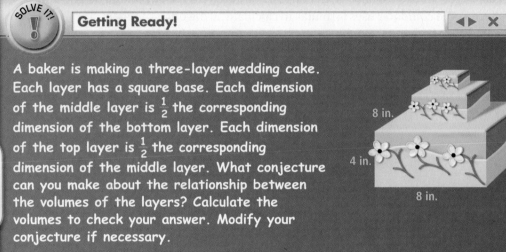

SOLVE IT!

Getting Ready!

A baker is making a three-layer wedding cake. Each layer has a square base. Each dimension of the middle layer is $\frac{1}{2}$ the corresponding dimension of the bottom layer. Each dimension of the top layer is $\frac{1}{2}$ the corresponding dimension of the middle layer. What conjecture can you make about the relationship between the volumes of the layers? Calculate the volumes to check your answer. Modify your conjecture if necessary.

8 in.

4 in.

8 in.

Will the bottom-to-middle ratio be the same as the middle-to-top ratio?

MATHEMATICAL PRACTICES

Essential Understanding You can use ratios to compare the areas and volumes of similar solids.

Similar solids have the same shape, and all their corresponding dimensions are proportional. The ratio of corresponding linear dimensions of two similar solids is the scale factor. Any two cubes are similar, as are any two spheres.

Plan

How do you check for similarity?
Check that the ratios of the corresponding dimensions are the same. A rectangular prism has three dimensions (ℓ, w, h), so you must check three ratios.

© Problem 1 **Identifying Similar Solids**

Are the two rectangular prisms similar? If so, what is the scale factor of the first figure to the second figure?

A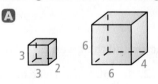

$$\frac{3}{6} = \frac{2}{4} = \frac{3}{6}$$

The prisms are similar because the corresponding linear dimensions are proportional.

The scale factor is $\frac{1}{2}$.

B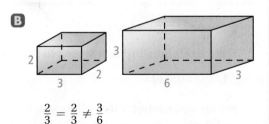

$$\frac{2}{3} = \frac{2}{3} \neq \frac{3}{6}$$

The prisms are not similar because the corresponding linear dimensions are not proportional.

Got It? **1.** Are the two cylinders similar? If so, what is the scale factor of the first figure to the second figure?

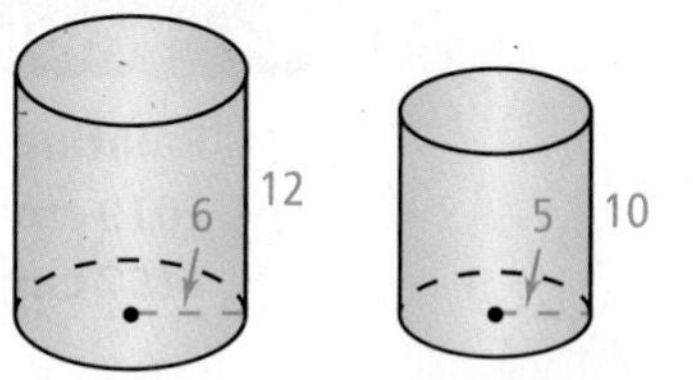

6 12   5 10

The two similar prisms shown here suggest two important relationships for similar solids.

The ratio of the side lengths is $1 : 2$.
The ratio of the surface areas is $22 : 88$, or $1 : 4$.
The ratio of the volumes is $6 : 48$, or $1 : 8$.

The ratio of the surface areas is the square of the scale factor. The ratio of the volumes is the cube of the scale factor. These two facts apply to all similar solids.

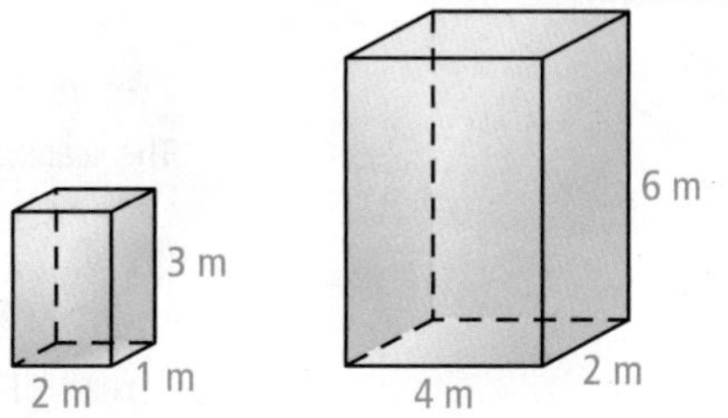

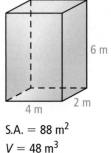

3 m
2 m 1 m
6 m
4 m 2 m

S.A. = 22 m²   S.A. = 88 m²
V = 6 m³   V = 48 m³

take note

Theorem 11-12   Areas and Volumes of Similar Solids

If the scale factor of two similar solids is $a : b$, then

- the ratio of their corresponding areas is $a^2 : b^2$
- the ratio of their volumes is $a^3 : b^3$

Plan

How can you use the given information?
You are given the volumes of two similar solids. Write a proportion using the ratio $a^3 : b^3$.

Problem 2   **Finding the Scale Factor**

The square prisms at the right are similar. What is the scale factor of the smaller prism to the larger prism?

$\dfrac{a^3}{b^3} = \dfrac{729}{1331}$ The ratio of the volumes is $a^3 : b^3$.

$\dfrac{a}{b} = \dfrac{9}{11}$ Take the cube root of each side.

The scale factor is $9 : 11$.

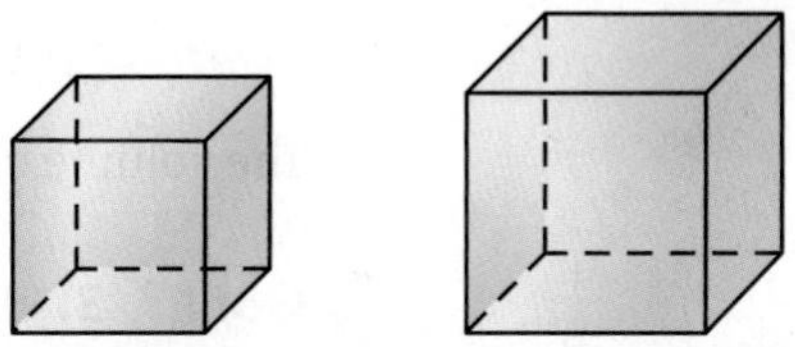

V = 729 cm³   V = 1331 cm³

 Got It? **2. a.** What is the scale factor of two similar prisms with surface areas 144 m² and 324 m²?
     b. Reasoning Are any two square prisms similar? Explain.

Problem 3 Using a Scale Factor

Painting The lateral areas of two similar paint cans are 1019 cm² and 425 cm². The volume of the smaller can is 1157 cm³. What is the volume of the larger can?

Know
- The lateral areas
- The volume of the smaller can

Need

The scale factor

Plan

Use the lateral areas to find the scale factor $a : b$. Then write and solve a proportion using the ratio $a^3 : b^3$ to find the volume of the larger can.

Step 1 Find the scale factor $a : b$.

$$\frac{a^2}{b^2} = \frac{1019}{425}$$ The ratio of the surface areas is $a^2 : b^2$.

$$\frac{a}{b} = \frac{\sqrt{1019}}{\sqrt{425}}$$ Take the positive square root of each side.

Step 2 Use the scale factor to find the volume.

Think

Does it matter how you set up the proportion?
Yes. The numerators should refer to the same paint can, and the denominators should refer to the other can.

$$\frac{V_{\text{large}}}{V_{\text{small}}} = \frac{\left(\sqrt{1019}\right)^3}{\left(\sqrt{425}\right)^3}$$ The ratio of the volumes is $a^3 : b^3$.

$$\frac{V_{\text{large}}}{1157} = \frac{\left(\sqrt{1019}\right)^3}{\left(\sqrt{425}\right)^3}$$ Substitute 1157 for V_{small}.

$$V_{\text{large}} = 1157 \cdot \frac{\left(\sqrt{1019}\right)^3}{\left(\sqrt{425}\right)^3}$$ Solve for V_{large}.

$$V_{\text{large}} \approx 4295.475437$$ Use a calculator.

The volume of the larger paint can is about 4295 cm³.

Got It? 3. The volumes of two similar solids are 128 m³ and 250 m³. The surface area of the larger solid is 250 m². What is the surface area of the smaller solid?

You can compare the capacities and weights of similar objects. The capacity of an object is the amount of fluid the object can hold. The capacities and weights of similar objects made of the same material are proportional to their volumes.

Problem 4 **Using a Scale Factor to Find Capacity** STEM

Containers A bottle that is 10 in. high holds 34 oz of milk. The sandwich shop shown at the right is shaped like a milk bottle. To the nearest thousand ounces how much milk could the building hold?

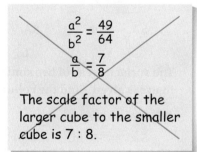

480 in.

Think

How does capacity relate to volume?
Since the capacities of similar objects are proportional to their volumes, the ratio of their capacities is equal to the ratio of their volumes.

The scale factor of the bottles is $1 : 48$.

The ratio of their volumes, and hence the ratio of their capacities, is $1^3 : 48^3$, or $1 : 110{,}592$.

$$\frac{1}{110{,}592} = \frac{34}{x}$$
Let $x =$ the capacity of the milk-bottle building.

$$x = 34 \cdot 110{,}592$$
Use the Cross Products Property.

$$x = 3{,}760{,}128$$
Simplify.

The milk-bottle building could hold about 3,760,000 oz.

Got It? **4.** A marble paperweight shaped like a pyramid weighs 0.15 lb. How much does a similarly shaped marble paperweight weigh if each dimension is three times as large?

Lesson Check

Do you know HOW?

1. Which two of the following cones are similar? What is their scale factor?

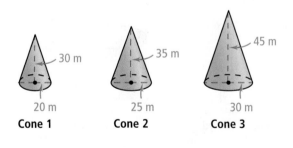

30 m   35 m   45 m

20 m   25 m   30 m

Cone 1   Cone 2   Cone 3

2. The volumes of two similar containers are 115 in.^{3} and 67 in.^{3}. The surface area of the smaller container is 108 in^2. What is the surface area of the larger container?

Do you UNDERSTAND? MATHEMATICAL PRACTICES

3. Vocabulary How are similar solids different from similar polygons? Explain.

4. Error Analysis Two cubes have surface areas 49 cm^2 and 64 cm^2. Your classmate tried to find the scale factor of the larger cube to the smaller cube. Explain and correct your classmate's error.

$$\frac{a^2}{b^2} = \frac{49}{64}$$

$$\frac{a}{b} = \frac{7}{8}$$

The scale factor of the larger cube to the smaller cube is 7 : 8.

Practice and Problem-Solving Exercises

MATHEMATICAL PRACTICES

A Practice

For Exercises 5–10, are the two figures similar? If so, give the scale factor of the first figure to the second figure.

See Problem 1.

5.

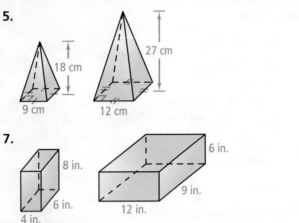

18 cm

27 cm

9 cm

12 cm

6.

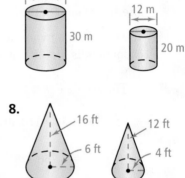

18 m

30 m

12 m

20 m

7.

8 in.

6 in.

4 in.

6 in.

9 in.

12 in.

8.

16 ft

6 ft

12 ft

4 ft

9. two cubes, one with 3-cm edges, the other with 4.5-cm edges

10. a cylinder and a square prism both with 3-in. radius and 1-in. height

Each pair of figures is similar. Use the given information to find the scale factor of the smaller figure to the larger figure.

See Problem 2.

11.

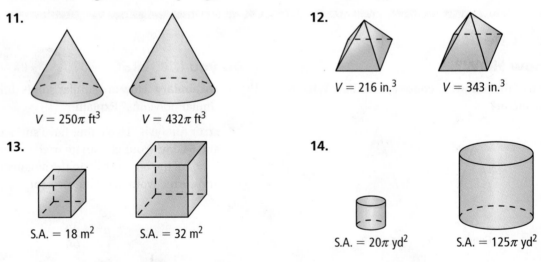

$V = 250\pi$ ft^3 $V = 432\pi$ ft^3

12.

$V = 216$ in.^{3} $V = 343$ in.^{3}

13.

S.A. $= 18$ m^2 S.A. $= 32$ m^2

14.

S.A. $= 20\pi$ yd^2 S.A. $= 125\pi$ yd^2

The surface areas of two similar figures are given. The volume of the larger figure is given. Find the volume of the smaller figure.

See Problem 3.

15. S.A. $= 248$ in.^{2}

S.A. $= 558$ in.^{2}

$V = 810$ in.^{3}

16. S.A. $= 192$ m^2

S.A. $= 1728$ m^2

$V = 4860$ m^3

17. S.A. $= 52$ ft^2

S.A. $= 208$ ft^2

$V = 192$ ft^3

The volumes of two similar figures are given. The surface area of the smaller figure is given. Find the surface area of the larger figure.

18. $V = 27$ in.^{3}
$V = 125$ in.^{3}
S.A. $= 63$ in.^{2}

19. $V = 27$ m^3
$V = 64$ m^3
S.A. $= 63$ m^2

20. $V = 2$ yd^3
$V = 250$ yd^3
S.A. $= 13$ yd^2

STEM 21. Packaging There are 750 toothpicks in a regular-sized box. If a jumbo box is made by doubling all the dimensions of the regular-sized box, how many toothpicks will the jumbo box hold? ◀ **See Problem 4.**

STEM 22. Packaging A cylinder with a 4-in. diameter and a 6-in. height holds 1 lb of oatmeal. To the nearest ounce, how much oatmeal will a similar 10-in.-high cylinder hold? (*Hint:* 1 lb $= 16$ oz)

Ⓒ 23. Compare and Contrast A regular pentagonal prism has 9-cm base edges. A larger, similar prism of the same material has 36-cm base edges. How does each indicated measurement for the larger prism compare to the same measurement for the smaller prism?
a. the volume   **b.** the weight

Ⓑ Apply

24. Two similar prisms have heights 4 cm and 10 cm.
a. What is their scale factor?
b. What is the ratio of their surface areas?
c. What is the ratio of their volumes?

Ⓒ 25. Think About a Plan A company announced that it had developed the technology to reduce the size of its atomic clock, which is used in electronic devices that transmit data. The company claims that the smaller clock will be similar to the existing clock made of the same material. The dimensions of the smaller clock will be $\frac{1}{10}$ the dimensions of the company's existing atomic clocks, and it will be $\frac{1}{100}$ the weight. Do these ratios make sense? Explain.
• What is the scale factor of the smaller clock to the larger clock?
• How are the weights of the two objects related to their scale factor?

Ⓒ 26. Reasoning Is there a value of x for which the rectangular prisms at the right are similar? Explain.

27. The volume of a spherical balloon with radius 3.1 cm is about 125 cm^3. Estimate the volume of a similar balloon with radius 6.2 cm.

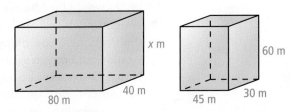

Ⓒ 28. Writing Are all spheres similar? Explain.

Ⓒ 29. Reasoning A carpenter is making a blanket chest based on an antique chest. Both chests have the shape of a rectangular prism. The length, width, and height of the new chest will all be 4 in. greater than the respective dimensions of the antique. Will the chests be similar? Explain.

30. Two similar pyramids have lateral area 20 ft^2 and 45 ft^2. The volume of the smaller pyramid is 8 ft^3. Find the volume of the larger pyramid.

31. The volumes of two spheres are 729 in.^{3} and 27 in.^{3}.
  a. Find the ratio of their radii.
  b. Find the ratio of their surface areas.

32. The volumes of two similar pyramids are 1331 cm^3 and 2744 cm^3.
  a. Find the ratio of their heights.
  b. Find the ratio of their surface areas.

33. A clown's face on a balloon is 4 in. tall when the balloon holds 108 in.^{3} of air. How much air must the balloon hold for the face to be 8 in. tall?

Copy and complete the table for the similar solids.

	Similarity Ratio	Ratio of Surface Areas	Ratio of Volumes
34.	1 : 2	■ : ■	■ : ■
35.	3 : 5	■ : ■	■ : ■
36.	■ : ■	49 : 81	■ : ■
37.	■ : ■	■ : ■	125 : 512

38. Literature In *Gulliver's Travels*, by Jonathan Swift, Gulliver first traveled to Lilliput. The Lilliputian average height was one twelfth of Gulliver's height.
  a. How many Lilliputian coats could be made from the material in Gulliver's coat? (*Hint:* Use the ratio of surface areas.)
  b. How many Lilliputian meals would be needed to make a meal for Gulliver? (*Hint:* Use the ratio of volumes.)

Challenge

39. Indirect Reasoning Some stories say that Paul Bunyan was ten times as tall as the average human. Assume that Paul Bunyan's bone structure was proportional to that of ordinary people.
  a. Strength of bones is proportional to the area of their cross section. How many times as strong as the average person's bones would Paul Bunyan's bones be?
  b. Weights of objects made of like material are proportional to their volumes. How many times the average person's weight would Paul Bunyan's weight be?
  c. Human leg bones can support about 6 times the average person's weight. Use your answers to parts (a) and (b) to explain why Paul Bunyan could not exist with a bone structure that was proportional to that of ordinary people.

40. Square pyramids *A* and *B* are similar. In pyramid *A*, each base edge is 12 cm. In pyramid *B*, each base edge is 3 cm and the volume is 6 cm^3.
  a. Find the volume of pyramid *A*.
  b. Find the ratio of the surface area of *A* to the surface area of *B*.
  c. Find the surface area of each pyramid.

41. A cone is cut by a plane parallel to its base. The small cone on top is similar to the large cone. The ratio of the slant heights of the cones is 1 : 2. Find each ratio. 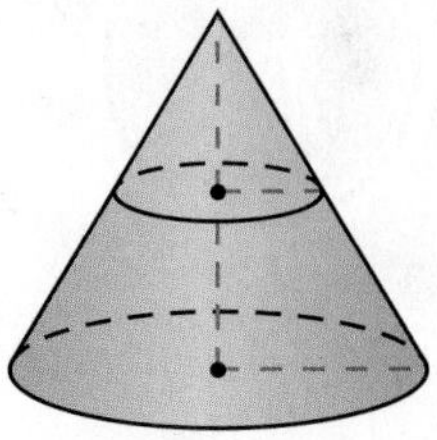
  a. the surface area of the large cone to the surface area of the small cone
  b. the volume of the large cone to the volume of the small cone
  c. the surface area of the frustum to the surface area of the large cone and to the surface area of the small cone
  d. the volume of the frustum to the volume of the large cone and to the volume of the small cone

Standardized Test Prep

GRIDDED RESPONSE

SAT/ACT

42. The slant heights of two similar pyramids are in the ratio 1 : 5. The volume of the smaller pyramid is 60 m³. What is the volume in cubic meters of the larger pyramid?

43. What is the value of x in the figure at the right?

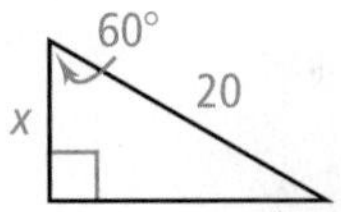

44. A dilation maps $\triangle JEN$ onto $\triangle J'E'N'$. If $JE = 4.5$ ft, $EN = 6$ ft, and $J'E' = 13.5$ ft, what is $E'N'$ in feet?

45. $\triangle CAR \cong \triangle BUS$, $m\angle C = 25$, and $m\angle R = 39$. What is $m\angle U$?

46. A regular pentagon has a radius of 5 in. What is the area of the pentagon to the nearest square inch?

Mixed Review

47. Sports Equipment The circumference of a regulation basketball is between 75 cm and 78 cm. What are the smallest and the largest surface areas that a basketball can have? Give your answers to the nearest whole unit.

See Lesson 11-6.

Find the volume of each sphere to the nearest tenth.

48. diameter = 6 in.   **49.** circumference = 2.5π m   **50.** radius = 6 in.

51. The altitude to the hypotenuse of right $\triangle ABC$ divides the hypotenuse into 12-mm and 16-mm segments. Find the length of each of the following.
  a. the altitude to the hypotenuse
  b. the shorter leg of $\triangle ABC$ **c.** the longer leg of $\triangle ABC$

See Lesson 7-4.

Get Ready! **To prepare for Lesson 12-1, do Exercises 52–54.**

Find the value of x.

See Lesson 8-1.

52.   **53.** 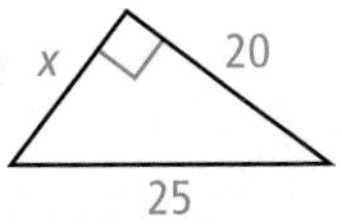   **54.**

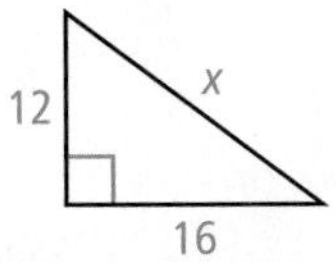

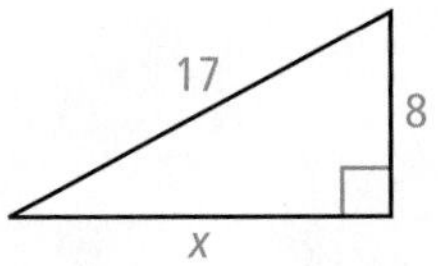

> To solve these problems, you will pull together many concepts and skills that you have studied about surface area and volume.

BIG idea Visualization

You can use a net to visualize a three-dimensional solid.

ⓒ **Performance Task 1**

The net of a composite space figure is shown below.

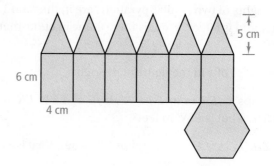

a. What figures make up the composite space figure?

b. What is the surface area of the composite space figure? Round your answer to the nearest square centimeter.

BIG idea Measurement

You can use formulas to find the volumes of various solids.

ⓒ **Performance Task 2**

Jack and Maureen are opening an ice hotel. The structure and furniture are made out of ice. They must determine the weight of some of the ice furniture to decide the size of the crane needed to lift each piece into place. Maureen knows that ice weighs approximately 62.4 pounds per cubic foot. Find the approximate weights of the figures in parts (a)–(c) below.

a. The stage in the banquet hall is shaped like a rectangular prism with measures 22 ft by 15 ft by 4 ft.

b. Two cylindrical pillars will flank the front door. Each cylinder has diameter 1 ft and height 9 ft.

c. A series of four spheres will decorate the reception area. The largest sphere has diameter 4 ft. The other three spheres have diameter 2 ft, 1 ft, and $\frac{1}{2}$ ft.

Connecting **BIG** ideas and Answering the Essential Questions

1 Visualization
You can determine the intersection of a solid and a plane by visualizing how the plane slices the solid to form a two-dimensional cross section.

Space Figures and Cross Sections (Lesson 11-1)
This vertical plane intersects the cylinder in a rectangular cross section.

Surface Areas and Volumes of Prisms, Cylinders, Pyramids, and Cones (Lessons 11-2 through 11-5)

2 Measurement
You can find the surface area or volume of a solid by first choosing a formula to use and then substituting the needed dimensions into the formula.

	Surface Area (S.A.)	Volume (V)
Prism	$ph + 2B$	Bh
Cylinder	$2\pi rh + 2B$	Bh
Pyramid	$\frac{1}{2}p\ell + B$	$\frac{1}{3}Bh$
Cone	$\pi r\ell + B$	$\frac{1}{3}Bh$

Surface Areas and Volumes of Spheres (Lesson 11-6)
$$\text{S.A.} = 4\pi r^2$$
$$V = \frac{4}{3}\pi r^3$$

3 Similarity
The surface areas of similar solids are proportional to the squares of their corresponding dimensions. The volumes are proportional to the cubes of their corresponding dimensions.

Areas and Volumes of Similar Solids (Lesson 11-7)
If the scale factor of two similar solids is $a : b$, then
• the ratio of their areas is $a^2 : b^2$
• the ratio of their volumes is $a^3 : b^3$

Chapter Vocabulary

- altitude (pp. 699, 701, 708, 711)
- center of a sphere (p. 733)
- cone (p. 711)
- cross section (p. 690)
- cylinder (p. 701)
- edge (p. 688)
- face (p. 688)
- great circle (p. 733)
- hemisphere (p. 733)
- lateral area (pp. 700, 702, 709, 711)
- lateral face (pp. 699, 708)
- polyhedron (p. 688)
- prism (p. 699)
- pyramid (p. 708)
- right cone (p. 711)
- right cylinder (p. 701)
- right prism (p. 699)
- slant height (pp. 708, 711)
- sphere (p. 733)
- surface area (pp. 700, 702, 709, 711)
- volume (p. 717)

Choose the correct term to complete each sentence.

1. A set of points in space equidistant from a given point is called a(n) __?__ .

2. A(n) __?__ is a polyhedron in which one face can be any polygon and the lateral faces are triangles that meet at a common vertex.

3. If you slice a prism with a plane, the intersection of the prism and the plane is a(n) __?__ of the prism.

11-1 Space Figures and Cross Sections

Quick Review

A **polyhedron** is a three-dimensional figure whose surfaces are polygons. The polygons are **faces** of the polyhedron. An **edge** is a segment that is the intersection of two faces. A **vertex** is a point where three or more edges intersect. A **cross section** is the intersection of a solid and a plane.

Example

How many faces and edges does the polyhedron have?

The polyhedron has 2 triangular bases and 3 rectangular faces for a total of 5 faces.

The 2 triangles have a total of 6 edges. The 3 rectangles have a total of 12 edges. The total number of edges in the polyhedron is one half the total of 18 edges, or 9.

Exercises

Draw a net for each three-dimensional figure.

4.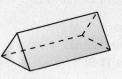

5.

Use Euler's Formula to find the missing number.

6. $F = 5$, $V = 5$, $E = $ ■   **7.** $F = 6$, $V = $ ■, $E = 12$

8. How many vertices are there in a solid with 4 triangular faces and 1 square base?

9. Describe the cross section in the figure at the right.

10. Sketch a cube with an equilateral triangle cross section.

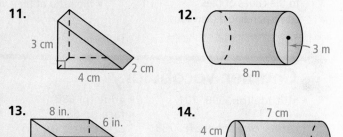

11-2 Surface Areas of Prisms and Cylinders

Quick Review

The **lateral area of a right prism** is the product of the perimeter of the base and the height. The **lateral area of a right cylinder** is the product of the circumference of the base and the height of the cylinder. The **surface area** of each solid is the sum of the lateral area and the areas of the bases.

Example

What is the surface area of a cylinder with radius 3 m and height 6 m? Leave your answer in terms of π.

$$\text{S.A.} = \text{L.A.} + 2B \qquad \text{Use the formula for surface area of a cylinder.}$$

$$= 2\pi rh + 2(\pi r^2) \qquad \text{Substitute formulas for lateral area and area of a circle.}$$

$$= 2\pi(3)(6) + 2\pi(3)^2 \qquad \text{Substitute 3 for } r \text{ and 6 for } h.$$

$$= 36\pi + 18\pi \qquad \text{Simplify.}$$

$$= 54\pi$$

The surface area of the cylinder is 54π m^2.

Exercises

Find the surface area of each figure. Leave your answers in terms of π where applicable.

11. 3 cm, 4 cm, 2 cm

12. 3 m, 8 m

13. 8 in., 6 in., 4 in.

14. 7 cm, 4 cm

15. A cylinder has radius 2.5 cm and lateral area 20π cm^2. What is the surface area of the cylinder in terms of π?

11-3 Surface Areas of Pyramids and Cones

Quick Review

The **lateral area of a regular pyramid** is half the product of the perimeter of the base and the slant height. The **lateral area of a right cone** is half the product of the circumference of the base and the slant height. The **surface area** of each solid is the sum of the lateral area and the area of the base.

Example

What is the surface area of a cone with radius 3 in. and slant height 10 in.? Leave your answer in terms of π.

S.A. = L.A. + B	Use the formula for surface area of a cone.
$= \pi r\ell + \pi r^2$	Substitute formulas for lateral area and area of a circle.
$= \pi(3)(10) + \pi(3)^2$	Substitute 3 for r and 10 for ℓ.
$= 30\pi + 9\pi$	Simplify.
$= 39\pi$	

The surface area of the cone is 39π in.^{2}.

Exercises

Find the surface area of each figure. Round your answers to the nearest tenth.

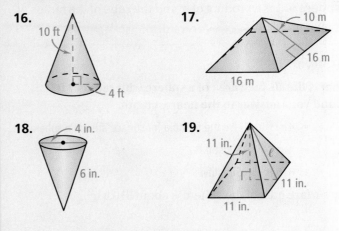

16.

17.

18.

19.

20. Find the formula for the base area of a prism in terms of surface area and lateral area.

11-4 and 11-5 Volumes of Prisms, Cylinders, Pyramids, and Cones

Quick Review

The **volume** of a space figure is the space that the figure occupies. Volume is measured in cubic units. The **volume of a prism** and the **volume of a cylinder** are the product of the area of a base and the height of the solid. The **volume of a pyramid** and the **volume of a cone** are one third the product of the area of the base and the height of the solid.

Example

What is the volume of a rectangular prism with base 3 cm by 4 cm and height 8 cm?

$V = Bh$	Use the formula for the volume of a prism.
$= (3 \cdot 4)(8)$	Substitute.
$= 96$	Simplify.

The volume of the prism is 96 cm^3.

Exercises

Find the volume of each figure. If necessary, round to the nearest tenth.

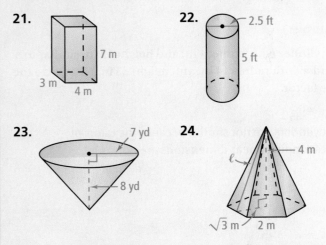

21.

22.

23.

24.

11-6 Surface Areas and Volumes of Spheres

Quick Review

The **surface area of a sphere** is four times the product of π and the square of the radius of the sphere. The **volume of a sphere** is $\frac{4}{3}$ the product of π and the cube of the radius of the sphere.

Example

What is the surface area of a sphere with radius 7 ft? Round your answer to the nearest tenth.

$\qquad$ S.A. $= 4\pi r^2$ $\qquad$ Use the formula for surface area of a sphere.

$\qquad\quad\; = 4\pi(7)^2$ $\qquad$ Substitute.

$\qquad\quad\; \approx 615.8$ $\qquad$ Simplify.

The surface area of the sphere is about 615.8 ft².

Exercises

Find the surface area and volume of a sphere with the given radius or diameter. Round your answers to the nearest tenth.

25. $r = 5$ in. $\qquad\qquad$ **26.** $d = 7$ cm

27. $d = 4$ ft $\qquad\qquad$ **28.** $r = 0.8$ ft

29. What is the volume of a sphere with a surface area of 452.39 cm²? Round your answer to the nearest hundredth.

30. What is the surface area of a sphere with a volume of 523.6 m³? Round your answer to the nearest square meter.

31. Sports Equipment The circumference of a lacrosse ball is 8 in. Find its volume to the nearest tenth of a cubic inch.

11-7 Areas and Volumes of Similar Solids

Quick Review

Similar solids have the same shape and all their corresponding dimensions are proportional.

If the scale factor of two similar solids is $a : b$, then the ratio of their corresponding surface areas is $a^2 : b^2$, and the ratio of their volumes is $a^3 : b^3$.

Example

Is a cylinder with radius 4 in. and height 12 in. similar to a cylinder with radius 14 in. and height 35 in.? If so, give the scale factor.

$$\frac{4}{14} \neq \frac{12}{35}$$

The cylinders are not similar because the ratios of corresponding linear dimensions are not equal.

Exercises

32. Open-Ended Sketch two similar solids whose surface areas are in the ratio $16 : 25$. Include dimensions.

For each pair of similar solids, find the ratio of the volume of the first figure to the volume of the second.

33. $\qquad\qquad\qquad\qquad$ **34.**

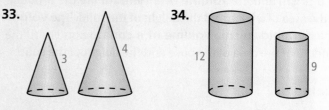

35. Packaging There are 12 pencils in a regular-sized box. If a jumbo box is made by tripling all the dimensions of the regular-sized box, how many pencils will the jumbo box hold?

11 Chapter Test

Do you know HOW?

Draw a net for each figure. Label the net with appropriate dimensions.

1.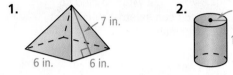
7 in.
6 in.   6 in.

2.
4 cm
10 cm

Use the polyhedron at the right for Exercises 3 and 4.

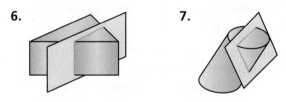

3. Verify Euler's Formula for the polyhedron.

4. Draw a net for the polyhedron. Verify $F + V = E + 1$ for the net.

5. What is the number of edges in a pyramid with seven faces?

Describe the cross section formed in each diagram.

6.

7.

8. Aviation The flight data recorders on commercial airlines are rectangular prisms. The base of a recorder is 15 in. by 8 in. Its height ranges from 15 in. to 22 in. What are the largest and smallest possible volumes for the recorder?

Find the volume and surface area of each figure to the nearest tenth.

9.
4 ft

10.
9 in.
8 in.
8 in.

11.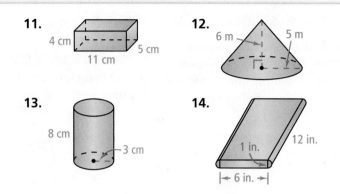
4 cm
5 cm
11 cm

12.
6 m   5 m

13.
8 cm
3 cm

14.
1 in.   12 in.
|← 6 in. →|

15. List these solids in order from the one with least volume to the one with the greatest volume.
A. a cube with edge 5 cm
B. a cylinder with radius 4 cm and height 4 cm
C. a square pyramid with base edges 6 cm and height 6 cm
D. a cone with radius 4 cm and height 9 cm
E. a rectangular prism with a 5 cm-by-5 cm base and height 6 cm

16. Painting The floor of a bedroom is 12 ft by 15 ft and the walls are 7 ft high. One gallon of paint covers about 450 ft^2. How many gallons of paint do you need to paint the walls of the bedroom?

Do you UNDERSTAND?

17. Reasoning What solid has a cross section that could either be a circle or a rectangle?

18. Visualization The triangle is revolved completely about the y-axis.
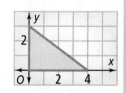
a. Describe the solid of revolution that is formed.
b. Find its lateral area and volume in terms of π.

19. Open-Ended Draw two different solids that have volume 100 in.^{3}. Label the dimensions of each solid.

TIPS FOR SUCCESS

Some questions ask you to use a formula to estimate volume. Read the sample question at the right. Then follow the tips to answer it.

The tank below is filled with gasoline. Which is closest to the volume of the tank in cubic feet?

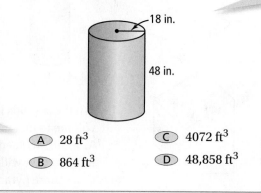

- A 28 ft³
- B 864 ft³
- C 4072 ft³
- D 48,858 ft³

TIP 2
Identify the given solid and what you want to find so you can select the correct formula.

TIP 1
Check whether the units in the problem match the units in the answer choices. You may need to convert measurements.

Think It Through
Convert the radius and the height given in the figure to feet: $r = 1.5$ ft and $h = 4$ ft. Substitute the values for r and h into the formula for the volume of a cylinder and simplify:
$$V = \pi r^2 h = \pi(1.5)^2(4) \approx 28.$$
The correct answer is A.

🔊 Vocabulary Builder

As you solve problems, you must understand the meanings of mathematical terms. Choose the correct term to complete each sentence.

A. The bases of a cylinder are (*circles, polygons*).

B. An arc of a circle that is larger than a semicircle is a (*major arc, minor arc*).

C. Each polygon of a polyhedron is called a(n) (*edge, face*).

D. A (*hemisphere, great circle*) is the intersection of a plane and a sphere through the center of the sphere.

E. The length of the altitude of a pyramid is the (*height, slant height*) of the pyramid.

F. The area of a net of a polyhedron is equal to the (*lateral area, surface area*) of the polyhedron.

G. In a parallelogram, one diagonal is always the (*perpendicular bisector, segment bisector*) of the other diagonal.

Multiple Choice

Read each question. Then write the letter of the correct answer on your paper.

1. A pyramid has a volume of 108 m³. A similar pyramid has base edges and a height that are $\frac{1}{3}$ those of the original pyramid. What is the volume of the second pyramid?
- A 3 m³
- B 4 m³
- C 12 m³
- D 36 m³

2. What is the volume of the triangular prism at the right?
- F 520 in.³
- G 560 in.³
- H 600 in.³
- I 1120 in.³

3. Which quadrilateral CANNOT have one diagonal that bisects the other?
- A square
- B trapezoid
- C kite
- D parallelogram

4. The diameter of the circle is 12 units. What is the minor arc length from point C to point D?

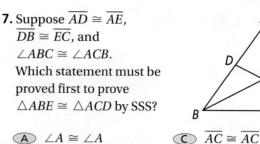

160°

- (F) $\frac{16}{3}\pi$ units
- (G) 8π units
- (H) 16π units
- (I) $\frac{160}{3}\pi$ units

5. Which is a net for a pentagonal pyramid?

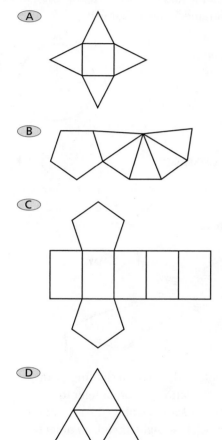

- (A)
- (B)
- (C)
- (D)

6. A frozen dinner is divided into three sections on a circular plate with a 12-in. diameter. What is the approximate arc length of the section containing green beans?

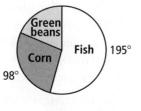

Green beans

Corn 98°

Fish 195°

- (F) 7 in.
- (G) 10 in.
- (H) 16 in.
- (I) 20 in.

7. Suppose $\overline{AD} \cong \overline{AE}$, $\overline{DB} \cong \overline{EC}$, and $\angle ABC \cong \angle ACB$. Which statement must be proved first to prove $\triangle ABE \cong \triangle ACD$ by SSS?

- (A) $\angle A \cong \angle A$
- (B) $\overline{AB} \cong \overline{BC}$
- (C) $\overline{AC} \cong \overline{AC}$
- (D) $\overline{DC} \cong \overline{EB}$

8. The formula for the volume of a cone is $V = \frac{1}{3}\pi r^2 h$. Which statement is true?

- (F) The volume depends on the mean of π, the radius, and the height.
- (G) The volume depends only on the square of the radius.
- (H) The radius depends on the product of the height and π.
- (I) The volume depends on both the height and the radius.

9. A wooden pole was broken during a windstorm. Before the pole broke, the height of the pole above the ground was 18 ft. When it broke, the top of the pole hit the ground 12 ft from the base.

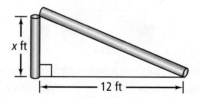

x ft

12 ft

How tall is the part of the pole that remained standing?

- (A) 5 ft
- (B) 10 ft
- (C) 13 ft
- (D) 15 ft

10. $\triangle ABC$ is a right isosceles triangle. Which statement CANNOT be true?

- (F) The length of the hypotenuse is $\sqrt{2} \cdot$ the length of a leg.
- (G) $AB = BC$
- (H) The hypotenuse is the shortest side of the triangle.
- (I) $m\angle B = 90$

11. The net of a cereal box is shown at the right. What is the surface area in square inches?

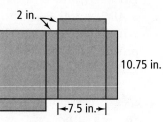

2 in.

10.75 in.

|←7.5 in.→|

12. A polyhedron has 10 vertices and 7 faces. How many edges does the polyhedron have?

13. The radius of a sphere with a volume of 2 m³ is quadrupled. What is the volume of the new sphere in cubic meters?

14. Molly stands at point *A* directly below a kite that Kat is flying from point *B*. Kat has let out 130 ft of the string. Molly stands 40 ft from Kat. To the nearest foot, how many feet from the ground is the kite?

130 ft

B
40 ft
A

15. The figure below shows a cylindrical tin of bath salts. To the nearest square centimeter, what is the surface area of the tin?

14 cm

4 cm

So Soft Bath Salts

16. A regular octagon has a side length of 8 m and an approximate area of 309 m². The side length of another regular octagon is 4 m. To the nearest square meter, what is the approximate area of the smaller octagon?

17. The lengths of the bases of an isosceles trapezoid are shown below. If the perimeter of this trapezoid is 44 units, what is its area in square units?

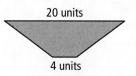

20 units

4 units

18. What is the surface area, in square centimeters, of a rectangular prism that is 9 cm by 8 cm by 10 cm?

Short Response

19. A great circle of a sphere has a circumference of 3π cm. What is the volume of the sphere in terms of π? Show your work.

20. Anna has cereal in a cylindrical container that has a diameter of 6 in. and a height of 12 in. She wants to store the cereal in a rectangular prism container. What is one possible set of dimensions for a rectangular prism container that would be close to the volume of the cylindrical container? Show your work.

21. What is the length of $\overset{\frown}{AB}$ in terms of π? Show your work.

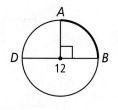

A

D
12
B

Extended Response

22. A log is cut lengthwise through its center. To the nearest square inch, what is the surface area of one of the halves? Explain your reasoning.

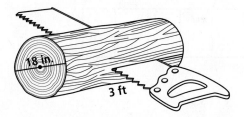

18 in.

3 ft

23. When airplane pilots make a visual sighting of an object outside the airplane, they often refer to the dial of a clock to help locate the object. For example, an object at 12 o'clock is straight ahead, an object at 3 o'clock is 90° to the right, and so on.

Suppose that two pilots flying two airplanes in the same direction spot the same object. Pilot A reports the object at 12 o'clock, and Pilot B reports the object at 2 o'clock. At the same time, Pilot A reports seeing the other airplane at 9 o'clock.

Draw a diagram showing the possible locations of the two planes and the object. If the planes are 800 ft apart, how far is Pilot A from the object? Show your work.

Get Ready!

CHAPTER 12

Skills Handbook, p. 893

◆ Solving Equations

Algebra Solve for x.

1. $\frac{1}{2}(x + 42) = 62$ **2.** $(5 + 3)8 = (4 + x)6$ **3.** $(9 + x)2 = (12 + 4)3$

Lesson 1-7

◆ Distance Formula

Find the distance between each pair of points.

4. $(13, 7), (6, 31)$ **5.** $(-4, 2,), (2, -4)$ **6.** $(-3, -1), (0, 3)$ **7.** $(2\sqrt{3}, 5), (-\sqrt{3}, 2)$

Lesson 4-5

◆ Isosceles and Equilateral Triangles

Algebra Find the value of x.

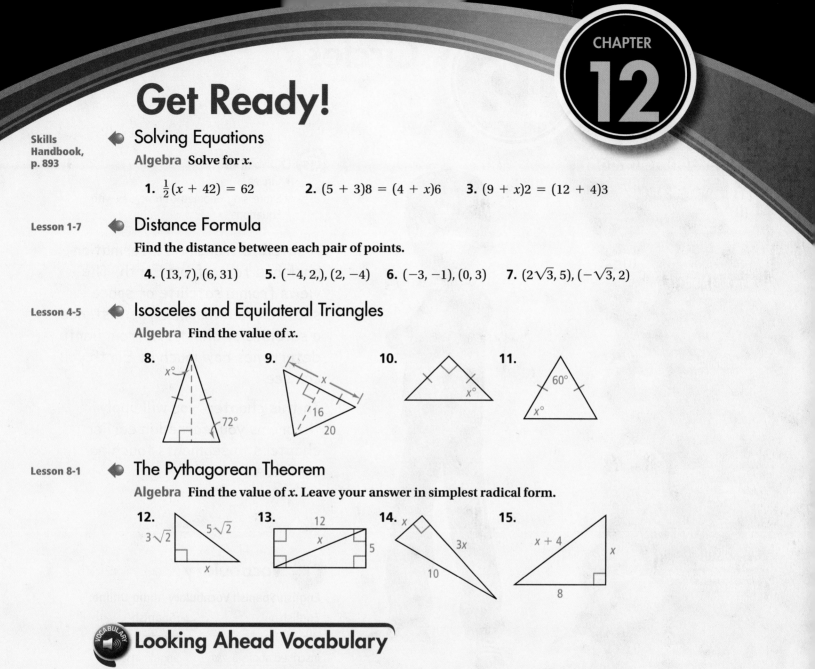

8. $x°$, $72°$

9. x, 16, 20

10. $x°$

11. $60°$, $x°$

Lesson 8-1

◆ The Pythagorean Theorem

Algebra Find the value of x. Leave your answer in simplest radical form.

12. $3\sqrt{2}$, $5\sqrt{2}$, x

13. 12, x, 5

14. x, $3x$, 10

15. $x + 4$, x, 8

Looking Ahead Vocabulary

16. When you are in a conversation and you go off on a *tangent*, you are leading the conversation away from the main topic. What do you think a line that is *tangent* to a circle might look like?

17. You learned how to *inscribe* a triangle in a circle in Chapter 5. What do you think an *inscribed* angle is?

18. A defensive player *intercepts* a pass when he catches the football before it reaches the intended receiver. On a circle, what might an *intercepted* arc of an angle be?

Circles

© **DOMAINS**

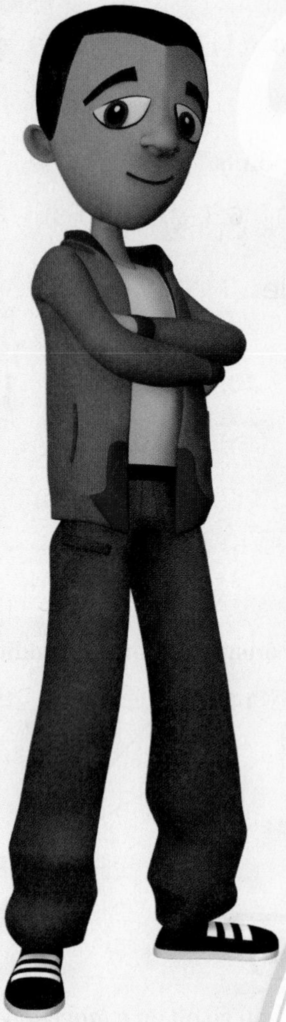

- Circles
- Expressing Geometric Properties with Equations

A satellite transmits information back and forth from Earth. The views from a satellite or space shuttle are awesome. The distance a satellite or shuttle is from Earth determines how much of Earth you can see.

In this chapter, you will apply theorems you learned in earlier chapters to segments touching circles.

 **Vocabulary**

English/Spanish Vocabulary Audio Online:

English	Spanish
chord, *p. 771*	cuerda
inscribed angle, *p. 780*	ángulo inscrito
intercepted arc, *p. 780*	arco interceptor
locus, *p. 804*	lugar geométrico
point of tangency, *p. 762*	punto de tangencia
secant, *p. 791*	secante
standard form of an equation of a circle, *p. 799*	forma normal de una ecuación lineal
tangent to a circle, *p. 762*	tagente de un círculo

My Math Video

`00:04:04`

VIDEO ▶

BIG ideas

1 Reasoning and Proof

Essential Question How can you prove relationships between angles and arcs in a circle?

2 Measurement

Essential Question When lines intersect a circle or within a circle, how do you find the measures of resulting angles, arcs, and segments?

3 Coordinate Geometry

Essential Question How do you find the equation of a circle in the coordinate plane?

Chapter Preview

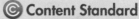
Content Standard

G.C.2 Identify and describe relationships among inscribed angles, radii, and chords . . . the radius of a circle is perpendicular to the tangent where the radius intersects the circle.

12-1 Tangent Lines

Objective To use properties of a tangent to a circle

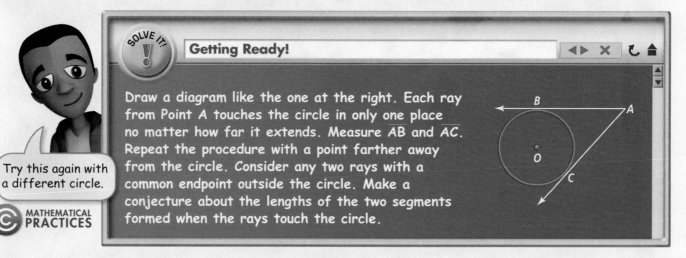

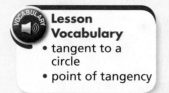

SOLVE IT!

Getting Ready!

Draw a diagram like the one at the right. Each ray from Point A touches the circle in only one place no matter how far it extends. Measure $\overline{AB}$ and $\overline{AC}$. Repeat the procedure with a point farther away from the circle. Consider any two rays with a common endpoint outside the circle. Make a conjecture about the lengths of the two segments formed when the rays touch the circle.

Try this again with a different circle.

MATHEMATICAL PRACTICES

In the Solve It, you drew lines that touch a circle at only one point. These lines are called tangents. This use of the word *tangent* is related to, but different from, the tangent ratio in right triangles that you studied in Chapter 8.

Lesson Vocabulary
• tangent to a circle
• point of tangency

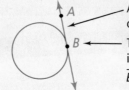

A **tangent to a circle** is a line in the plane of the circle that intersects the circle in exactly one point.

The point where a circle and a tangent intersect is the **point of tangency**.

$\overrightarrow{BA}$ is a tangent ray and $\overline{BA}$ is a tangent segment.

Essential Understanding A radius of a circle and the tangent that intersects the endpoint of the radius on the circle have a special relationship.

take note

Theorem 12-1

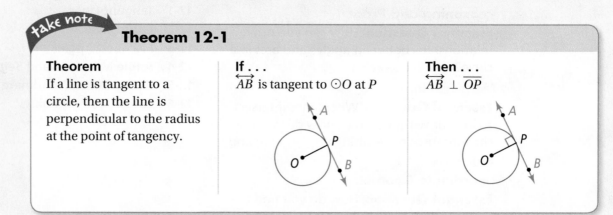

Theorem	**If . . .**	**Then . . .**
If a line is tangent to a circle, then the line is perpendicular to the radius at the point of tangency.	$\overleftrightarrow{AB}$ is tangent to $\odot O$ at P	$\overleftrightarrow{AB} \perp \overline{OP}$

Proof **Indirect Proof of Theorem 12-1**

Given: n is tangent to $\odot O$ at P.

Prove: $n \perp \overline{OP}$

Step 1 Assume that n is not perpendicular to $\overline{OP}$.

Step 2 If line n is not perpendicular to $\overline{OP}$, then, for some other point L on n, $\overline{OL}$ must be perpendicular to n. Also there is a point K on n such that $\overline{LK} \cong \overline{LP}$. $\angle OLK \cong \angle OLP$ because perpendicular lines form congruent adjacent angles. $\overline{OL} \cong \overline{OL}$. So, $\triangle OLK \cong \triangle OLP$ by SAS.

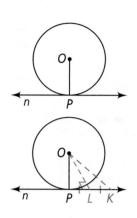

Since corresponding parts of congruent triangles are congruent, $\overline{OK} \cong \overline{OP}$. So K and P are both on $\odot O$ by the definition of a circle. For two points on n to also be on $\odot O$ contradicts the given fact that n is tangent to $\odot O$ at P. So the assumption that n is not perpendicular to $\overline{OP}$ must be false.

Step 3 Therefore, $n \perp \overline{OP}$ must be true.

© **Problem 1** **Finding Angle Measures**

Multiple Choice $\overline{ML}$ and $\overline{MN}$ are tangent to $\odot O$. What is the value of x?

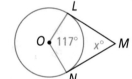

(A) 58

(C) 90

(B) 63

(D) 117

Think

What kind of angle is formed by a radius and a tangent?
The angle formed is a right angle, so the measure is 90.

Since $\overline{ML}$ and $\overline{MN}$ are tangent to $\odot O$, $\angle L$ and $\angle N$ are right angles. $LMNO$ is a quadrilateral. So the sum of the angle measures is 360.

$$m\angle L + m\angle M + m\angle N + m\angle O = 360$$
$$90 + m\angle M + 90 + 117 = 360 \quad \text{Substitute.}$$
$$297 + m\angle M = 360 \quad \text{Simplify.}$$
$$m\angle M = 63 \quad \text{Solve.}$$

The correct answer is B.

© ✓ **Got It?** **1. a.** $\overline{ED}$ is tangent to $\odot O$. What is the value of x?
   b. Reasoning Consider a quadrilateral like the one in Problem 1. Write a formula you could use to find the measure of any angle x formed by two tangents when you know the measure of the central angle c whose radii intersect the tangents.

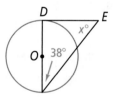

Problem 2 **Finding Distance** (STEM)

Earth Science The CN Tower in Toronto, Canada, has an observation deck 447 m above ground level. About how far is it from the observation deck to the horizon? Earth's radius is about 6400 km.

Step 1 Make a sketch. The length 447 m is about 0.45 km.

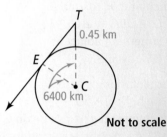

Not to scale

Plan

How does knowing Earth's radius help?
The radius forms a right angle with a tangent line from the observation deck to the horizon. So, you can use two radii, the tower's height, and the tangent to form a right triangle.

Step 2 Use the Pythagorean Theorem.

$$CT^2 = TE^2 + CE^2$$

$(6400 + 0.45)^2 = TE^2 + 6400^2$ Substitute.

$(6400.45)^2 = TE^2 + 6400^2$ Simplify.

$40{,}965{,}760.2025 = TE^2 + 40{,}960{,}000$ Use a calculator.

$5760.2025 = TE^2$ Subtract 40,960,000 from each side.

$76 \approx TE$ Take the positive square root of each side.

The distance from the CN Tower to the horizon is about 76 km.

Got It? **2.** What is the distance to the horizon that a person can see on a clear day from an airplane 2 mi above Earth? Earth's radius is about 4000 mi.

Theorem 12-2 is the converse of Theorem 12-1. You can use it to prove that a line or segment is tangent to a circle. You can also use it to construct a tangent to a circle.

take note

Theorem 12-2

Theorem	**If . . .**	**Then . . .**
If a line in the plane of a circle is perpendicular to a radius at its endpoint on the circle, then the line is tangent to the circle.	$\overleftrightarrow{AB} \perp \overline{OP}$ at P	$\overleftrightarrow{AB}$ is tangent to $\odot O$

You will prove Theorem 12-2 in Exercise 30.

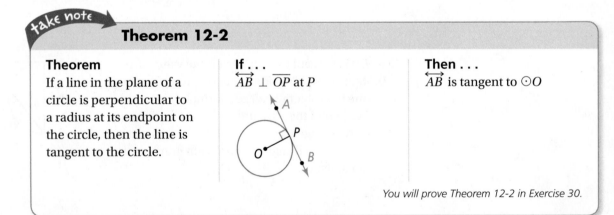

Problem 3 Finding a Radius

Think

Why does the value _x_ appear on each side of the equation?
The length of $\overline{AC}$, the hypotenuse, is the radius plus 8, which is on the left side of the equation. On the right side of the equation, the radius is one side of the triangle.

What is the radius of $\odot C$?

$$AC^2 = AB^2 + BC^2 \quad \text{Pythagorean Theorem}$$
$$(x + 8)^2 = 12^2 + x^2 \quad \text{Substitute.}$$
$$x^2 + 16x + 64 = 144 + x^2 \quad \text{Simplify.}$$
$$16x = 80 \quad \text{Subtract } x^2 \text{ and 64 from each side.}$$
$$x = 5 \quad \text{Divide each side by 16.}$$

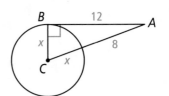

The radius is 5.

Got It? 3. What is the radius of $\odot O$?

Problem 4 Identifying a Tangent

Think

What information does the diagram give you?
• _LMN_ is a triangle.
• $NM = 25$, $LM = 24$, $NL = 7$
• $\overline{NL}$ is a radius.

Is $\overline{ML}$ tangent to $\odot N$ at _L_? Explain.

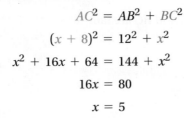

Know
The lengths of the sides of $\triangle LMN$

Need
To determine whether $\overline{ML}$ is tangent to $\odot O$

Plan
$\overline{ML}$ is a tangent if $\overline{ML} \perp \overline{NL}$. Use the Converse of the Pythagorean Theorem to determine whether $\triangle LMN$ is a right triangle.

$$NL^2 + ML^2 \stackrel{?}{=} NM^2$$
$$7^2 + 24^2 \stackrel{?}{=} 25^2 \quad \text{Substitute.}$$
$$625 = 625 \quad \text{Simplify.}$$

By the Converse of the Pythagorean Theorem, $\triangle LMN$ is a right triangle with $\overline{ML} \perp \overline{NL}$. So $\overline{ML}$ is tangent to $\odot N$ at _L_ because it is perpendicular to the radius at the point of tangency (Theorem 12-2).

Got It? 4. Use the diagram in Problem 4. If $NL = 4$, $ML = 7$, and $NM = 8$, is $\overline{ML}$ tangent to $\odot N$ at _L_? Explain.

In the Solve It, you made a conjecture about the lengths of two tangents from a common endpoint outside a circle. Your conjecture may be confirmed by the following theorem.

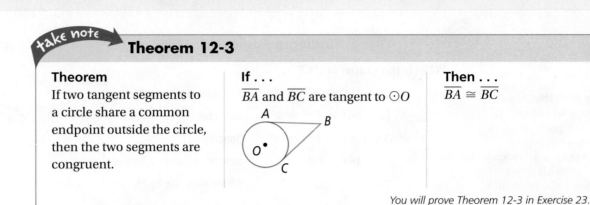

Theorem 12-3

Theorem	**If . . .**	**Then . . .**
If two tangent segments to a circle share a common endpoint outside the circle, then the two segments are congruent.	$\overline{BA}$ and $\overline{BC}$ are tangent to $\odot O$	$\overline{BA} \cong \overline{BC}$

You will prove Theorem 12-3 in Exercise 23.

In the figure at the right, the sides of the triangle are tangent to the circle. The circle is *inscribed in* the triangle. The triangle is *circumscribed about* the circle.

Problem 5 Circles Inscribed in Polygons

$\odot O$ is inscribed in $\triangle ABC$. What is the perimeter of $\triangle ABC$?

$AD = AF = 10$ cm 　 Two segments tangent to a circle from
$BD = BE = 15$ cm 　 a point outside the circle are congruent,
$CF = CE = 8$ cm 　 so they have the same length.

$$p = AB + BC + CA \qquad \text{Definition of perimeter } p$$
$$= AD + DB + BE + EC + CF + FA \qquad \text{Segment Addition Postulate}$$
$$= 10 + 15 + 15 + 8 + 8 + 10 \qquad \text{Substitute.}$$
$$= 66$$

The perimeter is 66 cm.

Got It? **5.** $\odot O$ is inscribed in $\triangle PQR$, which has a perimeter of 88 cm. What is the length of $\overline{QY}$?

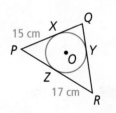

Plan

How can you find the length of $\overline{BC}$?
Find the segments congruent to $\overline{BE}$ and $\overline{EC}$. Then use segment addition.

Lesson Check

Do you know HOW?

1. If $m\angle A = 58$, what is $m\angle ACB$?

2. If $BC = 8$ and $DC = 4$, what is the radius?

3. If $AC = 12$ and $BC = 9$, what is the radius?

Do you UNDERSTAND?

4. Vocabulary How are the phrases *tangent ratio* and *tangent of a circle* used differently?

5. Error Analysis A classmate insists that $\overline{DF}$ is a tangent to $\odot E$. Explain how to show that your classmate is wrong.

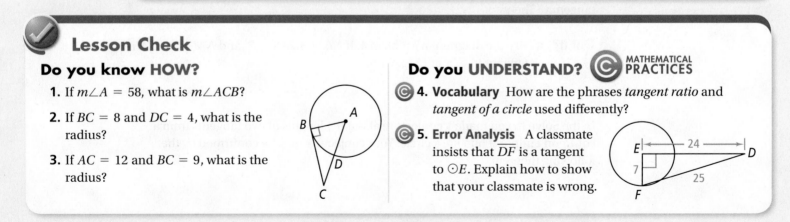

Ⓐ **Practice**

Algebra Lines that appear to be tangent are tangent. *O* is the center of each circle. What is the value of *x*?

◀ **See Problem 1.**

6.

7.

8.

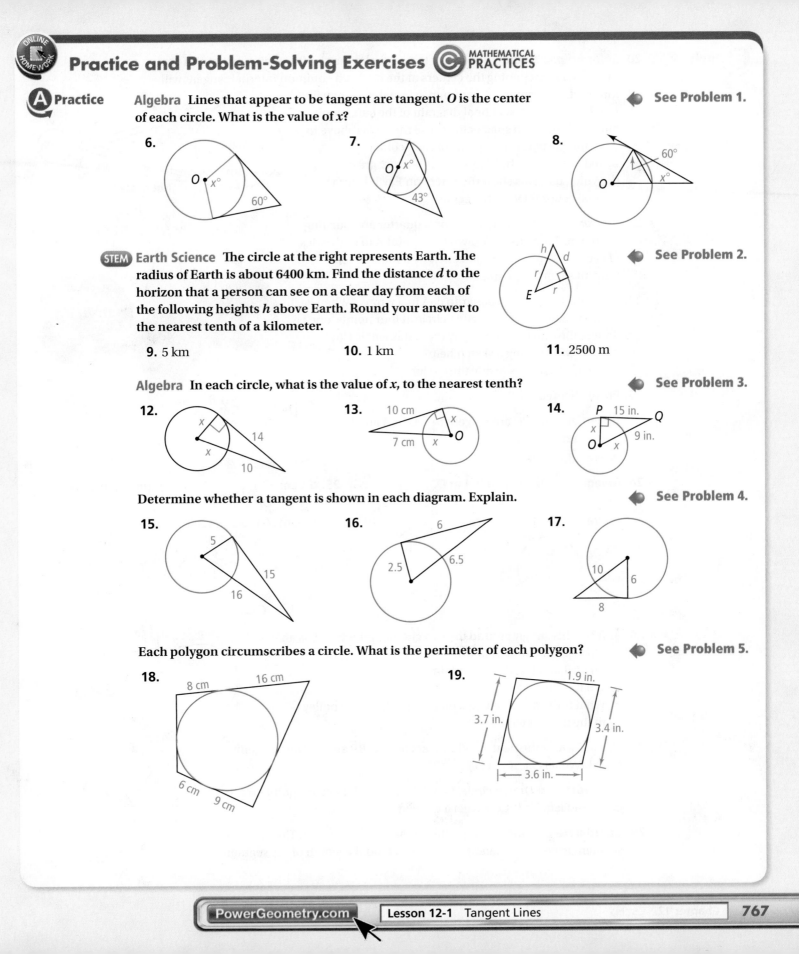

STEM **Earth Science** The circle at the right represents Earth. The radius of Earth is about 6400 km. Find the distance *d* to the horizon that a person can see on a clear day from each of the following heights *h* above Earth. Round your answer to the nearest tenth of a kilometer.

◀ **See Problem 2.**

9. 5 km

10. 1 km

11. 2500 m

Algebra In each circle, what is the value of *x*, to the nearest tenth?

◀ **See Problem 3.**

12.

13.

14.

Determine whether a tangent is shown in each diagram. Explain.

◀ **See Problem 4.**

15.

16.

17.

Each polygon circumscribes a circle. What is the perimeter of each polygon?

◀ **See Problem 5.**

18.

19.

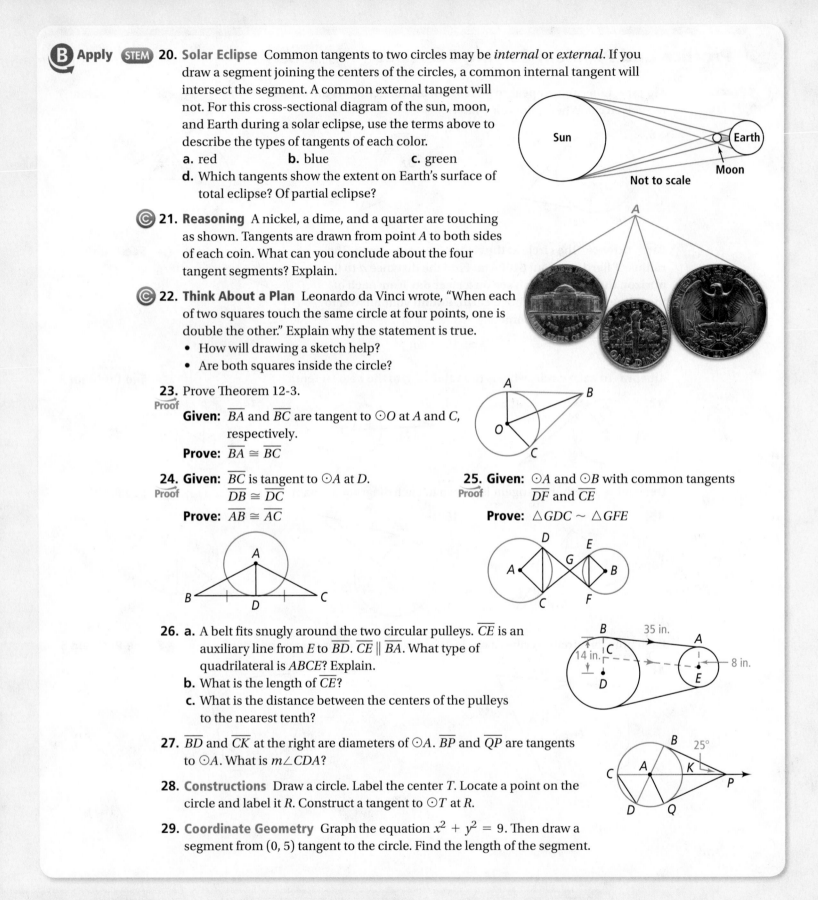

B Apply (STEM) **20. Solar Eclipse** Common tangents to two circles may be *internal* or *external*. If you draw a segment joining the centers of the circles, a common internal tangent will intersect the segment. A common external tangent will not. For this cross-sectional diagram of the sun, moon, and Earth during a solar eclipse, use the terms above to describe the types of tangents of each color.

  a. red   **b.** blue   **c.** green

  d. Which tangents show the extent on Earth's surface of total eclipse? Of partial eclipse?

21. Reasoning A nickel, a dime, and a quarter are touching as shown. Tangents are drawn from point *A* to both sides of each coin. What can you conclude about the four tangent segments? Explain.

22. Think About a Plan Leonardo da Vinci wrote, "When each of two squares touch the same circle at four points, one is double the other." Explain why the statement is true.

  • How will drawing a sketch help?

  • Are both squares inside the circle?

23. Prove Theorem 12-3.

Proof

  Given: $\overline{BA}$ and $\overline{BC}$ are tangent to $\odot O$ at *A* and *C*, respectively.

  Prove: $\overline{BA} \cong \overline{BC}$

24. Given: $\overline{BC}$ is tangent to $\odot A$ at *D*.

Proof $\overline{DB} \cong \overline{DC}$

  Prove: $\overline{AB} \cong \overline{AC}$

25. Given: $\odot A$ and $\odot B$ with common tangents

Proof $\overline{DF}$ and $\overline{CE}$

  Prove: $\triangle GDC \sim \triangle GFE$

26. a. A belt fits snugly around the two circular pulleys. $\overline{CE}$ is an auxiliary line from *E* to $\overline{BD}$. $\overline{CE} \parallel \overline{BA}$. What type of quadrilateral is *ABCE*? Explain.

  b. What is the length of $\overline{CE}$?

  c. What is the distance between the centers of the pulleys to the nearest tenth?

27. $\overline{BD}$ and $\overline{CK}$ at the right are diameters of $\odot A$. $\overline{BP}$ and $\overline{QP}$ are tangents to $\odot A$. What is $m\angle CDA$?

28. Constructions Draw a circle. Label the center *T*. Locate a point on the circle and label it *R*. Construct a tangent to $\odot T$ at *R*.

29. Coordinate Geometry Graph the equation $x^2 + y^2 = 9$. Then draw a segment from (0, 5) tangent to the circle. Find the length of the segment.

Challenge **30.** Write an indirect proof of Theorem 12-2.
Proof **Given:** $\overline{AB} \perp \overline{OP}$ at P.
Prove: $\overline{AB}$ is tangent to $\odot O$.

31. Two circles that have one point in common are *tangent circles*. Given any triangle, explain how to draw three circles that are centered at each vertex of the triangle and are tangent to each other.

Standardized Test Prep

GRIDDED RESPONSE

SAT/ACT

Lines in $\odot O$ that appear to be tangent are tangent. What is the value of x?

32.

33.

34. The diagram at the right shows the dimensions of a silo. What is the volume of the silo to the nearest cubic foot?

35. The perimeter of an equilateral triangle is 90 in. What is its area to the nearest square inch?

6 ft

15 ft

4 ft

Mixed Review

Two cubes have heights 6 in. and 8 in. Find each ratio.   ◀ **See Lesson 11-7.**

36. scale factor **37.** ratio of surface areas **38.** ratio of volumes

Algebra Find the value of x. Round answers to the nearest tenth.   ◀ **See Lesson 8-3.**

39.

5

$x°$

9

40.

15

$x°$

8

41.

3

$x°$

7.5

Get Ready!   **To prepare for Lesson 12-2, do Exercises 42–44.**

Find the value of each variable. Leave your answer in simplest radical form.   ◀ **See Lesson 8-2.**

42.

11

45°

a

43.

c

45°

$5\sqrt{2}$

44.

14

60°

b

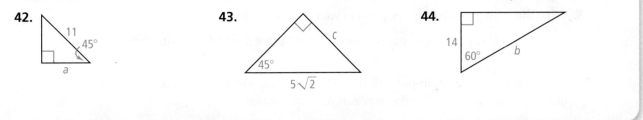

Paper Folding With Circles

© **Content Standard**
Prepares for G.C.2 Identify and describe relationships among inscribed angles, radii, and chords . . . the radius of a circle is perpendicular to the tangent where the radius intersects the circle.

A *chord* is a segment with endpoints on a circle. In these activities, you will explore some of the properties of chords.

Activity 1

Step 1 Use a compass. Draw a circle on tracing paper.

Step 2 Use a straightedge. Draw two radii.

Step 3 Set your compass to a distance shorter than the radii. Place its point at the center of the circle. Mark two congruent segments, one on each radius.

Step 4 Fold a line perpendicular to each radius at the point marked on the radius.

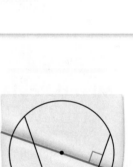

1. How do you measure the distance between a point and a line?

2. Each perpendicular contains a chord. Compare the lengths of the chords.

© **3. Make a Conjecture** What is the relationship among the lengths of the chords that are equidistant from the center of a circle?

Activity 2

Step 1 Use a compass. Draw a circle on tracing paper.

Step 2 Use a straightedge. Draw two chords that are not diameters.

Step 3 Fold the perpendicular bisector for each chord.

4. Where do the perpendicular bisectors appear to intersect?

5. Draw a third chord and fold its perpendicular bisector. Where does it appear to intersect the other two?

© **6. Make a Conjecture** What is true about the perpendicular bisector of a chord?

Exercises

7. Write a proof of your conjecture from Exercise 3 or give a counterexample.

8. What theorem provides a quick proof of your conjecture from Exercise 6?

© **9. Make a Conjecture** Suppose two chords have different lengths. How do their distances from the center of the circle compare?

10. You are building a circular patio table. You have to drill a hole through the center of the tabletop for an umbrella. How can you find the center?

Chords and Arcs

Content Standard
G.C.2 Identify and describe relationships among inscribed angles, radii, and chords.

Objectives To use congruent chords, arcs, and central angles
To use perpendicular bisectors to chords

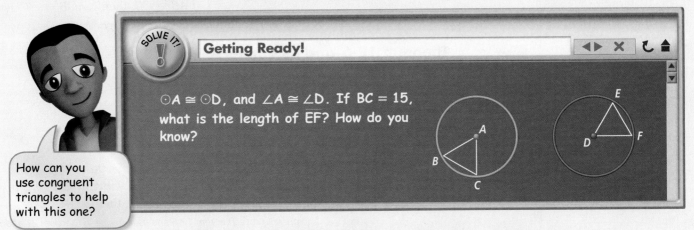

SOLVE IT!

Getting Ready!

⊙A ≅ ⊙D, and ∠A ≅ ∠D. If BC = 15, what is the length of EF? How do you know?

How can you use congruent triangles to help with this one?

MATHEMATICAL PRACTICES

In the Solve It, you found the length of a **chord,** which is a segment whose endpoints are on a circle. The diagram shows the chord $\overline{PQ}$ and its related arc, $\overparen{PQ}$.

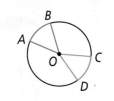

Essential Understanding You can use information about congruent parts of a circle (or congruent circles) to find information about other parts of the circle (or circles).

The following theorems and their converses confirm that if you know that chords, arcs, or central angles in a circle are congruent, then you know the other two parts are congruent.

Dynamic Activity
Chords and Arcs

Lesson Vocabulary
• chord

take note

Theorem 12-4 and Its Converse

Theorem
Within a circle or in congruent circles, congruent central angles have congruent arcs.

Converse
Within a circle or in congruent circles, congruent arcs have congruent central angles.

If ∠AOB ≅ ∠COD, then $\overparen{AB} ≅ \overparen{CD}$.
If $\overparen{AB} ≅ \overparen{CD}$, then ∠AOB ≅ ∠COD.

You will prove Theorem 12-4 and its converse in Exercises 19 and 35.

Theorem 12-5 and Its Converse

Theorem

Within a circle or in congruent circles, congruent central angles have congruent chords.

Converse

Within a circle or in congruent circles, congruent chords have congruent central angles.

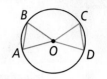

If $\angle AOB \cong \angle COD$, then $\overline{AB} \cong \overline{CD}$.

If $\overline{AB} \cong \overline{CD}$, then $\angle AOB \cong \angle COD$.

You will prove Theorem 12-5 and its converse in Exercises 20 and 36.

Theorem 12-6 and Its Converse

Theorem

Within a circle or in congruent circles, congruent chords have congruent arcs.

Converse

Within a circle or in congruent circles, congruent arcs have congruent chords.

If $\overline{AB} \cong \overline{CD}$, then $\overset{\frown}{AB} \cong \overset{\frown}{CD}$.

If $\overset{\frown}{AB} \cong \overset{\frown}{CD}$, then $\overline{AB} \cong \overline{CD}$.

You will prove Theorem 12-6 and its converse in Exercises 21 and 37.

Problem 1   **Using Congruent Chords**

Think

Why is it important that the circles are congruent?
Two circles may have central angles with congruent chords, but the central angles will not be congruent unless the circles are congruent.

In the diagram, $\odot O \cong \odot P$. Given that $\overline{BC} \cong \overline{DF}$, what can you conclude?

$\angle O \cong \angle P$ because, within congruent circles, congruent chords have congruent central angles (conv. of Thm. 12-5). $\overset{\frown}{BC} \cong \overset{\frown}{DF}$ because, within congruent circles, congruent chords have congruent arcs (Thm. 12-6).

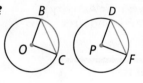

Got It? **1. Reasoning** Use the diagram in Problem 1. Suppose you are given $\odot O \cong \odot P$ and $\angle OBC \cong \angle PDF$. How can you show $\angle O \cong \angle P$? From this, what else can you conclude?

Theorem 12-7 and Its Converse

Theorem

Within a circle or in congruent circles, chords equidistant from the center or centers are congruent.

Converse

Within a circle or in congruent circles, congruent chords are equidistant from the center (or centers).

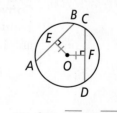

If $OE = OF$, then $\overline{AB} \cong \overline{CD}$.

If $\overline{AB} \cong \overline{CD}$, then $OE = OF$.

You will prove the converse of Theorem 12-7 in Exercise 38.

Proof **Proof of Theorem 12-7**

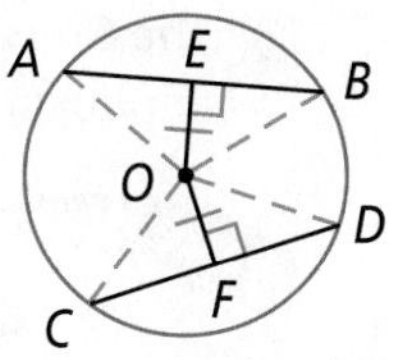

Given: $\odot O$, $\overline{OE} \cong \overline{OF}$, $\overline{OE} \perp \overline{AB}$, $\overline{OF} \perp \overline{CD}$

Prove: $\overline{AB} \cong \overline{CD}$

Statements	Reason
1) $\overline{OA} \cong \overline{OB} \cong \overline{OC} \cong \overline{OD}$	1) Radii of a circle are congruent.
2) $\overline{OE} \cong \overline{OF}$, $\overline{OE} \perp \overline{AB}$, $\overline{OF} \perp \overline{CD}$	2) Given
3) $\angle AEO$ and $\angle CFO$ are right angles.	3) Def. of perpendicular segments
4) $\triangle AEO \cong \triangle CFO$	4) HL Theorem
5) $\angle A \cong \angle C$	5) Corres. parts of $\cong$ ⚠ are $\cong$.
6) $\angle B \cong \angle A$, $\angle C \cong \angle D$	6) Isosceles Triangle Theorem
7) $\angle B \cong \angle D$	7) Transitive Property of Congruence
8) $\angle AOB \cong \angle COD$	8) If two ⚠ of a $\triangle$ are $\cong$ to two ⚠ of another $\triangle$, then the third ⚠ are $\cong$.
9) $\overline{AB} \cong \overline{CD}$	9) $\cong$ central angles have $\cong$ chords.

© **Problem 2** **Finding the Length of a Chord** **GRIDDED RESPONSE**

What is the length of $\overline{RS}$ in $\odot O$?

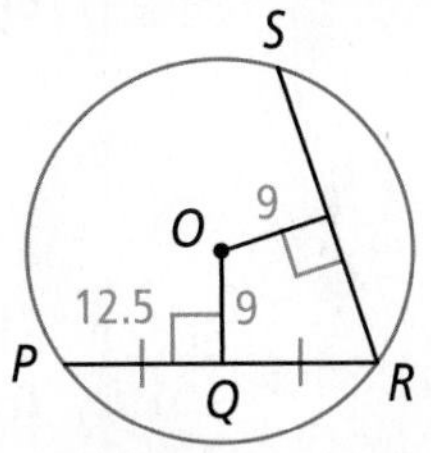

Know

The diagram indicates that $PQ = QR = 12.5$ and $\overline{PR}$ and $\overline{RS}$ are both 9 units from the center.

Need

The length of chord $\overline{RS}$

Plan

$\overline{PR} \cong \overline{RS}$, since they are the same distance from the center of the circle. So finding PR gives the length of $\overline{RS}$.

$PQ = QR = 12.5$	Given in the diagram
$PQ + QR = PR$	Segment Addition Postulate
$12.5 + 12.5 = PR$	Substitute.
$25 = PR$	Add.
$RS = PR$	Chords equidistant from the center of a circle are congruent.
$RS = 25$	Substitute.

✓ **Got It?** **2.** What is the value of x? Justify your answer.

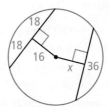

The Converse of the Perpendicular Bisector Theorem from Lesson 5-2 has special applications to a circle and its diameters, chords, and arcs.

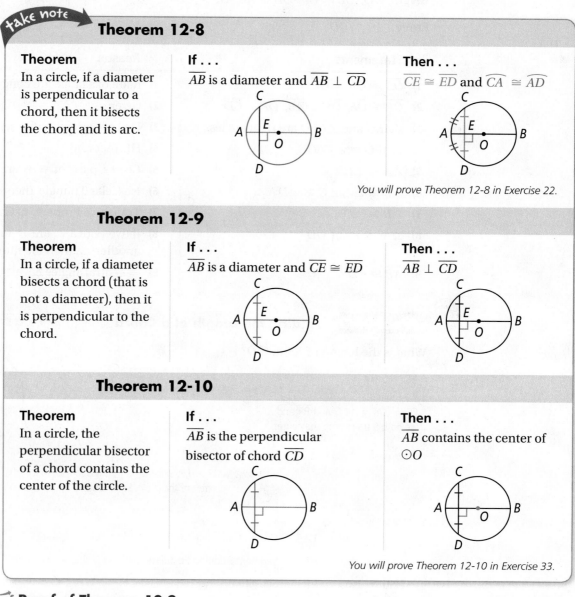

take note

Theorem 12-8

Theorem	**If . . .**	**Then . . .**
In a circle, if a diameter is perpendicular to a chord, then it bisects the chord and its arc.	$\overline{AB}$ is a diameter and $\overline{AB} \perp \overline{CD}$	$\overline{CE} \cong \overline{ED}$ and $\overset{\frown}{CA} \cong \overset{\frown}{AD}$

You will prove Theorem 12-8 in Exercise 22.

Theorem 12-9

Theorem	**If . . .**	**Then . . .**
In a circle, if a diameter bisects a chord (that is not a diameter), then it is perpendicular to the chord.	$\overline{AB}$ is a diameter and $\overline{CE} \cong \overline{ED}$	$\overline{AB} \perp \overline{CD}$

Theorem 12-10

Theorem	**If . . .**	**Then . . .**
In a circle, the perpendicular bisector of a chord contains the center of the circle.	$\overline{AB}$ is the perpendicular bisector of chord $\overline{CD}$	$\overline{AB}$ contains the center of $\odot O$

You will prove Theorem 12-10 in Exercise 33.

Proof **Proof of Theorem 12-9**

Given: $\odot O$ with diameter $\overline{AB}$ bisecting $\overline{CD}$ at E

Prove: $\overline{AB} \perp \overline{CD}$

Proof: $OC = OD$ because the radii of a circle are congruent. $CE = ED$ by the definition of bisect. Thus, O and E are both equidistant from C and D. By the Converse of the Perpendicular Bisector Theorem, both O and E are on the perpendicular bisector of $\overline{CD}$. Two points determine one line or segment, so $\overline{OE}$ is the perpendicular bisector of $\overline{CD}$. Since $\overline{OE}$ is part of $\overline{AB}$, $\overline{AB} \perp \overline{CD}$.

Archaeology An archaeologist found pieces of a jar. She wants to find the radius of the rim of the jar to help guide her as she reassembles the pieces. What is the radius of the rim?

Think

How does the construction help find the center?
The perpendicular bisectors contain diameters of the circle. Two diameters intersect at the circle's center.

Step 1 Trace a piece of the rim. Draw two chords and construct perpendicular bisectors.

Step 2 The center is the intersection of the perpendicular bisectors. Use the center to find the radius.

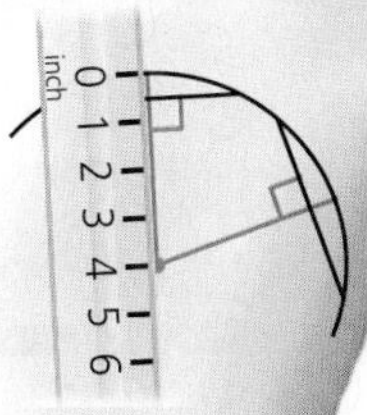

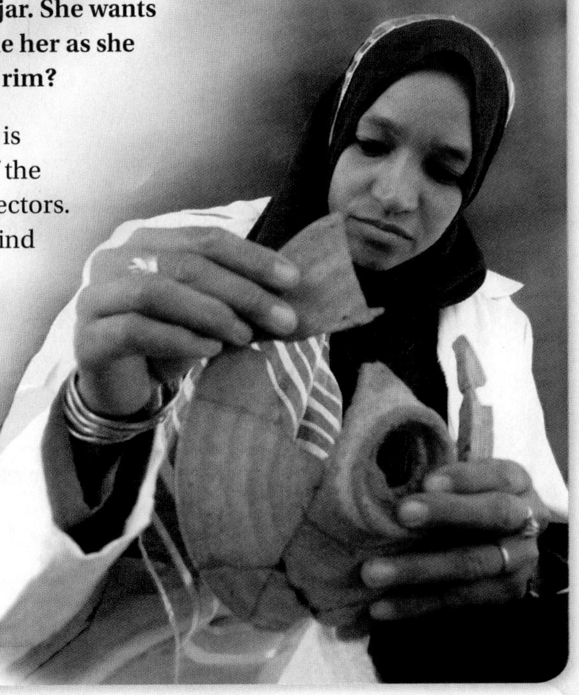

The radius is 4 in.

✓ **Got It?** **3.** Trace a coin. What is its radius?

 Problem 4 **Finding Measures in a Circle**

Plan

Find two sides of a right triangle. The third side is either the answer or leads to an answer.

Algebra What is the value of each variable to the nearest tenth?

Ⓐ

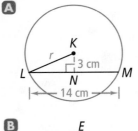

$LN = \frac{1}{2}(14) = 7$ A diameter $\perp$ to a chord bisects the chord.

$r^2 = 3^2 + 7^2$ Use the Pythagorean Theorem.

$r \approx 7.6$ Find the positive square root of each side.

Ⓑ

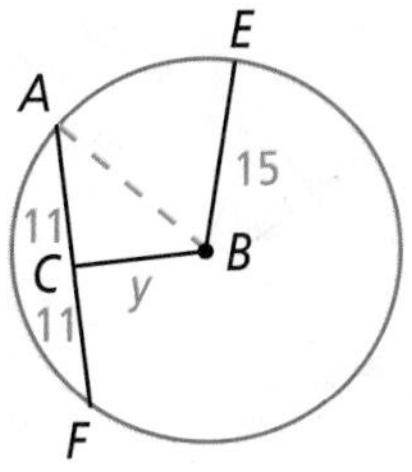

$\overline{BC} \perp \overline{AF}$ A diameter that bisects a chord that is not a diameter is $\perp$ to the chord.

$BA = BE = 15$ Draw an auxiliary $\overline{BA}$. The auxiliary $\overline{BA} \cong \overline{BE}$ because they are radii of the same circle.

$y^2 + 11^2 = 15^2$ Use the Pythagorean Theorem.

$y^2 = 104$ Solve for y^2.

$y \approx 10.2$ Find the positive square root of each side.

© ✓ **Got It?** **4. Reasoning** In part (b), how does the auxiliary $\overline{BA}$ make the problem simpler to solve?

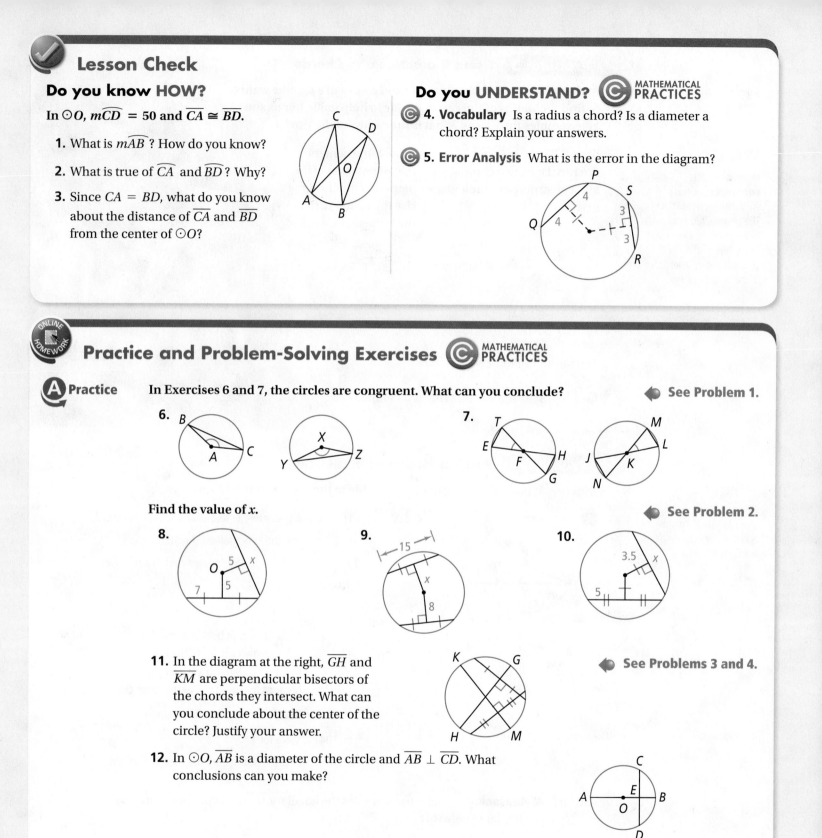

Lesson Check

Do you know HOW?

In $\odot O$, $m\widehat{CD} = 50$ and $\overline{CA} \cong \overline{BD}$.

1. What is $m\widehat{AB}$? How do you know?

2. What is true of $\widehat{CA}$ and $\widehat{BD}$? Why?

3. Since $CA = BD$, what do you know about the distance of $\overline{CA}$ and $\overline{BD}$ from the center of $\odot O$?

Do you UNDERSTAND? MATHEMATICAL PRACTICES

4. **Vocabulary** Is a radius a chord? Is a diameter a chord? Explain your answers.

5. **Error Analysis** What is the error in the diagram?

Practice and Problem-Solving Exercises MATHEMATICAL PRACTICES

A Practice

In Exercises 6 and 7, the circles are congruent. What can you conclude? ◀ **See Problem 1.**

6.

7.

Find the value of *x*. ◀ **See Problem 2.**

8.

9.

10.

11. In the diagram at the right, $\overline{GH}$ and $\overline{KM}$ are perpendicular bisectors of the chords they intersect. What can you conclude about the center of the circle? Justify your answer. ◀ **See Problems 3 and 4.**

12. In $\odot O$, $\overline{AB}$ is a diameter of the circle and $\overline{AB} \perp \overline{CD}$. What conclusions can you make?

Algebra Find the value of *x* to the nearest tenth.

13.

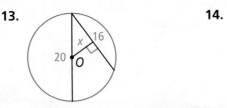

14.

15.

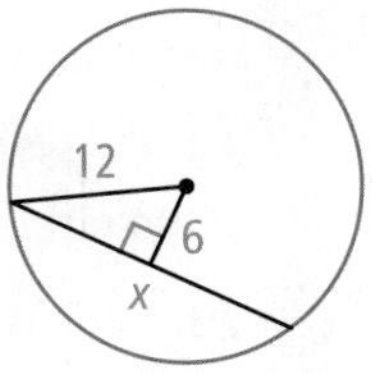

B Apply

16. Geometry in 3 Dimensions In the figure at the right, sphere *O* with radius 13 cm is intersected by a plane 5 cm from center *O*. Find the radius of the cross section ⊙*A*.

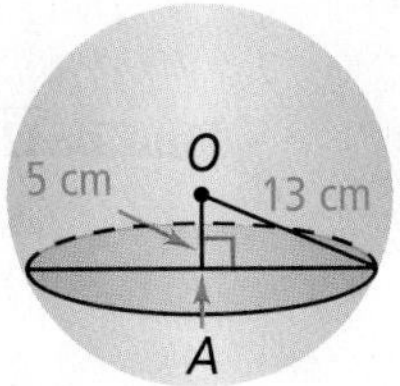

17. Geometry in 3 Dimensions A plane intersects a sphere that has radius 10 in., forming the cross section ⊙*B* with radius 8 in. How far is the plane from the center of the sphere?

C **18. Think About a Plan** Two concentric circles have radii of 4 cm and 8 cm. A segment tangent to the smaller circle is a chord of the larger circle. What is the length of the segment to the nearest tenth?
   - How will you start the diagram?
   - Where is the best place to position the radius of each circle?

19. Prove Theorem 12-4.
Proof **Given:** ⊙*O* with ∠*AOB* ≅ ∠*COD*
Prove: $\overset{\frown}{AB} \cong \overset{\frown}{CD}$

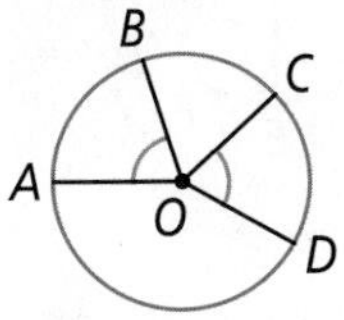

20. Prove Theorem 12-5.
Proof **Given:** ⊙*O* with ∠*AOB* ≅ ∠*COD*
Prove: $\overline{AB} \cong \overline{CD}$

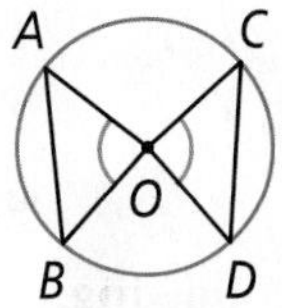

21. Prove Theorem 12-6.
Proof **Given:** ⊙*O* with $\overline{AB} \cong \overline{CD}$
Prove: $\overset{\frown}{AB} \cong \overset{\frown}{CD}$

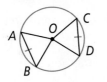

22. Prove Theorem 12-8.
Proof **Given:** ⊙*O* with diameter $\overline{ED} \perp \overline{AB}$ at *C*
Prove: $\overline{AC} \cong \overline{BC}$, $\overset{\frown}{AD} \cong \overset{\frown}{BD}$

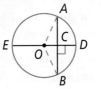

⊙*A* and ⊙*B* are congruent. $\overline{CD}$ is a chord of both circles.

23. If *AB* = 8 in. and *CD* = 6 in., how long is a radius?

24. If *AB* = 24 cm and a radius = 13 cm, how long is $\overline{CD}$?

25. If a radius = 13 ft and *CD* = 24 ft, how long is $\overline{AB}$?

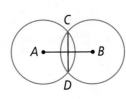

26. Construction Use Theorem 12-5 to construct a regular octagon.

27. In the diagram at the right, the endpoints of the chord are the points where the line $x = 2$ intersects the circle $x^2 + y^2 = 25$. What is the length of the chord? Round your answer to the nearest tenth.

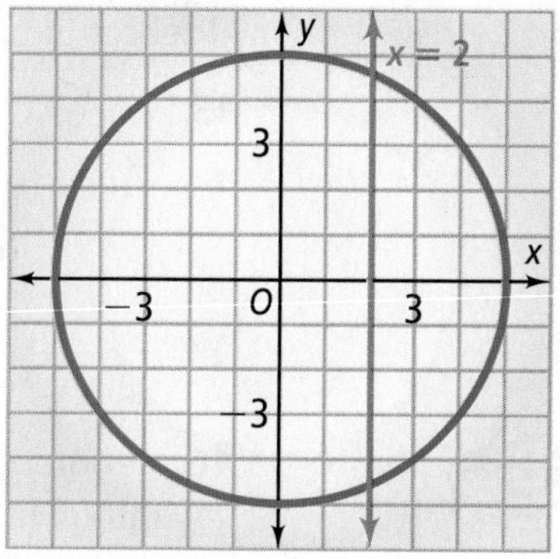

28. **Construction** Use a circular object such as a can or a saucer to draw a circle. Construct the center of the circle.

Ⓒ 29. **Writing** Theorems 12-4 and 12-5 both begin with the phrase, "within a circle or in congruent circles." Explain why the word *congruent* is essential for both theorems.

Find $m\widehat{AB}$. (*Hint:* You will need to use trigonometry in Exercise 32.)

30.

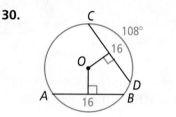

31.

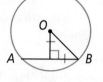

32.

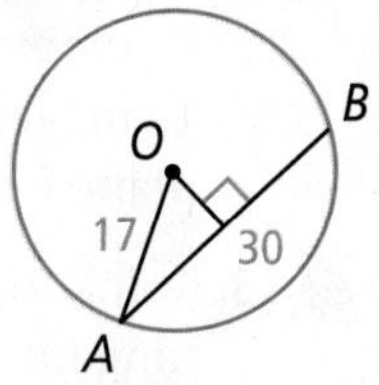

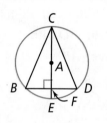

33. Prove Theorem 12-10.

Proof

Given: ℓ is the $\perp$ bisector of $\overline{WY}$.

Prove: ℓ contains the center of $\odot X$.

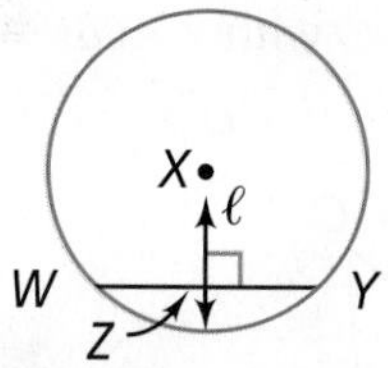

34. **Given:** $\odot A$ with $\overline{CE} \perp \overline{BD}$

Proof

Prove: $\widehat{BC} \cong \widehat{DC}$

Ⓒ **Challenge** Prove each of the following.

Proof

35. Converse of Theorem 12-4: Within a circle or in congruent circles, congruent arcs have congruent central angles.

36. Converse of Theorem 12-5: Within a circle or in congruent circles, congruent chords have congruent central angles.

37. Converse of Theorem 12-6: Within a circle or in congruent circles, congruent arcs have congruent chords.

38. Converse of Theorem 12-7: Within a circle or congruent circles, congruent chords are equidistant from the center (or centers).

39. If two circles are concentric and a chord of the larger circle is tangent to the smaller circle, prove that the point of tangency is the midpoint of the chord.

Proof

Standardized Test Prep

SAT/ACT

40. The diameter of a circle is 25 cm and a chord of the same circle is 16 cm. To the nearest tenth, what is the distance of the chord from the center of the circle?

   Ⓐ 9.0 cm   Ⓒ 18.0 cm

   Ⓑ 9.6 cm   Ⓓ 19.2 cm

41. The Smart Ball Company makes plastic balls for small children. The diameter of a ball is 8 cm. The cost for creating a ball is 2 cents per square centimeter. Which value is the most reasonable estimate for the cost of making 1000 balls?

   Ⓕ $2010   Ⓗ $16,080

   Ⓖ $4021   Ⓘ $42,900

42. From the top of a building you look down at an object on the ground. Your eyes are 50 ft above the ground and the angle of depression is 50°. Which distance is the best estimate of how far the object is from the base of the building?

   Ⓐ 42 ft   Ⓒ 65 ft

   Ⓑ 60 ft   Ⓓ 78 ft

Short Response

43. A bicycle tire has a diameter of 17 in. How many revolutions of the tire are necessary to travel 800 ft? Show your work.

Mixed Review

Assume that the lines that appear to be tangent are tangent. O is the center of each circle. Find the value of x to the nearest tenth.

◀ See Lesson 12-1.

44.

O 140° $x°$

45.

3.5   6.8 O x 2

46. The legs of a right triangle are 10 in. and 24 in. long. The bisector of the right angle cuts the hypotenuse into two segments. What is the length of each segment, rounded to the nearest tenth?

◀ See Lesson 7-5.

Get Ready!   **To prepare for Lesson 12-3, do Exercises 47–52.**

◀ See Lesson 10-6.

Identify the following in $\odot P$ at the right.

47. a semicircle   **48.** a minor arc   **49.** a major arc

Find the measure of each arc in $\odot P$.

50. $\overset{\frown}{ST}$ **51.** $\overset{\frown}{STQ}$ **52.** $\overset{\frown}{RT}$

T P Q 86°   145° S R

Inscribed Angles

© **Content Standards**
G.C.2 Identify and describe relationships among inscribed angles, radii, and chords.
G.C.3 . . . Prove properties of angles for a quadrilateral inscribed in a circle.
Also G.C.4

Objectives To find the measure of an inscribed angle
To find the measure of an angle formed by a tangent and a chord

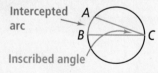

SOLVE IT!

Getting Ready!

Three high-school soccer players practice kicking goals from the points shown in the diagram. All three points are along an arc of a circle. Player A says she is in the best position because the angle of her kicks toward the goal is wider than the angle of the other players' kicks. Do you agree? Explain.

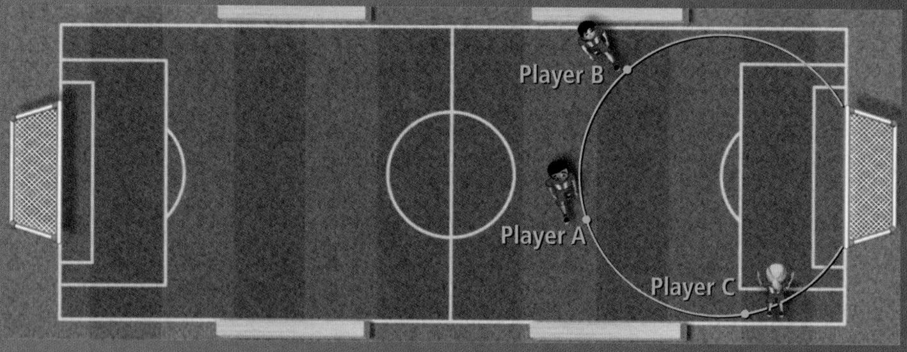

Draw a large diagram and draw the angle each point makes with the goal posts.

© **MATHEMATICAL PRACTICES**

Player B

Player A

Player C

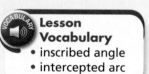

Dynamic Activity
Inscribed Angles

Lesson Vocabulary
• inscribed angle
• intercepted arc

An angle whose vertex is on the circle and whose sides are chords of the circle is an **inscribed angle.** An arc with endpoints on the sides of an inscribed angle, and its other points in the interior of the angle is an **intercepted arc.** In the diagram, inscribed $\angle C$ intercepts $\widehat{AB}$.

Intercepted arc

Inscribed angle

Essential Understanding Angles formed by intersecting lines have a special relationship to the arcs the intersecting lines intercept. In this lesson, you will study arcs formed by inscribed angles.

take note

Theorem 12-11   Inscribed Angle Theorem

The measure of an inscribed angle is half the measure of its intercepted arc.

$$m\angle B = \frac{1}{2}\, m\widehat{AC}$$

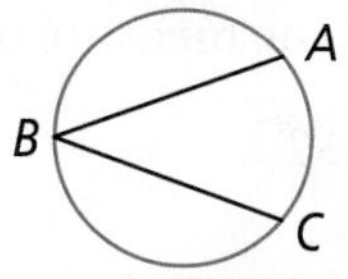

To prove Theorem 12-11, there are three cases to consider.

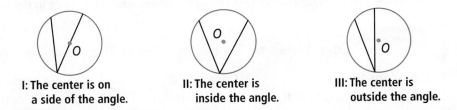

I: The center is on
a side of the angle.

II: The center is
inside the angle.

III: The center is
outside the angle.

Below is a proof of Case I. You will prove Case II and Case III in Exercises 26 and 27.

Proof **Proof of Theorem 12-11, Case I**

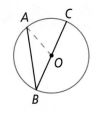

Given: $\odot O$ with inscribed $\angle B$ and diameter $\overline{BC}$

Prove: $m\angle B = \frac{1}{2}\,m\widehat{AC}$

Draw radius $\overline{OA}$ to form isosceles $\triangle AOB$ with $OA = OB$ and, hence, $m\angle A = m\angle B$ (Isosceles Triangle Theorem).

$m\angle AOC = m\angle A + m\angle B$	Triangle Exterior Angle Theorem
$m\widehat{AC} = m\angle AOC$	Definition of measure of an arc
$m\widehat{AC} = m\angle A + m\angle B$	Substitute.
$m\widehat{AC} = 2m\angle B$	Substitute and simplify.
$\frac{1}{2}\,m\widehat{AC} = m\angle B$	Divide each side by 2.

Ⓒ **Problem 1** **Using the Inscribed Angle Theorem**

Plan

Which variable should you solve for first?
You know the inscribed angle that intercepts $\widehat{PT}$, which has the measure a. You need a to find b. So find a first.

What are the values of a and b?

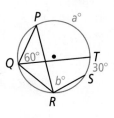

$m\angle PQT = \frac{1}{2}\,m\widehat{PT}$	Inscribed Angle Theorem
$60 = \frac{1}{2}\,a$	Substitute.
$120 = a$	Multiply each side by 2.
$m\angle PRS = \frac{1}{2}\,m\widehat{PS}$	Inscribed Angle Theorem
$m\angle PRS = \frac{1}{2}\,(m\widehat{PT} + m\widehat{TS})$	Arc Addition Postulate
$b = \frac{1}{2}\,(120 + 30)$	Substitute.
$b = 75$	Simplify.

✓ **Got It? 1. a.** In $\odot O$, what is $m\angle A$?

b. What are $m\angle A$, $m\angle B$, $m\angle C$, and $m\angle D$?

c. What do you notice about the sums of the measures of the opposite angles in the quadrilateral in part (b)?

You will use three corollaries to the Inscribed Angle Theorem to find measures of angles in circles. The first corollary may confirm an observation you made in the Solve It.

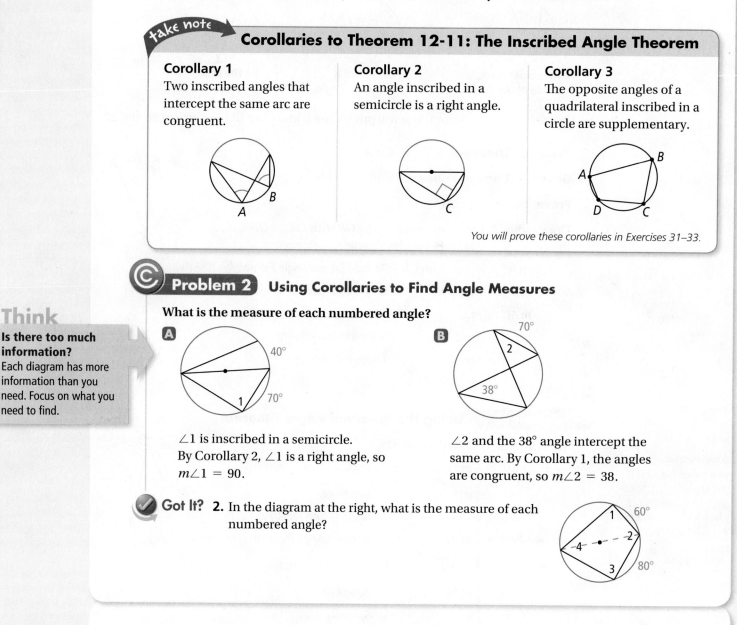

take note

Corollaries to Theorem 12-11: The Inscribed Angle Theorem

Corollary 1
Two inscribed angles that intercept the same arc are congruent.

Corollary 2
An angle inscribed in a semicircle is a right angle.

Corollary 3
The opposite angles of a quadrilateral inscribed in a circle are supplementary.

You will prove these corollaries in Exercises 31–33.

Problem 2 Using Corollaries to Find Angle Measures

Think

Is there too much information?
Each diagram has more information than you need. Focus on what you need to find.

What is the measure of each numbered angle?

A   40°   70°   1

∠1 is inscribed in a semicircle.
By Corollary 2, ∠1 is a right angle, so
$m\angle 1 = 90$.

B   70°   2   38°

∠2 and the 38° angle intercept the same arc. By Corollary 1, the angles are congruent, so $m\angle 2 = 38$.

Got It? **2.** In the diagram at the right, what is the measure of each numbered angle?

60° 80°

The following diagram shows point A moving along the circle until a tangent is formed. From the Inscribed Angle Theorem, you know that in the first three diagrams $m\angle A$ is $\frac{1}{2} m\widehat{BC}$. As the last diagram suggests, this is also true when A and C coincide.

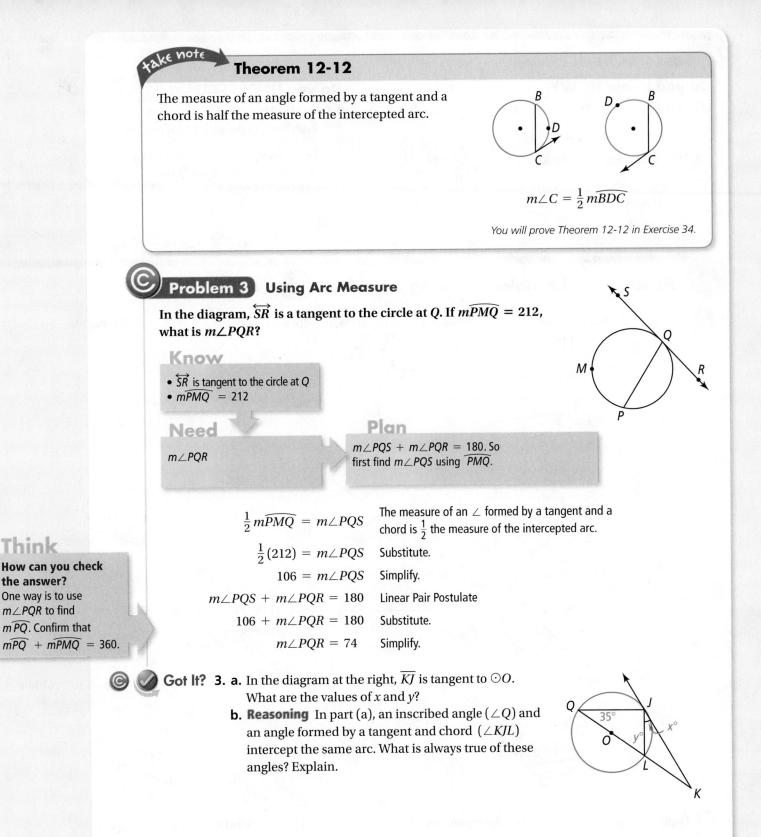

Theorem 12-12

The measure of an angle formed by a tangent and a chord is half the measure of the intercepted arc.

$$m\angle C = \tfrac{1}{2}\,m\widehat{BDC}$$

You will prove Theorem 12-12 in Exercise 34.

© **Problem 3** **Using Arc Measure**

In the diagram, $\overleftrightarrow{SR}$ is a tangent to the circle at Q. If $m\widehat{PMQ} = 212$, what is $m\angle PQR$?

Know

• $\overleftrightarrow{SR}$ is tangent to the circle at Q
• $m\widehat{PMQ} = 212$

Need

$m\angle PQR$

Plan

$m\angle PQS + m\angle PQR = 180$. So first find $m\angle PQS$ using $\widehat{PMQ}$.

$\tfrac{1}{2}\,m\widehat{PMQ} = m\angle PQS$	The measure of an $\angle$ formed by a tangent and a chord is $\tfrac{1}{2}$ the measure of the intercepted arc.
$\tfrac{1}{2}(212) = m\angle PQS$	Substitute.
$106 = m\angle PQS$	Simplify.
$m\angle PQS + m\angle PQR = 180$	Linear Pair Postulate
$106 + m\angle PQR = 180$	Substitute.
$m\angle PQR = 74$	Simplify.

Think

How can you check the answer?
One way is to use $m\angle PQR$ to find $m\widehat{PQ}$. Confirm that $m\widehat{PQ} + m\widehat{PMQ} = 360$.

© ✓ **Got It?** **3. a.** In the diagram at the right, $\overline{KJ}$ is tangent to $\odot O$. What are the values of x and y?

b. Reasoning In part (a), an inscribed angle ($\angle Q$) and an angle formed by a tangent and chord ($\angle KJL$) intercept the same arc. What is always true of these angles? Explain.

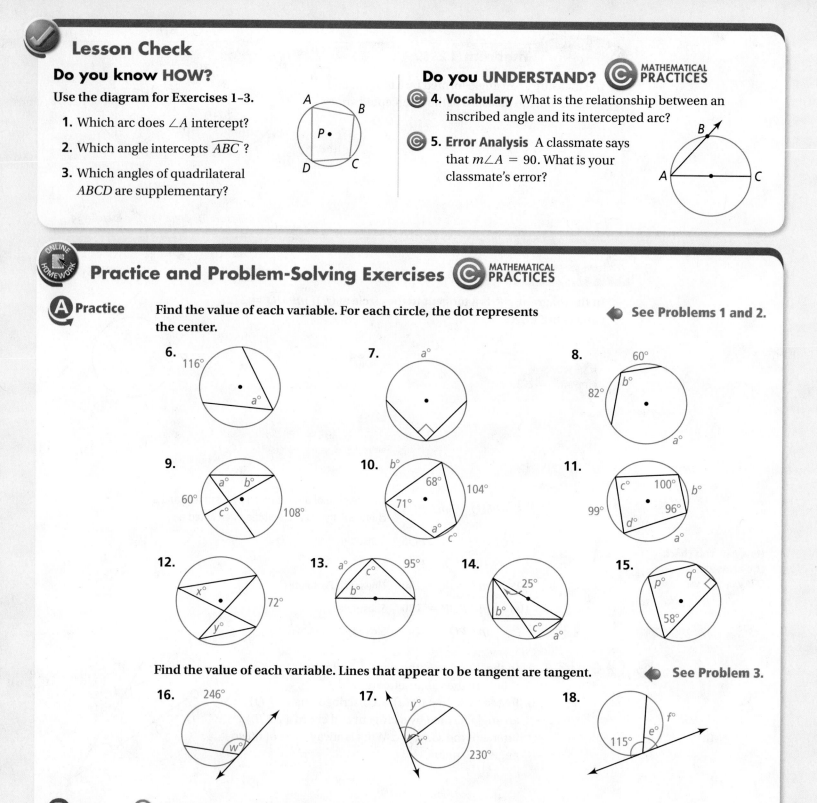

Lesson Check

Do you know HOW?

Use the diagram for Exercises 1–3.

1. Which arc does $\angle A$ intercept?

2. Which angle intercepts $\overset{\frown}{ABC}$?

3. Which angles of quadrilateral $ABCD$ are supplementary?

Do you UNDERSTAND? © MATHEMATICAL PRACTICES

© **4. Vocabulary** What is the relationship between an inscribed angle and its intercepted arc?

© **5. Error Analysis** A classmate says that $m\angle A = 90$. What is your classmate's error?

Practice and Problem-Solving Exercises © MATHEMATICAL PRACTICES

A Practice   Find the value of each variable. For each circle, the dot represents the center.

◀ See Problems 1 and 2.

6. 116° a°

7. a°

8. 60° b° 82° a°

9. a° b° 60° c° 108°

10. b° 68° 104° 71° a° c°

11. c° 100° b° 99° 96° d° a°

12. x° 72° y°

13. a° c° 95° b°

14. 25° b° c° a°

15. p° q° 58°

Find the value of each variable. Lines that appear to be tangent are tangent.   ◀ See Problem 3.

16. 246° w°

17. y° x° 230°

18. 115° e° f°

B Apply   © **19. Writing** A parallelogram inscribed in a circle must be what kind of parallelogram? Explain.

Find each indicated measure for ⊙O.

20.
a. $m\widehat{BC}$
b. $m\angle B$
c. $m\angle C$
d. $m\widehat{AB}$

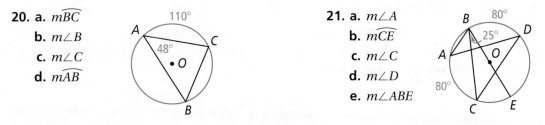

21.
a. $m\angle A$
b. $m\widehat{CE}$
c. $m\angle C$
d. $m\angle D$
e. $m\angle ABE$

22. Think About a Plan What kind of trapezoid can be inscribed in a circle? Justify your response.
- Draw several diagrams to make a conjecture.
- How can parallel lines help?

Find the value of each variable. For each circle, the dot represents the center.

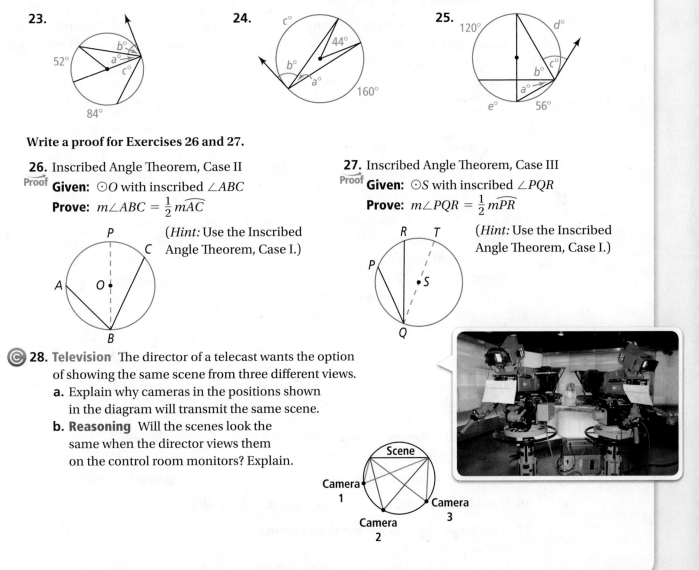

23.

24.

25.

Write a proof for Exercises 26 and 27.

26. Inscribed Angle Theorem, Case II
Proof **Given:** ⊙O with inscribed ∠ABC
Prove: $m\angle ABC = \frac{1}{2}\,m\widehat{AC}$

(*Hint:* Use the Inscribed Angle Theorem, Case I.)

27. Inscribed Angle Theorem, Case III
Proof **Given:** ⊙S with inscribed ∠PQR
Prove: $m\angle PQR = \frac{1}{2}\,m\widehat{PR}$

(*Hint:* Use the Inscribed Angle Theorem, Case I.)

28. Television The director of a telecast wants the option of showing the same scene from three different views.
a. Explain why cameras in the positions shown in the diagram will transmit the same scene.
b. **Reasoning** Will the scenes look the same when the director views them on the control room monitors? Explain.

29. Reasoning Can a rhombus that is not a square be inscribed in a circle? Justify your answer.

30. Constructions The diagrams below show the construction of a tangent to a circle from a point outside the circle. Explain why $\overleftrightarrow{BC}$ must be tangent to $\odot A$. (*Hint:* Copy the third diagram and draw $\overline{AC}$.)

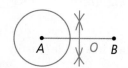

Given: $\odot A$ and point B
Construct the midpoint of $\overline{AB}$. Label the point O.

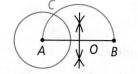

Construct a semicircle with radius OA and center O. Label its intersection with $\odot A$ as C.

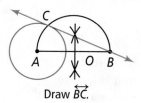

Draw $\overleftrightarrow{BC}$.

Write a proof for Exercises 31–34.

31. Inscribed Angle Theorem, Corollary 1
Proof **Given:** $\odot O$, $\angle A$ intercepts $\overset{\frown}{BC}$,
      $\angle D$ intercepts $\overset{\frown}{BC}$.
Prove: $\angle A \cong \angle D$

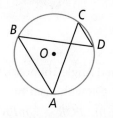

32. Inscribed Angle Theorem, Corollary 2
Proof **Given:** $\odot O$ with $\angle CAB$ inscribed
      in a semicircle
Prove: $\angle CAB$ is a right angle.

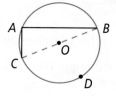

33. Inscribed Angle Theorem, Corollary 3
Proof **Given:** Quadrilateral $ABCD$
      inscribed in $\odot O$
Prove: $\angle A$ and $\angle C$ are supplementary.
      $\angle B$ and $\angle D$ are supplementary.

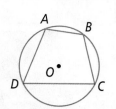

34. Theorem 12-12
Proof **Given:** $\overline{GH}$ and tangent ℓ
      intersecting $\odot E$ at H
Prove: $m\angle GHI = \frac{1}{2}m\overset{\frown}{GFH}$

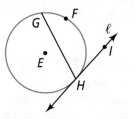

 Challenge

Reasoning Is the statement *true* or *false*? If it is true, give a convincing argument. If it is false, give a counterexample.

35. If two angles inscribed in a circle are congruent, then they intercept the same arc.

36. If an inscribed angle is a right angle, then it is inscribed in a semicircle.

37. A circle can always be circumscribed about a quadrilateral whose opposite angles are supplementary.

38. Prove that if two arcs of a circle are included between parallel chords, then the arcs
Proof   are congruent.

39. Constructions Draw two segments. Label their lengths x and y. Construct the
geometric mean of x and y. (*Hint:* Recall a theorem about a geometric mean.)

Standardized Test Prep

SAT/ACT

For Exercises 40 and 41, what is the value of each variable in $\odot O$?

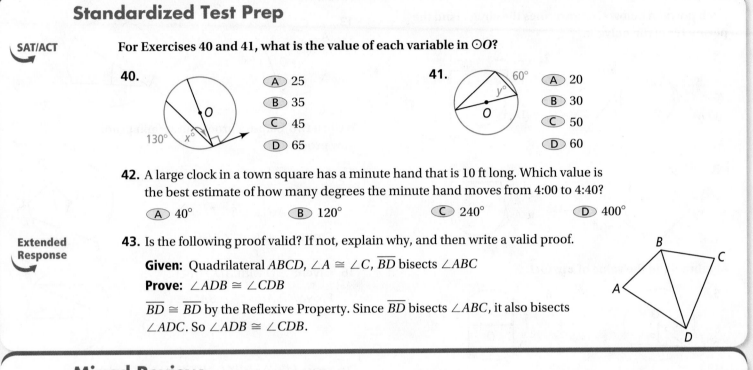

40.

- Ⓐ 25
- Ⓑ 35
- Ⓒ 45
- Ⓓ 65

130° $x°$

41.

60°
$y°$
O

- Ⓐ 20
- Ⓑ 30
- Ⓒ 50
- Ⓓ 60

42. A large clock in a town square has a minute hand that is 10 ft long. Which value is
the best estimate of how many degrees the minute hand moves from 4:00 to 4:40?

Ⓐ 40°   Ⓑ 120°   Ⓒ 240°   Ⓓ 400°

Extended
Response

43. Is the following proof valid? If not, explain why, and then write a valid proof.

Given: Quadrilateral $ABCD$, $\angle A \cong \angle C$, $\overline{BD}$ bisects $\angle ABC$

Prove: $\angle ADB \cong \angle CDB$

$\overline{BD} \cong \overline{BD}$ by the Reflexive Property. Since $\overline{BD}$ bisects $\angle ABC$, it also bisects
$\angle ADC$. So $\angle ADB \cong \angle CDB$.

Mixed Review

Algebra   Find the value of x in $\odot O$, to the nearest tenth.

See Lesson 12-2.

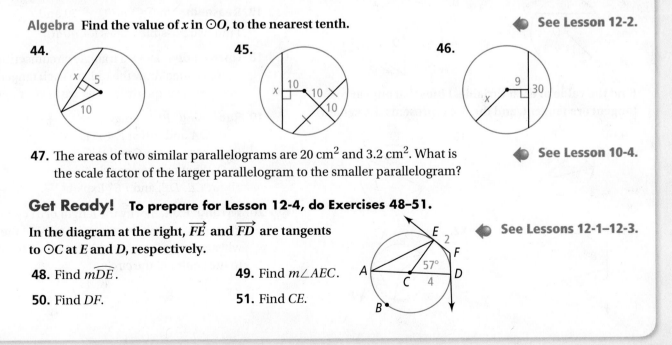

44.

x 5
10

45.

x 10
O 10
10

46.

9
x 30

47. The areas of two similar parallelograms are 20 cm² and 3.2 cm². What is
the scale factor of the larger parallelogram to the smaller parallelogram?

See Lesson 10-4.

Get Ready!   **To prepare for Lesson 12-4, do Exercises 48–51.**

In the diagram at the right, $\overrightarrow{FE}$ and $\overrightarrow{FD}$ are tangents
to $\odot C$ at E and D, respectively.

See Lessons 12-1–12-3.

48. Find $m\overset{\frown}{DE}$.   **49.** Find $m\angle AEC$.

50. Find DF.   **51.** Find CE.

E 2
F
A 57° D
C 4
B

MathXL® for School
Go to PowerGeometry.com

Do you know HOW?

Each polygon below circumscribes the circle. Find the perimeter of the polygon.

1.

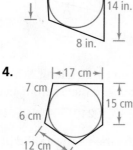

9 cm
13 cm   16 cm

2.

|← 13 in. →|
10 in.   14 in.
8 in.

3.

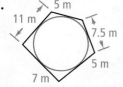

5 m
11 m   7.5 m
5 m
7 m

4.

|←17 cm→|
7 cm
6 cm   15 cm
12 cm

Algebra Find the value of *x* in ⊙O.

5.

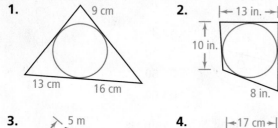

O
15   9
x

6.

5
O   x

7.

30
x   O
17

8.

230°
6   O
x   7
65°

Find the value of each variable. Lines that appear to be tangent are tangent, and the dot represents the center.

9.

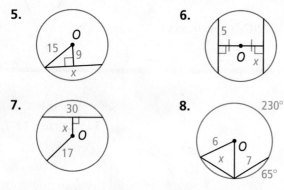

44°   w°   y°
x°
54°

10.

100°
60°   a°
b°
c°   d°
84°

11.

y°
w°
x°
150°

12.

125°
a°   b°
c°
140°

Find m⌢AB.

13.

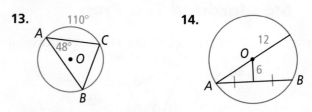

110°
A   C
48°
• O
B

14.

12
O
6
A   B

Write a two-column proof, paragraph proof, or flow proof.

15. Given: ⊙A with $\overline{BC} \cong \overline{DE}$,
$\overline{AF} \perp \overline{BC}$,
$\overline{AG} \perp \overline{DE}$

Prove: $\angle AFG \cong \angle AGF$

Proof

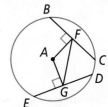

B
F
A   C
D
E   G

16. Given: ⊙O with $\overline{AD} \cong \overline{BC}$

Proof

Prove: $\triangle ABD \cong \triangle BAC$

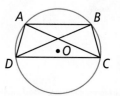

A   B
• O
D   C

Do you UNDERSTAND?

Ⓒ **17. Reasoning** In ⊙C, m⌢PQ = 50 and m⌢QR = 20. Find two possible values for m⌢PR.

Ⓒ **18. Open-Ended** Draw a triangle circumscribed about a circle. Then draw the radii to each tangent. How many convex quadrilaterals are in your figure?

Ⓒ **19. Reasoning** $\overline{EF}$ is tangent to both ⊙A and ⊙B at F. $\overline{CD}$ is tangent to ⊙A at C and to ⊙B at D. What can you conclude about $\overline{CE}$, $\overline{DE}$, and $\overline{FE}$? Explain.

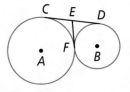

C   E   D
• F •
A   B

Ⓒ **20. Writing** Explain why the length of a segment tangent to a circle from a point outside the circle will always be less than the distance from the point to the center of the circle.

Concept Byte

Use With Lesson 12-4

TECHNOLOGY

Exploring Chords and Secants

© **Content Standard**
Extends G.C.2 Identify and describe relationships among inscribed angles, radii, and chords.

Activity 1

Construct ⊙A and two chords $\overline{BC}$ and $\overline{DE}$ that intersect at *F*.

1. Measure $\overline{BF}$, $\overline{FC}$, $\overline{EF}$, and $\overline{FD}$.

2. Use the calculator program of your software to find $BF \cdot FC$ and $EF \cdot FD$.

3. Manipulate the lines. What pattern do you observe in the products?

© **4. Make a Conjecture** What appears to be true for two intersecting chords?

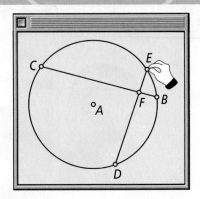

Activity 2

A *secant* is a line that intersects a circle in two points. A *secant segment* is a segment that contains a chord of the circle and has only one endpoint outside the circle. Construct a new circle and two secants $\overleftrightarrow{DG}$ and $\overleftrightarrow{DE}$ that intersect outside the circle at point *D*. Label the intersections with the circle as shown.

5. Measure $\overline{DG}$, $\overline{DF}$, $\overline{DE}$, and $\overline{DB}$.

6. Calculate the products $DG \cdot DF$ and $DE \cdot DB$.

7. Manipulate the lines. What pattern do you observe in the products?

© **8. Make a Conjecture** What appears to be true for two intersecting secants?

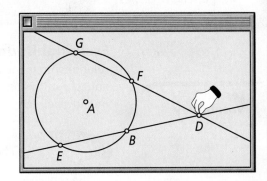

Activity 3

Construct ⊙A with tangent $\overline{DG}$ perpendicular to radius $\overline{AG}$ and secant $\overline{DE}$ that intersects the circle at *B* and *E*.

9. Measure $\overline{DG}$, $\overline{DE}$, and $\overline{DB}$.

10. Calculate the products $(DG)^2$ and $DE \cdot DB$.

11. Manipulate the lines. What pattern do you observe in the products?

© **12. Make a Conjecture** What appears to be true for the tangent segment and secant segment?

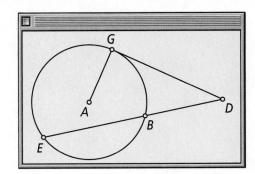

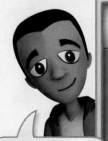

12-4 Angle Measures and Segment Lengths

Content Standard

Extends G.C.2 Identify and describe relationships among inscribed angles, radii, and chords.

Objectives To find measures of angles formed by chords, secants, and tangents
To find the lengths of segments associated with circles

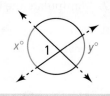

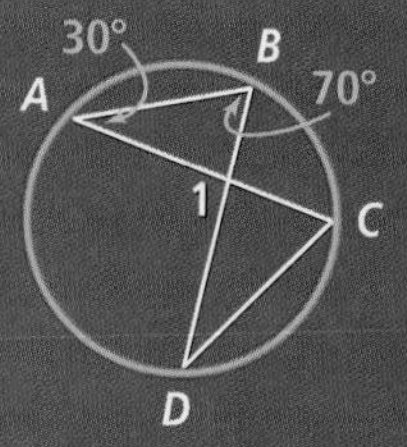

Think about how inscribed angles will help you out.

SOLVE IT! **Getting Ready!**

Find $m\angle 1$ and the sum of the measures of $\overset{\frown}{AD}$ and $\overset{\frown}{BC}$. What is the relationship between the measures? How do you know?

30°
B
70°
A
1
C
D

MATHEMATICAL PRACTICES

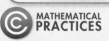

Essential Understanding Angles formed by intersecting lines have a special relationship to the related arcs formed when the lines intersect a circle. In this lesson, you will study angles and arcs formed by lines intersecting either within a circle or outside a circle.

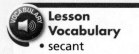

Lesson Vocabulary
- secant

take note

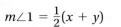

Theorem 12-13

The measure of an angle formed by two lines that intersect inside a circle is half the sum of the measures of the intercepted arcs.

$$m\angle 1 = \frac{1}{2}(x + y)$$

x°
1
y°

Theorem 12-14

The measure of an angle formed by two lines that intersect outside a circle is half the difference of the measures of the intercepted arcs.

x°
y°
1
x°
y°
1
x°
y°
1

$$m\angle 1 = \frac{1}{2}(x - y)$$

You will prove Theorem 12-14 in Exercises 35 and 36.

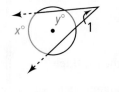

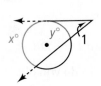

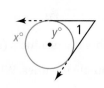

In Theorem 12-13, the lines from a point outside the circle going through the circle are called secants. A **secant** is a line that intersects a circle at two points. $\overleftrightarrow{AB}$ is a secant, $\overrightarrow{AB}$ is a secant ray, and $\overline{AB}$ is a secant segment. A chord is part of a secant.

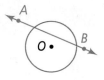

Proof **Proof of Theorem 12-13**

Given: $\odot O$ with intersecting chords $\overline{AC}$ and $\overline{BD}$

Prove: $m\angle 1 = \frac{1}{2}(m\widehat{AB} + m\widehat{CD})$

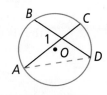

Begin by drawing auxiliary $\overline{AD}$ as shown in the diagram.

$m\angle BDA = \frac{1}{2}m\widehat{AB}$, and $m\angle CAD = \frac{1}{2}m\widehat{CD}$	$m\angle 1 = m\angle BDA + m\angle CAD$
Inscribed Angle Theorem	△Exterior Angle Theorem

$$m\angle 1 = \frac{1}{2}m\widehat{AB} + \frac{1}{2}m\widehat{CD}$$

Substitute.

$$m\angle 1 = \frac{1}{2}(m\widehat{AB} + m\widehat{CD})$$

Distributive Property

Problem 1 Finding Angle Measures

Algebra **What is the value of each variable?**

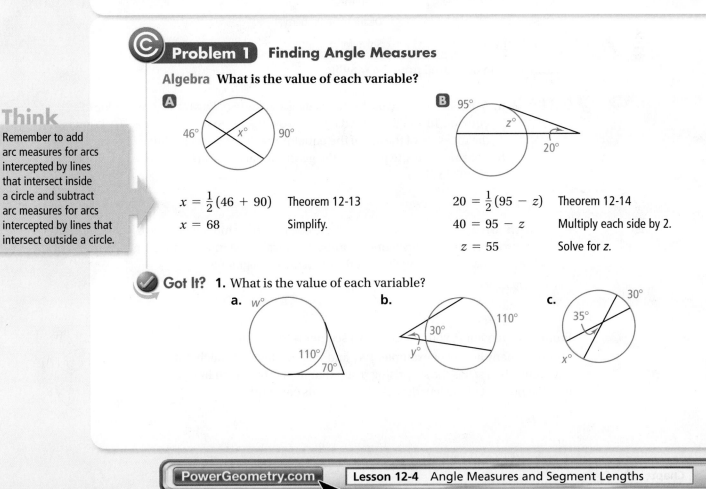

Think

Remember to add arc measures for arcs intercepted by lines that intersect inside a circle and subtract arc measures for arcs intercepted by lines that intersect outside a circle.

A

$x = \frac{1}{2}(46 + 90)$ Theorem 12-13

$x = 68$ Simplify.

B

$20 = \frac{1}{2}(95 - z)$ Theorem 12-14

$40 = 95 - z$ Multiply each side by 2.

$z = 55$ Solve for z.

Got It? **1.** What is the value of each variable?

a. b. c.

Problem 2 Finding an Arc Measure

Satellite A satellite in a geostationary orbit above Earth's equator has a viewing angle of Earth formed by the two tangents to the equator. The viewing angle is about 17.5°. What is the measure of the arc of Earth that is viewed from the satellite?

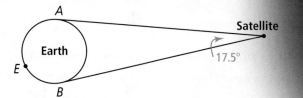

Think

How can you represent the measures of the arcs?
The sum of the measures of the arcs is 360°. If the measure of one arc is x, the measure of the other is $360 - x$.

Let $m\widehat{AB} = x$.

Then $m\widehat{AEB} = 360 - x$.

$17.5 = \frac{1}{2}\left(m\widehat{AEB} - m\widehat{AB}\right)$	Theorem 12-14
$17.5 = \frac{1}{2}\left[(360 - x) - x\right]$	Substitute.
$17.5 = \frac{1}{2}(360 - 2x)$	Simplify.
$17.5 = 180 - x$	Distributive Property
$x = 162.5$	Solve for x.

A 162.5° arc can be viewed from the satellite.

Got It? **2. a.** A departing space probe sends back a picture of Earth as it crosses Earth's equator. The angle formed by the two tangents to the equator is 20°. What is the measure of the arc of the equator that is visible to the space probe?

b. Reasoning Is the probe or the geostationary satellite in Problem 2 closer to Earth? Explain.

Essential Understanding There is a special relationship between two intersecting chords, two intersecting secants, or a secant that intersects a tangent. This relationship allows you to find the lengths of unknown segments.

From a given point P, you can draw two segments to a circle along infinitely many lines. For example, $\overline{PA_1}$ and $\overline{PB_1}$ lie along one such line. Theorem 12-15 states the surprising result that no matter which line you use, the product of the lengths $PA \cdot PB$ remains constant.

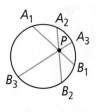

Theorem 12-15

For a given point and circle, the product of the lengths of the two segments from the point to the circle is constant along any line through the point and circle.

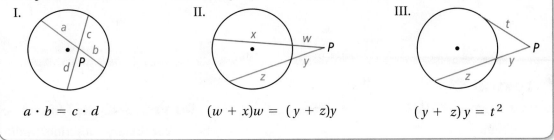

$$a \cdot b = c \cdot d \qquad (w + x)w = (y + z)y \qquad (y + z)y = t^2$$

As you use Theorem 12–15, remember the following.

- **Case I:** The products of the chord segments are equal.
- **Case II:** The products of the secants and their outer segments are equal.
- **Case III:** The product of a secant and its outer segment equals the square of the tangent.

Here is a proof for Case I. You will prove Case II and Case III in Exercises 37 and 38.

Proof **Proof of Theorem 12-15, Case I**

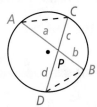

Given: A circle with chords $\overline{AB}$ and $\overline{CD}$ intersecting at P

Prove: $a \cdot b = c \cdot d$

Draw $\overline{AC}$ and $\overline{BD}$. $\angle A \cong \angle D$ and $\angle C \cong \angle B$ because each pair intercepts the same arc, and angles that intercept the same arc are congruent. $\triangle APC \sim \triangle DPB$ by the Angle-Angle Similarity Postulate. The lengths of corresponding sides of similar triangles are proportional, so $\frac{a}{d} = \frac{c}{b}$. Therefore, $a \cdot b = c \cdot d$.

Problem 3 **Finding Segment Lengths**

Algebra Find the value of the variable in $\odot N$.

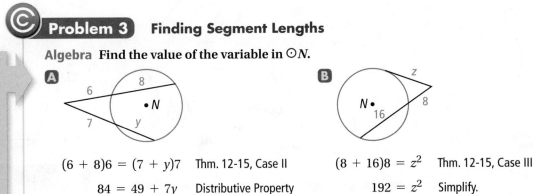

$(6 + 8)6 = (7 + y)7$ Thm. 12-15, Case II	$(8 + 16)8 = z^2$ Thm. 12-15, Case III
$84 = 49 + 7y$ Distributive Property	$192 = z^2$ Simplify.
$35 = 7y$	$13.9 \approx z$ Solve for z.
$5 = y$ Solve for y.	

Plan

How can you identify the segments needed to use Theorem 12-15? Find where segments intersect each other relative to the circle. The lengths of segments that are part of one line will be on the same side of an equation.

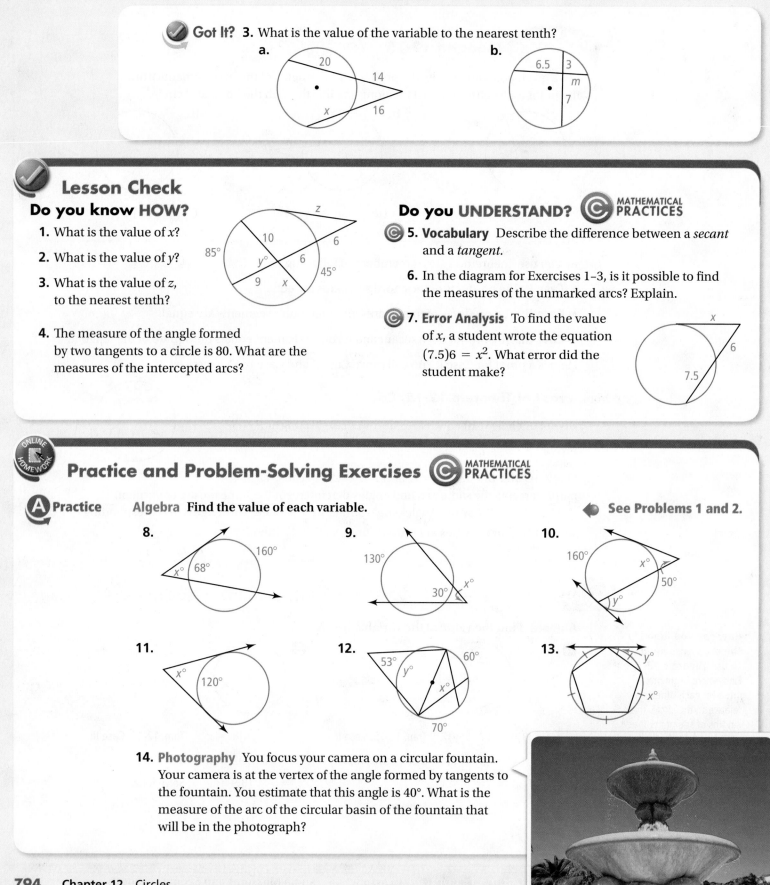

Got It? 3. What is the value of the variable to the nearest tenth?

a.

20

14

x   16

b.

6.5   3

m

7

Lesson Check

Do you know HOW?

1. What is the value of x?

2. What is the value of y?

3. What is the value of z, to the nearest tenth?

z

10   6

85° $y°$ 6

9   x 45°

4. The measure of the angle formed by two tangents to a circle is 80. What are the measures of the intercepted arcs?

Do you UNDERSTAND?   MATHEMATICAL PRACTICES

5. **Vocabulary** Describe the difference between a *secant* and a *tangent*.

6. In the diagram for Exercises 1–3, is it possible to find the measures of the unmarked arcs? Explain.

7. **Error Analysis** To find the value of x, a student wrote the equation $(7.5)6 = x^2$. What error did the student make?

x

6

7.5

Practice and Problem-Solving Exercises   MATHEMATICAL PRACTICES

A Practice   Algebra Find the value of each variable.   **See Problems 1 and 2.**

8.

160°

$x°$ 68°

9.

130°

30° $x°$

10.

160° $x°$

50°

$y°$

11.

$x°$

120°

12.

53° $y°$

60°

$x°$

70°

13.

$y°$

$x°$

$x°$

14. **Photography** You focus your camera on a circular fountain. Your camera is at the vertex of the angle formed by tangents to the fountain. You estimate that this angle is 40°. What is the measure of the arc of the circular basin of the fountain that will be in the photograph?

Algebra Find the value of each variable using the given chord, secant, and tangent lengths. If the answer is not a whole number, round to the nearest tenth.

◀ See Problem 3.

15.

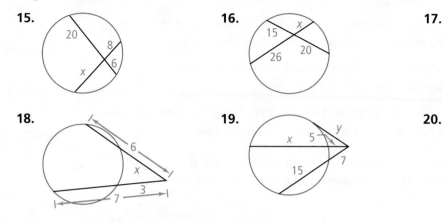

16.

17.

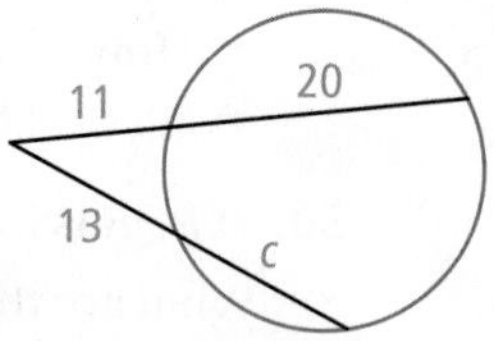

18.

19.

20.

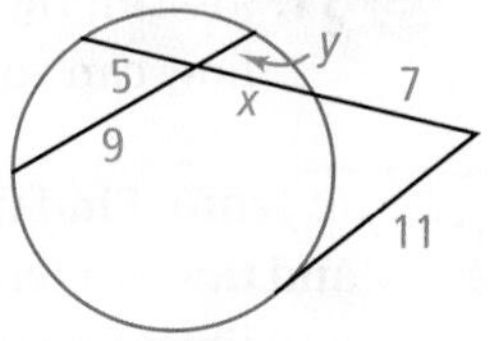

B Apply

Algebra $\overline{CA}$ and $\overline{CB}$ are tangents to $\odot O$. Write an expression for each arc or angle in terms of the given variable.

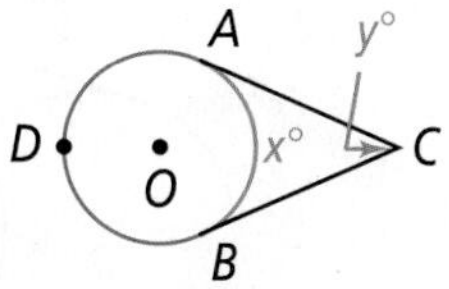

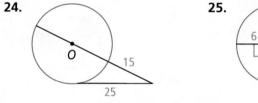

21. $m\widehat{ADB}$ using x 22. $m\angle C$ using x 23. $m\widehat{AB}$ using y

Find the diameter of $\odot O$. A line that appears to be tangent is tangent. If your answer is not a whole number, round it to the nearest tenth.

24.

25.

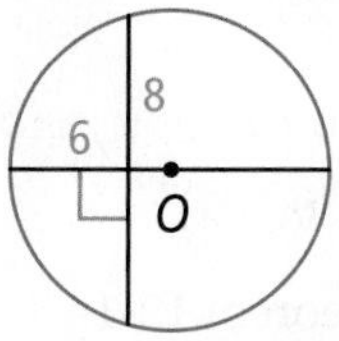

26.

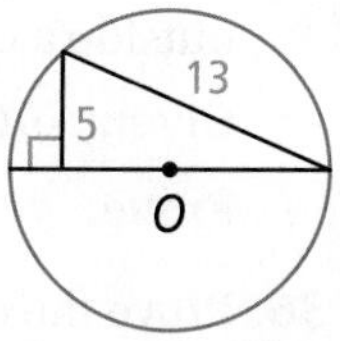

27. A circle is inscribed in a quadrilateral whose four angles have measures 85, 76, 94, and 105. Find the measures of the four arcs between consecutive points of tangency.

Wankel engine

STEM 28. **Engineering** The basis for the design of the Wankel rotary engine is an equilateral triangle. Each side of the triangle is a chord to an arc of a circle. The opposite vertex of the triangle is the center of the circle that forms the arc. In the diagram below, each side of the equilateral triangle is 8 in. long.
   a. Use what you know about equilateral triangles and find the value of x.

© b. **Reasoning** Copy the diagram and complete the circle with the given center. Then use Theorem 12-15 to find the value of x. Show that your answers to parts (a) and (b) are equal.

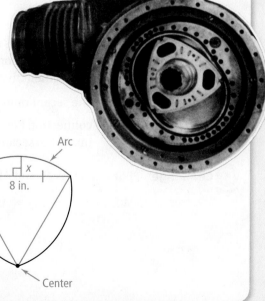

29. Think About a Plan In the diagram, the circles are concentric. What is a formula you could use to find the value of c in terms of a and b?
- How can you use the inscribed angle to find the value of c?
- What is the relationship of the inscribed angle to a and b?

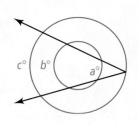

30. $\triangle PQR$ is inscribed in a circle with $m\angle P = 70$, $m\angle Q = 50$, and $m\angle R = 60$. What are the measures of $\widehat{PQ}$, $\widehat{QR}$, and $\widehat{PR}$?

31. Reasoning Use the diagram at the right. If you know the values of x and y, how can you find the measure of each numbered angle?

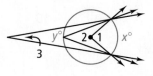

Algebra Find the values of x and y using the given chord, secant, and tangent lengths. If your answer is not a whole number, round it to the nearest tenth.

32.   **33.**   **34.**

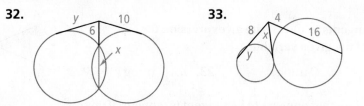

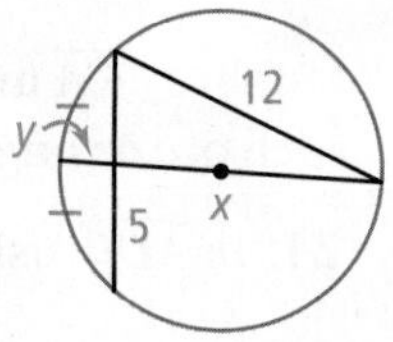

35. Prove Theorem 12-14 as it applies to two secants that intersect outside a circle.

Given: $\odot O$ with secants $\overline{CA}$ and $\overline{CE}$

Prove: $m\angle ACE = \frac{1}{2}(m\widehat{AE} - m\widehat{BD})$

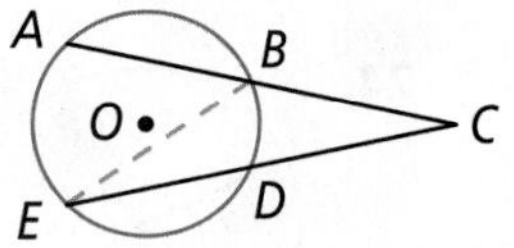

36. Prove the other two cases of Theorem 12-14. (See Exercise 35.)

For Exercises 37 and 38, write proofs that use similar triangles.

37. Prove Theorem 12-15, Case II.   **38.** Prove Theorem 12-15, Case III.

39. The diagram at the right shows a *unit circle*, a circle with radius 1.
- **a.** What triangle is similar to $\triangle ABE$?
- **b.** Describe the connection between the ratio for the tangent of $\angle A$ and the segment that is tangent to $\odot A$.
- **c.** The secant ratio is $\dfrac{\text{hypotenuse}}{\text{length of leg adjacent to an angle}}$. Describe the connection between the ratio for the secant of $\angle A$ and the segment that is the secant in the unit circle.

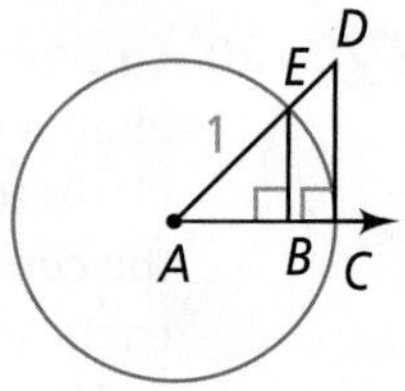

Challenge For Exercises 40 and 41, use the diagram at the right. Prove each statement.

40. $m\angle 1 + m\widehat{PQ} = 180$ 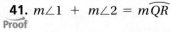 **41.** $m\angle 1 + m\angle 2 = m\widehat{QR}$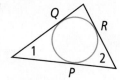

42. Use the diagram at the right and the theorems of this lesson to prove the Pythagorean Theorem.
Proof

43. If an equilateral triangle is inscribed in a circle, prove that the tangents to the circle at the vertices form an equilateral triangle.
Proof

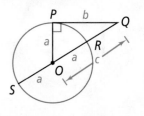

Standardized Test Prep

GRIDDED RESPONSE

SAT/ACT

For Exercises 44 and 45, use the diagram at the right.

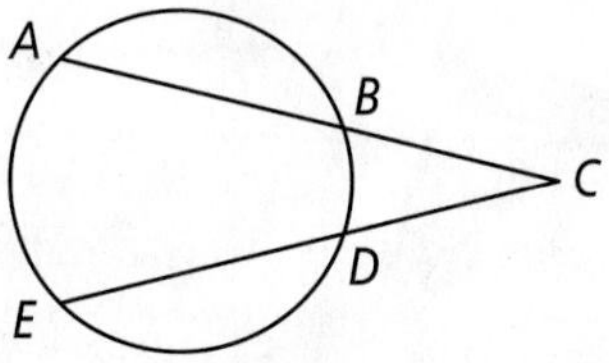

44. If $BC = 6$, $DC = 5$, and $CE = 12$, find AC.

45. If $m\angle C = 14$ and $m\widehat{AE} = 140$, find $m\widehat{BD}$.

46. The altitude to the hypotenuse of a right triangle divides the hypotenuse into segments of length 6 and 18. What is the length of the altitude to the nearest tenth?

47. A rectangular prism measures $3 \text{ cm} \times 4 \text{ cm} \times 5 \text{ cm}$. The length of the longest side of a similar rectangular prism is 12 cm. What is the volume of the larger prism to the nearest tenth?

Mixed Review

Find the value of each variable.

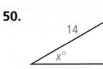 **See Lesson 12-3.**

48.

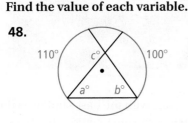

49.

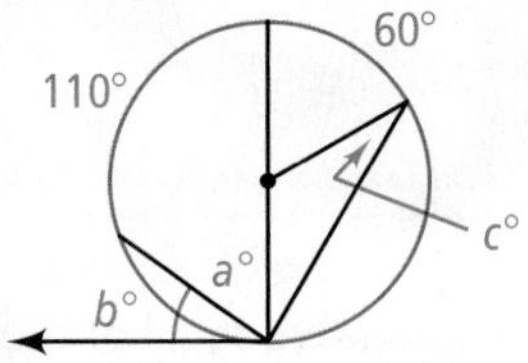

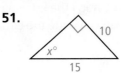

Find the value of x to the nearest whole number.

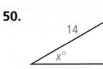 **See Lesson 8-3.**

50.

51.

52.

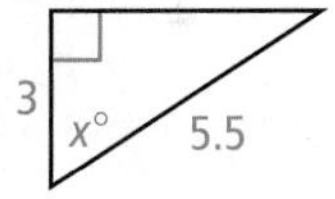

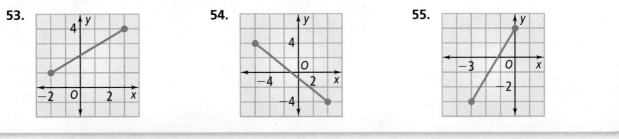

Get Ready! To prepare for Lesson 12-5, do Exercises 53–55.

Find the length of each segment to the nearest tenth.

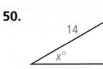 **See Lesson 1-7.**

53.

54.

55.

Circles in the Coordinate Plane

© **Content Standard**
G.GPE.1 Derive the equation of a circle given center and radius using the Pythagorean Theorem . . .

Objectives To write the equation of a circle
To find the center and radius of a circle

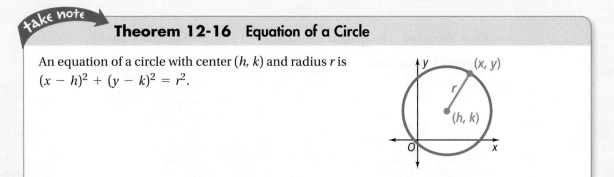

Getting Ready!

The owners of an outdoor adventure course want a way to communicate to all points on the course. They are considering purchasing walkie-talkies with a range of $\frac{1}{2}$ mi. A model of the course is at the right. Each grid unit represents $\frac{1}{8}$ mi. The base station is at (2, 4). Do you think the owners should buy the walkie-talkies? Why?

Do you need to check the distance to every part of the course?

© **MATHEMATICAL PRACTICES**

Dynamic Activity
Circles in the Coordinate Plane

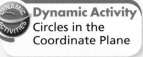
Lesson Vocabulary
• standard form of an equation of a circle

In the Solve It, all of the obstacles lie within or on a circle with the base station as the center. The information from the diagram is enough to write an equation for the circle.

Essential Understanding The information in the equation of a circle allows you to graph the circle. Also, you can write the equation of a circle if you know its center and radius.

take note

Theorem 12-16 Equation of a Circle

An equation of a circle with center (h, k) and radius r is
$(x - h)^2 + (y - k)^2 = r^2$.

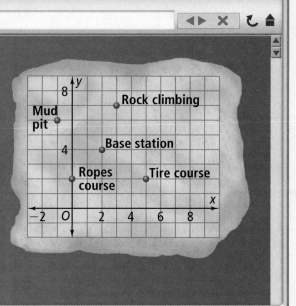

Here's Why It Works You can use the Distance Formula to find an equation of a circle with center (h, k) and radius r, which proves Theorem 12-16. Let (x, y) be any point on the circle. Then the radius r is the distance from (h, k) to (x, y).

$d = \sqrt{(x_2 - x_1)^2 + (y_2 - y_1)^2}$ Distance Formula

$r = \sqrt{(x - h)^2 + (y - k)^2}$ Substitute (x, y) for (x_2, y_2) and (h, k) for (x_1, y_1).

$r^2 = (x - h)^2 + (y - k)^2$ Square both sides.

The equation $(x - h)^2 + (y - k)^2 = r^2$ is the **standard form of an equation of a circle.** You may also call it the *standard equation* of a circle.

Ⓒ **Problem 1**   Writing the Equation of a Circle

Plan

What do you need to know to write the equation of a circle?
You need to know the values of h, k, and r; h is the x-coordinate of the center, k is the y-coordinate of the center, and r is the radius.

What is the standard equation of the circle with center $(5, -2)$ and radius 7?

$(x - h)^2 + (y - k)^2 = r^2$ Use the standard form of an equation of a circle.

$(x - 5)^2 + [y - (-2)]^2 = 7^2$ Substitute $(5, -2)$ for (h, k) and 7 for r.

$(x - 5)^2 + (y + 2)^2 = 49$ Simplify.

✓ **Got It?**   **1.** What is the standard equation of each circle?

     a. center $(3, 5)$; radius 6   **b.** center $(-2, -1)$; radius $\sqrt{2}$

Ⓒ **Problem 2**   Using the Center and a Point on a Circle

Think

How is this problem different from Problem 1?
In this problem, you don't know r. So the first step is to find r.

What is the standard equation of the circle with center $(1, -3)$ that passes through the point $(2, 2)$?

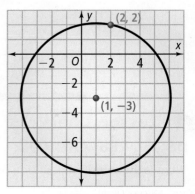

Step 1   Use the Distance Formula to find the radius.

$r = \sqrt{(x_2 - x_1)^2 + (y_2 - y_1)^2}$ Use the Distance Formula.

$\quad = \sqrt{(1 - 2)^2 + (-3 - 2)^2}$ Substitute $(1, -3)$ for (x_2, y_2) and $(2, 2)$ for (x_1, y_1).

$\quad = \sqrt{(-1)^2 + (-5)^2}$ Simplify.

$\quad = \sqrt{26}$

Step 2   Use the radius and the center to write an equation.

$(x - h)^2 + (y - k)^2 = r^2$ Use the standard form of an equation of a circle.

$(x - 1)^2 + [y - (-3)]^2 = (\sqrt{26})^2$ Substitute $(1, -3)$ for (h, k) and $\sqrt{26}$ for r.

$(x - 1)^2 + (y + 3)^2 = 26$ Simplify.

✓ **Got It?**   **2.** What is the standard equation of the circle with center $(4, 3)$ that passes through the point $(-1, 1)$?

If you know the standard equation of a circle, you can describe the circle by naming its center and radius. Then you can use this information to graph the circle.

 Problem 3 **Graphing a Circle Given Its Equation** STEM

Communications When you make a call on a cell phone, a tower receives and transmits the call. A way to monitor the range of a cell tower system is to use equations of circles. Suppose the equation $(x - 7)^2 + (y + 2)^2 = 64$ represents the position and the transmission range of a cell tower. What is the graph that shows the position and range of the tower?

Know	Need	Plan
The equation representing the cell tower's position and range	To draw a graph	Determine the values of (h, k) and r in the equation. Then draw a graph.

$(x - 7)^2 + (y + 2)^2 = 64$ Use the standard equation of a circle.

$(x - 7)^2 + [y - (-2)] = 8^2$ Rewrite to find h, k, and r.
    ↑   ↑   ↑
    h k r

The center is $(7, -2)$ and the radius is 8.

To graph the circle, place the compass point at the center $(7, -2)$ and draw a circle with radius 8.

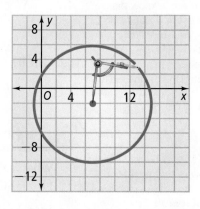

Got It? **3. a.** In Problem 3, what does the center of the circle represent? What does the radius represent?

b. What is the center and radius of the circle with equation $(x - 2)^2 + (y - 3)^2 = 100$? Graph the circle.

Lesson Check

Do you know HOW?

What is the standard equation of each circle?

1. center $(0, 0)$; $r = 4$

2. center $(1, -1)$; $r = \sqrt{5}$

What is the center and radius of each circle?

3. $(x - 8)^2 + y^2 = 9$

4. $(x + 2)^2 + (y - 4)^2 = 7$

Do you UNDERSTAND?  MATHEMATICAL PRACTICES

5. What is the least amount of information that you need to graph a circle? To write the equation of a circle?

6. Suppose you know the center of a circle and a point on the circle. How do you determine the equation of the circle?

7. Error Analysis A student says that the center of a circle with equation $(x - 2)^2 + (y + 3)^2 = 16$ is $(-2, 3)$. What is the student's error?

Ⓐ **Practice**   Write the standard equation of each circle.   ◀ **See Problem 1.**

8. center $(2, -8)$; $r = 9$ **9.** center $(0, 3)$; $r = 7$ **10.** center $(0.2, 1.1)$; $r = 0.4$

11. center $(5, -1)$; $r = 12$ **12.** center $(-6, 3)$; $r = 8$ **13.** center $(-9, -4)$; $r = \sqrt{5}$

14. center $(0, 0)$; $r = 4$ **15.** center $(-4, 0)$; $r = 3$ **16.** center $(-1, -1)$; $r = 1$

Write a standard equation for each circle in the diagram at the right.   ◀ **See Problem 2.**

17. ⊙P **18.** ⊙Q

Write the standard equation of the circle with the given center that passes through the given point.

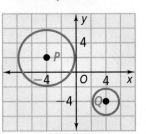

19. center $(-2, 6)$; point $(-2, 10)$ **20.** center $(1, 2)$; point $(0, 6)$

21. center $(7, -2)$; point $(1, -6)$ **22.** center $(-10, -5)$; point $(-5, 5)$

23. center $(6, 5)$; point $(0, 0)$ **24.** center $(-1, -4)$; point $(-4, 0)$

Find the center and radius of each circle. Then graph the circle.   ◀ **See Problem 3.**

25. $(x + 7)^2 + (y - 5)^2 = 16$ **26.** $(x - 3)^2 + (y + 8)^2 = 100$

27. $(x + 4)^2 + (y - 1)^2 = 25$ **28.** $x^2 + y^2 = 36$

Public Safety Each equation models the position and range of a tornado alert siren. Describe the position and range of each.

29. $(x - 5)^2 + (y - 7)^2 = 81$ **30.** $(x + 4)^2 + (y - 9)^2 = 144$

Ⓑ **Apply**   Write the standard equation of each circle.

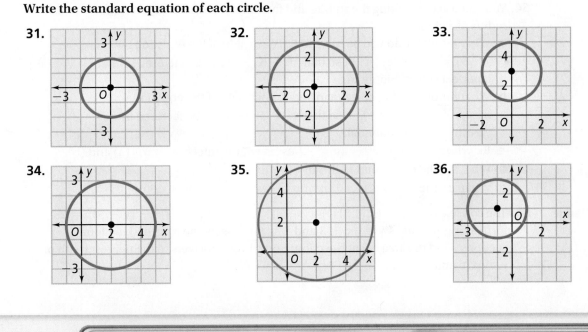

31.   **32.**   **33.**

34.   **35.**   **36.**

Write an equation of a circle with diameter $\overline{AB}$.

37. $A(0, 0)$, $B(8, 6)$

38. $A(3, 0)$, $B(7, 6)$

39. $A(1, 1)$, $B(5, 5)$

Ⓒ 40. Reasoning Describe the graph of $x^2 + y^2 = r^2$ when $r = 0$.

Determine whether each equation is the equation of a circle. Justify your answer.

41. $(x - 1)^2 + (y + 2)^2 = 9$

42. $x + y = 9$

43. $x + (y - 3)^2 = 9$

Ⓒ 44. Think About a Plan Find the circumference and area of the circle whose equation is $(x - 9)^2 + (y - 3)^2 = 64$. Leave your answers in terms of π.
- What essential information do you need?
- What formulas will you use?

45. Write an equation of a circle with area 36π and center $(4, 7)$.

46. What are the x- and y-intercepts of the line tangent to the circle $(x - 2)^2 + (y - 2)^2 = 5^2$ at the point $(5, 6)$?

47. For $(x - h)^2 + (y - k)^2 = r^2$, show that $y = \sqrt{r^2 - (x - h)^2} + k$ or $y = -\sqrt{r^2 - (x - h)^2} + k$.

Sketch the graphs of each equation. Find all points of intersection of each pair of graphs.

48. $x^2 + y^2 = 13$
$y = -x + 5$

49. $x^2 + y^2 = 17$
$y = -\frac{1}{4}x$

50. $x^2 + y^2 = 8$
$y = 2$

51. $x^2 + y^2 = 20$
$y = -\frac{1}{2}x + 5$

52. $(x + 1)^2 + (y - 1)^2 = 18$
$y = x + 8$

53. $(x - 2)^2 + (y - 2)^2 = 10$
$y = -\frac{1}{3}x + 6$

54. You can use completing the square and factoring to find the center and radius of a circle.
- **a.** What number c do you need to add to each side of the equation $x^2 + 6x + y^2 - 4y = -4$ so that $x^2 + 6x + c$ can be factored into a perfect square binomial?
- **b.** What number d do you need to add to each side of the equation $x^2 + 6x + y^2 - 4y = -4$ so that $y^2 - 4y + d$ can be factored into a perfect square binomial?
- **c.** Rewrite $x^2 + 6x + y^2 - 4y = -4$ using your results from parts (a) and (b).
- **d.** What are the center and radius of $x^2 + 6x + y^2 - 4y = -4$?
- **e.** What are the center and radius of $x^2 + 4x + y^2 - 20y + 100 = 0$?

Ⓒ Challenge

55. The concentric circles $(x - 3)^2 + (y - 5)^2 = 64$ and $(x - 3)^2 + (y - 5)^2 = 25$ form a ring. The lines $y = \frac{2}{3}x + 3$ and $y = 5$ intersect the ring, making four sections. Find the area of each section. Round your answers to the nearest tenth of a square unit.

56. Geometry in 3 Dimensions The equation of a sphere is similar to the equation of a circle. The equation of a sphere with center (h, j, k) and radius r is $(x - h)^2 + (y - j)^2 + (z - k)^2 = r^2$. $M(-1, 3, 2)$ is the center of a sphere passing through $T(0, 5, 1)$. What is the radius of the sphere? What is the equation of the sphere?

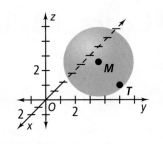

57. Nautical Distance A close estimate of the radius of Earth's equator is 3960 mi.
  a. Write the equation of the equator with the center of Earth as the origin.
  b. Find the length of a 1° arc on the equator to the nearest tenth of a mile.
  c. History Columbus planned his trip to the East by going west. He thought each 1° arc was 45 mi long. He estimated that the trip would take 21 days. Use your answer to part (b) to find a better estimate.

Standardized Test Prep

58. What is an equation of a circle with radius 16 and center $(2, -5)$?
  Ⓐ $(x - 2)^2 + (y + 5)^2 = 16$
  Ⓒ $(x + 2)^2 + (y - 5)^2 = 256$
  Ⓑ $(x + 2)^2 + (y - 5)^2 = 4$
  Ⓓ $(x - 2)^2 + (y + 5)^2 = 256$

59. What can you NOT conclude from the diagram at the right? 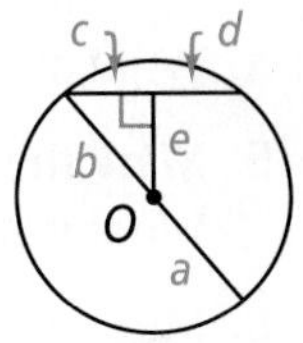
  Ⓕ $c = d$
  Ⓗ $a = b$
  Ⓖ $c^2 + e^2 = b^2$
  Ⓘ $e = d$

60. Are the following statements equivalent?
- In a circle, if two central angles are congruent, then they have congruent arcs.
- In a circle, if two arcs are congruent, then they have congruent central angles.

Mixed Review

Find the value of each variable.

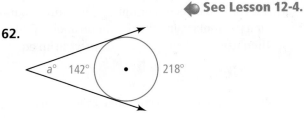

See Lesson 12-4.

61.

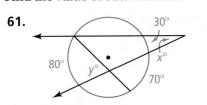

62.

Get Ready! **To prepare for Lesson 12-6, do Exercises 63–65.**   See Lessons 1-2 and 1-5.

Sketch each of the following.

63. the perpendicular bisector of $\overline{BC}$

64. line k parallel to line m and perpendicular to line w, all in plane N

65. $\angle EFG$ bisected by $\overrightarrow{FH}$

Equation of a Parabola

© **Content Standard**
G.GPE.2 Derive the equation of a parabola given a focus and directrix.

Recall from Algebra that a *parabola* is the graph of a quadratic function. In the following activities you will explore other properties of parabolas and how to translate between the geometric description of a parabola and its equation.

Activity 1

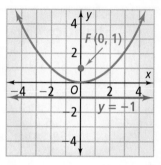

1. Copy the graph of $y = \frac{1}{4}x^2$. What is the vertex of this parabola?

2. Look at the point $F(0, 1)$ and the line $y = -1$ on the coordinate plane. What do you notice about the distance from F to the vertex and the distance from the line $y = -1$ to the vertex?

3. Mark the point $P(2, 1)$ on the parabola. Mark the point D directly below point P on the line $y = -1$. What are the coordinates of point D?

4. What are the lengths of $\overline{FP}$ and $\overline{PD}$?

5. Mark the point $Q(4, 4)$ on the parabola. What is the distance from this point to F? What is the vertical distance from $(4, 4)$ to the line $y = -1$?

6. Make a conjecture about any point on the parabola, the point $F(0, 1)$ and the line $y = -1$. Test your conjecture by using other points on the parabola.

A parabola can be defined as the set of all points that are equidistant from a fixed point, called the **focus of the parabola**, and a fixed line called the **directrix**. That means that any point on a parabola must be the same distance from its focus and directrix.

A focus and directrix uniquely determine a parabola. So, you can derive the equation for a parabola using the definition above. Call an arbitrary point on the parabola (x, y). Then use the distance formula to write an equation.

Activity 2

Refer to the graph to complete this activity.

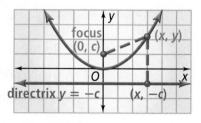

7. What is an expression for the distance between $(0, c)$ and (x, y)?

8. How can you write an expression for the distance between (x, y) and $(x, -c)$? What is the expression?

9. Write an equation for a parabola by setting the expressions you wrote in Exercises 7 and 8 equal to each other. Simplify the equation to get a quadratic equation in standard form.

10. What does the value of c represent in the equation?

11. What is an expression that represents the distance between the focus and the directrix?

Activity 3

In this Activity, you will derive the equation of a parabola with focus $F(2, 4)$ and directrix $y = -2$.

12. What are the coordinates of the vertex of the parabola? Explain how you know.

13. How can you use what you know to make a sketch of the parabola on the coordinate plane?

14. What do you know about the distance from any point (x, y) on the parabola to point $F(2, 4)$ to and the distance from (x, y) to the line $y = -2$?

15. Write an equation that represents the relationship described in Exercise 15.

16. What is the equation for the parabola in standard form?

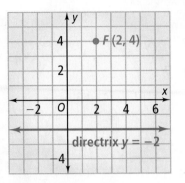

Exercises

Find the equation of the parabola with the focus and directrix shown.

17.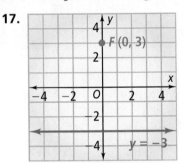

18.

19.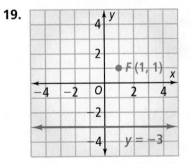

20.

21. What is the equation of a parabola with focus $F(3, -2)$ and directrix $y = -4$?

22. The vertex of a parabola is at $(3, -2)$ and its focus is at $(3, 8)$. What is the equation of its directrix?

STEM 23. **Solar Energy** Solar energy can be captured by using a solar disk. A cross section of the disk is the shape if a parabola. If you can model the cross section so that the vertex is at the origin, and the receiver (the focus) is at $(0, 7)$, what is the equation of this parabola?

Locus: A Set of Points

© **Content Standard**
G.GMD.4 . . . Identify three-dimensional objects generated by rotations of two-dimensional objects.

Objective To draw and describe a locus

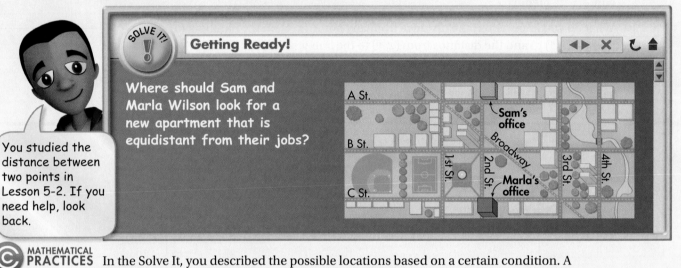

SOLVE IT!

Getting Ready!

Where should Sam and Marla Wilson look for a new apartment that is equidistant from their jobs?

You studied the distance between two points in Lesson 5-2. If you need help, look back.

© **MATHEMATICAL PRACTICES** In the Solve It, you described the possible locations based on a certain condition. A **locus** is a set of points, all of which meet a stated condition. *Loci* is the plural of locus.

Lesson Vocabulary
• locus

Essential Understanding You can use the description of a locus to sketch a geometric relationship.

© **Problem 1** **Describing a Locus in a Plane**

What is a sketch and description for each locus of points in a plane?

A the points 1 cm from a given point *C*

Draw a point *C*. Sketch several points 1 cm from *C*. Keep doing so until you see a pattern. Draw the figure the pattern suggests.

The locus is a circle with center *C* and radius 1 cm.

B the points 1 cm from $\overline{AB}$

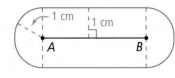

Draw $\overline{AB}$. Sketch several points on either side of $\overline{AB}$. Also sketch points 1 cm from point *A* and point *B*. Keep doing so until you see a pattern. Draw the figure the pattern suggests.

Think
Have you considered all possibilities?
Make sure that the endpoints as well as the segment are included in the sketch.

The locus is a pair of parallel segments, each 1 cm from $\overline{AB}$, and two semicircles with centers at *A* and *B*.

Got It? **1. Reasoning** If the question for part (b) asked for the locus of points in a plane 1 cm from $\overleftrightarrow{AB}$, how would the sketch change?

You can use locus descriptions for geometric terms.

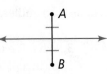

The locus of points in the interior of an angle that are equidistant from the sides of the angle is an angle bisector.

In a plane, the locus of points that are equidistant from a segment's endpoints is the perpendicular bisector of the segment.

Sometimes a locus is described by two conditions. You can draw the locus by first drawing the points that satisfy each condition. Then find their intersection.

Problem 2 **Drawing a Locus for Two Conditions**

What is a sketch of the locus of points in a plane that satisfy these conditions?

- **the points equidistant from intersecting lines *k* and *m***
- **the points 5 cm from the point where *k* and *m* intersect**

Know

Lines *k* and *m* intersect.

Need

Sketch that satisfies the given conditions

Plan

Make a sketch to satisfy the first condition. Then sketch the second condition. Look for the points in common.

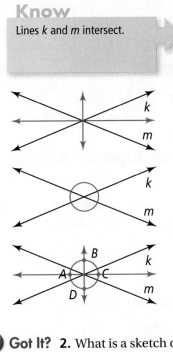

Sketch the points in a plane equidistant from lines *k* and *m*. These points form two lines that bisect the vertical angles formed by *k* and *m*.

Sketch the points in a plane 5 cm from the point where *k* and *m* intersect. These points form a circle.

Indicate the point or set of points that satisfies both conditions. This set of points is *A, B, C,* and *D*.

Got It? **2.** What is a sketch of the locus of points in a plane that satisfy these conditions?
- the points equidistant from two points *X* and *Y*
- the points 2 cm from the midpoint of $\overline{XY}$

Problem 3 Describing a Locus in Space

Think

How can making a sketch help?
Make a sketch of the points in a plane and then visualize what the figure would look like in three dimensions.

A What is the locus of points in space that are *c* units from a point *D*?

The locus is a sphere with center at point *D* and radius *c*.

B What is the locus of points in space that are 3 cm from a line ℓ?

The locus is an endless cylinder with radius 3 cm and centerline ℓ.

Got It? **3.** What is each locus of points?

a. in a plane, the points that are equidistant from two parallel lines

b. in space, the points that are equidistant from two parallel planes

Lesson Check

Do you know HOW?

What is a sketch and description for each locus of points in a plane?

1. points 4 cm from a point *X*

2. points 2 in. from $\overline{UV}$

3. points 3 mm from $\overleftrightarrow{LM}$

4. points 1 in. from a circle with radius 3 in.

Do you UNDERSTAND? MATHEMATICAL PRACTICES

5. Vocabulary How are the words *locus* and *location* related?

6. Compare and Contrast How are the descriptions of the locus of points for each situation alike? How are they different?

- in a plane, the points equidistant from points *J* and *K*
- in space, the points equidistant from points *J* and *K*

Practice and Problem-Solving Exercises MATHEMATICAL PRACTICES

A Practice Sketch and describe each locus of points in a plane.

See Problem 1.

7. points equidistant from the endpoints of $\overline{PQ}$

8. points in the interior of $\angle ABC$ and equidistant from the sides of $\angle ABC$

9. points equidistant from two perpendicular lines

10. midpoints of radii of a circle with radius 2 cm

For Exercises 11–15, sketch the locus of points in a plane that satisfy the given conditions.

See Problem 2.

11. equidistant from points *M* and *N* and on a circle with center *M* and radius $= \frac{1}{2} MN$

12. 3 cm from $\overline{GH}$ and 5 cm from *G*, where *GH* = 4.5 cm

13. equidistant from the sides of $\angle PQR$ and on a circle with center *P* and radius *PQ*

14. equidistant from both points
A and *B* and points *C* and *D*

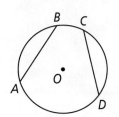

15. equidistant from the sides of
∠*JKL* and on ⊙*C*

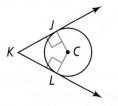

Describe each locus of points in space.

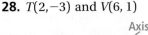 See Problem 3.

16. points 3 cm from a point *F*

17. points 4 cm from $\overleftrightarrow{DE}$

18. points 1 in. from plane *M*

19. points 5 mm from $\overrightarrow{PQ}$

B **Apply**

Describe the locus that each blue figure represents.

20.

21.

22.

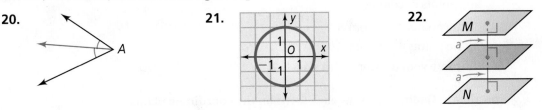

23. Open-Ended Give two examples of loci from everyday life, one in a plane and one in space.

24. Writing A classmate says that it is impossible to find a point equidistant from three collinear points. Is she correct? Explain.

25. Think About a Plan Write a locus description of the points highlighted in blue on the coordinate plane.
- How many conditions will be involved?
- What is the condition with respect to the origin?
- What are the conditions with respect to the *x*- and *y*-axes?

Coordinate Geometry **Write an equation for the locus of points in a plane equidistant from the two given points.**

26. *A*(0, 2) and *B*(2, 0)

27. *P*(1, 3) and *Q*(5, 1)

28. *T*(2, −3) and *V*(6, 1)

STEM **29. Meteorology** An anemometer measures wind speed and wind direction. In an anemometer, there are three cups mounted on an axis. Consider a point on the edge of one of the cups.

Axis

a. Describe the locus that this point traces as the cup spins in the wind.

b. Suppose the distance of the point from the axis of the anemometer is 2 in. Write an equation for the locus of part (a). Use the axis as the origin.

30. Landscaping The school board plans to construct a fountain in front of the school. What are all the possible locations for a fountain such that the fountain is 8 ft from the statue and 16 ft from the flagpole?

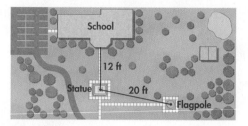

Make a drawing of each locus.

31. the path of a car as it turns to the right

32. the path of a doorknob as a door opens

33. the path of a knot in the middle of a jump-rope as it is being used

34. the path of the tip of your nose as you turn your head

35. the path of a fast-pitched softball

Ⓒ 36. Reasoning Points A and B are 5 cm apart. Do the following loci in a plane have any points in common?

  the points 3 cm from A

  the points 4 cm from $\overline{AB}$

Illustrate your answer with a sketch.

Ⓒ Coordinate Geometry Draw each locus on the coordinate plane.

37. all points 3 units from the origin

38. all points 2 units from $(-1, 3)$

39. all points 4 units from the y-axis

40. all points 5 units from $x = 2$

41. all points equidistant from $y = 3$ and $y = -1$

42. all points equidistant from $x = 4$ and $x = 5$

43. all points equidistant from the x- and y-axes

44. all points equidistant from $x = 3$ and $y = 2$

Ⓒ 45. a. Draw a segment to represent the base of an isosceles triangle. Locate three points that could be the vertex of the isosceles triangle.
  b. Describe the locus of possible vertices for the isosceles triangle.
  c. Writing Explain why points in the locus you described are the only possibilities for the vertex of the isosceles triangle.

46. Describe the locus of points in a plane 3 cm from the points on a circle with radius 8 cm.

47. Describe the locus of points in a plane 8 cm from the points on a circle with radius 3 cm.

48. Sketch the locus of points for the air valve on the tire of a bicycle as the bicycle moves down a straight path.

C Challenge

49. In the diagram, Moesha, Jan, and Leandra are seated at uniform distances around a circular table. Copy the diagram. Shade the points on the table that are closer to Moesha than to Jan or Leandra.

Playground Equipment Think about the path of a child on each piece of playground equipment. Draw the path from (a) a top view, (b) a front view, and (c) a side view.

50. a swing

51. a straight slide

52. a corkscrew slide

53. a merry-go-round

54. a firefighters' pole

Standardized Test Prep

SAT/ACT

55. What are the coordinates of the center of the circle whose equation is
$(x - 9)^2 + (y + 4)^2 = 1$?

Ⓐ $(3, -2)$ Ⓑ $(-3, 2)$ Ⓒ $(-9, 4)$ Ⓓ $(9, -4)$

56. A plane passes through two adjacent faces of a rectangular prism. The plane is perpendicular to the base of the prism. Which term is the most specific name for a figure formed by the cross section of the plane and the prism?

Ⓕ square   Ⓖ rectangle   Ⓗ parallelogram   Ⓘ kite

Short Response

57. Margie's cordless telephone can transmit up to 0.5 mi from her home. Carol's cordless telephone can transmit up to 0.25 mi from her home. Carol and Margie live 0.25 mi from each other. Can Carol's telephone work in a region that Margie's cannot? Sketch and label your diagram.

Mixed Review

Write an equation of the circle with center C and radius r.   ◀ See Lesson 12-5.

58. $C(6, -10), r = 5$ **59.** $C(1, 7), r = 6$ **60.** $C(-8, -1), r = \sqrt{13}$

Find the surface area of each figure to the nearest tenth.   ◀ See Lesson 11-2.

61.
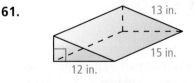
13 in.   15 in.   12 in.

62.
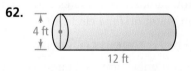
4 ft   12 ft

In ⊙O, find the area of sector AOB. Leave your answer in terms of π.   ◀ See Lesson 10-7.

63. $OA = 4, m\widehat{AB} = 90$ **64.** $OA = 8, m\widehat{AB} = 72$ **65.** $OA = 10, m\widehat{AB} = 36$

Pull It All Together © ASSESSMENT

To solve these problems you will pull together many concepts and skills that you have studied about relationships within circles.

BIG idea Reasoning and Proof

You can use triangle congruence theorems to prove relationships among tangents and secants.

© Performance Task 1

Four tangents are drawn from E to two concentric circles. A, B, C, and D are the points of tangency. Name as many pairs of congruent triangles as possible and tell how you can show each pair is congruent.

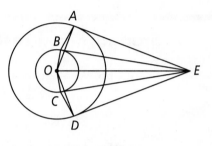

BIG idea Measurement

You can use facts about arcs and angle measures to solve real-world problems.

© Performance Task 2

The rocks near the shore between two lighthouses at points A and B make the waters unsafe. The measure of $\overset{\frown}{AXB}$ is 300. Waters inside this arc are unsafe. Suppose you are a navigator on a ship at sea. How can you use the lighthouses to keep the ship in safe waters? Explain your answer.

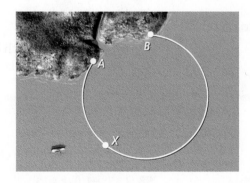

BIG idea Coordinate Geometry

You can use coordinates of the center of a circle and its radius to write an equation for a circle.

© Performance Task 3

A gardener wants the three rosebushes in her garden to be watered by a rotating water sprinkler. The gardener draws a diagram of the garden using a grid in which each unit represents 1 ft. The rosebushes are at (1, 3), (5, 11), and (11, 4). She wants to position the sprinkler at a point equidistant from each rosebush. Where should the gardener place the sprinkler? Show your work. What equation describes the boundary of the circular region that the sprinkler will cover?

Connecting **BIG** ideas and Answering the Essential Questions

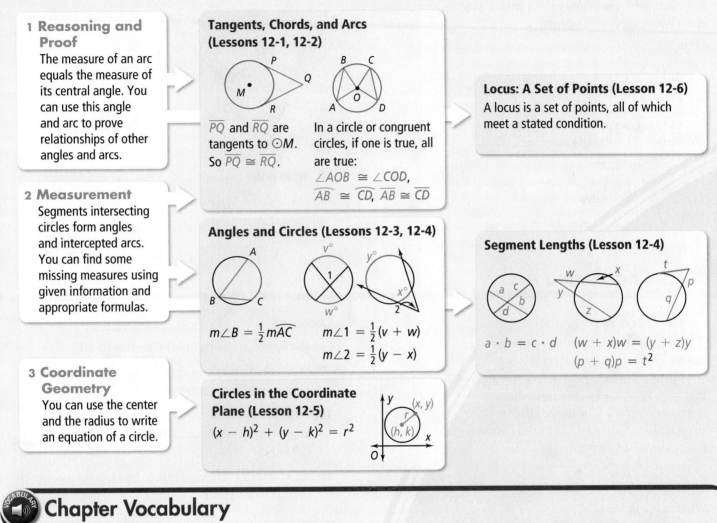

1 Reasoning and Proof

The measure of an arc equals the measure of its central angle. You can use this angle and arc to prove relationships of other angles and arcs.

Tangents, Chords, and Arcs (Lessons 12-1, 12-2)

$\overline{PQ}$ and $\overline{RQ}$ are tangents to $\odot M$. So $\overline{PQ} \cong \overline{RQ}$.

In a circle or congruent circles, if one is true, all are true:
$\angle AOB \cong \angle COD$,
$\overline{AB} \cong \overline{CD}$, $\overparen{AB} \cong \overparen{CD}$

Locus: A Set of Points (Lesson 12-6)

A locus is a set of points, all of which meet a stated condition.

2 Measurement

Segments intersecting circles form angles and intercepted arcs. You can find some missing measures using given information and appropriate formulas.

Angles and Circles (Lessons 12-3, 12-4)

$m\angle B = \frac{1}{2}m\overparen{AC}$ $m\angle 1 = \frac{1}{2}(v + w)$

$m\angle 2 = \frac{1}{2}(y - x)$

Segment Lengths (Lesson 12-4)

$a \cdot b = c \cdot d$ $(w + x)w = (y + z)y$

$(p + q)p = t^2$

3 Coordinate Geometry

You can use the center and the radius to write an equation of a circle.

Circles in the Coordinate Plane (Lesson 12-5)

$(x - h)^2 + (y - k)^2 = r^2$

Chapter Vocabulary

- chord (p. 771)
- inscribed angle (p. 780)
- intercepted arc (p. 780)
- locus (p. 804)
- point of tangency (p. 762)
- secant (p. 791)
- standard form of an equation of a circle (p. 799)
- tangent to a circle (p. 762)

Use the figure to choose the correct term to complete each sentence.

1. $\overleftrightarrow{EF}$ is (*a secant of, tangent to*) $\odot X$.

2. $\overline{DF}$ is a (*chord, locus*) of $\odot X$.

3. $\triangle ABC$ is made of (*chords in, tangents to*) $\odot X$.

4. $\angle DEF$ is an (*intercepted arc, inscribed angle*) of $\odot X$.

5. The set of all points equidistant from the endpoints of $\overline{CB}$ is a (*locus, tangent*).

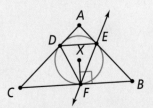

12-1 Tangent Lines

Quick Review

A **tangent** to a circle is a line that intersects the circle at exactly one point. The radius to that point is perpendicular to the tangent. From any point outside a circle, you can draw two segments tangent to a circle. Those segments are congruent.

Example

$\overrightarrow{PA}$ and $\overrightarrow{PB}$ are tangents. Find x.

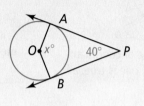

The radii are perpendicular to the tangents. Add the angle measures of the quadrilateral:

$$x + 90 + 90 + 40 = 360$$
$$x + 220 = 360$$
$$x = 140$$

Exercises

Use $\odot O$ for Exercises 6–8.

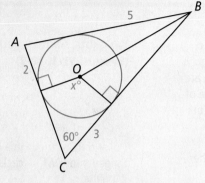

6. What is the perimeter of $\triangle ABC$?

7. $OB = \sqrt{28}$. What is the radius?

8. What is the value of x?

12-2 Chords and Arcs

Quick Review

A **chord** is a segment whose endpoints are on a circle. Congruent chords are equidistant from the center. A diameter that bisects a chord that is not a diameter is perpendicular to the chord. The perpendicular bisector of a chord contains the center of the circle.

Example

What is the value of d?

Since the chord is bisected, $m\angle ACB = 90$. The radius is 13 units. So an auxiliary segment from A to B is 13 units. Use the Pythagorean Theorem.

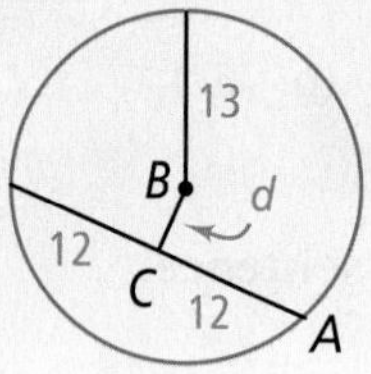

$$d^2 + 12^2 = 13^2$$
$$d^2 = 25$$
$$d = 5$$

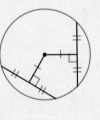

Exercises

Use the figure at the right for Exercises 9–11.

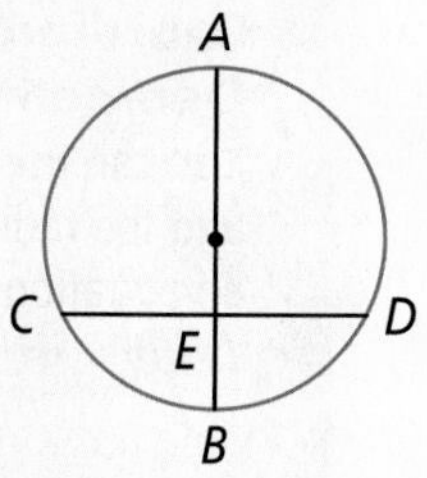

9. If $\overline{AB}$ is a diameter and $CE = ED$, then $m\angle AEC = \underline{\ ?\ }$.

10. If $\overline{AB}$ is a diameter and is at right angles to $\overline{CD}$, what is the ratio of CD to DE?

11. If $CE = \frac{1}{2}CD$ and $m\angle DEB = 90$, what is true of $\overline{AB}$?

Use the circle below for Exercises 12 and 13.

12. What is the value of x?

13. What is the value of y?

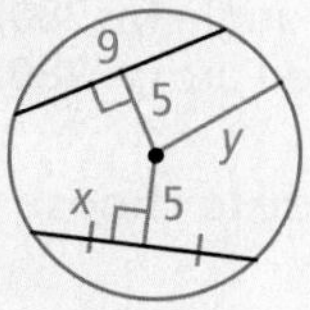

12-3 Inscribed Angles

Quick Review

An **inscribed angle** has its vertex on a circle and its sides are chords. An **intercepted arc** has its endpoints on the sides of an inscribed angle, and its other points in the interior of the angle. The measure of an inscribed angle is half the measure of its intercepted arc.

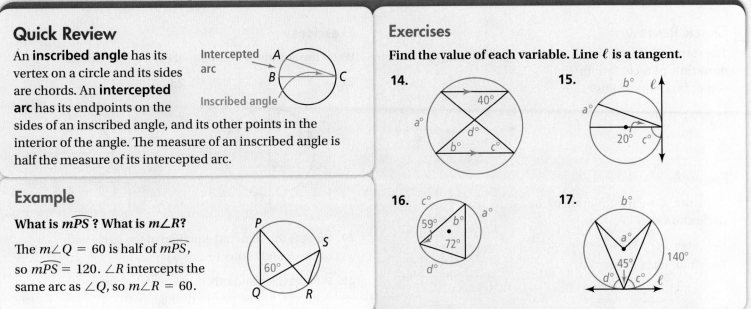

Intercepted arc

Inscribed angle

Example

What is $m\widehat{PS}$? What is $m\angle R$?

The $m\angle Q = 60$ is half of $m\widehat{PS}$, so $m\widehat{PS} = 120$. $\angle R$ intercepts the same arc as $\angle Q$, so $m\angle R = 60$.

Exercises

Find the value of each variable. Line ℓ is a tangent.

14.

15.

16.

17.

12-4 Angle Measures and Segment Lengths

Quick Review

A **secant** is a line that intersects a circle at two points. The following relationships are true:

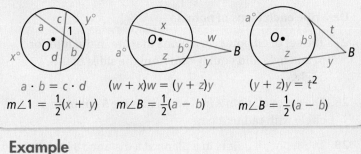

$a \cdot b = c \cdot d$

$m\angle 1 = \frac{1}{2}(x + y)$

$(w + x)w = (y + z)y$

$m\angle B = \frac{1}{2}(a - b)$

$(y + z)y = t^2$

$m\angle B = \frac{1}{2}(a - b)$

Example

What is the value of x?

$(x + 10)10 = (19 + 9)9$

$10x + 100 = 252$

$x = 15.2$

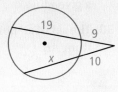

Exercises

Find the value of each variable.

18.

19.

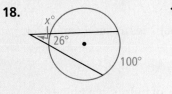

20.

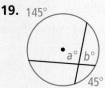

21.

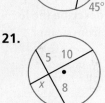

12-5 Circles in the Coordinate Plane

Quick Review

The **standard form of an equation of a circle** with center (h, k) and radius r is

$$(x - h)^2 + (y - k)^2 = r^2.$$

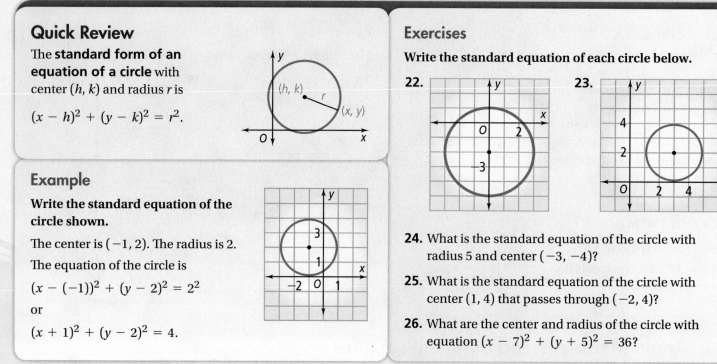

Example

Write the standard equation of the circle shown.

The center is $(-1, 2)$. The radius is 2.

The equation of the circle is

$$(x - (-1))^2 + (y - 2)^2 = 2^2$$

or

$$(x + 1)^2 + (y - 2)^2 = 4.$$

Exercises

Write the standard equation of each circle below.

22.

23.

24. What is the standard equation of the circle with radius 5 and center $(-3, -4)$?

25. What is the standard equation of the circle with center $(1, 4)$ that passes through $(-2, 4)$?

26. What are the center and radius of the circle with equation $(x - 7)^2 + (y + 5)^2 = 36$?

12-6 Locus: A Set of Points

Quick Review

A **locus** is a set of points that satisfies a stated condition.

Example

Sketch and describe the locus of points in a plane equidistant from points A and B.

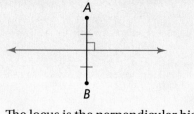

The locus is the perpendicular bisector of $\overline{AB}$.

Exercises

Describe each locus of points.

27. The set of all points in a plane that are in the interior of an angle and equidistant from the sides of the angle.

28. The set of all points in a plane that are 5 cm from a circle with radius 2 cm.

29. The set of all points in a plane at a distance 8 in. from a given line.

30. The set of all points in space that are a distance 6 in. from $\overline{AB}$.

Do you know HOW?

Algebra For Exercises 1–8, lines that appear tangent are tangent. Find the value of each variable. Round decimals to the nearest tenth.

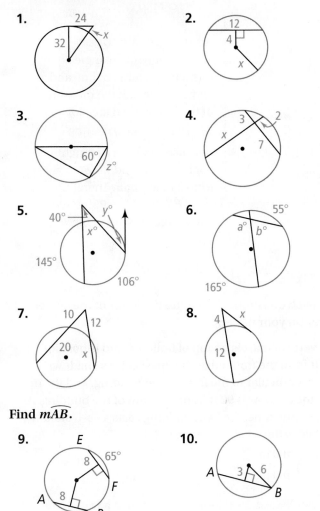

1.

2.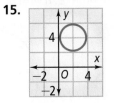

3.

4.

5.

6.

7.

8.

Find $m\widehat{AB}$.

9.

10.

11. Graph $(x + 3)^2 + (y - 2)^2 = 9$. Then label the center and radius.

12. Write an equation of the circle with center (3, 0) that passes through point (−2, −4).

13. What is the graph of $x^2 + y^2 = 0$?

14. Write an equation for the locus of points in the coordinate plane that are 4 units from (−5, 2).

Write the standard equation of each circle.

15.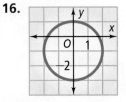

16.

Sketch each locus on a coordinate plane.

17. the set of all points 3 units from the line $y = -2$

18. the set of all points equidistant from the axes

Do you UNDERSTAND?

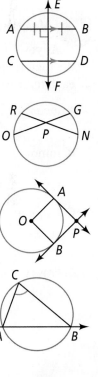

19. Writing What is special about a rhombus inscribed in a circle? Justify your answer.

20. Reasoning $\overleftrightarrow{EF}$ is the perpendicular bisector of chord $\overline{AB}$, and $\overline{CD} \parallel \overline{AB}$. Show that $\overleftrightarrow{EF}$ is the perpendicular bisector of chord $\overline{CD}$.

21. Error Analysis A student says that $\angle RPO \cong \angle NPG$ in this circle, since they are vertical angles, and thus $\widehat{RO} \cong \widehat{NG}$. Why is this incorrect?

22. Reasoning $\overleftrightarrow{PA}$ and $\overleftrightarrow{PB}$ are tangent to $\odot O$. PA is equal to the radius of the circle. What kind of quadrilateral is $PAOB$? Explain.

23. Reasoning A secant line passes through a circle at points A and B. Point C is also on the circle. Describe the locus of points P that satisfy these conditions: P and C are on the same side of the secant, and $m\angle APB = m\angle ACB$.

12 Cumulative Standards Review

ASSESSMENT

TIPS FOR SUCCESS

Some test questions require you to use the relationships between lines, segments, and circles. Read the sample question at the right. Then follow the tips to answer it.

In the figure below, $\overleftrightarrow{AC}$ is tangent to the circle at point B. If $m\widehat{BDE} = 156°$, what is $m\angle ABE$?

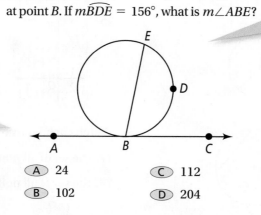

(A) 24
(B) 102
(C) 112
(D) 204

TIP 1

Because ∠ABE is an obtuse angle, you can eliminate choice A. Also, because the measure of a straight angle is 180° and choice D is greater than 180°, choice D can be eliminated.

TIP 2

Use the relationship between tangents, chords, angles, and intercepted arcs.

Think It Through

The measure of an angle formed by a tangent line and a chord is one half the measure of the intercepted arc. So, $m\angle EBC = \frac{1}{2}m\widehat{BDE} = 78°$. Because $\angle ABC$ is a straight angle, $\angle ABE$ and $\angle EBC$ form a linear pair. Therefore, $m\angle ABE = 180 - 78 = 102$. The correct answer is B.

Vocabulary Builder

As you solve test items, you must understand the meanings of mathematical terms. Match each term with its mathematical meaning.

A. locus	**I.** the intersection of a solid and a plane
B. minor arc	**II.** part of a circle that is smaller than a semicircle
C. major arc	**III.** a set of points, all of which meet a stated condition
D. cross section	**IV.** a part of a circle that is larger than a semicircle

Multiple Choice

Read each question. Then write the letter of the correct answer on your paper.

1. A vertical mast is on top of building and is positioned 6 ft from the front edge. The mast casts a shadow perpendicular to the front of the building, and the tip of the shadow is 90 ft from the front of the building. At the same time, the 24-ft building casts a 64-ft shadow. What is the height of the mast?

(A) 7 ft 6 in.
(B) 9 ft 9 in.
(C) 12 ft
(D) 33 ft

2. Javier leans a 20-ft long ladder against a wall. If the base of the ladder is positioned 5 feet from the wall, how high up the wall does the ladder reach? Round to the nearest tenth.

(F) 19.4 ft
(G) 18.8 ft
(H) 17.5 ft
(I) 15 ft

3. △*FGH* has the vertices shown below. If the triangle is rotated 90° counterclockwise about the origin, what are the coordinates of the rotated point *F′*?

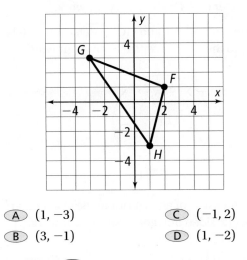

- (A) (1, −3)
- (B) (3, −1)
- (C) (−1, 2)
- (D) (1, −2)

4. What is $m\widehat{RT}$ in the figure at the right?

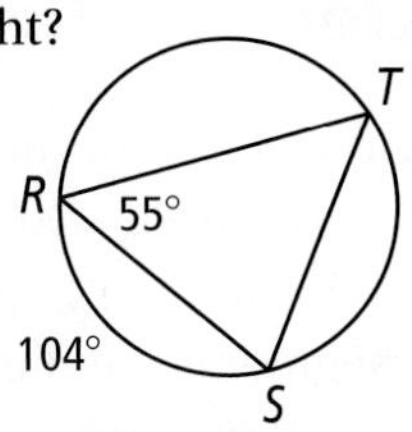

- (F) 162°
- (G) 146°
- (H) 110°
- (I) 73°

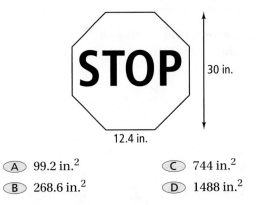

5. A rectangular prism has a volume of 204 cubic inches. A similar prism is created scaling each dimension by a factor of 2. What is the volume of the larger prism?

- (A) 102 in.^{3}
- (B) 408 in.^{3}
- (C) 816 in.^{3}
- (D) 1632 in.^{3}

6. What is the value of *x*?

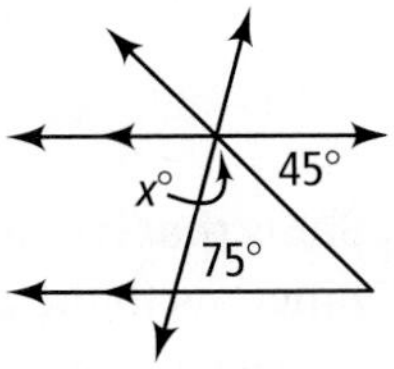

- (F) 45°
- (G) 75°
- (H) 60°
- (I) 105°

7. A bicycle tire has a radius of 14.5 inches. About how far does the bicycle travel if the tire makes 15 complete revolutions? Use 3.14 for π.

- (A) 57 ft
- (B) 114 ft
- (C) 825 ft
- (D) 1366 ft

8. What is the measure of *x* in the figure shown below? What is the value of *x*?

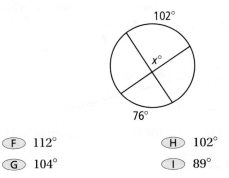

- (F) 112°
- (G) 104°
- (H) 102°
- (I) 89°

9. A stop sign is shaped like a regular octagon with the dimensions shown. What is the area of the stop sign?

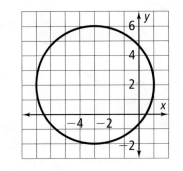

30 in.

12.4 in.

- (A) 99.2 in.^{2}
- (B) 268.6 in.^{2}
- (C) 744 in.^{2}
- (D) 1488 in.^{2}

10. What is the equation of the circle shown on the coordinate grid below?

- (F) $(x + 3)^2 + (y − 2)^2 = 4$
- (G) $(x + 3)^2 + (y − 2)^2 = 16$
- (H) $(x − 3)^2 + (y + 2)^2 = 4$
- (I) $(x − 3)^2 + (y + 2)^2 = 16$

11. A ski ramp on a lake has the dimensions shown below. To the nearest hundredth of a meter, what is the height h of the ramp?

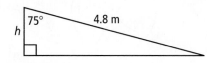

12. What is the value of n in the trapezoid shown below?

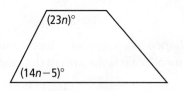

13. In the figure shown below, $\overline{MN} \parallel \overline{OP}$, $LM = 12$, $MN = 15$, and $MO = 6$. What is OP?

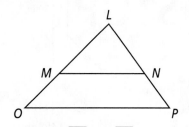

14. In the figure below, $\overline{XY}$ and $\overline{XZ}$ are tangent to $\odot O$ at points Y and Z, respectively. What is $m\angle YOZ$?

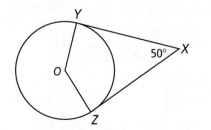

15. An aluminum juice can is shaped like a cylinder with a diameter of 7 centimeters and a height of 12 centimeters. To the nearest square centimeter, how much material was needed to make the can? Use 3.14 for π.

Short Response

16. $\triangle DEF$ has vertices $D(1, 1)$, $E(-2, 4)$, and $F(4, 7)$. What is the perimeter of $\triangle DEF$? Show your work.

17. In $\odot A$ below, $AE = 13.1$ and $\overline{AC} \perp \overline{BD}$. If $BC = 6.8$, what is AC, to the nearest tenth? Show your work.

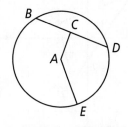

18. What is the measure of an exterior angle of a regular dodecagon (12-sided polygon)? Show your work.

Extended Response

19. The endpoints of $\overline{HK}$ are $H(-3, 7)$ and $K(6, 1)$, and the endpoints of $\overline{MN}$ are $M(-5, -8)$ and $N(7, 10)$. What are the slopes of the line segments? Are the line segments parallel, perpendicular, or neither? Explain how you know and show your work.

20. Suppose a square is inscribed in a circle such as square $HIJK$ shown below.

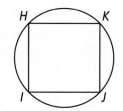

a. Show that if you form a new figure by connecting the tangents to the circle at H, I, J, and K, the new figure is also a square.

b. The inscribed square and the square formed by the tangents are similar. What is the scale factor of the similar figures?

c. Let a regular polygon with n-sides be inscribed in a circle. Do the tangent lines at the vertices of the polygon form another regular polygon with n-sides? Explain.

Get Ready!

CHAPTER 13

Skills Handbook, Page 895

Converting Between Percents, Fractions, and Decimals

Write two equivalent forms for each of the following.

1. 0.7875　　　　**2.** $\frac{7}{8}$　　　　**3.** 0.95　　　　**4.** 60%

Skills Handbook, Page 891

Simplifying Ratios

Simplify each ratio.

5. $\frac{45}{60}$　　　　**6.** $\frac{24}{36}$　　　　**7.** $\frac{7 \cdot 6}{4 \cdot 3}$

Skills Handbook, Page 890

Evaluating Expressions

Evaluate each expression for $x = 6$ and $y = -2$.

8. $9x - 5y + 8$　　　　**9.** $\frac{x}{y}$　　　　**10.** $\frac{2x + y}{y^2}$

Skills Handbook, Page 894

Solving Equations

Solve each equation.

11. $2n + 7 = 23$　　　　**12.** $3(n - 2) = 8$　　　　**13.** $\frac{5}{7}n - 12 = 15$

Lesson 10-8

Finding Geometric Probability

14. What is the probability that a point chosen at random is in the inner circle of the figure at the right?

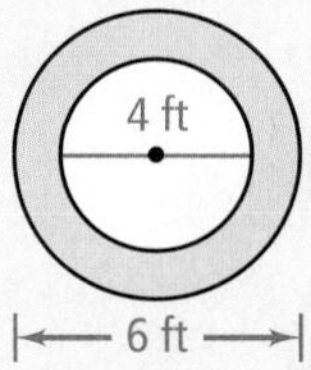

4 ft

|← 6 ft →|

Looking Ahead Vocabulary

15. When you describe the likelihood that a baseball player will get a hit, you are finding an *experimental probability*. What do you think is the experiment in this situation?

16. People training to be pilots sometimes use flight simulators, machines that create a simulation of actual flight. What do you suppose are some ways *simulations* could be used in a math class?

17. When the occurrence of one event does not affect the occurrence of another event, they are called *independent events*. Do you think the events *tossing heads with a penny* and *tossing tails with a quarter* are independent events? Explain.

Probability

© **DOMAINS**
- Conditional Probability and Rules of Probability
- Using Probability to Make Decisions

There are more seniors with cars than there are senior parking spaces. In this chapter you'll learn how to use probability to find ways to share the parking spaces fairly.

Vocabulary

English/Spanish Vocabulary Audio Online:

English	Spanish
combination, *p. 838*	combinación
conditional probability, *p. 851*	probabilidad condicional
dependent events, *p. 844*	sucesos dependientes
experimental probability, *p. 825*	probabilidad experimental
independent events, *p. 844*	sucesos independientes
mutually exclusive events, *p. 845*	sucesos mutuamente excluyentes
permutation, *p.837*	permutación
sample space, *p. 824*	espacio de muestral
theoretical probability, *p. 825*	probabilidad teórica

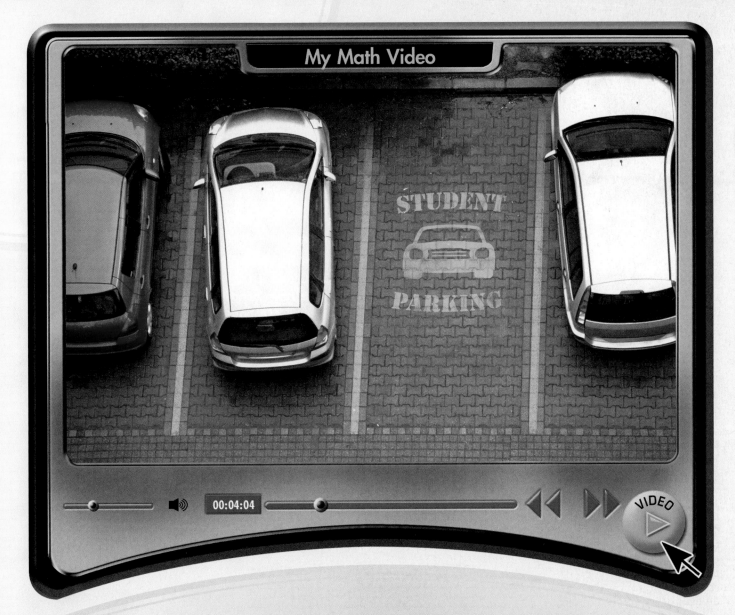

My Math Video

00:04:04

VIDEO

BIG ideas

1 **Probability**
Essential Question What is the difference between experimental probability and theoretical probability?

2 **Data Representation**
Essential Question What is a frequency table?

3 **Probability**
Essential Question What does it mean for an event to be random?

Chapter Preview

13-1 **Experimental and Theoretical Probability**

13-2 **Probability Distributions and Frequency Tables**

13-3 **Permutations and Combinations**

13-4 **Compound Probability**

13-5 **Probability Models**

13-6 **Conditional Probability Formulas**

13-7 **Modeling Randomness**

13-1 Experimental and Theoretical Probability

© **Content Standards**
S.CP.1 Describe events as subsets of a sample space using characteristics of the outcomes, or as unions, intersections, or complements of other events.
Also S.CP.4

Objectives To calculate experimental and theoretical probability

Getting Ready!

You find an old clock in your attic. The hour hand of the clock is broken off. Between 12:00 and 2:00, how many positions are possible for the hour hand? For a 12-hour period, how many positions are possible for the hour hand? What is the probability that the clock stopped some time between 12:00 and 2:00?

Visualize where the hour hand could be.

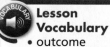 **MATHEMATICAL PRACTICES**

Lesson Vocabulary
• outcome
• event
• sample space
• probability
• experimental probability
• theoretical probability
• complement of an event

In the Solve It, you probably considered where the hour hand would be based on where the minute hand is. In the language of probability, this position would be a *favorable outcome*. An **outcome** is the possible result of a situation or experiment. An **event** may be a single outcome or a group of outcomes. For the clock, the hour hand being about halfway between 12 and 1 or about halfway between 1 and 2 are two favorable outcomes for the event of where the hour hand may stop between 12:00 and 2:00. The set of all possible outcomes is the **sample space**.

Essential Understanding Probability is a measure of the likelihood that an event will occur.

> **Key Concept** Probability
>
> **Definition**
> If the outcomes in a sample space are equally likely to occur, the **probability** of an event $P(\text{event})$ is a numerical value from 0 to 1 that measures the likelihood of an event.
>
> $$P(\text{event}) = \frac{\text{number of favorable outcomes}}{\text{number of possible outcomes}}$$

You can write the probability of an event as a ratio, decimal, or percent.

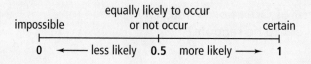

equally likely to occur
impossible   or not occur   certain
├──────────────────┼──────────────────┤
0   ← less likely   0.5   more likely →   1

You can find probabilities by using the results of an experiment or by reasoning mathematically.

take note

Key Concept   Experimental Probability

Experimental probability of an event measures the likelihood that the event occurs based on the actual results of an experiment.

$$P(\text{event}) = \frac{\text{number of times the event occurs}}{\text{number of times the experiment is done}} \quad \begin{array}{l} \leftarrow \text{number of favorable outcomes} \\ \leftarrow \text{number of possible outcomes} \end{array}$$

Problem 1   Calculating Experimental Probability

Quality Control A quality control inspector samples 500 LCD monitors and finds defects in three of them.

A **What is the experimental probability that a monitor selected at random will have a defect?**

$$P(\text{defect}) = \frac{\text{number of monitors with a defect}}{\text{number of monitors inspected}}$$

$$= \frac{3}{500}$$

$$= 0.006 \text{ or } 0.6\%$$

Think

You can report the result as a fraction, decimal, or percent. Use the form that will communicate most clearly.

The experimental probability that a monitor selected at random is defective is 0.6%.

B **If the company manufactures 15,240 monitors in a month, how many are likely to have a defect based on the quality inspector's results?**

$$\text{number of defective monitors} = P(\text{defect}) \cdot \text{total number of monitors}$$

$$= 0.006 \cdot 15,240$$

$$= 91.44$$

It is likely that approximately 91 monitors are defective.

Got It? **1.** A park has 538 trees. You choose 40 at random and determine that 25 are maple trees. What is the experimental probability that a tree chosen at random is a maple tree? About how many trees in the park are likely to be maple trees?

When a sample space consists of real data, you can find the experimental probability. **Theoretical probability** describes the likelihood of an event based on mathematical reasoning.

This chapter uses *standard number cubes* to illustrate probability. A standard number cube has 6 faces with a number from 1 to 6 on each face. No number is used twice.

Think

What results should you be looking for?
Any two cubes that result in the sum of 7, like a 1 and a 6.

What is the probability of rolling numbers that add to 7 when rolling two standard number cubes?

Step 1 Make a table of the possible results for the rolls of two number cubes. Circle the ones that sum to 7.

Step 2 Find the number of possible outcomes for the event that the sum of two cubes is 7.

Step 3 Find the probability.

$$P(\text{rolling a sum of 7}) = \frac{6}{36}$$

The probability of rolling numbers that add to 7 is $\frac{6}{36}$, or $\frac{1}{6}$.

	1	2	3	4	5	6
1	1,1	2,1	3,1	4,1	5,1	(6,1)
2	1,2	2,2	3,2	4,2	(5,2)	6,2
3	1,3	2,3	3,3	(4,3)	5,3	6,3
4	1,4	2,4	(3,4)	4,4	5,4	6,4
5	1,5	(2,5)	3,5	4,5	5,5	6,5
6	(1,6)	2,6	3,6	4,6	5,6	6,6

Got It? **2.** What is the probability of getting each sum when rolling two standard number cubes?

    a. 9   **b.** 2   **c.** 13

The **complement of an event** consists of all of the possible outcomes in the sample space that are not part of the event. For example, if you roll a standard number cube, the probability of rolling a number less than 3 is $P(\text{rolling} < 3) = \frac{2}{6}$, or $\frac{1}{3}$. The probability of *not* rolling a number less than 3 is $P(\text{not} < 3) = \frac{4}{6}$, or $\frac{2}{3}$.

Key Concept **Probability of a Complement**

The sum of the probability of an event and the probability of its complement is 1.

$$P(\text{event}) + P(\text{not event}) = 1 \qquad P(\text{not event}) = 1 - P(\text{event})$$

 **Problem 3** **Using Probabilities of Events and Their Complements**

Think

Can you find P(not green) another way?
You can find the total number of marbles that are not green, and then divide by the total number of marbles: $\frac{(10 + 5 + 6)}{29}$.

A jar contains 10 red marbles, 8 green marbles, 5 blue marbles, and 6 white marbles. What is the probability that a randomly selected marble is not green?

$$P(\text{not green}) = 1 - P(\text{green}) \qquad \text{Probability of the complement}$$
$$= 1 - \frac{8}{29} \qquad \text{Find } P(\text{green}).$$
$$= \frac{21}{29} \qquad \text{Simplify.}$$

The probability that the chosen marble is not green is $\frac{21}{29}$.

Got It? **3.** What is the probability that a randomly chosen marble is not red?

Lesson Check

Do you know HOW?

Use the spinner to find each theoretical probability.

1. P(an even number)

2. P(a number greater than 5)

3. P(a prime number)

Do you UNDERSTAND?  MATHEMATICAL PRACTICES

4. Vocabulary How are experimental and theoretical probability similar? How are they different?

5. Open-Ended Give an example of an impossible event.

6. Error Analysis Your friend says that the probability of rolling a number less than 7 on a standard number cube is 100. Explain your friend's error and find the correct probability.

Practice and Problem-Solving Exercises MATHEMATICAL PRACTICES

Ⓐ Practice

7. A baseball player got a hit 19 times of his last 64 times at bat.

   a. What is the experimental probability that the player got a hit?

   b. If the player comes up to bat 200 times in a season, about how many hits is he likely to get?

◀ **See Problem 1.**

8. A medical study tests a new cough medicine on 4,250 people. It is effective for 3982 people. What is the experimental probability that the medicine is effective? For a group of 9000 people, predict the approximate number of people for whom the medicine will be effective.

A bag contains letter tiles that spell the name of the state MISSISSIPPI. Find the theoretical probability of drawing one tile at random for each of the following.

◀ **See Problems 2 and 3.**

9. P(M)

10. P(I)

11. P(S)

12. P(P)

13. P(not M)

14. P(not I)

15. P(not S)

16. P(not P)

Ⓑ Apply

17. Think About a Plan Suppose that you flip 3 coins. What is the theoretical probability of getting at least 2 heads?

   • What is the sample space of possible outcomes?

   • What are the favorable outcomes?

18. Music A music collection includes 10 rock CDs, 8 country CDs, 5 classical CDs, and 7 hip hop CDs.

   a. What is the probability that a CD randomly selected from the collection is a classical CD?

   b. What is the probability that a CD randomly selected from the collection is not classical CD?

19. You are playing a board game with a standard number cube. It is your last turn and if you roll a number greater than 2, you will win the game. What is the probability that you will not win the game?

STEM **20. Weather** If there is a 70% chance of snow this weekend, what is the probability that it will not snow?

21. Quality Control From 15,000 graphing calculators produced by a manufacturer, an inspector selects a random sample of 450 calculators and finds 4 defective calculators. Estimate the total number of defective calculators out of the 15,000.

22. Suppose you choose a letter at random from the word shown below. What is the probability that you will not choose a B?

PROBABILITY

A student randomly selected 65 vehicles in the student parking lot and noted the color of each. She found that 9 were black, 10 were blue, 13 were brown, 7 were green, 12 were red, and 14 were a variety of other colors. What is each experimental probability?

23. $P(\text{red})$

24. $P(\text{black})$

25. $P(\text{not blue})$

26. $P(\text{not green})$

STEM **27. Genetics** Genetics was first studied by Gregor Mendel, who experimented with pea plants. He crossed pea plants having yellow, round seeds with pea plants having green, wrinkled seeds. The following are the probabilities for each type of new seed.

yellow, round: 56.25%   yellow, wrinkled: 18.75%

green, round: 18.75%   green, wrinkled: 6.25%

If 2014 seeds were produced, how many of each variety would you expect?

ⓒ **Reasoning** For Exercises 28–31, describe each of the following situations using one of the following probabilities. Explain your answer.

    I. 0   **II.** between 0 and 0.5   **III.** between 0.5 and 1   **IV.** 1

28. having school on Tuesday

29. two elephants in the city zoo having the same weight

30. getting your driver's license at the age of 10

31. turning on the TV while a commercial is playing

ⓒ **Challenge** **32.** The students in a math class took turns rolling a standard number cube. The results are shown in the table at the right.
    a. What is the theoretical probability of rolling the number 1 with the number cube?
    b. What was the experimental probability of rolling the number 1 for the experiment in class?

Number Cube Experiment						
Outcome	1	2	3	4	5	6
Times Rolled	39	40	47	42	38	44

33. Another way to express probability is with *odds*. Odds compare the number of favorable outcomes to the number of unfavorable outcomes. Odds in favor of an event are usually written as

number of favorable outcomes : number of unfavorable outcomes.

Suppose the probability of drawing a red marble from a bag of marbles is $\frac{3}{10}$.
   a. What are the odds in favor of drawing a red marble?
   b. What are the odds against drawing a red marble?

Standardized Test Prep

SAT/ACT

34. Which statement about quadrilaterals is NOT true?

   Ⓐ All squares are also rhombuses, rectangles, and parallelograms.

   Ⓑ A parallelogram with at least 2 right angles is a rectangle.

   Ⓒ A kite has two pairs of consecutive sides that are congruent.

   Ⓓ A rhombus has exactly two equal sides.

35. Which of the following could be the measures of the sides of a right triangle?
   Ⓐ 6 ft, 10 ft, 15 ft   Ⓒ 7 in., 24 in., 25 in.
   Ⓑ 26 mm, 24 mm, 8 mm   Ⓓ 9 m, 41 m, 42 m

36. An angle inscribed in a circle is a right angle. What is the measure of the angle's intercepted arc?
   Ⓐ 90   Ⓑ 120   Ⓒ 180   Ⓓ 270

Short Response

37. One side of a rectangle measures 6 in. The rectangle's diagonals measure $\frac{5}{6}$ ft. What is the area of the rectangle in square inches?

Mixed Review

Properly name each locus of points in a coordinate plane.

See Lesson 12-6.

38. the points exactly 5 units from the origin

39. the points equidistant from $y = x$ and $y = -x$

Get Ready!   **To Prepare for Lesson 13-2, do Exercises 40–42.**

Simplify each ratio.

See p. 891.

40. $\frac{21}{35}$ **41.** $\frac{18}{42}$ **42.** $\frac{9}{24}$

Probability Distributions and Frequency Tables

© **Content Standards**

S.CP.4 Construct and interpret two-way frequency tables of data when two categories are associated with each object being classified . . .

Also S.CP.5

Objective   To make and use frequency tables and probability distributions

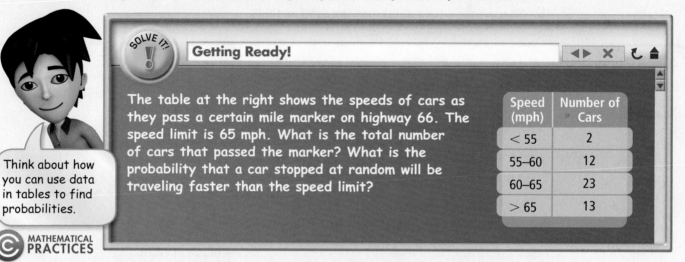

Think about how you can use data in tables to find probabilities.

© MATHEMATICAL PRACTICES

Getting Ready!

The table at the right shows the speeds of cars as they pass a certain mile marker on highway 66. The speed limit is 65 mph. What is the total number of cars that passed the marker? What is the probability that a car stopped at random will be traveling faster than the speed limit?

Speed (mph)	Number of Cars
< 55	2
55–60	12
60–65	23
> 65	13

In the Solve It, you used information from the table to calculate probability. A **frequency table** is a data display that shows how often an item appears in a category.

Essential Understanding   You can use data organized in tables that show frequencies to find probabilities.

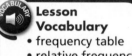

Lesson Vocabulary
- frequency table
- relative frequency
- probability distribution

take note

Key Concept   Relative Frequency

Relative frequency is the ratio of the frequency of the category to the total frequency.

© **Problem 1**   **Finding Relative Frequencies**

Plan

How do you find the denominator for a relative frequency?
Find the sum of the frequencies in the frequency table.

Surveys The results of a survey of students' music preferences are organized in this frequency table. What is the relative frequency of preference for rock music?

Use the frequency table to find the number of times rock music is chosen as the preference, and the total number of survey results.

$$\text{relative frequency} = \frac{\text{frequency of rock music preference}}{\text{total frequency}}$$
$$= \frac{10}{10 + 7 + 8 + 5 + 6 + 4} = \frac{10}{40} = \frac{1}{4}$$

The relative frequency of preference for rock music is $\frac{1}{4}$.

Type of Music Preferred	Frequency
Rock	10
Hip Hop	7
Country	8
Classical	5
Alternative	6
Other	4

Got It? 1. What is the relative frequency of preference for each type of music?

  a. classical

  b. hip hop

  c. country

  d. Critical Thinking Without calculating the rest of the relative frequencies, what is the sum of the relative frequencies for all the types of music? Explain how you know.

You can use relative frequency to approximate the probabilities of events.

Problem 2 **Calculating Probability by Using Relative Frequencies**

A student conducts a probability experiment by tossing 3 coins one after the other. Using the results below, what is the probability that exactly two heads will occur in the next three tosses?

Coin Toss Result	HHH	HHT	HTT	HTH	THH	THT	TTT	TTH
Frequency	5	7	9	6	2	9	10	2

Step 1 **Find the number of times a trial results in exactly two heads.**

The possible results that show exactly two heads are HHT, HTH, and THH.

The frequency of these results is $7 + 6 + 2 = 15$.

Step 2 **Find the total of all the frequencies.**

$5 + 7 + 9 + 6 + 2 + 9 + 10 + 2 = 50$

Think

How can you use the frequency table to find the probability?
The relative frequency is an approximation of the overall probability of the result.

Step 3 **Find the relative frequency of a trial with exactly two heads.**

$$\text{relative frequency} = \frac{\text{frequency of exactly two heads}}{\text{total of the frequencies}} = \frac{15}{50} = \frac{3}{10}$$

Based on the data collected, the probability that the next toss will be exactly two heads is $\frac{3}{10}$.

Got It? 2. A student conducts a probability experiment by spinning the spinner shown. Using the results in the frequency table, what is the probability of the spinner pointing at 4 on the next spin?

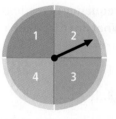

Spinner Result	1	2	3	4
Frequency	29	32	21	18

A **probability distribution** shows the probability of each possible outcome.
A probability distribution can be shown in a frequency table.

Finding a Probability Distribution

Archery In a recent competition, 50 archers shot 6 arrows each at a target. Three archers hit no bull's eyes; 5 hit one bull's eye; 7 hit two bull's eyes; 7 hit three bull's eyes; 11 hit four bull's eyes; 10 hit five bull's eyes; and 7 hit six bull's eyes. What is the probability distribution for the number of bull's eyes each archer hit?

Know	Need	Plan
The possible outcomes and the frequency of each outcome.	The probabilities of each outcome.	Make a frequency table and use relative frequencies to complete the probability distribution.

First, create a frequency table showing all of the possible outcomes: 0, 1, 2, 3, 4, 5, or 6 bull's eyes and the frequencies for each.

Next, use the table to find the relative frequencies for each number of bull's eyes. The relative frequencies are the probability distribution.

Probability Distribution of Bull's Eyes Hits							
Number of Bull's Eyes Hit	0	1	2	3	4	5	6
Frequency	3	5	7	7	11	10	7
Probability	$\frac{3}{50}$	$\frac{5}{50}$	$\frac{7}{50}$	$\frac{7}{50}$	$\frac{11}{50}$	$\frac{10}{50}$	$\frac{7}{50}$

Got It? **3.** On a math test, there were 10 scores between 90 and 100, 12 scores between 80 and 89, 15 scores between 70 and 79, 8 scores between 60 and 69, and 2 scores below 60. What is the probability distribution for the test scores?

Lesson Check

Do you know HOW?

Consumer Research The results of a survey show the frequency of responses for preferred music formats. What are the relative frequencies of the following formats?

1. CD   **2.** MP3

3. Blu-ray   **4.** Radio

5. What is the probability distribution for the music formats?

Preferred Format	Frequency
CD	54
Radio	50
Blu-ray	10
MP3	28

Do you UNDERSTAND? 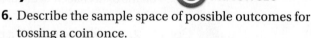 MATHEMATICAL PRACTICES

6. Describe the sample space of possible outcomes for tossing a coin once.

7. Error Analysis A friend says that her sweepstakes ticket will either win or not win, so the theoretical probability of it winning must be 50%. Explain your friend's error.

Practice and Problem-Solving Exercises

 MATHEMATICAL PRACTICES

A Practice

Blood Drive The honor society at a local high school sponsors a blood drive. High school juniors and seniors who weigh over 110 pounds may donate. The table at the right indicates the frequency of donor blood type.

◀ See Problem 1.

8. What is the relative frequency of blood type AB?

9. What is the relative frequency of blood type A?

10. Which blood type has the highest relative frequency? What is the relative frequency for this blood type?

11. The blood drive is extended for a second day, and the frequency doubles for each blood type. Do the relative frequencies for each blood type change? Explain.

Blood Drive Results

Blood Type	Frequency
O	30
A	25
B	6
AB	2

12. The data collected for new students enrolled at a community college show that 26 are under the age of eighteen, 395 are between the ages of eighteen and twenty-two, 253 are between the ages of twenty-three and twenty-seven, 139 are between the ages of twenty-eight and thirty-two, and 187 are over the age of thirty-two. What is the probability distribution for these data?

◀ See Problems 2 and 3.

Twenty-three preschoolers were asked what their favorite snacks are. The results are shown in the bar graph at the right.

13. What is the probability that a preschooler chosen at random chose popcorn as their favorite snack?

14. What is the probability that a preschooler chosen at random did not choose bananas as their favorite snack?

15. What is the probability distribution for the data in the graph?

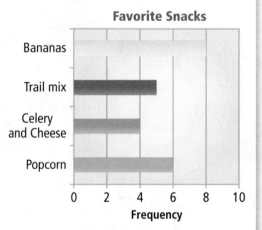

B Apply

© 16. **Reasoning** The possible outcomes for a spinner are 1, 2, 3, 4, 5, and 6.
The outcomes are equally likely. What is a probability distribution of the possible outcomes for the spinner?

17. **Text Messages** The table at the right shows number of text messages sent in one month by students at Metro High School is shown in the table at the right.
   a. If a student is chosen at random, what is the probability that the student sends 1500 or fewer text messages in one month?
   b. If a student is chosen at random, what is the probability that the student sends more than 1500 messages in a month?

Number of Text Messages t	Number of Students
$t \leq 500$	25
$500 < t \leq 1500$	120
$1500 < t \leq 2500$	300
$t > 2500$	538

18. Computers The results of a survey of 80 households in Westville are shown at the right.

Computer Survey

Number of Computers	Frequency
0	12
1	29
2	31
3	6
More than 3	2

   a. What is the probability distribution of the number of computers in Westville households?

   b. If Westville has 15,200 households, predict the number of households that will have exactly 3 computers.

   c. How many households will have either two or three computers?

19. Think About a Plan You can make a probability distribution table for theoretical probabilities. What is a probability distribution for four tosses of a single coin?

- What are all of the possible outcomes?
- What is the probability of each outcome?
- How can you display this in a table?

20. Employment The table below shows the number of people (in thousands) working in each occupational category, according to the U.S. Bureau of Labor Statistics.

U.S. Occupational Categories of Employed Workers

Occupational Category	Management, professional, and related	Service	Sales and office	Natural resources, construction, maintenance	Production, transportation, and material moving
Number (thousands)	9773	2271	1456	289	177

   a. Make a table showing the relative frequency for each occupational category.

   b. If an employed person is randomly selected, what is the likelihood the person works in the Service category? Explain how you know.

21. a. Make a probability distribution for the sum of the faces after rolling two standard number cubes.

   b. Are the probabilities theoretical or based on experimental results? Explain.

22. A student chose 100 letters at random from a page of a textbook. Partial results are in the table at the right. Make a probability distribution for the data. Include a category for all other letters not represented by the table.

23. Error Analysis The results of a survey about students' favorite type of pet are shown in the probability distribution below. What is wrong with this probability distribution? What information do you need to correct the probability distribution?

Letter	Tally
a	卌 III
e	卌 卌 IIII
i	卌 II
n	卌 I
o	卌 III
r	IIII
s	卌 I
t	卌 IIII

Favorite Pet	Dogs	Cats	Birds	Fish	Hamsters
Probability	0.36	0.34	0.15	0.08	0.12

24. A spinner has an equal number of even and odd numbers. The relative probability of getting an outcome of an even number on any spin is 50%. Make a probability distribution for even numbers for two spins, for three spins, and for four spins. How does the probability distribution change for each additional spin?

Standardized Test Prep

25. One hundred fifty students were asked the number of hours they spend on homework each night. Twenty students responded that they spend less than 1 hour on homework each night. What is the relative frequency of this response?

   Ⓐ 0.13   Ⓑ 0.87   Ⓒ 20   Ⓓ 130

26. Which pair of angles are supplementary?

   Ⓐ 110° and 70°   Ⓑ 25° and 65°

   Ⓒ 90° and 80°   Ⓓ 160° and 200°

27. What is the maximum number of right angles possible in a pentagon?

   Ⓐ 1   Ⓑ 2   Ⓒ 3   Ⓓ 4

28. What is the measure of an interior angle of a regular octagon?

Mixed Review

Determine whether each of the following is an experimental or theoretical probability. Explain your choice.

   ◀ **See Lesson 13-1.**

29. the probability of rolling two 5's when rolling two standard number cubes

30. the probability that a poll respondent chooses red as their favorite color

31. the probability of a number being even when randomly chosen from the whole numbers less than 11

The pairs of figures are similar. Find *x*.

   ◀ **See Lesson 7-1.**

32.

33.

Get Ready! **To Prepare for Lesson 13-3, do Exercises 34–36.**

   ◀ **See p. 891.**

Simplify each ratio.

34. $\dfrac{ab}{bc}$ **35.** $\dfrac{24 \cdot 15}{8 \cdot 3}$ **36.** $\dfrac{10 \cdot 9 \cdot 8 \cdot 7 \cdot 6 \cdot 5 \cdot 4 \cdot 3 \cdot 2 \cdot 1}{8 \cdot 7 \cdot 6 \cdot 5 \cdot 4 \cdot 3 \cdot 2 \cdot 1}$

13-3

Permutations and Combinations

© Content Standard
Prepares for S.CP.9 Use permutations and combinations to compute probabilities of compound events and solve problems.

Objectives To use permutations and combinations to solve problems

SOLVE IT!

Getting Ready! ◀▶ ✕ ↺ ⬆

In Chemistry class, you and your lab partner must add the samples in the test tubes to a mixture. The reactions that occur depend on the order in which you add them. How many different ways can you add the samples to the mixture?

Make a plan to find all of the possible ways to add the samples.

© MATHEMATICAL PRACTICES

In the Solve It, you may have drawn a diagram or listed all of the different possible ways to add the samples. Sometimes there are so many possibilities that listing them all is not practical.

Essential Understanding You can use counting techniques to find all of the possible ways to complete different tasks or choose items from a list.

VOCABULARY
Lesson Vocabulary
• Fundamental Counting Principle
• permutation
• n factorial
• combination

take note

Key Concept Fundamental Counting Principle

The **Fundamental Counting Principle** says that if event M occurs in m ways and event N occurs in n ways, then event M followed by event N can occur in $m \cdot n$ ways.

Example 4 entrees and 6 drinks gives $4 \cdot 6 = 24$ possible lunch specials

Here's Why It Works Suppose that you are choosing an outfit by first selecting blue jeans or black pants and then selecting a red, green, or blue shirt. The tree diagram shows the possible outfits you can choose. For m pant choices and n shirt choices, there are $m \cdot n$ or $2 \cdot 3 = 6$ possible outfits.

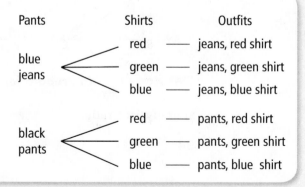

Pants	Shirts	Outfits
blue jeans	red	jeans, red shirt
	green	jeans, green shirt
	blue	jeans, blue shirt
black pants	red	pants, red shirt
	green	pants, green shirt
	blue	pants, blue shirt

You can use the Fundamental Counting Principle for situations that involve more than two events.

Plan

How can you find the number of possible lunch specials without listing each one?
Use the Fundamental Counting Principle. Multiply the choices for each category.

 **Problem 1** Using the Fundamental Counting Principle

Menus A deli offers a lunch special if you choose one from each of the following types of sandwiches, side items, and drink choices. How many different lunch specials are possible?

There are 6 possible sandwiches, 4 different side items, and 5 drink choices.

$$6 \cdot 4 \cdot 5 = 120 \qquad \text{Use the Fundamental Counting Principle.}$$

There are 120 different possible lunch specials.

Deli Menu

Sandwiches	Side Items	Drinks
ham & turkey	chips	juice
salami	potato salad	iced tea
tuna	fruit salad	lemonade
club	garden salad	milk
veggie		water
meatball		

 Got It? **1.** Suppose that a computer generates passwords that begin with a letter followed by 2 digits, like R38. The same digit can be used more than once. How many different passwords can the computer generate?

A **permutation** is an arrangement of items in which the order of the objects is important. You can use the Fundamental Counting Principle to find the total number of permutations. Suppose you want to find the number of ways to line up 4 friends. There are 4 ways to choose the first person, 3 ways to choose the second person, 2 ways to choose the third, and only 1 way to choose the last. So, there are $4 \cdot 3 \cdot 2 \cdot 1 = 24$ permutations, or ways to arrange your friends.

You can use *factorial notation* to write $4 \cdot 3 \cdot 2 \cdot 1$ as 4! (You say this as "4 factorial.") For any positive integer n, **n factorial** is $n! = n(n-1)(n-2) \cdot \ldots \cdot 3 \cdot 2 \cdot 1$. Zero factorial is defined to be equal to 1.

Think

Why use the Fundamental Counting Principle?
There are too many possibilities to make a list.

 **Problem 2** Finding the Number of Permutations

Music You download 8 songs on your music player. If you play the songs using the random shuffle option, how many different ways can the sequence of songs be played?

$$8! = 8 \cdot 7 \cdot 6 \cdot \ldots \cdot 2 \cdot 1 = 40{,}320 \qquad \text{Use the Fundamental Counting Principle.}$$

There are 40,320 different ways to randomly play the 8 songs.

 Got It? **2.** In how many ways can you arrange 12 books on a shelf?

Suppose you want 3 songs to play at a time from the 8 songs you have downloaded. There are 8 ways to select the first song, 7 ways to select the second song, and 6 ways to select the third song. By the Fundamental Counting Principle, this is $8 \cdot 7 \cdot 6$ ways, or 336 ways. You can express the result using factorials, as shown below.

Key Concept   Permutation Notation

The number of permutations of n items of a set arranged r items at a time is

$$_nP_r = \frac{n!}{(n-r)!} \text{ for } 0 \le r \le n.$$

Example $_8P_3 = \frac{8!}{(8-3)!} = \frac{8!}{5!} = \frac{8 \cdot 7 \cdot 6 \cdot 5 \cdot 4 \cdot 3 \cdot 2 \cdot 1}{5 \cdot 4 \cdot 3 \cdot 2 \cdot 1} = 8 \cdot 7 \cdot 6 = 336$

Problem 3   Finding a Permutation

The environmental club is electing a president, vice president, and treasurer. How many different ways can the officers be chosen from the 10 members?

Think

How do you know that the way the officers are chosen is a permutation?
This situation uses permutations because a different arrangement of the same 3 members is a different result. So the order of the choices matters.

Method 1   Use the formula for permutations.

There are 10 members, arranged 3 at a time. So $n = 10$ and $r = 3$.

$$_{10}P_3 = \frac{10!}{(10-3)!} \qquad \text{Substitute 10 for } n \text{ and 3 for } r.$$

$$= \frac{3,628,800}{5040} = 720 \qquad \text{Simplify.}$$

Method 2   Use a graphing calculator.

Press 1 0 (math) ◀ 2 3 (enter)

$_{10}P_3 = 720$

There are 720 ways that three of the ten members can be chosen as officers.

Got It?   3. Twelve swimmers compete in a race. In how many possible ways can the swimmers finish first, second, and third?

A **combination** is a selection of items in which order is *not* important. Suppose you select 3 different fruits to make a fruit salad. The order you select the fruits does not matter.

Key Concept   Combination Notation

The number of combinations of n items chosen r at a time is

$$_nC_r = \frac{n!}{r!(n-r)!} \text{ for } 0 \le r \le n.$$

Example $_9C_4 = \frac{9!}{4!(9-4)!} = \frac{9!}{4!\,5!} = \frac{9 \cdot 8 \cdot 7 \cdot 6 \cdot 5 \cdot 4 \cdot 3 \cdot 2 \cdot 1}{(4 \cdot 3 \cdot 2 \cdot 1)(5 \cdot 4 \cdot 3 \cdot 2 \cdot 1)} = \frac{9 \cdot 8 \cdot 7 \cdot 6}{4 \cdot 3 \cdot 2 \cdot 1} = 126$

Problem 4 **Using the Combination Formula**

Reading Suppose that you choose 4 books to read on summer vacation from a reading list of 12 books. How many different combinations of the books are possible?

Method 1 Use the formula for finding combinations.

There are 12 books chosen 4 at a time.

Think

If you are not using a calculator, how can you simplify the calculation?
Divide all common factors before multiplying.

$$_{12}C_4 = \frac{12!}{4!(12-4)!} = \frac{12!}{4!\,8!} = \frac{12 \cdot 11 \cdot 10 \cdot 9 \cdot 8 \cdot 7 \cdot 6 \cdot 5 \cdot 4 \cdot 3 \cdot 2 \cdot 1}{(4 \cdot 3 \cdot 2 \cdot 1)(8 \cdot 7 \cdot 6 \cdot 5 \cdot 4 \cdot 3 \cdot 2 \cdot 1)} = 495$$

Method 2 Use a calculator.

Press

$_{12}C_4 = 495$

There are 495 ways to choose 4 books from a reading list of 12 books.

Got It? **4.** A service club has 8 freshmen. Five of the freshmen are to be on the clean-up crew for the town's annual picnic. How many different ways are there to choose the 5 member clean-up crew?

To determine whether to use the permutation formula or the combination formula, you must decide whether order is important.

Problem 5 **Identifying Combinations and Permutations**

A A college student is choosing 3 classes to take during first, second, and third semseter from the 5 elective classes offered in his major. How many possible ways can the student schedule the three classes?

The order in which the classes are chosen does matter. Use a permutation.

$$_5P_3 = \frac{5!}{(5-3)!} = \frac{5!}{2!} = 60$$

There are 60 ways that the student can schedule the three classes.

Think

Why does order not matter?
After the selection takes place, the same 12 people are on the jury, regardless of the order in which they were chosen.

B A jury of 12 people is chosen from a pool of 35 potential jurors. How many different juries can be chosen?

The order in which the jurors are chosen is not important. Use a combination.

$$_{35}C_{12} = \frac{35!}{12!(35-12)!} = \frac{35!}{12!\,23!} = 834{,}451{,}800$$

There are 834,451,800 possible juries of 12 people.

Got It? **5.** A yogurt shop allows you to choose any 3 of the 10 possible mix-ins for a Just Right Smoothie. How many different Just Right Smoothies are possible?

You can use permutations and combinations to help you solve probability problems.

Problem 6 **Finding Probabilities**

Three pool balls are randomly chosen from a set numbered from 1 to 15. What is the probability that the pool balls chosen are numbered 5, 7, and 9?

Step 1 Use the probability formula.

$$P(\text{choosing 5, 7, and 9}) = \frac{\text{number of possible ways to choose 5, 7, and 9}}{\text{number of ways to choose 3 pool balls}}$$

Step 2 Find the numerator. Use the Fundamental Counting Principle to find the number of possible ways to choose 5, 7, and 9.

There are 3 ways to pick the first ball, 2 ways to pick the second ball, and 1 way to pick the last ball.

$$3 \cdot 2 \cdot 1 = 6$$

Think

What is the total number of outcomes?
Because the problem requires 3 pool balls chosen at random, the total number of outcomes is all the ways that you can choose 3 pool balls from a set of 15 pool balls.

Step 3 Find the sample space. Because choosing pool balls numbered 5, 7, and 9 is the same outcome as choosing pool balls numbered 9, 5, and 7, the order does not matter. Use the combination formula to find the total number of ways to choose 3 pool balls from 15 pool balls.

$$_{15}C_3 = \frac{15}{3!(15-3)!} = 455$$

Step 4 Find the probability.

$$P(\text{choosing 5, 7, and 9}) = \frac{6}{455} \approx 0.013$$

The probability of choosing the pool balls numbered 5, 7, and 9 is about 0.013, or 1.3%.

Got It? **6.** What is the probability of choosing first the number 1 ball, then the number 2 ball, and then the number 3 ball?

Lesson Check

Do you know HOW?

Evaluate each expression.

1. $3!$ **2.** $0!$ **3.** $_6P_2$

4. $_6P_3$ **5.** $_6C_2$ **6.** $_6C_3$

7. Sports How many ways can you choose 6 people to form a volleyball team out of a group of 10 players?

Do you UNDERSTAND?  MATHEMATICAL PRACTICES

8. Compare and Contrast How are combinations and permutations similar? How are they different?

9. Reasoning Your friend says that she can calculate any probability if she knows how many successful outcomes there are. Is there something else needed? Explain.

Practice and Problem-Solving Exercises ⓒ MATHEMATICAL PRACTICES

Ⓐ Practice

10. Telephones International calls require the use of a country code. Many country codes are 3-digit numbers. Country codes do not begin with a 0 or 1. There are no restrictions on the second and third digits. How many different 3-digit country codes are possible?

◀ See Problem 1.

11. Security To make an entry code, you need to first choose a single-digit number and then two letters, which can repeat. How many entry codes can you make?

Find the value of each expression.

◀ See Problems 2–4.

12. $6!$
13. $\dfrac{15!}{(15-10)!}$
14. $_{10}P_6$
15. $_{10}C_6$

16. Linguistics The Hawaiian alphabet has 12 letters. How many permutations are possible for each number of letters?
   a. 3 letters
   b. 5 letters

17. A class has 30 students. In how many ways can committees be formed using the following numbers of students?
   a. 3 students
   b. 5 students

For Exercises 18–19, determine whether to use a permutation or a combination. Then solve the problem.

◀ See Problem 5.

18. You and your friends pick up seven movies to watch over a holiday. You have time to watch only two. In how many ways can you select the two to watch?

19. Suppose that the math team at your school competes in a regional tournament. The math team has 12 members. Regional teams are made up of 4 people. How many different regional teams are possible?

20. You have a stack of 8 cards numbered 1–8. What is the probability that the first cards selected are 5 and 6?

◀ See Problem 6.

21. To win a lottery, 6 numbers are drawn at random from a pool of 50 numbers. Numbers cannot repeat. You have one lottery ticket. What is the probability you hold the winning ticket?

Ⓑ Apply

22. Entertainment Suppose that you and 4 friends go to a popular movie. You arrive late and cannot sit together, but you find 3 available seats in a row. How many possible ways can you and your friends sit in these seats?

ⓒ 23. Think About a Plan There are 8 online songs that you want to download. If you only have enough money to download 3 of the songs, how many different groups of songs can you buy?
   • Does the order in which you select the songs matter? Explain.
   • Should you use the permutation formula or combination formula?
   • What are the values of n and r?

24. Government The are 24 members of the U.S. Senate Committee on Finance. How many possible ways are there to choose a 13-member subcommittee to review current energy legislation?

25. Music What is the probability of the youngest four members of a 15-member choir being randomly selected for a quartet (a group of four singers)?

26. What is the probability of randomly choosing a specific set of 7 books off a bookshelf holding 12 books?

Ⓒ **27. Error Analysis** A friend says that there are 6720 different ways to combine 5 out of 8 ingredients to make a stew. Explain the error and find the correct answer.

Ⓒ **28. Reasoning** Can $_nC_r$ ever be equal to $_nP_r$? Explain.

Ⓒ **29.** A 4-digit code is needed to unlock a bicycle lock. The digits 0 through 9 can be used only once in the code.
  a. What is the probability that all of the digits are even?
  b. Writing These types of locks are usually called combination locks. What name might a mathematician prefer? Why?

Ⓒ **Challenge**

30. *Circular permutations* are arrangements of objects in a circle or a loop. For example, the number of different ways a group of friends can sit around a table represent circular permutations. Permutations that are equivalent after a rotation of the circle are considered the same. Make a table for the regular and circular permutations of 2 out of 2, 3 out of 3, and 4 out of 4. What formula do you think could be used to find the number of circular permutations for n out of n items?

Standardized Test Prep

GRIDDED RESPONSE

SAT/ACT

31. A cube has a volume of 64 cm^3. What is the total surface area of this cube in square centimeters?

32. The major arc of a circle measures 210°. What is the measure of the corresponding minor arc?

Mixed Review

33. The results of a survey on favorite movie genres are shown below in the frequency table below. What is the probability distribution of the data?

See Lesson 13-2.

Favorite Movie Genres					
Genre	Action	Comedy	Drama	Horror	Other
Frequency	9	8	3	6	4

Get Ready! **To Prepare for Lesson 13-4, do Exercises 34–36.**

Two standard number cubes are tossed. Find the following probabilities.

See Lesson 13-1.

34. P(two 5s)   **35.** P(sum of 10)   **36.** P(sum less than 9)

Do you know HOW?

A bag contains colored tiles. The tiles are randomly selected, and the results are recorded in the table below. Find the experimental probabilities.

Colored Tiles				
Tile	red	blue	green	yellow
Frequency	8	10	9	12

1. $P(\text{red})$ 2. $P(\text{green})$ 3. $P(\text{yellow})$

4. You toss two standard number cubes. What is the probability that you roll two odd numbers?

Suppose that you choose a number tile at random from the set {2, 5, 8, 10, 12, 19, 35}.

5. What is the probability that the number is even?

6. What is the probability that the number is not less than 10?

Students in the sophomore class at a local high school were asked whether they were left-handed or right-handed. The results of the survey are below.

Handedness	Frequency
Left-Handed	72
Right-Handed	478

7. What is the relative frequency of sophomores that are left-handed?

8. What is the probability distribution for this data?

Evaluate each expression.

9. $8!$

10. $3!$

11. $\frac{9!}{5!}$

12. $\frac{12!}{5!\,7!}$

13. $_8C_5$

14. $_{15}P_5$

15. $_4C_3$

16. $_{10}C_9$

In Exercises 17 and 18, determine whether each situation involves a permutation or a combination.

17. How many ways can presentation groups of 6 students be formed from a class of 24 students?

18. How many ways can you arrange 5 books on a shelf from a group of 9 books?

19. How many different security codes like A12 be formed using the digits 1–4 and the letters A, B, and N?

20. **Elections** A club has 16 members. All of the members are eligible for the offices of president, vice president, secretary, and treasurer. In how many different ways can the officers be elected?

21. The names of all 19 students in a class are written on slips of paper and placed in a basket. How many different ways can 5 names be randomly chosen from the basket?

Do you UNDERSTAND?

22. **Reasoning** The probability of an event is $\frac{4}{5}$. Is this event likely or unlikely to occur? Explain.

23. **Compare and Contrast** What is the same about experimental and theoretical probability? What is different?

24. **Open-Ended** Give an example of a situation that can be described using permutations and another situation that can be described using combinations.

25. **Error Analysis** The frequency table below shows the results of an experiment where a coin is tossed 100 times. Your friend says that the results show that the experimental probability of heads showing after a toss is $\frac{42}{58}$. Explain your friend's error.

Coin Shows	Frequency
Heads	42
Tails	58

13-4 Compound Probability

© Content Standards
S.CP.7 Apply the Addition Rule,
$P(A \text{ or } B) = P(A) + P(B) - P(A \text{ and } B) \ldots$
Also S.CP.8, S.CP.9

Objective To identify independent and dependent events
To find compound probabilities

SOLVE IT!

Getting Ready! ◄► X ↻ ⏏

Suppose you are traveling from Philadelphia, PA, to San Diego, CA. Do you think the probability of rain in Philadelphia affects the probability of rain in San Diego? Justify your reasoning.

Use what you know about the locations of the two cities.

© MATHEMATICAL PRACTICES

Philadelphia		San Diego	
Today	**Tomorrow**	**Today**	**Tomorrow**
Rain	Sunny	Cloudy	Partly Cloudy
43° F	58° F	70° F	75° F
Chance of rain: 80%	Chance of rain: 0%	Chance of rain: 50%	Chance of rain: 10%

If you were to find the probability of rain in both cities in the Solve It, you would be finding the probability of a *compound event*. A **compound event** is an event that is made up of two or more events.

Essential Understanding You can find the probability of compound events by using the probability of each part of the compound event.

If the occurrence of an event does not affect how another event occurs, the events are called **independent events**. If the occurence of an event does affect how another event occurs, the events are called **dependent events**. To calculate the probability of a compound event, first determine whether the events are independent or dependent.

Lesson Vocabulary
- compound event
- independent events
- dependent events
- mutually exclusive events
- overlapping events

Think

How can you tell that two events are independent?
Two events are independent if one does not affect the other.

© Problem 1 **Identifying Independent and Dependent Events**

Are the outcomes of each trial independent or dependent events?

Ⓐ Choose a number tile from 12 tiles. Then spin a spinner.

The choice of number tile does not affect the spinner result. The events are independent.

Ⓑ Pick one card from a set of 15 sequentially numbered cards. Then, without replacing the card, pick another card.

The first card chosen affects the possible outcomes of the second pick, so the events are dependent.

844   **Chapter 13** Probability

You can find the probability that two independent events will both occur by multiplying the probabilities of each event.

> **Key Concept** **Probability of A and B**
>
> If A and B are independent events, then $P(A \text{ and } B) = P(A) \cdot P(B)$.

Ⓒ **Problem 2** **Finding the Probability of Independent Events**

A desk drawer contains 5 red pens, 6 blue pens, 3 black pens, 24 silver paper clips, and 16 white paper clips. If you select a pen and a paper clip from the drawer without looking, what is the probability that you select a blue pen and a white paper clip?

Plan

Why are the events independent?
Selecting a blue pen has no affect on selecting a white paper clip.

Step 1 Let A = selecting a blue pen. Find the probability of A.

$P(A) = \frac{6}{14} = \frac{3}{7}$ 6 blue pens out of 14 pens

Step 2 Let B = selecting a white paper clip. Find the probability of B.

$P(B) = \frac{16}{40} = \frac{2}{5}$ 16 white paper clips out of 40 clips

Step 3 Find $P(A \text{ and } B)$.   Use the formula for the probability of independent events.

$P(A \text{ and } B) = P(A) \cdot P(B) = \frac{3}{7} \cdot \frac{2}{5} = \frac{6}{35} \approx 0.171$, or 17.1%

The probability that you select a blue pen and a white paper clip is about 17.1%.

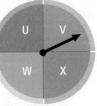

Got It? **2.** You roll a standard number cube and spin the spinner at the right. What is the probability that you roll a number less than 3 and the spinner lands on a vowel?

Events that cannot happen at the same time are called **mutually exclusive events**. For example, you cannot roll a 2 and a 5 on a standard number cube at the same time, so the events are mutually exclusive. If events A and B are mutually exclusive, then the probability of both A and B occurring is 0. The probability that either A or B occurs is the sum of the probability of A occurring and the probability of B occurring.

> **Key Concept** **Probability of Mutually Exclusive Events**
>
> If A and B are mutually exclusive events, then $P(A \text{ and } B) = 0$, and $P(A \text{ or } B) = P(A) + P(B)$.

Plan

Is there a way to simplify this problem?
You can model the probabilities with a simpler problem. Suppose there are 100 athletes. In the model 28 athletes will play basketball, and 24 will be on the swim team.

Problem 3 Finding the Probability of Mutually Exclusive Events

Athletics Student athletes at a local high school may participate in only one sport each season. During the fall season, 28% of student athletes play basketball and 24% are on the swim team. What is the probability that a randomly selected student athlete plays basketball or is on the swim team?

Because athletes participate in only one sport each season, the events are mutually exclusive. Use the formula $P(A \text{ or } B) = P(A) + P(B)$.

$P(\text{basketball or swim team}) = P(\text{basketball}) + P(\text{swim team})$

$= 28\% + 24\% = 52\%$ Substitute and Simplify.

The probability of an athlete either playing basketball or being on the swim team is 52%.

Got It? 3. In the Spring season, 15% of the athletes play baseball and 23% are on the track team. What is the probability of an athlete either playing baseball or being on the track team?

Overlapping events have outcomes in common. For example, for a standard number cube, the event of rolling an even number and the event of rolling a multiple of 3 overlap because a roll of 6 is a favorable outcome for both events.

take note

Key Concept Probability of Overlapping Events

If A and B are overlapping events, then $P(A \text{ or } B) = P(A) + P(B) - P(A \text{ and } B)$.

Here's Why It Works Suppose you have 7 index cards, each having one of the following letters written on it:

A B C D E F G

$P(\text{FACE})$, the probability of selecting a letter from the word FACE, is $\frac{4}{7}$.
$P(\text{CAB})$, the probability of selecting a letter from the word CAB, is $\frac{3}{7}$.

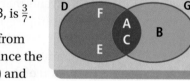

Consider $P(\text{FACE or CAB})$, the probability of choosing a letter from either the word FACE or the word CAB. These events overlap since the words have two letters in common. If you simply add $P(\text{FACE})$ and $P(\text{CAB})$, you get $\frac{4}{7} + \frac{3}{7} = \frac{4+3}{7}$. The value of the numerator should be the number of favorable outcomes, but there are only 5 distinct letters in the words FACE and CAB. The problem is that when you simply add, the letters A and C are counted twice, once in the favorable outcomes for the word FACE, and once for the favorable outcomes for the word CAB. You must subtract the number of letters that the two words have in common so they are only counted once.

$$P(\text{FACE or CAB}) = \frac{4+3-2}{7} = \frac{4}{7} + \frac{3}{7} - \frac{2}{7} = P(\text{FACE}) + P(\text{CAB}) - P(\text{AC})$$

Problem 4 Finding Probabilities of Overlapping Events

What is the probability of rolling either an even number or a multiple of 3 when rolling a standard number cube?

Know

You are rolling a standard number cube. The events are overlapping events because 6 is both even and a multiple of 3.

Need

You need the probability of rolling an even number and the probability of rolling a multiple of 3.

Plan

Find the probabilities and use the formula for probabilities of overlapping events.

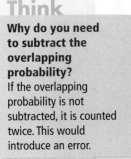

Think

Why do you need to subtract the overlapping probability?
If the overlapping probability is not subtracted, it is counted twice. This would introduce an error.

$P(\text{even or multiple of } 3) = P(\text{even}) + P(\text{multiple of } 3) - P(\text{even and multiple of } 3)$

$$= \frac{3}{6} + \frac{2}{6} - \frac{1}{6}$$

$$= \frac{4}{6}, \text{ or } \frac{2}{3}$$

The probability of rolling an even or a multiple of 3 is $\frac{2}{3}$.

Got It? **4.** What is the probability of rolling either an odd number or a number less than 4 when rolling a standard number cube?

Lesson Check

Do you know HOW?

1. Suppose A and B are independent events. What is $P(A \text{ and } B)$ if $P(A) = 50\%$ and $P(B) = 25\%$?

2. Suppose A and B are mutually exclusive events. What is $P(A \text{ or } B)$ if $P(A) = 0.6$ and $P(B) = 0.25$?

3. Suppose A and B are overlapping events. What is $P(A \text{ and } B)$ if $P(A) = \frac{1}{3}$; $P(B) = \frac{1}{2}$ and $P(A \text{ and } B) = \frac{1}{5}$?

Do you Understand? MATHEMATICAL PRACTICES

4. Open-Ended Give an example of independent events, and an example of dependent events. Describe how the examples differ.

5. Error Analysis Your brother says that the probability that it is cloudy tomorrow and the probability that it rains are independent events. Explain your brother's error.

Practice and Problem-Solving Exercises MATHEMATICAL PRACTICES

A Practice

Determine whether the outcomes of each trial are *independent* or *dependent* events.

See Problem 1.

6. You toss a coin and roll a number cube.

7. You draw a marble from a bag without looking. You do not replace it. You draw another marble from the bag.

8. A card is randomly chosen from a standard deck of cards, then replaced; another card is chosen at random.

9. Asking a student's age, and asking what year they expect to graduate.

You spin the spinner at the right and without looking, you choose a tile from a set of tiles numbered from 1 to 10. Find each probability.

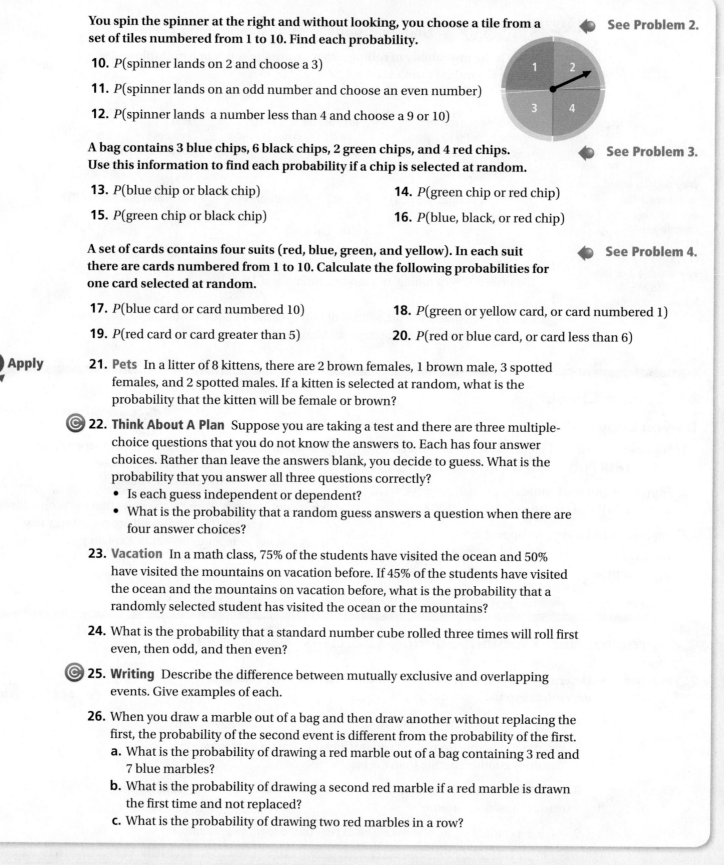

See Problem 2.

10. P(spinner lands on 2 and choose a 3)

11. P(spinner lands on an odd number and choose an even number)

12. P(spinner lands a number less than 4 and choose a 9 or 10)

A bag contains 3 blue chips, 6 black chips, 2 green chips, and 4 red chips. Use this information to find each probability if a chip is selected at random.

See Problem 3.

13. P(blue chip or black chip)

14. P(green chip or red chip)

15. P(green chip or black chip)

16. P(blue, black, or red chip)

A set of cards contains four suits (red, blue, green, and yellow). In each suit there are cards numbered from 1 to 10. Calculate the following probabilities for one card selected at random.

See Problem 4.

17. P(blue card or card numbered 10)

18. P(green or yellow card, or card numbered 1)

19. P(red card or card greater than 5)

20. P(red or blue card, or card less than 6)

B Apply

21. Pets In a litter of 8 kittens, there are 2 brown females, 1 brown male, 3 spotted females, and 2 spotted males. If a kitten is selected at random, what is the probability that the kitten will be female or brown?

22. Think About A Plan Suppose you are taking a test and there are three multiple-choice questions that you do not know the answers to. Each has four answer choices. Rather than leave the answers blank, you decide to guess. What is the probability that you answer all three questions correctly?
- Is each guess independent or dependent?
- What is the probability that a random guess answers a question when there are four answer choices?

23. Vacation In a math class, 75% of the students have visited the ocean and 50% have visited the mountains on vacation before. If 45% of the students have visited the ocean and the mountains on vacation before, what is the probability that a randomly selected student has visited the ocean or the mountains?

24. What is the probability that a standard number cube rolled three times will roll first even, then odd, and then even?

25. Writing Describe the difference between mutually exclusive and overlapping events. Give examples of each.

26. When you draw a marble out of a bag and then draw another without replacing the first, the probability of the second event is different from the probability of the first.
- **a.** What is the probability of drawing a red marble out of a bag containing 3 red and 7 blue marbles?
- **b.** What is the probability of drawing a second red marble if a red marble is drawn the first time and not replaced?
- **c.** What is the probability of drawing two red marbles in a row?

Challenge **Reasoning** For each set of probabilities, determine if the events A and B are mutually exclusive. Explain.

27. $P(A) = \frac{1}{2}$, $P(B) = \frac{1}{3}$, $P(A \text{ or } B) = \frac{2}{3}$

28. $P(A) = \frac{1}{6}$, $P(B) = \frac{3}{8}$, $P(A \text{ and } B) = 0$

ⓒ **29. Reasoning** Are mutually exclusive events dependent or independent? Explain.

Standardized Test Prep

SAT/ACT

30. Which of the following statements is NOT true?

Ⓐ The side lengths of an isosceles right triangle can be all whole numbers.

Ⓑ The side lengths of a right triangle can form a Pythagorean triple.

Ⓒ The side lengths of an equilateral triangle can be all whole numbers.

Ⓓ The angle measures of an equilateral triangle can be all whole numbers.

Short
Response

31. An arc of a circle measures 90° and is 10 cm long. How long is the circle's diameter?

32. You roll a standard number cube and then spin the spinner shown at the right. What is the probability that you will roll a 5 and spin a 3?

Mixed Review

Calculate the following permutations and combinations.

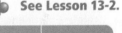 **See Lesson 13-3.**

33. The number of 3 letter sequences that can be made without reusing any letter.

34. The number of ways that 8 runners can finish a race, if there are no ties.

35. The number of ways a 5-member subcommittee can be formed from a 12-member student government.

Get Ready! **To Prepare for Lesson 13-5, do Exercises 36–38.**

See Lesson 13-2.

Students were asked about the number of siblings they have. The results of the survey are shown in the frequency table at the right. Find the following probabilities if a student is chosen at random from the respondents.

Number of Siblings	Frequency
0	5
1	12
2	15
3	7

36. $P(2 \text{ siblings})$

37. $P(\text{fewer than 3 siblings})$

38. $P(\text{more than 1 sibling})$

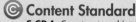

Content Standard
S.CP.4 Construct and interpret two-way frequency tables of data . . . Use the two-way table as a sample space to decide if events are independent and to approximate conditional probabilities.

Objective To construct and use probability models

Support your answer based on the data!

MATHEMATICAL PRACTICES

SOLVE IT!

Getting Ready!

The table at the right shows the number of students who passed their driving test as well as whether they took a driver's education class to prepare. What effect, if any, does taking the driver's education class have?

	Passed	Failed	Totals
Took the class	32	7	39
Do not take the class	18	23	41
Totals	50	30	80

In the Solve It, the data is displayed in a *two-way frequency table*. A **two-way frequency table**, or *contingency table*, displays the frequencies of data in two different categories.

Essential Understanding You can use two-way frequency tables to organize data and identify sample spaces to approximate probabilities.

Lesson Vocabulary
• two-way frequency table
• conditional probability

Think

What is the connection between relative frequency and probability?
You can use relative frequency to approximate a probability.

Problem 1 Using a Two-way Frequency Table

Activities The table shows data about student involvement in extracurricular activities at a local high school. What is the probability that a randomly chosen student is a female who is not involved in extracurricular activities?

Extracurricular Activities

	Involved in Activities	Not Involved in Activities	Totals
Male	112	145	257
Female	139	120	259
Totals	251	265	516

To find the probability, calculate the relative frequency.

relative frequency $= \frac{\text{females not involved}}{\text{total number of students}} = \frac{120}{516} \approx 0.233$

The probability that a student is a female who is involved in extracurricular activities is about 23.3%.

Got It? **1.** The two-way frequency table at the right shows the number of male and female students by grade level on the prom committee. What is the probability that a member of the prom committee is a male who is a junior?

	Male	Female	Totals
Juniors	3	4	7
Seniors	3	2	5
Totals	6	6	12

The probability that an event will occur, given that another event has already occurred is called a **conditional probability**. You can write the conditional probability of event *B*, given that event *A* has already occurred as $P(B \mid A)$. You read $P(B \mid A)$ as "the probability of event *B*, given event *A*."

 **Problem 2** **Finding Probability**

Opinion Polls Respondents of a poll were asked whether they were for, against, or had no opinion about a bill before the state legislature that would increase the minimum wage. What is the probability that a randomly selected person is over 60 years old, given that the person had no opinion on the state bill?

Age Group	For	Against	No Opinion	Totals
18–29	310	50	20	380
30–45	200	30	10	240
45–60	120	20	30	170
Over 60	150	20	40	210
Totals	780	120	100	1000

Plan

What part of the table do you need to use?
Since the group you are interested in is the one with no opinion, you only need to look at that column.

The condition that the person selected has no opinion on the minimum-wage bill limits the total outcomes to the 100 people who had no opinion. Of those 100 people, 40 respondents were over 60 years old.

$$P(\text{over 60} \mid \text{no opinion}) = \frac{40}{100} = 0.4$$

 Got It? **2. a.** What is the probability that a randomly selected person is 30–45 years old, given that the person is in favor of the minimum-wage bill?
   b. Reasoning What is the probability that a randomly selected person is not 18–29, given that the person is in favor of the minimum wage bill?

Problem 3 Using Relative Frequencies

Business A company has 150 sales representatives. Two months after a sales seminar, the company vice-president made the table based on sales results. What is the probability that someone who attended the seminar had an increase in sales?

	Attended Seminar	Did not Attend Seminar	Totals
Increased Sales	0.48	0.02	0.5
No Increase in Sales	0.32	0.18	0.5
Totals	0.8	0.2	1

Think

Given the relative frequencies, how do you find the frequencies?
Multiply the number or sales representatives times the relative frequency for a category.

Method 1 Find frequencies first.

Find the number of people who attended the seminar and had increased sales: $0.48 \cdot 150 = 72$

Find the number of people who attended the seminar: $0.8 \cdot 150 = 120$

Find P(increased sales | sales seminar): $\frac{72}{150} = 0.6$, or 60%

Method 2 Use relative frequencies.

P(increased sales | sales seminar):

$$= \frac{\text{relative frequency of attend seminar and increased sales}}{\text{total frequency of increased sales}} = \frac{0.48}{0.8} = 0.6$$

The probability that some who attended the seminar had an increase in sales is 0.6, or 60%.

Got It? 3. What is the probability that a randomly selected sales representative, who did not attend the seminar, did not see an increase in sales?

Lesson Check

Do you know HOW?

Use the two-way frequency table to find the probabilities.

	Supports the Issue	Does not supports the issue	Totals
Democrat	24	36	60
Republican	27	33	60
Totals	51	69	120

1. P(democrat and supports the issue)

2. P(democrat | supports the issue)

Do you UNDERSTAND?  MATHEMATICAL PRACTICES

3. Error Analysis Using the table at the left, a student calculated the relative frequency of those who do not support the issue, given that they are Republican, as $\frac{33}{33 + 36} \approx 0.478$. What error did the student make?

4. Vocabulary What is a two-way frequency table?

5. Suppose A is a female student and B is a student who plays sports. What does $P(B \mid A)$ mean?

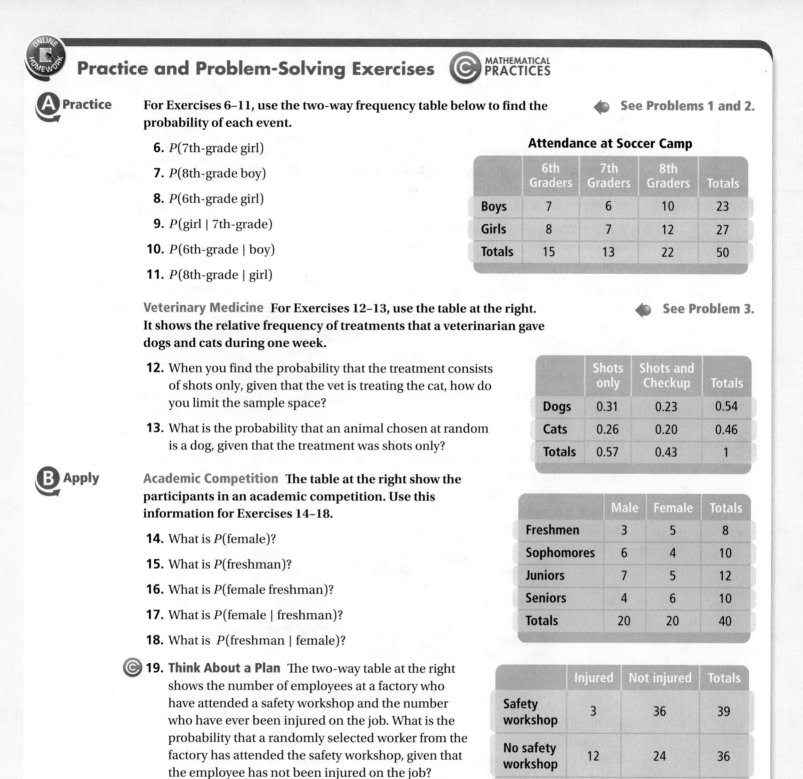

Practice and Problem-Solving Exercises

MATHEMATICAL PRACTICES

A Practice

For Exercises 6–11, use the two-way frequency table below to find the probability of each event.

See Problems 1 and 2.

6. P(7th-grade girl)

7. P(8th-grade boy)

8. P(6th-grade girl)

9. P(girl | 7th-grade)

10. P(6th-grade | boy)

11. P(8th-grade | girl)

Attendance at Soccer Camp

	6th Graders	7th Graders	8th Graders	Totals
Boys	7	6	10	23
Girls	8	7	12	27
Totals	15	13	22	50

Veterinary Medicine For Exercises 12–13, use the table at the right. It shows the relative frequency of treatments that a veterinarian gave dogs and cats during one week.

See Problem 3.

12. When you find the probability that the treatment consists of shots only, given that the vet is treating the cat, how do you limit the sample space?

13. What is the probability that an animal chosen at random is a dog, given that the treatment was shots only?

	Shots only	Shots and Checkup	Totals
Dogs	0.31	0.23	0.54
Cats	0.26	0.20	0.46
Totals	0.57	0.43	1

B Apply

Academic Competition The table at the right show the participants in an academic competition. Use this information for Exercises 14–18.

14. What is P(female)?

15. What is P(freshman)?

16. What is P(female freshman)?

17. What is P(female | freshman)?

18. What is P(freshman | female)?

	Male	Female	Totals
Freshmen	3	5	8
Sophomores	6	4	10
Juniors	7	5	12
Seniors	4	6	10
Totals	20	20	40

19. **Think About a Plan** The two-way table at the right shows the number of employees at a factory who have attended a safety workshop and the number who have ever been injured on the job. What is the probability that a randomly selected worker from the factory has attended the safety workshop, given that the employee has not been injured on the job?
   - Which column is needed for solving this problem?
   - Do you need the total number of workers or the total not injured?

	Injured	Not injured	Totals
Safety workshop	3	36	39
No safety workshop	12	24	36
Totals	15	60	75

20. Reasoning Recall that two events are independent when the occurrence of one has no effect the other. The table at the right is a frequency table.

	B	D	F	Total
A	7	5	4	16
C	3	4	5	12
E	11	7	2	28
Total	21	16	11	48

   a. Calculate $P(B)$.
   b. Calculate $P(B \mid A)$.
   c. Does the occurrence of A have any effect on the probability of B? What can you say about events A and B?
   d. Are events C and D independent? Explain.
   e. Are events C and F independent? Explain.

21. Writing What is the sum of the probabilities of each outcome in a probability experiment? How does this relate to the relative frequencies in a contingency table? Explain and give an example.

22. Healthcare The table at the right is a relative frequency distribution for healthy people under the age of 65.

Flu Vaccines

	Got the Flu	Did not Get the Flu	Totals
Vaccinated	■	54%	60%
Not Vaccinated	■	■	■
Totals	15%	■	100%

   a. Copy and complete the table.
   b. What is the probability of getting the flu, given you are vaccinated?
   c. What is the probability of getting the flu, given you have not been vaccinated?

Challenge

23. When you construct a two-way frequency table, you add across rows and down columns to find values for the totals row and column. The values in the totals column and totals row of the table are called *marginal frequencies*, and the values in the interior of the table are called *joint frequencies*. In the table to the right, the values 9, 3, 5, and 8 represent the joint frequencies, and the values 14, 11, 12, 13, and 25 represent the marginal frequencies.

	Exam Score ≥ 85%	Exam Score < 85%	Totals
Studied more than 4 hours	9	3	12
Studied less than 4 hours	5	8	13
Totals	14	11	25

   a. Why do you think the values in the interior of the table are called joint frequencies?
   b. What do the marginal frequencies of the table represent?
   c. If you replace the joint and marginal frequencies in a two-way frequency with the respective relative frequencies, what do the values in the totals row and column represent? Use the given table to provide an example.

Standardized Test Prep

SAT/ACT

24. The two-way frequency table shows the number of males and females that either support or are against the building of a new mall. What is the probability that a randomly selected person is a female, given that the person supports the new mall? Round to the nearest hundredth.

	For	Against	Totals
Male	62	48	110
Female	78	32	110
Totals	140	80	220

25. The area of a kite is 150 in.^{2}. The length of one diagonal is 50 in. What is the length, in inches, of the other diagonal?

26. What is the x-coordinate of the midpoint of $\overline{AB}$ for $A(-3, 9)$ and $B(-5, -3)$?

27. The segment with endpoints $A(1, 5)$ and $B(2,1)$ is reflected over $x = 3$ and translated up 2 units and to the right 3 units. What is the x-coordinate of the midpoint of $\overline{A'B'}$?

Mixed Review

Find the sum of the interior angle measures of each polygon.

See Lesson 6-1.

28. quadrilateral

29. dodecagon

30. 20-gon

31. A cylindrical carton with a radius of 2 in. is 7 in. tall. Assuming no surfaces overlap, what is the surface area of the carton? Round your answer to the nearest square inch.

See Lesson 11-2.

32. A cube with edges 8 cm fits within the sphere, as shown at the right. The diagonal of the cube is the diameter of the sphere.
  a. Find the radius of the sphere. Leave your answer in simplest radical form.
  b. What the volume of the sphere?

See Lesson 11-6.

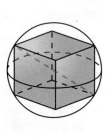

Get Ready! **To prepare for Lesson 13-6, do Exercises 33–36.**

Find the probability of each event.

33. $P(\text{large} \mid \text{red})$

34. $P(\text{red} \mid \text{large})$

35. $P(\text{small} \mid \text{blue})$

36. $P(\text{large} \mid \text{blue})$

	Large	Small	Totals
Blue	17	3	20
Red	8	12	20
Totals	25	15	40

Conditional Probability Formulas

Content Standards
S.CP.3 Understand the conditional probability of *A* given *B* as *P(A* and *B)/P(B)* . . .
S.CP.5 Recognize and explain the concepts of conditional probability and independence in everyday language and everyday situations.
Also **S.CP.2, S.CP.6**

Objective To understand and calculate conditional probabilities

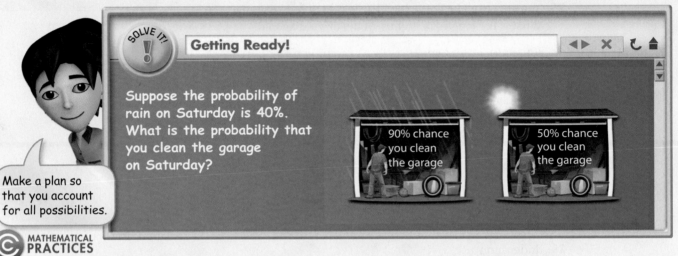

SOLVE IT!

Getting Ready!

Suppose the probability of rain on Saturday is 40%. What is the probability that you clean the garage on Saturday?

90% chance you clean the garage

50% chance you clean the garage

Make a plan so that you account for all possibilities.

MATHEMATICAL PRACTICES

In the Solve It, you may have calculated the probability that it rains and you clean the garage and the probability that it does not rain and you clean the garage and then added the probabilities together.

Essential Understanding You can find conditional probabilities using a formula.

In the previous lesson, you found probabilities using frequency tables. For example, in the relative frequency table at the right, the probability that a sales representative had an increase in sales, given that he attended the training seminar, is $\frac{0.48}{0.8} = 0.6$.

	Attended Seminar	Did not Attend Seminar	Totals
Increased Sales	0.48	0.02	0.5
No Increase in Sales	0.32	0.18	0.5
Totals	0.8	0.2	1

With respect to probability, this would be

$$P(\text{increased sales} \mid \text{attend seminar}) = \frac{P(\text{attend seminar and had increased sales})}{P(\text{attended semiar})}.$$

This suggests an algebraic formula for finding conditional probability.

take note

Conditional Probability Formula

For any two events *A* and *B*, the probability of *B* occurring, given that event *A* has occurred, is

$$P(B \mid A) = \frac{P(A \text{ and } B)}{P(A)}, \text{ where } P(A) \neq 0.$$

 **Problem 1** **Using Conditional Probabilities**

Pharmaceutical Testing In a study designed to test the effectiveness of a new drug, half of the volunteers received the drug. The other half of the volunteers received a placebo, a tablet or pill containing no medication. The probability of a volunteer receiving the drug and getting well was 45%. What is the probability of someone getting well, given that he receives the drug?

Think

How can you find $P(A)$?
A is the event that represents the sample space. In this case, it is the volunteers that received the drug. Since this is half of the volunteers, $P(A)$ is $\frac{1}{2}$.

Step 1 Identify the probabilities.
$P(B \mid A) = P(\text{getting well, given taking the new drug})$
$P(A) = P(\text{taking the new drug}) = \frac{1}{2} = 0.5$
$P(A \text{ and } B) = P(\text{taking the new drug and getting well}) = 45\%, \text{ or } 0.45$

Step 2 Find $P(B \mid A)$.
$P(B \mid A) = \frac{0.45}{0.5} = 0.9, \text{ or } 90\%$ Use the conditional probability formula.

The probability of getting well if someone received the drug is 90%.

Got It? **1.** The probability of a volunteer receiving the placebo and having his or her health improve was 20%. What is the conditional probability of a volunteer's health improving, given that they received the placebo?

Conditional probabilities are usually not reversible. $P(A \mid B) \neq P(B \mid A)$.

Problem 2 **Comparing Conditional Probabilities**

Pets In a survey of pet owners, 45% own a dog, 27% own a cat, and 12% own both a dog and a cat. What is the conditional probability that a dog owner also owns a cat? What is the conditional probability that a cat owner also owns a dog?

Think

How are the sample spaces limited in each of the conditional probabilities?
In $P(\text{cat} \mid \text{dog})$, the sample space is limited to the probability that a pet owner owns a dog. Similarly, $P(\text{dog} \mid \text{cat})$, the sample space is the probability that an owner owns a cat.

$P(\text{cat} \mid \text{dog}) = \frac{P(\text{owns cat and dog})}{P(\text{owns dog})}$ Definition $P(\text{dog} \mid \text{cat}) = \frac{P(\text{owns cat and dog})}{P(\text{owns cat})}$

$= \frac{0.12}{0.45}$ Substitute. $= \frac{0.12}{0.27}$

$\approx 0.267, \text{ or } 26.7\%$ Simplify. $\approx 0.444, \text{ or } 44.4\%$

The conditional probability that a dog owner also owns a cat is about 26.7%.
The conditional probability that a cat owner also owns a dog is about 44.4%.

Got It? **2.** The same survey showed that 5% of the pet owners own a dog, a cat, and at least one other type of pet.
  a. What is the conditional probability that a pet owner owns a cat and some other type of pet, given that they own a dog?
  b. What is the conditional probability that a pet owner owns a dog and some other type of pet, given that they own a cat?
  c. Critical Thinking Can you calculate the conditional probability of owning another pet for a pet owner owning a cat and no dogs? Explain.

Because $P(B \mid A) = \frac{P(A \text{ and } B)}{P(A)}$, then $P(A \text{ and } B) = P(A) \cdot P(B \mid A)$.

You can use this form of the conditional rule when you know the conditional probability. You can also combine conditional probabilities to find the probability of an event that can happen in more than one way.

Problem 3   Using a Tree Diagram

Graduation Rate   A college reported the following based on their graduation data.
- **70% of freshmen had attended public schools**
- **60% of freshmen who had attended public schools graduated within 5 years**
- **80% of other freshmen graduated within 5 years**

What percent of freshmen graduated within 5 years?

You can use a tree diagram to organize the information.

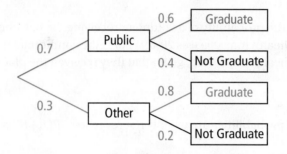

$P(\text{Public and Graduate}) = P(\text{Graduate, given Public}) \cdot P(\text{Public})$
$$= 0.6 \cdot 0.7$$
$$= 0.42$$

$P(\text{Other and Graduate}) = P(\text{Graduate, given Other}) \cdot P(\text{Other})$
$$= 0.8 \cdot 0.3$$
$$= 0.24$$

Plan

Why do you add the probabilities here?
You add the probabilities because the events are mutually exclusive.

$P(\text{Graduate}) = P(\text{Public and Graduate}) + P(\text{Other and Graduate})$
$$= 0.42 + 0.24$$
$$= 0.66$$

66% of the freshmen graduate within 5 years.

Got It?   **3. a.** A soccer team wins 65% of its games on muddy fields and 30% of their games on dry fields. The probability of the field being muddy for their next game is 70%. What is the probability that the team will win their next game?

      b. Critical Thinking   If the probability of the field being muddy increases, how will that influence the probability of the soccer team winning their next game? Explain.

Lesson Check

Do you know HOW?

A jar contains 10 large red marbles, 4 small red marbles, 6 large blue marbles, and 5 small blue marbles. Calculate the following conditional probabilities for choosing a marble at random.

1. $P(\text{large} \mid \text{red})$ **2.** $P(\text{large} \mid \text{blue})$

3. For the same jar of marbles, which of the conditional probabilities is larger, $P(\text{red} \mid \text{small})$ or $P(\text{small} \mid \text{red})$? Explain.

Do you UNDERSTAND? MATHEMATICAL PRACTICES

4. Error Analysis Your friend says that the conditional probability of one event is 0% if it is independent of another given event. Explain your friend's error.

5. Compare and Contrast How is finding a conditional probability like finding a compound probability? How are they different?

Practice and Problem-Solving Exercises MATHEMATICAL PRACTICES

A Practice

6. Allowance Suppose that 62% of children are given a weekly allowance, and 38% of children do household chores to earn an allowance. What is the probability that a child does household chores, given that the child gets an allowance?

See Problems 1 and 2.

7. You roll two standard number cubes. What is the probability that the sum is even, given that one number cube shows a 2?

Softball Suppose that your softball team has a 75% chance of making the playoffs. Your cross-town rivals have an 80% chance of making the playoffs. Teams that make the playoffs have a 25% chance of making the finals. Use this information to find the following probabilities.

See Problem 3.

8. $P(\text{your team makes the playoffs and the finals})$

9. $P(\text{cross-town rivals make the playoffs and the finals})$

B Apply

10. Think About a Plan Suppose there are two stop lights between your home and school. On the many times you have taken this route, you have determined that 70% of the time you are stopped on the first light and 40% of the time you are stopped on both lights. If you are not stopped on the first light, there is 50% chance you are stopped on the second light. What is the probability that you make it to school without having to stop at a stoplight?
- What conditional probability are you looking for?
- How can a tree diagram help?

11. **Sports** There is a 40% chance that a school's basketball team will make the playoffs this year. If they make the playoffs, there is a 15% chance that they will win the championship. There is also a 30% chance that the same school's volleyball team will make the playoffs this year. If the volleyball team makes the playoffs, there is a 30% chance that they will win the championship. What is the probability that at least one of these teams will win a championship this year?

STEM **Science** In a research study, one third of the volunteers received drug A, one third received drug B, and one third received a placebo. Out of all the volunteers, 10% received drug A and got better, 8% recieved drug B and got better, and 12% received the placebo and got better.

12. What is the conditional probability of a volunteer getting better if they were given drug A?

13. What is the conditional probability of a volunteer getting better if they were given drug B?

14. What is the conditional probability of a volunteer getting better if they were given the placebo?

STEM 15. **Chemistry** A scientist discovered that a certain element was present in 35% of the samples she studied. In 15% of all samples, the element was found in a special compound. What is the probability that the compound is in a sample that contains the element?

16. **Music** Three students compared the music on their MP3 players. Find the probability that a randomly selected song is country, given that it is not on Student A's MP3 player. Round to the nearest hundredth.

Student	Rock	Country	R & B
A	0.10	0.02	0.03
B	0.13	0.05	0.23
C	0.10	0.01	0.33

17. **Fire Drill** A fire drill will begin at a randomly-chosen time between 8:30 A.M. and 3:30 P.M. You have a math test planned for 2:05 P.M. to 3:00 P.M. If the fire drill is in the afternoon, what is the probability that it will start during the test?

C **Challenge** 18. The table below shows the average standardized test scores for a group of students. If 35% of the students are seniors, 40% are juniors, and 25% are freshman, what is the probability that a student chosen at random will have a score at least 125?

	Average Score < 125	Average Score between 125 and 145	Average Score > 145
Freshman	48%	37%	15%
Junior	36%	52%	12%
Senior	13%	69%	18%

Standardized Test Prep

SAT/ACT

19. Which of the following statements is true?

⒜ A kite has two pairs of congruent angles.

Ⓑ The measure of an inscribed angle equals the measure of the arc it intersects.

Ⓒ A rhombus has two consecutive angles congruent if and only if the rhombus is a square.

Ⓓ Two equilateral triangles can be combined to form a square.

20. If $CD = 3$ and $AD = 12$, what is BD?

⒜ 2   Ⓒ 6

Ⓑ 4   Ⓓ 8

Short Response

21. The sides of a rectangle are 5 cm and 12 cm long. What is the sum of the lengths of the rectangle's diagonals?

Mixed Review

Let A and B be independent events, C and D be mutually exclusive events, and F and G be overlapping events. Calculate the following probabilities if $P(A) = 25\%$, $P(B) = 25\%$, $P(C) = 10\%$, $P(D) = 40\%$, $P(F) = 80\%$, $P(G) = 50\%$, and $P(F$ and $G) = 40\%$.

See Lesson 13-4.

22. $P(A$ and $B)$

23. $P(\text{not } A)$

24. $P(C$ and $D)$

25. $P(C$ or $D)$

26. $P(F$ or $G)$

27. $P((\text{not } F)$ and $(\text{not } G))$

Get Ready! **To Prepare for Lesson 13-7, do Exercises 28–29.**

See Lesson 13-1.

28. You roll a standard number cube. What is the probability that a number less than 5 is showing?

29. Your friend tosses a coin 58 times. Heads shows on the coin 25 times. What is the experimental probability that the coin shows tails?

13-7 Modeling Randomness

© **Content Standards**
S.MD.6 Use probabilities to make fair decisions (e.g., drawing by lots, using a random number generator).
S.MD.7 Analyze decisions and strategies using probability concepts (e.g., product testing, medical testing, pulling a hockey goalie at the end of a game).

Objective To understand random numbers
To use probabilities in decision-making

Getting Ready!

A class of 25 students wants to choose 4 students at random to bring food for a class party to be used in the platter shown. Any set of 4 students should have an equal chance of being chosen. What are some ways that the class could determine a random selection of 4 students?

Think of as many ways as you can.

MATHEMATICAL PRACTICES

In the Solve It, you thought of ways to select 4 people. In this lesson, you will learn ways to make a random selection and use probabilities to make decisions and solve real world problems.

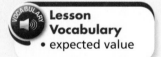

Lesson Vocabulary
• expected value

Essential Understanding You can use probability to make choices and to help make decisions based on prior experience.

A random event has no predetermined pattern or bias toward one outcome or another. You can use random number tables or randomly generated numbers using graphing calculators or computer software to help you make fair decisions. In Reference 5 of this book, you will find a random number table to use with this lesson.

Think

How can a single-digit number be represented by two digits?
A zero and a number between one and nine can represent a single-digit number. For example, 09 can represent 9.

© **Problem 1** **Making Random Selections**

There are 28 students in a homeroom. Four students are chosen at random to represent the homeroom on a student committee. How can a random number table be used to fairly choose the students?

Step 1 Select a line from a random number table.

18823 18160 93593 67294 09632 62617 86779

Step 2 Group the line from the table into two digit numbers.

18 82 31 81 60 93 59 36 72 94 09 63 26 26 17 86 77 9

Step 3 Match the first four numbers less than 28 with the position of the students' names on a list. Duplicates and numbers greater than 28 are discarded because they don't correspond to any student on the list.

18 82 31 81 60 93 59 36 72 94 09 63 26 26 17 86 77 9

The students listed 18th, 9th, 26th, and 17th on the list are chosen fairly.

 Got It? **1.** A teacher wishes to choose three students from a class of 25 students to raise the school's flag. What numbers should the teacher choose based on this line from a random number table?

65358 70469 87149 89509 72176 18103 55169 79954 72002 20582

Problem 2 **Making a Simulation**

A cereal company is having a promotion in which 1 of 6 different prizes is given away with each box. The prizes are equally and randomly distributed in the boxes of cereal. On average, how many boxes of cereal will a customer need to buy in order to get all 6 prizes?

Plan

How do you start to model a situation?
The first step in modeling a situation is determining the possible outcomes and assigning a number to each outcome.

Step 1 Let the digits from 1 to 6 represent the six prizes.

Step 2 Use a graphing calculator and enter the function randInt(1,6) to generate integers from 1 to 6 to simulate getting each prize. One trial is completed when all 6 digits have appeared. The circled numbers represent the first appearance of each number.

Step 3 Count how many boxes of cereal will be bought before all the digits 1 through 6 have appeared.

In this trial, all six prizes were collected after the purchase of the 16th box of cereal.

Step 4 Conduct additional trials. For 19 more simulations the results were: 17, 12, 21, 17, 8, 11, 14, 10, 16, 23, 19, 15, 9, 27, 20, 10, 18, 12, 13. For the 20 simulations, it will take, on average, $\frac{308}{20} = 15.4$ boxes of cereal to get all 6 prizes.

Got It? **2.** Suppose that to win a game, you must roll a standard number cube until all of the sides show at least one time. On average, how many times will you have to roll the number cube before each side shows at least one time? Use a random number table or a random number generator.

In Problem 2, you figured out how many boxes of cereal on average you would expect to have to buy to collect all 6 prizes. You can use this information to make a decision about whether it is worth trying to collect all 6 prizes. Similarly, you can use the *expected value* of a situation that involves uncertainty to make a decision. *Expected value* uses theoretical probability to tell you what you can expect in the long run. If you know what *should* happen mathematically, you make better decisions in problem situations. The **expected value** is the sum of each outcome's value multiplied by its probability.

take note

Key Concept Calculating Expected Value

If A is an event that includes outcomes $A_1, A_2, A_3, \ldots$ and $\text{Value}(A_n)$ is a quantitative value associated with each outcome, the expected value of A is given by
$$\text{Value}(A) = P(A_1) \cdot \text{Value}(A_1) + P(A_2) \cdot \text{Value}(A_2) + \ldots$$

Problem 3 Calculating an Expected Value

Suppose you are at a carnival and are throwing darts at a board like the one at the right. There is an equally likely chance that your dart lands anywhere on the board. You receive 20 points if your dart lands in the white area, 10 points if it lands in the red area, and –5 points if it lands in the blue area. How many points can you expect to get given that the areas for each region are white, 36 in.^{2}; red, 108 in.^{2}; and blue, 432 in.^{2}? The total area is 576 in.^{2}.

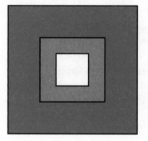

Value (points)

$= P(\text{white area})(\text{white points}) + P(\text{red area})(\text{red points}) + P(\text{blue area})(\text{blue points})$

$= \frac{36}{576} \cdot 20 + \frac{108}{576} \cdot 10 + \frac{432}{576} \cdot (-5)$ Substitute the ratio of each area to the total area and the points for each section.

$= 1.25 + 1.875 + (-3.75)$ Multiply.

$= -0.625$ Add and subtract.

You can expect to get -0.625 points.

✓ **Got It? 3. a.** How many points can you expect to get for the dart board shown in Example 3 if you receive these points: 30 points for white, 15 points for red, and −10 points for blue?

　　b. For part (a), what number of points could be set for the blue area that would result in an expected value of 0?

Expected values can be used to make data-driven decisions. You can calculate the expected values of events of interest, and compare them to decide which is most favorable.

Think

How can a coach know the probability for his team's kicking a field goal?
He can know the experimental probability based on how many successful field goals divided by the number of attempts in similar situations.

Problem 4   Making Decisions Based on Expected Values

Football   On the opening drive, a A football coach must decide whether to kick a field goal (FG) or go for a touchdown (TD). The probabilities for each choice based on his team's experience are shown on the page from his playbook at the right. A field goal will give his team 3 points. His team will get 7 points if the touchdown is successful. Which play should he choose?

	Pts.	Prob.
FG	3	90%
TD	7	35%

Step 1   Calculate the expected value of both plays.

Field Goal: 90% · 3 = 2.7 points

Touchdown: 35% · 7 = 2.45 points

Step 2   Choose the play with the greater expected value.

2.7 > 2.45

The coach should choose the field goal.

Got It?   **4. a.** Suppose the probability for the field goal was 80% and the probability for a touchdown was 30%. Which play should the coach choose?

    b. Are there situations where the coach should choose a play that doesn't have the greatest expected value? Explain.

Lesson Check

Do you know HOW?

1. What are the first four numbers between 1 and 45 which would be chosen on the basis of the following line from a random number table?

81638   36566   42709   33717   59943   12027   46547

2. A basketball player can either attempt a 3-point shot (with a 25% probability of scoring) or pass to a teammate with a 50% probability of scoring 2 points. What are the expected values for each choice? Which choice should he make?

Do you UNDERSTAND?   MATHEMATICAL PRACTICES

3. Reasoning   A friend says that using a random number table to pick a person at random isn't as fair as throwing a dart at the list of names. Explain your friend's error.

4. Vocabulary   Explain the meaning of expected value. Include an example.

5. Writing   Describe how you can use a random number generator to model the results of tossing a coin a hundred times.

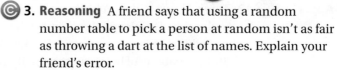

Practice and Problem-Solving Exercises

 MATHEMATICAL PRACTICES

A Practice

Use the lines from a random number table to select numbers to use in each problem.

See Problem 1.

6. Choose 3 students from a list of 45 volunteers.

72749 13347 65030 26128 49067 27904 49953 74674 94617 13317

7. Choose 5 families to survey from a phone directory page with 950 names.

11873 57196 32209 67663 07990 12288 59245 83638 23642 61715

For Exercises 8 and 9, describe how to use random numbers to do a simulation for each situation.

See Problem 2.

8. A teacher assigns students new seats randomly each week. On average, how long will it be before a student is assigned the same seat for two weeks in a row?

9. Sports A basketball player makes 80% of her free throws attempts. Find the average number of free throw attempts needed in order to make 3 free throws in a row.

10. Games In a game show, a contestant receives a prize that has a 5% probability of being worth $1000 and a 95% probability of being worth $1. What is the expected value of winning a prize?

See Problems 3 and 4.

11. Video Games In a video game, a player has a 80% probability of winning 1000 points if he attacks a certain monster. There is a 20% probability that he will lose 25,000 points. What is the expected value of points earned? Should the player attack the monster? Explain.

B Apply

© 12. Think About a Plan A business owner is choosing which of two potential products to develop. Developing Product A will cost $10,000, and the product has a 60% probability of being successful. Product B will cost $15,000 to develop, and it has a 30% probability of being successful. If Product A is successful, the business will gain $200,000. If product B is successful, the business will gain $450,000. Which product should the business owner choose?
- How can the expected values of Product A and Product B help the business owner make the decision?
- Compare the expected values. Recommend the business owner choose the product with the greater expected value.

13. Business A company is deciding whether to invest in a new business opportunity. There is a 40% chance that the company will lose $25,000, a 25% chance that the company will break even, and a 35% chance that the company will make $40,000. Should company executives decide to invest in the opportunity? Explain.

14. Finance A stock has a 25% probability of increasing by $10 per share over the next month. The stock has a 75% probability of decreasing by $5 per share over the same period. What is the expected value of the stock's increase or decrease? Based on this information, should you buy the stock? Explain.

15. Test Taking You earn 1 point for each correct response on a multiple choice test and lose 0.5 point for an incorrect response. Each question has four answer choices. If a student does not know the correct answer and guesses, what is the expected value of the guess?

16. Games A bag contains 10 marbles; 4 are red, 5 are yellow, and 1 is blue. You draw one marble from the bag without looking. If you draw a blue marble, you win $10. If you draw a red marble, you win $5, and if you draw a yellow marble, you win $3. What is the expected value of drawing one marble? Would you play this game for $5? Explain.

Ⓒ **17. Writing** A polling company has been hired to survey 500 households out of a possible 2500 households. How can the 500 households be randomly selected for the survey?

Ⓒ **Challenge**

18. Stocks An investor is choosing between three stocks she might purchase. The potential outcomes are listed in the table.
   a. Calculate the expected value of the gain or loss for each stock.
   b. Set up a model and use a random number table to model the investment in each stock.
   c. How do the results of the random number table model compare with the expected values?
   d. Does the expected value tell you anything about the riskiness of each stock?

Potential Outcomes

	Lose $25 per share	Gain $5 per share	Gain $45 per share
Stock ABC	40%	15%	45%
Stock JKL	15%	65%	20%
Stock MNO	5%	80%	15%

Standardized Test Prep

SAT/ACT

19. What is the sum of the interior angles of an octagon?
   Ⓐ 135   Ⓑ 360   Ⓒ 480   Ⓓ 1080

20. A kite has an 80° angle and a 50° angle. Which of the following might be the measure of the remaining angles?
   Ⓐ 80 and 50   Ⓑ 50 and 180   Ⓒ 80 and 150   Ⓓ 50 and 150

Short Response

21. What are the lengths of the two legs of a right triangle with a 60° angle and a hypotenuse 1 meter long? Round your answer to the nearest centimeter.

Mixed Review

Calculate the following conditional probabilities.   ◀ See Lesson 13-5.

22. The conditional probability $P(A \mid B)$ when $P(B) = 30\%$ and $P(A \text{ and } B) = 20\%$

23. The conditional probability $P(A \mid B)$ when $P(B) = 75\%$ and $P(A \text{ and } B) = 50\%$

What is the standard equation of each circle?   ◀ See Lesson 12-5.

24. center $(3, -5)$; radius, 9

25. center $(-2, 8)$; radius, 5

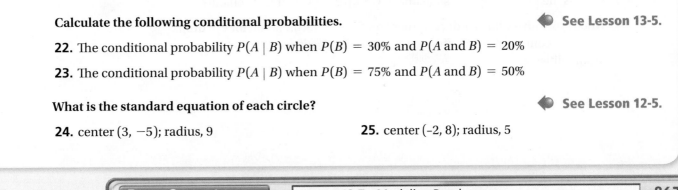

Probability and Decision Making

© Content Standard
S.MD.7 Analyze decisions and strategies using probability concepts (e.g., product testing, medical testing, pulling a hockey goalie at the end of a game).

Understanding probability is important in areas from game playing to testing new pharmacuetical drugs.

Activity 1

Your older sister has devised a game to play with your younger brother. She says that your brother can choose who goes first. Here are the rules:

> Roll a standard number cube.
>
> Player 1 wins a point if the number cube shows an even number or a 5.
>
> Player 2 wins a point if the number cube is a factor of 24.

1. a. In each turn, what is the probability that Player 1 scores a point? What is the probability that Player 2 scores a point?

   b. What advice would you give your brother? Should he choose to be Player 1 or Player 2? Explain.

Activity 2

A pharmaceutical company tested a new formulation for eye drops on 160 people who suffered from very dry eyes. Two groups of test subjects were randomly selected. The first group received a placebo and the second group received the new eye drops. Below are the results of the testing.

Eye Drop Test Results

	Improvement	No Improvement
Group 1	48	32
Group 2	39	41

2. What is the probability that someone in Group 1 showed improvement?

3. What is the probability that someone in Group 2 showed improvement?

4. Would you advise further development on the new formulation for eye drops or for the company to drop this new medication? Justify your answer using probabilities.

BIG idea Probability

You can find theoretical and experimental probabilities to make decisions or predictions about future events.

> To solve these problems, you will pull together concepts and skills related to probability.

© Performance Task 1

Use a random number table or generate random numbers to simulate the experiment of rolling a standard number cube 60 times.

   a. What is the experimental probability of rolling each number?

   b. What is the theoretical probability of rolling each number? In 60 rolls, how many times would you expect to roll each number?

   c. Do the results of your experiment match the expected number of rolls of each number in 60 trials? If not, explain why there might be differences.

© Performance Task 2

Take a survey of at least 20 people. Record whether they are male or female and whether or not their first names end with a vowel (include y as a vowel for this survey).

   a. Make a tree diagram of the data.

   b. Find P(female | ends in vowel) and P(ends in vowel | male).

   c. Is there any difference in how a name ends based on gender?

© Performance Task 3

In a carnival game, you get 5 darts to throw at the dartboard at the right. For each dart you throw, you score the number of points shown. At the end of the game, you add all your points and get one dollar for each point in your total. If the number of total points is negative, you get nothing.

   a. What is the expected value for the number of points when throwing one dart? (Assume that your dart lands at random on the dartboard.)

   b. If it costs $5 to play this game, do you expect to make money or lose money at the end of the game? Explain.

-5
0
10
20

Width of each ring = *r*.

r

Connecting **BIG** ideas and Answering the Essential Questions

1 Probability
Theoretical probability is based on what should happen mathematically. Experimental probability is based on observations.

Experimental and Theoretical Probability (Lesson 13-1)

Experimental Probability:
$$P(\text{event}) = \frac{\text{number of times the event occurs}}{\text{number of times the experiment is done}}$$

Theoretical Probability:
$$P(\text{event}) = \frac{\text{number of favorable outcomes}}{\text{number of possible outcomes}}$$

Compound Probability (Lessons 13–4)

For independent events A and B,
$P(A \text{ and } B) = P(A) \cdot P(B)$.

For mutually exclusive events A and B,
$P(A \text{ or } B) = P(A) + P(B)$.

For overlapping events A and B,
$P(A \text{ or } B) = P(A) + P(B) - P(A \text{ and } B)$.

2 Data Representation
A frequency table represents data from one category. A two-way frequency table represents data from two categories.

Probability Distributions and Frequency Tables and Probability Models (Lessons 13-2 and 13-5)

Tables are used to organize data, and can be helpful when finding probabilities.

Permutations and Combinations (Lesson 13–3)

$_nP_r = \dfrac{n!}{(n-r)!}$ for $0 \le r \le n$

$_nC_r = \dfrac{n!}{r!(n-r)!}$ for $0 \le r \le n$

Conditional Probability Formulas (Lesson 13–6)

You can use two-way frequency tables to calculate conditional probability. For dependent events A and B,
$P(A \text{ and } B) = P(A) \cdot P(B \mid A)$.

3 Probability
A random event has no bias or inclination toward any outcome.

Modeling Randomness (Lesson 13–7)

You can use random number tables and graphing calculators to model random events and make fair decisions.

Chapter Vocabulary

- combination (p. 838)
- complement of an event (p. 826)
- conditional probability (p. 851)
- dependent events (p. 844)
- event (p. 824)
- experimental probability (p. 825)
- expected value (p. 864)
- frequency table (p. 830)
- independent events (p. 844)
- mutually exclusive events (p. 845)
- permutation (p. 837)
- probability (p. 824)
- sample space (p. 824)
- theoretical probability (p. 825)
- two-way frequency table (p. 850)

Choose the correct term to complete each sentence.

1. A ___?___ is a grouping of items in which order is important.

2. ___?___ is based on what should happen mathematically.

3. The outcomes of ___?___ events affect each other.

4. A ___?___ should be used to show the results of a survey on students' favorite teacher.

13-1 Experimental and Theoretical Probability

Quick Review

Experimental probability is based on how many successes are observed in repeated trials of an event. **Theoretical probability** describes the likelihood of an event based on what should happen mathematically. The **sample space** is the set of all possible outcomes of an experiment.

Example

There are 14 boys and 11 girls in a homeroom. If a student is selected at random, what is the probability that the student is a girl?

There are 25 possible outcomes in the sample space. Of these, 11 are girls. The probability that the student is a girl is $P(\text{girl}) = \frac{11}{25} = 0.44$.

Exercises

There are 12 math teachers, 9 science teachers, 3 music teachers, and 6 social studies teachers at a conference. If a teacher is selected at random, find each probability.

5. $P(\text{science})$

6. $P(\text{not music})$

7. $P(\text{reading})$

8. $P(\text{not math})$

9. Bowling Suppose you rolled 5 strikes and 7 non-strikes in a bowling game. Based on these results, what is the experimental probability that you will roll a strike?

13-2 Probability Distributions and Frequency Tables

Quick Review

A **frequency table** shows how often an item appears in a category. The **relative frequency** of an item is the ratio of the frequency of the category to the total frequency. A **probability distribution** shows the probabilities of every possible outcome of an experiment.

Example

What is the relative frequency of sedans?

There are 30 cars in all, and 12 of them are sedans. So, the relative frequency of sedans is $\frac{12}{30} = 0.4$.

Type of Car	Frequency
Truck	3
Minivan	6
Sedan	12
Sports	2
Other	7

Exercises

Use the frequency table below to estimate each probability.

Test Grade	Frequency
A	7
B	11
C	6
D	3
F	1

10. $P(A)$

11. $P(B)$

12. $P(B \text{ or } C)$

13. $P(\text{grade below } C)$

14. $P(\text{grade above } C)$

15. $P(\text{grade above } D)$

13-3 Permutations and Computations

Quick Review

According to the **Fundamental Counting Principle**, if event M can occur in m ways and event N can occur in n ways, then event M and event N can occur in $m \cdot n$ ways.

n factorial: $n! = n \cdot (n - 1) \cdot (n - 2) \cdot \ldots \cdot 3 \cdot 2 \cdot 1$

Permutation formula (order matters):

$$_nP_r = \frac{n!}{(n - r)!}$$

Combination formula (order is not important):

$$_nC_r = \frac{n!}{r!(n - r)!}$$

Example

How many ways can a teacher choose a group of 4 students from 15 if order does not matter?

$$_{15}C_4 = \frac{15!}{4!(15 - 4)!} = 1365 \text{ ways}$$

Exercises

Evaluate each of the following.

16. $5!$

17. $8!$

18. $\frac{7!}{3!}$

19. $\frac{12!}{9!3!}$

20. $_8P_5$

21. $_{10}P_2$

22. $_9C_7$

23. $_{20}C_3$

24. License Plate A license plate is made up of 3 letters followed by 4 digits. How many different license plates are possible if none of the letters or digits can be repeated?

25. Music A student wants to put 3 songs on her MP3 player from a collection of 25 songs. How many different ways can she do this?

13-4 Compound Probability

Quick Review

For independent events A and B,
$P(A \text{ and } B) = P(A) \cdot P(B)$. For mutually exclusive events A and B, $P(A \text{ or } B) = P(A) + P(B)$. For overlapping events A and B, $P(A \text{ or } B) = P(A) + P(B) - P(A \text{ and } B)$.

Example

If a classmate rolls a standard number cube, what is the probability that she rolls a prime number or an even number?

$$P(\text{prime}) = \frac{3}{6} = \frac{1}{2} \qquad \text{2, 3, and 5 are prime.}$$

$$P(\text{even}) = \frac{3}{6} = \frac{1}{2} \qquad \text{2, 4, and 6 are even.}$$

$$P(\text{even and prime}) = \frac{1}{6} \qquad \text{2 is even and prime.}$$

$$P(\text{even or prime}) = \frac{1}{2} + \frac{1}{2} - \frac{1}{6} = \frac{5}{6}$$

Exercises

Suppose A and B are independent events, $P(A) = 0.46$, and $P(B) = 0.25$. Find each probability.

26. $P(A \text{ and } B)$

27. $P(A \text{ or } B)$

28. A bag contains 3 red marbles, 4 green marbles, 5 blue marbles, and 4 yellow marbles. If a marble is selected at random, what is the probability that it is red or green?

29. Writing Explain why the formula $P(A \text{ or } B) = P(A) + P(B) - P(A \text{ and } B)$ can be used for any two events A and B.

13-5 Probability Models

Quick Review

A **two-way frequency** table shows data that fall into two different categories. You can use two-way frequency tables to find relative frequencies of items in different categories.

Example

What is the relative frequency of female juniors on the student council?

	Male	Female	Totals
Juniors	2	2	4
Seniors	1	3	4
Totals	3	5	8

There are 8 members and 2 of them are female juniors. The relative frequency is $\frac{2}{8} = 0.25$.

Exercises

Studying Use the two-way frequency table to answer each question.

	Studied	Did not study	Totals
Passed	19	7	26
Failed	1	4	5
Totals	20	11	31

30. Did more students choose to study or to not study for the quiz?

31. What is the relative frequency of students who studied for the quiz and passed?

32. What is the relative frequency of students who did not study and failed?

13-6 Conditional Probability Formulas

Quick Review

To find the conditional probability that dependent events A and B occur, multiply the probability that event A occurs by the probability that event B occurs, given that event A has already occurred.

$$P(A \text{ and } B) = P(A) \cdot P(B \mid A)$$

For any two events A and B with $P(A) \neq 0$,

$$P(B \mid A) = \frac{P(A \text{ and } B)}{P(A)}.$$

Example

A coach's desk drawer contains 3 blue pens, 5 black pens, and 4 red pens. What is the probability that he selects a black pen followed by a red pen, if the first pen is not replaced?

$$P(\text{black, then red}) = P(\text{black}) \cdot P(\text{red, given black})$$

$$= \frac{5}{12} \cdot \frac{4}{11} = \frac{5}{33}$$

Exercises

Solve each problem.

33. **Recreation** The ball bin in the gym contains 8 soccer balls, 12 basketballs, and 10 kickballs. If Coach Meyers selects 2 balls at random and does not replace the first, what is the probability that she selects 2 basketballs?

34. **Tests** Suppose your teacher has given 2 pop quizzes this week. Seventy percent of her class passed the first quiz, and 50% passed both quizzes. If a student is selected at random, what is the probability that he or she passed the second quiz, given that the student passed the first quiz?

35. **Shopping** At a shoe store yesterday, 74% of customers bought shoes, 38% bought accessories, and 55% bought both shoes and accessories. If a customer is selected at random, what is the probability that he bought accessories, given that he bought shoes?

13-7 Modeling Randomness

Quick Review

Random events have no predetermined pattern or bias toward one outcome or the other. You can use random number tables and graphing calculators to help you model random events and make fair decisions. A simulation is a probability model that can be used to make predictions about real world situations.

Example

Twenty students are assigned 2-digit numbers from 01 to 20. Use the excerpt from the random number table below to randomly select the first 3 students.

    80261 53267 85048 53091 18479 06286 38323

Reading from left to right, the 2-digit numbers are:

    80, 26, 15, 32, 67, 85, 04, 85, 30, 91, 18, . . .

The first 3 students are indicated by the numbers 15, 04, and 18.

Exercises

Golf On average, Raymond shoots lower than 80 in 3 out of 4 rounds of golf. He conducts a simulation to predict how many times he will shoot lower than 80 in his next 5 rounds. Let 1, 2, and 3 represent shooting lower than 80, and let 4 represent shooting 80 or higher.

22113	22413	34141	12131	34432
34112	34142	24241	44231	41122
23132	13314	14133	11432	41214
34423	13332	12422	11422	33312

36. How many trials of the simulation did Raymond conduct?

37. In how many trials of the simulation did Raymond shoot lower than 80 in at least 4 of the 5 rounds?

38. According to this simulation, what is the experimental probability that Raymond will shoot lower than 80 in at least 4 of his next 5 rounds?

Do you know HOW?

You select a number from 1 to 10 at random. Find each probability.

1. $P(\text{odd number})$
2. $P(1 \text{ or } 4)$
3. $P(\text{less than } 8)$
4. $P(\text{greater than } 10)$

For Exercises 5–8, use the frequency table to find each relative frequency.

Trees Planted	Frequency
Oak	7
Maple	12
Birch	9
Pear	17

5. oak trees
6. maple trees
7. birch trees
8. pear trees

Evaluate each expression.

9. $8!$
10. $\frac{9!}{5!}$
11. $\frac{6!}{4!2!}$
12. $_7C_4$
13. $_8P_5$
14. $_6C_6$

If A and B are independent events, find $P(A \text{ and } B)$ and $P(A \text{ or } B)$.

15. $P(A) = 0.2, P(B) = 0.4$
16. $P(A) = \frac{2}{3}, P(B) = \frac{1}{4}$

For Exercises 17 and 18, find each conditional probability, given that $P(A) = \frac{11}{20}$, $P(B) = \frac{17}{25}$, and $P(A \text{ and } B) = \frac{3}{25}$.

17. $P(A \mid B)$
18. $P(B \mid A)$

Use the excerpt from a random number table below to find each of the following in Exercises 19 and 20.

59717 37738 07632 84854 38223 65202 50822

19. the first three 2-digit numbers from 01 to 30
20. the first two 3-digit numbers from 100 to 299

Do you UNDERSTAND?

For Exercises 21 and 22, determine whether each pair of events is independent or dependent.

21. playing a stringed instrument and playing a percussion instrument
22. selecting a letter tile, then selecting a second letter tile without replacing the first

23. **Writing** If four friends are lining up for a photo, does the situation describe a combination or a permutation? Explain. How many different ways are there for four friends to line up for a photo?

24. **Weather** The probability of school being cancelled because of snow on any given day in January is 7.5%. Describe a simulation you could conduct to predict how many snow days you will have in January if there are 21 school days during the month.

25. A and B are independent events. Find two probabilities, $P(A)$ and $P(B)$, such that $P(A \text{ and } B) = \frac{1}{8}$ and $P(A \text{ or } B) = \frac{5}{8}$.

26. **Writing** Describe how to find relative frequencies of items in different categories in a contingency table.

End-of-Course Assessment © ASSESSMENT

Multiple Choice

Read each question. Then write the letter of the correct answer on your paper.

1. Which term is defined by the following phrase?

a set of points in a plane a given distance from a point

- (A) segment
- (C) sector
- (B) circle
- (D) angle

2. The United States has a land area of about 3,536,278 mi^2. Illinois has a land area of about 57,918 mi^2. What is the probability that a location in the United States chosen at random is not in Illinois?

- (F) 1.6%
- (H) 16.3%
- (G) 9.8%
- (I) 98.4%

3. $\overleftrightarrow{AB}$ is tangent to $\odot C$ at point B. Which of the following can you NOT conclude is true?

- (A) $m\angle CAB < m\angle ACB$
- (B) $AB^2 + BC^2 = AC^2$
- (C) $\angle CAB$ and $\angle ACB$ are complements.
- (D) $\overleftrightarrow{AB} \perp \overleftrightarrow{BC}$

4. What is the relationship between the sine of an angle and the cosine of the angle's complement?

- (F) They are inverses.
- (H) They are the same.
- (G) Their sum is 1.
- (I) They are opposites.

5. What is the area of $\triangle RST$ in square feet?

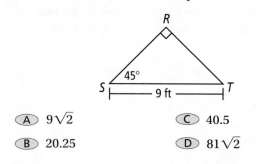

- (A) $9\sqrt{2}$
- (C) 40.5
- (B) 20.25
- (D) $81\sqrt{2}$

6. Which is an equation of the circle below?

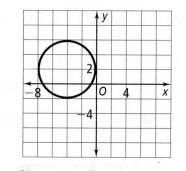

- (F) $(x + 4)^2 + (y - 2)^2 = 16$
- (G) $(x - 4)^2 + (y - 2)^2 = 4$
- (H) $(x - 2)^2 + (y + 1)^2 = 4$
- (I) $(x + 2)^2 + (y - 1)^2 = 2$

7. What is the area of the shaded sector? Use 3.14 for π.

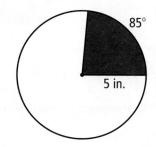

- (A) about 18.5 in.^{2}
- (B) about 66.8.5 in.^{2}
- (B) about 78.5 in.^{2}
- (D) about 85 in.^{2}

8. Which describes the transformation of $\triangle ABC$ to $\triangle A'B'C'$?

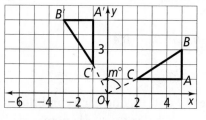

- (F) $R_{y=x}(\triangle ABC) = \triangle A'B'C'$
- (G) $T_{<2,5>}(\triangle ABC) = \triangle A'B'C'$
- (H) $r_{(m°, O)}(\triangle ABC) = \triangle A'B'C'$
- (I) $R_{y=x} + T_{<2,0>}(\triangle ABC) = \triangle A'B'C'$

9. What are the values of a and b in the figure below?

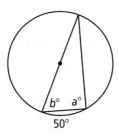

- **A** $a = 90$ and $b = 130$
- **B** $a = 50$ and $b = 130$
- **C** $a = 90$ and $b = 65$
- **D** $a = 65$ and $b = 65$

10. What is the locus of points in space that are 5 cm from segment m?

- **F** a cylinder with radius 5 cm capped at each end with a half sphere
- **G** a cylinder with radius 5 cm
- **H** a sphere with radius 5 cm
- **I** a cylinder with radius 5 cm capped at each end with a filled circle

11. Figure $QRST$ is shown in the coordinate plane. Which transformation creates an image with a vertex at the point $(-2, 1)$?

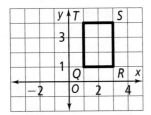

- **A** Rotate figure $QRST$ by 90° about R.
- **B** Reflect figure $QRST$ across the line $y = 1$.
- **C** Reflect figure $QRST$ across the line $x = 1$.
- **D** Rotate figure $QRST$ by 90° about Q.

12. Suppose you roll two number cubes, one red and one blue. What is the probability that you will roll 3 on the red cube and an even number on the blue cube?

- **F** $\frac{1}{4}$
- **G** $\frac{1}{3}$
- **H** $\frac{1}{6}$
- **I** $\frac{1}{12}$

13. Amy is trying to prove that $a \cdot b = c \cdot d$ using $\odot O$ at the right by first proving that $\triangle APC \sim \triangle DPB$. Which similarity theorem or postulate can Amy use?

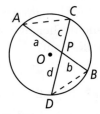

- **A** Side-Side-Side Similarity Theorem
- **B** Side-Angle-Side Similarity Theorem
- **C** Angle-Angle Similarity Postulate
- **D** None of these

14. All four angles of a quadrilateral have the same measure. Which statement is true?

- **F** All four sides of the quadrilateral must have the same length.
- **G** All four angles of the quadrilateral are acute.
- **H** Opposite sides of the quadrilateral are parallel.
- **I** The quadrilateral must be a square.

15. A triangular park is bordered by three streets, as shown in the map below.

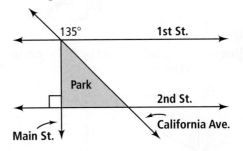

If 1st Street and 2nd Street are parallel, what are the measures of the three angles of the park?

- **A** 90, 45, 45
- **B** 90, 35, 55
- **C** 90, 25, 65
- **D** 135, 25, 10

16. $\overline{AB}$ is dilated by a scale factor of 2 with a center of dilation C, which is not on $\overline{AB}$. Which best describes $\overline{A'B'}$ and its length?

I. $A'B' = 2AB$

II. $\overline{A'B'} \parallel \overline{AB}$

- **F** I only
- **G** II only
- **H** I and II
- **I** neither I nor II

17. Which equation can be used to find the height h of the triangle at the right?

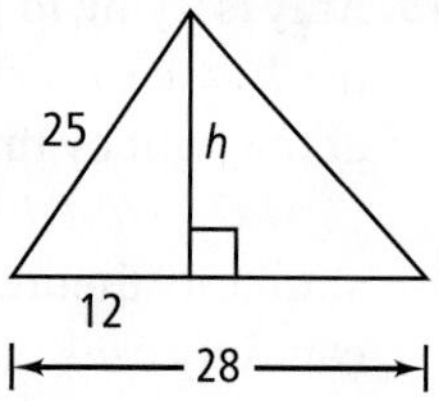

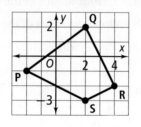

- Ⓐ $h^2 = 28^2 - 25^2$
- Ⓑ $h^2 = 25^2 + 12^2$
- Ⓒ $25^2 = 12^2 + h^2$
- Ⓓ $25^2 = 12^2 - h^2$

18. In the figure at the right, what is the length of $\overline{PS}$?

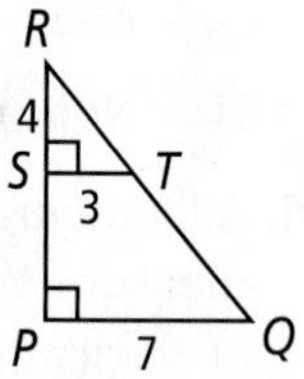

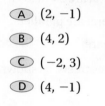

- Ⓕ 3
- Ⓗ 4
- Ⓖ $3\sqrt{2}$
- Ⓘ $5\frac{1}{3}$

19. Quadrilateral $PQRS$ is reflected across the line $x = 1$. Its image is $P'Q'R'S'$. What are the coordinates of P'?

- Ⓐ $(2, -1)$
- Ⓑ $(4, 2)$
- Ⓒ $(-2, 3)$
- Ⓓ $(4, -1)$

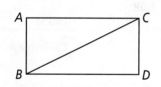

20. Which of the following facts would be sufficient to prove $\triangle ACB \cong \triangle DBC$?

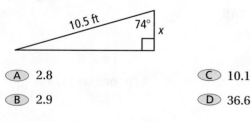

- Ⓕ $\angle A$ is a right angle.
- Ⓖ $\overline{BC} \cong \overline{BC}$
- Ⓗ $\angle ABC$ and $\angle DCB$ are acute.
- Ⓘ $\overline{AB} \parallel \overline{CD}$ and $\overline{AC} \parallel \overline{BD}$.

21. To the nearest tenth of a foot, what is the value of x in the figure below?

- Ⓐ 2.8
- Ⓑ 2.9
- Ⓒ 10.1
- Ⓓ 36.6

22. Bob walked diagonally across a rectangular field that measured 240 ft by 320 ft. Which expression could be used to determine how far Bob walked?

- Ⓕ $2(240 + 320)$
- Ⓖ $\sqrt{240} + \sqrt{320}$
- Ⓗ $\dfrac{240 + 320}{2}$
- Ⓘ $\sqrt{240^2 + 320^2}$

23. Which triangle is drawn with its medians?

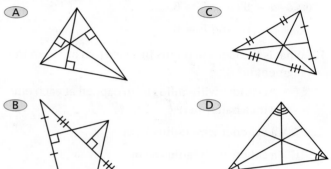

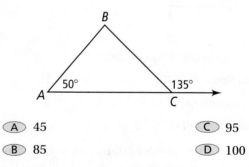

24. Which statement is true for both a rhombus and a kite?

- Ⓕ The diagonals are congruent.
- Ⓖ Opposite sides are congruent.
- Ⓗ The diagonals are perpendicular.
- Ⓘ Opposite sides are parallel.

25. Given $\triangle ABC$ below, what is $m\angle B$?

- Ⓐ 45
- Ⓑ 85
- Ⓒ 95
- Ⓓ 100

26. How can you determine that a point lies on the perpendicular bisector of $\overline{PQ}$ with endpoints $P(-3, -6)$ and $Q(-3, 4)$?

   (F) The point has x-coordinate -3.

   (G) The point has y-coordinate -3.

   (H) The point lies on the line $x = -1$.

   (I) The point lies on the line $y = -1$.

27. The table shows the number of degree recipients in a recent year. What is the probability that the recipient was a female getting a bachelor's degree?

Number of Degree Recipients (thousands)

Degree	Male	Female
Associates	245	433
Bachelors	598	853

   (A) 36%   (C) 59%

   (B) 40%   (D) 64%

28. What is the volume of the square pyramid?

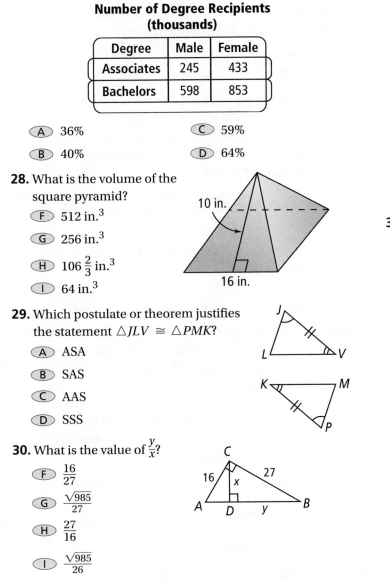

   (F) 512 in.^{3}

   (G) 256 in.^{3}

   (H) $106\frac{2}{3}$ in.^{3}

   (I) 64 in.^{3}

29. Which postulate or theorem justifies the statement $\triangle JLV \cong \triangle PMK$?

   (A) ASA

   (B) SAS

   (C) AAS

   (D) SSS

30. What is the value of $\frac{y}{x}$?

   (F) $\frac{16}{27}$

   (G) $\frac{\sqrt{985}}{27}$

   (H) $\frac{27}{16}$

   (I) $\frac{\sqrt{985}}{26}$

31. For the triangle at the right, what is $m\angle E$?

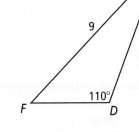

   (A) 17°

   (B) 23°

   (C) 47°

   (D) 53°

32. In the figure below, $\overline{PQ}$ is parallel to $\overline{RS}$, and $\overline{PS}$ and $\overline{QR}$ intersect at point T. What is the length of $\overline{PS}$?

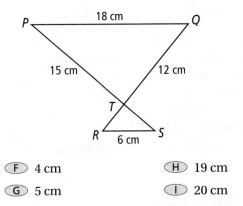

   (F) 4 cm   (H) 19 cm

   (G) 5 cm   (I) 20 cm

33. A hiker is traveling north through Anza Borrego State Park. When he is 0.8 mi from his destination, he veers off course for 1 mi. Use the diagram below to determine how many miles x the hiker is from his destination.

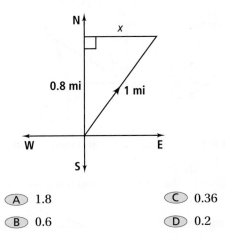

   (A) 1.8   (C) 0.36

   (B) 0.6   (D) 0.2

34. A school library has 85 books on history. It also has 240 books of fiction, of which 25 are historical fiction. If the library holds 1000 books, what is the probability that it is history or fiction but not historical fiction?

(F) $\frac{1}{14}$ (H) $\frac{3}{10}$

(G) $\frac{2}{5}$ (I) $\frac{13}{14}$

35. What values of x and y make the quadrilateral a parallelogram?

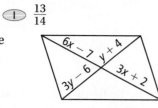

(A) $x = 3$, $y = 3$ (C) $x = 5$, $y = 3$

(B) $x = 3$, $y = 5$ (D) $x = 5$, $y = 7$

36. On a globe, lines of latitude form circles, as shown at the right. The arc from the equator to the Artic Circle is 66.5°. The approximate radius of Earth is 6471 km. What is the circumference of the Artic Circle?

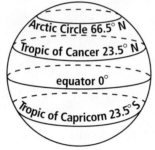

(F) about 2693 km

(G) about 16,213 km

(H) about 2,662,006 km

(I) about 22,775,168 km

37. Which sequence of transformations maps $\triangle ABC$ onto $\triangle DEF$?

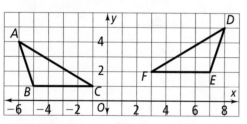

(A) reflection across the line $x = 0$ and translation right 2 units

(B) reflection across the line $x = 1$ and translation up 1 unit

(C) 180° rotation about the origin and reflection over the line $x = \frac{1}{2}$

(D) 180° rotation about the origin and translation right 2 units up 1 unit

38. Read this excerpt from a news article.

The Casco Bay Bridge, a double-leaf drawbridge in Maine, opened in 1997. The bridge replaced the old Million Dollar Bridge over the Fore River. The old bridge had a clearance of 24 feet between the water and the closed bridge. The new bridge has a clearance of 65 feet when closed, so it does not need to be opened as often as the old bridge. Each leaf of the new bridge is approximately 143 feet long and opens up to an angle of 78°. The new bridge may need to be opened less often, but it takes about 6 minutes longer to open and close than the old bridge.

How high off the water must the tip of each leaf be when the Casco Bay Bridge is open?

(F) 65 ft   (H) 140 ft

(G) 95 ft   (I) 205 ft

39. Which method is a fair way to randomly select 12 students from a class of 30 students?

(A) Assign each student a number from 01 to 30. Use a random number table, selecting the first 12 pairs of digits from 01 to 12.

(B) Assign each student a number from 01 to 30. Use a random number table, selecting the first 12 pairs of digits from 01 to 30.

(C) Roll red and blue number cubes. The red cube represents the first digit and the second represents the second digit. Roll 12 times.

(D) Have the 30 students elect the 12 student.

40. Derek lives close to both his school and the library, as shown below. After school, Derek walked 5 min to the library. He then walked 8 min to his home when he realized that he left a book at school. Which time is the best estimate of how long it will take Derek to walk directly from his home to his school if the angle formed by the School-Library-Derek's home is 87°? (Assume Derek walks at the same rate each time.)

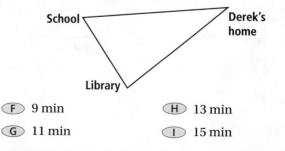

(F) 9 min   (H) 13 min

(G) 11 min   (I) 15 min

41. What is the value of x in the circle at the right?

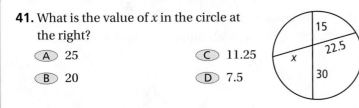

- Ⓐ 25
- Ⓒ 11.25
- Ⓑ 20
- Ⓓ 7.5

42. Which of the following could be the side lengths of a right triangle?

- Ⓕ 4.1, 6.2, 7.3
- Ⓖ 40, 60, 72
- Ⓗ 3.2, 5.4, 6.2
- Ⓘ 33, 56, 65

43. In the figure at the right, what is the length of $\overline{CD}$?

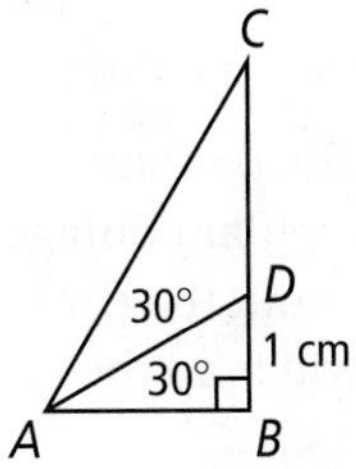

- Ⓐ $2\sqrt{3}$ cm
- Ⓑ 3 cm
- Ⓒ 2 cm
- Ⓓ $\sqrt{3}$ cm

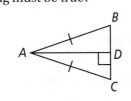

44. A manufacturer is comparing two packages for a new product. Package A is a rectangular prism that is 9 in. by 5 in. by 8 in. Package B is a triangular prism with height 15 in. Its bases are right triangles with 6-in. and 8-in. legs. Which statement best describes the relationship between the two prisms?

- Ⓕ The triangular prism has $\frac{1}{2}$ the volume of the rectangular prism.
- Ⓖ The rectangular prism has the greater volume.
- Ⓗ The volumes are equal.
- Ⓘ not enough information

45. Which of the following must be true?

I. $\angle BAC \cong \angle B$

II. $\angle B \cong \angle C$

III. $\overline{AD} \cong \overline{AB}$

IV. $\overline{BD} \cong \overline{CD}$

- Ⓐ I and II only
- Ⓒ II and IV only
- Ⓑ I and III only
- Ⓓ III and IV only

46. What are the values of x and y?

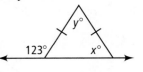

- Ⓕ $x = 46, y = 67$
- Ⓖ $x = 67, y = 46$
- Ⓗ $x = 57, y = 66$
- Ⓘ $x = 66, y = 57$

47. What is the volume of the cylinder to the nearest cubic inch?

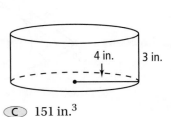

- Ⓐ 603 in.^{3}
- Ⓒ 151 in.^{3}
- Ⓑ 226 in.^{3}
- Ⓓ 113 in.^{3}

48. A large box of laundry detergent has the shape of a rectangular prism. A similar box has length, width, and height that are one half of the large box. How many times the volume of the small box is the volume of the large box?

- Ⓕ 4
- Ⓗ 64
- Ⓖ 8
- Ⓘ 512

49. In the figure at the right, $\overline{AB}$ is tangent to $\odot O$, $AB = 15$ cm, and $BC = 9$ cm. What is the radius of $\odot O$?

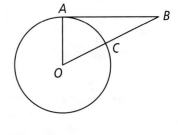

- Ⓐ 7 cm
- Ⓒ 9 cm
- Ⓑ 8 cm
- Ⓓ 16 cm

50. Myra multiplies the length of each side of a triangle by $\frac{1}{5}$. By what factor can she multiply the perimeter of the original triangle to find the perimeter of the new triangle?

- Ⓕ 0.008
- Ⓗ 0.2
- Ⓖ 0.04
- Ⓘ 5

51. What is the area of a regular pentagon with side length 4 cm? Round your answer to the nearest tenth.

- Ⓐ 16.2 cm^2
- Ⓒ 41.5 cm^2
- Ⓑ 27.5 cm^2
- Ⓓ 55.0 cm^2

52. The triangle circumscribes the circle. What is the perimeter of the triangle?

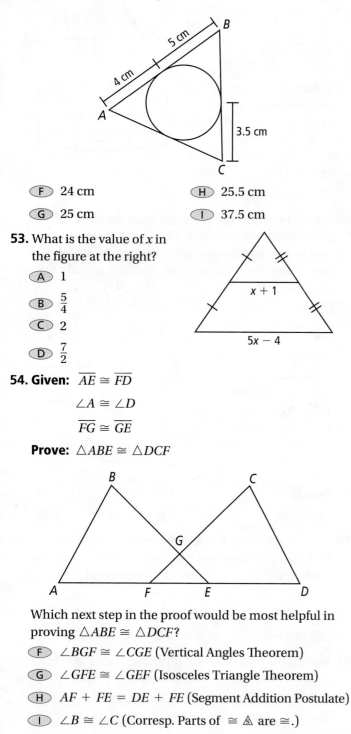

- (F) 24 cm
- (G) 25 cm
- (H) 25.5 cm
- (I) 37.5 cm

53. What is the value of x in the figure at the right?

- (A) 1
- (B) $\frac{5}{4}$
- (C) 2
- (D) $\frac{7}{2}$

$x + 1$

$5x - 4$

54. Given: $\overline{AE} \cong \overline{FD}$

$\angle A \cong \angle D$

$\overline{FG} \cong \overline{GE}$

Prove: $\triangle ABE \cong \triangle DCF$

Which next step in the proof would be most helpful in proving $\triangle ABE \cong \triangle DCF$?

- (F) $\angle BGF \cong \angle CGE$ (Vertical Angles Theorem)
- (G) $\angle GFE \cong \angle GEF$ (Isosceles Triangle Theorem)
- (H) $AF + FE = DE + FE$ (Segment Addition Postulate)
- (I) $\angle B \cong \angle C$ (Corresp. Parts of $\cong$ ▲ are $\cong$.)

55. The diagram below shows a standard construction with straightedge and compass. What has been constructed?

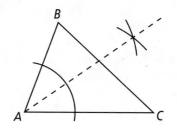

- (A) a median
- (B) an altitude
- (C) a perpendicular bisector
- (D) an angle bisector

56. The pentagon at the right is a regular pentagon. Which rotation about the center of the pentagon will map the pentagon onto itself?

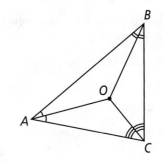

- (F) 58°
- (G) 90°
- (H) 144°
- (I) 180°

57. In $\triangle ABC$ below, point O has been constructed.

What is point O?

- (A) centroid
- (B) center of the inscribed circle
- (C) center of the circumscribed circle
- (D) none of the above

58. What is the surface area of the sphere?

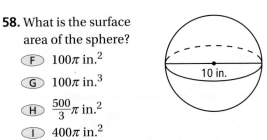

- (F) 100π in.^{2}
- (G) 100π in.^{3}
- (H) $\frac{500}{3}\pi$ in.^{2}
- (I) 400π in.^{2}

59. Which is the cross section of a cylinder that is intersected by a plane that is perpendicular to the two bases?

- (A) circle
- (C) trapezoid
- (B) rectangle
- (D) triangle

60. If O is the center of the circle, what can you conclude from the diagram?

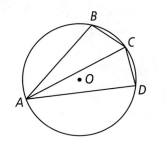

- (F) $AB > AD$
- (G) $AB = AD$
- (H) $AB < AD$
- (I) There is not enough information to compare AB and AD.

61. The height of the cone below is 5.

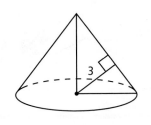

What is the radius of the cone?

- (A) 3
- (C) 3.75
- (B) $\sqrt{10}$
- (D) 4

Short Response

62. How do you construct a $\triangle DEF$ congruent to $\triangle ACB$?

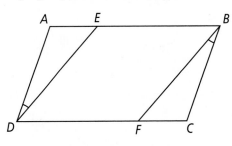

63. Rieko is trying to prove the following theorem:

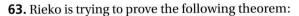

If $ABCD$ is a parallelogram, and $\angle ADE \cong \angle CBF$, then $DEBF$ is a parallelogram.

One strategy is to show that both pairs of opposite angles are congruent. How would you show that $\angle EDF \cong \angle EBF$?

Extended Response

64. In one game on a reality TV show, players must dig for a prize hidden somewhere within a sand-filled circular area. There are 18 posts equally spaced around the circle and clues are given so that players narrow the location by crossing two ropes between the posts. What is the angle measure w formed by the ropes in the diagram? Explain your reasoning.

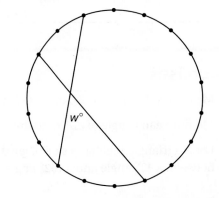

Skills Handbook

Using a Ruler and Protractor

Knowing how to use a ruler and protractor is crucial for success in geometry.

Example

Draw a triangle that has a 28° angle between sides of length 5.2 cm and 3.0 cm.

Step 1 Use a ruler to draw a segment 5.2 cm long.

Step 2 Place the hole of a protractor at one endpoint of the segment. Make a small mark at the 28° position along the protractor.

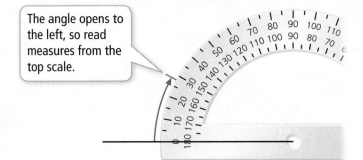

The angle opens to the left, so read measures from the top scale.

Step 3 Align the ruler along the small mark and the same endpoint. Place the zero point of the ruler at the endpoint. Draw a segment 3.0 cm long.

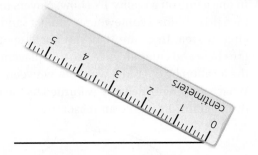

Step 4 Complete the triangle by connecting the endpoints of the first and second segments.

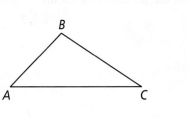

Exercises

1. Measure sides $\overline{AB}$ and $\overline{BC}$ of $\triangle ABC$ to the nearest millimeter.

2. Measure each angle of $\triangle ABC$ to the nearest degree.

3. Draw a triangle that has a side of length 2.4 cm between a 43° angle and a 102° angle.

Classifying Triangles

You can classify a triangle by its angles and sides.

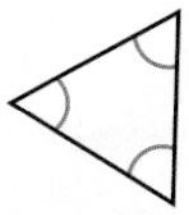

Equiangular
all angles congruent

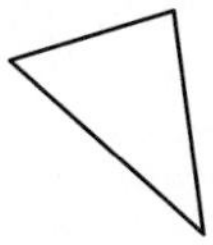

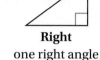

Acute
all angles acute

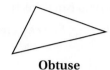

Right
one right angle

Obtuse
one obtuse angle

Equilateral
all sides congruent

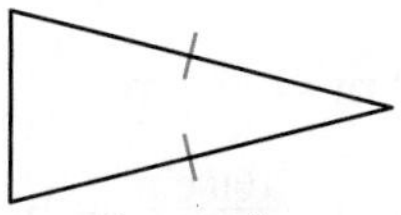

Isosceles
at least two sides congruent

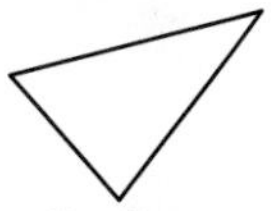

Scalene
no sides congruent

Example

What type of triangle is shown below?

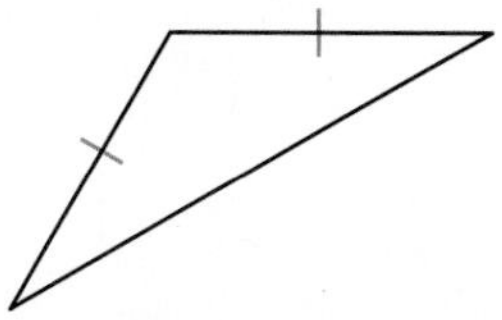

At least two sides are congruent, so the triangle is isosceles. One angle is obtuse, so the triangle is obtuse. The triangle is an obtuse isosceles triangle.

Exercises

Classify each triangle by its sides and angles.

1.
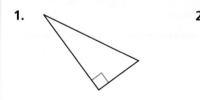

2.

3.
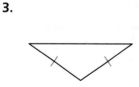

If possible, draw a triangle to fit each description. Mark the triangle to show known information. If you cannot draw the triangle, write *not possible* and explain why.

4. acute equilateral

5. right equilateral

6. obtuse scalene

7. acute isosceles

8. right isosceles

9. acute scalene

Measurement Conversions

To convert from one unit of measure to another, you multiply by a conversion factor in the form of a fraction. The numerator and denominator are in different units, but they represent the same amount. So you can think of this as multiplying by 1.

An example of a conversion factor is $\frac{1 \text{ ft}}{12 \text{ in.}}$. You can create other conversion factors using the table on page 837.

Example 1

Complete each statement.

a. 88 in. = ■ ft

$$88 \text{ in.} \cdot \frac{1 \text{ ft}}{12 \text{ in.}} = \frac{88}{12} \text{ ft} = 7\tfrac{1}{3} \text{ ft}$$

b. 5.3 m = ■ cm

$$5.3 \text{ m} \cdot \frac{100 \text{ cm}}{1 \text{ m}} = 5.3(100) \text{ cm} = 530 \text{ cm}$$

Area is always in square units, and volume is always in cubic units.

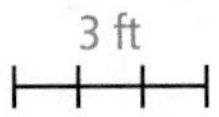

1 yd = 3 ft

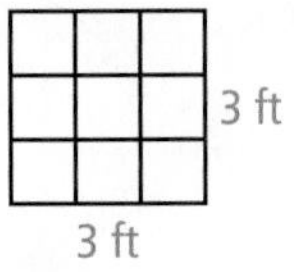

$1 \text{ yd}^2 = 9 \text{ ft}^2$

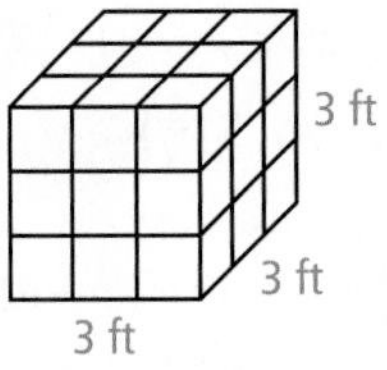

$1 \text{ yd}^3 = 27 \text{ ft}^3$

Example 2

Complete each statement.

a. $300 \text{ in.}^2 = ■ \text{ ft}^2$

$1 \text{ ft} = 12 \text{ in., so } 1 \text{ ft}^2 = (12 \text{ in.})^2 = 144 \text{ in.}^2$.

$$300 \text{ in.}^2 \cdot \frac{1 \text{ ft}^2}{144 \text{ in.}^2} = 2\frac{1}{12} \text{ ft}^2$$

b. $200{,}000 \text{ cm}^3 = ■ \text{ m}^3$

$1 \text{ m} = 100 \text{ cm, so } 1 \text{ m}^3 = (100 \text{ cm})^3 = 1{,}000{,}000 \text{ cm}^3$.

$$200{,}000 \text{ cm}^3 \cdot \frac{1 \text{ m}^3}{1{,}000{,}000 \text{ cm}^3} = 0.2 \text{ m}^3$$

Exercises

Complete each statement.

1. 40 cm = ■ m

2. 1.5 kg = ■ g

3. 60 cm = ■ mm

4. 200 in. = ■ ft

5. 28 yd = ■ in.

6. 1.5 mi = ■ ft

7. 15 g = ■ mg

8. 430 mg = ■ g

9. 34 L = ■ mL

10. 1.2 m = ■ cm

11. 43 mm = ■ cm

12. 3600 s = ■ min

13. 14 gal = ■ qt

14. 4500 lb = ■ t

15. 234 min = ■ h

16. $3 \text{ ft}^2 = ■ \text{ in.}^2$

17. $108 \text{ m}^2 = ■ \text{ cm}^2$

18. $21 \text{ cm}^2 = ■ \text{ mm}^2$

19. $1.4 \text{ yd}^2 = ■ \text{ ft}^2$

20. $0.45 \text{ km}^2 = ■ \text{ m}^2$

21. $1300 \text{ ft}^2 = ■ \text{ yd}^2$

22. $1030 \text{ in.}^2 = ■ \text{ ft}^2$

23. $20{,}000{,}000 \text{ ft}^2 = ■ \text{ mi}^2$

24. $1000 \text{ cm}^3 = ■ \text{ m}^3$

Measurement, Rounding Error, and Reasonableness

There is no such thing as an *exact* measurement. Measurements are always approximate. No matter how precise it is, a measurement actually represents a range of values.

Example 1

Chris's height, to the nearest inch, is 5 ft 8 in. What range of values does this measurement represent?

The height is given to the nearest inch, so the error is $\frac{1}{2}$ in. Chris's height, then, is between 5 ft $7\frac{1}{2}$ in. and 5 ft $8\frac{1}{2}$ in., or 5 ft 8 in. $\pm \frac{1}{2}$ in. Within this range are all the measures that, when rounded to the nearest inch, equal 5 ft 8 in.

As you calculate with measurements, errors can accumulate.

Example 2

Jean drives 18 km to work each day. This distance is given to the nearest kilometer. What is the range of values for the round-trip distance?

The driving distance is between 17.5 and 18.5 km, or 18 ± 0.5 km. Double the lower limit, 17.5, and the upper limit, 18.5. Thus, the round trip can be anywhere between 35 and 37 km, or 36 ± 1 km. Notice that the error for the round trip is double the error for a single leg of the trip.

So that your answers will be reasonable, keep precision and error in mind as you calculate. For example, in finding AB, the length of the hypotenuse of $\triangle ABC$, it would be inappropriate to give the answer as 8.6533 if the sides are given to the nearest tenth. Round your answer to 8.7.

Exercises

Each measurement is followed by its unit of greatest precision. Find the range of values that each measurement represents.

1. 24 ft (ft)

2. 124 cm (cm)

3. 340 mL (mL)

4. $5\frac{1}{2}$ mi. $\left(\frac{1}{2}\text{ mi}\right)$

5. 73.2 mm (0.1 mm)

6. 34 yd^2 (yd^2)

7. The lengths of the sides of *TJCM* are given to the nearest tenth of a centimeter. What is the range of values for the figure's perimeter?

8. To the nearest degree, two angles of a triangle are 49° and 73°. What is the range of values for the measure of the third angle?

9. The lengths of the legs of a right triangle are measured as 131 m and 162 m. You use a calculator to find the length of the hypotenuse. The calculator display reads 208.33867. What should your answer be?

The Effect of Measurement Errors on Calculations

Measurements are always approximate, and calculations with these measurements produce error. Percent error is a measure of accuracy of a measurement or calculation. It is the ratio of the greatest possible error to the measurement.

$$\text{percent error} = \frac{\text{greatest possible error}}{\text{measurement}}$$

Example

The dimensions of a box are measured as 18 in., 12 in., and 9 in. What is the percent error in calculating the box's volume?

The measurements are to the nearest inch, so the greatest possible error is 0.5 in.

Volume:

as measured	maximum value	minimum value
$V = \ell \cdot w \cdot h$	$V = \ell \cdot w \cdot h$	$V = \ell \cdot w \cdot h$
$= 18 \cdot 12 \cdot 9$	$= 18.5 \cdot 12.5 \cdot 9.5$	$= 17.5 \cdot 11.5 \cdot 8.5$
$= 1944 \text{ in.}^3$	$\approx 2196.9 \text{ in.}^3$	$\approx 1710.6 \text{ in.}^3$

Possible Error:

maximum value − measured   measured − minimum value

$2196.9 - 1944 = 252.9$ $1944 - 1710.6 = 233.4$

$$\text{percent error} = \frac{\text{greatest possible error}}{\text{measurement}}$$

$$= \frac{252.9}{1944}$$

$$\approx 0.1300926$$

The percent error is about 13%.

Exercises

Find the percent error in calculating the volume of each box given its dimensions. Round to the nearest percent.

1. 10 cm by 5 cm by 20 cm

2. 1.2 mm by 5.7 mm by 2.0 mm

3. 1.24 cm by 4.45 cm by 5.58 cm

4. $8\frac{1}{4}$ in. by $17\frac{1}{2}$ in. by 5 in.

Find the percent error in calculating the perimeter of each figure.

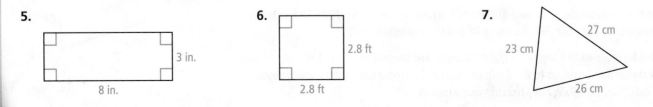

5. 8 in. by 3 in.

6. 2.8 ft by 2.8 ft

7. 27 cm, 23 cm, 26 cm

Squaring Numbers and Finding Square Roots

Skills Handbook

The square of a number is found by multiplying the number by itself. An exponent of 2 is used to indicate that a number is being squared.

Example 1

Simplify.

a. 5^2

$5^2 = 5 \cdot 5$

$\quad = 25$

b. $(-3.5)^2$

$(-3.5)^2 = (-3.5) \cdot (-3.5)$

$\quad = 12.25$

c. $\left(\frac{2}{7}\right)^2$

$\left(\frac{2}{7}\right)^2 = \frac{2}{7} \cdot \frac{2}{7}$

$\quad = \frac{4}{49}$

The square root of a number is itself a number that, when squared, results in the original number. A radical symbol ($\sqrt{}$) is used to represent the positive square root of a number.

Example 2

Simplify. Round to the nearest tenth if necessary.

a. $\sqrt{36}$

$\sqrt{36} = 6$, since $6^2 = 36$.

b. $\sqrt{174}$

$\sqrt{174} \approx 13.2$, since $13.2^2 \approx 174$.

You can solve equations that include squared numbers.

Example 3

Algebra Solve.

a. $x^2 = 144$

$x = 12$ or -12

b. $a^2 + 3^2 = 5^2$

$a^2 + 9 = 25$

$a^2 = 16$

$a = 4$ or -4

Exercises

Simplify.

1. 11^2

2. $(-14)^2$

3. 5.1^2

4. $\left(\frac{8}{5}\right)^2$

5. -6^2

6. $\left(-\frac{3}{7}\right)^2$

Simplify. Round to the nearest tenth if necessary.

7. $\sqrt{100}$

8. $\sqrt{169}$

9. $\sqrt{74}$

10. $\sqrt{50}$

11. $\sqrt{\frac{4}{9}}$

12. $\sqrt{\frac{49}{81}}$

Algebra Solve. Round to the nearest tenth if necessary.

13. $x^2 = 49$

14. $a^2 = 9$

15. $y^2 + 7 = 8$

16. $5 + x^2 = 11$

17. $8^2 + b^2 = 10^2$

18. $5^2 + 4^2 = c^2$

19. $p^2 + 12^2 = 13^2$

20. $20^2 = 15^2 + a^2$

Evaluating and Simplifying Expressions

To evaluate an expression with variables, substitute a number for each variable. Then simplify the expression using the order of operations. Be especially careful with exponents and negative signs. For example, the expression $-x^2$ always yields a negative or zero value, and $(-x)^2$ is always positive or zero.

Order of Operations
1. Perform any operation(s) inside grouping symbols.
2. Simplify any term with exponents.
3. Multiply and divide in order from left to right.
4. Add and subtract in order from left to right.

Example 1

Algebra Evaluate each expression for $r = 4$.

a. $-r^2$

$$-r^2 = -4^2 = -16$$

b. $-3r^2$

$$-3r^2 = -3(4^2) = -3(16) = -48$$

c. $(r + 2)^2$

$$(r + 2)^2 = (4 + 2)^2 = (6)^2 = 36$$

To simplify an expression, you eliminate any parentheses and combine like terms.

Example 2

Algebra Simplify each expression.

a. $5r - 2r + 1$

Combine like terms.
$$5r - 2r + 1 = 3r + 1$$

b. $\pi(3r - 1)$

Use the Distributive Property.
$$\pi(3r - 1) = 3\pi r - \pi$$

c. $(r + \pi)(r - \pi)$

Multiply polynomials.
$$(r + \pi)(r - \pi) = r^2 - \pi^2$$

Exercises

Algebra Evaluate each expression for $x = 5$ and $y = -3$.

1. $-2x^2$

2. $-y + x$

3. $-xy$

4. $(x + 5y) \div x$

5. $x + 5y \div x$

6. $(-2y)^2$

7. $(2y)^2$

8. $(x - y)^2$

9. $\dfrac{x + 1}{y}$

10. $y - (x - y)$

11. $-y^x$

12. $\dfrac{2(1 - x)}{y - x}$

13. $x \cdot y - x$

14. $x - y \cdot x$

15. $\dfrac{y^3 - x}{x - y}$

16. $-y(x - 3)^2$

Algebra Simplify.

17. $6x - 4x + 8 - 5$

18. $2(\ell + w)$

19. $-(4x + 7)$

20. $y(4 - y)$

21. $-4x(x - 2)$

22. $3x - (5 + 2x)$

23. $2t^2 + 4t - 5t^2$

24. $(r - 1)^2$

25. $(1 - r)^2$

26. $(y + 1)(y - 3)$

27. $4h + 3h - 4 + 3$

28. $\pi r - (1 + \pi r)$

29. $(x + 4)(2x - 1)$

30. $2\pi h(1 - r)^2$

31. $3y^2 - (y^2 + 3y)$

32. $-(x + 4)^2$

Simplifying Ratios

The ratio of the length of the shorter leg to the length of the longer leg for this right triangle is 4 to 6. This ratio can be written in three ways.

4 to 6 $\dfrac{4}{6}$ $4:6$

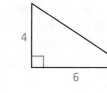

Example

Algebra Simplify each ratio.

a. **4 to 6**

$4 \text{ to } 6 = \dfrac{4}{6}$

$\qquad = \dfrac{2 \cdot 2}{2 \cdot 3}$ Find and remove the common factor.

$\qquad = \dfrac{2}{3}$

b. $3ab : 27ab$

$3ab : 27ab = \dfrac{3ab}{27ab}$

$\qquad\qquad = \dfrac{3ab \cdot 1}{3ab \cdot 9}$

$\qquad\qquad = \dfrac{1}{9}$

c. $\dfrac{4a + 4b}{a + b}$

$\dfrac{4a + 4b}{a + b} = \dfrac{4(a + b)}{a + b}$ Factor the numerator. The denominator cannot be factored. Remove the common factor $(a + b)$.

$\qquad\qquad = 4$

Exercises

Algebra Simplify each ratio.

1. 25 to 15

2. $6 : 9$

3. $\dfrac{36}{54}$

4. 0.8 to 2.4

5. $\dfrac{7}{14x}$

6. $\dfrac{12c}{14c}$

7. $22x^2$ to $35x$

8. $0.5ab : 8ab$

9. $\dfrac{4xy}{0.25x}$

10. $1\frac{1}{2}x$ to $5x$

11. $\dfrac{x^2 + x}{2x}$

12. $\frac{1}{4}r^2$ to $6r$

13. $0.72t : 7.2t^2$

14. $(2x - 6) : (6x - 4)$

15. $12xy : 8x$

16. $(9x - 9y)$ to $(x - y)$

17. $\dfrac{\pi r}{r^2 + \pi r}$

18. $\dfrac{8ab}{32xy}$

Express each ratio in simplest form.

19. shorter leg : longer leg

20. hypotenuse to shorter leg

21. $\dfrac{\text{shorter leg}}{\text{hypotenuse}}$

22. $\dfrac{\text{longer leg}}{\text{hypotenuse}}$

23. longer leg to shorter leg

24. hypotenuse : longer leg

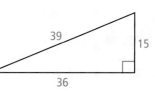

Absolute Value

Absolute value is used to represent the distance of a number from 0 on a number line. Since distance is always referred to as a nonnegative number, the absolute value of an expression is nonnegative.

On the number line at the right, both 4 and -4 are four units from zero. Therefore, $|4|$ and $|-4|$ are both equal to four.

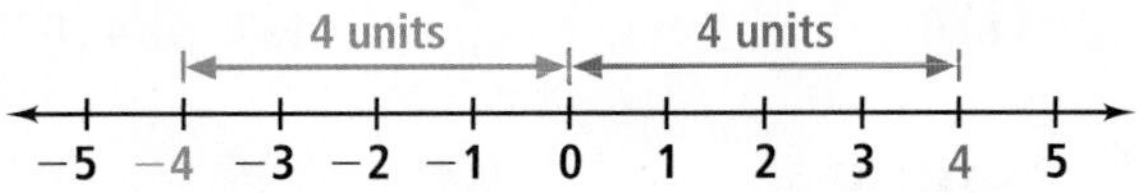

When working with more complicated expressions, always remember to simplify within absolute value symbols first.

Example 1

Simplify each expression.

a. $|4| + |-19|$

$|4| + |-19| = 4 + 19$

$= 23$

b. $|4 - 8|$

$|4 - 8| = |-4|$

$= 4$

c. $-3|-7 - 4|$

$-3|-7 - 4| = -3|-11|$

$= -3 \cdot 11$

$= -33$

To solve the absolute value equation $|x| = a$, find all the values x that are a units from 0 on a number line.

Example 2

Algebra Solve.

a. $|x| = 7$

$x = 7 \text{ or } -7$

b. $|x| - 3 = 22$

$|x| - 3 = 22$

$|x| = 25$

$x = 25 \text{ or } -25$

Exercises

Simplify each expression.

1. $|-8|$

2. $|11|$

3. $|-7| + |15|$

4. $|-12| - |-12|$

5. $|-5| - |10|$

6. $|-4| + |-2|$

7. $10 - |-20|$

8. $|-9| - 15$

9. $|4 - 17|$

10. $|-9 - 11|$

11. $2|-21 + 16|$

12. $-8|-9 + 4|$

Algebra Solve.

13. $|x| = 16$

14. $1 = |x|$

15. $|x| + 7 = 27$

16. $|x| - 9 = 15$

The Coordinate Plane

Two number lines that intersect at right angles form a coordinate plane. The horizontal axis is the *x*-axis and the vertical axis is the *y*-axis. The axes intersect at the origin and divide the coordinate plane into four sections called quadrants.

An ordered pair of numbers names the location of a point in the plane. These numbers are the coordinates of the point. Point *B* has coordinates $(-3, 4)$.

| The first coordinate is the *x*-coordinate. | $(-3, 4)$ | The second coordinate is the *y*-coordinate. |

You use the *x*-coordinate to tell how far to move right (positive) or left (negative) from the origin. You then use the *y*-coordinate to tell how far to move up (positive) or down (negative) to reach the point (x, y).

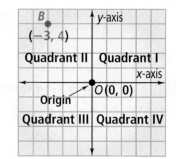

Example 1

Graph each point in the coordinate plane. In which quadrant or on which axis would you find each point?

a. Graph point $A(-2, 3)$ in the coordinate plane.

To graph $A(-2, 3)$, move 2 units to the left of the origin. Then move 3 units up. Since the *x*-coordinate is negative and the *y*-coordinate is positive, point *A* is in Quadrant II.

b. Graph point $B(2, 0)$ in the coordinate plane.

To graph $B(2, 0)$, move 2 units to the right of the origin. Since the *y*-coordinate is 0, point *B* is on the *x*-axis.

Exercises

Name the coordinates of each point in the coordinate plane at the right.

1. S **2.** T **3.** U **4.** V

Graph each ordered pair in the same coordinate plane.

5. $(0, -5)$ **6.** $(4, -1)$ **7.** $(-2, -2)$ **8.** $\left(-1\frac{1}{2}, 4\right)$

In which quadrant or on which axis would you find each point?

9. $(0, 10)$ **10.** $\left(1\frac{1}{2}, -3\right)$ **11.** $(-5, 0)$ **12.** $(-9, -2)$

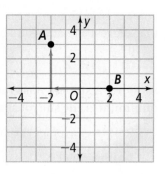

Solving and Writing Linear Equations

To solve a linear equation, use the properties of equality and properties of real numbers to find the value of the variable that satisfies the equation.

Example 1

Algebra Solve each equation.

a. $5x - 3 = 2$

$5x - 3 = 2$

$\quad 5x = 5$ Add 3 to each side.

$\quad\quad x = 1$ Divide each side by 5.

b. $1 - 2(x + 1) = x$

$1 - 2(x + 1) = x$

$\quad 1 - 2x - 2 = x$ Use the Distributive Property.

$\quad\quad -1 - 2x = x$ Simplify the left side.

$\quad\quad\quad -1 = 3x$ Add $2x$ to each side.

$\quad\quad -\frac{1}{3} = x$ Divide each side by 3.

You will sometimes need to translate word problems into equations. Look for words that suggest a relationship or some type of mathematical operation.

Example 2

Algebra A student has grades of 80, 65, 78, and 92 on four tests. What is the minimum grade she must earn on her next test to ensure an average of 80?

Relate   average of 80, 65, 78, 92, and next test is 80   Pull out the key words and numbers.

Define   Let x = the grade on the next test.   Let a variable represent what you are looking for.

Write $\dfrac{80 + 65 + 78 + 92 + x}{5} = 80$ Write an equation.

$\quad\quad\quad \dfrac{315 + x}{5} = 80$ Combine like terms.

$\quad\quad\quad 315 + x = 400$ Multiply each side by 5.

$\quad\quad\quad\quad\quad x = 85$ Subtract 315 from each side.

The student must earn 85 on the next test for an average of 80.

Exercises

Algebra Solve each equation.

1. $3n + 2 = 17$

2. $5a - 2 = -12$

3. $2x + 4 = 10$

4. $3(n - 4) = 15$

5. $4 + 2y = 8y$

6. $-6z + 1 = 13 - 3z$

7. $6 - (3t + 4) = t$

8. $7 = -2(4n - 4.5)$

9. $(w + 5) - 5 = (2w + 5)$

10. $\frac{5}{7}p - 10 = 30$

11. $\frac{m}{-3} - 3 = 1$

12. $5k + 2(k + 1) = 23$

13. Twice a number subtracted from 35 is 9. What is the number?

14. The Johnsons pay $9.95 a month plus $.035 per min for local phone service. Last month, they paid $12.75. How many minutes of local calls did they make?

Percents

A percent is a ratio in which a number is compared to 100. For example, the expression *60 percent* means "60 out of 100." The symbol % stands for "percent."

A percent can be written in decimal form by first writing it in ratio form, and then writing the ratio as a decimal. For example, 25% is equal to the ratio $\frac{25}{100}$ or $\frac{1}{4}$. As a decimal, $\frac{1}{4}$ is equal to 0.25. Note that 25% can also be written directly as a decimal by moving the decimal point two places to the left.

Example 1

Convert each percent to a decimal.

  a. 42%   **b.** 157%   **c.** 12.4%   **d.** 4%

    42% = 0.42   157% = 1.57   12.4% = 0.124   4% = 0.04

To calculate a percent of a number, write the percent as a decimal and multiply.

Example 2

Simplify. Where necessary, round to the nearest tenth.

  a. 30% of 242   **b.** 7% of 38

    30% of 242 = $0.3 \cdot 242$ 7% of 38 = $0.07 \cdot 38$

              = 72.6   = 2.66 ≈ 2.7

For a percent problem, it is a good idea to check that your answer is reasonable by estimating it.

Example 3

Estimate 23% of 96.

23% ≈ 25% and 96 ≈ 100. So 25% $\left(\text{or } \frac{1}{4}\right)$ of 100 = 25.
A reasonable estimate is 25.

Exercises

Convert each percent to a decimal.

  1. 50%   **2.** 27%   **3.** 6%   **4.** 84.6%   **5.** 109%   **6.** 2.5%

Simplify. Where necessary, round to the nearest tenth.

  7. 21% of 40   **8.** 45% of 200   **9.** 6% of 120   **10.** 23.8% of 176

Estimate.

  11. 12% of 70   **12.** 48% of 87   **13.** 73% of 64   **14.** 77% of 42

Reference

Table 1 **Measures**

	United States Customary	Metric
Length	12 inches (in.) = 1 foot (ft) 36 in. = 1 yard (yd) 3 ft = 1 yard 5280 ft = 1 mile (mi) 1760 yd = 1 mile	10 millimeters (mm) = 1 centimeter (cm) 100 cm = 1 meter (m) 1000 mm = 1 meter 1000 m = 1 kilometer (km)
Area	144 square inches $(in.^2)$ = 1 square foot (ft^2) 9 ft^2 = 1 square yard (yd^2) 43,560 ft^2 = 1 acre (a) 4840 yd^2 = 1 acre	100 square millimeters (mm^2) = 1 square centimeter (cm^2) 10,000 cm^2 = 1 square meter (m^2) 10,000 m^2 = 1 hectare (ha)
Volume	1728 cubic inches $(in.^3)$ = 1 cubic foot (ft^3) 27 ft^3 = 1 cubic yard (yd^3)	1000 cubic millimeters (mm^3) = 1 cubic centimeter (cm^3) 1,000,000 cm^3 = 1 cubic meter (m^3)
Liquid Capacity	8 fluid ounces (fl oz) = 1 cup (c) 2 c = 1 pint (pt) 2 pt = 1 quart (qt) 4 qt = 1 gallon (gal)	1000 milliliters (mL) = 1 liter (L) 1000 L = 1 kiloliter (kL)
Weight or Mass	16 ounces (oz) = 1 pound (lb) 2000 pounds = 1 ton (t)	1000 milligrams (mg) = 1 gram (g) 1000 g = 1 kilogram (kg) 1000 kg = 1 metric ton
Temperature	32°F = freezing point of water 98.6°F = normal human body temperature 212°F = boiling point of water	0°C = freezing point of water 37°C = normal human body temperature 100°C = boiling point of water

Customary Units and Metric Units	
Length	1 in. = 2.54 cm 1 mi ≈ 1.61 km 1 ft ≈ 0.305 m
Capacity	1 qt ≈ 0.946 L
Weight and Mass	1 oz ≈ 28.4 g 1 lb ≈ 0.454 kg

Time		
60 seconds (s) = 1 minute (min) 60 minutes = 1 hour (h) 24 hours = 1 day (d) 7 days = 1 week (wk)	4 weeks (approx.) = 1 month (mo) 365 days = 1 year (yr) 52 weeks (approx.) = 1 year	12 months = 1 year 10 years = 1 decade 100 years = 1 century

Table 2 **Formulas**

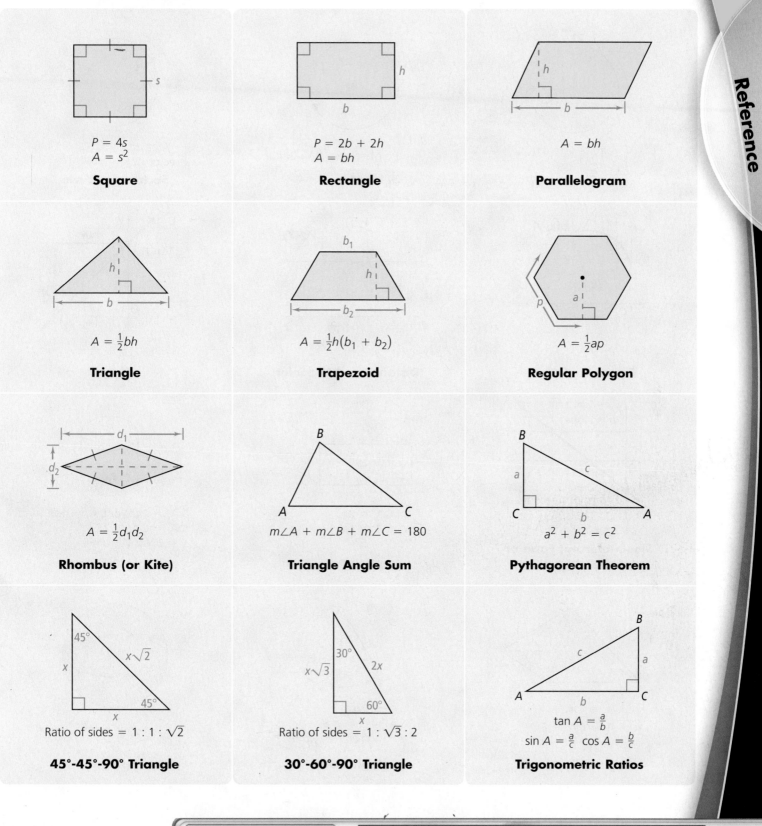

$P = 4s$
$A = s^2$

Square

$P = 2b + 2h$
$A = bh$

Rectangle

$A = bh$

Parallelogram

$A = \frac{1}{2}bh$

Triangle

$A = \frac{1}{2}h(b_1 + b_2)$

Trapezoid

$A = \frac{1}{2}ap$

Regular Polygon

$A = \frac{1}{2}d_1d_2$

Rhombus (or Kite)

$m\angle A + m\angle B + m\angle C = 180$

Triangle Angle Sum

$a^2 + b^2 = c^2$

Pythagorean Theorem

Ratio of sides = $1 : 1 : \sqrt{2}$

45°-45°-90° Triangle

Ratio of sides = $1 : \sqrt{3} : 2$

30°-60°-90° Triangle

$\tan A = \frac{a}{b}$
$\sin A = \frac{a}{c}$ $\cos A = \frac{b}{c}$

Trigonometric Ratios

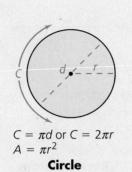

$C = \pi d$ or $C = 2\pi r$
$A = \pi r^2$

Circle

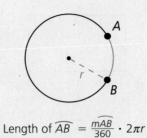

Length of $\overset{\frown}{AB} = \frac{m\overset{\frown}{AB}}{360} \cdot 2\pi r$

Arc

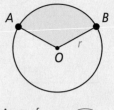

Area of sector $AOB = \frac{m\overset{\frown}{AB}}{360} \cdot \pi r^2$

Sector of a Circle

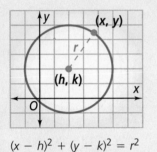

$(x - h)^2 + (y - k)^2 = r^2$

Equation of Circle

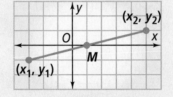
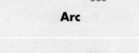

$d = \sqrt{(x_2 - x_1)^2 + (y_2 - y_1)^2}$
$M = \left(\frac{x_1 + x_2}{2}, \frac{y_1 + y_2}{2}\right)$

Distance and Midpoint

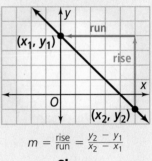

$m = \frac{\text{rise}}{\text{run}} = \frac{y_2 - y_1}{x_2 - x_1}$

Slope

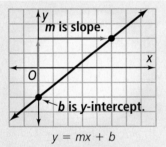
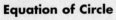

$y = mx + b$

**Slope-Intercept Form of
a Linear Equation**

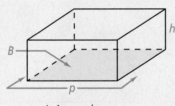
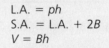

L.A. = ph
S.A. = L.A. + $2B$
$V = Bh$

Right Prism

L.A. = $2\pi rh$ or L.A. = πdh
S.A. = L.A. + $2B$
$V = Bh$ or $V = \pi r^2 h$

Right Cylinder

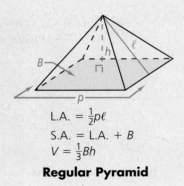

L.A. = $\frac{1}{2}p\ell$
S.A. = L.A. + B
$V = \frac{1}{3}Bh$

Regular Pyramid

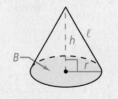

L.A. = $\pi r\ell$
S.A. = L.A. + B
$V = \frac{1}{3}Bh$ or $V = \frac{1}{3}\pi r^2 h$

Right Cone

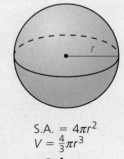

S.A. = $4\pi r^2$
$V = \frac{4}{3}\pi r^3$

Sphere

Table 3 **Reading Math Symbols**

Symbols	Words
...	and so on
=	is equal to, equality
≈	is approximately equal to
≠	is not equal to
>	is greater than
<	is less than
≥	is greater than or equal to
≤	is less than or equal to
≯	is not greater than
≮	is not less than
±	plus or minus
%	percent
$\lvert a \rvert$	absolute value of a
( ), [ ]	parentheses and brackets for grouping
$p \to q$	if p, then q
$p \leftrightarrow q$	p if and only if q
$\sim p$	not p
→	maps to
d	distance
M	midpoint
°	degree(s)
$\overleftrightarrow{AB}$	line through points A and B
$\overline{AB}$	segment with endpoints A and B
$\overrightarrow{AB}$	ray with endpoint A and through point B
AB	length of $\overline{AB}$
$\angle A$	angle with vertex A
$\angle ABC$	angle with sides $\overrightarrow{BA}$ and $\overrightarrow{BC}$
$m\angle A$	measure of angle A
$\angle\!\!\!\angle$	angles
$\triangle ABC$	triangle with vertices A, B, and C
⌐	right angle symbol
$\triangle\!\!\!\triangle$	triangles
≅	is congruent to
≇	is not congruent to
~	is similar to
$\stackrel{?}{=}$	Is this statement true?
$\square ABCD$	parallelogram with vertices A, B, C, and D
$\square\!\!\!\square$	parallelograms
A'	image of A, A prime
A	area
s	length of a side
b	base length
h	height, length of an altitude

Symbols	Words
d	diameter
r	radius
P	perimeter
π	pi, ratio of the circumference of a circle to its diameter
C	circumference
b_1, b_2	bases of a trapezoid
d_1, d_2	lengths of diagonals
a	apothem
B	area of a base
L.A.	lateral area
S.A.	surface area
ℓ	slant height
V	volume
n-gon	polygon with n sides
$\odot A$	circle with center A
$\overset{\frown}{AB}$	arc with endpoints A and B
$\overset{\frown}{ABC}$	arc with endpoints A and C and containing B
$m\overset{\frown}{AB}$	measure of $\overset{\frown}{AB}$
∥	is parallel to
⊥	is perpendicular to
m	slope of a linear function
b	y-intercept of a linear function
$a : b, \frac{a}{b}$	ratio of a to b
$\tan A$	tangent of $\angle A$
$\sin A$	sine of $\angle A$
$\cos A$	cosine of $\angle A$
(a, b)	ordered pair with x-coordinate a and y-coordinate b
$T_{<x,\,y>}(A)$	translation of A by x units horizontally and y units vertically
$R_\ell(A)$	reflection of A across the line ℓ
$r_{(x°,\,P)}(A)$	rotation of A by $x°$ about the point P
$D_x(A)$	dilation of A in the coordinate plane by scale factor x
$\begin{bmatrix} 1 & 2 \\ 3 & 4 \end{bmatrix}$	matrix
$P(\text{event})$	probability of an event
$P(B \mid A)$	conditional probability of B given A
$n!$	n factorial
$_nP_r$	permutation
$_nC_r$	combination

Table 4 **Properties of Real Numbers**

Unless otherwise stated, a, b, c, and d represent real numbers.

Identity Properties

Addition $a + 0 = a$ and $0 + a = a$

Multiplication $a \cdot 1 = a$ and $1 \cdot a = a$

Commutative Properties

Addition $a + b = b + a$

Multiplication $a \cdot b = b \cdot a$

Associative Properties

Addition $(a + b) + c = a + (b + c)$

Multiplication $(a \cdot b) \cdot c = a \cdot (b \cdot c)$

Inverse Properties

Addition The sum of a number and its *opposite*, or *additive inverse*, is zero.

$a + (-a) = 0$ and $-a + a = 0$

Multiplication The *reciprocal*, or *multiplicative inverse*, of a rational number $\frac{a}{b}$ is $\frac{b}{a}$ (a, $b \neq 0$).

$a \cdot \frac{1}{a} = 1$ and $\frac{1}{a} \cdot a = 1$ ($a \neq 0$)

Distributive Properties

$a(b + c) = ab + ac \qquad (b + c)a = ba + ca$

$a(b - c) = ab - ac \qquad (b - c)a = ba - ca$

Properties of Equality

Addition	If $a = b$, then $a + c = b + c$.
Subtraction	If $a = b$, then $a - c = b - c$.
Multiplication	If $a = b$, then $a \cdot c = b \cdot c$.
Division	If $a = b$ and $c \neq 0$, then $\frac{a}{c} = \frac{b}{c}$.
Substitution	If $a = b$, then b can replace a in any expression.
Reflexive	$a = a$
Symmetric	If $a = b$, then $b = a$.
Transitive	If $a = b$ and $b = c$, then $a = c$.

Properties of Proportions

$\frac{a}{b} = \frac{c}{d}$ (a, b, c, $d \neq 0$) is equivalent to

(1) $ad = bc$ 　　(2) $\frac{b}{a} = \frac{d}{c}$

(3) $\frac{a}{c} = \frac{b}{d}$ 　　(4) $\frac{a + b}{b} = \frac{c + d}{d}$

Zero-Product Property

If $ab = 0$, then $a = 0$ or $b = 0$.

Properties of Inequality

Addition	If $a > b$ and $c \geq d$, then $a + c > b + d$.
Multiplication	If $a > b$ and $c > 0$, then $ac > bc$. If $a > b$ and $c < 0$, then $ac < bc$.
Transitive	If $a > b$ and $b > c$, then $a > c$.
Comparison	If $a = b + c$, and $c > 0$, then $a > b$.

Properties of Exponents

For any nonzero numbers a and b, any positive number c, and any integers m and n,

Zero Exponent	$a^0 = 1$
Negative Exponent	$a^{-n} = \frac{1}{a^n}$
Product of Powers	$a^m \cdot a^n = a^{m+n}$
Quotient of Powers	$\frac{a^m}{a^n} = a^{m-n}$
Power to a Power	$(c^m)^n = c^{mn}$
Product to a Power	$(ab)^n = a^n b^n$
Quotient to a Power	$\left(\frac{a}{b}\right)^n = \frac{a^n}{b^n}$

Properties of Square Roots

For any nonnegative numbers a and b, and any positive number c,

Product of Square Roots	$\sqrt{a} \cdot \sqrt{b} = \sqrt{ab}$
Quotient of Square Roots	$\frac{\sqrt{a}}{\sqrt{c}} = \sqrt{\frac{a}{c}}$

Table 5 **Random Numbers**

18823	18160	93593	67294	19632	62617	86779	74024
65358	70469	87149	89509	72176	18103	55169	79954
72002	20582	25325	22451	22445	62132	81638	36566
42709	33717	59943	12027	46547	72749	13347	65030
26128	49067	27904	49953	74674	94617	13317	58697
31973	06303	94202	62287	56164	79157	98375	24558
99241	38449	46438	91579	01907	72146	05764	22400
94490	49833	09258	11873	57196	32209	67663	07990
12288	59245	83638	23642	61715	35483	84563	79956
88618	54619	24853	59783	47537	88822	47227	09262
25041	57862	19203	86103	02800	23198	70639	43757
52064	75820	50994	31050	67304	16730	29373	96700
07845	69584	70548	52973	72302	97594	92241	15204
42665	29990	57260	75846	01152	30141	35982	96088
04003	36893	51639	65625	28426	90634	32979	05449
32959	06776	72420	55622	81422	67587	93193	67479
29041	35939	80920	31801	38638	87905	37617	53135
63364	20495	50868	54130	32625	30799	94255	03514
27838	19139	82031	46143	93922	32001	05378	42457
94248	29387	32682	86235	35805	66529	00886	25875
40156	92636	95648	79767	16307	71133	15714	44142
44293	19195	30569	41277	01417	34656	80207	33362
71878	31767	40056	52582	30766	70264	86253	07179
24757	57502	51033	16551	66731	87844	41420	10084
55529	68560	50069	50652	76104	42086	48720	96632
39724	50318	91370	68016	06222	26806	86726	52832
80950	27135	14110	92292	17049	60257	01638	04460
21694	79570	74409	95087	75424	57042	27349	16229
06930	85441	37191	75134	12845	67868	51500	97761
18740	35448	56096	37910	35485	19640	07689	31027
40657	14875	70695	92569	40703	69318	95070	01541
52249	56515	59058	34509	35791	22150	56558	75286
86570	07303	40560	57856	22009	67712	19435	90250
62962	66253	93288	01838	68388	55481	00336	19271
78066	09117	62350	58972	80778	46458	83677	16125
89106	30219	30068	54030	49295	48985	01624	72881
88310	18172	89450	04987	02781	37935	76222	93595
20942	90911	57643	34009	20728	88785	81212	08214
93926	66687	58252	18674	18501	22362	37319	33201
88294	55814	67443	77285	36229	26886	66782	89931
29751	08485	49910	83844	56013	26596	20875	34568
11169	15529	33241	83594	01727	86595	65723	82322
06062	54400	80649	70749	50395	48993	77447	24862
87445	17139	43278	55031	79971	18515	61850	49101
39283	22821	44330	82225	53534	77235	42973	60190

Postulates, Theorems, and Constructions

Chapter 1 Tools of Geometry

Postulate 1-1
Through any two points there is exactly one line. (p. 13)

Postulate 1-2
If two distinct lines intersect, then they intersect in exactly one point. (p. 13)

Postulate 1-3
If two distinct planes intersect, then they intersect in exactly one line. (p. 14)

Postulate 1-4
Through any three noncollinear points there is exactly one plane. (p. 15)

Postulate 1-5
Ruler Postulate
Every point on a line can be paired with a real number. This makes a one-to-one correspondence between the points on the line and the real numbers. (p. 20)

Postulate 1-6
Segment Addition Postulate
If three points A, B, and C are collinear and B is between A and C, then $AB + BC = AC$. (p. 21)

Postulate 1-7
Protractor Postulate
Consider $\overrightarrow{OB}$ and a point A on one side of $\overrightarrow{OB}$. Every ray of the form $\overrightarrow{OA}$ can be paired one to one with a real number from 0 to 180. (p. 28)

Postulate 1-8
Angle Addition Postulate
If point B is in the interior of $\angle AOC$, then $m\angle AOB + m\angle BOC = m\angle AOC$. (p. 30)

Postulate 1-9
Linear Pair Postulate
If two angles form a linear pair, then they are supplementary. (p. 36)

The Midpoint Formulas
On a Number Line
The coordinate of the midpoint M of $\overline{AB}$ is $\frac{a + b}{2}$.

In the Coordinate Plane
Given $\overline{AB}$ where $A(x_1, y_1)$ and $B(x_2, y_2)$, the coordinates of the midpoint of $\overline{AB}$ are $M\left(\frac{x_1 + x_2}{2}, \frac{y_1 + y_2}{2}\right)$. (p. 50)

The Distance Formula
The distance between two points $A(x_1, y_1)$ and $B(x_2, y_2)$ is $d = \sqrt{(x_2 - x_1)^2 + (y_2 - y_1)^2}$. (p. 52)
- Proof on p. 497, Exercise 35

The Distance Formula (Three Dimensions)
In a three-dimensional coordinate system, the distance between two points (x_1, y_1, z_1) and (x_2, y_2, z_2) can be found with this extension of the Distance Formula.
$$d = \sqrt{(x_2 - x_1)^2 + (y_2 - y_1)^2 + (z_2 - z_1)^2}$$ (p. 56)

Postulate 1-10
Area Addition Postulate
The area of a region is the sum of the areas of its nonoverlapping parts. (p. 63)

Chapter 2 Reasoning and Proof

Law of Detachment
If the hypothesis of a true conditional is true, then the conclusion is true. In symbolic form:
If $p \rightarrow q$ is true and p is true, then q is true. (p. 106)

Law of Syllogism
If $p \rightarrow q$ is true and $q \rightarrow r$ is true, then $p \rightarrow r$ is true. (p. 108)

Properties of Congruence
Reflexive Property
$\overline{AB} \cong \overline{AB}$ and $\angle A \cong \angle A$

Symmetric Property
If $\overline{AB} \cong \overline{CD}$, then $\overline{CD} \cong \overline{AB}$.
If $\angle A \cong \angle B$, then $\angle B \cong \angle A$.

Transitive Property
If $\overline{AB} \cong \overline{CD}$, and $\overline{CD} \cong \overline{EF}$, then $\overline{AB} \cong \overline{EF}$.
If $\angle A \cong \angle B$, and $\angle B \cong \angle C$, then $\angle A \cong \angle C$.
If $\angle B \cong \angle A$, and $\angle B \cong \angle C$, then $\angle A \cong \angle C$. (p. 114)

Theorem 2-1
Vertical Angles Theorem
Vertical angles are congruent. (p. 120)
- Proof on p. 121

Theorem 2-2
Congruent Supplements Theorem
If two angles are supplements of the same angle (or of congruent angles), then the two angles are congruent. (p. 122)
- Proof on p. 123, Problem 3

Theorem 2-3

Congruent Complements Theorem
If two angles are complements of the same angle (or of congruent angles), then the two angles are congruent. (p. 123)
- Proof on p. 125, Exercise 13

Theorem 2-4

All right angles are congruent. (p. 123)
- Proof on p. 125, Exercise 18

Theorem 2-5

If two angles are congruent and supplementary, then each is a right angle. (p. 123)
- Proof on p. 126, Exercise 23

Chapter 3 Parallel and Perpendicular Lines

Theorem 3-2

Corresponding Angles Theorem
If a transversal intersects two parallel lines, then corresponding angles are congruent. (p. 149)
- Proof on p. 155, Exercise 25

Theorem 3-1

Alternate Interior Angles Theorem
If a transversal intersects two parallel lines, then alternate interior angles are congruent. (p. 149)
- Proof on p. 150

Postulate 3-1

Same-Side Interior Angles Postulate
If a transversal intersects two parallel lines, then same-side interior angles are supplementary. (p. 148)

Theorem 3-3

Alternate Exterior Angles Theorem
If a transversal intersects two parallel lines, then alternate exterior angles are congruent. (p. 151)
- Proof on p. 150, Got It 2

Theorem 3-4

Converse of the Corresponding Angles Theorem
If two lines and a transversal form corresponding angles that are congruent, then the lines are parallel. (p. 156)
- Proof on p. 161, Exercise 29

Theorem 3-5

Converse of the Alternate Interior Angles Theorem
If two lines and a transversal form alternate interior angles that are congruent, then the two lines are parallel. (p. 157)
- Proof on p. 158

Theorem 3-6

Converse of the Same-Side Interior Angles Postulate
If two lines and a transversal form same-side interior angles that are supplementary, then the two lines are parallel. (p. 157)
- Proof on p. 158, Got It 2

Theorem 3-7

Converse of the Alternate Exterior Angles Theorem
If two lines and a transversal form alternate exterior angles that are congruent, then the two lines are parallel. (p. 157)
- Proof on p. 158, Problem 2

Theorem 3-8

If two lines are parallel to the same line, then they are parallel to each other. (p. 164)
- Proof on p. 167, Exercise 7

Theorem 3-9

In a plane, if two lines are perpendicular to the same line, then they are parallel to each other. (p. 165)
- Proof on p. 165

Theorem 3-10

Perpendicular Transversal Theorem
In a plane, if a line is perpendicular to one of two parallel lines, then it is perpendicular to the other. (p. 166)
- Proof on p. 168, Exercise 10

Postulate 3-2

Parallel Postulate
Through a point not on a line, there is one and only one line parallel to the given line. (p. 171)

Theorem 3-11

Triangle Angle-Sum Theorem
The sum of the measures of the angles of a triangle is 180. (p. 172)
- Proof on p. 172

Theorem 3-12

Triangle Exterior Angle Theorem
The measure of each exterior angle of a triangle equals the sum of the measures of its two remote interior angles. (p. 173)
- Proof on p. 177, Exercise 33

Corollary
The measure of an exterior angle of a triangle is greater than the measure of each of its remote interior angles. (p. 325)
- Proof on p. 325

Spherical Geometry Parallel Postulate

Through a point not on a line, there is no line parallel to the given line. (p. 179)

Postulate 3-3

Perpendicular Postulate
Through a point not on a line, there is one and only one line perpendicular to the given line. (p. 184)

Slopes of Parallel Lines

If two nonvertical lines are parallel, then their slopes are equal. If the slopes of two distinct nonvertical lines are equal, then the lines are parallel. Any two vertical lines or horizontal lines are parallel. (p. 197)

- Proofs on p. 457, Exercises 33, 34

Slopes of Perpendicular Lines

If two nonvertical lines are perpendicular, then the product of their slopes is −1. If the slopes of two lines have a product of −1, then the lines are perpendicular. Any horizontal line and vertical line are perpendicular. (p. 198)

- Proofs on p. 418, Exercise 28; p. 497, Exercise 51; p. 466, Exercise 44

Chapter 4 Congruent Triangles

Theorem 4-1

Third Angles Theorem

If the two angles of one triangle are congruent to two angles of another triangle, then the third angles are congruent. (p. 220)

- Proof on p. 220

Postulate 4-1

Side-Side-Side (SSS) Postulate

If the three sides of one triangle are congruent to the three sides of another triangle, then the two triangles are congruent. (p. 227)

Postulate 4-2

Side-Angle-Side (SAS) Postulate

If two sides and the included angle of one triangle are congruent to two sides and the included angle of another triangle, then the two triangles are congruent. (p. 228)

Postulate 4-3

Angle-Side-Angle (ASA) Postulate

If two angles and the included side of one triangle are congruent to two angles and the included side of another triangle, then the two triangles are congruent. (p. 234)

Theorem 4-2

Angle-Angle-Side (AAS) Theorem

If two angles and a nonincluded side of one triangle are congruent to two angles and the corresponding nonincluded side of another triangle, then the triangles are congruent. (p. 236)

- Proof on p. 236

Theorem 4-3

Isosceles Triangle Theorem

If two sides of a triangle are congruent, then the angles opposite those sides are congruent. (p. 250)

- Proofs on p. 251; p. 255, Exercise 22

Corollary

If a triangle is equilateral, then the triangle is equiangular. (p. 252)

- Proof on p. 255, Exercise 24

Theorem 4-4

Converse of the Isosceles Triangle Theorem

If two angles of a triangle are congruent, then the sides opposite the angles are congruent. (p. 251)

- Proof on p. 255, Exercise 23

Corollary

If a triangle is equiangular, then the triangle is equilateral. (p. 252)

- Proof on p. 255, Exercise 24

Theorem 4-5

If a line bisects the vertex angle of an isosceles triangle, then the line is also the perpendicular bisector of the base. (p. 252)

- Proof on p. 255, Exercise 26

Theorem 4-6

Hypotenuse-Leg (HL) Theorem

If the hypotenuse and a leg of one right triangle are congruent to the hypotenuse and a leg of another right triangle, then the triangles are congruent. (p. 259)

- Proof on p. 259

Chapter 5 Relationships Within Triangles

Theorem 5-1

Triangle Midsegment Theorem

If a segment joins the midpoints of two sides of a triangle, then the segment is parallel to the third side and is half as long. (p. 285)

- Proof on p. 415, Got It 2

Theorem 5-2

Perpendicular Bisector Theorem

If a point is on the perpendicular bisector of a segment, then it is equidistant from the endpoints of the segment. (p. 293)

- Proof on p. 298, Exercise 32

Theorem 5-3

Converse of the Perpendicular Bisector Theorem

If a point is equidistant from the endpoints of a segment, then it is on the perpendicular bisector of the segment. (p. 293)

- Proof on p. 298, Exercise 33

Theorem 5-4

Angle Bisector Theorem

If a point is on the bisector of an angle, then the point is equidistant from the sides of the angle. (p. 295)

- Proof on p. 298, Exercise 34

Theorem 5-5

Converse of the Angle Bisector Theorem

If a point in the interior of an angle is equidistant from the sides of the angle, then the point is on the angle bisector. (p. 295)

- Proof on p. 298, Exercise 35

Theorem 5-6

Concurrency of Perpendicular Bisectors Theorem
The perpendicular bisectors of the sides of a triangle are concurrent at a point equidistant from the vertices. (p. 301)
- Proof on p. 302

Theorem 5-7

Concurrency of Angle Bisectors Theorem
The bisectors of the angles of a triangle are concurrent at a point equidistant from the sides of the triangle. (p. 303)
- Proof on p. 306, Exercise 24

Theorem 5-8

Concurrency of Medians Theorem
The medians of a triangle are concurrent at a point that is two-thirds the distance from each vertex to the midpoint of the opposite side. (p. 309)
- Proof on p. 417, Exercise 25

Theorem 5-9

Concurrency of Altitudes Theorem
The lines that contain the altitudes of a triangle are concurrent. (p. 310)
- Proof on p. 417, Exercise 26

Comparison Property of Inequality
If $a = b + c$ and $c > 0$, then $a > b$. (p. 324)
- Proof on p. 324

Theorem 5-10
If two sides of a triangle are not congruent, then the larger angle lies opposite the longer side. (p. 325)
- Proof on p. 330, Exercise 40

Theorem 5-11
If two angles of a triangle are not congruent, then the longer side lies opposite the larger angle. (p. 326)
- Proof on p. 326

Theorem 5-12

Triangle Inequality Theorem
The sum of the lengths of any two sides of a triangle is greater than the length of the third side. (p. 327)
- Proof on p. 331, Exercise 45

Theorem 5-13

The Hinge Theorem (SAS Inequality Theorem)
If two sides of one triangle are congruent to two sides of another triangle and the included angles are not congruent, then the longer third side is opposite the larger included angle. (p. 332)
- Proof on p. 338, Exercise 25

Theorem 5-14

Converse of the Hinge Theorem (SSS Inequality)
If two sides of one triangle are congruent to two sides of another triangle and the third sides are not congruent, then the larger included angle is opposite the longer third side. (p. 334)
- Proof on p. 334

Chapter 6 Polygons and Quadrilaterals

Theorem 6-1

Polygon Angle-Sum Theorem
The sum of the measures of the angles of an n-gon is $(n - 2)180$. (p. 353)
- Proof on p. 357, Exercise 40

Corollary
The measure of each angle of a regular n-gon is $\frac{(n - 2)180}{n}$. (p. 354)
- Proof on p. 358, Exercise 43

Theorem 6-2

Polygon Exterior Angle-Sum Theorem
The sum of the measures of the exterior angles of a polygon, one at each vertex, is 360. (p. 355)
- Proofs on p. 352 (using a computer); p. 357, Exercise 39

Theorem 6-3
If a quadrilateral is a parallelogram, then its opposite sides are congruent. (p. 359)
- Proof on p. 360

Theorem 6-4
If a quadrilateral is a parallelogram, then its consecutive angles are supplementary. (p. 360)
- Proof on p. 365, Exercise 32

Theorem 6-5
If a quadrilateral is a parallelogram, then its opposite angles are congruent. (p. 361)
- Proof on p. 361, Problem 2

Theorem 6-6
If a quadrilateral is a parallelogram, then its diagonals bisect each other. (p. 362)
- Proof on p. 364, Exercise 13

Theorem 6-7
If three (or more) parallel lines cut off congruent segments on one transversal, then they cut off congruent segments on every transversal. (p. 363)
- Proof on p. 366, Exercise 43

Theorem 6-8
If both pairs of opposite sides of a quadrilateral are congruent, then the quadrilateral is a parallelogram. (p. 367)
- Proof on p. 373, Exercise 20

Theorem 6-9
If an angle of a quadrilateral is supplementary to both of its consecutive angles, then the quadrilateral is a parallelogram. (p. 368)
- Proof on p. 373, Exercise 21

Theorem 6-10
If both pairs of opposite angles of a quadrilateral are congruent, then the quadrilateral is a parallelogram. (p. 368)
- Proof on p. 373, Exercise 18

Reference

Theorem 6-11

If the diagonals of a quadrilateral bisect each other, then the quadrilateral is a parallelogram. (p. 369)
- Proof on p. 369

Theorem 6-12

If one pair of opposite sides of a quadrilateral is both congruent and parallel, then the quadrilateral is a parallelogram. (p. 370)
- Proof on p. 373, Exercise 19

Theorem 6-13

If a parallelogram is a rhombus, then its diagonals are perpendicular. (p. 376)
- Proof on p. 377

Theorem 6-14

If a parallelogram is a rhombus, then each diagonal bisects a pair of opposite angles. (p. 376)
- Proof on p. 381, Exercise 45

Theorem 6-15

If a parallelogram is a rectangle, then its diagonals are congruent. (p. 378)
- Proof on p. 381, Exercise 41

Theorem 6-16

If the diagonals of a parallelogram are perpendicular, then the parallelogram is a rhombus. (p. 383)
- Proof on p. 383

Theorem 6-17

If one diagonal of a parallelogram bisects a pair of opposite angles, then the parallelogram is a rhombus. (p. 384)
- Proof on p. 387, Exercise 23

Theorem 6-18

If the diagonals of a parallelogram are congruent, then the parallelogram is a rectangle. (p. 384)
- Proof on p. 387, Exercise 24

Theorem 6-19

If a quadrilateral is an isosceles trapezoid, then each pair of base angles is congruent. (p. 389)
- Proof on p. 396, Exercise 45

Theorem 6-20

If a quadrilateral is an isosceles trapezoid, then its diagonals are congruent. (p. 391)
- Proof on p. 396, Exercise 53

Theorem 6-21

Trapezoid Midsegment Theorem

If a quadrilateral is a trapezoid, then
(1) the midsegment is parallel to the bases, and
(2) the length of the midsegment is half the sum of the lengths of the bases. (p. 391)
- Proofs on p. 409, Problem 3; p. 415, Problem 2

Theorem 6-22

If a quadrilateral is a kite, then its diagonals are perpendicular. (p. 392)
- Proof on p. 392

Chapter 7 Similarity

Postulate 7-1

Angle-Angle Similarity (AA ~) Postulate

If two angles of one triangle are congruent to two angles of another triangle, then the triangles are similar. (p. 450)

Theorem 7-1

Side-Angle-Side Similarity (SAS ~) Theorem

If an angle of one triangle is congruent to an angle of a second triangle, and the sides that include the two angles are proportional, then the triangles are similar. (p. 451)
- Proof on p. 457, Exercise 35

Theorem 7-2

Side-Side-Side Similarity (SSS ~) Theorem

If the corresponding sides of two triangles are proportional, then the triangles are similar. (p. 451)
- Proof on p. 458, Exercise 36

Theorem 7-3

The altitude to the hypotenuse of a right triangle divides the triangle into two triangles that are similar to the original triangle and to each other. (p. 460)
- Proof on p. 461

Corollary 1

The length of the altitude to the hypotenuse of a right triangle is the geometric mean of the lengths of the segments of the hypotenuse. (p. 462)
- Proof on p. 466, Exercise 42

Corollary 2

The altitude to the hypotenuse of a right triangle separates the hypotenuse so that the length of each leg of the triangle is the geometric mean of the length of the hypotenuse and the length of the segment of the hypotenuse adjacent to the leg. (p. 463)
- Proof on p. 466, Exercise 43

Theorem 7-4

Side-Splitter Theorem

If a line is parallel to one side of a triangle and intersects the other two sides, then it divides those sides proportionally. (p. 471)
- Proof on p. 472

Converse

If a line divides two sides of a triangle proportionally, then it is parallel to the third side.
- Proof on p. 476, Exercise 37

Corollary

If three parallel lines intersect two transversals, then the segments intercepted on the transversals are proportional. (p. 473)
- Proof on p. 477, Exercise 46

Theorem 7-5

Triangle-Angle-Bisector Theorem
If a ray bisects an angle of a triangle, then it divides the opposite side into two segments that are proportional to the other two sides of the triangle. (p. 473)
- Proof on p. 477, Exercise 47

Chapter 8 Right Triangles and Trigonometry

Theorem 8-1

Pythagorean Theorem
If a triangle is a right triangle, then the sum of the squares of the lengths of the legs is equal to the square of the length of the hypotenuse.
$a^2 + b^2 = c^2$ (p. 491)
- Proof on p. 497, Exercise 49

Theorem 8-2

Converse of the Pythagorean Theorem
If the sum of the squares of the lengths of two sides of a triangle is equal to the square of the length of the third side, then the triangle is a right triangle. (p. 493)
- Proof on p. 498, Exercise 52

Theorem 8-3

If the square of the length of the longest side of a triangle is greater than the sum of the squares of the lengths of the other two sides, then the triangle is obtuse. (p. 494)
- Proof on p. 498, Exercise 53

Theorem 8-4

If the square of the length of the longest side of a triangle is less than the sum of the squares of the lengths of the other two sides, then the triangle is acute. (p. 494)
- Proof on p. 498, Exercise 54

Theorem 8-5

45°-45°-90° Triangle Theorem
In a 45°-45°-90° triangle, both legs are congruent and the length of the hypotenuse is $\sqrt{2}$ times the length of a leg.
hypotenuse = $\sqrt{2} \cdot$ leg (p. 499)
- Proof on p. 499

Theorem 8-6

30°-60°-90° Triangle Theorem
In a 30°-60°-90° triangle, the length of the hypotenuse is twice the length of the shorter leg. The length of the longer leg is $\sqrt{3}$ times the length of the shorter leg.
hypotenuse = $2 \cdot$ shorter leg
longer leg = $\sqrt{3} \cdot$ shorter leg (p. 501)
- Proof on p. 501

Law of Sines
$\frac{\sin A}{a} = \frac{\sin B}{b} = \frac{\sin C}{c}$ (p. 522)
- Proof on p. 522

Law of Cosines
$a^2 = b^2 + c^2 - 2bc \cos A$
$b^2 = a^2 + c^2 - 2ac \cos B$
$c^2 = a^2 + b^2 - 2ab \cos C$ (p. 527)
- Proof on p. 527

Chapter 9 Transformations

Theorem 9-1
The composition of two or more isometries is an isometry. (p. 570)

Theorem 9-2

Reflections Across Parallel Lines
A composition of reflections across two parallel lines is a translation. (p. 571)

Theorem 9-3

Reflections Across Intersecting Lines
A composition of reflections across two intersecting lines is a rotation. (p. 572)

Chapter 10 Area

Theorem 10-1

Area of a Rectangle
The area of a rectangle is the product of its base and height.
$A = bh$ (p. 616)

Theorem 10-2

Area of a Parallelogram
The area of a parallelogram is the product of a base and the corresponding height.
$A = bh$ (p. 616)

Theorem 10-3

Area of a Triangle
The area of a triangle is half the product of a base and the corresponding height.
$A = \frac{1}{2}bh$ (p. 618)

Theorem 10-4

Area of a Trapezoid
The area of a trapezoid is half the product of the height and the sum of the bases.
$A = \frac{1}{2}h(b_1 + b_2)$ (p. 623)

Theorem 10-5

Area of a Rhombus or a Kite
The area of a rhombus or a kite is half the product of the lengths of its diagonals.
$A = \frac{1}{2}d_1d_2$ (p. 624)

Reference

Postulate 10-1

If two figures are congruent, then their areas are equal. (p. 630)

Theorem 10-6

Area of a Regular Polygon

The area of a regular polygon is half the product of the apothem and the perimeter.

$A = \frac{1}{2}ap$ (p. 630)

- Proof on p. 630

Theorem 10-7

Perimeters and Areas of Similar Figures

If the scale factor of two similar figures is $\frac{a}{b}$, then

(1) the ratio of their perimeters is $\frac{a}{b}$ and

(2) the ratio of their areas is $\frac{a^2}{b^2}$. (p. 635)

Theorem 10-8

Area of a Triangle Given SAS

The area of a triangle is half the product of the lengths of two sides and the sine of the included angle.

Area of $\triangle ABC = \frac{1}{2}bc(\sin A)$ (p. 645)

- Proof on p. 645

Postulate 10-2

Arc Addition Postulate

The measure of the arc formed by two adjacent arcs is the sum of the measures of the two arcs.

$m\widehat{ABC} = m\widehat{AB} + m\widehat{BC}$ (p. 650)

Theorem 10-9

Circumference of a Circle

The circumference of a circle is π times the diameter.

$C = \pi d$ or $C = 2\pi r$ (p. 651)

Theorem 10-10

Arc Length

The length of an arc of a circle is the product of the ratio $\frac{\text{measure of the arc}}{360}$ and the circumference of the circle.

length of $\widehat{AB} = \frac{m\widehat{AB}}{360} \cdot 2\pi r$ or

length of $\widehat{AB} = \frac{m\widehat{AB}}{360} \cdot \pi d$ (p. 653)

Theorem 10-11

Area of a Circle

The area of a circle is the product of π and the square of the radius.

$A = \pi r^2$ (p. 660)

Theorem 10-12

Area of a Sector of a Circle

The area of a sector of a circle is the product of the ratio $\frac{\text{measure of the arc}}{360}$ and the area of the circle.

Area of sector $AOB = \frac{m\widehat{AB}}{360} \cdot \pi r^2$ (p. 661)

Chapter 11 Surface Area and Volume

Theorem 11-1

Lateral and Surface Areas of a Prism

The lateral area of a right prism is the product of the perimeter of the base and the height of the prism.

L.A. $= ph$

The surface area of a right prism is the sum of the lateral area and the areas of the two bases.

S.A. $=$ L.A. $+ 2B$ (p. 700)

Theorem 11-2

Lateral and Surface Areas of a Cylinder

The lateral area of a right cylinder is the product of the circumference of the base and the height of the cylinder.

L.A. $= 2\pi rh$, or L.A. $= \pi dh$

The surface area of a right cylinder is the sum of the lateral area and areas of the two bases.

S.A. $=$ L.A. $+ 2B$, or S.A. $= 2\pi rh + 2\pi r^2$ (p. 702)

Theorem 11-3

Lateral and Surface Areas of a Pyramid

The lateral area of a regular pyramid is half the product of the perimeter p of the base and the slant height ℓ of the pyramid.

L.A. $= \frac{1}{2}p\ell$

The surface area of a regular pyramid is the sum of the lateral area and the area B of the base.

S.A. $=$ L.A. $+ B$ (p. 709)

Theorem 11-4

Lateral and Surface Areas of a Cone

The lateral area of a right cone is half the product of the circumference of the base and the slant height of the cone.

L.A. $= \frac{1}{2} \cdot 2\pi r\ell$, or L.A. $= \pi r\ell$

The surface area of a right cone is the sum of the lateral area and the area of the base.

S.A. $=$ L.A. $+ B$ (p. 711)

Theorem 11-5

Cavalieri's Principle

If two space figures have the same height and the same cross-sectional area at every level, then they have the same volume. (p. 718)

Theorem 11-6

Volume of a Prism

The volume of a prism is the product of the area of the base and the height of the prism.

$V = Bh$ (p. 718)

Theorem 11-7

Volume of a Cylinder

The volume of a cylinder is the product of the area of the base and the height of the cylinder.

$V = Bh$, or $V = \pi r^2 h$ (p. 719)

Theorem 11-8

Volume of a Pyramid
The volume of a pyramid is one third the product of the area of the base and the height of the pyramid.

$V = \frac{1}{3}Bh$ (p. 726)

Theorem 11-9

Volume of a Cone
The volume of a cone is one third the product of the area of the base and the height of the cone.

$V = \frac{1}{3}Bh$, or $V = \frac{1}{3}\pi r^2 h$ (p. 728)

Theorem 11-10

Surface Area of a Sphere
The surface area of a sphere is four times the product of π and the square of the radius of the sphere.

S.A. $= 4\pi r^2$ (p. 734)

Theorem 11-11

Volume of a Sphere
The volume of a sphere is four thirds the product of π and the cube of the radius of the sphere.

$V = \frac{4}{3}\pi r^3$ (p. 735)

Theorem 11-12

Areas and Volumes of Similar Solids
If the scale factor of two similar solids is $a : b$, then
- the ratio of their corresponding areas is $a^2 : b^2$, and
- the ratio of their volumes is $a^3 : b^3$. (p. 743)

Chapter 12 Circles

Theorem 12-1
If a line is tangent to a circle, then the line is perpendicular to the radius at the point of tangency. (p. 762)
- Proof on p. 763

Theorem 12-2
If a line in the plane of a circle is perpendicular to a radius at its endpoint on the circle, then the line is tangent to the circle. (p. 764)
- Proof on p. 769, Exercise 30

Theorem 12-3
If two segments are tangent to a circle from a point outside the circle, then the two segments are congruent. (p. 766)
- Proof on p. 768, Exercise 23

Theorem 12-4
Within a circle or in congruent circles, congruent central angles have congruent arcs. (p. 771)
- Proof on p. 777, Exercise 19

Converse
Within a circle or in congruent circles, congruent arcs have congruent central angles. (p. 771)
- Proof on p. 778, Exercise 35

Theorem 12-5
Within a circle or in congruent circles, congruent central angles have congruent chords. (p. 772)
- Proof on p. 777, Exercise 20

Converse
Within a circle or in congruent circles, congruent chords have congruent central angles. (p. 772)
- Proof on p. 778, Exercise 36

Theorem 12-6
Within a circle or in congruent circles, congruent chords have congruent arcs. (p. 772)
- Proof on p. 777, Exercise 21

Converse
Within a circle or in congruent circles, congruent arcs have congruent chords. (p. 772)
- Proof on p. 778, Exercise 37

Theorem 12-7
Within a circle or in congruent circles, chords equidistant from the center (or centers) are congruent. (p. 772)
- Proof on p. 773

Converse
Within a circle or in congruent circles, congruent chords are equidistant from the center (or centers). (p. 772)
- Proof on p. 778, Exercise 38

Theorem 12-8
In a circle, if a diameter is perpendicular to a chord, it bisects the chord and its arc. (p. 774)
- Proof on p. 777, Exercise 22

Theorem 12-9
In a circle, if a diameter bisects a chord (that is not a diameter), it is perpendicular to the chord. (p. 774)
- Proof on p. 774

Theorem 12-10
In a circle, the perpendicular bisector of a chord contains the center of the circle. (p. 774)
- Proof on p. 778, Exercise 33

Theorem 12-11

Inscribed Angle Theorem
The measure of an inscribed angle is half the measure of its intercepted arc. (p. 780)
- Proofs on p. 781; p. 785, Exercises 26, 27

Corollary 1
Two inscribed angles that intercept the same arc are congruent. (p. 782)
- Proof on p. 786, Exercise 31

Corollary 2
An angle inscribed in a semicircle is a right angle. (p. 782)
- Proof on p. 786, Exercise 32

Corollary 3
The opposite angles of a quadrilateral inscribed in a circle are supplementary. (p. 782)
- Proof on p. 786, Exercise 33

Theorem 12-12
The measure of an angle formed by a tangent and a chord is half the measure of the intercepted arc. (p. 783)
- Proof on p. 786, Exercise 34

Theorem 12-13
The measure of an angle formed by two lines that intersect inside a circle is half the sum of the measures of the intercepted arcs. (p. 790)
- Proof on p. 791

Theorem 12-14
The measure of an angle formed by two lines that intersect outside a circle is half the difference of the measures of the intercepted arcs. (p. 790)
- Proofs on p. 796, Exercises 35, 36

Theorem 12-15
For a given point and circle, the product of the lengths of the two segments from the point to the circle is constant along any line through the point and circle. (p. 793)
- Proofs on p. 793; p. 796, Exercises 37, 38

Theorem 12-16
An equation of a circle with center (h, k) and radius r is $(x - h)^2 + (y - k)^2 = r^2$. (p. 798)
- Proof on p. 799

Constructions

Construction 1
Congruent Segments
Construct a segment congruent to a given segment. (p. 43)

Construction 2
Congruent Angles
Construct an angle congruent to a given angle. (p. 44)

Construction 3
Perpendicular Bisector
Construct the perpendicular bisector of a segment. (p. 45)

Construction 4
Angle Bisector
Construct the bisector of an angle. (p. 45)

Construction 5
Parallel Through a Point Not on a Line
Construct the line parallel to a given line and through a given point that is not on the line. (p. 182)

Construction 6
Quadrilateral With Parallel Sides
Construct a quadrilateral with one pair of parallel sides of lengths a and b. (p. 183)

Construction 7
Perpendicular Through a Point on a Line
Construct the perpendicular to a given line at a given point on the line. (p. 184)

Construction 8
Perpendicular Through a Point Not on a Line
Construct the perpendicular to a given line through a given point not on the line. (p. 185)

Visual **Glossary**

English

A

Spanish

Acute angle (p. 29)   An acute angle is an angle whose measure is between 0 and 90.

Ángulo agudo (p. 29)   Un ángulo agudo es un ángulo que mide entre 0 y 90 grados.

Example

17°

Acute triangle (p. 885)   An acute triangle has three acute angles.

Triángulo acutángulo (p. 885)   Un triángulo acutángulo tiene los tres ángulos agudos.

Example

75°
60°   45°

Adjacent angles (p. 34)   Adjacent angles are two coplanar angles that have a common side and a common vertex but no common interior points.

Ángulos adyacentes (p. 34)   Los ángulos adyacentes son dos ángulos coplanarios que tienen un lado común y el mismo vértice, pero no tienen puntos interiores comunes.

Example

1
2

3
4

∠1 and ∠2 are   ∠3 and ∠4 are
adjacent.   *not* adjacent.

Adjacent arcs (p. 650)   Adjacent arcs are on the same circle and have exactly one point in common.

Arcos adyacentes (p. 650)   Los arcos adyacentes están en el mismo círculo y tienen exactamente un punto en común.

Example

B
C
A

$\overarc{AB}$ and $\overarc{BC}$ are
adjacent arcs.

Alternate interior (exterior) angles (p. 142)   Alternate interior (exterior) angles are nonadjacent interior (exterior) angles that lie on opposite sides of the transversal.

Ángulos alternos internos (externos) (p. 142)   Los ángulos alternos internos (externos) son ángulos internos (externos) no adyacentes situados en lados opuestos de la transversal.

Example

5 *t*
1 / 3   ℓ
4 / 2
6   *m*

∠1 and ∠2 are alternate interior angles,
as are ∠3 and ∠4. ∠5 and ∠6 are
alternate exterior angles.

English

Spanish

Altitude *See* **cone; cylinder; parallelogram; prism; pyramid; trapezoid; triangle.**

Altura *Ver* **cone; cylinder; parallelogram; prism; pyramid; trapezoid; triangle.**

Altitude of a triangle (p. 310) An altitude of a triangle is the perpendicular segment from a vertex to the line containing the side opposite that vertex.

Altura de un triángulo (p. 310) Una altura de un triángulo es el segmento perpendicular que va desde un vértice hasta la recta que contiene el lado opuesto a ese vértice.

Example

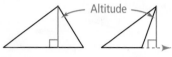

Angle (p. 27) An angle is formed by two rays with the same endpoint. The rays are the *sides* of the angle and the common endpoint is the *vertex* of the angle.

Ángulo (p. 27) Un ángulo está formado por dos semirrectas que convergen en un mismo extremo. Las semirrectas son los *lados* del ángulo y los extremos en común son el *vértice*.

Example

This angle could be named $\angle A$, $\angle BAC$, or $\angle CAB$.

Angle bisector (p. 37) An angle bisector is a ray that divides an angle into two congruent angles.

Bisectriz de un ángulo (p. 37) La bisectriz de un ángulo es una semirrecta que divide al ángulo en dos ángulos congruentes.

Example

$\overrightarrow{LN}$ bisects $\angle KLM$.
$\angle KLN \cong \angle NLM$.

Angle of elevation or depression (p. 516) An angle of elevation (depression) is the angle formed by a horizontal line and the line of sight to an object above (below) the horizontal line.

Ángulo de elevación o depresión (p. 516) Un ángulo de elevación (depresión) es el ángulo formado por una línea horizontal y la recta que va de esa línea a un objeto situado arriba (debajo) de ella.

Example

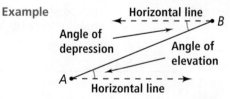

Angle of rotation (p. 561) *See* **rotation.**

Ángulo de rotación (p. 561) *Ver* **rotation.**

Apothem (p. 629) The apothem of a regular polygon is the distance from the center to a side.

Apotema (p. 629) La apotema de un polígono regular es la distancia desde el centro hasta un lado.

Example

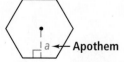

English

Arc   *See* **major arc; minor arc.** *See also* **arc length; measure of an arc.**

Arc length (p. 653)   The length of an arc of a circle is the product of the ratio $\frac{\text{measure of the arc}}{360}$ and the circumference of the circle.

Example

$$\text{Length of } \overset{\frown}{DE} = \frac{60}{360} \cdot 2\pi(5) = \frac{5\pi}{3}$$

Area (p. 59)   The area of a plane figure is the number of square units enclosed by the figure. A list of area formulas is on pp. 897–898.

Example

The area of the rectangle is 12 square units, or 12 units².

Auxiliary line (p. 172)   An auxiliary line is a line that is added to a diagram to help explain relationships in proofs.

Example

Axes (p. 893)   *See* **coordinate plane.**

Axiom (p. 13)   *See* **postulate.**

B

Base(s)   *See* **cone; cylinder; isosceles triangle; parallelogram; prism; pyramid; trapezoid; triangle.**

Base angles   *See* **trapezoid; isosceles triangle.**

Biconditional (p. 98)   A biconditional statement is the combination of a conditional statement and its converse. A biconditional contains the words "if and only if."

Example   This biconditional statement is true:
Two angles are congruent *if and only if* they have the same measure.

Bisector   *See* **segment bisector; angle bisector.**

Spanish

Arco   *Ver* **major arc; minor arc.** *Ver también* **arc length; measure of an arc.**

Longitud de un arco (p. 653)   La longitud del arco de un círculo es el producto del cociente $\frac{\text{medida del arco}}{360}$ por la circunferencia del círculo.

Área (p. 59)   El área de una figura plana es la cantidad de unidades cuadradas que contiene la figura. Una lista de fórmulas para calcular áreas está en las págs. 897–898.

Línea auxiliar (p. 172)   Una línea auxiliar es aquella que se le agrega a un diagrama para explicar la relación entre pruebas.

Ejes (p. 893)   *Ver* **coordinate plane.**

Axioma (p. 13)   *Ver* **postulate.**

Base(s)   *Ver* **cone; cylinder; isosceles triangle; parallelogram; prism; pyramid; trapezoid; triangle.**

Ángulos de base   *Ver* **trapezoid; isosceles triangle.**

Bicondicional (p. 98)   Un enunciado bicondicional es la combinación de un enunciado condicional y su recíproco. El enunciado bicondicional incluye las palabras "si y solo si".

Bisectriz   *Ver* **segment bisector; angle bisector.**

Center *See* **circle; dilation; regular polygon; rotation; sphere.**

Centro *Ver* **circle; dilation; regular polygon; rotation; sphere.**

Central angle of a circle (p. 649) A central angle of a circle is an angle whose vertex is the center of the circle.

Ángulo central de un círculo (p. 649) Un ángulo central de un círculo es un ángulo cuyo vértice es el centro del círculo.

Example

$\angle ROK$ is a central angle of $\odot O$.

Centroid of a triangle (p. 309) The centroid of a triangle is the point of concurrency of the medians of the triangle.

Centroide de un triángulo (p. 309) El centroide de un triángulo es el punto de intersección de sus medianas.

Example *P* is the centroid of △*ABC*.

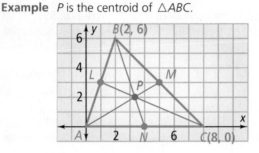

Chord (p. 771) A chord of a circle is a segment whose endpoints are on the circle.

Cuerda (p. 771) Una cuerda de un círculo es un segmento cuyos extremos son dos puntos del círculo.

Example

$\overline{HD}$ and $\overline{HR}$ are chords of $\odot C$.

Circle (pp. 649, 798) A circle is the set of all points in a plane that are a given distance, the *radius*, from a given point, the *center*. The standard form for an equation of a circle with center (h, k) and radius r is $(x - h)^2 + (y - k)^2 = r^2$.

Círculo (pp. 649, 798) Un círculo es el conjunto de todos los puntos de un plano situados a una distancia dada, el *radio*, de un punto dado, el *centro*. La fórmula normal de la ecuación de un círculo con centro (h, k) y radio r es $(x - h)^2 + (y - k)^2 = r^2$.

Example

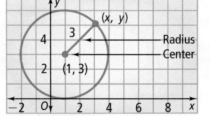

The equation of the circle whose center is $(1, 3)$ and whose radius is 3 is $(x - 1)^2 + (y - 3)^2 = 9$.

English

Spanish

Circumcenter of a triangle (p. 301)   The circumcenter of a triangle is the point of concurrency of the perpendicular bisectors of the sides of the triangle.

Circuncentro de un triángulo (p. 301)   El circuncentro de un triángulo es el punto de intersección de las bisectrices perpendiculares de los lados del triángulo.

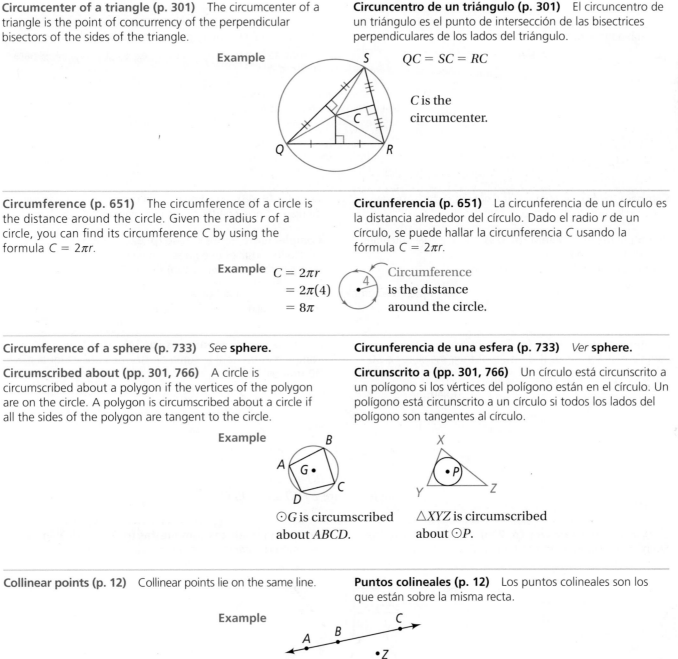

Example

$QC = SC = RC$

C is the circumcenter.

Circumference (p. 651)   The circumference of a circle is the distance around the circle. Given the radius r of a circle, you can find its circumference C by using the formula $C = 2\pi r$.

Circunferencia (p. 651)   La circunferencia de un círculo es la distancia alrededor del círculo. Dado el radio r de un círculo, se puede hallar la circunferencia C usando la fórmula $C = 2\pi r$.

Example

$$C = 2\pi r$$
$$= 2\pi(4)$$
$$= 8\pi$$

Circumference is the distance around the circle.

Circumference of a sphere (p. 733)   *See* **sphere**.

Circunferencia de una esfera (p. 733)   *Ver* **sphere**.

Circumscribed about (pp. 301, 766)   A circle is circumscribed about a polygon if the vertices of the polygon are on the circle. A polygon is circumscribed about a circle if all the sides of the polygon are tangent to the circle.

Circunscrito a (pp. 301, 766)   Un círculo está circunscrito a un polígono si los vértices del polígono están en el círculo. Un polígono está circunscrito a un círculo si todos los lados del polígono son tangentes al círculo.

Example

$\odot G$ is circumscribed about $ABCD$.

$\triangle XYZ$ is circumscribed about $\odot P$.

Collinear points (p. 12)   Collinear points lie on the same line.

Puntos colineales (p. 12)   Los puntos colineales son los que están sobre la misma recta.

Example

Points A, B, and C are collinear, but points A, B, and Z are noncollinear.

English

Spanish

Combination (p. 838) Any unordered selection of r objects from a set of n objects is a combination. The number of combinations of n objects taken r at a time is $_nC_r = \frac{n!}{r!(n-r)!}$ for $0 \le r \le n$.

Combinación (p. 838) Cualquier selección no ordenada de r objetos tomados de un conjunto de n objetos es una combinación. El número de combinaciones de n objetos, cuando se toman r objetos cada vez, es $_nC_r = \frac{n!}{r!(n-r)!}$ para $0 \le r \le n$.

Example The number of combinations of seven items taken four at a time is
$$_7C_4 = \frac{7!}{4!(7-4)!} = 35.$$
There are 35 ways to choose four items from seven items without regard to order.

Compass (p. 43) A compass is a geometric tool used to draw circles and parts of circles, called arcs.

Compás (p. 43) El compás es un instrumento usado para dibujar círculos y partes de círculos, llamados arcos.

Complement of an event (p. 826) All possible outcomes that are not in the event.
$P(\text{complement of event}) = 1 - P(\text{event})$

Complemento de un suceso (p. 826) Todos los resultados posibles que no se dan en el suceso.
$P(\text{complemento de un suceso}) = 1 - P(\text{suceso})$

Example The complement of rolling a 1 or a 2 on a standard number cube is rolling a 3, 4, 5, or 6.

Complementary angles (p. 34) Two angles are complementary angles if the sum of their measures is 90.

Ángulos complementarios (p. 34) Dos ángulos son complementarios si la suma de sus medidas es igual a 90 grados.

Example

$\angle HKI$ and $\angle IKJ$ are complementary angles, as are $\angle HKI$ and $\angle EFG$.

Composite space figures (p. 720) A composite space figure is the combination of two or more figures into one object.

Figuras geométricas compuestas (p. 720) Una figura geométrica compuesta es la combinación de dos o más figuras en un mismo objeto.

Example

English

Composition of transformations (p. 548)   A composition of two transformations is a transformation in which a second transformation is performed on the image of a first transformation.

Example

If you reflect $\triangle ABC$ across line m to get $\triangle A'B'C'$ and then reflect $\triangle A'B'C'$ across line n to get $\triangle A''B''C''$, you perform a composition of transformations.

Compound event (p. 844)   An event that consists of two or more events linked by the word *and* or the word *or*.

Examples   Rolling a 5 on a standard number cube and then rolling a 4 is a compound event.

Compound statement (p. 96)   A compound statement is a statement formed by combining two or more statements.

Example   A square is a rectangle *and* it is a rhombus.
You will walk to school *or* you will take the bus.

Concave polygon (p. 58)   *See* **polygon.**

Concentric circles (p. 651)   Concentric circles lie in the same plane and have the same center.

Example

The two circles both have center D and are therefore concentric.

Conclusion (p. 89)   The conclusion is the part of an *if-then* statement (conditional) that follows *then*.

Example   In the statement, "If it rains, then I will go outside," the conclusion is "I will go outside."

Spanish

Composición de transformaciones (p. 548)   Una composición de dos transformaciones es una transformación en la cual una segunda transformación se realiza a partir de la imagen de la primera.

Suceso compuesto (p. 844)   Suceso que consiste en dos o más sucesos unidos por medio de la palabra *y* o la palabra *o*.

Enunciado compuesto (p. 96)   Un enunciado compuesto es un enunciado que combina dos o más enunciados.

Polígono cóncavo (p. 58)   *Ver* **polygon.**

Círculos concéntricos (p. 651)   Los círculos concéntricos están en el mismo plano y tienen el mismo centro.

Conclusión (p. 89)   La conclusión es lo que sigue a la palabra *entonces* en un enunciado (condicional), *si . . ., entonces. . . .*

Concurrent lines (p. 301)   Concurrent lines are three or more lines that meet in one point. The point at which they meet is the *point of concurrency.*

Rectas concurrentes (p. 301)   Las rectas concurrentes son tres o más rectas que se unen en un punto. El punto en que se unen es el *punto de concurrencia.*

Example

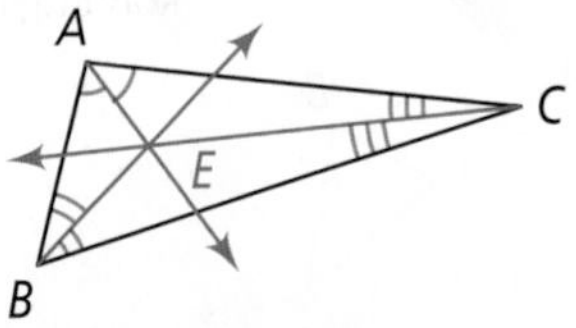

Point *E* is the point of concurrency of the bisectors of the angles of △*ABC.* The bisectors are concurrent.

Conditional (p. 89)   A conditional is an *if-then* statement.

Condicional (p. 89)   Un enunciado condicional es del tipo *si . . ., entonces. . . .*

Example *If* you act politely, *then* you will earn respect.

Conditional probability (p. 851)   A conditional probability contains a condition that may limit the sample space for an event. The notation $P(B|A)$ is read "the probability of event B, given event A." For any two events A and B in the sample space, $P(B|A) = \dfrac{P(A \text{ and } B)}{P(A)}$.

Probabilidad condicional (p. 851)   Una probabilidad condicional contiene una condición que puede limitar el espacio muestral de un suceso. La notación $P(B|A)$ se lee "la probabilidad del suceso B, dado el suceso A". Para dos sucesos cualesquiera A y B en el espacio muestral,

$$P(B|A) = \dfrac{P(A \text{ y } B)}{P(A)}.$$

Example $= \dfrac{P(\text{departs and arrives on time})}{P(\text{departs on time})}$

$= \dfrac{0.75}{0.83}$

≈ 0.9

Cone (p. 711)   A cone is a three-dimensional figure that has a circular *base*, a *vertex* not in the plane of the circle, and a curved lateral surface, as shown in the diagram. The *altitude* of a cone is the perpendicular segment from the vertex to the plane of the base. The *height* is the length of the altitude. In a *right cone*, the altitude contains the center of the base. The *slant height* of a right cone is the distance from the vertex to the edge of the base.

Cono (p. 711)   Un cono es una figura tridimensional que tiene una *base* circular, un *vértice* que no está en el plano del círculo y una superficie lateral curvada (indicada en el diagrama). La *altura* de un cono es el segmento perpendicular desde el vértice hasta el plano de la base. La *altura*, por extensión, es la longitud de la altura. Un *cono recto* es un cono cuya altura contiene el centro de la base. La *longitud de la generatriz* de un cono recto es la distancia desde el vértice hasta el borde de la base.

Example

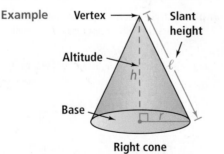

Right cone

English

Congruence transformation (p. 580)   *See* **isometry.**

Congruent angles (p. 29)   Congruent angles are angles that have the same measure.

Example

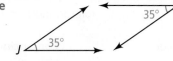

$m\angle J = m\angle K,$ so $\angle J \cong \angle K.$

Congruent arcs (p. 771)   Congruent arcs are arcs that have the same measure and are in the same circle or congruent circles.

Example

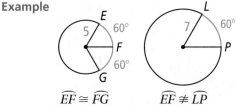

$\overset{\frown}{EF} \cong \overset{\frown}{FG}$ $\overset{\frown}{EF} \not\cong \overset{\frown}{LP}$

Congruent circles (p. 649)   Congruent circles are circles whose radii are congruent.

Example

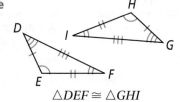

$\odot A$ and $\odot B$ have the same radius, so $\odot A \cong \odot B.$

Congruent polygons (p. 219)   Congruent polygons are polygons that have corresponding sides congruent and corresponding angles congruent.

Example

$\triangle DEF \cong \triangle GHI$

Congruent segments (p. 22)   Congruent segments are segments that have the same length.

Example

$\overline{AB} \cong \overline{CD}$

Spanish

Transformación de congruencia (p. 580)   *Ver* **isometry.**

Ángulos congruentes (p. 29)   Los ángulos congruentes son ángulos que tienen la misma medida.

Arcos congruentes (p. 771)   Arcos congruentes son arcos que tienen la misma medida y están en el mismo círculo o en círculos congruentes.

Círculos congruentes (p. 649)   Los círculos congruentes son círculos cuyos radios son congruentes.

Polígonos congruentes (p. 219)   Los polígonos congruentes son polígonos cuyos lados correspondientes son congruentes y cuyos ángulos correspondientes son congruentes.

Segmentos congruentes (p. 22)   Los segmentos congruentes son segmentos que tienen la misma longitud.

English

Spanish

Conjecture (p. 83)   A conjecture is a conclusion reached by using inductive reasoning.

Conjetura (p. 83)   Una conjetura es una conclusión obtenida usando el razonamiento inductivo.

Example   As you walk down the street, you see many people holding unopened umbrellas. You make the conjecture that the forecast must call for rain.

Conjunction (p. 96)   A conjunction is a compound statement formed by connecting two or more statements with the word *and*.

Conjunción (p. 96)   Una conjunción es un enunciado compuesto que conecta dos o más enunciados por medio de la palabra *y*.

Example   The sky is blue *and* the grass is green.

Consecutive angles (p. 360)   Consecutive angles of a polygon share a common side.

Ángulos consecutivos (p. 360)   Los ángulos consecutivos de un polígono tienen un lado común.

Example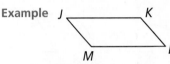

In □*JKLM*, ∠*J* and ∠*M* are consecutive angles, as are ∠*J* and ∠*K*. ∠*J* and ∠*L* are *not* consecutive.

Construction (p. 43)   A construction is a geometric figure made with only a straightedge and compass.

Construcción (p. 43)   Una construcción es una figura geométrica trazada solamente con una regla sin graduación y un compás.

Example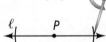

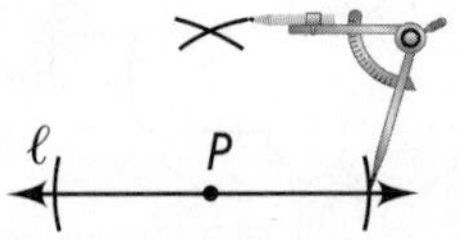

The diagram shows the construction (in progress) of a line perpendicular to a line ℓ through a point *P* on ℓ.

Contrapositive (p. 91)   The contrapositive of the conditional "if *p*, then *q*" is the conditional "if not *q*, then not *p*." A conditional and its contrapositive always have the same truth value.

Contrapositivo (p. 91)   El contrapositivo del condicional "si *p*, entonces *q*" es el condicional "si no *q*, entonces no *p*". Un condicional y su contrapositivo siempre tienen el mismo valor verdadero.

Example   **Conditional:** If a figure is a triangle, then it is a polygon.
Contrapositive: If a figure is not a polygon, then it is not a triangle.

Converse (p. 91)   The statement obtained by reversing the hypothesis and conclusion of a conditional.

Expresión recíproca (p. 91)   Enunciado que se obtiene al intercambiar la hipótesis y la conclusión de una situación condicional.

Example   The converse of "If I was born in Houston, then I am a Texan" would be "If I am a Texan, then I am born in Houston."

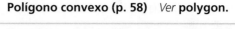
Convex polygon (p. 58)   *See* **polygon.**

Polígono convexo (p. 58)   *Ver* **polygon.**

Coordinate(s) of a point (pp. 20, 893)   The coordinate of a point is its distance and direction from the origin of a number line. The coordinates of a point on a coordinate plane are in the form (x, y), where x is the x-coordinate and y is the y-coordinate.

Coordenada(s) de un punto (pp. 20, 893)   La coordenada de un punto es su distancia y dirección desde el origen en una recta numérica. Las coordenadas de un punto en un plano de coordenadas se expresan como (x, y), donde x es la coordenada x, e y es la coordenada y.

Example

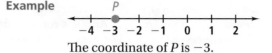

The coordinate of P is -3.

The coordinates of T are $(-4, 3)$.

Coordinate plane (p. 893)   The coordinate plane is formed by two number lines, called the *axes*, intersecting at right angles. The x-axis is the horizontal axis, and the y-axis is the vertical axis. The two axes meet at the *origin*, $O(0, 0)$. The axes divide the plane into four *quadrants*.

Plano de coordenadas (p. 893)   El plano de coordenadas se forma con dos rectas numéricas, llamadas *ejes*, que se cortan en ángulos rectos. El eje x es el eje horizontal y el eje y es el eje vertical. Los dos ejes se unen en el *origen*, $O(0, 0)$. Los ejes dividen el plano de coordenadas en cuatro *cuadrantes*.

Example

	y-axis	
Quadrant II		Quadrant I
	Origin	
	O	*x*-axis
Quadrant III		Quadrant IV

Coordinate proof (p. 408)   *See* **proof.**

Prueba de coordenadas (p. 408)   *Ver* **proof.**

Coplanar figures (p. 12)   Coplanar figures are figures in the same plane.

Figuras coplanarias (p. 12)   Las figuras coplanarias son las figuras que están localizadas en el mismo plano.

Example

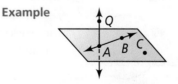

Point C and $\overleftrightarrow{AB}$ are coplanar but points A, B, C, and Q are noncoplanar.

Corollary (p. 252) A corollary is a theorem that can be proved easily using another theorem.

Corolario (p. 252) Un corolario es un teorema que se puede probar fácilmente usando otro teorema.

Example **Theorem:** If two sides of a triangle are congruent, then the angles opposite those sides are congruent.
Corollary: If a triangle is equilateral, then it is equiangular.

Corresponding angles (p. 142) Corresponding angles lie on the same side of the transversal t and in corresponding positions relative to ℓ and m.

Ángulos correspondientes (p. 142) Los ángulos correspondientes están en el mismo lado de la transversal t y en las correspondientes posiciones relativas a ℓ y m.

Example

$\angle 1$ and $\angle 2$ are corresponding angles, as are $\angle 3$ and $\angle 4$, $\angle 5$ and $\angle 6$, and $\angle 7$ and $\angle 8$.

Cosine ratio (p. 507) *See* **trigonometric ratios.**

Razón coseno (p. 507) *Ver* **trigonometric ratios.**

Counterexample (pp. 84, 90) An example showing that a statement is false.

Contraejemplo (pp. 84, 90) Ejemplo que demuestra que un enunciado es falso.

Example **Statement:** All apples are red.
Counterexample: A Granny Smith Apple is green.

Cross Products Property (p. 434) The product of the extremes of a proportion is equal to the product of the means.

Propiedad de los productos cruzados (p. 434) El producto de los extremos de una proporción es igual al producto de los medios.

Example If $\frac{x}{3} = \frac{12}{21}$, then $21x = 3 \cdot 12$.

Cross section (p. 690) A cross section is the intersection of a solid and a plane.

Sección de corte (p. 690) Una sección de corte es la intersección de un plano y un cuerpo geométrico.

Example

The cross section is a circle.

Cube (p. 691) A cube is a polyhedron with six faces, each of which is a square.

Cubo (p. 691) Un cubo es un poliedro de seis caras, cada una de las caras es un cuadrado.

Example

English

Spanish

Cylinder (p. 701)   A cylinder is a three-dimensional figure with two congruent circular *bases* that lie in parallel planes. An *altitude* of a cylinder is a perpendicular segment that joins the planes of the bases. Its length is the *height* of the cylinder. In a *right cylinder*, the segment joining the centers of the bases is an altitude. In an *oblique cylinder*, the segment joining the centers of the bases is not perpendicular to the planes containing the bases.

Cilindro (p. 701)   Un cilindro es una figura tridimensional con dos *bases* congruentes circulares en planos paralelos. Una *altura* de un cilindro es un segmento perpendicular que une los planos de las bases. Su longitud es, por extensión, la *altura* del cilindro. En un *cilindro recto*, el segmento que une los centros de las bases es una altura. En un *cilindro oblicuo*, el segmento que une los centros de las bases no es perpendicular a los planos que contienen las bases.

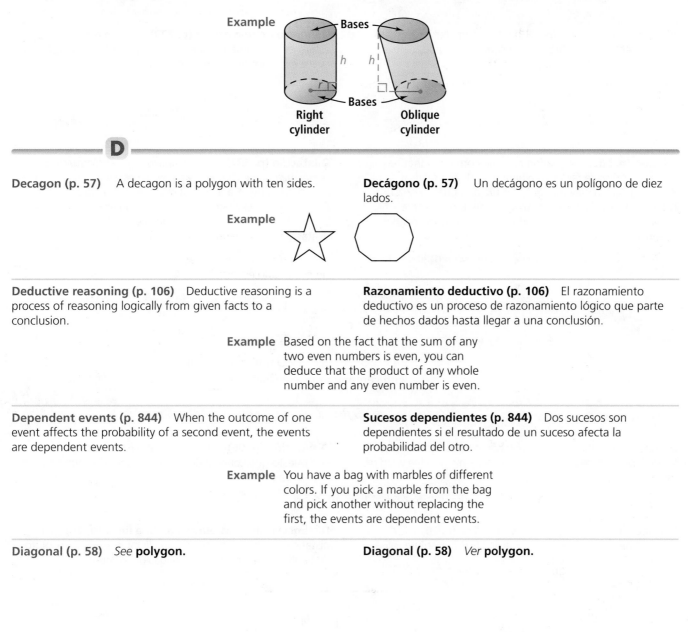

Example

Bases

Bases

Right cylinder

Oblique cylinder

D

Decagon (p. 57)   A decagon is a polygon with ten sides.

Decágono (p. 57)   Un decágono es un polígono de diez lados.

Example

Deductive reasoning (p. 106)   Deductive reasoning is a process of reasoning logically from given facts to a conclusion.

Razonamiento deductivo (p. 106)   El razonamiento deductivo es un proceso de razonamiento lógico que parte de hechos dados hasta llegar a una conclusión.

Example   Based on the fact that the sum of any two even numbers is even, you can deduce that the product of any whole number and any even number is even.

Dependent events (p. 844)   When the outcome of one event affects the probability of a second event, the events are dependent events.

Sucesos dependientes (p. 844)   Dos sucesos son dependientes si el resultado de un suceso afecta la probabilidad del otro.

Example   You have a bag with marbles of different colors. If you pick a marble from the bag and pick another without replacing the first, the events are dependent events.

Diagonal (p. 58)   *See* **polygon.**

Diagonal (p. 58)   *Ver* **polygon.**

Visual Glossary

Diameter of a circle (p. 649) A diameter of a circle is a segment that contains the center of the circle and whose endpoints are on the circle. The term *diameter* can also mean the length of this segment.

Diámetro de un círculo (p. 649) Un diámetro de un círculo es un segmento que contiene el centro del círculo y cuyos extremos están en el círculo. El término *diámetro* también puede referirse a la longitud de este segmento.

Example

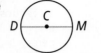

$\overline{DM}$ is a diameter of $\odot C$.

Diameter of a sphere (p. 733) The diameter of a sphere is a segment passing through the center, with endpoints on the sphere.

Diámetro de una esfera (p. 733) El diámetro de una esfera es un segmento que contiene el centro de la esfera y cuyos extremos están en la esfera.

Example

Dilation (p. 587) A dilation is a transformation that has *center C* and *scale factor n*, where $n > 0$, and maps a point R to R' in such a way that R' is on $\overrightarrow{CR}$ and $CR' = n \cdot CR$. The center of a dilation is its own image. If $n > 1$, the dilation is an *enlargement*, and if $0 < n < 1$, the dilation is a *reduction*.

Dilatación (p. 587) Una dilatación, o *transformación de semejanza*, tiene *centro C* y *factor de escala n* para $n > 0$, y asocia un punto R a R' de tal modo que R' está en $\overrightarrow{CR}$ y $CR' = n \cdot CR$. El centro de una dilatación es su propia imagen. Si $n > 1$, la dilatación es un *aumento*, y si $0 < n < 1$, la dilatación es una *reducción*.

Example

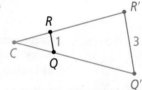

$\overline{R'Q'}$ is the image of $\overline{RQ}$ under a dilation with center C and scale factor 3.

Directrix (p. 804) *See* **parabola.**

Directriz (p. 804) *Ver* **parabola.**

Disjunction (p. 96) A disjunction is a compound statement formed by connecting two or more statements with the word *or*.

Disyunción (p. 96) Una disyunción es un enunciado compuesto que conecta dos o más enunciados por medio de la palabra *o*.

Example x is less than 10 *or* x is greater than 2.

Distance between two points on a line (p. 20) The distance between two points on a line is the absolute value of the difference of the coordinates of the points.

Distancia entre dos puntos de una línea (p. 20) La distancia entre dos puntos de una línea es el valor absoluto de la diferencia de las coordenadas de los puntos.

Example

$$AB = |a - b|$$

English

Spanish

Distance from a point to a line (p. 294) The distance from a point to a line is the length of the perpendicular segment from the point to the line.

Distancia desde un punto hasta una recta (p. 294) La distancia desde un punto hasta una recta es la longitud del segmento perpendicular que va desde el punto hasta la recta.

Example

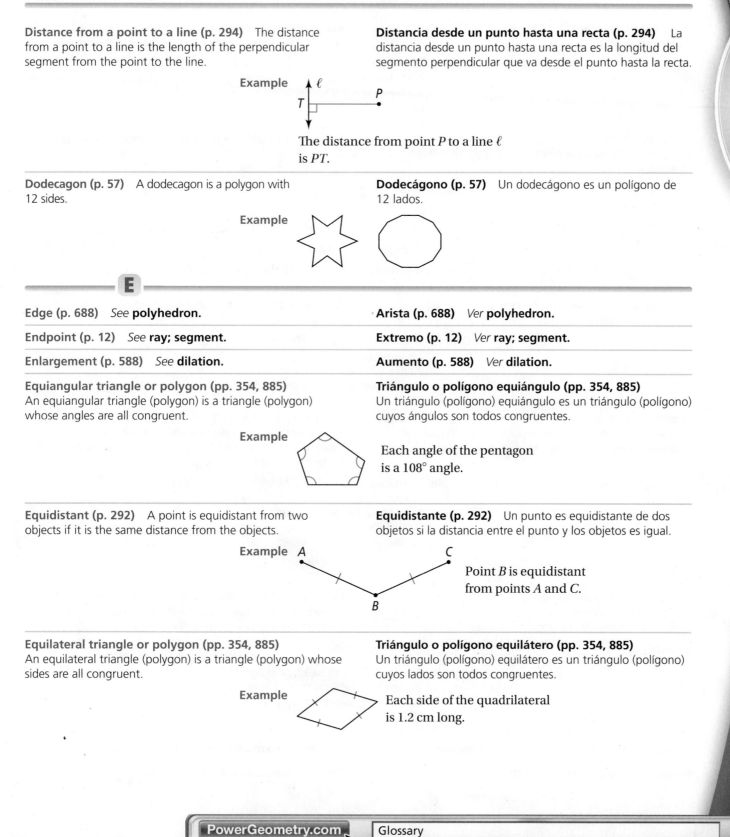

The distance from point *P* to a line ℓ is *PT*.

Dodecagon (p. 57) A dodecagon is a polygon with 12 sides.

Dodecágono (p. 57) Un dodecágono es un polígono de 12 lados.

Example

E

Edge (p. 688) *See* **polyhedron.**

Arista (p. 688) *Ver* **polyhedron.**

Endpoint (p. 12) *See* **ray; segment.**

Extremo (p. 12) *Ver* **ray; segment.**

Enlargement (p. 588) *See* **dilation.**

Aumento (p. 588) *Ver* **dilation.**

Equiangular triangle or polygon (pp. 354, 885)
An equiangular triangle (polygon) is a triangle (polygon) whose angles are all congruent.

Triángulo o polígono equiángulo (pp. 354, 885)
Un triángulo (polígono) equiángulo es un triángulo (polígono) cuyos ángulos son todos congruentes.

Example

Each angle of the pentagon is a 108° angle.

Equidistant (p. 292) A point is equidistant from two objects if it is the same distance from the objects.

Equidistante (p. 292) Un punto es equidistante de dos objetos si la distancia entre el punto y los objetos es igual.

Example

Point *B* is equidistant from points *A* and *C*.

Equilateral triangle or polygon (pp. 354, 885)
An equilateral triangle (polygon) is a triangle (polygon) whose sides are all congruent.

Triángulo o polígono equilátero (pp. 354, 885)
Un triángulo (polígono) equilátero es un triángulo (polígono) cuyos lados son todos congruentes.

Example

Each side of the quadrilateral is 1.2 cm long.

English

Spanish

Equivalent statements (p. 91)   Equivalent statements are statements with the same truth value.

Enunciados equivalentes (p. 91)   Los enunciados equivalentes son enunciados con el mismo valor verdadero.

Example   The following statements are equivalent:
If a figure is a square, then it is a rectangle.
If a figure is not a rectangle, then it is not a square.

Euclidean geometry (p. 179)   Euclidean geometry is a geometry of the plane in which Euclid's Parallel Postulate is accepted as true.

Geometría euclidiana (p.179)   La geometría euclidiana es una geometría del plano en donde el postulado paralelo de Euclides es verdadero.

Example

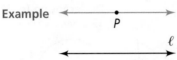

In Euclidean geometry, there is exactly one line parallel to line ℓ through point P.

Event (p. 824)   Any group of outcomes in a situation involving probability.

Suceso (p. 824)   En la probabilidad, cualquier grupo de resultados.

Example   When rolling a number cube, there are six possible outcomes. Rolling an even number is an event with three possible outcomes, 2, 4, and 6.

Expected value (p. 864)   The average value you can expect for a large number of trials of an experiment; the sum of each outcome's value multiplied by its probability.

Valor esperado (p. 864)   El valor promedio que se puede esperar para una cantidad grande de pruebas en un experimento; la suma de los valores de los resultados multiplicados cada uno por su probabilidad.

Example   In a game, a player has a 25% probability of earning 10 points by spinning an even number and a 75% probability of earning 5 points by spinning an odd number.

expected value = 0.25(10) + 0.75(5) = 6.25

Experimental probability (p. 825)   The ratio of the number of times an event actually happens to the number of times the experiment is done.

$$P(\text{event}) = \frac{\text{number of times an event happens}}{\text{number of times the experiment is done}}$$

Probabilidad experimental (p. 825)   La razón entre el número de veces que un suceso sucede en la realidad y el número de veces que se hace el experimento.

$$P(\text{suceso}) = \frac{\text{número de veces que sucede un suceso}}{\text{número de veces que se hace el experimento}}$$

Example   A baseball player's batting average shows how likely it is that a player will get a hit, based on previous times at bat.

Extended proportion (p. 440)   *See* **proportion.**

Proporción extendida (p. 440)   *Ver* **proportion.**

Extended ratio (p. 433)   *See* **ratio.**

Razón extendida (p. 433)   *Ver* **ratio.**

English

Exterior angle of a polygon (p. 173)   An exterior angle of a polygon is an angle formed by a side and an extension of an adjacent side.

Example

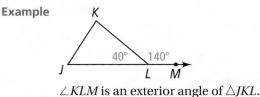

$\angle KLM$ is an exterior angle of $\triangle JKL$.

Extremes of a proportion (p. 434)   In the proportion $\frac{a}{b} = \frac{c}{d}$, a and d are the extremes.

Example   The product of the extremes of $\frac{x}{4} = \frac{x + 3}{2}$ is $2x$.

F

Face (p. 688)   *See* **polyhedron.**

Fibonacci sequence (p. 468)   The Fibonacci sequence is the infinite sequence of numbers beginning with 1, 1, . . . such that each term is the sum of the two previous terms.

Example   1, 1, 2, 3, 5, 8, 13, 21, . . .

Flow proof (p. 158)   *See* **proof.**

Focus of a parabola (p. 804)   *See* **parabola.**

Formula (p. 59)   A formula is a rule that shows the relationship between two or more quantities.

Example   The formula $P = 2\ell + 2w$ gives the perimeter of a rectangle in terms of the length and width.

Frequency table (p. 830)   A table that groups a set of data values into intervals and shows the frequency for each interval.

Example

Interval	Frequency
0–9	5
10–19	8
20–29	4

Spanish

Ángulo exterior de un polígono (p. 173)   El ángulo exterior de un polígono es un ángulo formado por un lado y una extensión de un lado adyacente.

Valores extremos de una proporción (p. 434)   En la proporción $\frac{a}{b} = \frac{c}{d}$, a y d son los valores extremos.

Cara (p. 688)   *Ver* **polyhedron.**

Sucesión de Fibonacci (p. 468)   La sucesión de Fibonacci es la sucesión infinita de números que comienza con 1, 1, . . . de forma tal que cada término es la suma de los dos términos anteriores.

Prueba de flujo (p. 158)   *Ver* **proof.**

Foco de una parábola (p. 804)   *Ver* **parabola.**

Fórmula (p. 59)   Una fórmula es una regla que muestra la relación entre dos o más cantidades.

Tabla de frecuencias (p. 830)   Tabla que agrupa un conjunto de datos en intervalos y muestra la frecuencia de cada intervalo.

English	Spanish

Fundamental Counting Principle (p. 836)   If there are m ways to make the first selection and n ways to make the second selection, then there are $m \cdot n$ ways to make the two selections.

Principio fundamental de Conteo (p. 836)   Si hay m maneras de hacer la primera selección y n maneras de hacer la segunda selección, quiere decir que hay $m \cdot n$ maneras de hacer las dos selecciones.

Example For 5 shirts and 8 pairs of shorts, the number of possible outfits is
$$5 \cdot 8 = 40.$$

G

Geometric mean (p. 462)   The geometric mean is the number x such that $\frac{a}{x} = \frac{x}{b}$, where a, b, and x are positive numbers.

Media geométrica (p. 462)   La media geométrica es el número x tanto que $\frac{a}{x} = \frac{x}{b}$, donde a, b y x son números positivos.

Example The geometric mean of 6 and 24 is 12.
$$\frac{6}{x} = \frac{x}{24}$$
$$x^2 = 144$$
$$x = 12$$

Geometric probability (p. 668)   Geometric probability is a probability that uses a geometric model in which points represent outcomes.

Probabilidad geométrica (p. 668)   La probabilidad geométrica es una probabilidad que utiliza un modelo geométrico donde se usan puntos para representar resultados.

Example

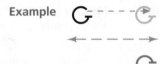

$$P(H \text{ on } \overline{BC}) = \frac{BC}{AD}$$

Glide reflection (p. 572)   A glide reflection is the composition of a translation followed by a reflection across a line parallel to the direction of translation.

Reflexión deslizada (p. 572)   Una reflexión por deslizamiento es la composición de una traslación seguida por una reflexión a través de una línea paralela a la dirección de traslación.

Example

The blue G in the diagram is a glide reflection image of the black G.

Golden rectangle, golden ratio (p. 468)   A *golden rectangle* is a rectangle that can be divided into a square and a rectangle that is similar to the original rectangle. The *golden ratio* is the ratio of the length of a golden rectangle to its width. The value of the golden ratio is $\frac{1 + \sqrt{5}}{2}$, or about 1.62.

Rectángulo áureo, razón áurea (p. 468)   Un *rectángulo áureo* es un rectángulo que se puede dividir en un cuadrado y un rectángulo semejante al rectángulo original. La *razón áurea* es la razón de la longitud de un rectángulo áureo en relación a su ancho. El valor de la razón áurea es $\frac{1 + \sqrt{5}}{2}$ o aproximadamente 1.62.

Example

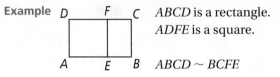

$ABCD$ is a rectangle.
$ADFE$ is a square.

$ABCD \sim BCFE$

English

Spanish

Great circle (p. 733) A great circle is the intersection of a sphere and a plane containing the center of the sphere. A great circle divides a sphere into two *hemispheres*.

Círculo máximo (p. 733) Un círculo máximo es la intersección de una esfera y un plano que contiene el centro de la esfera. Un círculo máximo divide una esfera en dos *hemisferios*.

Example **Hemispheres**   **Great circle**

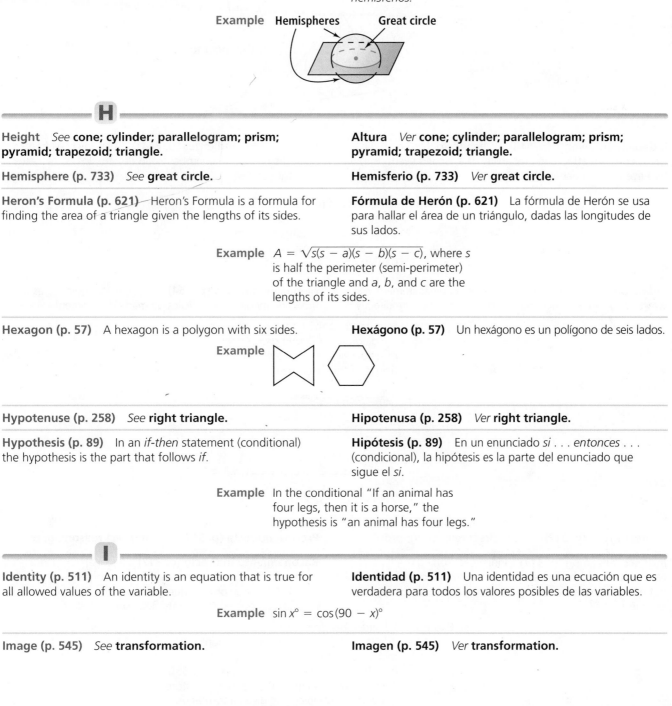

H

Height *See* **cone; cylinder; parallelogram; prism; pyramid; trapezoid; triangle.**

Altura *Ver* **cone; cylinder; parallelogram; prism; pyramid; trapezoid; triangle.**

Hemisphere (p. 733) *See* **great circle.**

Hemisferio (p. 733) *Ver* **great circle.**

Heron's Formula (p. 621) Heron's Formula is a formula for finding the area of a triangle given the lengths of its sides.

Fórmula de Herón (p. 621) La fórmula de Herón se usa para hallar el área de un triángulo, dadas las longitudes de sus lados.

Example $A = \sqrt{s(s-a)(s-b)(s-c)}$, where s is half the perimeter (semi-perimeter) of the triangle and a, b, and c are the lengths of its sides.

Hexagon (p. 57) A hexagon is a polygon with six sides.

Hexágono (p. 57) Un hexágono es un polígono de seis lados.

Example

Hypotenuse (p. 258) *See* **right triangle.**

Hipotenusa (p. 258) *Ver* **right triangle.**

Hypothesis (p. 89) In an *if-then* statement (conditional) the hypothesis is the part that follows *if*.

Hipótesis (p. 89) En un enunciado *si . . . entonces . . .* (condicional), la hipótesis es la parte del enunciado que sigue el *si*.

Example In the conditional "If an animal has four legs, then it is a horse," the hypothesis is "an animal has four legs."

I

Identity (p. 511) An identity is an equation that is true for all allowed values of the variable.

Identidad (p. 511) Una identidad es una ecuación que es verdadera para todos los valores posibles de las variables.

Example $\sin x° = \cos(90 - x)°$

Image (p. 545) *See* **transformation.**

Imagen (p. 545) *Ver* **transformation.**

Incenter of a triangle (p. 303)   The incenter of a triangle is the point of concurrency of the angle bisectors of the triangle.

Incentro de un triángulo (p. 303)   El incentro de un triángulo es el punto donde concurren las tres bisectrices de los ángulos del triángulo.

Example

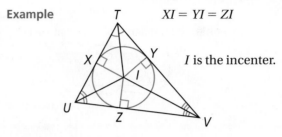

$XI = YI = ZI$

I is the incenter.

Independent events (p. 844)   When the outcome of one event does not affect the probability of a second event, the two events are independent.

Sucesos independientes (p. 844)   Cuando el resultado de un suceso no altera la probabilidad de otro, los dos sucesos son independientes.

Example   The results of two rolls of a number cube are independent. Getting a 5 on the first roll does not change the probability of getting a 5 on the second roll.

Indirect measurement (p. 454)   Indirect measurement is a way of measuring things that are difficult to measure directly.

Medición indirecta (p. 454)   La medición indirecta es un modo de medir cosas difíciles de medir directamente.

Example   By measuring the distances shown in the diagram and using proportions of similar figures, you can find the height of the taller tower.

$$\frac{196}{540} = \frac{x}{1300} \rightarrow x \approx 472 \text{ ft}$$

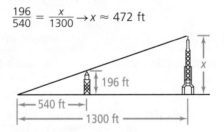

Indirect proof (p. 317)   *See* **indirect reasoning; proof.**

Prueba indirecta (p. 317)   *Ver* **indirect reasoning; proof.**

Indirect reasoning (p. 317)   Indirect reasoning is a type of reasoning in which all possiblities are considered and then all but one are proved false. The remaining possibility must be true.

Razonamiento indirecto (p. 317)   Razonamiento indirecto es un tipo de razonamiento en el que se consideran todas las posibilidades y se prueba que todas son falsas, a excepción de una. La posibilidad restante debe ser verdadera.

Example   Eduardo spent more than $60 on two books at a store. Prove that at least one book costs more than $30.
Proof: Suppose neither costs more than $30. Then he spent no more than $60 at the store. Since this contradicts the given information, at least one book costs more than $30.

English

Spanish

Inductive reasoning (p. 82)   Inductive reasoning is a type of reasoning that reaches conclusions based on a pattern of specific examples or past events.

Razonamiento inductivo (p. 82)   El razonamiento inductivo es un tipo de razonamiento en el cual se llega a conclusiones con base en un patrón de ejemplos específicos o sucesos pasados.

Example   You see four people walk into a building. Each person emerges with a small bag containing food. You use inductive reasoning to conclude that this building contains a restaurant.

Inscribed angle (p. 780)   An angle is inscribed in a circle if the vertex of the angle is on the circle and the sides of the angle are chords of the circle.

Ángulo inscrito (p. 780)   Un ángulo está inscrito en un círculo si el vértice del ángulo está en el círculo y los lados del ángulo son cuerdas del círculo.

Example

$\angle C$ is inscribed in $\odot M$.

Inscribed in (pp. 303, 766)   A circle is inscribed in a polygon if the sides of the polygon are tangent to the circle. A polygon is inscribed in a circle if the vertices of the polygon are on the circle.

Inscrito en (pp. 303, 766)   Un círculo está inscrito en un polígono si los lados del polígono son tangentes al círculo. Un polígono está inscrito en un círculo si los vértices del polígono están en el círculo.

Example

$\odot T$ is inscribed in $\triangle XYZ$.

$ABCD$ is inscribed in $\odot J$.

Intercepted arc (p. 780)   An intercepted arc is an arc of a circle having endpoints on the sides of an inscribed angle, and its other points in the interior of the angle.

Arco interceptor (p. 780)   Un arco interceptor es un arco de un círculo cuyos extremos están en los lados de un ángulo inscrito y los punto restantes están en el interior del ángulo.

Example

$\overset{\frown}{UV}$ is the intercepted arc of inscribed $\angle T$.

Intersection (p. 13)   The intersection of two or more geometric figures is the set of points the figures have in common.

Intersección (p. 13)   La intersección de dos o más figuras geométricas es el conjunto de puntos que las figuras tienen en común.

Example

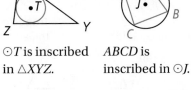

The intersection of lines r and s is point P.

English

Spanish

Inverse (p. 91)   The inverse of the conditional "if *p*, then *q*" is the conditional "if not *p*, then not *q*."

Inverso (p. 91)   El inverso del condicional "si *p*, entonces *q*" es el condicional "si no *p*, entonces no *q*".

Example   **Conditional:** If a figure is a square, then it is a parallelogram.
Inverse: If a figure is not a square, then it is not a parallelogram.

Isometric drawing (p. 5)   An isometric drawing shows a corner view of a three-dimensional figure. It is usually drawn on isometric dot paper. An isometric drawing allows you to see the top, front, and side of an object in the same drawing.

Dibujo isométrico (p. 5)   Un dibujo isométrico muestra la perspectiva de una esquina de una figura tridimensional. Generalmente se dibuja en papel punteado isométrico. Un dibujo isométrico permite ver la cima, el frente, y el lado de un objeto en el mismo dibujo.

Example

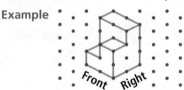

Isometry (p. 570)   An isometry, also known as a *congruence transformation,* is a transformation in which an original figure and its image are congruent.

Isometría (p. 570)   Una isometría, conocida también como una *transformación de congruencia,* es una transformación en donde una figura original y su imagen son congruentes.

Example   The four isometries are reflections, rotations, translations, and glide reflections.

Isosceles trapezoid (p. 389)   An isosceles trapezoid is a trapezoid whose nonparallel opposite sides are congruent.

Trapecio isósceles (p. 389)   Un trapecio isosceles es un trapecio cuyos lados opuestos no paralelos son congruentes.

Example

Isosceles triangle (p. 885)   An isosceles triangle is a triangle that has at least two congruent sides. If there are two congruent sides, they are called *legs*. The *vertex angle* is between them. The third side is called the *base* and the other two angles are called the *base angles*.

Triángulo isósceles (p. 885)   Un triángulo isósceles es un triángulo que tiene por lo menos dos lados congruentes. Si tiene dos lados congruentes, éstos se llaman *catetos*. Entre ellos se encuentra el *ángulo del vértice*. El tercer lado se llama *base* y los otros dos ángulos se llaman *ángulos de base*.

Example

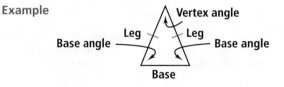

English

Kite (p. 392) A kite is a quadrilateral with two pairs of consecutive sides congruent and no opposite sides congruent.

Example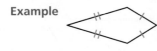

Lateral area (pp. 700, 702, 709, 711) The lateral area of a prism or pyramid is the sum of the areas of the lateral faces. The lateral area of a cylinder or cone is the area of the curved surface. A list of lateral area formulas is on p. 839.

Example

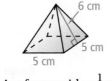
6 cm

5 cm

5 cm

$$\text{L.A. of pyramid} = \tfrac{1}{2}p\ell$$
$$= \tfrac{1}{2}(20)(6)$$
$$= 60 \text{ cm}^2$$

Lateral face *See* **prism; pyramid.**

Law of Cosines (p. 527) In $\triangle ABC$, let a, b, and c represent the lengths of the sides opposite $\angle A$, $\angle B$, and $\angle C$, respectively. Then
$a^2 = b^2 + c^2 - 2bc \cos A,$
$b^2 = a^2 + c^2 - 2ac \cos B,$ and
$c^2 = a^2 + b^2 - 2ab \cos C$

Example

11.41 L

18°

K ———— M
8.72

$LM^2 = 11.41^2 + 8.72^2 - 2(11.42)(8.72) \cos 18°$
$LM^2 = 16.9754$
$LM = 4.12$

Spanish

Cometa (p. 392) Una cometa es un cuadrilátero con dos pares de lados congruentes consecutivos y sin lados opuestos congruentes.

Área lateral (pp. 700, 702, 709, 711) El área lateral de un prisma o pirámide es la suma de las áreas de sus caras laterals. El área lateral de un cilindro o de un cono es el área de la superficie curvada. Una lista de las fórmulas de áreas laterales está en la p. 839.

Cara lateral *Ver* **prism; pyramid.**

Ley de cosenos (p. 527) En $\triangle ABC$, sean a, b y c las longitudes de los lados opuestos a $\angle A$, $\angle B$ y $\angle C$, respectivamente. Entonces
$a^2 = b^2 + c^2 - 2bc \cos A,$
$b^2 = a^2 + c^2 - 2ac \cos B$ y
$c^2 = a^2 + b^2 - 2ab \cos C$

English

Spanish

Law of Sines (p. 522) In $\triangle ABC$, let a, b, and c represent the lengths of the sides opposite $\angle A$, $\angle B$, and $\angle C$, respectively. Then $\frac{\sin A}{a} = \frac{\sin B}{b} = \frac{\sin C}{c}$.

Ley de senos (p. 522) En $\triangle ABC$, sean a, b y c las longitudes de los lados opuestos a $\angle A$, $\angle B$ y $\angle C$, respectivamente. Entonces $\frac{\text{sen } A}{a} = \frac{\text{sen } B}{b} = \frac{\text{sen } C}{c}$.

Example

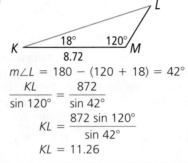

$$m\angle L = 180 - (120 + 18) = 42°$$
$$\frac{KL}{\sin 120°} = \frac{872}{\sin 42°}$$
$$KL = \frac{872 \sin 120°}{\sin 42°}$$
$$KL = 11.26$$

Leg *See* **isosceles triangle; right triangle; trapezoid.**

Cateto *Ver* **isosceles triangle; right triangle; trapezoid.**

Line (pp. 11, 179) In Euclidean geometry, a line is undefined. You can think of a line as a straight path that extends in two opposite directions without end and has no thickness. A line contains infinitely many points. In spherical geometry, you can think of a line as a great circle of a sphere.

Recta (pp. 11, 179) En la geometría euclidiana, una recta es indefinida. Se puede pensar en una recta como un camino derecho que se extiende en direcciones opuestas sin fin ni grosor. Una recta tiene un número infinito de puntos. En la geometría esférica, se puede pensar en una recta como un gran círculo de una esfera.

Example

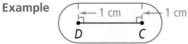

Linear pair (p. 36) A linear pair is a pair of adjacent angles whose noncommon sides are opposite rays.

Par lineal (p. 36) Un par lineal es un par de ángulos adjuntos cuyos lados no comunes son semirrectas opuestas.

Example

$\angle 1$ and $\angle 2$ are a linear pair.

Line of reflection (p. 554) *See* **reflection.**

Eje de reflexión (p. 554) *Ver* **reflection.**

Line of symmetry (p. 568) *See* **reflectional symmetry.**

Eje de simetría (p. 568) *Ver* **reflectional symmetry.**

Line symmetry (p. 568) *See* **reflectional symmetry.**

Simetría axial (p. 568) *Ver* **reflectional symmetry.**

Locus (p. 806) A locus is a set of points, all of which meet a stated condition.

Lugar geométrico (p. 806) Un lugar geométrico es un conjunto de puntos que cumplen una condición dada.

Example

1 cm   1 cm

D   C

The points in blue are the locus of points in a plane 1 cm from $\overline{DC}$.

Visual **Glossary**

English

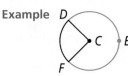

Major arc (p. 649) A major arc of a circle is an arc that is larger than a semicircle.

Example

$\widehat{DEF}$ is a major arc of $\odot C$.

Map (p. 545) *See* **transformation.**

Means of a proportion (p. 434) In the proportion $\frac{a}{b} = \frac{c}{d}$, b and c are the means.

Example The product of the means of $\frac{x}{4} = \frac{x + 3}{2}$ is $4(x + 3)$ or $4x + 12$.

Measure of an angle (p. 28) Consider $\overrightarrow{OD}$ and a point C on one side of $\overrightarrow{OD}$. Every ray of the form $\overrightarrow{OC}$ can be paired one to one with a real number from 0 to 180. The measure of $\angle COD$ is the absolute value of the difference of the real numbers paired with $\overrightarrow{OC}$ and $\overrightarrow{OD}$.

Spanish

Arco mayor (p. 649) Un arco mayor de un círculo es cualquier arco más grande que un semicírculo.

Trazar (p. 545) *Ver* **transformation.**

Valores medios de una proporción (p. 434) En la proporción $\frac{a}{b} = \frac{c}{d}$, b y c son los valores medios.

Medida de un ángulo (p. 28) Toma en cuenta $\overrightarrow{OD}$ y un punto C a un lado de $\overrightarrow{OD}$. Cada semirrecta de la forma $\overrightarrow{OC}$ puede ser emparejada exactamente con un número real de 0 a 180. La medida de $\angle COD$ es el valor absoluto de la diferencia de los números reales emparejados con $\overrightarrow{OC}$ y $\overrightarrow{OD}$.

Example

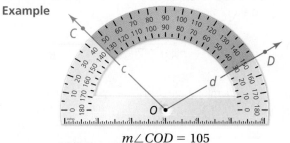

$m\angle COD = 105$

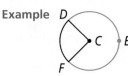 Visual **Glossary**

English	Spanish

Measure of an arc (p. 650)   The measure of a minor arc is the measure of its central angle. The measure of a major arc is 360 minus the measure of its related minor arc.

Medida de un arco (p. 650)   La medida de un arco menor es la medida de su ángulo central. La medida de un arco mayor es 360 menos la medida en grados de su arco menor correspondiente.

Example

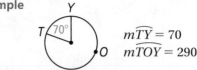

$$m\overset{\frown}{TY} = 70$$
$$m\overset{\frown}{TOY} = 290$$

Median of a triangle (p. 309)   A median of a triangle is a segment that has as its endpoints a vertex of the triangle and the midpoint of the opposite side.

Mediana de un triángulo (p. 309)   Una mediana de un triángulo es un segmento que tiene en sus extremos el vértice del triángulo y el punto medio del lado opuesto.

Example

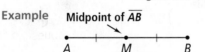

Midpoint of a segment (p. 22)   A midpoint of a segment is the point that divides the segment into two congruent segments.

Punto medio de un segmento (p. 22)   El punto medio de un segmento es el punto que divide el segmento en dos segmentos congruentes.

Example   **Midpoint of $\overline{AB}$**

Midsegment of a trapezoid (p. 391)   The midsegment of a trapezoid is the segment that joins the midpoints of the nonparallel opposite sides of a trapezoid.

Segmento medio de un trapecio (p. 391)   El segmento medio de trapecio es el segmento que une los puntos medios de los lados no paralelos de un trapecio.

Example   Midsegment

Midsegment of a triangle (p. 285)   A midsegment of a triangle is a segment that joins the midpoints of two sides of the triangle.

Segmento medio de un triángulo (p. 285)   Un segmento medio de un triángulo es un segmento que une los puntos medios de dos lados del triángulo.

Example   Midsegment

Minor arc (p. 649)   A minor arc is an arc that is smaller than a semicircle.

Arco menor (p. 649)   Un arco menor de un círculo es un arco más corto que un semicírculo.

Example

$\overset{\frown}{KC}$ is a minor arc of $\odot S$.

Mutually exclusive events (p. 845)   When two events cannot happen at the same time, the events are mutually exclusive. If A and B are mutually exclusive events, then $P(A \text{ or } B) = P(A) + P(B)$.

Sucesos mutuamente excluyentes (p. 845)   Cuando dos sucesos no pueden ocurrir al mismo tiempo, son mutuamente excluyentes. Si A y B son sucesos mutuamente excluyentes, entonces $P(A \text{ o } B) = P(A) + P(B)$.

Example   Rolling an even number E and rolling a multiple of five M on a standard number cube are mutually exclusive events.

$$P(E \text{ or } M) = P(E) + P(M)$$
$$= \frac{3}{6} + \frac{1}{6}$$
$$= \frac{4}{6}$$
$$= \frac{2}{3}$$

N

n factorial (p. 837)   The product of the integers from n down to 1, for any positive integer n. You write n factorial as $n!$. The value of 0! is defined to be 1.

n factorial (p. 837)   Producto de todos los enteros desde n hasta 1, de cualquier entero positivo n. El factorial de n se escribe $n!$. El valor de 0! se define como 1.

Example $4! = 4 \cdot 3 \cdot 2 \cdot 1 = 24$

Negation (p. 91)   The negation of a statement has the opposite meaning of the original statement.

Negación (p. 91)   La negación de un enunciado tiene el sentido opuesto del enunciado original.

Example   **Statement:** The angle is obtuse.
Negation: The angle is not obtuse.

Net (p. 4)   A net is a two-dimensional pattern that you can fold to form a three-dimensional figure.

Plantilla (p. 4)   Una plantilla es una figura bidimensional que se puede doblar para formar una figura tridimensional.

Example

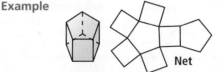

Net

The net shown can be folded into a prism with pentagonal bases.

n-gon (p. 57)   An n-gon is a polygon with n sides.

n-ágono (p. 57)   Un n-ágono es un polígono de n lados.

Example   A polygon with 25 sides is a 25-gon.

Nonagon (p. 57)   A nonagon is a polygon with nine sides.

Nonágono (p. 57)   Un nonágono es un polígono de nueve lados.

Example

Oblique cylinder or prism   *See* **cylinder; prism.**

Cilindro oblicuo o prisma   *Ver* **cylinder; prism.**

Obtuse angle (p. 29)   An obtuse angle is an angle whose measure is between 90 and 180.

Ángulo obtuso (p. 29)   Un ángulo obtuso es un ángulo que mide entre 90 y 180 grados.

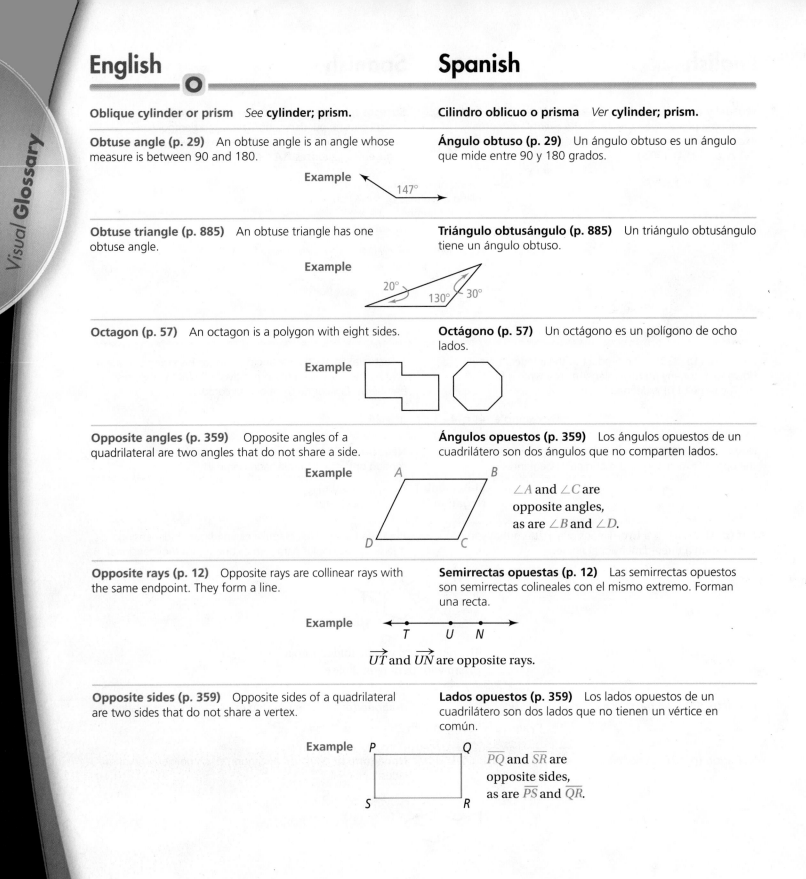

Example   147°

Obtuse triangle (p. 885)   An obtuse triangle has one obtuse angle.

Triángulo obtusángulo (p. 885)   Un triángulo obtusángulo tiene un ángulo obtuso.

Example   20°   130°   30°

Octagon (p. 57)   An octagon is a polygon with eight sides.

Octágono (p. 57)   Un octágono es un polígono de ocho lados.

Example

Opposite angles (p. 359)   Opposite angles of a quadrilateral are two angles that do not share a side.

Ángulos opuestos (p. 359)   Los ángulos opuestos de un cuadrilátero son dos ángulos que no comparten lados.

Example A B

$\angle A$ and $\angle C$ are opposite angles, as are $\angle B$ and $\angle D$.

D C

Opposite rays (p. 12)   Opposite rays are collinear rays with the same endpoint. They form a line.

Semirrectas opuestas (p. 12)   Las semirrectas opuestos son semirrectas colineales con el mismo extremo. Forman una recta.

Example T U N

$\overrightarrow{UT}$ and $\overrightarrow{UN}$ are opposite rays.

Opposite sides (p. 359)   Opposite sides of a quadrilateral are two sides that do not share a vertex.

Lados opuestos (p. 359)   Los lados opuestos de un cuadrilátero son dos lados que no tienen un vértice en común.

Example P Q

$\overline{PQ}$ and $\overline{SR}$ are opposite sides, as are $\overline{PS}$ and $\overline{QR}$.

S R

English

Spanish

Orientation (p. 554) Two congruent figures have *opposite* orientation if a reflection is needed to map one onto the other. If a reflection is not needed to map one figure onto the other, the figures have the same orientation.

Orientación (p. 554) Dos figuras congruentes tienen orientación opuesta si una reflexión es necesaria para trazar una sobre la otra. Si una reflexión no es necesaria para trazar una figura sobre la otra, las figuras tiene la misma orientación.

Example R Я The two R's have opposite orientation.

Origin (p. 893) *See* **coordinate plane.**

Origen (p. 893) *Ver* **coordinate plane.**

Orthocenter of a triangle (p. 311) The orthocenter of a triangle is the point of concurrency of the lines containing the altitudes of the triangle.

Ortocentro de un triángulo (p. 311) El ortocentro de un triángulo es el punto donde se intersecan las alturas de un triángulo.

Example *D* is the orthocenter.

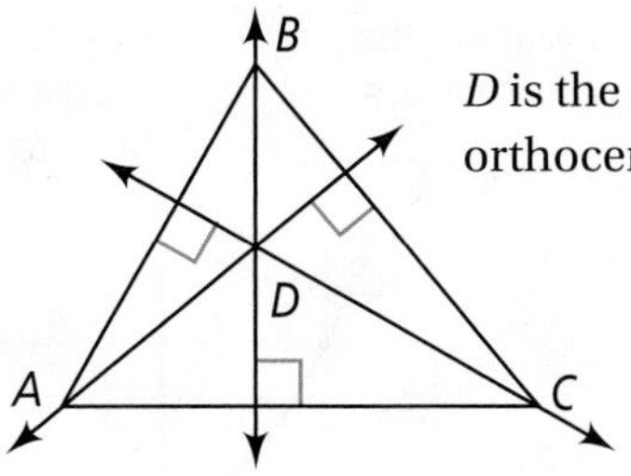

Orthographic drawing (p. 6) An orthographic drawing is the top view, front view, and right-side view of a three-dimensional figure.

Dibujo ortográfico (p. 6) Un dibujo ortográfico es la vista desde arriba, la vista de frente y la vista del lado derecho de una figura tridimensional.

Example The diagram shows an isometric drawing (upper right) and the three views that make up an orthographic drawing.

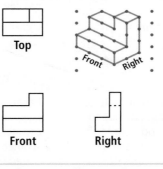

Top

Front Right

Front

Right

Outcome (p. 824) The result of a single trial in a probability experiment.

Resultado (p. 824) Lo que se obtiene al hacer una sola prueba en un experimento de probabilidad.

Example The outcomes of rolling a number cube are 1, 2, 3, 4, 5, and 6.

English	Spanish

Overlapping events (p. 846)   Events that have at least one common outcome. If A and B are overlapping events, then $P(A \text{ or } B) = P(A) + P(B) - P(A \text{ and } B)$.

Sucesos traslapados (p. 846)   Sucesos que tienen por lo menos un resultado en común. Si A y B son sucesos traslapados, entonces $P(A \text{ ó } B) = P(A) + P(B) - P(A \text{ y } B)$.

Example   Rolling a multiple of 3 and rolling an odd number on a number cube are overlapping events.

$$P(\text{multiple of 3 or odd}) = P(\text{multiple of 3}) + P(\text{odd}) - P(\text{multiple of 3 and odd})$$
$$= \frac{1}{3} + \frac{1}{2} - \frac{1}{6}$$
$$= \frac{2}{3}$$

P

Parabola (p. 804)   A parabola is the graph of a quadratic function. It is the set of all points P in a plane that are the same distance from a fixed point F, the focus, as they are from a line d, the directrix.

Parábola (p. 804)   La parábola es la gráfica de una función cuadrática. Es el conjunto de todos los puntos P situados en un plano a la misma distancia de un punto fijo F, o foco, y de la recta d, o directriz.

Example

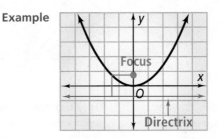

Paragraph proof (p. 122)   *See* **proof.**

Prueba de párrafo (p. 122)   *Ver* **proof.**

Parallel lines (p. 140)   Two lines are parallel if they lie in the same plane and do not intersect. The symbol ‖ means "is parallel to."

Rectas paralelas (p. 140)   Dos rectas son paralelas si están en el mismo plano y no se cortan. El símbolo ‖ significa "es paralelo a".

Example $\ell \parallel m$

The red symbols indicate parallel lines.

Parallelogram (p. 359)   A parallelogram is a quadrilateral with two pairs of parallel sides. You can choose any side to be the *base*. An *altitude* is any segment perpendicular to the line containing the base drawn from the side opposite the base. The *height* is the length of an altitude.

Paralelogramo (p. 359)   Un paralelogramo es un cuadrilátero con dos pares de lados paralelos. Se puede escoger cualquier lado como la *base*. Una *altura* es un segmento perpendicular a la recta que contiene la base, trazada desde el lado opuesto a la base. La *altura*, por extensión, es la longitud de una altura.

Example

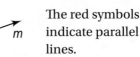

English

Parallel planes (p. 140) Parallel planes are planes that do not intersect.

Example

Planes Y and Z are parallel.

Pentagon (p. 57) A pentagon is a polygon with five sides.

Example

Perimeter of a polygon (p. 59) The perimeter of a polygon is the sum of the lengths of its sides.

Example

$$P = 4 + 4 + 5 + 3$$
$$= 16 \text{ in.}$$

4 in.

4 in.

3 in.

5 in.

Permutation (p. 837) An arrangement of some or all of a set of objects in a specific order. You can use the notation $_nP_r$ to express the number of permutations, where n equals the number of objects available and r equals the number of selections to make.

Example How many ways can you arrange 5 objects 3 at a time?

$$_5P_3 = \frac{5!}{(5-3)!} = \frac{5!}{2!} = \frac{5 \cdot 4 \cdot 3 \cdot 2 \cdot 1}{2 \cdot 1} = 60$$

There are 60 ways to arrange 5 objects 3 at a time.

Perpendicular bisector (p. 44) The perpendicular bisector of a segment is a line, segment, or ray that is perpendicular to the segment at its midpoint.

Example

$\overleftrightarrow{YZ}$ is the perpendicular bisector of $\overline{AB}$. It is perpendicular to $\overline{AB}$ and intersects $\overline{AB}$ at midpoint M.

Spanish

Planos paralelos (p. 140) Planos paralelos son planos que no se cortan.

Pentágono (p. 57) Un pentágono es un polígono de cinco lados.

Perímetro de un polígono (p. 59) El perímetro de un polígono es la suma de las longitudes de sus lados

Permutación (p. 837) Disposición de algunos o de todos los objetos de un conjunto en un orden determinado. El número de permutaciones se puede expresar con la notación $_nP_r$, donde n es igual al número total de objetos y r es igual al número de selecciones que han de hacerse.

Mediatriz (p. 44) La mediatriz de un segmento es una recta, segmento o semirrecta que es perpendicular al segmento en su punto medio.

English

Spanish

Perpendicular lines (p. 44)   Perpendicular lines are lines that intersect and form right angles. The symbol ⊥ means "is perpendicular to."

Rectas perpendiculares (p. 44)   Las rectas perpendiculares son rectas que se cortan y forman ángulos rectos. El símbolo ⊥ significa "es perpendicular a".

Example

$m \perp n$

Perspective drawing (p. 696)   Perspective drawing is a way of drawing objects on a flat surface so that they look the same way as they appear to the eye. In *one-point perspective*, there is one *vanishing point*. In *two-point perspective*, there are two vanishing points.

Dibujar en perspectiva (p. 696)   Dibujar en perspectiva es una manera de dibujar objetos en una superficie plana de modo que se vean como los percibe el ojo humano. En la *perspectiva de un punto* hay un *punto de fuga*. En la *perspectiva de dos puntos* hay dos puntos de fuga.

Example

One-point perspective

Two-point perspective

Pi (p. 651)   Pi (π) is the ratio of the circumference of any circle to its diameter. The number π is irrational and is approximately 3.14159.

Pi (p. 651)   Pi (π) es la razón de la circunferencia de cualquier círculo a su diámetro. El número π es irracional y se aproxima a $\pi \approx 3.14159$.

Example

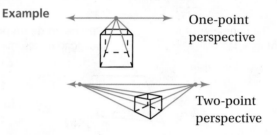

$\pi = \dfrac{C}{d}$

Plane (p. 11)   In Euclidean geometry, a plane is undefined. You can think of a plane as a flat surface that extends without end and has no thickness. A plane contains infinitely many lines.

Plano (p. 11)   En la geometría euclidiana, un plano es indefinido. Se puede pensar en un plano como una superficie plana sin fin, ni grosor. Un plano tiene un número infinito de rectas.

Example

Plane *ABC* or plane *Z*

Point (p. 11)   In Euclidean geometry, a point is undefined. You can think of a point as a location. A point has no size.

Punto (p. 11)   En la geometría euclidiana, un punto es indefinido. Puedes imaginarte a un punto como un lugar. Un punto no tiene dimensión.

Example   • *P*

Point of concurrency (p. 301)   *See* **concurrent lines.**

Punto de concurrencia (p. 301)   *Ver* **concurrent lines.**

Point of tangency (p. 762)   *See* **tangent to a circle.**

Punto de tangencia (p. 762)   *Ver* **tangent to a circle.**

Point-slope form (p. 190)   The point-slope form for a nonvertical line with slope m and through point (x_1, y_1) is $y - y_1 = m(x - x_1)$.

Forma punto-pendiente (p. 190)   La forma punto-pendiente para una recta no vertical con pendiente m y que pasa por el punto (x_1, y_1) es $y - y_1 = m(x - x_1)$.

Example $y + 1 = 3(x - 4)$

In this equation, the slope is 3 and (x_1, y_1) is $(4, -1)$.

Point symmetry (p. 568)   Point symmetry is the type of symmetry for which there is a rotation of 180° that maps a figure onto itself.

Simetría central (p. 568)   La simetría central es un tipo de simetría en la que una figura se ha rotado 180° sobre sí misma.

Example

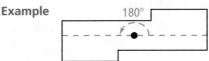

Polygon (p. 57)   A polygon is a closed plane figure formed by three or more segments. Each segment intersects exactly two other segments, but only at their endpoints, and no two segments with a common endpoint are collinear. The *vertices* of the polygon are the endpoints of the sides. A *diagonal* is a segment that connects two nonconsecutive vertices. A polygon is *convex* if no diagonal contains points outside the polygon. A polygon is *concave* if a diagonal contains points outside the polygon.

Polígono (p. 57)   Un polígono es una figura plana compuesta por tres o más segmentos. Cada segmento interseca los otros dos segmentos exactamente, pero únicamente en sus puntos extremos y ningúno de los segmentos con extremos comunes son colineales. Los *vértices* del polígono son los extremos de los lados. Una *diagonal* es un segmento que conecta dos vértices no consecutivos. Un polígono es *convexo* si ninguna diagonal tiene puntos fuera del polígono. Un polígono es *cóncavo* si una diagonal tiene puntos fuera del polígono.

Example

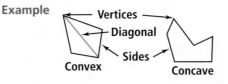

Polyhedron (p. 688)   A polyhedron is a three-dimensional figure whose surfaces, or *faces*, are polygons. The vertices of the polygons are the *vertices* of the polyhedron. The intersections of the faces are the *edges* of the polyhedron.

Poliedro (p. 688)   Un poliedro es una figura tridimensional cuyas superficies, o *caras*, son polígonos. Los vértices de los polígonos son los *vértices* del poliedro. Las intersecciones de las caras son las *aristas* del poliedro.

Example

Postulate (p. 13)   A postulate, or *axiom*, is an accepted statement of fact.

Postulado (p. 13)   Un postulado, o *axioma*, es un enunciado que se acepta como un hecho.

Example   **Postulate:** Through any two points there is exactly one line.

Preimage (p. 545)   *See* **transformation.**

Preimagen (p. 545)   *Ver* **transformation.**

Prime notation (p. 546)   *See* **transformation.**

Notación prima (p. 546)   *Ver* **transformation.**

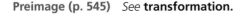

English

Spanish

Prism (p. 699) A prism is a polyhedron with two congruent and parallel faces, which are called the *bases*. The other faces, which are parallelograms, are called the *lateral faces*. An *altitude* of a prism is a perpendicular segment that joins the planes of the bases. Its length is the *height* of the prism. A *right prism* is one whose lateral faces are rectangular regions and a lateral edge is an altitude. In an *oblique prism*, some or all of the lateral faces are nonrectangular.

Prisma (p. 699) Un prisma es un poliedro con dos caras congruentes paralelas llamadas *bases*. Las otras caras son paralelogramos llamados *caras laterales*. La *altura* de un prisma es un segmento perpendicular que une los planos de las bases. Su longitud es también la *altura* del prisma. En un *prisma rectangular*, las caras laterales son rectangulares y una de las aristas laterales es la altura. En un *prisma oblicuo*, algunas o todas las caras laterales no son rectangulares.

Example

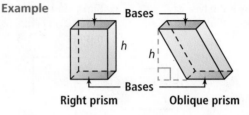

Bases

h *h*

Bases

Right prism   Oblique prism

Probability (p. 824) How likely it is that an event will occur (written formally as *P*(event)).

Probabilidad (p. 824) La posibilidad de que un suceso ocurra, escrita formalmente *P*(suceso).

Example You have 4 red marbles and 3 white marbles. The probability that you select one red marble, and then, without replacing it, randomly select another red marble is $P(\text{red}) = \frac{4}{7} \cdot \frac{3}{6} = \frac{2}{7}$.

Probability distribution (p. 831) A probability distribution is a function that tells the probability of each outcome in a sample space.

Distribución de probabilidades (p. 831) Una distribución de probabilidades es una función que señala la probabilidad de que cada resultado ocurra en un espacio muestral.

Example

Roll	Fr.	Prob.
1	5	0.125
2	9	0.225
3	7	0.175
4	8	0.2
5	8	0.2
6	3	0.075

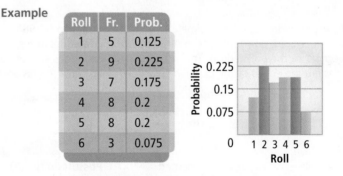

The table and graph both show the experimental probability distribution for the outcomes of 40 rolls of a standard number cube.

Visual **Glossary**

English

Proof (pp. 115, 122, 158, 317, 408)   A proof is a convincing argument that uses deductive reasoning. A proof can be written in many forms. In a two-column proof, the statements and reasons are aligned in columns. In a paragraph proof, the statements and reasons are connected in sentences. In a flow proof, arrows show the logical connections between the statements. In a coordinate proof, a figure is drawn on a coordinate plane and the formulas for slope, midpoint, and distance are used to prove properties of the figure. An indirect proof involves the use of indirect reasoning.

Spanish

Prueba (pp. 115, 122, 158, 317, 408)   Una prueba es un argumento convincente en el cual se usa el razonamiento deductivo. Una prueba se puede escribir de varias maneras. En una *prueba de dos columnas*, los enunciados y las razones se alinean en columnas. En una *prueba de párrafo*, los enunciados y razones están unidos en oraciones. En una *prueba de flujo*, hay flechas que indican las conexiones lógicas entre enunciados. En una *prueba de coordenadas*, se dibuja una figura en un plano de coordenadas y se usan las fórmulas de la pendiente, punto medio y distancia para probar las propiedades de la figura. Una *prueba indirecta* incluye el uso de razonamiento indirecto.

Example

Given: $\triangle EFG$, with right angle $\angle F$
Prove: $\angle E$ and $\angle G$ are complementary.

Paragraph Proof: Because $\angle F$ is a right angle, $m\angle F = 90$. By the Triangle Angle-Sum Theorem, $m\angle E + m\angle F + m\angle G = 180$. By substitution, $m\angle E + 90 + m\angle G = 180$. Subtracting 90 from each side yields $m\angle E + m\angle G = 90$. $\angle E$ and $\angle G$ are complementary by definition.

Proportion (p. 434)   A proportion is a statement that two ratios are equal. An *extended proportion* is a statement that three or more ratios are equal.

Proporción (p. 434)   Una proporción es un enunciado en el cual dos razones son iguales. Una *proporción extendida* es un enunciado que dice que tres razones o más son iguales.

Example $\frac{x}{5} = \frac{3}{4}$ is a proportion.
$\frac{9}{27} = \frac{3}{9} = \frac{1}{3}$ is an extended proportion.

Pyramid (p. 708) A pyramid is a polyhedron in which one face, the *base*, is a polygon and the other faces, the *lateral faces*, are triangles with a common vertex, called the *vertex* of the pyramid. An *altitude* of a pyramid is the perpendicular segment from the vertex to the plane of the base. Its length is the *height* of the pyramid. A *regular pyramid* is a pyramid whose base is a regular polygon and whose lateral faces are congruent isosceles triangles. The *slant height* of a regular pyramid is the length of an altitude of a lateral face.

Pirámide (p. 708) Una pirámide es un poliedro en donde una cara, la *base*, es un polígono y las otras caras, las *caras laterales*, son triángulos con un vértice común, llamado el *vértice* de la pirámide. Una *altura* de una pirámide es el segmento perpendicular que va del vértice hasta el plano de la base. Su longitud es, por extensión, la *altura* de la pirámide. Una *pirámide regular* es una pirámide cuya base es un polígono regular y cuyas caras laterales son triángulos isósceles congruentes. La *apotema* de una pirámide regular es la longitud de la altura de la cara lateral.

Example

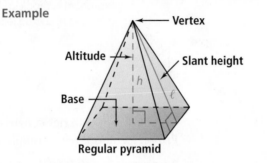

Vertex

Altitude

Slant height

Base

Regular pyramid

Pythagorean triple (p. 492) A Pythagorean triple is a set of three nonzero whole numbers a, b, and c, that satisfy the equation $a^2 + b^2 = c^2$.

Tripleta de Pitágoras (p. 492) Una tripleta de Pitágoras es un conjunto de tres números enteros positivos a, b, and c que satisfacen la ecuación $a^2 + b^2 = c^2$.

Example The numbers 5, 12, and 13 form a
Pythagorean triple because
$5^2 + 12^2 = 13^2 = 169$.

Q

Quadrant (p. 893) *See* **coordinate plane.**

Cuadrante (p. 893) *Ver* **coordinate plane.**

Quadrilateral (p. 57) A quadrilateral is a polygon with four sides.

Cuadrilátero (p. 57) Un cuadrilátero es un polígono de cuatro lados.

Example

R

Radius of a circle (p. 649) A radius of a circle is any segment with one endpoint on the circle and the other endpoint at the center of the circle. *Radius* can also mean the length of this segment.

Radio de un círculo (p. 649) Un radio de un círculo es cualquier segmento con extremo en el círculo y el otro extremo en el centro del círculo. *Radio* también se refiere a la longitud de este segmento.

Example

D E $\overline{DE}$ is a radius of $\odot D$.

Visual **Glossary**

English	Spanish

Radius of a regular polygon (p. 629) The radius of a regular polygon is the distance from the center to a vertex.

Radio de un polígono regular (p. 629) El radio de un polígono regular es la distancia desde el centro hasta un vértice.

Example

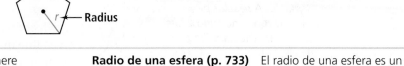

Radius

Radius of a sphere (p. 733) The radius of a sphere is a segment that has one endpoint at the center and the other endpoint on the sphere.

Radio de una esfera (p. 733) El radio de una esfera es un segmento con un extremo en el centro y otro en la esfera.

Example

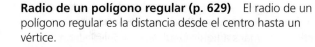

Ratio (p. 432) A ratio is a comparison of two quantities by division. An *extended ratio* is a comparison of three or more quantities by division.

Razón (p. 432) Una razón es una comparación de dos cantidades usando la división. Una *razón extendida* es una comparación de tres o más cantidades usando la división.

Example 5 to 7, 5 : 7, and $\frac{5}{7}$ are ratios.
3 : 5 : 6 is an extended ratio.

Ray (p. 12) A ray is the part of a line that consists of one *endpoint* and all the points of the line on one side of the endpoint.

Semirrecta (p. 12) Una semirrecta es la parte de una recta que tiene un *extremo* de donde parten todos los puntos de la recta.

Example

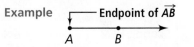

Endpoint of $\overrightarrow{AB}$

A   B

Rectangle (p. 375) A rectangle is a parallelogram with four right angles.

Rectángulo (p. 375) Un rectángulo es un paralelogramo con cuatro ángulos rectos.

Example

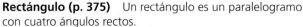

Reduction (p. 588) *See* **dilation.**

Reducción (p. 588) *Ver* **dilation.**

Reflection (p. 554) A reflection (*flip*) across line r, called the *line of reflection*, is a transformation such that if a point A is on line r, then the image of A is itself, and if a point B is not on line r, then its image B' is the point such that r is the perpendicular bisector of $\overline{BB'}$.

Reflexión (p. 554) Una reflexión (*inversión*) a través de una línea r, llamada el *eje de reflexión*, es una transformación en la que si un punto A es parte de la línea r, la imagen de A es sí misma, y si un punto B no está en la línea r, su imagen B' es el punto en el cual la línea r es la bisectriz perpendicular de $\overline{BB'}$.

Example

B

$A = A'$ r

B'

English	Spanish

Reflectional symmetry (p. 568) Reflectional symmetry, or *line symmetry*, is the type of symmetry for which there is a reflection that maps a figure onto itself. The reflection line is the *line of symmetry*. The line of symmetry divides a figure with reflectional symmetry into two congruent halves.

Simetría reflexiva (p. 568) Simetría reflexiva, o *simetría lineal*, es el tipo de simetría donde hay una reflexión que ubica una figura en sí misma. El eje de reflexión es el *eje de simetría*. El eje de simetría divide una figura con simetría reflexiva en dos mitades congruentes.

Example

A reflection across the given line maps the figure onto itself.

Regular polygon (p. 354) A regular polygon is a polygon that is both equilateral and equiangular. Its *center* is the point that is equidistant from its vertices.

Polígono regular (p. 354) Un polígono regular es un polígono que es equilateral y equiangular. Su *centro* es el punto equidistante de sus vértices.

Example

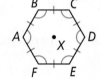

ABCDEF is a regular hexagon. Point *X* is its center.

Regular pyramid (p. 708) *See* **pyramid.**

Pirámide regular (p. 708) *Ver* **pyramid.**

Relative frequency (p. 830) The ratio of the number of times an event occurs to the total number of events in the sample space.

Frecuencia relativa (p. 830) La razón del número de veces que ocurre un evento al número de eventos en el espacio muestral.

Example

Archery Results				
Scoring Region	Yellow	Red	Blue	Black	White
Arrow Strikes	52	25	10	8	5

$$\text{Relative frequency of spinning } 1 = \frac{\text{frequency of spinning } 1}{\text{total frequencies}}$$

$$= \frac{29}{100}$$

Remote interior angles (p. 173) Remote interior angles are the two nonadjacent interior angles corresponding to each exterior angle of a triangle.

Ángulos interiores remotos (p. 173) Los ángulos interiores remotos son los dos ángulos interiores no adyacentes que corresponden a cada ángulo exterior de un triángulo.

Example

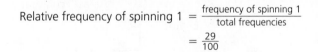

$\angle 1$ and $\angle 2$ are remote interior angles of $\angle 3$.

Rhombus (p. 375) A rhombus is a parallelogram with four congruent sides.

Rombo (p. 375) Un rombo es un paralelogramo de cuatro lados congruentes.

Example

Right angle (p. 29) A right angle is an angle whose measure is 90.

Ángulo recto (p. 29) Un ángulo recto es un ángulo que mide 90.

Example

90°

This symbol indicates a right angle.

Right cone (p. 711) *See* **cone.**

Cono recto (p. 711) *Ver* **cone.**

Right cylinder (p. 701) *See* **cylinder.**

Cilindro recto (p. 701) *Ver* **cylinder.**

Right prism (p. 699) *See* **prism.**

Prisma rectangular (p. 699) *Ver* **prism.**

Right triangle (pp. 258, 885) A right triangle contains one right angle. The side opposite the right angle is the *hypotenuse* and the other two sides are the *legs*.

Triángulo rectángulo (pp. 258, 885) Un triángulo rectángulo contiene un ángulo recto. El lado opuesto del ángulo recto es la *hipotenusa* y los otros dos lados son los *catetos*.

Example

Leg | Hypotenuse

Leg

Rigid motion (p. 545) A transformation in the plane that preserves distance and angle measure.

Movimiento rígido (p. 545) Una transformación en el plano que no cambia la distancia ni la medida del ángulo.

Example Translations, reflections, and rotations are rigid motions.

Rotation (p. 561) A rotation (*turn*) of $x°$ about a point R, called the *center of rotation*, is a transformation such that for any point V, its image is the point V', where $RV = RV'$ and $m\angle VRV' = x$. The image of R is itself. The positive number of degrees x that a figure rotates is the *angle of rotation*.

Rotación (p. 561) Una rotación (*giro*) de $x°$ sobre un punto R, llamado el *centro de rotación*, es una transformación en la que para cualquier punto V, su imagen es el punto V', donde $RV = RV'$ y $m\angle VRV' = x$. La imagen de R es sí misma. El número positivo de grados x que una figura rota es el *ángulo de rotación*.

Example

R' | V'

135°

R

V

Rotational symmetry (p. 568) Rotational symmetry is the type of symmetry for which there is a rotation of 180° or less that maps a figure onto itself.

Simetría rotacional (p. 568) La simetría rotacional es un tipo de simetría en la que una rotación de 180° o menos vuelve a trazar una figura sobre sí misma.

Example

120°

The figure has 120° rotational symmetry.

English

S

Same-side interior angles (p. 142) Same-side interior angles lie on the same side of the transversal t and between ℓ and m.

Example

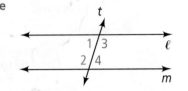

$\angle 1$ and $\angle 2$ are same-side interior angles, as are $\angle 3$ and $\angle 4$.

Sample space (p. 824) The part of a population that is surveyed.

Example Let the set of all males between the ages of 19 and 34 be the population. A random selection of 900 males between those ages would be a sample of the population.

Scale (p. 443) A scale is the ratio of any length in a scale drawing to the corresponding actual length. The lengths may be in different units.

Example 1 cm to 1 ft
1 cm = 1 ft
1 cm : 1 ft

Scale drawing (p. 443) A scale drawing is a drawing in which all lengths are proportional to corresponding actual lengths.

Example

Living room	Bedroom
	Bath

Scale:
1 in. = 30 ft

Spanish

Ángulos internos del mismo lado (p. 142) Los ángulos internos del mismo lado están situados en el mismo lado de la transversal t y dentro de ℓ y m.

Muestra (p. 824) Porción que se estudia de una población.

Escala (p. 443) Una escala es la razón de cualquier longitud en un dibujo a escala en relación a la longitud verdadera correspondiente. Las longitudes pueden expresarse en distintas unidades.

Dibujo a escala (p. 443) Un dibujo a escala es un dibujo en el que todas las longitudes son proporcionales a las longitudes verdaderas correspondientes.

English

Spanish

Scale factor (pp. 440, 742)   A scale factor is the ratio of corresponding linear measurements of two similar figures.

Factor de escala (pp. 440, 742)   El factor de escala es la razón de las medidas lineales correspondientes de dos figuras semejantes.

Example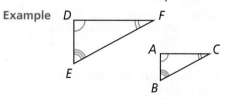

$$\triangle ABC \sim \triangle DEF$$
$$\frac{AB}{DE} = \frac{BC}{EF} = \frac{CA}{FD}$$

Scale factor of a dilation (p. 587)   The scale factor of a dilation is the ratio of the distances from the center of dilation to an image point and to its preimage point.

Factor de escala de dilatación (p. 587)   El factor de escala de dilatación es la razón de las distancias desde el centro de dilatación hasta un punto de la imagen y hasta un punto de la preimagen.

Example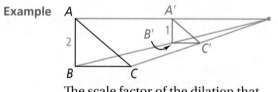

The scale factor of the dilation that maps $\triangle ABC$ to $\triangle A'B'C'$ is $\frac{1}{2}$.

Scalene triangle (p. 885)   A scalene triangle has no congruent sides.

Triángulo escaleno (p. 885)   Un triángulo escaleno no tiene lados congruentes.

Example

Secant (p. 791)   A secant is a line, ray, or segment that intersects a circle at two points.

Secante (p. 791)   Una secante es una recta, semirrecta o segmento que corta un círculo en dos puntos.

Example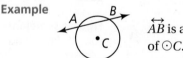

$\overleftrightarrow{AB}$ is a secant of $\odot C$.

Sector of a circle (p. 661)   A sector of a circle is the region bounded by two radii and their intercepted arc.

Sector de un círculo (p. 661)   Un sector de un círculo es la región limitada por dos radios y el arco abarcado por ellos.

Example

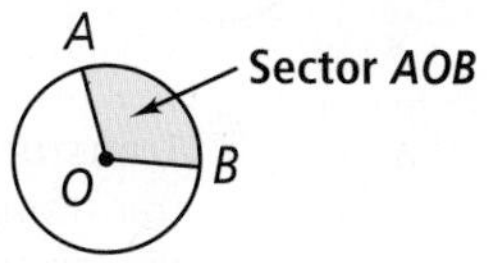

Sector *AOB*

English

Spanish

Segment (p. 12)   A segment is the part of a line that consists of two points, called *endpoints*, and all points between them.

Example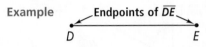

Segmento (p. 12)   Un segmento es la parte de una recta que tiene dos puntos, llamados *extremos*, entre los cuales están todos los puntos de esa recta.

Segment bisector (p. 22)   A segment bisector is a line, segment, ray, or plane that intersects a segment at its midpoint.

Example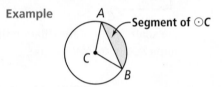

ℓ bisects $\overline{KJ}$.

Bisectriz de un segmento (p. 22)   La bisectriz de un segmento es una recta, segmento, semirrecta o plano que corta un segmento en su punto medio.

Segment of a circle (p. 662)   A segment of a circle is the part of a circle bounded by an arc and the segment joining its endpoints.

Example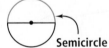

Segment of ⊙C

Segmento de un círculo (p. 662)   Un segmento de un círculo es la parte de un círculo bordeada por un arco y el segmento que une sus extremos.

Semicircle (p. 649)   A semicircle is half a circle.

Example

Semicircle

Semicírculo (p. 649)   Un semicírculo es la mitad de un círculo.

Side   *See* **angle; polygon.**

Lado   *Ver* **angle; polygon.**

Similar figures (pp. 440, 596)   Similar figures are two figures that have the same shape, but not necessarily the same size.

Example

Figuras semejantes (pp. 440, 596)   Los figuras semejantes son dos figuras que tienen la misma forma pero no necesariamente el mismo tamaño.

Similarity transformation (p. 596)   A composition of a rigid motion and a dilation.

Transformación de semejanza (p. 596)   Una transfomación que contiene un movimiento rígido y una dilatación.

English

Spanish

Similar polygons (p. 440)   Similar polygons are polygons having corresponding angles congruent and the lengths of corresponding sides proportional. You denote similarity by ∼.

Polígonos semejantes (p. 440)   Los polígonos semejantes son polígonos cuyos ángulos correspondientes son congruentes y las longitudes de los lados correspondientes son proporcionales. El símbolo ∼ significa "es semejante a".

Example

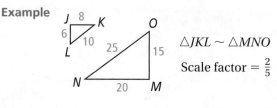

$\triangle JKL \sim \triangle MNO$

Scale factor $= \frac{2}{5}$

Similar solids (p. 742)   Similar solids have the same shape and have all their corresponding dimensions proportional.

Cuerpos geométricos semejantes (p. 742)   Los cuerpos geométricos semejantes tienen la misma forma y todas sus dimensiones correspondientes son proporcionales.

Example

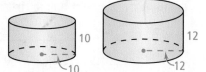

Sine ratio (p. 507)   *See* **trigonometric ratios.**

Razón seno (p. 507)   *Ver* **trigonometric ratios.**

Skew lines (p. 140)   Skew lines are lines that do not lie in the same plane.

Rectas cruzadas (p. 140)   Las rectas cruzadas son rectas que no están en el mismo plano.

Example

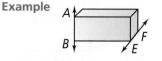

$\overleftrightarrow{AB}$ and $\overleftrightarrow{EF}$ are skew.

Slant height   *See* **cone; pyramid.**

Generatriz (cono) o apotema (pirámide)   *Ver* **cone; pyramid.**

Slope-intercept form (p. 190)   The slope-intercept form of a linear equation is $y = mx + b$, where m is the slope of the line and b is the y-intercept.

Forma pendiente-intercepto (p. 190)   La forma pendiente-intercepto es la ecuación lineal $y = mx + b$, en la que m es la pendiente de la recta y b es el punto de intersección de esa recta con el eje y.

Example $y = \frac{1}{2}x - 3$

In this equation, the slope is $\frac{1}{2}$ and the y-intercept is -3.

Slope of a line (p. 189)   The slope of a line is the ratio of its vertical change in the coordinate plane to the corresponding horizontal change. If (x_1, y_1) and (x_2, y_2) are points on a nonvertical line, then the slope is $\frac{y_2 - y_1}{x_2 - x_1}$. The slope of a horizontal line is 0 and the slope of a vertical line is undefined.

Pendiente de una recta (p. 189)   La pendiente de una recta es la razón del cambio vertical en el plano de coordenadas en relación al cambio horizontal correspondiente. Si (x_1, y_1) y (x_2, y_2) son puntos en una recta no vertical, entonces la pendiente es $\frac{y_2 - y_1}{x_2 - x_1}$. La pendiente de una recta horizontal es 0, y la pendiente de una recta vertical es indefinida.

Example

The line containing $P(-1, -1)$ and $Q(1, -2)$
has slope $\frac{-2 - (-1)}{1 - (-1)} = \frac{-1}{2} = -\frac{1}{2}$.

Space (p. 12)   Space is the set of all points.

Espacio (p. 12)   El espacio es el conjunto de todos los puntos.

Sphere (p. 733)   A sphere is the set of all points in space that are a given distance r, the *radius*, from a given point C, the *center*. A *great circle* is the intersection of a sphere with a plane containing the center of the sphere. The *circumference* of a sphere is the circumference of any great circle of the sphere.

Esfera (p. 733)   Una esfera es el conjunto de los puntos del espacio que están a una distancia dada r, el *radio*, de un punto dado C, el *centro*. Un *círculo máximo* es la intersección de una esfera y un plano que contiene el centro de la esfera. La *circunferencia* de una esfera es la circunferencia de cualquier círculo máximo de la esfera.

Example

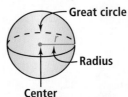

Great circle

Radius

Center

Visual **Glossary**

English

Spanish

Spherical geometry (p. 179)   In spherical geometry, a plane is considered to be the surface of a sphere and a line is considered to be a great circle of the sphere. In spherical geometry, through a point not on a given line there is no line parallel to the given line.

Geometría esférica (p. 179)   En la geometría esférica, un plano es la superficie de una esfera y una recta es un círculo máximo de la esfera. En la geometría esférica, a través de un punto que no está en una recta dada, no hay recta paralela a la recta dada.

Example   In spherical geometry, lines are represented by great circles of a sphere.

Square (p. 375)   A square is a parallelogram with four congruent sides and four right angles.

Cuadrado (p. 375)   Un cuadrado es un paralelogramo con cuatro lados congruentes y cuatro ángulos rectos.

Example

Standard form of an equation of a circle (p. 799)   The standard form of an equation of a circle with center (h, k) and radius r is $(x - h)^2 + (y - k)^2 = r^2$.

Forma normal de la ecuación de un círculo (p. 799)   La forma normal de la ecuación de un círculo con un centro (h, k) y un radio r es $(x - h)^2 + (y - k)^2 = r^2$.

Example   In $(x + 5)^2 + (y + 2)^2 = 48$, $(-5, -2)$ is the center of the circle.

Straight angle (p. 29)   A straight angle is an angle whose measure is 180.

Ángulo llano (p. 29)   Un ángulo llano es un ángulo que mide 180.

Example

$m\angle AOB = 180$

Straightedge (p. 43)   A straightedge is a ruler with no markings on it.

Regla sin graduación (p. 43)   Una regla sin graduación no tiene marcas.

Supplementary angles (p. 34)   Two angles are supplementary if the sum of their measures is 180.

Ángulos suplementarios (p. 34)   Dos ángulos son suplementarios cuando sus medidas suman 180.

Example

$\angle MNP$ and $\angle ONP$ are supplementary, as are $\angle MNP$ and $\angle QRS$.

Visual **Glossary**

Surface area (pp. 700, 702, 709, 711, 734) The surface area of a prism, cylinder, pyramid, or cone is the sum of the lateral area and the areas of the bases. The surface area of a sphere is four times the area of a great circle. A list of surface area formulas is on p. 839.

Área (pp. 700, 702, 709, 711, 734) El área de un prisma, pirámide, cilindro o cono es la suma del área lateral y las áreas de las bases. El área de una esfera es igual a cuatro veces el área de un círculo máximo. Una lista de fórmulas de áreas está en la p. 839.

Example

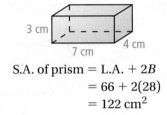

$$\text{S.A. of prism} = \text{L.A.} + 2B$$
$$= 66 + 2(28)$$
$$= 122 \text{ cm}^2$$

Symmetry (p. 568) A figure has symmetry if there is an isometry that maps the figure onto itself. *See also* **point symmetry; reflectional symmetry; rotational symmetry.**

Simetría (p. 568) Una figura tiene simetría si hay una isometría que traza la figura sobre sí misma. *Ver también* **point symmetry; reflectional symmetry; rotational symmetry.**

Example

A regular pentagon has reflectional symmetry and 72° rotational symmetry.

T

Tangent ratio (p. 507) *See* **trigonometric ratios.**

Razón tangente (p. 507) *Ver* **trigonometric ratios.**

Tangent to a circle (p. 762) A tangent to a circle is a line, segment, or ray in the plane of the circle that intersects the circle in exactly one point. That point is the *point of tangency.*

Tangente de un círculo (p. 762) Una tangente de un círculo es una recta, segmento o semirrecta en el plano del círculo que corta el círculo en exactamente un punto. Ese punto es el *punto de tangencia.*

Example

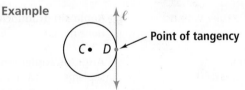

Line ℓ is tangent to $\odot C$. Point D is the point of tangency.

Theorem (p. 120) A theorem is a conjecture that is proven.

Teorema (p. 120) Un teorema es una conjetura que se demuestra.

Example The theorem "Vertical angles are congruent" can be proven by using postulates, definitions, properties, and previously stated theorems.

English

Theoretical probability (p. 825) The ratio of the number of favorable outcomes to the number of possible outcomes if all outcomes have the same chance of happening.

$$P(\text{event}) = \frac{\text{number of favorable outcomes}}{\text{number of possible outcomes}}$$

Example In tossing a coin, the events of getting heads or tails are equally likely. The likelihood of getting heads is $P(\text{heads}) = \frac{1}{2}$.

Transformation (p. 545) A transformation is a change in the position, size, or shape of a geometric figure. The given figure is called the *preimage* and the resulting figure is called the *image*. A transformation *maps* a figure onto its image. *Prime notation* is sometimes used to identify image points. In the diagram, X' (read "X prime") is the image of X.

Translation (p. 547) A translation (*slide*) is a transformation that moves points the same distance and in the same direction.

Transversal (p. 141) A transversal is a line that intersects two or more lines at distinct points.

Spanish

Probabilidad teórica (p. 825) Si cada resultado tiene la misma probabilidad de darse, la probabilidad teórica de un suceso se calcula como la razón del número de resultados favorables al número de resultados posibles.

$$P(\text{suceso}) = \frac{\text{numero de resultados favorables}}{\text{numero de resultados posibles}}$$

Transformación (p. 545) Una transformación es un cambio en la posición, tamaño o forma de una figura. La figura dada se llama la *preimagen* y la figura resultante se llama la *imagen*. Una transformación *traza* la figura sobre su propia imagen. La *notación prima* a veces se utilize para identificar los puntos de la imagen. En el diagrama de la derecha, X' (leído X prima) es la imagen de X.

Example

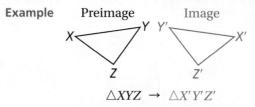

$$\triangle XYZ \; \rightarrow \; \triangle X'Y'Z'$$

Traslación (p. 547) Una traslación (*desplazamiento*) es una transformación en la que se mueven puntos la misma distancia en la misma dirección.

Example

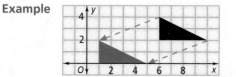

The blue triangle is the image of the black triangle under the translation $\langle -5, -2 \rangle$.

Transversal (p. 141) Una transversal es una línea que interseca dos o más líneas en puntos precisos.

Example

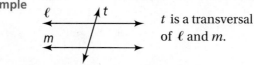

t is a transversal of ℓ and m.

English

Spanish

Trapezoid (p. 389)   A trapezoid is a quadrilateral with exactly one pair of parallel sides, the *bases*. The nonparallel sides are called the *legs* of the trapezoid. Each pair of angles adjacent to a base are *base angles* of the trapezoid. An *altitude* of a trapezoid is a perpendicular segment from one base to the line containing the other base. Its length is called the *height* of the trapezoid.

Trapecio (p. 389)   Un trapecio es un cuadrilátero con exactamente un par de lados paralelos, las *bases*. Los lados no paralelos se llaman los *catetos* del trapecio. Cada par de ángulos adyacentes a la base son los *ángulos de base* del trapecio. Una *altura* del trapecio es un segmento perpendicular que va de una base a la recta que contiene la otra base. Su longitud se llama, por extensión, la *altura* del trapecio.

Example

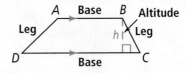

In trapezoid *ABCD*, ∠*ADC* and ∠*BCD* are one pair of base angles, and ∠*DAB* and ∠*ABC* are the other.

Triangle (pp. 57, 618)   A triangle is a polygon with three sides. You can choose any side to be a *base*. The *height* is the length of the altitude drawn to the line containing that base.

Triángulo (pp. 57, 618)   Un triángulo es un polígono con tres lados. Se puede escoger cualquier lado como *base*. La *altura*, entonces, es la longitud de la altura trazada hasta la recta que contiene la base.

Example

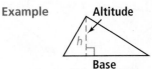

Trigonometric ratios (p. 507)   In right △*ABC* with acute ∠*A*,

$$\text{sine } \angle A = \sin A = \frac{\text{leg opposite } \angle A}{\text{hypotenuse}}$$

$$\text{cosine } \angle A = \cos A = \frac{\text{leg adjacent to } \angle A}{\text{hypotenuse}}$$

$$\text{tangent } \angle A = \tan A = \frac{\text{leg opposite } \angle A}{\text{leg adjacent to } \angle A}$$

Razones trigonométricas (p. 507)   En un triángulo rectángulo △*ABC* con ángulo agudo ∠*A*,

$$\text{seno } \angle A = \text{sen } A = \frac{\text{cateto opuesto a } \angle A}{\text{hipotenusa}}$$

$$\text{coseno } \angle A = \cos A = \frac{\text{cateto adyecente a } \angle A}{\text{hipotenusa}}$$

$$\text{tangente } \angle A = \tan A = \frac{\text{cateto opuesto a } \angle A}{\text{cateto adyecente a } \angle A}$$

Example

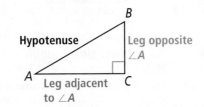

English

Truth table (p. 97) A truth table is a table that lists all the possible combinations of truth values for two or more statements.

Example

p	q	$p \rightarrow q$
T	T	T
T	F	F
F	T	T
F	F	T

Truth value (p. 90) The truth value of a statement is "true" or "false" according to whether the statement is true or false, respectively.

Two-column proof (p. 115) *See* **proof.**

Two-way frequency table (p. 850) A table that displays frequencies in two different categories.

Example

	Male	Female	Totals
Juniors	3	4	7
Seniors	3	2	5
Totals	6	6	12

The last column shows a total of 7 juniors and 5 seniors.
The last row shows a total of 6 males and 6 females.

Spanish

Tabla de verdad (p. 97) Una tabla de verdad es una tabla que muestra todas las combinaciones posibles de valores de verdad de dos o más enunciados.

Valor verdadero (p. 90) El valor verdadero de un enunciado es "verdadero" o "falso" según el enunciado sea *verdadero* o falso, respectivamente.

Prueba de dos columnas (p. 115) *Ver* **proof.**

Tabla de frecuencias de doble entrada (p. 850) Una tabla de frecuencias que contiene dos categorías de datos.

V

Vertex *See* **angle; cone; polygon; polyhedron; pyramid.** The plural form of *vertex* is *vertices*.

Vértice *Ver* **angle; cone; polygon; polyhedron; pyramid.**

Vertex angle (p. 250) *See* **isosceles triangle.**

Ángulo del vértice (p. 250) *Ver* **isosceles triangle.**

Vertical angles (p. 34) Vertical angles are two angles whose sides form two pairs of opposite rays.

Ángulos opuestos por el vértice (p. 34) Dos ángulos son ángulos opuestos por el vértice si sus lados son semirrectas opuestas.

Example 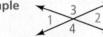 ∠1 and ∠2 are vertical angles, as are ∠3 and ∠4.

Volume (p. 717) Volume is a measure of the space a figure occupies. A list of volume formulas is on p. 898.

Volumen (p. 717) El volumen es una medida del espacio que ocupa una figura. Una lista de las fórmulas de volumen está en la p. 898.

Selected Answers

Chapter 1

Get Ready! p. 1

1. 9 **2.** 16 **3.** 121 **4.** 37 **5.** 78.5 **6.** 13 **7.** 1 **8.** $-\frac{3}{5}$
9. 5 **10.** 8 **11.** 4 **12.** 3 **13.** 3 **14.** 6 **15.** 1
16. Answers may vary. Sample: building or making a geometric object, possibly involving several steps
17. Answers may vary. Sample: a point that falls exactly in the middle of a geometric object **18.** Answers may vary. Sample: a type of line that has a source and no ending point **19.** Answers may vary. Sample: part of the same line

Lesson 1-1 pp. 4–10

Got It? 1. E, C
2a. Answers may vary. Sample:

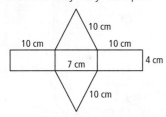

b. Yes; answers may vary. Sample:

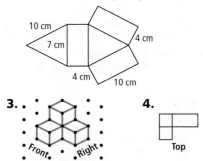

3.

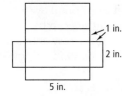

4.

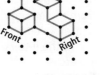

Lesson Check

1. Answers may vary. Sample:

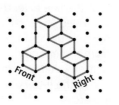

2.

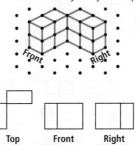

3.

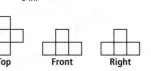

4a. net **b.** orthographic **c.** isometric **d.** none
5. Answers may vary. Sample: In an isometric drawing, you see three sides of a figure from one corner view. In an orthographic drawing, you see three separate views of the figure. In both drawings, you see the same three sides of the figure (top, front, and right). Also, both drawings represent a three-dimensional object in two dimensions.

Exercises 7. A

9. Answers may vary. Sample:

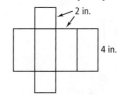

11. Answers may vary. Sample:

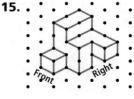

13.

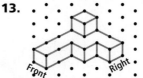

15.

17.

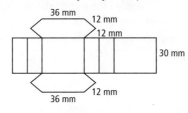

19.

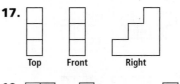

21a. Answers may vary. Sample:

b.

23. Answers may vary. Sample: Dürer may have thought that the printed pattern resembled a fishing net. **25.** C
27. Miquela
29.

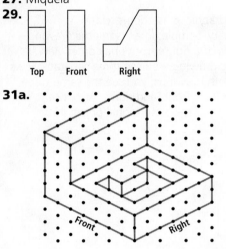

Top
Front
Right

31a.

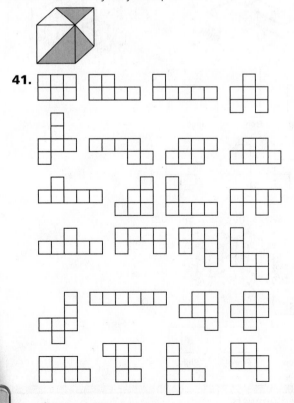

Front
Right

b.

Top
Front
Right

33. Answers may vary. Sample: for a tourist map showing locations of attractions **35.** green
37. purple

39. Answers may vary. Sample:

41.

43. C
45.

y
2
(−3, 0)
−4
O
2
x
(0, −1.5)
x + 2y = −3
−4

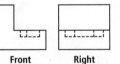

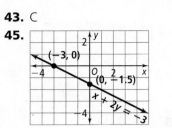

46. $DE = 31$ mm, $EF = 41$ mm **47.** $m\angle D = 60$, $m\angle E = 80$, $m\angle F = 40$
48. Answers may vary. Sample:

5 cm
6 cm

49.

y
4
2
−2
O
2
x
−2

50.

y
4
2
x
−4
−2
O
2

51.

y
x
−4
−2
O
2
4
−2
−4

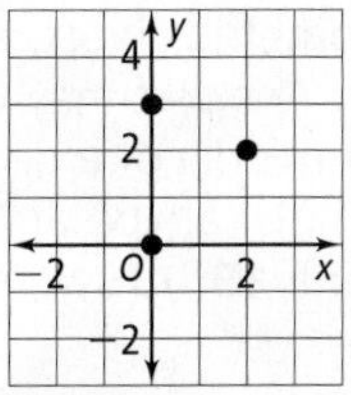

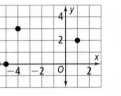

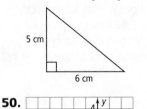

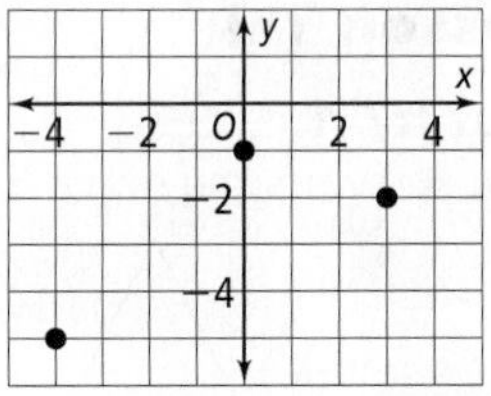

Lesson 1-2 pp. 11–19

Got It? 1a. Answers may vary. Sample: $\overleftrightarrow{RQ}$, $\overleftrightarrow{QS}$
b. Answers may vary. Sample: plane *RVS*, plane *VQS*
c. *N*, *Q*, *T* **d.** *N*, *T* **2.** No; they do not have the same endpoint. **3a.** Answers may vary. Sample: plane *BFE*, plane *BFG* **b.** Postulate 1-3 says that two distinct planes intersect in exactly one line, so you only need two points to name the line of intersection, by Postulate 1-1.
4a.

M
L
J
Q
N
P

b. Answers may vary. Sample: $\overleftrightarrow{JM}$

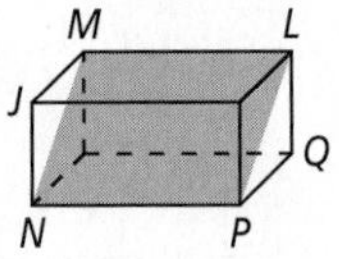

Lesson Check 1. Answers may vary. Sample: $\overleftrightarrow{XR}$, $\overleftrightarrow{RY}$
2. $\overrightarrow{RX}$, $\overrightarrow{RY}$ **3.** $\overleftrightarrow{RS}$ **4.** $\overline{RS}$, $\overline{SR}$ **5.** No; they have different endpoints and extend in opposite directions. **6.** to show that the line extends in both directions **7.** To name both, you need to identify two points on the ray or line. For a ray, you use a single-sided arrow that must point away from the endpoint. For a line, the two letters can be written in either order and a double-sided arrow appears above the letters. A line can also be named with a single lowercase letter, but a ray cannot.
Exercises 9. Answers may vary. Sample: plane *EBG*, plane *BFG* **11.** *E*, *B*, *F*, *G* **13.** $\overrightarrow{RS}$, $\overrightarrow{SR}$, $\overrightarrow{ST}$, $\overrightarrow{TS}$, $\overrightarrow{TW}$, $\overrightarrow{WT}$, $\overrightarrow{TR}$, $\overrightarrow{RT}$, $\overrightarrow{WR}$, $\overrightarrow{RW}$, $\overrightarrow{WS}$, $\overrightarrow{SW}$ **15.** $\overleftrightarrow{RS}$ **17.** $\overleftrightarrow{UV}$ **19.** plane *QUX*, plane *QUV* **21.** plane *XTQ*, plane *XTS*

23.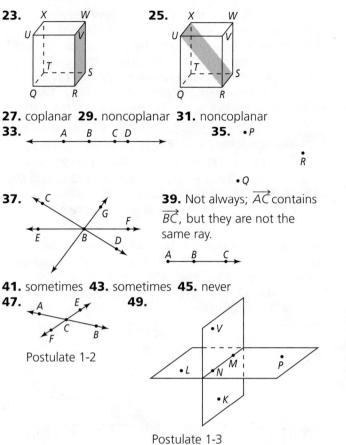

25.

27. coplanar **29.** noncoplanar **31.** noncoplanar

33.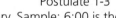

A B C D

35. • P

• R

• Q

37.

39. Not always; $\overleftrightarrow{AC}$ contains $\overrightarrow{BC}$, but they are not the same ray.

A   B   C

41. sometimes **43.** sometimes **45.** never

47.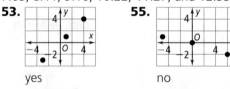

49.

Postulate 1-2

Postulate 1-3

51. Answers may vary. Sample: 6:00 is the only "exact" time. Other times are about 1:38, 2:43, 3:49, 4:54, 5:59, 7:05, 8:11, 9:16, 10:22, 11:27, and 12:33.

53. yes

55. no

57. yes

59. Infinitely many; answers may vary. Sample: The three collinear points are contained in one line. There are infinitely many planes that can intersect in that line.

61a. Answers may vary. Sample: Since the plane is flat, the line would have to curve in order to contain the two points and not lie in the plane, but lines are straight, so the line must also be in plane P.

b. One; points A, B, and C are noncollinear. By Postulate 1-4, they are coplanar. Thus, by part (a), $\overleftrightarrow{AB}$ and $\overleftrightarrow{BC}$ are coplanar.

63. $\frac{1}{4}$ **65.** D **67.** A19

69.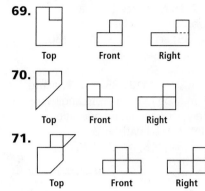

Top   Front   Right

70.

Top   Front   Right

71.

Top   Front   Right

72. 5 to 2 **73.** $\frac{3}{7}$ **74.** $\frac{n+1}{4}$ **75.** 6 **76.** 3.5 **77.** 3 **78.** 4
79. 9 **80.** $\frac{1}{3}$

Lesson 1-3   pp. 20–26

Got It? 1. $UV = 4$, $SV = 18$ **2.** $JK = 42$, $KL = 78$
3a. no **b.** yes; $|5 - (-2)| = |7| = 7$ **4a.** No; since $PQ = QR$, when you solve and get PQ, you know QR.
b. $TU = 35$, $UV = 35$, $TV = 70$
Lesson Check 1. B **2.** A, G **3.** 0 **4.** Answers may vary.
Sample: $\overline{BD}$ **5.** line ℓ, point Q **6.** Answers may vary.
Sample: You would use "congruent" when you are referring to a segment, for example, when describing the trusses of a bridge. You would use "equal length" when you are referring to the measurement of a segment, for example, when describing the distance between two buildings. **7.** Answers may vary. Sample: Distance is always a nonnegative measure because it is the absolute value of the difference of two values.
Exercises 9. 9 **11.** 6 **13.** 25 **15.** no **17.** yes **19a.** 9
b. $AY = 9$, $XY = 18$ **21.** 34 **23.** $XY = 4$, $ZW = 4$;
congruent **25.** $YZ = 4$, $XW = 12$; not congruent
27. −3.5 or 3.5 **29.** −2 or 8 **35.** about 1 h, 21 min
37. The distance is $|65 - 80|$, or 15 mi. The driver added the values instead of subtracting them. **39.** $y = 15$;
$AC = 24$, $DC = 12$ **41.** Not always; the Segment Addition Postulate can be used only if P, Q, and R are collinear points.
43a. $(2x + 3) - x + (4x - 3)$, or $5x$
b. $GH = 9$, $JK = 15$ **45.** G
47.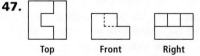

Top   Front   Right

48. always **49.** always **50.** never **51.** always **52a.** yes
b. no **c.** no **d.** yes **53.** 14 **54.** 6.5 **55.** −3 **56.** 12.8

Lesson 1-4   pp. 27–33

Got It? 1a. ∠LMK, ∠2 **b.** No; since there are three ∡ that have vertex M, it would not be clear which one you intended. **2.** m∠LKH = 35, acute; m∠HKN = 180, straight; m∠MKH = 145, obtuse **3.** 49
4. m∠DEC = 142, m∠CEF = 38
Lesson Check 1. ∠ABC, ∠CBA **2.** 85 − x **3.** acute
4. 0 or 1; congruent ∡ may be two separate angles, or they may have the same vertex and share one side.
5. No; the diagram is not marked with ≅ ∡.
Exercises 7. ∠ABC, ∠CBA, ∠B, or ∠1 **9.** 70, acute
11. 110, obtuse **13.** 85, acute
15. Answers may vary. Sample: **17.**

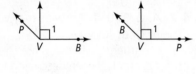

19. ∠BJA **21.** 130 **23.** m∠RQS = 43, m∠TQS = 137
25. about 90°; right **27.** about 88°; acute **29.** x = 8; m∠AOB = 30, m∠BOC = 50, m∠COD = 30
31. A **33.** 180 **35.** 30 **37.** 40
39a. yes;

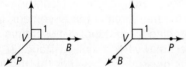

41. B **43a.** 4 **b.** 52 **44.** 47 **45.** x = 8, EF = 19, FG = 30 **46.** 2x + 4 = 28; 12 **47.** 90 − x = 3x; 22.5
48. x + 5x = 180; 30, 150 **49.** 65

Lesson 1-5   pp. 34–40

Got It? 1a. Yes; ∠AFE and ∠CFD are formed by opposite rays $\overrightarrow{FA}$, $\overrightarrow{FD}$, $\overrightarrow{FC}$, and $\overrightarrow{FE}$. **b.** No; m∠BFC = 28 and m∠DFE = 118, so 28 + 118 ≠ 180. **c.** Yes; ∠BFD and ∠AFB share $\overrightarrow{FB}$, and they have no common interior points. **2a.** Yes; they have corresponding ≅ tick marks.
b. No; they do not have corresponding ≅ tick marks.
c. No; it (or its supplements) do not have a right angle symbol. **d.** No; $\overline{PW}$ and $\overline{WQ}$ do not have corresponding ≅ tick marks. **3a.** Adding the measures of both ∡ should give 180. **b.** m∠ADB = 77, m∠BDC = 103 **4.** 36
Lesson Check 1–3. Answers may vary. Samples are given.
1. ∠AFE and ∠CFD (or ∠AFC and ∠EFD) **2.** ∠AEF and ∠DEF (or ∠AEC and ∠DEC) **3.** ∠BCE and ∠ECD (or any two adjacent ∡ with common vertex F) **4.** 20
5. Answers may vary. Sample: The angles combine to

form a line. **6.** Since the ∡ are complementary, the sum of the two measures should be 90, not 180. So, x = 15.
Exercises 7. Yes; the angles share a common side and vertex, and have no interior points in common. **9.** No; they are supplementary. **11.** ∠DOC, ∠AOB **13.** ∠EOC
15. Answers may vary. Sample: ∠AOB, ∠DOC **17.** No; they are not marked as ≅. **19.** Yes. Answers may vary. Sample: The two ∡ form a linear pair. **21.** No; $\overline{JC}$ and $\overline{CD}$ are not marked as ≅. **23.** Yes; they are formed by $\overleftrightarrow{JF}$ and $\overleftrightarrow{ED}$. **25.** m∠EFG = 69, m∠GFH = 111 **27.** x = 5, m∠ABC = 50 **29.** x = 11, m∠ABC = 56 **31.** 120; 60
33. 90 **35.** 155 **37a.** 19.5 **b.** m∠RQS = 43, m∠TQS = 137 **c.** Answers may vary. Sample: 43 + 137 = 180 **39.** Both are correct; if you multiply both sides of the equation m∠ABX = $\frac{1}{2}$m∠ABC by 2, you get 2m∠ABX = m∠ABC. **41.** The four vertical angles are all right angles. **43.** ∠KML **45.** ∠PMR, ∠KML, ∠KMQ, ∠MQP **47.** 30 **49.** l **51.** ∠WXY **52.** ∠WXZ, ∠YXZ
53. 39 **54–59.** Answers may vary. Samples are given.

54. G ⟶ H

55. C — D

56. ⟵ A — B ⟶

57. (figure with points A, B, C)

58. P, S, T (figure)

59. ⟵ X — Y — Z ⟶

Lesson 1-6   pp. 43–48

Got It?
1. X •——• Y

R —|—— S

2a. (figures with B, F)

b. Answers may vary. Sample: You use a compass setting to copy a distance.
3. (figure S, T)   **4.** (figure X, P, Y, Z)

Lesson Check
1. PQ (figure X, Y)   **2.** (figure P, Q)

3.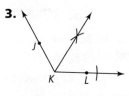

4. compass, straightedge
5. Answers may vary. Sample: When you sketch a figure, it does not require accurate measurements for angles and sides. When you draw a figure with a ruler and protractor, you use measurements to determine the lengths of sides or the sizes of angles. When you construct a figure, the only tools you use are a compass and straightedge. **6.** Since $\overleftrightarrow{XY}$ is ⊥ to and contains the midpoint of $\overline{AB}$, then $\overleftrightarrow{XY}$ is the ⊥ bis. of $\overline{AB}$, not the other way around.

Exercises

7.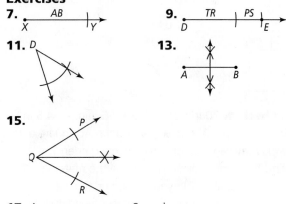

9.

11. D

13.

15.

17. Answers may vary. Sample:

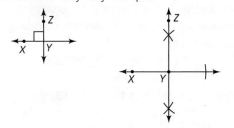

Find a segment on $\overleftrightarrow{XY}$ so that you can construct $\overleftrightarrow{YZ}$ as its perpendicular bisector.
19. Answers may vary. Sample: Both constructions involve drawing arcs with the same radius from two different points, and using the point(s) of intersection of those arcs. Arcs must intersect at two points for the ⊥ bis., but only one point for the ∠ bis. **21a.** A segment has exactly one midpoint; using the Ruler Postulate (Post. 1-5), each point corresponds with exactly one number, and exactly one number represents half the length of a segment.
b. A segment has infinitely many bisectors because many lines can be drawn through the midpoint. **c.** In the plane with the segment, there is one ⊥ bis. because only one line in that plane can be drawn through the midpoint so that it forms a right angle with the given segment.
d. Consider the plane that is the ⊥ bis. of the segment. Any line in that plane that contains the midpoint of the segment is a ⊥ bis. of the segment, and there are infinitely many such lines.

23.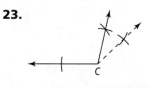

25a. With *P* as center, draw an arc with radius slightly more than $\frac{1}{2}PQ$. Keeping that radius, draw an arc with *Q* as center. Those two arcs meet at 2 points; the line through those 2 points intersects $\overline{PQ}$ at its midpoint. **b.** Follow the steps in part (a) to find the midpoint *C* of $\overline{PQ}$. Then repeat the process for segments $\overline{PC}$ and $\overline{CQ}$.

27. possible

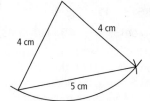

29. Not possible; the two 2-cm sides do not meet.
31a. 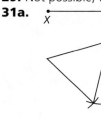 **b.** The measure of each angle is 60°. **c.** Draw an angle congruent to one of the angles of the triangle from part (a) to get a 60° ∠. Then construct its angle bisector to get two 30° ▵.
33. ⊥; contains the intersection of that line with the plane

35.

In the angle bisector construction, $\overline{AB} \cong \overline{AC}$, $\overline{BD} \cong \overline{CD}$, and $\overline{AD} \cong \overline{AD}$. Using the statement that two triangles are ≅ if three pairs of sides are ≅, then $\triangle ABD \cong \triangle ACD$. Since the ▵ are ≅, each ∠ of one △ is ≅ to an ∠ of the other △. So, ∠*BAD* ≅ ∠*CAD* and $\overrightarrow{AD}$ is the ∠ bisector of ∠*BAC*.
37. I **39.** 116 **40.** yes; $m\angle TUV + m\angle VUW = 180$
41. 6 **42.** 10 **43.** 4 **44.** 3 **45.** 196 **46.** 10 **47.** −1

Lesson 1-7 pp. 50–56

Got It? 1a. −4 **b.** (4, −2) **2.** (11, −13) **3a.** 15.8
b. Yes; the diff. of the coordinates are opposite, but their squares are the same. $VU = \sqrt{(-11)^2 + 8^2} = \sqrt{185} = 13.6$ **4.** $\sqrt{1325}$ m, or about 36.4 m

Lesson Check 1. (0.5, 5.5), or $\left(\frac{1}{2}, \frac{11}{2}\right)$ **2.** (7, −8)
3. $\sqrt{73}$, or about 8.5 units **4.** Answers may vary. Sample: For two different points, the expression $(x_2 - x_1)^2 + (y_2 - y_1)^2$ in the Distance Formula is always positive. So the positive square root of a positive number is

positive. **5.** He did not keep the *x*-value and *y*-values together; so, $d = \sqrt{(1-3)^2 + (5-8)^2} = \sqrt{4+9} = \sqrt{13}$ units.

Exercises 7. −1.5, or $-\frac{3}{2}$ **9.** −10 **11.** (3, 1) **13.** (6, 1) **15.** $\left(3\frac{7}{8}, -3\right)$ **17.** (5, −1) **19.** (12, −24) **21.** (5.5, −13.5) **23.** 18 **25.** 9 **27.** 10 **29.** 12.2 **31.** 8.2 **33.** 8.5 **35.** Everett, Charleston, Brookline, Fairfield, Davenport **37a.** 5.8 **b.** $\left(\frac{3}{2}, \frac{1}{2}\right)$, or (1.5, 0.5) **39a.** 5.4 **b.** $\left(-\frac{5}{2}, 3\right)$, or (−2.5, 3) **41a.** 2.8 **b.** (−4, −4) **43a.** 5.4 **b.** $\left(3, \frac{1}{2}\right)$, or (3, 0.5) **45.** 165 units; flying *T* to *V* then to *U* is shortest distance. **47a.** Answers may vary. Sample: Distance Formula (Find *KP*, then divide it by 2.) **b.** Answers may vary. Sample: Distance Formula (If *M* is the given midpoint, find *KM* and then multiply it by 2.) **49a.** 10.7 **b.** (3, −4)

51a.

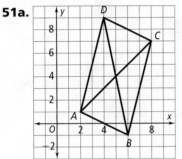

The midpoints are the same, (5, 4).
b. Answers may vary. Sample: The diagonals bisect each other. **53.** 7 mi **55.** 3.2 mi **57a.** Answers may vary. Sample: (0, 2) and (4, 2); (2, 0) and (2, 4); (0, 4) and (4, 0); (0, 0) and (4, 4) **b.** Infinitely many; draw a circle with center (2, 2) and radius 4. Any diameter of that circle has length 8 and midpoint (2, 2).
59. *A*(0, 0, 0), *B*(6, 0, 0), *C*(6, −3, 0), **61.** 11.7 units **63.** F
D(0, −3, 0), *E*(0, 0, 9), *F*(6, 0, 9),
G(0, −3, 9)

65.

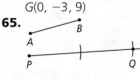

66.

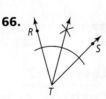

67. ∠*PQR*, ∠*RQP* **68.** 150 **69.** $10\frac{5}{6}$ **70.** 504 **71.** 9 **72.** 10,560

Review   pp. 57–58

1. yes **3.** no; not a plane figure **5.** Sample: *FBWMX*; sides are $\overline{FB}, \overline{BW}, \overline{WM}, \overline{MX}, \overline{XF}$; angles are ∠*F*, ∠*B*, ∠*W*, ∠*M*, ∠*X*. **7.** Sample: *AGNHEPT*; sides are $\overline{AG}, \overline{GN}, \overline{NH}, \overline{HE}, \overline{EP}, \overline{PT}, \overline{TA}$; angles are ∠*A*, ∠*G*, ∠*N*, ∠*H*, ∠*E*, ∠*P*, ∠*T*. **9.** nonagon or enneagon, convex

Lesson 1-8   pp. 59–67

Got It? 1a. 24 in. **b.** 32 in. **2a.** 48π m **b.** 75.4 m

3.

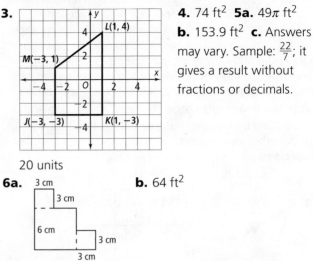

4. 74 ft² **5a.** 49π ft²
b. 153.9 ft² **c.** Answers may vary. Sample: $\frac{22}{7}$; it gives a result without fractions or decimals.

20 units

6a.

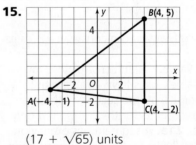

b. 64 ft²

Lesson Check 1. 20 in.; 21 in.² **2a.** 56.5 in.; 254.5 in.²
b. 22.9 m; 41.9 m² **3.** $(12 + 2\sqrt{2})$ units; 10 square units
4. Answers may vary. Sample: To fence a garden you would find the perimeter; to determine the material needed to make a tablecloth you would find the area.
5. Answers may vary. Sample: Remind your friend that 2π*r* has only one variable, so it must compute the circumference. π*r*² has one variable squared, and square units indicate area.
6. The classmate seems to have forgotten to multiply *r*² by π. The correct answer is $A = \pi r^2 = \pi(30)^2 = 900\pi \approx 2827.4$ in.².
Exercises 7. 22 in. **9.** 38 ft **11.** 10π ft **13.** $\frac{\pi}{2}$ m

15.

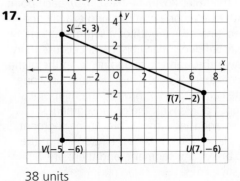

$(17 + \sqrt{65})$ units

17.

38 units

19. 4320 in.², or $3\frac{1}{3}$ yd² **21.** 8000 cm², or 0.8 m²
23. 400π m² **25.** $\frac{3969}{400}\pi$ ft² **27.** 153.9 ft²
29. 452.4 cm² **31.** 310 m² **33.** 208 ft² **35.** Perimeter;
the crown molding must fit the edges of the ceiling.
37. Area; the floor is a surface. **39a.** 144 in.²; 1 ft²
b. 144 **41.** 16 cm **43.** 96 cm² **45.** 27 in.² **47a.** Yes;
substitute s for each of a and b to get perimeter,
$P = 2s + 2s$ or $P = 4s$. **b.** No; we need to know the
length and width of a rectangle to find its perimeter.
c. $A = \frac{p^2}{16}$ **49.** $\frac{25}{4}\pi$ units²
51.

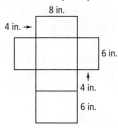

10 units, 4 square units
53a. Answers may vary. Sample:

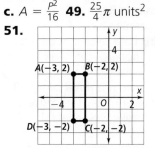

b. 208 in.²; 208 in.² **55.** \$35.70 **57.** $\frac{3a}{20}$ square units
59. $(10x^2 + \frac{7}{2}xy - 3y^2)$ square units **61.** 1104 ft²
63. 27.9 **64a.** 8.5 units **b.** $\left(\frac{11}{2}, 5\right)$, or (5.5, 5)
65a. 5.8 units **b.** $\left(-\frac{3}{2}, \frac{11}{2}\right)$, or (−1.5, 5.5) **66a.** 6.7 units
b. $\left(-\frac{5}{2}, -2\right)$, or (−2.5, −2) **67.** 90 **68.** $\overline{WK}$, $\overline{KR}$
69a. $1^2 = 1$, $2^2 = 4$, $3^2 = 9$, $4^2 = 16$, $5^2 = 25$,
$6^2 = 36$, $7^2 = 49$, $8^2 = 64$, $9^2 = 81$, $10^2 = 100$
b. It is odd.

Chapter Review pp. 70–74

1. angle bisector **2.** perpendicular lines **3.** net
4. complementary angles **5.** 4, 6, 11
6.

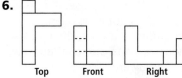

Top   Front   Right

7. Answers may vary. Sample: $\overleftrightarrow{QA}$ and $\overleftrightarrow{AB}$ **8.** $\overleftrightarrow{QR}$
9. Answers may vary. Sample: A, B, C **10.** True;

Postulate 1-1 states, "Through any two points, there is
exactly one line." **11.** False; they have different
endpoints. **12.** −7, 3 **13.** $\frac{1}{2}$ or 0.5 **14.** 15 **15.** $XY = 21$,
$YZ = 29$ **16.** acute **17.** right **18.** 36 **19.** 14 **20–24.**
Answers may vary. Samples are given. **20.** $\angle ADB$ and
$\angle BDC$ **21.** $\angle ADB$ and $\angle BDF$ **22.** $\angle ADC$ and
$\angle EDF$ **23.** $\angle ADC$ and $\angle ADE$ **24.** 31 **25.** 15

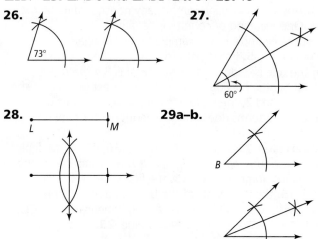

30. 1.4 units **31.** 7.6 units **32.** 14.4 units **33.** (0, 0)
34. 7.2 units **35.** (6, −2) **36.** (1, 1) **37.** (−6, −7)
38. 32 cm; 64 cm² **39.** 32 in.; 40 in.² **40.** 6π in.;
9π in.² **41.** 15π m; $\frac{225}{4}\pi$ m²

Chapter 2

Get Ready! p. 79

1. 50 **2.** −3 **3.** 25.5 **4.** 10.5 **5.** 15 **6.** 11 **7.** 7 **8.** 5
9. 6 **10.** 20 **11.** 18 **12.** $\angle ACD$, $\angle DCA$ **13.** 3
14. $m\angle 1 = 48$, $m\angle 2 = 42$ **15.** $\angle ADC$ and $\angle CDB$
16. $\angle 1$ and $\angle 2$ **17.** $\angle ADB$ or $\angle BDA$ **18.** Answers may
vary. Sample: Similar: They are both statements you start
with. Different: In geometry you do not try to prove the
hypothesis of a statement. **19.** Answers may vary.
Sample: A conclusion in geometry answers questions
raised by the hypothesis. **20.** Answers may vary. Sample:
In geometry you use deductive reasoning to draw
conclusions from other information.

Lesson 2-1 pp. 82–88

Got It? 1a. 25, 20 **b.**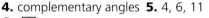

2. Every 3rd term is B, so the 21st term will be B. **3.** The
sum of the first 30 odd numbers is 30^2, or 900. **4a.** Sales
will be about 500 fewer than 8000, or 7500. **b.** No; sales
may increase because students may want backpacks for
school. **5a–c.** Answers may vary. Samples are given. **a.** A

carnation can be red, and it is not a rose. **b.** When three points are collinear, the number of planes that can be drawn through them is infinite. **c.** When you multiply 5 (or any odd number) by 3, the product is not divisible by 6.

Lesson Check 1. 31, 37 **2.**

G	P		B	G
B	R		R	P

3. Answers may vary. Sample: any nonsquare rectangle
4. One meaning of *counter* is "against," so a counterexample is an example that goes against a statement. **5.** In the pattern 2, 4, . . . , the next term is 6 if the rule is "add 2"; the next term is 8 if the rule is "double the previous term"; and the next term is 7 if the rule is "add 2, then add 3, then add 4, . . ." Just giving the first 2 terms does not give enough information to describe the pattern.
Exercises 7. Find the next square; 36, 49. **9.** Multiply the previous term by $\frac{1}{2}$; $\frac{1}{16}$, $\frac{1}{32}$. **11.** Subtract 3 from the previous number; 3, 0. **13.** the first letter of the months; J, J **15.** the Presidents of the U.S.; Madison, Monroe
17. state postal abbreviations in alphabetical order; CO, CT **19.** **21.** blue **23.** blue

25–30. Answers may vary. Samples are given. **25.** The sum of the first 100 positive odd numbers is 100^2, or 10,000. **27.** The sum of two odd numbers is even.
29. The product of two even numbers is even.
31. 1 mi **33–37.** Answers may vary. Samples are given. **33.** two right angles **35.** −2 and −3 **37.** −2 and −3 **39.** Add 1 then add 3; add 1 then add 3; . . . ; 10, 13. **41.** Multiply by 3, add 1; multiply by 3, add 1; . . . ; 201, 202. **43.** Add $\frac{1}{2}$, add $\frac{1}{4}$, add $\frac{1}{8}$ . . . ; $\frac{31}{32}$, $\frac{63}{64}$.
45. 123,454,321 **47.** **49.** 102 cm

51a. sì-shí-sān; lìu-shí-qī; bā-shí-sì **b.** Yes; the second part of the number repeats each ten numbers. **53.** His conjecture is probably false because most people's growth slows by 18 until they stop growing sometime between 18 and 22 years. **55.** Answers may vary.
57. 1 × 1: 64 squares; 2 × 2: 49 squares;
  3 × 3: 36 squares; 4 × 4: 25 squares;
  5 × 5: 16 squares; 6 × 6: 9 squares;
  7 × 7: 4 squares; 8 × 8: 1 square; total number of squares: 204 **59.** C
61. No; since all four variables are positive, the two ordered pairs must be in Quadrant I, so the midpoint of the segment will also be in Quadrant I. **62.** 16π in.^{2}
63. 20 m **64.** 2 **65.** True; explanations may vary. Sample: If the two even numbers are $2a$ and $2b$, the sum is $2a + 2b = 2(a + b)$, which is the form of an even

number. **66.** True; explanations may vary. Sample: if the three odd numbers are $2a + 1$, $2b + 1$, and $2c + 1$, the sum is $2(a + b + c) + 2 + 1 = 2(a + b + c + 1) + 1$, which is the form of an odd number.

Lesson 2-2 pp. 89–95

Got It? 1. Hypothesis: An angle measures 130. Conclusion: The angle is obtuse. **2.** If an animal is a dolphin, then it is a mammal. **3a.** False; January has 28 days, plus 3 more. **b.** True; the sum of the measures of two angles that form a linear pair is 180.
4. Counterexamples may vary. Samples are given. Converse: If a vegetable contains beta carotene, then it is a carrot. Inverse: If a vegetable is not a carrot, then it does not contain beta carotene. Contrapositive: If a vegetable does not contain beta carotene, then it is not a carrot. The conditional and the contrapositive are true. The converse and inverse are false; counterexample: any vegetable, such as spinach, that contains beta carotene.
Lesson Check 1. Hypothesis: Someone is a resident of Key West. Conclusion: The person lives in Florida. Conditional: If someone is a resident of Key West, then that person lives in Florida. **2.** Converse: If a figure has a perimeter of 10 cm, then it is a rectangle with sides 2 cm and 3 cm. Inverse: If a figure is not a rectangle with sides 2 cm and 3 cm, then it does not have a perimeter of 10 cm. Contrapositive: If a figure does not have a perimeter of 10 cm, then it is not a rectangle with sides 2 cm and 3 cm. The original conditional and the contrapositive are true. **3.** The hypothesis and conclusion were exchanged. The conditional should be "If it is Sunday, then you jog." **4.** Both are true because a conditional and its contrapositive have the same truth value, and a converse and an inverse have the same truth value.
Exercises 5. Hypothesis: You are an American citizen. Conclusion: You have the right to vote. **7.** Hypothesis: You want to be healthy. Conclusion: You should eat vegetables. **9.** If $3x − 7 = 14$, then $3x = 21$. **11.** If an object or example is a counterexample for a conjecture, then the object or example shows that the conjecture is false. **13.** If something is blue, then it has a color.
15. If something is wheat, then it is a grain. **17.** false; Mexico **19.** true **21.** Conditional: If a person is a pianist, then that person is a musician. Converse: If a person is a musician, then that person is a pianist. Inverse: If a person is not a pianist, then that person is not a musician. Contrapositive: If a person is not a musician, then that person is not a pianist. The conditional and contrapositive are true. The converse and inverse are false; counterexample: a percussionist is a musician.
23. Conditional: If a number is an odd natural number

Selected Answers

less than 8, then the number is prime. Converse: If a number is prime, then it is an odd natural number less than 8. Inverse: If a number is not an odd natural number less than 8, then the number is not prime. Contrapositive: If a number is not prime, then it is not an odd natural number less than 8. All four statements are false; counterexamples: 1 and 11. **25.** If a group is half the people, then that group should make up half the Congress. **27.** If an event has a probability of 1, then that event is certain to occur. **29.** Answers may vary. Sample: If an angle is acute, its measure is less than 90; if the measure of an angle is 85, then it is acute. **31.** Natalie is correct because a conditional statement and its contrapositive have the same truth value.

33.

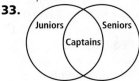

35. If $|x| = 6$, then $x = -6$; false, $x = 6$ is a counterexample.

37. If $x^3 < 0$, then $x < 0$; true. **39.** If you wear Snazzy sneakers, then you will look cool. **41.** If two figures are congruent, then they have equal areas. **43.** All integers divisible by 8 are divisible by 2. **45.** Some musicians are students. **47.** A **49.** D **51.** Answers may vary. Sample: 4 collinear points **52.** Answers may vary. Sample: 0.5 **53.** 36 in. **54.** 21 cm **55.** 4.5 yd or 162 in. **56.** 23.2 m or 2320 cm **57.** If tomorrow is October 1, then today is September 30; both the statement and the converse are true. **58.** If $\overline{AB}$ and $\overline{CD}$ are perpendicular, then $\overline{AB}$ is the perpendicular bisector of $\overline{CD}$; the statement is true and the converse is false.

Lesson 2-3   pp. 98–104

Got It? 1. Converse: If two angles are congruent, then the angles have equal measure; true. Biconditional: Two angles have equal measure if and only if the angles are congruent. **2.** If two numbers are reciprocals, then their product is 1. If the product of two numbers is 1, then the numbers are reciprocals. **3.** Yes, it is reversible; an angle is a straight angle if and only if its measure is 180. **4a.** No, it is not precise; a rectangle is also a figure with four right angles. **b.** Answers may vary. Sample: Obtuse angles have measures between 90 and 180.

Lesson Check 1. If points are collinear, then they lie on the same line. If points lie on the same line, then they are collinear. **2.** This month is June if and only if next month is July. **3.** Two angles are vertical angles if and only if their sides are opposite rays. **4.** The prefix *bi-* means "two." **5.** The word *gigantic* is not precise. **6.** The second statement is a better definition. A counterexample for the first statement is any two nonadjacent right angles.

Exercises 7. Converse: If two segments are congruent, then they have the same length; true. Biconditional: Two segments have the same length if and only if they are congruent. **9.** Converse: If a number is even, then it is divisible by 20; false. **11.** Converse: If it is Independence Day in the United States, then it is July 4; true. Biconditional: In the United States, it is July 4 if and only if it is Independence Day. **13.** If a line bisects a segment, then it intersects the segment only at its midpoint. If a line intersects a segment only at its midpoint, then the line bisects the segment. **15.** If you live in Washington, D.C., then you live in the capital of the United States. If you live in the capital of the United States, then you live in Washington, D.C. **17.** If an angle is a right angle, then it measures 90. If an angle measures 90, then it is a right angle. **19.** A line, segment, or ray is a perpendicular bisector of a segment if and only if it is perpendicular to the segment at its midpoint. **21.** A person is a Tarheel if and only if the person was born in North Carolina. **23.** not reversible **25.** No, it is not reversible; some endangered animals are not red wolves. **27.** No, it is not precise; straightedges and protractors are geometric tools. **29.** yes **31.** No; a straight angle has a measure greater than 90, but it is not an obtuse angle. **33.** That statement, as a biconditional, is "an angle is a right angle if and only if it is greater than an acute angle." Counterexamples to that statement are obtuse angles and straight angles. **35.** A point is in Quadrant III if and only if it has two negative coordinates. **37.** A number is a whole number if and only if it is a nonnegative integer. **39.** good definition **41.** good definition **43.** If ∠A and ∠B are a linear pair, then they are supplementary. **45.** If ∠A and ∠B are a linear pair, then they are adjacent, supplementary angles. **47.** Answers may vary. Sample: A line is a circle on the sphere formed by the intersection of the sphere and a plane containing the center of the sphere. **49.** D **51a.** If you go to the store, then you want to buy milk; false. **b.** Answers may vary. Sample: A counterexample is going to the store because you want to buy juice. **52.** If your grades suffer, then you do not get enough sleep. **53.** If you have a good voice, then you are in the school chorus.
54. true **55.** 60, 50 **56.** 4, $\frac{4}{5}$ **57.** 4, −2

Lesson 2-4   pp. 106–112

Got It? 1a. Marla is not safe out in the open. **b.** No conclusion is possible. **2a.** If a whole number ends in 0, then it is divisible by 5; Law of Syllogism. **b.** No conclusion is possible. **3a.** The Nile is the longest river in the world; Law of Syllogism and Law of Detachment. **b.** Yes; if you use the Law of Detachment first, then you must use it again to reach the same conclusion. The Law of Syllogism is not used.

Lesson Check **1.** No conclusion is possible. **2.** Figure *ABC* is a triangle; Law of Detachment. **3.** If it is Saturday, then you wear sneakers; Law of Syllogism. **4.** The Law of Detachment cannot be applied because the hypothesis is not satisfied. **5.** Answers may vary. Sample: Deductive reasoning uses logic to reach conclusions, while inductive reasoning bases conclusions on unproved (but possibly true) conjectures.

Exercises **7.** No conclusion is possible; the hypothesis has not been satisfied. **9.** No conclusion is possible; the hypothesis has not been satisfied. **11.** If an animal is a Florida panther, then it is endangered. **13.** If a line intersects a segment at its midpoint, then it divides the segment into two congruent segments. **15.** Alaska's Mount McKinley is the highest mountain in the U.S. **17.** If you are studying botany, then you are studying a science. (Law of Syllogism only) No conclusion can be made about Shanti. **19.** Must be true; by E and A, it is breakfast time; by D, Julio is drinking juice. **21.** May be true; by E and A, it is breakfast time. You don't know what Kira drinks at breakfast. **23.** May be true; by E, Maria is drinking juice. You don't know if she also drinks water. **25.** strange **27.** If a figure is a square, then it is a rectangle; *ABCD* is a rectangle. **29.** If a person is a high school student, then the person likes art; no conclusion is possible because the hypothesis is not satisfied.

31a. The result is two more than the chosen integer.

  b. $x + 2$

  c. The expression in part (b) is equivalent to the conjecture in part (a). In part (a) inductive reasoning was used to make a conjecture based on a pattern. In part (b) deductive reasoning was used in order to write and simplify an expression.

33. B **35.** A type of reasoning is called inductive if and only if it is based on patterns that you observe. **36.** $\angle AOB$, $\angle BOA$ **37.** $\angle BOC$, $\angle COB$ **38.** $\overrightarrow{OB}$ **39.** acute

Lesson 2-5   pp. 113–119

Got It? **1.** 75; $x = 2x - 75$ (Def. of an $\angle$ bis.); $x + 75 = 2x$ (Add. Prop. of Eq.); $75 = 2x - x$ (Subtr. Prop. of Eq.); $75 = x$ **2a.** Sym. Prop. of $\cong$ **b.** Distr. Prop. **c.** Mult. Prop. of Eq. **d.** Refl. Prop. of Eq. **3a.** Answers may vary. Sample: $\overline{AB} \cong \overline{CD}$ (Given); $AB = CD$ ($\cong$ segments have = length.); $BC = BC$ (Refl. Prop. of Eq.); $AB + BC = BC + CD$ (Add. Prop. of =); $AB + BC = AC$, $BC + CD = BD$ (Seg. Add. Post.); $AC = BD$ (Trans. Prop. of Eq.); $\overline{AC} \cong \overline{BD}$ (Segments with = length are $\cong$.) **b.** Answers may vary. Sample: You need to establish equality in order to add the same quantity ($m\angle 2$) to each side of the equation in Statement 3.

Lesson Check **1.** Trans. Prop. of Eq. **2.** Distr. Prop. **3.** Subtr. Prop of Eq. **4a.** Given **b.** Subtr. Prop. of Eq. **c.** Div. Prop. of Eq.

Exercises **5a.** Mult. Prop. of Eq. **b.** Distr. Prop. **c.** Add. Prop. of Eq. **7a.** def. of suppl. $\angle$ **b.** Subst. Prop. **c.** Distr. Prop. **d.** Subtr. Prop. of Eq. **e.** Div. Prop. of Eq. **9.** Distr. Prop. **11.** Sym. Prop. of $\cong$ **13a.** Given **b.** A midpt. divides a seg. into two $\cong$ segments. **c.** Substitution **d.** $2x = 12$ **e.** Div. Prop. of Eq. **15.** $\angle K$ **17.** 3 **19.** $\angle XYZ \cong \angle WYT$ **21.** Since $\overline{LR}$ and $\overline{RL}$ are two ways to name the same segment and $\angle CBA$ and $\angle ABC$ are two ways to name the same $\angle$, then both statements are examples of saying that something is $\cong$ to itself. **23.** $KM = 35$ (Given); $KL + LM = KM$ (Seg. Add. Post.); $(2x - 5) + 2x = 35$ (Subst. Prop.); $4x - 5 = 35$ (Distr. Prop.); $4x = 40$ (Add. Prop. of Eq.); $x = 10$ (Div. Prop. of Eq.); $KL = 2x - 5$ (Given); $KL = 2(10) - 5$ (Subst. Prop.); $KL = 15$ (Simplify) **25.** The error is in the 5th step when both sides of the equation are divided by $b - a$, which is 0, and division by 0 is not defined. **27.** Transitive only; A cannot be taller than A; if A is taller than B, then B is not taller than A. **29.** 58.5 **31.** 153.86 **33.** 58 **34.** Walt's science teacher is concerned. **35.** 80 **36.** 65 **37.** 125 **38.** 90 **39.** 50 **40.** 90 **41.** 35

Lesson 2-6   pp. 120–127

Got It? **1.** 40 **2a.** $\angle 1 \cong \angle 2$ (Given); $\angle 1 \cong \angle 3$, $\angle 2 \cong \angle 4$ (Vert. $\angle$ are $\cong$); $\angle 1 \cong \angle 4$, $\angle 2 \cong \angle 3$ (Trans. Prop. of $\cong$); $\angle 1 \cong \angle 2 \cong \angle 3 \cong 4$ (Trans. Prop. of $\cong$) **b.** Answers may vary. Sample: $m\angle 1 + m\angle 2 = 180$ because they form a linear pair. So $m\angle 1 = 90$ and $m\angle 2 = 90$ because $\angle 1 \cong \angle 2$. Then, using the relationship that $m\angle 2 + m\angle 3 = 180$ and $m\angle 1 + m\angle 4 = 180$, you can show that $m\angle 3 + m\angle 4 = 90$ by the Subtr. Prop. of Eq. Then $\angle 1 \cong \angle 2 \cong \angle 3 \cong \angle 4$ because their measures are =. **3.** Answers may vary. Sample: $\angle 1$ and $\angle 3$ are vert. $\angle$ because it is given. $\angle 1$ and $\angle 2$ are suppl. and $\angle 2$ and $\angle 3$ are suppl. because $\angle$ that form a linear pair are suppl. So, $m\angle 1 + m\angle 2 = 180$ and $m\angle 2 + m\angle 3 = 180$ by the def. of suppl. $\angle$. By the Trans. Prop. of Eq., $m\angle 1 + m\angle 2 = m\angle 2 + m\angle 3$. By the Subtr. Prop. of Eq., $m\angle 1 = m\angle 3$. So, $\angle 1 \cong \angle 3$ because $\angle$ with the same measure are $\cong$.

Lesson Check **1.** $m\angle 1 = 90$, $m\angle 2 = 50$, $m\angle 3 = 40$ **2.** B **3.** $\angle B \cong \angle C$ because both are suppl. to $\angle A$ and if two $\angle$ are suppl. to the same $\angle$, then they are $\cong$. **4.** He used the Trans. Prop. of $\cong$, which does not apply here. $\angle 2$ and $\angle 3$ are $\cong$, not compl. If two $\angle$ are compl. to the same $\angle$, then they are $\cong$ to each other. **5.** Answers may vary. Sample: A postulate is a statement that is assumed to be true, while a theorem is a statement that is proved to be true.

Exercises 7. $x = 38$, $y = 104$ **9.** 60, 60 **11.** 120, 120 **13a.** 90 **b.** 90 **c.** $m\angle 3$ **d.** $\cong$ **15.** Answers may vary. Sample: scissors **17.** $x = 14$, $y = 15$; $3x + 8 = 50$, $5x - 20 = 50$, $5x + 4y = 130$ **19.** $x = 50$, $y = 50$ **21.** $\angle EIG \cong \angle FIH$ because all rt. $\angle$s are $\cong$; $\angle EIF \cong \angle HIG$ because each one is compl. to $\angle FIG$ and compl. of the same $\angle$ are $\cong$. **23a.** It is given. **b.** $m\angle V$ **c.** 180 **d.** Division **e.** right **25.** By Theorem 2-5: If two $\angle$s are $\cong$ and suppl., then each is a right $\angle$. **27.** $m\angle A = 30$, $m\angle B = 60$ **29.** $m\angle A = 90$, $m\angle B = 90$ **31.** Answers may vary. Sample: $(-5, -1)$ **33.** $x = 30$, $y = 90$; 60, 120, 60 **34.** $x = 35$, $y = 70$; 70, 110, 70 **37.** 20 **39.** 60 **40.** Subtr. Prop. of Eq. **41.** Div. Prop. of Eq. **42.** Trans. Prop. of $\cong$ **43.** points F, I, H, B **44.** no **45.** yes **46.** line r (or $\overleftrightarrow{EG}$, $\overleftrightarrow{GH}$, $\overleftrightarrow{HC}$, and so on) **47.** any three of $\overleftrightarrow{FI}$ (or $\overleftrightarrow{IF}$), $\overrightarrow{FH}$, $\overrightarrow{FB}$, $\overrightarrow{IH}$, $\overrightarrow{IB}$, $\overrightarrow{HB}$ **48.** point H

Chapter Review   pp. 129–132

1. conclusion **2.** deductive reasoning **3.** truth value **4.** converse **5.** biconditional **6.** theorem **7.** hypothesis **8.** Divide the previous term by 10; 1, $\frac{1}{10}$. **9.** Multiply the previous term by -1; 5, -5. **10.** Subtract 7 from the previous term; 6, -1. **11.** Multiply the previous term by 4; 1536, 6144. **12.** Answers may vary. Sample: $-1 \cdot 2 = -2$, and -2 is not greater than 2 **13.** Answers may vary. Sample: Portland, Maine **14.** If a person is a motorcyclist, then that person wears a helmet. **15.** If two nonparallel lines intersect, then they intersect in one point. **16.** If two $\angle$s form a linear pair, then the $\angle$s are supplementary. **17.** If today is one of a certain group of holidays, then school is closed. **18.** Converse: If the measure of an $\angle$ is greater than 90 and less than 180, then the $\angle$ is obtuse. Inverse: If an angle is not obtuse, then it is not true that its measure is greater than 90 and less than 180. Contrapositive: If it is not true that the measure of an $\angle$ is greater than 90 and less than 180, then the $\angle$ is not obtuse. All four statements are true. **19.** Converse: If a figure has four sides, then the figure is a square. Inverse: If a figure is not a square, then it does not have four sides. Contrapositive: If a figure does not have four sides, then it is not a square. The conditional and the contrapositive are true. The converse and inverse are false. **20.** Converse: If you play an instrument, then you play the tuba. Inverse: If you do not play the tuba, then you do not play an instrument. Contrapositive: If you do not play an instrument, then you do not play the tuba. The conditional and the contrapositive are true. The converse and inverse are false. **21.** Converse: If you are busy on Saturday night, then you baby-sit. Inverse: If you do not baby-sit, then you are not busy on Saturday night. Contrapositive: If you are not busy on Saturday night, then you do not baby-sit. The conditional and the contrapositive are true. The converse and inverse are

false. **22.** No; it is not reversible; a magazine is a counterexample. **23.** yes **24.** No; it is not reversible; a line is a counterexample. **25.** A phrase is an oxymoron if and only if it contains contradictory terms. **26.** If two $\angle$s are complementary, then the sum of their measures is 90; if the sum of the measures of two $\angle$s is 90, then the $\angle$s are complementary. **27.** Colin will become a better player. **28.** $m\angle 1 + m\angle 2 = 180$ **29.** If two angles are vertical, then their measures are equal. **30.** If your father buys new gardening gloves, then he will plant tomatoes. **31a.** Given **b.** Seg. Add. Post. **c.** Subst. Prop. **d.** Distr. Prop. **e.** Subtr. Prop. of Eq. **f.** Div. Prop. of Eq. **32.** BY **33.** $p - 2q$ **34.** 18 **35.** 74 **36.** 74 **37.** 106 **38.** $\angle 1$ is compl. to $\angle 2$, $\angle 3$ is compl. to $\angle 4$, and $\angle 2 \cong \angle 4$ are all given. $m\angle 2 = m\angle 4$ by the def. of $\cong$. $\angle 1$ and $\angle 4$ are compl. by the Subst. Post. $\angle 1 \cong \angle 3$ by the $\cong$ Compl. Thm.

Chapter 3

Get Ready!   p. 137

1. $\angle 1$ and $\angle 5$, $\angle 5$ and $\angle 2$ **2.** $\angle 3$ and $\angle 4$ **3.** $\angle 1$ and $\angle 2$ **4.** $\angle 1$ and $\angle 5$, $\angle 5$ and $\angle 2$ **5.** Div. Prop. of $=$ **6.** Trans. Prop. of $\cong$ **7.** 4 **8.** 61 **9.** 15 **10.** 5 **11.** $2\sqrt{17}$ **12.** $\sqrt{17}$ **13.** Answers may vary. Sample: A figure divides a plane or space into three parts: the figure itself, the region inside the figure—called its interior—and the region outside the figure—called its exterior. **14.** Answers may vary. Sample: *Trans-* means "cross"; a transversal crosses other lines. **15.** Answers may vary. Sample: A flow proof shows the individual steps of the proof and how each step is related to the other steps.

Lesson 3-1   pp. 140–146

Got It? 1a. $\overline{EH}$, $\overline{BC}$, $\overline{FG}$ **b.** Sample: They are both in plane $FEDC$, so they are coplanar. **c.** plane BCG $\parallel$ plane ADH **d.** any two of $\overline{AB}$, $\overline{BF}$, $\overline{EF}$, and $\overline{AE}$ **2.** any three of $\angle 1$ and $\angle 3$, $\angle 2$ and $\angle 4$, $\angle 8$ and $\angle 6$, $\angle 7$ and $\angle 5$ **3.** corresp. $\angle$s

Lesson Check 1–7. Answers may vary. Samples are given. **1.** $\overline{EF}$ and $\overline{HG}$ **2.** $\overline{EF}$ and $\overline{GC}$ **3.** plane ABF $\parallel$ plane DCG **4.** $\angle 8$ and $\angle 6$ **5.** $\angle 3$ and $\angle 8$ **6.** $\angle 1$ and $\angle 3$ **7.** $\angle 1$ and $\angle 4$ **8.** Although lines that are not coplanar do not intersect, they are not parallel. **9.** Alt. int. $\angle$s are $\angle$s between two lines on opposite sides of a transversal. **10.** Carly; the lines are coplanar since they are both in plane ABH, so $\overline{AB} \parallel \overline{HG}$.

Exercises 11. plane JCD $\parallel$ plane ELH **13.** $\overleftrightarrow{GB}$, $\overleftrightarrow{JE}$, $\overleftrightarrow{CL}$, $\overleftrightarrow{FA}$ **15.** $\overleftrightarrow{GB}$, $\overleftrightarrow{DH}$, $\overleftrightarrow{CL}$ **17.** $\angle 7$ and $\angle 6$ (lines a and b with transversal d), $\angle 2$ and $\angle 5$ (lines b and c with transversal e)

19. ∠5 and ∠6 (lines *d* and *e* with transversal *b*); ∠2 and ∠4 (lines *b* and *e* with transversal *c*) **21.** ∠1 and ∠2 are corresp. ⦞; ∠3 and ∠4 are alt. int. ⦞; ∠5 and ∠6 are corresp. ⦞. **23.** ∠1 and ∠2 are corresp. ⦞; ∠3 and ∠4 are same-side int. ⦞; ∠5 and ∠6 are alt. int. ⦞. **25.** 2 pairs **27.** 2 pairs **29.** Skew; answers may vary. Sample: Since the paths are not coplanar, they are skew. **31.** False; $\overleftrightarrow{ED}$ and $\overleftrightarrow{HG}$ are skew. **33.** False; the planes intersect. **35.** False; both lines are in plane *ABC*. **37.** always **39.** always **41.** sometimes **43a.** Lines may be intersecting, parallel, or skew. **b.** Answers may vary. Sample: In a classroom, two adjacent edges of the floor are intersecting, two opposite edges of the floor are parallel, and one edge of the floor is skew to each of the vertical edges of the opposite wall. **45a.** The lines of intersection are ‖. **b.** Sample: The lines of intersection of a wall with the ceiling and floor (or the lines of intersection of any of the 6 planes with two different, opposite faces)

47. Yes; **49.** B **51.** C

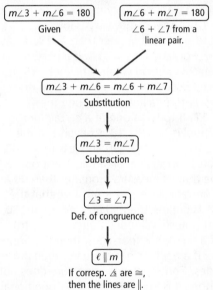

53. 121 **54.** 59 **55.** 29.5 **56.** 16, −32 **57.** corresp. ⦞ **58.** alt. int. ⦞ **59.** alt. ext. ⦞ **60.** same-side int. ⦞

Lesson 3-2   pp. 148–155

Got It? 1a. Yes, if you have the measure of at least one angle. **2.** (1) *a* ‖ *b* (Given) (2) ∠1 ≅ ∠5 (If lines are ‖, then corresp. ⦞ are ≅.) (3) ∠5 ≅ ∠7 (Vert. ⦞ are ≅.) (4) ∠1 ≅ ∠7 (Trans. Prop. of ≅) **3a.** 75; $m\angle 1 = m\angle 4$ by the Alt. Int. ⦞ Thm. **b.** 75; $m\angle 2 = m\angle 4$ by the Corresp. ⦞ Thm. **c.** 105; $m\angle 5 = 105$ by the Corresp. ⦞ Thm. **d.** 105; Alt. Int. ⦞ Thm. **e.** 105; Vert. ⦞. Thm. **f.** 105; ∠8 ≅ ∠6 by the Corresp. ⦞ Thm. **4a.** $x = 64$, $y = 40$ **b.** Clockwise from the bottom left, the measures are 52, 128, 120, 60.

Lesson Check 1–2. Answer may vary. Samples are given. **1.** ∠4 and ∠5, ∠2 and ∠6, ∠3 and ∠7, ∠4 and ∠8 **2.** ∠2 and ∠5, ∠4 and ∠7 **3.** 70 **4.** 55 **5.** Alike: Two parallel lines are cut by a tra nsversal and the angles are congruent; different: The int. ⦞ are between the two parallel lines, while the ext. ⦞ are not between the two parallel lines. **6.** same-side ext. ⦞, because they are ext. ⦞ on the same side of the transversal

Exercises 7. ∠1 (vert. ⦞), ∠7 (alt. int. ⦞), ∠4 (corresp. ⦞) **9.** ∠3 (alt. int. ⦞), ∠1 (corresp. ⦞) **11.** (1) *a* ‖ *b*; *c* ‖ *d* (Given) (2) ∠1 ≅ ∠4 (Alt. int. ⦞ are ≅.) (3) ∠4 ≅ ∠3 (Corresp. ⦞ are ≅.) (4) ∠1 ≅ ∠3 (Trans. Prop. of ≅) **13.** $m\angle 1 = 120$ because corresp. ⦞ are ≅;

$m\angle 2 = 60$ because ∠2 forms a linear pair with the given ∠. **15.** $x = 115$, $x - 50 = 65$ **17.** 20; $5x = 100$, $4x = 80$ **19.** $x = 135$, $y = 45$ **21.** 90; all the ⦞ are ≅ because each pair form vert. ⦞, corresp. ⦞, or suppl. ⦞. **23a.** 117 **b.** same-side int. ⦞
25. (1) ℓ ‖ *m* (Given)
  (2) $m\angle 2 + m\angle 3 = 180$ (⦞ that form a linear pair are suppl.)
  (3) $m\angle 3 + m\angle 6 = 180$ (Same-side interior angles are suppl.)
  (4) $m\angle 2 + m\angle 3 = m\angle 3 + m\angle 6$ (Substitution)
  (5) $m\angle 2 = m\angle 6$ (Subtraction)
  (6) ∠2 ≅ ∠6 (Def. of Congruence)
27. $m\angle 1 = 48$, $m\angle 2 = 132$ **29.** 65 **31.** 60
33. never **34.** never **35.** never **36.** sometimes **37.** If a △ has a 90° angle, then it is a right △; true. **38.** If two ⦞ are ≅, then they are vert. ⦞; false. **39.** If two ⦞ are suppl., then they are same-side int. ⦞; false.

Lesson 3-3   pp. 156–163

Got It? 1. ℓ ‖ *m* by the Converse of the Corresp. ⦞ Thm.
2. Answers may vary. Sample:

| $m\angle 3 + m\angle 6 = 180$ | $m\angle 6 + m\angle 7 = 180$ |
| Given | ∠6 + ∠7 from a linear pair. |

$m\angle 3 + m\angle 6 = m\angle 6 + m\angle 7$
Substitution

$m\angle 3 = m\angle 7$
Subtraction

∠3 ≅ ∠7
Def. of congruence

ℓ ‖ *m*
If corresp. ⦞ are ≅, then the lines are ‖.

3. ∠2 ≅ ∠3 (Vert. ⦞ are ≅.), so ∠1 ≅ ∠3 (Trans. Prop. of ≅). So *r* ‖ *s* by the Converse of the Corresp. ⦞ Thm.
4. 19
Lesson Check 1. Conv. of Corresp. ⦞ Thm. **2.** Conv. of Alt. Int. ⦞ Thm. **3.** 115 **4.** If you want to prove that alt. int. ⦞ are ≅, use the Alt. Int. ⦞ Thm.; if you want to prove that two lines are parallel, use the Converse of the Alt. Int. ⦞ Thm. **5.** Alike: Both give statements and reasons; different: The proofs use different formats. **6.** $\overleftrightarrow{DC}$ is the transversal, so the two same-side int. ⦞ show that $\overleftrightarrow{AD}$ and $\overleftrightarrow{BC}$ are parallel.

Exercises **7.** $\overleftrightarrow{BE} \parallel \overleftrightarrow{CG}$ by the Converse of the Corresp. ⓢ Thm. **9.** $\overleftrightarrow{CA} \parallel \overrightarrow{HR}$ by the Converse of the Corresp. ⓢ Thm. **11a.** Given **b.** ∠1 and ∠2 form a linear pair. **c.** ⓢ that form a linear pair are suppl. **d.** ∠2 ≅ ∠3 **e.** If corresp. ⓢ are ≅, then lines are ∥. **13.** 30 **15.** 59 **17.** $a \parallel b$; if same-side int. ⓢ are suppl., then the lines are ∥. **19.** $a \parallel b$; if same-side int. ⓢ are suppl., then the lines are ∥. **21.** none **23.** $a \parallel b$ (Conv. of the Alt. Ext. ⓢ Thm.) **25.** none **27.** 5 **29.** ∠3 ≅ ∠7 (Given), $m\angle3 = m\angle7$ (Def. of congruence), and $m\angle6 = m\angle7 = 180$ (∠6 and ∠7 form a linear pair). Then $m\angle6 = m\angle3 = 180$ (substitution). Therefore $\ell \parallel m$ (Thm. 3-6) **31.** $x = 10$; $m\angle1 = m\angle2 = 70$ **33.** $x = 2.5$; $m\angle1 = m\angle2 = 30$ **35.** Answers may vary. Sample: If ∠3 ≅ ∠5, then $\ell \parallel n$ by the Converse of Corresp. ⓢ Thm. **37.** Answers may vary. Sample: If ∠5 ≅ ∠3, then $j \parallel k$ by the Converse of the Corresp. ⓢ Thm. **39.** If alt. ext. ⓢ are ≅, then the lines are ∥.

41. Answers may vary. Sample:

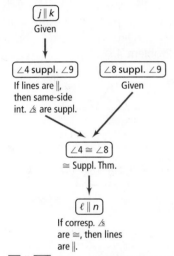

43. $\overline{PL} \parallel \overline{NA}$; if same-side int. ⓢ are suppl., then the lines are ∥. **45.** $\overline{PN} \parallel \overline{LA}$; if same-side int. ⓢ are suppl., then the lines are ∥. **47.** A **49.** C **51.** A sketch of a closed plane figure consisting of 5 segments (sides); the pentagon is convex because all diagonals are inside the pentagon OR the pentagon is concave because at least one diagonal has points outside the pentagon. **52.** $m\angle1 = 70$ (∠1 is suppl. to a 110° ∠.); $m\angle2 = 110$ (∠2 is suppl. to ∠1, which is a 70° angle.) **53.** $m\angle1 = 66$ (Alt. int. ⓢ are ≅.); $m\angle2 = 86$ (∠2 is suppl. to a 94° angle.) **54.** always **55.** sometimes **56.** sometimes **57.** never

Lesson 3-4 pp. 164–169

Got It? **1.** Yes; place the pieces with 60° ⓢ opposite each other and place the pieces with 30° ⓢ opposite each other. All four corners will be 90°, so opposite sides will be ∥. **2.** Yes; $a \parallel b$ because they are both ⊥ to d, and in a plane, two lines ⊥ to the same line are ∥.
Lesson Check **1.** They are ⊥; using Main Street as a transversal, Avenue B ⊥ Main Street by Thm. 3-10. **2.** $a \parallel b$; in a plane, if two lines are ⊥ to the same line, then they are ∥. **3.** Sample: Even if the 3 lines are not in the same plane, each line is parallel to the other 2 lines. **4.** Thm. 3-9 uses the Converse of the Corresp. ⓢ Thm.; the ⊥ Trans. Thm. uses the Corresp. ⓢ Thm. **5.** The diagram should show that m and r are ⊥.
Exercises **7a.** corresp. ⓢ **b.** ∠1 **c.** ∠3 **d.** Converse of Corresp. ⓢ Thm. **9.** Measure any three int. ⓢ to be rt. ⓢ and opp. walls will be ∥ because two walls ⊥ to the same wall are ∥. **11.** The rungs are ∥ to each other because they are all ⊥ to the same side. **13.** The rungs are ⊥ to both sides. The rungs are ⊥ to one of two ∥ sides, so they are ⊥ to both sides. **15.** The rungs are ∥ because they are all ⊥ to one side. **17.** Sample: Using the diagram underneath Thm. 3-10, $\overleftrightarrow{EC}$ and $\overleftrightarrow{AB}$ are both ⊥ to $\overleftrightarrow{AC}$, but $\overleftrightarrow{EC}$ and $\overleftrightarrow{AB}$ are skew, so they cannot be ∥. **19.** $a \parallel d$ by Thm. 3-8 **21.** $a \perp d$ by Thm. 3-10 **23.** $a \parallel d$ by Thms. 3-8 and 3-9 **25.** Reflexive: $a \parallel a$; false; every line intersects itself. Symmetric: If $a \parallel b$ then $b \parallel a$; true; lines a and b are coplanar and do not meet. Transitive: If $a \parallel b$ and $b \parallel c$, then $a \parallel c$; true; that is Thm. 3-8. **27.** A **29.** C **31.** 53 **32.** 46 **33.** right **34.** obtuse **35.** acute **36.** 60 **37.** 20 **38.** 40 **39.** 58

Lesson 3-5 pp. 171–178

Got It? **1.** 29 **2.** 127, 127, 106 **3.** Yes; answers may vary. Sample: $m\angle ACB$ must = 100, so by the △ ∠-Sum Thm., $m\angle A + 30 + 100 = 180$, and $m\angle A = 50$.
Lesson Check **1.** 58 **2.** 45 **3.** 68 **4.** $130 - x$ **5.** $m\angle1 = 130$ **6.** $m\angle3 = 38$ **7.** Answers may vary. Sample: Consider the int. ∠A of △ABC. By the △ ∠-Sum Thm., the sum of the measures of angles A, B, and C is 180°. ∠A is suppl. to its ext. ∠. So the sum of the measures of angles B and C is equal to the measure of the ext. ∠ of ∠A. **8.** A; all 3 ⓢ are int. ⓢ, so the solution should use the △ ∠-Sum Thm.
Exercises **9.** 30 **11.** 90 **13.** $x = y = 80$ **15a.** ∠5, ∠6, ∠8 **b.** For ∠5: ∠1 and ∠3; for ∠6: ∠1 and ∠2; for ∠8: ∠1 and ∠2 **c.** ∠6 ≅ ∠8 **17.** 123 **19.** $m\angle3 = 92$, $m\angle4 = 88$ **21.** 114 **23.** 60, 80 **25.** 102, 65, 13 **27.** 60; answers may vary. Sample: $180 \div 3 = 60$, so each ∠ is 60. **29.** $x = 37$; $m\angle P = 65$, $m\angle Q = 78$, $m\angle R = 37$ **31.** $a = 67$, $b = 58$, $c = 125$, $d = 23$,

e = 90 **33.** ∠1 is an ext. ∠ of the △. (Given); ∠1 and ∠4 are suppl. (∠ that form a straight ∠ are suppl.); m∠1 + m∠4 = 180 (Def. of suppl.); m∠2 + m∠3 + m∠4 = 180 (△∠-Sum Thm.); m∠1 + m∠4 = m∠2 + m∠3 + m∠4 (Subst. Prop.); m∠1 = m∠2 + m∠3 (Subtr. Prop. of =) **35.** 40, 50 **37.** $\frac{1}{7}$ **39.** $\frac{1}{19}$

41. 115 **43.** A **45.** C **47.** a ∥ c; if 2 same-side ext. ∠ are suppl., then the lines are ∥. **48.** a ∥ b; if 2 lines are ∥ to the same line, they are ∥ to each other.

49. 32 **50.** m∠1 = m∠2 = 90; sample: If the sum of two equal numbers is 180, then each number is 90.

51.
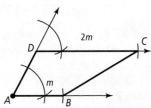

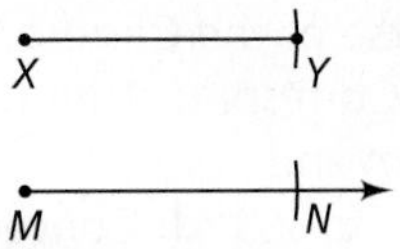

52. 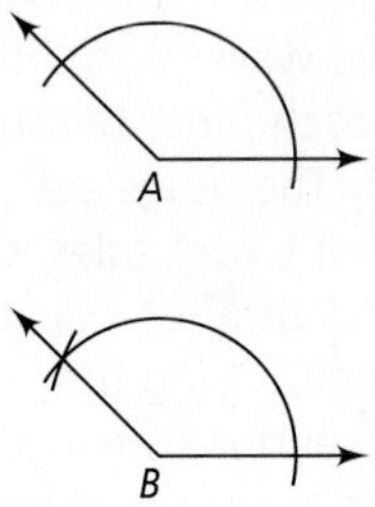 **53.**
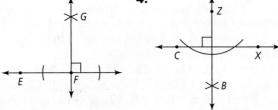

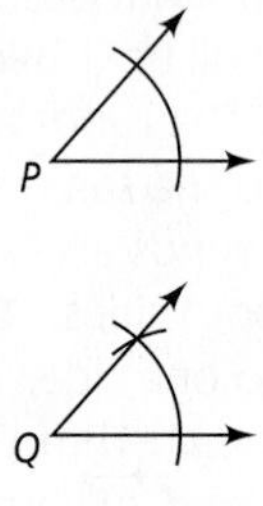

Lesson 3-6 pp. 182–188

Got It? 1. ∠1 and ∠NHJ are corresp. ∠ for lines m and ℓ. Since ∠1 ≅ ∠NHJ, then m ∥ ℓ.
2a. Answers may vary. Sample:

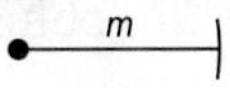

b. No; the length of $\overline{AB}$ and m∠A are not determined.
3. **4.**

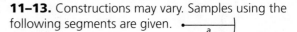

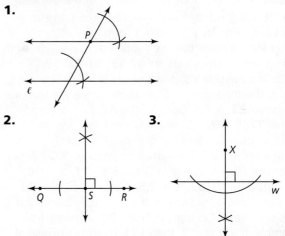

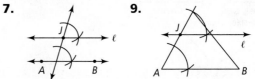

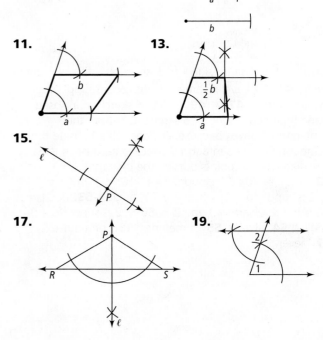

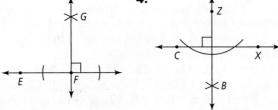

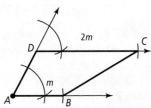

Lesson Check
1.

2. **3.**

4. Yes; the same compass opening is used to draw the arcs at C. **5.** No; points E and F would have been further apart, but the new point G would determine the same line $\overleftrightarrow{RG}$ as in Step 4. **6.** Similar: You are constructing a line ⊥ to a given line through a given point. Different: The given point is on the given line in Problem 3 and is not on the given line in Problem 4.

Exercises
7. **9.**

11–13. Constructions may vary. Samples using the following segments are given.

11. **13.**

15.

17. **19.**

974

21. Construct a ≅ alt. int. ∠, then draw the ∥ line.

23. **25.**

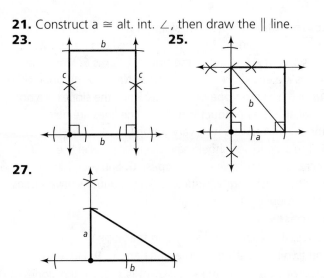

27.

29a. II, IV, III, I **b.** III: points C and G; I: the intersection of $\overleftrightarrow{GC}$ with the arcs from Step III

31–39. Constructions may vary. Samples are given.

31. **33.**

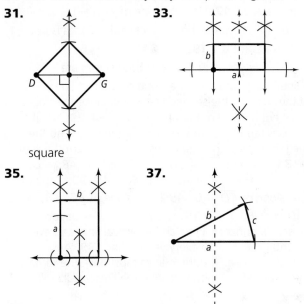

square

35. **37.**

39. Not possible; the shorter sides would meet at a point on the longer side, forming a segment. **41.** I

43. $3y = 120$, $(y - 15) = 25$ **44.** $x = 104$, $(x - 28) = 76$, $y = 35$, $(2y - 1) = 69$ **45.** $\frac{1}{2}$ **46.** 1

47. -2

Lesson 3-7 pp. 189–196

Got It? 1a. $\frac{4}{3}$ **b.** 0

2a. **b.**

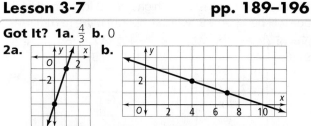

3a. $y = -\frac{1}{2}x + 2$ **b.** $y - 4 = -3(x + 1)$
4a. $y + 1 = \frac{6}{5}(x + 2)$ **b.** $y = \frac{6}{5}x + \frac{7}{5}$; $y = \frac{6}{5}x + \frac{7}{5}$; they represent the same line. **5a.** horizontal: $y = -3$; vertical: $x = 4$ **b.** No; the slope is undefined for a vertical line, so you cannot use the slope-intercept form because that requires a value for the slope.

Lesson Check 1. 5 **2.** -2 **3.** $y = 8x + 10$
4. $y - 3 = 4(x - 3)$ or $y - 7 = 4(x - 4)$ **5.** Answers may vary. Sample: The slope-intercept form $y = mx + b$ uses the slope m and the y-intercept b; the point-slope form $y - y_1 = m(x - x_1)$ uses a point (x_1, y_1) on the line and the slope m. **6.** The lines have the same y-int., but one line has a steep positive slope and the other has a less steep negative slope. **7.** Your classmate switched the x- and y-values in the formula for slope. The slope of the line is undefined.

Exercises 9. $-\frac{5}{6}$ **11.** $-\frac{3}{2}$ **13.** -8 **15.** undefined

17. **19.**

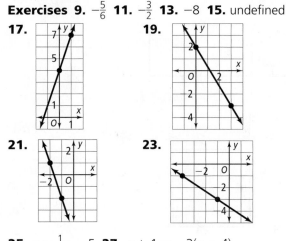

21. **23.**

25. $y = \frac{1}{2}x - 5$ **27.** $y + 1 = -3(x - 4)$
29. $y - 6 = -(x + 2)$ or $y - 3 = -(x - 1)$
31. $y - 2 = -\frac{1}{2}(x - 6)$ or $y - 4 = -\frac{1}{2}(x - 2)$
33. $y = \frac{1}{2}(x + 1)$ or $y + 1 = \frac{1}{2}(x + 3)$ **35.** horizontal: $y = -2$; vertical: $x = 3$ **37.** horizontal: $y = 4$; vertical: $x = 6$ **39.** **41.**

43. Yes; if the ramp is 24 in. high and 72 in. long, the slope will be $\frac{24}{72} = 0.\overline{3}$, which is less than the maximum slope of $\frac{4}{11} = 0.\overline{36}$. **45.** $y = -x + 2$ **47.** $y = -\frac{3}{2}x + 5$ **49.** (6, −4) **51.** (−1, 3)

53. No; answers may vary. Sample: $\frac{1}{12} < \frac{3}{10}$ so the ramp would need to zigzag to comply with the law.

55a. Undefined; the y-axis is a vertical line, and the slope of a vertical line is undefined. **b.** $x = 0$ **57a.** $y = \frac{5}{2}x$ **b.** $y - 5 = -\frac{5}{2}(x - 2)$ or $y = -\frac{5}{2}x + 10$ **c.** The abs. value of the slopes is the same, but one slope is pos. and the other is neg. One y-int. is 0 and the other is 10.

59. No; the slope of the line through the first two points is $-\frac{1}{3}$ and the slope of the line through the last two points is −1, so the points do not lie on the same line.

61. 9 **63.** $\frac{1}{6}$ **65.** G **67.** G

69.

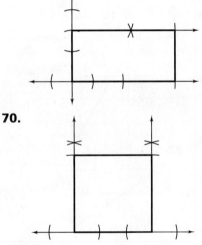

70.

71. Distr. Prop. **72.** Substitution **73.** Reflexive Prop. of ≅ **74.** Symmetric Prop. of ≅ **75.** $\frac{1}{2}$ **76.** $\frac{5}{2}$ **77.** −5

Lesson 3-8   pp. 197–204

Got It? 1. No; the slope of ℓ_3 is $\frac{6 - 2}{-13 - (-1)} = \frac{4}{-12} = -\frac{1}{3}$ And the slope of ℓ_4 is $\frac{6 - 7}{3 - 6} = \frac{-1}{-3} = \frac{1}{3}$. The slopes are not equal. **2.** $y - 3 = -(x + 5)$ **3.** No; the slope of ℓ_3 is $\frac{7 - (-1)}{2 - 3} = \frac{8}{-1} = -8$ and the slope of ℓ_4 is $\frac{6 - 7}{-2 - 8} = \frac{-1}{-10} = \frac{1}{10}$. Since the product of the slopes is not −1, the Lines are not ⊥. **4.** $y - 7 = \frac{1}{3}(x + 3)$ **5.** $y - 40 = \frac{1}{3}(x - 90)$

Lesson Check 1. ⊥; the slope of $\overleftrightarrow{AB}$ is 2 and the slope of $\overleftrightarrow{CD}$ is $-\frac{1}{2}$. Since $(2)\left(-\frac{1}{2}\right) = -1$, the lines are ⊥. **2.** ∥; the slope of $\overleftrightarrow{AB}$ is 6 and the slope of $\overleftrightarrow{CD}$ is 6. Since the slopes are equal, the lines are ∥. **3.** Neither; the slope of $\overleftrightarrow{AB}$ is 0 and the slope of $\overleftrightarrow{CD}$ is 1. Since the slopes are not equal and their product is not −1, the lines are neither ∥ nor ⊥. **4.** Answers may vary. Sample: $y + 3 = \frac{1}{4}(x - 2)$ **5.** The second line should say "slope of parallel line = 3" because ∥ lines have equal slopes. **6.** Sample: ∥ line equations have equal slopes. ⊥ line equations have slopes with product −1.

Exercises

7. Yes; the slope of ℓ_1 is $-\frac{1}{2}$ and the slope of ℓ_2 is $-\frac{1}{2}$, and two lines with the same slope are ∥. **9.** No; the slope of ℓ_1 is $\frac{3}{2}$ and the slope of ℓ_2 is 2. Since the slopes are not equal the lines are not ∥. **11.** $y = -2x + 3$ **13.** $y - 4 = \frac{1}{2}(x + 2)$ **15.** Yes; the slope of ℓ_1 is $-\frac{1}{2}$ and the slope of ℓ_2 is 2. Since the product of the slopes is −1, the lines are ⊥. **17.** No; the slope of ℓ_1 is −1 and the slope of ℓ_2 is $\frac{4}{5}$. Since the product of the slopes is not −1, the lines are not ⊥. **19.** $y - 6 = -\frac{3}{2}(x - 6)$ **21.** $y - 4 = \frac{1}{2}(x - 4)$ **23.** Yes; both slopes are −1 so the lines are ∥. **25.** No; the slope of the first line is $-\frac{3}{4}$ and the slope of the second line is −3. Since the slopes are not equal, the lines are not ∥. **27.** −4 **29.** No; if two equations represent lines with the same slope and the same y-intercept, the equations must represent the same line. **31.** slope of $\overline{AB}$ = slope of $\overline{CD} = -\frac{3}{4}$, $\overline{AB} \parallel \overline{CD}$; slope of $\overline{BC}$ = slope of $\overline{AD}$ = 1, $\overline{BC} \parallel \overline{AD}$ **33.** slope of $\overline{AB}$ = slope of $\overline{CD}$ = 0, $\overline{AB} \parallel \overline{CD}$; slope of $\overline{BC}$ = 3, slope of $\overline{AD} = \frac{3}{2}$, $\overline{BC} \nparallel \overline{AD}$ **35.** A **37.** Yes; the equations represent a horizontal line and a vertical line, and every horizontal line is ⊥ to every vertical line. **39.** Answers may vary. Sample: The three lines must have the same slope or undefined slope, so all three lines are ∥.

41a. $y = -\frac{1}{2}x + 100$ **b.** (100, 50) **c.** 112 yd

43. Slope of $\overline{AB}$ is $-\frac{1}{8}$; slope of $\overline{CD}$ is 8; the lines are ⊥. **45.** The slope of $\overline{AC}$ is $\frac{10}{-2} = -5$ and the slope of $\overline{BD}$ is $\frac{2}{10} = \frac{1}{5}$. Since the product of the slopes is −1, the diagonals are ⊥. The midpoint of $\overline{AC}$ is $\left(\frac{7 + 9}{2}, \frac{11 + 1}{2}\right) = (8, 6)$ and the midpoint of $\overline{BD}$ is $\left(\frac{13 + 3}{2}, \frac{7 + 5}{2}\right) = (8, 6)$. Since the two diagonals have the same midpoint, they bisect each other.

47. $y - 5 = \frac{1}{3}(x - 4)$ **49.** 6 **51.** 7 **53.** $y = -\frac{1}{2}x + 3$ **54.** $y - 2 = \frac{5}{3}(x + 4)$ or $y - 7 = \frac{5}{3}(x + 1)$ **55.** $y + 2 = \frac{3}{4}(x - 3)$ or $y + 8 = \frac{3}{4}(x + 5)$ **56.** Reflexive Prop. of ≅ **57.** Mult.

Prop. of Equality **58.** Distr. Prop. **59.** Symmetric Prop. of ≅
60. Yes; ∠1 and ∠2 are vert. ∡, and vert. ∡ are ≅.
61. Yes; ∠1 and ∠2 are both rt. ∡, and all rt. ∡ are ≅.
62. No; m∠1 = 54 (Given) and m∠2 = 90 − 54 = 36 (because ∠1 and ∠2 are compl.).

Chapter Review   pp. 206–210

1. transversal **2.** ext. ∠ **3.** point-slope **4.** alt. int. ∡
5. skew lines **6.** slope-intercept **7.** ∠2 and ∠7, a and b, transversal d; ∠3 and ∠6, c and d, transversal e; ∠3 and ∠8; b and e; transversal c **8.** ∠5 and ∠8, lines a and b, transversal c; ∠2 and ∠6; a and e; transversal d **9.** ∠1 and ∠4, lines c and d, transversal b; ∠2 and ∠4, lines a and b, transversal d; ∠2 and ∠5, lines c and d, transversal a; ∠1 and ∠5, lines a and b, transversal c; ∠3 and ∠4; b and c; transversal e **10.** ∠1 and ∠7, lines c and d, transversal b **11.** corresp. ∡ **12.** alt. int. ∡
13. m∠1 = 120 because corresp. ∡ are ≅; m∠2 = 120 because ∠1 and ∠2 are vert. ∡. **14.** m∠1 = 75 because same-side int. ∡ are suppl.; m∠2 = 105 because alt. int. ∡ are ≅. **15.** x = 118, y = 37 **16.** 20
17. 20 **18.** n ∥ p; if corresp. ∡ are ≅, then the lines are ∥. **19.** none; ∠3 and ∠6 form a linear pair. **20.** ℓ ∥ m; if same-side int. ∡ are suppl., then the lines are ∥.
21. n ∥ p; if alt. int. ∡ are ≅, then the lines are ∥.
22. ∥ **23.** a **24.** 1st Street and 3rd Street are ∥ because they are both ⊥ to Morris Avenue. Since 1st Street and 5th street are both ∥ to 3rd Street, 1st Street and 5th Street are ∥ to each other. **25.** x = 60, y = 60
26. x = 45, y = 45 **27.** 30 **28.** 55 **29.** 3
30.   **31.**

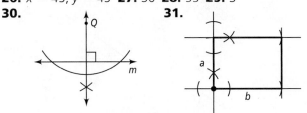

32.   **33.**

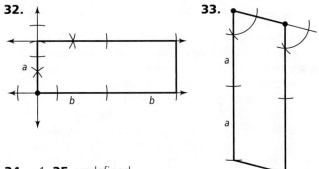

34. −1 **35.** undefined

36. slope: 2;
y-intercept: −1

37. slope: −2;
point: (−5, 3)

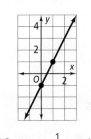 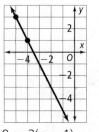

38. $y = -\frac{1}{2}x + 12$ **39.** $y + 9 = 3(x - 1)$
40. $y - 2 = 4(x - 4)$ or $y + 2 = 4(x - 3)$
41. neither **42.** ∥ **43.** ⊥ **44.** ∥
45. $y - 2 = 8(x + 6)$ **46.** $y + 3 = -6(x - 3)$

Chapter 4

Get Ready!   p. 215

1. AB = 4, BC = 3, AC = 5 **2.** AB = 8, BC = $\sqrt{265}$, AC = $\sqrt{137}$ **3.** AB = $\sqrt{58}$, BC = $\sqrt{32}$, AC = $\sqrt{58}$
4. ∠J ≅ ∠L **5.** m∠M = m∠N = 90 **6.** ∠B is a rt. ∠.
7. ∠AFB ≅ ∠CFD **8.** ∠B ≅ ∠C, ∠A ≅ ∠D, ∠AEB ≅ ∠CED **9.** ∠DAC ≅ ∠BCA, ∠DCA ≅ ∠BAC, ∠DAB ≅ ∠BCD, ∠B ≅ ∠D **10.** m∠A = 21, m∠B = 71, m∠C = 88 **11-13.** Answers may vary. Sample: **11.** The base is the side that meets each of the two ≅ sides of the △. **12.** The legs are the ≅ sides of an isosc. △.
13. Corresp. parts are the sides or ∡ that are in the same relative position in each figure.

Lesson 4-1   pp. 218–224

Got It? 1. $\overline{WY} \cong \overline{MK}$, $\overline{YS} \cong \overline{KV}$, $\overline{WS} \cong \overline{MV}$, ∠W ≅ ∠M, ∠Y ≅ ∠K, ∠S ≅ ∠V **2.** m∠V = 83; ∠W ≅ ∠M and ∠Y ≅ ∠K because they are corresp. parts of ≅ △. By the Triangle Angle-Sum Theorem, m∠M + m∠K + m∠V = 180. By substitution, 62 + 35 + m∠V = 180. So by subtraction, m∠V = 83. **3.** Answers may vary. Sample: You know that $\overline{AD} \cong \overline{CD}$ (Given) and $\overline{BD} \cong \overline{BD}$ (Reflexive Prop. of ≅), but you have no other information about the sides and ∡ of the △, so you cannot conclude that △ABD ≅ △CBD. **4.** ∠A ≅ ∠D (Given), and ∠ABE ≅ ∠DBC because vertical ∡ are ≅. Also, ∠AEB ≅ ∠DCB (Third ∡ Theorem). The three pairs of sides are ≅ (Given), so △AEB ≅ △DCB by the def. of ≅ △.
Lesson Check 1a. $\overline{NY}$ **b.** ∠X **2a.** $\overline{RO}$
b. ∠T **3a.** ∠A **b.** $\overline{KL}$ **c.** CKLU **4a.** ∠M ≅ ∠T
b. 92 **5.** Answers may vary. Sample: finding the correct top for a food container **6.** No; the ∡ could be the same shape but not necessarily the same size. **7.** He has not shown that corresp. ∡ are ≅.

Exercises 9. $\overline{EF} \cong \overline{HI}$, $\overline{FG} \cong \overline{IJ}$, $\overline{EG} \cong \overline{HJ}$, $\angle EFG \cong \angle HIJ$, $\angle FGE \cong \angle IJH$, $\angle FEG \cong \angle IHJ$ **11.** $\overline{CM}$ **13.** $\angle B$ **15.** $\angle J$ **17.** $\triangle CLM$ **19.** $\triangle MCL$ **21.** $\angle P \cong \angle S$, $\angle O \cong \angle I$, $\angle L \cong \angle D$, $\angle Y \cong \angle E$ **23.** 45 ft **25.** 52 **27.** 280 ft **29.** 128 **31.** No; there are not three pairs of $\cong$ corresp. sides. **33.** C **35.** $m\angle A = m\angle D = 20$ **37.** $BC = EF = 8$ **39.** 43 **41.** 5 **43.** Answers may vary. Sample: If $\triangle PQR \cong \triangle XYZ$, then $\overline{PQ} \cong \overline{XY}$, $\overline{QR} \cong \overline{YZ}$, $\overline{PR} \cong \overline{XZ}$, $\angle P \cong \angle X$, $\angle Q \cong \angle Y$, and $\angle R \cong \angle Z$. **45.** Two pairs of $\cong$ sides are given, and the third pair of sides are $\cong$ because $\overline{PQ}$ bisects $\overline{RT}$, so $\overline{TS} \cong \overline{RS}$. $\overline{PR} \parallel \overline{TQ}$, so $\angle P \cong \angle Q$ and $\angle R \cong \angle T$ because they are alt. int. $\triangle$; the third pair of $\triangle$ are vertical $\triangle$, so they are $\cong$. Thus $\triangle PRS \cong \triangle QTS$ by the def. of $\cong \triangle$. **47.** $KL = 4$, $LM = 3$, $KM = 5$ **49a.** 15 quadrilaterals

b. 11 convex, 4 concave
51. 50 **53.** 28 **54.** $y = -\frac{2}{3}x + \frac{25}{3}$ **55.** $y = -\frac{1}{4}x + \frac{5}{4}$
56. 5 **57.** 18 **58.** 10 **59.** $\overline{AB} \cong \overline{DE}$, $\angle C \cong \angle F$
60. $\angle Q \cong \angle S$, $\angle QPR \cong \angle SRP$, $\angle QPR$ and $\angle SPR$ are adjacent, $\angle QRP$ and $\angle SRP$ are adjacent.
61. $\angle M \cong \angle U$, $\overline{TO} \cong \overline{NV}$, $\overline{TV} \cong \overline{NO}$, $\angle MOT$ and $\angle MON$ are adjacent and suppl., $\angle UVT$ and $\angle UVN$ are adjacent and suppl.

Lesson 4-2   pp. 226–233

Got It? 1. Two pairs of sides are given as $\cong$, and $\overline{BD} \cong \overline{BD}$ by the Refl. Prop. of $\cong$. So $\triangle BCD \cong \triangle BFD$ by SSS. **2.** $\overline{LE} \cong \overline{BN}$ **3.** SSS; three pairs of corresp. sides are $\cong$.
Lesson Check 1a. $\angle PEN$ (or $\angle E$) **b.** $\angle NPE$ (or $\angle P$) **2a.** $\overline{HA}$ and $\overline{HT}$ **b.** $\overline{TH}$ and $\overline{TA}$ **3.** SAS **4.** SSS **5.** Answers may vary. Sample: Alike: Both use three pairs of $\cong$ parts to prove $\triangle \cong$. Different: SSS uses three pairs of $\cong$ sides, while SAS uses two pairs of $\cong$ sides and their $\cong$ included $\triangle$. **6.** No; the $\cong$ $\triangle$ are not included between the pairs of $\cong$ sides. **7.** No; the $\triangle$ have the same perimeter, but the three side lengths of one $\triangle$ are not necessarily $=$ to the three side lengths of the other

$\triangle$, so you cannot use SSS. There is no information about the $\triangle$ of the $\triangle$, so you cannot use SAS.
Exercises
9. F is the midpt. of $\overline{GI}$ (Given), so $\overline{IF} \cong \overline{GF}$ because a midpt. divides a segment into two $\cong$ segments. The other two pairs of sides are given as $\cong$, so $\triangle EFI \cong \triangle HFG$ by SSS. **11.** You need to know $\overline{LG} \cong \overline{MN}$; the diagram shows that $\overline{LT} \cong \overline{MQ}$ and $\angle L \cong \angle M$. $\angle L$ is included between $\overline{LG}$ and $\overline{LT}$, and $\angle M$ is included between $\overline{MN}$ and $\overline{MQ}$. **13.** Not enough information; the congruent vertical angles TQP and RQS are not included by the pairs of $\cong$ sides. **15.** If the 40° $\angle$ is *always* included between the two 5-in. sides, then all the $\triangle$ will be $\cong$ by SAS. If the 40° $\angle$ is *never* included between the two 5-in. sides, then the angles of the $\triangle$ will be 40°, 40°, and 100°, with the 100° angle included between the 5-in. sides, so all the $\triangle$ will be $\cong$ by SAS. But a $\triangle$ with the 40° angle included between the 5-in. sides will not be $\cong$ to a $\triangle$ with the 40° angle not included between the 5-in. sides. **17.** X is the midpt. of $\overline{AG}$ and $\overline{NR}$ (Given), so $\overline{AX} \cong \overline{GX}$ and $\overline{NX} \cong \overline{RX}$ by the def. of midpt. Also, $\angle AXN \cong \angle GXR$ because they are vertical $\triangle$, so $\triangle ANX \cong \triangle GRX$ by SAS. **19.** $AB = \sqrt{25 + 16} = \sqrt{41}$ and $DE = \sqrt{16 + 36} = \sqrt{52}$, so $\triangle ABC \not\cong \triangle DEF$.
21. Answers may vary. Sample: roof trusses for a house, sections of a ferris wheel, sawhorses used by a carpenter; explanations will vary.
23a. Answers may vary.   **b.** Answers may vary.
     Sample: Sample:

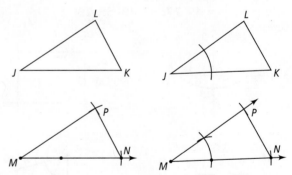

25. Not enough information; you need $\overline{DY} \cong \overline{TK}$ to show the $\triangle$ are $\cong$ by SSS, or you need $\angle H \cong \angle P$ to show the $\triangle$ are $\cong$ by SAS. **27.** Not necessarily; the $\cong$ $\triangle$ are not included between the pairs of $\cong$ sides. **29.** $\overline{AE}$ and $\overline{BD}$ bisect each other (Given), so $\overline{AC} \cong \overline{EC}$ and $\overline{DC} \cong \overline{BC}$ (Def. of bisector). $\angle ACB \cong \angle ECD$ (Vert. $\triangle$ are $\cong$.), so $\triangle ACB \cong \triangle ECD$ by SAS. **31.** Given the $\perp$ segments, $\angle B \cong \angle CMA$ because all rt. $\triangle$ are $\cong$. M is the midpt. of $\overline{AB}$ (Given), so $\overline{AM} \cong \overline{MB}$ by the def. of midpt. Since $\overline{DB} \cong \overline{CM}$ (Given), then $\triangle AMC \cong \triangle MBD$ by SAS. **33.** Answers may vary. Sample: $\angle N \cong \angle L$, $\overline{MN} \cong \overline{OL}$, and $\overline{NO} \cong \overline{LM}$ (Given), so $\triangle MNO \cong \triangle OLM$ by SAS. $\angle NMO \cong \angle LOM$ (Corresp. parts of $\cong$ $\triangle$ are $\cong$.) So

$\overline{MN} \parallel \overline{OL}$ because if alt. int. $\&$ are $\cong$, then the lines are $\parallel$.
35. A **37.** D **39.** $\angle E$ **40.** $\overline{AB}$ **41.** $\overline{FG}$ **42.** $\angle C$ **43.** If $2x = 6$, then $x = 3$; both are true. **44.** If $x^2 = 9$, then $x = 3$; the statement is true and its converse is false.
45. $\overline{JH}$ **46.** $\angle MNL$ (or $\angle N$)

Lesson 4-3   pp. 234–241

Got It? 1. $\triangle HGO \cong \triangle ACT$ because $\overline{HG} \cong \overline{AC}$ and the $\cong$ segments are included between two pairs of $\cong$ $\&$.
2. $\angle B \cong \angle E$ because all rt. $\&$ are $\cong$. $\overline{AB} \cong \overline{AE}$ and $\angle CAB \cong \angle DAE$ (Given), so $\triangle ABC \cong \triangle AED$ by ASA.
3a. $\overline{RP}$ bisects $\angle SRQ$ (Given), so $\angle SRP \cong \angle QRP$ by the def. of $\angle$ bisector. $\angle S \cong \angle Q$ (Given) and $\overline{RP} \cong \overline{RP}$ (Refl. Prop. of $\cong$), so $\triangle SRP \cong \triangle QRP$ by AAS. **b.** After Step 3 in the proof, state that $\angle MRW \cong \angle KWR$ by the Third $\&$ Theorem and write Step 4, so $\triangle WMR \cong \triangle RKW$ by ASA. **4.** Yes; $\overline{PR} \cong \overline{SR}$ and $\angle A \cong \angle I$ (Given). $\angle ARP \cong \angle IRS$ (Vert. $\&$ are $\cong$.), so $\triangle PAR \cong \triangle SIR$ by AAS.
Lesson Check 1. $\overline{RS}$ **2.** $\angle N$, $\angle O$ **3.** ASA
4. AAS **5.** Answers may vary. Sample: Alike: Both postulates use three pairs of $\cong$ corresp. parts. Different: To use the ASA Postulate, the sides must be included between the pairs of corresp. $\&$, while to use the SAS Postulate, the $\&$ must be included between the pairs of corresp. sides. **6.** $\overline{LM}$ is not included between the pairs of $\cong$ corresp. $\&$. **7.** $\angle F \cong \angle G$; $\angle D \cong \angle H$
Exercises
9. $\triangle ABC \cong \triangle EDF$ **11.** $\overline{AC} \perp \overline{BD}$ (Given), so $\angle ACB \cong \angle ACD$ because $\perp$ lines form rt. $\&$, and all rt. $\&$ are $\cong$. $\angle BAC \cong \angle DAC$ (Given) and $\overline{AC} \cong \overline{AC}$ (Refl. Prop. of $\cong$), so $\triangle ABC \cong \triangle ADC$ by ASA. **13a.** Vert. $\&$ are $\cong$. **b.** Given **c.** $\overline{TQ} \cong \overline{RQ}$ **d.** AAS **15.** Given the $\perp$ segments, $\angle Q \cong \angle S$ because $\perp$ lines form rt. $\&$, and all rt. $\&$ are $\cong$. It is given that T is the midpt. of $\overline{PR}$, so $\overline{PT} \cong \overline{RT}$ by the def. of midpt. $\angle PTQ \cong \angle RTS$ because vert. $\&$ are $\cong$, so $\triangle PQT \cong \triangle RST$ by AAS.
17. $\triangle UST \cong \triangle RTS$ by AAS. **19.** It is given that $\angle N \cong \angle P$ and $\overline{MO} \cong \overline{QO}$. Also, $\angle MON \cong \angle QOP$ because vert. $\&$ are $\cong$. So $\triangle MON \cong \triangle QOP$ by AAS.
21. Answers may vary. Sample: Yes; ASA guarantees a unique triangle with vertices at the oak tree, the maple tree, and the time capsule. **23.** No; the common side is included between the two $\cong$ $\&$ in one $\triangle$, but it is not included between the $\cong$ $\&$ in the other $\triangle$. **25.** $\overline{AE} \parallel \overline{BD}$ (Given), so $\angle A \cong \angle DBC$ (If $\parallel$ lines, corresp. $\&$ are $\cong$.). Since $\angle E \cong \angle D$ and $\overline{AE} \cong \overline{BD}$ (Given), then $\triangle AEB \cong \triangle BDC$ by ASA.

27. Answers may vary. Sample:

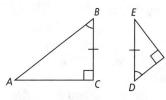

29. $\triangle EAB \cong \triangle ECD$, $\triangle EBC \cong \triangle EDA$, $\triangle ABD \cong \triangle CDB$, $\triangle ABC \cong \triangle CDA$
31. $\frac{13}{20}$ **33.** I
35. Converse: If you are too young to vote in the United States, then you are less than 18 years old; true. **36.** SSS; you are given two pairs of $\cong$ sides. Another pair of sides are $\cong$ by the Refl. Prop. of $\cong$. **37.** SAS; you are given two pairs of $\cong$ sides. The pair of included angles are congruent because they are vertical angles.
38. $\angle T \cong \angle L$, $\angle I \cong \angle O$, $\angle C \cong \angle K$
39. $\overline{TI} \cong \overline{LO}$, $\overline{IC} \cong \overline{OK}$, $\overline{TC} \cong \overline{LK}$

Lesson 4-4   pp. 244–248

Got It? 1. $\overline{BA} \cong \overline{DA}$ and $\overline{CA} \cong \overline{ED}$ (Given). $\angle CAB \cong \angle EAD$ (Vert. $\&$ are $\cong$.) So $\triangle ABC \cong \triangle ADE$ by SAS and $\angle C \cong \angle E$ because corresp. parts of $\cong$ $\triangle$ are $\cong$.
2a. It is given that M is the midpt. of $\overline{BC}$, so $\overline{BM} \cong \overline{CM}$ by the def. of midpt. $\overline{AB} \cong \overline{AC}$ (Given) and $\overline{AM} \cong \overline{AM}$ (Refl. Prop. of $\cong$), so $\triangle AMB \cong \triangle AMC$ by SSS. Thus $\angle AMB \cong \angle AMC$ because corresp. parts of $\cong$ $\triangle$ are $\cong$.
b. No; while $\overline{TR} \perp \overline{RS}$, if point L is not at sea level, then $\overline{TR}$ would not be $\perp$ to $\overline{RL}$.
Lesson Check 1. SAS; so $\overline{EA} \cong \overline{MA}$ because corresp. parts of $\cong$ $\triangle$ are $\cong$. **2.** SSS; so $\angle U \cong \angle E$ because corresp. parts of $\cong$ $\triangle$ are $\cong$. **3.** "Corresp. parts of $\cong$ $\triangle$ are $\cong$" is a short version of the def. of $\cong$ $\triangle$.
4. $\triangle KHL \cong \triangle NHM$ by AAS Thm.
Exercises 5. $\triangle KLJ \cong \triangle OMN$ by SAS; $\overline{KJ} \cong \overline{ON}$, $\angle K \cong \angle O$, $\angle J \cong \angle N$. **7.** $\overline{OM} \cong \overline{EB}$ and $\overline{ME} \cong \overline{RO}$ (Given). $\overline{OE} \cong \overline{OE}$ by the Refl. Prop. of $\cong$. $\triangle MOE \cong \triangle REO$ by SSS, so $\angle M \cong \angle R$ because corresp. parts of $\cong$ $\triangle$ are $\cong$. **9.** A pair of $\cong$ sides and a pair of $\cong$ $\&$ are given. Since $\overline{PT} \cong \overline{PT}$ (Refl. Prop. of $\cong$), then $\triangle STP \cong \triangle OTP$ by SAS. $\angle S \cong \angle O$ because corresp. parts of $\cong$ $\triangle$ are $\cong$. **11.** $\overline{KL}$ bisects $\angle PKQ$, so $\angle PKL \cong \angle QKL$. $\overline{KL} \cong \overline{KL}$ by Refl. Prop. of $\cong$. $\triangle PKL \cong \triangle QKL$ by SAS, so $\angle P \cong \angle Q$ because corresp. parts of $\cong$ $\triangle$ are $\cong$. **13.** $\angle PLK \cong \angle QLK$ because $\perp$ lines form rt. $\&$, and all rt. $\&$ are $\cong$. From the def. of $\angle$ bisector, $\angle PKL \cong \angle QKL$. So with $\overline{KL} \cong \overline{KL}$ by the Refl. Prop. of $\cong$, $\triangle PKL \cong \triangle QKL$ by ASA and $\angle P \cong \angle Q$ because corresp. parts of $\cong$ $\triangle$ are $\cong$.
15. $\overline{BA} \cong \overline{BC}$ (Given) and $\overline{BD}$ bisects $\angle ABC$ (Given) $\angle ABD \cong \angle CBD$ (Def. of $\angle$ bisector). $\overline{BD} \cong \overline{BD}$ (Refl. Prop. of $\cong$), so $\triangle ABD \cong \triangle CBD$ by SAS. $\angle ADB \cong \angle CDB$

(Corresp. parts of ≅ ▲ are ≅.) and ∠ADB and ∠CDB are suppl. so they must be rt. ▲. By def. of ⊥ lines, $\overline{BD} \perp \overline{AC}$. $\overline{AD} \cong \overline{CD}$ (Corresp. parts of ≅ ▲ are ≅.), so $\overline{BD}$ bisects $\overline{AC}$ (Def. of seg. bisector). **17.** The construction makes $\overline{AC} \cong \overline{BE}$, $\overline{AD} \cong \overline{BF}$, and $\overline{CD} \cong \overline{EF}$. So △ACD ≅ △BEF by SSS. Thus ∠A ≅ ∠B because corresp. parts of ≅ ▲ are ≅. **19.** It is given that $\overline{JK} \parallel \overline{QP}$, so ∠K ≅ ∠Q and ∠J ≅ ∠P because they are alt. int. ▲. With $\overline{JK} \cong \overline{PQ}$ (Given), △KJM ≅ △QPM by ASA and then $\overline{JM} \cong \overline{PM}$ because corresp. parts of ≅ ▲ are ≅. Thus M is the midpt. of $\overline{JP}$ by def. of midpt. So $\overline{KQ}$, which contains point M, bisects $\overline{JP}$ by the def. of segment bisector. **21.** Using the given information and $\overline{AE} \cong \overline{AE}$ (Refl. Prop. of ≅), △AKE ≅ △ABE by SSS. Thus ∠KAS ≅ ∠BAS because corresp. parts of ≅ ▲ are ≅. In △KAS and △BAS, $\overline{AK} \cong \overline{AB}$ (Given) and $\overline{AS} \cong \overline{AS}$ (Refl. Prop. of ≅), so △KAS ≅ △BAS by SAS. Thus $\overline{KS} \cong \overline{BS}$ because corresp. parts of ≅ ▲ are ≅, and S is the midpt. of $\overline{BK}$ by the def. of midpt. **23.** 3.5 **25.** 38 **27.** ASA **28.** AAS **29.** $\overline{AC}$ **30.** ∠C **31.** ∠A **32.** 105

Lesson 4-5   pp. 250–256

Got It? 1a. Yes; since $\overline{WV} \cong \overline{WS}$, ∠WVS ≅ ∠S by the Isosc. △ Thm.; yes; since ∠WVS ≅ ∠S, and ∠R ≅ ∠WVS (Given), ∠R ≅ ∠S (Trans. Prop. of ≅). Therefore, $\overline{TR} \cong \overline{TS}$ by the Converse of isosc. △ Thm. **b.** No; there is not enough information about the sides or ▲ of △RUV. **2.** 63 **3.** m∠A = 61, m∠BCD = 119 **Lesson Check 1a.** 70 **b.** 53 **2a.** 75 **b.** 48 **3.** 23, 134 **4a.** The ▲ opposite the ≅ sides are ≅. **b.** All three ▲ have measure 60, and all three sides are ≅. **5.** The ≅ ▲ should be opposite the ≅ sides.
Exercises 7. $\overline{UW}$; Converse of Isosc. △ Thm. **9.** Answers may vary. Sample: ∠VUY; Isosc. △ Thm. **11.** x = 38, y = 4 **13.** 108 **15.** 45 and 45; the sum of the measures of the acute ▲ must be 90, so the measure of each acute ∠ must be half of 90. **17.** 2.5 **19.** 35 **21.** 20, 80, 80 or 50, 50, 80 **23a.** $\overline{RS}$ **b.** $\overline{RS}$; Proof: $\overline{RS} \cong \overline{RS}$ (Refl. Prop. of ≅) and ∠PRS ≅ ∠QRS (def. of ∠ bisector). Also, ∠P ≅ ∠Q (Given). So △PRS ≅ △QRS by AAS. $\overline{PR} \cong \overline{QR}$ because corresp. parts of ≅ ▲ are ≅. **25.** $\overline{AE} \cong \overline{DE}$ (Given), so ∠A ≅ ∠D by the Isosc. △ Thm. Since $\overline{AB} \cong \overline{DC}$ (Given), then △ABE ≅ △DCE by SAS. **27a.** isosc. ▲ **b.** 900 ft; 1100 ft **c.** The tower is the ⊥ bisector of each base.
29.

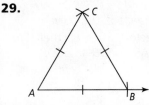

Draw $\overline{AB}$. Using AB as a radius, draw arcs with centers A and B. The intersection of these arcs is C. **31.** m = 36, n = 27

33. (−4, 0), (0, 0), (0, −4), (4, 4), (4, 8), (8, 4) **35.** (−1, 6), (2, 6), (2, 9), (5, 0), (5, 3), (8, 3) **37.** B **39.** C **41.** RC = GV; there are three pairs of ≅ ▲ and one pair of ≅ sides, so △TRC ≅ △HGV by AAS or ASA, and $\overline{RC} \cong \overline{GV}$ because corresp. parts of ≅ ▲ are ≅. **42.** The letters are the first letters of the days of the week; S, S. **43.** Yes; the ▲ share a common side, so they are ≅ by SAS. **44.** Yes; the vertical ▲ are ≅, so the ▲ are ≅ by SAS.

Algebra Review   p. 257

1. (−3, −7) **3.** no solution **5.** infinitely many solutions **7.** infinitely many solutions

Lesson 4-6   pp. 258–264

Got It? 1a. △PRS and △RPQ are rt. ▲ with ≅ hypotenuses ($\overline{SP} \cong \overline{QR}$) and ≅ legs ($\overline{PR} \cong \overline{PR}$). So △PRS ≅ △RPQ by HL. **b.** Yes; the two ▲ satisfy the three conditions of the HL Thm., so they are ≅. **2.** It is given that $\overline{AD}$ is the ⊥ bisector of $\overline{CE}$, so △CBD and △EBA are rt. ▲ and $\overline{CB} \cong \overline{EB}$ by the def. of ⊥ bisector. Also, $\overline{CD} \cong \overline{EA}$ (Given), so △CBD ≅ △EBA by HL.
Lesson Check 1. yes; △BCA ≅ △EFD **2.** yes; △MPL ≅ △MNO **3.** no **4.** yes; △XVR ≅ △TVR **5.** 13 cm; the hypotenuse is the longest side of a rt. △. **6.** Answers may vary. Sample: Alike: They both require information on two pairs of sides and one pair of ▲. Different: For HL, the rt. ▲ are NOT included between the two pairs of ≅ sides, while for SAS the ▲ ARE included between the two pairs of ≅ sides. **7.** No; △LMJ and △JKL are rt. ▲ with ≅ hypotenuses ($\overline{MJ} \cong \overline{KL}$) and ≅ legs ($\overline{LJ} \cong \overline{LJ}$), so △LMJ ≅ △JKL by HL.
Exercises 9a. △ABE and △DEB are rt. ▲. **b.** $\overline{BE} \cong \overline{EB}$ **c.** $\overline{AB} \cong \overline{DE}$ **d.** HL **11.** From the given information about ⊥ segments, △PTM and △RMJ are rt. ▲. $\overline{PM} \cong \overline{RJ}$ (Given), and since M is the midpt. of $\overline{TJ}$, $\overline{TM} \cong \overline{JM}$. Thus △PTM ≅ △RMJ by HL. **13.** x = −1, y = 3 **15.** Yes; the two ▲ are rt. ▲ with ≅ hypotenuses and one pair of ≅ legs, so the two ▲ are ≅ by HL. Then $\overline{RQ} \cong \overline{CB}$ because corresp. parts of ≅ ▲ are ≅. **17.** Using the information about ⊥ segments, △RST and △TUV are rt. ▲. $\overline{RS} \cong \overline{TU}$ (Given), and T is the midpt. of $\overline{RV}$ (Given), so $\overline{RT} \cong \overline{TV}$ (Def. of midpt.). Thus △RST ≅ △TUV by HL.
19.

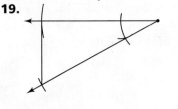

21.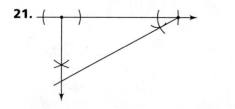

23. From the given information about an isosc. △, rt. ∡, and midpt., you can conclude that $\overline{KG} \cong \overline{KE}$ (Def. of isosc. △), △LKG and △DKE are rt.∡ (Def. of rt. △), and $\overline{LK} \cong \overline{DK}$ (Def. of midpt.). So △LKG ≅ △DKE by HL, and $\overline{LG} \cong \overline{DE}$ because corresp. parts of ≅ ▲ are ≅. **25.** No, the triangles are not ≅. Explanations may vary. Sample: $\overline{DF}$ is the hypotenuse of △DEF, so it is the longest side of the triangle. Therefore, it is greater than 5 and greater than 13 because it is longer than either of the legs. So $\overline{DF}$ cannot be congruent to $\overline{AC}$, which is the hypotenuse of △ABC and has length 13. **27.** △AEB and △CEB are rt. ▲ because the given information includes $\overline{BE} \perp \overline{EA}$ and $\overline{BE} \perp \overline{EC}$. △ABC is equilateral (Given), so $\overline{AB} \cong \overline{CB}$ by the def. of equilateral. Also, $\overline{BE} \cong \overline{BE}$ by the Refl. Prop. of ≅. So △AEB ≅ △CEB by HL. **29.** C **31.** △XRY ≅ △XRZ by HL, △YQX ≅ △YQZ by HL, △ZPX ≅ △ZPY by HL, and △XPS ≅ △YPS by SAS OR other correct pairs and explanations. **32.** △STU is isosceles. $\overline{ST} \cong \overline{UT}$ because corresp. parts of ≅ ▲ are ≅. **33.** △STU is equilateral. $\overline{ST} \cong \overline{US}$, $\overline{TU} \cong \overline{ST}$, and $\overline{US} \cong \overline{TU}$ because corresp. parts of ≅ ▲ are ≅. **34.** Yes; △ABC ≅ △LMN by HL. **35.** No; △LMN and △HJK have one pair of ≅ sides and one pair of ≅ ▲, but that is not enough to conclude that they are ≅. **36.** No; the hypotenuse of rt. △ABC is ≅ to a leg of rt. △RST, so the ▲ cannot be ≅.

Lesson 4-7   pp. 265–271

Got It? 1a. $\overline{AD}$ **b.** $\overline{AB}$ **2.** It is given that △ACD ≅ △BDC, so ∠ADC ≅ ∠BCD because corresp. parts of ≅ ▲ are ≅. Therefore, $\overline{CE} \cong \overline{DE}$ by the Converse of the Isosc. △ Thm. **3.** △PSQ ≅ △RSQ by SAS because $\overline{PS} \cong \overline{RS}$ (Given), ∠PSQ ≅ ∠RSQ (Given), and $\overline{SQ} \cong \overline{SQ}$ (Refl. Prop. of ≅). So $\overline{PQ} \cong \overline{RQ}$ and ∠PQT ≅ ∠RQT (Corresp. parts of ≅ ▲ are ≅.). Also, $\overline{QT} \cong \overline{QT}$ (Refl. Prop. of ≅), so △QPT ≅ △QRT by SAS. **4.** Using $\overline{AD} \cong \overline{AD}$ (Refl. Prop. of ≅) and the two given pairs of ≅ ▲, △ACD ≅ △AED by AAS. Then $\overline{CD} \cong \overline{ED}$ (Corresp. parts of ≅ ▲ are ≅.) and ∠BDC ≅ ∠FDE (Vert. ▲ are ≅.). Therefore, △BDC ≅ △FDE by ASA, and $\overline{BD} \cong \overline{FD}$ because corresp. parts of ≅ ▲ are ≅.

Lesson Check 1. $\overline{JK}$ **2.** ∠D

3.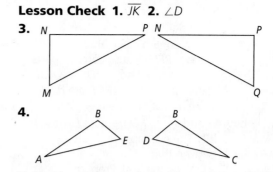

4.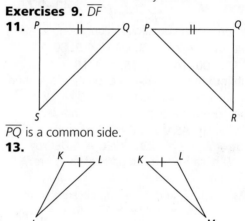

5. No; there are several ▲ with vertex J and several ▲ with vertex K, and a different ∠ at each vertex is in each △. **6.** Answers may vary. Sample: Based on the given statement that △PSY ≅ △SPL, $\overline{PL} \cong \overline{SY}$, and ∠L ≅ ∠Y because corresp. parts of ≅ ▲ are ≅. ∠PRL ≅ ∠SRY because vert. ▲ are ≅. So △PRL ≅ △SRY by AAS. **7.** Answers may vary. Sample: Prove △AEB ≅ △CED (by SAS) to get $\overline{AB} \cong \overline{CD}$ and ∠BAE ≅ ∠DCE. Use those ≅ segments and ≅ angles, along with rt. ▲ ADC and ABC, to show △ACD ≅ △CAB by ASA.

Exercises 9. $\overline{DF}$

11.

$\overline{PQ}$ is a common side.

13.

$\overline{KL}$ is a common side.

15. $\overline{RS} \cong \overline{UT}$ and $\overline{RT} \cong \overline{US}$ (Given), and $\overline{ST} \cong \overline{ST}$ (Refl. Prop. of ≅), so △RST ≅ △UTS by SSS. **17.** ∠1 ≅ ∠2 and ∠3 ≅ ∠4 (Given), and $\overline{QB} \cong \overline{QB}$ by the Refl. Prop. of ≅. So △QTB ≅ △QUB by ASA. Thus $\overline{QT} \cong \overline{QU}$ (Corresp. parts of ≅ ▲ are ≅.). $\overline{QE} \cong \overline{QE}$ (Refl. Prop. of ≅), so △QET ≅ △QEU by SAS. **19.** Since VT = VU + UT = UT + TS = US, $\overline{VT} \cong \overline{US}$. Therefore, △QVT ≅ △PSU by SAS. **21.** It is given that $\overline{AC} \cong \overline{EC}$ and $\overline{CD} \cong \overline{CB}$, and ∠C ≅ ∠C by the Refl. Prop. of ≅. So △ACD ≅ △ECB by SAS, and ∠A ≅ ∠E because corresp. parts of ≅ ▲ are ≅. **23.** Answers may vary. Sample:

25. $\overline{TE} \cong \overline{RI}$ and $\overline{TI} \cong \overline{RE}$ (Given) and $\overline{EI} \cong \overline{EI}$ (Refl. Prop. of $\cong$), so $\triangle TEI \cong \triangle RIE$ by SSS. Thus $\angle TIE \cong \angle REI$ because corresp. parts of $\cong$ $\triangle$ are $\cong$. Also, $\angle TDI \cong \angle ROE$ because $\angle TDI$ and $\angle ROE$ are rt. $\triangle$ (Given) and all rt. $\triangle$ are $\cong$. So $\triangle TDI \cong \triangle ROE$ by AAS and $\overline{TD} \cong \overline{RO}$ because corresp. parts of $\cong$ $\triangle$ are $\cong$. **27.** The overlapping $\triangle$ are $\triangle CAE$ and $\triangle CBD$. It is given that $\overline{AC} \cong \overline{BC}$ and $\angle A \cong \angle B$. Also, $\angle C \cong \angle C$ by the Refl. Prop. of $\cong$. So $\triangle CAE \cong \triangle CBD$ by ASA. **29.** D **31.** C **33a.** right **b.** right **c.** Reflexive **d.** HL
34.

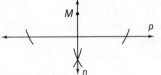

35. (1, 2) **36.** (1.5, 5.5) **37.** (1, 1)

Chapter Review   pp. 273–276

1. legs **2.** hypotenuse **3.** corollary **4.** congruent polygons **5.** $\overline{ML}$ **6.** $\angle U$ **7.** $\overline{ST}$ **8.** *ONMLK* **9.** 80 **10.** 3 **11.** 5 **12.** 35 **13.** 100 **14.** 145 **15.** $\angle D$ **16.** $\overline{MR}$ **17.** not enough information **18.** not enough information **19.** SAS **20.** AAS or ASA **21.** $\triangle TVY \cong \triangle YWX$ by AAS, so $\overline{TV} \cong \overline{YW}$ because corresp. parts of $\cong$ $\triangle$ are $\cong$. **22.** $\triangle BEC \cong \triangle DEC$ by ASA, so $\overline{BE} \cong \overline{DE}$ because corresp. parts of $\cong$ $\triangle$ are $\cong$. **23.** $\triangle BEC \cong \triangle DEC$ by SSS, so $\angle B \cong \angle D$ because corresp. parts of $\cong$ $\triangle$ are $\cong$. **24.** If $\parallel$ lines, alt. int. $\triangle$ are $\cong$, so $\angle LKM \cong \angle NMK$. Then $\triangle LKM \cong \triangle NMK$ by SAS, and $\overline{KN} \cong \overline{ML}$ because corresp. parts of $\cong$ $\triangle$ are $\cong$. **25.** $x = 4, y = 65$ **26.** $x = 55$, $y = 62.5$ **27.** $x = 65, y = 90$ **28.** $x = 7, y = 60$ **29.** $\overline{LN} \perp \overline{KM}$ (Given), so $\triangle KLN$ and $\triangle MLN$ are rt. $\triangle$. $\overline{KL} \cong \overline{ML}$ (Given) and $\overline{LN} \cong \overline{LN}$ (Refl. Prop. of $\cong$), so $\triangle KLN \cong \triangle MLN$ by HL. **30.** The given information on $\perp$ segments means $\triangle PSQ$ and $\triangle RQS$ are rt. $\triangle$. You know $\overline{PQ} \cong \overline{RS}$ (Given) and $\overline{QS} \cong \overline{QS}$ (Refl. Prop. of $\cong$). So $\triangle PSQ \cong \triangle RQS$ by HL. **31.** $\triangle AEC \cong \triangle ABD$ by SAS or ASA or AAS. **32.** $\triangle FIH \cong \triangle GHI$ by SAS. **33.** $\triangle TAR \cong \triangle TSP$ by ASA.

Chapter 5

Get Ready!   p. 281

1.   **2.**

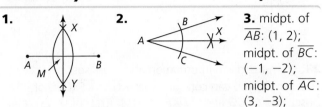

3. midpt. of $\overline{AB}$: (1, 2); midpt. of $\overline{BC}$: (−1, −2); midpt. of $\overline{AC}$: (3, −3); $AB = 2\sqrt{17}$; $BC = 2\sqrt{29}$; $AC = 4\sqrt{5}$ **4.** midpt. of $\overline{AB}$: (4, 2); midpt. of $\overline{BC}$: (4, 5); midpt. of $\overline{AC}$: (−1, 5); $AB = 10$; $BC = 2\sqrt{34}$; $AC = 6$ **5.** midpt. of $\overline{AB}$: (0, −3); midpt. of $\overline{BC}$: (1, 0); midpt. of $\overline{AC}$: (−1, 0); $AB = 4$; $BC = 2\sqrt{10}$; $AC = 2\sqrt{10}$ **6.** The team did not win. **7.** It is too late. **8.** $m\angle R \le 60$ **9.** −6 **10.** $-\frac{8}{3}$ **11.** undefined **12.** the length of a segment from a vertex to the opposite side **13.** the length of a $\perp$ segment from the point to the line **14.** a segment that connects the midpts. of 2 sides of the $\triangle$ **15.** The lines intersect at one point, or the lines have exactly one point in common.

Lesson 5-1   pp. 285–291

Got It? 1a. $\overline{AC} \parallel \overline{YZ}$, $\overline{CB} \parallel \overline{XY}$, $\overline{AB} \parallel \overline{XZ}$ **b.** 65; $\overline{UV}$ is a midsegment of $\triangle NOM$, so by the $\triangle$ Midseg. Thm., $\overline{UV} \parallel \overline{NM}$. Then $m\angle VUO = m\angle N = 65$ because corresp. $\triangle$ of $\parallel$ lines are $\cong$. **2.** $DC = 6$; $AC = 12$; $EF = 6$; $AB = 15$ **3.** 1320 ft
Lesson Check 1. $\overline{NO}$ **2.** 23 **3.** 4 **4.** A midsegment is a segment whose endpoints are the midpts. of two sides of a triangle. **5.** The segments are $\parallel$. **6.** The student is assuming that *L* is the midpt. of $\overline{OT}$, which is not given.
Exercises 7. $\overline{UY} \parallel \overline{XV}$, $\overline{UW} \parallel \overline{TX}$, $\overline{YW} \parallel \overline{TV}$ **9.** $\overline{FE}$ **11.** $\overline{AB}$ **13.** $\overline{AC}$ **15.** 40 **17.** 160 **19.** 13 **21.** 6 **23.** 17 **25.** 156 m **27.** 114 ft 9 in.; because the red segments divide the legs into four $\cong$ parts, the white segment divides each leg into two $\cong$ parts. The white segment is a midsegment of the triangular face of the building, so its length is one half the length of the base. **29.** 40; $\overline{ST}$ is a midsegment of $\triangle PQR$, so by the $\triangle$ Midseg. Thm., $\overline{ST} \parallel \overline{PR}$. Then $m\angle QPR = m\angle QST$ because corresp. $\triangle$ of $\parallel$ lines are $\cong$. **31.** 60 **33.** 100 **35.** 18.5 **37.** C **39.** 50 **41.** $x = 6$; $y = 6.5$ **43.** 24 **45.** Draw $\overrightarrow{CA}$. Find *P* on $\overrightarrow{CA}$ such that $CA = AP$. Draw $\overline{PD}$. Construct the $\perp$ bisector of $\overline{PD}$. Label the intersection point *B*. Draw $\overline{AB}$. This is a midsegment of $\triangle CPD$. According to the $\triangle$ Midsegment Thm., $\overline{AB} \parallel \overline{CD}$ and $AB = \frac{1}{2}CD$.
47. *G*(4, 4); *H*(0, 2); *J*(8, 0) **49.** 1.8 **51.** 80

53. △FBD ≅ △FCE, △BAE ≅ △CAD, △DAF ≅ △EAF, △ABF ≅ △ACF **54.** Answers may vary. Sample: ∠BFD ≅ ∠CFE because they are vertical ⦞. ∠1 ≅ ∠2 is given. By the ∠ Addition Post., it follows that ∠BFA ≅ ∠CFA. $\overline{BF}$ ≅ $\overline{CF}$ is given, and $\overline{FA}$ ≅ $\overline{FA}$ by the Refl. Prop. Therefore, △BFA ≅ △CFA by SAS. $\overline{AB}$ ≅ $\overline{AC}$ because corresp. parts of ≅ △ are ≅. **55.** 6 **56.** 68 **57.**

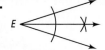

Lesson 5-2   pp. 292–299

Got It? 1. 8 **2a.** Any point on the ⊥ bis. of $\overline{PS}$ **b.** At the intersection point of ℓ and the perpendicular bisector of $\overline{PS}$; let X be the intersection point of ℓ and the perpendicular bisector of $\overline{PS}$. By the ⊥ Bis. Thm., XR = XS and XS = XP, so XR = XS = XP. Thus, X is equidistant from R, S, and P. **3.** 21

Lesson Check 1. $\overline{AC}$ is the ⊥ bisector of $\overline{DB}$. **2.** 15 **3.** 18 **4.** Answers may vary. Sample: **5.** Draw the ⊥ seg-

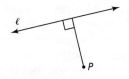

ments that join P to $\overrightarrow{OL}$ and $\overrightarrow{OX}$. Use a ruler to determine if OL = OX. If OL = OX, then P is on the bisector of ∠LOX.

Exercises 7. 3 **9.** Coleman School; it is on 6th Ave., which is (approximately) the ⊥ bisector of 14th St. between 8th Ave. and Union Square. **11.** Draw $\overline{HS}$ and find its midpt., M. Through M, construct the line ⊥ to $\overline{HS}$. Any point on this line will be equidistant from H and S. **13.** $\overrightarrow{HL}$ bisects ∠KHF; point L is equidistant from the sides of the ∠, so L is on the bisector of ∠KHF by the Converse of the ∠ Bisector Thm. **15.** 54; 54 **17.** y = 3, ST = 15, TU = 15 **19.** 10 **21.** isosc., because TW = ZW **23.** At the point on $\overline{XY}$ that lies on the bisector of ∠GPL; the goalie does not know to which side of her the player will aim his shot, so she should keep herself equidistant from the sides of ∠GPL. Points on the bisector of ∠GPL are equidistant from $\overline{PG}$ and $\overline{PL}$. If she moves to a point on the ⊥ bisector of $\overline{GL}$, she will be closer to $\overline{PL}$ than to $\overline{PG}$. **25a.**

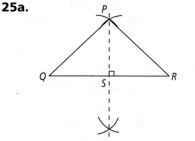

b. Answers may vary. Sample: Since P is on the perpendicular bisector of QR, it is equidistant from Q and R by the perpendicular bisector theorem. **27.** A pt. is on the bisector of an ∠ if and only if it is equidistant from the sides of the ∠. **29.** No; A is not equidistant from the sides of ∠TXR. **31.** Yes; A is equidistant from the sides of ∠TXR. **33.** $\overline{PA}$ ≅ $\overline{PB}$ (Given) and ∠AMP ≅ ∠BMP because all rt. ⦞ are ≅. Also, $\overline{PM}$ ≅ $\overline{PM}$ by the Refl. Prop. of ≅. So rt. △PMA ≅ rt. △PAB by HL and $\overline{AM}$ ≅ $\overline{BM}$ because corresp. parts of ≅ △ are ≅. Therefore $\overleftrightarrow{PM}$ is the ⊥ bisector of $\overline{AB}$, by the def. of ⊥ bisector. **35.** In rt. △SPQ and rt. △SRQ, $\overline{SP}$ ≅ $\overline{SR}$ (Given) and $\overline{QS}$ ≅ $\overline{QS}$ (Refl. Prop. of ≅), so △SPQ ≅ △SRQ by HL. Thus ∠PQS ≅ ∠RQS because corresp. parts of ≅ △ are ≅, and $\overrightarrow{QS}$ bisects ∠PQR by the def. of ∠ bisector. **37.** Line ℓ through the midpts. of two sides of △ABC is equidistant from A, B, and C. This is because △1 ≅ △2 and △3 ≅ △4 by ASA. $\overline{AD}$ ≅ $\overline{BE}$ and $\overline{BE}$ ≅ $\overline{CF}$ because corresp. parts of ≅ △ are ≅. By the Trans. Prop. of ≅,

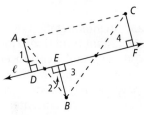

$\overline{AD}$ ≅ $\overline{BE}$ ≅ $\overline{CF}$. By the def. of ≅, AD = BE = CF, so points A, B, and C are equidistant from line ℓ. **39.** C **41.** C **43.** 6 **44.** 120 **45.** $\frac{1}{3}$ **46.** undefined **47.** Answers may vary. Sample: It is a vertical line that contains the point (5, 0).

Lesson 5-3   pp. 301–307

Got It? 1. (6, 5) **2.** at the circumcenter of the △ whose vertices are the three trees

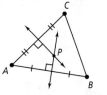

3a. 61 **b.** No; answers may vary. Sample: The distance from Q to $\overline{KL}$ is QN, the length of the shortest segment from Q to $\overline{KL}$. From part (a), QN = 61, so QP > 61.

Lesson Check 1. (3, 2.5) **2.** 6 **3.** obtuse △ **4.** Since the three ⊥ bisectors of a △ are concurrent, the third ⊥ bisector goes through the pt. of intersection of the other two ⊥ bisectors. **5.** Answers may vary. Sample: The diagram does not show that $\overline{QC}$ bisects ∠SQR, so you cannot conclude that point C is equidistant from the sides of ∠SQR. **6.** Each one is a point of concurrency of bisectors of parts of a △, each is equidistant from three parts of the △, and each is the center of a ⊙ that contains three points of the △. The circumcenter is equidistant from three points, while the incenter is equidistant from three segments. The △ is inside the ⊙ centered at the circumcenter and outside the ⊙ centered at the incenter.

Exercises 7. (−2, −3) **9.** (1.5, 1) **11.** (−3, 1.5) **13.** (3.5, 3) **15.** C **17.** 2 **19.** Isosceles; SR = ST,

so ∠SRT ≅ ∠STR (Isosc. △ Thm.). Since P is the incenter of △RST, $\overline{PR}$ and $\overline{PT}$ are ∠ bisectors. So m∠PRT = $\frac{1}{2}$m∠SRT = $\frac{1}{2}$m∠STR = m∠PTR. Thus PR = PT by the Converse of the Isosc. △ Thm.
21. Same method as for Exercise 20.

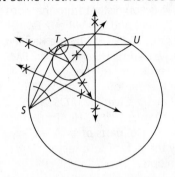

23. An interpretation of the passage is that the treasure is equidistant from three Norway pines. To find the treasure, Karl can find the circumcenter of the △ whose vertices are the three pines. **25.** P; the markings in the diagram show that P is the incenter of the triangular station and C is the circumcenter. If you stand at P, you will be equidistant from the three sides along which the buses are parked. If you move away from P, you will move closer to some of the buses. **27.** true

29.

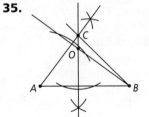

As the diagram shows, circle C is circumscribed about both △PQR and △PQS, so points R and S do not have to coincide.

31. Never; if you have three ∥ lines ℓ, m, and n, with m in between ℓ and n, then a point equidistant from ℓ and m would be (midway) between them. A point equidistant from m and n would be (midway) between those two lines. A point equidistant from all three would therefore have to be on both sides of m! This is impossible.
33. B **35.** C **37.** 4 **38.** 17 **39.** (3, 8) **40.** (5, 3.5)

Lesson 5-4   pp. 309–315

Got It? 1a. 13.5 **b.** 2 : 1; ZA = $\frac{2}{3}$CZ and AC = $\frac{1}{3}$CZ, so ZA : AC = $\frac{2}{3}$: $\frac{1}{3}$ = 2 : 1. **2a.** A median; it connects a vertex of △ABC and the midpt. of the opposite side. **b.** Neither; E is a midpt. of △ABC, but G is not a vertex of △ABC. **c.** An altitude; it extends from a vertex of △ABC and is ⊥ to the opposite side. **3.** (1, 2)
Lesson Check 1. median **2.** 6 **3.** 7.5 **4.** $\overline{AB}$, $\overline{AC}$
5. $\overline{HJ}$ does not contain a vertex of △ABC, so it is not an altitude of △ABC. **6.** No; any pair of altitudes meet at the orthocenter of the △. **7.** They are ⊥; since A is the orthocenter of △ABC, A lies on the altitude from B to $\overline{AC}$.

B also lies on this altitude, so the altitude from B to $\overline{AC}$ must be $\overline{BA}$. Therefore, $\overline{BC}$ ⊥ $\overline{AC}$.
Exercises 9. ZY = 4.5, ZU = 13.5 **11.** Median; it connects a vertex of △ABC and the midpt. of the opposite side. **13.** Altitude; it extends from a vertex of △ABC and is ⊥ to the opposite side. **15.** (6, 4)
17. H **19.** J **21.** 125
23.

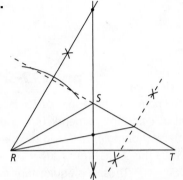

25. $\overline{BD}$ **27.** $\overline{OD}$ **29.** The folds should show the ⊥ bisectors of the sides to identify the midpt. of each side, and also show the fold through each vertex and the midpt. of the opposite side. **31.** C **33.** Answers may vary. Sample: The ∠ bisector of the vertex ∠ forms two △ that are ≅ by SAS. Therefore the 2 segments formed on the base are ≅ (so the ∠ bisector contains a median), and the two △ formed by the ∠ bisector and the base are rt. △ (so the ∠ bisector contains an altitude). Thus the median and the altitude are the same.
35.

Draw $\overleftrightarrow{AB}$. Construct the ⊥ to $\overleftrightarrow{AB}$ through O. Draw $\overleftrightarrow{BO}$. Construct the ⊥ to $\overleftrightarrow{BO}$ through A. The two ⊥s intersect at C. Draw $\overline{BC}$.
37. A is the intersection of the altitudes, so it is the orthocenter; B is the intersection of the ∠ bisectors, so it is the incenter; C is the intersection of the medians, so it is the centroid; D is the intersection of the ⊥ bisectors of the sides, so it is the circumcenter. **39.** incenter **41.** H
43. Both; the markings show directly that $\overline{XY}$ is a ⊥ bisector. The two △ formed are congruent by SAS, so the two △ at top are ≅. Therefore, $\overline{XY}$ is also an ∠ bisector. **44.** Neither; $\overline{XY}$ connects vertex X and the midpt., Y, of side $\overline{PQ}$, so $\overline{XY}$ is a median. **45.** Two angles are not congruent. **46.** You are 16 years old.
47. m∠A ≥ 90

Lesson 5-5   pp. 317–322

Got It? 1a. Assume temporarily that $\triangle BOX$ is acute.
b. Assume temporarily that no pair of shoes you bought cost more than $25. **2a.** II and III **b.** No; if $\triangle ABC$ is an isosc., nonequilateral $\triangle$, then Statement III is true but Statement II is not true. Therefore, Statements II and III are not equivalent. **3.** Assume temporarily that $y = 6$. Then $7(x + 6) = 70$; divide each side by 7 to get $x + 6 = 10$ and so $x = 4$. But this contradicts the given statement that $x \neq 4$. The temporary assumption that $y = 6$ led to a contradiction, so we can conclude that $y \neq 6$.

Lesson Check 1. Assume temporarily that at least one $\angle$ in quadrilateral $ABCD$ is not a rt. $\angle$.
2. Lines a and b meet at P.

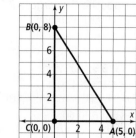

3. The negation of "$\angle A$ is obtuse" is "$\angle A$ is an acute or a rt. $\angle$."

Exercises 5. Assume temporarily that $\angle J$ is a rt. $\angle$.
7. Assume temporarily that no $\angle$ is obtuse. **9.** Assume temporarily that $m\angle 2 \leq 90$. **11.** I and III **13.** I and II **15a.** rt. $\angle$ **b.** rt. $\angle s$ **c.** 90 **d.** 180 **e.** 90 **f.** 90 **g.** 0 **h.** more than one rt. $\angle$ **i.** at most one rt. $\angle$ **17.** Assume temporarily $\ell \parallel p$. Then $\angle 1 \cong \angle 2$ because if lines are $\parallel$ then corresp. $\angle s$ are $\cong$. But this contradicts the given statement that $\angle 1 \not\cong \angle 2$. Therefore the temporary assumption is false, and we can conclude that $\ell \nparallel p$.
19. Assume temporarily that $\overline{XB} \not\cong \overline{XA}$. **21.** I and III
23. Assume temporarily that at least one base $\angle$ is a rt. $\angle$. Then both base $\angle s$ must be rt. $\angle s$, by the Isosc. $\triangle$ Thm. But this contradicts the fact that a $\triangle$ is formed, because in a plane two lines $\perp$ to the same line are $\parallel$. Therefore the temporary assumption is false that at least one base $\angle$ is a rt. $\angle$, and we can conclude that neither base $\angle$ is a rt. $\angle$.
25. Assume temporarily that an obtuse $\triangle$ can contain a rt. $\angle$. Then the measure of the obtuse $\angle$ plus the measure of the rt. $\angle$ must be greater than $90 + 90 = 180$. This contradicts the $\triangle$ Angle-Sum Thm., so the temporary assumption that an obtuse $\triangle$ can contain a rt. $\angle$ is incorrect. We can conclude that an obtuse $\triangle$ cannot contain a rt. $\angle$. **27.** The culprit entered the room through a hole in the roof; all the other possibilities were ruled out. **29.** Assume temporarily $\overline{XB} \perp \overline{AC}$. Then $\angle BXA \cong \angle BXC$ (All rt. $\angle s$ are $\cong$.), $\angle ABX \cong \angle CBX$ (Given), and $\overline{BX} \cong \overline{BX}$ (Reflexive Prop. of $\cong$), so $\triangle BXA \cong \triangle BXC$ by ASA and $BA = BC$ because corresp. parts of $\cong$ $\angle s$ are $\cong$. But this contradicts the given statement that $\triangle ABC$ is scalene. Therefore the temporary assumption that $\overline{XB} \perp \overline{AC}$ is wrong, and we can conclude that $\overline{XB}$ is not $\perp$ to $\overline{AC}$. **31.** D **33.** C **35.** 24 cm

36. 30 and 120 **37.** Law of Syllogism
38. $\overline{AC}$, $\overline{BC}$, $\overline{AB}$ **39.** $\overline{CA}$, $\overline{BC}$, $\overline{BA}$

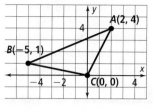

40. $\overline{AB}$, $\overline{AC} \cong \overline{BC}$

Algebra Review   p. 323

1. $x \leq -1$ **3.** $x > -10.5$ **5.** $a \leq 90$ **7.** $z > 0.5$
9. $n \leq -\frac{7}{2}$ **11.** $x > -5$ **13.** $x \leq -1$ **15.** $x > -8$
17. $x \geq -362$

Lesson 5-6   pp. 324–331

Got It? 1. $\angle 5$ is an ext. $\angle$ of $\triangle ACD$, so by the Corollary to the $\triangle$ Ext. $\angle$ Thm., $m\angle 5 > m\angle C$.
2. Holingsworth Rd. and MLK Blvd. **3.** $\overline{OX}$; $m\angle X = 180 - (130 + 24) = 26$, so $m\angle O > m\angle X > m\angle S$. By Theorem 5-11, $SX > OS > OX$. **4a.** No; $2 + 6 \not> 9$. **b.** Yes; the sum of the lengths of any two sides is greater than the length of the third side. **5.** 3 in. $< x < 11$ in.

Lesson Check 1. $\overline{BC}$ **2.** $\angle C$ **3.** No; $5 + 4 \not> 10$.
4. If the perimeter is 16 and the length of one side is 8, then the sum of the lengths of the other two sides is $16 - 8 = 8$. However, the $\triangle$ Inequality Thm. tells you that if the length of one side is 8, then the sum of the lengths of the other two sides is greater than 8. So the friend is incorrect. **5.** No; the adjacent interior $\angle$ would measure 92. Then, because a second $\angle$ of the $\triangle$ measures 90, the sum of the $\angle$ measures would exceed 180, which contradicts the $\triangle$ Angle-Sum Thm.

Exercises 7. This is true by the Corollary to the $\triangle$ Ext. $\angle$ Thm. **9.** $\angle M$, $\angle L$, $\angle K$ **11.** $\angle G$, $\angle H$, $\angle J$ **13.** $\angle E$, $\angle F$, $\angle D$ **15.** $\overline{MN}$, $\overline{NO}$, $\overline{MO}$ **17.** $\overline{TU}$, $\overline{UV}$, $\overline{TV}$ **19.** $\overline{EF}$, $\overline{DE}$, $\overline{DF}$
21. No; $2 + 3 \not> 6$. **23.** No; $8 + 10 \not> 19$. **25.** Yes; $2 + 9 > 10$, $9 + 10 > 2$, and $2 + 10 > 9$.
27. 4 ft $< x < 20$ ft **29.** 0 cm $< x < 12$ cm
31. 3 yd $< x < 11$ yd **33.** Place the computer at the

corner that forms a rt. ∠; place the bookshelf along the wall opposite the rt. ∠. In a rt. △ the rt. ∠ is the largest ∠, and the longest side of a △ is opposite the largest ∠. **35.** The dashed red line and the courtyard walkway determine three sides of a △, so by the △ Inequality Thm., the path that follows the dashed red line is longer than the courtyard walkway. **37.** $\overline{RS}$ **39.** $\overline{XY}$ **41.** Answers may vary. Sample: The sum of the ∠ measures of a △ is 180, so $m\angle T + m\angle P + m\angle A = 180$. Since $m\angle T = 90$, $m\angle P + m\angle A = 90$ and so $m\angle T > m\angle A$ (Comparison Prop. of Inequality). Therefore $PA > PT$ by Thm. 5-11. **43.** (2, 4), (2, 5), (2, 6), (3, 3), (3, 4), (3, 5), (3, 6), (3, 7), (4, 3), (4, 4), (4, 5), (4, 6), (4, 7), (4, 8) **45.** Answers may vary. Sample:

Find point D on $\overrightarrow{BC}$ such that $DC = AC$. $m\angle D = m\angle CAD$ by the Isosc. △ Thm. Now $m\angle DAB > m\angle DAC$ by the Comparison Prop. of Inequality, and so $m\angle DAB > m\angle D$ by substitution. Thus $DB > AB$ by Thm. 5-11. We know $DC + CB = DB$ by the Segment Add. Post., so $DC + CB > AB$ (Substitution) and $AC + CB > AB$ (Substitution). **47.** 5 **49.** 2 **50.** Assume temporarily that the side is less than 2 ft long. **51.** Assume temporarily that no two ∠ of △PQR are ≅. **52.** SSS **53.** 40 **54.** 25 **55.** no

Lesson 5-7   pp. 332–339

Got It? 1a. $LN > OQ$ **b.** Assume temporarily that $m\angle P \not> m\angle A$. If $m\angle P = m\angle A$, then $\triangle ABC \cong \triangle PQR$ (SAS), but this contradicts the fact that $BC \neq QR$. If $m\angle P < m\angle A$, then by the Hinge Thm., $QR < BC$. This contradicts the fact that $QR > BC$. Therefore, $m\angle P > m\angle A$. **2.** The 40° opening; the lengths of the blades do not change as the scissors open. The included angle between the blades of the 40° opening is greater than the included angle of the 35° opening, so by the Hinge Thm., the distance between the blades is greater for the 40° opening. **3.** $-6 < x < 24$ **4.** From the given information, $LO = ON$ (def. of midpt.) and $m\angle MOL = 100$ (suppl. ∠ to $\angle MON$). Since $\overline{MO} \cong \overline{MO}$, and $m\angle MOL > m\angle MON$, the Hinge Thm. yields $LM > MN$.

Lesson Check 1. $FD > BC$ **2.** $m\angle UST > m\angle VST$ **3.** Answers may vary. Sample: As a door opens, and the angle between the door and doorway increases, the distance between the door jamb and the nonhinge vertical edge of the door increases. **4.** The two ∠ that are formed by ≅ sides are $\angle ABD$ and $\angle CDB$. Since the side opposite $\angle ABD$ is longer than the side opposite $\angle CDB$, the correct conclusion is $m\angle ABD > m\angle CDB$. **5.** Answers may vary. Sample: Both deal with a pair of ▲ that have two pairs of ≅ corresponding sides along with a relationship between the ∠ formed by those sides.

Exercises 7. $PR < RT$ **9.** no conclusion **11.** $6 < x < 38$ **13.** $3.5 < y < 17.5$ **15a.** Converse of Isosc. △ Thm. **b.** Given **c.** Def. of midpt. **d.** $BC = CD$ **e.** Given **f.** Hinge Theorem **17.** $m\angle QTR > m\angle RTS$; $m\angle PTQ + m\angle QTR + m\angle RTS = 180$, so $m\angle PTQ + m\angle RTS = 88$. Thus $m\angle RTS < 88$ by the Comparison Prop. of Inequality, so $m\angle QTR > m\angle RTS$ by the Transitive Prop. of Inequality. **19a.** The two labeled ∠ are formed by ≅ corresp. sides of the two ▲, so the side opposite the 94° ∠ should be longer than the side opposite the 91° ∠, by the Hinge Thm. Thus the side labeled "13" must be longer than the side labeled "14." **b.** Answers may vary. Sample: Switch the angle labels 91° and 94°. **21.** A **23.** $\triangle ABE \cong \triangle CBD$ (Given) so $\triangle ABE$ and $\triangle CBD$ are isosc. with $AB = EB = DB = CB$. Since $m\angle EBD > m\angle ABE$ (Given), $ED > AE$ by the Hinge Thm. **25.** A Using the diagram in the Plan for Proof, $BC = YZ$, $BD = YX$, and $m\angle ZYX = m\angle CBD$, so $\triangle DBC \cong \triangle XYZ$ by SAS. $\angle FBA \cong \angle FBD$ (Def. of ∠ bisector), $\overline{BD} \cong \overline{BA}$ (because each is ≅ to $\overline{XY}$), and $\overline{BF} \cong \overline{BF}$, so $\triangle ABF \cong \triangle DBF$ by SAS. $\overline{AF} \cong \overline{DF}$, because corresp. parts of ≅ ▲ are ≅. $AF + FC = AC$ (Segment Addition Post.), so $DF + FC = AC$. Using the △ Inequality Thm. in $\triangle FDC$, $DF + FC > DC$. Now $AC > DC$ by substitution. Since $DC = XZ$ (Corresp. parts of ≅ ▲ are ≅.), it follows that $AC > XZ$ by substitution. **27.** I **29.** I **31.** $\angle T$, $\angle F$, $\angle R$ **32.** $\angle M$, $\angle L$, $\angle K$ **33.** 4 cm $< x <$ 34 cm **34.** 5 ft $< x <$ 17 ft **35.** 0 in. $< x <$ 6 in. **36.** $\overline{GH}$ **37.** $\frac{3}{4}$ **38.** $-\frac{8}{3}$ **39.** 1

Chapter Review   pp. 341–344

1. median **2.** distance from a point to a line **3.** incenter **4.** 15 **5.** 11 **6.** $L\left(\frac{5}{2}, -\frac{1}{2}\right)$; $M\left(\frac{7}{2}, \frac{1}{2}\right)$; slope of $\overline{AB} = 1$ and slope of $\overline{LM} = 1$, so $\overline{LM} \parallel \overline{AB}$; $AB = 2\sqrt{2}$ and $LM = \sqrt{2}$, so $LM = \frac{1}{2}AB$. **7.** Let point S be second base and point T be third base. Find the midpt. M of $\overline{ST}$ and then through M construct the line $\ell \perp$ to $\overline{ST}$. Points of the baseball field that are on line ℓ are equidistant from second and third base. **8.** 40 **9.** 40 **10.** 6 **11.** 11 **12.** 33 **13.** 33 **14.** (0, 0) **15.** (3, 2) **16.** (4, 4) **17.** (5, 1) **18.** 45 **19.** 40 **20.** 25 **21.** $\overline{AB}$ is an altitude; it is a segment from a vertex that is $\perp$ to the opposite side. **22.** $\overline{AB}$ is a median; it is a segment from a vertex to the midpt. of the opposite side. **23.** $QZ = 8$, $QM = 12$ **24.** (0, −1) **25.** (2, −3) **26.** Assume temporarily that neither of the two numbers is even. That means each number is odd, so the product of the two numbers must be odd. That contradicts the statement that the product of the two numbers is even. Thus the temporary assumption is false, and we can conclude that at least one of the numbers must be even. **27.** Assume temporarily that the third line intersects neither of the first two. Then it is ∥ to both of them. Since the first two lines are ∥ to the same line, they are ∥ to each other. This contradicts the given information. Therefore the temporary assumption is false, and the third line must intersect at least one of the two others. **28.** Assume temporarily that there is a △ with two obtuse ⦞. Then the sum of the measures of those two ⦞ is greater than 180, which contradicts the △ Angle-Sum Thm. Therefore the temporary assumption is false, and a △ can have at most one obtuse ∠. **29.** Assume temporarily that an equilateral △ has an obtuse ∠. Since all the ⦞ are ≅ in an equilateral △, then all three ⦞ must be obtuse. But we showed in Ex. 28 that a △ can have at most one obtuse ∠. Therefore the temporary assumption is false, and an equilateral △ cannot have an obtuse ∠. **30.** Assume temporarily that each of the three integers is less than or equal to 3. Then the sum of the three integers must be less than or equal to 3 · 3, or 9. This contradicts the given statement that the sum of the three integers is greater than 9. Therefore the temporary assumption is false, and you can conclude that one of the integers must be greater than 3. **31.** $\overline{RS}$, $\overline{ST}$, $\overline{RT}$ **32.** No; $5 + 8 \not> 15$. **33.** Yes; $10 + 12 > 20$, $10 + 20 > 12$, and $12 + 20 > 10$. **34.** 1 ft $< x <$ 25 ft **35.** $<$ **36.** $>$ **37.** $<$

Chapter 6

Get Ready! p. 349

1. 30 **2.** 42 **3.** 22 **4.** yes **5.** no **6.** yes **7.** ∥ **8.** ⊥ **9.** neither **10.** ASA **11.** SAS **12.** AAS **13.** Answers may vary. Sample: polygon in which all the ⦞ are ≅ **14.** Answers may vary. Sample: four-sided figure formed by joining two isosc. ⦞. **15.** Answers may vary. Sample: Angles that follow one right after the other.

Lesson 6-1 pp. 353–358

Got It? 1a. 2700 **b.** Answers may vary. Sample: Divide 1980 by 180, and then add 2. **2.** 140 **3.** 102 **4.** 40 **Lesson Check 1.** 1620 **2.** 360 **3.** 144, 36 **4.** Yes; explanations may vary. Sample: rectangle that is not square **5.** ∠2 and ∠4; their measures are equal; answers may vary. Sample: Two ⦞ suppl. to the same ∠ must be ≅. **6.** Answers may vary. Sample: ext. ∠ would measure 50, which is not a factor of 360.
Exercises 7. 900 **9.** 2160 **11.** 180,000 **13.** 150 **15.** 60, 120, 120, 60 **17.** 145 **19.** 10 **21.** 3.6 **23.** 8 **25.** 18 **27.** octagon; $m\angle 1 = 135$, $m\angle 2 = 45$ **29.** $y = 103$, $z = 70$ **31.** 36 **33.** 144; 10 **35.** 150; 12 **37.** 45, 45, 90 **39a.** $180n$ **b.** $(n - 2) \cdot 180$ **c.** $180n - [(n - 2) \cdot 180] = 360$ **d.** Polygon Ext. ∠ Sum Theorem **41.** octagon **43a.** Answers may vary. Sample: The sum of the interior ∠

measures $= (n - 2)180$. All ⦞ of a regular n-gon are ≅. So each interior ∠ measure $= \frac{180(n - 2)}{n}$, and $\frac{180(n - 2)}{n} = \frac{180n - 360}{n} = 180 - \frac{360}{n}$. **b.** As n gets larger, $\frac{360}{n}$ gets smaller. The interior angle measure gets closer to 180. The polygon becomes more like a circle. **45.** 225 **47.** 79 **49.** $\overline{CD}$; the longer side is opposite the larger ∠. **50.** Distr. Prop. **51.** Refl. Prop. of ≅ **52.** Sym. Prop. of ≅ **53.** ASA **54a.** ∠HGE **b.** ∠GHE **c.** ∠HEG **d.** $\overline{GH}$ **e.** $\overline{HE}$ **f.** $\overline{EG}$

Lesson 6-2 pp. 359–366

Got It? 1. 94
2. 1. ABCD is a ▱ and $\overline{AK} \cong \overline{MK}$. (Given)
   2. ∠A ≅ ∠BCD (Opp. ⦞ of a ▱ are ≅.)
   3. ∠A ≅ ∠CMD (Isosc. △ Theorem)
   4. ∠BCD ≅ ∠CMD (Transitive Prop. of ≅)
3a. $x = 4$, $y = 5$, $PR = 16$, $SQ = 10$ **b.** No; answers may vary. Sample: Solutions to a system of equations do not depend on the method used to solve it. **4.** 5
Lesson Check 1. 53 **2.** 127 **3.** 5 **4.** 7 **5.** $ED = 12$, $FD = 24$ **6.** Answers may vary. Sample: The ∠ opposite the given ∠ is congruent to it. The other two ⦞ and the given ∠ are consecutive ⦞, so they are supplements of the given ∠. **7.** A quad. and a ▱ both have four sides, but if both pairs of opp. sides are ∥, then the figure is a ▱. **8.** It is not given that $\overleftrightarrow{PQ}$, $\overleftrightarrow{RS}$, and $\overleftrightarrow{TV}$ are ∥.
Exercises 9. 127 **11.** 100 **13a.** Def. of ▱ **b.** If lines are ∥, then alt. int. ⦞ are ≅. **c.** Opp. sides of a ▱ are ≅. **d.** △ABE ≅ △CDE **e.** Corresp. parts of ≅ ⦞ are ≅. **f.** $\overline{AC}$ and $\overline{BD}$ bisect each other at E. **15.** $x = 5$, $y = 7$ **17.** 3 **19.** 9 **21.** 2.25 **23.** 4.5 **25.** 20 **27.** $x = 12$, $y = 4$ **29.** 22, $AB = 23.6$, $BC = 18.5$, $CD = 23.6$, $AD = 18.5$ **31a.** 2.5 ft **b.** 129 **c.** Answers may vary. Sample: As $m\angle E$ increases, $m\angle D$ decreases. ∠E and ∠D are suppl.
33. Answers may vary. Sample:
   1. ▱ LENS and NGTH (Given)
   2. ∠L ≅ ∠ENS and ∠GNH ≅ ∠T. (Opp. ⦞ of a ▱ are ≅.)
   3. ∠ENS ≅ ∠GNH (Vert. ⦞ are ≅.)
   4. ∠L ≅ ∠T (Transitive Prop. of ≅)
35. Answers may vary. Sample:
   1. ▱ LENS and NGTH (Given)
   2. ∠E is suppl. to ∠ENS. (Consecutive ⦞ in a ▱ are suppl.)
   3. ∠GNH ≅ ∠ENS (Vert. ⦞ are ≅.)
   4. ∠GNH ≅ ∠T (Opp. ⦞ of a ▱ are ≅.)
   5. ∠ENS ≅ ∠T (Transitive Prop. of ≅)
   6. ∠E is suppl. to ∠T. (Substitution Prop.)
37. 1. ▱ RSTW and XYTZ (Given)
   2. $\overline{XY} \parallel \overline{TZ}$ and $\overline{TZ} \parallel \overline{RS}$. (Def. of ▱)
   3. $\overline{XY} \parallel \overline{RS}$ (If two lines are ∥ to the same line, then they are ∥ to each other.)

39. $m\angle 1 = 71$, $m\angle 2 = 28$, $m\angle 3 = 81$

41. $AB = CD = 13$, $BC = AD = 33$

43. Answers may vary. Sample:
1. $\overline{AB} \parallel \overline{CD}$, $\overline{CD} \parallel \overline{EF}$ (Given)
2. $\overline{BG} \parallel \overline{AC}$, $\overline{DH} \parallel \overline{CE}$ (Construction)
3. $ABGC$ and $CDHE$ are ▱. (Def. of ▱)
4. $\overline{AC} \cong \overline{BG}$, $\overline{CE} \cong \overline{DH}$ (Opp. sides of a ▱ are ≅.)
5. $\overline{AC} \cong \overline{CE}$ (Given)
6. $\overline{BG} \cong \overline{DH}$ (Trans. Prop. of ≅)
7. $\overline{BG} \parallel \overline{DH}$ (If two lines are ∥ to the same line, then they are ∥ to each other.)
8. $\angle 3 \cong \angle 6$ and $\angle GBD \cong \angle HDF$. (If lines are ∥, corresp. ≰ are ≅.)
9. $\triangle GBD \cong \triangle HDF$ (AAS)
10. $\overline{BD} \cong \overline{DF}$ (Corresp. parts of ≅ ▲ are ≅.)

45. D **47.** C **49.** 1440 **50.** 2520

51. 4140 **52.** 6840 **53.** $\overline{AC} \perp \overline{DB}$ (or $\angle ACD$ and $\angle ACB$ are rt. ≰) **54.** 42

Lesson 6-3   pp. 367–374

Got It? 1. $x = 10$, $y = 43$ **2a.** No; $DEFG$ could be an isosc. trapezoid. (One pair of sides must be both ≅ and ∥.)
b. yes
1. $\angle ALN \cong \angle DNL$; $\angle ANL \cong \angle DLN$ (Given)
2. $\overline{AN} \parallel \overline{LD}$ and $\overline{AL} \parallel \overline{ND}$. (If alt. int. ≰ are ≅, then lines are ∥.)
3. $LAND$ is a ▱. (Def. of ▱)

3. 6 ft; explanations may vary. Sample: The maximum height occurs when $\overline{QP}$ is vertical.

Lesson Check 1. 112 **2.** Yes; opp. ≰ are ≅. **3.** No; the diagonals may not bisect each other. **4.** because Thm. 6-3 and its converse are both true **5.** Thm. 6-11 and Thm. 6-6 are converses of each other. Use Thm. 6-11 if you need to show the figure is a ▱. Use Thm. 6-6 if it is given that the figure is a ▱. **6.** It is a ▱ only if the same pair of opp. sides are ≅ and ∥.

Exercises 7. 5 **9.** $x = 21$, $y = 39$ **11.** 5 **13.** Yes; both pairs of opp. sides are ≅. **15.** Yes; both pairs of opp. ≰ are ≅. **17.** A quad. is a ▱ if and only if its opp. sides are ≅; a quad. is a ▱ if and only if its consecutive ≰ are suppl.; a quad. is a ▱ if and only if its opp. ≰ are ≅; a quad. is a ▱ if and only if its diagonals bisect each other.

19. Answers may vary. Sample:
1. Draw $\overline{BD}$. (Construction)
2. $\angle CBD \cong \angle ADB$ (Alt. int. ≰ are ≅.)
3. $\overline{BC} \cong \overline{DA}$ (Given)
4. $\overline{BD} \cong \overline{BD}$ (Refl. Prop. of ≅)
5. $\triangle BCD \cong \triangle DAB$ (SAS)
6. $\angle BDC \cong \angle DBA$ (Corresp. parts of ≅ ▲ are ≅.)
7. $\overline{AB} \parallel \overline{CD}$ (If alt. int. ≰ are ≅, then lines are ∥.)
8. $ABCD$ is a ▱. (Def. of ▱)

21. Answers may vary. Sample:
1. $\angle A$ is suppl. to $\angle B$. (Given)
2. $\overline{BC} \parallel \overline{AD}$ (Converse of Corresp. ≰ Postulate)
3. $\angle A$ is suppl. to $\angle D$. (Given)
4. $\overline{AB} \parallel \overline{DC}$ (Converse of Corresp. ≰ Postulate)
5. $ABCD$ is a ▱. (Def. of ▱)

23. $x = 3$, $y = 11$

25. Answers may vary. Sample:
1. $\triangle TRS \cong \triangle RTW$ (Given)
2. $\overline{SR} \cong \overline{WT}$ and $\overline{ST} \cong \overline{WR}$. (Corresp. parts of ≅ ▲ are ≅.)
3. $RSTW$ is a ▱. (If both pairs of opp. sides of a quad. are ≅, then the quad. is a ▱.)

27. Answers may vary. Sample:
1. $\overline{AB} \cong \overline{CD}$, $\overline{AC} \cong \overline{BD}$ (Construction)
2. ▱$ABCD$ (Opp. sides of a ▱ are ≅.)
3. M is the midpt. of $\overline{BC}$. (Diagonals of a ▱ bisect each other.)
4. $\overline{AM}$ is a median. (Def. of median)

29. D

31a. $7x - 11 = 6x$, $x = 11$ **b.** Yes; $m\angle F = 66$ and $m\angle FED = 114$. So $m\angle F + m\angle FED = 66 + 114 = 180$, and $\overline{AF} \parallel \overline{DE}$. (Converse of Corresp. ≰ Postulate)
c. Yes; $\overline{BD} \parallel \overline{FE}$ (Given) and $\overline{BF} \parallel \overline{DE}$ from part (b). So $BDEF$ is a ▱. (Def. of ▱)

32. $a = 8$, $h = 30$, $k = 120$ **33.** $m = 9.5$, $x = 15$

34. $c = 204$, $e = 13$, $f = 11$

35. 1. $\overline{AD} \cong \overline{BC}$, $\angle DAB \cong \angle CBA$ (Given)
2. $\overline{AB} \cong \overline{AB}$ (Refl. Prop. of ≅)
3. $\triangle ACB \cong \triangle BDA$ (SAS)
4. $\overline{AC} \cong \overline{BD}$ (Corresp. parts of ≅ ▲ are ≅.)

36. 7.47 **37.** 7.47 **38.** 7.47 **39.** 3.5 **40.** 13.2 **41.** 124

42. 56 **43.** 56 **44.** 28

Lesson 6-4   pp. 375–382

Got It? 1. Rhombus; opp. sides of a ▱ are ≅, so all sides of $EFGH$ are ≅, and there are no rt. ≰.
2. $m\angle 1 = m\angle 2 = m\angle 3 = m\angle 4 = 38$ **3a.** 43
b. Isosc.; diagonals of a rectangle are ≅ and bisect each other.

Lesson Check 1. Square; it is a rectangle because of the rt. $\angle$, and a rhombus because it has 4 ≅ sides.
2. Rhombus; it has 4 ≅ sides, and no rt. ≰. **3.** $m\angle 1 = 40$, $m\angle 2 = 90$, $m\angle 3 = 50$ **4.** 4, 4 **5.** rectangle and square; rhombus and square **6.** The first step should be $2x + 8 + 9x - 6 = 90$.

Exercises 7. Rectangle; the ▱ has 4 rt. ≰ and does not have 4 ≅ sides. **9.** $m\angle 1 = m\angle 2 = m\angle 3 = m\angle 4 = 37$ **11.** $m\angle 1 = 118$, $m\angle 2 = m\angle 3 = 31$ **13.** $m\angle 1 = 32$, $m\angle 2 = 90$, $m\angle 3 = 58$, $m\angle 4 = 32$ **15.** $m\angle 1 = 55$, $m\angle 2 = 35$, $m\angle 3 = 55$, $m\angle 4 = 90$ **17.** $m\angle 1 = 90$,

$m\angle 2 = 55$, $m\angle 3 = 90$ **19.** $x = 3$; $LN = MP = 7$
21. $x = 9$; $LN = MP = 67$ **23.** $x = 2.5$; $LN = MP = 12.5$ **25.** ▱ **27.** rectangle **29.** ▱, rhombus, rectangle, square **31.** ▱, rhombus, rectangle, square **33.** ▱, rhombus, rectangle, square **35.** rectangle, square **37.** rhombus, square **39.** $x = 5$, $y = 4$; all sides are 3. **41a.** Given **b.** Def. of rectangle **c.** Refl. Prop. of ≅ **d.** Def. of rectangle **e.** $\overline{AB} \cong \overline{DC}$ **f.** $\triangle ABC \cong \triangle DCB$ **g.** All rt. ⩟ are ≅. **h.** Corresp. parts of ≅ ⩟ are ≅.
43. $x = 5$, $y = 32$, $z = 7.5$
45. Answers may vary. Sample:
1. $ABCD$ is a rhombus. (Given)
2. $\overline{AB} \cong \overline{AD}$ and $\overline{CB} \cong \overline{CD}$. (Def. of rhombus)
3. $\overline{AC} \cong \overline{AC}$ (Refl. Prop. of ≅)
4. $\triangle ABC \cong \triangle ADC$ (SSS)
5. $\angle 3 \cong \angle 4$ and $\angle 2 \cong \angle 1$. (Corresp. parts of ≅ ⩟ are ≅.)
6. $\overline{AC}$ bisects $\angle BAD$ and $\angle BCD$. (Def. of ∠ bisector)
47. $m\angle H = m\angle J = 58$, $m\angle K = m\angle G = 122$, $HK = KJ = JG = GH = 6$ **49.** $AC = BD = 16$
51. $AC = BD = 1$ **53.** 2 **55.** D **57.** A **59.** Yes; both pairs of opp. sides of the quad. are ≅. **60.** No; two opp. sides are ≅ and two opp. sides are ∥, but not the same pair of opp. sides.
61. Yes; diagonals of the quad. bisect each other. **62.** 6
63. 16 **64.** 5 **65.** $\overline{RQ}$ **66.** $\overline{PR}$ **67.** $\overline{ST}$
68. Answers may vary. Sample:

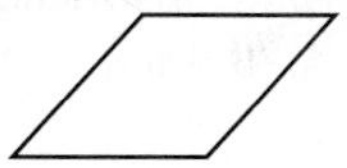

69. Answers may vary. Sample:

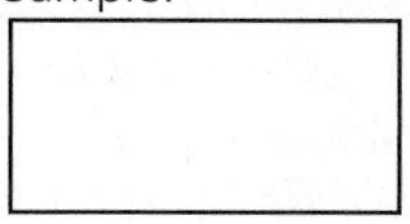

Lesson 6-5   pp. 383–388

Got It? 1a. The ▱ is not a rectangle or a square because ⩟ are not rt. ⩟. It might be a rhombus. **b.** No; the fact that the diagonals bisect each other is true of all ▱. **2.** 4 **3.** Yes; make diagonals ⊥. The result will be a rectangle and a rhombus, so it is square.
Lesson Check 1. Rectangle; diagonals are ≅.
2. Rhombus; diagonals are ⊥. **3.** 2 **4.** 3 **5a.** rhombus, square **b.** rectangle, square **c.** rhombus, square **d.** rectangle, rhombus, square **e.** rhombus, square **6.** The only ▱ with ⊥ diagonals are rhombuses and squares. **7.** Rectangle; diagonals are ≅.
Exercises 9. Rhombus; diagonals are ⊥. **11.** 12 **13.** 10 **15.** Answers may vary. Sample: Measure the lengths of the frame's diagonals. If they are ≅, then the frame has the shape of a rectangle, and therefore a parallelogram; measure the two pairs of alt. int. ⩟ formed by the turnbuckle (the transversal). If both pairs of ⩟ are ≅, then both pairs of opposite sides of the frame are ∥.

17. 11 **19.** 16
21. Rhombus; answers may vary. Sample:

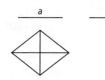

23. Answers may vary. Sample:
1. $\overline{AB}$ bisects $\angle BAD$ and $\angle BCD$. (Given)
2. $\angle 1 \cong \angle 2$ and $\angle 3 \cong \angle 4$. (Def. of bisect)
3. $\overline{AC} \cong \overline{AC}$ (Refl. Prop. of ≅)
4. $\triangle ABC \cong \triangle ADC$ (ASA)
5. $\overline{AB} \cong \overline{AD}$ and $\overline{BC} \cong \overline{CD}$. (Corresp. parts of ≅ ⩟ are ≅.)
6. $\overline{AB} \cong \overline{CD}$ and $\overline{BC} \cong \overline{AD}$. (Opp. sides of a ▱ are ≅.)
7. $\overline{AB} \cong \overline{AD} \cong \overline{BC} \cong \overline{CD}$ (Trans. Prop. of ≅)
8. $ABCD$ is a rhombus. (Def. of rhombus)
25. Construct the midpt. of each diagonal. Copy the diagonals so the two midpts. coincide. Connect the endpoints of the diagonals. **27.** Construct the midpts. of each diagonal. Construct two ⊥ lines, and mark off diagonal lengths on the ⊥ lines. Connect the endpoints of the diagonals. **29.** Yes; ≅ diagonals in a ▱ mean it can be a rectangle with 2 opp. sides 2 cm long. **31.** "If one diagonal of a ▱ bisects one ∠, then the ▱ is a rhombus." The new statement is true. If $\overline{AC}$ bisects $\angle BCD$, then $\angle BCA \cong \angle DCA$. (Def. of ∠ bisector) $ABCD$ is a ▱, so $\angle B \cong \angle D$. (Opp. ⩟ of a ▱ are ≅.) $\overline{AC} \cong \overline{AC}$ (Reflexive Prop. of ≅), so $\triangle BCA \cong \triangle DCA$ (AAS) and $\overline{BC} \cong \overline{DC}$. (Corresp. parts of ≅ ⩟ are ≅.) Since opp. sides of a ▱ are ≅, $\overline{AB} \cong \overline{CD}$ and $\overline{BC} \cong \overline{DA}$. So $\overline{AB} \cong \overline{BC} \cong \overline{CD} \cong \overline{DA}$, and ▱$ABCD$ is a rhombus. (Def. of rhombus) **33.** I
35. $\left(\dfrac{-7 + x}{2}, \dfrac{10 + y}{2}\right) = (-1, 4)$. $-7 + x = -2$, $x = 5$, and $10 + y = 8$, $y = -2$, so $Q(5, -2)$.
36. $m\angle 1 = 128$, $m\angle 2 = 26$, $m\angle 3 = 26$
37. $m\angle 1 = 57$, $m\angle 2 = 57$, $m\angle 3 = 66$ **38.** $m\angle 1 = 90$, $m\angle 2 = 58$, $m\angle 3 = 90$ **39.** A ▱ is a rhombus if and only if its diagonals are ⊥. **40.** A ▱ is a rectangle if and only if its diagonals are ≅. **41.** $a = 5.6$, $b = 6.8$; 4.5, 4.5, 4.2, 4.2 **42.** 3; 18, 4.8, 18, 16.4 **43.** $m = 5$, $n = 15$; 15, 15, 21, 21

Lesson 6-6   pp. 389–397

Got It? 1a. $m\angle P = m\angle Q = 74$, $m\angle S = 106$ **b.** Yes; $\overline{DE} \parallel \overline{CF}$ so same-side int. ⩟ are suppl. **2.** obtuse ∠ measure: 102; acute ∠ measure: 78 **3a.** 6; 23 **b.** 3; 1; A △ has 3 midsegments joining any pair of side midpts. A trapezoid has 1 midsegment joining the midpts. of the two legs. **4.** $m\angle 1 = 90$, $m\angle 2 = 54$, $m\angle 3 = 36$

Lesson Check 1. $m\angle 1 = 78$, $m\angle 2 = 90$, $m\angle 3 = 12$
2. $m\angle 1 = 94$, $m\angle 2 = 132$ **3.** 20 **4.** No; a kite's opp. sides are not $\cong$ or $\parallel$. **5.** Answers may vary. Sample: Similar: diagonals are $\perp$, consecutive sides $\cong$. Different: one diagonal of a kite bisects opp. $\angle$s but the other diagonal does not; all sides of a rhombus are $\cong$. **6.** Def. of trapezoid is a quad. with exactly one pair of $\parallel$ sides. A $\square$ has two pairs of $\parallel$ sides, so a $\square$ is not a trapezoid.
Exercises 7. $m\angle 1 = 77$, $m\angle 2 = 103$, $m\angle 3 = 103$
9. $m\angle 1 = 49$, $m\angle 2 = 131$, $m\angle 3 = 131$
11. $m\angle 1 = m\angle 2 = 115$, $m\angle 3 = 65$ **13.** 9 **15.** 9
17. $m\angle 1 = 90$, $m\angle 2 = 45$, $m\angle 3 = 45$ **19.** $m\angle 1 = 90$, $m\angle 2 = 26$, $m\angle 3 = 90$ **21.** $m\angle 1 = 90$, $m\angle 2 = 55$, $m\angle 3 = 90$, $m\angle 4 = 55$, $m\angle 5 = 35$ **23.** $m\angle 1 = 90$, $m\angle 2 = 90$, $m\angle 3 = 90$, $m\angle 4 = 90$, $m\angle 5 = 46$, $m\angle 6 = 34$, $m\angle 7 = 56$, $m\angle 8 = 44$, $m\angle 9 = 56$, $m\angle 10 = 44$
25. Answers may vary. Sample:

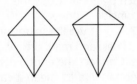

27. No; explanations may vary. Sample: Assume $\overline{KM}$ bisects both $\angle$s. Then $\angle MKL \cong \angle MKN \cong \angle KML \cong \angle KMN$. Both pairs of sides of $KLMN$ would be $\parallel$, and $KLMN$ would be a $\square$. It is impossible for an isosc. trap. to also be a $\square$, so $\overline{KM}$ cannot bis. $\angle LMN$ and $\angle LKN$.
29. 15 **31.** $AD = 4$, $EF = 9$, $BC = 14$ **33.** $HG = 2$, $CD = 5$, $EF = 8$ **35.** $x = 35$, $y = 30$ **37.** Isosc. trapezoid; $\overline{AB} \parallel \overline{DC}$ (If alt. int. $\angle$s are $\cong$, then lines are $\parallel$) and $\overline{AD} \cong \overline{BC}$. (Corresp. parts of $\cong$ $\triangle$s are $\cong$.) **39.** Yes; the $\cong$ $\angle$s can be obtuse. **41.** Yes; if two $\cong$ $\angle$s are rt. $\angle$s, they are suppl. The other two $\angle$s are also suppl. **43.** Yes; the $\cong$ $\angle$s each have measure 45.
45. Answers may vary. Sample:
1. Draw $\overline{AE} \parallel \overline{DC}$. (Construction)
2. $AECD$ is a $\square$. (Def. of $\square$)
3. $\overline{AE} \cong \overline{DC}$ (Opp. sides of a $\square$ are $\cong$.)
4. $\angle 1 \cong \angle C$ (If $\parallel$ lines, corresp. $\angle$s are $\cong$.)
5. $\angle B \cong \angle 1$ (Isosc. $\triangle$ Thm.)
6. $\angle B \cong \angle C$ (Transitive Prop. of $\cong$)
7. $\angle D$ and $\angle C$ are suppl. (If $\parallel$ lines, same-side int. $\angle$s are suppl.)
8. $\angle BAD$ and $\angle B$ are suppl. (If $\parallel$ lines, same-side int. $\angle$s are suppl.)
9. $\angle BAD \cong \angle D$ ($\angle$s suppl. to $\cong$ $\angle$s are $\cong$.)
47. Isosc. trapezoid; answers may vary. Sample:

49. Rectangle, square; answers may vary. Sample:

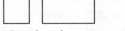

51. Kite, rhombus, square; answers may vary. Sample:

53. Answers may vary. Sample:
1. $\overline{AB} \cong \overline{DC}$ (Given)
2. $\angle BAD \cong \angle CDA$ (Base $\angle$s of an isosc. trapezoid are $\cong$.)
3. $\overline{AD} \cong \overline{AD}$ (Refl. Prop. of $\cong$)
4. $\triangle BAD \cong \triangle CDA$ (SAS)
5. $\overline{BD} \cong \overline{CA}$ (Corresp. parts of $\cong$ $\triangle$s are $\cong$.)
55. Answers may vary. Sample:
1. Draw $\overline{TA}$ and $\overline{PR}$. (Construction)
2. $\overline{TR} \cong \overline{PA}$ (Given)
3. $\angle TRA \cong \angle PAR$ (Base $\angle$s of an isosc. trapezoid are $\cong$.)
4. $\overline{RA} \cong \overline{RA}$ (Refl. Prop. of $\cong$)
5. $\triangle TRA \cong \triangle PAR$ (SAS)
6. $\angle RTA \cong \angle APR$ (Corresp. parts of $\cong$ $\triangle$s are $\cong$.)
57. True; a square is a $\square$ with 4 rt. $\angle$s. **59.** False; a rhombus has 4 $\cong$ sides, and a kite does not. **61.** False; counterexample: kites and trapezoids are not $\square$.
63. Answers may vary. Sample:
1. $\overline{RT}$ and $\overline{PA}$ are not $\parallel$. (Def. of trapezoid)
2. Extend $\overline{RT}$ and $\overline{PA}$ to meet at M. (Construction)
3. $\angle MTP \cong \angle R$ and $\angle MPT \cong \angle A$. (If $\parallel$ lines, then corresp. $\angle$s are $\cong$.)
4. $\angle MTP \cong \angle MPT$ (Trans. Prop. of $\cong$)
5. $\overline{MT} \cong \overline{MP}$ (Converse of Isosc. $\triangle$ Thm.)
6. $\angle MIT$ and $\angle MIP$ are rt. $\angle$s. (A line $\perp$ to one of two $\parallel$ lines is also $\perp$ to the other line.)
7. $\overline{MI} \cong \overline{MI}$ (Refl. Prop. of $\cong$)
8. $\triangle MIT \cong \triangle MIP$ (HL)
9. $\overline{TI} \cong \overline{PI}$ (Corresp. parts of $\cong$ $\triangle$s are $\cong$.)
10. $\overline{BI}$ is the $\perp$ bis. of $\overline{TP}$. (Def. of $\perp$ bis.)
65. half the difference of the bases; $\triangle$ Midsegment Thm. **67.** B **69.** D **71.** 61 **72.** 27 **73.** 12 **74.** 89
75. (1, 3); $\sqrt{200}$ or $10\sqrt{2}$ **76.** $-\frac{1}{4}$

Algebra Review   p. 399

1. $5\sqrt{2}$ **3.** 8 **5.** $4\sqrt{3}$ **7.** $6\sqrt{2}$ **9.** 6 **11.** $7\sqrt{2}$ **13.** $2\sqrt{6}$
15. $\frac{3\sqrt{10}}{2}$

Lesson 6-7   pp. 400–405

Got It? 1. scalene **2a.** Yes; slope of $\overline{MN}$ = slope of $\overline{PQ} = -3$ and slope of $\overline{NP}$ = slope of $\overline{MQ} = \frac{1}{3}$, so opp. sides are $\parallel$. The product of slopes is -1, so sides are $\perp$.

b. Yes; $MN = PQ = NP = MQ = \sqrt{10}$. **c.** Yes; slope of $\overline{AB} = \frac{3}{4}$ and slope of $\overline{BC} = -\frac{4}{3}$, so the product of their slopes is -1. Therefore, $\overline{AB} \perp \overline{BC}$ and $\angle B$ is a rt. $\angle$. So $\triangle ABC$ is a rt. $\triangle$ by def. of rt. $\triangle$. **3.** rhombus (The length of each side is $\sqrt{13}$.)

Lesson Check 1. isosceles **2.** No; explanations may vary. Sample: The diagonal lengths ($\sqrt{29}$ and 5) are not equal. **3.** Find the coordinates and use the Distance Formula to compare lengths. **4.** Answers may vary. Sample: *DEFG* is not a $\square$.

Exercises 5. Scalene; side lengths are 4, 5, and $\sqrt{17}$. **7.** Isosceles; side lengths are $2\sqrt{2}$, $\sqrt{34}$, and $\sqrt{34}$. **9.** Rhombus; explanations may vary. Sample: All four sides are $\cong$ (with length $\sqrt{5}$), and diagonals are not $\cong$ (with lengths 2 and 4). **11.** None; explanations may vary. Sample: Consecutive sides are not $\cong$ or $\perp$. **13.** Rhombus; explanations may vary. Sample: All sides are $\cong$ and consecutive sides are not $\perp$.

15. rhombus

17.

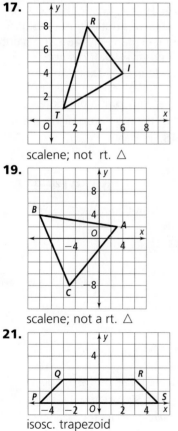

scalene; not rt. $\triangle$

19.

scalene; not a rt. $\triangle$

21.

isosc. trapezoid

23.

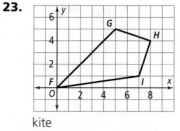

kite

25.

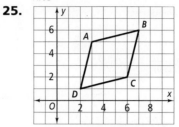

rhombus

27.

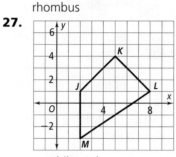

quadrilateral

29.

kite

31. Yes; $PR = SW = 4$, $PQ = ST = \sqrt{10}$, $QR = TW = 3\sqrt{2}$, so $\triangle PQR \cong \triangle STW$ by SSS. **33.** $\square$; 24 units^2 **35.** slope of $\overline{DE} = 2$; slope of $\overline{AB} = 2$; $DE = \frac{1}{2}\sqrt{5}$; $AB = \sqrt{5}$. So $\overline{DE} \parallel \overline{AB}$ and $DE = \frac{1}{2}AB$. **37.** Answers may vary. Sample: Chairs are not at vertices of a $\square$. Move right-most chair down by 1 grid unit. **39.** $G(-4, 1)$, $H(1, 3)$ **41.** $\left(-1, 6\frac{2}{3}\right)$, $\left(1, 8\frac{1}{3}\right)$, $(3, 10)$, $\left(5, 11\frac{2}{3}\right)$, $\left(7, 13\frac{1}{3}\right)$ **43.** $(-2.76, 5.2)$, $(-2.52, 5.4)$, $(-2.28, 5.6)$, $\ldots$, $(8.52, 14.6)$, $(8.76, 14.8)$ **45.** D **47.** A **49.** $m\angle 1 = 62$, $m\angle 2 = m\angle 3 = 118$, $x = 2.5$ **50.** $(3, 2)$ **51.** $(-3, -4)$ **52.** -1 **53.** 0 **54.** $\frac{b}{c + d - a}$

Lesson 6-8   pp. 406–412

Got It? 1a. $R(-b, 0)$, $E(-b, a)$, $C(b, a)$, $T(b, 0)$
b. $K(-b, 0)$, $I(0, a)$, $T(c, 0)$, $E(0, -a)$ **2a.** Answers may vary. Sample: x-coordinate of B is $2a$ more than x-coordinate of C. **b.** yes; $TR = AP = \sqrt{a^2 - 2ab + b^2 + c^2}$

3.

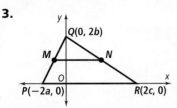

Given: $\triangle PQR$, midpoints M and N

Prove: $\overline{MN} \parallel \overline{PR}$ and $MN = \frac{1}{2}PR$

- First, use the Midpoint Formula to find the coordinates of M and N.
- Then, use the Slope Formula to determine whether the slopes of $\overline{MN}$ and $\overline{PR}$ are equal. If they are, then $\overline{MN} \parallel \overline{PR}$.
- Finally, use the Distance Formula to find and compare the lengths of $\overline{MN}$ and $\overline{PR}$.

Lesson Check 1. $K(2b, c)$, $M(2a, 0)$ **2.** The slope of $\overline{KM}$ is $\frac{c}{2b - 2a}$, and the slope of $\overline{OL}$ is $\frac{c}{2a + 2b}$.
3. $\left(a + b, \frac{c}{2}\right)$ **4.** Answers may vary. Sample: Using variables allows the figure to represent all possibilities.
5. rectangle **6.** Answers may vary. Sample: Classmate ignored the coefficient 2 in the coordinates. The endpoints are (b, c) and $(a + d, c)$.
Exercises 7. $O(0, 0)$, $S(0, h)$, $T(b, h)$, $W(b, 0)$
9. $S\left(-\frac{b}{2}, -\frac{b}{2}\right)$, $T\left(-\frac{b}{2}, \frac{b}{2}\right)$, $W\left(\frac{b}{2}, \frac{b}{2}\right)$, $Z\left(\frac{b}{2}, -\frac{b}{2}\right)$
11. $W(r, 0)$, $T(0, t)$, $S(-r, 0)$, $Z(0, -t)$ **13.** Yes, $ABCD$ is a rhombus. The slope of $\overline{AC} = -1$, and the slope of $\overline{BD} = 1$, so the diagonals are $\perp$.
15. Answers may vary. Sample:

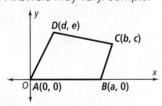

17. $P(c - a, b)$ **19.** $P(-b, 0)$
21a. Answers may vary. Sample:
    b. Answers may vary. Sample:

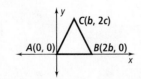
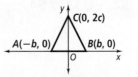

c. $\sqrt{b^2 + 4c^2}$, $\sqrt{b^2 + 4c^2}$ **d.** $\sqrt{b^2 + 4c^2}$, $\sqrt{b^2 + 4c^2}$
e. The results are the same. **23.** Answers may vary. Sample: Place vertices at $A(0, 0)$, $B(a, 0)$, $C(a + b, 0)$, and $D(b, c)$. Use the Distance Formula to find the lengths of opp. sides. **25.** Answers may vary. Sample: Place vertices at $A(0, 0)$, $B(0, a)$, $C(a, a)$, and $D(a, 0)$. Use the fact that a horizontal line is $\perp$ to a vertical line. **27.** isosc. trapezoid
29. square
31. Answers may vary. Sample:

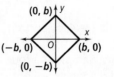

33. Answers may vary. Sample: B, D, H, F **35.** Answers may vary. Sample: A, C, G, E **37.** Answers may vary. Sample: A, D, G, F **39.** G **41.** 1, 2, 3, 4 **42.** No; product of slopes is not -1, so there are no rt. $\angle$s. **43a.** If $x \neq 51$, then $2x \neq 102$. **b.** If $2x \neq 102$, then $x \neq 51$.
44a. If $a \neq 5$, then $a^2 \neq 25$. **b.** If $a^2 \neq 25$, then $a \neq 5$.
45a. If b not less than -4, then b is not negative. **b.** If b is not negative, then b is not less than -4. **46a.** If c is not greater than 0, then c is not positive. **b.** If c is not positive, then c is not greater than 0. **47a.** If the sum of the measures of the interior $\angle$s of a polygon is 360, then the polygon is a quadrilateral. **b.** If a polygon is a quadrilateral, then the sum of the measures of the interior $\angle$s of the polygon is 360. **48.** $y = \frac{5}{4}x$
49. $y - q = \frac{a}{b}(x - p)$

Lesson 6-9   pp. 414–418

Got It? 1. The factor 2 avoids fractions.
2. Answers may vary. Sample:

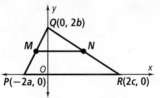

Given: $\triangle PQR$, midpoints M and N

Prove: $\overline{MN} \parallel \overline{PR}$, $MN = \frac{1}{2}PR$
By the Midpoint Formula, coordinates of the midpoints are $M(-a, b)$ and $N(c, b)$. By the Slope Formula, slope of $\overline{MN} =$ slope of $\overline{PR} = 0$, so $\overline{MN} \parallel \overline{PR}$. By the Distance Formula, $MN = \sqrt{(c + a)^2}$ and $PR = 2\sqrt{(c + a)^2}$, so $MN = \frac{1}{2}PR$.

Lesson Check

1a.

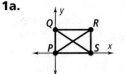

b. $(0, b)$, (a, b), and $(a, 0)$
c. Given: Rectangle $PQRS$
Prove: $\overline{PR} \cong \overline{SQ}$ **d.** Answers may vary. Sample: By the Distance Formula,

$PR = \sqrt{(0 - a)^2 + (0 - b)^2} = \sqrt{a^2 + b^2}$ and $SQ = \sqrt{(0 - a)^2 + (b - 0)^2} = \sqrt{a^2 + b^2}$. So $\overline{PR} \cong \overline{SQ}$.

2. Answers may vary. Sample: Place the vertices on the x- and y-axes so that the axes are the diagonals of the rhombus. **3.** Your classmate assumes $PQRO$ is an isosc. trapezoid.

Exercises 5a.
$M(-a, b)$, $N(a, b)$ **b.** $PN = \sqrt{9a^2 + b^2}$, $RM = \sqrt{9a^2 + b^2}$ **c.** The Distance Formula shows that $\overline{PN}$ and $\overline{RM}$ are the same length. **7.** Yes; use Slope Formula. **9.** Yes; use Midpoint Formula. **11.** No; you need $\angle$ measures. **13.** Yes; use Distance Formula. **15.** Yes; answers may vary. Sample: Show four sides have the same length or show diagonals $\perp$. **17.** No; you need $\angle$ measures.
19. Answers may vary. Sample:

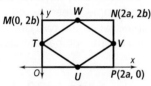

Given: $MNPO$ is a rectangle.
   T, W, V, U are midpoints of its sides.
Prove: $TWVU$ is a rhombus.

By the Midpoint Formula, the coordinates of the midpoints are $T(0, b)$, $W(a, 2b)$, $V(2a, b)$, and $U(a, 0)$. By the Slope Formula,

$\quad$ slope of $\overline{TW} = \dfrac{2b - b}{a - 0} = \dfrac{b}{a}$

$\quad$ slope of $\overline{WV} = \dfrac{2b - b}{a - 2a} = -\dfrac{b}{a}$

$\quad$ slope of $\overline{VU} = \dfrac{b - 0}{2a - a} = \dfrac{b}{a}$

$\quad$ slope of $\overline{UT} = \dfrac{b - 0}{0 - a} = -\dfrac{b}{a}$

So $\overline{TW} \parallel \overline{VU}$ and $\overline{WV} \parallel \overline{UT}$. Therefore, $TWVU$ is a $\square$. By the Slope Formula, slope of $\overline{TV} = 0$, and slope of $\overline{WU}$ is undefined. $\overline{TV} \perp \overline{WU}$ because horiz. and vert. lines are $\perp$. Since the diagonals of $\square TWVU$ are $\perp$, it must be a rhombus.
21. Answers may vary. Sample:

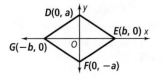

Given: $DEFG$ is a parallelogram.
Prove: $\overline{GE} \perp \overline{DF}$
By the Slope Formula, slope of $\overline{GE} = \dfrac{0 - 0}{b - (-b)} = 0$, and slope of $\overline{DF} = \dfrac{a - (-a)}{0 - 0}$, which is undefined. So $\overline{GE}$ must be horizontal and $\overline{DF}$ must be vertical. Therefore, $\overline{GE} \perp \overline{DF}$ because horiz. and vert. lines are $\perp$.
23. Answers may vary. Sample:

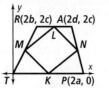

Given: Trapezoid $TRAP$, M, L, N, and K are midpoints of its sides
Prove: $MLNK$ is a $\square$.
By the Midpoint Formula, the coordinates of the midpoints are $M(b, c)$, $L(b + d, 2c)$, $N(a + d, c)$, and $K(a, 0)$. By the Slope Formula, the slope of $\overline{ML} = \dfrac{c}{d}$, the slope of $\overline{LN} = \dfrac{c}{b - a}$, the slope of $\overline{NK} = \dfrac{c}{d}$, and the slope of $\overline{KM} = \dfrac{c}{b - a}$. Since slopes are $=$, $\overline{ML} \parallel \overline{NK}$ and $\overline{LN} \parallel \overline{KM}$. Therefore, $MLNK$ is a $\square$ by def. of $\square$.
25a. $L(3q, 3r)$, $M(3p + 3q, 3r)$, $N(3p, 0)$

b. equation of $\overleftrightarrow{AM}$: $y = \dfrac{r}{p + q}x$

$\quad$ equation of $\overleftrightarrow{BN}$: $y = \dfrac{2r}{2q - p}(x - 3p)$

$\quad$ equation of $\overleftrightarrow{CL}$: $y = \dfrac{r}{q - 2p}(x - 6p)$

c. $P(2p + 2q, 2r)$
d. The coordinates of P satisfy the equation for $\overleftrightarrow{CL}$: $y = \dfrac{r}{q - 2p}(x - 6p)$.

$\quad 2r = \dfrac{r}{q - 2p}(2p + 2q - 6p)$

$\quad 2r = \dfrac{r}{q - 2p}(2q - 4p)$

$\quad 2r = 2r$

e. $AM = \sqrt{(3p + 3q - 0)^2 + (3r - 0)^2} = \sqrt{(3p + 3q)^2 + (3r)^2}$;

$\quad \dfrac{2}{3}AM = \dfrac{2}{3}\sqrt{(3p + 3q)^2 + (3r)^2} = $

$\quad \sqrt{\dfrac{4}{9}\left[(3p + 3q)^2 + (3r)^2\right]} = $

$\quad \sqrt{\left[\dfrac{2}{3}(3p + 3q)^2\right] + \left[\dfrac{2}{3}(3r)^2\right]} = $

$\quad \sqrt{(2p + 2q)^2 + (2r)^2}$;

$\quad AP = \sqrt{(2p + 2q - 0)^2 + (2r - 0)^2} = \sqrt{(2p + 2q)^2 + (2r)^2}$

So $AP = \frac{2}{3}AM$. You can find the other two distances similarly.

27a. Answers may vary. Sample: The area of a $\triangle$ with base b and height c is $\frac{1}{2}bc$. The area of a $\triangle$ with base d and height a is $\frac{1}{2}ad$. In both cases, the remaining area of the triangle has base $(b - d)$ and height a. Therefore $\frac{1}{2}ad = \frac{1}{2}bc$ by the Transitive Prop. of Eq. So $ad = bc$. **b.** Slope of $\ell = \frac{a}{b}$ or $\frac{c}{d}$. So $\frac{a}{b} = \frac{c}{d}$ and $ad = bc$. **29.** 50 **31.** 19 **33.** $(a, -b)$ **34.** Answers may vary. Sample: $\angle A \cong \angle C$, $\angle ADB \cong \angle CDB$, and $\overline{AD} \cong \overline{CD}$ (Given), so $\triangle ABD \cong \triangle CBD$ by ASA. Then $\overline{AB} \cong \overline{CB}$ because corresp. parts of $\cong \triangle$ are $\cong$. **35.** Answers may vary. Sample: $\overline{HE} \cong \overline{FG}$, $\overline{EF} \cong \overline{GH}$, (Given) and $\overline{HF} \cong \overline{HF}$ (Reflexive Prop. of $\cong$), so $\triangle HEF \cong \triangle FGH$ by SSS. Then $\angle 1 \cong \angle 2$ because corresp. parts of $\cong \triangle$ are $\cong$. **36.** $\overline{KN} \cong \overline{ML}$ (Given), $\angle KNL \cong \angle MLN$ (All rt. $\triangle$ are $\cong$.), and $\overline{NL} \cong \overline{NL}$ (Reflexive Prop. of $\cong$). Then $\triangle KNL \cong \triangle MLN$ by SAS, and $\angle K \cong \angle M$ because corresp. parts of $\cong \triangle$ are $\cong$. **37.** 12, -12 **38.** 8, -8 **39.** 5, -5 **40.** 16.6, -16.6

Chapter Review pp. 420–424

1. rhombus **2.** equiangular polygon **3.** consecutive angles **4.** trapezoid **5.** 120, 60 **6.** 157.5, 22.5 **7.** 108, 72 **8.** 360, 360, 360 **9.** 159 **10.** 69 **11.** $m\angle 1 = 38$, $m\angle 2 = 43$, $m\angle 3 = 99$ **12.** $m\angle 1 = 101$, $m\angle 2 = 79$, $m\angle 3 = 101$ **13.** $m\angle 1 = 37$, $m\angle 2 = 26$, $m\angle 3 = 26$ **14.** $m\angle 1 = 45$, $m\angle 2 = 45$, $m\angle 3 = 45$ **15.** $x = 3$, $y = 7$ **16.** $x = 2$, $y = 5$ **17.** no **18.** yes **19.** $x = 29$, $y = 28$ **20.** $x = 4$, $y = 5$ **21.** $m\angle 1 = 58$, $m\angle 2 = 32$, $m\angle 3 = 90$ **22.** $m\angle 1 = 124$, $m\angle 2 = 28$, $m\angle 3 = 62$ **23.** sometimes **24.** always **25.** sometimes **26.** sometimes **27.** sometimes **28.** always **29.** No; two sides are $\parallel$ in all $\square$. **30.** Yes; the $\square$ is a rhombus and a rectangle so it must be a square. **31.** $x = 18$; a diagonal bisects a pair of $\triangle$ in a rhombus. **32.** $x = 4$; a rectangle has $\cong$ diagonals that bisect each other. **33.** $m\angle 1 = 135$, $m\angle 2 = 135$, $m\angle 3 = 45$ **34.** $m\angle 1 = 80$, $m\angle 2 = 100$, $m\angle 3 = 100$ **35.** $m\angle 1 = 90$, $m\angle 2 = 25$ **36.** $m\angle 1 = 52$, $m\angle 2 = 52$ **37.** 2 **38.** scalene **39.** isosceles **40.** parallelogram **41.** kite **42.** rhombus **43.** isosc. trapezoid **44.** $F(0, 2b)$, $L(a, 0)$, $P(0, -2b)$, $S(-a, 0)$ **45.** $(a - b, c)$ **46.** Answers may vary. Sample:

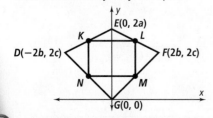

Given: Kite $DEFG$, K, L, M, N are midpoints of sides
Prove: $KLMN$ is a rectangle.
By the Midpoint Formula, coordinates of midpoints are $K(-b, a + c)$, $L(b, a + c)$, $M(b, c)$, and $N(-b, c)$. By the Slope Formula, slope of $\overline{KL}$ = slope of $\overline{NM}$ = 0, and slope of $\overline{KN}$ and slope of $\overline{LM}$ are undefined. $\overline{KL} \parallel \overline{NM}$ and $\overline{KN} \parallel \overline{LM}$ so $KLMN$ is a $\square$. $\overline{KL} \perp \overline{LM}$, $\overline{LM} \perp \overline{NM}$, $\overline{KN} \perp \overline{NM}$, and $\overline{KN} \perp \overline{KL}$ so $KLMN$ is a rectangle.

Chapter 7

Get Ready! p. 429

1. 70; if lines are $\parallel$, same-side int. $\triangle$ are suppl. **2.** 110; if lines are $\parallel$, corresponding $\triangle$ are $\cong$. **3.** 70; adjacent angles forming a straight $\angle$ are suppl. **4.** 70; it is a vert. $\angle$ with $\angle 1$; vert. $\triangle$ are $\cong$. **5.** $\overline{DL}$ **6.** $\angle A$ **7.** $\angle DLH$ **8.** $\triangle APC$ **9.** $\triangle KNP \cong \triangle LNM$ by SAS. **10.** $\triangle BAC \cong \triangle BED$ by AAS. **11.** $\triangle UGH \cong \triangle UGB$ by SSS. **12.** 6, 6 **13.** 4.7, 9.4 **14.** Answers may vary. Sample: The relative sizes of the body parts in the drawing are the same as those of a real person. **15.** Answers may vary. Sample: They might be similar if they have the same shape. **16.** Answers may vary. Sample: Measure the number of inches on the map between the two cities, and multiply that number of inches by the number of miles represented by 1 in.

Lesson 7-1 pp. 432–438

Got It? 1. $3 : 4$ **2.** 36, 144 **3.** 12 cm, 21 cm, 27 cm **4a.** 63 **b.** 0.25 **5a.** $\frac{7}{y}$; Prop. of Proportions (1) **b.** $\frac{x + 6}{6}$; Prop. of Proportions (3) **c.** The proportion is equivalent to $\frac{x - 6}{6} = \frac{y - 7}{7}$ by Prop. of Proportions (1). Then by Prop. of Proportions (3), $\frac{x - 6 + 6}{6} = \frac{x - 7 + 7}{7}$, which simplifies to $\frac{x}{6} = \frac{y}{7}$.

Lesson Check 1. $23 : 42$ **2.** $5x$, $9x$ **3.** 12 **4a.** $\frac{a}{13} = \frac{7}{b}$ **b.** $\frac{a - 7}{7} = \frac{13 - b}{b}$ **c.** $\frac{7}{a} = \frac{b}{13}$ **5.** A ratio is a single comparison, while a proportion is a statement that two ratios are equal. **6.** Answers may vary. Sample: 3 in., 6 in., 7 in.; or 6 in., 12 in., 14 in. **7.** The second line should equate the product of the means and the product of the extremes: $7x = 12$. Then the third line would be $x = \frac{12}{7}$. **8.** $\frac{9}{6} = \frac{18}{12}$, $\frac{9}{18} = \frac{6}{12}$, $\frac{12}{6} = \frac{18}{9}$, or $\frac{12}{18} = \frac{6}{9}$

Exercises 9. $\frac{14}{5}$ or $14 : 5$ **11.** $\frac{10}{17}$ or $10 : 17$ **13.** won 110, lost 44 **15.** 24 cm, 28 cm, 36 cm **17.** 4 **19.** $\frac{36}{5}$ **21.** 32 **23.** 7 **25.** 6 **27.** $\frac{4}{3}$; Prop. of Proportions (1) **29.** $\frac{a}{3}$; Prop. of Proportions (2) **31.** $\frac{7}{4}$; Prop. of Proportions (3) **33.** 1 **35.** 4 **37.** length: 15 in.; width: 10 in. **39a.** 12 in. **b.** 1.5 in. **41.** 1.5 **43.** 0.2

45.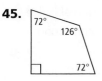

47. The product of the means is $26 \cdot 16 = 416$, and the product of the extremes is $10 \cdot 42 = 420$. Since $416 \neq 420$, it is not a valid proportion. **49.** $\frac{9}{4}$; divide each side by $4n$. **51.** $\frac{b}{2}$; Prop. of Proportions (3) **53.** $\frac{c}{d}$; Prop. of Proportions (2), then (3), then (2) **55.** $\frac{a}{b} = \frac{c}{d}$ (given); $\frac{a}{b}(bd) = \frac{c}{d}(bd)$ (Mult. Prop. of =); $ad = bc$ (simplify and Commutative Prop. of Mult.); $bc = ad$ (Sym. Prop. of =); $\frac{bc}{ac} = \frac{ad}{ac}$ (Div. Prop. of =); $\frac{b}{a} = \frac{d}{c}$ (simplify) **57.** $\frac{a}{b} = \frac{c}{d}$ (given); $\frac{a}{b} + 1 = \frac{c}{d} + 1$ (Add. Prop. of Eq.); $\frac{a}{b} + \frac{b}{b} = \frac{c}{d} + \frac{d}{d}$ (Subst. Prop. of Eq.); $\frac{a + b}{b} = \frac{c + d}{d}$ (simplify) **59.** $-\frac{5}{6}, \frac{1}{2}$ **61.** 3 **63.** 70 **65.** 13 **66.** Use the coordinates $A(0, 0)$, $B(a, 0)$, $C(a, a)$, and $D(0, a)$ for square $ABCD$. The slope of diagonal $\overline{AC}$ is $\frac{a}{a} = 1$ and the slope of diagonal $\overline{BD}$ is $\frac{a}{-a} = -1$. The slopes are negative reciprocals, so $\overline{AC} \perp \overline{BD}$. **67.** I and III **68.** II and III **69.** $\angle A \cong \angle H$, $\angle B \cong \angle I$, $\angle C \cong \angle J$, $\overline{AB} \cong \overline{HI}$, $\overline{BC} \cong \overline{IJ}$, $\overline{AC} \cong \overline{HJ}$

Algebra Review   p. 439

1. $-7, 2$ **3.** $-3, -\frac{1}{2}$ **5.** $\frac{5 + \sqrt{3}}{2}, \frac{5 - \sqrt{3}}{2}$; 3.37, 1.63 **7.** $-4, \frac{5}{2}$ **9.** 0, 4 **11.** $\frac{-5 + \sqrt{55}}{6}, \frac{-5 - \sqrt{55}}{6}$; 0.40, -2.07

Lesson 7-2   pp. 440–447

Got It? 1a. $\angle D \cong \angle H$, $\angle E \cong \angle J$, $\angle F \cong \angle K$, $\angle G \cong \angle L$ **b.** $\frac{DE}{HJ} = \frac{EF}{JK} = \frac{FG}{KL} = \frac{GD}{LH}$ **2a.** not similar **b.** $ABCDE \sim SRVUT$ or $ABCDE \sim UVRST$; 2 : 1 **3.** $\frac{10}{3}$ **4.** 28.8 in. high by 48 in. wide **5a.** Using 0.8 cm as the height of the towers, then $\frac{1}{200} = \frac{0.8}{h}$ and $h = 160$ m. **b.** No; using a scale of 1 in. = 50 ft, the paper must be more than 12 in. long.
Lesson Check 1. $\angle H$ **2.** JT **3.** yes; $DEGH \sim PLQR$; 3 : 2 **4.** 6 **5.** Answers may vary. Sample: The scale indicates how many units of length of the actual object are represented by each unit of length in the drawing. **6.** A is incorrect. Sample explanation: In the diagram, $\angle T$ corresp. to $\angle P$ (or to $\angle U$), but in the similarity statement $TRUV \sim NPQV$, $\angle T$ corresp. to $\angle N$. **7.** Every figure is $\sim$ to itself, so similarity is reflexive. If figure 1 $\sim$ figure 2 and figure 2 $\sim$ figure 3, then figure 1 $\sim$ figure 3, so similarity is transitive. If figure 1 $\sim$ figure 2, then figure 2 $\sim$ figure 1, so the similarity is symmetric. **8.** any three of the following: $\triangle ABS \sim \triangle PRS$, $\triangle ASB \sim \triangle PSR$, $\triangle SAB \sim \triangle SPR$, $\triangle SBA \sim \triangle SRP$, $\triangle BAS \sim \triangle RPS$, $\triangle BSA \sim \triangle RSP$
Exercises 9. $\angle R \cong \angle D$, $\angle S \cong \angle E$, $\angle T \cong \angle F$, $\angle V \cong \angle G$; $\frac{RS}{DE} = \frac{ST}{EF} = \frac{TV}{FG} = \frac{VR}{GD}$

11. $\angle K \cong \angle H$, $\angle L \cong \angle G$, $\angle M \cong \angle F$, $\angle N \cong \angle D$, $\angle P \cong \angle C$; $\frac{KL}{HG} = \frac{LM}{GF} = \frac{MN}{FD} = \frac{NP}{DC} = \frac{PK}{CH}$ **13.** $ABDC \sim FEDG$ (or $ABDC \sim FGDE$, $ABDC \sim DEFG$, $ABDC \sim DGFE$); scale factor is 2 : 3. **15.** Not similar; sample explanation: The ratio of the longer sides is $\frac{12}{9}$ or $\frac{4}{3}$, and the ratio of the shorter sides is $\frac{10}{8}$ or $\frac{5}{4}$. Since $\frac{4}{3} \neq \frac{5}{4}$, the corresp. sides are not proportional and the figures are not $\sim$. **17.** Not similar; sample explanation: The $\angle$ measures are not the same. **19.** $x = 8$, $y = 9$, $z = 5.25$ **21.** 120 pixels wide by 90 pixels high **23.** 5 in. **25.** 3 : 5 **27.** 5 : 3 **29.** 25 **31a.** The slope of $\overline{AB}$, $\overline{CD}$, $\overline{AE}$, and $\overline{FG}$ is -2. The slope of $\overline{BC}$, $\overline{AD}$, $\overline{EF}$, and $\overline{AG}$ is $\frac{1}{2}$. For each pair of consecutive sides of $ABCD$, the slopes are negative reciprocals, so $ABCD$ has four rt. $\angle$. Similarly, $AEFG$ has four rt. $\angle$. The measure of $\angle A$, $\angle ABC$, $\angle BCD$, $\angle CDA$, $\angle E$, $\angle F$, and $\angle G$, is 90. **b.** By the Distance Formula, $AB = BC = CD = AD = \sqrt{5}$ and $AE = EF = FG = AG = 2\sqrt{5}$. **c.** All the angles of $AEFG$ and $ABCD$ are $\cong$. $\frac{AB}{AE} = \frac{BC}{EF} = \frac{CD}{FG} = \frac{AD}{AG} = \frac{\sqrt{5}}{2\sqrt{5}} = \frac{1}{2}$ The corresp. sides are proportional, so $AEFG \sim ABCD$. **33.** No; for polygons with more than 3 sides, you also need to know that corresp. $\angle$ are $\cong$ in order to state that the polygons are $\sim$. **35.** 1 : 3 **37.** $x = 10$; 2 : 1 **43.** always **45.** sometimes **47.** 21 ft by 40 ft **49.** All $\angle$ in any rectangle are right $\angle$, so all corresp. $\angle$ are $\cong$. The ratio of two pair of consecutive sides for each rectangle is the same. Since opposite sides of a parallelogram are equal, the other two pair of sides will also have the same ratio. So corresp. sides form equal ratios and are proportional. So $BCEG \sim LJAW$. **51.** A **53.** D **55.** $7y$ **56.** $\frac{7}{9}$ **57.** $\frac{y + 9}{9}$ **58.** $\triangle BDC$, $\triangle AEC$, $\triangle FED$ **59.** $\overline{BD}$, $\overline{AF}$ **60.** 8 **61.** 69 **62.** SSS **63.** SAS **64.** ASA

Lesson 7-3   pp. 450–458

Got It? 1a. The measures of the two acute $\angle$ in each $\triangle$ are 39 and 51, so the $\triangle$ are $\sim$ by the AA $\sim$ Post. **b.** Each of the base $\angle$ in the $\triangle$ at the left measures 68, while each of the base $\angle$ in the $\triangle$ at the right measures $\frac{1}{2}(180 - 62) = 59$; the $\triangle$ are not $\sim$. **2a.** The ratio for each of the three pairs of corresp. sides is 3 : 4, so $\triangle ABC \sim \triangle EFG$ by SSS $\sim$. **b.** $\angle A$ is in each $\triangle$ and $\frac{AL}{AC} = \frac{AW}{AE} = \frac{1}{2}$, so $\triangle ALW \sim \triangle ACE$ by SAS $\sim$.
3a. $\overline{MP} \parallel \overline{AC}$ (given), so $\angle A \cong \angle P$ and $\angle C \cong \angle M$ because if two lines are $\parallel$, then alt. int. $\angle$ are $\cong$. So $\triangle ABC \sim \triangle PBM$ by AA $\sim$. **b.** No; the $\cong$ vertical angles are not included by the proportional sides, so it is not possible to prove that the triangles are similar. **4.** The

triangles formed will not be similar unless both Darius and the cliff form right angles with the ground.

Lesson Check 1. Yes; $m\angle R = 180 - (35 + 45) = 100$, and $\angle AEZ \cong \angle REB$ (Vert. $\angle$ are $\cong$.), so $\triangle AEZ \sim \triangle REB$ by AA$\sim$. **2.** Yes; the ratios of corresp. sides are all $2 : 3$, so $\triangle ABC \sim \triangle FED$ by SSS$\sim$. **3.** Yes; $\angle G \cong \angle E$ and $\frac{UG}{FE} = \frac{AG}{BE} = \frac{4}{5}$, so $\triangle GUA \sim \triangle EFB$ by SAS $\sim$. **4.** Answers may vary. Sample: Measure your shadow and the flagpole's shadow. Use the proportion $\frac{\text{your shadow}}{\text{flagpole's shadow}} = \frac{\text{your height}}{\text{flagpole's height}}$

5. Method A is not correct because the ratio, $\frac{4}{8}$ does not use corresp. sides. **6a.** Answers may vary. Sample: Both use two pairs of corresp. sides and the $\angle$ included by those sides, but SAS$\sim$ uses pairs of equal ratios, while SAS $\cong$ uses pairs of $\cong$ sides. **b.** Both involve all three sides of a $\triangle$, but corresp. sides are proportional for SSS $\sim$ and $\cong$ for SSS $\cong$.

Exercises 7. $\triangle FGH \sim \triangle KJH$; AA$\sim$ **9.** $\triangle RST \sim \triangle PSQ$; SAS$\sim$ **11.** Not $\sim$; $m\angle U = 180 - (25 + 35) = 120$, while $m\angle A = 110$. **13.** $\angle A \cong \angle A$ (Refl. Prop. of $\cong$) and $\angle ABC \cong \angle ACD$ (given), so $\triangle ABC \sim \triangle ACD$ by AA$\sim$.
15. There are a pair of $\cong$ vert. $\angle$ and a pair of $\cong$ rt. $\angle$, so the $\angle$ are $\sim$ by AA$\sim$; 180 ft **17.** about 169.2 m
19. $\triangle LMN \sim \triangle SMT$ by AA$\sim$. **21a.** No; the ratios of the sides that form the vertex $\angle$ are $=$, but the vertex $\angle$ may not be $\cong$. **b.** Yes; sample explanation: An isosc. rt. $\triangle$ has two $\angle$ 45°, so any two isosc. rt. $\triangle$ are $\sim$ by AA$\sim$.
23. 180 ft **25.** 20 **27.** In $\triangle PQR$ and $\triangle STV$, $\angle Q \cong \angle T$ because $\perp$ lines form rt. $\angle$, which are $\cong$. The sides that contain the $\angle$ are proportional (given). So $\triangle PQR \sim \triangle STV$ by SAS$\sim$, and $\angle KRV \cong \angle KVR$ because corresp. $\angle$ of $\sim \triangle$ are $\cong$. Thus $\triangle VKR$ is isosc. by the Converse of Isosc. $\triangle$ Thm. **29.** Yes; the two $\parallel$ lines and the two sides determine two pairs of $\cong$ corr. $\angle$, so the two $\triangle$ are $\sim$ by AA$\sim$. **31.** $4 : 3$; sample explanation: Since $\angle P \cong \angle S$ and $\angle PQM \cong \angle STR$, $\triangle PQM \sim \triangle STR$ by AA$\sim$. So the ratio $\frac{MQ}{RT} = \frac{PM}{SR} = $ the ratio of corresp. sides in $\triangle PMN$ and $\triangle SRW$ namely, $4 : 3$. **33.** It is given that $\ell_1 \parallel \ell_2$, so $\angle BAC \cong \angle EDF$ because if lines are $\parallel$, then corresponding $\angle$ are $\cong$. The given $\perp$ lines mean $\angle ACB \cong \angle DFE$ because $\perp$ lines form rt. $\angle$, which are $\cong$. So $\triangle ABC \sim \triangle DEF$ by AA$\sim$, and $\frac{BC}{EF} = \frac{AC}{DF}$ because corresp. sides of $\sim \triangle$ are proportional. Then Prop. of Proportions (2) lets us conclude that $\frac{BC}{AC} = \frac{EF}{DF}$.

35.

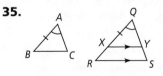

Choose point X on $\overline{QR}$ so that $QX = AB$. Then draw $\overline{XY} \parallel \overline{RS}$ (Through a point not on a line, there is exactly one line $\parallel$ to the given line.). $\angle A \cong \angle Q$ (Given) and $\angle QXY \cong \angle R$ (If two lines are $\parallel$, then corresp. $\angle$ are $\cong$.), so

$\triangle QXY \sim \triangle QRS$ by AA$\sim$. Therefore, $\frac{QX}{QR} = \frac{XY}{RS} = \frac{QY}{QS}$

because corresp. sides of $\sim \triangle$ are proportional. Since $QX = AB$, substitute QX for AB in the given proportion $\frac{AB}{QR} = \frac{AC}{QS}$ to get $\frac{QX}{QR} = \frac{AC}{QS}$. Therefore, $\frac{QX}{QR} = \frac{QY}{QS} = \frac{AC}{QS}$, and $QY = AC$. So $\triangle ABC \cong \triangle QXY$ by SAS. $\angle B \cong \angle QXY$ (Corresp. parts of $\cong \triangle$ are $\cong$.) and $\angle B \cong \angle R$ by the Transitive Prop. of $\cong$. Therefore, $\triangle ABC \sim \triangle QRS$ by AA$\sim$.
37. C **39.** C **41.** $2 : 3$ **42.** 135 **43.** 12 **44.** $\frac{3}{2}$ **45.** 125; obtuse **46.** 88; acute **47.** 180; straight **48.** 110; obtuse **49.** 8, 18; x, 24; 6 **50.** m, 18; 12, 20; $\frac{40}{3}$ or $13\frac{1}{3}$ **51.** $x + 2$, 9; 15, x; 3 **52.** $x + 4$, 5; $x - 3$, 9; $\frac{47}{4}$ or 11.75

Lesson 7-4   pp. 460–467

Got It? 1a. $\triangle PRQ \sim \triangle SPQ \sim \triangle SRP$ **b.** $\frac{SR}{SP} = \frac{SP}{SQ}$, $\frac{SR}{SP} = \frac{PR}{QP}$ **2.** $6\sqrt{2}$ **3.** $x = 6$, $y = 2\sqrt{5}$ **4.** 12 in.
Lesson Check 1. 6 **2.** $\sqrt{48}$ or $4\sqrt{3}$ **3.** h **4.** g **5.** j, h or h, j **6.** d, d **7a.** $\overline{RT}$ **b.** $\overline{RP}$, $\overline{PT}$ **c.** $\overline{PT}$ **8.** The length 8 is the entire hypotenuse, so the segments of the hypotenuse have lengths 3 and 5. The correct proportion is $\frac{3}{x} = \frac{x}{5}$.

Exercises 9. Answers may vary. Sample: $\triangle KJL \sim \triangle NJK \sim \triangle NKL$ **11.** Answers may vary. Sample: $\triangle OMN \sim \triangle PMO \sim \triangle PON$ **13.** 12 **15.** $\sqrt{63}$ or $3\sqrt{7}$ **17.** 14 **19.** $x = 20$, $y = 10\sqrt{5}$ **21.** $x = 3\sqrt{7}$, $y = 12$ **23a.** 4 cm

b.

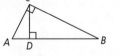

4 cm
2 cm   8 cm

c. Answers may vary. Sample: Draw a 10-cm segment. Construct a $\perp$ of length 4 cm that is 2 cm from one endpoint; connect to form a $\triangle$. **25.** 2.5 **27.** 1 **29.** Yes; the proportion $\frac{a}{\sqrt{ab}} = \frac{\sqrt{ab}}{b}$ is true by the Cross Products Prop. and satisfies the definition of the geometric mean.
31. 8.50 m **33.** $\ell_1 = \sqrt{2}$, $\ell_2 = \sqrt{2}$, $a = 1$, $s_2 = 1$ **35.** $\ell_2 = 2\sqrt{3}$, $h = 4$, $a = \sqrt{3}$, $s_1 = 1$
37. $(-2, 6)$, $(10, 6)$ **39.** 4 **41.** 5 **43.** $\triangle ABC \sim \triangle ACD$ and $\triangle ABC \sim \triangle CBD$ by Thm. 7-3. Then $\frac{AB}{AC} = \frac{AC}{AD}$ and $\frac{AB}{CB} = \frac{BC}{BD}$ because corresp. sides of $\sim \triangle$ are proportional.
45a.

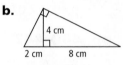

Given: $\overline{AC} \perp \overline{BC}$, $\overline{AB} \perp \overline{CD}$

Prove: $AC \cdot BC = AB \cdot CD$

b. The conjecture is true. You can express the area of $\triangle ABC$ as $\frac{1}{2}(AC)(BC)$ or as $\frac{1}{2}(AB)(CD)$, so $AC \cdot BC = AB \cdot CD$.
47. Answers may vary. Sample: $\triangle ABC \sim \triangle DEC$ (AA $\sim$ Post.), so $\frac{AC}{DC} = \frac{BC}{EC}$ (Corr. sides of $\sim \triangle$ are in prop.). By the Subtraction Property of $=$, $\frac{AC - DC}{DC} = \frac{BC - EC}{EC}$, or $\frac{AD}{DC} = \frac{BE}{EC}$. **49.** H **51.** $\angle R \cong \angle P$ (given) and $\angle RNM \cong \angle PNQ$ (Vert. $\angle$ are $\cong$.), so $\triangle NRM \sim \triangle NPQ$

by AA ∼. **52.** $x = 5$, $y = 8$ **53.** $x = 3$, $y = 4$
54. 28 cm **55.** 9.8 in. **56.** $\frac{24}{7}$ mm or $3\frac{3}{7}$ mm

Lesson 7-5   pp. 471–478

Got It? 1a. 8 **b.** $RS = \frac{1}{2}XZ$ (Midsegment Thm.)
2. 5.76 yd **3.** 14.4
Lesson Check 1. d **2.** c **3.** d **4.** 5 **5.** 15 **6.** Answers
may vary. Sample: The corollary to the Side-Splitter Thm.
takes the same three (or more) ‖ lines as in Thm. 6-7, but
instead of cutting off ≅ segments it allows the segments
to be proportional. **7.** Answers may vary. Sample: Alike:
Both involve a △ and a seg. from one vertex to the
opposite side of the △. Different: In Corollary 1 to
Thm. 7-3, the △ is a rt. △ and the seg. is an alt., while in
the △-∠-Bis. Thm. the △ does not have to be a rt. △
and the seg. is an ∠ bis. **8.** The Side-Splitter Thm.
involves only the segments formed on the two sides
intersected by the ‖ line. (To find x, you can use a
proportion involving the two ∼ △.)
Exercises 9. 7.5 **11.** 10 **13.** 8 mm **15.** 7.5 **17.** $3\frac{5}{13}$
19. 6 **21.** 35 **23.** Use the Side-Splitter Thm. to write
the proportion $\frac{AB}{BD} = \frac{AC}{CE}$, then find the values of BD, AC,
and CE to calculate the unknown length AB. **25.** KS by
the △-∠-Bis. Thm. **27.** JP by the Side-Splitter Thm.
29. KM by the △-∠-Bis. Thm. **31.** 575 ft **33.** 20
35. $\frac{2}{7}$ or 3 **37.** $\frac{XR}{RQ} = \frac{YS}{SQ}$ (Given); $\frac{XR + RQ}{RQ} = \frac{YS + SQ}{SQ}$
(Prop. of Proportions (3)); $XQ = XR + RQ$,
$YQ = YS + SQ$ (Seg. Add. Post.); $\frac{XQ}{RQ} = \frac{YQ}{SQ}$ (Subst.);
∠Q ≅ ∠Q (Refl. Prop. of ≅); △XQY ∼ △RQS (SAS ∼
Post.); ∠1 ≅ ∠2 (Corresp. ∡ of ∼ △ are ≅.); $\overline{RS} \parallel \overline{XY}$
(If corresp. ∡ are ≅, the lines are ‖.) **39.** no;
$\frac{28}{12} \neq \frac{24}{10}$ **41.** 12.5 cm or 4.5 cm **43.** Isosc.; $AC : BC$ is
1 : 1 by the △-∠-Bis. Thm. **45.** 5.2 **47.** By the Side-
Splitter Thm., $\frac{CD}{DB} = \frac{CA}{AF}$. By the Corresp. ∡ Post.,
∠3 ≅ ∠1. Since $\overrightarrow{AD}$ bisects ∠CAB, ∠1 ≅ ∠2. By the
Alt. Int. ∡ Thm., ∠2 ≅ ∠4. So, ∠3 ≅ ∠4 by the Trans.
Prop. of ≅. By the Converse of the Isosc. △ Thm.,
$BA = AF$. Substituting BA for AF, $\frac{CD}{DB} = \frac{CA}{BA}$.
49. Use the diagram with Ex. 47, with $\overline{AD} \parallel \overline{EB}$. It is given
that $\frac{CD}{DB} = \frac{CA}{BA}$, and you want to prove that ∠1 ≅ ∠2. By
the Side-Splitter Thm., $\frac{CA}{AF} = \frac{CD}{DB}$. So $\frac{CA}{BA} = \frac{CA}{AF}$ by the
Transitive Prop. of =, and $BA = AF$. Therefore, ∠3 ≅ ∠4
by the Isosc. △ Thm. Using properties of ‖ lines,
∠1 ≅ ∠3 and ∠2 ≅ ∠4. So ∠1 ≅ ∠2 by the Transitive
Prop. of ≅, and $\overrightarrow{AD}$ bisects ∠CAB by the def. of ∠ bis.
51. 20 **53.** 118 **55.** m **56.** m **57.** c **58.** h **59.** (3, −3)
60. (0, 2) **61.** (1.5, 2.5) **62.** $(3 \text{ m})^2 = 9 \text{ m}^2$,
$(4 \text{ m})^2 = 16 \text{ m}^2$, $(5 \text{ m})^2 = 25 \text{ m}^2$ **63.** $(5 \text{ in.})^2 = 25 \text{ in.}^2$,
$(12 \text{ in.})^2 = 144 \text{ in.}^2$, $(13 \text{ in.})^2 = 169 \text{ in.}^2$
64. $(4 \text{ m})^2 = 16 \text{ m}^2$, $(4\sqrt{2} \text{ m})^2 = 32 \text{ m}^2$

Chapter Review   pp. 480–482

1. similar **2.** proportion **3.** scale factor **4.** means,
extremes (in either order) **5.** 1 : 116 or $\frac{1}{116}$ **6.** 36
7. 6 **8.** $\frac{55}{4}$ or $13\frac{3}{4}$ **9.** 6 **10.** 7 **11.** $JEHN \sim JKLP$; 3 : 4
12. △PQR ∼ △XYZ; 3 : 2 **13.** 120 ft **14.** 45 ft **15.** The
ratios of each pair of corresp. sides is 2 : 1, so
△AMY ∼ △ECD by SSS∼. **16.** If lines are ‖, then
corresp. ∡ are ≅, so △RPT ∼ △SGT by AA∼.
17. 12 **18.** $2\sqrt{15}$ **19.** $x = 6\sqrt{2}$, $y = 6\sqrt{6}$ **20.** $\sqrt{35}$
21. $x = 2\sqrt{21}$; $y = 4\sqrt{3}$ **22.** $x = 12$, $y = 4\sqrt{5}$
23. 7.5 **24.** 3.6 **25.** 22.5 **26.** 12 **27.** 17.5 **28.** 77

Chapter 8

Get Ready!   p. 487

1. 4.648 **2.** 40.970 **3.** 6149.090
4. −5 **5.** AA ∼ **6.** SSS ∼ **7.** SAS ∼ **8.** 12 **9.** 8
10. $2\sqrt{13}$ **11.** 9 **12.** Answers may vary. Sample: When
something is "elevated" you look up to see it, so an ∠ of
elevation is formed by a horizontal line and the line of
sight. **13.** Answers may vary. Sample: The prefix *tri-*
means 3; triangles are associated with trigonometric
ratios. **14.** Answers may vary. Sample: Because the Law
of Cosines can be derived from the Pythagorean Theorem,
and you can use the Law of Cosines to find angle
measures, the Law of Cosines can probably be used to
find side lengths and angle measures.

Lesson 8-1   pp. 491–498

Got It? 1a. 26 **b.** Yes; 10, 24, and 26 are whole
numbers that satisfy $a^2 + b^2 = c^2$. **2.** $6\sqrt{3}$ **3.** 15.5 in.
4a. No; $16^2 + 48^2 \neq 50^2$. **b.** No; $a^2 + b^2 = b^2 + a^2$
for any values of a and b. **5.** acute
Lesson Check 1. 37 **2.** $\sqrt{130}$ **3.** 4 **4.** $4\sqrt{3}$ **5.** The
three numbers a, b, and c must be whole numbers that
satisfy $a^2 + b^2 = c^2$. **6.** The longest side is 34, so the
student should have tested $16^2 + 30^2 \overset{?}{=} 34^2$.
Exercises 7. 10 **9.** 34 **11.** 97 **13.** no; $4^2 + 5^2 \neq 6^2$
15. yes; $15^2 + 20^2 = 25^2$ **17.** $\sqrt{33}$ **19.** $\sqrt{105}$
21. $5\sqrt{3}$ **23.** 17 m **25.** no; $8^2 + 24^2 \neq 25^2$
27. acute **29.** acute **31.** right **33.** 4.2 in. **35a.** $|x_2 - x_1|$;
$|y_2 - y_1|$ **b.** $PQ^2 = (x_2 - x_1)^2 + (y_2 - y_1)^2$
c. $PQ = \sqrt{(x_2 - x_1)^2 + (y_2 - y_1)^2}$ **37.** $8\sqrt{5}$ **39.** 29
41. 84 **43–48.** Answers may vary. Samples are given.
43a. 6 **b.** 7 **45a.** 8 **b.** 11 **47a.** 8 **b.** 10 **49.** $\frac{q}{b} = \frac{b}{c}$ and
$\frac{r}{a} = \frac{a}{c}$ because each leg is the geometric mean of the
adj. hypotenuse segment and the hypotenuse. By the
Cross Products Property, $b^2 = qc$ and $a^2 = rc$. Then
$a^2 + b^2 = qc + rc = c(q + r)$. Substituting c for $q + r$

gives $a^2 + b^2 = c^2$. **51a.** Horiz. lines have slope 0, and vert. lines have undef. slope. Neither could be mult. to get -1. **b.** Assume the lines do not intersect. Then they have the same slope m. Then $m \cdot m = m^2 = -1$, which is impossible. So the lines must intersect. **c.** Let ℓ, be $y = \frac{b}{a}x$ and ℓ_2 be $y = -\frac{a}{b}x$. Define $C(a,b)$, $A(0,0)$, and $B(a, -\frac{a^2}{b})$.

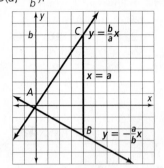

Using the Distance Formula, $AC = \sqrt{a^2 + b^2}$, $BA = \sqrt{a^2 + \frac{a^4}{b^2}}$, and $CB = b + \frac{a^2}{b}$. Then $AC^2 + BA^2 = CB^2$ and $m\angle A = 90$ by the Conv. of the Pythagorean Thm. So $\ell_1 \perp \ell_2$.

53. Draw right $\triangle FDE$ with legs $\overline{DE}$ of length a and $\overline{EF}$ of length b, and hypotenuse of length x. By the Pythagorean Thm., $a^2 + b^2 = x^2$. $\triangle ABC$ has sides of length a, b, and c, where $c^2 > a^2 + b^2 \cdot c^2 > x^2$ and $c > x$ by Prop. of Inequalities. If $c > x$, then $m\angle C > m\angle E$ by the Converse of the Hinge Thm. An angle with measure > 90 is obtuse, so $\triangle ABC$ is an obtuse $\triangle$. **55.** 4 **57.** 61

59. 4, 5 **60.** $\sqrt{3}$ **61.** $15\sqrt{2}$ **62.** $\frac{16\sqrt{3}}{3}$

Lesson 8-2   pp. 499–505

Got It? 1. $5\sqrt{6}$ **2a.** $5\sqrt{2}$ **b.** $\frac{\sqrt{2}}{\sqrt{2}} = 1$, so multiplying by $\frac{\sqrt{2}}{\sqrt{2}}$ is the same as multiplying by 1. **3.** 141 ft

4. $\frac{10\sqrt{3}}{3}$ **5.** 15.6 mm

Lesson Check 1. $7\sqrt{2}$ **2.** 3 **3.** $4\sqrt{2}$ **4.** $6\sqrt{3}$ **5.** Rika; 5 should be opposite the 30° $\angle$ and $5\sqrt{3}$ should be opposite the 60° $\angle$. **6.** Answers may vary. Sample: The $\triangle$ is isosc. The length of each leg is the same. Use the Pythagorean Thm. to find the hypotenuse; 6, $6\sqrt{2}$.

Exercises 7. $x = 8$, $y = 8\sqrt{2}$ **9.** $60\sqrt{2}$ **11.** $5\sqrt{2}$
13. 14.1 cm **15.** $x = 20$, $y = 20\sqrt{3}$ **17.** $x = 5$, $y = 5\sqrt{3}$ **19.** $x = 4$, $y = 2$ **21.** 50 ft **23.** $a = 7$, $b = 14$, $c = 7$, $d = 7\sqrt{3}$ **25.** $a = 10\sqrt{3}$, $b = 5\sqrt{3}$, $c = 15$, $d = 5$ **27.** $a = 3$, $b = 7$ **29.** 14.4 s

31. Answers may vary. Sample: A ramp up to a door is 12 ft long. The ramp forms a 30° $\angle$ with the ground. How high off the ground is the door? 6 ft
33a. $\sqrt{3}$ units **b.** $2\sqrt{3}$ units **c.** $s\sqrt{3}$ units **35.** I

37. $AC = 6$; $\frac{3}{AC} = \frac{AC}{12}$, $AC^2 = 36$
38. $\sqrt{11}$ in. **39.** $4\sqrt{21}$ cm **40.** $\frac{12}{7}$ **41.** $\frac{54}{11}$ **42.** $\frac{15}{2}$ **43.** $\frac{60}{7}$

Lesson 8-3   pp. 507–513

Got It? 1. $\frac{15}{17}$; $\frac{8}{17}$; $\frac{15}{8}$ **2a.** 13.8 **b.** 1.9 **c.** 3.8 **d.** 44 ft
3a. 68 **b.** No; you can use any of the three trigonometric ratios as long as you identify the appropriate leg that is opp. or adj. to each acute $\angle$.

Lesson Check 1. $\frac{8}{10}$ or $\frac{4}{5}$ **2.** $\frac{6}{10}$ or $\frac{3}{5}$ **3.** $\frac{8}{6}$ or $\frac{4}{3}$
4. $\frac{6}{10}$ or $\frac{3}{5}$ **5.** $\frac{8}{10}$ or $\frac{4}{5}$ **6.** $\frac{6}{8}$ or $\frac{3}{4}$ **7.** 12.1 **8.** 57.5

9. The word is made up of the first letters of each ratio: $S = \frac{O}{H}$, $C = \frac{A}{H}$, and $T = \frac{O}{A}$. **10.** No; $\sin X = \frac{YZ}{YX}$, $\sin A = \frac{BC}{BA}$, and $\triangle XYZ \sim \triangle ABC$ by AA $\sim$, so $\frac{YZ}{YX} = \frac{BC}{BA}$ because corresp. sides of $\sim \triangle$ are proportional. Therefore, $\sin X = \sin A$.

Exercises 11. $\frac{7}{25}$; $\frac{24}{25}$; $\frac{7}{24}$ **13.** $\frac{\sqrt{3}}{2}$; $\frac{1}{2}$; $\sqrt{3}$ **15.** 8.3
17. 17.0 **19.** 21.4 **21.** 1085 ft **23.** 58 **25.** 59
27. 66 **29.** about 17 ft 8 in.
31. $\cos X \cdot \tan X = \frac{\text{adjacent}}{\text{hypotenuse}} \cdot \frac{\text{opposite}}{\text{adjacent}} = \frac{\text{opposite}}{\text{hypotenuse}} = \sin X$
33. $w = 3$, $x \approx 41$ **35.** $w \approx 68.3$, $x \approx 151.6$
37a. They are equal; yes; sine and cosine of compl. $\triangle$ are $=$. **b.** $\angle B$; $\angle A$ **c.** Sample: The cosine is the complement's sine.
39a.

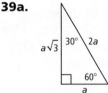

Using the ratio of sides $1 : \sqrt{3} : 2$ for a 30°-60°-90° $\triangle$, $\tan 60° = \frac{\sqrt{3}}{1} = \sqrt{3}$.

b. Answers may vary. Sample: $\sin 60° = \sqrt{3} \cdot \cos 60°$
41. $\frac{15}{9}$ or $\frac{5}{3}$ **43.** $\frac{15}{9}$ or $\frac{5}{3}$ **45.** $\frac{12}{9}$ or $\frac{4}{3}$
47a. No; answers may vary. Sample: $\tan 45° + \tan 30° = 1 + \frac{\sqrt{3}}{3} \approx 1.6$, but $\tan 75° \approx 3.7$.
b. No; assume $\tan A - \tan B = \tan (A - B)$; $\tan A = \tan B + \tan (A - B)$ by the Add. Prop. of $=$; let $A = B + C$, then $\tan (B + C) = \tan B + \tan C$ by the Subst. Prop.; part (a) proved this false; this contradicts the assumption, so $\tan A - \tan B \neq \tan (A - B)$.
49. $(\sin B)^2 + (\cos B)^2 = \left(\frac{b}{c}\right)^2 + \left(\frac{a}{c}\right)^2 = \frac{b^2}{c^2} + \frac{a^2}{c^2} = \frac{b^2 + a^2}{c^2} = \frac{c^2}{c^2} = 1$
51. $\frac{1}{(\sin A)^2} - \frac{1}{(\tan A)^2} = \frac{1}{\left(\frac{a}{c}\right)^2} - \frac{1}{\left(\frac{a}{b}\right)^2} = \frac{c^2}{a^2} - \frac{b^2}{a^2} = \frac{c^2 - b^2}{a^2} = \frac{a^2}{a^2} = 1$

53a. 1.5 AU **b.** 5.2 AU **55.** G **57.** G **57.** 4, $4\sqrt{3}$
58. $5\sqrt{2}$ units **59.** ∠7 **60.** ∠11 **61.** ∠6 **62.** 90

Lesson 8-4   pp. 516–521

Got It? 1a. ∠ of elevation from the person in the hot-air balloon to bird **b.** ∠ of depression from the person in the hot-air balloon to base of mountain **2.** about 631 ft **3.** about 6.2 km
Lesson Check 1. ∠ of elevation from C to A **2.** ∠ of depression from A to C **3.** ∠ of elevation from A to D **4.** ∠ of elevation from A to B **5.** ∠ of depression from B to A **6.** ∠1 ≅ ∠2 (alt. int. ∠); ∠4 ≅ ∠5 (alt. int. ∠) **7.** Answers may vary. Sample: An ∠ of elevation is formed by two rays with a common endpoint when one ray is horizontal and the other ray is above the horizontal ray. **8.** Answers may vary. Sample: The ∠ labeled in the sketch is the complement of the ∠ of depression.
Exercises 9. ∠ of elevation from sub to boat **11.** ∠ of elevation from boat to tree **13.** ∠ of elevation from Max to top of waterfall **15.** ∠ of depression from top of waterfall to Max **17.** 34.2 ft **19.** 986 m **21.** 0.6 km **23.** 64° **25.** 72, 72 **27.** 27, 27 **29a.** length of any guy wire = distance on the ground from the tower to the guy wire div. by the cosine of the ∠ formed by the guy wire and the ground **b.** height of attachment = distance on the ground from the tower to the guy wire times the tangent of the ∠ formed by the guy wire and the ground **31.** about 2.8 **33.** 3300 m **35.** Answers may vary. **37.** G **39.** 85.2 m **40.** 38.2 ft **41.** 45 **42.** $2\sqrt{17} \approx 8.2$ **43.** $\sqrt{229} \approx 15.1$ **44.** $2\sqrt{37} \approx 12.2$

Lesson 8-5   pp. 522–526

Got It? 1. 9.5 units **2.** 38.4 **3.** 40.6 ft
Lesson Check 1. 43.6 **2.** about 18.1 **3.** about 9.1 **4.** No, you need to know at least one angle measure to use the Law of Sines. **5.** ∠R is opposite side PQ, so the proportion should be written as $\frac{\sin 75°}{4} = \frac{\sin P}{3}$.
Exercises 7. 54.2 **9.** $x \approx 2.1$, $y \approx 3.6$ **11.** $x \approx 12.7$, $y \approx 9.4$ **13.** 259.4 yd **15.** about 97.4 mi **17.** 36.9° **19.** 7.05 in. **21.** 0.6 **22.** 4492 ft **23.** 8 **24.** 0.53 **25.** 54.02 **26.** 0.65 **27.** 51.32

Lesson 8-6   pp. 527–532

Got It? 1. 61.8 **2.** 76.4° **3.** 2.1 mi
Lesson Check 1. 11.2 **2.** 81.5 **3.** 26.0 sq units **4.** $m\angle X \approx 86.4$, $m\angle Y \approx 58.8$, $m\angle Z \approx 34.8$ **5.** The variable c should be squared, so the actual value of c is about 7.98. **6.** Use the Law of Cosines to solve for the measure of the angle that is opposite the longest side of the triangle.

Exercises 7. 10.0 **9.** 3.9 **11.** $x \approx 46.8$, $y \approx 35.0$ **13.** $x \approx 54.1$, $y \approx 72.0$ **15.** 54.7 **17.** 115 ft **19.** Law of Sines; 59.0 **21.** Law of Sines; 17.5 **23.** 18.4 ft **25.** 26.9 **27.** 73.7, 53.1, 53.1 **29.** 3.8 **31.** 12 **33.** 74.6 **34.** Yes, the slopes of sides AB and CD are both 0, and the slopes of sides BC and AD are both $\frac{8}{3}$. **35.** $\overline{EF}$ **36.** $\overline{AC}$ **37.** $\overline{BC}$ **38.** ∠G **39.** ∠A **40.** ∠F

Chapter Review   pp. 534–536

1. Trigonometric ratios **2.** ∠ of elevation **3.** Pythagorean triple **4.** $2\sqrt{113}$ **5.** 17 **6.** $12\sqrt{2}$ **7.** $9\sqrt{3}$ **8.** $x = 7$, $y = 7\sqrt{2}$ **9.** $5\sqrt{2}$ **10.** $x = 6\sqrt{3}$, $y = 12$ **11.** $x = 7$, $y = 7\sqrt{3}$ **12.** 70.7 ft **13.** $\frac{2\sqrt{19}}{20}$ or $\frac{\sqrt{19}}{10}$; $\frac{18}{20}$ or $\frac{9}{10}$; $\frac{2\sqrt{19}}{18}$ or $\frac{\sqrt{19}}{9}$ **14.** $\frac{16}{20}$ or $\frac{4}{5}$; $\frac{12}{20}$ or $\frac{3}{5}$; $\frac{16}{12}$ or $\frac{4}{3}$ **15.** 16.5 **16.** 33.1 **17.** 38.2 ft **18.** $x = 14.7$ cm **19.** 29.9° **20.** $m\angle D = 82.1$ **21.** $m\angle N = 86.0$

Chapter 9

Get Ready!   p. 541

1. △ADC **2.** △LJK **3.** △RTS **4.** △LHC **5.** 108 **6.** 135 **7.** 144 **8.** 160 **9.** always **10.** never **11.** sometimes **12.** always **13.** 50 ft; Because 1 in. = 20 ft, 2.5 in. = 2.5(20 ft) = 50 ft. **14.** 0.25 in.; Because 20 ft = 1 in. and 5 ft = $\frac{1}{4}$(20 ft), by substitution, 5 ft = $\frac{1}{4}$(1 in.) = 0.25 in. **15.** left hand; 4 ft **16.** the point at the center of the clock **17.** Answers may vary. Sample: When you dilate a geometric figure, you change its size.

Lesson 9-1   pp. 545–552

Got It? 1a. Yes; the distance between the vertices and the angle measures of the image are the same as the preimage. **b.** Yes; the distances between the vertices and the angle measures of the image are the same as the preimage.
2a. ∠U; P **b.** $\overline{NI}$ and $\overline{SU}$, $\overline{ID}$ and $\overline{UP}$, $\overline{DN}$ and $\overline{PS}$
3a–b. Graph:

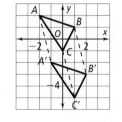

a. $A'(-1, -2)$, $B'(2, -3)$, $C'(1, -5)$ **b.** $AA' = BB' = CC'$ and $\overline{AA'} \parallel \overline{BB'} \parallel \overline{CC'}$; because rigid motions preserve distance and the slope of each segment is −4. **4.** $T_{<7, -1>}(\triangle LMN)$ **5.** 3 squares right and 5 squares down

Lesson Check 1. P'; $\overline{T'J'}$

2.

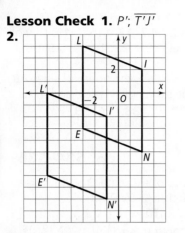

3. $T_{<-12, 4>}(H)$

4. Answers may vary. There are points in the image that are not the same distance from each other as the corresponding

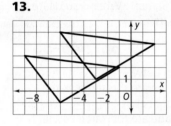

points in the preimage. **5.** The transformation that maps $\triangle ABC$ to $\triangle PQR$ maps A to P and C to R, so it is a reflection, not a translation. The transformation that maps $\triangle ABC$ onto $\triangle RQP$ is a translation. **6.** $T_{<1, 0>}(x, y)$ followed by $T_{<0, -3>}(x, y)$

Exercises 7. Yes; distances between corresponding pairs of points are equal. **9.** No; distances between corresponding pairs of points are not equal.

11a. Answers may vary. Sample: $\angle R \rightarrow \angle R'$ **b.** $\overline{RP}$ and $\overline{R'P'}$; $\overline{PT}$ and $\overline{P'T'}$; $\overline{RT}$ and $\overline{R'T'}$

13.   **15.**

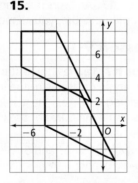

17. $T_{<1, -1>}(x, y)$ **19.** 1 block west and 7 blocks north
21. $T_{<-3, 1>}(x, y)$ **23.** The vertices of $P'L'A'T'$ are $P'(0, -3)$, $L'(1, -2)$, $A'(2, -2)$, and $T'(1, -3)$. Slope of $\overline{PP'}$ = slope of $\overline{LL'}$ = slope of $\overline{AA'}$ = slope of $\overline{TT'}$ = $-\frac{3}{2}$, so $\overline{PP'} \parallel \overline{LL'} \parallel \overline{AA'} \parallel \overline{TT'}$.

25.   **27.** Answers may vary.
29. at least 5 ft east and 10 ft north **31.** Sample answer:

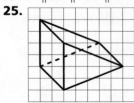

$T_{<4, -1>}(\triangle JKL)$;
$T_{<2, -1>}(\triangle JKL)$;
$T_{<4, -4>}(\triangle JKL)$

33. $T_{<13, -2.5>}(x, y)$ **35.** Translate a line segment in some other direction than along the segment. Then connect the endpoints of the line segment and its image to form a parallelogram. **37.** F **39a.** $(-5, 2)$ **b.** Yes; answers may vary. Sample: The slope of $\overline{DB}$ = 1 and the slope of $\overline{AC}$ = -1, so $\overline{DB} \perp \overline{AC}$. Since $ABCD$ is a $\square$ with $\perp$ diagonals, $ABCD$ is a rhombus.
40. about 431.7 km **41.** $\overline{BC} \cong \overline{EF}$ and $\overline{BC} \parallel \overline{EF}$ (Given), so $\angle BCA \cong \angle F$ (Corresp. $\angle$ of $\parallel$ lines are $\cong$). $\overline{AD} \cong \overline{DC} \cong \overline{CF}$ (Given), so $AC = AD + DC = DC + CF = DF$ (Segment Addition Post., Trans. Prop. of Equality). So $\triangle BCA \cong \triangle EFD$ by SAS, and $\overline{AB} \cong \overline{DE}$ (Corresp. parts of $\cong$ $\triangle$ are $\cong$). **42.** $y = -2$ **43.** $x = -1$
44. $y = -x + 1$

Lesson 9-2   pp. 554–560

Got It? 1. $(-1, 4)$

2.

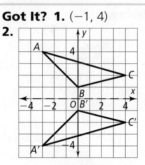

3. R_m(Triangle 3) = Triangle 1
4. No; it is not possible to prove that $\triangle GHJ$ is equilateral since you cannot prove that $HJ = HG$ or $HJ = GJ$.

Lesson Check 1. $(-4, -3)$ **2.** $(4, 2)$

3.

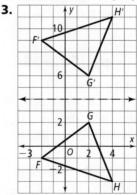

4. The line of reflection is the $\perp$ bis. of any seg. whose endpts. are corresp. pts. of the preimage and image.
5. $\overline{AA'}$ should be $\perp$ to r.

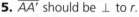

6. $R_{y\text{-axis}}(x, y) = (-x, y)$; $R_{x\text{-axis}}(x, y) = (x, -y)$

Exercises 7. $(-1, -2)$ **9.** $(-3, 2)$ **11.** $(-5, -3)$
13. $J'(1, -4)$, $A'(3, -5)$, **15.** $J'(1, 0)$, $A'(3, -1)$, $G'(2, 3)$
$G'(2, -1)$

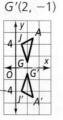

17. $J'(-3, 4)$, $A'(-5, 5)$, $G'(-4, 1)$

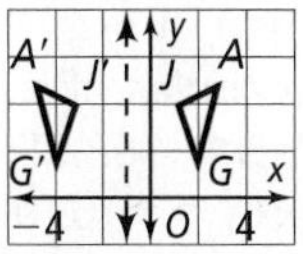

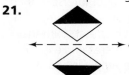

19a. Figure 3 = R_j (Figure 1) because the line j is the perpendicular bisector of the line segments between corresponding vertices of Figures 1 and 3. **b.** Figure 2 = R_n (Figure 4) because line n is the perpendicular bisector of the line segments between corresponding vertices of Figures 2 and 4. **c.** Figure 4 = R_n (Figure 2) because line n is the perpendicular bisector of the line segments between corresponding vertices of Figures 4 and 2.

21. 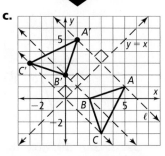   **23a.** -1 **b.** $B'(0, 2)$; $C'(-3, 3)$

c. 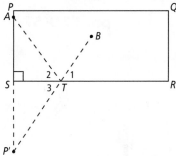 **d.** The coordinates of P' will be (b, a); the x- and y-coordinates will switch.

25. Reflect P across $\overline{SR}$ to P'. Because the pool table is a rectangle, $\overline{PS} \perp \overline{SR}$, and thus P' is collinear with S and P. The ball should bounce off the point T that is the intersection of $\overline{BP'}$ and $\overline{SR}$. Let A be the point on $\overline{SP}$ that the ball rolls to after it bounces off $\overline{SR}$. To see why A is the same point as P, look at $\triangle AST$ and $\triangle P'ST$.

Since the ball bounces off $\overline{SR}$ so that $\angle 1 \cong \angle 2$ and $\angle 1 \cong \angle 3$ (vertical $\angle$s), $\angle 2 \cong \angle 3$ by the Trans. Prop. of $\cong$. Right $\triangle$s AST and $P'ST$ are $\cong$ and $\overline{TS} \cong \overline{TS}$, so $\triangle ATS \cong \triangle P'TS$ by ASA. Then $\overline{AS} \cong \overline{P'S}$ because corresp. parts of $\cong$ $\triangle$s are $\cong$. But $\overline{P'S} \cong \overline{PS}$ by the definition of reflection across a line, so A and P must be the same point.
27a. $(3, 5)$, $(1.5, 3.5)$ **b.** $y = x + 2$ **c.** $R_{y=x+2}(ABCDE) = A'B'C'D'E'$

29.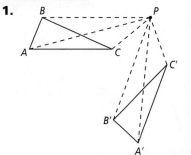

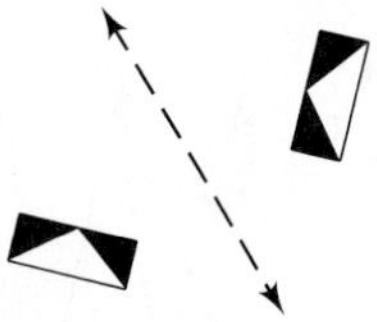

31. Answers may vary. Sample: scissors, baseball glove, golf clubs **33.** $(4, 0)$ **35.** No; each point moves a distance equal to twice the point's distance from the line of reflection.
37a. $(3, 1)$ **b.** $(-1, -3)$ **c.** $(-3, -1)$ **d.** $(1, 3)$ **e.** They are the same point. **39.** Yes; follow the steps of Exercise 38 using one leg of an isosc. $\triangle$ to first form a $\square$. Then reflect the original $\triangle$ across the $\perp$ bis. of the base of the second $\triangle$ to form an isosc. trapezoid. **41.** Yes; reflect an isosc. $\triangle$ across its base. **43.** Yes; reflect an isosc. rt. $\triangle$ across its hyp. **45.** D **47.** A **49.** $T_{<4, -2>}(x, y) = (x + 4, y - 2)$ **50.** $T_{<5, 1>}(x, y) = (x + 5, y + 1)$
51. 277.5 km **52–57.** Drawings to be checked by teacher.

Lesson 9-3   pp. 561–567

Got It?
1.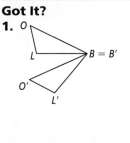

2.

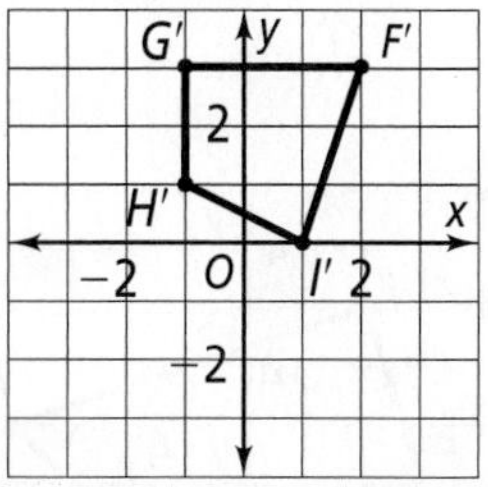

3. No; there is not enough information about $WXYZ$ to know that there is a rotation that maps $\overline{XW}$ to $\overline{WZ}$.
Lesson Check
1.

2. R **3.** $\overline{SE}$ **4.** Because A is the center of $SQRE$, $\overline{SR}$ rotated 90° clockwise about point A maps to $\overline{QE}$. Because rotations preserve distance, $SR = QE$. **5.** Draw $\overline{AO}$ and $\overline{A'O}$ and then measure $\angle AOA'$. **6.** The diagram shows a reflection, not a rotation. R' is a 115° clockwise rotation of R. All points of $\triangle PQR$ must be rotated counterclockwise.

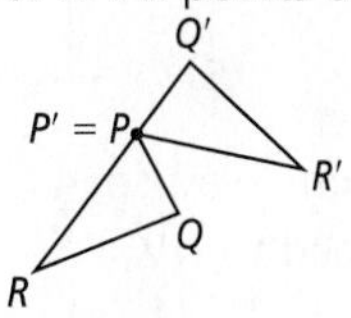

7. Both are rigid motions. A reflection reverses orientation. A rotation has the same orientation. **8.** $(-x, -y)$; Sample: The coordinates are the same as a single rotation of 180° since $135° + 45° = 180°$.

Exercises

9.

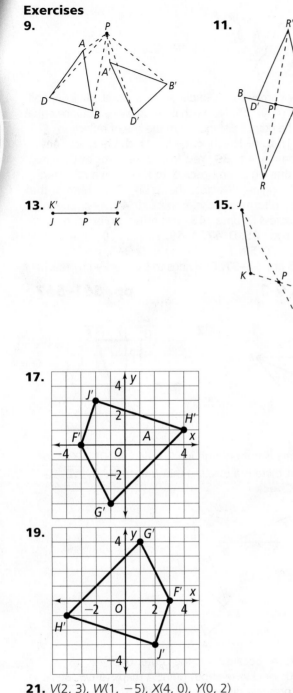

11.

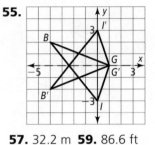

13.

K' ———•——— J'
J   P   K

15.

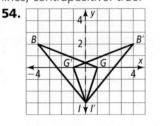

17.

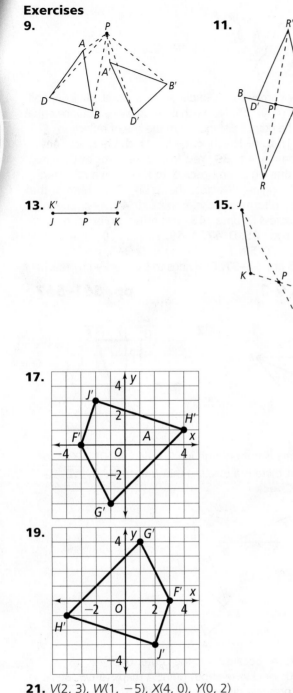

19.

21. V(2, 3), W(1, −5), X(4, 0), Y(0, 2)

23. Answers may vary. Sample answer: Because M is the midpoint of the diagonals, VM = QM and TM = NM. Since V and Q are equidistant from M, and the m∠VMQ = 180, you know that $r_{(180°, M)}(V) = Q$. Similarly, $r_{(180°, M)}(T) = N$.

Every point on $\overline{VT}$ can be rotated in this same way, so $r_{(180°, T)}(\overline{VT}) = \overline{QN}$. Also, $\overline{TQ}$ can be mapped to $\overline{NV}$, so $r_{(180°, M)}(\overline{TQ}) = \overline{NV}$.

Because rotations are rigid motions and preserve distance, TV = NQ and TQ = VN. **25.** No. In general, the diagonals of a rectangle do not bisect the angles of a rectangle. **27.** a 180° rotation **29.** 110° **31.** 168.75° **33.** any two rotations of a° and b° if a > 0, b > 0, and a + b = 360 **35.** 280° **37.** The image of $\overline{ED}$ is $\overline{BA}$, not $\overline{AB}$. **39.** M **41.** C **43.** A **45.** K **47.** J **49.** Draw two segments connecting preimage points A and B to image points A' and B'. If $\overline{AA'}$ and $\overline{BB'}$ are not collinear, construct the ⊥ bis. of $\overline{AA'}$ and $\overline{BB'}$ to find C, the center of rotation. m∠ACA' is the ∠ of rotation. If $\overline{AA'}$ and $\overline{BB'}$ are collinear, then the midpoint of $\overline{AA'}$ is the center of rotation, and 180° is the angle of rotation. **50.** A **51.** F **52.** C **53a.** Converse: If two lines do not intersect, then they are ∥; inverse: If two lines are not ∥, then they intersect; contrapositive: If two lines intersect, then they are not ∥. **b.** Converse: false; a counterexample is two skew lines; inverse: false; a counterexample is two skew lines; contrapositive: true.

54.

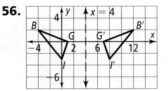

55.

56.

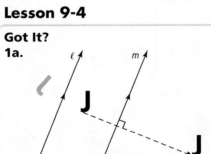

57. 32.2 m **59.** 86.6 ft
59. (2, −3) **60.** (3, 0)
61. (−2, −3)

Lesson 9-4   pp. 570–576

Got It?

1a.

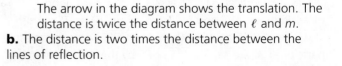

The arrow in the diagram shows the translation. The distance is twice the distance between ℓ and m.
b. The distance is two times the distance between the lines of reflection.

2a.

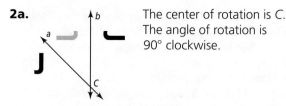

The center of rotation is *C*. The angle of rotation is 90° clockwise.

b. The center of rotation is the intersection of the lines of reflection; the ∠ of rotation is two times the measure of the acute or right ∠ formed by the lines of reflection.

3.

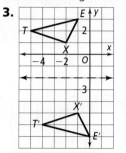

Lesson Check

1.

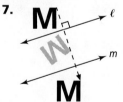

2.

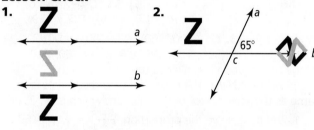

3. *P'*(3, −8), *Q'*(8, −6), *R'*(6, −4)   **4.** parallel
5. Answers may vary. Sample: He assumed that reflections are commutative, when, in general, they are not.

Exercises

7.

A translation; the arrow in the diagram shows the direction, determined by a line perpendicular to ℓ and *m*. The distance is twice the distance between ℓ and *m*.

9–11. A rotation; the center of rotation is *C*.

9.

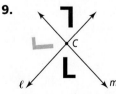

The ∠ of rotation is 170° clockwise.

11.

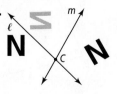

The ∠ of rotation is 150° clockwise.

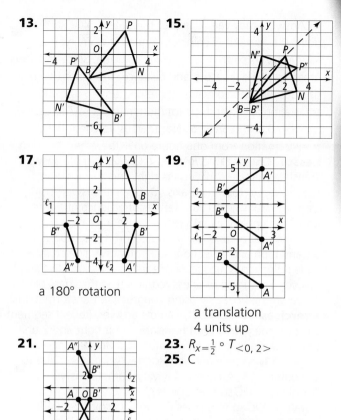

13.

15.

17.

a 180° rotation

19.

a translation 4 units up

21.

a translation 4 units up

23. $R_{x=\frac{1}{2}} \circ T_{<0,\,2>}$
25. C

27. Answers may vary.   **29.** rotation; center *C*, ∠ of rotation 180°   **31.** translation; $T_{<-9,\,0>}(x, y)$
33. reflection; *x* = 4   **35.** rotation; center (3, 0), ∠ of rotation 180°   **37.** translation; $T_{<-11,\,-4>}(x, y)$
39. Answers may vary. Sample: Translate the black R so that one point moves to its corresponding point on the blue R. Then reflect across a line passing through that point and the point halfway between two other corresponding points.   **41.** A   **43.** C   **45.** *A'*(−4, 0), *B'*(0, 0), *C'*(1, −3)   **46.** *A'*(−4, −2), *B'*(−2, −8), *C'*(−8, 0)   **47.** I and II   **48.** I and III   **49.** yes; ASA   **50.** yes; Pythagorean Theorem and SSS or HL   **51.** Not enough information; you need to show one pair of the corresponding sides are congruent to use ASA or AAS.

Lesson 9-5   pp. 578–585

Got It? 1. *m∠A* = *m∠X*, *m∠B* = *m∠Y*, *m∠C* = *m∠Z*; *AB* = *XY*, *BC* = *YZ*, *AC* = *XZ*   **2.** Answers may vary. Sample: △*UVW* can be mapped to △*QNM* by a translation 6 units left followed by a reflection over the *x*-axis. Parallelogram *ABCD* can be mapped to *HIJK* by a translation 6 units right and 5 units down.   **3.** Sample answer: translation 5 units left, reflection across the *x*-axis.

4. Answers may vary. Sample: Translate △YDT so that point D and point N coincide. Since $\overline{TD} \cong \overline{EN}$, you can rotate △YDT so that $\overline{TD}$ and $\overline{EN}$ coincide. Since rotations preserve angle measure and distance, the other two pairs of sides will also coincide. Therefore, this composition of a translation followed by a rotation maps △YDT to △SNE, and △YDT ≅ △SNE. **5.** No, there is no congruence transformation from one figure onto the other.

Lesson Check 1. △RAV ≅ △QSI **2.** Sample answer: Translate triangle RAV 5 units right and 1 unit down; then reflect across the x-axis. **3.** Sample answer: Using transformations, you can define congruence of figures other than polygons. **4.** Yes, because a rotation is a rigid motion and a glide reflection is a composition of a translation and reflection, so a rotation followed by a glide reflection is a congruence transformation. **5.** Sample answer: The game of chess requires that the chess pieces move on the board by using congruence transformations.

Exercises 7. $\overline{GC} \cong \overline{FD}$; Sample answer: Reflect segment GC over the y-axis; then translate 1 unit right and 2 units down. **9.** Sample answer: Rotate △LMN 180° about the origin. **11.** Sample answer: Rotate △LMN 180° about the origin. **13.** Answers may vary. Sample answer: Translate △IQC so that points Q and Z coincide. Rotate △IQC so that sides QC and NZ coincide. Reflect △IQC across side NZ so that the triangles overlap. △IQC ≅ △VZN. **15.** no, there is no congruence transformation **17.** translation **19.** reflection **21.** Answers may vary. Sample: Translate the top triangle down 6 units; reflect across the x-axis; rotate the bottom triangle 180° about the point (−3, 0), then reflect across the line x = −3; reflect the bottom triangle across the line x = −4, then rotate 180° about the point (−4, 0). **23a.** rotations and glide reflections **b.** translations and glide reflections **25a.** Congruence transformations preserve distances and angle measures. **b.** Use SAS, proven in Problem 4. **27.** Sample answer: Draw and label the midpoint of $\overline{GH}$ as point M. Draw $\overline{FM}$. Using the identity mapping, $\overline{FM} \cong \overline{FM}$. It is given that $\overline{FG} \cong \overline{FH}$, and $\overline{GM} \cong \overline{MH}$ by the definition of midpoint. Therefore, if triangle FHM is reflected across $\overline{FM}$, it will overlap triangle FGM. Because there is an isometry mapping triangle FHM onto triangle FGM, △FHM ≅ △FGM. Therefore, ∠G ≅ ∠H because corresp. parts of ≅ △ are ≅. **29.** 14.14 **31.** 6.4 **33.** A'(−3, 3), B'(−4, −2), C'(−4, 4) **34.** greater than 10 in. and less than 42 in. **35.** greater than 1 ft and less than 40 ft **36.** greater than 0 m and less than 18 m **37.** greater than 3.5 yd and less than 12.5 yd **38.** 3 in. by 4 in. **39.** 2 in. by 2.5 in. **40.** 1.5 in. by 2.25 in.

Lesson 9-6   pp. 587–593

Got It? 1. reduction; $n = \frac{1}{2}$ **2a.** P'(−0.5, 0), Z'(−1, 0.5),

G'(0, −1) **b.** Answers may vary. Sample: Use the Distance Formula to find the lengths of the sides of △P'Z'G' and △PZG. Then show that the corresp. sides are proportional, so the △ are ~ by SSS ~ Thm. **3.** 5.1 cm

Lesson Check 1. enlargement; 1.5 **2.** D'(2, −10) **3.** T'(0, 2) **4.** M'(0, 0) **5.** a number between 0 and 1 **6a.** The student used 6, instead of 2 + 6 = 8, as the preimage length in the denominator; the correct scale factor is $n = \frac{2}{2 + 6} = \frac{1}{4}$. **b.** The student did not write the scale factor with the image length in the numerator; the correct scale factor is $n = \frac{1}{4}$.

Exercises 7. enlargement; $\frac{3}{2}$ **9.** enlargement; $\frac{3}{2}$ **11.** reduction; $\frac{1}{3}$ **13.** reduction; $\frac{1}{2}$ **15.** enlargement; $\frac{3}{2}$ **17.** P'(−50, 10), Q'(−30, 30), R'(10, −30)

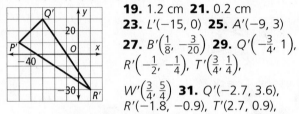

19. 1.2 cm **21.** 0.2 cm **23.** L'(−15, 0) **25.** A'(−9, 3) **27.** $B'\left(\frac{1}{8}, -\frac{3}{20}\right)$ **29.** $Q'\left(-\frac{3}{4}, 1\right)$, $R'\left(-\frac{1}{2}, -\frac{1}{4}\right)$, $T'\left(\frac{3}{4}, \frac{1}{4}\right)$, $W'\left(\frac{3}{4}, \frac{5}{4}\right)$ **31.** Q'(−2.7, 3.6), R'(−1.8, −0.9), T'(2.7, 0.9), W'(2.7, 4.5) **33.** Q'(−300, 400), R'(−200, −100), T'(300, 100), W'(300, 500) **35.** x = 3, y = 60; the image of a dilation is similar to the preimage, so △L'N'M' ~ △LNM. The ratio of the corresp. sides is the same as the scale factor of the dilation, which is 4 : 2, or 2 : 1. To find x, solve the proportion $\frac{x + 3}{x} = \frac{2}{1}$. y = 60 because corresponding angles of ~ figures are ≅.

37.

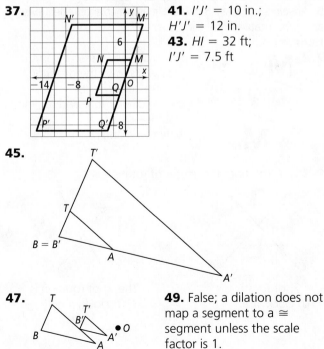

41. I'J' = 10 in.; H'J' = 12 in. **43.** HI = 32 ft; I'J' = 7.5 ft

45.

47.

49. False; a dilation does not map a segment to a ≅ segment unless the scale factor is 1.

51. True; the image and preimage are ~, so the corresp. ∠s are ≅.

53.

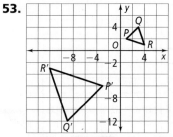

55. 1 ft **57.** G **59a.** no **b.** yes; 1
60. $J'(25, 1)$, $K'(6, 2)$, $L'(3, -20)$ **61.** 4 **62.** 7.5 **63.** 40

Lesson 9-7   pp. 594–600

Got It? 1. $L''(0, 2)$, $M''(0.5, -0.5)$, $N''(1.5, 1.5)$
2. $D_{0.5} \circ R_{y\text{-axis}}$ **3.** No, there is no similarity transformation that maps one triangle to the other. The side lengths are not all proportional. **4.** Yes, there is a similarity transformation: rotation, translation, and then dilation.
Lesson Check 1. Answers may vary. Sample:
$D_{1.5} \circ R_{y\text{-axis}}$ **2.** $R''(\frac{3}{4}, -\frac{1}{2})$, $S''(\frac{1}{4}, -\frac{1}{4})$, $T''(\frac{1}{2}, -1)$
3. Sample answer: The pupils of your eyes dilate when you go from dark to bright locations or from bright to dark. The pupils are reduced or enlarged proportionally to form similar pupils. **4.** Answers may vary.

Exercises

5.

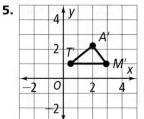

7.

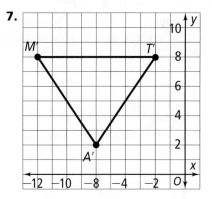

9. $D_{1.5} \circ r_{(180°, O)}$ **11.** Sample answer: △AVS is similar to △RGI. Translate △RGI so that points R and A coincide. Rotate by 180°. Then dilate with center A and scale factor 1.5. **13.** Sample answer: The similarity transformation is a rotation about point A followed by a dilation with respect to point A. △SCA is similar to △ELA. **15.** Sample answer: Yes, there is a similarity transformation between the two figures: a translation and a rotation followed by a dilation. **17.** Yes, the triangles are similar because there is a similarity transformation that maps △AJL to △ABC.
19. sometimes **21.** always **23a.** Yes, the triangles are similar because there is a similarity transformation between them: rotation followed by a dilation. **b.** 330 m
25. Sample answer: Yes, a rigid motion is a similarity transformation with a scale factor of 1. The preimage and image of a rigid motion are congruent, so they are also similar. **27.** Sample answer: No, there is not a similarity transformation. To create △NOP, △ABC is reflected across the x-axis. Then the x-coordinates are scaled by a factor of 5 and the y-coordinates are scaled by a factor of 4. Because the reflected triangle is 5 times as wide as △ABC but only 4 times as tall, the figures are not similar. Therefore, there is no similarity transformation between them. **29a.** true **b.** true **c.** false **31.** 6 **33.** 7.5
34. H, 180°; I, 180°; N, 180°; O, any rotation; S, 180°; X, 180°; Z, 180° **35.** (−2, −2) or (7, 1) **36.** 25 cm²
37. 28 in.² **38.** 11.5 m² **39.** 1.5 ft²

Chapter Review   pp. 602–606

1. transformation **2.** similarity transformation
3. translation **4.** image **5a.** No; the distances between corresponding points in the image and preimage are not the same. **b.** $\overline{LA}$, W
6. $R'(-4, 3)$, $S'(-6, 6)$, $T'(-10, 8)$

7. $T_{<-5, 10>}(x, y)$

8. $T_{<-2, 7>}(x, y)$

9. $A'(6, -4)$, $B'(-2, -1)$, $C'(5, 0)$

10. $A'(2, 4)$, $B'(10, 1)$, $C'(3, 0)$

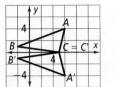

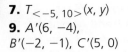

11. $A'(4, 6)$, $B'(1, -2)$, $C'(0, 5)$

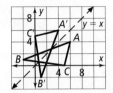

12.

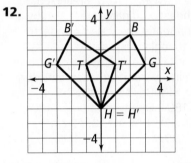

13.

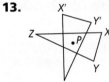

14. $P'(4, -1)$

15. $W'(-1, -3)$, $X'(2, -5)$, $Y'(8, 0)$, and $Z'(-1, -2)$

16. E is translated right, twice the distance between ℓ and m.

17. same; rotation **18.** same; translation **19.** opposite; glide reflection **20.** $\triangle T'A'M'$ with vertices $T'(-4, -9)$, $A'(0, -5)$, $M'(-1, -10)$ **21.** $(r_{(90°, O)} \circ R_{x\text{-axis}})(\triangle XYZ)$
22. Answers may vary. Sample: Yes, the letters are congruent. The p can be mapped to the d with a composition of a translation followed by a rotation.
23. enlargement; 2 **24.** $M'(-15, 20)$, $A'(-30, -5)$, $T'(0, 0)$, $H'(15, 10)$
25.

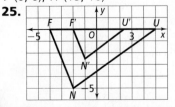

26. $L'N' = 6.5$ ft, $M'N' = 11.25$ ft
27.

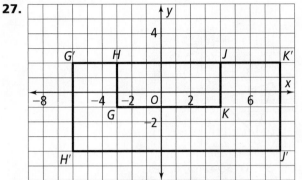

28. No. The side lengths are not proportional. **29.** No, because all of the dimensions of the airplane must dilate by the same scale factor for the figures to be similar.
30. Answers may vary. Sample answer: The figures are similar because a composition of a translation, rotation, and a dilation maps p to d.

Chapter 10
Get Ready! p. 611

1. 9 **2.** 64 **3.** 144 **4.** 225 **5.** 4 **6.** 8 **7.** 10 **8.** 13 **9.** ± 8
10. ± 15 **11.** ± 12 **12.** $2\sqrt{2}$ **13.** $3\sqrt{3}$ **14.** $5\sqrt{3}$
15. $24\sqrt{2}$ **16.** $\frac{2}{7}$ **17.** 5 bags **18.** rhombus
19. parallelogram **20.** rhombus **21–23.** Answers may vary. Samples are given. **21.** half of a circle **22.** more than half a circle **23.** arcs that are next to each other

Lesson 10-1 pp. 616–622

Got It? 1. 108 m² **2.** 7.5 cm **3.** 30 in.² or $\frac{5}{24}$ ft²
4. The area is doubled.
Lesson Check 1. 200 m² **2.** 64 ft² **3.** 96 cm²
4. 36 in.² **5.** No; two altitudes of an obtuse $\triangle$ lie outside the $\triangle$. The legs of a right $\triangle$ are two altitudes of the $\triangle$.
6. Answers may vary. Sample: You can cut and paste a section of the $\square$ to make a rectangle that is $\cong$ to the given rectangle. **7.** The area of $\triangle ABC$ is half the area of the $\square$.

Exercises 9. 26.79 in.² **11.** 11.2 units **13.** $16\frac{8}{13}$ units
15. 13.5 yd² **17a.** 1390 ft² **b.** Find the entire area and subtract the areas for flowers.

c. $(50)(31) - 2\left[\frac{1}{2}(10)(16)\right] = 1550 - 160 = 1390$ ft²

19. B **21.** 18 in.; 12 in. **25.** 6 units² **27.** 12 units²
29. 3 units² **31.** The area is tripled; explanations may vary.
Sample: If $A = \frac{1}{2}b \cdot h$, then $\frac{1}{2}(b \cdot 3h) = 3 \cdot \frac{1}{2}(b \cdot h) = 3A$.

33a. 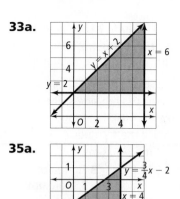 **b.** 18 units^2

35a. **b.** 6 units^2

37. 60 units^2 **39.** 20 units^2 **41.** 312.5 ft^2
43. 12,800 m^2 **45.** 126 m^2 **47.** B

49. No; the sum of the two shorter legs is 6 + 4. By the
△ Inequality Thm., that sum must be greater than the
length of the third side of the △. Since 6 + 4 < 11,
a △ with sides 6, 4, and 11 is not possible. **50.** Similar;
the similarity transformation is a reflection followed by a
dilation with scale factor $\frac{3}{4}$. **51.** Similar; the similarity
transformation is a 90° rotation followed by a dilation
with scale factor $\frac{2}{5}$. **52.** 108 **53.** 72 **54.** 72 **55.** 36
56. 36 **57.** 36 **58.** $A = bh$ **59.** $A = \frac{1}{2}bh$
60. 9 units^2 **61.** 7 units^2 **62.** 12 units^2

Lesson 10-2   pp. 623–628

Got It? 1. 94.5 cm^2 **2.** 12 m^2 **3.** 54 in.^{2} **4.** 96 cm^2
Lesson Check 1. 42 m^2 **2.** 378 in.^{2} **3.** 30 ft^2
4. 288 in.^{2} **5.** 300 m^2 **6.** 8 cm^2 **7.** No; in the formula
for the area of a trapezoid, half the sum of the bases
would have to equal the length of the base of the
parallelogram in order for the areas to be the same. This is
not possible since the other base of the trapezoid will be
longer or shorter than the given base. **8.** No; if you know
the height, then you need only the lengths of the bases,
but not the legs, to find the area. **9.** No; unless the
rhombus is a square, you cannot calculate the area
without knowing the lengths of the diagonals. **10.** No;
you can calculate the area of a kite from the lengths of
the diagonals, without knowing the lengths of the sides.
Exercises 11. 472 in.^{2} **13.** 108 ft^2 **15.** $\frac{5}{6}$ ft^2
17. 30 ft^2 **19.** 72 m^2 **21.** 18 m^2 **23.** 1200 ft^2
25. 24 m^2 **27.** about 35.4 cm^2 **29.** 11.3 cm^2
31. 1.8 m^2 **33.** 15 units^2 **35.** C **37.** 18 cm^2
39. $\frac{128\sqrt{3}}{3}$ in.^{2} **41a.** $A = \frac{1}{2}b_1h$, $A = \frac{1}{2}b_2h$ **b.** Add the
areas of the △ to get the area of the trapezoid: Area of
trapezoid $= \frac{1}{2}b_1h + \frac{1}{2}b_2h = \frac{1}{2}h(b_1 + b_2)$.
43. 1.5m^2 **45.** A

47.

48. 72 cm^2 **49.** 15 ft **50.** 140 **51.** 25$\sqrt{3}$ cm^2
52. 50 ft^2 **53.** $\frac{100\sqrt{3}}{3}$ m^2

Lesson 10-3   pp. 629–634

Got It? 1. $m\angle 1 = 45$, $m\angle 2 = 22.5$, $m\angle 3 = 67.5$
2a. 232 cm^2 **b.** It is reduced by half; explanations may
vary. Sample: The perimeter of the original polygon is
$n \cdot s$. If the side is reduced to half its length, the new
perimeter is $n \cdot \frac{1}{2}s$, or $\frac{1}{2}ns$. **3.** 665 ft^2
Lesson Check 1. 100.0 in^2 **2.** 23.4 ft^2 **3.** 5.2 m^2
4. 166.3 units^2 **5.** A radius is the distance from the
center to a vertex, while the apothem is the perpendicular
distance from the center to a side. **6a.** $s = 2a$
b. $s = \frac{2\sqrt{3}}{3}a$ **c.** $s = 2\sqrt{3}a$ **7.** Special △ have ⦟ of
30°, 60°, 90° or 45°, 45°, 90° and are found in equilateral
△, squares, and regular hexagons.
Exercises 9. $m\angle 4 = 90$, $m\angle 5 = 45$, $m\angle 6 = 45$
11. 2144.475 cm^2 **13.** 12,080 in.^{2} **15.** 1168.5 m^2
17. 841.8 ft^2 **19.** 93.5 m^2 **21.** 72 cm^2 **23.** 162$\sqrt{3}$ m^2
25. 12$\sqrt{3}$ in.^{2} **27a.** 45 **b.** 67.5 **29a.** 30 **b.** 75
31. 9.7 ft **33a.** 9.1 in. **b.** 6 in. **c.** 3.7 in. **d.** Answers
may vary. Sample: About 4 in.; the length of a side of a
pentagon should be between 3.7 in. and 6 in.
35. The apothem is one leg of a rt. △ and the radius is
the hypotenuse. **37.** 17.0; 18 **39.** 51.0; 187.1 **41.** The
apothem is ⊥ to a side of the pentagon. Two right △ are
formed with the radii of the pentagon. The △ are ≅ by
HL. So, the ⦟ formed by the apothem and radii are ≅
because corresp. parts of ≅△ are ≅. Therefore, the
apothem bisects the vertex ∠. **43a.** (2.8, 2.8)
b. 5.6 units^2 **c.** 45 units^2 **45.** F **47.** (2, 2$\sqrt{3}$) and
(2, −2$\sqrt{3}$), or equivalent decimal approximations
(2, 3.464) and (2, −3.464); the length of each side of the
△ is 4 units. The third vertex must lie on the altitude of
the triangle, which is a point on the line $x = 2$ and
has x-coordinate 2. Using the Distance Formula,
$\sqrt{(2-0)^2 + (y-0)^2} = 4$; $\sqrt{4 + y^2} = 4$;
$4 + y^2 = 16$; $y^2 = 12$; $y = \pm\sqrt{12}$; $y = \pm 2\sqrt{3}$

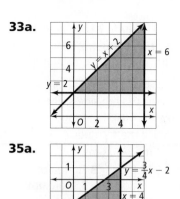

48. 46 m² **49.** 8 m **50.** $P = 28$ in.; $A = 49$ in.²
51. $P = 24$ m; $A = 32$ m² **52.** $P = 24$ cm; $A = 24$ cm²

Lesson 10-4   pp. 635–641

Got It? 1a. 7 : 5 **b.** 49 : 25 **2.** 54 in.² **3a.** $6.94
b. In order for the two plots to be $\sim$, the pairs of corresp. sides must have the same ratio. **4.** $5\sqrt{5}$: 3
Lesson Check 1. 2 : 3; 4 : 9 **2.** 4 : 3; 16 : 9 **3.** 69.3 ft²
4. $\sqrt{6}$: 4 **5.** For two $\sim$ figures, the ratio of their areas is the square of the ratio of the perimeters. **6.** $\sqrt{2}$: 1; the ratio of the areas is 2 : 1, so the ratio of the perimeters is the square root of that ratio, which is $\sqrt{2}$: 1. **7.** Answers may vary. Sample: The ratios of perimeters and areas of $\sim$ figures are not = (unless the figures are $\cong$, in which case each ratio is 1). **8.** The ratio of the areas of two $\cong$ figures is 1, while the ratio of the areas of two $\sim$ figures is the square of the scale factor.
Exercises 9. 1 : 2; 1 : 4 **11.** 2 : 3; 4 : 9 **13.** 24 in.²
15. 59 ft² **17.** $384 **19.** 1 : 2; 1 : 2 **21.** 7 : 3; 7 : 3
23. 4 : 1; 4 : 1 **25.** 3 : 1; 9 : 1 **27.** 2 : 3; 4 : 9 **29.** 6 : 1; 36 : 1 **31.** While the ratio of lengths is 2 : 1, the ratio of areas is 4 : 1. **33.** 252 m² **35.** $x = 2\sqrt{2}$ cm, $y = 3\sqrt{2}$ cm **37.** $x = \frac{8\sqrt{3}}{3}$ cm, $y = 4\sqrt{3}$ cm
39. $x = 8$ cm, $y = 12$ cm **43a.** 8 : 3 **b.** 64 : 9
45a. $6\sqrt{3}$ cm² **b.** $54\sqrt{3}$ cm²; $13.5\sqrt{3}$ cm²; $96\sqrt{3}$ cm² **47a–c.** Answers may vary. Samples are given.
a.

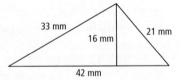

33 mm   16 mm   21 mm   42 mm

b. 96 mm; 336 mm² **c.** 457 yd; 7619 yd² **49.** Sometimes; a 1 unit-by-8 unit rectangle and a 2 unit-by-4 unit rectangle have the same area, but they are not $\sim$.
51. Sometimes; if they are $\cong$, they are $\sim$ and have = areas. **53.** $\frac{26}{3}$ **55.** 155 **56.** 50 cm² **57.** 690 units²
58. 480 units² **59.** $5\frac{1}{3}$ cm, 12 cm **60.** 36 m²
61. 4536 in.² **62.** 168 ft²

Lesson 10-5   pp. 643–648

Got It? 1. 28 in.² **2a.** 265 in.² **b.** The area is quadrupled; explanations may vary. Sample: Both the apothem and the side length are doubled if the radius is doubled. **3.** 45 in.²
Lesson Check 1. 41.6 m² **2.** 277.0 cm² **3.** 22 in.²
4. Yes; the diagonal of a regular hexagon is two times the side, and you have several ways to find the area of a regular hexagon with 6-cm sides. **5.** He set up the wrong ratio. The correct ratio is $\frac{4}{a} = \tan 36°$.

Exercises 7. 123.1 yd² **9.** 141.7 in.² **11.** 12.4 mm²
13. 2540.5 cm² **15.** 18.0 ft² **17.** 311.3 km² **19.** 0.8 ft²
21. Multiply the formula for the area of an equilateral $\triangle$, $A = \frac{s^2\sqrt{3}}{4}$, by 6 to get $\frac{3s^2\sqrt{3}}{2}$; use a 30°-60°-90° $\triangle$ to find the height of one equilateral $\triangle$ with side s, then multiply the area of that $\triangle$ by 6; or use the tangent ratio to find the apothem and then use the formula $A = \frac{1}{2}ap$.
23. 20.8 m, 20.8 m² **25.** 61.2 m, 282.8 m²
27. 1,459,000 ft² **29.** about 925.8 cm² **31.** area of Pentagon A $\approx$ 1.53 · (area of Pentagon B) **33.** area of Octagon B $\approx$ 1.17 · (area of Octagon A) **35.** $162\sqrt{3}$ ft² or about 280.6 ft² **37.** about 48.2 cm² **39.** 320 ft
41. 17 **43.** 47.2 **45a.** 2 : 3 **b.** 173.8 in.² **46.** $x = 16.1$ m; $y = 20$ m **47.** $x = 29.7°$; $y = 6$ cm **48.** 14 cm
49. 2.5 in. **50.** 3.2 m

Lesson 10-6   pp. 649–657

Got It? 1a. $\overset{\frown}{SP}$, $\overset{\frown}{SQ}$, $\overset{\frown}{PQ}$, $\overset{\frown}{QR}$, $\overset{\frown}{RS}$ **b.** $\overset{\frown}{RSP}$, $\overset{\frown}{RQP}$
c. $\overset{\frown}{PQS}$, $\overset{\frown}{PSQ}$, $\overset{\frown}{SPR}$, $\overset{\frown}{QRS}$, $\overset{\frown}{RSQ}$ **2a.** 77 **b.** 103 **c.** 208
d. 283 **3a.** about 29.5 ft **b.** 2 : 1; if the radius of $\odot A$ is r, then its circumference is $2\pi r$. $\odot B$ will have a circumference of πr. The ratio of their circumferences is $\frac{2\pi r}{\pi r} = \frac{2}{1}$, or 2 : 1. **4.** 1.3π m
Lesson Check 1–3. Answers may vary. Samples are given. **1.** $\overset{\frown}{AB}$ **2.** $\overset{\frown}{DAB}$ **3.** $\overset{\frown}{CAB}$ **4.** 81 **5.** 18π cm
6. $\frac{23\pi}{4}$ cm **7.** The measure of an arc corresponds to the measure of a central angle; an arc length is a fraction of the circle's circumference. **8.** The student substituted the diameter into the formula that requires the radius.
Exercises 9. $\overset{\frown}{BC}$, $\overset{\frown}{BD}$, $\overset{\frown}{CD}$, $\overset{\frown}{CE}$, $\overset{\frown}{DE}$, $\overset{\frown}{DF}$, $\overset{\frown}{EF}$, $\overset{\frown}{FB}$
11. $\overset{\frown}{BCE}$, $\overset{\frown}{BFE}$, $\overset{\frown}{CBF}$, $\overset{\frown}{CDF}$ **13.** 180 **15.** 270 **17.** 308
19. 90 **21.** 90 **23.** 270 **25.** 6π ft **27.** 14π in. **29.** 19 in.
31. 8π ft **33.** 33π in. **35.** $\frac{5\pi}{4}$ m **37.** 70 **39.** 110
41. 235 **43.** about 183.3 ft **45.** Find the measure of the major arc, then use Thm. 10-10; or find the length of the minor arc using Thm. 10-10, then subtract that length from the circumference of the circle. **47.** 38 **49.** 31 m
51. 3 : 4 **53.** 2.6π in. **55.** 7.9 units **57.** Since $\overline{AR} \cong \overline{RW}$ and $AR + RW = AW$ by the Seg. Add. Post., $AW = 2 \cdot AR$. So the radius of the outer circle is twice the radius of the inner circle. Because $\angle QAR$ and $\angle SAU$ are vertical $\angle$s, and $m\angle SAT = \frac{1}{2} m\angle SAU$, $m\angle QAR = 2 \cdot m\angle SAT$. The length of $\overset{\frown}{ST} = \frac{m\angle SAT}{360} \cdot 2\pi(2 \cdot AR) = \frac{m\angle SAT}{90} \cdot \pi(AR)$ and the length of $\overset{\frown}{QR} = \frac{m\angle QAR}{360} \cdot 2\pi(AR) = \frac{2 \cdot m\angle SAT}{360} \cdot 2\pi(AR) = \frac{m\angle SAT}{90} \cdot \pi(AR)$. Therefore the length of $\overset{\frown}{ST}$ = the length of $\overset{\frown}{QR}$ by the

Trans. Prop. of Eq. **59.** 325.7 yd, 333.5 yd, 341.4 yd, 349.2 yd, 357.1 yd, 365.0 yd, 372.8 yd, 380.6 yd **61.** F
63. Using the Distance Formula, $AB = CD = 3$, $BC = AD = 5$ and $RS = TV = 6$, $ST = RV = 10$. The slopes of $\overline{AB}$ and $\overline{CD} = 0$ and the slopes of $\overline{BC}$ and $\overline{AD}$ are undefined. So both $\overline{AB}$ and $\overline{CD}$ are $\perp$ to $\overline{BC}$ and $\overline{AD}$. Therefore, $ABCD$ is a rectangle and $\angle$ A, B, C, and D are rt. $\angle$. The slopes of $\overline{RS}$ and $\overline{TV}$ are undefined and the slopes of $\overline{ST}$ and $\overline{RV} = 0$. So, $RSTV$ is a rectangle and $\angle$ R, S, T, and V are rt. $\angle$. Since all rt. $\angle$ are $=$, the pairs of corresponding $\angle$ are $\cong$. The short sides of the two rectangles are 3 and 6, and the long sides are 5 and 10. Since $\frac{3}{6} = \frac{5}{10} = \frac{1}{2}$, the corresp. sides are proportional. Therefore, $ABCD \sim RSTV$ by the def. of $\sim$ polygons.
64. $m\angle 1 = 30$, $m\angle 2 = 15$, $m\angle 3 = 75$, $m\angle 4 = 30$
65. 18.6 mm **66.** Answers may vary slightly. Samples: 120 mm; 1116 mm^2 **67.** No; it could be an isosc. trapezoid.
68. Yes; the diagonals bis. each other, so it is a $\square$.
69. Yes; one pair of sides is both $\cong$ and $\parallel$, so it is a $\square$.
70. 17π in. or about 53.4 in. **71.** 3π cm or about 9.4 cm

Lesson 10-7   pp. 660–666

Got It? 1a. about 1385 ft^2 **b.** The area is $\frac{1}{4}$ the original area; explanations may vary. Sample: half the radius is $\frac{r}{2}$. So, if $A = \pi r^2$, then $\pi\left(\frac{r}{2}\right)^2 = \frac{1}{4}\pi r^2 = \frac{1}{4}A$.
2. 2π in.^{2} **3.** 4.6 m^2

Lesson Check 1. 64π in.^{2} **2.** $\frac{135}{8}\pi$ in.^{2}, or 16.875π in.^{2} **3.** $\left(\frac{4}{3}\pi - \sqrt{3}\right)$ m^2 **4.** A sector of a circle is a region bounded by an arc and the two radii to the endpoints of the arc. A segment is a part of a circle bounded by an arc and the seg. joining the arc's endpoints. **5.** No; the central $\angle$ corresponding to the arcs and the radii of the circles may be different. Circles with different radii do not have the same area. **6.** 6^2 was incorrectly evaluated as $6 \cdot 2$.

Exercises 7. 9π m^2 **9.** 0.7225π ft^2 **11.** about 282,743 ft^2 **13.** 40.5π yd^2 **15.** $\frac{169\pi}{6}$ m^2 **17.** 12π ft^2
19. $\frac{25\pi}{4}$ m^2 **21.** 24π in.^{2} **23.** 22.1 cm^2 **25.** 3.3 m^2
27. $(54\pi + 20.25\sqrt{3})$ cm^2 **29.** $(4 - \pi)$ ft^2
31. $(784 - 196\pi)$ in.^{2} **33.** 314 ft^2 **35.** 116 mm^2
37. 22.6 mm^2 **39.** 12 in. **41a.** Answers may vary. Sample: Subtract the minor arc segment from the area of the circle; or add the areas of the major sector and the $\triangle$ that is part of the minor arc sector.
b. $(25\pi - 50)$ units^2; $(75\pi + 50)$ units^2 **43.** 4.4 m^2

45. $\left(\frac{5\pi}{6} - 2 \cdot \sin 75°\right)$ ft^2, or
$$\left[\frac{5\pi}{6} - 4(\sin 37.5°)(\cos 37.5°)\right]$$
47. $(200 - 50\pi)$ m^2 **49.** Blue region: Let $AB = 2$. Area of blue $= 4 - \pi$; area of yellow $= \pi - 2$, and $4 - \pi < \pi - 2$. **51.** B **53.** B **55.** 10π cm **56.** 2π m
57. 28π in. **58.** $11\frac{1}{4}$ in., $11\frac{1}{4}$ in., $11\frac{1}{4}$ in., $15\frac{1}{4}$ in.
59. $4 : 9$ **60.** $\frac{4}{9}$

Lesson 10-8   pp. 668–674

Got It? 1. $\frac{1}{2}$ or 50% **2.** $\frac{1}{5}$ or 20% **3.** $\frac{1}{2}$ or 50%
4a. 0.04, or 4% **b.** The black zone; the area of the black zone is greater than the area of the red zone, so P(black zone) $> P$(red zone).

Lesson Check 1. $\frac{3}{7}$ **2.** $\frac{6}{7}$ **3.** $\frac{4}{7}$ **4.** $\frac{3}{7}$ **5.** about 0.09, or 9% **6.** $\frac{2}{3}$; explanations may vary. Sample: Since $\frac{SQ}{QT} = \frac{1}{2}$, you can let $SQ = x$ and $QT = 2x$, where x is not 0. Then $ST = 3x$ and the ratio $\frac{QT}{ST} = \frac{2x}{3x} = \frac{2}{3}$. **7.** The numerator should be (area of square $-$ area of semicircles); the favorable region is the shaded region and its area is the area left when the areas of the semicircles are subtracted from the area of the square.

Exercises 9. $\frac{1}{10}$ **11.** $\frac{2}{5}$ **13.** $\frac{2}{5}$ **15.** $\frac{2}{5}$, or 40% **17.** $\frac{2}{9}$, or about 22% **19.** $\frac{5}{9}$, or about 56% **21.** $\frac{1}{49}$, or about 2%
23. $\frac{24}{49}$, or about 49% **25.** $\frac{3}{10}$, or 30% **27.** $\frac{3}{20}$, or 15%
29. $\frac{9}{19}$, or about 47% **31.** $\frac{1}{4}$; $m\widehat{AB} = 90$, so the length of $\widehat{AB} = \frac{90}{360} \cdot 2\pi r = \frac{1}{4} \cdot 2\pi r$. The ratio of the length of $\widehat{AB}$ to the circumference is $\frac{1}{4}$. **33.** $\frac{3}{5}$ **35.** $\frac{1}{10}$
37. $\frac{1}{40}$ **39.** $\frac{\pi}{20}$, or about 16% **41.** 36 s **43a.** about 8.7% **b.** about 19.6% **45.** 50% **47.** G **49.** G
51. 100π ft^2 **52.** 12π cm^2 **53.** $A'''(-2, -2)$; $B'''(1, -1)$; $C'''(-1, 1)$ **54.** $A'''(-5, 6)$; $B'''(-4, 3)$; $C'''(-2, 5)$
55. $A'''(-2, 1)$; $B'''(2, -1)$; $C'''(1, -2)$
56.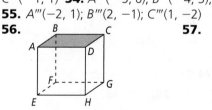
57.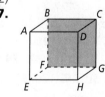
58.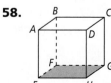
59.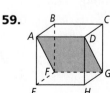

60.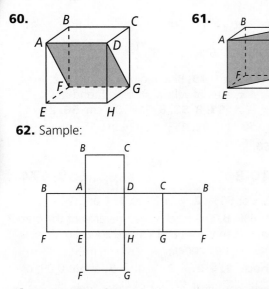

61.

62. Sample:

Chapter Review   pp. 676–680

1. base **2.** sector **3.** radius **4.** adjacent arcs **5.** 10 m²
6. 90 in.² **7.** 30 ft² **8.** 160 ft² **9.** 30 ft²
10. 96√3 mm² **11.** 96 ft² **12.** 117 cm² **13.** 256 ft²
14. 54 m² **15.** 9√3 in.² **16.** 28 m² **17.** 2400√3 cm²
18. 112.5 m²

19.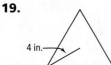

4 in.

20.8 in.²

20.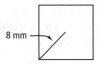

8 mm

128 mm²

21.

7 cm

127.3 cm²

22. 4 : 9 **23.** 9 : 4 **24.** 1 : 4 **25.** 4 : 1 **26.** 2√2 : 5
27. 73.5 ft² **28.** 232.5 cm² **29.** 124.7 in.² **30.** 8 m²
31. 331.4 ft² **32.** 24.6 ft² **33.** 100.8 cm² **34.** 70.4 m²
35. 30 **36.** 120 **37.** 330 **38.** 120 **39.** $\frac{22\pi}{9}$ in.
40. π mm **41.** $\frac{25\pi}{9}$ m **42.** 4π m **43.** 144π in.²
44. $\frac{49\pi}{4}$ ft² **45.** 41.0 cm² **46.** 18.3 m² **47.** 36.2 cm²
48. $\frac{1}{2}$, or 50% **49.** $\frac{3}{8}$, or 37.5% **50.** $\frac{1}{6}$, or about 16.7%
51. $\frac{1}{2}$, or 50% **52.** $\frac{1}{2}$, or 50%

Chapter 11

Get Ready!   p. 685

1. 17 **2.** 8√2 **3.** 6 **4.** 4√5
5. 6√2 **6.** 4√3 **7.** 44 units² **8.** 14√3 units²
9. 234 units² **10.** 54√3 units² **11.** 24 **12.** 2√2 : 5
13. the ⊥ segment from one base to a parallel base or a
vertex to the base **14.** the sum of the areas of each side
(face) of a figure **15.** An Egyptian pyramid has 4 sides
that are triangles and a bottom (base) that is a square.

Lesson 11-1   pp. 688–695

Got It? 1a. 6 vertices: R, S, T, U, V, W; 9 edges: $\overline{SR}$, $\overline{ST}$,
$\overline{UR}$, $\overline{UV}$, $\overline{UT}$, $\overline{RV}$, $\overline{SW}$, $\overline{VW}$, $\overline{TW}$; 5 faces: $\triangle URV$, $\triangle STW$,
quadrilateral $RSTU$, quadrilateral $RSWV$, quadrilateral
$TWVU$ **b.** No; an edge is a segment formed by the
intersection of two faces. $\overline{TV}$ is a segment that is
contained in only one face, so it is not an edge.
2a. 12 **b.** 30 **3a.** 6 + 8 = 12 + 2
b.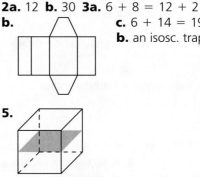
c. 6 + 14 = 19 + 1 **4a.** a circle
b. an isosc. trapezoid

5.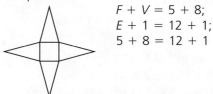

a square

Lesson Check
1. 5 faces: $\triangle ABC$, $\triangle ACD$, $\triangle ADE$, $\triangle AEB$, quadrilateral
$BCDE$; 8 edges: $\overline{AB}$, $\overline{AC}$, $\overline{AD}$, $\overline{AE}$, $\overline{BC}$, $\overline{CD}$, $\overline{DE}$, $\overline{EB}$;
5 vertices: A, B, C, D, E
2. Sample:

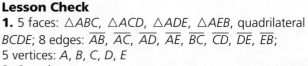

$F + V = 5 + 8$;
$E + 1 = 12 + 1$;
$5 + 8 = 12 + 1$

3. a rectangle **4.** 24 edges: There are 8 edges on each of
the two octagonal bases, and there are 8 edges that
connect pairs of vertices of the bases. **5.** A cylinder is not
a polyhedron because its faces are not polygons.
Exercises
7. 8 vertices: A, B, C, D, E, F, G, H; 12 edges: $\overline{AB}$, $\overline{BC}$, $\overline{CD}$,
$\overline{DA}$, $\overline{EF}$, $\overline{FG}$, $\overline{GH}$, $\overline{HE}$, $\overline{AE}$, $\overline{BF}$, $\overline{CG}$, $\overline{DH}$; 6 faces:
quadrilaterals $ABCD$, $EFGH$, $ABFE$, $BCGF$, $DCGH$, $ADHE$

9. 8 **11.** 12 **13.** 5
15. 5 + 6 = 9 + 2; answers may vary. Sample:

17. 7 + 7 = 12 + 2; answers may vary. Sample:

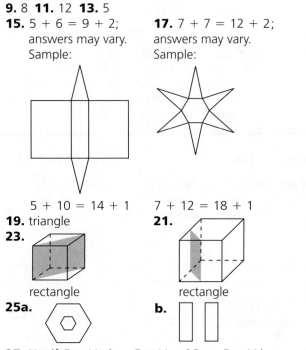

5 + 10 = 14 + 1   7 + 12 = 18 + 1
19. triangle   **21.**
23.

rectangle   rectangle
25a.   **b.**

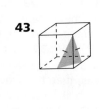

27. No; if $F = V$, then $F + V = 2F$, so $F + V$ is even. So $E \neq 9$ because $E + 2$ must be even.
29.

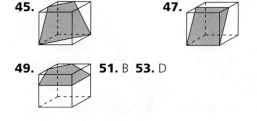

rectangle

31. cone **33.** a cylinder attached to a cone **35.** 4 + 6 = 9 + 1
37. 5 + 5 = 9 + 1 **39.** 6 in.

41.   **43.**

45.   **47.**

49.   **51.** B **53.** D

55. (5, 2), (−1, 0), or (3, −4); the fourth vertex lies on a line parallel to an opposite side such that the length of the side is equal to the length of the opposite side.
56.

0 4 8 12 16 20

60%

57. 25% **58.** 4.7 **59.** 8.3
60. 96 cm² **61.** 40π cm²
62. 9√3 m²

Review   p. 698

1. $r = \frac{C}{2\pi}$ **3.** $r = \sqrt{\frac{A}{\pi}}$ **5.** $y = x \tan A$; $x = \frac{y}{\tan A}$
7. $C = 2\sqrt{\pi A}$ **9.** $a = \frac{\sqrt{6A\sqrt{3}}}{6}$ or $a = \frac{\sqrt{6A} \cdot \sqrt[4]{3}}{6}$

Lesson 11-2   pp. 699–707

Got It? 1. 216 cm² **2a.** 432 m² **b.** 54√3 m²
c. 619 m² **3.** 380π cm² **4a.** 11.8 in.² **b.** $\frac{1}{4}$; $\frac{1}{4}$

Lesson Check 1. 130 in.² **2.** (133 + 42√2) ft² or about 192.4 ft² **3.** 48π cm² or about 150.8 cm²
4. 170π m² or about 534.1 m² **5.** lateral faces: *BFGC*, *DCGH*, *ADHE*, *EFBA*; bases: *ABCD*, *EFGH* **6.** The diameter of the circular bases does not match the length of the rectangle. If the diameter is 2 cm, then the length must be 2π cm, or if the length is 4 cm, then the diameter should be $\frac{4}{\pi}$ cm, or about 1.3 cm.
Exercises
7. 1726 cm²

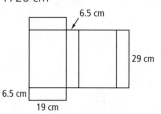

9. (80 + 32√2) in.², or about 125.3 in.²

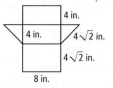

11. 220 ft² **13.** 1121 cm² **15.** 170 m² **17.** 40π cm²
19. 101.5π in.² **21.** 20 cm **23.** 4080 mm²
25a. 94 units² **b.** 376 units² **c.** 4 : 1 **d.** 438 units²; 1752 units²; 4 : 1 **e.** The surface area is multiplied by 4.
27. just under 150 cm² **29.** 110 in.² **31a.** 7 units
b. 196π units² **33.** cylinder of radius 4 and height 2; 48π units² **35.** cylinder of radius 2 and height 4; 24π units² **37a.** The lateral area is doubled.
b. The surface area is more than doubled. **c.** If r doubles, S.A. $= 2\pi(2r)^2 + 2\pi(2r)h = 8\pi r^2 + 4\pi rh = 2(4\pi r^2 + 2\pi rh)$. So the surface area $2\pi r^2 + 2\pi rh$ is more than doubled. **39.** (182π + 232) cm²
41. (220 − 8π) in.² **43.** $h = 6$ m; $r = 3$ m
45. 13 **47.** 68

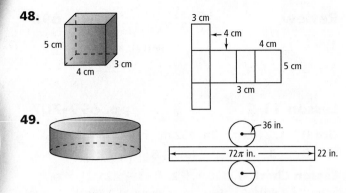

48.

49.

50. 37.7 cm² **51.** 22.1 cm² **52a.** P(a, −b), Q(a, b), R(c, b), S(c, −b) **b.** | a − c |; | a − c |; 2b; 2b **c.** Both pairs of opposite sides are ≅. **53.** √233 in. **54.** √130 m **55.** √313 cm

Lesson 11-3   pp. 708–715

Got It? 1a. 55 m² **b.** The L.A. will double. Sample explanation: Since L.A. = $\frac{1}{2}p\ell$, then replacing ℓ with 2ℓ gives $\frac{1}{2}p(2\ell) = 2(\frac{1}{2}p\ell) = 2 \cdot$ L.A. **2a.** 5649 ft² **b.** The slant height is the hypotenuse of a rt. △ with a leg of length equal to the height of the pyramid, so the slant height is greater than the height. **3.** 704π m²
4a. 934 in.² **b.** The L.A. will be halved. Sample explanation: Since L.A. = $\pi r\ell$, then replacing r with $\frac{r}{2}$ gives $\pi(\frac{r}{2})\ell = \frac{1}{2}(\pi r\ell) = \frac{1}{2} \cdot$ L.A.
Lesson Check 1. 60 m² **2.** 85 m² **3.** 2π√29 ft², or about 33.8 ft² **4.** (2π√29 + 4π) ft², or about 46.4 ft²
5. The height is the distance from the vertex to the center of the base, while the slant height is the distance from the vertex to the midpoint of an edge of the base.
6. Alike: Both are the sum of a lateral area and the areas of the bases. Different: For a prism the area includes two bases, while for a pyramid the surface area includes just one base. **7.** 5; 6; n **8.** The height 7 is not the slant height. The slant height is √(7² + 3²) = √58, so L.A. = πrℓ = π(3)(√58) = 3π√58 units².
Exercises 9. 408 in.² **11.** 179 in.² **13.** 354 cm²
15. 834,308 ft² **17.** 31 m² **19.** 144π cm² **21.** 119π
23. 4 in. **25.** 8 ft **27.** 471 ft²
29. Answers may vary. Sample:

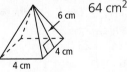

64 cm²

31. Cylinder; the L.A. of 2 cones is 30π in.², and the L.A. of the cylinder is 48π in.². **33a.** $\ell = \frac{S.A.}{\pi r} - r$

b. $r = \frac{-\pi\ell + \sqrt{\pi^2\ell^2 + 4\pi \cdot S.A.}}{2\pi}$ **35.** s = 12 m, L.A. = 240 m²; S.A. = 384 m² **37.** cone with r = 4 and h = 3; 36π **39.** cylinder with cone-shaped hole; 60π units² **41.** 129.6 **43.** L.A. = 25π√5 cm²; S.A. = (25π√5 + 25π) cm² **45.** H **47.** Yes; if the legs of each isosc. △ are two consecutive sides of the ▱, then the ▱ is a rhombus. **49.** 76 ft² **50.** 127 in.²
51. 26 in. **52.** 4 cm² **53.** 176.7 in.²

Lesson 11-4   pp. 717–724

Got It? 1a. 60 ft³ **b.** No; explanations may vary. Sample: The volume is the product of the three dimensions, and multiplication is commutative. **2a.** 150 m³
b. The volume is doubled. Using V = B · h and replacing h with 2h gives B · (2h) = 2 · B · h = 2 · V.
3a. 3π m³ **b.** The volume is $\frac{1}{4}$ the volume of the cylinder in part (a). Using V = πr²h and replacing r with $\frac{r}{2}$ gives π($\frac{r}{2}$)²h = $\frac{1}{4}$πr²h = $\frac{1}{4}$ · V. **4.** 501 in.³
Lesson Check 1. 54 ft³ **2.** 339 in.³ **3.** Yes; it is a combination of a cylinder and a cone. **4.** Alike: Both are the product of the base area and the height. Different: For a prism the base is a polygon, while for a cylinder the base is a circle. **5.** The volumes are the same, 24 m³, because multiplication is commutative.
Exercises 7. 80 in.³ **9.** 14 cm³ **11.** 22.5 ft³
13. 22.5 in.³ **15.** 40π cm³; 125.7 cm³ **17.** π yd³; 3.1 yd³
19. 144 cm³ **21.** 1747 lb **23.** 40 cm **25.** 6 ft **27.** 96 ft³
29. Volume is 27 times greater. Using V = B · h = ℓ · w · h for a rectangular prism, (3ℓ) · (3w) · (3h) = 27 · ℓ · w · h = 27 · V. **31.** Answers may vary. Sample: If two plane figures have the same height and the same width at every level, then they have the same area.
33. 80 units³ **35.** bulk; cost of bags = $1167.50, cost of bulk ≈ $1164 **37.** 125.7 cm³ **39.** cylinder with r = 2 and h = 4; 16π units³ **41.** cylinder with r = 2 and h = 4; 16π units³ **43a.** C = 8.5 in. and h = 11 in.: V ≈ 63.2 in.³; C = 11 in. and h = 8.5 in.: V ≈ 81.8 in.³; the cylinder with the greater circumference has the greater volume. **45.** The volume of B is twice the volume of A. **47.** H **49.** L.A. = 2πrh and A = bh; 2πr is the length of the base when the cylinder is unwrapped.
50. 204.2 mm² **51.** 469.2 ft² **52.** 37 cm **53.** 240 ft

Lesson 11-5   pp. 726–732

Got It? 1. 32,100,000 ft³ **2.** 960 m³ **3a.** 77 ft³
b. The volume of the original tepee is 8 times the volume of the child's tepee. **4a.** 144π m³; 452 m³

Selected Answers

b. They are equal because both cones have the same base and same height.

Lesson Check 1. 96 in.^{3} **2.** 3.1 cm^3 **3.** Alike: Both formulas are $\frac{1}{3}$ the area of the base times height. Different: Because the bases are different figures, the base area will require different formulas. **4.** The areas of the bases are not equal; the area of the base of the pyramid is $13^2 = 169$ ft^2, but the area of the base of the cone is $\pi(6.5)^2 \approx 132.7$ ft^2.

Exercises 5. 200 cm^3 **7.** 50 m^3 **9.** 443.7 cm^3
11. 2048 m^3 **13.** 3714.5 mm^3 **15.** about 66.4 cm^3
17. $\frac{16}{3}\pi$ ft^3; 17 ft^3 **19.** 4π m^3; 13 m^3 **21.** Volume is halved; $V = \frac{1}{3}Bh$, so if h is replaced with $\frac{h}{2}$, then the volume is $\frac{1}{3}B\left(\frac{h}{2}\right) = \frac{1}{2}\left[\frac{1}{3}Bh\right]$. **23.** 123 in.^{3} **25.** 10,368 ft^3
27a. 79,000 m^3 **b.** $20\frac{2}{3}$ m, or about 20.7 m
29a. 120π ft^3 **b.** 60π ft^3 **c.** 240π ft^3 **31.** 3 **33.** cone with $r = 4$ and $h = 3$; 16π **35.** cylinder with $r = 4$, $h = 3$, with a cone of $r = 4$, $h = 3$ removed from it; 32π **37a.** The frustum has volume that is the difference of the volumes of the entire cone and the small cone. The frustum has volume $V = \frac{1}{3}\pi R^2 H - \frac{1}{3}\pi r^2 h$ or $\frac{1}{3}\pi(R^2 H - r^2 h)$. **b.** about 784.6 in.^{3} **39.** A **41.** B **43.** 3600 cm^3 **44.** $JC > KN$ **45.** 7.1 in.^{2} **46.** 13 cm

Lesson 11-6   pp. 733–740

Got It? 1. 196π in.^{2}; 616 in.^{2} **2.** 100 in.^{2}
3a. 113,097 in.^{3} **b.** The volume is $\left(\frac{1}{2}\right)^3 = \frac{1}{8}$ of the original volume. Using $V = \frac{4}{3}\pi r^3$, replacing r with $\frac{r}{2}$ gives $V = \frac{4}{3}\pi\left(\frac{r}{2}\right)^3 = \frac{1}{8}\left(\frac{4}{3}\pi r^3\right)$. **4.** 1258.9 ft^2

Lesson Check 1. 144π ft^2 **2.** 904.8 ft^3 **3.** 193 cm^2
4. 1 : 4 **5.** The surface area will quadruple, but the volume will be 8 times the original volume. $V = \frac{4}{3}\pi(2r)^3 = 8\left(\frac{4}{3}\pi r^3\right)$

Exercises 7. 400π in.^{2} **9.** $40,000\pi$ yd^2 **11.** 441π cm^2
13. 62 cm^2 **15.** 20 cm^2 **17.** $\frac{500}{3}\pi$ ft^3; 524 ft^3
19. $\frac{1125}{2}\pi$ in.^{3}; 1767 in.^{3} **21.** 2304π yd^3; 7238 yd^3
23. 451 in.^{2} **25.** 130 cm^2 **27.** Answers may vary. Sample: sphere with $r = 3$ in., cylinder with $r = 3$ in. and $h = 4$ in.
29. 0.9 in. **31.** 1.7 lb **33.** An infinite number of planes pass through the center of a sphere, so there are an infinite number of great circles. **35.** 36π in.^{3} **37.** $\frac{500}{3}\pi$ mm^3
39. 288π cm^3 **41.** $\frac{1125}{2}\pi$ mi^3 **43a.** about 8.9 in.^{2}
b. The answer is less than the actual surface area since the dimples on the golf ball add to the surface area.

45a. on **b.** inside **c.** outside **47.** 38,792.4 ft^3
49. 22π cm^2; $\frac{46}{3}\pi$ cm^3 **51.** 22π cm^2; $\frac{14}{3}\pi$ cm^3
53. Answers may vary. Sample: You could lift the small ball because it weighs about 75 lb. The big ball would be much harder to lift since it weighs about 253 lb. **55.** 3 m
57a. Cube; explanations may vary. Sample: If $s^3 = \frac{4}{3}\pi r^3$, then $s = r \cdot \sqrt[3]{\frac{4\pi}{3}}$. So $6s^2 = 6\left(r \cdot \sqrt[3]{\frac{4\pi}{3}}\right)^2 \approx 15.6r^2 > 4\pi r^2$ (which is about $12.6r^2$). **b.** Answers may vary. Sample: Spheres are difficult to stack in a display or on a shelf.
59. 2 : 3 **61.** G **63.** I **65.** 16 m^3 **66.** 19 in.^{3}
67. 19,396 mm^3 **68.** 35; 55 **69.** 109, 71, 109, 71
70. yes; 3 : 1 **71.** yes; 3 : $\sqrt{2}$ or $3\sqrt{2}$: 2

Lesson 11-7   pp. 742–749

Got It? 1. yes; 6 : 5 or $\frac{6}{5}$ **2a.** 2 : 3 **b.** No; the bases are similar but the heights may not be in the same ratio as the edges of the bases. **3.** 160 m^2 **4.** 4.05 lb
Lesson Check 1. Cone 1 and Cone 3 are similar; 2 : 3.
2. about 155 in.^{2} **3.** Answers may vary. Sample: There are many relationships that must be true for the solids to be similar: all corresponding angles must be $\cong$; the corresponding faces must be similar and all corresponding edges and heights proportional. **4.** Your classmate found the scale factor of the smaller cube to the larger cube. The scale factor should be 8 : 7.
Exercises 5. no **7.** yes; 2 : 3 **9.** yes; 2 : 3 **11.** 5 : 6
13. 3 : 4 **15.** 240 in.^{3} **17.** 24 ft^3 **19.** 112 m^2
21. 6000 toothpicks **23a.** It is 64 times the volume of the smaller prism. **b.** It is 64 times the weight of the smaller prism. **25.** No; explanations may vary. Sample: If the scale factor is $\frac{1}{10}$, then the weight of the smaller clock should be $\frac{1}{1000}$ the weight of the existing clock.
27. about 1000 cm^3 **29.** No; the same increase to all the dimensions does not result in proportional ratios unless the original prism is a cube. **31a.** 3 : 1 **b.** 9 : 1
33. 864 in.^{3} **35.** 9 : 25; 27 : 125 **37.** 5 : 8; 25 : 64
39a. 100 times **b.** 1000 times **c.** His weight is 1000 times the weight of an average person, but his bones can support only 600 times the weight of an average person. **41a.** 4 : 1 **b.** 8 : 1 **c.** $(3\ell + 5r) : (4\ell + 4r)$; $(3\ell + 5r) : (\ell + r)$, where r is the radius and ℓ is the slant height of the small cone. **d.** 7 : 8 and 7 : 1 **43.** 10
45. 116 **47.** about 1790 cm^2 and 1937 cm^2
48. 113.1 in.^{3} **49.** 8.2 m^3 **50.** 904.8 in.^{3} **51a.** $8\sqrt{3}$ mm, or about 13.9 mm **b.** $4\sqrt{21}$ mm, or about 18.3 mm
c. $8\sqrt{7}$ mm, or about 21.2 mm **52.** 20 **53.** 15 **54.** 15

Chapter Review   pp. 751–754

1. sphere **2.** pyramid **3.** cross section **4–5.** Answers may vary. Samples are given.

4.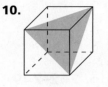

5.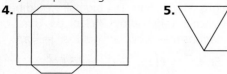

6. 8 **7.** 8 **8.** 5 **9.** a circle

10.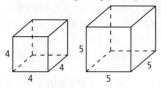

11. 36 cm² **12.** 66π m²
13. 208 in.² **14.** 36π cm²
15. 32.5π cm² **16.** 185.6 ft²
17. 576 m² **18.** 50.3 in.²
19. 391.6 in.²

20. $B = \frac{S.A. - L.A.}{2}$ **21.** 84 m³ **22.** 24.5 ft³
23. 410.5 yd³ **24.** 13.9 m³ **25.** S.A. = 314.2 in.²;
V = 523.6 in.³ **26.** S.A. = 153.9 cm²; V = 179.6 cm³
27. S.A. = 50.3 ft²; V = 33.5 ft³ **28.** S.A. = 8.0 ft²;
V = 2.1 ft³ **29.** 904.78 cm³ **30.** 314 m² **31.** 8.6 in.³
32. Answers may vary. Sample:

33. 27 : 64 **34.** 64 : 27 **35.** 324 pencils

Chapter 12

Get Ready!   p. 759

1. 82 **2.** $6\frac{2}{3}$ **3.** 15 **4.** 25 **5.** $6\sqrt{2}$ **6.** 5 **7.** 6 **8.** 18
9. 24 **10.** 45 **11.** 60 **12.** $4\sqrt{2}$ **13.** 13 **14.** $\sqrt{10}$
15. 6 **16.** Answers may vary. Sample: A tangent touches a circle at one point. **17.** Answers may vary. Sample: An inscribed ∠ has its vertex on a circle and its sides are inside the circle. **18.** Answers may vary. Sample: An intercepted arc is the part of a circle that lies in the interior of an ∠.

Lesson 12-1   pp. 762–769

Got It? 1a. 52 **b.** $x = 180 - c$ **2.** about 127 mi **3.** $5\frac{1}{3}$
4. no; $4^2 + 7^2 = 65 \neq 8^2$ **5.** 12 cm
Lesson Check 1. 32 **2.** 6 units **3.** $\sqrt{63} \approx 7.9$ units
4. Answers may vary. Sample: *Tangent ratio* refers to a ratio of the lengths of two sides of a rt. △, while *tangent to a circle* refers to a line or part of a line that is in the plane of a circle and touches the circle in exactly one point. **5.** If $\overline{DF}$ is tangent to ⊙E, then $\overline{DF} \perp \overline{EF}$. That

would mean that △DEF contains two rt. ∡, which is impossible. So $\overline{DF}$ is not a tangent to ⊙E.
Exercises 7. 47 **9.** 253.0 km **11.** 178.9 km
13. 3.6 cm **15.** no; $5^2 + 15^2 \neq 16^2$ **17.** yes;
$6^2 + 8^2 = 10^2$ **19.** 14.2 in. **21.** All 4 are ≅; the two tangents to each coin from A are ≅, so by the Transitive Prop. of ≅, all the tangents are ≅. **23.** 1. $\overline{BA}$ and $\overline{BC}$ are tangent to ⊙O at A and C. (Given) 2. $\overline{AB} \perp \overline{OA}$ and $\overline{BC}$ $\perp \overline{OC}$ (If a line is tan. to a ⊙, it is ⊥ to the radius.) 3. △BAO and △BCO are rt. ∡. (Def. of rt. △) 4. $\overline{AO} \cong \overline{OC}$ (Radii of a circle are ≅.) 5. $\overline{BO} \cong \overline{BO}$ (Refl. Prop. of ≅) 6. △$BAO \cong △BCO$ (HL) 7. $\overline{BA} \cong \overline{BC}$ (Corresp. parts of ≅ ∡ are ≅.) **25.** 1. ⊙A and ⊙B with common tangents $\overline{DF}$ and $\overline{CE}$ (Given) 2. $GD = GC$ and $GE = GF$ (Two tan. segments from a pt. to a ⊙ are ≅.) 3. $\frac{GD}{GC} = 1$, $\frac{GF}{GE} = 1$ (Div. Prop. of =) 4. $\frac{GD}{GC} = \frac{GF}{GE}$ (Trans. Prop. of =) 5. ∠$DGC \cong \angle EGF$ (Vert. ∡ are ≅.) 6. △$GDC \sim △GFE$ (SAS ~ Thm.) **27.** 57.5

29.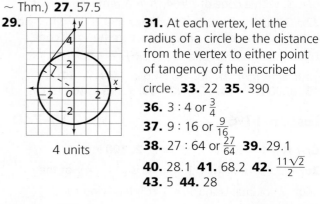

4 units

31. At each vertex, let the radius of a circle be the distance from the vertex to either point of tangency of the inscribed circle. **33.** 22 **35.** 390
36. 3 : 4 or $\frac{3}{4}$
37. 9 : 16 or $\frac{9}{16}$
38. 27 : 64 or $\frac{27}{64}$ **39.** 29.1
40. 28.1 **41.** 68.2 **42.** $\frac{11\sqrt{2}}{2}$
43. 5 **44.** 28

Lesson 12-2   pp. 771–779

Got It? 1. Since the circles are ≅, their radii are = and ∡BOC and DPF are isosceles. So $\overline{OB} \cong \overline{OC} \cong \overline{PD} \cong \overline{DF}$. Since ∠$B \cong \angle D$ and the ∡ are isosceles, ∠$B \cong \angle C \cong \angle D \cong \angle F$. So △$BOC \cong △DPF$ by AAS. So ∠$O \cong \angle P$. Therefore, $\overline{BC} \cong \overline{DF}$ (either by corresp. parts of ≅ ∡ are ≅ or by within ≅ circles, ≅ central ∡ have ≅ chords) and $\overset{\frown}{BC} \cong \overset{\frown}{DF}$ (within ≅ circles, ≅ central ∡ have ≅ arcs).
2. 16; ≅ chords are equidistant from the center **4.** $\overline{BA}$ is the hypotenuse of rt. △BAC, so the Pythagorean Theorem can be used.
Lesson Check 1. 50; ∠$COD \cong \angle AOB$ (Vert. ∡ are ≅), so $\overset{\frown}{CD} \cong \overset{\frown}{AB}$ because ≅ central ∡ have ≅ arcs. Therefore, $m\overset{\frown}{CD} = m\overset{\frown}{AB}$. **2.** $\overset{\frown}{CA} \cong \overset{\frown}{BD}$ because in a circle ≅ chords have ≅ arcs. **3.** The distances are equal because in a circle ≅ chords are equidistant from the center. **4.** A radius is *not* a chord because one of its endpoints is not on the circle. A diameter *is* a chord because both of its endpoints

are on the circle. **5.** Chords $\overline{SR}$ and $\overline{QP}$ are equidistant from the center, so their lengths must be equal.

Exercises 7. Answers may vary. Sample: $\overarc{ET} \cong \overarc{GH} \cong \overarc{JN} \cong \overarc{ML}$; $\overline{ET} \cong \overline{GH} \cong \overline{JN} \cong \overline{ML}$; $\angle TFE \cong \angle HFG$; $\angle JKN \cong \angle MKL$ **9.** 8 **11.** The center is at the intersection of $\overline{GH}$ and $\overline{KM}$, because if a chord is the $\perp$ bis. of another chord, then the first chord is a diameter; two diameters intersect at the center of a circle. **13.** 6 **15.** 20.8 **17.** 6 in. **19.** Since $\angle AOB \cong \angle COD$, it follows that $m\angle AOB = m\angle COD$. Now $m\angle AOB = m\overarc{AB}$ and $m\angle COD = m\overarc{CD}$ (Definition of arc measure). So $m\overarc{AB} = m\overarc{CD}$ (Substitution). Therefore, $\overarc{AB} \cong \overarc{CD}$ (Definition of $\cong$ arcs). **21.** $\odot O$ with $\overline{AB} \cong \overline{CD}$ (Given); $\overline{AO} \cong \overline{BO} \cong \overline{CO} \cong \overline{DO}$ (All radii of a $\odot$ are $\cong$); $\triangle AOB \cong \triangle COD$ (SSS); $\angle AOB \cong \angle COD$ (Corresp. parts of $\cong \triangle$ are $\cong$.); $\overarc{AB} \cong \overarc{CD}$ ($\cong$ central $\angle$ have $\cong$ arcs.). **23.** 5 in. **25.** 10 ft **27.** 9.2 units **29.** The length of a chord or an arc is determined not only by the measure of the central $\angle$, but also by the radius of the $\odot$. **31.** 90 **33.** $\overline{XW} \cong \overline{XY}$ (All radii of a circle are $\cong$); X is on the $\perp$ bis. of $\overline{WY}$ (Converse of $\perp$ Bis. Thm.); ℓ is the $\perp$ bis. of $\overline{WY}$ (Given); X is on ℓ (Subst. Prop.), so ℓ contains the center of $\odot X$.

35.

Given: $\odot O$ with $\overarc{AB} \cong \overarc{CD}$
Prove: $\angle AOB \cong \angle COD$
Proof: $m\angle AOB = m\overarc{AB}$ and $m\angle COD = m\overarc{CD}$ (definition of arc measure). $\overarc{AB} \cong \overarc{CD}$ (given), so $m\overarc{AB} = m\overarc{CD}$ (Def. of $\cong$ arcs). Therefore, $m\angle AOB = m\angle COD$ (Substitution). Hence $\angle AOB \cong \angle COD$ (Def. of $\cong \angle$).

37.

Given: $\odot O$ with $\overarc{AB} \cong \overarc{CD}$
Prove: $\overline{AB} \cong \overline{CD}$
Proof: It is given that $\overarc{AB} \cong \overarc{CD}$, so $\angle AOB \cong \angle COD$ (if arcs are $\cong$ then their central $\angle$ are $\cong$). Also, $AO = BO = CO = DO$ (radii of a $\odot$ are $\cong$), so $\triangle AOB \cong \triangle COD$ (SAS), and $\overline{AB} \cong \overline{CD}$ (corresp. parts of $\cong \triangle$ are $\cong$).

39.
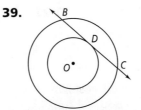

Given: Concentric circles, $\overline{BC}$ is tangent to the smaller circle at D
Prove: D is the midpt. of $\overline{BC}$
Proof: It is given that $\overline{BC}$ is tangent to the smaller circle, so $\overline{BC} \perp \overline{OD}$ (a tangent is $\perp$ to a radius at the point of tangency). $\overline{OD}$ is part of a diameter of the larger circle, so $\overline{BD} \cong \overline{CD}$ (if a diameter is $\perp$ to a chord, it bisects the chord). D is the midpt. of $\overline{BC}$ (Def. of midpt.)

41. G
43. During one revolution the bicycle moves $C = \pi d = \pi(17) = 53.4$ in., or about 4.45 ft. So the number of revolutions needed to travel 800 ft is $\frac{800}{4.45} \approx 180$ revolutions. **44.** 40 **45.** 5.5 **46.** 7.6 in. and 18.4 in. **47-49.** Answers may vary. Samples are given. **47.** $\overarc{STQ}$ **48.** $\overarc{ST}$ **49.** $\overarc{STR}$ **50.** 86 **51.** 180 **52.** 121

Lesson 12-3   pp. 780–787

Got It? 1a. 90 **b.** $m\angle A = 95$, $m\angle B = 77$, $m\angle C = 85$, and $m\angle D = 103$ **c.** The sum of the measures of opposite $\angle$ is 180. **2.** $m\angle 1 = 90$, $m\angle 2 = 110$, $m\angle 3 = 90$, $m\angle 4 = 70$ **3a.** $x = 35$, $y = 55$ **b.** An inscribed $\angle$, and an $\angle$ formed by a tangent and chord, are both equal to half the measure of the intercepted arc. Since the $\angle$ intercept the same arc, their measures are $=$ and they are $\cong$.

Lesson Check 1. $\overarc{BD}$ **2.** $\angle D$ **3.** $\angle A$ and $\angle C$ are suppl., and $\angle B$ and $\angle D$ are suppl. **4.** Sample answer: For inscribed $\angle ABC$, B is the vertex and A, B, and C are points on the circle. The intercepted arc of $\angle ABC$ consists of points A, C, and all the points on the circle in the interior of $\angle ABC$. **5.** $\angle A$ is not inscribed in a semicircle.

Exercises 7. 180 **9.** $a = 54$, $b = 30$, $c = 96$ **11.** $a = 101$, $b = 67$, $c = 84$, $d = 80$ **13.** $a = 85$, $b = 47.5$, $c = 90$ **15.** $p = 90$, $q = 122$ **17.** $x = 65$, $y = 130$ **19.** Rectangle; opposite $\angle$ are $\cong$ (because figure is $\square$) and suppl. (because opp. $\angle$ intercept arcs whose measures sum to 360). $\cong$ suppl. $\angle$ are rt. $\angle$, so the inscribed $\square$ must be a rectangle. **21a.** 40 **b.** 50 **c.** 40 **d.** 40 **e.** 65 **23.** $a = 26$, $b = 64$, $c = 42$

25. $a = 30$, $b = 60$, $c = 62$, $d = 124$, $e = 60$ **27.** $\odot S$ with inscribed $\angle PQR$ (Given); $m\angle PQT = \frac{1}{2}m\widehat{PT}$ (Inscribed $\angle$ Thm., Case I); $m\angle RQT = \frac{1}{2}m\widehat{RT}$ (Inscribed $\angle$ Thm., Case I); $m\widehat{PR} = m\widehat{PT} - m\widehat{RT}$ (Arc Add. Post.); $m\angle PQR = m\angle PQT - m\angle RQT$ ($\angle$ Add. Post.); $m\angle PQR = \frac{1}{2}m\widehat{PT} - \frac{1}{2}m\widehat{RT}$ (Subst. Prop.); $m\angle PQR = \frac{1}{2}m\widehat{PR}$ (Subst. Prop.) **29.** No; since opposite $\angle$s of a quadrilateral inscribed in a circle must be supplementary, the only rhombus that meets the criteria is a square. **31.** $\odot O$, $\angle A$ intercepts $\widehat{BC}$, and $\angle D$ intercepts $\widehat{BC}$ (Given); $m\angle A = \frac{1}{2}m\widehat{BC}$ and $m\angle D = \frac{1}{2}m\widehat{BC}$ (Inscribed $\angle$ Thm.); $m\angle A = m\angle D$ (Subst. Prop.); $\angle A \cong \angle D$ (Def. of $\cong$ $\angle$s). **33.** Quadrilateral $ABCD$ inscribed in $\odot O$ (Given); $m\angle A = \frac{1}{2}m\widehat{BCD}$ and $m\angle C = \frac{1}{2}m\widehat{BAD}$ (Inscribed $\angle$ Thm.); $m\angle A + m\angle C = \frac{1}{2}m\widehat{BCD} + \frac{1}{2}m\widehat{BAD}$ (Add. Prop.); $m\widehat{BCD} + m\widehat{BAD} = 360$ (Arc measure of circle is 360); $\frac{1}{2}m\widehat{BCD} + \frac{1}{2}m\widehat{BAD} = 180$ (Mult. Prop.) $m\angle A + m\angle C = 180$ (Subst. Prop.); $\angle A$ and $\angle C$ are suppl. (Def. of suppl.); $m\angle B = \frac{1}{2}m\widehat{ADC}$ and $m\angle D = \frac{1}{2}m\widehat{ABC}$ (Inscribed $\angle$ Thm.); $m\angle B + m\angle D = \frac{1}{2}m\widehat{ADC} + \frac{1}{2}m\widehat{ABC}$ (Add. Prop.); $m\widehat{ADC} + m\widehat{ABC} = 360$ (Arc measure of circle is 360); $\frac{1}{2}m\widehat{ADC} + \frac{1}{2}m\widehat{ABC} = 180$ (Mult. Prop.) $m\angle B + m\angle D = 180$ (Subst. Prop.); $\angle B$ and $\angle D$ are suppl. (Def. of suppl. $\angle$s). **35.** false **37.** True; opposite $\angle$s in an inscribed

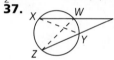

quadrilateral intercept nonoverlapping arcs totaling 360 and inscribed $\angle$s have half the measure of the intercepted arcs, so the opposite $\angle$s are suppl.

41. D

43. No; answers may vary: Sample: Although $\overline{BD}$ bisects $\angle ABC$, you cannot assume that it also bisects $\angle ADC$. This assumption leads to the invalid conclusion that the $\angle$s are $\cong$. Proof: From the given information, $\angle A \cong \angle C$ and $\overline{BD}$ bisects $\angle ABC$. By the def. of $\angle$ bisector, $\angle ABD \cong \angle CBD$. $\overline{BD} \cong \overline{BD}$ by the Refl. Prop. of $\cong$ $\triangle ADB \cong \triangle CDB$ by AAS. So, $\angle ADB \cong \angle CDB$ because corresp. parts of $\cong$ $\triangle$s are $\cong$. **44.** 17.3 **45.** 34.6 **46.** 17.5 **47.** 5 : 2 or $\frac{5}{2}$ **48.** 57 **49.** 28.5 **50.** 2 **51.** 4

Lesson 12-4   pp. 790–797

Got It? 1a. 250 **b.** 40 **c.** 40 **2a.** 160 **b.** The probe is closer; as an observer moves away from Earth, the viewing angle decreases and the measure of the arc of Earth that is viewed gets larger and approaches 180. **3a.** 13.8 **b.** 3.2

Lesson Check 1. 5.4 **2.** 65 **3.** 11.2 **4.** 100, 260 **5.** A secant is a line that intersects a circle at two points; a tangent is a line that intersects a circle at one point. **6.** No; we can find the sum of the measures of the two arcs (in this situation, that sum is 230), but there is not enough information to find the measure of each arc. **7.** The student forgot to multiply by the length of the entire secant seg.; the equation should be $(13.5)(6) = x^2$.

Exercises 9. 50 **11.** 60 **13.** $x = 72$, $y = 36$ **15.** 15 **17.** 13.2 **19.** $x = 25.8$, $y \approx 12.4$ **21.** $360 - x$ **23.** $180 - y$ **25.** 16.7 **27.** 95, 104, 86, 75 **29.** $c = b - a$ **31.** $\angle 1$ is a central $\angle$, so $m\angle 1 = x$; $\angle 2$ is an inscribed $\angle$, so $m\angle 2 = \frac{1}{2}x$; $\angle 3$ is formed by the secants, so $m\angle 3 = \frac{1}{2}(x - y)$. **33.** $x \approx 8.9$, $y = 2$ **35.** 1. $\odot O$ with secants $\overline{CA}$ and $\overline{CE}$ (Given) 2. Draw $\overline{BE}$ (2 pts. determine a line.) 3. $m\angle BEC = \frac{1}{2}m\widehat{BD}$ and $m\angle ABE = \frac{1}{2}m\widehat{AE}$ (The measure of an inscribed $\angle$ is half the measure of its intercepted arc.) 4. $m\angle BEC + m\angle BCE = m\angle ABE$ (Ext. $\angle$ Thm.) 5. $\frac{1}{2}m\widehat{BD} + m\angle BCE = \frac{1}{2}m\widehat{AE}$ (Subst. Prop. of =) 6. $m\angle BCE = \frac{1}{2}m\widehat{AE} - \frac{1}{2}m\widehat{BD}$ (Subst. Prop. of =) 7. $m\angle BCE = \frac{1}{2}(m\widehat{AE} - m\widehat{BD})$ (Distr. Prop.) 8. $\angle BCE \cong \angle ACE$ (Refl. Prop. of $\cong$) 9. $m\angle ACE = \frac{1}{2}(m\widehat{AE} - m\widehat{BD})$ (Subst. Prop. of =)

37.

Given: A $\odot$ with secant segments $\overline{XV}$ and $\overline{ZV}$
Prove: $XV \cdot WV = ZV \cdot YV$.
Proof: Draw $\overline{XY}$ and $\overline{ZW}$ (2 pts. determine a line); $\angle XVY \cong \angle ZVW$ (Refl. Prop. of $\cong$); $\angle VXY \cong \angle WZV$ (2 inscribed $\angle$s that intercept the same arc are $\cong$); $\triangle XVY \sim \triangle ZVW$ (AA$\sim$); $\frac{XV}{ZV} = \frac{YV}{WV}$ (In similar figures, corresp. sides are proportional); $XV \cdot WV = ZV \cdot YV$ (Prop. of Proportion) **39a.** $\triangle ACD$ **b.** $\tan A = \frac{DC}{AC} = \frac{DC}{1} = DC$, length of tangent seg. **c.** $\sec A = \frac{AD}{AC} = \frac{AD}{1} = AD$, length of secant seg. **41.** $m\angle 1 = \frac{1}{2}m\widehat{QRP} - \frac{1}{2}m\widehat{PQ}$ and $m\angle 2 = \frac{1}{2}m\widehat{RQP} - \frac{1}{2}m\widehat{RP}$ (vertex outside $\odot$, $m\angle$ = half difference of intercepted arcs); $m\angle 1 + m\angle 2 = \frac{1}{2}m\widehat{QRP} + \frac{1}{2}m\widehat{RQP} - \frac{1}{2}m\widehat{PQ} - \frac{1}{2}m\widehat{RP}$ (Subst. Prop. of =); $m\angle 1 + m\angle 2 = \frac{1}{2}m\widehat{QR} + \frac{1}{2}m\widehat{RP} + \frac{1}{2}m\widehat{QR} + \frac{1}{2}m\widehat{PQ} - \frac{1}{2}m\widehat{PQ} - \frac{1}{2}m\widehat{RP}$ (Arc Add. Postulate and Distr. Prop.); $m\angle 1 + m\angle 2 = m\widehat{QR}$ (Distr. Prop.).

43.

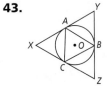

Given: Equilateral $\triangle ABC$ is inscribed in $\odot O$; $\overline{XY}$, $\overline{YZ}$, and $\overline{XZ}$ are tangents to $\odot O$
Prove: $\triangle XYZ$ is equilateral.
Proof: $m\widehat{AB} = m\widehat{BC} = m\widehat{AC} = 120$, since chords $\overline{AB}$, $\overline{BC}$, and $\overline{CA}$ are all $\cong$. So the measures of $\angle X$, $\angle Y$, and $\angle Z$ are $\frac{1}{2}(240 - 120) = 60$, and $\triangle XYZ$ is equiangular, so it is also equilateral.

45. 112 **47.** 829.4 cm^3 **48.** $a = 50$, $b = 55$, $c = 105$
49. $a = 55$, $b = 35$, $c = 30$ **50.** 30 **51.** 42 **52.** 57
53. 5.8 **54.** 12.8 **55.** 5.8

Lesson 12-5   pp. 798–803

Got It? 1a. $(x - 3)^2 + (y - 5)^2 = 36$ **b.** $(x + 2)^2 + (y + 1)^2 = 2$ **2.** $(x - 4)^2 + (y - 3)^2 = 29$
3a. The center of the circle represents the cell tower's position. The radius represents the cell tower's transmission range.
b. center (2, 3); radius 10

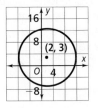

Lesson Check 1. $x^2 + y^2 = 16$
2. $(x - 1)^2 + (y + 1)^2 = 5$ **3.** center (8, 0); radius 3
4. center (−2, 4); radius $\sqrt{7}$ **5.** its center and its radius; its center and its radius **6.** Using the two known points, use the Distance Formula to find the distance between them; that is the radius. Then use the center and the radius to write the standard equation for the circle.
7. Sample explanation: The student should have rewritten the equation as $(x - 2)^2 + (y - (-3))^2 = 16$ to realize that the center is (2, −3).
Exercises 9. $x^2 + (y - 3)^2 = 49$
11. $(x - 5)^2 + (y + 1)^2 = 144$
13. $(x + 9)^2 + (y + 4)^2 = 5$
15. $(x + 4)^2 + y^2 = 9$
17. $(x + 4)^2 + (y - 2)^2 = 16$
19. $(x + 2)^2 + (y - 6)^2 = 16$

21. $(x - 7)^2 + (y + 2)^2 = 52$
23. $(x - 6)^2 + (y - 5)^2 = 61$
25. center (−7, 5);   **27.** center (−4, 1);
radius 4   radius 5

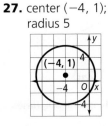

29. position (5, 7); range 9 **31.** $x^2 + y^2 = 4$
33. $x^2 + (y - 3)^2 = 4$ **35.** $(x - 2)^2 + (y - 2)^2 = 16$
37. $(x - 4)^2 + (y - 3)^2 = 25$
39. $(x - 3)^2 + (y - 3)^2 = 8$ **41.** Yes; it is a circle with center (1, −2) and radius 3. **43.** No; the x term is not squared. **45.** $(x - 4)^2 + (y - 7)^2 = 36$
47. $(x - h)^2 + (y - k)^2 = r^2$
$(y - k)^2 = r^2 - (x - h)^2$
$y - k = \pm\sqrt{r^2 - (x - h)^2}$
$y = \pm\sqrt{r^2 - (x - h)^2} + k$

49.

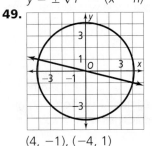

(4, −1), (−4, 1)

51.   **53.**

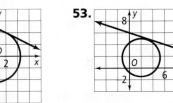

(2, 4)   (3, 5)
55. about 11.5, 11.5, 49.8, and 49.8 units^2.
57a. $x^2 + y^2 = 15,\!681,\!600$ **b.** 69.1 mi
c. about 32 days **59.** I **61.** $x = 25$, $y = 75$ **62.** 38
63.   **64.**

65.

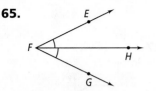

Lesson 12-6   pp. 806–811

Got It? 1. a pair of ∥ lines, each 1 cm from $\overleftrightarrow{AB}$

2.

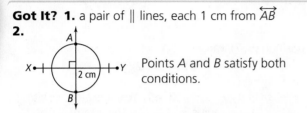

Points A and B satisfy both conditions.

3a. The locus is the line ∥ to and equidistant from the given ∥ lines (midway between them). **b.** The locus is a plane ∥ to and equidistant from the given ∥ planes (midway between them).

Lesson Check

1.

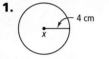

The locus is a circle with center x and radius 4 cm.

2.

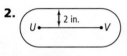

The locus is a pair of ∥ segments, each segment 2 in. from $\overline{UV}$, and two semicircles with radius 2 in. and centers U and V.

3.

The locus is a pair of ∥ lines, each 3 mm from $\overleftrightarrow{LM}$.

4.

The locus is two circles concentric with the original circle; the smaller circle has radius 2 in. and the larger circle has radius 4 in.

5. Answers may vary. Sample: A *locus* is a set of points, and a *location* can be thought of as a description of a single point. **6.** The locus in a plane is a line (the ⊥ bis. of $\overline{JK}$) and the locus in space is a plane (it contains the midpt. of $\overline{JK}$ and is ⊥ to it).

Exercises

7. The locus is the ⊥ bis. of $\overline{PQ}$.

9. The locus is the two lines that bis. the rt. ∠s.

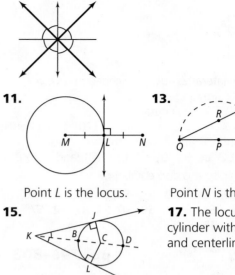

11.   **13.**

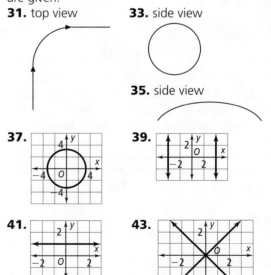

Point L is the locus.   Point N is the locus.

15.   **17.** The locus is an endless cylinder with radius 4 cm and centerline $\overleftrightarrow{DE}$.

The locus is points B and D.

19. The locus is an endless cylinder with radius 5 mm and centerline $\overrightarrow{PQ}$, and a hemisphere of radius 5 mm centered at P, "capping off" the cylinder. **21.** The locus is the set of all points 2 units from the origin. **25.** The locus will be points in the plane that are 1 unit from the x-axis and 2 units from the origin. **27.** $y = 2x - 4$ **29a.** a circle **b.** $x^2 + y^2 = 4$ **31–35.** Answers may vary. Samples are given.

31. top view   **33.** side view

35. side view

37.   **39.**

41.   **43.**

45a. Sample: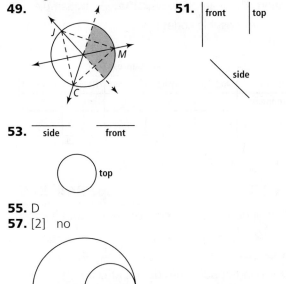
b. The locus is the ⊥ bis. of the base except for the midpt. of the base. **c.** Sample explanation: The vertex of the isosc. △ must be equidistant from the endpoints of the base, and all the points (in a plane) that are equidistant from two points lie on the ⊥ bis. of the segment whose endpoints are the two given points.
47. The locus is a circle of radius 11 cm, concentric with the original.

49.

51. front | top

side

53. side   front

top

55. D
57. [2] no

CM = 0.25 mi

[1] incorrect answer OR incorrect diagram/ explanation

58. $(x - 6)^2 + (y + 10)^2 = 25$
59. $(x - 1)^2 + (y - 7)^2 = 36$
60. $(x + 8)^2 + (y + 1)^2 = 13$ **61.** 510 in.^{2}
62. 175.9 ft^2 **63.** 4π units^2 **64.** $\frac{64\pi}{5}$ units^2
65. 10π units^2

Chapter Review   pp. 811–814

1. secant of **2.** chord **3.** tangents to **4.** inscribed ∠
5. locus **6.** 20 units **7.** $\sqrt{3}$ **8.** 120 **9.** 90 **10.** 2 : 1 or $\frac{2}{1}$
11. $\overline{AB}$ is a diameter of the circle. **12.** 4.5 **13.** $\frac{\sqrt{181}}{2} \approx 6.7$
14. $a = 80$, $b = 40$, $c = 40$, $d = 100$ **15.** $a = 40$, $b = 140$, $c = 90$ **16.** $a = 118$, $b = 49$, $c = 144$, $d = 98$ **17.** $a = 90$, $b = 90$, $c = 70$, $d = 65$
18. 37 **19.** $a = 95$, $b = 85$ **20.** 6.5 **21.** 4
22. $x^2 + (y + 2)^2 = 9$ **23.** $(x - 3)^2 + (y - 2)^2 = 4$
24. $(x + 3)^2 + (y + 4)^2 = 25$ **25.** $(x - 1)^2 + (y - 4)^2 = 9$ **26.** center (7, −5); radius 6 **27.** The locus

is the ray that bisects the ∠. **28.** The locus is a circle, concentric with the given circle, with radius 7 cm.
29. The locus is two lines, one on each side of the given line and ∥ to it, each at a distance of 8 in. from the given line. **30.** The locus consists of a cylinder with radius 6 in. that has $\overline{AB}$ as its centerline, along with two hemispheres with centers A and B, each with radius 6 in.

Chapter 13

Get Ready!   p. 821

1. $\frac{63}{80}$, 78.75% **2.** 0.875, 87.5% **3.** $\frac{19}{20}$, 95% **4.** $\frac{3}{5}$, 0.6

5. $\frac{3}{4}$ **6.** $\frac{2}{3}$ **7.** $\frac{7}{2}$ **8.** 72 **9.** −3 **10.** $\frac{5}{2}$ **11.** 8 **12.** $\frac{14}{3}$ **13.** $\frac{189}{5}$

14. $\frac{4}{9}$ **15.** the batter's at-bats **16.** Answers may vary. Sample answer: A simulation can be used to model situations that are difficult or impractical to conduct. **17.** Yes; how the penny lands does not affect how the quarter lands.

Lesson 13-1   pp. 824–829

Got It? 1. 0.625 or 62.5%; about 336 **2a.** $\frac{4}{36}$ or $\frac{1}{9}$
b. $\frac{1}{36}$ **c.** 0 **3.** $\frac{19}{29}$
Lesson Check 1. $\frac{4}{8}$ or $\frac{1}{2}$ **2.** $\frac{3}{8}$ **3.** $\frac{4}{8}$ **4.** Sample answer: The computation of each is similar, but experimental probability is based on the results of trials. Theoretical probability is based on what should happen mathematically. **5.** Sample answer: Rolling a 9 with a standard number cube. **6.** Probabilities are expressed as numbers from 0 to 1 or from 0% to 100%. The probability is 1 or 100%.
Exercises 7a. about 0.297 or 29.7% **b.** about 59 times
9. $\frac{1}{11}$ **11.** $\frac{4}{11}$ **13.** $\frac{10}{11}$ **15.** $\frac{7}{11}$ **17.** $\frac{4}{8}$ or $\frac{1}{2}$ **19.** $\frac{2}{6}$ or $\frac{1}{3}$
21. about 133 **23.** $\frac{12}{65}$ **25.** $\frac{55}{65}$ or $\frac{11}{13}$ **27.** yellow, round: about 1133; yellow, wrinkled: about 378; green, round: about 378; green, wrinkled: about 126 **29.** Answers will vary. **31.** Answers will vary. **33a.** 3 : 7 **b.** 7 : 3 **35.** C
37. 48 in.^{2} **38.** circle with radius 5 centered at the origin
39. all points on either axis **40.** $\frac{3}{5}$ **41.** $\frac{3}{7}$ **42.** $\frac{3}{8}$

Lesson 13-2   pp. 830–835

Got It? 1a. $\frac{1}{8}$ **b.** $\frac{7}{40}$ **c.** $\frac{1}{5}$ **d.** $\frac{40}{40}$ or 1; the sum of the frequencies includes all of the events in the sample space. **2.** $\frac{18}{100}$ or $\frac{9}{50}$

3.

Math Test Scores				
Score	90–100	80–89	70–79	60–69	0–59
Frequency	10	12	15	8	2
Probability	$\frac{10}{47}$	$\frac{12}{47}$	$\frac{15}{47}$	$\frac{8}{47}$	$\frac{2}{47}$

Lesson Check **1.** $\frac{54}{142}$ or $\frac{37}{71}$ **2.** $\frac{28}{142}$ or $\frac{14}{71}$ **3.** $\frac{10}{142}$ or $\frac{5}{71}$
4. $\frac{50}{142}$ or $\frac{25}{71}$
5.

Preferred Music Format

Result	CD	Radio	Blu-ray	MP3
Frequency	54	50	10	28
Probability	$\frac{37}{71}$	$\frac{25}{71}$	$\frac{5}{71}$	$\frac{14}{71}$

6. The possible outcomes are heads or tails. **7.** There are many ways she will not win but only one way she will win, so both the probability of winning is much smaller.

Exercises **9.** $\frac{25}{63}$ **11.** The relative frequencies do not change because both the numerator and denominator are multiplied by 2. **13.** $\frac{6}{23}$

15.

Favorite Snacks

Snack	Bananas	Trail Mix	C&C	Popcorn
Frequency	8	5	4	6
Probability	$\frac{8}{23}$	$\frac{5}{23}$	$\frac{4}{23}$	$\frac{6}{23}$

17a. $\frac{145}{983}$ **b.** $\frac{838}{983}$

19. Answers may vary. Sample answer:

Coin Toss

Result	4 Heads	3 Heads 1 Tail	2 Heads 1 Tail	1 Head 3 Tails	4 Tails
Frequency	8	5	4	6	1
Probability	$\frac{1}{16}$	$\frac{1}{4}$	$\frac{3}{8}$	$\frac{1}{4}$	$\frac{1}{16}$

21a.

Sum	2	3	4	5	6	7	8	9	10	11	12
Frequency	1	1	2	4	5	6	5	4	3	2	1
Probability	$\frac{1}{36}$	$\frac{1}{18}$	$\frac{1}{12}$	$\frac{1}{9}$	$\frac{5}{36}$	$\frac{1}{6}$	$\frac{5}{36}$	$\frac{1}{9}$	$\frac{1}{12}$	$\frac{1}{18}$	$\frac{1}{36}$

b. theoretical; the results are based on each face of one number cube having a probability of $\frac{1}{6}$.
23. The probabilities do not add to 1; they add to 1.05. You need to know the frequency distribution of the favorite pets. **25.** A **27.** C **29.** theoretical; though experimental results can be found, the more you do the experiment, the closer the results will be to the theoretical value.
30. experimental; no information is available to find results until an experiment is done. **31.** theoretical; all of the data are available to find the probability without an experiment.
32. 1.5 in. **33.** 20.25 mm **34.** $\frac{a}{c}$ **35.** 15 **36.** 90

Lesson 13-3 pp. 836–842

Got It? **1.** 2600 **2.** 479,001,600 **3.** 1320 **4.** 56
5. 120 **6.** $\frac{1}{2730}$

Lesson Check **1.** 6 **2.** 1 **3.** 30 **4.** 120 **5.** 15
6. 20 **7.** 210 **8.** Both are methods of counting. With permutations order is important, but with combinations order is not important. **9.** Yes, she also needs to know the total number of possible outcomes in the sample space.

Exercises **11.** 6760 **13.** 10,897,286,400 **15.** 210
17a. 4060 **b.** 142,506 **19.** 495 **21.** $\frac{1}{15,890,700}$ **23.** 56
25. $\frac{1}{1365}$ **27.** The friend used a permutation instead of a combination. There are 56 ways. **29a.** $\frac{1}{42}$ **b.** Sample: The number of possible codes is actually a permutation.
31. 96 cm^2

33.

Favorite Movie Genres

Genres	Action	Comedy	Drama	Horror	Other
Frequency	9	8	3	6	4
Probability	$\frac{3}{10}$	$\frac{4}{15}$	$\frac{1}{10}$	$\frac{1}{5}$	$\frac{2}{15}$

34. $\frac{1}{36}$ **35.** $\frac{3}{36}$ **36.** $\frac{26}{36}$

Lesson 13-4 pp. 844–849

Got It? **1.** independent; the outcomes do not affect each other **2.** $\frac{1}{12}$ **3.** 38% **4.** $\frac{2}{3}$
Lesson Check **1.** 12.5% **2.** 0.85 **3.** $\frac{19}{30}$ **4.** Answers will vary. When the outcomes do not affect each other, the events are independent. If the outcome of one event affects the outcome of another event, they are dependent events. **5.** Sample answer: The events are dependent because it needs to be cloudy in order for it to rain.

Exercises **7.** dependent **9.** dependent **11.** $\frac{1}{4}$ **13.** $\frac{3}{5}$
15. $\frac{8}{15}$ **17.** $\frac{13}{40}$ **19.** $\frac{5}{8}$ **21.** $\frac{3}{4}$ **23.** 80% **25.** Mutually exclusive events cannot happen at the same time, but overlapping events can occur at the same time.

27. No; Answers may vary. Sample: $\frac{1}{2} + \frac{1}{3} \neq \frac{2}{3}$

29. dependent events; If one event occurs, then the other event cannot occur because they are mutually exclusive. So, the outcome of the second event depends on the outcome of the first event. **31.** $\frac{41}{\pi}$ cm **33.** 15,600
34. 40,320 **35.** 792 **36.** $\frac{5}{13}$ **37.** $\frac{32}{39}$ **38.** $\frac{22}{39}$

Lesson 13-5 pp. 850–855

Got It? **1.** $\frac{1}{4}$ **2a.** about 0.26 **b.** about 0.76
3. 0.90

Lesson Check **1.** 0.2 **2.** about 0.47 **3.** He divided by the number of people who do not support the issue instead of by the total number of Republicans.

4. A two-way frequency table displays the frequencies of data in two different categories. **5.** The probability that a student plays sports, given that the student is a female.

Exercises 7. 0.2 **9.** $\frac{7}{13}$ **11.** about 0.444 **13.** about 0.54 **15.** $\frac{1}{5}$ **17.** $\frac{5}{8}$ **19.** $\frac{3}{5}$ **21.** Answers may vary: Sample: The sum of the probabilities is 1. The sum of the relative frequencies

is also 1 because they represent the experimental probabilities of each possible outcome. **23a.** Each cell contains the frequency of the joint event described by its row and column. **b.** The marginal frequencies are the total frequencies for each category in the two-way table. **c.** The sum of the joint probabilities in any row or column of a joint probability distribution is the same as the marginal probability associated with that row or column. **25.** 6 **27.** 7.5 **28.** 360° **29.** 1800° **30.** 3240° **31.** 113 in.² **32a.** $4\sqrt{3}$ cm **b.** about 1393 cm³ **33.** $\frac{2}{5}$ **34.** $\frac{8}{25}$ **35.** $\frac{3}{20}$ **36.** $\frac{17}{20}$

Lesson 13-6   pp. 856–861

Got It? 1. 40% **2a.** $\frac{1}{9}$ **b.** $\frac{5}{27}$ **c.** No, you don't know P(other pet) or P(cat and no dogs). **3a.** 54.5% **b.** Their chances of winning will increase because they have a better chance of winning on a muddy field.

Lesson Check 1. $\frac{5}{7}$ **2.** $\frac{6}{11}$ **3.** P(red | small); P(red | small) $= \frac{4}{9}$, P(small | red) $= \frac{2}{7}$ **4.** No. If two events are independent, then $P(A \text{ and } B) = P(A) \cdot P(B)$, which is not necessarily zero. The conditional probability would be 0 for mutually exclusive events. **5.** Sample answer: When finding a conditional probability you divide, but when you find a compound probability you often multiply.

Exercises 7. $\frac{1}{2}$ **9.** 20% **11.** 14.46% **13.** 24% **15.** $\frac{3}{7}$ **17.** $\frac{11}{42}$ **19.** C **21.** 26 cm **22.** 6.25% **23.** 75% **24.** 0 **25.** 50% **26.** 90% **27.** 10% **28.** $\frac{2}{3}$ **29.** $\frac{33}{58}$

Lesson 13-7   pp. 862–867

Got It? 1. 04, 14, 09 **2.** Answers may vary. Sample: about 15 **3a.** -2.8125 **b.** -6.25 **4a.** field goal **b.** Sample answer: Yes. If it is late in the game and they are down by more than 3 points, the touchdown might be a better option.

Lesson Check 1. 42, 37, 17, 31 **2.** 0.75; 1; pass to the teammate **3.** With the dart, there is some control over which name will be chosen. The table is more random. **4.** Expected value is the average value of a trial based on the probabilities and the values of the outcomes. **5.** Sample answer: Let 0–4 represent heads and 5–9

represent tails.

Exercises 7. 118, 735, 719, 632, 209 **9.** Answers may vary. Sample: Use a random number generator to produce numbers from 1 to 5. Count how many trials it takes before producing 8 numbers in a row that are not 1. Conduct the simulations many times and find the average value. **11.** –4200; No, the player should not continue because the expected value is negative. **13.** Yes, the expected value is $4000. **15.** -0.125 **17.** Answers may vary. Sample: Assign numbers from 1 to 2500 to each household. Use a calculator to generate random numbers between 1 and 2500. Select the first 500 unique numbers that the calculator generates. **19.** D **21.** 50 cm and 87 cm **22.** $\frac{2}{3}$ **23.** $\frac{2}{3}$ **24.** $(x - 3)^2 + (y + 5)^2 = 81$ **25.** $(x + 2)^2 + (y - 8)^2 = 25$

Chapter Review   pp. 870–874

1. permutation **2.** Theoretical probability **3.** dependent **4.** frequency table **5.** $\frac{3}{10}$ **6.** $\frac{9}{10}$ **7.** 0 **8.** $\frac{3}{5}$ **9.** $\frac{5}{12}$ **10.** $\frac{1}{4}$ **11.** $\frac{11}{28}$ **12.** $\frac{17}{28}$ **13.** $\frac{1}{7}$ **14.** $\frac{9}{14}$ **15.** $\frac{6}{7}$ **16.** 120 **17.** 40,320 **18.** 840 **19.** 220 **20.** 6720 **21.** 90 **22.** 36 **23.** 1140 **24.** 78,624,000 **25.** 2300 **26.** 0.115 **27.** 0.595 **28.** $\frac{7}{16}$ **29.** Sample answer: If the events are mutually exclusive, then $P(A \text{ and } B) = 0$ and the formula becomes $P(A \text{ or } B) = P(A) + P(B)$. Otherwise, the formula is used as stated. **30.** More students chose to study. **31.** 0.95 **32.** about 0.364 **33.** $\frac{22}{145}$ **34.** $\frac{5}{7}$ or about 71.4% **35.** $\frac{55}{74}$ or about 74.3% **36.** 20 **37.** 13 **38.** $\frac{13}{20}$

Skills Handbook

p. 884 1. Answer may vary slightly due to measuring method. Sample: 20 mm; 25 mm

3.

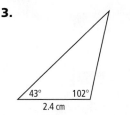

p. 885 1. right, scalene **3.** obtuse, isosceles **5.** Not possible; a rt. △ will always have one longest side opposite the rt. ∠.

7. 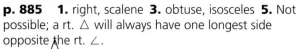 **9.**

p. 886   **1.** 0.4 **3.** 600 **5.** 1008 **7.** 15,000
9. 34,000 **11.** 4.3 **13.** 56 **15.** 3.9 **17.** 1,080,000
19. 12.6 **21.** $144\frac{4}{9}$ **23.** $\frac{3125}{4356}$

p. 887   **1.** $23\frac{1}{2}$ ft to $24\frac{1}{2}$ ft **3.** $339\frac{1}{2}$ mL to $340\frac{1}{2}$ mL
5. 73.15 mm to 73.25 mm **7.** 10.8 cm to 11.2 cm
9. 208 cm

p. 888   **1.** 18% **3.** 1% **5.** $\approx$ 9% **7.** $\approx$ 2%

p. 889   **1.** 121 **3.** 26.01 **5.** -36 **7.** 10 **9.** 8.6
11. $\frac{2}{3}$ **13.** ±7 **15.** ±1 **17.** ±6 **19.** ±5

p. 890   **1.** -50 **3.** 15 **5.** 2 **7.** 36 **9.** -2 **11.** 243
13. -20 **15.** -4 **17.** $2x + 3$ **19.** $-4x - 7$
21. $-4x^2 + 8x$ **23.** $-3t^2 + 4t$ **25.** $1 - 2r + r^2$
27. $7h - 1$ **29.** $2x^2 + 7x - 4$ **31.** $2y^2 - 3y$

p. 891   **1.** $\frac{5}{3}$ **3.** $\frac{2}{3}$ **5.** $\frac{1}{2x}$ **7.** $\frac{22x}{35}$ **9.** $16y$ **11.** $\frac{x + 1}{2}$
13. $\frac{1}{10t}$ **15.** $\frac{3y}{2}$ **17.** $\frac{\pi}{r + \pi}$ **19.** $\frac{5}{12}$ **21.** $\frac{5}{13}$ **23.** $\frac{12}{5}$

p. 892   **1.** 8 **3.** 22 **5.** -5 **7.** -10 **9.** 13 **11.** 10
13. -16 or 16 **15.** -20 or 20

p. 893   **1.** $(0, -3)$ **3.** $(4, 3)$

5–8. **9.** y-axis **11.** x-axis

p. 894   **1.** 5 **3.** 3 **5.** $\frac{2}{3}$ **7.** $\frac{1}{2}$ **9.** -5 **11.** -12
13. $35 - 2x = 9$; 13

p. 895   **1.** 0.5 **3.** 0.06 **5.** 1.09 **7.** 8.4 **9.** 7.2
11–14. Answers may vary. Samples are given. **11.** 7
13. 45

Index

A

AAS (Angle-Angle-Side) Theorem. *See* Angle-Angle-Side (AAS) Theorem

AA ~ (Angle-Angle Similarity) Postulate. *See* Angle-Angle Similarity (AA ~) Postulate

absolute value, 20, 28, 892

activity. *See also* Extensions; Technology
Building Congruent Triangles, 225
Circle Graphs, 658
Compass Designs, 42
Dynamic. *See* Dynamic Activity
Exploring Spherical Geometry, 179–180
Exploring the Area of a Circle, 659
Finding Volume, 725
The Golden Ratio, 468–469
Logic and Truth Tables, 96–97
Measuring From Afar, 515
Paper Folding and Reflections, 553
Paper Folding Bisectors, 300
Paper-Folding Conjectures, 249
Paper Folding With Circles, 770
Perpendicular Lines and Planes, 170
Probability and Decision Making, 868
Pythagorean Theorem, 490
Symmetry, 568
Tracing Paper Transformations, 544
Transforming to Find Area, 614–615

acute angle, 70

acute triangle, 494–495

Addition Property, 113, 323
addition
of angles, 30
of arcs, 650
of segments, 21

adjacent angles, 34–35, 73

adjacent arcs, 611, 649, 650, 676

algebra. *See also* Algebra Review
exercises, 21, 23, 24, 25, 26, 30, 31, 32, 33, 36, 39, 41, 48, 66, 67, 75, 79, 88, 93, 94, 101, 114, 117, 125, 126, 127, 132, 137, 153, 154, 155, 159, 161, 162, 169, 173, 176, 177, 204, 211, 223, 254, 255, 256, 262, 264, 275, 277, 288–289, 290, 293, 295, 297, 304, 306, 316, 323, 328, 329, 335, 337, 339, 342, 345, 349, 356, 357, 364, 365, 372, 373, 374, 380, 381, 385, 387, 388, 391–392, 395, 397, 398, 405, 410, 419, 425, 434, 436, 438, 439, 442, 445, 446, 457, 458, 459, 463, 465, 467, 475, 476, 477, 481, 482, 483, 487, 492, 495, 496, 497, 502, 504, 505, 514, 520, 537, 620, 628, 640, 656, 673, 698, 706, 716, 723, 731, 759, 767, 769, 775, 777, 787, 788, 791, 793, 794, 795, 815
reasoning in, 113–116, 129–132

algebraic expressions, 23, 32, 79

Algebra Review
Literal Equations, 698
Simplifying Radicals, 399
Solving Inequalities, 323
Solving Quadratic Equations, 439
Systems of Linear Equations, 257

alternate exterior angles, 140, 142–143, 151–152, 206
theorem, 151–152

Alternate Exterior Angles Theorem, 151–152

alternate interior angles, 140, 142–143, 206
theorem, 149–152

Alternate Interior Angles Theorem, 149–152

altitude
of cone, 711, 751
of cylinder, 699, 701, 751
of parallelogram, 616, 676
of prism, 699
of pyramid, 708, 751
of triangle, 281, 308–312, 341, 343, 460–461

angle(s)
acute, 29, 70
addition of, 30
adjacent, 34–35, 73
alternate exterior, 140, 142–143, 151–152, 206
alternate interior, 140, 142–143, 206
base, 249, 250, 389
bisector of, 37, 70, 294–296, 300–304, 341, 342–343, 805
central, 649, 676, 771–772
classifying, 29, 72
complementary, 34–35, 70, 73, 123
congruence of, 72, 120–123, 132, 220–221, 361–362, 389, 578
consecutive, 349, 360–361, 420
constructing, 44
corresponding, 140, 142–143, 148–152, 206
of depression, 516–518, 534
of elevation, 487, 516–518, 534, 536
exterior, 173–174, 206, 209, 325–326, 352, 355
formed by a tangent and a chord, 782–783

inequalities involving angles of triangles, 325–326
inscribed, 780–783, 811, 813
lines and, 140–143, 207
measuring, 27–31, 72, 253, 353–355, 389–390, 392, 419, 420, 629
obtuse, 29, 70
opposite, 359–360, 420, 421
pairs of, 34–35, 70, 73, 140, 142–143
remote interior, 173–174, 206, 209
right, 29, 70
same-side interior, 140, 142–143, 148–152, 206
straight, 29, 70
supplementary, 34–35, 70, 73, 122–123, 360–361
vertex, 250, 273
vertical, 34–35, 70, 73, 120–123

Angle Addition Postulate, 30

Angle-Angle-Side (AAS) Theorem, 236–237, 274, 583

Angle-Angle Similarity (AA ~) Postulate, 450–451

angle bisector, 37, 70, 294–296, 300–304, 341, 342–343, 805
constructing, 45
Triangle-Angle-Bisector Theorem, 473–474, 480, 482

Angle Bisector Theorem, 295, 342

angle of rotation, 561

Angle-Side-Angle (ASA) Postulate, 234, 235–236, 274, 583

Angle-Sum Theorems, 353–355, 421

apothem, 629–631, 643–645, 676, 678

applications
academic competition, 853
accessibility, 195
activities, 850
advertising, 94
aerial views, 9, 144, 520
agriculture, 637, 663
airplane, 531
allowance, 859
archaeology, 412, 466, 775
archery, 670, 832
architecture, 9, 10, 143, 254, 289, 445, 446, 465, 504, 511, 647, 714, 727, 728, 729, 731
art, 437, 442, 445, 469, 584, 595, 632
astronomy, 497, 513, 672, 739, 768
athletics, 846
automobiles, 467
aviation, 436, 503, 520, 755
baking, 584

for lateral area of cylinder, 702
for lateral area of prism, 700
for lateral area of pyramid, 709
literal equations as, 894
for midpoint, 50, 74, 281, 286, 400
for perimeter, 59
polygon angle-sum, 353
quadratic, 439
for slope, 400
for surface area of cone, 711
for surface area of cylinder, 702
for surface area of prism, 700
for surface area of pyramid, 709
for surface area of sphere, 709
for volume of cone, 729
for volume of cylinder, 719
for volume of prism, 718
for volume of pyramid, 726
for volume of sphere, 735
table of, 897–898

Fractals, 448–449

frequency table, 830

frustum, of cone, 732

function notation, 547, 555

Fundamental Counting Principle, 836

G

Gauss, Karl, 88

geography, 521, 623, 626, 734, 739

geology, 287

geometric mean, 460, 462–463, 480

geometric probability, 668–670, 676, 680

geometric solid. *See* polyhedron(s); solid(s)

geometry, 437, 445
reasoning in, 113–116, 129–132
spherical, 179–180

Geometry in 3 Dimensions, 56, 264,
505, 551, 706, 738, 777, 802

Get Ready!
for chapters, 1, 79, 137, 215, 281,
349, 429, 487, 541, 611, 685, 759
for lessons, 10, 19, 26, 33, 40, 48, 56,
67, 88, 95, 104, 112, 119, 127,
146, 155, 163, 169, 178, 188, 196,
204, 224, 233, 241, 248, 256, 264,
271, 291, 299, 307, 315, 322, 331,
339, 358, 366, 374, 382, 388, 397,
405, 412, 418, 438, 447, 458, 498,
505, 513, 521, 526, 532, 552, 560,
567, 576, 593, 600, 622, 628, 634,
641, 648, 657, 666, 674, 695, 707,
715, 724, 732, 740, 749, 769, 779,
787, 797, 803, 821, 829, 835, 842,
849, 855, 86?

glide reflection, 570, 572, 595, 602, 604

Glossary, 911–960

Golden Gate Bridge
scale drawing, 443

golden ratio, 468–469

Got It? *See* assessment, Got It?

graph(s)
circle, 658, 798–800
of circle given its equation, 800
of line in coordinate plane, 190–191
of linear equation, 191
of system of equations, 257
reflection images, 556

graphing calculator. *See also* calculator
maximum/minimum values, 68
probability, 673
slope, 203
tangent lines, 802
trigonometric ratios, 512

great circle, 179–180, 733

Gridded Response
exercises, 67, 78, 119, 127, 136, 155,
204, 214, 224, 248, 280, 291, 331,
348, 358, 418, 428, 438, 478, 486,
498, 526, 532, 540, 585, 600, 610,
641, 648, 684, 707, 749, 758, 769,
797, 842, 855
problems, 52, 121, 152, 252, 310,
368, 472, 518, 644, 727, 773

guess and check, 84, 245

H

height
of cone, 711
of cylinder, 699, 701
of parallelogram, 616, 676, 677
of prism, 699
of pyramid, 708
slant, 708, 711, 727, 751
of trapezoid, 623, 676, 677
of triangle, 616

hemisphere, 733

Here's Why It Works
Cross Products Property, 434
equation of a circle, 799
Fundamental Counting Principle, 836
Probability of Overlapping Events, 846
Triangle Midsegment Theorem, 286

Heron's Formula, 621

hexagon, 57

**Hinge Theorem (SAS Inequality
Theorem),** 332–334, 344

history, 8, 88, 315, 320, 559, 621, 739,
803

horizon line, 696

horizontal line, 193

hypotenuse, 258, 273. *See also* right
triangle(s)

Hypotenuse-Leg (HL) Theorem,
259–260, 276
proof of, 259

hypothesis, 79, 89–92, 129

I

identity
trigonometric ratio as, 511

If-then statements. *See* conditional
statements.

images in transformations, 545, 554,
562, 587, 602
dilation, 586, 587
glide reflection, 572
preimage, 545, 554, 561, 573, 587
reflection, 554
rotation, 561
translation, 547

incenter of triangle, 301, 303–304, 341

inclinometer, 515

independent events, 844

indirect measurement, 454, 456, 480,
481, 483, 519, 599

indirect proof, 317–319, 326, 334, 341,
344, 763

indirect reasoning, 317–319, 326–327,
341, 748

inductive reasoning, 82–84, 129–132

inequalities
involving angles of triangles, 324–327,
332–334, 341, 344
involving sides of triangles, 327,
334–336, 341, 344
properties of, 323
solving, 323
in triangles, 324–328, 332–336, 341,
344

inscribed angle, 780–783, 811, 813

Inscribed Angle Theorem, 780

inscribed figure, 759
circle, 766
polygon, 667
triangle, 303, 667

intercepted arc, 759, 780, 811, 813

interdisciplinary
archaeology, 412, 466, 775
architecture, 9, 10, 143, 254, 289,
445, 446, 465, 504, 511, 647, 714,
727, 728, 729, 731
art, 437, 442, 445, 469, 595, 632
astronomy, 497, 513, 672, 739, 768
biology, 87, 112, 354, 475, 589, 631
chemistry, 321, 712, 730
earth science, 764, 767
ecology, 110
engineering, 395, 722, 795

Index

742, 762, 771, 780, 790, 798, 804, 824, 830, 836, 844, 850, 856, 862

Think About a Plan exercises, 8, 17, 25, 32, 39, 47, 54, 65, 87, 94, 102, 111, 118, 125, 145, 154, 162, 168, 176, 187, 195, 202, 223, 231, 240, 247, 254, 263, 269, 289, 297, 306, 313, 320, 329, 338, 357, 365, 373, 380, 387, 395, 404, 411, 417, 437, 446, 456, 465, 476, 496, 504, 511, 520, 525, 530, 551, 559, 566, 574, 591, 599, 620, 626, 633, 640, 647, 655, 664, 672, 693, 705, 714, 722, 730, 738, 747, 768, 777, 785, 796, 802, 807, 827, 834, 841, 848, 853, 859, 866

problem-solving strategies

draw a diagram, 36, 37, 46, 54, 60, 94, 146, 176, 240, 297, 303, 313, 314, 402, 409, 631, 669

guess and check, 84, 245

look for a pattern, 82, 84

make a chart (table), 83, 308, 506, 635, 668, 858

solve a simpler problem, 83, 623, 720

work backwards, 159, 494, 574, 637

write an equation/inequality, 21, 51, 121, 287, 295, 298, 310, 311, 328, 360, 378

product

Cross Products Property, 434, 480, 481

proof(s)

of Angle-Angle-Side (AAS) Theorem, 236

of Angle-Angle Side (AAS) Theorem using transformations, 583

of Angle-Side-Angle (ASA) Postulate using transformations, 583

choice of, 236

of Comparison Property of Inequality, 324

congruent circles and chord theorems, 773, 774, 791

congruent triangles, 226–229, 234–237, 244–245, 265, 273–274, 419

of Concurrency of Perpendicular Bisectors Theorem, 302

coordinate proof, 406, 408–409, 414–415, 420, 424

of Corollary to the Triangle Exterior Angle Theorem, 325

in exercises, 117, 118, 121–123, 126, 150, 153, 155, 158, 161, 162, 165–166, 167, 168, 172, 176, 177, 203, 211, 220–221, 223, 224, 227, 230, 231, 232, 233, 237, 238, 239, 240, 244, 245, 246, 247, 248, 255, 259, 262, 263, 264, 270, 271, 291, 298, 306, 319, 320–321, 322, 330, 331, 335, 338, 364, 366, 373, 377, 379, 387–388, 392, 396, 398, 414–415,

416, 417, 419, 452, 456, 457, 458, 466–467, 476, 477, 497, 498, 501, 512, 583, 584, 633, 634, 656, 768, 769, 777, 778, 785, 786–787, 796–797

flow, 137, 158, 206

of Hypotenuse-Leg (HL) Theorem, 259

indirect, 317–319, 326, 334, 341, 763

of Inscribed Angle Theorem, 781

of Isosceles Triangle Theorem using transformations, 585

of kite theorem, 392

overlapping triangles in, 265–267

paragraph, 120–123, 129

of parallel lines, 156–159

parallelograms and, 361, 369, 377, 383–385

of Pythagorean Theorem, 490, 497

related to segment(s) and circle(s) theorems, 793

of Side-Angle-Side Congruence Postulate using transformations, 581

of Side-Angle-Side Similarity (SAS ~) Theorem, 452

of Side-Splitter Theorem, 472

of theorems about similar triangles, 453, 480

of tangent to a circle theorem, 763

of Triangle Inequality Theorem, 331

two-column, 113, 115–116, 129

using coordinate geometry, 414–415

Properties

Comparison Property of Inequality, 324

of congruence, 114–116

Cross Products, 434, 480, 481

Distributive, 114

of equality, 113–116, 435

of inequality, 323

of kites, 392–393, 423

of midsegments, 284–291, 391–392

of parallel lines, 148–152, 207

of parallelograms, 359–363, 375–378, 383–385

of proportions, 435

of real numbers, 900

Reflexive, 113–114

Symmetric, 113

of tangents, 762–766

Transitive, 113

of trapezoids, 389–392, 420, 423

Zero-Product Property, 439

proportions. *See also* scale

extended, 440–441, 480

properties of, 435

and ratios, 432–435, 480, 481

and similarity, 440

in triangle(s), 471–474, 480, 482

protractor

Protractor Postulate, 28

using, 884

Pull It All Together, 69, 128, 205, 272, 340, 419, 479, 533, 601, 675, 750, 810, 869

pyramid

altitude of, 708

base of, 708

defined, 708

height of, 708

lateral area of, 709

slant height of, 708

surface area of, 709

vertex of, 708

volume of, 726–727, 751, 753

Pythagorean Theorem, 491

applying, 491–495, 534

Converse of, 493, 535

proof of, 490, 497

with special triangles, 499–501

statement of, 491, 535

Pythagorean triple, 492, 534

Q

quadratic equations, 439

quadratic formula, 439

quadrilateral(s)

classifying, 402

constructing, 183

diagonals of, 367, 369, 371, 391

identifying, 367–371, 421

quadrilaterals in, 413

relationships among, 393

special, 183, 389–393

Quick Review, 71–74, 130–132, 207–210, 274–276, 342–344, 421–424, 481–482, 535–536, 603–606, 677–680, 752–754, 812–814 , 871–874

R

radical expressions, 399

radius

of circle, 649, 676, 798

of cylinder, 702

defined, 649

radius-tangent relationship, 762

finding area with, 643–644, 678

of regular polygon, 629–631, 676, 678

of sphere, 733–734, 751

random number table, 901

ratios

cosine, 507

defined, 432

extended, 432, 433, 480

golden, 468–469

and proportions, 432–435, 440, 480, 481, 675

W

X

Y

Z

Acknowledgments

Staff Credits

The people who made up the High School Mathematics team—representing composition services, core design digital and multimedia production services, digital product development, editorial, editorial services, manufacturing, marketing, and production management—are listed below.

Dan Anderson, Scott Andrews, Christopher Anton, Carolyn Artin, Michael Avidon, Margaret Banker, Charlie Bink, Niki Birbilis, Suzanne Biron, Beth Blumberg, Kyla Brown, Rebekah Brown, Judith Buice, Sylvia Bullock, Stacie Cartwright, Carolyn Chappo, Christia Clarke, Mary Ellen Cole, Tom Columbus, Andrew Coppola, AnnMarie Coyne, Bob Craton, Nicholas Cronin, Patrick Culleton, Damaris Curran, Steven Cushing, Sheila DeFazio, Cathie Dillender, Emily Dumas, Patty Fagan, Frederick Fellows, Jorgensen Fernandez, Mandy Figueroa, Suzanne Finn, Sara Freund, Matt Frueh, Jon Fuhrer, Andy Gaus, Mark Geyer, Mircea Goia, Andrew Gorlin, Shelby Gragg, Ellen Granter, Gerard Grasso, Lisa Gustafson, Toni Haluga, Greg Ham, Marc Hamilton, Chris Handorf, Angie Hanks, Scott Harris, Cynthia Harvey, Phil Hazur, Thane Heninger, Aun Holland, Amanda House, Chuck Jann, Linda Johnson, Blair Jones, Marian Jones, Tim Jones, Gillian Kahn, Matthew Keefer, Brian Keegan, Jonathan Kier, Jennifer King, Tamara King, Elizabeth Krieble, Meytal Kotik, Brian Kubota, Roshni Kutty, Mary Landry, Christopher Langley, Christine Lee, Sara Levendusky, Lisa Lin, Wendy Marberry, Dominique Mariano, Clay Martin, Rich McMahon, Eve Melnechuk, Cynthia Metallides, Hope Morley, Christine Nevola, Michael O'Donnell, Michael Oster, Ameer Padshah, Stephen Patrias, Jeffrey Paulhus, Jonathan Penyack, Valerie Perkins, Brian Reardon, Wendy Rock, Marcy Rose, Carol Roy, Irene Rubin, Hugh Rutledge, Vicky Shen, Jewel Simmons, Ted Smykal, Emily Soltanoff, William Speiser, Jayne Stevenson, Richard Sullivan, Dan Tanguay, Dennis Tarwood, Susan Tauer, Tiffany Taylor-Sullivan, Catherine Terwilliger, Maria Torti, Mark Tricca, Leonid Tunik, Ilana Van Veen, Lauren Van Wart, John Vaughan, Laura Vivenzio, Samuel Voigt, Kathy Warfel, Don Weide, Laura Wheel, Eric Whitfield, Sequoia Wild, Joseph Will, Kristin Winters, Allison Wyss, Dina Zolotusky

Additional Credits: Michele Cardin, Robert Carlson, Kate Dalton-Hoffman, Dana Guterman, Narae Maybeth, Carolyn McGuire, Manjula Nair, Rachel Terino, Steve Thomas

Illustration

Stephen Durke: 4, 5, 8, 9, 11, 18, 20, 25, 27, 32, 43, 50, 53, 54, 59, 60, 61, 66, 67, 69, 140, 148, 156, 161, 164, 168, 171, 174, 179, 182,189, 195, 197, 203, 218, 219, 222, 226, 228, 232, 234, 235, 240, 244, 245, 248, 250, 255, 265, 270, 284, 289, 291, 293, 295, 296, 300, 302, 304, 305, 308, 313, 316, 322, 324, 328, 330, 331, 332, 334, 338, 466, 504, 511, 515, 516, 519, 520, 521, 548, 620, 641, 655, 656, 657, 665, 811, 812; **Jeff Grunwald represented by Wilkinson Studios, Inc.:** 82, 83, 89, 94, 98, 106, 109, 111, 113, 120, 125, 128, 200, 357, 359, 371, 375, 383, 385, 389, 400, 404, 406, 417, 437, 440, 443, 446, 450, 454, 456, 464, 465, 466, 471, 473, 475, 476, 479, 499, 516, 533, 639.; **Phil Guzy:** 34, 320, 330, 491, 496, 507, 513, 544, 549, 554, 559, 569, 570, 616, 623, 643, 649, 652, 655, 672, 673, 688, 697, 699, 703, 705, 706, 708, 717, 720, 726, 728, 733, 735, 739, 742, 748, 758, 762, 770, 771, 780, 790, 792, 798; **Rob Schuster:** 522; **XNR Productions:** 526, 623, 806, 810.

Technical Illustration

Datagrafix, Inc.

Photography

All photographs not listed are the property of Pearson Education.

Back Cover: Bon Appetit/Almay

Chapter 1: Page 3, Julian Smith/Corbis; **15,** Kelly Redinger/Design Pics Inc./Alamy; **30,** Pete Saloutos/zefa/Corbis; **32,** Stuart Melvin/Alamy; **39,** Richard Menga/Fundamental Photographs; **62,** Charles O. Cecil/Alamy.

Chapter 2: Page 81, Hoge Noorden/epa/Corbis; **126,** IPS Co., Ltd./Beateworks/Corbis; **103,** www.Lifeprint.com.

Chapter 3: Page 139, Laurent Gillieron/epa/Corbis; **144,** Kevin Fleming/Corbis; **154,** photo courtesy of Frank Adelstein, Ithaca, NY; **143,** Peter Cade/Iconica/Getty Images; **176,** Bill Brooks/Alamy; **159,** Robert Slade/Manor Photography/Alamy; **162,** Robert Llewellyn/Corbis.

Chapter 4: Page 217, Robin Utrecht/Staff/AFP/Getty Images; **227,** Stan Honda/Staff/AFP/Getty Images; **246,** Viktor Kitaykin/iStockphoto; **253,** John Wells/Photo Researchers, Inc; **263,** Image Source Black/Jupiterimages; **258,** Tony Freeman/PhotoEdit; **260,** Paul Jones/Iconica/Getty Images; **260,** Llado, M./Plainpicture Photography/Veer; **266,** Photo by Pearson Education, created by JC Nolan, folded by Sara Adams.

Chapter 5: Page 281, Floto + Warner/Arcaid/Alamy; **286,** Momatiuk - Eastcott/Corbis; **288,** Joseph Sohm/Visions of America, LLC/Alamy; **331,** Gunter Marx Photography/Corbis.

Chapter 6: Page 351, White Star/Friedrichsmeier/B. Nasner/imagebroker/Alamy; **354 l,** Robert Harding Picture Library Ltd/Alamy Inc.; **354 r,** Anthony Bannister/Gallo Images/Corbis; **356 l,** Inger Hogstrom/DanitaDelimont.com; **356 c,** Laurie Strachan/Alamy; **356 r,** BestShot/iStockphoto; **360,** Eric Hood/iStockphoto; **365,** Esa Hiltula/Alamy; **367,** Victor Fraile/Reuters; **376,** Kirsty McLaren/Alamy; **379 l,** Claro Cortes IV/Reuters/Landov; **379 r,** Michael Jenner/Alamy; **387,** Rodney Raschke; **395,** Colin Underhill/Alamy.

Chapter 7: Page 431, Michel Setboun/Corbis; **432,** Chuck Eckert/Alamy; **443,** Ron Watts/Corbis; **466,** James L. Amos/Corbis; **469 l,** Mark A. Johnson/Corbis; **469 c,** Corey Hochachka/Design Pics/Fotosearch; **469 r,** Photodisc/Fotosearch; **475,** Victor R. Boswell Jr./Contributor/National Geographic/Getty Images.

Chapter 8: Page 489, Tim Woodcock/Alamy; **493,** Petra Wegner/Alamy; **502,** Hemera Technologies/Jupiterimages; **508,** Steve Vidler/ImageState; **517,** Dave Reede/All Canada Photos/Alamy.

Chapter 9: Page 543, I G Kinoshita/amana images/Getty Images; **544,** RubberBall/Alamy; **559,** North Wind Picture Archives/Alamy; **564,** Alan Copson/City Pictures/Alamy; **600 tl,** Owaki - Kulla/Corbis; **600 tr,** D. Hurst/Alamy; **584 l,** M.C. Escher's

"Symmetry E56" © 2009 The M.C. Escher Company-Holland. All rights reserved. www.mcescher.com; **584 r,** M.C. Escher's "Symmetry E18" © 2009 The M.C. Escher Company-Holland. All rights reserved. www.mcescher.com; **587,** Martin William Allen/ Alamy; **589 l,** Keith Leighton/Alamy; **589 r,** Reven T.C. Wurman/ Alamy; Keith Leighton/Alamy; **601,** Douglas Kirkland/Corbis.

Chapter 10: Page 613, Jim West/Alamy; **618,** Bob Gates/Alamy; **627,** artpartner-images.com/Alamy; **625,** Joe Sohm/Visions of America, LLC/Alamy; **633,** Dennis Marsico/Corbis; **644,** Alan Schein Photography/Corbis; **661,** Matthias Tunger/Photonica/ Getty Images; **668,** Clive Streeter/Dorling Kindersley; **670,** amana images inc./Alamy.

Chapter 11: Page 685, Associated Press; **693,** Sports Bokeh/ Alamy; **705,** Ron Chapple Stock/Alamy; **710,** Adam Eastland/ Alamy; **727,** age fotostock/SuperStock; **728,** John E Marriott/ Alamy; **733,** D. Hurst/Alamy; **737 l,** Andrew Paterson/Alamy; **737 c,** Stockbyte/Getty Images; **737 r,** Image Source/Getty Images; **738,** Stephen Sweet/Alamy; **744,** Jupiterimages/ BananaStock/Alamy; **745,** Andre Jenny/Alamy.

Chapter 12: 761, NASA and STScI; **764 t,** T. Pohling/Alamy; **764 b,** NASA Marshall Space Flight Center Collection; **768,** Clive Streeter/Dorling Kindersley; **775,** Cris Bouroncle/Staff/AFP/Getty Images; **785,** Vario Images GmbH & Co.KG/Alamy; **792,** NASA Marshall Space Flight Center Collection; **794,** Melvyn Longhurst/ Alamy; **795,** dpa/Corbis; **809,** matthiasengelien.com/Alamy.

Chapter 13: 823, Jokerpro/Shutterstock.

Acknowledgments